Conversion Factors (cont.)

Volume	$1\ cm^3 = 10^{-6}\ m^3 = 3.53 \times 10^{-5}\ ft^3 = 6.10 \times 10^{-2}\ in.^3$
	$1\ m^3 = 10^6\ cm^3 = 10^3\ L = 35.3\ ft^3 = 6.10 = 10^3\ in.^3 = 264\ gal$
	$1\ liter = 10^3\ cm^3 = 10^{-3}\ m^3 = 1.056\ qt = 0.264\ gal$
	$1\ in.^3 = 5.79 \times 10^{-4}\ ft^3 = 16.4\ cm^3 = 1.64 \times 10^{-5}\ m^3$
	$1\ ft^3 = 1728\ in.^3 = 7.48\ gal = 0.0283\ m^3 = 28.3\ L$
	$1\ qt = 2\ pt = 946.5\ cm^3 = 0.946\ L$
	$1\ gal = 4\ qt = 231\ in.^3 = 3.785\ L$
Time	$1\ h = 60\ min = 3600\ s$
	$1\ day = 24\ h = 1440\ min = 8.64 \times 10^4\ s$
	$1\ year = 365\ days = 8.76 \times 10^3\ h = 5.26 \times 10^5\ min = 3.16 \times 10^7\ s$
Angle	$360° = 2\pi\ rad$
	$180° = \pi\ rad$
	$90° = \pi/2\ rad$
	$60° = \pi/3\ rad$
	$45° = \pi/4\ rad$
	$30° = \pi/6\ rad$
	$1\ rad = 57.3°$
	$1° = 0.0175\ rad$
Speed	$1\ m/s = 3.6\ km/h = 3.28\ ft/s = 2.24\ mi/h$
	$1\ km/h = 0.278\ m/s = 0.621\ mi/h = 0.911\ ft/s$
	$1\ ft/s = 0.682\ mi/h = 0.305\ m/s = 1.10\ km/h$
	$1\ mi/h = 1.467\ ft/s = 1.609\ km/h = 0.447\ m/s$
	$60\ mi/h = 88\ ft/s$
Force	$1\ N = 10^5\ dynes = 0.225\ lb$
	$1\ dyne = 10^{-5}\ N = 2.25 \times 10^{-6}\ lb$
	$1\ lb = 4.45 \times 10^5\ dynes = 4.45\ N$
	Equivalent weight of 1-kg mass = 2.2 lb = 9.8 N
Pressure	$1\ Pa\ (N/m^2) = 1.45 \times 10^{-4}\ lb/in.^2 = 7.5 \times 10^{-3}\ torr\ (mm\ Hg) = 10\ dynes/cm^2$
	$1\ torr\ (mm\ Hg) = 133\ Pa\ (N/m^2) = 0.02\ lb/in.^2 = 1333\ dynes/cm^2$
	$1\ atm = 14.7\ lb/in.^2 = 1.013 \times 10^5\ N/m^2 = 1.013 \times 10^6\ dynes/cm^2$
	$= 30\ in.\ Hg = 76\ cm\ Hg$
	$1\ bar = 10^6\ dynes/cm^2 = 10^5\ Pa$
	$1\ millibar = 10^3\ dynes/cm^2 = 10^2\ Pa$
Energy	$1\ J = 10^7\ ergs = 0.738\ ft\text{-}lb = 0.239\ cal = 9.48 \times 10^{-4}\ Btu = 6.24 \times 10^{18}\ eV$
	$1\ kcal = 4186\ J = 4.186 \times 10^{10}\ ergs = 3.968\ Btu$
	$1\ Btu = 1055\ J = 1.055 \times 10^{10}\ ergs = 778\ ft\text{-}lb = 0.252\ kcal$
	$1\ cal = 4.186\ J = 3.97 \times 10^{-3}\ Btu = 3.09\ ft\text{-}lb$
	$1\ ft\text{-}lb = 2.69 \times 10^6\ J = 1.36 \times 10^7\ ergs = 1.29 \times 10^{-3}\ Btu$
	$1\ eV = 1.60 \times 10^{-19}\ J = 1.60 \times 10^{-12}\ ergs$
	$1\ kWh = 3.6 \times 10^6\ J$
Power	$1\ W = 0.738\ ft\text{-}lb/s = 1.34 \times 10^{-3}\ hp = 3.41\ Btu/h$
	$1\ ft\text{-}lb/s = 1.36\ W = 1.82 \times 10^{-3}\ hp$
	$1\ hp = 550\ ft\text{-}lb/s = 745.7\ W = 2545\ Btu/h$
Mass-Energy Equivalents (at rest)	$1\ u = 1.66 \times 10^{-27}\ kg \leftrightarrow 931.5\ MeV$
	$1\ electron\ mass = 9.11 \times 10^{-31}\ kg = 5.49 \times 10^4\ u \leftrightarrow 0.511\ MeV$
	$1\ proton\ mass = 1.672 \times 10^{-27}\ kg = 1.007276\ u \leftrightarrow 938.28\ MeV$
	$1\ neutron\ mass = 1.674 \times 10^{-27}\ kg = 1.008665\ u \leftrightarrow 939.57\ MeV$

College Physics

College Physics

Jerry D. Wilson

Lander College

Allyn and Bacon

Boston   London   Sydney   Toronto

Series Editor: Nancy Forsyth
Developmental Editor: Judith Hauck
Senior Editorial Assistant: Ann Guth
Senior Production Administrator: Elaine Ober
Production Administrator: Mary Beth Finch
Copy Editor: Jane Hoover
Editorial-Production Services: Mary Hill
Composition Buyer: Linda Cox
Manufacturing Buyer: Tamara Johnson
Cover Administrator: Linda Dickinson
Cover Designer: Lynda Fishbourne

Copyright © 1990 by Allyn and Bacon
A Division of Simon & Schuster, Inc.
160 Gould Street
Needham Heights, Massachusetts 02194

Library of Congress Cataloging-in-Publication Data

Wilson, Jerry D.
    College physics / Jerry D. Wilson.
    p.   cm.
    ISBN 0-205-12110-1
    1. Physics.   I. Title.
QC21.2.W548 1990   89-27632
                              CIP
Printed in the United States of America
10 9 8 7 6 5 4 3 2 1   93 92 91 90

CREDITS
Credits begin on page 876, which constitutes a continuation of the copyright page.

Dedicated to James T. Shipman,
friend and colleague

Contents

Preface   *xvii*

1   Units and Problem Solving   *1*

1.1   SI Units for Length, Mass and Time   *2*
1.2   More about the Metric System   *6*
1.3   Dimensional Analysis   *9*
1.4   Unit Conversion   *11*
1.5   Significant Figures   *13*
1.6   Problem Solving   *15*
Important Formulas
Questions   *18*
Problems   *19*

2   Kinematics: The Description of Motion   *23*

2.1   A Change of Position   *23*
2.2   Speed and Velocity   *25*
2.3   Acceleration   *32*
2.4   Kinematic Equations   *35*
2.5   Free Fall   *37*
Insight: Galileo Galilei   *38*
Important Formulas   *45*
Questions   *45*
Problems   *46*

3 Motion in Two Dimensions   *52*

3.1 Components of Motion   *52*
3.2 Vector Addition and Subtraction   *57*
3.3 Projectile Motion   *63*
3.4 Uniform Ciruclar Motion and Centripetal Acceleration   *70*
Important Formulas   *74*
Question   *75*
Problems   *75*

4 Force and Motion   *82*

4.1 Newton's First Law of Motion   *83*
Insight: Isaac Newton   *84*
4.2 The Concept of Force   *86*
4.3 Newton's Second Law of Motion   *87*
Demonstration 1   Newton's First Law and Inertia   *87*
4.4 Applications of Newton's Second Law   *91*
4.5 Newton's Third Law of Motion   *97*
Demonstration 2   Tension in a String   *100*
4.6 Friction   *100*
Important Formulas   *108*
Questions   *109*
Problems   *110*

5 Work and Energy   *116*

5.1 The Definition of Work   *116*
5.2 Work Done by Variable Forces   *120*
5.3 The Work-Energy Theorem: Kinetic Energy   *124*
5.4 Potential Energy   *126*
5.5 The Conservation of Energy   *129*
Demonstration 3   High Road/Low Road   *135*
5.6 Power   *136*
Important Formulas   *138*
Questions   *139*
Problems   *139*

6 Momentum and Collisions   *146*

6.1 Linear Momentum   *146*
6.2 The Conservation of Linear Momentum   *150*
Demonstration 4   Internal Forces (and Conservation of Momentum?)   *152*
6.3 Impulse and Collisions   *154*
6.4 Center of Mass   *163*

6.5 Jet Propulsion   *167*
Demonstration 5   Jet Propulsion   **169**
Important Formulas   *170*
Questions   *170*
Problems   *171*

7   Circular Motion and Gravitation   *177*

7.1 Angular Measure   *177*
7.2 Angular Speed and Velocity   *181*
Demonstration 6   Constant Angular Velocity, but not Tangential Velocity   **183**
7.3 Angular Acceleration   *184*
7.4 Newton's Law of Gravitation   *188*
Insight: Gravitational and Inertial Mass   *191*
7.5 Kepler's and Earth Satellites   *196*
7.6 Frames of Reference and Inertial Forces   *201*
Demonstration 7   Accelerated Reference Frame   **202**
Insight: The Coriolis Force and the Foucault Pendulum   *206*
Important Formulas   *208*
Questions   *208*
Problems   *210*

8   Rotational Motion   *216*

8.1 Rigid Bodies, Translations, and Rotations   *216*
8.2 Torque and Moment of Inertia   *220*
8.3 Applications of Rotational Dynamics   *223*
8.4 Rotational Work and Kinetic Energy   *227*
Insight: Slide or Roll to a Stop?   *231*
8.5 Angular Momentum   *232*
Important Formulas   *237*
Questions   *237*
Problems   *239*

9   Mechanical Properties   *245*

9.1 Mechanical Equilibrium   *245*
9.2 Stability and the Center of Gravity   *250*
Demonstration 8   Stability and Center of Gravity   **254**
9.3 Mechanical Advantage   *254*
9.4 Elastic Moduli   *259*
Insight: Feat of Strength or Knowledge of Materials?   *261*
Demonstration 9   Stress, Strain, and Elasticity   **263**
Important Formulas   *265*
Questions   *266*
Problems   *267*

10

Fluids: Liquids and Gases   *274*

10.1   Pressure and Pascal's Principle   *274*
10.2   Bouyancy and Archimedes' Principle   *281*
Demonstration 10   Buoyancy and Density   *285*
10.3   Surface Tension and Capillary Action   *287*
10.4   Fluid Dynamics and Bernoulli's Equation   *292*
Demonstration 11   Bernoulli-trapped Ball   *296*
Insight: Blood Pressure   *300*
10.5   Viscosity, Poiseuille's Law, and Reynolds Number   *297*
Important Formulas   *302*
Questions   *302*
Problems   *305*

11

Temperature   *310*

11.1   Temperature and Heat   *310*
11.2   The Celsius and Fahrenheit Temperature Scales   *311*
Insight: Daniel Gabriel Fahrenheit and Anders Celsius   *315*
11.3   Gas Laws and Absolute Temperature   *315*
Demonstration 12   Boyle's Shaving Cream   *318*
11.4   Thermal Expansion   *319*
11.5   The Kinetic Theory of Gases   *325*
Important Formulas   *328*
Questions   *329*
Problems   *331*

12

Heat   *334*

12.1   Units of Heat   *334*
Insight: Heat of Combustion   *336*
12.2   Specific Heat   *338*
12.3   Phase Changes and Latent Heat   *342*
12.4   Heat Transfer   *347*
Insight: Heat Transfer by Radiation: The Microwave Oven   *352*
12.5   Evaporation and Relative Humidity   *354*
Important Formulas   *356*
Questions   *356*
Problems   *358*

13

Thermodynamics   *362*

13.1   Thermodynamic Systems, States, and Processes   *362*
13.2   The First Law of Thermodynamics   *364*
13.3   The Second Law of Thermodynamics and Entropy   *369*
Insight: Time's Arrow and the Heat Death of the Universe   *372*
13.4   Heat Engines and Heat Pumps   *374*

13.5   The Carnot Cycle and Ideal Heat Engines   *379*
Important Formulas   *381*
Questions   *381*
Problems   *383*

14 Vibrations and Waves   *388*

14.1   Simple Harmonic Motion   *388*
14.2   Equations of Motion   *392*
Demonstration 13   Simple Harmonic Motion (SHM) and Sinusoidal Wave   ***398***
14.3   Wave Motion   *400*
Insight: Earthquakes and Seismology   *404*
14.4   Wave Phenomena   *404*
14.5   Standing Waves and Resonance   *408*
Demonstration 14   Flame Standing Wave Pattern   ***410***
Insight: Unwanted Resonance   *412*
Important Formulas   *413*
Questions   *414*
Problems   *414*

15 Sound   *419*

15.1   Sound Waves   *419*
15.2   The Speed of Sound   *421*
15.3   Sound Intensity   *422*
15.4   Sound Phenomena   *426*
15.5   Sound Characteristics   *434*
Important Formulas   *437*
Questions   *437*
Problems   *438*

16 Electric Charge, Force, and Energy   *443*

16.1   Electric Charge   *443*
16.2   Electrostatic Charging   *446*
16.3   Electric Force and Electric Field   *449*
Insight: Xerography and Electrostatic Copiers   *450*
16.4   Electrical Energy and Electric Potential   *456*
16.5   Capacitance and Dielectrics   *461*
Important Formulas   *468*
Questions   *468*
Problems   *470*

17 Electric Current and Resistance   *475*

17.1   Batteries and Direct Current   *475*
17.2   Current and Drift Velocity   *477*

17.3   Ohm's Law and Resistance   *479*
17.4   Electric Power   *484*
Insight: Superconductivity   *485*
Important Formulas   *489*
Questions   *489*
Problems   *490*

18   Electric Circuits   *495*

18.1   Resistances in Series, Parallel, and Series-Parallel Combinations   *495*
18.2   Multiloop Circuits and Kirchoff's Rules   *502*
18.3   RC Circuits   *506*
18.4   Ammeters and Voltmeters   *509*
Insight: Electricity and Personal Safety   *516*
18.5   Household Circuits and Electrical Safety   *512*
Important Formulas   *517*
Questions   *517*
Problems   *518*

19   Magnetism   *525*

19.1   Magnets and Magnetic Poles   *525*
19.2   Electromagnetism and the Source of Magnetic Fields   *527*
19.3   Magnetic Materials   *533*
19.4   Magnetic Forces on Current-Carrying Wires   *535*
19.5   Applications of Electromagnetism   *539*
19.6   The Earth's Magnetic Field   *543*
Important Formulas   *546*
Questions   *546*
Problems   *548*

20   Electromagnetic Induction   *553*

20.1   Induced emf's: Faraday's Law and Lenz's Law   *553*
Demonstration 15   Magnetic Levitation and Lenz's Law   *557*
Insight: Electromagnetic Induction in Action   *558*
20.2   Generators and Back emf   *560*
20.3   Transformers and Power Transmission   *565*
20.4   Electromagnetic Waves   *569*
Insight: CAT Scan   *574*
Important Formulas   *575*
Questions   *575*
Problems   *577*

21 AC Circuits   581

21.1 Resistance in an AC Circuit   581
21.2 Capacitive Reactance   584
21.3 Inductive Reactance   585
21.4 Impedance: RLC Circuits   587
21.5 Circuit Resonance   591
Important Formulas   593
Questions   594
Problems   594

22 Geometrical Optics: Reflection and Refraction of Light   598

22.1 Wave Fronts and Rays   598
22.2 Reflection   599
22.3 Refraction   600
Demonstration 16 Refraction in Action   602
22.4 Total Internal Reflection and Fiber Optics   607
22.5 Dispersion   611
Important Formulas   613
Questions   613
Problems   614

23 Mirrors and Lenses   618

23.1 Plane Mirrors   618
Demonstration 17 A Candle Burning Underwater?   619
23.2 Spherical Mirrors   621
Demonstration 18 Candle Burning at Both Ends   625
23.3 Lenses   629
Demonstration 19 Inverted Image   633
Insight: Fresnel Lenses   636
23.4 Lens Aberrations   636
235. The Lens Maker's Equation   638
Important Formulas   641
Questions   642
Problems   643

24 Physical Optics: The Wave Nature of Light   647

24.1 Young's Double-Slit Experiment   647
24.2 Thin-Film Interference   651
Insight: Nonreflecting Lenses   653
Demonstration 20 Thin-Film Interference   654
24.3 Diffraction   656
24.4 Polarization   662

Insight: LCD's and Polarized Light   *666*
24.5   Atmospheric Scattering of Light   *668*
Important Formulas   *671*
Questions   *671*
Problems   *672*

25   Optical Instruments   *676*

25.1   The Human Eye   *676*
25.2   Microscopes   *683*
25.3   Telescopes   *687*
Insight: Telescopes Using Nonvisible Radiation   *692*
25.4   Diffraction and Resolution   *693*
25.5   Color and Optical Illusions   *695*
Demonstration 21   Geometrical Optical Illusions—What Do You See?   *699*
Important Formulas   *700*
Questions   *700*
Problems   *701*

26   Relativity   *705*

26.1   Classical Relativity   *705*
26.2   The Michelson-Morley Experiment   *708*
26.3   The Special Theory of Relativity   *711*
Insight: Albert Einstein   *712*
26.4   Relativistic Mass and Energy   *720*
26.5   The General Theory of Relativity   *724*
Important Formulas   *727*
Questions   *728*
Problems   *728*

27   Quantum Physics   *732*

27.1   Quantization: Planck's Hypothesis   *732*
27.2   Quanta of Light: Photons and the Photoelectric Effect   *735*
27.3   Quantum "Particles": The Compton Effect   *739*
27.4   The Bohr Theory of the Hydrogen Atom   *742*
27.5   A Quantum Success: The Laser   *747*
Insight: The Compact Disc   *751*
Important Formulas   *752*
Questions   *752*
Problems   *753*

28   Quantum Mechanics   *756*

28.1   Matter Waves: The de Broglie Hypothesis   *757*
Insight: The Electron Microscope   *760*

28.2 The Schrödinger Wave Equation   *762*
28.3 The Heisenberg Uncertainty Principle   *766*
28.4 Particles and Antiparticles   *769*
Important Formulas   *772*
Questions   *772*
Problems   *773*

29 The Nucleus   *775*

29.1 Nuclear Stability and the Nuclear Force   *775*
29.2 Radioactivity   *779*
29.3 Decay Rate and Half-Life   *783*
29.4 Nuclear Stability and Binding Energy   *788*
29.5 Radiation Detection and Applications   *793*
Important Formulas   *801*
Questions   *801*
Problems   *802*

30 Nuclear Reactions and Elementary Particles   *806*

30.1 Nuclear Reactions   *806*
30.2 Nuclear Energy: Fission and Fusion   *811*
Demonstration 22   A Simulated Chain Reaction (without Fission)   *814*
30.3 Beta Decay and the Neutrino   *821*
30.4 Fundamental Forces and Exchange Particles   *823*
30.5 Elementary Particles   *826*
Important Formulas   *829*
Questions   *829*
Problems   *830*

Appendix I   Tables   *834*
   Table 1   Physical Constants   *834*
   Table 2   Conversion Factors   *835*
   Table 3   Trigonometric Relationships   *836*
   Table 4   Trigonometric Tables   *837*
   Table 5   Areas and Volumes of Some Common Shapes   *838*
   Table 6   Planetary Data   *838*
Appendix II   Kinetic Theory of Gases   *839*
Appendix III   Einstein's Letter to President Roosevelt   *841*
Appendix IV   Alphabetical Listing of the Chemical Elements   *843*
Appendix V   Properties of Selected Isotopes   *845*

Answers to Odd-numbered Problems   *849*
Index   *866*

Preface

Welcome to physics! You might be wondering why this text was selected as your introductory, college-level physics text. One reason is that it was written with you, the student, in mind. Students often view physics as a "difficult" subject. However, when properly presented and explained, physical principles are easily learned and their consistency and logic can be appreciated. A knowledge of physics will help you better understand the world around you—to know why things happen the way they do and how they "work."

In this text, the emphases are on "the basics" and on problem solving. The latter is probably where physics gets its "difficult" label. Physics is a problem-solving science—there's no getting around it. What this text does to help you learn and understand problem solving is to address the topic right up front. In the first chapter, a complete section is devoted to problem solving, along with other tricks of the trade, such as dimensional analysis. SI units and unit conversion are also considered.

The help with problem solving continues throughout the text. There are many solved examples, and *Problem Solving Hints* are given where appropriate to help you avoid common mistakes.

Please don't think that in College Physics all we do is solve problems. There are many descriptive applications and special demonstration photos to illustrate various physical principles. Also, I've included chapter *Insights* on special interest topics. Let me summarize a few of the features of the text which should help you in studying from it:

- Important terms are in boldface type in the text material for emphasis and easy reference.
- Colors enhance illustrations and features, with about 80 photographs of classical physics demonstrations and over 400 line drawings.
- A variety of chapter *Insights* cover special interest or related topics (for example, compact dics—do you know how these work?).
- End-of-chapter summaries of important formulas are provided for easy reference when working assigned problems.
- Answers to odd-numbered problems are given at the back of the book—triple checked for accuracy.

And there's more. A *Student Study Guide* by John Kinard and David Cordes is available separately. It contains a chapter-by-chapter review, key terms, and solved problems (about 20% of these are end-of-chapter problems *from the text* and the rest are worked problems *not* found in the text for you to use in studying and practicing your problem solving techniques.)

So, you see that I have tried to make your study of physics an enjoyable and rewarding experience. During the course of your study, if you think of something that might help other students learn physics, please let me know. Perhaps we can use it in a future edition. I look forward to hearing your comments.

Acknowledgements

My sincere thanks to the colleagues listed below. These professors read and commented upon various portions of the manuscript as I wrote them. Their suggestions were uniformly thoughtful and constructive, and this book has benefitted greatly from them.

Philip Gash
Cal State University

Richard Gehrenback
Rhode Island College

Groves E. Herrick
Maine Maritime Academy

Joe Pifer
Rutgers University

Joseph Chalmers
University of Louisville

John Anderson
University of Pittsburgh

Lewis Ford
Texas A&M University

Kirby Kemper
Florida State University

David Markowitz
University of Connecticut

William Riley
Ohio State University

J. P. Davidson
University of Kansas

Charles Bacon
Ferris State College

Dean Zollman
Kansas State University

Robert Graham
University of Nebraska

Rubin Laudan
Oregon State University

Bennet Brabson
Indiana University

George Rainey
Cal State Polytechnic University

Richard Graham
Ricks College

Special thanks goes to John Kinard for his overall assistance and ever diligent work on the *Annotated Instructor's Edition* and *Solutions Manuals,* and to John Smith, who took many photographs.

The supplements authors also contributed greatly to the overall teaching/learning

Michael Browne
University of Idaho

David M. Cordes
Belleville Area Community College

Thanks also to Ian Avruch, Alex Azima, Robert Coakley, Jeff Conn, Geoffrey Dixon, and Frank Lomanno for conducting independent checks of the solutions and answers to the end-of-chapter problems and to Martha Pond and Scott Goelzer for their work on the end-of-chapter questions.

Special thanks to the Physics Department at Boston University, Boston University Video Unit (part of the BU School of Communications), Vishnu Patel, and William Alston for their cooperation and excellent work in producing a high quality demonstration video.

At Allyn & Bacon, the editorial staff was particularly helpful. Special recognition goes to the "Science Team" for their enthusiasm, creativity, and commitment to this project. The team consists of Nancy Forsyth, Senior Series Editor; Ann Guth, Senior Editorial Assistant; Judy Hauck, Developmental Editor; Carol Kennedy, Advertising Manager; Virginia Lanigan, Marketing Manager; Elaine Ober, Senior Production Administrator; Mary Beth Finch, Production Administrator; Lisa Feder, Production Assistant; Tamara Johnson, Manufacturing Buyer. Thanks also to Mary Hill, Jane Hoover, Kathleen Smith, Glenna Collett, Louise Gelinas, and Barbara Willette.

Jerry Wilson
Division of Science and Mathematics
Lander College
Greenwood, SC 29646

Accompanying Study Aids
Study Guide/Student Solutions Manual.

Accompanying Teaching Aids
Instructor's Solutions Manual
Transparencies for College Physics
Test Bank
Computerized Test Bank
Video Demonstrations for College Physics

Units and Problem Solving

*Weights and measures may be ranked among the necessities of life
to every individual of human society.... The knowledge of them,
as established in use, is among the first elements of education,...*
John Quincy Adams, *Report to Congress*, 1821

Physics is concerned with the description of nature, and one way to describe things is to measure them. For example, you might measure the length of a table, and when you tell someone the time, you are actually measuring how much of the day has passed. There is a special significance to some physical descriptions and measurements, however. For instance, color is perceived because of a physiological response of the eye to light waves. But color perception is subjective and may vary from person to person. Objectively, the color of light is described and measured in terms of wavelengths or frequencies of the light waves, which are the same for all observers. Thus, physics describes nature in a verifiable, reproducible, and universally communicable manner.

Measurements are expressed in unit values, or units, and as you are probably aware, a large variety of units are used to express measured values. Some of the earliest units of measurement were referenced to parts of the human body; for instance, the hand is still used as a unit to measure the heights of horses. If a unit becomes officially accepted, it is called a **standard unit**. Traditionally, a head of state or a government establishes standard units.

A group of standard units and their combinations is called a **system of units**. Two major systems of units are in use today—the metric system and the British system. The latter is still widely used in the United States but will probably be phased out in favor of the metric system, which is used predominantly throughout the world. Different units in the same system or units from different systems can be used to describe the same thing. For example, your height can be expressed in inches, feet, centimeters, or meters. It is always possible to convert from one unit to another, however, and this is sometimes necessary.

Measured quantities are frequently used in calculations. Through the application of physical principles and mathematical analyses, physicists and engineers seek to solve problems and to further knowledge and understanding. For example, measuring the intensity of sunlight received by the Earth allows computation of the daily total energy output of the Sun, which can then be converted to the equivalent amount of matter or mass lost in the energy-producing process.

In a sense, mathematics is the language of physics. Principles can be expressed in words but are stated most clearly in mathematical terms. One equation is worth a thousand words, so to speak. But, in order to use equations to do calculations, you must have the proper background and understanding. Toward this end, you will begin your study of physics with a look at some basic quantities and units and an approach to problem solving. A firm understanding of these will benefit you throughout this entire course of study.

1.1 SI Units for Length, Mass, and Time

Length, mass, and time are fundamental physical quantities that describe a great many objects and phenomena. In fact, the topics of mechanics (the study of motion and force) covered in the first part of this book require *only* these physical quantities. The system of units used by scientists to represent these and other quantities is based on the metric system.

Historically, the metric system was the outgrowth of proposals for a more uniform system of weights and measures in France during the seventeenth and eighteenth centuries. In 1790, during the French Revolution, the Academy of Sciences was asked to develop an invariable standard for all measures and weights. Some of the units developed at that time were used in 1960 when an international committee adopted a set of definitions and standards for fundamental quantities. The newly established system involved extensive revision, simplification, and modernization of the metric system and is called the **International System of Units**, officially abbreviated as the **SI system** (from the French, *Le Système International des Unités*). The SI system includes base quantities and derived quantities, which are described by **base units** and **derived units**, respectively. Base units such as the meter and kilogram are easily represented by standards. Other quantities may be expressed in terms of combinations of base units and are called derived units. (As a familiar analogy, think of how we commonly measure the length of a trip in miles and express how fast we travel in the derived unit of miles/hour, which represents distance/time.)

One of the refinements of the SI system was the adoption of new standard references for some base units, including those for length, mass, and time.

Length

Length is the base quantity used to measure space. We commonly say that length is the distance between two points. But the distance between any two points depends on how space is traversed, which may be in a straight or a curved line.

The SI unit of length is the **meter (m)**. The meter was originally defined as 1/10,000,000 of the distance from the North Pole to the Equator along a meridian running through Paris [Fig. 1.1(a)]. A portion of this meridian between Dunkirk, France, and Barcelona, Spain, was surveyed to establish the standard length, which was assigned the name *metre* (from the Greek word *metron*,

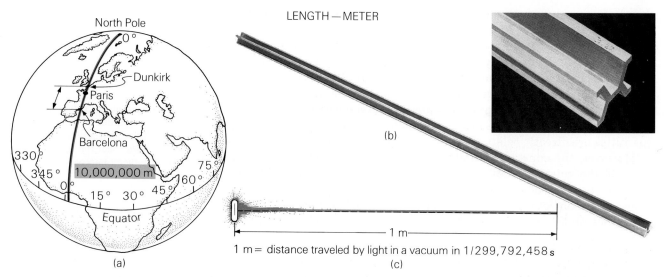

North Pole

0°

Dunkirk

Paris

Barcelona

330°

345°

10,000,000 m

0°

15° 30° 45° 60° 75°

Equator

(a)

(b)

1 m

1 m = distance traveled by light in a vacuum in 1/299,792,458 s

(c)

Figure 1.1 **The SI length standard—the meter** (a) The meter was originally defined as 1/10,000,000 (10^{-7}) of the distance from the North Pole to the Equator along a meridian running through Paris. A portion of this meridian between Dunkirk and Barcelona was surveyed to establish the standard length of the meter. Based on these results, a metal bar (called the Meter of the Archives) was constructed. (b) Prototype Meter No. 27 is the United States' copy of the original standard meter bar. It was sent from France in 1890. (c) The meter is currently defined in terms of the speed of light.

meaning "a measure"; the English spelling is *meter*). A meter is slightly longer than a yard (3.37 inches longer).

The length of the meter was initially preserved in the form of a material standard: the distance between two marks on a metal bar (made of a platinum-iridium alloy) that was stored under controlled conditions [Fig. 1.1(b)]. However, it is not desirable to have a reference standard which changes with external conditions, such as temperature. Research provided more accurate standards, and new definitions of the meter were adopted in 1960 and again in 1983. In 1960, the meter was defined in terms of the wavelength of light from krypton-86, an atomic property not affected by external influences.

In 1983, the meter was redefined in terms of another property of light: the length of the path traveled by light in a vacuum during an interval of 1/299,792,458 of a second [Fig. 1.1(c)]. Another way of saying this is that light travels 299,792,458 meters in a second, or that the speed of light in a vacuum is defined to be 299,792,458 meters per second. Note that the length standard is referenced to time, which is one of the most accurate measurements.

Mass

Mass is the base quantity used to describe amounts of matter. A common definition of mass is that it is the amount of matter of which an object is composed. The more massive an object, the more matter it contains. In general, you can think of mass as representing the number of atoms or molecules in an object.

The SI unit of mass is the **kilogram (kg)**. The kilogram was originally defined in terms of a specific volume of water (see Section 1.2), but is now referenced to a specific material standard: the mass of a prototype platinum-iridium cylinder kept at the International Bureau of Weights and Measures in Sèvres, France. The United States has a duplicate of the prototype cylinder, which serves as the standard mass for this country (Fig. 1.2).

You may have noticed that the phrase "weights and measures" is generally used instead of "masses and measures." This may be a reflection of the fact that there is commonly some confusion about the concepts of mass and weight. In the familiar British system, weight units are generally used instead of mass units, for example, pound (weight) instead of kilogram (mass). By definition, a weight is a measure of the gravitational attraction a celestial body, such as Earth, has for an object. To measure an object's mass, you would first weigh it to find its weight (w) and then compute its mass (m), since these quantities are directly related by a constant (g); that is, $w = mg$. This is somewhat like measuring the circumference (c) of a circle and then computing its diameter (d), since these quantities are directly related by the constant π ($c = \pi d$).

However, the constant g (and therefore weight) is directly related to the mass of the celestial body that supplies the gravitational attraction. For example, the value of g is less on the moon than on Earth (only about one-sixth as great). An object with a certain amount of mass has a particular weight on Earth, but on the moon, where the gravitational attraction is less, the same amount of mass will weigh less (only about one-sixth as much). *Weight is related to mass, but they are not the same. Mass is the base quantity.* Base quantities remain the same regardless of where they are measured, under normal or stationary conditions.*

The distinction between mass and weight will be fully explained in a later chapter. The discussion meanwhile will be chiefly concerned with mass.

Time

Time is a difficult concept to define. (Try it.) A common definition is that time is the forward flow of events. This is not so much a definition as an observation that time has never been known to run backwards (as it might appear to do when you view a film run backwards in a projector). Time is sometimes said to be a fourth dimension, accompanying the three dimensions of space. Thus, if something exists in space, it also exists in time. In any case, events can be used to mark time measurements. The events are analogous to the marks on a meterstick used for length measurements.

The SI unit of time is the **second (s)**. The solar "clock" was originally used to define the second. A solar day is the interval of time that elapses between two successive crossings of the same meridian by the Sun, at its highest point in the sky at that meridian [Fig. 1.3(a)]. A second was fixed as 1/86,400 of this apparent solar day (1 day = 24 h = 1440 min = 86,400 s). However, the elliptical path of the Earth's motion around the Sun causes apparent solar days to vary in length.

As a more precise standard, an average, or mean, solar day was computed from the lengths of the apparent solar days during a solar year. In 1956, the second was referenced to this mean solar day. But the mean solar day is not exactly the same for each yearly period because of minor variations in the Earth's motions and a steady slowing of its rate of rotation due to tidal friction. So scientists kept looking for something better.

In 1967, an atomic standard was adopted as a more precise reference. Currently, the second is defined as the duration of 9,192,631,770 cycles (periods) of the radiation associated with a particular transition of the cesium-133 atom. The atomic "clock" used to maintain this standard doesn't look much like an ordinary clock [Fig. 1.3(b), (c)].

* There are relativistic effects when objects move at high speeds, approaching the speed of light (see Chapter 26).

The constant g is called the acceleration due to gravity and has a relatively constant value of 9.8 m/s^2 near the Earth's surface (see Chapter 3).

MASS — KILOGRAM

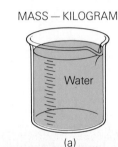

Water

(a)

(b)

Figure 1.2 **The SI mass standard—the kilogram** (a) Originally defined in terms of the mass of a specific volume of water, the standard kilogram is now defined by a metal cylinder. (b) Prototype Kilogram No. 20 is the national standard of mass for the United States and is kept under a double bell jar in a vault at the National Institute of Standards and Technology (formerly the National Bureau of Standards) in Washington, D.C.

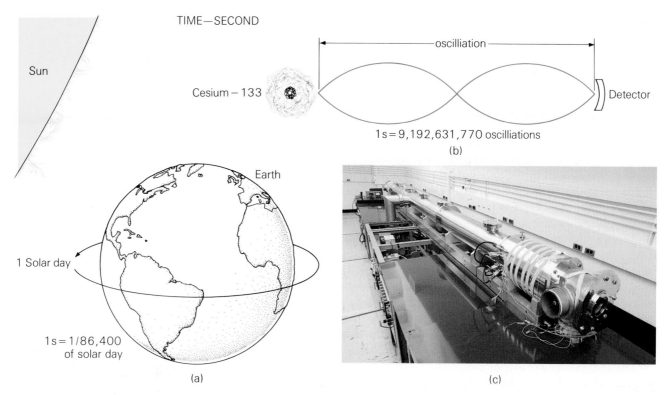

TIME—SECOND

Sun

oscilliation

Cesium – 133

Detector

1 s = 9,192,631,770 oscilliations

(b)

Earth

1 Solar day

1 s = 1/86,400
of solar day

(a)

(c)

Figure 1.3 **The SI time standard—the second**
(a) The second was originally defined in terms of the average solar day. (b) It is currently defined in terms of the frequency of the radiation associated with an atomic transition. (c) This atomic "clock" is the primary frequency standard at the National Institute of Standards and Technology. The device keeps time with an accuracy of about three millionths of a second per year.

SI Base Units

The SI system has the seven base quantities and units listed in Table 1.1. This is thought to be the smallest number of base quantities needed for a full description of everything observed or measured in nature. Also included in the table are two supplemental units for angular measure: the two-dimensional plane angle and the three-dimensional solid angle. There has not been general agreement about whether these geometric units are base or derived.

Table 1.1
SI Base and Supplemental Units

Quantity	Unit (and Abbreviation)
Base Units	
length	meter (m)
mass	kilogram (kg)
time	second (s)
electric current	ampere (A)
temperature	kelvin (K)
amount of substance	mole (mol)
luminous intensity	candela (cd)
Supplemental Units	
plane angle	radian (rad)
solid angle	steradian (sr)

1.2 More about the Metric System

The metric system involving the standard units of length, mass, and time now incorporated in the SI system was once called the **mks system** (for *meter-kilogram-second*). Another metric system that has been used in dealing with relatively small quantities is the **cgs system** (for *centimeter-gram-second*). In the United States, the system still generally in use is the British (or English) engineering system, in which the standard units of length, mass, and time are foot, slug, and second, respectively. You may not have heard of the slug, because gravitational force, or weight, is commonly used instead of mass (pounds instead of slugs) to describe quantities of matter. As a result, the British system is sometimes called the **fps system** (*foot-pound-second*).

The metric system is predominant throughout the world and is coming into increasing use in the United States. Primarily because of its mathematical simplicity, it is the preferred system of units for science and technology. SI units are used throughout most of this book. All quantities can be expressed in SI units. However, some units from other systems are accepted for limited use as a matter of practicality, for example, the time unit of hour and the temperature unit of degree Celsius. British units will sometimes be used in the early chapters for comparison purposes, since these units are still employed in everyday activities and for many practical applications.

The increasing worldwide use of the metric system means that you should be familiar with it. One of the greatest advantages of the metric system is that it is

Table 1.2
Multiples and Prefixes for Metric Units*

Multiple	Prefix (and Abbreviation)	Pronunciation
10^{18}	exa- (E)	ex´a (*a* as in *about*)
10^{15}	peta- (P)	pet´a (as in *petal*)
10^{12}	tera- (T)	ter´a (as in *terrace*)
10^{9}	giga- (G)	ji´ga (*a* as in *about*)
10^{6}	mega- (M)	meg´a (as in *megaphone*)
10^{3}	kilo- (k)	kil´o (as in *kilowatt*)
10^{2}	hecto- (h)	hek´to (*heck-toe*)
10	deka- (da)	dek´a (*deck* plus *a* as in *about*)
10^{-1}	deci- (d)	des´i (as in *decimal*)
10^{-2}	centi- (c)	sen´ti (as in *sentimental*)
10^{-3}	milli- (m)	mil´li (as in *military*)
10^{-6}	micro- (μ)	mi´kro (as in *microphone*)
10^{-9}	nano- (n)	nan´oh (*an* as in *ant*)
10^{-12}	pico- (p)	pe´ko (*peek-oh*)
10^{-15}	femto- (f)	fem´toe (*fem* as in *feminine*)
10^{-18}	atto- (a)	at´to (as in *anatomy*)

* For example, 1 gram (g) multiplied by 1000 (10^3) is 1 kilogram (kg), or multiplied by 1/1000 (10^{-3}) is 1 milligram (mg).

a decimal, or base-10, system. This means that larger or smaller units are obtained by multiplying or dividing a base unit by multiples of 10. A list of the various multiples and corresponding prefixes is given in Table 1.2. In decimal measurements, the prefixes milli-, centi-, and kilo- are the most commonly used ones. The decimal characteristic makes it convenient to change measurements from one size metric unit to another. With the familiar British system, different factors must be used, such as 16 for converting pounds to ounces and 12 for converting feet to inches.

You are already familiar with a base-10 system—U.S. currency. Just as a meter can be divided into 10 decimeters, 100 centimeters, or 1000 millimeters, the "base unit" of the dollar can be broken down into 10 "decidollars" (dimes), 100 "centidollars" (cents), or 1000 "millidollars" (tenths of a cent, or mils, used in figuring property taxes and bond levies). Since the metric prefixes are all in magnitudes of 10, there are no metric analogues for quarters or nickels.

The official metric prefixes can help eliminate confusion. In the United States, a billion is a thousand million (10^9); in Great Britain, a billion is a million million (10^{12}). The use of metric prefixes eliminates any confusion, since giga- indicates 10^9 and tera- stands for 10^{12}.

In the SI system, the standard unit of volume is the cubic meter (m^3). Since this is a rather large quantity, it is often more convenient to use the nonstandard unit of volume (or capacity) of a cube 1 dm (decimeter), or 10 cm (centimeters) on a side (Fig. 1.4). This volume unit was given the name litre, which is spelled liter in English. The volume of a **liter (L)** is 1000 cm^3 (10 cm × 10 cm × 10 cm), *and* since 1 L = 1000 mL (milliliters), it follows that 1 mL = 1 cm^3 (cubic centimeter is sometimes abbreviated cc, particularly in

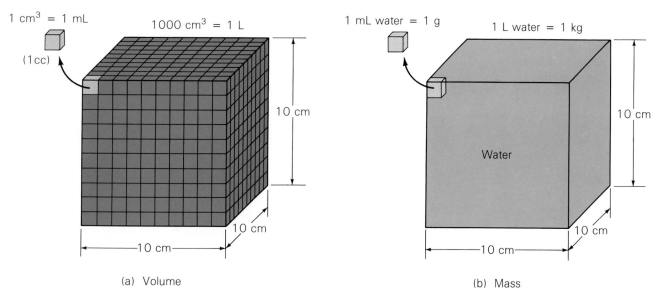

(a) Volume (b) Mass

Figure 1.4 **The liter and the kilogram**
Other metric units were derived from the meter. (a) A unit of volume (capacity) was taken to be a cube 1 dm (10 cm) on a side and was given the name liter. (b) The mass of a liter of water (at maximum density) was defined to be 1 kg. Note that the decimeter cube contains 1000 cm^3, or 1000 mL. Thus, 1 cm^3, or 1 mL, of water has a mass of 1 g.

Figure 1.5 **One, two, three, and one-half**
The metric liter is a common unit for soft drinks.

chemistry and biology.) The use of the liter is becoming quite common in the United States, as Fig. 1.5 indicates.

The standard unit of mass, the kilogram, was originally defined to be the mass of 1 liter of water at its maximum density (at $\cong 4\,°C$). Since 1 kg = 1000 g and 1 L = 1000 cm^3, then *1 cm^3 (or 1 mL) of water has a mass of 1 g*. Large quantities of mass are sometimes expressed in metric tons. A metric ton (or *tonne*) is 1000 kg.

Because the metric system is coming into increasing use in this country, you may find it helpful to have an idea of how metric and British units compare. The relative sizes of some units are illustrated in Fig. 1.6. The mathematical conversion from one unit to another will be discussed shortly.

Note that for the comparison of the kilogram and pound in Fig. 1.6, "approximately equal to" signs are used to reflect that these two units measure different things, that is, that this is a comparison of mass and equivalent weight.

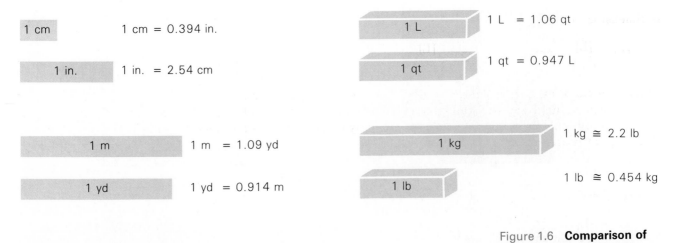

1 cm	1 cm = 0.394 in.
1 in.	1 in. = 2.54 cm
1 m	1 m = 1.09 yd
1 yd	1 yd = 0.914 m
1 km	1 km = 0.621 m
1 mi	1 mi = 1.61 km

1 L	1 L = 1.06 qt
1 qt	1 qt = 0.947 L
1 kg	1 kg $\cong$ 2.2 lb
1 lb	1 lb $\cong$ 0.454 kg

Figure 1.6 **Comparison of some SI and British units**
The bars illustrate the relative magnitudes of each pair of units. The comparison scales differ in each case.

1.3 Dimensional Analysis

The fundamental, or base, quantities used in physical descriptions are called dimensions. For example, length, mass, and time are dimensions. You could measure the distance between two points and express it in units of meters, centimeters, or feet. In any case, the quantity would have the dimension of length.

It is common to express dimensional quantities by bracketed symbols, such as [L], [M], and [T] for length, mass, and time, respectively. Derived quantities are combinations of dimensions; for example, velocity (v) has the dimensions [L]/[T] (think of miles/hour or meters/second), and volume (V) has the dimensions [L] × [L] × [L], or [L³]. Addition and subtraction can only be done with quantities that have the same dimensions, for example, $10\,\text{s} + 20\,\text{s} = 30\,\text{s}$, or [T] + [T] = [T].

Dimensional analysis is a procedure by which the dimensional consistency of any equation may be checked. You have used equations and know that an equation is a mathematical equality. Since physical quantities used in equations have dimensions, *the two sides of an equation must be equal not only in numerical magnitude, but also in dimensions*. And *dimensions can be treated as algebraic quantities*. That is, they can be multiplied or divided. For example, suppose the quantities in an equation have the dimension of length: $3\,\text{m} \times 4\,\text{m} = 12\,\text{m}^2$. This is [L] × [L] = [L²] dimensionally, and both sides of the equation are numerically and dimensionally equal.

One of the major advantages of dimensional analysis is its usefulness for checking whether an equation that has been derived or is being used in solving a problem has the correct form. For example, suppose that you think you can solve a problem about the distance an object travels using the equation $x = at$, where x represents distance, or length, and has the dimension [L], a represents acceleration and has the dimensions [L]/[T²], and t represents time and has the dimension [T]. First, before trying to use numbers with the equation, you can check to see if it is dimensionally correct:

$$x = at$$

is dimensionally

$$[\text{L}] = \frac{[\text{L}]}{[\text{T}^2]} \times [\text{T}] \qquad \text{or} \qquad [\text{L}] = \frac{[\text{L}]}{[\text{T}]}$$

which is obviously not true. Thus, $x = at$ *cannot* be a correct equation.

Dimensional analysis will tell you if an equation is incorrect, but a dimensionally consistent equation may *not* correctly express the real relationship of quantities. For example,

$$x = at^2$$

is dimensionally

$$[\text{L}] = \frac{[\text{L}]}{[\text{T}^2]} \times [\text{T}^2] \qquad \text{or} \qquad [\text{L}] = [\text{L}]$$

This equation is dimensionally correct. But, as you will see in Chapter 2, it is not physically correct. The correct form of the equation—both dimensionally and physically—is $x = \frac{1}{2}at^2$. The fraction $\frac{1}{2}$ has no dimensions; it is a dimensionless constant, like the constant π. [For a circle, the circumference (c) is related to the diameter (d) by $c = \pi d$, or $\pi = c/d$. Do you see why π is dimensionless?]

Using symbols for dimensions is fine, but in actual practice it is often more convenient to use units, which are often written into an equation along with the numerical magnitudes of the quantities. Units can also be treated as algebraic quantities in a form of dimensional analysis that can be called **unit analysis**. The following example demonstrates unit analysis.

Example 1.1 Unit Analysis

Analyze the following equations dimensionally; x is distance in meters (m), v is velocity, or speed, in meters/second (m/s), a is acceleration in (meter/second)/second, or meter/second^2 (m/s^2), and t is time in seconds (s): (a) $v = v_o + at$ (where v_o is the initial velocity) and (b) $x = v/2a$.

Solution

(a) The equation is

$$v = v_o + at$$

Putting the units into it gives

$$\frac{m}{s} = \frac{m}{s} + \left(\frac{m}{s^2} \times s\right) \qquad \text{or} \qquad \frac{m}{s} = \frac{m}{s} + \frac{m}{s}$$

The equation is dimensionally correct, since the only units on each side are meters per second. If an incorrect term had been used in the equation, the unit analysis would have shown the error.

(b) The equation

$$x = \frac{v}{2a}$$

is dimensionally

$$m = \frac{\left(\dfrac{m}{s}\right)}{\left(\dfrac{m}{s^2}\right)} = \frac{m}{s} \times \frac{s^2}{m} \qquad \text{or} \qquad m = s$$

which is *untrue*. In this case, the equation is incorrect. A correct equation for some situations is $x = v^2/2a$, which is dimensionally correct. (Do you agree?) ∎

Mixed Units

Unit analysis also allows you to check for mixed units. In general, in working problems, you should always use the same unit for a given dimension throughout a problem. For example, suppose that you wanted to measure the floor area of a rectangular room in order to buy a new carpet to fit it. To compute the area, you would measure the lengths of two adjacent sides of the room and multiply them. Would you measure the sides in different units and express the area in ft-yd or ft-m? Probably not. Such mixed units are not usually very useful.

Suppose that you used centimeters as the unit for x in this equation:

$$v^2 = v_o^2 + 2ax$$

Dimensionally, this would give

$$\left(\frac{m}{s}\right)^2 = \left(\frac{m}{s}\right)^2 + \left(\frac{m}{s^2} \times cm\right)$$

or $\quad \dfrac{m^2}{s^2} = \dfrac{m^2}{s^2} + \dfrac{m - cm}{s^2}$

which is dimensionally correct. That is, it is equivalent to

$$\frac{[L^2]}{[T^2]} = \frac{[L^2]}{[T^2]} + \frac{[L^2]}{[T^2]}$$

But the units are mixed (m and cm). The terms on the right-hand side could not be added together (and make sense).

Determining the Units of Quantities

Another aspect of unit analysis that is very important in physics is the determination of the units of quantities from defining equations. For example, density (ρ) is defined by the equation

$$\rho = \frac{m}{V} \tag{1.1}$$

where m is mass and V is volume. (Density is the mass per unit of volume and is a measure of the compactness of the mass of an object or substance.) What are the units of density? Since mass is measured in kilograms and volume in cubic meters in SI units, the defining equation

$$\rho = \frac{m}{V} \left(= \frac{kg}{m^3} \right)$$

gives the derived SI unit for density as kilograms per cubic meter (kg/m^3).

What are the units of π? Since the relationship of the circumference and the diameter of a circle is given by the equation $c = \pi d$, then $\pi = c/d$, and if length is measured in meters,

$$\pi = \frac{c}{d} \left(= \frac{m}{m} \right)$$

Thus, the constant π has no units because they cancel out. It is unitless.

1.4 Unit Conversions

Because different units in the same system or in different systems can express the same quantity, it is sometimes necessary to convert the units of a quantity from one unit to another, for example, from feet to yards or from inches to centimeters. You already know how to do many unit conversions. For example, if a room is 12 feet long, what is its length in yards? Your immediate answer is 4 yards.

What you do in this operation is to use a **conversion factor**. You know that 1 yd = 3 ft, and this can be written as a ratio: 1 yd/3 ft or 3 ft/1 yd. (The one is often omitted in the denominator of such ratios, for example, 3 ft/yd.) Note that dividing the equation 1 yd = 3 ft by 3 ft on both sides gives

$$\frac{1 \text{ yd}}{3 \text{ ft}} = \frac{3 \text{ ft}}{3 \text{ ft}} = 1$$

This illustrates that a conversion factor (ratio) always has a magnitude of 1. You don't change the magnitude, or size, of a quantity by multiplying it by 1. A conversion factor simply lets you express a quantity in terms of other units.

What is done in converting 12 feet to yards may be expressed mathematically as follows:

$$12 \text{ ft} \times \frac{1 \text{ yd}}{3 \text{ ft}} = 4 \text{ yd}$$

The feet cancel, giving yd = yd.

Suppose you are asked to convert 12 inches to centimeters. You may not know the conversion factor in this case, but you can get it from a table (such as Table 2 in Appendix I or the table inside the front cover of this book): 1 in. = 2.54 cm or 1 cm = 0.394 in. It makes no difference which form of the conversion factor is used. The question is whether to divide or multiply the given quantity to make the conversion. *In doing unit conversions, you should take advantage of unit analysis.*

Note that you may use either 1 in./2.54 cm or 2.54 cm/in. for the conversion factor in ratio form. The unit analysis tells you that in this case the second form works more easily:

$$12 \text{ in.} \times \frac{2.54 \text{ cm}}{\text{in.}} = 30.5 \text{ cm}$$

Alternatively, you can express the conversion factor as follows:

$$12 \text{ in.} \times \frac{1 \text{ cm}}{0.394 \text{ in.}} = 30.5 \text{ cm}$$

A few commonly used conversion factors are not dimensionally correct; for example, 1 kg = 2.2 lb is used for conversions, but the kilogram is a unit of mass and the pound is a unit of weight. What the conversion factor really means is that 1 kilogram is equivalent to 2.2 pounds, or a 1-kilogram mass has a weight of 2.2 pounds. In any case, 1 kg/2.2 lb is a convenient conversion factor for converting kilograms of mass to pounds of weight, and vice versa. (Mass and weight are related by a constant, and this is taken into account in the ratio given as the conversion factor.)

The inverse ratio gives 12 in. × (1 in./2.54 cm) = 4.7 in.^{2}/cm, which is dimensionally correct but is not a conversion to centimeters. The units in.^{2}/cm for length are nonstandard and a bit confusing. To use this ratio properly, the quantity should be divided by the conversion factor.

Example 1.2 Unit Conversion
Do the following unit conversions: (a) 15 meters to feet, (b) 30 days to seconds, and (c) 50 miles/hour to meters/second.

Solution
(a) From the conversion table, 1 m = 3.28 ft, so

$$15 \text{ m} \times \frac{3.28 \text{ ft}}{\text{m}} = 49.2 \text{ ft}$$

(b) The conversion factor for days and seconds is available from the table (1 day = 86,400 s), but you may not always have a table handy, so instead use several better-known conversion factors to get the result:

$$30 \text{ days} \times \frac{24 \text{ h}}{\text{day}} \times \frac{60 \text{ min}}{\text{h}} \times \frac{60 \text{ s}}{\text{min}} = 2.6 \times 10^6 \text{ s}$$

Note how unit analysis checks the conversion factors for you. The rest is simple arithmetic.

(c) In this case, from the conversion table, 1 mi = 1609 m and 1 h = 3600 s. (The latter is easily computed.) These ratios are used to cancel the units that are to be changed, leaving behind the ones that are wanted:

$$\frac{50 \text{ mi}}{\text{h}} \times \frac{1609 \text{ m}}{\text{mi}} \times \frac{1 \text{ h}}{3600 \text{ s}} = 22.3 \text{ m/s} \quad \blacksquare$$

Example 1.3 Conversion of Units of Area

Convert 2.5 m^2 to square centimeters (cm^2).

Solution

A common error in such conversions is the use of incorrect conversion factors. It is sometimes assumed that 1 m^2 = 100 cm^2, which is *wrong*. Unit analysis quickly reveals this:

$$2.5 \text{ m}^2 \times \left(\frac{10^2 \text{ cm}}{1 \text{ m}}\right)^2 = 2.5 \text{ m}^2 \times \left(\frac{10^4 \text{ cm}^2}{\text{m}^2}\right) = 2.5 \times 10^4 \text{ cm}^2$$

(Similarly, converting 2.5 m^3 to cubic centimeters gives 2.5×10^6 cm^3. The advantage of using the decimal metric system in calculations is evident.) $\quad \blacksquare$

1.5 Significant Figures

> *"What is the meaning of it all, Mr. Holmes?" "Ah. I have no data. I cannot tell,"* he said.
>
> Arthur Conan Doyle,
> *The Adventure of the Copper Beeches*, 1892

Unlike Sherlock Holmes, you will be given numerical data when asked to solve a problem. In general, such data are either exact numbers or measured numbers (quantities). **Exact numbers** include factors such as 100 as used to calculate a percentage and 2 in the equation $r = d/2$ relating the radius and diameter of a circle. **Measured numbers** are those obtained from measurement processes and generally have some uncertainty or error.

When calculations are done with measured numbers, the error of measurement is propagated, or carried along, by the mathematical operations. A question of how to report a result arises. For example, suppose that you are asked to find time (t) from the formula $x = vt$, and you are given that $x = 5.3$ m and $v = 1.67$ m/s. Then

$$t = \frac{x}{v} = \frac{5.3 \text{ m}}{1.67 \text{ m/s}} = ?$$

Doing the division operation on a hand calculator yields a result such as 3.173652695 (see Fig. 1.7). How many figures, or digits, should you report in the answer?

The error or uncertainty of the result of a mathematical operation may be computed by statistical methods. A simpler, widely used procedure for making an estimate of this uncertainty involves the use of **significant figures (sf)**, sometimes called significant digits. The degree of accuracy of a measured quantity depends on how finely divided the measuring scale of the instrument is. For example, you might measure the length of an object as 2.5 cm with one instrument and 2.54 cm with another; the second instrument provides more significant figures and a greater degree of accuracy. In general, *the number of significant figures of a numerical quantity is the number of reliably known digits it contains*. For a measured quantity, this is usually defined as all of the digits that can be read directly from the instrument used in making the measurement plus one uncertain digit that is obtained by estimating the fraction of the smallest division of the instrument's scale.

Of the measured quantities mentioned above, 2.5 cm has two significant figures and 2.54 cm has three significant figures. Other examples are 0.074 (2 sf), 0.0704 (3 sf), and 1.074 (4 sf). Note that in the first two of these examples, the two leftmost zeros simply locate the decimal point and aren't significant figures. Intermediate zeros (those between two other digits) are measured numbers and therefore are significant.

A typical problem in determining the number of significant figures occurs with numbers having one or more zeros at the rightmost end. For example, a car may have a reported mass of 2500 kg. Since the zeros may or may not be significant, a question arises about the number of significant figures in this quantity. This ambiguity is removed by using powers of 10, or scientific notation. If the mass of the car is reported as 2.5×10^3 kg, then the quantity has two significant figures; if 2.50×10^3 kg is reported, then the number of significant figures is three.

The following general rule tells how to determine the number of significant figures in the result of a multiplication or division:

> The final result of a multiplication or division operation should have the same number of significant figures as the quantity with the least number of significant figures that was used in the calculation.

What this means is that the result of a calculation can be no more accurate than the least accurate quantity used; that is, you cannot gain accuracy in performing mathematical operations. Thus, the result that should be reported for the division operation discussed above is

$$\frac{5.3 \text{ m}}{1.67 \text{ m/s}} = 3.2 \text{ s}$$

The result is rounded off to two significant figures (see Fig. 1.7).

For multiple operations, rounding off to the proper number of significant figures should not be done at each step since rounding errors may accumulate. It is usually suggested that one or two insignificant figures be carried along, or if a calculator is being used, rounding off may be done on the final result of the multiple calculations.

A calculator may have a floating decimal point and list up to ten figures or digits. The number of digits displayed on some calculators can be preset.

Figure 1.7 **Significant and insignificant figures**

For the division operation of 5.3/1.67, a calculator with a floating decimal point gives an ample number of figures. A calculated quantity can be no more accurate than the least accurate quantity involved in the calculation, so this result should be rounded off to two significant figures, that is, 3.2.

Rounding off is done like this: If the next digit after the last significant figure is 5 or greater, the last significant figure is increased by 1. If the next digit is less than 5, the last significant figure is left unchanged.

There is also a general rule for determining significant figures for the results of addition and subtraction:

The final result of the addition or subtraction of numbers should have the same number of decimal places as the quantity with the least number of decimal places that was used in the calculation.

This rule is applied by rounding off numbers to the least number of decimal places before adding or subtracting, for example:

$$23.1 \text{ m} + 0.546 \text{ m} = 23.1 \text{ m} + 0.5 \text{ m} = 23.6 \text{ m}$$

and

$$157 \text{ kg} - 5.5 \text{ kg} = 157 \text{ kg} - 6 \text{ kg} = 151 \text{ kg}$$

Basically, the number of digits reported in a result depends on the number of digits in the given data. The rules for rounding off will generally be observed in the solutions given for the examples in this book. Rounding off results is a good practice to follow when you work problems. In general, the examples and problems will involve quantities with two or three significant figures.

You should keep in mind, however, that the ultimate criterion for rounding off an answer is relative uncertainty and not the number of significant figures in the data. The significant figure approximation is convenient for most purposes, but there are situations where it is practical to assume that some numbers have equal relative uncertainties. For example, suppose that the dimensions of a tabletop are measured as 0.97 m and 1.08 m with equal accuracy. The area of the tabletop is then

$$0.97 \text{ m} \times 1.08 \text{ m} = 1.0476 \text{ m}^2 = 1.05 \text{ m}^2$$

Rigorous application of the results for significant figures would give a result of 1.0 m^2. However, as in this case, it is sometimes clearer to retain the "insignificant" digit. This assumes a greater number of significant figures than are given. Such an assumption is convenient when rounding off would eliminate numbers on one side of the decimal point; for example, $5.09 \times 6.2 = 31.558 = 31.6$ (or 32 by the rounding rule for multiplication). Retaining an extra digit will be done in examples when scientific notation is inappropriate and a calculation is done in steps.

Rounding off of results of intermediate steps of multiple operations will be done in some examples for instructional purposes.

1.6 Problem Solving

One of the most important aspects of physics is problem solving. In general, this involves the application of physical principles and equations to given data in order to find some unknown or wanted quantity. There is no set method for approaching the solution of problems. In fact, more than one method may be appropriate for solving a particular problem.

One of the most important aids in problem solving is having a thorough understanding of the problem. All too often a student will start working on a problem without really understanding the situation described or knowing what is to be found. (Basically, if you don't know exactly what you want, it is very difficult to find it.) Often coupled with a lack of understanding of what is wanted in a problem is the misapplication of physical principles or equations. Principles and equations generally are subject to certain conditions or are limited in their applicability to physical situations.

The following stepwise procedure will assist you in solving problems.

Suggested Problem-Solving Procedure

1. Read the problem carefully and analyze it. *Write down the given data and what it is you are to find.* [Some data may not be given explicitly in numerical form. For example, if a car "starts from rest," its initial speed is zero ($v_o = 0$). In some instances, you may be expected to know certain quantities or to look them up in tables.]
2. Decide what principles and equations are applicable to obtaining the desired result (that is, getting from the information given to what is to be found). *It may be helpful to draw a diagram as an aid in visualizing and analyzing the physical situation of the problem.*
3. Check units before doing calculations. Make unit conversions if necessary so that all units are in the same system and quantities with the same dimensions have the same units (preferably standard units). This avoids mixed units and is helpful in unit analysis. (Unit checking and conversions are often done in Step 1 when writing the data.)
4. Substitute given quantities into equation(s) and perform calculations. *Report the result with proper units and number of significant figures.* Consider whether the result is reasonable, that is, whether it has an appropriate magnitude. (This means "in the right ball park.") For example, if a person's calculated height turns out to be 5.2 m, the result should be questioned since 5.2 m is approximately 17 ft.

The following examples illustrate the procedure. The steps are numbered to help you follow along.

Example 1.4 Finding the Area of a Rectangle
Two students measure the lengths of adjacent sides of their dorm room. One reports 15 ft, 18 in., and the other reports 4.25 m. What is the area of the room in square meters?

Solution
1. *Given:* Length $= l = 15$ ft, 8 in.   *Find:* Area (in m^2)
   Width $= w = 4.25$ m

2. It is well known that $A = l \times w$.

3. It is obvious that a unit change is necessary. Using commonly known conversion factors gives

$$15 \text{ ft} + 8 \text{ in.} = \left(15 \text{ ft} \times \frac{12 \text{ in.}}{\text{ft}} \right) + 8 \text{ in.} = 188 \text{ in.}$$

and

$$188 \text{ in.} \times \frac{2.54 \text{ cm}}{\text{in.}} = 478 \text{ cm} = 4.78 \text{ m}$$

4. Then

$$A = l \times w = 4.78 \text{ m} \times 4.25 \text{ m} = 20.315 \text{ m}^2 = 20.3 \text{ m}^2$$

5. Rounded to the proper number of significant figures, the result is 20.3 m². [Suppose that you inadvertently punched 47.8 instead of 4.78 on your calculator. The result would be $A = 47.8\,\text{m} \times 4.25\,\text{m} = 203\,\text{m}^2$. A room with an area of about 200 m² would have to have dimensions of about 10 m by 20 m, which is roughly 30 ft by 65 ft. Since this is not the size of a typical dorm room, the magnitude of the result would make you suspect there is an error.] ∎

Example 1.5   Finding the Length of One Side of a Triangle
If $x = 0.54$ m and $y = 0.49$ m as shown in Fig. 1.8, what is the length of r?

Solution
1. *Given*: $x = 0.54$ m   *Find*: r (in m)
   $y = 0.49$ m
   values for angles shown in figure

2. Glancing hastily at the figure, you might think that r can be found readily by applying the Pythagorean theorem; that is, $r = \sqrt{x^2 + y^2}$. But, since the triangle is not a right triangle, that theorem does not apply. Therefore, the law of sines or the law of cosines should be used (see Table 3 in Appendix I for a review of these laws). Note that the angle opposite r is $\theta = 180° - (40° + 45°) = 95°$. The law of sines is

$$\frac{b}{\sin \beta} = \frac{c}{\sin \gamma}$$

For this problem, $b = y$, $c = r$, $\beta = 40°$, and $\gamma = \theta = 95°$. Thus,

$$\frac{y}{\sin 40°} = \frac{r}{\sin 95°}$$

3. Since the sine of any angle is a dimensionless number and y is given in meters, no unit conversions are needed to get r in meters.

4. $$r = \frac{y \sin 95°}{\sin 40°}$$

$$= \frac{(0.49\ \text{m})(0.996)}{0.643}$$

5. $$= 0.76\ \text{m} \quad ∎$$

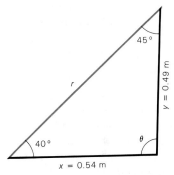

Figure 1.8 **The length of a triangle's side**
See Example 1.5.

Example 1.6   Finding the Capacity of a Cylinder
A cylindrical container with a height of 45.5 cm and an inside diameter of 12.0 cm is filled with water. What is the mass of the water in kilograms?

Solution
1. *Given*: $h = 45.5$ cm   *Find*: m (in kg)
   $d = 12.0$ cm

2. The problem involves mass and volume, so you should think of density: $\rho = m/V$, or $m = \rho V$. The density of water is not given; however, you can look it up in a table, or you may remember from the definition of the liter that $\rho_{\text{H}_2\text{O}} = 1.00$ g/cm³ (at maximum density, which will be assumed here). Therefore, you need to compute the volume of the inside of the cylinder to solve the problem. If you can't remember the formula for the volume of a cylinder,

Given: *Find:*

h = 45.5 cm m (in kg)
d = 12.0 cm

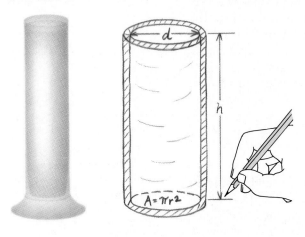

Figure 1.9 **The capacity of
a cylinder**
See Example 1.6.

drawing a diagram like the one in Fig. 1.9 may be helpful. (Formulas for areas
and volumes of common shapes can be found in Table 5 of Appendix I.) The area
of the circular end of the cylindrical volume is $A = \pi r^2 = \pi d^2/4$ (since $r = d/2$),
and the volume is this area times the height of the cylinder, or $V = Ah$. Thus,

$$m = \rho V = \rho Ah = \rho\left(\frac{\pi d^2}{4}\right)h$$

3 and 4. $= \left(\frac{1.00\ \text{g}}{\text{cm}^3}\right)\left(\frac{\pi(12.0\ \text{cm})^2}{4}\right)(45.5\ \text{cm}) = 5.15 \times 10^3\ \text{g}$

5. $= 5.15\ \text{kg}$ ■

Important Formula

Density: $\rho = \dfrac{m}{V}$

Questions

Units and the Metric System

1. Suppose some quantity cannot be described by a
derived unit. What does this imply?

2. Why is mass and not weight a base quantity?

3. Why is time sometimes called the fourth dimension?

4. Explain what is meant by the acronyms mks, cgs,
and fps.

5. When a portion of the Paris meridian was measured
to determine the length of the meter, the result was slightly
different from the length determined by more recent
measurements. What effect does this have on the metric
length standard? Explain.

6. How was the original metric mass standard related
to the length standard?

7. Using the appropriate metric prefixes, express a metric ton in (a) kilograms and (b) grams.

Dimensional Analysis

8. Can dimensional analysis tell you whether you have used the correct equation in solving a problem? Explain.

9. Is the unit equation kg-cm-s/g-min^2 = m/s dimensionally correct? If so, explain how this can be with so many different units.

10. How many dimensions are there in the SI system? Explain.

11. Explain the difference in dimensions between a conversion factor expressed in equation form and in ratio form. What is the numerical value of a conversion factor in ratio form?

Unit Conversions

12. Which is longer, a 50-yd dash or a 50-m dash? How much longer, expressed as a percentage?

13. The width of the nail on your little finger is about the size of a common metric unit. Which one?

14. If you wanted to express your height as a larger number, which unit would you use: (a) meter or yard, (b) decimeter or foot, (c) centimeter or inch?

15. Would the gas consumption of an automobile appear lower if it were expressed in mi/gal or km/gal? Explain.

Significant Figures

16. If a measured length is reported as 25.483 cm, could this result have been measured with an ordinary meterstick whose smallest division is millimeters? Discuss in terms of significant figures.

17. Explain why rounding off should be done on the final result of a calculation involving multiple steps rather than on the results of each intermediate step.

18. Accuracy is the difference between a measured quantity and the true (or accepted) value. Strictly speaking, true values are never known except when discrete objects are counted and for defined quantities. Explain why, and give an example of each of the exceptions.

19. Why should results of calculations be reported with not only the proper number of significant figures but also the units? Give some practical examples where not doing so might lead to serious difficulties.

Problems

1.3 Dimensional Analysis*

■1. Show that the equation $A = 4\pi r^2$, where A is the surface area and r is the radius of a sphere, is dimensionally correct.

■2. Show that the equation $x = x_0 + vt$, where v is velocity and x and x_0 are lengths, is dimensionally correct.

■3. Is the equation $V = \pi d^2/4$ dimensionally correct? (V is volume and d is diameter.) If not, how could it be made so, assuming that its right-hand side is correct?

■4. Is the equation $t = 2x/v$, where t is time, x is length, and v is velocity, dimensionally correct?

■5. If $x = gt^2/2$, where x is length and t is time, is dimensionally correct, what are the units of the constant g? Is g a particular derived quantity? If so, identify it.

■■6. Show that the equation $x = x_0 + v_0t + \frac{1}{2}at^2$ is dimensionally correct (a is acceleration, v_0 is velocity, and x and x_0 are lengths).

■■7. Show that the equation $v = (2y/t) - v_0$ is dimensionally correct (v and v_0 are velocities, y is length, and t is time).

■■8. Show that the equation $a = (v^2 - v_0^2)/2x$ is dimensionally correct (a is acceleration, v and v_0 are velocities, and x is length).

■■9. Using unit analysis, determine whether the following equations are dimensionally correct (t is time, x is length, a is acceleration, and v is velocity): (a) $t = \sqrt{2x/a}$, and (b) $v = \sqrt{2ax}$.

■■10. Is the equation for the area of a trapezoid, $A = \frac{1}{2}a(b_1 + b_2)$, where a is the altitude and b_1 and b_2 are the bases, dimensionally correct? (See Fig. 1.10.) If not, how should it be changed to correct it?

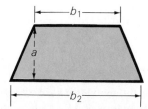

Figure 1.10 **The area of a trapezoid**
See Problem 10.

* Dimensions and/or units of velocity and acceleration are given in the chapter.

■■11. (a) Is the equation $V = 4\pi r^3$, where V is the volume and r is the radius of a sphere, dimensionally correct? (b) Is this the correct equation for the volume of a sphere? Explain.

■■12. Using unit analysis, determine if the following equation is dimensionally correct: $v = \sqrt{2ax - v_0 t}$, where a is acceleration, v and v_0 are velocities, and x is length.

■■13. The formulas for a straight line, a parabola, and a hyperbola are commonly given as (a) $y = mx$, (b) $y = bx^2$, and (c) $y = c/x$, respectively, where m, b, and c are constants. What are the units of each of the constants if x and y are lengths?

■■14. In the equation $mg = kx$, g is acceleration and k is a constant called the spring constant. What are the SI units of the spring constant?

■■15. The equation for the distance traveled by an object is given by $x = v_0 t + bt^2$, where v_0 is velocity, t is time, and b is a constant. What are the SI units of b?

■■16. (a) Show that the equation $A = 2(\pi r^2) + 2\pi rL$, where r is a radius and L is a length, is dimensionally correct. (b) Prove that this formula gives the total area of a solid circular cylinder.

■■17. Newton's second law of motion is expressed by the equation $F = ma$, where F represents force, m is mass, and a is acceleration. (a) The SI unit of force is, appropriately, the newton (N). What are the units of the newton in terms of base quantities? (b) Using the result of part (a), show that the equation $Ft = mv$, where v is velocity and t is time, is dimensionally correct.

■■18. Newton's law of gravitation is expressed by the equation $F = Gm_1 m_2 / r^2$, where F is force, G is a constant (called the universal gravitational constant), m_1 and m_2 are masses, and r is a distance. (a) What are the units of the constant G when the force is in newtons? (b) As given in Problem 17, $F = ma$, where m is mass and a is acceleration. What are the units of G in SI base quantities?

1.4   Unit Conversions*

■19. What is your height in centimeters?

■20. How many meters are there in (a) 25 ft and (b) 5280 ft?

■21. Do the following conversions: (a) 15 m to feet, (b) 12 in. to centimeters, and (c) 30 days to seconds.

---

* Conversion factors are listed in Table 2 of Appendix I.

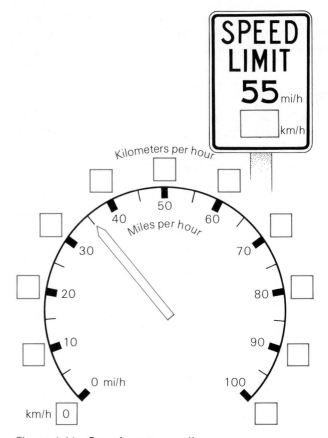

Figure 1.11   **Speedometer readings**
See Problem 23.

■22. A road sign indicates that a town is 50 km away. How far away is it in miles?

■■23. An automobile speedometer is shown in Fig. 1.11. (a) What would the equivalent scale readings be in km/h? (b) What would a 55 mi/h speed limit be in km/h?

■■24. Which is longer and by how many centimeters, a 100-m dash or a 100-yd dash?

■■25. Suppose that a 16-oz soda and a 500-mL soda sell for the same price. Which would you choose to get the most for your money and how much more (in milliliters) would you get? (Hint: 1 pint = 16 oz.)

■■26. A professor regularly buys 10 gal of gas, but the gas station got new pumps that deliver liters. How many liters of gas (rounded off to a whole number) should she ask for?

■■27. (a) What is the mass in kilograms of a 165-lb person? (b) What is your mass in kilograms?

■■**28.** Some common product labels are shown in Fig. 1.12. From the units given on the labels, form a set of conversion factors.

Figure 1.12 Conversion factors
See Problem 28. (Notice the difference in the gram amounts on the cans. This is due to a rounding error. Which is correct? 1 lb = 0.4535924 kg.)

■■**29.** (a) A football field is 300 ft long and 160 ft wide. What are the field's dimensions in meters and its area in square centimeters? (b) A football is 11 to $11\frac{1}{4}$ in. long. What is its length in centimeters?

■■**30.** The width and length of a room are 3.5 yd and 4.2 yd. (a) What is the area of the floor in square meters? (b) If the height of the room is 8.0 ft, what is the volume of the room in (1) cubic meters and (2) cubic feet?

■■**31.** In the Bible, Noah is instructed to build an ark 300 cubits long, 50.0 cubits wide, and 30.0 cubits high. (A cubit was a unit of length based on the length of the forearm and equal to half of a yard.) What would the dimensions of the ark be in meters? What would its volume be in cubic meters? Assume that the ark was rectangular.

■■■**32.** Suppose that when the United States goes completely metric, the dimensions of a football field are established as 100 m by 54 m. What would be the difference between the area of the metric football field and that of a current football field (see Problem 29)?

■■■**33.** The metal mercury has a density of 13.6 g/cm³. How much mercury would be required to fill a 0.500-L container? Compare this to the mass of the same volume of a soft drink. (Use the density of water.)

■■■**34.** Engineers often express the density of a substance as a weight density, for example, in lb/ft³. (a) What is the weight density of water? (b) What is the weight of a gallon of water?

1.5 Significant Figures

■**35.** Determine the number of significant figures in the following measured numbers: (a) 2.05 cm, (b) 810.2 kg, (c) 0.033 μs (microseconds), (d) 167.043 m.

■**36.** Express each of the numbers in Problem 35 with two significant figures.

■**37.** Express each of the following numbers with three significant figures: (a) 17.48 m, (b) 608.4 km, (c) 0.0013025 kg, (d) 93,000,000 mi.

■■**38.** Express the length 64,018 μm (micrometers) in centimeters, decimeters, and meters to three significant figures.

■■**39.** A circular flower bed has a radius of 5.45 m. Compute its (a) circumference and (b) area.

■■**40.** The parallel bases of a trapezoid are 8.5 cm and 5.68 cm, respectively, and the altitude is 3.25 cm. What is the area of the trapezoid? (Hint: see Problem 10.)

■■**41.** A spherical communications satellite has a diameter of 1.56 m. How much space does the satellite occupy?

■■**42.** The outside dimensions of a cylindrical soda can are reported as 6.744 cm for the diameter and 12.3 cm for the height. What is the total outside area of the can?

■■**43.** Assume that the radius of the Earth is 4.00×10^3 mi and that it is a perfect sphere. How many cubic meters of space does the Earth occupy?

1.6 Problem Solving

■**44.** In checking a car's gas consumption over a period of time, the owner finds that 32.0 gal of gasoline were used to travel 756 mi. How many miles per gallon (mpg) does the car get?

■**45.** The mass of the Earth is 6.0×10^{24} kg, and it has a volume of 1.1×10^{21} m³. What is the Earth's average density?

■**46.** Determine whether each of the following reported measurements has an appropriate magnitude. (a) A subcompact car has a mass of 2050 kg. (b) A person is 183 cm tall. (c) A runner runs a mile in 3.00×10^8 μs. (d) A normal highway driving speed is 25 m/s. (e) A dachshund has a tail 280 mm long and stands 5.0 dm off the ground. (f) A professor has a mass of 98 kg and a 50-cm waist.

■■**47.** A car travels with a constant speed of 55 km/h. How many kilometers will it travel in 4 h, 45 min? 260 km

■■**48.** The thickness of a textbook, not including the covers, is measured as 3.54 cm. If the last page of the book is numbered 826, what is the average thickness of a page?

■■**49.** A light year is a unit of distance corresponding to the distance light can travel in a vacuum in 1 year. If the speed of light is 3.0×10^8 m/s, what is the length of a light year in kilometers?

■■**50.** Joe's Pizza Palace sells a 9-in. (diameter) pizza for $7.95 and a 12-in. pizza for $13.50. Which is the better buy?

■■■**51.** In 1986, the experimental aircraft *Voyager* flew around the world without refueling. It used approximately 1.2×10^3 gallons of fuel to travel 2.5×10^4 miles in 9.1 days. (a) What was *Voyager*'s average mileage rating in mi/gal and km/L? (b) What was its average fuel consumption in gal/h and L/h? (c) What was its average speed in mi/h and km/h?

■■■**52.** A flat porch roof measuring 4.0 m by 9.0 m collapsed when snow had accumulated on it to a depth of 46 cm. If the average density of the snow was 15 kg/m³, what was the weight in pounds of the snow on the roof when it collapsed?

Additional Problems

53. Is the equation $v^2 = (4x^2/t^2) - v_o^2$, where v and v_o are velocity, x is length, and t is time, dimensionally correct?

54. A thick-walled spherical metal shell has an inner diameter of 20 cm and an outer diameter of 25 cm. What is the volume occupied by the shell itself?

55. Which holds more soda and how much more, a half-gallon bottle or a two-liter bottle?

56. A fathom is a nautical measure of length or depth and is equal to 6 ft. A furlong is a unit of distance used in horseracing and is equal to $\frac{1}{8}$ mi. What are the values of each of these distances expressed in a multiple of the meter that allows the numerical part of the result to be as close as possible to 1?

57. The top of a rectangular table measures 1.450 m by 0.855 m. (a) What is the smallest division on the scale of the measurement instrument? (b) What is the area of the tabletop?

58. A basketball team from the United States has a center who is 6 ft 7 in. tall and weighs 190 lb. If the team plays exhibition games in Europe, what values will be listed for the center's height and weight on programs for fans there?

59. A cylindrical drinking glass has a diameter of 8.0 cm and a depth of 10 cm. If a person drinks a completely full glass of water, how much of a liter will be consumed?

60. When computing the average speed of a cross-country runner, a student gets 20 m/s. Is this a reasonable result? Justify your answer.

61. Using the conversion factor 1 in. = 2.54 cm and other generally known factors, convert 6.67×10^3 mi/h to m/s.

62. The average density of the moon is 3.3 g/cm³, and it has a diameter of 2160 mi. What is the total mass of the moon?

63. Using unit analysis, show that the equation $t = x/(v_o + at/2)$ (where v_o is velocity, a is acceleration, x is length, and t is time) is dimensionally correct.

64. A solid sphere has a radius of 10 cm. What is its surface area in (a) square centimeters and (b) square meters?

65. Assuming that the Earth is a sphere and that the original definition of the meter is accurate, what is the diameter of the Earth?

66. The general equation for a parabola is $y = ax^2 + bx + c$, where a, b, and c are constants. What are the units of each constant if y is in meters?

67. A rectangular block has the dimensions 4.8 cm, 6.52 cm, and 15.51 cm. What is the volume of the block in cubic centimeters?

68. Which is greater and by how much, a metric ton or a U.S. ton (2000 lb)?

Kinematics: The Description of Motion

2

The description of motion involves the representation of a restless world. You sit, apparently at rest, but your blood flows, and air moves into and out of your lungs. The air is composed of gas molecules moving at different speeds and directions. Also, while you experience stillness, you, your chair, the building, and the air you breathe are all revolving through space with the Earth, part of a solar system in a spiraling galaxy in an expanding universe.

The branch of physics concerned with the study of motion and what produces and affects it is called **mechanics**. The roots of mechanics and of human interest in motion go back to early civilizations. The study of the motions of heavenly bodies, or celestial mechanics, grew out of the need to measure time and location. Several early Greek scientists, notably Aristotle, put forth theories of motion that were useful descriptions but were later proved to be inaccurate. Currently accepted concepts of motion were formulated in large part by Galileo (1564–1642) and Isaac Newton (1642–1727).

Mechanics is generally divided into two parts: kinematics and dynamics. **Kinematics** deals with the description of the motion of objects without consideration of what causes the motion. **Dynamics** takes into account, or includes, the causes of motion. This chapter covers kinematics and reduces the description of motion to its simplest terms by considering the simple case of motion in a straight line, or linear motion, which is motion in one dimension (of space). Chapter 3 focuses on motion in two dimensions (which can easily be extended to three dimensions). Chapter 4 investigates dynamics to show what causes motion.

2.1 A Change of Position

What is motion? This seems a simple question, but you might have some difficulty giving an immediate answer. After a little thought, you should conclude that **motion** is the changing of position.

For straight-line, or linear, motion, it is convenient to specify position using the familiar Cartesian coordinate system with x and y axes at right angles. The straight-line path may be in any direction, but for simplicity the motion is usually described as being along the x or y axis.

Motion is described by specifying how far something travels in changing its position. This may be expressed in terms of distance or displacement. The distinction between these terms is important in distinguishing speed and velocity. **Distance** is simply the total path length traveled in moving from one location or point to another. For example, in Fig. 2.1(a), the distance between points x_1 and x_2 on the graph is 8.0 m. The distance between these points is the same whether the motion is from x_1 to x_2 *or* from x_2 to x_1. Distance is a **scalar** quantity, that is, one with magnitude, or numerical value, only (expressed in units).

However, the direction of motion may also be specified. Suppose that the person moves from x_1 to x_2, toward the physics lab door, as shown in Fig. 2.1(b). The **displacement** is the straight-line distance between two points *plus* direction. Displacement is a **vector** quantity, or one with both magnitude and

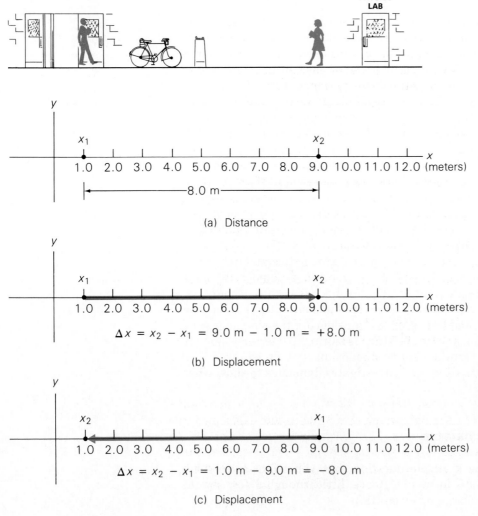

(a) Distance

$\Delta x = x_2 - x_1 = 9.0\ \text{m} - 1.0\ \text{m} = +8.0\ \text{m}$

(b) Displacement

$\Delta x = x_2 - x_1 = 1.0\ \text{m} - 9.0\ \text{m} = -8.0\ \text{m}$

(c) Displacement

Figure 2.1 **Distance and displacement**
(a) The distance (actual path length) between x_1 and x_2 is 8.0 m and is a scalar quantity.
(b) The displacement is $+8.0$ m (in the positive x direction) and is a vector quantity.
(c) The displacement is -8.0 m (in the negative x direction). Note how the sign gives the direction for straight-line motion.

24   CH. 2   Kinematics: The Description of Motion

direction. Given x_1 as the initial position and x_2 as the final position, the displacement (Δx) is

$$\Delta x = x_2 - x_1$$
$$= 9.0 \text{ m} - 1.0 \text{ m} = +8.0 \text{ m}$$

In this case, the displacement (magnitude and direction) is 8.0 m in the positive x direction, as indicated by the plus sign of the result; that is, the motion is in the positive x direction. (The plus sign of a result is commonly omitted, being taken as understood.)

Note in Fig. 2.1(b) that the displacement vector is conveniently represented by an arrow. The length of the arrow is proportional to the magnitude of the displacement, and the arrowhead indicates its direction.

Suppose that the other person in Fig. 2.1 moves from x_1 to $x_2 = 1.0$ m, as indicated in Fig. 2.1(c). In this case, the displacement is

$$\Delta x = x_2 - x_1$$
$$= 1.0 \text{ m} - 9.0 \text{ m} = -8.0 \text{ m}$$

The minus sign indicates that the direction of the displacement is in the negative x direction.

It is customary to indicate vector quantities in boldface type, for example, $\Delta \mathbf{x}$ (or by writing an arrow over the quantity, $\Delta \vec{x}$). However, in doing computations concerning linear motion, these notations do not have to be used if the directions of the vector quantities are specified as positive and negative values. (Chapter 3 considers vectors in more detail.)

2.2 Speed and Velocity

When something is in motion, its position changes with time. That is, it moves a certain distance in a given time. Both length and time are therefore important considerations in describing motion. For example, imagine a car and a pedestrian moving down a street and traveling a distance (length) of one block. In general, the car travels faster, or covers the distance in a shorter time, than the person does. This can be expressed by using length and time to give the time rate of change in position for each.

Average speed is the distance traveled divided by the total time elapsed in traveling that distance:

$$\text{average speed} = \frac{\text{distance traveled}}{\text{total time to travel that distance}}$$

or

$$\bar{v} = \frac{\Delta d}{\Delta t} \tag{2.1}$$

where the symbol d is used for distance, the actual path length, which does not have to be in a straight line. For example, you may have used the distance your car traveled on a trip to calculate the average speed you attained; you get this distance from the starting and ending odometer readings, and it reflects the actual path length. The standard units of speed are m/s (length/time), although km/h and mi/h are used in everyday applications. (The British standard unit is ft/s.)

Note that the symbol for speed is written with an overbar ($\bar{v}$) in Eq. 2.1 to indicate that it is in general an average over a time interval (Δt). Suppose that a car travels 100 m in 25 s. Then its average speed is $\bar{v} = 100$ m/25 s $= 4.0$ m/s. The car does not necessarily travel 4.0 m each second. It may have started from rest and would then have to travel at speeds greater than 4.0 m/s during some seconds to have that be the average value. The average speed gives only a general description of motion and is analogous to an average test score for a class.

If the time interval (Δt) considered becomes smaller and smaller and approaches zero, the speed calculation is an **instantaneous speed**. This is how fast something is moving at an instant of time. The speedometer of a car gives an approximate instantaneous speed (see Fig. 2.2). If something moves with a constant, or uniform, speed, the average and instantaneous speeds are equal. (Do you agree? Think about the average test score analogy.)

Like the distance it incorporates, speed is a scalar quantity (having only magnitude). Another quantity used to describe motion is velocity. Speed and velocity are often used synonymously, but they have quite different meanings in physics. Velocity is a vector quantity and therefore has magnitude and direction.

The **average velocity** is the displacement divided by the total travel time:

$$\text{average velocity} = \frac{\text{displacement}}{\text{total travel time}}$$

$$\bar{v} = \frac{\Delta x}{\Delta t} = \frac{x - x_\text{o}}{t - t_\text{o}} \tag{2.2}$$

The vector difference, $\Delta x = x - x_\text{o}$, is simply the displacement between the initial and final positions. In Fig. 2.1, x_2 and x_1 denote the positions, but x and x_o are used in Eq. 2.2. These are the more general symbols for position and are more convenient because they involve the use of fewer subscripts. The subscripts on x_o and t_o stand for *original* and indicate that these quantities refer to original, or initial, position and time. The symbols x and t represent position and time at some arbitrary later time and are sometimes called the final position and time.

As discussed in Section 2.1, for motion in one dimension, it is convenient to use plus and minus signs ($+$ and $-$) to indicate the directions of the displacement and velocity along the positive and negative axes, for example, $+x$ and $+v$ (or simply x and v) and $-x$ and $-v$. Also, it is common to take $x_\text{o} = 0$ and $t_\text{o} = 0$, so Eq. 2.2 becomes

$$\bar{v} = \frac{x}{t} \quad \text{or} \quad x = \bar{v}t \tag{2.3}$$

As can be readily seen from this equation, the standard units for velocity are the same as those for speed: m/s or ft/s.

You might be wondering whether there is a relationship between average speed and average velocity. A quick look at Fig. 2.1 will show you that *if* the motion is in one direction, the distance is equal to the magnitude of the displacement, and the average speed is the magnitude of the average velocity. However, be careful. This is not true if there is motion in both directions, as the following example shows.

Figure 2.2 **Instantaneous speed**
The speedometer of a car gives the speed over a very short interval of time, so its reading approaches the instantaneous speed. Add the direction of the car, and you have a good approximation of the instantaneous velocity.

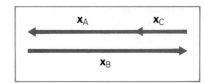

Figure 2.3 **Average speed and velocity**
The jogger starts at A, jogs to B, and then back to C, and finally back to A. What are the average speeds and velocities for the segments of the lap? See Example 2.1.

Example 2.1   Average Speed and Velocity

A jogger jogs from one end to the other of a straight 300-m track (from point A to point B in Fig. 2.3) in 2.50 min and then turns around and jogs 100 m back toward the starting point (to point C) in another 1.00 min. What are the jogger's average speeds and velocities in going (a) from A to B and (b) from A to C?

Solution

Given: $x_B = +300$ m (from A to B)   *Find*: Average speeds and average
$\quad\quad x_C = -100$ m (from B to C) $\quad\quad\quad$ velocities
$\quad\quad t_B = (2.50\ \text{min})(60\ \text{s/min}) = 150$ s
$\quad\quad t_C = (1.00\ \text{min})(60\ \text{s/min}) = 60$ s

(a) The jogger's average speed in going from A to B is easily computed:

$$\bar{v}_B = \frac{\Delta d}{\Delta t} = \frac{\Delta x}{\Delta t}$$

$$= \frac{300\ \text{m}}{150\ \text{s}} = 2.00\ \text{m/s}$$

The average velocity in going from A to B with $x_o = 0$ is also easily found, but direction must be indicated:

$$\bar{v}_B = \frac{x_B}{t_B}$$

$$= \frac{+300\ \text{m}}{150\ \text{s}} = +2.00\ \text{m/s}$$

The positive direction is to the right in Fig. 2.3. Note that the average speed is equal to the magnitude of the average velocity in this instance.

(b) The average speed in going from A to C involves the total distance traveled, so

$$\bar{v}_C = \frac{\Delta x}{\Delta t}$$

$$= \frac{300\ \text{m} + 100\ \text{m}}{150\ \text{s} + 60\ \text{s}} = 1.90\ \text{m/s}$$

where the directional signs are omitted. (Why?)

The average velocity, on the other hand, involves the sum of the vector displacements:

$$\bar{v}_{\text{C}} = \frac{x_{\text{B}} + x_{\text{C}}}{t_{\text{B}} + t_{\text{C}}}$$

$$= \frac{+300 \text{ m} - 100 \text{ m}}{150 \text{ s} + 60 \text{ s}} = +0.95 \text{ m/s}$$

Note that direction makes a difference; the average speed is *not* equal to the magnitude of the average velocity in this case. This is because the effective displacement is from the initial starting point (A) to the stopping point (C), or +200 m.

Suppose that next the jogger sprints from point C back to the starting point (A) in 0.50 min. What are the average speed and average velocity for the whole lap? The additional data are $x_{\text{A}} = -200$ m (Fig. 2.3) and $t_{\text{A}} = 30$ s, so, for the round trip, the average speed is

$$\bar{v}_{\text{A}} = \frac{\Delta x}{\Delta t}$$

$$= \frac{300 \text{ m} + 100 \text{ m} + 200 \text{ m}}{150 \text{ s} + 60 \text{ s} + 30 \text{ s}} = 2.50 \text{ m/s}$$

The average velocity is

$$\bar{v}_{\text{A}} = \frac{x_{\text{B}} + x_{\text{C}} + x_{\text{A}}}{t_{\text{B}} + t_{\text{C}} + t_{\text{A}}}$$

$$= \frac{300 \text{ m} - 100 \text{ m} - 200 \text{ m}}{240 \text{ s}} = 0 \text{ m/s}$$

The average velocity in this case is zero!

The total displacement is measured from the initial starting point to the final stopping point. When the jogger comes back to the starting point, the displacement is zero and so is the velocity. This is true for any round trip. (Note from the vector diagram in Fig. 2.3 how displacement vectors in opposite directions tend to cancel each other out, making the vector sum zero.) ∎

As Example 2.1 shows, average velocity is not very useful as a general descriptive quantity for motion. One thing that can be done to obtain a more useful measure is to look at smaller segments of the motion, that is, to let the time of observation (Δt) become smaller. As with speed, this leads to **instantaneous velocity** (when Δt approaches zero). Then $x = vt$, since the average bar is no longer needed. The instantaneous velocity is the instantaneous speed (magnitude) plus the direction of the motion at that particular instant.

In some instances, the motion of an object may be uniform, which means that the velocity or speed (or both) is constant. For example, the car in Fig. 2.4(a) has a uniform velocity (as well as a uniform speed). It travels the same distance in equal time intervals (50 km each hour), and the direction of its motion does not change.

Graphical analysis is often helpful in understanding motion and its related quantities. In this case, the car's motion may be represented on a plot of displacement versus time, or x versus t. As can be seen from Fig. 2.4(a), a straight line is obtained for a uniform, or constant, velocity on such a graph (and for uniform speed on a plot of distance versus time). Since the slope of the line is $\Delta x/\Delta t$ (analogous to $\Delta y/\Delta x$ on a plot using Cartesian coordinates), the slope is equal to the average velocity ($\bar{v} = \Delta x/\Delta t$). The positive slope indicates that x increases with time, and the motion can be seen to be in the positive x direction.

Suppose that the line had a negative slope. What would this indicate? In this case, the straight line would slant from upper left to lower right on the graph. The position values (x values) would get smaller with time, indicating that the car was traveling in uniform motion in the negative x direction.

For uniform motion, the average and instantaneous velocities are equal; that is, $v = \bar{v}$. (Why?)

In general, the motion of an object is usually nonuniform, meaning that different distances are covered in equal intervals of time. An x versus t plot for such motion in one dimension is a curved line, as illustrated in Fig. 2.4(b). The average velocity for a particular interval of time is the slope of a straight line between the two points on the curve corresponding to the starting and ending times of the interval. In the figure, the average velocity for the total trip is the slope of the straight line joining the beginning and end points of the curve

Figure 2.4 Graphical analysis of motion—displacement versus time
(a) For uniform (constant) velocity, an x versus t plot is a straight line, since equal displacements are covered in equal times. The slope of the line is equal to the magnitude of the velocity. (Average velocity equals instantaneous velocity in this case.) (b) For a nonuniform velocity, an x versus t plot is a curved line. The slope of a line between two positions is the average velocity between those positions, and the instantaneous velocity is that of a line tangent to the curve at any point (two tangent lines are shown in the figure).

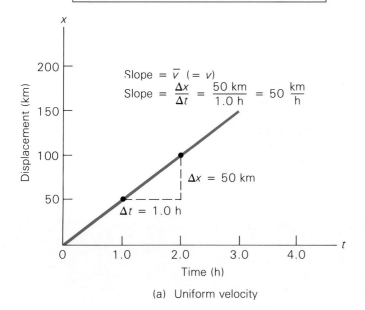

x (km)	t (h)	x/t
50	1	50 km/1 h = 50 km/h
100	2	100 km/2 h = 50 km/h
150	3	150 km/3 h = 50 km/h

(a) Uniform velocity

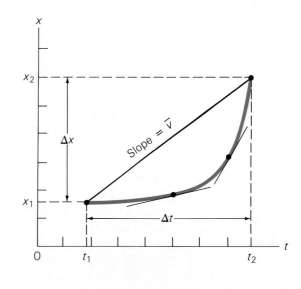

(b) Nonuniform velocity

(t_1 and t_2). The instantaneous velocity is equal to the slope of a straight line tangent to the curve at a specific point. As can be seen from the two tangent lines shown in Fig. 2.4(b), this slope varies, indicating that the velocity (or speed) is changing and that the motion is nonuniform.

Relative Velocity ◆

Since velocity involves displacement, it too must be measured from a fixed reference point, usually designated as the origin of a coordinate system in which the displacement is measured. Where you designate the origin of the coordinate axes to be is arbitrary and entirely a matter of choice. For example, you may "attach" the coordinate system to the road or the ground and then measure the velocity of the car relative to the axes.

Once the coordinate system has been designated for a situation, it is considered fixed. However, you may look at the motion of an object from another frame of reference that is moving relative to the original one. (For example, a car moves along a highway relative to the ground, but the Earth itself is moving relative to the Sun.) In any case, in analyzing motion, you do not change the physical situation or what is taking place, but only the point of view from which you choose to describe it. This choice is similar to the arbitrary choice of units. An object has a particular physical length regardless of whether you describe it in meters, feet, or light years (an astronomical unit, but still one of length).

As an illustration, consider cars moving at constant velocities along a straight, level highway, as shown in Fig. 2.5(a). The velocities shown in the figure are *relative* to a stationary observer (in parked car A), or to a set of coordinate axes fixed to the car or the ground at that point. The relative velocity between two objects is given by the velocity (vector) difference. For example, the velocity of car B relative to car A is given by

$$\mathbf{v}_{BA} = \mathbf{v}_B - \mathbf{v}_A = +90 \text{ km/h} - 0 = +90 \text{ km/h}$$

A person sitting in car A would see car B move away at a speed of 90 km/h (in the positive x direction). Boldface vector symbols are used here to remind you that the directional signs of the velocities must be supplied (in addition to the minus sign in the formula).

Similarly,

$$\mathbf{v}_{CA} = \mathbf{v}_C - \mathbf{v}_A = -75 \text{ km/h} - 0 = -75 \text{ km/h}.$$

That is, a person in car A would see car C approaching with a speed of 75 km/h (in the negative x direction).

But suppose that you want to know the velocities of the other cars *relative* to car B, that is, from the point of view of an observer in car B, or relative to a set of coordinate axes with the origin at car B [Fig. 2.5(b)]. Relative to those axes, car B is not moving and acts as the fixed reference point. The other cars are moving relative to it at relative velocities of, for car C,

$$\mathbf{v}_{CB} = \mathbf{v}_C - \mathbf{v}_B = -75 \text{ km/h} - (90 \text{ km/h}) = -165 \text{ km/h}$$

and, for car A,

$$\mathbf{v}_{AB} = \mathbf{v}_A - \mathbf{v}_B = 0 - (90 \text{ km/h}) = -90 \text{ km/h}$$

Relative to B, the other cars are moving in the negative x direction. That is, C is

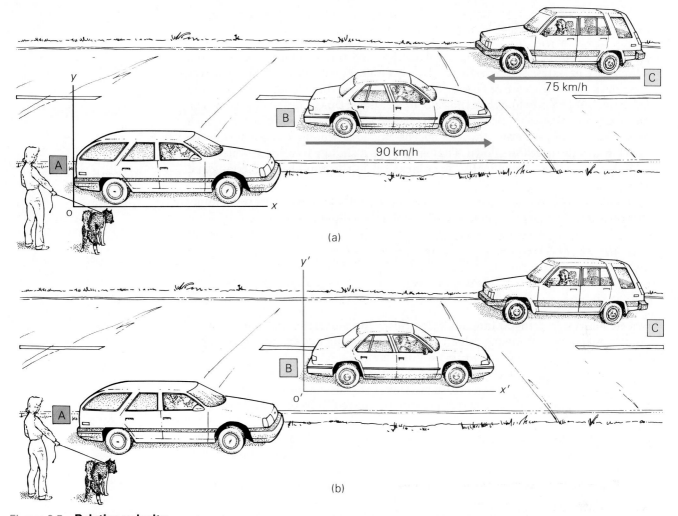

75 km/h

C

B

90 km/h

A

(a)

y'

C

B

o'

x'

A

(b)

Figure 2.5 **Relative velocity**
The observed velocity of a car depends on, or is relative to, the frame of reference.
(See the text for a specific description.)

approaching B with a velocity of 165 km/h in the negative x direction and A appears to be receding from B with a velocity of 90 km/h in the negative x direction. (Imagine yourself in car B, and take that position as stationary. Car C would appear to be coming toward you at a high rate of speed, and car A would be getting farther and farther away, as though it were moving backward relative to you.) In general,

$$\mathbf{v}_{AB} = -\mathbf{v}_{BA}$$

What about the velocities relative to C? From the point of view (or reference point) of C, both A and B would appear to be moving in the positive x direction. For B relative to C,

$$\mathbf{v}_{BC} = \mathbf{v}_B - \mathbf{v}_C = 90 \text{ km/h} - (-75 \text{ km/h}) = +165 \text{ km/h}$$

Can you show that $\mathbf{v}_{AC} = +75$ km/h?

2.3 Acceleration

The basic description of motion involves the time rate of change of position, which may be expressed by velocity. However, going one step further involves considering how the rate of change changes. Suppose that something is moving at a constant velocity and then the velocity changes; this is an acceleration. The gas pedal on an automobile is commonly called the accelerator. When you push down on the accelerator, the car speeds up; and when you let up on the accelerator, the car slows down. That is, there is a change in velocity with time, or an acceleration. Specifically, **acceleration** is the time rate of change of velocity.

Analogous to average velocity is the **average acceleration**, or the change in velocity divided by the time taken to make the change:

$$\text{average acceleration} = \frac{\text{change in velocity}}{\text{time to make the change}}$$

or

$$\bar{\mathbf{a}} = \frac{\Delta \mathbf{v}}{\Delta t} = \frac{\mathbf{v} - \mathbf{v}_o}{t - t_o} \tag{2.4}$$

where $\mathbf{v}$ and $\mathbf{v}_o$ are instantaneous velocities—the velocities at times t and t_o. Analogous to instantaneous velocity is **instantaneous acceleration**, which is the acceleration at a particular instant of time.

The dimensions of acceleration are (length/time)/time (as is obvious from $\Delta v/\Delta t$). The SI units for acceleration are therefore (m/s)/s, or m/s-s, commonly written m/s^2 (and read as "meters per second squared"). In the British system, the units are ft/s^2.

Since velocity is a vector quantity, an acceleration may result from two possible kinds of changes in velocity. A vector has both magnitude and direction. Therefore, it can be changed by changing its magnitude or its direction. Velocity, then, can be changed by varying the speed (the magnitude) or the direction of the motion, or both, as illustrated in Fig. 2.6.

For straight-line, or linear, motion, as for displacement, plus and minus signs can be used to indicate the direction. Thus, Eq. 2.4 can be written as

$$\bar{a} = \frac{v - v_o}{t} \tag{2.5}$$

where t_o is taken to be zero (v_o may not be zero, so it cannot generally be omitted).

Example 2.2   Average Acceleration
An automobile traveling on a straight road at 90 km/h slows down to 40 km/h in 5.0 s. What is its average acceleration?

Solution

Given: $v_o = (90 \text{ km/h})\left(\dfrac{0.278 \text{ m/s}}{1 \text{ km/h}}\right)$ *Find:* $\bar{a}$

$\qquad\quad = 25 \text{ m/s}$

$\qquad\quad v = (40 \text{ km/h})\left(\dfrac{0.278 \text{ m/s}}{1 \text{ km/h}}\right)$

$\qquad\quad = 11 \text{ m/s}$

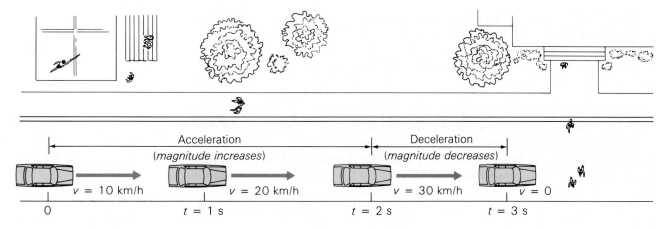

(a) Change in magnitude but *not* direction

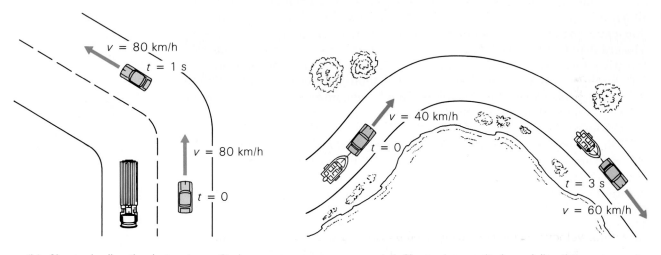

(b) Change in direction but *not* magnitude

(c) Change in magnitude *and* direction

Figure 2.6 **Acceleration—the time rate of change of velocity**
Since velocity is a vector quantity with magnitude and direction, an acceleration
can occur when there is (a) a change in magnitude but not direction or (b) a change in
direction but not magnitude or (c) a change in both magnitude and direction.

(Positive direction taken to be the direction of the initial motion)

$$t = 5.0 \text{ s}$$

Using Eq. 2.5 gives

$$\bar{a} = \frac{v - v_\text{o}}{t}$$

$$= \frac{11 \text{ m/s} - 25 \text{ m/s}}{5.0 \text{ s}} = -2.8 \text{ m/s}^2$$

The minus sign indicates the direction of the (vector) acceleration. In this case,
the acceleration is opposite to the direction of the initial motion ($+v_\text{o}$), and it
slows the car. Sometimes a slowing acceleration is called a deceleration.

A negative acceleration doesn't necessarily slow an object down. If the velocity and acceleration are both in the *same* direction, the car will speed up. For example, if the car is initially traveling in the negative x direction, a negative acceleration will speed it up in that direction. ∎

Although acceleration can vary with time, this study of motion will be restricted to constant accelerations for simplicity. (An important constant acceleration is the acceleration due to gravity near the Earth's surface, which will be considered in the next section.) Since for a constant acceleration the average is equal to the constant value ($\bar{a} = a$), the bar over the acceleration in Eq. 2.5 may be omitted. Thus, the equation is commonly written like this:

$$v = v_0 + at \qquad (2.6)$$

Uniform or constant accelerations are easy to represent graphically. A v versus t plot is a straight line whose slope is equal to the acceleration, as illustrated by Fig. 2.7(a). [Note that Eq. 2.6 can be written $v = at + v_0$, which, as you may recognize, has the form of an equation for a straight line, $y = mx + b$.] The graph in Fig. 2.7(b) might represent a car uniformly accelerating from rest (positive slope means positive acceleration), then moving at a constant velocity for a while (zero slope, zero acceleration), and finally uniformly decelerating to a stop (negative slope). You should be able to compute each acceleration, or slope. For nonuniform acceleration, the v versus t plot would not be a straight line. (Why?)

When an object moves with a constant acceleration, its velocity changes by the same amount in each time unit. For example, if the acceleration is 10 m/s^2, the object's velocity increases by 10 m/s in each second. Suppose that the object has an initial velocity (v_0) of 20 m/s at $t_0 = 0$. Then, for $t = 1, 2, 3,$ and 4 s, the velocities are 30, 40, 50, and 60 m/s, respectively. The average velocity over the 4-s interval is $\bar{v} = 40$ m/s. [The uniformly increasing series of numbers 20, 30, 40, 50, and 60 has an average of 40, which you may immediately recognize as the median (midway) value or compute in the regular manner. Also, note that in this particular case, the average of the extreme (initial and final) values gives the average of the series: $(20 + 60)/2 = 40$.]

When the velocity changes at a uniform rate because of a constant acceleration, $\bar{v}$ will be the median, or midway point, between the initial and final velocities:

$$\bar{v} = \frac{v + v_0}{2} \qquad \text{(for a constant acceleration only)} \qquad (2.7)$$

Example 2.3   A Constantly Accelerating Motorboat

A motorboat starting from rest on a lake accelerates in a straight line at a constant rate of 3.0 m/s^2 for 8.0 s. How far does the boat travel during this time?

Solution

Given: $v_0 = 0$ *Find:* x
        $a = 3.0 \text{ m/s}^2$
        $t = 8.0 \text{ s}$

The velocity of the boat at the end of 8.0 s is

$$v = v_0 + at = 0 + (3.0 \text{ m/s}^2)(8.0 \text{ s}) = 24 \text{ m/s}$$

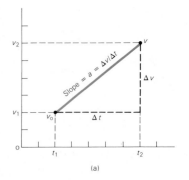

(a)

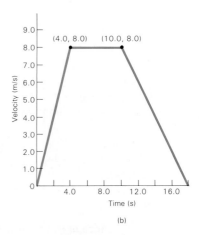

(b)

Figure 2.7 **Graphical analysis of motion—velocity versus time**
(a) For a uniformly changing velocity, a v versus t plot is a straight line. The slope of the line is equal to the magnitude of the acceleration, which is constant. (b) This graphical representation could reflect the motion of a car accelerating uniformly from rest, moving with a constant velocity for a time, and then decelerating uniformly to a stop.

The average velocity over that time interval is

$$\bar{v} = \frac{v + v_\mathrm{o}}{2} = \frac{24 \text{ m/s} + 0}{2} = 12 \text{ m/s}$$

Finally, the distance traveled is

$$x = \bar{v}t$$

$$= (12 \text{ m/s})(8.0 \text{ s}) = 96 \text{ m} \quad \blacksquare$$

2.4 Kinematic Equations

The description of motion in one dimension with constant acceleration requires only three basic equations. From previous sections, these are

$$x = \bar{v}t \qquad (2.3)$$

$$\bar{v} = \frac{v + v_\mathrm{o}}{2} \qquad (2.7)$$

$$v = v_\mathrm{o} + at \qquad (2.6)$$

(Keep in mind that the first equation is general and is not limited to situations where there is constant acceleration as are the latter two.)

However, as Example 2.3 showed, the description of motion in some instances requires multiple applications of these equations, which may not be obvious at first. It would be helpful if there were a way to reduce the number of operations in solving kinematic problems, and there is—combining equations algebraically.

For instance, combining the above equations involves first substituting for $\bar{v}$ from Eq. 2.7 into 2.3:

$$x = \bar{v}t = \left(\frac{v + v_\mathrm{o}}{2}\right)t$$

Then substituting for v from Eq. 2.6 gives

$$x = \left(\frac{v + v_\mathrm{o}}{2}\right)t = \left[\frac{(v_\mathrm{o} + at) + v_\mathrm{o}}{2}\right]t$$

Simplifying gives

$$x = v_\mathrm{o}t + \tfrac{1}{2}at^2 \qquad (2.8)$$

Essentially, this was done in Example 2.3. This combined equation allows the distance traveled by the motorboat in that example to be computed directly:

$$x = v_\mathrm{o}t + \tfrac{1}{2}at^2 = 0 + \tfrac{1}{2}(3.0 \text{ m/s}^2)(8.0 \text{ s})^2 = 96 \text{ m}$$

Another possibility is to use Eq. 2.6 to eliminate time (t), rather than the final velocity (v), by writing that equation in the form $t = (v - v_\mathrm{o})/a$. Then, as before, substituting for $\bar{v}$ in Eq. 2.3 from 2.7 gives

$$x = \bar{v}t = \left(\frac{v + v_\mathrm{o}}{2}\right)t$$

But then substituting for t gives

$$x = \left(\frac{v + v_o}{2}\right)t = \left(\frac{v + v_o}{2}\right)\left(\frac{v - v_o}{a}\right)$$

Simplifying gives

$$v^2 = v_o^2 + 2ax \qquad (2.9)$$

Example 2.4   A Dropped Ball

A ball is dropped from the top of a building. If it accelerates toward the ground at a rate of 9.80 m/s², what is its velocity when it has fallen 4.00 m?

Solution

Given: $v_o = 0$ $\qquad\qquad\qquad$ *Find*: v
$\qquad\quad a = -9.80$ m/s²
$\qquad\quad x = -4.00$ m

Here upward is taken to be the positive direction, and downward the negative direction. The velocity may be obtained directly from Eq. 2.9:

$$v^2 = v_o^2 + 2ax$$
$$= 0 + 2(-9.80 \text{ m/s}^2)(-4.00 \text{ m}) = 78.4 \text{ m}^2/\text{s}^2$$

and

$$v = \sqrt{78.4 \text{ m}^2/\text{s}^2} = -8.85 \text{ m/s}$$

where the negative root is taken because the motion is downward.

Using the combined forms of the equations can often save you a lot of steps and calculations. ∎

The equations of motion for linear motion with constant acceleration are summarized in Table 2.1 for your convenience. Note that only the first three are basic equations; the last two are convenient combinations.

Example 2.5   Stopping Distance

The stopping distance for a vehicle after the brakes have been applied is an important factor in highway safety. This distance depends on the initial velocity (v_o) and the braking capacity, or deceleration ($-a$, which is assumed to be constant). [Recall that the negative acceleration indicates that the acceleration is in the opposite direction from the velocity.] Express the stopping distance x in terms of these quantities.

Solution

The stopping distance may be found from Eq. 2.9. Expressing the negative acceleration explicitly gives

$$v^2 = v_o^2 - 2ax$$

Because the vehicle is to come to a stop, $v = 0$, and

$$x = \frac{v_o^2}{2a}$$

Note that the stopping distance is proportional to the square of the initial

Table 2.1

Equations for Linear Motion with Constant Acceleration*

$x = \bar{v}t$	(2.3)[†]
$\bar{v} = \dfrac{v + v_o}{2}$	(2.7)
$v = v_o + at$	(2.6)
$x = v_o t + \frac{1}{2}at^2$	(2.8)
$v^2 = v_o^2 + 2ax$	(2.9)

* It is assumed that $x_o = 0$ and the velocity v_o is that at $t_o = 0$. The initial position and time, x_o and t_o, may be included for general cases, for example, $x - x_o = \bar{v}(t - t_o)$.
[†] Note that Eq. 2.3 is not limited to constant acceleration but applies generally.

velocity of the vehicle. Doubling the initial velocity ($2v_o$) therefore increases the stopping distance by a factor of 4 (for the same deceleration). That is,

$$x_1 \propto v_o^2 \quad \text{and} \quad x_2 \propto (2v_o)^2 = 4v_o^2$$

Do you think this is an important consideration in setting speed limits, for example, in school zones? (The driver's reaction time should also be considered. A method to approximate a person's reaction time is given in the next section.) ∎

2.5 Free Fall

One of the more familiar cases of constant acceleration is that due to gravity near the Earth's surface. When an object is dropped, its initial velocity is zero (at the instant it is released), but at a later time, it has speeded up. There has been a change in velocity and, by definition, an acceleration. This **acceleration due to gravity** (g) has an approximate value (magnitude) of

$$g = 9.80 \text{ m/s}^2$$

or 980 cm/s² and is directed downward (toward the center of the Earth). In British units, the value of g is about 32 ft/s².

The values given here for g are only approximate because the acceleration due to gravity varies slightly at different locations as a result of differences in elevation and regional average mass density of the Earth. These small variations will be ignored in this book unless otherwise noted. (Gravitation is studied in more detail in Chapter 7.) Air resistance is another factor that affects the acceleration of a falling object. But for relatively dense objects and over the short distances of fall commonly encountered, air resistance produces only a small effect, which will also be ignored here for simplicity. (The frictional effect of air resistance will be considered in Chapter 4.)

Objects in motion solely under the influence of gravity are said to be in **free fall**. The acceleration due to gravity is the *constant* acceleration for all free-falling objects, regardless of their mass or weight. It was once thought that heavier bodies fell faster than lighter bodies. This was part of Aristotle's theory of motion. You can easily observe that a coin falls faster than a sheet of paper when dropped simultaneously from the same height. But in this case air resistance plays a noticeable role. If the paper is crumpled into a compact ball, it gives the coin a better race. Similarly, a feather "floats" down much more slowly than a coin falls. However, in a good partial vacuum (Fig. 2.8), the feather and the coin fall with the same acceleration—the acceleration due to gravity.

Astronaut David Scott performed a similar experiment on the moon in 1971 by simultaneously dropping a feather and a hammer from the same height. Of course, he did not need a vacuum pump since the moon has no atmosphere. The hammer and the feather reached the lunar surface together, but fell at a slower rate than on Earth. The acceleration due to gravity near the moon's surface is approximately one-sixth of that near the Earth's surface ($g_m \cong \frac{1}{6}g$).

Currently accepted ideas about the motion of falling bodies are due in large part to Galileo. He challenged Aristotle's theory and experimentally investigated the motion of objects. Galileo made measurements on balls rolling

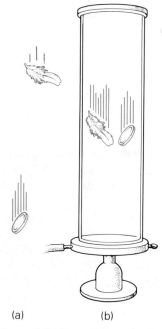

(a)   (b)

Figure 2.8 **Free fall and air resistance**
(a) When dropped simultaneously from the same height, a feather falls more slowly to the ground than a coin because of air resistance. (b) When placed in a partial vacuum, the same feather and coin fall together, at a constant acceleration.

Galileo Galilei (1564–1642)

Galileo Galilei was born in Pisa, Italy, in 1564 during the Renaissance. Today, he is known throughout the world by his first name and often referred to as the father of modern science or the father of modern mechanics and experimental physics, which attests to the magnitude of his scientific contributions (Fig. 1).

Galileo pursued a variety of scientific endeavors. These include the study of time, motion, machines, floating bodies, temperature, and heat and the invention of various measurement instruments. Although he did not invent the telescope, he made improvements that increased its image quality and magnification, and he was one of the first to use it to study the stars and planets. Some of his astronomical observations and discoveries include the mountains of the moon, the phases of Venus, sunspots, the moons of Jupiter, and the rings of Saturn (the latter he observed but did not distinguish as rings).

Celestial observations lead Galileo to support the Copernican theory, which held that the Earth and other planets revolved about the Sun. At that time, it was the official view of the Catholic Church that the Earth was the center of the universe, which accorded with Aristotelian theory. To believe otherwise was heresy. Galileo was tried by the Inquisition, forced to recant his views, and spent the last eight years of his life under house arrest (he was finally "rehabilitated" by Pope John Paul II in 1984). He died early in 1642, the year another great scientist, Isaac Newton, was born.

One of Galileo's greatest contributions to science was the establishment of the scientific method, that is, investigation through experiment. In contrast, Aristotle's approach was based on logical deduction. By the scientific method, for a theory to be valid it must predict or agree with experimental results. If it doesn't, it is invalid or requires modification. Galileo said, "I think that in the discussion of natural problems we ought not to begin at the authority of places of Scripture, but at sensible experiments and necessary demonstrations."[*]

Probably the most popular and well-known legend about Galileo is that he performed experiments with falling bodies by dropping objects from the Leaning Tower of Pisa (see Fig. 2.). There is some doubt as to whether Galileo actually did this, but there is little doubt that he questioned Aristotle's view on the motion of falling objects. In 1638, Galileo wrote:

> Aristotle says that an iron ball of one hundred pounds falling from a height of one hundred cubits reaches the ground before a one-pound ball has fallen a single cubit. I say that they arrive at the same time. You find, on making the experiment, that the larger outstrips the smaller by two finger-breadths, that is, when the larger has reached the ground, the other is short of it by two finger-breadths; now you would not hide behind these two fingers the ninety-nine cubits of Aristotle.[†]

This and other writings show that Galileo was aware of the effect of air resistance.

The experiments at the Tower of Pisa were supposed to have taken place around 1590. In his writings of about that time, Galileo mentions dropping objects from a high tower, but never specifically names the Tower of Pisa. A letter

Figure 1 **Galileo**
Galileo is alleged to have performed free-fall experiments by dropping objects off the Leaning Tower of Pisa.

[*] From *Growth of Biological Thought: Diversity, Evolution & Inheritance*, by F. Meyr (Cambridge, MA: Harvard University Press, 1982).
[†] This and the next quotation are both from *Aristotle, Galileo, and the Tower of Pisa*, by L. Cooper (Ithaca, NY: Cornell University Press, 1935).

down inclined planes in his studies of constantly accelerated motion. Also, legend has it that he studied the accelerations of falling bodies by dropping objects of different weights from the top of the Leaning Tower of Pisa (see the Insight feature).

The words "free fall" bring to mind dropped objects that are moving downward under the influence of gravity ($g = 9.80$ m/s^2 in the absence of air written to Galileo from another scientist in 1641 describes the dropping of a cannon ball and a musket ball from the Tower of Pisa. The first account of Galileo doing a similar experiment was written a dozen years after his death by Vincenzo Viviani, his last pupil and first biographer. Viviani related that the falling bodies "all moved at the same speed" and that Galileo demonstrated "this with repeated experiments from the height of the Campanile [Tower] of Pisa in the presence of the other teachers and philosophers, and the whole assembly of students." However, there is no other record of this event, which seems odd given that a crowd of people were supposedly witnessing it.

It is not known whether Galileo told this story to Viviani in his declining years or Viviani created this picture of his former teacher.

The important point is that Galileo recognized (and probably experimentally showed) that free-falling objects fall at the same rate regardless of their mass or weight (see Fig. 3). Galileo gave no reason why objects in free fall have the same acceleration, but Newton did, as you will learn in a later chapter.

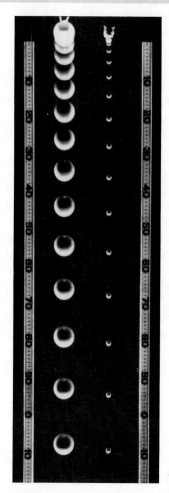

Figure 3 **Free fall**
A multiflash photograph shows a baseball and a golf ball released simultaneously, falling together. All freely falling bodies fall at the same rate, regardless of their weight or mass.

Figure 2 **Leaning Tower of Pisa**

resistance). However, the term can be applied in general to any vertical motion under the influence of gravity. An object with an initial velocity, directed either up or down, may be thought of as being projected in one dimension and having an acceleration equal to g. (Even when an object projected upward is traveling upward, it is accelerating downward.) Thus, the set of equations for motion in one dimension (in Table 2.1) can be used to describe generalized free fall.

It is customary to use y to represent the vertical direction and to take upward as positive. Since the acceleration due to gravity is always downward, it is in the negative direction ($a = -g$), which may be expressed explicitly in the equations for linear motion:

$$y = \bar{v}t \qquad (2.3')$$

$$\bar{v} = \frac{v + v_o}{2} \qquad (2.7')$$

$$v = v_o - gt \qquad (2.6')$$

$$y = v_o t - \tfrac{1}{2}gt^2 \qquad (2.8')$$

$$v^2 = v_o^2 - 2gy \qquad (2.9')$$

The equations can be written with $a = g$, for example, $v = v_o + gt$, and the minus sign associated with the value of g, for example, $g = -9.80$ m/s^2. The choice of the sign of the direction is arbitrary. Since upward is generally taken to be positive (+y axis on a graph), writing $a = -g$ explicitly reminds you of directional differences.

The origin ($y = 0$) of the reference frame is usually taken to be at the initial position of the object.

Note that you have to be explicit about the directions of vector quantities when using these equations. The displacement y and the velocities v and v_o may be positive or negative, depending on the direction of motion. The use of these equations and the sign convention is illustrated in the following examples.

Example 2.6 An Object Thrown Downward

A boy on a bridge throws a stone vertically downward toward the river below with an initial velocity of 14.7 m/s. If the stone hits the water 2.00 s later, what is the height of the bridge above the water?

Solution

Given: $v_o = -14.7$ m/s   *Find*: v
            Downward taken as the
            negative direction
            $t = 2.00$ s
            $g = 9.80$ m/s^2

The distance the stone travels in 2.00 s is given by Eq. 2.8':

$$y = v_o t - \tfrac{1}{2}gt^2 = (-14.7 \text{ m/s})(2.00 \text{ s}) - \tfrac{1}{2}(9.80 \text{ m/s}^2)(2.00)^2$$

$$= -29.4 \text{ m} - 19.6 \text{ m} = -49.0 \text{ m}$$

The minus sign indicates that the displacement is downward, which agrees with what you know from the statement of the problem. (Could you find how long it would take the stone to reach the river if the boy had dropped it rather than thrown it?) ∎

Example 2.7 Reaction Time

Reaction time is how long it takes a person to notice, think, and act in response to a situation, for example, the time between first observing and then responding to something happening on the road ahead while driving an automobile. Reaction time varies with the complexity of the situation (and the individual). In general, the largest component of a person's reaction time is spent thinking, but practice in dealing with a given situation can reduce this time.

A person's reaction time for a simple situation may be measured by having another person drop a ruler (without warning) through the thumb and forefinger as shown in Fig. 2.9. The falling ruler is grasped by the first person as quickly as possible, and the length of the ruler below the top of the finger is noted. If the ruler descends 18 cm on the average before it is grasped, what is the person's average reaction time?

Solution

Given: $v_0 = 0$ *Find:* t
$\qquad y = -18 \text{ cm} = -0.18 \text{ m}$
$\qquad g = 9.80 \text{ m/s}^2$

Eq. 2.8' is

$$y = v_0 t - \tfrac{1}{2} g t^2$$

or $y = -\tfrac{1}{2} g t^2$ (with $v_0 = 0$)

Solving for t gives

$$t = \sqrt{\frac{2y}{-g}}$$

$$= \sqrt{\frac{2(-0.18 \text{ m})}{-9.80 \text{ m/s}^2}} = 0.19 \text{ s}$$

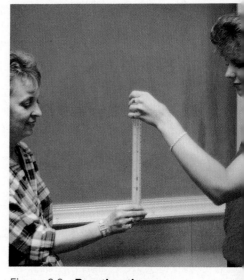

Figure 2.9 **Reaction time**
A person's reaction time can be measured by having her (or him) grasp a dropped ruler. See Example 2.7.

Try this yourself. Why should another person drop the ruler without warning? Substitute a dollar bill for the ruler. Are you fast enough to catch it? ∎

Example 2.8   Free Fall Up and Down

A worker on a scaffold on a billboard throws a ball straight up. It has an initial velocity of 11.2 m/s when it leaves his hand at the top of the billboard (Fig. 2.10). (a) What is the maximum height the ball reaches relative to the top of the billboard? (b) How long does it take for the ball to reach this height? (c) What is the position of the ball at $t = 2.00$ s? (d) What is the total time for the ball to return to its starting point? (e) What velocity does the ball have when it returns to its starting point? (f) What is the position of the ball at $t = 3.00$ s? (All aspects of free fall are illustrated by this example for completeness and comparison. Study and understand each part before going on to the next. Note in Fig. 2.10 that the g vector is constant and v changes.)

Solution

Given: $v_0 = 11.2 \text{ m/s}$ *Find:* (a) y_{max}
$\qquad g = 9.80 \text{ m/s}^2$ (b) t_u
$\qquad\qquad\qquad\qquad\qquad\qquad$ (c) y (at $t = 2.00$ s)
$\qquad\qquad\qquad\qquad\qquad\qquad$ (d) t (at $y = 0$ on return)
$\qquad\qquad\qquad\qquad\qquad\qquad$ (e) v (at $y = 0$ on return)
$\qquad\qquad\qquad\qquad\qquad\qquad$ (f) y (at $t = 3.00$ s)

(a) At the maximum height, y_{max}, the velocity of the ball is zero ($v = 0$). Then, Eq. 2.9' is

$$v^2 = 0 = v_0^2 - 2g y_{max}$$

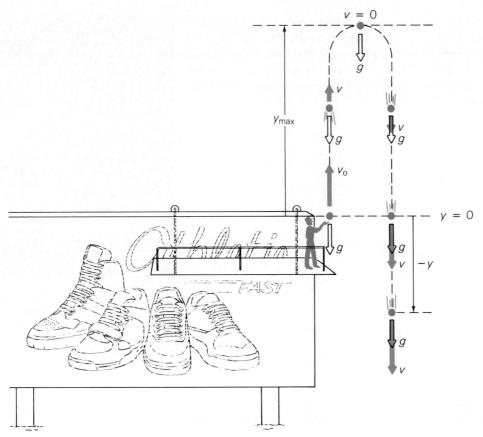

Figure 2.10 **Free fall up and down**
(The ball is horizontally displaced for illustration.) Note the lengths of the velocity and acceleration vectors at different times. See Example 2.8.

and

$$y_{max} = \frac{v_o^2}{2g}$$

$$= \frac{(11.2 \text{ m/s})^2}{2(9.80 \text{ m/s}^2)} = 6.40 \text{ m}$$

relative to $y = 0$, or the top of billboard (see Fig. 2.10).

(b) With $v = 0$ at the maximum height, the time elapsed for upward flight (t_u) is found using Eq. 2.6′:

$$v = 0 = v_o - gt_u$$

and

$$t_u = \frac{v_o}{g}$$

$$= \frac{11.2 \text{ m/s}}{9.80 \text{ m/s}^2} = 1.14 \text{ s}$$

(c) The height of the ball at $t = 2.00$ s is given directly by Eq. 2.8′:

$$y = v_o t - \tfrac{1}{2}gt^2$$

$$= (11.2 \text{ m/s})(2.00 \text{ s}) - \tfrac{1}{2}(9.80 \text{ m/s}^2)(2.00 \text{ s})^2 = 22.4 \text{ m} - 19.6 \text{ m} = 2.8 \text{ m}$$

Note that this is 2.8 m above $(+y)$, or measured upward from, the reference point $(y = 0)$.

Considered from another reference point, this situation is like dropping a ball from a height of y_{max} above the top of the billboard with $v_o = 0$ and asking how far it falls in a time $t = 2.00 \text{ s} - t_u = 2.00 \text{ s} - 1.14 \text{ s} = 0.86 \text{ s}$. The answer is

$$y = v_o t - \tfrac{1}{2}gt^2$$
$$= 0 - \tfrac{1}{2}(9.80 \text{ m/s}^2)(0.86 \text{ s})^2 = -3.6 \text{ m}$$

This is the same as the position found above but is measured with respect to the maximum height as the reference point; that is,

$$y_{max} - 3.6 \text{ m} = 6.4 \text{ m} - 3.6 \text{ m} = 2.8 \text{ m}$$

(d) The time to reach the maximum height is $t_u = 1.14 \text{ s}$. The time for the return trip downward, t_d, is the time it takes the ball to fall 6.4 m (the maximum height) from rest ($v = 0$ at maximum height). This situation is like dropping the ball from its maximum height. With $v_o = 0$ in this case, Eq. 2.8' is

$$y = -\tfrac{1}{2}gt_u^2$$

and

$$t_d = \sqrt{\frac{2y}{-g}}$$
$$= \sqrt{\frac{2(-6.40 \text{ m})}{-9.80 \text{ m/s}^2}} = 1.14 \text{ s}$$

Adding this to the upward time from part (b) gives the total time:

$$t = t_u + t_d$$
$$= 1.14 \text{ s} + 1.14 \text{ s} = 2.28 \text{ s}$$

Note that the time to return to the starting point is the same as that to reach the maximum height. This is a general result.

Another way to calculate the total time is to note that when the ball returns to its original starting point, it is again at $y = 0$, and then use Eq. 2.8' to find the total time:

$$y = v_o t - \tfrac{1}{2}gt^2 = (v_o - \tfrac{1}{2}gt)t = 0$$

where v_o is the initial upward velocity. The roots of this equation are $t = 0$ (which is when the ball is initially at $y = 0$), and $v_o - \tfrac{1}{2}gt = 0$, which is when the ball is again at $y = 0$ (on the return trip). From the latter,

$$t = \frac{2v_o}{g}$$
$$= \frac{2(11.2 \text{ m/s})}{9.80 \text{ m/s}^2} = 2.28 \text{ s}$$

Note that since $t = t_u + t_d$ and $t_u = 1.14 \text{ s}$ from part (b), it follows that the time upward from the original starting point is the same as the time back down to it $(t_u = t_d)$.

(e) The velocity of the ball when it returns to its starting point may be found using Eq. 2.9′:

$$v^2 = v_0^2 - 2gy = v_0^2 - 0 \quad \text{or} \quad v^2 = v_0^2$$

The negative root is physically significant (to show the proper sign, or direction—see the Problem-Solving Hint):

$$v = -v_0 = -11.2 \text{ m/s}$$

This velocity can also be found by considering the velocity of the ball at time t_d, or 1.14 s after falling from its maximum height. Eq. 2.6′ with $v_0 = 0$ gives

$$v = v_0 - gt_d$$

$$= 0 - (9.80 \text{ m/s}^2)(1.14 \text{ s}) = -11.2 \text{ m/s}$$

Note that the ball returns to the starting point with the same speed it had initially.

(f) At $t = 3.00$ s, the ball has fallen below its starting point, but its position is still given by Eq. 2.8′:

$$y = v_0 t - \tfrac{1}{2}gt^2$$

$$= (11.2 \text{ m/s})(3.00 \text{ s}) - \tfrac{1}{2}(9.80 \text{ m/s}^2)(3.00 \text{ s})^2 = 33.6 \text{ m} - 44.1 \text{ m} = -10.5 \text{ m}$$

The minus sign indicates that the ball has fallen 10.5 m below its original starting point ($y = 0$), assuming that the top of the billboard is higher than 10.5 m above the ground. (Could you find when the ball will strike the ground and how fast it would be going if you knew the height of the top of the billboard?) ∎

PROBLEM-SOLVING HINT

Physical Significance of Roots   When solving problems, you may run into quadratic equations, which have two roots, or solutions. This occurred in part (d) of Example 2.8, where the equation $y = v_0 t - \tfrac{1}{2}gt^2$ was solved for t. Since y was equal to zero, the roots of that quadratic equation were conveniently found by factoring. However, in general, such roots are found using the quadratic formula:

$$x = \frac{-b \pm \sqrt{b^2 - 4ac}}{2a} \quad \text{where } ax^2 + bx + c = 0$$

When you obtain the two roots from a quadratic equation, you must determine their physical significance, or meaning, to be able to select the desired answer. Both, one, or neither of the roots may have physical significance. Basically, you must decide whether the numerical answers have meaning. For example, if you solve an equation for time (t) and one root has a negative value, it is not physically significant, since time cannot be negative. Also, if the quantity under the square root sign in the quadratic formula is negative, the roots are imaginary numbers, and neither one has physical significance. (See Problem 81.)

In the case of Example 2.8 (d), the quadratic equation could have been avoided by using a series of the kinematic equations from which $y = v_0 t - \tfrac{1}{2}gt^2$ was derived. However, this method can be complicated and time-consuming, and a direct quadratic approach is more convenient. When solving problems, use the mathematical tools available to you—don't let them use you.

Important Formulas

Kinematic equations for linear motion with constant acceleration:

$x = \bar{v}t$ (general, not limited to constant acceleration)

$$\bar{v} = \frac{v + v_o}{2}$$

$v = v_o + at$

$x = v_o t + \frac{1}{2}at^2$

$v^2 = v_o^2 + 2ax$

Acceleration due to gravity:

$g = 9.80 \text{ m/s}^2 = 980 \text{ cm/s}^2 = 32 \text{ ft/s}^2$

Kinematic equations applied to free fall: (with downward taken as the negative direction and g expressed explicitly):

$y = \bar{v}t$

$$\bar{v} = \frac{v + v_o}{2}$$

$v = v_o - gt$

$y = v_o t - \frac{1}{2}gt^2$

$v^2 = v_o^2 - 2gy$

Questions

A Change of Position

1. What is the difference between kinematics and dynamics?

2. Can there be a change in position without motion? Explain.

3. What is needed to specify an object's position?

4. Distinguish between a scalar quantity and a vector quantity. Can a scalar and a vector be related? Explain.

5. Is the distance between two points always equal to the magnitude of the displacement? Explain.

6. Can a displacement be zero and the corresponding distance nonzero? Can a distance be zero and the corresponding displacement nonzero? Explain.

Speed and Velocity

7. When is speed equal to the magnitude of velocity? Give two instances.

8. Which unit is greater, km/h or mi/h? Give an estimate of how much greater.

9. When is the average velocity equal to the instantaneous velocity?

10. What is the physical meaning of a horizontal line on an x versus t plot?

11. How is motion in the $-x$ direction represented on an x versus t plot? Is there a negative horizontal axis? Explain.

12. Discuss (a) the motions of the Earth relative to the Sun and (b) the apparent motions of the Sun and the moon relative to Earth.

Acceleration and Free Fall

13. The gas pedal of an automobile is commonly called the accelerator. Could (a) the steering wheel, (b) the brakes, (c) the headlights, or (d) the clutch be called an accelerator? Explain.

14. What is the physical meaning of a horizontal line on a v versus t plot?

15. Describe the general motions of the two objects that have the v versus t plots shown in Fig. 2.11.

16. Explain how the value of the acceleration due to gravity could be determined from the photo of the falling balls in Fig. 3 in the Insight feature on Galileo.

17. Suppose that the equation $y = v_o t + \frac{1}{2}gt^2$ is used to describe an object projected vertically upward without indicating that v_o and g are in opposite directions. What

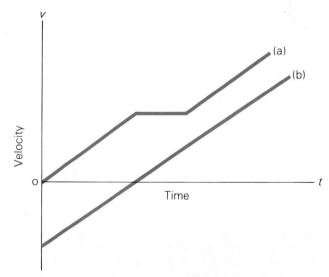

Figure 2.11 **Description of motion**
See Question 15.

would be the physical implication? (Hint: what would happen as time went on?)

18. Suppose that you are asked to find the time it takes an object to fall a given distance after being dropped from rest. The equation $y = v_o t - \frac{1}{2}gt^2$ with $v_o = 0$ yields $t = \sqrt{-2g/y}$, with a negative quantity under the square root sign. Does this mean there is no real root or physically significant answer? Explain.

19. Compare the motions of (a) a dropped object and (b) an object thrown vertically upward both on the moon and on Earth. (Do you want to exclude air resistance?)

20. Sketch the general forms of the plots of (a) v versus t and (b) y versus t for a dropped object in free fall.

21. Sketch the general forms of the plots of (a) v versus t and (b) y versus t for an object projected upward.

Problems

2.2   Speed and Velocity

■**1.** A student walks 0.25 km to class in 5.0 min. What is the student's average speed in m/s?

■**2.** A small airplane flies in a straight line at a speed of 140 km/h. How long does it take the plane to fly 360 km?

■**3.** A bus travels on an interstate highway at an average speed of 95 km/h. How far does the bus travel in 20 min on the average?

■**4.** A motorist drives 150 km from one city to another in 2.0 h, but makes the return trip in only 1.5 h. What are the average speeds for (a) each half of the round trip and (b) the total trip?

■**5.** If the velocity of car C in Fig. 2.5 were in the opposite direction, what would the velocities of the other cars be (a) relative to car C and (b) relative to car B?

■■**6.** A jogger makes four complete laps around a 0.50-km track in 15 min. What are the jogger's (a) average speed and (b) average velocity?

■■**7.** Is the average speed of the jogger in Example 2.1 going from A to C equal to the average of the average speeds in going from A to B and from B to C? Justify your answer.

■■**8.** A plot of displacement versus time is shown in Fig. 2.12 for an object in linear motion. (a) What are the average velocities for the segments AB, BC, CD, DE, EF, FG, and BG? (b) State whether the motion is uniform or nonuniform in each case. (c) What is the instantaneous velocity at point D?

■■**9.** In demonstrating a dance step, a person moves in one dimension as shown in Fig. 2.13. What are (a) the

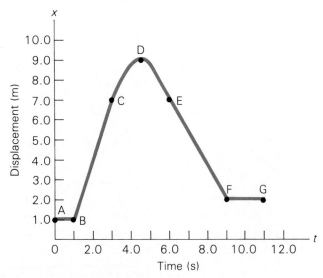

Figure 2.12   **Displacement versus time**
See Problem 8.

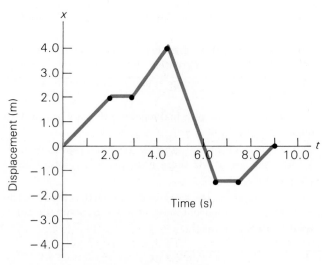

Figure 2.13   **Displacement versus time**
See Problem 9.

average speed and (b) the average velocity for each phase of the motion? (c) What are the instantaneous velocities at $t = 1.0$ s, 2.5 s, 4.5 s, and 6.0 s? (d) What is the average velocity for the interval between $t = 4.5$ s and $t = 9.0$ s? (Hint: recall that the effective displacement is the displacement between the starting point and the stopping point.)

■■**10.** The Indianapolis 500 was first run in 1911 in a time of 6 h, 42 min, and 8 s. In 1987, the race was run in a time of 3 h, 4 min, and 59 s. (a) What are the average speeds for the Indy 500 for these years? (b) What is the percentage change in the average speed from 1911 to 1987?

■■**11.** On a walk in the country, a couple goes 1.80 km east along a straight road in 15.0 min and then 2.40 km directly north in 20.0 min. (a) What is the average velocity for each segment of their walk? (b) What is the average speed for the total distance walked? (c) If they walk along a straight-line path back to their original starting place in 25.0 min, what is the average speed and average velocity for the total trip?

■■**12.** According to the theory of continental drift and plate tectonics, the continents were once parts of a single giant supercontinent, called Pangaea, which broke up. Recent measurements of sea-floor spreading along mid-oceanic ridges show the continental drift to be on the order of 4 cm per year. Assume that this rate has been constant throughout the past, and take the current distance between South America and Africa to be 5000 mi. Approximately how long ago did Pangaea break up?

■■**13.** A boat travels on a river with a constant speed of 50.0 km/h, as observed by a person on the bank. If the river has a flow rate (current) of 2.5 km/h, what is the boat's speed relative to the water if it is traveling directly (a) upstream or (b) downstream?

■■■**14.** On a cross-country trip, the average speeds and distances traveled on each of three days were 95.0 km/h for 800 km on day 1, 87.5 km/h for 640 km on day 2, and 92.8 km/h for 710 km on day 3. What was the average speed for the whole trip?

■■■**15.** Two runners approaching each other on a straight track have constant velocities of $+5.0$ m/s and -4.0 m/s, respectively, when they are 100 m apart. How long will it take for the runners to meet and at what position will this occur?

■■■**16.** Two motorcyclists race against the clock on a 35-km cross-country route. The first cyclist travels the route with an average speed of 55 km/h. The second cyclist starts 3.0 min after the first but crosses the finish line at the same time. What is the average speed of the second cyclist?

2.3 Acceleration

■**17.** An automobile traveling at 30.0 km/h along a straight road accelerates to 60.0 km/h in 4.00 s. What is the average acceleration?

■**18.** A car moving with a velocity of 15 m/s on a one-way street must be brought to a stop in 3.0 s. What is the required average acceleration?

■■**19.** A drag racer starting from rest reaches a speed of 200 km/h in 8.85 s along a straight track. (a) What is the average acceleration of the racer? (b) Assuming that the acceleration is constant, what is the average velocity of the racer?

■■**20.** If an automobile moving with a velocity of 36 km/h along a straight road is braked uniformly to rest in 5.0 s, by how much must the velocity change each second?

■■**21.** Compute the accelerations represented on the graph in Fig. 2.7(b).

■■**22.** After landing, a jet plane comes uniformly to rest along a straight runway with an average velocity of -40 km/h. If this takes 8.0 s, what is the plane's acceleration?

■■**23.** (Work this problem and Problem 24 as a pair.) Show that the area under the curve of a v versus t plot for a constant acceleration is equal to the displacement. Do this for the cases where (a) $a = 0$, (b) $v_o = 0$, and (c) $v_o \neq 0$ [as in Fig. 2.7(a)]. (Hint: the area of a triangle is $\frac{1}{2}ab$, or one-half the altitude times the base.)

■■**24.** Compute the distance traveled for the motion represented by Fig. 2.7(b).

■■■**25.** A uniform acceleration of -4.0 m/s^2 is applied to a moving object so that it is brought to rest with an average velocity of 10 m/s. Is the deceleration adequate to bring the object to rest in 4.0 s? (Justify your answer.) If it is not adequate, what is the required deceleration?

■■■**26.** (Work this problem and Problem 27 as a pair.) The area under the curve of a v versus t plot for a constant acceleration is equal to the displacement (see Problem 23). (a) Show that this is the case for a positive acceleration and for a negative acceleration. (b) Is there a relationship between the areas under these curves and the displacement?

■■■**27.** Fig. 2.14 shows velocity versus time for an object in linear motion. (a) Compute the acceleration for each phase of the motion. (b) Describe how the object moves during the 11.0 s represented by the graph. (c) Compute the

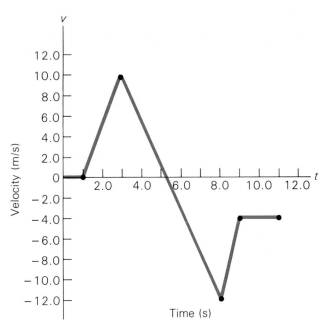

Figure 2.14 **Velocity versus time**
See Problem 27.

final displacement of the object. (d) Compute the total distance the object travels.

2.4 Kinematic Equations

■28. A sprinter accelerates from rest at a uniform rate of 1.8 m/s^2 for 4.0 s to reach his maximum speed. What is that speed in km/h and mi/h?

■29. A car accelerates uniformly from rest at a rate of 6.50 m/s^2. (a) How far does the car travel in 5.00 s? (b) What is the speed of the car at that time?

■30. A motorboat traveling on a straight course slows down uniformly from 72 km/h to 36 km/h in a distance of 50 m. What is the acceleration?

■31. A car initially at rest rolls down a hill with a uniform acceleration of 5.0 m/s^2. How long will it take the car to travel 100 m, the distance to the bottom of the hill?

■■32. The driver of a pickup truck going 90.0 km/h applies the brakes, giving the truck a uniform deceleration of 6.50 m/s^2 while it travels 20.0 m. (a) What is the velocity of the truck in km/h at the end of this distance? (b) How much time has elapsed?

■■33. A motorboat is traveling due north in calm water at a speed of 6.0 m/s. The operator puts the motor in reverse and the boat receives an acceleration of −1.5 m/s^2. (a) How

long after the boat is put into reverse does it momentarily come to a halt? (b) What is the position of the boat when it momentarily comes to a halt relative to the point where the motor was put into reverse?

■■34. A bullet traveling horizontally with a speed of 30 m/s strikes a tree and penetrates to a depth of 8.0 cm before coming to rest. Assuming a constant acceleration, how long after hitting the tree does the bullet come to a stop?

■■35. A bullet traveling horizontally with a speed of 35.0 m/s hits a board perpendicular to the surface, passes through it, and emerges on the other side with a speed of 21.0 m/s. If the board is 4.00 cm thick, how long does the bullet take to pass through it?

■■36. The speed limit in a school zone is 40 km/h (about 25 mi/h). A driver traveling at this speed sees a child run into the road 17 m ahead of his car. He applies the brakes, and the car decelerates at a rate of 8.0 m/s^2. If the driver's reaction time and the brake application time total 0.75 s, will the car stop before hitting the child?

■■37. Assume that the driver's reaction time and brake application time total 0.75 s and construct a table of the stopping distances for an automobile traveling at initial velocities of 36 km/h, 72 km/h, and 90 km/h with decelerations of (a) 8.0 m/s^2 (on a dry pavement) and (b) 4.0 m/s^2 (on a wet pavement). Also list the initial velocities in mi/h and the stopping distances in ft in parentheses.

■■38. An object moves in the positive x direction with a speed of 50 m/s. As it passes through the origin, it starts to experience a constant acceleration of 2.0 m/s^2 in the negative x direction. How much time elapses before the object returns to the origin?

■■■39. A drag racer accelerates uniformly from rest on a straight track at 12.0 m/s^2 for 5.50 s, maintains the speed it has reached for 3.00 s, and then ejects a parachute which uniformly decelerates it to a stop at a rate of 14.0 m/s^2. (a) What is the total distance traveled by the racer? (b) How long was it in motion?

■■■40. A car traveling with a velocity of 35 km/h along a straight road accelerates uniformly at a rate of 4.0 m/s^2 for 2.5 s. How far does the car travel during this time?

■■■41. An object moves in the positive x direction with a constant acceleration. At $x = 5.0$ m, its speed is 10 m/s. In 2.5 s, the object is at $x = 65$ m. What is the acceleration of the object?

■■■42. A car and a motorcycle start from rest at the same

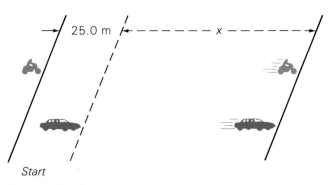

25.0 m ← → x

Start

Figure 2.15 **A tie race**
See Problem 42.

time on a straight track, but the motorcycle is 25.0 m behind the car (Fig. 2.15). The car accelerates at a uniform rate of 3.70 m/s², and the motorcycle at a uniform rate of 4.40 m/s². (a) How much time elapses until the motorcycle overtakes the car? (b) How far will each have traveled during that time? (c) How far ahead of the car will the motorcycle be 2.00 s later? (Both vehicles are still accelerating.)

■■■**43.** A pitcher throws a fastball toward home plate 60.5 ft away with an initial velocity of 90 mi/h in the horizontal direction. (a) If the batter's combined reaction and swing times total 0.28 s, how long can the batter watch the ball after it has left the pitcher's hand before making a decision to swing? (A major reason why some players illegally use hollowed out bats with cork or superball filling is that such bats are lighter, which decreases the swing time, allowing more time to follow the ball and make a decision to swing.) (b) How far does the ball drop from its original horizontal line in traveling to the plate? (Hint: consider the motions in the x and y directions independently.)

2.5 Free Fall

■**44.** If a dropped object falls 19.6 m in 2.00 s, how far will it fall in 4.00 s?

■**45.** A stone is dropped from a height of 20.0 m. (a) How long does it take for the stone to hit the ground? (b) What velocity does it have when it hits the ground?

■**46.** An object dropped from the top of a cliff takes 2.40 s to hit the water in the lake below. What is the height of the cliff above the water?

■**47.** With what velocity must an object be projected vertically upward for it to reach a maximum height of 15.0 m above its starting point?

■**48.** The ceiling of a classroom is 4.50 m above the floor. A student tosses an apple vertically upward, releas-

ing it 0.75 m above the floor. What is the maximum magnitude of the initial velocity that can be given to the apple if it is not to touch the ceiling?

■**49.** A ball is projected vertically downward with a velocity of 2.5 m/s. (a) How far does the ball travel in 1.8 s? (b) What is the velocity of the ball at that time?

■**50.** A baseball is thrown vertically upward with a velocity of 22.5 m/s. (a) How high above its point of release does the ball go? (b) How long does it take for the ball to return to the point of release? (c) What velocity does it have when it returns to that point?

■**51.** A stone is thrown vertically downward with an initial velocity of 15.0 m/s from a height of 65.0 m above the ground. (a) How far does the stone travel in 2.00 s? (b) What is its velocity when it hits the ground?

■**52.** How long would it take a dropped object to reach a speed of 55 mi/h? (Make a guess before you do the calculations.)

■■**53.** A baseball thrown vertically upward is caught at the same height 3.20 s later. What are (a) the initial velocity of the ball and (b) its maximum height above its starting point?

■■**54.** A certain person can jump a vertical distance of 0.85 m. (a) What is the total time the person is off the ground? (b) With what velocity does the person hit the ground?

■■**55.** A ball is thrown upward with an initial velocity of 8.0 m/s by someone on the top of a building 40 m tall who is leaning over the edge so that the ball will not strike the building on the return trip. (a) How far above the ground will the ball be at the end of 1.0 s? (b) What is the ball's velocity at that time? (c) When and with what speed will the ball strike the ground?

■■**56.** For an object dropped from rest, what is the distance it will fall *during* the second second (that is, between t = 1.00 s and t = 2.00 s)? Does this distance double during the fourth second of fall? Explain.

■■**57.** Draw plots of y versus t and v versus t for the following: (a) an object dropped from rest from a height of 30 m above the ground, and (b) an object projected vertically upward with an initial velocity of 34.3 m/s and its return to the same point. (c) Draw plots of a versus t for these motions.

■■**58.** Someone throws an object vertically upward with a velocity of 6.15 m/s from the top of a tall building; the thrower leans over the edge so that the object will not strike the building on the return trip. (a) What is the velocity of the object when it has traveled 30.0 m? (b) How long does it take to travel this distance?

■■**59.** Suppose that the boy on the bridge in Example 2.6 throws the stone vertically upward instead of downward. How long will the stone be in flight before it hits the river?

■■**60.** A photographer in a helicopter ascending vertically at a constant rate of 2.50 m/s accidentally drops a camera when the helicopter is 40.0 m above the ground. (a) How long will it take the camera to reach the ground? (b) What will its speed be when it hits?

■■**61.** The acceleration due to gravity on the moon is one-sixth of that on Earth. (a) If an object were dropped from the same height on the moon and on the Earth, how much longer (by what factor) would it take it to hit the surface of the moon? (b) For a projectile with an initial velocity of 23.5 m/s upward, what would be the maximum height and the total time of flight on the moon and on the Earth?

■■■**62.** A student at a window on the second floor of a dorm sees his math professor coming along the walkway beside the building. He drops a water balloon from 18.0 m above the ground when the prof is 1.00 m from the point directly beneath the window. If the prof is 170 cm tall and walks at a rate of 0.45 m/s, does the balloon hit his head? Does it hit him at all?

■■■**63.** A student drops a stone from the top of a building 26.0 m tall. Another student simultaneously throws a second stone downward, and the thrown stone hits the ground 0.30 s before the dropped stone does. What was the initial velocity of the thrown stone?

■■■**64.** A dropped object passes a window that is 1.35 m tall in 0.21 s. From what height above the top of the window was the object released?

■■■**65.** (This is an old one.) In order to find the depth of the water surface in a well, a person drops a stone from the top of the well and simultaneously starts a stopwatch. The watch is stopped when the splash is heard, giving a reading of 3.65 s. The speed of sound is 340 m/s. Estimate a range for the depth of the water surface below the top of the well. Take the person's reaction time for stopping the watch to be 0.25 s. (How could the solution be obtained directly? Is there a problem with doing this?)

Additional Problems

66. A drag racer traveling on a straight track with a speed of 160 km/h ejects a parachute and slows uniformly to a speed of 20 km/h in 12 s. (a) What is the racer's acceleration? (b) How far does it travel in the 12-s interval?

67. On a cross-country trip, a couple drives 450 mi in 10 h on the first day, 380 mi in 8.0 h on the second day, and 600 mi in 12 h on the third day. What was their average speed on the trip?

68. An aluminum ball with a mass of 4.0 kg and an iron ball of the same size with a mass of 11.6 kg are dropped simultaneously from a height of 49 m. (a) Neglecting air resistance, how long does it take the aluminum ball to fall to the ground? (b) How much later does the heavier iron ball strike the ground?

69. A car going 90 km/h on a straight road is brought uniformly to a stop in 10 s. How far does the car travel during that time?

70. An object initially at rest experiences an acceleration of 2.0 m/s² for 5.0 s and then travels at that constant velocity for another 10 s. What is the object's average velocity over the 15-s interval?

71. An arrow shot from a bow vertically upward has an initial velocity of 45 m/s. (a) What is the maximum height of the arrow above its launch point? (b) How long does it take for the arrow to return to its launch point? (Neglect air resistance and assume that the arrow travels in a straight line.)

72. An automobile is braked to a stop with a uniform deceleration in a time of t_s. Show that the distance traveled during this time is given by $x = v_o^2/a + \frac{1}{2}at_s^2$, where the positive direction is taken to be in the direction of the initial motion.

73. The air speed of a plane is the speed at which the air travels over its wings, or the speed of the plane relative to the air. (Air moving over the wings is important in providing lift for the plane.) In still air, the air speed of a plane is equal to its ground speed (speed relative to the ground). If an airplane has a ground speed of 200 km/h, what is its air speed with (a) a tail wind of 25 km/h and (b) a head wind of 35 km/h?

74. When checking reaction times as described in Example 2.7, some students decide to calibrate the ruler in time units so they can obtain a quick approximation without having to do a calculation each time the test is made. They want to calibrate the ruler in time intervals of 20 ms. Show on a sketch of a ruler like that shown in Fig. 2.16 where the time gradations would be for the first 20 cm.

75. A vertically moving projectile reaches a maximum

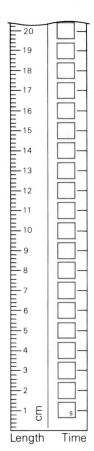

<figure>
Figure 2.16 **Time calibration**
See Problem 74.
</figure>

Length   Time

height of 16.5 m above its starting position. (a) What was the projectile's initial velocity? (b) What is its height above the starting point at $t = 2.10$ s?

76. Given that the speed of sound is 340 m/s, how much time will elapse between seeing a lightning flash and hearing the resulting thunder if the lightning strikes 1.75 mi away? (The speed of light is 3.0×10^8 m/s, or

about 186,000 mi/s, so the lightning flash is seen instantaneously.)

77. A spring-loaded gun shoots a 0.0050-kg bullet vertically upward with an initial velocity of 21 m/s. (a) What is the height of the bullet 4.0 s after firing? (b) At what times is the bullet 12 m above the muzzle of the gun?

78. A student driving home for the holidays starts at 8:00 a.m. to make the 675-km trip, practically all of which is on nonurban interstate highway. If she wants to arrive home no later than 3:00 p.m., what must her average speed be, at minimum? Will she have to exceed the 65 mi/h speed limit?

79. A rocket car travels with a constant velocity of 200 km/h on a salt flat. The driver gives the car a reverse thrust, and it experiences a continuous and constant deceleration of 8.60 m/s². How much time elapses until the car is 175 m from the point where the reverse thrust is applied?

80. An object starting from rest and traveling in a straight line has velocities of 5.0 m/s, 10 m/s, 15 m/s, 20 m/s, and 25 m/s at the ends of the first, second, third, fourth, and fifth seconds, respectively. (a) What is the acceleration of the object? (b) What is its average velocity for the 5-s interval? (c) Sketch a graph of v versus t. (d) Sketch a graph of x versus t.

81. A projectile is given an initial velocity of 8.0 m/s directly upward. Use the quadratic formula (with $g = 10$ m/s² to simplify the calculations) to determine (a) the time when projectile is 1.4 m above its starting point and (b) the time when it is 3.5 m above its starting point. (c) Suppose that the projectile is instead given an initial velocity of 8.0 m/s directly downward. At what time is the projectile a distance of 1.8 m below its starting point? (Comment on both roots of the quadratic equation in each case.)

Motion in Two Dimensions

3

Chapter 2 presented the simplest possible scenario and considered straight-line, or linear, motion in one dimension. The next logical step is to extend the study of motion to two dimensions, to consider motion in a plane. This includes curved paths of motion, such as that of a high fly ball or a car going around a circular race track. These are examples of **curvilinear motion**. The analysis and understanding of curved-path motions will eventually lead you to the study of more dramatic things, such as planetary and atomic orbits.

Curvilinear motion is quite easy to analyze using rectangular components of motion. Essentially, you break down the motion into rectangular (x and y) components and look at the straight-line motion in each dimension. To those you can apply the kinematic equations introduced in Chapter 2. To find the position of an object moving in a curved path, you simply find x and y at any time; then the object's position is given as the point (x, y).

Components of motion can be conveniently represented in vector notation: a displacement vector contains the position components (x and y); a velocity vector has the velocity components (v_x and v_y); an acceleration vector has the acceleration components (a_x and a_y). In some instances, aspects of motion can be analyzed by adding vectors directly. Because every vector has both magnitude and direction, the scalar addition of numbers you learned in grade school does not apply. In this chapter, you'll learn how to add vectors, an addition operation that takes direction into account.

3.1 Components of Motion

An object moving in a straight line was considered in Chapter 2 to be moving along one of the Cartesian axes (x or y). However, since the orientation of a set of axes is arbitrary, straight-line motion may also be described in two dimensions. For example, consider a ball moving uniformly across a tabletop as shown in Fig. 3.1. (Considering the components of the ball's straight-line motion will help you understand components of curvilinear motion.)

The ball has a constant velocity (**v**) in a direction at an angle θ relative to the x axis. Note how the velocity vector (arrow) can be resolved, or broken down, into components in the x and y directions, that is, $\mathbf{v}_x$ and $\mathbf{v}_y$, whose magnitudes are

$$v_x = v \cos \theta \qquad\qquad\qquad (3.1a)$$

$$v_y = v \sin \theta \qquad\qquad\qquad (3.1b)$$

The position of the ball, (x, y), or the distance traveled in each of the component directions at time t, is given by

$$x = v_x t \qquad\qquad\qquad (3.2a)$$

$$y = v_y t \qquad\qquad\qquad (3.2b)$$

(with $x_o = y_o = 0$ at $t = 0$). Thus, you may think of the ball moving in the x direction and at the same time moving in the y direction. The distance the ball travels along its actual path is given by $d = \sqrt{x^2 + y^2}$.

Example 3.1   Components of Motion
If the ball in Fig. 3.1 has a velocity of 0.50 m/s at an angle of 37° relative to the x axis, how far does it travel in 3.0 s?

Solution
Given: $v = 0.50$ m/s   *Find*: d (distance)
    $\theta = 37°$
    $t = 3.0$ s

First compute the velocity components v_x and v_y:

$$v_x = v \cos 37° = (0.50 \text{ m/s})(0.80) = 0.40 \text{ m/s}$$

$$v_y = v \sin 37° = (0.50 \text{ m/s})(0.60) = 0.30 \text{ m/s}$$

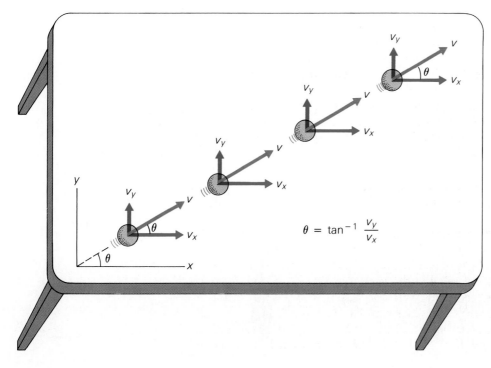

$$\theta = \tan^{-1} \frac{v_y}{v_x}$$

Figure 3.1   **Components of motion**
Even for uniform straight-line motion, the velocity (and displacement) may have x and y components because of the chosen orientation of the coordinate axes. Since the motion is uniform, the ratio v_y/v_x (and therefore θ) is constant.

Then compute the component distances:

$$x = v_x t = (0.40 \text{ m/s})(3.0 \text{ s}) = 1.2 \text{ m}$$

$$y = v_y t = (0.30 \text{ m/s})(3.0 \text{ s}) = 0.90 \text{ m}$$

Thus, the actual path distance is

$$d = \sqrt{x^2 + y^2} = \sqrt{(1.2 \text{ m})^2 + (0.90 \text{ m})^2} = 1.5 \text{ m}$$

Note that this can also be obtained directly from

$$d = vt = (0.50 \text{ m/s})(3.0 \text{ s}) = 1.5 \text{ m} \quad \blacksquare$$

Example 3.2   Across and Down the River

The current of a river that is 200 m wide has a flow rate of 2.5 km/h. A motorboat traveling through the water with a speed of 30 km/h crosses the river (see Fig. 3.2). (a) If the boat is headed directly toward the opposite bank, how far will it travel across *and* down the river in 9.0 s? (b) How long will it take the boat to get to the opposite bank, and how far downstream will its landing point be from the point directly opposite its starting point? (Landing at the constant velocity isn't very realistic, but assume that the boat comes instantaneously to rest on a sandy shore.)

Solution

Given: $y_{max} = 200$ m   *Find:* (a) y and x
$\quad\quad\quad v_x = v_r = 2.5$ km/h = 0.70 m/s   (b) x_{max}
$\quad\quad\quad v_y = v_b = 30$ km/h = 8.3 m/s
$\quad\quad\quad t = 9.0$ s

Note that as the boat moves toward the opposite bank, it is also carried downstream by the current. These velocity components would be apparent to

Figure 3.2   Velocity components
As the boat moves across the river, it is carried downstream by the current. See Example 3.2.

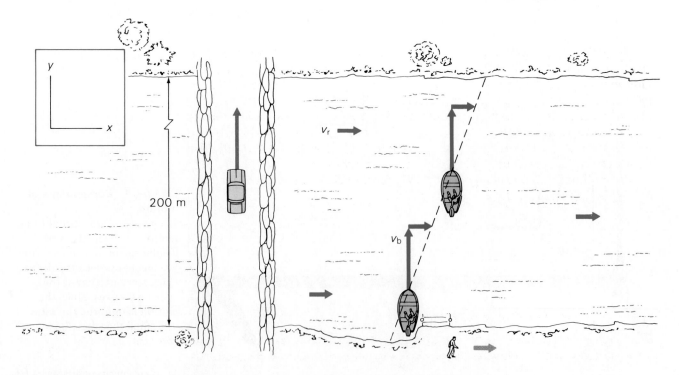

the driver of the car moving across the bridge and the person walking downstream along the straight river bank in Fig. 3.2. If both of these stay even with the boat, the velocity of each will match the respective component of the boat's velocity. Since the velocity components are constant, the boat travels in a straight line, although that line is at an angle to the bridge and the bank.

(a) At 9.0 s, the y component of the boat's displacement is

$$y = v_y t = (8.3 \text{ m/s})(9.0 \text{ s}) = 75 \text{ m}$$

and the current has carried the boat downstream a distance of

$$x = v_x t = (0.70 \text{ m/s})(9.0 \text{ s}) = 6.3 \text{ m}$$

The actual displacement during this time is $d = \sqrt{x^2 + y^2}$ at an angle of $\theta = \tan^{-1}(x/y) = \tan^{-1}(6.3/75) = 4.8°$ relative to the direction of the boat's velocity. [Why does $\tan \theta$ equal x/y instead of y/x as it is usually written? Note that $\theta = \tan^{-1}(v_r/v_b)$ also. As you will learn in the next section, the boat has a net, or effective, velocity in this direction with a magnitude of $v = \sqrt{v_b^2 + v_r^2}$.]
(b) The time to travel 200 m in the y direction is

$$t = \frac{y}{v_y} = \frac{200 \text{ m}}{8.3 \text{ m/s}} = 24 \text{ s}$$

During this time, the boat travels downstream a distance

$$x = v_x t = (0.70 \text{ m/s})(24 \text{ s}) = 17 \text{ m}$$

Checking this result by trigonometry gives

$$x = y \tan \theta = y \left(\frac{v_r}{v_b} \right)$$

$$= (200 \text{ m}) \left(\frac{0.70 \text{ m/s}}{8.3 \text{ m/s}} \right) = 17 \text{ m} \quad \blacksquare$$

The preceding examples involve two-dimensional motion in a plane. And since the velocity is constant, that is, has constant components v_x and v_y, the motion is in a straight line. But motion can also be along a curved path. For the motion of an object to be curvilinear, that is, to vary from a straight-line path, an acceleration is required. Also, this acceleration (a vector) must be at some angle to the velocity (also a vector) that is different from 0° or 180°. If the acceleration were in the direction of the velocity (0°) or opposite to it (180°), the object would just speed up or slow down in a straight line.

For motion in a plane *with a constant acceleration* having components a_x and a_y, the displacement and velocity components are given by

$$x = v_{x_o} t + \tfrac{1}{2} a_x t^2 \tag{3.3a}$$

$$y = v_{y_o} t + \tfrac{1}{2} a_y t^2 \tag{3.3b}$$

$$v_x = v_{x_o} + a_x t \tag{3.3c}$$

$$v_y = v_{y_o} + a_y t \tag{3.3d}$$

with constant acceleration only

A curved path is the result of a deflecting acceleration, which means that the ratio of the velocity components varies with time, or that $\tan \theta = v_y/v_x$ is not constant and the motion is not in a straight line. Consider a ball initially moving along the x axis as illustrated in Fig. 3.3. Assume that starting at time $t_o = 0$ it receives a constant acceleration a_y in the y direction. (You'll learn what produces an acceleration in the next chapter.) The x component of the ball's displacement is given by $x = v_{x_o}t$, since there is no acceleration in the x direction. Prior to t_o, the motion is in a straight line along the x axis. But after t_o, there is a y displacement of $y = \frac{1}{2}a_y t^2$ (with $v_{y_o} = 0$). The result is a curved path.

Note the change in the length of the velocity component, v_y, with time. The total velocity vector is said to be tangent to the curve. It is at an angle θ with the x axis, whose tangent is found from the velocity components; that is, $\tan\theta = v_y / v_x$.

Example 3.3 Curvilinear Motion

Suppose that the ball in Fig. 3.3 has an initial velocity of 1.50 m/s along the x axis and starting at t_o receives an acceleration of 2.80 m/s^2 in the y direction. (a) What is the position of the ball 3.00 s after t_o? (b) What is the velocity of the ball at that time?

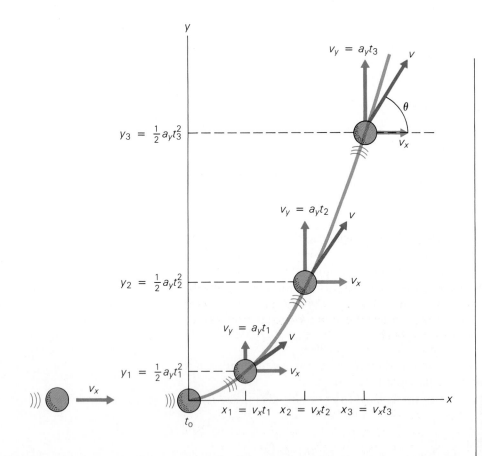

Figure 3.3 **Curvilinear motion**
An acceleration not parallel to the instantaneous motion produces a curved path. Here an acceleration ($\mathbf{a}_y$) is applied at t_o to a ball moving with a constant velocity ($\mathbf{v}_x$). The result is a curved path with the components of motion shown.

Solution

Given: $v_{x_o} = 1.50$ m/s

$\quad\quad\;\; v_{y_o} = 0$

$\quad\quad\;\; a_x = 0$

$\quad\quad\;\; a_y = 2.80$ m/s^2

$\quad\quad\;\; t = 3.00$ s

Find: (a) (x, y)

$\quad\quad$ (b) v

(a) At 3 s after $t = 0$, the ball has traveled the following distances from the origin in the x and y directions:

$$x = v_{x_o}t + \tfrac{1}{2}a_x t^2 = (1.50 \text{ m/s})(3.00 \text{ s}) + 0 = 4.50 \text{ m}$$

$$y = v_{y_o}t + \tfrac{1}{2}a_y t^2 = 0 + \tfrac{1}{2}(2.80 \text{ m/s}^2)(3.00)^2 = 12.6 \text{ m}$$

Thus, its position is $(x, y) = (4.50$ m, 12.6 m$)$. If you computed the distance, $d = \sqrt{x^2 + y^2}$, what would this be?

(b) The x component of the velocity is given by

$$v_x = v_{x_o} + a_x t = v_{x_o} + 0 = 1.50 \text{ m/s}$$

(This is constant since there is no acceleration in the x direction.) The y component of the velocity is

$$v_y = v_{y_o} + a_y t = 0 + (2.80 \text{ m/s}^2)(3.00 \text{ s}) = 8.40 \text{ m/s}$$

The velocity therefore has a magnitude of

$$v = \sqrt{v_x^2 + v_y^2} = \sqrt{(1.50 \text{ m/s})^2 + (8.40 \text{ m/s})^2} = 8.53 \text{ m/s}$$

and its direction relative to the x axis is

$$\theta = \tan^{-1}\left(\frac{v_y}{v_x}\right) = \tan^{-1}\left(\frac{8.40 \text{ m/s}}{1.50 \text{ m/s}}\right) = 79.9° \quad\blacksquare$$

3.2 Vector Addition and Subtraction

Many physical quantities are vectors. You have worked with a few associated with motion and will encounter more throughout this course of study. Therefore, it is important for you to know how to add and subtract vector quantities. These operations are not the same as the analogous scalar ones since both magnitudes and directions are involved.

Vector Addition

In general, there are geometric (graphic) methods and analytical (calculation) methods of **vector addition**. The geometric methods are instructive about the concepts of vector addition because they help you visualize them. But analytical methods are more commonly used because they are faster and more precise. One thing that scalar addition and vector addition have in common is that the quantities being added must have the same units. You cannot meaningfully add a displacement vector to a velocity vector, just as you cannot add the scalars of distance and speed and obtain a result with any physical meaning.

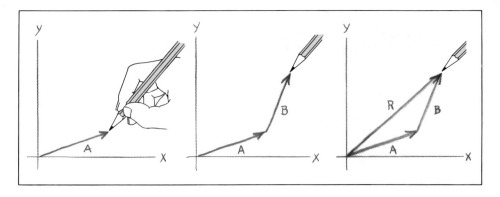

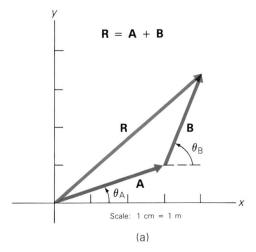

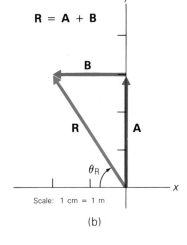

Figure 3.4 **Triangle method of vector addition**
The vectors **A** and **B** are placed tip to tail. The vector from the tail of **A** to the tip of **B**, forming the third side of the triangle, is the resultant, $\mathbf{R} = \mathbf{A} + \mathbf{B}$.

Triangle Method   To add two vectors, for example, to add **B** to **A** (that is, to find **A** + **B**) by the **triangle method**, you first draw **A** on a sheet of graph paper to some scale. For example, if **A** is a displacement in meters, a convenient scale is 1 cm:1 m, or 1 cm of vector length on the graph corresponding to 1 meter of displacement. As shown in Fig. 3.4(a), the direction of the vector is specified as being at an angle (θ) relative to a coordinate axis, usually the x axis. Next draw **B** with its tail starting at the tip of **A**. (Thus, this method is also called the tip-to-tail method.) The vector from the tail of **A** to the tip of **B** is then the vector sum , or the resultant of the two vectors: $\mathbf{R} = \mathbf{A} + \mathbf{B}$.

The physical significance of **R** is that it gives the same displacement as **A** followed by **B**; this is the net, or effective, displacement. Also, since the lengths of two sides of the obtuse triangle are known, an interior angle can be found from the angles of the vector orientations. The length (magnitude) and the orientation of **R** can then be computed using the law of sines or the law of cosines. In the case of a right triangle, such as is illustrated in Fig. 3.4(b), the Pythagorean theorem can be used to find the length of the resultant (the hypotenuse) and the angle of its direction is given by the appropriate inverse trigonometric function.

Parallelogram Method   Another graphical method of vector addition similar to the triangle method is the **parallelogram method**. In Fig. 3.5, **A** and **B** are drawn tail to tail, and a parallelogram is formed as shown. Basically, **B**

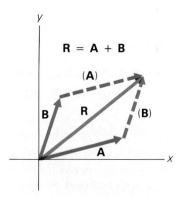

Figure 3.5 **Parallelogram method of vector addition**
The diagonal of a parallelogram formed after **A** and **B** are drawn with their tails at the origin is the resultant, $\mathbf{R} = \mathbf{A} + \mathbf{B}$.

could be moved over to the other side of the parallelogram to form the **A** + **B** triangle. In general, a vector (arrow) can be moved around in vector addition methods. As long as you don't change its length (magnitude) or direction, you don't change the vector. In Fig. 3.5, the shifting of vector arrows shows that **A** + **B** = **B** + **A**.

Polygon Method For the addition of more than two vectors, the tip-to-tail method can be extended to include any number of vectors. The method is then called the **polygon method** because the resulting graphical figure is a polygon.

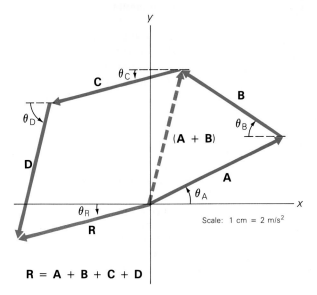

R = A + B + C + D

Figure 3.6 **Polygon method of vector addition**
The vectors to be added are placed tip to tail. The resultant **R** is the vector from the tail of the first vector (**A**) to the tip of the last vector (**D**) to complete a polygon. This method is essentially multiple application of the triangle method: (**A** + **B**) + **C**, and so on.

This is illustrated for four vectors in Fig. 3.6, where **R** = **A** + **B** + **C** + **D**. Note that this addition is essentially three applications of the triangle method. The length and direction of the resultant could be found analytically by successive applications of the laws of sines and cosines, but a much easier analytical method, the component method, is described below.

Vector Subtraction

Vector subtraction is a special case of vector addition:

$$\mathbf{A} - \mathbf{B} = \mathbf{A} + (-\mathbf{B})$$

That is, to subtract **B** from **A**, a *negative* **B** is added to **A**. In Chapter 2, you learned that a minus sign simply means that the direction of a vector is reversed. (The magnitude of a vector is simply a number without a sign, or an absolute value.) The vector −**B** has the same magnitude as the vector **B**, but is in the opposite direction. Figure 3.7 shows the graphical representations of both **A** + **B** and **A** − **B**.

Vector Components and the Analytical Component Method

You have already had an introduction to vector components in the earlier discussion of components of motion. Suppose that **A** and **B**, two vectors at right angles, are added, as illustrated in Fig. 3.8(a). The right angle makes things

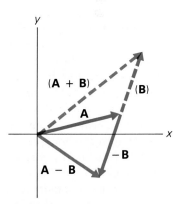

Figure 3.7 **Vector subtraction**
Vector subtraction is a special case of vector addition; that is, **A** − **B** = **A** + (−**B**), where −**B** has the same magnitude as **B** but is in the opposite direction.

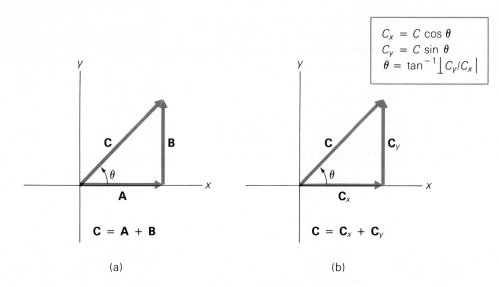

$$C_x = C \cos \theta$$
$$C_y = C \sin \theta$$
$$\theta = \tan^{-1}\lfloor C_y/C_x \rfloor$$

C = A + B

(a)

C = C$_x$ + C$_y$

(b)

Figure 3.8 **Vector components**
(a) The vectors **A** and **B** along the x and y axes, respectively, add to give **C**. (b) A vector **C** may be resolved into rectangular components $\mathbf{C}_x$ and $\mathbf{C}_y$.

easy. The magnitude of **C** is given by the Pythagorean theorem:

$$C = \sqrt{A^2 + B^2}$$

The orientation of **C** to the x axis is given by the angle

$$\theta = \tan^{-1}\left(\frac{B}{A}\right)$$

This is how a resultant is expressed in **magnitude-angle form**.

Fig. 3.8(b) illustrates the opposite situation, showing how a given vector **C** is the sum of its rectangular components, $\mathbf{C}_x$ and $\mathbf{C}_y$. The vector is resolved into these components, which have magnitudes

$$C_x = C \cos \theta \qquad (3.4a)$$

$$C_y = C \sin \theta \qquad (3.4b)$$

(similar to $v_x = v \cos \theta$ and $v_y = v \sin \theta$ in Example 3.1). The angle of direction for **C** can also be expressed in terms of the components since $\tan \theta = C_y/C_x$, or

$$\theta = \tan^{-1}\left(\frac{C_y}{C_x}\right) \qquad (3.5)$$

The angle could also be found using the sine or cosine since $\sin \theta = C_y/C$ and $\cos \theta = C_x/C$.

A vector may be expressed in **component notation**, for example, $\mathbf{C} = C_x\mathbf{x} + C_y\mathbf{y}$, where **x** and **y** are called unit vectors. That is, they have a magnitude of unity, or one, and so simply indicate the components' directions. For example, the ball's displacement from the origin in Example 3.3 could be written as $\mathbf{d} = (4.50\ \text{m})\mathbf{x} + (12.6\ \text{m})\mathbf{y}$. Note that a negative sign associated with a vector component is part of, or goes with, the unit vector that indicates the component's direction, for example, $-C_x\mathbf{x} = C_x(-\mathbf{x})$, or a magnitude of C_x in the negative x direction.

The **analytical component method** of vector addition involves resolving the vectors into components and adding the components. This is illustrated graphically in Fig. 3.9 for two vectors, $\mathbf{F}_1$ and $\mathbf{F}_2$. Note that the sums of the x and y components of the vectors being added are equal to the corresponding components of the resultant vector.

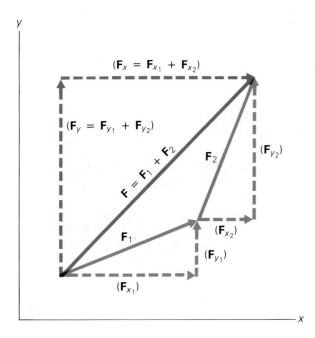

Figure 3.9 **Component addition**
Note that the sums of the x and y components of $\mathbf{F}_1$ and $\mathbf{F}_2$ are the components of the resultant $\mathbf{F}$; that is, $\mathbf{F}_x = \mathbf{F}_{x_1} + \mathbf{F}_{x_2}$ and $\mathbf{F}_y = \mathbf{F}_{y_1} + \mathbf{F}_{y_2}$.

Suppose that you are given three (or more) vectors to add. You could find the resultant by applying the graphical tip-to-tail method as illustrated in Fig. 3.10(a). However, this involves drawing the vectors to scale and using a protractor to measure angles, which is time-consuming. Note that the magnitude of the directional angle θ is not obvious in the figure and would have to be measured (or computed trigonometrically). If you use the component method, you do not have to draw the vectors tip to tail. In fact, it is usually more convenient to put all of the tails together at the origin, as shown in Fig. 3.10(b). Also, the vectors do not have to be drawn to scale, since the sketch is just a visual aid in applying the analytical method.

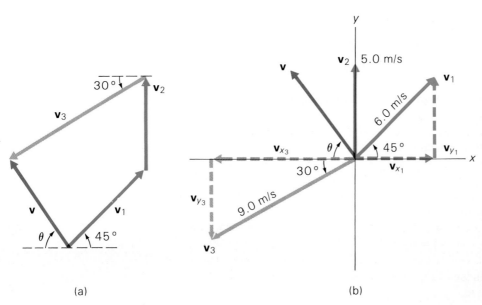

Figure 3.10 **Component method of vector addition**
(a) The vectors to be added are placed tip to tail, and the resultant (**v**) forms the polygon. (b) The vectors to be added are instead extended from the origin so that they may be easily resolved into rectangular components. The summations of the x and y components give the resultant (**v**). See Example 3.4.

Procedure for the Component Method The recommended procedure for adding vectors analytically is as follows:

1. Resolve the vectors to be added into their x and y components. Use the acute angles (those less than 90°) between the vectors and the x axis and indicate the components' directions by plus and minus signs.

2. Add all of the x components together and all of the y components together vectorially to obtain the x and y components of the resultant, or the vector sum.

3. Find the magnitude of the resultant using the Pythagorean formula, and find the angle of direction (relative to the x axis) by taking the arctangent of the ratio of the x and y components. The signs of the components will indicate the quadrant in which the resultant lies, and the arctangent of the absolute value of the component ratio gives the angle relative to the closer part of the x axis, the positive or the negative portion. [Taking the absolute value means to ignore any minus sign: $|-3| = 3$. This is done to avoid negative angles and angles greater than 90°.]

This eliminates working with trigonometric functions for angles greater than 90°. For example, if a vector **v** is in the second quadrant at an angle of 150° relative to the positive x axis, then $v_x = v\cos 150°$ and $v_y = v\sin 150°$. But using the acute angle of 30° between the vector and the negative x axis yields $v_x = -v\cos 30°$ and $v_y = v\sin 30°$. The minus sign indicates that the x component of **v** is in the negative x direction.

Example 3.4   Analytical Component Method

The steps of the analytical method can be applied to the addition of the vectors in Fig. 3.10(b). The vectors with units of m/s represent velocities.

Solution

1. The rectangular components of the vectors are shown in the figure.

2. Summing these components gives

$$\mathbf{v} = (v_{x_1} + v_{x_2} + v_{x_3})\mathbf{x} + (v_{y_1} + v_{y_2} + v_{y_3})\mathbf{y}$$

where

$$v_x = v_{x_1} + v_{x_2} + v_{x_3} = v_1 \cos 45° + v_2 \sin 90° - v_3 \cos 30°$$

$$= (6.0 \text{ m/s})(0.707) + 0 - (9.0 \text{ m/s})(0.866) = -3.6 \text{ m/s}$$

$$v_y = v_{y_1} + v_{y_2} + v_{y_3} = v_1 \sin 45° + v_2 \sin 90° - v_3 \sin 30°$$

$$= (6.0 \text{ m/s})(0.707) + 5.0 \text{ m/s} - (9.0 \text{ m/s})(0.50) = 4.7 \text{ m/s}$$

Note that the directions of the components are indicated by signs and that v_2 has no x component.

   In component notation, the resultant vector may be written

$$\mathbf{v} = (-3.6 \text{ m/s})\mathbf{x} + (4.7 \text{ m/s})\mathbf{y}$$

3. Then the resultant velocity has a magnitude of

$$v = \sqrt{(v_x^2) + (v_y^2)} = \sqrt{(-3.6 \text{ m/s})^2 + (4.7 \text{ m/s})^2} = 5.9 \text{ m/s}$$

Since the x component is negative and the y component is positive, the resultant lies in the second quadrant at an angle of

$$\theta = \tan^{-1}\left|\frac{v_y}{v_x}\right| = \tan^{-1}\left(\frac{4.7 \text{ m/s}}{3.6 \text{ m/s}}\right) = 53°$$

relative to the negative x axis [see Fig. 3.10(b)]. ∎

Although this discussion is limited to motion in two dimensions (in a plane), the component method is easily extended to three dimensions. For a velocity in three dimensions, the vector has x, y, and z components; that is, $\mathbf{v} = v_x\mathbf{x} + v_y\mathbf{y} + v_z\mathbf{z}$.

3.3 Projectile Motion

A familiar example of two-dimensional, curvilinear motion is the motion of objects that are thrown or projected by some means. The motion of a stone thrown across a stream or a golf ball driven off a tee is **projectile motion**. Such motion is analyzed from the instant of projection, with the projectile having a specified initial velocity. A special case of projectile motion in one dimension is free fall when an object is projected vertically. Air resistance will be neglected here, so the projectile motion is free-fall motion.

Projectile motion is easily analyzed using vector components. Essentially, you break up the motion and look at its individual one-dimensional components.

Horizontal Projections

It is worthwhile to first analyze the special case of the motion of an object projected horizontally, or parallel to the Earth's level surface. Suppose that you throw an object horizontally with an initial velocity v_{x_o} [see Fig. 3.11(a) on page 64]. Once the object is released (after time $t = 0$), the horizontal acceleration is zero. Thus, $a_x = 0$ for $t > 0$, and the horizontal velocity is constant, or $v_x = v_{x_o}$.

According to the equation $x = v_x t$ (Eq. 3.2a), the object should continue to travel in the horizontal direction indefinitely. However, you know that this is not what really happens. As soon as the object is released, it is in free fall in the vertical direction, with $v_{y_o} = 0$ (as though it were dropped) and $a_y = -g$. The time of flight of the object horizontally is equal to the time that it takes the object to fall vertically to the ground.

Essentially, the object travels at a uniform velocity in the horizontal direction while *at the same time* it accelerates downward under the influence of gravity (free fall). The result is a curved path. If there were no initial horizontal motion, the object would simply drop to the ground in a straight line.

Note the components of the velocity vector in Fig. 3.11(a). The length of the horizontal component of the velocity vector remains the same, but the length of the vertical component increases with time. What is the instantaneous velocity at any point along the path? (Think in terms of vector addition, covered in the last section.) The photo in Fig. 3.11(b) shows the actual motions of a horizontally projected golf ball and one that is simultaneously dropped from rest. The horizontal reference lines show that the balls fall vertically at the same rate. The horizontally projected ball also travels to the right as it falls.

Example 3.5 Horizontal Projection

Suppose that the ball in Fig. 3.11(a) is projected from a height of 25.0 m above the ground and is thrown with an initial horizontal velocity of 8.25 m/s. (a) How long is the ball in flight before striking the ground? (b) How far from the building does the ball strike the ground?

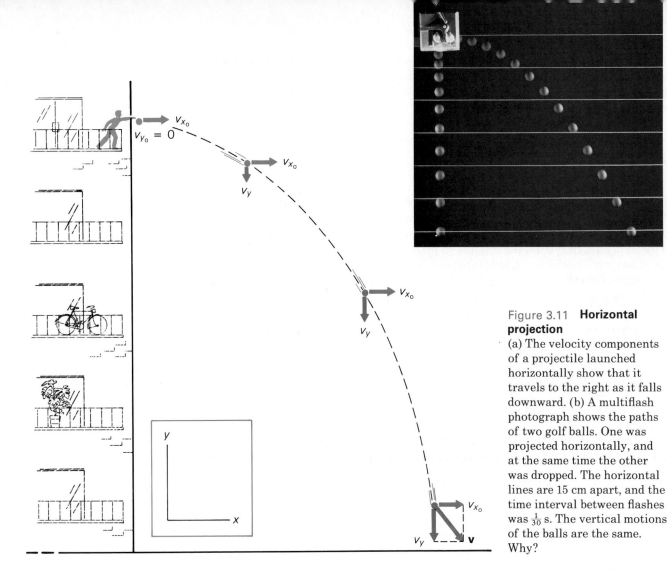

Figure 3.11 **Horizontal projection**
(a) The velocity components of a projectile launched horizontally show that it travels to the right as it falls downward. (b) A multiflash photograph shows the paths of two golf balls. One was projected horizontally, and at the same time the other was dropped. The horizontal lines are 15 cm apart, and the time interval between flashes was $\frac{1}{30}$ s. The vertical motions of the balls are the same. Why?

Solution

Given: $y = -25.0$ m   *Find:* (a) t
$\qquad v_{x_0} = 8.25$ m/s   (b) x_{max}
$\qquad a_y = -g$
$\qquad v_{y_0} = 0$

Downward taken as the negative direction

(a) The time of flight is the same as the time it takes for the ball to fall vertically to the ground. This may be found using the equation $y = v_{y_0}t - \frac{1}{2}gt^2$, in which the negative direction of g is expressed explicitly as was done in Chapter 2. With $v_{y_0} = 0$,

$$y = -\tfrac{1}{2}gt^2$$

and

$$t = \sqrt{\frac{2y}{-g}} = \sqrt{\frac{2(-25.0 \text{ m})}{-9.80 \text{ m/s}^2}} = 2.26 \text{ s}$$

(b) The ball travels in the x direction for the same amount of time it travels in the y direction (that is, 2.26 s); so, with $a_x = 0$,

$$x = v_{x_o}t = (8.25 \text{ m/s})(2.26 \text{ s}) = 18.6 \text{ m} \quad \blacksquare$$

Projections at Arbitrary Angles

The general case of projectile motion is when an object is projected at an arbitrary angle θ relative to the horizontal, for example, when a golf ball is hit (Fig. 3.12). This motion is also analyzed using its components. As before, upward is taken as the positive direction and downward as the negative direction.

First, the initial velocity v_o is resolved into rectangular components:

$$v_{x_o} = v_o \cos \theta \tag{3.6a}$$

$$v_{y_o} = v_o \sin \theta \tag{3.6b}$$

Since there is no horizontal acceleration and gravity acts in the negative y direction, the x component of the velocity is constant and the y component varies with time (see Eq. 3.3):

$$v_x = v_{x_o} = v_o \cos \theta \tag{3.7a}$$

$$v_y = v_{y_o} - gt = (v_o \sin \theta) - gt \tag{3.7b}$$

The components of the instantaneous velocity at various times are illustrated in Fig. 3.12. Note that the instantaneous velocity is tangent to the curved path of the ball at any point.

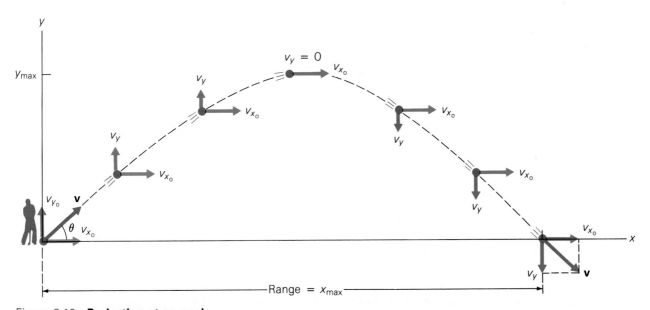

Figure 3.12 **Projection at an angle**
The velocity components are shown for various times. Note that $v_y = 0$ at the top of the arc, or at y_{max}. The range (R) is the maximum horizontal distance, or x_{max}. Essentially, the ball travels up and then down as it travels to the right.

Similarly, the displacement components are given by

$$x = v_{x_o}t = (v_o \cos \theta)t \qquad\qquad (3.8a)$$

$$y = v_{y_o}t - \tfrac{1}{2}gt^2 = (v_o \sin \theta)t - \tfrac{1}{2}gt^2 \qquad\qquad (3.8b)$$

The curve produced by these equations, or the path of motion of the projectile, is called a **parabola**. The path of projectile motion is often referred to as a parabolic arc.

During projectile motion, the object travels up and down while traveling horizontally with a constant velocity. As in the case of horizontal projection, *time is the common feature shared by the components of motion.* Aspects of projectile motion that may be of interest in various situations include the time of flight, the maximum height reached, and the **range**, which is the horizontal distance traveled.

Example 3.6   Projection at an Angle

The golf ball in Fig. 3.12 leaves the tee with an initial velocity of 30.0 m/s at an angle of 37° to the horizontal. (a) What is the maximum height reached by the ball? (b) What is its range?

Solution

Given: $v_o = 30.0$ m/s   *Find:* (a) y_{max}
$\qquad\quad \theta = 37°$ (b) $R = x_{max}$
$\qquad\quad a_y = -g$

Computing v_{x_o} and v_{y_o} explicitly eliminates the sine and cosine terms in the kinematic equations:

$$v_{x_o} = v_o \cos 37° = (30.0 \text{ m/s})(0.799) = 24.0 \text{ m/s}$$

$$v_{y_o} = v_o \sin 37° = (30.0 \text{ m/s})(0.602) = 18.1 \text{ m/s}$$

(a) Just as for an object thrown vertically upward, $v_y = 0$ at the maximum height (y_{max}). Thus, the time to reach the maximum height (t_m) is given by Eq. 3.7b:

$$v_y = 0 = v_{y_o} - gt_m \qquad \text{and} \qquad t_m = \frac{v_{y_o}}{g} = \frac{18.1 \text{ m/s}}{9.80 \text{ m/s}^2} = 1.85 \text{ s}$$

As in the case of vertical projection, the time in going up is equal to the time in coming down, so the total time of flight is $t = 2t_m$ (to return to the elevation from which the object was projected).

The maximum height (y_{max}) is then obtained by substituting t_m into Eq. 3.8b:

$$y_{max} = v_{y_o}t_m - \tfrac{1}{2}gt_m^2 = (18.1 \text{ m/s})(1.85 \text{ s}) - \tfrac{1}{2}(9.80 \text{ m/s}^2)(1.85 \text{ s})^2 = 16.7 \text{ m}$$

The maximum height could also be obtained directly from $v_y^2 = v_{y_o}^2 - 2gy$ with $y = y_{max}$ and $v_y = 0$. However, the method of solution used here illustrates how the time of flight is obtained.

(b) The range (R) is equal to the horizontal distance traveled (x_{max}), which is easily found by substituting the total time of flight $t = 2t_m = 2(1.85 \text{ s}) = 3.70$ s into Eq. 3.8a:

$$R = x_{max} = v_{x_o}t = v_{x_o}(2t_m) = (24.0 \text{ m/s})(3.70 \text{ s}) = 88.8 \text{ m} \quad \blacksquare$$

The range of a projectile is an important consideration in various applications. This is particularly true in those sports where a maximum range is desired, such as when driving a golf ball or throwing a javelin.

At what angle should an object be projected so as to attain its maximum range? To answer this requires looking more closely at the equation used in Example 3.6 to calculate the range, $R = v_x t$. In the example, $v_x = v_{x_o} = v_o \cos \theta$ and $t = 2t_m$, where $t_m = v_{y_o}/g = (v_o \sin \theta)/g$. Then

$$R = v_x t = (v_o \cos \theta)(2t_m) = (v_o \cos \theta)\left(\frac{2v_o \sin \theta}{g}\right) = \frac{2v_o^2 \sin \theta \cos \theta}{g}$$

Using the trigonometric identity $\sin 2\theta = 2 \cos \theta \sin \theta$ gives

$$R = \frac{v_o^2 \sin 2\theta}{g} \tag{3.9}$$

Note that the range depends on the initial velocity (v_o) and the angle of projection (θ). The range for a particular initial velocity (g is assumed to be the same in each case) is a maximum when $\sin 2\theta$ is a maximum, that is, is equal to 1. In that case,

$$R_{\text{max}} = \frac{v_o^2}{g} \tag{3.10}$$

Since $\sin 90° = 1$, if $2\theta = 90°$, then $\theta = 45°$ *for the maximum range for a given initial velocity* (when the projectile returns to the elevation from which it was projected). At a greater or smaller angle, for a projectile with the same initial velocity, the range will be less, as illustrated in Fig. 3.13. Also, the range is the same at equal angles above and below 45°. (Can you prove this?)

In actual situations involving projectiles, other considerations such as air resistance and spin may affect the range. The angle of projection for the maximum range is less than 45° with air resistance. A reverse spin on a driven golf ball provides lift, and the projection angle for the maximum range may then be considerably less than 45°. Also, when an object is projected from a height above the ground (or above its landing elevation), the angle of projection for

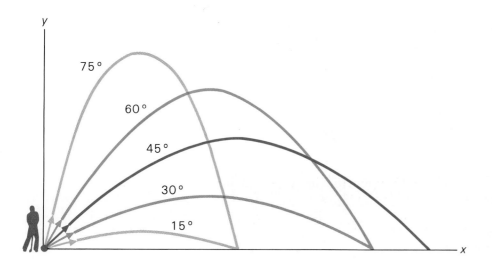

Figure 3.13 **Range**
For a projectile with a given initial velocity, the maximum range is attained with a projection angle of 45°. For angles above and below 45°, the range is shorter and equal for equal differences from 45°.

maximum range is less than 45°. However, when a baseball must be thrown directly in from the outfield to cut a runner off at home plate, a good throw may be one made at a relatively small angle of projection, to cause the ball to hop and minimize its time of flight, rather than maximizing the range.

Example 3.7 Extended Range of a Projectile

A stone thrown from a bridge 20 m above a river has an initial speed of 12 m/s at an angle of 45° to the horizontal (Fig. 3.14). What is the range of the stone?

Solution

Given: $y = -20$ m   *Find:* $R = x_{max}$
$\quad\quad\quad v_o = 12$ m/s
$\quad\quad\quad \theta = 45°$
$\quad\quad\quad a_y = -g$

Eq. 3.9 is not applicable here, since it was derived for $t = 2t_m$. At that time, $y = 0$ and the x coordinate is the range for a symmetrical projectile path (the top part of the arc in the figure). The extended range of the projectile in this case may be found in two ways; both involve finding the total time of flight.

The velocity components are

$$v_{x_o} = v_o \cos 45° = (12 \text{ m/s})(0.707) = 8.5 \text{ m/s}$$

$$v_{y_o} = v_o \sin 45° = (12 \text{ m/s})(0.707) = 8.5 \text{ m/s}$$

The stone travels horizontally and vertically for the same amount of time. This time may be obtained by analyzing the vertical motion. As in Example 3.6, $v_y = 0$ at the maximum height, and the time to reach this height is

$$t_m = \frac{v_{y_o}}{g} = \frac{8.5 \text{ m/s}}{9.8 \text{ m/s}^2} = 0.87 \text{ s}$$

Then

$$y_{max} = v_{y_o} t_m - \tfrac{1}{2} g t_m^2 = (8.5 \text{ m/s})(0.87 \text{ s}) - \tfrac{1}{2}(9.8 \text{ m/s}^2)(0.87 \text{ s})^2 = 3.7 \text{ m}$$

which is the maximum height relative to the bridge (or 23.7 m above the water surface).

After reaching the maximum height, the stone falls downward 23.7 m (as though released from rest) for a time t_d. Since $y = -\tfrac{1}{2} g t^2$,

$$t_d = \sqrt{\frac{2y}{-g}} = \sqrt{\frac{2(-23.7 \text{ m})}{-9.80 \text{ m/s}^2}} = 2.20 \text{ s}$$

Thus, the total time of flight is

$$t = t_m + t_d = 0.87 \text{ s} + 2.20 \text{ s} = 3.07 \text{ s}$$

During this time, the stone travels a horizontal distance of

$$R = x_{max} = v_{x_o} t = (8.5 \text{ m/s})(3.07 \text{ s}) = 26 \text{ m}$$

The other way to find the total time of flight is to find the time when the stone is at $y = -20$ m using the equation

$$y = v_{y_o} t - \tfrac{1}{2} g t^2$$

$$-20 \text{ m} = (8.5 \text{ m/s})t - \tfrac{1}{2}(9.8 \text{ m/s}^2)t^2$$

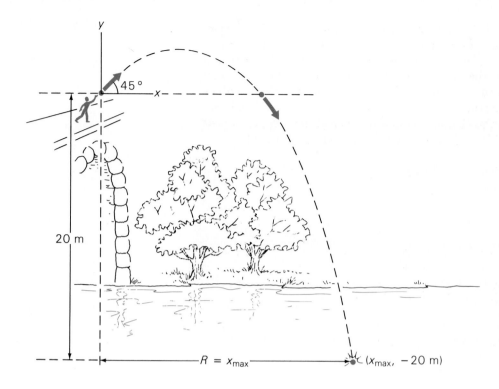

45°

20 m

$R = x_{\text{max}}$

$(x_{\text{max}}, -20 \text{ m})$

Figure 3.14 **Extended range of a projectile**
See Example 3.7.

This is a quadratic equation $(at^2 + bt^2 + c = 0)$. You can use the quadratic formula to find the roots; in this case, the positive root gives the total time (why?). The range may then be found as before. (The quadratic equation for t can always be avoided. Here you could solve $v_y^2 = v_{y_o}^2 - 2gy$ for v_y and then find t from $v_y = v_{y_o} - gt$.) ∎

The preceding example shows that there are a variety of approaches for solving problems involving projectile motion.

Another special case of projectile motion is downward projection at an angle. In Fig. 3.14, this is essentially the stone's motion for the last part of its flight, after it passes below $y = 0$. Since the stone returns to $y = 0$ with the same vertical speed and the horizontal speed is unchanged, the magnitude of the velocity is the same as it was initially and directed at an angle of $-45°$ relative to the horizontal (which you should easily be able to prove).

Example 3.8   Downward Projection at an Angle
Suppose that the boy in Fig. 3.14 throws another stone with an initial speed of 12 m/s at an angle of $-45°$ relative to the horizontal (that is, 45° below the x axis). What is the range of the stone?

Solution
Given: $y = -20$ m   *Find*: $R = x_{\text{max}}$
$v_o = 12$ m/s
$\theta = 45°$ (below horizontal)

The velocity components are found as in the previous example:

$$v_{x_o} = v_o \cos 45° = 8.5 \text{ m/s}$$

$$v_{y_o} = -v_o \sin 45° = -8.5 \text{ m/s}$$

Note that the y component of the initial velocity is negative in this case. You can use $y = v_{y_o}t - \frac{1}{2}gt^2$ to find the time of fall. As noted previously, this involves solving a quadratic equation. You can avoid that by adding an extra step. First, you find the y component of the velocity with which the stone hits the water:

$$v_y^2 = v_{y_o}^2 - 2gy$$

$$= (-8.5 \text{ m/s})^2 - 2(9.8 \text{ m/s}^2)(-20 \text{ m})$$

$$= 72.3 \text{ m}^2/s^2 + 392 \text{ m}^2/s^2 = 464.3 \text{ m}^2/s^2$$

and $\quad v_y = \sqrt{464.3 \text{ m}^2/s^2} = -21.5 \text{ m/s}$

where the negative root is the significant one because of the directional sign.

Then the time of flight (or fall) is found by solving $v_y = v_{y_o} - gt$ for t:

$$t = \frac{v_{y_o} - v}{g} = \frac{-8.5 \text{ m/s} - (-21.5 \text{ m/s})}{9.8 \text{ m/s}^2} = 1.3 \text{ s}$$

Note that this is the difference between the time for the stone to fall from the maximum height in Example 3.7 and the time to reach that height: 2.20 s − 0.87 s = 1.3 s.

The range is then

$$R = x_{\max} = v_{x_o}t = (8.5 \text{ m/s})(1.3 \text{ s}) = 11 \text{ m} \quad \blacksquare$$

3.4 Uniform Circular Motion and Centripetal Acceleration

Another simple, but important, type of motion in a plane is **uniform circular motion**, which occurs when an object moves at a constant speed in a circular path. An example of this is a car going around a circular track (Fig. 3.15). The motion of the moon around the Earth and of some electrons around the nucleus of an atom are approximated by uniform circular motion. Such motion is curvilinear, so you know from the earlier discussion that there must be an acceleration. But what is its magnitude and direction?

Obviously, the acceleration is not in the same direction as the instantaneous velocity (which is tangent to the circular path at any point). If it were, the object would speed up and the motion wouldn't be uniform. Recall that acceleration is the time rate of change of velocity, and velocity has both magnitude and direction. In uniform circular motion, the direction of the velocity is continually changing, which is a clue to the direction of the acceleration.

The velocity vectors at the beginning and end of a time interval give the change in velocity, or $\Delta \mathbf{v}$, via vector addition (or subtraction). As illustrated in Fig. 3.16 (on page 72), since $\Delta \mathbf{v}$ is not zero, there is an acceleration ($\mathbf{a} = \Delta \mathbf{v}/\Delta t$). The vector triangles for several instantaneous velocities have been drawn in the figure. All of the instantaneous velocity vectors have the same magnitude or length (constant speed), but differ in direction.

Note that as Δt (or $\Delta \theta$) becomes smaller, $\Delta \mathbf{v}$ becomes more nearly perpendicular to the instantaneous velocity vectors. As Δt becomes very small,

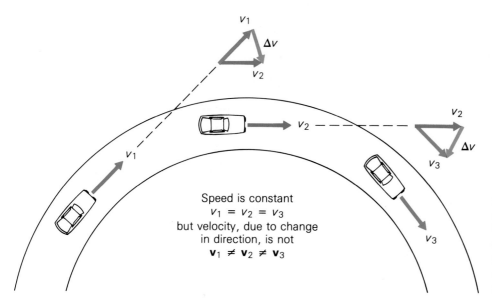

Speed is constant
$v_1 = v_2 = v_3$
but velocity, due to change
in direction, is not
$\mathbf{v}_1 \neq \mathbf{v}_2 \neq \mathbf{v}_3$

Figure 3.15 **Uniform circular motion**
The speed is constant for uniform circular motion, but the velocity changes because of changes in direction. Thus, there is an acceleration.

those velocity vectors become almost parallel and $\Delta\mathbf{v}$ will be approximately perpendicular to them. As Δt approaches zero, the instantaneous change in the velocity, and therefore the acceleration, is toward the center of the circle. As a result, the acceleration is called **centripetal acceleration**, which means center-seeking acceleration (in Latin *centri* means "center" and *petere* means "to move toward").

Without the centripetal acceleration, the motion would not be in a curved path, but in a straight line. Also, the centripetal acceleration must be directed radially inward, that is, with no component in the direction of the (tangential) velocity, or else the magnitude of that velocity will change.

The magnitude of the centripetal acceleration may be deduced from the small triangles in Fig. 3.16(a). (For very short time intervals, the arc length of Δs is almost a straight line.) These two triangles are similar (have the same angle $\Delta\theta$ and the same ratio of lengths of sides). Thus, Δv is to v as Δs is to r, which may be written as

$$\frac{\Delta v}{v} = \frac{\Delta s}{r}$$

For the approximation using short time intervals, $\Delta s \cong v\,\Delta t$, so

$$\frac{\Delta v}{v} \cong \frac{v\,\Delta t}{r}$$

and

$$\frac{\Delta v}{\Delta t} \cong \frac{v^2}{r}$$

Then, as Δt approaches zero, $\Delta v/\Delta t = a_{\mathrm{c}}$, the magnitude of the instantaneous centripetal acceleration, and

$$a_{\mathrm{c}} = \frac{v^2}{r} \tag{3.11}$$

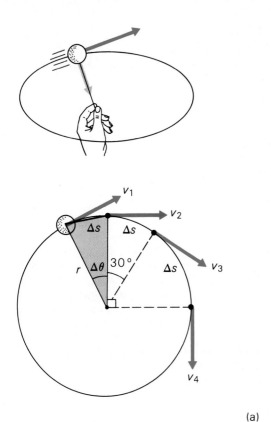

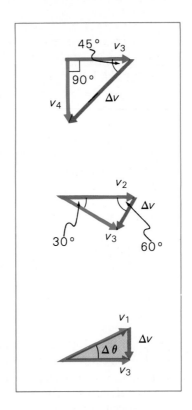

(a)

Figure 3.16 **Analysis of centripetal acceleration** (a) The velocity vector of an object in uniform circular motion is continually changing direction. As Δt (or $\Delta\theta$) is taken smaller and smaller, the vector $\Delta\mathbf{v}$ becomes more nearly perpendicular to the velocity vectors. (b) As Δt approaches zero, $\Delta\mathbf{v}$ (and therefore the acceleration) is directed toward the center of the circle. Analysis shows that this centripetal, or center-seeking, acceleration has a magnitude of $a_c = v^2/r$.

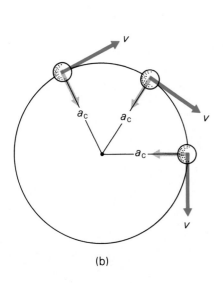

(b)

As Fig. 3.16(b) shows, for an object in uniform circular motion, the direction of the centripetal acceleration is continually changing. In terms of x and y components, a_x and a_y are not constant, and this motion is different from projectile motion.

The concept of centripetal acceleration will be clearer if you look at the displacement components of uniform circular motion for a short time interval,

Δt (Fig. 3.17). The rectangular components in this case are directed tangentially and radially. An object travels a distance $v\,\Delta t$ tangent to the circle, and at the same time it is displaced a distance $\frac{1}{2}a_c(\Delta t^2)$ radially toward the center of the circle. As Δt approaches zero, a circular path is described. Because of its centripetal acceleration, the moon in a sense is continually falling toward the Earth. Fortunately, it never gets here because it is traveling tangentially at the same time.

Suppose that an automobile is moving into a circular curve. To negotiate the curve, the car must have a centripetal acceleration determined by the equation $a_c = v^2/r$. Chapter 4 will describe how this is supplied by friction (a force) between the tires and the road. However, this friction has a maximum limiting value. If the speed of the car is high enough, the friction will not be sufficient to supply the necessary centripetal acceleration, and the car will skid outward from the center of the curve. Or the friction between the tires and the road can be reduced if the car moves onto a wet or icy spot; this produces the same result.

Passengers may consider the car to be sliding toward the outside of the curve, but what is really happening is that the car is not being sufficiently accelerated radially toward the center. If the car hit a wet or icy spot and the friction and centripetal acceleration became essentially zero, the car would skid tangentially to the curve in a straight line. (Imagine swinging a ball on a string in a horizontal circle above your head. If the string broke or you let go, which way would the ball fly off; that is, in what direction would its initial velocity be? Obviously, v_o would be tangential to the circular path. At the instant the string breaks, you'd have a horizontal projectile. Think about it.)

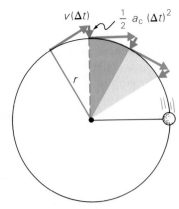

Figure 3.17 **Components of uniform circular motion** In a short time, Δt, an object travels a distance $v\,\Delta t$ tangentially, and a distance $\frac{1}{2}a_c(\Delta t)^2$ radially. Thus, it is essentially falling toward the center of the circle. As Δt approaches zero, these components describe a circular path.

Example 3.9 Centripetal Acceleration
A communications satellite is in a circular orbit about the Earth at an altitude of 5.0×10^2 km. If the satellite makes one revolution every 90 min, what are its (a) orbital speed and (b) centripetal acceleration?

Solution
Given: $h = 5.0 \times 10^2$ km   *Find*: (a) v
$\qquad\quad = 5.0 \times 10^5$ m   (b) a_c
$\qquad\quad t = 90$ min $= 5.4 \times 10^3$ s

(a) The radius of the circular orbit is not h, but $R_e + h$, where R_e is the radius of the Earth (6.4×10^6 m, see Table 6 in Appendix I):

$$r = R_e + h = (6.4 \times 10^6 \text{ m}) + (0.50 \times 10^6 \text{ m}) = 6.9 \times 10^6 \text{ m}$$

The satellite travels the circumference of its circular orbit ($2\pi r$) in the given time, so the tangential speed is

$$v = \frac{2\pi r}{t} = \frac{2\pi(6.9 \times 10^6 \text{ m})}{5.4 \times 10^3 \text{ s}} = 8.0 \times 10^3 \text{ m/s}$$

(b) Then the centripetal acceleration is

$$a_c = \frac{v^2}{r} = \frac{(8.0 \times 10^3 \text{ m/s})^2}{6.9 \times 10^6 \text{ m}} = 9.3 \text{ m/s}^2$$

directed toward the center of the orbit. ∎

Thus, an object in uniform circular motion has a centripetal acceleration of constant magnitude. What if the circular motion is not uniform? In that case, the tangential speed will vary, and there will be a tangential acceleration as well as a centripetal acceleration (Fig. 3.18). The resultant acceleration is the vector sum of these rectangular components: $\mathbf{a} = \mathbf{a}_t + \mathbf{a}_c$. In this case, $\mathbf{a}$, $\mathbf{a}_t$, and $\mathbf{a}_c$ all change *in magnitude and direction*. When the object speeds up (v increases), the centripetal acceleration (v^2/r) must increase to keep the object on a circular path of radius r. (What would happen if that acceleration didn't increase?)

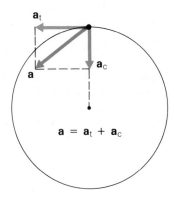

Figure 3.18 **Tangential and centripetal accelerations** In general, an object in circular motion may have both a tangential acceleration ($\mathbf{a}_t$) and a centripetal acceleration ($\mathbf{a}_c$). The total instantaneous acceleration is the vector sum ($\mathbf{a} = \mathbf{a}_t + \mathbf{a}_c$).

Important Formulas

Components of initial velocity:

$$v_x = v_o \cos \theta$$
$$v_y = v_o \sin \theta$$

Components of displacement (constant acceleration only):

$$x = v_{x_o}t + \tfrac{1}{2}a_x t^2$$
$$y = v_{y_o}t + \tfrac{1}{2}a_y t^2$$

Components of velocity:

$$v_x = v_{x_o} + a_x t$$
$$v_y = v_{y_o} + a_y t$$

Component method of vector addition:

$$\mathbf{v} = \mathbf{v}_1 + \mathbf{v}_2 + \mathbf{v}_3 + \cdots$$
$$\mathbf{v}_x = \Sigma \, \mathbf{v}_{x_i} \quad \text{and} \quad \mathbf{v}_y = \Sigma \, \mathbf{v}_{y_i}$$
$$v = \sqrt{v_x^2 + v_y^2}$$
$$\theta = \tan^{-1} \left| \frac{v_y}{v_x} \right|$$

Projectile range ($y_{\text{initial}} = y_{\text{final}}$ only):

$$R = \frac{v_o^2 \sin 2\theta}{g}$$

Centripetal acceleration:

$$a_c = \frac{v^2}{r}$$

Questions

Components of Motion

1. Do you think problems involving motion in three dimensions could be solved using rectangular components? If so, what other equations would be needed?

2. In Example 3.2, what is the actual path of the boat?

3. Comment on the difference in the ratios v_y/v_x and y/x for rectilinear and curvilinear motions.

4. When you are sitting in a stationary car and it is raining (with no wind), the rain falls straight down relative to the car and ground. But when you're driving, the rain appears to hit the windshield at an angle, which is increased as the velocity of the car increases. Explain this effect in terms of velocity components. (Hint: consider the velocity of the rain relative to you. Move your head to the left. Which way does the book appear to move relative to you?)

5. Describe the resulting motion if an object moving in a straight line receives both an acceleration in the direction of motion and a deflecting acceleration. How about an acceleration opposite to the direction of motion and a deflecting acceleration?

Vector Addition and Subtraction

6. What is one thing that scalar addition and vector addition have in common?

7. Show that the parallelogram method of vector addition is essentially the same as successive applications of the triangle method.

8. If **B** is at an angle θ relative to **A**, at what angle is $-\mathbf{B}$ relative to **A**?

9. Is there any limit to the number of components into which a vector may be resolved? Consider (a) rectangular components and (b) components with no restriction on the angles between them.

10. The procedure given for the component method recommends that the angle for the resultant be found by taking the arctangent of the absolute value of the ratio of the component sums. What might happen if the absolute value were not used? (See Example 3.4.) Suppose all angles were measured counterclockwise from the positive x axis. Would this complicate matters?

11. Describe the motion of an object that is initially traveling with a constant velocity and then receives an acceleration (a) in a direction parallel to the initial velocity, (b) in a direction perpendicular to the initial velocity, and (c) that is always perpendicular to the instantaneous velocity.

Projectile Motion

12. Is it possible for a projectile to travel in a circular arc rather than a parabolic one? Explain.

13. A projectile has an initial velocity v_o at an angle θ relative to a horizontal surface. What will be the angle of the final instantaneous velocity (just before the projectile hits the surface) relative to the horizontal?

14. Air resistance was neglected in analyzing projectile motion. What would be the general effects on the path of motion if air resistance were taken into account?

15. Describe the general effect of each of the following on the angle of projection for maximum range of a projectile: (a) launch height above the landing height and (b) air resistance.

Uniform Circular Motion and Centripetal Acceleration

16. What is the average acceleration for an object in uniform circular motion during one revolution? Explain.

17. A ball attached to a string is swung with uniform circular motion in a horizontal circle about a person's head with the string making a small angle relative to the horizontal. (a) Identify the accelerations acting on the ball. (You'll learn in a later chapter that it is impossible for the string to be exactly horizontal. Can you guess why?) (b) Describe the subsequent motion of the ball if the string breaks.

18. A level curve on an interstate highway is a segment of a circle and has a speed limit of 65 mi/h. A cloverleaf exit which feeds directly into an interstate highway is also a circular segment but has a speed limit of only 25 mi/h. Why the difference?

19. Does an object in (a) uniform or (b) nonuniform circular motion have constant tangential and centripetal accelerations? Explain.

Problems

3.1 Components of Motion

■**1.** An object moves with a velocity of 5.0 m/s at an angle of 30° to the x axis. What are the x and y components of the velocity?

■**2.** The y component of a velocity that has an angle of 25° to the x axis has a magnitude of 4.4 m/s. (a) What is the magnitude of the velocity? (b) What is the magnitude of the x component of the velocity?

■3. (a) An object moving in a straight line with a speed of 10 m/s has a velocity with an x component of 8.0 m/s. In what direction is the object moving? (b) What is the value of the y component of the velocity?

■■4. The displacement vector of a moving object initially at the origin has a magnitude of 8.85 cm and an angle of 120° with the positive x axis at a particular instant. What are the coordinates of the object at that instant?

■■5. An airplane flies directly north with a speed of 200 km/h while a west wind blows with a constant speed of 40 km/h. How far does the plane travel during the 2 h of flight?

■■6. A swimmer can maintain a speed of 0.80 m/s when swimming directly toward the opposite shore of a straight river with a current that flows at 3.5 m/s. (a) How far downstream is the swimmer carried in 1.5 min? (b) What is the velocity of the swimmer relative to an observer on the shore?

■■7. After being hit from a corner of the rink, a hockey puck has a constant velocity of 30.0 m/s at an angle of 30° to the side wall along the length of the rink. (Neglect friction.) (a) Using the wall as an axis, what are the components of the puck's velocity? (b) What are the displacement components for the puck 2.5 s after it is hit? (Let the direction along the wall be the x axis.)

■■8. For the situation described in Example 3.2, (a) in what direction should the boat be headed in order to land it at a point directly across the river from its starting point? (b) How long would such a crossing take?

■■9. A ball moving along the y axis with a speed of 3.0 m/s experiences an acceleration of 0.75 m/s² in the x direction when it is at $(x, y) = (0, 1.0 \text{ m})$. What are (a) the velocity components and (b) the position of the ball 4.0 s after the acceleration is applied?

■■10. A ball moves diagonally across a table top from corner to corner with a constant speed of 0.24 m/s. The table top is 3.0 m wide and 4.0 m long. Another ball, starting at the same time as the first one, moves along the longer edge of the table. What constant speed must the second ball have in order to reach the corner at the same time as the ball traveling diagonally?

■■11. A pontoon boat that can move at 3.50 m/s in the water crosses a straight river 200 m wide and arrives at a point 25.0 m upriver from the point directly across from its starting point. If the boat is directed upriver at an angle of 40° (relative to a line directly across the river), what is the speed of the current?

■■12. A jet plane takes off with a constant speed of 140 km/h at an angle of 35°. In 3.00 s, (a) how high is the plane above the ground, and (b) what distance has it traveled from the lift-off point?

■■■13. A ball moving along the x axis with a speed of 2.5 m/s experiences an acceleration of 1.0 m/s² at an angle of 37° to the x axis, starting when the ball is at the origin ($t = 0$) and continuing without interruption. What are the coordinates of the ball at $t = 3.0$ s?

■■■14. A boat that travels at a speed of 10.0 m/s goes directly across a river and back (Fig. 3.19). The current flows at 2.50 m/s. (a) At what angle must the boat be steered? (b) How long does it take to make a round trip? (Assume that the boat's speed is constant at all times and neglect turnaround time.)

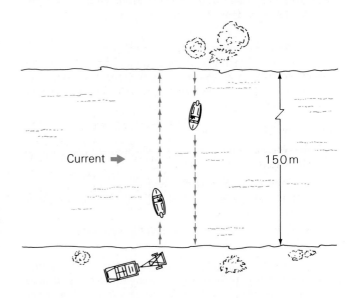

Figure 3.19 **A boat race**
See Problem 14.

3.2 Vector Addition and Subtraction

■15. Using the triangle method, show graphically that (a) $\mathbf{A} + \mathbf{B} = \mathbf{B} + \mathbf{A}$, and (b) if $\mathbf{A} - \mathbf{B} = \mathbf{C}$, then $\mathbf{A} = \mathbf{B} + \mathbf{C}$.

■16. In Chapter 2, the displacement between two points on the x axis was written $\Delta x = x_2 - x_1$. Show what this means graphically in terms of vector subtraction.

■17. Find the rectangular components of a vector $\mathbf{C}$ that has a magnitude of 12.0 m/s and is oriented at an angle of 60° to the positive x axis.

■**18.** The resultant of two rectangular component vectors has a magnitude of 4.5 m and an angle of 37° relative to the positive x axis in the fourth quadrant. What are the component vectors?

■**19.** For each of the following vectors, give a vector that when added to it yields a null vector (a vector with a magnitude of zero): (a) $\mathbf{A} = (3.0 \text{ cm})\mathbf{x}$, (b) $\mathbf{B} = (-10 \text{ cm})\mathbf{y}$, (c) $\mathbf{C} = 8.0$ cm at an angle of 30° relative to the negative x axis in the second quadrant.

■■**20.** For the two vectors $\mathbf{x}_1 = (20 \text{ m})\mathbf{x}$ and $\mathbf{x}_2 = (15 \text{ m})\mathbf{x}$, compute and show graphically (a) $\mathbf{x}_1 + \mathbf{x}_2$, (b) $\mathbf{x}_1 - \mathbf{x}_2$, and (c) $\mathbf{x}_2 - \mathbf{x}_1$.

■■**21.** For three arbitrary vectors, $\mathbf{F}_1$, $\mathbf{F}_2$, and $\mathbf{F}_3$, show graphically that the resultant $\mathbf{F}$ is the same regardless of the order in which the vectors are added. (How many possibilities are there?)

■■**22.** A block rests on an inclined plane. Its weight is a force directed vertically downward, as illustrated in Fig. 3.20. Find the components of the force along the surface of the plane and perpendicular to it.

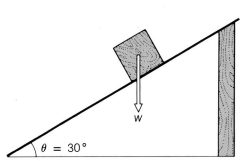

Figure 3.20 **Block on an inclined plane**
See Problem 22.

■■**23.** Find the resultant (or sum) of the vectors in Fig. 3.21.

■■**24.** If **A** in Fig. 3.21 were at an angle of 30° with the x axis, instead of 37°, what would be the resultant (or sum) **A** and **B**?

■■**25.** For the vectors in Fig. 3.21, determine (a) $\mathbf{C}_1 = \mathbf{A} - \mathbf{B}$ and (b) $\mathbf{C}_2 = \mathbf{B} - \mathbf{A}$.

■■**26.** A student works four different problems involving the addition of different vectors $\mathbf{F}_1$ and $\mathbf{F}_2$. He states that the magnitudes of the four resultants are given by

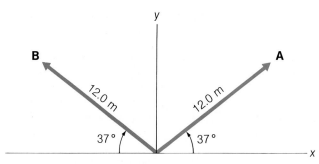

Figure 3.21 **Vector addition**
See Problems 23–25.

(a) $F_1 + F_2$, (b) $F_1 - F_2$, (c) $F = \sqrt{F_1^2 + F_2^2}$, and (d) $F_1 + F_2 = F_1 - F_2$. Is this possible? If so, describe the vectors in each case.

■■**27.** For the vectors in Fig. 3.22, find (a) $\mathbf{v}_1 + \mathbf{v}_2$, (b) $\mathbf{v}_1 + \mathbf{v}_3$, and (c) $\mathbf{v}_2 - \mathbf{v}_1$.

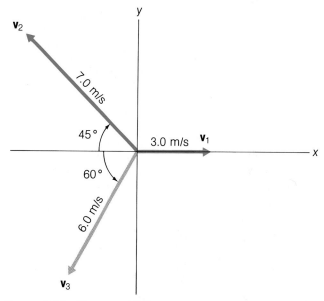

Figure 3.22 **Vector addition**
See Problems 27–29.

■■**28.** Find the resultant (or sum) of the three vectors in Fig. 3.22. (Express in both magnitude-angle form and component notation.)

■■**29.** For the vectors in Fig. 3.22, find the magnitude and orientation of $\mathbf{v}_1 - \mathbf{v}_2 - \mathbf{v}_3$.

■■30. Find the resultant of the displacement vectors $\mathbf{d}_1 = (3.0 \text{ m})\mathbf{x} + (2.0 \text{ m})\mathbf{y}$, $\mathbf{d}_2 = (2.5 \text{ m})\mathbf{x} - (6.0 \text{ m})\mathbf{y}$, and $\mathbf{d}_3 = -(8.0 \text{ m})\mathbf{x} + (1.5 \text{ m})\mathbf{y}$; that is, find $\mathbf{d}_1 + \mathbf{d}_2 + \mathbf{d}_3$. (Express in both component notation and magnitude-angle form.)

■■■31. Find the resultant (or sum) of the four vectors in Fig. 3.23.

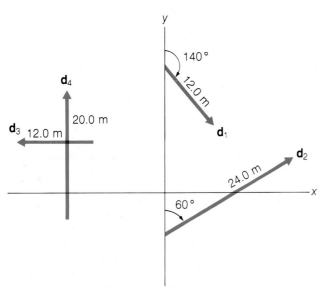

Figure 3.23 **Vector addition**
See Problems 31 and 32.

■■■32. For the vectors in Fig. 3.23, find $(\mathbf{d}_2 - \mathbf{d}_4) + (-\mathbf{d}_1 + \mathbf{d}_3)$.

■■■33. When sitting in a parked car in a drizzling rain with no wind, you note that the rain comes straight down. However, when driving, the rain appears to come toward you at an angle, which increases the faster you go. Show vectorially why the direction of the rain changes relative to you.

3.3 Projectile Motion*

■34. A ball with a horizontal speed of 1.5 m/s rolls off a table 1.0 m tall. (a) How long will it take the ball to reach the floor? (b) How far from a point on the floor directly below the edge of the table will it land?

■35. A spring-loaded gun can project a marble with an initial speed of 4.5 m/s. With the gun lying horizontally on

* Neglect air resistance in solving these problems.

78   CH. 3   Motion in Two Dimensions

a level table 1.0 m above the floor, what is the range of a marble fired from it?

■■36. While warming up, a pitcher throws a fast ball horizontally with a speed of 25.0 m/s. How far will the ball have dropped from its straight-line path in a distance of 18.5 m (the distance from the pitcher's mound to home plate)? Is the answer realistic? Explain.

■■37. A stunt in a movie calls for driving a car at a speed of 100 km/h off a level cliff top 35.5 m above a lake. (a) How long does the stunt driver have to get the door open before the car hits the water? (b) How far from the base of the cliff does the car hit the water?

■■38. A ball is thrown horizontally from the top of a building at a height of 32.5 m above the ground and hits the level ground 56.0 m from the base of the building. (a) What is the initial velocity of the ball? (b) What is the velocity of the ball just before it hits the ground?

■■39. A package is to be dropped from an airplane to hit the ground at a designated spot near some campers. The airplane approaches the spot at an altitude of 245 m above level ground, moving horizontally with a constant velocity of 162 km/h. Having the designated point in sight, the pilot prepares to drop the package. (a) What should the angle between the horizontal and the pilot's line of sight be when the package is released? (b) What is the location of the plane when the package hits the ground?

■■40. A student throws a softball horizontally from a dorm window 15 m above the ground. Another student standing 10 m away from the dorm catches the ball at a height of 1.5 m above the ground. What is the initial velocity of the ball?

■■41. A monster climbs to the top of a building 40 m above the ground and hurls a boulder downward with a velocity of 20 m/s at an angle of 30° below the horizontal. How far from the building does the boulder land?

■■42. A football resting on the ground is kicked at an angle of 40° with an initial speed of 23.0 m/s. (a) What is the maximum height reached by the ball? (b) What is its range?

■■43. The shells fired from an artillery piece have a muzzle velocity of 1.80×10^2 m/s, and the target is 3.00 km away. (a) At what angle relative to the horizontal should the gun be aimed? (b) Could the gun hit a target 3.50 km away?

■■44. Show that the path of a projectile is a parabolic arc. The general form of the equation for a parabola is

$y = ax - bx^2$, where a and b are constants. (Hint: use the appropriate forms of Eq. 3.3a or 3.3b for displacements.)

■■**45.** For projectile motion in general at an angle θ relative to a level surface, sketch graphs of (a) y versus t, (b) x versus t, (c) v_x versus t, (d) v_y versus t, and (e) v versus t.

■■**46.** An artillery shell with a muzzle velocity of 125 m/s is fired at an angle of 35° to the horizontal. If the shell explodes 10.0 s after being projected, where does the blast occur?

■■**47.** A javelin is thrown at angles of 35° and 60° to the horizontal from the same height with the same speed in each case. For which throw does the javelin go farther and how many times farther? (Assume that the landing place is at the same height as the launching.)

■■**48.** An astronaut on the moon fires a projectile from a launcher sitting on a level surface so as to get the maximum range. If the launcher gives the projectile a muzzle velocity of 36 m/s, what is the range of the projectile? (Hint: don't forget where the launcher is and gravitational effects.)

■■**49.** A firefighter holding the nozzle of a hose 50.0 m from a flaming building is just able to reach a third-story window with a stream at a height of 11.0 m above the ground. If the firefighter is holding the nozzle 0.50 m above the ground, what is the speed of the water coming out of it?

■■**50.** At a track and field meet, the best long jump is measured as 8.15 m. The jumper took off at an angle of 30° to the horizontal. (a) What was the jumper's initial speed? (b) If there were another meet on the moon and the same jumper could attain only half of the initial earthly speed, what would be the maximum jump there? [Air resistance does not have to be neglected in part (b). Why?]

■■■**51.** Determine the initial velocity of the horizontally projected golf ball in Fig. 3.11(b).

■■■**52.** A field goal is attempted with the football resting at the center of the field 40 yd from the goalposts. If the kicker gives the ball a velocity of 70.0 ft/s toward the goalposts at an angle of 45° to the horizontal, will the kick be good? (The crossbar of the goalposts is 10 ft above the ground, and the ball must be higher than this when it reaches the goalposts for the field goal to be good.)

■■■**53.** A diver goes off a 4.0-m high springboard with a velocity of 9.0 m/s at an angle of 20° above the horizontal. (a) What is the maximum height of the diver above the water? (b) How far from a point directly below the edge of the board will the diver hit the water?

■■■**54.** An extremely difficult hole on a golf course is one where the golfers have to hit the ball upward to the green. If the front edge of the green is located 100 m from the tee and 15 m above it, at what angle should the ball be hit with an initial speed of 20 m/s so it just lands on the front edge of the green? (Hint: helpful trigonometric identities are $\sin\theta/\cos\theta = \tan\theta$, $1/\cos^2\theta = \sec^2\theta$, and $\sec^2\theta = 1 + \tan^2\theta$.)

■■■**55.** Three identical balls are thrown or projected from a cliff at a particular height (h) above a horizontal plain. One ball is projected at an angle of 45° to the horizontal, the second ball is thrown horizontally, and the third is thrown at an angle of 30° below the horizontal (that is, downward). If all of the balls have the same initial speed, which one will have the greatest speed when it hits the ground? (Justify your answer mathematically.)

■■■**56.** A quarterback passes a football with a velocity of 50 ft/s at an angle of 40° to the horizontal toward an intended receiver 30 yd downfield. The pass is released 5.0 ft above the ground. Assume that the receiver is stationary and that he will catch the ball if it comes to him. Will the pass be completed?

■■■**57.** The apparatus for a popular lecture demonstration is shown in Fig. 3.24. The gun is aimed directly at the can, and when the gun is fired, the can is simultaneously released. This so-called monkey gun won't miss as long as the initial velocity of the bullet is sufficient to reach the falling target before it hits the floor. Verify this, using the figure. (Hint: note that $y_o = x\tan\theta$.)

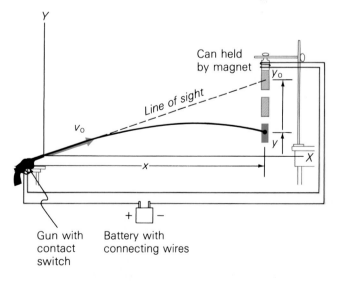

Figure 3.24 **A sure shot**
See Problem 57.

■58. For an object to move uniformly in a circular path
with a radius of 0.45 m at a speed of 0.45 m/s, what is the
necessary centripetal acceleration?

■59. A race car goes around a level, circular track with
a radius of 1.0 km at a speed of 1.2×10^2 km/h. What is the
centripetal acceleration of the car?

■60. A ball is swung at a uniform speed in a horizontal
circle with a radius of 0.75 m. If the ball has a centripetal
acceleration of 5.3 m/s^2, what is its orbital speed?

■■61. A car enters a circular curve with a radius of
curvature of 0.400 km at a constant speed of 83.0 km/h. If
the friction between the road and the car's tires can supply
a centripetal acceleration of 1.25 m/s^2, does the car
negotiate the curve smoothly? Justify your answer.

■■62. Suppose that you swing an object attached to the
end of a string at a constant speed in a horizontal circle
with a radius of 1.5 m. It takes 1.6 s for the object to make
one revolution. (a) What is the magnitude of the tangential
velocity of the object? (b) What centripetal acceleration
are you imparting to the object via the string?

■■63. A ball at the end of a rope being swung in a
horizontal circle has a centripetal acceleration of 12 m/s^2.
If the circular path has a radius of 1.5 m, how long does
it take for one revolution?

■■64. The moon revolves around the Earth in 29.5 days
in a nearly circular orbit with a radius of 3.8×10^5 km.
Assume that the moon's motion is uniform. With what acceleration is it falling toward the Earth?

■■65. The Earth revolves around the Sun in a nearly
circular orbit; its average distance from the Sun is about
1.5×10^8 km. Assume that the Earth's motion is uniform.
(a) Compute the orbital speed in SI units. (Convert the
answer to mi/h to get a more familiar measure of how fast
you are traveling through space.) (b) Compute the centripetal acceleration of the Earth. What supplies this
acceleration?

■■66. A particle traveling with uniform circular motion
at a speed of 0.480 m/s in a path with a radius of 1.75 m
experiences a tangential acceleration of 0.766 m/s^2. If the
particle stays on the same circular path, (a) what is its
centripetal acceleration and (b) its total acceleration at the
end of 3.00 s?

■■■67. Compute the centripetal acceleration of a person

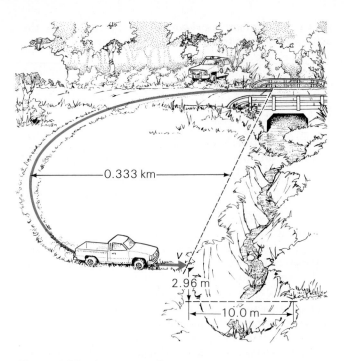

Figure 3.25 **Over the gully**
See Problem 69.

on the rotating Earth (a) at the Equator, (b) at a latitude of
40° N, and (c) at the North Pole.

■■■68. A ball on a rope is swung in a horizontal circle
2.0 m above the ground. The ball has a centripetal acceleration of 24 m/s^2, and the radius of its circular path is 1.5 m.
If the ball suddenly breaks free of the rope, where will it hit
the level ground?

■■■69. For a scene in a movie, a stunt driver drives a
1500-kg pickup truck with a length of 4.25 m around a
circular curve with a radius of curvature of 0.333 km
(Fig. 3.25). The truck is to curve off the road, jump across a
gully, and land on the other side 2.96 m below and 10.0 m
away. What is the minimum centripetal acceleration the
truck must have going around the circular curve so that it
will clear the gully and land on the other side?

Additional Problems

70. Find the resultant (or sum) of the three vectors
$\mathbf{A} = (4.0 \text{ m})\mathbf{x} + (2.0 \text{ m})\mathbf{y}$, $\mathbf{B} = (-6.0 \text{ m})\mathbf{x} + (3.5 \text{ m})\mathbf{y}$, and
$\mathbf{C} = (-5.5 \text{ m})\mathbf{y}$.

71. An object is projected with an initial speed of
24.0 m/s at an angle less than 45°. Its range at this angle is

51.0 m. At what angle greater than 45° could the object be projected at the same initial speed so the range would be the same?

72. A pilot flies a jet plane in a vertical circular arc with a radius of curvature of 3.0 km. (a) What uniform speed would be required for the centripetal acceleration to be equal to the acceleration due to gravity? (b) At the maximum height (the top of the circular arc), what supplies this centripetal acceleration? (c) Speculate on the sensation experienced by the pilot at this point. (Hint: think about being in an elevator with a large downward acceleration.)

73. A ball rolling on a table has velocity with rectangular components $v_x = 0.30$ m/s and $v_y = 0.40$ m/s. What is the displacement of the ball in an interval of 2.5 s?

74. A student walks two blocks west and one block south. If the length of a block is 60 m, what displacement will bring the student back to the starting point?

75. A ditch 2.5 m wide crosses a trailbike path. An upward incline of 15° has been built up on the approach so that the end of the incline is level with the top of the ditch. What is the minimum speed at which trailbike must be moving to clear the ditch? (Add 1.4 m to the range so the back of the bike clears the ditch safely.)

76. A diver cleaves the water at an angle of 45°, and the water is 25.0 m deep. Assume that she maintains a constant underwater velocity with a magnitude of 0.85 m/s. Will she reach the bottom of the pool in 2.00 s?

77. A pickup truck travels on a straight, level highway with a speed of 90.0 km/h. If a person riding in the back of the truck throws an object toward the side of the highway with a speed of 12.5 m/s at an angle of 25° relative to the horizontal, where does the object land? Assume that the object is released 2.20 m above ground level and neglect air resistance.

78. An object is projected from a level surface with an initial velocity of 18.0 m/s at an angle of 30° relative to the surface. (a) What is the maximum height of the projectile? (b) What is its range?

79. If the flow rate of the current in a straight river is greater than the speed of a boat in the water, the boat cannot make a trip directly across the river. Prove this statement.

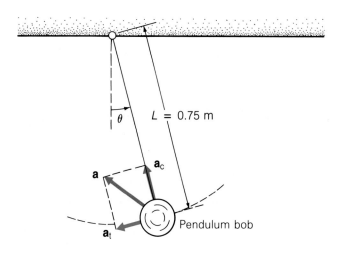

Figure 3.26 **Pendulum swinging**
See Problem 81.

80. A gardener holds a hose nozzle 1.00 m above the ground. If water comes out of the nozzle with a speed of 7.50 m/s, is it possible for the gardener to direct a stream of water on a spot 6.67 m away? If so, at what angle to the horizontal should the nozzle be held? Is there more than one answer to that question? Explain.

81. A pendulum swinging in a circular arc under the influence of gravity, as shown in Fig. 3.26, has both centripetal and tangential components of acceleration. (a) If the pendulum bob has a speed of 2.7 m/s when the cord makes an angle of $\theta = 15°$ with the vertical, what are the magnitudes of the components at this time? (Hint: consider the acceleration due to gravity.) (b) When is the centripetal acceleration a maximum? What is the value of the tangential acceleration at that time?

82. A soccer player kicks the ball, giving it a velocity of 20.0 m/s at an angle of 45° to the horizontal. (a) What is the maximum height reached by the ball? (b) What is its range? (c) How could the range be increased?

83. One ball rolls on a table 1.0 m tall at a speed of 2.0 m/s, and another ball rolls on the floor directly under the first ball and with the same speed and direction. Will the balls collide after the first ball rolls off the table? If so, how far from the point directly below the edge of the table will they be when they strike each other?

84. How are the vectors $\mathbf{A} - \mathbf{B}$ and $\mathbf{B} - \mathbf{A}$ related? Write an equation and also show graphically.

Force and Motion

4

The relationship between force and changes in motion is the focus of this chapter. You know how to analyze motion in terms of kinematics, but you need to know more about what causes motion and changes in motion, that is, to consider the dynamics of motion. If an object is at rest, a shove or a push may be used to set it into motion. That is, a force is applied to the object. Similarly, a moving object may be speeded up or slowed down by applying a force.

There are a variety of types of forces. For example, the gravitational force that attracts people and objects to the Earth gives rise to weight, or weight force. But this force varies a great deal, as you know from picking up different objects (and setting them into motion). There is also the ever present force of friction, which tends to oppose motion. For simplicity, frictional effects are sometimes ignored (as they were in Chapter 2 when air resistance was neglected in analyzing free fall). However, in many instances, friction cannot be ignored and must be included in the analysis of motion. In general, it is important to know that a force is acting on an object and to be able to calculate the effects of the force.

Before learning more about the nature of forces, you need to be familiar with some basic force terminology. A force is a vector quantity and therefore has both direction and magnitude. When several forces act on an object, you will often be interested in their combined effect, or the net force. The **net force** is the vector sum, or resultant, of all the forces acting on an object or system. As illustrated in Fig. 4.1, the net force is zero when forces of equal magnitudes act in opposite directions, and such forces are said to be **balanced forces**. A nonzero net force is referred to as an **unbalanced force**. In this case, the situation can be analyzed as though only this single force were acting.

The study of force and motion occupied many early scientists. It was Isaac Newton who summarized the various relationships and principles into three statements, or laws, which not surprisingly are known as Newton's laws of motion. These laws sum up the concepts of dynamics. In this chapter, you'll learn what Newton had to say about force and motion. (An Insight feature gives you some background on this great scientist.)

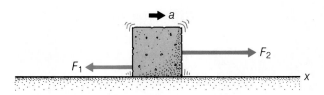

(a) Zero net force (balanced forces)

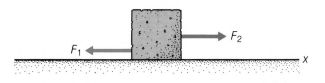

$$F_{net} = F_2 - F_1 = 0$$

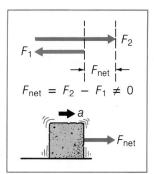

$$F_{net} = F_2 - F_1 \neq 0$$

(b) Nonzero net force (unbalanced forces)

Figure 4.1 Net force
(a) The vector resultant, or net force acting on the block in the x direction, is zero. The forces acting on the block are balanced. (b) The vector resultant is not zero. A nonzero net force, or unbalanced force, acts on the block. This gives a motional effect (which, as you will learn in this chapter, is an acceleration).

4.1 Newton's First Law of Motion

The groundwork for Newton's first law of motion was laid by Galileo. In his experimental investigations, Galileo dropped objects to observe motion under the influence of gravity. However, the relatively large acceleration due to gravity causes dropped objects to move quite fast and quite far in a short time. From the kinematic equations in Chapter 2, you can see that 3 s after being dropped, an object in free fall has a speed of about 29 m/s (64 mi/h) and has fallen a distance of 44 m (about 48 yd, or almost half the length of a football field). Thus, experimental measurements of free-fall distance versus time were particularly difficult to make with the instrumentation available in Galileo's time.

To slow things down so that he could study motion, Galileo used balls rolling on inclined planes. He allowed a ball to roll down one inclined plane and then up other planes, each having a different degree of incline (Fig. 4.2). Galileo noted that the ball rolled to approximately the same height in every case, but it rolled farther in the horizontal direction when the angle of incline was smaller. When allowed to roll onto a horizontal surface, the ball traveled a considerable

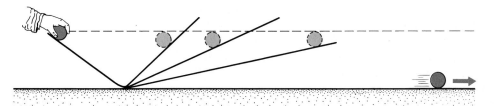

Figure 4.2 Galileo's experiment
A ball rolls further along the upward incline as its angle of incline is decreased. On a smooth, horizontal surface, the ball rolls a great distance before coming to rest. How far would the ball travel on an ideal, perfectly smooth surface?

Isaac Newton (1642–1727)

Isaac Newton, who lived three centuries ago, is thought by many to be the greatest scientist ever (Fig. 1). Newton was definitely the foremost scientific intellect of his time and made many outstanding contributions to physics. Not content to be only a scientist, he was also an administrator, a theologian, and, to some degree, a mystic. So renowned was he that his contemporary, poet Alexander Pope, wrote:

> Nature and Nature's laws lay hid in night:
> God said, *Let Newton be*! and all was light.

Newton's scientific achievements were almost all realized within a relatively short period of his life. In only 22 years (from 1665 to 1687), he produced or formulated the foundation for virtually all his important works.

Newton was born on Christmas Day in 1642 in the village of Woolsthorpe in Lincolnshire, England. The household was in mourning for his father, who had died about two months earlier. In his early years, Isaac showed no particular genius in school and no desire to work on the small farm his mother inherited through a second marriage. Fortunately, the headmaster of his school encouraged Isaac to pursue his education, and he entered Trinity College at Cambridge in 1661. His mother could not provide sufficient financial support, so he entered the university as a subsizar, a subsidized student who had to help pay for his education by doing various tasks.

In 1665, Newton received his B.A. degree and ordinarily would have gone on for a master's degree. However, an epidemic of bubonic plague broke out and the university was closed. Newton returned to Woolsthorpe, and in the following two years he laid the groundwork for many of his contributions to physics, mathematics, and astronomy. In his own recollection of this time, he said, "I was in the prime of my age for invention, and minded mathematics and philosophy [science] more than at any time since."

Figure 1 **Isaac Newton**
Shown here from a painting by Kneller, he is considered by many to be one of the greatest physicists of all time.

Two decades later, at the age of 44, Newton published his famous treatise *Philosophiae Naturalis Principia Mathematica* (Mathematical Principles of Natural Philosophy), in which he set forth his laws of motion and the theory of gravitation (Fig. 2). The publication of *Principia* was financed by Newton's friend Edmund Halley. (Halley used Newton's theories to predict the return of the comet that bears his name.) Although reportedly a shy man, Newton did not hesi-

distance and went even farther when the surface was made smoother. Galileo wondered how far the ball would roll if the horizontal surface could be made perfectly smooth (frictionless). Since this situation was impossible to attain experimentally, Galileo reasoned that in this ideal case (with an infinitely long surface), the ball would continue to travel indefinitely with straight-line, uniform motion, since there would be nothing (no force) to cause its motion to change.

According to Aristotle's theory of motion, which had been accepted for about 1500 years prior to Galileo's time, the normal state of a body was to be at rest (with the exception of celestial bodies, which were naturally in motion). Aristotle probably observed that objects moving on a surface tend to slow down and come to rest, so this conclusion would have seemed logical to him. However,

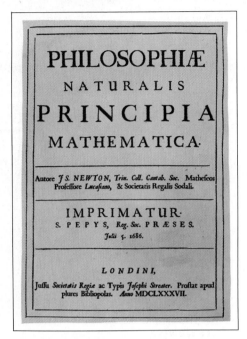

Figure 2 Flyleaf of Newton's *Principia*
This is the title page. Can you read the date of publication at the lower right?

against him. Common speculation held that it was the death (in 1689) of his mother, to whom he was very close, that led to the nervous collapse. However, some evidence seems to indicate that his behavior may have been due to chemical experiments he had carried out with the toxic metals antimony and mercury. Those experiments may have had to do with the production of mirrored surfaces for the reflecting telescope he invented.

Significantly, it is recorded that Newton had a disregard for personal safety and often tasted the products of experiments. He even joked in later years that his hair had turned gray when he was 30 because of his experiments with quicksilver (mercury). Interestingly, a modern analysis of four of Newton's hairs (recovered from personal effects supplied by his descendants) showed abnormally high concentrations of metals, particularly mercury. The symptoms of his madness are similar to those of mercury poisoning, which affects the nervous system; so perhaps that is what he had.*

In any case, the symptoms abated and Newton was again able to pursue a variety of activities. Although his scientific endeavors never equaled those of his earlier days, he was elected to Parliament and served as Warden of the Mint and later Master of the Mint, supervising the enormous task of a complete recoinage program for the country. He was knighted by Queen Anne in 1705.

Sir Isaac Newton died in 1727 at the age of 85. The austere bachelor and giant of science, who explored what he called "the great ocean of truth," was buried with honors in Westminster Abbey in London.

---

tate to defend his theories vigorously. His disputes with contemporaries Robert Hooke (on the nature of light) and Gottfried Leibniz (on the development of the calculus) are well documented.

Records of the time indicate that Newton apparently went mad for a short period in the middle of his career, around 1693. He broke off various relationships with colleagues, could not sleep, and accused friends of plotting

* Mercury compounds were used to treat beaver pelts in the hatting industry in the United States in the nineteenth century, and the early symptoms of mercury poisoning (giddiness, silliness, and strange behavior to the degree of madness) were common among workers. This fact probably gave rise to the phrase "mad as a hatter," as well as the character of the Mad Hatter in Lewis Carroll's *Alice in Wonderland.*

---

from his experiments, Galileo concluded that bodies in motion exhibit the behavior of maintaining that motion, and that if an object is initially at rest, it will remain so, unless something causes it to move.

Galileo called this tendency of an object to maintain its initial state of motion **inertia**. That is,

> Inertia is the property of matter that describes its resistance to changes in motion.

For example, if you've ever tried to stop a slowly rolling automobile by pushing on it, you felt its resistance to a change in motion, to slowing down. Physicists describe the property of inertia in terms of observed behavior, as is the case for all physical phenomena. A comparative example of inertia is illustrated in

In Chapter 1, mass was identified as a base quantity. However, the mass of a given quantity of matter is not totally independent of conditions. The mass of an object depends on its speed and even on its temperature. But these effects are normally too small to detect, and a quantity of matter is generally acceptable as a definition of mass.

Fig. 4.3. You may be quick to note that the large punching bag is heavier than the smaller one and to correlate weight with inertia. You're on the right track, but remember that mass is the fundamental property.

Newton related the concept of inertia to mass. Originally, he called mass a quantity of matter, but later redefined it as follows:

Mass is a measure of inertia.

That is, a large (massive) object has more inertia, or resistance to a change in motion, than a small (less massive) object does. For example, a car has more inertia than a bicycle.

Newton's first law of motion, sometimes called the law of inertia, summarizes these observations:

In the absence of an unbalanced applied force, a body at rest remains at rest, and a body already in motion remains in motion with a constant velocity.*

(See Demonstration 1.)

4.2 The Concept of Force

An operational definition of force is based on observed effects. That is, a force is described in terms of what it does. Either from your own experience or from Newton's first law, you know that a force can produce a change in motion, that is, a change in velocity (speed and/or direction), or an acceleration. Therefore, an observed change in motion, including motion starting from rest, is evidence of a force. A common definition of force is

A force is something capable of changing an object's state of motion.

The word "capable" is very significant here. It takes into account the fact that a force may be balanced by another force or forces acting so that the net force on the object is zero; thus, a force does not necessarily produce a change in motion. An unbalanced force, however, does *produce a change in motion, as evidenced by an acceleration*. In some instances, an applied force may deform an object, that is, change its size or shape. (See Chapter 9.)

Forces are sometimes divided into two classes. The more familiar of these consists of **contact forces**. Such forces arise because of physical contact between objects. For example, when you push on a door to open it or throw or kick a ball, you exert a contact force on the door or ball.

The other class of forces is called **action-at-a-distance forces**. Examples of these forces are gravity, the electrical force between two charges, and the magnetic force between two magnets. The moon is attracted to the Earth by a gravitational force (which supplies the necessary centripetal acceleration to keep the moon in orbit), but there seems to be nothing physically transmitting that force. (Later in this book, you will learn that it is generally believed that something called exchange particles are transmitted between objects experiencing action-at-a-distance forces.)

* What Newton actually said was this: "Every body preserves its state of rest, or of uniform motion in a right [straight] line unless it is compelled to change that state by forces impressed thereon." From *A Source Book in Physics*, by W. F. Magie (Cambridge, MA: Harvard University Press, 1963).

Figure 4.3 **A difference in inertia**
The larger punching bag has more mass, and hence more inertia or resistance to a change in motion.

Newton's First Law and Inertia

According to Newton's first law, an object remains at rest or in motion with a constant velocity unless acted upon by an unbalanced force.

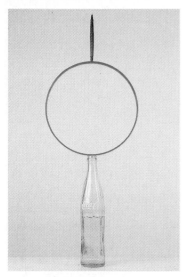

(a) A pen is at rest on an embroidery hoop on top of a bottle.

(b) The hoop is struck sharply, accelerating horizontally. Because the friction between the pen and hoop is small *and* only acts for a short time, the pen does not move appreciably in the horizontal direction. However, there is now an unbalanced force acting vertically upon it—gravity.

(c) The pen falls into the bottle.

4.3 Newton's Second Law of Motion

Since a change in motion, or an acceleration, is evidence of a force, it seems logical that the acceleration is directly proportional to the force, that is, that $a \propto F$. However, as Newton recognized, the inertia or mass of the object also plays a role. For a given force, the more massive the object, the less its acceleration will be; that is, the acceleration is inversely proportional to the mass (m) of the object. Thus,

$$a \propto \frac{F}{m}$$

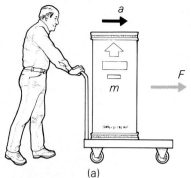

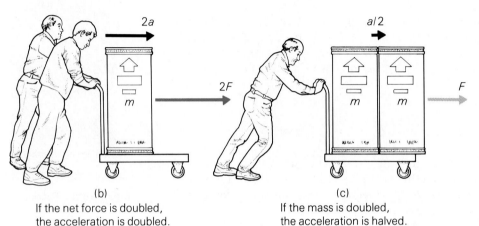

(a)

A net force accelerates the crate $a \propto F/m$

(b)

If the net force is doubled, the acceleration is doubled.

(c)

If the mass is doubled, the acceleration is halved.

Figure 4.4 Newton's second law
The relationships among force, acceleration, and mass shown here are expressed by Newton's second law of motion.

See Fig. 4.4 for illustrations of this relationship.

Newton's second law of motion is commonly expressed in this form:

$$\mathbf{F} = \mathbf{ma} \tag{4.1}$$

In other words, the force is equal to the mass times the acceleration.

> force = mass × acceleration

Thus, if the net force acting on an object is zero, its acceleration is zero, and it remains at rest or in uniform motion, which is consistent with the first law. For a nonzero net force (an unbalanced force), the resulting acceleration is in the same direction as the force.

The SI unit of force is, appropriately, the **newton (N)**. As Eq. 4.1 shows, the base units of the newton are kilograms times meters per second squared:

$$F = ma$$

$$N \equiv (kg)(m/s^2) = kg\text{-}m/s^2$$

That is, a force of 1 N gives a mass of 1 kg an acceleration of 1 m/s^2. The British unit of force is the pound (lb), and 1 lb is equivalent to about 4.5 N (actually 4.448 N). The average apple weighs about 1 N.

To change a proportion into an equation requires the use of a constant of proportionality, generally k. Thus, $F \propto ma$ should be expressed as $F = kma$. As an analogy, the amount of deposit money ($) received for returnable bottles depends on the number (n) returned, $\$ \propto n$. But if you know that you get a nickel per bottle, you can write $\$ = kn$, where $k = \$0.05/\text{bottle}$. The k in $F = kma$ has been assigned a value of one, that is, $1 \text{ N}/(\text{kg-m/s}^2)$, and can therefore be left out.

Eq. 4.1 thus gives the relationship between different quantities of mass and weight. The **weight** of an object is a measure of the gravitational force acting on it, most commonly the gravitational attraction of the Earth. This force is easily demonstrated. When you drop an object, it falls (accelerates) toward the Earth.

The magnitude of the weight of an object on the surface of the Earth is commonly written $w = mg$, where g is the acceleration due to gravity. Note that this is a special case of Newton's second law:

$$F = ma$$

and $w = mg$ (4.2)

The weight of 1.0 kg of mass is then $w = mg = (1.0 \text{ kg})(9.8 \text{ m/s}^2) = 9.8$ N. In places where the metric system is predominant, the weights of objects are commonly expressed in kilograms (or kilos for short), which are really units of mass. Recall that 1.0 kg of mass has an equivalent weight of approximately 2.2 lb (or 9.8 N). Keep in mind that mass is the fundamental property; it doesn't depend on the value of g, but weight does. Although g is taken to have the constant value of 9.80 m/s^2, it actually has slightly different values at different locations near the surface of the Earth. So an object can have slightly different weights at different locations—but its mass is always the same.

The equation $w = mg$ also indicates why all objects in free fall have the same acceleration near the surface of the Earth, where g is considered to be constant. For any object, $w/m = g$. Therefore, an object with twice as much mass as another has twice the weight, or twice the gravitational force acting on it. But the more massive object also has twice the inertia, so twice as much force is needed to accelerate it at the same rate.

The second law applies to any acceleration resulting from an unbalanced force. For example, the centripetal acceleration (a_c) you learned about in Chapter 3 can be said to result from a **centripetal force**, which, in terms of the second law, may be expressed as follows:

$$F_c = ma_c = \frac{mv^2}{r}$$ (4.3)

Note that the F in the equation form of Newton's second law stands for a general force. That is, the equation applies to *any* force that produces an acceleration. Similarly, F_c, the centripetal force in Eq. 4.3, stands for any real force that produces a centripetal acceleration. This might be a gravitational force, an electrical force, or a tension force in a string.

Newton's second law allows you to analyze dynamic situations. In using this equation, you should keep in mind that F is the *magnitude of the net force* and m is the *total mass of the system*. The system is composed of all the objects or masses involved in the given situation. The limits of a system are defined by real or imaginary boundaries. For example, a class might be considered a system, and the total mass of the system is all of the students in it. A system could also be the gas molecules in an arbitrary cubic meter of air. Thus, a dynamic system may consist of one or more discrete masses. You can often isolate a given mass within a system in applications of Newton's second law. Thus, Newton's second law applies to a total system or to any part of it, as the following example illustrates.

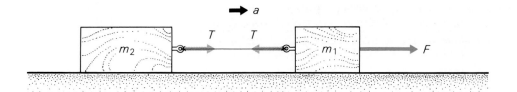

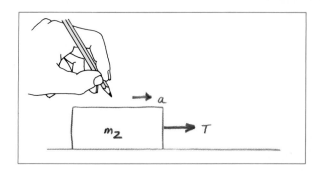

Figure 4.5 **An accelerated system**
See Example 4.1.

Example 4.1   Newton's Second Law—All or Part of the System

Two masses, $m_1 = 2.5$ kg and $m_2 = 3.5$ kg, rest on a frictionless surface and are connected by a light string (Fig. 4.5). A horizontal force of 12.0 N is applied to m_1 as shown in the figure. (a) What is the magnitude of the acceleration of the masses (the system)? (b) What is the magnitude of the tension force (**T**) in the string?

Solution

Given: $m_1 = 2.5$ kg   Find: (a) a
       $m_2 = 3.5$ kg   (b) T
       $F = 12.0$ N

(a) You can find the magnitude of the acceleration directly from the second law:

$$a = \frac{F}{m} = \frac{F}{m_1 + m_2} = \frac{12.0 \text{ N}}{2.5 \text{ kg} + 3.5 \text{ kg}} = 2.0 \text{ m/s}^2$$

The acceleration is in the direction of the applied force, as the figure indicates. Note that m is the *total* mass of the system, or all the mass that is in motion or accelerating. (The mass of the string is small enough to be ignored.)

(b) The tension force in the string may be found by applying the second law to either of the isolated masses, which are also systems. *Newton's second law applies to the total system or to any part of it.* Thus, isolating m_2, as shown in the figure, gives

$$T = m_2 a = (3.5 \text{ kg})(2.0 \text{ m/s}^2) = 7.0 \text{ N}$$

The tension force is transmitted *undiminished* through the string. That is, the magnitude of **T** acting on m_2 is the same as that acting on m_1. (This is true only if the string is assumed to have zero mass. If the mass of the string were taken

When an object is described as being "light," you can ignore its mass in analyzing the problem situation. That is, the mass is negligible relative to the other masses.

into account, the magnitude of **T** would be different for m_1 and m_2; the difference is the net force required to accelerate or move the mass of the string.)

Sketch an isolated m_1 and apply the second law to find T. You should find that

$$F - T = m_1 a$$

or $\quad T = F - m_1 a$

$$= 12.0\ \text{N} - (2.5\ \text{kg})(2.0\ \text{m/s}^2) = 12.0\ \text{N} - 5.0\ \text{N} = 7.0\ \text{N} \quad \blacksquare$$

Not only does Newton's second law hold for any part of a system, it also holds for all the components of motion. Thus, a force may be expressed in component notation for two dimensions:

$$\mathbf{F} = F_x \mathbf{x} + F_y \mathbf{y} = m(a_x \mathbf{x} + a_y \mathbf{y})$$

and $\quad F_x = ma_x$ (4.4a)

$\qquad F_y = ma_y$ (4.4b)

(Also, $F_z = ma_z$ for three dimensions.) An example of how the second law is applied to components of motion is given in the following section, which presents several applications of Newton's second law.

4.4 Applications of Newton's Second Law

The simple relationship expressed by Newton's second law, $F = ma$, allows the quantitative analysis of force and motion. This relationship can be thought of as a cause-and-effect one, with force being the cause and acceleration being the motional effect.

In general, you will be concerned with applications involving constant forces. Constant forces result in constant accelerations for moving objects and allow the use of the kinematic equations from Chapter 2 in analyzing the motion. When there is a variable force, Newton's second law holds for the *instantaneous* force and acceleration, but the acceleration will vary with time. This course of study will generally be limited to average and constant accelerations and forces, as expressed by $F = ma$.

This section presents several examples of applications of Newton's second law so that you may become familiar with its use. This small but powerful equation will be used again and again throughout this course of study. The generality of the dynamic equation is emphasized by the first example, which concerns centripetal force.

Example 4.2 Newton's Second Law and Centripetal Force

Suppose that two masses, $m_1 = 2.5$ kg and $m_2 = 3.5$ kg, are in uniform circular motion on a frictionless surface as illustrated in Fig. 4.6, where $r_1 = 1.0$ m and $r_2 = 1.3$ m. The forces acting on the masses are similar to those in Example 4.1, but in this case $F = 4.5$ N and $T = 2.9$ N. Find (a) the centripetal accelerations and (b) the magnitudes of the tangential velocities (tangential speeds) of the masses.

Solution

Given: $r_1 = 1.0$ m and $r_2 = 1.3$ m   *Find:* (a) $\mathbf{a}_{c_1}$ and $\mathbf{a}_{c_2}$
        $m_1 = 2.5$ kg and $m_2 = 3.5$ kg   (b) v_1 and v_2
        $F = 4.5$ N
        $T = 2.9$ N

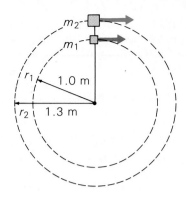

By isolating m_2 in the figure, you can see that the centripetal force is provided by the tension in the string ($\mathbf{T}$ is the only force acting on m_2 toward the center of its circular path). Thus,

$$T = m_2 a_{c_2}$$

and

$$a_{c_2} = \frac{T}{m_2} = \frac{2.9 \text{ N}}{3.5 \text{ kg}} = 0.83 \text{ m/s}^2$$

The acceleration is toward the center of the circle. You can find the tangential speed of m_2 from $a_c = v^2/r$:

$$v = \sqrt{a_{c_2} r_2} = \sqrt{(0.83 \text{ m/s}^2)(1.3 \text{ m})} = 1.0 \text{ m/s}$$

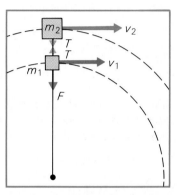

The situation is a bit different for m_1. In this case, the centripetal force is *not* simply $\mathbf{F}$, because Newton's second law requires that you use the net force. The tension force $\mathbf{T}$ also acts radially on m_1, and

$$F_{\text{net}} = F - T = m_1 a_{c_1} = \frac{m_1 v_1^2}{r_1}$$

where the radial direction (toward the center of the circular path) is taken to be positive. Then

$$a_{c_1} = \frac{F - T}{m_1} = \frac{4.5 \text{ N} - 2.9 \text{ N}}{2.5 \text{ kg}} = 0.64 \text{ m/s}^2$$

and

$$v_1 = \sqrt{a_{c_1} r_1} = \sqrt{(0.64 \text{ m/s}^2)(1.0 \text{ m})} = 0.80 \text{ m/s}$$

Figure 4.6 **Newton's second law and centripetal force** See Example 4.2.

Note that the centripetal accelerations are different; that of m_2 (the outer mass) is greater. This is to be expected since it must travel a greater radial distance to maintain a circular orbit. Also note that v_2 is greater than v_1. This is to be expected as well, since m_2 must travel a longer distance than m_1 in the same amount of time (the circular path of m_2 has a larger radius than that of m_1). ∎

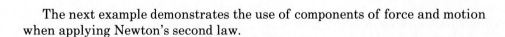

The next example demonstrates the use of components of force and motion when applying Newton's second law.

Example 4.3   Force and Motion

A 5.0-N force is applied at an angle of $30°$ to the horizontal on a 2.0-kg block resting on a frictionless surface (Fig. 4.7). (a) What is the acceleration of the block? (b) If the force is applied for 3.0 s, how far does the block travel in that time? (c) What happens after 3.0 s, if the force is no longer applied?

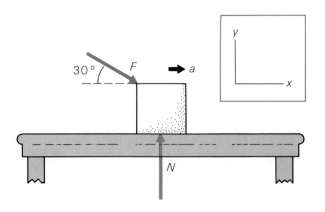

Figure 4.7 **Application of Newton's second law**
See Example 4.3.

Solution

Given: $F = 5.0$ N   *Find*: (a) a
       $m = 2.0$ kg   (b) x
       $t = 3.0$ s   (c) Analyze situation when
       $v_o = 0$ force is no longer applied

(a) The acceleration is given by Newton's second law, but note that *only* a component of the force acts in the direction of the motion. With the coordinate axes shown in Fig. 4.7,

$$F_x = F\cos 30° = ma_x$$

and

$$a_x = \frac{F\cos 30°}{m} = \frac{(5.0 \text{ N})(0.866)}{2.0 \text{ kg}} = 2.2 \text{ m/s}^2$$

(b) The distance traveled in 3.0 s is given by

$$x = v_o t + \tfrac{1}{2}at_x^2 = 0 + \tfrac{1}{2}(2.2 \text{ m/s}^2)(3.0 \text{ s})^2 = 9.9 \text{ m}$$

where $v_o = 0$ since the block is initially at rest.

(c) When the force stops acting, the block will be traveling at a velocity of

$$v = v_o + at = 0 + (2.2 \text{ m/s}^2)(3.0 \text{ s}) = 6.6 \text{ m/s}$$

(in the x direction). By Newton's first law, the block will continue to travel with this velocity until acted on by another force (generally friction for real surfaces).

There is no acceleration in the vertical direction, so $F_y = 0$. This means that the downward y component of **F** *and* the weight force on the block are balanced by the upward force **N** that the table exerts on the block (**N** is commonly called the normal force because it is normal, or perpendicular, to the surface). Thus,

$$F_y = N - F\sin 30° - mg = 0$$

and

$$N = F\sin 30° + mg$$

$$= (5.0 \text{ N})(0.50) + (2.0 \text{ kg})(9.8 \text{ m/s}^2) = 22 \text{ N}$$

The table exerts a force of 22 N on the block. Note that there are three forces acting on the block: **F**, mg (force due to gravity), and **N**. ∎

You can also use Newton's second law to find the force from the motional effects, as the following example shows.

Example 4.4   Finding a Force from Motional Effects

A car traveling at 72.0 km/h along a straight road is brought uniformly to a stop in a distance of 40.0 m. If the car weighs 8.80×10^3 N, what is the braking force?

Solution

Given: $v_0 = 72.0$ km/h $= 20.0$ m/s   Find: F
$\qquad v = 0$
$\qquad x = 40.0$ m
$\qquad w = 8.80 \times 10^3$ N

The constant acceleration may be found from the equation $v^2 = v_0^2 + 2ax$:

$$a = \frac{v^2 - v_0^2}{2x} = \frac{0 - (20.0 \text{ m})^2}{2(40.0 \text{ m})} = -5.00 \text{ m/s}^2$$

The minus sign indicates that the acceleration is opposite to v_0, as expected for a braking force, which slows the car.

The mass of the car is obtained from its weight ($m = w/g$), giving

$$F = ma = \left(\frac{w}{g}\right)a$$

$$= \left(\frac{8.80 \times 10^3 \text{ N}}{9.80 \text{ m/s}^2}\right)(-5.00 \text{ m/s}^2) = -4.49 \times 10^3 \text{ N} \quad \blacksquare$$

Example 4.5   The Atwood Machine

The Atwood machine consists of two masses suspended from a fixed pulley, as shown in Fig. 4.8(a). If $m_1 = 0.55$ kg and $m_2 = 0.80$ kg, what is the acceleration of the system? (Consider the pulley to be frictionless and the masses of the string and the pulley to be negligible.)

Solution

Given: $m_1 = 0.55$ kg   Find: a
$\qquad m_2 = 0.80$ kg

Since m_2 is greater than m_1, it is obvious that m_2 will fall and m_1 will rise (with the same magnitude of acceleration). The pulley is simply a direction changer, and the horizontal situation shown in Fig. 4.8(b) is equivalent to that in (a). Applying Newton's second law to the system as a whole gives

$$F_2 - F_1 = m_2 g - m_1 g$$

$$= (m_1 + m_2)a$$

net force = total mass × acceleration

Solving for a gives

$$a = \frac{(m_2 - m_1)g}{m_1 + m_2}$$

$$= \frac{(0.80 \text{ kg} - 0.55 \text{ kg})(9.8 \text{ m/s}^2)}{0.55 \text{ kg} + 0.80 \text{ kg}} = 1.8 \text{ m/s}^2$$

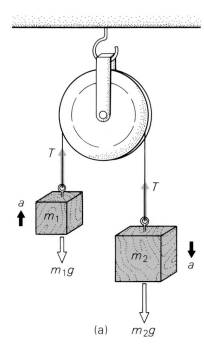

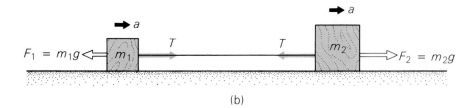

Figure 4.8 **The Atwood machine**
See Example 4.5. (a) A fixed pulley is simply a direction changer. (b) An equivalent horizontal situation (frictionless surface).

(a)   (b)

Note that the tension forces in the string cancel out and need not be considered in finding the solution. However, the problem may also be worked by applying Newton's second law to each isolated mass on which the tension acts. This gives two equations:

$$T - m_1g = m_1a$$

$$m_2g - T = m_2a$$

where the direction of the acceleration for each mass (upward for m_1 and downward for m_2) is taken as positive so as to avoid a minus sign for a. The magnitudes of the accelerations of the masses are equal. Eliminating T from the two equations gives the equation derived above:

$$m_2g - m_1g = m_2a + m_1a = (m_1 + m_2)a$$

Solving the problem this way gives you an equation containing T (actually, two of them) in case you need to find its value, for example, to see whether it exceeds the tensile strength of the string (the force that would cause the string to break). Once the acceleration is found, you simply use it in either of the above two equations, solved for T. Using the first equation gives

$$T = m_1a + m_1g = m_1(a + g)$$

$$= (0.55 \text{ kg})(1.8 \text{ m/s}^2 + 9.8 \text{ m/s}^2) = 6.4 \text{ N} \quad \blacksquare$$

Incidentally, the Atwood machine is named after George Atwood (1746–1807), who used the arrangement to study motion and measure the value of g. Obviously, the masses can be chosen to minimize the acceleration, making it easier to measure the time of fall. The next example concerns a variation of Atwood's machine, where one of the masses is on an inclined plane.

Example 4.6 Motion on a Frictionless Inclined Plane

In Fig. 4.9, two masses are connected by a light string running over a light, frictionless pulley. One mass ($m_1 = 5.0$ kg) is on a frictionless 20° inclined plane, and the other ($m_2 = 1.5$ kg) is freely suspended. What is the acceleration of the system?

Solution

Given: $m_1 = 5.0$ kg   *Find*: a
$\quad\quad\quad m_2 = 1.5$ kg
$\quad\quad\quad \theta = 20°$

Assume that m_1 moves up the plane, as indicated in the figure. [Note that the weight force ($m_1 g$) has a component that acts down the plane. The other component acts perpendicularly to the plane and is balanced by the normal force of the plane on the block.] Applying the second law to the system gives

$$m_2 g - m_1 g \sin 20° = (m_1 + m_2)a$$

or $\quad a = \dfrac{m_2 g - m_1 g \sin 20°}{m_1 + m_2}$

$$= \frac{(1.5 \text{ kg})(9.8 \text{ m/s}^2) - (5.0 \text{ kg})(9.8 \text{ m/s}^2)(0.34)}{5.0 \text{ kg} + 1.5 \text{ kg}}$$

$$= -0.31 \text{ m/s}^2$$

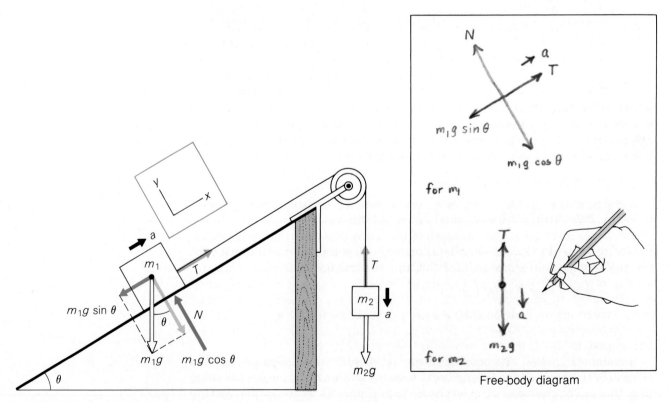

Figure 4.9 **Application of Newton's second law**
See Example 4.6.

Free-body diagram

The minus sign indicates that the acceleration is opposite to the assumed direction. That is, m_1 actually moves down the plane and m_2 rises. As this example shows, if you assume the motion to be in the wrong direction, the sign on the result will tell you. ∎

4.5 Newton's Third Law of Motion

Newton formulated a third law, which concerns the forces involved in seat belt safety. When the brakes are suddenly applied in a moving car, you continue to move forward (the frictional force on the seat of your pants is not enough to stop you). In doing so, you exert a force on the seat belt and shoulder strap and it exerts a force on you, and you slow down with the car. If you haven't buckled up, you may keep on going (Newton's first law) until another applied force, such as that applied by the dashboard or windshield, slows you down.

We commonly think of single forces. However, Newton recognized that it is impossible to have a single force. He observed that in any application of force, there is always a mutual interaction, and forces always occur in pairs. An example given by Newton was this: if you press on a stone with a finger, then the finger is also pressed by, or receives a force from, the stone.

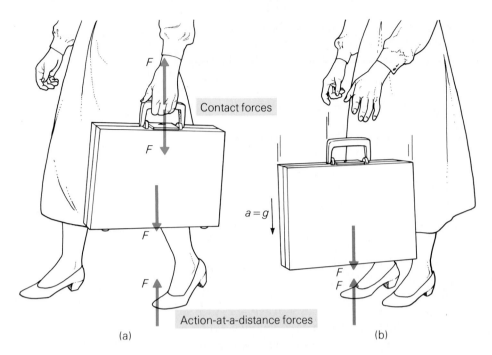

Figure 4.10 **Force pairs of
Newton's third law**
(a) When the person holds the
briefcase, there are two force
pairs: a contact pair and an
action-at-a-distance pair.
Note that the net force acting
on the briefcase is zero. The
upward contact force
balances the downward
weight force. (b) When the
briefcase is falling, there is an
unbalanced force acting on
the case (its weight force),
and it accelerates downward
(at *g* for free fall).

Newton termed the paired forces action and reaction, and **Newton's third
law of motion** is

> For every action (force), there is an equal and opposite reaction
> (force).

$$F_{12} = -F_{21}$$

That is, if object 1 exerts a force on object 2 (F_{12}), then object 2 exerts an
equal and opposite force on object 1 ($-F_{21}$, where the opposite direction is
indicated by the minus sign). Which force is the action or reaction is arbitrary.

The third law may seem to contradict the second law. If there are always
equal and opposite forces, how can there be a nonzero net force? *An important
thing to remember about a force pair of the third law is that the opposing forces act
on different objects.* The second law is concerned with force(s) acting on a
particular object (or system). The opposing forces of the third law act on
different objects. For example, in Fig. 4.10(a), two force pairs are acting when
the person is holding the briefcase. There is a pair of contact forces: the person's
hand exerts an upward force on the handle, and the handle exerts an equal
downward force on the person's hand. There is also a pair of action-at-a-distance
forces associated with gravitational attraction: the Earth attracts the briefcase
(its weight force), and the briefcase attracts the Earth. The contact force on the
person's hand is equal to the action-at-a-distance weight force, but they are not
the same force.

On the isolated stationary briefcase, the net force is zero, as is required by
Newton's first and second laws. However, when the person drops the briefcase
[Fig. 4.10(b)], there is then a nonzero net force on the case and it accelerates. In

this situation, there is still a pair of third law forces *acting on different objects*—the briefcase and the Earth.

Fig. 4.11, on the other hand, illustrates two non–third law forces acting when the briefcase is left on a table. The gravitational situation here is similar. The Earth attracts the briefcase (**F** is the weight of the briefcase), and the briefcase attracts the Earth (the vector arrow is not shown). But, the case exerts a (contact) force on the table, and the table exerts an upward force on the case. This force on the case is called a **normal force** because it is *normal, or perpendicular, to the surface.* ["Normal" refers only to the orientation of the force relative to the surface.] The normal force is equal and opposite to the weight force for an object on a horizontal surface, but it is *not* the partner of the weight force in a third law pair. Note that the weight force (**F**) and the normal force (**N**) both act on the *same* object, the briefcase. The normal force is an important consideration when frictional effects are included in the analysis of forces.

In most instances when you are applying Newton's second law, you need to consider only those forces acting on an object. But the reaction forces of the third law are always there. For example, for the arrangement in Fig. 4.12(a), you immediately perceive that the weight is pulling on the wall. But the wall also pulls on the weight (via the string), as you would realize if you were substituted for the wall in an equivalent situation [Fig. 4.12(b)]. For another look at Newton's third law see Demonstration 2.

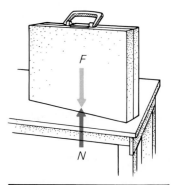

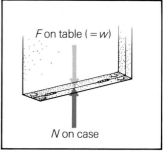

Figure 4.11 **Normal force**
The force exerted by a surface on an object (**N**) is called the normal force because it is perpendicular, or normal, to the surface. Note that **N** is equal and opposite to **F**, but not its partner in a third law force pair since the forces act on the same object. With a horizontal surface, *F* is equal in magnitude to the weight force (*w*), and thus $N = w = mg$.

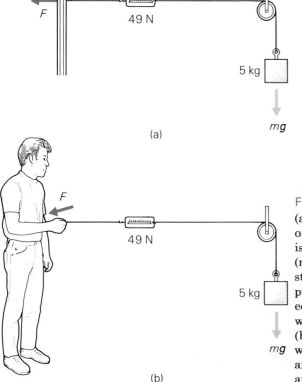

(a)

(b)

Figure 4.12 **Third law forces**
(a) The object exerts a force on the wall via the string that is equal to the object's weight (neglecting the mass of the string and friction of the pulley). The wall exerts an equal and opposite force, which may not be obvious. (b) However, replace the wall with yourself, and the equal and opposite force becomes apparent.

Tension in a String

A demonstration showing action and reaction forces.

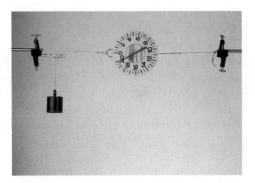

(a) Two 2-kg masses are suspended on each side of a scale (calibrated in newtons). The total suspended weight is $w = mg = (4.0 \text{ kg})(9.8 \text{ m/s}^2) = 39.2$ N, yet, the scale reads about 20 N. Is something wrong with the scale?

(b) No, think of it in this manner. The weight of one mass supplies the reaction force that keeps the scale stationary, while the other mass stretches the scale spring giving a reading of 20 N [or $w = mg = (2.0 \text{ kg})(9.8 \text{ m/s}^2) = 19.6$ N]. The reaction force can be equivalently supplied by a fixed support.

(c) A fixed pulley merely changes the direction of the force—the fixed support or reaction force can be vertical with the same effect. In all cases, the tension in the string is 19.6 N.

4.6 Friction

Friction refers to the ever present resistance to motion between materials, or media. This resistance occurs with all types of media—solids, liquids, and gases—and is characterized as the **force of friction**. Until this point, examples and problems have ignored friction (and air resistance) for simplicity. Now that you know how to describe motion, you are ready to consider situations that are more realistic in that the effects of friction are included.

In some real situations, we want to increase friction, for example, by putting sand on an icy road or sidewalk to improve traction. This might seem contradictory, since an increase in friction presumably would increase the resistance to motion. However, if you analyze the forces illustrated in Fig. 4.13, you'll see that the direction of the force of friction prevents slipping (or the spinning of an automobile wheel). In other situations, we try to reduce friction by lubricating moving machine parts, to lessen wear and expenditure of energy. Automobiles would not run without friction-reducing oils and greases.

This section is concerned chiefly with friction between solid surfaces. All surfaces are microscopically rough, no matter how smooth they appear or feel. It was originally thought that friction was primarily due to the mechanical interlocking of surface irregularities, or asperities (high spots). However, research has shown that the friction between the contacting surfaces of ordinary solids (metals in particular) is mostly due to local adhesion. When

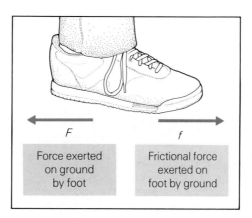

Figure 4.13 **Friction and walking**
Note that the force of friction is shown in the direction of the walking motion. This may seem wrong at first glance, but it's not. The force of friction *resists* the motion of the foot in slipping backwards, so it will stay in place while the other foot is brought forward. The frictional force balances internal forces.

Force exerted on ground by foot

Frictional force exerted on foot by ground

surfaces are pressed together, local welding or bonding occurs in a few small patches where the largest asperities make contact. To overcome this local adhesion, a force great enough to pull apart the bonded regions must be applied. Once contacting surfaces are in relative motion, another form of friction may result when the asperities of a harder material dig into a softer material, with a "plowing" effect.

Friction between solids is generally classified into three types: static, sliding (kinetic), and rolling. **Static friction** includes all cases where the frictional force is sufficient to prevent relative motion between surfaces. **Sliding friction**, or **kinetic friction**, occurs where there is relative tangential motion at the interface of the surfaces in contact. **Rolling friction** exists when one surface rotates and does not slip or slide at the point or area of contact with another surface. Rolling friction, such as occurs between a train wheel and a rail, is attributed to local deformations in the contact region. This type of friction will be considered further in Chapter 8, along with rotational motion.

This section will consider the forces of friction on stationary and sliding objects. These are called the force of static friction and the force of kinetic (or sliding) friction. Experimentally, it is found that the force of friction depends on both the nature of the two surfaces and the load, or the force with which the surfaces are pressed together. For an object on a horizontal surface, this force is equal to the object's weight. However, as shown in Fig. 4.9, on an inclined plane only a component of the weight force contributes to the load. Thus, to avoid confusion, you should remember that the force of friction is proportional to the normal force ($f \propto N$), that is, the force perpendicular to the surface acting on an object (the force exerted by the surface on the object). In the absence of other perpendicular forces, the normal force is equal in magnitude to the component of the weight force acting perpendicular to the surface, by the second law.

The force of static friction (f_s) between parallel surfaces in contact is in the direction that opposes relative motion between the surfaces. The magnitude has different values such that

$$f_s \leq \mu_s N \tag{4.5}$$

where μ_s is the **coefficient of static friction**. (Note that it is a dimensionless constant. Why?) You might wonder how the force of static friction can have more than one value, or different magnitudes. In Fig. 4.14(a), one person pushes on a file cabinet and it doesn't move. With no motion, the net force on the cabinet is zero, and $F - f_s = 0$, or $F = f_s$. Suppose that a second person also

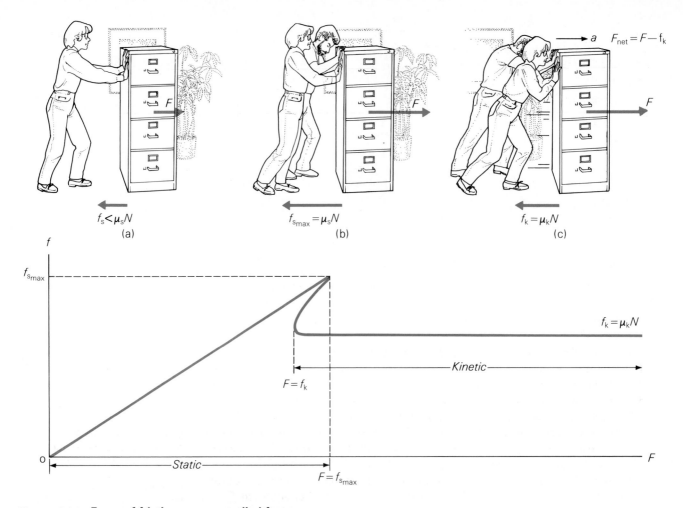

Figure 4.14 **Force of friction versus applied force**
(a) In the static region of the graph, as the applied force F increases, so does f; that is, $f = F$ and $f_s < \mu_s N$. (b) When the applied force exceeds $f_{s_{max}} = \mu_s N$, the file cabinet is set into motion. With $F = f_k$, the cabinet moves with a constant velocity. Note that $f_k < f_s$. (c) When $F > f_k$, there is a net force, and the cabinet is accelerated.

pushes and the file cabinet still doesn't budge [Fig. 4.14(b)]. Then f_s still equals F but has increased, since the applied force has been increased. Finally, if the applied force is made large enough to overcome the static friction, motion occurs. The greatest, or maximum, force of static friction is exerted just before the cabinet starts to slide, and for this case Eq. 4.5 can be written with an equals sign:

$$f_{s_{max}} = \mu_s N \qquad (4.6)$$

Once an object is in motion, or sliding, there is a force of kinetic friction in the direction opposite to the motion and having a magnitude of

$$f_k = \mu_k N \qquad (4.7)$$

where μ_k is the **coefficient of kinetic friction** (sometimes called the coefficient of sliding friction). Generally, the coefficient of kinetic friction is less than the coefficient of static friction ($\mu_k < \mu_s$) for two surfaces, which means that the

Table 4.1
**Approximate Values for Coefficients of
Static and Kinetic Friction between
Certain Surfaces.**

Friction between Materials	μ_s	μ_k
aluminum on aluminum	1.05	1.40
glass on glass	0.94	0.35
rubber on concrete		
dry	1.20	0.85
wet	0.80	0.60
steel on aluminum	0.61	0.47
steel on steel		
dry	0.75	0.48
lubricated	0.12	0.07
Teflon on steel	0.04	0.04
Teflon on Teflon	0.04	0.04
waxed wood on snow	0.05	0.03
wood on wood	0.58	0.40

force of kinetic friction is less than $f_{s_{max}}$ as illustrated in Fig. 4.14. The coefficients of friction between some common materials are listed in Table 4.1.

Note that the force of static friction (f_s) exists in response to an applied force. The magnitude of f_s and its direction depend on the magnitude and direction of the applied force. Up to its maximum value, the force of static friction is equal and opposite to the applied force (F), since there is no motion ($F - f_s = 0 = ma$). Note that if the person in Fig. 4.14(a) pushed on the cabinet in the opposite direction, f_s would be in the opposite direction. If there were no applied force F, then f_s would not exist. When F exceeds $f_{s_{max}}$, the block slides and kinetic friction comes into effect, with $f_k = \mu_k N$. If F is equal to f_k, the cabinet will slide with a constant velocity; and if F is greater than f_k, it will accelerate.

Also, it has been experimentally determined that the coefficients of friction (and therefore the forces of friction) are nearly independent of the size of the contact area between metal surfaces. This means that the force of friction between a brick-shaped metal block and a metal surface is the same regardless of whether the block is lying on a larger side or a smaller side. This observation is not generally valid for other surfaces, such as wood, and does not apply to plastic or polymer surfaces. The lack of dependence of friction on contact area is related to pressure, which is force per unit of area ($p = F/A$). If the smaller side of a metal block has an area that is half as large as the area of a larger side, it will have, on the average, only half as many asperities for local welding as the larger side. However, the pressure producing the welding will be twice as great on the smaller side (the same weight acting over half the area).

Finally, you should keep in mind that although the equation $f = \mu N$ holds in general for frictional forces, friction may not be linear over a wide range. That is, μ is not always constant. For example, the coefficient of kinetic friction varies somewhat with the relative speed of the surfaces. However, for speeds up to several meters per second the coefficients are relatively constant. Thus, this discussion will neglect any variations due to speed, and the forces of static and kinetic friction will depend only on the load and the nature of the materials (as expressed in the given coefficients of friction).

Frictional forces that obey the general equation $f = \mu N$ are said to represent Coulomb friction, and the equation is sometimes called Coulomb's law of friction. The French scientist Charles A. de Coulomb studied friction during the latter half of the sixteenth century. Although Coulomb gets the credit, the relationship was actually formulated earlier by Leonardo da Vinci.

Example 4.7   Forces of Friction

(a) If the coefficient of static friction between the 40.0-kg crate in Fig. 4.15(a) and the floor is 0.650, with what horizontal force must the worker pull to move the crate? (b) If the worker maintains that force once the crate starts to move and the coefficient of kinetic friction between the surfaces is 0.500, what is the acceleration of the crate? (c) If the applied force is at an angle of 30° to the horizontal, as shown in Fig. 4.15(b), what will its maximum magnitude have to be in order for the crate to move? (Before looking at the solution, in which case would you expect the worker to have to apply a greater force to move the crate?)

Solution

Given: $m = 40.0$ kg   Find: (a) F
       $\mu_s = 0.650$ (b) a
       $\theta = 30°$ (c) F
       $\mu_k = 0.500$

(a) The weight of the crate is equal to the magnitude of the normal force in this case, and the maximum force of static friction is

$$f_{s_{max}} = \mu_s N = \mu_s(mg)$$

$$= (0.650)(40.0 \text{ kg})(9.80 \text{ m/s}^2)$$

$$= 255 \text{ N} \quad \text{(about 57 lb)}$$

The crate moves if the applied force exceeds this maximum force.

(b) With $F = f_{s_{max}}$, there is a net force on the crate, and by Newton's second law,

$$F - f_k = F - \mu_k N = ma$$

or $a = \dfrac{F - \mu_k N}{m} = \dfrac{F - \mu_k(mg)}{m}$

$$= \frac{255 \text{ N} - (0.500)(40.0 \text{ kg})(9.80 \text{ m/s}^2)}{40.0 \text{ kg}}$$

$$= 1.48 \text{ m/s}^2$$

(c) In this case, the horizontal component of the applied force ($F \cos 30°$) acts against the frictional force, and when the latter is at a maximum,

$$F \cos 30° = f_{s_{max}} = \mu_s N$$

However, the magnitude of the normal force is *not* equal to that of the weight of the crate because of the upward component of the applied force. By the second law,

$$N + F \sin 30° - mg = 0$$

or $N = mg - F \sin 30°$

In effect, the applied force partially supports the weight of the crate. Substituting this expression for N into the preceding equation gives

$$F \cos 30° = \mu_s(mg - F \sin 30°)$$

Solving for F gives

$$F = \frac{mg}{(\cos 30°/\mu_s) + \sin 30°}$$

$$= \frac{(40.0 \text{ kg})(9.80 \text{ m/s}^2)}{(0.866/0.650) + 0.500}$$

$$= 214 \text{ N} \qquad (\text{about 48 lb})$$

Thus, less applied force is needed in this case since the frictional force is less because of the reduced load, or reduced normal force. As the angle between the applied force and the horizontal increases, the normal force gets smaller and so does $f_{s_{max}}$. ∎

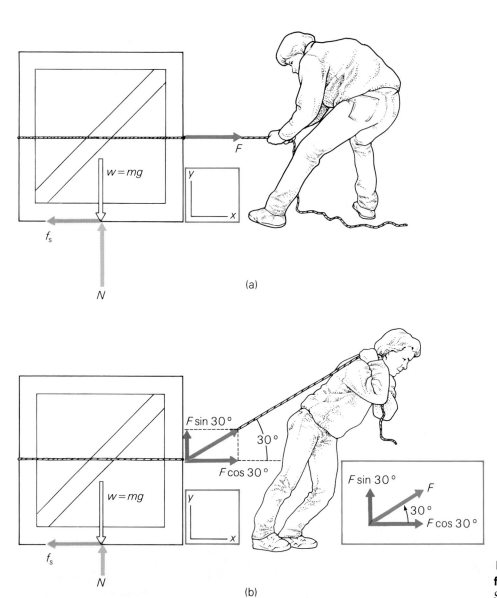

(a)

(b)

Figure 4.15 **Force of static friction**
See Example 4.7.

Example 4.8 Experimental Determination of Coefficients of Friction

A block slides with a constant velocity down a plane inclined at 37° to the horizontal (Fig. 4.16). What is the coefficient of kinetic friction between the block and the plane?

Solution

Given: $a = 0$ (because v is constant)   *Find:* μ_k
$\theta = 37°$

(a) Since $a = 0$, there is no net force on the block, and

$$F_x = 0 = mg \sin \theta - f_k$$

$$F_y = 0 = N - mg \cos \theta$$

or $f_k = mg \sin \theta$

$N = mg \cos \theta$

Then, since $f_k = \mu_k N$,

$$\mu_k = \frac{f_k}{N} = \frac{mg \sin \theta}{mg \cos \theta} = \tan \theta = \tan 37° = 0.75$$

Thus, adjusting the angle of incline until the velocity of the block sliding down the plane is constant allows μ_k to be determined experimentally from the angle of incline.

Suppose that μ_s between the plane and the block is 0.90. Can you determine the angle of incline at which the block will start to move down the plane? It will be an angle just slightly greater than that at which the component of the weight force down the plane equals the maximum force of static friction. That is,

$$mg \sin \theta = f_{s_{max}} = \mu_s N = \mu_s(mg \cos \theta)$$

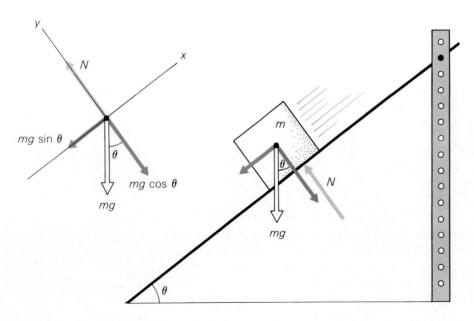

Figure 4.16 **Determination of coefficients of friction** See Example 4.8.

and

$$\frac{\sin \theta}{\cos \theta} = \tan \theta = \mu_{\mathrm{s}}$$

Thus,

$$\theta = \tan^{-1} \mu_{\mathrm{s}} = \tan^{-1}(0.90) = 42°$$

Therefore, the block will move if the angle of incline exceeds 42°. Adjusting the angle of incline until the block just starts to slide down the plane is an experimental way of approximating μ_{s}. (This critical angle is called the angle of repose.) ■

Example 4.9   Frictional Centripetal Force

A car approaches a level, circular curve with a radius of 45.0 m. If the concrete pavement is dry, what is the maximum speed at which the car can negotiate the curve?

Solution

Given: $r = 45.0$ m   *Find*: v
$\quad\quad\quad \mu_{\mathrm{s}} = 1.20$ (from Table 4.1)

To go around the curve at a particular speed, the car must have a centripetal acceleration or a centripetal force, $F_{\mathrm{c}} = mv^2/r$, acting on it. This is supplied by the friction between the tires and the road, and at the maximum velocity,

$$f_{\mathrm{s}_{\max}} = \mu_{\mathrm{s}}N = \mu_{\mathrm{s}}mg = \frac{mv^2}{r}$$

So $v = \sqrt{\mu_{\mathrm{s}}rg}$

$$= \sqrt{(1.20)(45.0 \text{ m})(9.80 \text{ m/s}^2)} = 23.0 \text{ m/s} \quad\quad \text{(about 83 km/h or 52 mi/h)}$$

Note that the speed is independent of the mass (the m's cancel out), so the equation holds for any vehicle. (Is the centripetal force the same for all vehicles?) ■

Air Resistance

In analyses of free fall, you can generally ignore the effect of air resistance and still get valid approximations for objects falling relatively short distances. However, for longer distances, air resistance cannot be ignored.

Air resistance is essentially an effect produced when a moving object bumps into air molecules, and therefore it depends on the object's shape and size (exposed area) and speed. The larger the object and the faster it falls, the more air molecules it will collide with. (Air density is also a factor, but this can be assumed to be constant near the Earth's surface.) Since air resistance is dependent on velocity or speed, as a falling object gains speed (from the acceleration due to gravity), the retarding force of air resistance increases [Fig. 4.17(a)]. This continues until this resistant force equals the object's weight force. The net force on the object is then zero and the object falls with a constant velocity, which is called the **terminal velocity**.

For a skydiver with an unopened parachute, the terminal velocity is about

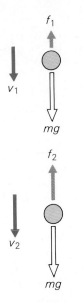

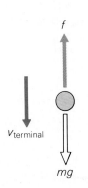

(a) As v increases, so does f.

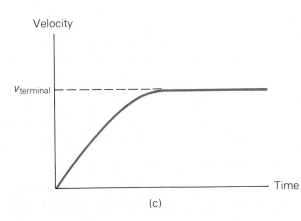

(b) When $f = mg$, the object falls with a constant (terminal) velocity.

Velocity

$v_{terminal}$

Time

(c)

Figure 4.17 **Air resistance and terminal velocity**
(a) As the speed of a falling object increases, so does the frictional force of air resistance. (b) When this force of friction equals the weight of the object, the net force is zero and the object falls with a constant (terminal) velocity. (c) This is a plot of velocity versus time showing these conditions.

Figure 4.18 **Terminal velocity**
Skydivers assume a spread-eagle position in order to reach terminal velocity more quickly and prolong the length of a fall.

200 km/h (125 mi/h). To reduce the terminal velocity, so it can be reached sooner and the time of fall be extended, a skydiver will try to increase exposed body area to a maximum by assuming a spread-eagle position (Fig. 4.18). Doing this takes advantage of the shape and size dependence of air resistance. Once the parachute is open (giving a larger exposed area and a shape that catches the air), the additional air resistance slows the diver down to about 40 km/h (25 mi/h), which is preferable for landing.

Important Formulas

Newton's second law:

$F = ma$

Weight:

$w = mg$

Centripetal force:

$$F_c = ma_c = \frac{mv^2}{r}$$

Force of static friction:

$$f_s \leq \mu_s N$$

$$f_{s_{max}} = \mu_s N$$

Force of kinetic friction:

$$f_k = \mu_k N$$

Questions

Newton's Laws of Motion

1. Can Newton's first law be demonstrated experimentally? Explain.

2. An object weighs 300 N on Earth and 50 N on the moon. Does the object also have less inertia on the moon?

3. What does the expression "a person has a lot of inertia" mean literally?

4. Explain the principle underlying each of the following: (a) tightening a loose hammerhead by striking the base of the handle on a hard surface; (b) the old parlor trick of pulling a table cloth from beneath a setting of plates and glasses.

5. Analyze each of the following in terms of Newton's first law: (a) being pushed back in a car seat when accelerating quickly from rest; (b) the purpose of shoulder straps and seat belts in cars; (c) whiplash injuries when a car is struck from behind.

6. Is Newton's first law a special case of his second law, or vice versa? Explain.

7. Consider an air-bubble level sitting on a horizontal surface (Fig. 4.19). (a) If you push on the level with a horizontal force so as to accelerate it, would the bubble move forward as shown here, or is this a trick? Explain. Which way would it move when the force is removed and the level comes to rest? (b) Such a level is sometimes used as an "accelerometer" to indicate the *direction* of the acceleration of an applied force. Explain the principle involved. (c) If the level is positioned with its long axis radially on a turntable that is rotated with a constant speed, what will be the location of the bubble?

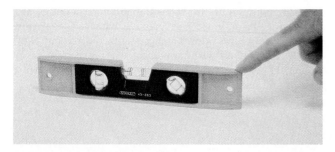

Figure 4.19 **An accelerometer**
See Question 7.

8. If an Atwood machine is used to measure g experimentally, how is the acceleration of the masses determined?

9. In general, this chapter considered forces that were being applied to objects of constant mass. What would be the situation if mass were added or lost by an object while a force was being applied? Give examples of situations where this might happen.

10. Identify all of the forces in Fig. 4.7 in terms of Newton's third law. How many forces are not explicitly illustrated in the figure?

11. A tug-of-war takes place between two college clubs. Analyze the situation in terms of Newton's second and third laws when the teams pull with opposing forces that are (a) equal and (b) unequal.

Friction

12. Note in Fig. 4.13 that the force of friction is in the same direction as the motion, that is, the direction in which the person is walking. How can this be if friction always resists motion?

13. Why is the normal force used in the equations for the force of friction rather than the weight force?

14. In general, μ_k is less than μ_s, but in Table 4.1 there is a noticeable exception. What might be a reason for this exception?

15. For the situation where a metal block rests on one side (of area A) on a metal surface, is the pressure on *each* asperity equal to mg/A? Explain. (Hint: think in terms of vertical forces.)

16. It is a common practice on ships to wrap a rope once or twice around a rail so that a large load on the other end can be held or gradually released with a minimum of effort. Explain why doing this gives an advantage in handling a load.

17. Answer the question at the end of Example 4.9: "Is the centripetal force the same for all vehicles?"

18. Automobiles are streamlined so that they will get better gas mileage. Explain this in terms of air resistance.

19. If air resistance is not ignored, does a falling object fall with a constant acceleration? Explain.

20. If the air is in motion as well as a falling object, how does this affect air resistance? Comment on the effects of (a) an updraft and (b) a downdraft on the terminal velocities of falling objects.

21. From a high altitude, a balloonist simultaneously drops two balls of identical size but appreciably different weights. If the altitude is such that both balls reach terminal velocity during the fall, will they strike the ground at the same time? Justify your answer.

Problems

4.1 Newton's First Law of Motion

■1. A large car weighs 4800 lb, and a compact car weighs 1850 lb. Which car has the greater inertia, and how many times greater?

■■2. Which has more inertia, 10 cm³ of gold or 20 cm³ of iron, and how much more? (See Table 10.1.)

■■3. Three forces act on an object that is moving in a straight line with a constant speed. If two of the forces are $\mathbf{F}_1 = (5.0 \text{ N})\mathbf{x} - (1.5 \text{ N})\mathbf{y}$ and $\mathbf{F}_2 = (-3.5 \text{ N})\mathbf{x} - (1.0 \text{ N})\mathbf{y}$, what is the third force?

■■4. A particle with a mass of 5.0×10^{-6} kg moves in the positive x direction with a constant velocity. If applied forces of $F_1 = 2.0$ N at an angle of 30° to the positive x axis and $F_2 = 3.0$ N at an angle of 30° relative to the negative x axis (third quadrant) act on the particle, what other force, applied simultaneously, will allow the particle to maintain a constant velocity?

■■■5. A 5.0-kg block at rest on a frictionless surface is acted on by forces $F_1 = 6.0$ N and $F_2 = 4.0$ N, as illustrated in Fig. 4.20. What additional applied force will keep the block at rest?

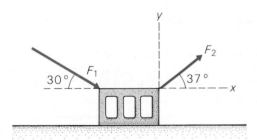

Figure 4.20 **Two applied forces**
See Problems 5 and 64.

■■■6. A 1.5-kg object moves up the y axis with a constant velocity. When the object reaches the origin, the forces $\mathbf{F}_1 = 4.0$ N at 30° relative to the positive x axis (first quadrant), $\mathbf{F}_2 = (2.5 \text{ N})\mathbf{x}$, $\mathbf{F}_3 = 3.5$ N at 45° relative to the negative x axis (third quadrant), and $\mathbf{F}_4 = (-1.5 \text{ N})\mathbf{y}$ are applied to it. (a) Will the object maintain a path along the y axis? (b) If not, what simultaneously applied force will keep the object moving along the y axis with a constant speed?

4.3 and 4.4 Newton's Second Law of Motion*

■7. Determine the net force required to give a 4.50-kg object an acceleration of 1.25 m/s².

■8. A stalled 1800-kg automobile is towed by another car with a horizontal rope, and the cars accelerate at a rate of 0.75 m/s². Determine the tension force in the rope.

■9. A net force of 20 N acts on a container containing 1 L of water. Determine the resulting acceleration. (Neglect the mass of the container.)

■10. A worker pushes on a crate, and it experiences a net force of 100 N. If the crate moves with an acceleration of 0.855 m/s², what is its weight?

■11. (a) What are the mass in kilograms and the weight in newtons of a 167-lb person? (b) What are your mass and weight in these units?

■12. A force acts on a mass. If the force is tripled and the mass is halved, how is the acceleration affected? (Give a factor of change, for example, two times as great or one-half as great.)

■13. What centripetal force is required to keep a 0.10-kg object in a circular orbit of radius 0.55 m with a constant speed of 2.8 m/s?

■■14. At a frat party, 18 students lift a sports car. While holding the car off the ground, each student is exerting a force of 414 N. (a) What is the mass of the car in kilograms? (b) What is its weight in pounds?

■■15. A 70-kg gymnast hangs vertically from a set of parallel rings. (a) If the ropes supporting the rings are attached to the ceiling directly above, what is the tension in the ropes? (b) If the ropes are supported so that they make an angle of 45° with the ceiling, what is the tension in the ropes? (c) Analyze the tension as the angle becomes smaller and smaller. (Why are telephone and electric lines allowed to sag between the poles? Wouldn't it take less

---

* In problems with strings and pulleys, "ideal conditions" means that the masses of the string(s) and pulley(s) should be neglected, as well as the friction of the pulley.

wire and be more economical if they were strung tightly?)

■■16. What is the force acting on a 0.56-kg object in free fall?

■■17. A professor's car, which has a weight of 1.47×10^4 N, moves with a constant velocity of 55.0 km/h. Determine the unbalanced force acting on the car. $\Sigma F = 0$

■■18. A horizontal force of 7.5 N acts on an object resting on a level, frictionless surface on the moon, where the object has a weight of 98 N. (a) What is the acceleration of the object? (b) What would be the acceleration of the object in a similar situation on Earth?

■■19. When a horizontal force of 250 N is applied to a 7.50-kg crate, it slides on a level floor opposed by a force of kinematic friction of 60.0 N. What is the acceleration of the crate?

■■20. A 1.75-kg block slides down a 37.0° frictionless inclined plane. What is the acceleration of the block?

■■21. A loaded jet plane with a weight of 2.75×10^6 N is ready for take-off. If its engines supply 6.35×10^6 N of net thrust, how long a runway will the plane need to reach its minimum take-off speed of 285 km/h?

■■22. Two blocks of ice, weighing 90 N and 60 N, sit side by side in contact with each other on a horizontal surface. (a) If a constant horizontal force of 45 N is applied to one of the blocks in the direction of the other block, what is the resulting acceleration? (Neglect friction.) (b) If the force is applied to the block at an angle of 20° below the horizontal, what would be the acceleration in this case?

■■23. A 1500-kg car traveling at 90 km/h on a straight level road is brought uniformly to rest. What are the magnitude and the direction of the braking force if this is done in (a) a time of 5.0 s or (b) a distance of 50 m?

■■24. A motorboat on a lake traveling with an initial velocity of 40 km/h is slowed nonuniformly to a velocity of 10 km/h in 3.0 s. If the motorboat has a mass of 65 kg, what is the net average force acting on the boat?

■■25. In catching a baseball traveling horizontally with a speed of 15.0 m/s, a player moves the glove straight backward 20.0 cm from the time of contact to the time the ball comes to rest. If the ball has a mass of 0.075 kg, what is the average force acting on the ball during that interval?

■■26. A 1.25-kg object moves with uniform circular motion at a speed of 3.8 m/s. If the centripetal force acting on the object is doubled and the object stays in the same circular path, what will be the effect on its speed?

■■27. An Atwood machine (see Fig. 4.8) has suspended masses of 0.25 kg and 0.20 kg. What will be the acceleration of the smaller mass with ideal conditions?

■■28. One mass, $m_1 = 155$ g, of an ideal Atwood machine rests on the floor 1.25 m below the other mass, $m_2 = 175$ g. If the masses are released from rest, how long does it take m_2 to reach the floor?

■■29. For the arrangement in Fig. 4.9, with ideal conditions, what will be the acceleration of the system if $m_1 = 2.00$ kg, $m_2 = 1.65$ kg, and $\theta = 30.0°$?

■■30. Suppose that $\theta = 37°$ and $m_2 = 1.5$ kg for the arrangement in Fig. 4.9. What should m_1 be if it is to move with an acceleration of (a) 0.25 m/s² down the plane, or (b) 0.10 m/s² down the plane?

■■31. Three blocks are pulled along a frictionless surface by a horizontal force, as shown in Fig. 4.21. (a) What is the acceleration of the system? (b) What are the tension forces in the light strings? (Hint: can T_1 equal T_2? Investigate by isolating m_2.)

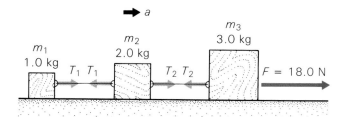

Figure 4.21 **Three blocks**
See Problems 31 and 63.

■■32. Assume ideal conditions for the arrangement illustrated in Fig. 4.22. What is the acceleration of the system if (a) $m_1 = 0.25$ kg, $m_2 = 0.50$ kg, and $m_3 = 0.25$ kg, and (b) $m_1 = 0.35$ kg, $m_2 = 0.15$ kg, and $m_3 = 0.50$ kg?

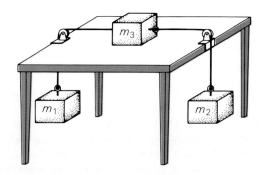

Figure 4.22 **Which way will they go?**
See Problems 32, 66, and 67.

■■33. A hoist is designed to give a maximum acceleration of 0.60 m/s^2 to a maximum load of 9.5×10^2 kg. What is the tension in the support cable when this load travels (a) upward and (b) downward? Are these the maximum and minimum tensions? Explain.

■■34. A horizontal net force of 50 N acting on a block on a frictionless level surface produces an acceleration of 2.5 m/s^2. A second block with a mass of 5.0 kg is dropped onto the first. What is the acceleration of the combination if the same force continues to act? (Assume that the second block does not slide on the first block.)

■■35. A centripetal force of 12 N keeps a 0.025-kg object in uniform circular motion at a speed of 8.8 m/s. How far does the object travel in making one revolution?

■■■36. A double Atwood machine is illustrated in Fig. 4.23. Assuming ideal conditions and letting $m_3 = 4.0$ kg and $m_1 = m_2 = 3.0$ kg, (a) what is the acceleration of m_3? (b) What are the magnitudes of the tensions in the strings?

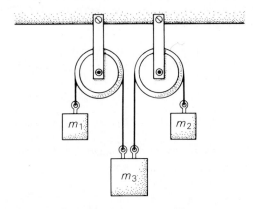

Figure 4.23 **A double Atwood machine**
See Problems 36–38 and 44.

37. If the masses for a double Atwood machine similar to the one shown in Fig. 4.23 are $m_1 = 2.0$ kg, $m_2 = 3.0$ kg, and $m_3 = 8.0$ kg, what is the acceleration of the system? (Assume ideal conditions.)

■■■38. What is the acceleration of the system if the masses for a double Atwood machine similar to that in Fig. 4.23 are (a) $m_3 = 3.0$ kg, $m_2 = 4.0$ kg, and $m_1 = 3.0$ kg, or (b) $m_3 = 2.5$ kg, $m_2 = 3.5$ kg, and $m_1 = 4.0$ kg? (Assume ideal conditions.)

■■■39. Two blocks are on a frictionless double inclined plane as illustrated in Fig. 4.24. If one block (m_1) has a mass of 15 kg and the other (m_2) has a mass of 20 kg, what is the acceleration of the system? (Assume ideal conditions.)

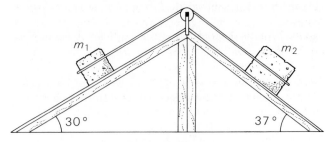

Figure 4.24 **Double inclined plane**
See Problems 39, 40, and 69.

■■■40. If the blocks on the double frictionless incline in Fig. 4.24 are at rest or move with a constant speed, (a) which block is more massive? (b) How many times more massive?

■■■41. A 65.0-kg person stands on a scale in an elevator (Fig. 4.25). What does the scale read if the elevator is (a) at rest on the second floor, (b) moving upward with an upward acceleration of 0.525 m/s^2, (c) moving downward with a downward acceleration of 0.525 m/s^2?

■■■42. The force of gravity supplies the centripetal force to keep the moon and the Earth in their nearly circular orbits. Assuming orbital motions to be uniform, compute the approximate magnitudes of the centripetal forces attracting (a) the moon to the Earth and (b) the Earth to the Sun. (The period of the moon is 29.5 days.)

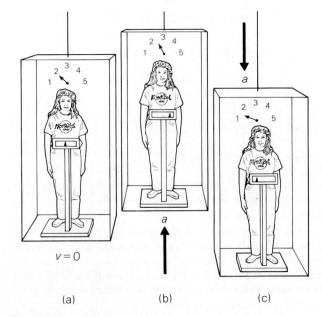

Figure 4.25 **Scale readings**
See Problem 41.

■■■43. A ball attached to the end of a string of length L is swung in a circle over a person's head such that the string makes a small angle θ with the horizontal. (a) Analyze the forces on the ball and show that its speed of uniform circular motion is given by $v = \sqrt{gL\cos^2\theta/\sin\theta}$. (b) Discuss the situations where $\theta = 90°$ and $\theta = 0°$. Can the ball be swung in a circle with the string perfectly horizontal?

4.5 Newton's Third Law of Motion

■44. Analyze the forces acting in the system of the double Atwood machine of Fig. 4.23 in terms of Newton's third law of motion.

■45. Three blocks with masses of 1.0 kg, 2.0 kg, and 3.0 kg are stacked on a table with the smallest on top and the largest on the bottom. Make a sketch and analyze this system in terms of the force pairs of Newton's third law.

■46. A rifle weighs 50.0 N, and its barrel is 0.750 m long. It shoots a 35-g bullet, which leaves the barrel with a speed of 300 m/s (muzzle velocity) after being uniformly accelerated. What is the reaction force on the rifle?

4.6 Friction*

■47. A 50-kg crate is at rest on a level surface. If the coefficient of static friction between the crate and the surface is 0.85, what horizontal force is required to move the crate?

■48. In moving a 35.0-kg desk from one side of a classroom to the other, a professor finds that a horizontal force of 255 N is necessary to set the desk in motion and a force of 175 N is necessary to keep it in motion with a constant speed. What are the coefficients of (a) static and (b) kinetic friction between the desk and the floor?

■49. When doing a lab experiment, a student finds that on the average a 1.0-kg metal weight just starts to slide down an adjustable inclined plane at an angle of incline of 22°. What is the coefficient of static friction between the weight and the plane surface?

■50. A skier on waxed skis coasts down a small slope with a constant velocity. What is the angle of the slope? (Hint: see Table 4.1.)

■51. A wooden block is placed on an adjustable wooden inclined plane. (a) What is the angle of incline above which the block will start to slide down the plane? (b) Once in motion, at what angle of incline will the block slide down the plane at a constant speed?

* Neglect air resistance unless problem states otherwise.

■52. A glass paperweight with a weight of 3.0 N is placed on a pane of glass measuring 0.75 m × 0.75 m and elevated at one end to act as an inclined plane. What is the elevation of the higher end of the pane when the paperweight slides down with a constant velocity once it is in motion?

■53. A packing crate is placed on a 20° incline plane. If the coefficient of static friction between the crate and the plane is 0.65, will the crate slide down the plane? (Justify your answer.)

■■54. The coefficient of static friction between a 9.0-kg object and a horizontal surface is 0.45. (a) Would a force of 35 N applied horizontally cause the object to move from rest? (b) If not, could this force applied in another manner cause the object to move? Justify your answer.

■■55. In the normal operation of a machine, a 5.0-kg steel part moves on a horizontal steel surface. The applied force to the part is downward at an angle of 30° from the horizontal. (a) What is the magnitude of the applied force required to set the part in motion if the surface is dry? (b) If the surface is lubricated, by what factor is that necessary applied force reduced?

■■56. Consider a system as in Fig. 4.9. (a) If the coefficient of static friction between $m_1 = 10$ kg and the plane surface is 0.80, what suspended mass (m_2) will just set the system into motion? (b) What suspended mass is required to keep m_1 moving up the plane at a constant velocity if $\mu_k = 0.60$? (Hint: the incline angle is not the same.) (c) What suspended mass is required to keep m_1 moving down the plane with a constant velocity? (Assume ideal conditions for the string and pulley.)

■■57. A 30-kg block is pushed up an inclined plane ($\theta = 25°$) with a constant velocity of 4.0 m/s by a force of 280 N parallel to the plane. What is the coefficient of kinetic friction between the block and the plane?

■■58. A 200-kg block of wood on a wooden table is connected to a vertically suspended mass of 100 g (similar to Fig. 4.9 with $\theta = 0°$). If the block is released from rest, how long will it take it to slide 1.5 m across the table if the coefficient of kinetic friction between it and the table is 0.30?

■■59. A crate containing machine parts sits unrestrained on the back of a flatbed truck traveling along a straight road at a speed of 70 km/h. The driver applies a constant braking force and comes to a stop in a distance of 20 m. What is the minimum coefficient of friction between the crate and the truck bed if the crate is not to slide forward?

■■60. A 1500-kg automobile travels at a speed of 90 km/h along a straight concrete highway. Faced with an emergency situation, the driver jams on the brakes and skids to a stop. What will the stopping distance be for (a) dry pavement and (b) wet pavement?

■■61. A school bus pulls into an intersection as a car approaches on a icy street at a speed of 40 km/h. Seeing the bus from 25 m away, the driver of the car locks the brakes causing the car to slide toward the intersection. If the coefficient of kinetic friction between the car's tires and the icy road is 0.25, does the car hit the bus? (You will learn in Chapter 8 why shorter stopping distances result from pumping the brakes.)

■■62. What is the maximum speed at which a car can round a level circular curve with a radius of curvature of 50.0 m without skidding outward on the concrete road (a) on a dry day and (b) on a wet day?

■■63. For the system illustrated in Fig. 4.21, if $\mu_s = 0.45$ and $\mu_k = 0.35$ between the blocks and the surface, what applied forces will (a) set the blocks in motion and (b) move the blocks with a constant velocity?

■■64. For the situation shown in Fig. 4.20, what is the minimum coefficient of static friction between the block and the surface that will keep the block from moving? ($F_1 = 6.0$ N and $F_2 = 4.0$ N.)

■■■65. A ramp 135 m long is to be built for a ski jump. If a skier starting from rest at the top is to have a speed of 24 m/s at the bottom, should the angle of the incline be greater than 20°? Justify your answer.

■■■66. For the arrangement in Fig. 4.22, what is the minimum value of the coefficient of static friction between the block (m_3) and the table that would keep the system at rest if $m_1 = 0.25$ kg, $m_2 = 0.50$ kg, and $m_3 = 0.35$ kg? (Assume ideal conditions for the string and pulleys.)

■■■67. If the coefficient of kinetic friction between the block and the table in Fig. 4.22 is 0.56, and $m_1 = 0.15$ kg and $m_2 = 0.25$ kg, (a) what should m_3 be if the system is to move with a constant speed? (b) If $m_3 = 0.10$ kg, what is the acceleration of the system? (Assume ideal conditions for the string and pulleys.)

■■■68. For an Atwood machine with suspended masses of 0.30 kg and 0.40 kg, the acceleration of the masses is measured as 0.95 m/s². What is the effective force of friction for the system?

■■■69. For the double inclined plane illustrated in Fig. 4.24, if $m_1 = 1.5$ kg and $m_2 = 2.7$ kg, what is the mini-mum value of the coefficient of static friction between the blocks and the planes that will keep the system at rest? (Assume the same coefficient for both surfaces and ideal conditions for the string and pulleys.)

■■■70. The coefficients of kinetic friction between the blocks and the inclined surfaces in Fig. 4.24 are $\mu_{k_1} = 0.46$ and $\mu_{k_2} = 0.38$, respectively, for m_1 and m_2. (a) If $m_1 = 0.50$ kg, what should m_2 be if the system is to move with a constant speed? (b) If $m_2 = 1.7$ kg, what is the acceleration of the system? (Assume ideal conditions for the string and pulley.)

Additional Problems

71. A person pushes on a block of wood that has been placed against a wall. Make a sketch, and analyze this situation in terms of Newton's third law.

72. A 0.45-kg shuffleboard puck is given an initial velocity of 4.5 m/s down the playing surface. If the coefficient of sliding friction between the puck and the surface is 0.25, how far will the puck slide before coming to rest?

73. An object is acted on by $F_1 = 4.0$ N at an angle of 37° relative to the positive x axis and $F_2 = 6.0$ N along the negative x axis. (a) What single additional force will ensure that the particle has zero acceleration? (b) Describe the particle's situation in this case.

74. Two blocks initially at rest on a horizontal frictionless surface are each acted on by a constant force of 20 N. One block has a mass of 4.0 kg, and the other a mass of 5.0 kg. If the identical forces act on the blocks for 3.0 s, which block will have traveled the greater distance at the end of this time, and how far ahead of the other block will it be?

75. For a system with two masses like that in Fig. 4.9, if m_1 and m_2 are 4.0 kg and 2.0 kg, respectively, at what angle of incline will the system have zero acceleration?

76. A 5.0-kg block initially at rest at a height of 2.5 m on a frictionless inclined plane slides 3.0 m to the bottom of the plane and out on to a level surface. If the block experiences a constant frictional force of 2.5 N on the level surface, how far from the bottom of the inclined plane does the block travel before coming to rest?

77. Example 4.9 illustrated how important a role friction plays when a car goes around a level curve. However, because friction can vary with road conditions and therefore cannot be relied on, curves are usually banked so that the normal force of the road on a car has a component directed horizontally toward the center of the curve. Show that a car can negotiate a curve with a radius of curvature r at a speed v *without* friction if the road is banked at an angle θ given by $\tan \theta = v^2/gr$.

78. Prove that it is impossible to swing a ball on a string in a circle over one's head with the string exactly horizontal.

79. Does it require more force to start a block of steel moving on a horizontal steel surface or on a horizontal aluminum surface? How many times as much?

80. The maximum load that can safely be supported by a rope in an overhead hoist is 200 N. What is the maximum acceleration that can safely be given to a 15-kg object being hoisted vertically upward?

81. A hockey player hits a puck with his stick, giving it an initial velocity of 5.0 m/s. If the puck slows uniformly and comes to rest in a distance of 25 m, what is the coefficient of kinetic friction between the ice and the puck?

82. In a cheerleading act, two cheerleaders with masses of 60 kg and 65 kg, respectively, are thrown from the same height vertically upward with the same initial speed of 3.6 m/s. What is the maximum height above the point of release reached by each?

83. A daredevil stunt involves riding a motorcycle around the vertical inside wall of a cylindrical structure (Fig. 4.26). Assume that the cylinder has a radius of 15 m and that the coefficient of static friction between the motorcycle tires and the wall is 1.1. (a) What is the minimum speed that will keep the motorcycle and the rider from sliding down the wall? Express this speed in km/h and mi/h. (b) Would the speed be different for a bigger (heavier) motorcycle? Explain.

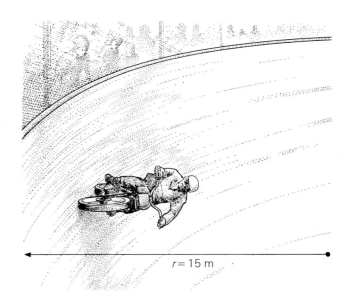

$r = 15$ m

Figure 4.26 **A daredevil stunt**
See Problem 83.

84. When mowing a lawn, a person pushes on the handle of a 25.0-kg mower with a force of 90.0 N at an angle of 35° to the horizontal. What is the normal force on the mower?

85. A crate weighing 9.80×10^3 N is pulled 6.00 m along (up) an incline by a pulley arrangement. If the coefficient of kinetic friction between the crate and the surface of the plane is 0.750, what is the magnitude of the applied force (parallel to the plane) required to move the crate with a constant velocity? (Assume ideal conditions for the pulley arrangement.)

Work and Energy

5

Two important concepts in both physics and everyday life are work and energy. We usually think of work as being associated with doing or accomplishing something. Because work makes us physically (and sometimes mentally) tired, we have invented machines and use them to decrease the amount of work we have to do personally. In any case, energy is required to do work.

It is surprisingly difficult to define energy precisely, although we use the term frequently. The burning of fuels supplies energy used for the generation of electricity, for transportation, for heating homes and offices, and for many other purposes. Food is the fuel that supplies the energy our bodies need to carry out life processes. We say that we are full of energy when ready (and willing) to do something. Since energy is used to perform tasks, one common definition or observation is that when something possesses energy, it has the ability to do work. That is, *energy is the ability to do work.*

In this chapter, you will study the scientific relationship between work and energy as well as one of the cornerstones of physics, the conservation of energy. To begin, you need to know the technical definition of work.

5.1 The Definition of Work

Work is done in lifting an object a certain distance. Such mechanical work involves force *and motion.* Twice as much work is done in lifting an object twice as heavy through the same distance. That is, you must exert twice as much force through the same distance. Holding a heavy object is commonly considered work, since a force is exerted and the effort is physically tiring. However, no mechanical work is done in this case.

What then is a definition of work? Mechanical work involves force and displacement. In the simplest case, a *constant* force is involved:

Internal work is done, however, by the minute contractions and relaxations of muscle fibers and the expenditure of chemical energy.

The work done by a constant force in moving an object is equal to the product of the magnitudes of the displacement and the component of the force parallel to the displacement.

Basically, work is done when a force moves through a distance and some component of the force is along the line of motion (see Fig. 5.1). For a constant force **F** acting in the same direction as the displacement **d**, the work (W) is simply

$$W = Fd \tag{5.1}$$

However, if the force is at an angle θ to the displacement [Fig. 5.1(c)], then $F \cos \theta$ is the component parallel to the displacement, giving this more general equation:

$$W = (F \cos \theta)d = Fd \cos \theta \tag{5.2}$$

Note that if $\theta = 0°$, Eq. 5.2 reduces to Eq. 5.1.

Actually, θ is the angle *between* the force and displacement vectors. The force may be in the opposite direction to the displacement, for example, a braking force that tends to slow down, or decelerate, an object. In this case, $\theta = 180°$ and $\cos 180° = -1$. Then the work is negative: $W = Fd \cos 180° = -Fd$.

Work is a scalar quantity and has SI units of newton-meter (N-m), since $W = Fd$ and force is given in newtons and displacement in meters.

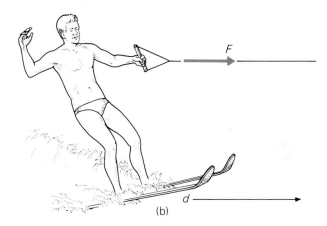
(b)

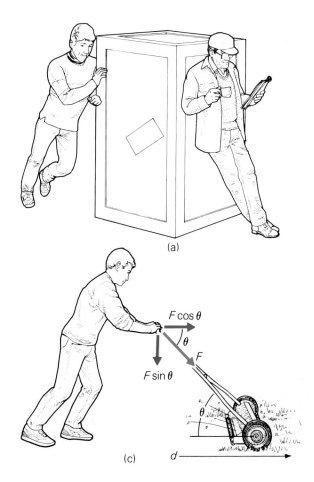
(a)

(c)

Figure 5.1 **Work—the product of the magnitudes of the parallel component of a force and the displacement** (a) If there is no motion, no work is done: $W = 0$. (b) For a constant force in the same direction as the displacement, $W = Fd$. (c) For a constant force at an angle to the displacement, $W = (F \cos \theta)d = Fd \cos \theta$.

$$Fd = W$$

N-m ≡ joule (J)

The newton-meter has been given the name **joule (J)** (pronounced "jool") in honor of James Prescott Joule (1818–1889), a British scientist who investigated work and energy. The British standard unit for work is the foot-pound (ft-lb).

Example 5.1 Mechanical Work Requires Motion

A student holds his psychology textbook, which has a mass of 1.5 kg, out of a second-story dormitory window and then releases it. (a) How much work is done by the student in simply holding the book out the window? (b) How much work will have been done by the force of gravity during the time in which the book falls 3.0 m?

Solution

Given: $v_o = 0$ *Find*: (a) W
        $m = 1.5$ kg   (b) W
        $d = 3.0$ m

(a) Despite what you may think, the work done by the student in holding the book stationary is zero. The student exerts an upward force on the book (equal in magnitude to its weight), but the displacement is zero in this case ($d = 0$). Thus, $W = Fd = F \times 0 = 0$.

(b) While the book is falling, the net force acting on it is the force of gravity, $F = mg$ (neglecting air resistance). For a displacement of $d = y = 3.0$ m,

$$W = Fd = mgy = (1.5 \text{ kg})(9.8 \text{ m/s}^2)(3.0 \text{ m}) = 44 \text{ J} \quad \blacksquare$$

Example 5.2 Parallel Component of a Force

If the person in Fig. 5.1(c) pushes on the lawn mower with a constant force of 90.0 N at an angle of 40° to the horizontal, how much work does he do in pushing it a horizontal distance of 7.50 m?

Solution

Given: $F = 90.0$ N   *Find*: W
        $\theta = 40°$
        $d = 7.50$ m

Work is done by the horizontal component of the applied force, or $F \cos \theta$, so Eq. 5.2 applies:

$$W = Fd \cos 40° = (90.0 \text{ N})(7.50 \text{ m})(0.766) = 517 \text{ J}$$

The vertical component of the force does no work. Why? ∎

The preceding examples involve work done by a single constant force. If more than one force acts on an object, the work done by each may be calculated separately. The total, or net, work is then the scalar sum of those quantities of work. Alternatively, you may find the vector sum of the forces and use this net force to compute the total, or net, work. If the resultant force is zero, the net work is zero. Balanced forces work against each other, so to speak, and produce no net work. In this case, each force does work, but equal positive and negative work is done.

In general, you should specify what is doing work on what. For example, the force of gravity does work on a falling object. When you lift an object, you are doing work on the object and *against* the force of gravity because it is acting in the opposite direction to the lift force you are applying.

Example 5.3 Total or Net Work

A 0.75-kg block slides down a 20° inclined plane with a uniform velocity (Fig. 5.2). (a) How much work is done by the force of friction on the block as it slides the total length of the plane? (b) What is the net work done on the block? (c) Discuss the work done if the angle of incline is adjusted so that the block accelerates down the plane.

Solution

Given: $m = 0.75$ kg   Find: (a) W_f
       $\theta = 20°$ (b) W_{net}
       $x = 1.2$ m   (c) W with acceleration

(a) The work done by the force of friction is

$$W_f = f_k d \cos 180° = -f_k d = -\mu_k N d$$

The angle of $180°$ indicates that the force and displacement are in opposite directions. (It is common in such cases to write $W_f = -f_k d$ directly, since friction typically opposes motion.) You know that

$$N = mg \cos \theta$$

And you can use trigonometry to write

$$d = \frac{x}{\cos \theta}$$

But what is μ_k? It doesn't seem to be given (nor are the types of surfaces specified). This is an example where a key piece of data is given to you indirectly in a problem. Note that the block slides down the plane with a uniform (constant) velocity. In Chapter 4, you learned that $\mu_k = \tan \theta$ for uniform motion down a plane. Thus,

$$W_f = -\mu_k N d = -(\tan \theta)(mg \cos \theta)\left(\frac{x}{\cos \theta}\right) = -mgx \tan 20°$$

$$= -(0.75 \text{ kg})(9.8 \text{ m/s}^2)(1.2 \text{ m})(0.364) = -3.2 \text{ J}$$

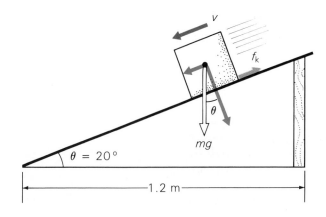

Figure 5.2 **Total or net work**
See Example 5.3.

[Note how the expression for W_f was easily simplified by using the algebraic forms of N and d instead of computing these quantities initially. It is a good rule of thumb not to put numbers into an equation until you have to. How to simplify an equation through cancellation is easier to see with symbols, and computation time is saved.]

(b) The net work is zero since the net force on the block is zero (constant velocity). Computing the work done by the force of gravity (W_g) will help prove this:

$$W_g = Fd = (mg \sin \theta)\left(\frac{x}{\cos \theta}\right) = mgx \tan 20° = 3.2 \text{ J}$$

Then

$$W_{net} = W_g + W_f = 3.2 \text{ J} - 3.2 \text{ J} = 0$$

Remember that work is a scalar quantity, so scalar addition is used to find net work.

(c) If the block accelerates down the plane, then $F = mg \sin \theta$ is greater than f_k, or $F > f_k$. This means that the force of gravity does more work than the force of kinetic friction, and that net work is done on the block. You might be wondering what the effect of nonzero net work is. As you will learn shortly, it causes a change in the amount of energy an object has. ∎

5.2 Work Done by Variable Forces ◆

The discussion in the preceding section was limited to work done by constant forces. In general, however, forces are variable; that is, they change with time and/or position. For example, a force applied to an object to overcome the force of static friction may increase with time until it exceeds $f_{s_{max}}$ (see Section 4.6). However, the force of static friction does no work. (Why?)

Examples of variable forces doing work are illustrated in Fig. 5.3. As a spring is stretched (or compressed) further and further, more force is required. The force is directly proportional to the displacement of the mass, or the change in the length of the spring. This is expressed as an equation:

$$F = k \Delta x = k(x - x_o)$$

or, with $x_o = 0$,

$$F = kx \tag{5.3}$$

The k in this equation is a constant of proportionality and is commonly called the **spring, or force, constant**. The greater the value of k, the stiffer or stronger is the spring. As you should be able to prove to yourself, the units of k are newtons per meter (N/m).

The relationship expressed by Eq. 5.3 holds only for ideal springs. Real springs approximate this linear relationship between force and displacement within certain limits. For example, if a spring is stretched beyond a certain point, called its elastic limit, it will be pulled out of shape and $F = kx$ will no longer apply.

Note that a spring exerts an equal and opposite force, $F_s = -k \Delta x$, when its linear length is changed by an amount Δx. The minus sign indicates that this spring force is in the direction opposite to the displacement whether the spring

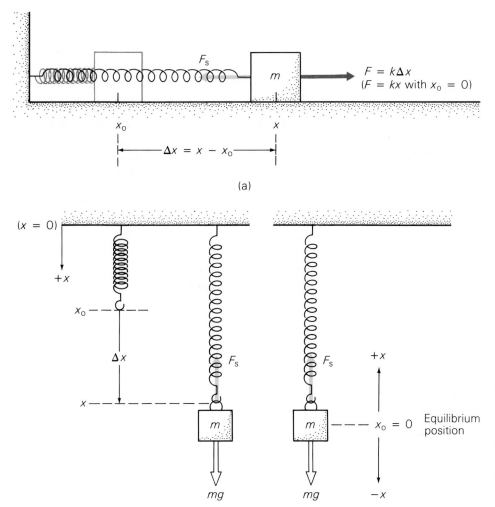

$F = k\Delta x$
($F = kx$ with $x_o = 0$)

$\Delta x = x - x_o$

(a)

($x = 0$)

$+x$

x_o

Δx

F_s

x

m

mg

F_s

m

$+x$

$x_o = 0$ Equilibrium position

$-x$

mg

Figure 5.3 **Spring force**
(a) The force exerted by a spring depends on the change in the spring's length. This is often referenced to the position of a mass on the end of the spring. (b) When the net force on the mass is zero, it is at its equilibrium position, which is often taken as $x_o = 0$.

is stretched or compressed. This equation is a form of what is known as Hooke's law, after Robert Hooke, a contemporary of Newton.

PROBLEM-SOLVING HINT

The reference position x_o for the change in length of a spring is arbitrary and is usually chosen for convenience. *The important quantity is the displacement difference Δx, or the net change in the length of the spring.* As shown in Fig. 5.3(b) for a mass suspended on a spring, x_o can be referenced to the unloaded length of the spring or to the loaded position, which may be taken as zero for convenience. When the net force on the mass parallel to the length of the spring is zero, the mass is at its equilibrium position ($x = x_o = 0$). When the mass is displaced from the equilibrium position, work is done (against the spring force by an applied force, $F = kx$).

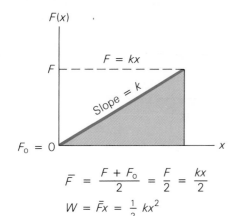 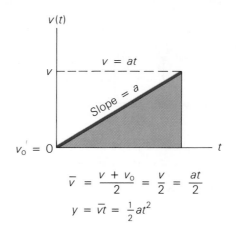

$$\bar{F} = \frac{F + F_o}{2} = \frac{F}{2} = \frac{kx}{2}$$

$$W = \bar{F}x = \frac{1}{2}kx^2$$

$$\bar{v} = \frac{v + v_0}{2} = \frac{v}{2} = \frac{at}{2}$$

$$y = \bar{v}t = \frac{1}{2}at^2$$

Figure 5.4 Work done by a uniformly varying force The work done by a uniformly varying force of the form $F = kx$ is $W = \frac{1}{2}kx^2$. This special case of a variable force is graphically analogous to a uniformly varying velocity: $v = at$. The work (W) and distance (y) are equal to the areas under the respective lines.

This course of study will generally be limited to constant or average force considerations. As you learned in studying kinematics (Chapter 2), an average value may not be very useful except in special cases. One of these cases is the average velocity for a constant acceleration: $\bar{v} = (v + v_o)/2$. A special case of a variable force that can be analyzed without calculus is the spring force, where the spring constant is k. A plot of F versus x is shown in Fig. 5.4 (which also shows the analogous case of the plot of v versus t for a constant acceleration). The slope of the line is equal to k, and F increases uniformly with x. The average force is then

$$\bar{F} = \frac{F + F_o}{2}$$

or, if $F_o = 0$,

$$\bar{F} = \frac{F}{2}$$

Thus, the work done in stretching or compressing the spring is

$$W = \bar{F}x = \frac{Fx}{2}$$

Since $F = kx$, the work done is

$$W = \frac{1}{2}kx^2 \tag{5.4}$$

Note that the work is the area under the curve in Fig. 5.4.

Example 5.4 Determining the Spring Constant and the Work Done in Stretching a Spring

A 0.15-kg mass is suspended from a vertical spring and descends a distance of 4.6 cm, after which it hangs at rest (Fig. 5.5). An additional 0.50-kg mass is then suspended from the first. (a) What is the total extension of the spring? (b) How much work is done in stretching the spring?

Solution

Given: $m_1 = 0.15$ kg
$\quad\quad y_1 = 4.6$ cm $= 0.046$ m
$\quad\quad m_2 = 0.50$ kg

Find: (a) y
$\quad\quad$ (b) W

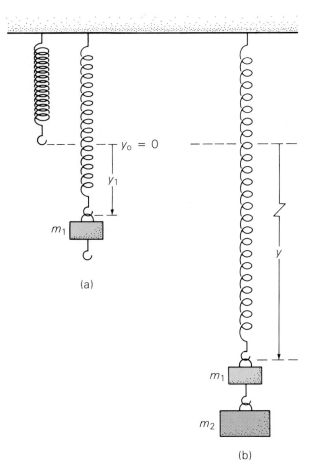

$y_0 = 0$

y_1

m_1

(a)

y

m_1

m_2

(b)

Figure 5.5 **Determining the spring constant and work done in stretching a spring** See Example 5.4.

Here the displacement is taken to be in the y direction. The spring constant must be known to find the total extension and work. It can be found using the result from suspending m_1 and measuring the resulting displacement (y_1):

$$F = m_1 g = k y_1$$

and

$$k = \frac{m_1 g}{y_1} = \frac{(0.15 \text{ kg})(9.8 \text{ m/s}^2)}{0.046 \text{ m}} = 32 \text{ N/m}$$

(a) The total extension of the spring is then found from

$$F = (m_1 + m_2)g = ky$$

Thus,

$$y = \frac{(m_1 + m_2)g}{k} = \frac{(0.15 \text{ kg} + 0.50 \text{ kg})(9.8 \text{ m/s}^2)}{32 \text{ N/m}} = 0.20 \text{ m} \qquad \text{(or 20 cm)}$$

(b) The total work done is

$$W = \tfrac{1}{2}ky^2 = \tfrac{1}{2}(32 \text{ N/m})(0.20 \text{ m})^2 = 0.64 \text{ J} \quad \blacksquare$$

5.3 The Work-Energy Theorem: Kinetic Energy

Now that you know an operational definition of work, you are ready to look at how it is related to energy. One form of energy that is closely associated with work is kinetic energy. (Another form of energy, potential energy, will be considered in the next section.) **Kinetic energy** is often called the energy of motion and is defined mathematically as one-half of the product of the mass and the square of the (instantaneous) velocity of a moving object:

$$K = \tfrac{1}{2}mv^2 \tag{5.5}$$

For a constant net force doing work on an object as illustrated in Fig. 5.6, the force does an amount of work, $W = Fx$. But what are the kinematic effects? The force causes the object to accelerate, and from the equation $v^2 = v_o^2 + 2ax$,

$$a = \frac{v^2 - v_o^2}{2x}$$

where v_o may or may not be zero. Writing the force in its Newton's second law form ($F = ma$) and then substituting for a from the above equation gives

$$W = Fx = max = m\left(\frac{v^2 - v_o^2}{2x}\right)x$$

$$= \tfrac{1}{2}mv^2 - \tfrac{1}{2}mv_o^2$$

In terms of kinetic energy, then,

$$W = \tfrac{1}{2}mv^2 - \tfrac{1}{2}mv_o^2 = K - K_o = \Delta K$$

or $W_{\text{net}} = \Delta K$ 　　　　　　　　　　　　　　　　　　　(5.6)

This equation is called the **work-energy theorem** and relates the net work done on an object to the change in its kinetic energy. That is, the net work done is equal to the change in kinetic energy. In general, you might say that

Work is a measure of the transfer of energy.

Keep in mind that the work-energy theorem is true in general and not just for the special case considered in deriving Eq. 5.6. As a concrete example, recall that in Example 5.1 the force of gravity did 44 J of work on a book that fell from rest through a distance of $y = 3.0$ m. At this position and instant, the falling book has 44 J of kinetic energy. This can be easily shown. Since $v_o = 0$ in this case, $v^2 = 2gy$, and

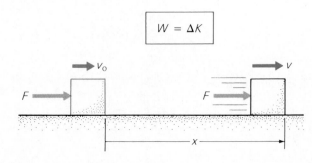

Figure 5.6 **The relationship of work and energy**
The work done on the block is equal to the change in kinetic energy, or $W = \Delta K$.

$$W = Fd = mgy = \frac{mv^2}{2} = K = 44 \text{ J}$$

where $K_o = 0$.

As you can see from the work-energy theorem, kinetic energy has the same units as work. You should also note that, like work, kinetic energy is a scalar quantity.

What the work-energy theorem tells you is that when work is done, there is a change or transfer of energy, and vice versa. A force doing work on an object and causing it to speed up gives rise to an increase in the object's kinetic energy. Work done by the force of kinetic friction, on the other hand, may cause a moving object to slow down and decrease its kinetic energy. Note that for a change in kinetic energy there must be a net force and net work. (Why?)

When an object is in motion, it possesses kinetic energy and has the ability to do work. For example, a moving automobile has kinetic energy and can do work in crumpling a fender—not *useful* work in that case, but still work. In the case illustrated in Fig. 5.6, there is no indication of what supplied the force or did the work. You may not know this in some situations where something expended energy to do work (and energy was transferred). In any case, you can think of work as the process of energy transfer.

Example 5.5 Work and Kinetic Energy

A shuffleboard player pushes a 0.25-kg puck in a way that causes a horizontal constant force of 6.0 N to act on it through a distance of 0.50 m. (a) What are the kinetic energy and the speed of the puck when the force is removed? (b) How much work would be required to bring the puck to rest?

Solution

Given: $m = 0.25$ kg   Find: (a) K and v
          $F = 6.0$ N   (b) W
          $d = 0.50$ m
          $v_o = 0$

(a) The work done on the puck is

$$W = Fd = (6.0 \text{ N})(0.50 \text{ m}) = 3.0 \text{ J}$$

By the work-energy theorem,

$$W = \Delta K = K - K_o = 3.0 \text{ J}$$

And $K_o = \frac{1}{2}mv_o^2 = 0$ since $v_o = 0$, so

$$K = 3.0 \text{ J}$$

The speed of the puck may be found from the kinetic energy. Since $K = \frac{1}{2}mv^2$,

$$v = \sqrt{\frac{2K}{m}} = \sqrt{\frac{2(3.0 \text{ J})}{0.25 \text{ kg}}} = 4.9 \text{ m/s}$$

(b) As you might guess, the work required to bring the puck to rest is equal to its kinetic energy (the amount of energy that must be "removed" from the puck to bring it to rest). In this case, $v_o = 4.9$ m/s and $v = 0$, so

$$W = K - K_o = 0 - K_o = -\tfrac{1}{2}mv_o^2$$

$$= -\tfrac{1}{2}(0.25 \text{ kg})(4.9 \text{ m/s})^2 = -3.0 \text{ J}$$

The minus sign indicates that the puck loses this energy as it slows down (similarly, an acceleration and a velocity of opposite signs result in the slowing down of an object). The work is done *against* the motion of the puck; that is, the applied force is opposite to the direction of motion. The force that does the work is friction, of course. ∎

Note that the work in the work-energy theorem is equal to the change in the kinetic energy, which is *not* given by the change in velocity. Suppose that a moving object experiences an increase in velocity from 2.0 m/s to 5.0 m/s. The change in velocity is $\Delta v = v - v_0 = 5.0$ m/s $- 2.0$ m/s $= 3.0$ m/s. But the change in kinetic energy is *not* equal to $\frac{1}{2}m(\Delta v)^2$. That is,

$$(v - v_0)^2 \neq v^2 - v_0^2$$

Note that

$$(v - v_0)^2 = (5.0 \text{ m/s} - 2.0 \text{ m/s})^2 = 9.0 \text{ m}^2/\text{s}^2$$

and

$$v^2 - v_0^2 = (5.0 \text{ m/s})^2 - (2.0 \text{ m/s})^2 = 21 \text{ m}^2/\text{s}^2$$

In terms of kinetic energy,

$$W = \Delta K = K - K_0 = \tfrac{1}{2}mv^2 - \tfrac{1}{2}mv_0^2 = \tfrac{1}{2}m(v^2 - v_0^2)$$

What this means is that you must compute the kinetic energy of an object at one point or time (using the instantaneous velocity and getting the instantaneous kinetic energy) and also at another point or time, and then subtract the second quantity from the first to find the change in kinetic energy, or the work. Or, alternatively, you can find the difference of the squares of the velocities, as indicated in the preceding equation.

5.4 Potential Energy

An object in motion has kinetic energy. However, whether an object is in motion or not, it may have another form of energy—potential energy. Potential energy is associated with location, or position (or configuration), because a force is required and work is done in moving an object from one position to another (when lifting something or deforming an object, such as a spring). Consequently, **potential energy** is often called the energy of position.

As the name implies, potential energy can be thought of as stored work. You have already seen an example of potential energy in Section 5.2 when work was done in compressing a spring. Recall that the work done in such a case is $W = \frac{1}{2}kx^2$ (with $x_0 = 0$). Note that the amount of work done depends on position (x). Because work is done, there is a *change* of position and a *change* in potential energy (ΔU), which is equal to the work done *by* the applied force in compressing the spring:

$$W = \Delta U = U - U_{\text{o}} = \tfrac{1}{2}kx^2 - \tfrac{1}{2}kx_{\text{o}}^2$$

Thus, with $x_{\text{o}} = 0$, as it is commonly taken to be, the **potential energy of a spring** is

$$U = \tfrac{1}{2}kx^2 \qquad \text{(for a spring)} \tag{5.7}$$

[Since the potential energy has x^2 as one term, the previous problem-solving hint applies when $x_{\text{o}} \neq 0$; that is, $x^2 - x_{\text{o}}^2 \neq (x - x_{\text{o}})^2$.]

Perhaps the most common type of potential energy is **gravitational potential energy**. In this case, position refers to the height of an object above some reference point such as the floor or the ground. Suppose that an object of mass m is lifted a distance Δh (Fig. 5.7). Work is done against the force of gravity, and the minimum applied force is $F = mg$. The work done by the applied force F on the object (which is the negative of the work done by the force of gravity) is then equal to the change in potential energy:

$$W = F\,\Delta h = \Delta U = U - U_{\text{o}} = mgh - mgh_{\text{o}}$$

And, with $h_{\text{o}} = 0$, the gravitational potential energy is

$$U = mgh \tag{5.8}$$

(gravitational potential energy)

> Alternatively, the change in potential energy of the spring is equal to the negative work done by the spring force. That work is negative since the spring force and the displacement are in opposite directions.

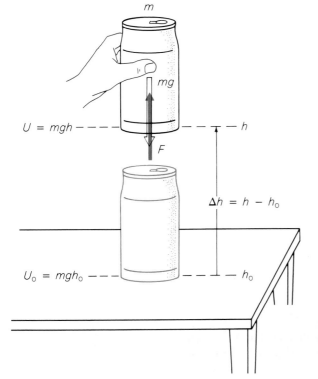

$$W = \Delta U = U - U_{\text{o}} = mgh - mgh_{\text{o}}$$

Figure 5.7 Gravitational potential energy
The work done in lifting an object is equal to the change in gravitational potential energy, or $W = F\,\Delta h = mg(h - h_{\text{o}})$.

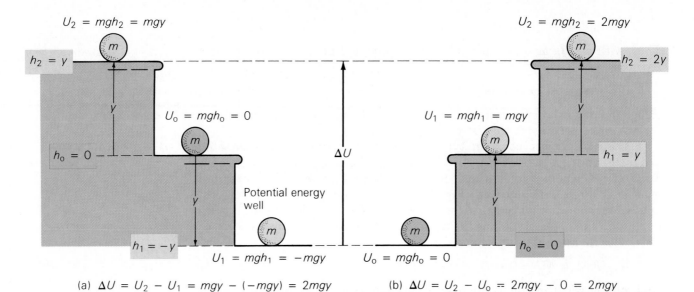

(a) $\Delta U = U_2 - U_1 = mgy - (-mgy) = 2mgy$ (b) $\Delta U = U_2 - U_o = 2mgy - 0 = 2mgy$

Example 5.6   Kinetic and Potential Energies

A 0.50-kg ball is projected vertically upward from ground level with an initial velocity of 10 m/s. (a) What is the change in the ball's kinetic energy in going from ground level to its maximum height? (b) What is the ball's potential energy at its maximum height? (c) What are the total changes in the ball's kinetic energy (ΔK_T) and potential energy (ΔU_T) when it has returned to its starting point? (Neglect air resistance.)

Solution

Given: $m = 0.50$ kg   *Find*: (a) K
$\qquad v_o = 10$ m/s   (b) U
$\qquad a = g$ (c) ΔK_T and ΔU_T
$\qquad v = 0$, at y_{max}

(a) At the maximum height, $v = 0$, and therefore $K = 0$. Thus,

$$\Delta K = K - K_o = 0 - K_o = -\tfrac{1}{2}mv_o^2 = -\tfrac{1}{2}(0.50 \text{ kg})(10 \text{ m/s})^2 = -25 \text{ J}$$

That is, the ball loses 25 J of kinetic energy, as negative work is done on it by the force of gravity.

(b) Using $v^2 = v_o^2 - 2gy$, with $v = 0$ at the maximum height (y_{max}), gives

$$h = y_{max} = \frac{v_o^2}{2g} = \frac{(10 \text{ m/s})^2}{2(9.8 \text{ m/s}^2)} = 5.1 \text{ m}$$

Then, since $h_o = 0$ at ground level,

$$U = mgh = (0.50 \text{ kg})(9.8 \text{ m/s}^2)(5.1 \text{ m}) = 25 \text{ J}$$

The potential increases by 25 J, as might be expected.

(c) The ball returns to its starting point with the same speed as it had initially, or $v = v_o$. Therefore, $K = K_o$ and $\Delta K_T = K - K_o = 0$ (that is, the ball gains 25 J of kinetic energy on the return trip—positive work is done on the ball by the

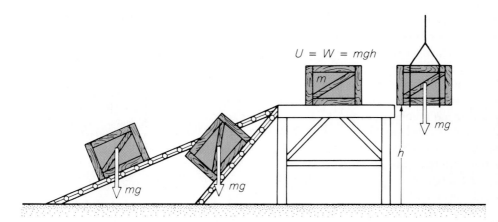

$U = W = mgh$

Figure 5.9 **Path-
independence of the change
in gravitational potential
energy**
When the crate is resting on
the loading dock, it has the
same potential energy
regardless of how it got there.
Only the vertical component
of the force that moves the
crate does work against
gravity.

force of gravity). It should be evident that the ball loses 25 J of potential energy on the return trip and $\Delta U_\mathrm{T} = 0$ for the round trip (because $mg\,\Delta h = 0$ since $\Delta h = 0$). ■

Both spring and gravitational potential energies are a function of position. Note that the potential energy at a particular position (U) is referenced to the potential energy at some other position (U_o). The reference position is arbitrary ("dealer's choice," so to speak, like the origin of a set of coordinate axes for analyzing a system). It is usually chosen with convenience in mind, for example, $x_\mathrm{o} = 0$ or $h_\mathrm{o} = 0$. The value of the potential energy at a particular position depends on the reference position used. However, the *difference, or change, in potential energy associated with two positions is the same regardless of the reference position*, as illustrated in Fig. 5.8.

Note that the potential energy can be negative. When an object has a negative potential energy, it is in a potential energy well, which is something like being in an actual well. Work is needed to raise the object to a higher position in the well or to get it out of the well.

Also, for gravitational potential energy, the path by which an object is raised (or lowered) makes no difference (see Fig. 5.9). That is, *the change in gravitational potential energy is independent of path*. As illustrated in the figure, an object raised to a height h has a change in potential energy of $\Delta U = mgh$ whether it is lifted vertically or moved along an inclined plane (or any other path). This is because the force of gravity always acts downward and only a vertical displacement (or vertical component) is involved in doing work against gravity and changing the potential energy.

5.5 The Conservation of Energy

When physicists say that something is conserved, they mean that it is constant or has a constant value. Because things continually change in everyday processes, conserved quantities are helpful for understanding and describing natural phenomena. Such quantities are given special status by being made the subject of conservation laws.

One of the most important conservation laws is that concerning conservation of energy. (You may have seen this coming in Example 5.6.) A familiar statement is that the total energy of the universe is conserved. This is true if the whole universe is taken to be a system. In effect, the universe is the largest possible isolated, or closed, system, that is, one whose particles may interact with each other but have absolutely no interaction with anything outside. Another way of saying this is that no mechanical work is done on or by the system and no energy is transmitted to or from the system (including thermal energy and radiation). Thus, *the total energy of an isolated system is always conserved.* Within such a system, energy may be converted from one form to another, but the total amount of energy is constant, or unchanged.

A general distinction among systems may be made by considering two general categories of forces that may act within them: conservative and nonconservative forces. You have already been introduced to a couple of conservative forces: the force due to gravity (see Fig. 5.9) and the spring force. A classic nonconservative force, friction, was considered in Chapter 4. Here's the distinction:

> **A force is said to be conservative if the work done by it in moving an object is independent of the object's path.**

This means that the work done or the energy transferred depends only on the initial and final coordinates of an object, or its position. Consequently, potential energy is associated with only conservative forces, and a change in potential energy may be defined as the work done by a conservative force.

It follows that *a force is conservative if the work done by it in moving an object through a round trip is zero.* A book resting on a table has gravitational potential energy, $U = mgh$, relative to $h_o = 0$. You could pick up the book, carry it around with you all day, and then place it back at its original position (a round trip). Its gravitational potential energy will be the same. Thus, the change in its potential energy or the work done by the conservative force of gravity is $\Delta U = W = 0$. However, suppose that you pushed the book around on the tabletop and eventually back to its original position. In this case, the work done against the nonconservative force of friction would depend on the path the book took (it could be a long path, or a short path).

In general, both conservative and nonconservative forces may act on an object and do work. The work-energy theorem in this case is written

$$W_{net} = W_c + W_{nc} = \Delta K \qquad (5.9)$$

where W_c is all the work done by conservative forces and W_{nc} is all the work done by nonconservative forces.

But the work done by a conservative force is the *negative* change in the potential energy. That is, $W_c = -\Delta U$, as you have seen for gravitational and spring potential energy. Thus, Eq. 5.9 may be written as

$$W_{nc} = \Delta K - W_c = \Delta K + \Delta U$$

That is,

$$W_{nc} = \Delta K + \Delta U \qquad (5.10)$$

There are a variety of instances in which frictional forces and other nonconservative forces are not acting, or at least do negligible work, while an object is

Recall that a system is a physical situation with real or imaginary boundaries. A classroom might be considered a system and so might an arbitrary cubic meter of air.

in motion. When this is the case ($W_{nc} \cong 0$) and only conservative forces act, then

$$\Delta K + \Delta U = 0$$

or $\quad (K - K_o) + (U + U_o) = 0$

Thus,

$$K + U = K_o + U_o \qquad (5.11)$$

where U is the potential energy associated with one or more conservative forces. The sum of the kinetic and potential energies is called the **total mechanical energy**:

$$E = K + U$$

For a **conservative system** (one in which only conservative forces act), Eq. 5.11 says that the mechanical energy is constant, or conserved; that is,

$$E = K + U = K_o + U_o = E_o$$

or $\quad \frac{1}{2}mv^2 + U = \frac{1}{2}mv_o^2 + U_o \qquad (5.12)$

This is a mathematical statement of the **conservation of mechanical energy**:

> **In a conservative system, the sum of the kinetic and potential energies is constant and equals the total mechanical energy of the system.**

The kinetic and potential energies in a conservative system may change, but their sum is always constant. This is illustrated by the diagram in Fig. 5.10. Note that if there is a nonconservative force, the mechanical energy is not conserved (more about this shortly). If the mechanical energy is conserved, the total mechanical energy of the system can be determined, usually in a convenient situation for which K or U is zero. Then the constant value of the mechanical energy can be used to analyze the system at a later time, as the following examples show.

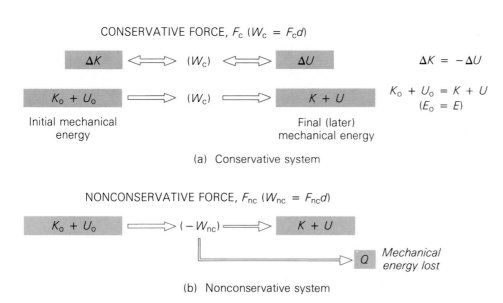

(a) Conservative system

(b) Nonconservative system

Figure 5.10 **Conservative and nonconservative systems**
(a) The work done by a conservative force (F_c) exchanges energy between kinetic and potential forms; that is, $\Delta K = -\Delta U$. The total mechanical energy is conserved. (b) Part of the work done by a nonconservative force (F_{nc}) does not go into mechanical energy, and the mechanical energy is not conserved.

Example 5.7 Conservation of Mechanical Energy

A painter on a scaffold accidentally drops a 1.50-kg can of paint from a height of 6.00 m. (a) What is the kinetic energy of the can when it is at a height of 4.00 m? (b) With what speed will the can hit the ground? (Neglect air resistance.)

Solution

Given: $m = 1.50$ kg
$\quad\quad\ h_o = 6.00$ m
$\quad\quad\ h = 4.00$ m

Find: (a) K
$\quad\quad$ (b) v

(a) Before it is dropped, the can's total mechanical energy is equal to its potential energy ($v_o = 0$):

$$E = K + U = 0 + mgh_o = (1.50 \text{ kg})(9.80 \text{ m/s}^2)(6.00 \text{ m}) = 88.2 \text{ J}$$

In falling 2.00 m (a third of the distance), $\frac{1}{3}$ of the original potential energy is converted into kinetic energy, and

$$K = \tfrac{1}{3}E = \tfrac{1}{3}(88.2 \text{ J}) = 29.4 \text{ J}$$

Computing explicitly with the can at $h = 4.00$ m gives the same result:

$$K = E - V = E - mgh = 88.2 \text{ J} - (1.50 \text{ kg})(9.80 \text{ m/s}^2)(4.00 \text{ m}) = 29.4 \text{ J}$$

(b) Just before the can strikes the ground ($h = 0$, $U = 0$), the total mechanical energy is all kinetic, or

$$E = K = \tfrac{1}{2}mv^2$$

Thus,

$$v = \sqrt{\frac{2E}{m}} = \sqrt{\frac{2(88.2 \text{ J})}{1.50 \text{ kg}}} = 10.8 \text{ m/s}$$

Basically, all of the potential energy of a free-falling object released from some height h is converted into kinetic energy just before it hits the ground, so

$$K = U$$

$$\tfrac{1}{2}mv^2 = mgh$$

or $\quad v = \sqrt{2gh}$

This result is also obtained from the kinematic equation $v^2 = v_o^2 + 2gy$, with $v_o = 0$ and $y = h$. ∎

Example 5.8 Conservative Force and Mechanical Energy of a Spring

A 0.30-kg mass sliding on a horizontal frictionless surface with a speed of 2.5 m/s, as depicted in Fig. 5.11, strikes a spring, which has a spring constant of 3.0×10^3 N/m. (a) What is the total mechanical energy of the system? (b) What is the kinetic energy (K_1) of the mass when the spring is compressed a distance $x_1 = 1.0$ cm? (c) How far will the spring be compressed when the mass comes to a stop? (Assume that no energy is lost in the collision.)

Solution

Given: $m = 0.30$ kg
$\quad\quad\ v_o = 2.5$ m/s
$\quad\quad\ k = 3.0 \times 10^3$ N/m
$\quad\quad\ x_1 = 1.0$ cm $= 0.010$ m

Find: (a) E
$\quad\quad$ (b) K_1
$\quad\quad$ (c) x_{max}

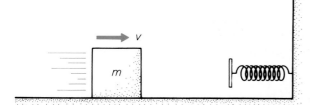

Figure 5.11 **Conservative force and mechanical energy of a spring** See Example 5.8.

(a) Before the mass makes contact with the spring, the total mechanical energy of the system is

$$E = K = \tfrac{1}{2}mv^2 = \tfrac{1}{2}(0.30 \text{ kg})(2.5 \text{ m/s})^2 = 0.94 \text{ J}$$

Since the system is conservative, this is the total energy at any time.

(b) When the spring is compressed a distance x_1,

$$E = K_1 + \tfrac{1}{2}kx_1^2$$

or $K_1 = E - \tfrac{1}{2}kx_1^2$

$$= 0.94 \text{ J} - \tfrac{1}{2}(3.0 \times 10^3)(0.010 \text{ m})^2 = 0.94 \text{ J} - 0.15 \text{ J} = 0.79 \text{ J}$$

(c) When the mass stops and the spring is compressed a distance x_{max},

$$E = \tfrac{1}{2}kx_{max}^2 = 0.94 \text{ J}$$

Thus,

$$x_{max} = \sqrt{\frac{2(0.94 \text{ J})}{k}} = \sqrt{\frac{2(0.94 \text{ J})}{3.0 \times 10^3 \text{ N/m}}} = 0.025 \text{ m} \quad \blacksquare$$

In the preceding examples, the force of friction, which is no doubt the most common nonconservative force, was ignored. If a nonconservative force acts on a moving object, the mechanical energy is not conserved. For example, when an object is thrown vertically upward, air resistance retards the motion in both directions, and the object returns to the starting position with less kinetic energy than it had initially. In this case, Eq. 5.10 may be written as

$$W_{nc} = \Delta K + \Delta U$$

and

$$K_o + U_o = K + U - W_{nc} \tag{5.13}$$

Thus, an amount of mechanical energy (Q) is lost. This is equal to the negative of the work done by the nonconservative force:

$$Q = -W_{nc}$$

However, the total energy is conserved, and you can sometimes use this to find Q. That is, for a nonconservative system, initial total energy equals final total energy, or

$$K_o + U_o = K + U + Q$$

That is,

$$E_o = E + Q$$

and

$$Q = E_o - E$$

Example 5.9 Work Done by a Nonconservative Force

A skier with a mass of 80 kg starts from rest and skis down a slope from an elevation of 110 m (Fig. 5.12). The speed of the skier at the bottom of the slope is 20 m/s. (a) Show that the system is nonconservative. (b) How much work is done by the nonconservative force?

Solution

Given: $m = 80$ kg   Find: (a) Show that E is not
  $v_o = 0$ constant
  $v = 20$ m/s   (b) $Q = -W_{nc}$
  $h = 110$ m

(a) If the system is conservative, the total mechanical energy is constant. Initially,

$$E_o = U_o = mgh = (80 \text{ kg})(9.8 \text{ m/s}^2)(110 \text{ m}) = 8.6 \times 10^4 \text{ J}$$

At the bottom of the slope,

$$E = K = \tfrac{1}{2}mv^2 = \tfrac{1}{2}(80 \text{ kg})(20 \text{ m/s})^2 = 1.6 \times 10^4 \text{ J}$$

Therefore, this system is not conservative (as you might expect).

(b) The work done by the nonconservative force of friction is equal in magnitude to the change in the mechanical energy, or to the amount of energy lost:

$$Q = E_o - E = (8.6 \times 10^4 \text{ J}) - (1.6 \times 10^4 \text{ J}) = 7.0 \times 10^4 \text{ J}$$

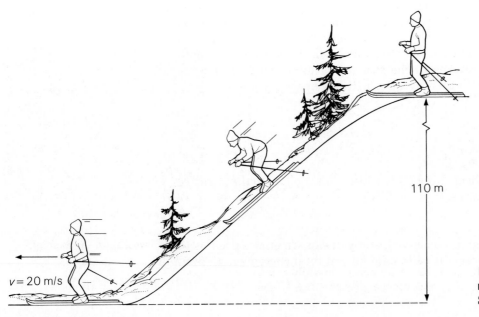

$v = 20$ m/s

110 m

Figure 5.12 **Work done by a nonconservative force**
See Example 5.9.

and

$$W_{nc} = -Q = -7.0 \times 10^4 \text{ J}$$

This is over 80% of the initial energy. (Where did this energy actually go?) ∎

Note that in a nonconservative system the total energy is conserved, but it is not all available for mechanical work. For a conservative system, you get back what you put in, so to speak. That is, if you do work on the system, the transferred energy is generally available to do work. Of course, conservative systems are idealizations, because all real systems are nonconservative to some

Demonstration 3

High Road/Low Road

A demonstration of potential energy and kinetic energy conversion.

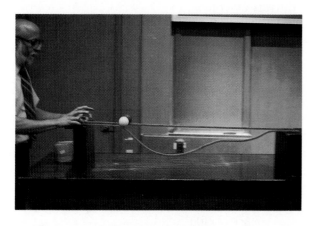

(a) They're off! Two billiard balls are released together. Which will make it to the end of its track first? Note that the yellow ball on the low road must roll a longer distance, and that the change in initial and final heights and gravitational potential energy difference is the same for both balls.

(b) The red ball actually pulls ahead, but not for long.

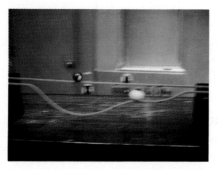

(c) Although the low-road ball must travel farther, it gains a greater speed by giving up more gravitational potential energy for kinetic energy. This allows it to cover the greater distance in a shorter time.

(d) By the time the low-road ball rises to the height of the end of the track, it is well ahead of the high-road ball.

degree. However, working with ideal conservative systems gives an understanding of the conservation of energy. (See Demonstration 3.)

During this course of study, you will be concerned with other forms of energy, such as thermal, electrical, nuclear, and chemical energy. In general, on the microscopic and submicroscopic levels, these forms of energy can be described in terms of kinetic energy and potential energy. Also, you will learn that mass is a form of energy and that the law of the conservation of energy must take this into account to be applied to the analysis of nuclear reactions.

5.6 Power

A particular task may require a certain amount of work, but it might be done over different lengths of time or at different rates. For example, suppose that you have to mow a lawn. This takes a certain amount of work, but you might do the job in a half hour, or you might take an hour or two. There's a practical distinction to be made here. There is usually not only an interest in the amount of work done, but also an interest in how fast it is done, that is, the rate at which it is done. **Power** is the rate of doing work.

The average power is the work done divided by the time it takes to do the work, or work per unit of time:

$$\bar{P} = \frac{\Delta W}{\Delta t} \tag{5.14}$$

If W_o and t_o are taken to be zero,

$$\bar{P} = \frac{W}{t}$$

If the work is done by a constant force,

$$\bar{P} = \frac{W}{t} = \frac{Fd}{t} = F\bar{v} \tag{5.15}$$

where it is assumed that F is in the direction of the displacement [otherwise, it is the component of the force in that direction, and $\bar{P} = (Fd\cos\theta)/t$] and $\bar{v}$ is the average velocity. If the velocity is constant, $\bar{P} = P = Fv$. The instantaneous power when v is the instantaneous velocity is $P = Fv$.

As you can see from Eq. 5.14, the SI units of power are joules per second (J/s).

$$\text{J/s} \equiv \text{watt (W)}$$

The SI power unit is called a **watt (W)** in honor of James Watt (1736–1819), a Scottish engineer who developed one of the first practical steam engines. The British units are foot-pounds per second (ft-lb/s). However, a larger unit, the **horsepower (hp)** is more commonly used:

$$1 \text{ hp} \equiv 550 \text{ ft-lb/s} = 746 \text{ W}$$

Power tells you how fast work is being done *or* how fast energy is transferred. For example, motors have power ratings (commonly given in horsepower). A 1-hp motor can do a given amount of work in half the time that a $\frac{1}{2}$-hp motor would take, or twice the work in the same amount of time; that is, a 1-hp motor is twice as powerful as a $\frac{1}{2}$-hp motor (see Example 5.11).

This unit was originated by James Watt. At the time, steam engines were replacing horses for work in mines and mills. To characterize the performance of his new engine, which was more efficient than other existing ones, Watt used the average rate at which a horse could do work as a unit—a horsepower.

Example 5.10 Power delivery

A hoist crane like those shown in Fig. 5.13 lifts a load of 1.0 metric ton a vertical distance of 25 m in 9.0 s at a constant velocity. (a) How much work is done by the hoist? (b) What is the minimum horsepower delivered by the hoist motor?

Solution

Given: $m = 1.0$ metric ton
$\quad\quad\quad = 1.0 \times 10^3$ kg
$\quad h = 25$ m
$\quad t = 9.0$ s

Find: (a) W
$\quad\quad\;$ (b) P (in hp)

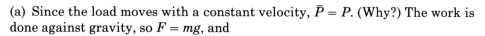

Figure 5.13 **Power delivery**
See Example 5.10.

(a) Since the load moves with a constant velocity, $\bar{P} = P$. (Why?) The work is done against gravity, so $F = mg$, and

$$P = \frac{W}{t} = \frac{Fd}{t} = \frac{mgh}{t}$$

$$= \frac{(1.0 \times 10^3 \text{ kg})(9.8 \text{ m/s}^2)(25 \text{ m})}{9.0 \text{ s}}$$

$$= 2.7 \times 10^4 \text{ W} \quad \text{(or 27 kW)}$$

Note that the velocity is $v = d/t = 25$ m$/9.0$s $= 2.8$ m/s, and the power is $P = Fv$.
(b) The minimum power refers to useful work output (neglecting frictional and other losses), and this was computed in part (a). In horsepower, it is

$$P = (2.7 \times 10^4 \text{ W})(1 \text{ hp}/746 \text{ W}) = 36 \text{ hp} \quad \blacksquare$$

Example 5.11 Work and Time

Two motors have net power outputs of 1.00 hp and 0.500 hp. (a) How much work in joules can each motor do in 3.00 min? (b) How long does it take for each motor to do 56.0 kJ of work?

Solution

Given: $P_1 = 1.00$ hp $= 746$ W
$\quad\quad\;\; P_2 = 0.500$ hp $= 373$ W
$\quad\quad\;\; t = 3.0$ min $= 180$ s
$\quad\quad\;\; W = 56.0$ kJ $= 56.0 \times 10^3$ J

Find: (a) W
$\quad\quad\;$ (b) t

(a) Since $P = W/t$,

$$W_1 = P_1 t = (746 \text{ W})(180 \text{ s}) = 1.34 \times 10^5 \text{ J}$$

$$W_2 = P_2 t = (373 \text{ W})(180 \text{ s}) = 0.67 \times 10^5 \text{ J}$$

Note that the smaller motor does half the work the larger one does in the same time, as you would expect.
(b) The times are given by $t = W/P$, and for the same amount of work,

$$t_1 = \frac{W}{P_1} = \frac{56.0 \times 10^3 \text{ J}}{746 \text{ W}} = 75 \text{ s}$$

$$t_2 = \frac{W}{P_2} = \frac{56.0 \times 10^3 \text{ J}}{373 \text{ W}} = 150 \text{ s}$$

Note that the smaller motor takes twice as long as the larger one to do the same amount of work. $\quad \blacksquare$

Efficiency

Machines and motors are quite common, and we often talk about their efficiency. Efficiency involves work, energy, and/or power. Simple and complex machines that do work have mechanical parts that move, so some energy is always lost because of friction or some other cause (perhaps in the form of sound). Thus, not all of the input energy goes into doing useful work.

Mechanical efficiency is essentially a measure of what you get out for what you put in, that is, the useful work output for the energy input. **Efficiency (ε)** is given as a fraction (or percentage):

$$\varepsilon = \frac{\text{work output}}{\text{energy input}} (\times 100\%)$$

$$= \frac{W_{out}}{E_{in}} (\times 100\%) \tag{5.16}$$

For example, if a machine has a 100-J (energy) input and a 50-J (work) output, then its efficiency is

$$\varepsilon = \frac{W_{out}}{E_{in}} = \frac{50 \text{ J}}{100 \text{ J}} = 0.50 \, (\times 100\%) = 50\%$$

An efficiency of 0.50 or 50% means that half of the energy input is lost because of friction or some other cause and doesn't go into its intended purpose.

Note that you can also write efficiency in terms of power (P):

$$\varepsilon = \frac{P_{out}}{P_{in}} (\times 100\%) \tag{5.17}$$

If both terms of the ratio in Eq. 5.16 are divided by time (t), $W_{out}/t = P_{out}$ and $E_{in}/t = P_{in}$.

Important Formulas

Work:

$W = Fd \cos \theta$

Hooke's law:

$F = -k(x - x_o)$

$F = -kx \qquad$ (with $x_o = 0$)

Work done (or potential energy change) in stretching or compressing a spring:

$W = \frac{1}{2}k(x^2 - x_o^2)$

$W = \frac{1}{2}kx^2 \qquad$ (with $x_o = 0$)

Kinetic energy:

$K = \frac{1}{2}mv^2$

Work-energy theorem:

$W_{ret} = K - K_o = \Delta K$

Elastic (spring) potential energy:

$U = \frac{1}{2}kx^2 \qquad$ (with $x_o = 0$)

Gravitational potential energy:

$U = mgh \qquad$ (with $h_o = 0$)

Conservation of mechanical energy:

$\frac{1}{2}mv^2 + U = \frac{1}{2}mv_o^2 + U_o$

Conservation of energy (with a nonconservative force):

$K + U = K_o + U_o + Q$

Work done by a nonconservative force:

$W_{nc} = -Q = \Delta K + \Delta U$

Power:

$\bar{P} = W/t = Fd/t = F\bar{v}$
(constant force in direction of d and v, otherwise $F \cos \theta$)

Questions

Work

1. Can work be done if there is no motion? Explain.

2. In Example 5.3, is it possible for W_f to be greater than W_g? If this were the case, what would it mean?

3. In one dimension, $W = Fx$, and the spring force is $F = kx$ (without the minus sign). Therefore, you might think that the work done in compressing a spring would be $W = Fx = (kx)x = kx^2$, but this is wrong. Why?

4. Could you find the work done by a constant force from a plot of F versus x? Explain in detail how this would be done.

5. How would *negative* work be represented on a graph of force versus displacement?

6. Does the centripetal force acting on an object moving with uniform circular motion do work on the object? Explain.

7. A spring is compressed and a mass on its end is moved from $x_0 = 2.0$ cm to $x = 6.0$ cm. Is the work done equal to $\frac{1}{2}k(\Delta x)^2$, where $\Delta x = x - x_0 = 4.0$ cm? Explain your answer.

Energy

8. How is the work-energy theorem stated in words? Is the converse statement true?

9. Describe the mechanical energy of an ideal simple pendulum at various positions in its swing (look ahead to Fig. 5.18).

10. (a) Explain why spring force is a conservative force. Are there any restrictions? (b) The force of friction is a nonconservative force, yet the force of static friction does not violate the conservation of mechanical energy. Why?

11. Describe how you would make and calibrate a scale for measuring force (weight) using a linear spring.

12. Imagine that you are at the bottom of a (dry) well and throw a stone vertically upward from a distance h below the ground. What is the minimum kinetic energy the stone would have to have to make it to the top of the well? How much work would you have to do on the stone and when is this done?

Power

13. Two students start at the same ground floor location and at the same time go to a classroom on the third floor by different routes. Who does more work against gravity? If they arrive at different times, which student will have expended more power? Explain.

14. What is actually purchased from a power company in buying units of kilowatt-hours?

15. A factory worker may be paid either at an hourly rate or for piecework on various jobs. Are any power considerations incorporated in these methods of paying for work?

Problems

5.1 Work

■1. A tractor exerts a constant force of 4.5×10^3 N on a horizontal chain while moving a load a horizontal distance of 25 cm. How much work is done by the tractor?

■2. A crane lifts a 500-kg load a vertical distance of 30.0 m. If the speed of the load is constant, how much work is done in lifting it?

■3. Two constant horizontal forces have magnitudes of 50 N and 75 N. The 75-N force does 300 J of work. Through what horizontal distance would the 50-N force have to act to do the same work?

■■4. The formula for work is sometimes written $W = F_\parallel d$, where $F_\parallel$ is the component of the force parallel to the displacement. Show that work is also given by $Fd_\parallel$, where $d_\parallel$ is the component of the displacement parallel to the force.

■■5. A person pushing a lawnmower on a level lawn applies a constant force of 150 N at an angle of 37.0° to the horizontal. How far does the person push the mower in doing 1.44×10^3 J of work?

■■6. A 2.0-kg block slides down a frictionless plane inclined 20° to the horizontal. If the length of the plane's surface is 1.5 m, how much work is done and by what force?

■■7. In planing a piece of wood 30 cm long, a carpenter applies a force of 40 N to a jack plane at an angle of 25° to the horizontal. How much work is done by the carpenter?

■■8. A helicopter with a mass of 1500 kg accelerates vertically upward at a constant rate of 2.2 m/s² for a distance of 50 m. (a) What is the net work done on the helicopter? (b) How much work is done by the downward force produced by the helicopter motor and rotor?

■■9. A 120-kg sleigh is pulled by one horse at a constant velocity for a distance of 0.64 km on a level snowy surface. The coefficient of kinetic friction between the sleigh runners and the snow is 0.20. (a) Calculate the work done by the horse. (b) Calculate the work done by friction.

■■10. A 50-kg crate slides down a 5.5-m loading ramp that is inclined at an angle of 20° to the horizontal. A worker pushes on the crate parallel to the ramp surface so that it slides down with a constant velocity. If the coefficient of kinetic friction between the crate and the ramp is 0.30, how much work is done by (a) the worker, (b) the force of friction, and (c) the force of gravity? (d) What is the net work?

■■11. The sprocket arm on a five-speed bicycle is 7.0 in. long. When moving at a constant velocity, a cyclist exerts an average downward force of 5.0 lb on a pedal during each downward stroke (a half cycle). How much work is done by the cyclist on the pedal during each cycle in (a) foot-pounds and (b) joules?

■■■12. A father pulls his young daughter on a sled with a constant velocity on a level surface through a distance of 10 m, as illustrated in Fig. 5.14(a). If the total mass of the sled and the girl is 35 kg and the coefficient of kinetic friction between the sled runners and the snow is 0.15, how much work does the father do?

■■■13. A father pushes horizontally on his daughter's sled to move it up a snowy incline, as illustrated in Fig. 5.14(b). If the sled goes up the hill with a constant acceleration of 0.25 m/s², how much work is done by the father in getting it from the bottom to the top of the hill? (Some necessary data are given in Problem 12.)

5.2 Work Done by Variable Forces ◆

■14. A spring has a spring constant of 45 N/m. How much work is required to stretch the spring 3.0 cm?

■15. Compressing a spring 5.0 cm requires 0.20 J of work. What is the force constant of the spring?

■■16. A particular force is described by the equation $\mathbf{F} = (75 \text{ N})\mathbf{x}$. How much work is done when this force acts through the distance between (a) $x_o = 0$ and $x = 0.25$ m, and (b) $x_o = 0.15$ m and $x = 0.35$ m?

■■17. When a 50-g mass is suspended from a vertical spring, the spring is stretched from a length of 4.0 cm to a length of 6.0 cm. If the mass is pulled downward an additional 10 cm, what is the *total* work done against the spring force?

■■18. Compute the work done by the force in the F versus x graph in Fig. 5.15.

■■19. A force of $\mathbf{F} = (-50 \text{ N})\mathbf{x}$ stretches a spring from $x_o = 0$ to $x_1 = 10$ cm and then from $x_1 = 10$ cm to $x_2 = 20$ cm. Is more work required for the second stretching than for the first? Justify your answer.

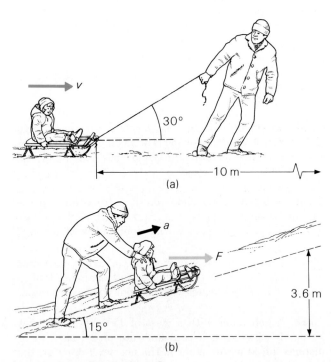

(a)

(b)

Figure 5.14 **Fun and work**
See Problems 12 and 13.

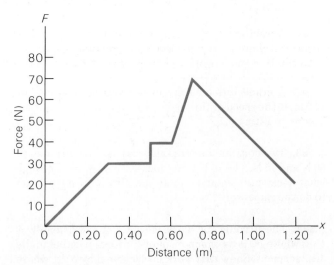

Figure 5.15 **Work done by a variable force**
See Problem 18.

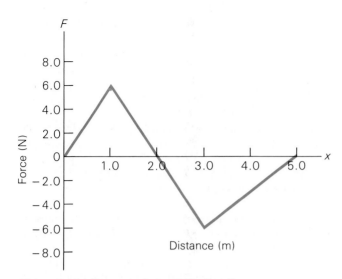

Figure 5.16 **How much work is done?**
See Problem 20.

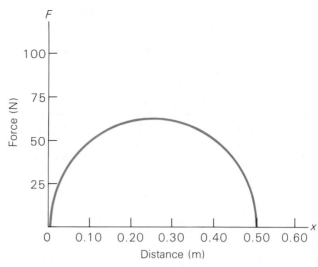

Figure 5.17 **Definitely a varying force**
See Problem 22.

■■■20. Compute the work done by the force in the F versus x graph in Fig. 5.16.

■■■21. A spring force (in newtons) is given by $F = -\pi(x - 0.10)$. (a) How much work is done in stretching the spring from its equilibrium position ($x = 0$) to $x = 0.25$ m? (b) How much work is done if the position of the free end of the spring is changed from $x_1 = 0.05$ m to $x_2 = 0.15$ m?

■■■22. (a) Compute the approximate work done by the force in the graph of F versus x in Fig. 5.17 using two triangles. (b) Compute the approximate work done using straight line segments for symmetric intervals with $\Delta x = 0.10$ m and $\Delta x = 0.15$ m. (c) Explain how a better approximation can be obtained.

5.3 The Work-Energy Theorem: Kinetic Energy

■23. An automobile with a mass of 1.0×10^3 kg travels at a speed of 90 km/h. (a) What is its kinetic energy? (b) What is the minimum work that would be required to bring it to rest?

■24. What is the minimum work required to give a 15-kg object a velocity of 2.0 m/s?

■25. A constant net force of 75 N acts on an object initially at rest and acts through a parallel distance of 0.60 m. (a) What is the final kinetic energy of the object? (b) If the object has a mass of 0.20 kg, what is its final speed?

■■26. A 3.0-g bullet traveling at 350 m/s hits a tree and penetrates a distance of 10 cm. What was the average force exerted on the bullet in bringing it to rest?

■■27. An electron ($m = 9.11 \times 10^{-31}$ kg) has 8.00×10^{-17} J of kinetic energy. What is the speed of the electron?

■■28. (Work this problem and the next one as a pair.) The stopping distance of a vehicle is an important safety factor. Assuming a constant braking force, show that a vehicle's stopping distance is proportional to the square of its instantaneous speed or velocity.

■■29. An automobile traveling at 45 km/h is brought to a stop in 50 m. Assume that the same conditions (same braking force, driver reaction time, and so on) hold for all cases. (a) What would be the stopping distance for an initial speed of 90 km/h? (b) What would be the initial speed for a stopping distance of 100 m?

■■30. A 0.16-kg puck moves on a (frictionless) air hockey table in a straight line with a speed of 2.0 m/s. How much work would be required to increase its speed to 4.0 m/s?

■■■31. A 150-g block sliding on a frictionless surface moves directly toward a fixed, horizontal spring with a speed of 4.75 m/s. The spring has a spring constant of 200 N/m. (a) After the block makes contact with the spring, by how much will the spring be compressed when the block comes to rest? (b) What will the speed of the block be when the spring has been compressed 6.00 cm? (c) What will the kinetic energy and the speed of the block be when it recoils free of the spring? (Assume that no energy is lost in the collision.)

■■■32. A sports car weighs a third as much as a large luxury car. (a) If the sports car is traveling at a speed of 90 km/h, at what speed would the larger car have to travel to have the same kinetic energy? (b) Suppose that the large car is traveling at a speed that gives it half the kinetic energy of the sports car. What is the speed of the large car?

5.4 Potential Energy

■33. How much more gravitational potential energy does a 1.0-kg hammer have when it is on a shelf 1.5 m high than when it is on a shelf 1.2 m high?

■34. To store exactly 1 J of potential energy in a spring with $k = 65$ N/m, how much would the spring have to be stretched beyond its equilibrium length?

■35. Sketch a plot of U versus x for a mass oscillating on a spring between the limits of $-x_{max}$ and $+x_{max}$.

■■36. The floor of the basement of a house is 3.5 m below ground level and the floor of the attic is 4.5 m above ground level. (a) What are the potential energies of 1.5-kg objects on each floor, relative to ground level? (b) If the object in the attic were brought to ground level and the object in the basement were taken to the attic, what would be the change in potential energy for each?

■■37. A 0.20-kg stone is projected vertically upward with an initial velocity of 8.0 m/s, from a starting point 1.2 m above the ground. (a) What is the potential energy of the stone at its maximum height relative to the ground? (b) What is the change in the potential energy of the stone?

■■38. A spring compressed from $x_0 = 4.5$ cm to $x = 9.5$ cm has a change in potential energy of 0.63 J. What is the spring constant?

■■39. The force constant of one spring is twice that of another spring, but has only $\frac{2}{3}$ as much potential energy as the other spring. Which spring is compressed further from equilibrium and by how many times more?

■■■40. A student has six textbooks, each with a thickness of 4.0 cm and a weight of 25 N. What is the minimum work the student would have to do to stack the books on top of one another?

■■■■41. A 0.10-kg ball is dropped from a height of 3.0 m above the top of a fixed, vertical spring with a spring constant of 250 N/m. The ball lands on the spring and compresses it. How far has the spring been compressed when the ball comes to rest?

5.5 The Conservation of Energy

■42. A 0.50-kg ball thrown vertically upward has an initial kinetic energy of 60 J. (a) What are its kinetic and potential energies when it has traveled $\frac{3}{4}$ of the distance to its maximum height? (b) What is its velocity at this point? (c) What is its potential energy at the maximum height?

■43. An automobile is traveling at 90 km/h on level ground and then coasts up a hill. (a) Neglecting friction, how high above the level ground will the car go before stopping? (b) Assuming that 40% of the energy is lost to friction, how high will the car go?

■44. A person standing on a bridge at a height of 120 m above a river drops a 0.250-kg rock. (a) What is the rock's mechanical energy at the time of release relative to the river? (b) What are the rock's kinetic and potential energies after it has fallen 75.0 m? (c) What are the rock's velocity and total mechanical energy just before hitting the water? (d) Answer parts (a)–(c) taking the reference point ($h = 0$) at the elevation where the rock is released.

■■45. A girl swings back and forth on a swing whose ropes are 4.8 m long. The maximum height she reaches is 2.50 m above the ground. At the lowest point of the swing, she is 0.50 m above the ground. What is the girl's maximum speed, and where does she attain it?

■■46. A roller coaster starts from rest at a point 50 m above the bottom of a dip. Neglecting friction, what will be the speed of the roller coaster at the top of the next slope, which is 30 m above the bottom of the dip?

■■47. A simple pendulum has a length of 0.85 m and a bob whose mass is 0.20 kg. The bob is released from an angle of 25° relative to a vertical reference line (Fig. 5.18). (a) Show that the vertical height of the bob when released is $h = L(1 - \cos 25°)$. (b) What is the kinetic energy of the bob when the string is at an angle of 9.0°? (c) What is the speed of the bob at the bottom of the swing? (Neglect friction and the mass of the string.)

■■48. The grade of a hill is given by the rise (Δy) over the run (Δx), which is the slope of the incline, or the tangent of the angle of incline ($\tan \theta$), usually expressed as a percentage. If a car starts up a hill with a 3.0% grade traveling at 90 km/h in neutral gear, what distance up the hill will it coast before coming to a stop? Assume a frictional energy loss of 40%.

■■49. A 0.25-kg projectile is given an initial velocity of 15 m/s at an angle of 37° to the horizontal. (a) What are the projectile's potential energy and kinetic energy at $t = 0.50$ s? (b) What is its kinetic energy at $t = 1.1$ s?

■■50. On a snowy day when school has been canceled,

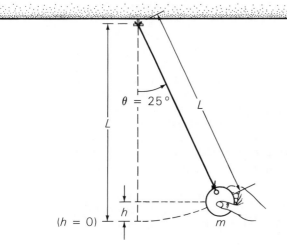

Figure 5.18 A pendulum swings
See Problem 47.

a student on a sled starts from rest at a vertical height of 25 m above the horizontal base of the hill and slides down. If the sled and the student have a speed of 20 m/s at the bottom of the hill, is mechanical energy conserved? Justify your answer.

■■**51.** Suppose that at the base of the hill the student on the sled in Problem 51 slides onto a rough horizontal surface with a coefficient of kinetic friction of $\mu_k = 1.5$. How far will the sled travel from the base of the hill before coming to rest?

■■**52.** A 0.35-kg block on a frictionless surface is pushed against a horizontal spring with a force constant of 200 N/m, and the spring is compressed 10 cm. The spring is released and the block is propelled directly toward another horizontal spring with a force constant of 265 N/m. (a) What is the speed of the block when it has traveled 5.0 cm from its point of release? (b) What is the speed of the block when it is not in contact with either spring? (c) How far does the block compress the second spring? (Neglect energy loss due to the collision.)

■■■**53.** A ball with a mass of 0.36 kg is dropped from a height of 1.2 m above the top of a fixed vertical spring, whose force constant is 300 N/m. (a) What is the maximum distance the spring is compressed by the ball? (Neglect energy loss due to the collision.) (b) What is the speed of the ball when the spring has been compressed 5.0 cm?

■■■**54.** A water slide has a height of 3.7 m, and the people coming down the slide shoot out horizontally at the bottom, which is a distance of 1.5 m above the water in the swimming pool. If a person starts down the slide from rest, and there are no frictional losses, how far from a point directly below the bottom of the slide does the person land?

Does it make any difference whether the person is a small child or an adult?

■■■**55.** When a certain rubber ball is dropped from a height of 1.25 m onto a hard surface, it loses 18.0% of its mechanical energy in the collision. (a) How high will the ball bounce on the first bounce? (b) How high will it bounce on the second bounce? (c) With what speed would the ball have to be thrown downward to make it reach its original height on the first bounce?

■■■**56.** Three identical balls are thrown or projected from a cliff at a height h above level ground. The first ball is thrown at an angle of 45° to the horizontal, the second is thrown horizontally, and the third is thrown at an angle of −30° to the horizontal (that is, downward). If all the balls are given the same initial speed, which will strike the ground with the greatest speed? (Hint: use energy considerations. Also see Problem 55 in Chapter 3. Can you work the problem using methods in that chapter? Which way is easier?)

■■■**57.** A hiker plans to swing across a ravine in the mountains on a rope, as illustrated in Fig. 5.19, and to drop when he is just above the far edge. (a) With what horizontal speed should he be moving when starting to swing? (b) Will he be in any danger of hitting the edge and falling into the ravine? Explain. (Can you compute how far from the edge

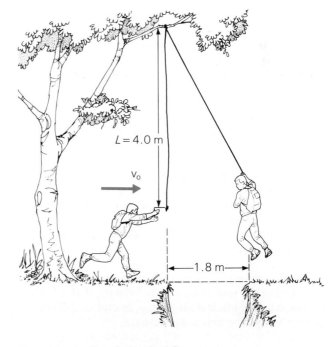

Figure 5.19 Can he make it?
See Problem 57.

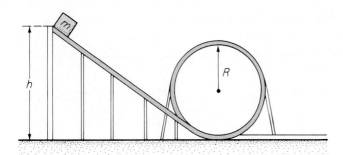

Figure 5.20 **Loop-the-loop**
See Problem 58.

of the ravine he will land? It's rather involved, but you have the background to do it.)

■■■**58.** A block of mass m slides down an inclined plane into a loop-the-loop of radius R (Fig. 5.20). (a) Neglecting friction, what is the minimum speed the block must have at the highest point of the loop to stay in the loop? (Hint: what force must act on the block at the top of the loop to keep it on a circular path?) (b) At what vertical height on the inclined plane (in terms of the radius of the loop) must the block be released if it is to have the required minimum speed at the top of the loop?

5.6 Power

■**59.** What is the horsepower rating of a 100-W light bulb?

■**60.** (a) How many joules of electrical energy are used by a 1200-W hair dryer in 10 min? (b) If the dryer is operated a total of 4.5 h during one month, how much will this contribute to the electrical bill if the billing rate is 9¢/kWh? (See Question 14.)

■**61.** A constant force of 50 N moves an object 4.0 m in 5.0 s. (a) How much energy is expended? (b) What is the average power developed?

■**62.** A 65-kg student who is late for a class runs up two flights of stairs whose combined vertical height is 7.0 m in 10 s (and then silently enters the classroom while the professor is writing on the blackboard). Compute the student's power output in doing work against gravity in (a) watts and (b) horsepower.

■■**63.** A tractor pulls a wagon with a constant force of 700 N, giving the wagon a constant velocity of 20.0 km/h. (a) How much work is done by the tractor in 3.50 min? (b) What is the tractor's power output?

■■**64.** A self-propelled lawn mower with a power delivery

of 3.5 hp travels at a constant rate of 2.0 ft/s. (a) What is the magnitude of the propelling force? (b) What is the magnitude of the frictional force? (c) How much work is done by each force in 6.0 s?

■■**65.** A 1.0-kg stone is dropped and free falls a distance of 5.0 m. What is the average power developed by gravity?

■■**66.** A 1500-kg elevator accelerates upward at a constant rate of 0.50 m/s². How much power is developed during the time the elevator's speed goes from 0.25 m/s to 1.75 m/s?

■■**67.** What is the average power developed or expended by each of the forces in the graphs in Figs. 5.15 and 5.16 if the forces act for 8.0 s in each case?

■■**68.** An electric motor with a 2.0-hp output drives a machine with an efficiency of 45%. What is the energy output of the machine per second?

■■■**69.** A pump delivers water from a 100-ft well to a house at a rate of 10 gal/min. If 25% of the power is lost to friction and other causes, what is the minimum power rating of the pump motor in horsepower?

■■■**70.** A 3.25×10^3-kg commercial aircraft takes 12.5 min to achieve its cruising altitude of 10.0 km and cruising speed of 850 km/h. If the plane's engines deliver, on the average, 1500 hp of power during this time, what is the efficiency of the aircraft?

■■■**71.** A large electric motor with an efficiency of 75% has a power output of 1.5 hp. If the motor is run steadily for 2.0 h and the cost of electricity is 10¢/kWh, how much does it cost to run the motor for this period? (See Question 14.)

Additional Problems

72. A student suspends a 0.45-kg mass from a spring, which stretches a distance of 1.5 cm as a result. The student then pulls down on the suspended mass, doing 4.0 J of work in further stretching the spring. How far is the spring stretched as a result of the work done by the student?

73. A constant force of 6.0 N acts on a 1.0-kg object initially at rest on a horizontal surface and gives it 50 J of kinetic energy. How much power was expended in doing this?

74. A 0.50-kg projectile is given an initial speed of 24 m/s at an angle of 37° to a horizontal surface and lands a certain distance (range) from its launch point. How much work is done on the projectile after it is launched? (Neglect air resistance.)

75. A fossilized dinosaur bone with a mass of 5.0 kg is dug out from a depth of 2.8 m below the ground surface. The bone is taken to the second story of a nearby building where a lab has been set up; it is placed on a table at a height of 9.0 m above ground level. (a) When the bone is in the dig hole, what is its potential energy relative to ground level? (b) What is the change in the bone's potential energy when it is taken from the dig hole to the lab table? (c) When the bone is on the table, what is its potential energy relative to the bottom of the hole? (Hint: make a sketch of the situation.)

76. A 2.0-kg block with an initial velocity of 3.5 m/s slides to a stop on a horizontal surface. If the coefficient of kinetic friction between the block and the surface is 0.42, how much work is done by friction? Compute this two ways using work and energy considerations.

77. How much work does a 1000-W motor do in 5.0 min if it has an efficiency of 90%?

78. A 1.0-kg object with an initial velocity of 4.0 m/s in the positive x direction is acted on by a force in the direction of the motion. The force does 2.0 J of work. (a) What is the final velocity of the object? (b) What would its final velocity be if the applied force were in the direction opposite to the motion?

79. A 0.50-kg block slides 1.2 m down a 25° inclined plane with a constant acceleration of 0.30 m/s². (a) How much work is done by the force of gravity? (b) What is the coefficient of kinetic friction between the block and plane?

80. A motorboat traveling on a lake with a constant speed of 2.50 m/s has a loss of 12.0 hp due to friction between the hull and the water. What is the average force of friction?

81. A 1.0-kg block slides on a level, frictionless surface with a velocity of 3.0 m/s and runs into a horizontal spring. (a) If the spring has a force constant of 720 N/m, what is its maximum compression distance after the block collides with it? (Neglect any energy loss due to the collision.) (b) As the spring recoils and the block moves outward, what will the speed of the block be when the spring is still compressed 5.0 cm?

82. A 0.40-kg block attached to a fixed, horizontal spring that has a force constant of 6.5×10^2 N/m can slide back and forth on a tabletop. There is a coefficient of kinetic friction of 0.34 between the surfaces. (a) If the block

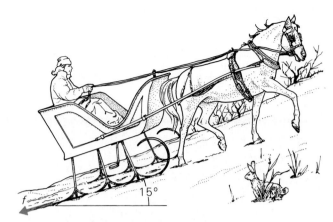

Figure 5.21 **A one-horse open sleigh**
See Problem 84.

is displaced and the spring is stretched 10 cm from equilibrium, how far will the spring be compressed when the block is released and travels to the other side of its equilibrium position? (b) How far will the spring be stretched on the return trip? (c) Is the same amount of energy lost to friction during each round trip, or cycle? Explain.

83. A student does 6.0×10^3 J of work in pushing a stalled car a distance of 20 m. What is the average force exerted by the student?

84. A horse pulls a sled whose total mass is 120 kg up a hill with a 15° incline, as illustrated in Fig. 5.21. (a) If the overall retarding frictional force is 950 N and the sled moves up the hill with a constant velocity of 5.0 km/h, what is the power output of the horse? (Express in horsepower, of course. Note the magnitude of your answer and explain.) (b) Suppose that in a spurt of energy the horse accelerates the sled uniformly from 5.0 km/h to 20 km/h in 5.0 s. What is the horse's maximum instantaneous power output? Assume the same force of friction.

85. A block of mass m may be put on a table of height h by three routes: by lifting it vertically and by sliding it up 30° and 60° frictionless inclined planes. Show that the work done against gravity is the same in all cases. What does this tell you about the gravitational force?

86. A 10-g bullet is fired vertically upward with an initial speed of 200 m/s. If the bullet reaches a maximum height of 1.2 km, what percentage of mechanical energy is lost to air resistance?

Momentum and Collisions

6

Another quantity that is particularly useful in describing and analyzing motion is called momentum. This word may bring to mind a football player running down the field with a lot of momentum, knocking down other players who are trying to stop him. Or you might have heard someone say that a team lost its momentum (and lost the game). Such everyday usages give some insight into the meaning of momentum. The idea of mass in motion, or inertia, is implied.

We tend to think of heavy or massive objects as having a great deal of momentum, even if they move very slowly. However, as you will learn from the technical definition of momentum given below, a light object can have just as much momentum as a heavier one, and sometimes more. Also, like energy, momentum is conserved under certain conditions. This makes it particularly useful in analyzing various situations, especially collisions.

6.1 Linear Momentum

In his *Principia*, Newton referred to what modern physicists know as **linear momentum** as "the quantity of motion... arising from velocity and the quantity of matter conjointly." In other words,

> The linear momentum of an object is the product of its mass and velocity:

$$\mathbf{p} = m\mathbf{v} \qquad (6.1)$$

It is common to refer to linear momentum as simply momentum. From Eq. 6.1, you can see that the SI units for momentum are kg-m/s. Momentum is a vector quantity having the same direction as the velocity. Like velocity, it can be resolved into rectangular components: $\mathbf{p}_x = m\mathbf{v}_x$ and $\mathbf{p}_y = m\mathbf{v}_y$.

Eq. 6.1 expresses the momentum of a single object or particle. For a system of more than one particle, the total linear momentum of the system is the vector sum of the momenta (plural of momentum) of the individual particles:

$$\mathbf{P} = \mathbf{p}_1 + \mathbf{p}_2 + \mathbf{p}_3 + \cdots \qquad (6.2)$$

Example 6.1 Comparison of Momenta

A 100-kg football player runs straight down the field with a velocity of 4.0 m/s. A 1.0-kg shell leaves the barrel of a gun with a muzzle velocity of 500 m/s. Which has the greater momentum?

Solution
Given: $m_p = 100$ kg   *Find*: p_p and p_s
$\quad\quad\quad v_p = 4.0$ m/s
$\quad\quad\quad m_s = 1.0$ kg
$\quad\quad\quad v_s = 500$ m/s

The magnitude of the momentum of the player is

$$p_p = m_p v_p = (100 \text{ kg})(4.0 \text{ m/s}) = 400 \text{ kg-m/s}$$

and that of the shell is

$$p_s = m_s v_s = (1.0 \text{ kg})(500 \text{ m/s}) = 500 \text{ kg-m/s}$$

The much lighter, or less massive, shell has the greater momentum. Keep in mind that the magnitude of momentum depends on *both* mass and velocity. ∎

Example 6.2 Total Momentum

What is the total momentum for each of the systems of particles illustrated in Fig. 6.1?

Solution
Given: Magnitudes and directions   *Find*: (a) Total momentum (**P**)
$\quad\quad\quad$ of momenta from Fig. 6.1 $\quad\quad\quad$ (b) Total momentum (**P**)

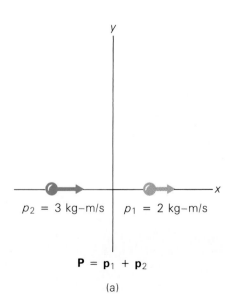

$p_2 = 3$ kg–m/s $p_1 = 2$ kg–m/s

$\mathbf{P} = \mathbf{p}_1 + \mathbf{p}_2$

(a)

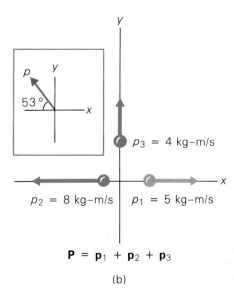

$p_3 = 4$ kg–m/s

$p_2 = 8$ kg–m/s $p_1 = 5$ kg–m/s

$\mathbf{P} = \mathbf{p}_1 + \mathbf{p}_2 + \mathbf{p}_3$

(b)

Figure 6.1 **Total momentum** The total momentum of a system of particles is the vector sum of the individual momenta. See Example 6.2.

(a) The total momentum of a system is the vector sum of the momenta of the individual particles, so

$$\mathbf{P} = \mathbf{p}_1 + \mathbf{p}_2 = 2 \text{ kg-m/s} + 3 \text{ kg-m/s} = 5 \text{ kg-m/s} \qquad (+x \text{ direction})$$

(Remember that plus and minus signs are used to indicate directions when working with vector quantities in one dimension.)

(b) Computing the total momenta in the x and y directions gives

$$\mathbf{P}_x = \mathbf{p}_1 + \mathbf{p}_2 = 5 \text{ kg-m/s} - 8 \text{ kg-m/s}$$
$$= -3 \text{ kg-m/s} \quad (-x \text{ direction})$$

$$\mathbf{P}_y = \mathbf{p}_3 = 4 \text{ kg-m/s} \quad (+y \text{ direction})$$

Then

$$\mathbf{P} = \mathbf{P}_x + \mathbf{P}_y = (-3 \text{ kg-m/s})\mathbf{x} + (4 \text{ kg-m/s})\mathbf{y}$$

Expressed in magnitude-angle form, the total momentum is

$$P = \sqrt{P_x^2 + P_y^2} = \sqrt{(3 \text{ kg-m/s})^2 + (4 \text{ kg-m/s})^2}$$
$$= \sqrt{25 \text{ kg}^2\text{-m}^2/\text{s}^2} = 5 \text{ kg-m/s}$$

and

$$\theta = \tan^{-1}\left|\frac{P_y}{P_x}\right| = \tan^{-1}\left|\frac{4}{3}\right| = 53°$$

relative to the negative x axis in the second quadrant. See Fig. 6.1(b). [Note: if the motion of one (or more) of the particles is not along a coordinate axis, its momentum is broken up, or resolved, into rectangular components and added to the component sums—just as you learned to do with force components in Chapter 4.] ■

A change in momentum can result from a change in magnitude and/or direction, since momentum is a vector quantity. Examples of changes in the momenta of particles due to changes of direction on collision are illustrated in Fig. 6.2. In the figure, the magnitude of a particle's momentum is taken to be the same before and after the collision. Note that the change in momentum ($\Delta\mathbf{p}$) is the vector difference.

As you know, a change in velocity (an acceleration) requires a force. Similarly, since momentum is directly related to velocity, a change in momentum also requires a force. In fact, Newton originally expressed his second law of motion in terms of momentum rather than acceleration. The force-momentum relationship may be seen by starting with $\mathbf{F} = m\mathbf{a}$ and using $\mathbf{a} = (\mathbf{v} - \mathbf{v}_o)/\Delta t$:

$$\mathbf{F} = m\mathbf{a} = \frac{m(\mathbf{v} - \mathbf{v}_o)}{\Delta t} = \frac{m\mathbf{v} - m\mathbf{v}_o}{\Delta t} = \frac{\mathbf{p} - \mathbf{p}_o}{\Delta t} = \frac{\Delta\mathbf{p}}{\Delta t}$$

or $\quad \mathbf{F} = \dfrac{\Delta\mathbf{p}}{\Delta t}$ \hfill (6.3)

where $\mathbf{F}$ is the average force if the acceleration is not constant or the instantaneous force if Δt goes to zero.

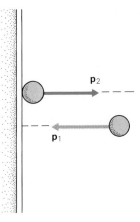

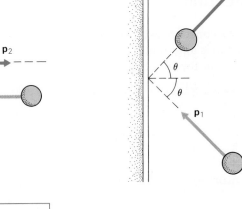

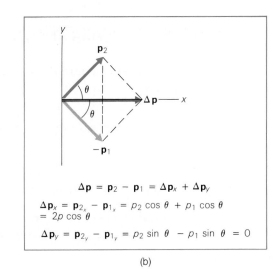

$$\Delta \mathbf{p} = \mathbf{p}_2 - \mathbf{p}_1 = \Delta \mathbf{p}_x + \Delta \mathbf{p}_y$$

$$\Delta \mathbf{p}_x = \mathbf{p}_{2_x} - \mathbf{p}_{1_x} = p_2 \cos \theta + p_1 \cos \theta$$
$$= 2p \cos \theta$$

$$\Delta \mathbf{p}_y = \mathbf{p}_{2_y} - \mathbf{p}_{1_y} = p_2 \sin \theta - p_1 \sin \theta = 0$$

(b)

$$\Delta \mathbf{p} = \mathbf{p}_2 - \mathbf{p}_1$$
$$= mv - (-mv) = 2mv$$

(a)

Figure 6.2 **Change in momentum**
The change in momentum is given by the *difference* between the momentum vectors. (a) Here the vector sum is zero, but the vector difference, or the change in momentum, is not. (b) Here the change in momentum is found using the change in the components. As shown in the diagram, vector subtraction is a special case of vector addition; that is, $\Delta \mathbf{p} = \mathbf{p}_2 + (-\mathbf{p}_1)$.

Expressed in this form, Newton's second law is that the net force acting on an object is equal to the time rate of change of the object's momentum. You can see that the equations $\mathbf{F} = m\mathbf{a}$ and $\mathbf{F} = \Delta \mathbf{p}/\Delta t$ are equivalent if the mass is constant.

Just as the equation $\mathbf{F} = m\mathbf{a}$ indicates that an acceleration is evidence of a force, the equation $\mathbf{F} = \Delta \mathbf{p}/\Delta t$ indicates that *a change in momentum is evidence of a force*. For example, as illustrated in Fig. 6.3, the momentum of a projectile is tangential to the projectile's parabolic path and changes in both magnitude and direction. The change in momentum indicates that there is a force acting on the projectile, which you know is the force of gravity. [In Fig. 6.2, what are the forces acting on the particles to cause the changes in momentum? Can you identify the reaction forces of the wall on the particles as the cause?]

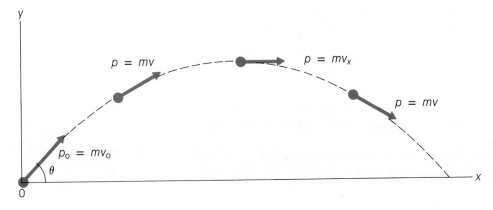

Figure 6.3 **Change in momentum of a projectile**
The total momentum vector of a projectile is tangential to its path (as is the velocity) and changes in magnitude and direction because of the action of an external force (gravity). The *x* component of the momentum is constant. Why?

6.2 The Conservation of Linear Momentum

For the linear momentum of a particle or an object to be conserved (to remain constant with time), a condition must hold. This condition is apparent from the momentum form of Newton's second law (Eq. 6.2). If the force acting on a particle is zero, that is,

$$\mathbf{F} = \frac{\Delta \mathbf{p}}{\Delta t} = 0$$

then

$$\Delta \mathbf{p} = 0 = \mathbf{p} - \mathbf{p}_o$$

where $\mathbf{p}_o$ is the initial momentum and $\mathbf{p}$ is the momentum at some later time. Since they are equal, the momentum is conserved:

$$\mathbf{p} = \mathbf{p}_o$$

$$\text{or} \quad m\mathbf{v} = m\mathbf{v}_o \tag{6.4}$$

Note that this is consistent with Newton's first law: an object remains at rest ($\mathbf{p} = 0$) or in motion with a *uniform* velocity (constant $\mathbf{p}$), unless acted on by a net external force.

The conservation of momentum may be extended easily to a system of particles by writing Newton's second law in terms of the sums (resultants) of the forces acting on the system and of the momenta of the particles: $\mathbf{F} = \Sigma \mathbf{F}_i$ and $\mathbf{P} = \Sigma \mathbf{p}_i = \Sigma m\mathbf{v}_i$. As stated in the development of Eq. 6.4, if there is no net force acting on the system, then $\mathbf{P} = \mathbf{P}_o$. This generalized condition is referred to as the **law of the conservation of linear momentum**:

> The total linear momentum of a system is conserved if the net external force acting on the system is zero.

There are other ways to express this. For example, recall that an isolated system is one on which no work is done. Therefore there is no net external force acting on the system, and the total linear momentum of an isolated system is conserved.

Within a system, internal forces may be acting, for example, when particles collide. These are force pairs of Newton's third law. However, there is a good reason why such forces are not explicitly referred to in the condition for the conservation of momentum. By Newton's third law, these internal forces are equal and opposite and vectorially cancel each other. Thus, the net internal force of a system is zero.

An important point to understand is that the momenta of individual particles or objects within a system may change. But in the absence of an external force, the vector sum of the momenta remains the same. If the objects are initially at rest (total momentum is zero) and then set in motion as the result of internal forces, the total momentum must still add up to zero. An example of this is illustrated in Fig. 6.4 and analyzed in Example 6.3. Objects in an isolated system may initially be in motion, and the motion may be changed by internal forces, but the total momentum after the changes must add up to the initial value (see Demonstration 4).

The conservation of momentum is often a convenient tool for analyzing situations involving motion. Its application is illustrated in the following examples.

Example 6.3   Zero Total Momentum

Two masses, $m_1 = 1.0$ kg and $m_2 = 3.0$ kg, are held on either side of a compressed spring by a light string joining them, as shown in Fig. 6.4. The string is burned (negligible external force), and the masses move apart on the frictionless surface, with m_1 having a velocity of 1.8 m/s to the left. What is the velocity of m_2?

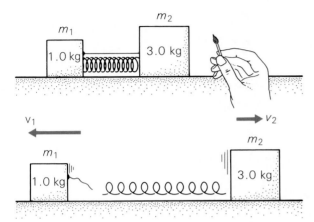

Figure 6.4   **An internal force and the conservation of momentum**
The spring force is an internal force, and so the momentum of the system is conserved. See Example 6.3.

Solution

Given: $m_1 = 1.0$ kg   *Find:* v_2
       $m_2 = 3.0$ kg
       $v_1 = 1.8$ m/s

Here the system consists of the two masses and the spring. Since the spring force is internal to the system, the momentum is conserved. It should be apparent that the initial momentum of the system ($\mathbf{P_o}$) is zero:

$$\mathbf{P_o} = \mathbf{P} = 0 \quad \text{and} \quad \mathbf{P} = \mathbf{p}_1 + \mathbf{p}_2 = 0$$

Thus,

$$\mathbf{p}_2 = -\mathbf{p}_1$$

which means that the momenta of m_1 and m_2 are equal and opposite. Using directional signs (with $+x$ to the right in figure) gives

$$p_2 - p_1 = m_2 v_2 - m_1 v_1 = 0$$

or $m_2 v_2 = m_1 v_1$

and

$$v_2 = \left(\frac{m_1}{m_2}\right) v_1 = \left(\frac{1.0 \text{ kg}}{3.0 \text{ kg}}\right)(1.8 \text{ m/s}) = 0.60 \text{ m/s}$$

Thus, the velocity of m_2 is 0.60 m/s in the positive x direction, or to the right in the figure.

Note that the speeds are related by the ratio of the masses in this case; that is, $v_2/v_1 = m_1/m_2$. Suppose that one object were twice as massive as the other. What would the relationship of the velocities be then? ∎

Demonstration 4

Internal Forces (and Conservation of Momentum?)

(a) Getting ready to push off. Note that the forces are internal to the system.

(b) Play ball! Here a ball is used to exchange momentum between parts of the system.

In both cases, the initial total momentum is zero. What would happen after the actions (and reactions) begin? Actually there is an external force, so the momentum is not conserved and the motion is limited.

Example 6.4   Components of Momentum
A moving shuffleboard puck has a glancing collision with a stationary one of the same mass, as shown in Fig. 6.5. Considering friction to be negligible, what are the velocities of the pucks after the collision?

Solution
Given: $v_{1_o} = 0.95$ m/s $\qquad\qquad$ *Find:* v_1 and v_2
$\qquad\quad \theta_1 = 50°$
$\qquad\quad \theta_2 = 40°$
$\qquad\quad m_1 = m_2$

The collision forces of the two-puck system are internal forces, so the total momentum is conserved (the external force of friction is negligible). This means that the rectangular components of momentum are also conserved. That is, $\mathbf{P} = \mathbf{P}_x + \mathbf{P}_y$ is a constant, so $\mathbf{P}_x$ and $\mathbf{P}_y$ must be constant. For the x component of the momentum,

$\qquad$ before $\quad$ after

$$p_{x_o} = p_{x_1} + p_{x_2}$$

or $\quad m_1 v_{1_o} = m_1 v_1 \cos 50° + m_2 v_2 \cos 40° \qquad\qquad\qquad (1)$

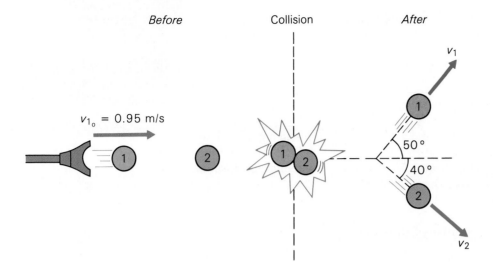

$v_{1_o} = 0.95$ m/s

v_1

$50°$

$40°$

v_2

Figure 6.5 **A glancing collision**
Momentum is conserved in an isolated system and the motion in two dimensions may be analyzed in terms of the components of momentum. See Example 6.4.

For the y component,

$$0 = p_{y_1} - p_{y_2}$$

or $0 = m_1 v_1 \sin 50° - m_2 v_2 \sin 40°$ (2)

where the minus sign indicates that the component is in the negative y direction. With two equations and two unknowns, the rest is simple algebra. The masses cancel in Eq. 2 since they are equal ($m_1 = m_2$), and

$$v_2 = \left(\frac{\sin 50°}{\sin 40°}\right)v_1 = \left(\frac{0.766}{0.643}\right)v_1 = 1.19v_1 \tag{3}$$

Substituting for v_2 in Eq. 1 and canceling the masses (as above) gives

$$v_{1_o} = v_1 \cos 50° + (1.19v_1)(\cos 40°)$$

and

$$v_1 = \frac{v_{1_o}}{\cos 50° + (1.19)(\cos 40°)} = \frac{0.95 \text{ m/s}}{0.643 + (1.19)(0.766)} = 0.61 \text{ m/s}$$

Using this value in Eq. 3 gives

$$v_2 = 1.19v_1 = (1.19)(0.61 \text{ m/s}) = 0.73 \text{ m/s}$$

The directions of v_1 and v_2 are given by θ_1 and θ_2, respectively. ∎

The conservation of momentum is one of the most important principles in physics. It is used to analyze the collisions of objects ranging from subatomic particles to automobiles in traffic accidents. In many instances, external forces may be acting on the objects, which means that the momentum is not conserved. But, as you will learn in the next section, the conservation of momentum often allows a good approximation *over the short time of a collision* because the internal forces, for which momentum is conserved, are much greater than the external forces. For example, external forces such as gravity and friction also act on colliding objects, but are usually relatively small compared to the internal forces. Therefore, if the objects interact for only a brief time, the external forces are usually negligible compared to those of the large internal forces.

(a)

6.3 Impulse and Collisions

When two objects—such as a hammer and a nail, a golf club and a golf ball, or even two automobiles—collide, they exert very large forces on one another for a short period of time (see Fig. 6.6). Newton's second law in momentum form is useful for analyzing such situations. Since $\bar{\mathbf{F}} = \Delta\mathbf{p}/\Delta t$,

$$\bar{\mathbf{F}}\,\Delta t = \Delta\mathbf{p} = \mathbf{p} - \mathbf{p}_\text{o} \tag{6.5}$$

The term $\bar{\mathbf{F}}\,\Delta t$ is known as the **impulse** of the force,

$$\text{Impulse} = \bar{\mathbf{F}}\,\Delta t = \Delta\mathbf{p} = m\mathbf{v} - m\mathbf{v}_\text{o} \tag{6.6}$$

The impulse is equal to the change in momentum. This is referred to as the **impulse-momentum theorem.** Impulse obviously has units of newton-seconds (N-s), which are also units for momentum by Eq. 6.5.

In Chapter 5, you learned that the area under an F versus x curve is equal to the work or the change in kinetic energy (by the work-energy theorem, $W = F\Delta x = \Delta K$). Similarly, the area under an F versus t curve is equal to the impulse or the change in momentum (Fig. 6.7). An impulse force varies with time and is not a constant force as implied by Eq. 6.5. However, in general, it is convenient to talk about the equivalent *constant* average force $\bar{F}$ acting over a time interval Δt to give the same impulse (same area under the force versus time curve) as shown in Fig. 6.7.

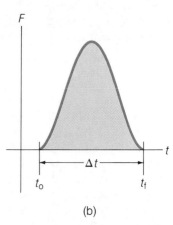
(b)

Figure 6.6 Collision impulse
(a) Collision impulse causes this golf ball to be deformed. (b) The impulse is the area under the curve of an F versus t graph. Note that the impulse force acting on an object first increases and then decreases.

Example 6.5 Impulse and Force
A golfer drives a 0.10-kg ball from an elevated tee, giving it an initial horizontal speed of 40 m/s (about 90 mi/h). The club and the ball are in contact for 1.0 ms (millisecond). What is the average force exerted by the club on the ball during this time?

Solution
Given: $m = 0.10$ kg   *Find*: $\bar{F}$
$\qquad\quad v = 40$ m/s
$\qquad\quad v_\text{o} = 0$
$\qquad\quad \Delta t = 1.0$ ms $= 1.0 \times 10^{-3}$ s

The average force may be computed using the impulse-momentum theorem:

$$\bar{F}\Delta t = p - p_\text{o} = mv - mv_\text{o}$$

Thus,

$$\bar{F} = \frac{mv - mv_\text{o}}{\Delta t} = \frac{(0.10\text{ kg})(40\text{ m/s}) - 0}{1.0 \times 10^{-3}\text{ s}}$$

$$= 4000\text{ N} \quad \text{(about 900 lb)}$$

The force is in the direction of the motion and is the *average* force. The instantaneous force is even greater near the midpoint of time interval of the collision (Δt in Fig. 6.7). ∎

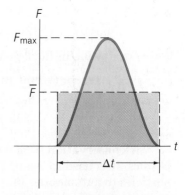

Figure 6.7 Average impulse force
The area under the average force line ($\bar{F}\Delta t$) is the same as the area under the F versus t curve, which is usually difficult to evaluate.

Example 6.5 illustrates the large magnitudes of the internal forces that colliding objects exert on one another during short time intervals.

The word "impulse" implies that the impulse force acts only briefly, and this is true in many instances. However, the definition of impulse places no limit on the time interval over which the force may act. Technically, a comet at its closest approach to the Sun is involved in a collision, because in physics collision forces do not have to be contact forces. Basically, a **collision** is an interaction of objects in which there is an exchange of momentum and energy. As you might expect from the work-energy theorem and the impulse-momentum theorem, momentum and kinetic energy are directly related. Kinetic energy can be expressed in terms of the magnitude of momentum:

$$K = \tfrac{1}{2}mv^2 = \frac{(mv)^2}{2m} = \frac{p^2}{2m} \qquad (6.7)$$

You have probably tried to minimize impulse on occasion. When jumping from a height onto a hard surface, you try not to land stiff-legged. The abrupt stop (small Δt) would apply a large impulse force to your leg bones and joints and could do damage. Instead, you bend your knees on landing. Since the impulse is constant in this case ($\bar{F}\Delta t = \Delta p = -mv_0$, with the final velocity zero), increasing the time interval Δt makes the impulse force smaller.

In other cases, the contact time (Δt) may be purposefully increased to get a greater impulse or change in momentum. This is the principle of following through in sports, for example, when hitting a ball with a bat or driving a golf ball. In the latter case [see Fig. 6.8(a)], assuming that the golfer supplies the same average force with each swing, the longer the contact time, the greater will be the impulse or the momentum the ball receives. That is, with $\bar{F}\Delta t = mv$ (since $v_0 = 0$), the greater the value of Δt, the greater will be the velocity of the ball. As you learned in Chapter 3, a greater projection velocity gives a projectile a greater range. (What angle of projection should a golfer try to achieve when driving the ball on a level fairway?)

Another application where the contact time is increased is shown in Fig. 6.8(b). In this instance, the purpose is to decrease the impulse force, as is the case in the following example.

Figure 6.8 **Increasing the contact time**
(a) A golfer follows through on his swing to increase the contact time so the ball receives greater impulse and momentum. (b) An automobile airbag increases the contact time and decreases the impulse force that could cause injury.

Example 6.6   Increasing the Contact Time to Reduce the Sting

A boy catches a 0.16-kg baseball coming directly toward him at a speed of 25 m/s in his bare hands with his arms rigidly extended. He emits an audible "ouch!" because the ball stings his hands. He learns quickly that if he moves his hands with the ball as he catches it, the sting is reduced. If the contact time for the collision is increased from 0.25 s to 0.75 s in this way, how do the average impulse forces compare? (Assume that the ball has the same initial velocity for each catch.)

Solution

Given: $m = 0.16$ kg   *Find*: $\bar{F}_1$ and $\bar{F}_2$
        $v_0 = 25$ m/s
        $t_1 = 0.25$ s
        $t_2 = 0.75$ s

The impulse is equal to the same change in momentum for each catch, and the change in momentum is equal in magnitude to the initial momentum ($v = 0$).

Thus,

$$\bar{F}_1 \Delta t_1 = \bar{F}_2 \Delta t_2 = mv_o = (0.16 \text{ kg})(25 \text{ m/s}) = 4.0 \text{ N-s}$$

and

$$\bar{F}_2 = \left(\frac{\Delta t_1}{\Delta t_2}\right)\bar{F}_1 = \left(\frac{0.25 \text{ s}}{0.75 \text{ s}}\right)\bar{F}_1 = \tfrac{1}{3}\bar{F}_1$$

The average force is reduced to a third of its original value. The magnitudes of the forces may be computed directly:

$$\bar{F}_1 = \frac{mv_o}{\Delta t_1} = \frac{4.0 \text{ N-s}}{0.25 \text{ s}} = 16 \text{ N}$$

$$\bar{F}_2 = \frac{mv_o}{\Delta t_2} = \frac{4.0 \text{ N-s}}{0.75 \text{ s}} = 5.3 \text{ N}$$

A force of 16 N (3.6 lb) may not seem like much, but this is the average force. The maximum instantaneous force is considerably larger. ∎

Elastic and Inelastic Collisions

Taking a closer look at collisions in terms of the conservation of momentum is simpler if an isolated system is considered, such as a system of particles (or balls) involved in head-on collisions. These collisions can also be analyzed in terms of the conservation of energy. The total kinetic energy is considered for two types of collisions: elastic and inelastic.

In an **elastic collision**, the total kinetic energy is conserved. That is, the total kinetic energy of all the objects of the system after the collision is the same as their total kinetic energy before the collision:

before   after

$$K_i \ = K_f \tag{6.8}$$

condition for an elastic collision

During the collision, some of the initial kinetic energy is temporarily converted to potential energy as the objects are deformed. But, after the maximum deformations, the objects elastically spring back to their original shapes, and the system has the same total kinetic energy as it did initially. For example, two steel balls or two billiard balls may have a nearly elastic collision with each ball having the same shape afterwards as before; that is, there is no permanent deformation.

An elastic collision between two objects can be represented on a graph of force versus displacement as shown in Fig. 6.9. Since the areas bounded by the curves and the x axis are equal in magnitude, they cancel each other out and no net work is done. As you might guess, objects interact via conservative forces in elastic collisions.

In an **inelastic collision**, the total kinetic energy is *not* conserved. One or more of the colliding objects does not spring back to its original shape, so work is done and some kinetic energy is lost:

$$K_f < K_i \tag{6.9}$$

condition for an inelastic collision

In reality, only atoms and subatomic particles have truly elastic collisions, but some larger hard objects have nearly elastic collisions in which the kinetic energy is approximately conserved.

For example, a hollow aluminum ball that collides with a solid steel ball may be dented and permanently deformed. Inelastic collisions involve nonconservative forces.

For isolated systems, momentum is conserved in both elastic and inelastic collisions. For an inelastic collision, only an amount of kinetic energy consistent with the conservation of momentum may be lost. It may seem strange that energy can be lost and momentum still be conserved, but this fact provides insight into the difference between vector and scalar quantities.

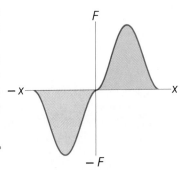

Figure 6.9 **Work in an elastic collision**
When two object collide in an elastic collision, the opposing reaction forces do an equal amount of work. Since the net work is equal to zero, the change in kinetic energy is zero ($W = \Delta K = 0$), and the kinetic energy is conserved.

PROBLEM-SOLVING HINT

To see how momentum can remain constant while the kinetic energy changes (decreases) in inelastic collisions, consider the examples illustrated on the next page. In Fig. 6.10(a), the total momentum before the collision is vectorially zero, but the scalar kinetic energy is not zero. After the collision, the balls are stuck together and stationary, so the total momentum is unchanged and still zero. The total kinetic energy, however, has decreased to zero. In this case, the kinetic energy went into the work done in permanently deforming the balls or the work done against friction or was lost in some other way (for example, in producing sound). [Could the balls still be moving after the collision and the momentum be conserved (still equal to zero) and the kinetic energy not conserved, but not zero? If the balls recoil in opposite directions, the momentum vectors could sum to zero.]

In Fig. 6.10(b), the balls stick together after collision, but are still in motion. Both of these cases are examples of a **completely inelastic collision**, in which the objects stick together and have the same velocity after colliding.

Assume that the balls in Fig. 6.10(b) have different masses. Since the momentum is conserved even in inelastic collisions,

before   *after*

$$m_1 v_0 = (m_1 + m_2)v$$

and

$$v = \frac{m_1 v_0}{m_1 + m_2} \qquad (1)$$

Thus, v is less than v_0, since $m_1/(m_1 + m_2)$ must be less than 1. Some kinetic energy must have been lost. How much? Initially, $K_i = \frac{1}{2}m_1 v_0^2$, and finally, or after the collision,

$$K_f = \frac{1}{2}(m_1 + m_2)v^2 = \frac{1}{2}(m_1 + m_1)\left(\frac{m_1 v_0}{m_1 + m_2}\right)^2$$

where Eq. 1 was used to substitute for v. Then

$$K_f = \frac{\frac{1}{2}m_1^2 v_0^2}{m_1 + m_2} = \left(\frac{m_1}{m_1 + m_2}\right)\frac{1}{2}m_1 v_0^2 = \left(\frac{m_1}{m_1 + m_2}\right)K_i$$

or

$$\frac{K_f}{K_i} = \frac{m_1}{m_1 + m_2} \qquad (2)$$

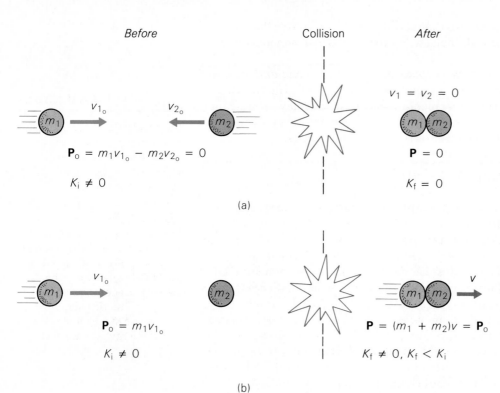

Before　　　　　　　　　Collision　　　　　After

$v_1 = v_2 = 0$

$\mathbf{P}_0 = m_1 v_{1_0} - m_2 v_{2_0} = 0$

$K_i \neq 0$

$\mathbf{P} = 0$

$K_f = 0$

(a)

v_{1_0}

$\mathbf{P}_0 = m_1 v_{1_0}$

$K_i \neq 0$

v

$\mathbf{P} = (m_1 + m_2)v = \mathbf{P}_0$

$K_f \neq 0,\ K_f < K_i$

(b)

Figure 6.10 **Inelastic collisions**
In inelastic collisions, the momentum is conserved but the kinetic energy is not. Collisions like the one shown here, in which the objects stick together afterward, are called completely inelastic collisions. The maximum amount of kinetic energy lost is consistent with the conservation of momentum.

Eq. 2 gives the fractional amount of the initial kinetic energy that the system has after the inelastic collision. For example, if the masses of the balls are equal ($m_1 = m_2$), then $m_1/(m_1 + m_2) = 1/2$ and $K_f/K_i = 1/2$ or $K_f = K_i/2$. That is, half of the initial kinetic energy is lost (consistent with the conservation of momentum).

Note that all the kinetic energy cannot be lost in this case no matter what the masses of the balls are. Why? Eq. 2 doesn't help much. According to this equation, for K_f to be zero, m_1 would have to be zero or m_2 would have to be infinitely large. The momentum (velocity) after collision cannot be zero, since it was not zero initially. Thus, the balls must be moving and must have kinetic energy.

For a general elastic collision of two objects, the conditions are as follows:

	before	*after*

Conservation of $\mathbf{P}$:　$m_1\mathbf{v}_{1_0} + m_2\mathbf{v}_{2_0} = m_1\mathbf{v}_1 + m_2\mathbf{v}_2$　　　　(6.10a)

Conservation of K:　$\frac{1}{2}m_1 v_{1_0}^2 + \frac{1}{2}m_2 v_{2_0}^2 = \frac{1}{2}m_1 v_1^2 + \frac{1}{2}m_2 v_2^2$　　　　(6.10b)

Knowing the masses of the objects and the initial velocities allows you to find the final velocities (there are two equations and two unknowns).

A common collision situation is that in which one of the objects is initially stationary (Fig. 6.11). Assume that the motion of both of the balls after the collision is to the right, or in the positive direction, as shown in the figure. If this is not the case, the math will tell you.

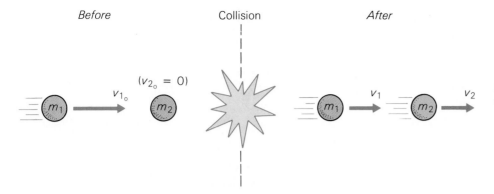

Figure 6.11 **General elastic collision**
For an elastic collision between two bodies, one of which is initially at rest, the velocities after the collision depend on the relative masses.

The conditions for an elastic collision in this situation are

$$m_1 v_{1_o} = m_1 v_1 + m_2 v_2 \qquad (6.11\text{a})$$

and

$$\tfrac{1}{2} m_1 v_{1_o}^2 = \tfrac{1}{2} m_1 v_1^2 + \tfrac{1}{2} m_2 v_2^2 \qquad (6.11\text{b})$$

Rearranging these equations gives

$$m_1(v_{1_o} - v_1) = m_2 v_2$$

and

$$m_1(v_{1_o}^2 - v_1^2) = m_2 v_2^2$$

Using the relationship $x^2 - y^2 = (x + y)(x - y)$ and dividing the second equation by the first gives

$$v_{1_o} + v_1 = v_2 \qquad (6.12)$$

Eq. 6.12 can be used to eliminate v_1 or v_2 from either of the preceding two equations. Thus, each of these final velocities may be expressed in terms of the initial velocity (v_{1_o}):

$$v_1 = \left(\frac{m_1 - m_2}{m_1 + m_2} \right) v_{1_o} \qquad (6.13)$$

and

$$v_2 = \left(\frac{2m_1}{m_1 + m_2} \right) v_{1_o} \qquad (6.14)$$

Note that the final velocities depend on the masses of the objects. Looking at Eq. 6.13, you can see that if m_1 is greater than m_2, then v_1 is positive, or in the same direction as the initial velocity of the incoming ball, as was assumed in Fig. 6.11. However, if m_2 is greater than m_1, then v_1 is negative, and the incoming ball recoils in the opposite direction after collision. Eq. 6.14 shows that v_2 is always in the same direction as the velocity of the incoming ball.

You can also get some general ideas about what happens after such a collision by considering three possible situations for the relative masses of the objects, illustrated in Fig. 6.12. For the case in Fig. 6.12(a), where the masses of the two balls are equal, from Eqs. 6.13 and 6.14,

$$v_1 = 0 \qquad \text{and} \qquad v_2 = v_{1_o}$$

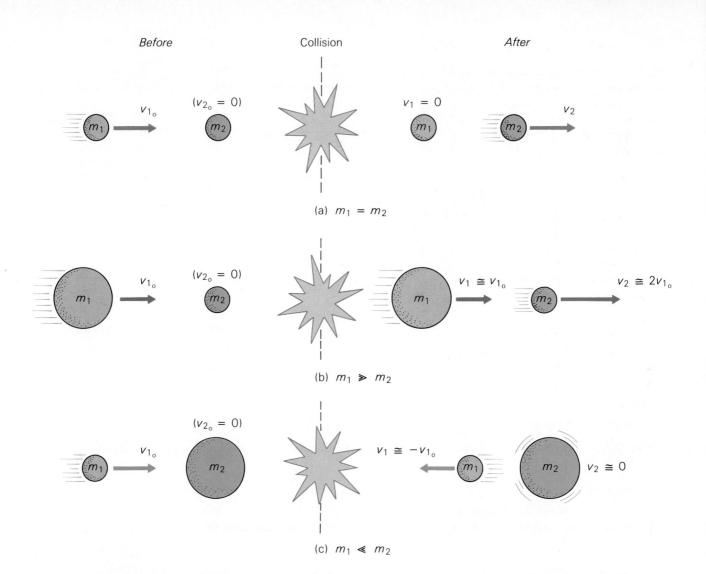

Before   Collision   *After*

v_{1_o} $(v_{2_o} = 0)$ $v_1 = 0$ v_2

(a) $m_1 = m_2$

v_{1_o} $(v_{2_o} = 0)$ $v_1 \cong v_{1_o}$ $v_2 \cong 2v_{1_o}$

(b) $m_1 \gg m_2$

v_{1_o} $(v_{2_o} = 0)$ $v_1 \cong -v_{1_o}$ $v_2 \cong 0$

(c) $m_1 \ll m_2$

Figure 6.12 **Special cases of elastic collisions**
(a) When a moving particle collides elastically with a stationary particle of equal mass, there is a complete exchange of momentum and energy. (b) When a large moving mass collides elastically with a much lighter stationary mass (a particle), the large mass continues to move as before, essentially, and the particle is given a velocity almost twice the initial velocity of the large mass. (c) When a moving particle collides elastically with a much larger stationary mass, the particle recoils at approximately the same speed (in the opposite direction) and the large mass remains essentially stationary.

That is, if the masses of the colliding objects are equal, the objects simply exchange momentum and energy. The incoming ball is stopped on collision, and the originally stationary ball moves off with the velocity that the incoming ball had.

The second situation, where m_1 is *very* much greater than m_2, is shown in Fig. 6.12(b). In this case, m_2 can be ignored in the addition and subtraction with m_1 in Eqs. 6.13 and 6.14, and

$$v_1 \cong v_{1_o} \qquad \text{and} \qquad v_2 \cong 2v_{1_o}$$

This tells you that if a very massive object collides with a stationary light object, the massive object is slowed down only slightly by the collision, and the light object is knocked away with a velocity almost twice that of the initial velocity of the massive object.

When m_1 is *very* much less than m_2, as in Fig. 6.12(c), m_1 can be ignored in the addition and subtraction with m_2 in Eqs. 6.13 and 6.14, and

$$v_1 \cong -v_{1_o} \qquad \text{and} \qquad v_2 \cong 0$$

(In the second equation, the approximation $m_1/m_2 \cong 0$ is made.)

Thus, if a light object collides with a massive stationary one, the massive object remains *almost* stationary, and the light object recoils backward with approximately the same speed it had before collision. An extreme case of this type seems similar to a particle striking a solid, immovable wall (see Fig. 6.2). But, for the case in Fig. 6.12(c), the massive ball must move a bit after the collision since the momentum is conserved.

Example 6.7   Elastic Collision

A 0.30-kg object with a speed of 2.0 m/s in the positive x direction has a head-on elastic collision with a stationary 0.75-kg object located at $x = 0$. What is the distance separating the objects 2.5 s after the collision?

Solution

Given: $m_1 = 0.30$ kg   Find: $\Delta x = x_2 - x_1$
$\quad\quad v_{1_o} = 2.0$ m/s
$\quad\quad m_2 = 0.70$ kg
$\quad\quad t = 2.5$ s

From Eqs. 6.13 and 6.14, the velocities after collision are

$$v_1 = \left(\frac{m_1 - m_2}{m_1 + m_2}\right)v_{1_o} = \left(\frac{0.30 \text{ kg} - 0.70 \text{ kg}}{0.30 \text{ kg} + 0.70 \text{ kg}}\right)(2.0 \text{ m/s}) = -0.80 \text{ m/s}$$

$$v_2 = \left(\frac{2m_1}{m_1 + m_2}\right)v_{1_o} = \left[\frac{2(0.30 \text{ kg})}{0.30 \text{ kg} + 0.70 \text{ kg}}\right](2.0 \text{ m/s}) = 1.2 \text{ m/s}$$

Here m_1 is less than m_2, but not *very much* less, as it was in case (c) of Fig. 6.12.
   The positions of the objects 2.5 s after collision are

$$x_1 = v_1 t = (-0.80 \text{ m/s})(2.5 \text{ s}) = -2.0 \text{ m}$$

$$x_2 = v_2 t = (1.2 \text{ m/s})(2.5 \text{ s}) = 3.0 \text{ m}$$

and

$$\Delta x = x_2 - x_1 = 3.0 \text{ m} - (-2.0 \text{ m}) = 5.0 \text{ m}$$

The objects are 5.0 m apart at that time.   ∎

Example 6.8   Multiple Elastic Collisions

A popular novelty item is the apparatus shown in Fig. 6.13(a). One or more of the identical steel balls are allowed to swing in at one end, and an equal number of them swing out at the other end. (a) Explain why this happens. (b) If two balls swing in, why doesn't one ball swing out with twice the velocity?

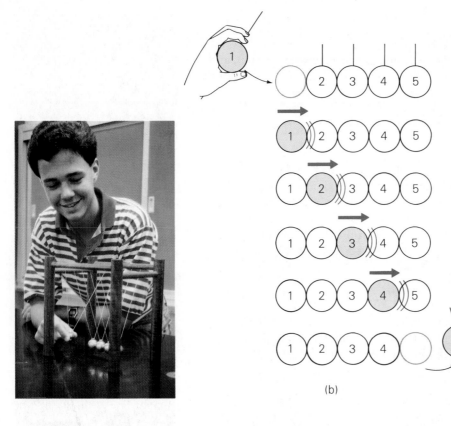

Figure 6.13 **Multiple collisions**
(a) This novelty apparatus has five steel balls of equal mass. (b) In the collision of an incoming ball with a stationary ball of equal mass, there is a complete transfer of momentum and kinetic energy. Multiple collisions transfer momentum and energy down the line. See Example 6.8.

(b)

Solution

(a) As the balls swing in, an external force (gravity) acts on them and momentum is not conserved. However, when the ball is moving horizontally, in the collision region, the downward gravitational weight force is balanced by the tension in the string, and momentum is conserved over the contact time. Also, the hard steel balls have nearly elastic collisions, so the kinetic energy is conserved. What is occurring is multiple elastic collisions.

Since the balls have equal masses, the situation is like that in Fig. 6.12(a); that is, $m_1 = m_2$. So, when one ball swings in, it collides with the next stationary ball, transferring all of its momentum and kinetic energy. The first ball comes to a stop, and the second ball moves away with the same initial momentum and kinetic energy. But the balls are in contact, so the second ball immediately collides with the third ball, and so on, and the momentum and energy are transferred down the line. The final ball receives it and swings out [see Fig. 6.13(b)]. When it does so, the momentum is no longer conserved, and the kinetic energy is converted into gravitational potential energy.

When two or more balls swing in, the leading ball collides with a stationary ball, transfers all of its momentum, and comes to rest. It is immediately hit by the incoming ball behind it and again collides with the next ball in line, which is again stationary, having transferred the momentum it received to the next ball in line. In this manner, pulses of momentum are transferred down the line of balls, and the end ball swings out when a pulse is received. All of this happens so quickly that you see a number of balls swing in and the same number swing out.

(b) You might think that if two balls of mass m swing in with velocity v, then one ball could swing out with velocity $2v$. Certainly the momentum would be conserved:

before   after

$$(2m)v \;=\; m(2v)$$

But, for an elastic collision, kinetic energy must also be conserved. The initial and final kinetic energies would be

before   after

$$K_i = \tfrac{1}{2}(2m)v^2 \qquad K_f = \tfrac{1}{2}m(2v)^2$$
$$= 2[\tfrac{1}{2}(2m)v^2]$$
$$= 2K_i$$

So, for two balls to swing in and one out as described would require that $K_f = 2K_i$, which means energy would have to be created since there is no internal energy source. It just doesn't happen that way. If it did, the law of the conservation of energy would be violated. ∎

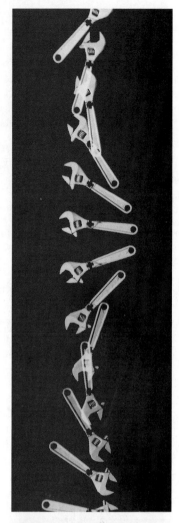

6.4 Center of Mass

If no resultant external force acts on a particle, its linear momentum is constant. Similarly, if no net external force acts on a system of particles, the linear momentum of the system is constant. This similarity implies that a system of particles might be represented by an *equivalent* single particle. Moving objects, such as balls, automobiles, and so forth, are essentially systems of particles and can be effectively represented by equivalent single particles when analyzing motion. Such representation is done through the concept of the **center of mass (CM)**.

> The center of mass is the point at which all of the mass of an object or system may be considered to be concentrated in representing the object or system as a particle.

Even if the object is rotating, the center of mass moves as though it were a particle (Fig. 6.14). The center of mass is sometimes descriptively said to be at the balance point of an object. For example, if you balance a meterstick on your finger, the center of mass of the stick is located directly above your finger and all of the mass (or weight) seems to be concentrated there.

Newton's second law applies to a system when the center of mass is used:

$$\mathbf{F} = M\mathbf{A}_{CM} \tag{6.15}$$

Where **F** is the net external force, M is the total mass of the system or the sum of the masses of the particles of the system ($M = m_1 + m_2 + m_3 + \cdots + m_n$), where the system has n particles), and $\mathbf{A}_{CM}$ is the acceleration of the center of mass. In words, Eq. 6.15 says that the center of mass of a system of particles moves as though all the mass of the system were concentrated there and acted on by the resultant of the external forces.

Figure 6.14 **Center of mass** The center of mass of a rotating object moves as though it were a particle. Note the X on the wrench that marks the center of mass. (b) Even after exploding, the center of mass of a fireworks projectile follows a parabolic path.

Also, if the net external force on a system of particles is zero, the total linear momentum of the center of mass is conserved (stays constant) since $\mathbf{F} = M \Delta \mathbf{V}_{CM}/\Delta t$ as for a particle. This means that the center of mass either moves with a constant velocity or remains at rest. Although you may more readily visualize the center of mass of a solid object, the concept of the center of mass applies to any system of particles or objects, even a quantity of gas.

For a system of n particles arranged in one dimension, along the x axis, the location of the center of mass is given by

$$X_{CM} = \frac{m_1\mathbf{x}_1 + m_2\mathbf{x}_2 + m_3\mathbf{x}_3 + \cdots + m_n\mathbf{x}_n}{m_1 + m_2 + m_3 + \cdots + m_n} \qquad (6.16)$$

That is, X_{CM} is the x coordinate of the center of mass of a system of particles. In shorthand notation (using signs to indicate vector directions),

$$X_{CM} = \frac{\sum_i m_i x_i}{M} \qquad (6.17)$$

where Σ_i indicates the summation of the products $m_i x_i$ for i particles ($i = 1, 2, 3, \ldots, n$). If $\Sigma_i m_i x_i = 0$, then $X_{CM} = 0$, and the center of mass of the one-dimensional system is located at the origin.

Other coordinates of the center of mass for systems of particles are similarly defined. For a two-dimensional distribution of masses, the coordinates of the center of mass are (X_{CM}, Y_{CM}).

Example 6.9   Center of Mass
Three masses, 2.0 kg, 3.0 kg, and 6.0 kg, are located at positions $(3.0, 0)$, $(6.0, 0)$, and $(-4.0, 0)$, respectively, in meters from the origin. Where is the center of mass of this system?

Solution
Given: $m_1 = 2.0$ kg   Find: X_{CM}
  $m_2 = 3.0$ kg
  $m_3 = 6.0$ kg
  $x_1 = 3.0$ m
  $x_2 = 6.0$ m
  $x_3 = -4.0$ m

$$X_{CM} = \frac{\sum_i x_i x_i}{M}$$

$$= \frac{(2.0 \text{ kg})(3.0 \text{ m}) + (3.0 \text{ kg})(6.0 \text{ m}) + (6.0 \text{ kg})(-4.0 \text{ m})}{2.0 \text{ kg} + 3.0 \text{ kg} + 6.0 \text{ kg}}$$

$$= 0$$

The center of mass is at the origin. ∎

Example 6.10   Center of Mass and Frame of Reference
A dumbbell (Fig. 6.15) has a connecting bar of negligible mass. Find the location of the center of mass (a) if m_1 and m_2 are each 5.0 kg, and (b) if m_1 is 5.0 kg and m_2 is 10.0 kg.

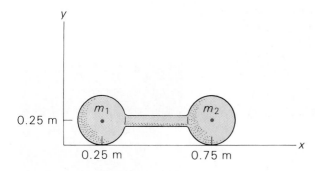

Figure 6.15 **Location of the center of mass**
See Example 6.10.

Solution

Given: (a) $m_1 = m_2 = 5.0$ kg
    $x_1 = 0.25$ m
    $x_2 = 0.75$ m
    $y_1 = y_2 = 0.25$ m
   (b) $m_1 = 5.0$ kg
    $m_2 = 10.0$ kg

Find: (a) (X_{CM}, Y_{CM})
    (b) (X_{CM}, Y_{CM})

Note that each mass is considered to be a particle located at the center of the sphere (its center of mass).

(a) Finding X_{CM} gives

$$X_{CM} = \frac{m_1 x_1 + m_2 x_2}{m_1 + m_2}$$

$$= \frac{(5.0 \text{ kg})(0.25 \text{ m}) + (5.0 \text{ kg})(0.75 \text{ m})}{5.0 \text{ kg} + 5.0 \text{ kg}} = 0.50 \text{ m}$$

Similarly, it is easy to find that $Y_{CM} = 0.25$ m. (You might have seen this right away, since each center of mass is at this height.) The center of mass of the dumbbell is located at $(X_{CM}, Y_{CM}) = (0.50 \text{ m}, 0.25 \text{ m})$, or midway between the end masses.

(b) With $m_2 = 10.0$ kg,

$$X_{CM} = \frac{m_1 x_1 + m_2 x_2}{m_1 + m_2}$$

$$= \frac{(5.0 \text{ kg})(0.25 \text{ m}) + (10.0 \text{ kg})(0.75 \text{ m})}{5.0 \text{ kg} + 10.0 \text{ kg}} = 0.58 \text{ m}$$

which is $\frac{1}{3}$ of the length of the bar away from m_1.

 That the location of the center of mass does not depend on the frame of reference can be demonstrated by putting the origin at the point of contact of the 5.0-kg mass. In this case, $x_1 = 0$ and $x_2 = 0.50$ m, and

$$X_{CM} = \frac{(5.0 \text{ kg})(0) + (10.0 \text{ kg})(0.50 \text{ m})}{5.0 \text{ kg} + 10.0 \text{ kg}} = 0.33 \text{ m}$$

The y coordinate of the center of mass is again $Y_{CM} = 0.25$ m, as you can easily prove for yourself. ■

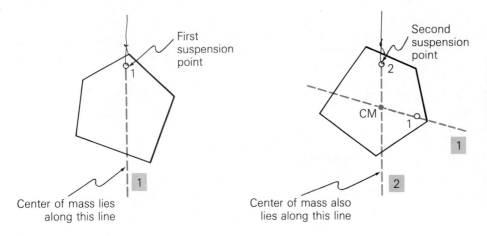

Figure 6.16 **Location of the center of mass by suspension**
The center of mass of a flat, irregularly shaped object may be found by suspending
the object from two or more points. The center of mass (or center of gravity) lies on
a vertical line under any point of suspension, so the intersection of two such lines
locates it.

In Example 6.10, when the value of one of the masses changed, the x
coordinate of the center of mass changed. You might have expected the y
coordinate to change also. However, the centers of the end masses were still at
the same height, and Y_{CM} remained the same. To increase Y_{CM}, one or both of the
end masses would have had to rise, which would have required work against
gravity.

Closely associated with the center of mass is the **center of gravity (CG)**,
the point where all of the weight of an object may be considered to be
concentrated in representing the object as a particle. Taking the acceleration
due to gravity as constant, as it is generally taken to be near the Earth's surface,
allows Eq. 6.17 to be rewritten as

$$MgX_{CM} = \sum_i m_i g x_i \tag{6.18}$$

All of the weight, Mg, is concentrated at X_{CM}, and the center of mass and the
center of gravity coincide. As you may have noticed, the location of the center
of gravity was implied in some figures where the vector arrows for weight ($m\mathbf{g}$)
were drawn from a point at or near the center of an object.

In some cases, the center of mass or the center of gravity of an object may be
located by symmetry. For example, for a spherical object which is homogeneous
(the mass is distributed evenly throughout), the center of mass is at the
geometrical center (or center of symmetry). In Example 6.10(a), where the end
masses of the dumbbell were equal, it was probably apparent that the center of
mass was midway between them.

The location of the center of mass of an irregularly shaped object is not so
evident and is usually difficult to calculate (even with advanced mathematical
methods that are beyond the scope of this book). In some instances, the center of
mass may be located experimentally. For example, the center of mass of a flat,
irregularly shaped object may be determined experimentally by suspending it

freely from different points (Fig. 6.16). A moment's thought should convince you that the center of mass (or center of gravity) always lies vertically below the point of suspension. Since the center of mass is defined as a particle, this is analogous to a particle of mass suspended from a string. Suspending the object from two or more points and marking the vertical lines on which the center of mass must lie locates the center of mass as the intersection of the lines.

The center of mass of an object may lie outside the body of the object. Examples are shown in Fig. 6.17. The center of mass of a homogeneous ring is at its center. The mass in any section of the ring is canceled out by the mass in an equivalent section directly across the ring, and by symmetry the center of mass is at the center. For an L-shaped object with equal legs, the center of mass lies on a line that makes a 45° angle with both legs. Its location can easily be determined by suspending the L from a point on one of the legs and noting where a vertical line from that point intersects the diagonal line.

Keep in mind that the location of the center of mass or center of gravity of an object depends on the distribution of mass. Therefore, for a flexible object such as the human body, the position of the center of gravity changes as the object changes configuration (mass distribution). For example, when a person raises both arms overhead, his or her center of gravity is raised several centimeters. For a high jumper going over a bar, the center of gravity lies outside the body because it is arched (Fig. 6.18). In fact, the center of gravity passes *beneath* the bar. This is done purposefully because work or effort is required to raise the center of gravity, and only the jumper's body has to clear the bar.

6.5 Jet Propulsion and Rockets ◆

The word "jet" is sometimes used to refer to a stream of liquid or gas emitted at a high speed, for example, a jet of water from a fountain or a jet of air from an automobile tire. **Jet propulsion** is the application of such jets to the production of motion. This usually brings to mind jet planes and rockets, but squids and octopuses propel themselves by squirting jets of water. You have probably tried the simple application of blowing up a balloon and releasing it. Lacking any guidance or rigid exhaust system, the balloon zig-zags around, propelled by the escaping air. The air is forced out, and an equal and opposite force is exerted on the balloon.

Jet propulsion is explained by Newton's third law. In the absence of external forces, the conservation of momentum also applies. You may understand this better by considering the recoil of a rifle, taking the rifle as an isolated system (Fig. 6.19). Initially, the total momentum of this system is zero. When the rifle is fired (by remote control to avoid external forces), the expansion of the gases from the exploding charge (an internal force) accelerates the bullet down the barrel. The reaction force on the rifle produces the recoil (which gives rise to the kick experienced by a person firing a rifle). At any instant, the momenta of rifle and bullet are equal and opposite, so the total momentum is zero. After the bullet leaves the barrel, there is no propelling force, so the bullet and the rifle move with constant velocities.

Similarly, the thrust for a rocket is created by exhausting the gas from burning fuel out the rear of the rocket. The rocket exerts a force on the exhaust

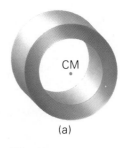
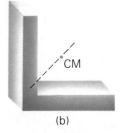

Figure 6.17 Center of mass located outside the body of an object
The center of mass may lie outside the body of an object. (a) For a uniform ring, the center of mass is at the center. (b) For an L-shaped object, if the mass distribution is symmetrical and the legs are of equal length, the center of mass lies on the diagonal line between the legs.

Figure 6.18 Center of gravity
By arching his body, the center of gravity of this high jumper passes beneath the bar. Work or effort is required to raise the center of gravity, and only the jumper's body need clear the bar.

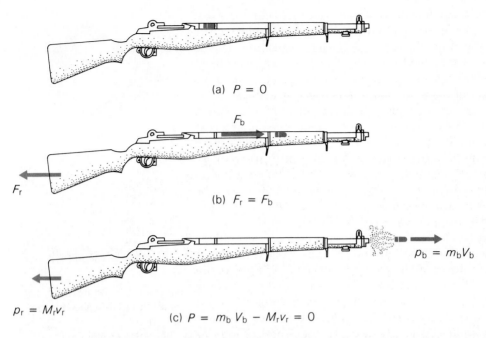

(a) $P = 0$

F_b

F_r

(b) $F_r = F_b$

$p_b = m_b V_b$

$p_r = M_r v_r$

(c) $P = m_b V_b - M_r v_r = 0$

Figure 6.19 **Conservation of momentum**
(a) Before being fired, the total momentum of the rifle (as an isolated system) is zero. (b) During firing, there are equal and opposite internal forces, and the instantaneous total momentum is zero (neglecting external forces). (c) When the bullet leaves the barrel, the total momentum is still zero.

gas, and the gas exerts an equal and opposite force on the rocket. The latter force propels the rocket forward (Fig. 6.20). If the rocket is at rest when the engines are turned on and there are no external forces (as in space, where friction is zero and gravitational forces are negligible), then the instantaneous momentum of the exhaust gas is equal and opposite to that of the rocket. The numerous exhaust gas molecules have small masses and high velocities, and the rocket has a much larger mass and a smaller velocity. (See Demonstration 5.)

Unlike a rifle firing a single shot, a rocket engine continually loses mass when burning fuel (it is more like a machine gun). Thus, the rocket is a system for which the mass is not constant. As the mass of the rocket decreases, it accelerates more easily. Multistage rockets take advantage of this fact: the hull of a burnt-out stage is jettisoned to give an in-flight reduction in mass (Fig. 6.21). The payload is actually a very small part of the initial mass of rockets for space flights.

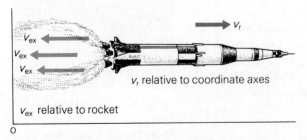

v_r

v_{ex}

v_{ex}

v_{ex}

v_r relative to coordinate axes

v_{ex} relative to rocket

O

Figure 6.20 **Jet propulsion and mass reduction**
A rocket burning fuel is continually losing mass, and so becomes easier to accelerate. The force on (or thrust of) the rocket resulting from this can be shown to depend on the product of the rate of change of its mass with time and the velocity of the exhaust gas: $(\Delta m/\Delta t)\mathbf{v_{ex}}$. Since the mass is decreasing, $(\Delta m/\Delta t)$ is negative and the thrust is opposite $\mathbf{v_{ex}}$.

Jet Propulsion

A demonstration of Newton's third law and conservation of momentum. For a rocket or jet engine, thrust is obtained by burning fuel and exhausting the gas out the rear of the engine. Here a fire extinguisher exhausting CO_2 takes the place of the engine.

(a) Ignition and blast off!

(b) Away we go—equal and opposite forces.

Suppose that the purpose of a space flight is to land a payload on the moon. At some point on the journey, the gravitational attraction of the moon will become greater than that of Earth, and the spacecraft will accelerate toward the moon. A soft landing is desirable, so the spacecraft must be slowed down enough to go into orbit about the moon. This is done by using the rocket engines to apply a reverse thrust, or braking thrust. The spacecraft is maneuvered through a 180° angle, or turned around, which is quite easy to do in space. The rocket engines are then fired, expelling the exhaust gas toward the moon and supplying a braking action.

You have experienced a reverse thrust effect if you have flown in a commercial jet. (In this instance, however, the craft is not turned around.) After landing, the jet engines are revved up and a braking action can be felt.

Figure 6.21 **Multistage rockets**
(a) An artist's depiction of the first- and second-stage separation of a Saturn V rocket. (b) A photo of actual separation after 148 seconds of burn time.

Ordinarily, revving up the engines accelerates the plane forward. The reverse thrust is accomplished by placing something in the stream of exhaust gas that deflects the gas forward. The gas experiences an impulse force and a change in momentum in the forward direction [see Fig. 6.2(b)], and the engine and the aircraft have an equal and opposite momentum change and braking impulse force. The deflection of the gas is usually accomplished by adjustable deflector doors within the engine of large aircraft. However, on some smaller jets, the deflector doors may be external, forming part of the engine casing or cowling. Many jet planes use reverse thrust not only for braking on landing, but also to back away from the terminal loading dock. Large jet planes are often towed when they have to back up, but when reverse thrust is used, a panel on the side of the engine slides back to allow the deflected exhaust gases to escape in the forward direction.

Important Formulas

Linear momentum:

$\mathbf{p} = m\mathbf{v}$

Total linear momentum of a system:

$\mathbf{P} = \mathbf{p}_1 + \mathbf{p}_2 + \mathbf{p}_3 + \cdots$

Newton's second law in terms of momentum:

$\mathbf{F} = \dfrac{\Delta \mathbf{p}}{\Delta t}$

Impulse-momentum theorem:

$\text{impulse} = \bar{\mathbf{F}}\,\Delta t = \Delta \mathbf{p} = m\mathbf{v} - m\mathbf{v}_\text{o}$

Condition for an elastic collision:

$K_\text{f} = K_\text{i}$

Condition for an inelastic collision:

$K_\text{f} < K_\text{i}$

Coordinate of the center of mass:

$X_\text{CM} = \dfrac{\displaystyle\sum_i m_i x_i}{M}$

Questions

Linear Momentum and Its Conservation

1. Explain why a change in momentum is evidence of a force.

2. Why is the momentum of a projectile tangential to its path?

3. The momenta of the particles of an isolated system can continually change, yet the total momentum of the system will remain constant. Explain why this is so.

4. If the string in Example 6.3 had an appreciable mass, would the momentum of the system still be conserved? Explain.

5. The component of an object's momentum in one direction may be constant, and the component in another direction is not, for example, for a projectile. What does this imply? Is the total momentum conserved?

6. Explain how a sailing ship is propelled by the wind.

7. A fan boat of the type used in swampy and marshy areas is shown in Fig. 6.22(a). Explain the principle of its propulsion.

8. A sailboat that glides on an air track is equipped with a battery-operated fan [Fig. 6.22(b)]. There are two interchangeable sails, one made of soft cloth and the other of hard plastic sheet. Assuming that all of the air molecules propelled by the fan strike the sail, explain the motion or lack of it in the following cases. (a) When the fan is removed from the boat and directed toward the cloth sail, the sailboat moves away from the fan. But when the fan is on the boat, the boat is *nearly* stationary on the track (the air track blower is operating, of course). (b) When the plastic sail is used with the fan on the boat, the boat moves forward (in the direction the fan is blowing). (c) If the fan is on the boat without a sail, will the boat move? Explain. Would the direction the fan is blowing make a difference?

Impulse and Collisions

9. Explain the difference for each of the following pairs of actions in terms of impulse: (a) a golfer's drive and chip shots, (b) a boxer's jab and knock-out punches, and (c) a baseball player's bunting action and home-run swing.

10. Explain the principle behind (a) using Styrofoam

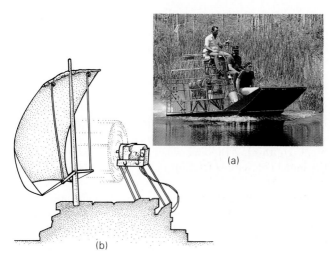

(a)

(b)

Figure 6.22 **Fan boat and air track sailboat**
See Questions 7 and 8.

packing to prevent breakage and (b) padding dashboards in cars to prevent injurious bumps.

11. What is a collision? Must the colliding objects come into contact? Give some examples.

12. If $K = p^2/2m$, how can the momentum be constant for an inelastic collision and the kinetic energy change (decrease)?

13. Is it possible for a nonconservative force to act during a collision and still have the total momentum be conserved? Explain.

14. Is the mechanical energy conserved in (a) an elastic collision and (b) an inelastic collision? Explain.

15. If a collision is completely inelastic, is the final kinetic energy zero? Explain.

16. Two balls move toward each other and collide; the impact sets off a small explosive charge on one of the balls. Is this an elastic collision?

17. For an apparatus like that shown in Fig. 6.13, suppose that four balls swing in toward a single stationary one. Explain why four balls swing out.

18. A neutron (a subatomic particle) moving at a high velocity collides with a stationary (a) carbon atom and (b) uranium atom. Describe what happens in each case. (Hint: find the masses of these particles in Table 1 of Appendix I.)

Center of Mass

19. Explain how the location of the center of mass of an object depends on the distribution of its mass.

20. The acceleration due to gravity decreases with altitude. Thus, the center of gravity of a tall building or mountain does not exactly coincide with the center of mass. Is the center of gravity above or below the center of mass in such cases?

21. Could the suspension method be used to find the center of mass of a three-dimensional object? If so, explain how this might be done. (Hint: a plumb bob could be hung from the bottom of the suspended object.)

22. A spacecraft is initially at rest in space, and then its rocket engines are fired. Describe the motion of the center of mass of the system.

Jet Propulsion and Rockets ◆

23. Bazookas or rocket launchers are essentially open tubes used to fire projectiles. They are hand-held and have very little recoil. Explain why.

24. Is it correct to use the conservation of momentum to analyze the motion of a rocket on blast-off? Explain.

Problems

6.1 Linear Momentum

■1. Find the magnitude of the momentum of (a) a 0.50-kg ball traveling at 8.0 m/s and (b) a 1500-kg automobile traveling at 100 km/h.

■2. What is the linear momentum of a 7.3-kg bowling ball going down the alley with a speed of 22 m/s?

■3. The magnitude of the instantaneous momentum of a runner who is moving at 20.0 km/h is 479 kg-m/s. What is the runner's mass?

■■4. The fastest time for the straight 200-m dash is about 19.8 s. What is the average momentum of a 70-kg runner who finishes the dash in that time?

■■5. How fast would an automobile having a mass of 900 kg have to be going to have the same linear momentum as a truck having a mass of 2 metric tons and traveling along a straight road with a constant speed of 30 km/h?

■■6. If a 0.25-kg ball is dropped from a height of 50 m, what is the momentum of the ball (a) 1.8 s after being released and (b) just before it hits the ground?

■■7. What is the total momentum of a system consisting of two particles, $m_1 = 0.015$ kg and $m_2 = 0.025$ kg, if (a) $v_1 = 8.5$ m/s and $v_2 = 9.0$ m/s in the positive x direction, and (b) $v_1 = -6.4$ m/s and $v_2 = 7.5$ m/s along the y axis?

■■8. A 0.15-kg baseball traveling with a horizontal speed of 3.40 m/s is hit by a bat and is then moving with a speed of 44.7 m/s in the opposite direction. What is the change in the ball's momentum?

■■9. Two balls of equal mass (0.25 kg) approach the origin along the positive x and y axes at the same speed (3.8 m/s). (a) What is the total momentum of the system? (b) Will the balls necessarily collide at the origin? If not, what is the total momentum of the system after they have both passed through the origin?

■■10. A 0.20-kg billiard ball traveling at a speed of 15 m/s strikes the side rail of the table at an angle of 60° (Fig. 6.23). If the ball rebounds at the same speed and angle, what is the change in its momentum?

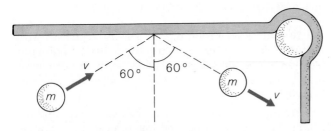

Figure 6.23 **Glancing collision**
See Problems 10, 12, and 40.

■■■11. At a basketball game, a 120-lb cheerleader is tossed vertically upward with a speed of 3.8 m/s by another cheerleader. (a) What is the cheerleader's change in momentum from the time of release to just before being caught if she is caught at the same height? (b) Would there by any difference if she were caught 0.25 m below the point of release? If so, what is it?

■■■12. Suppose that the billiard ball in Fig. 6.23 approaches the rail at a speed of 15 m/s and an angle of 60° as shown, but rebounds at a speed of 12 m/s and an angle of 50°. What is the change in momentum in this case? (Hint: use components.)

■■■13. Four particles have masses $m_1 = 10$ g, $m_2 = 15$ g, $m_3 = 13$ g, and $m_4 = 18$ g and velocities $v_1 = (2.0 \text{ m/s})\mathbf{x}$, $v_2 = (-0.50 \text{ m/s})\mathbf{x}$, $v_3 = (0.70 \text{ m/s})\mathbf{y}$, and $v_4 = (-1.9 \text{ m/s})\mathbf{y}$. What is the total momentum of the four-particle system?

6.2 The Conservation of Linear Momentum

■14. Two identical billiard balls approach each other at the same speed (2.0 m/s). At what speeds do they rebound after a head-on collision?

■15. A 80-kg hunter jumps to shore from a stationary 35-kg canoe. The hunter has a horizontal speed of 1.5 m/s (relative to the shore). At what speed does the canoe initially move away from the shore?

■16. A 70-kg man and his 40-kg daughter on skates stand together on a frozen lake. If they push apart and the father has a velocity of 1.5 m/s eastward, what is the velocity of the daughter?

■■17. A 100-g bullet is fired horizontally into a 14.9-kg block of wood resting on a horizontal surface, and the bullet becomes embedded in the block. If the muzzle velocity of the bullet is 250 m/s, what is the velocity of the block containing the embedded bullet immediately after the impact? (Neglect friction.)

■■18. A 1500-kg empty hopper car rolls under a loading bin with a speed of 2.5 m/s, and a 3000-kg load is deposited in the car. What is the speed of the car immediately after being loaded?

■■19. For a movie scene, a 90-kg stunt man drops from a tree onto a 20-kg sled moving with a velocity of 5.0 m/s toward the shore on a frozen lake. (a) What is the velocity of the sled after the stunt man is on board? (Neglect friction.) (b) If the sled hits the bank and stops but the stunt man keeps on going, with what velocity does he leave the sled?

■■20. A 70-kg athlete achieves a height of 1.55 m in a standing high jump. Considering the jumper and the Earth as an isolated system, with what speed does the Earth initially move?

■■21. A 90-kg astronaut is stranded in space at a point 12 m from his spaceship. In order to get back, he throws a 0.50-kg piece of equipment so that it moves at a speed of 4.0 m/s directly away from the spaceship. How long will it take the astronaut to reach the ship?

■■22. A toy gun fired vertically sends a 0.10-kg ball into a 0.050-kg holder sitting on the open end of the barrel. The ball lodges in the holder, and both rise to a height of 0.76 m above the end of the barrel. What was the muzzle velocity of the ball, the velocity just before it hits the holder?

■■23. Identical railroad freight cars hit each other and couple together. In each of the following cases, what are the velocities of the cars immediately after coupling? (a) A moving car approaches a stationary one with a velocity of +10 km/h. (b) Two cars approach each other with velocities of +20 km/h and -15 km/h, respectively. (c) Two cars travel in the same direction with velocities of +20 km/h and +15 km/h.

■■24. An isolated 3.0-kg object initially at rest explodes and splits into three fragments. One fragment has a mass of 0.50 kg and flies off along the negative x axis at a speed of 2.8 m/s, and another has a mass of 1.3 kg and flies off along the negative y axis at a speed of 1.5 m/s. What is the speed and direction of the third fragment?

■■■25. Suppose that the 3.0-kg object in Problem 24 is initially traveling at a speed of 2.5 m/s in the positive x direction. What will the speed and direction of the third fragment be in this case?

■■■26. After dumping their loads (through bottom doors) at a terminal, coal cars roll from rest down an inclined ramp from a vertical height of 2.5 m onto a level track. One of the empty cars is inadvertently stopped at the bottom of the ramp. The next car that comes down couples to it, and both cars move off together. Finally, the last two cars to be unloaded come down the ramp and successively catch up to and couple with the already joined pairs. Neglecting external forces, what is the final velocity of the four-car system?

■■■27. A projectile that is fired from a gun has an initial velocity of 75 m/s at an angle of 60° to the horizontal. When the projectile is at the top of its trajectory, an internal explosion causes it to separate into two fragments that have equal masses. One of the fragments falls straight downward as though it had been released from rest. How far does the other fragment land from the gun?

■■■28. A ballistic pendulum is a device used to measure the velocity of a projectile, for example, the muzzle velocity of a rifle bullet. The projectile is shot horizontally into and becomes embedded in the bob of a pendulum as illustrated in Fig. 6.24. The pendulum swings upward to some height (h), which is measured. The masses of the block and the bullet are known. Using the laws of momentum and energy, show that the initial velocity of the projectile is given by $v_o = \sqrt{(m + M)/m](2gh)}$.

6.3 Impulse and Collisions

■29. Show that the units for impulse (N-s) are equivalent to those for momentum (kg-m/s).

■30. When thrown upward and hit, a 0.20-kg softball receives an impulse of 10 N-s. With what speed does the ball move away from the bat?

■31. An automobile with a linear momentum of 3.2×10^4 kg-m/s is brought to a stop in 2.0 s. What is the average braking force?

■32. One billiard ball travels toward another at rest with a velocity of 5.0 m/s. If the balls collide head-on, what are their velocities afterward?

■■33. During a snowball fight, a 0.15-kg snowball traveling at a speed of 12 m/s hits a student in the back of the head. (a) What is the impulse? (Is this an elastic collision?) (b) If the contact time is 0.10 s, what is the average impulse force on the student's head?

■■34. A basketball with a mass of 0.30 kg is thrown horizontally against a wall with a velocity of 20 m/s. If the ball rebounds with a velocity of 15 m/s, what is the impulse of the collision? (Was the collision elastic?)

■■35. (a) Sketch graphs of force versus time for a two-body elastic collision and a two-body inelastic collision. Show the forces acting on both bodies on the same graph, as in Fig. 6.9. (b) Sketch a graph of force versus displacement for an inelastic collision.

■■36. Fig. 6.25 is an approximate plot of force versus time for a case where a 0.25-kg cue ball is hit by a cue stick in a pool game. (a) What is the impulse delivered to the ball? (b) What is the average force exerted on the ball? (c) What is the speed of the ball immediately after the collision?

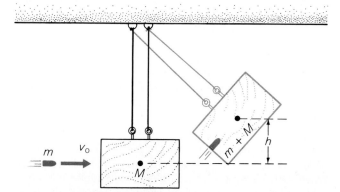

Figure 6.24 A ballistic pendulum
See Problems 28, 51 and 55.

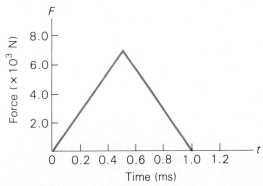

Figure 6.25 Force versus time graph
See Problem 36.

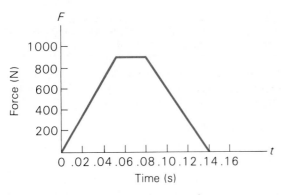

Figure 6.26 **Force versus time graph**
See Problem 37.

■■37. A one-dimensional impulse force acts on a 1.5-kg object as diagrammed in the graph of Fig. 6.26. Find (a) the impulse given to the object, (b) the average force, and (c) the final velocity if the object had an initial velocity of −6.0 m/s.

■■38. A 1600-kg automobile is traveling at 25 m/s along a straight, level road, and then its speed is reduced to 15 m/s in 4.0 s. What is the magnitude of the average braking force? Work the problem two ways: first, using impulse, and second, finding the average acceleration.

■■39. A 0.25-kg piece of putty is dropped from a height of 1.5 m above a flat surface. When the putty hits the surface, it comes to rest in 0.10 s. What is the average force exerted on the putty by the surface?

■■40. If the billiard ball in Fig. 6.23 is in contact with the rail for 0.01 s, what is the magnitude of the average force exerted on the ball? (See Problem 10.)

■■41. For a typical drive, the golf club and the ball are in contact for about 0.50 ms, and the ball leaves the tee with a speed of 70 m/s. What is the average force exerted by the club on the ball? Express your final answer in pounds. (The official weight of a golf ball is 1.620 oz.)

■■42. A 4.0-kg ball with a velocity of 3.0 m/s has a head-on elastic collision with a stationary 2.0-kg ball. Describe the motions of the balls after the collision.

■■43. A ball with a mass of 100 g is traveling with a velocity of 50 cm/s and collides head-on with a 5.0-kg ball that was at rest. Find the velocities of the balls after the collision; assume that it is elastic.

■■44. For the apparatus in Fig. 6.13, show that one ball swinging in with velocity v_o will not cause two balls to swing out with velocity $v_o/2$.

■■45. (a) If the balls described in Problem 9 collide at the origin in a completely inelastic collision, what will the velocity be immediately afterward? (b) If the particles in Problem 13 collide at the origin in a completely inelastic collision, what will the velocity be immediately afterward?

■■46. In nuclear reactors, subatomic particles called neutrons are slowed down by allowing them to collide with the atoms of a moderator material, such as carbon. (a) How much energy is lost by a neutron in a head-on, elastic collision with a carbon atom? (A carbon atom is about 12 times more massive than a neutron.) (b) If the neutron has an initial speed of 1.5×10^7 m/s, by what percentage will that be reduced in the collision?

■■47. Two balls with masses of 2.0 kg and 6.0 kg travel toward each other at speeds of 12 m/s and 4.0 m/s, respectively. If the balls have a head-on, inelastic collision and the 2.0-kg ball recoils with a speed of 8.0 m/s, how much kinetic energy is lost in the collision?

■■48. A freight car with a mass of 2.0×10^4 kg rolls down an inclined track through a vertical distance of 3.2 m. At the bottom of the incline, on a level track, the car collides and couples with an identical freight car that was at rest. What percentage of the initial kinetic energy is lost in the collision?

■■49. A gondola car has a mass of 5.0 metric tons when empty. While it is moving at a speed of 2.0 m/s along a level track under a grain elevator, 8.0 metric tons of wheat are loaded into it from directly above. Is this an elastic collision? If not, where did the kinetic energy go?

■■50. Suppose that the block and surface in Problem 17 have a coefficient of kinetic friction of 0.35. After collision, how far will the block with the embedded bullet slide before coming to rest?

■■51. A 10-g bullet is fired horizontally into and becomes embedded in a suspended block of wood whose mass is 0.890 kg (see Fig. 6.24). (a) What is the velocity of the block with the embedded bullet immediately after the collision in terms of the initial velocity (v_o)? (b) If the block with the embedded bullet swings upward and its center of mass is raised 0.40 m, what was the initial velocity of the bullet? (c) Was the collision elastic? If not, what percentage of the initial kinetic energy was lost?

■■52. A 15,000-N automobile travels at a speed of 45 km/h northward along a street, and a 7500-N sports car travels at a speed of 60 km/h eastward along an intersecting street. (a) If neither driver brakes and the cars collide at the intersection and stick together, what will the velocity of the cars be immediately after the collision?

(b) How much of the initial kinetic energy will be lost in the collision?

■■53. A 75-g hard rubber ball is dropped from a height of 1.50 m onto a horizontal surface, and the ball rebounds to a height of 1.25 m. If the average impulse force on the ball during collision is 80 N, how long is the ball in contact with the surface?

■■■54. Show that the amount of kinetic energy lost in the collision in Fig. 6.10(b) is equal to $m_2/(m_1 + m_2)$.

■■■55. Show that the amount of kinetic energy lost in a ballistic pendulum collision (as in Fig. 6.24) is equal to $M/(m + M)$.

■■■56. A moving billiard ball collides with an identical stationary one, and the incoming ball is deflected at an angle of 45° from its original direction. Show that if the collision is elastic, both balls will have the same speed afterward and will move at a right angle (90°) relative to each other.

■■■57. (a) For an elastic, two-body collision, show that in general $v_2 - v_1 = -(v_{2_o} - v_{1_o})$. That is, the relative speed of recession after the collision is the same as the relative speed of approach before it. (b) In general, a collision is either completely inelastic or completely elastic, or somewhere in between. The degree of elasticity is sometimes expressed as the *coefficient of restitution (e)*, which is defined as the ratio of the relative velocities of recession and approach: $v_2 - v_1 = -e(v_{2_o} - v_{1_o})$. What are the values of e for an elastic collision and a completely inelastic collision?

■■■58. The coefficient of restitution (see Problem 57) for steel colliding with steel is 0.95. If a steel ball is dropped from a height h above a steel plate, to what height will the ball rebound?

6.4 Center of Mass

■59. The center of mass of a system consisting of two 100-g particles is located at the origin. If one of the particles is at $(0.40 \text{ m}, 0)$ where is the other?

■60. The centers of a 5.0-kg sphere and a 8.0-kg sphere are separated by a distance of 2.0 m. Where is the center of mass of the system?

■■61. (a) Find the center of mass of the Earth-moon system. (Hint: use data from Appendix I and consider the distance between the two to be measured from their centers.) (b) Where is that center of mass relative to the surface of the Earth?

■■62. Find the center of mass of a system composed of three spherical objects with masses of 3.0 kg, 2.0 kg, and 4.0 kg and centers located at $(-6.0 \text{ m}, 0)$, $(1.0 \text{ m}, 0)$, and $(3.0 \text{ m}, 0)$, respectively.

■■63. For the system described in Problem 62, where would a fourth sphere with a mass of 0.50 kg have to be located for the center of mass of the system to be at the origin?

■■64. Suppose that the mass of dumbbell bar in Fig. 6.15 is not negligible, that the bar has a uniform mass of 1.0 kg. How does this affect the location of the center of mass of the dumbbell?

■■65. Three particles, each with a mass of 0.25 kg, are located at $(-4.0 \text{ m}, 0)$, $(2.0 \text{ m}, 0)$, and $(0, 3.0 \text{ m})$ and are acted on by forces $\mathbf{F}_1 = (-3.0 \text{ N})\mathbf{y}$, $\mathbf{F}_2 = (5.0 \text{ N})\mathbf{y}$, and $\mathbf{F}_3 = (4.0 \text{ N})\mathbf{x}$, respectively. Find the acceleration (magnitude and direction) of the center of mass. (Hint: consider the components of that acceleration.)

■■66. A uniform, flat piece of metal is shaped like an equilateral triangle with sides that are 30 cm long. What are the coordinates of the center of mass in the xy plane if one apex is at the origin and one side along the y axis?

■■■67. Two skaters with masses of 90 kg and 60 kg, respectively, stand 8.0 m apart; each holds one end of a piece of rope. (a) If they pull themselves along the rope until they meet, how far does each skater travel? (Neglect friction.) (b) If only the 60-kg skater pulls herself along the rope until she meets her friend on the opposite end (who just holds onto the rope), how far does each skater travel?

■■■68. A system of two masses has a center of mass given by X_{CM_1}. Another system of three masses has a center of mass given by X_{CM_2}. Show that if all five masses are considered to be one system, the center of mass of that combined system is *not* $X_{CM} = X_{CM_1} + X_{CM_2}$.

■■■69. Once in a while, there is a so-called grand alignment of planets; that is, all nine of the (known) planets are located along a straight line running through the Sun. Assuming that there is such an alignment, with all the planets on one side of the Sun, compute the approximate location of the center of mass of the planets system at that time. (Obtain the necessary data from an outside source. Any introductory astronomy book will have what you need.) Where is that center of mass relative to the surface of the Sun? Does the center of mass lie far from the ecliptic plane? When the planets move out of alignment, what effect does this have on the location of the center of mass?

Additional Problems

70. A truck with a mass of 25 metric tons travels at a constant speed of 90 km/h. (a) What is the magnitude of the truck's instantaneous linear momentum? (b) What average force would be required to stop the truck in 4.0 s?

71. A pool player imparts an impulse of 6.0 N-s to a 0.25-kg cue ball with a cue stick. What is the initial speed of the ball?

72. A piece of uniform sheet metal measures 25 cm by 25 cm. If a circular piece with a radius of 5.0 cm is cut from the center of the sheet, where is the center of mass of the resulting shape?

73. A 3.5-kg block sliding on a frictionless horizontal surface with a velocity of 8.0 m/s approaches a stationary 6.5-kg block. (a) If the blocks have a completely inelastic collision, what is their velocity after it? (b) How much mechanical energy is lost in the completely inelastic collision?

74. A constant force gives a 2.5-kg block that is initially at rest on a horizontal surface a velocity of 8.0 m/s in 5.0 s. What is the magnitude of the force?

75. Another way of describing an elastic collision between two objects is to say that their relative velocities are the same before and after collision. Apply this statement to the collision between a tennis racket and ball when the ball is being served (the ball is thrown upward and hit just at its maximum height). Show that the ball leaves the racket with a speed approximately twice that of the initial speed of the racket.

76. A 2.0-kg block of wood at rest on a horizontal frictionless surface is hit by a 100-g bullet traveling at a speed of 300 m/s parallel to the surface. If the bullet passes through the block and emerges with a velocity of 250 m/s in the initial direction, what will the block's velocity be?

77. A 75-kg man stands in the far end of a 50-kg boat 100 m from the shore, as pictured in Fig. 6.27. If he walks to the other end of the 6.0-m boat, how far is the man from the shore? Neglect friction and assume that the center of mass of the boat is at its center. (Hint: the answer is *not* 94 m, because the boat moves as the man walks. Why? With no external force, the acceleration of the center of mass of the man-boat system is zero, and thus it is stationary. Calculate the location of the center of mass initially, and then apply to the final configuration.)

78. A 0.010-kg bullet leaves a rifle barrel with a muzzle velocity of 150 m/s. The rifle has a mass of 4.0 kg and a barrel length of 0.80 m. (a) What is the average force on the bullet while it is in the barrel? (b) What is the total impulse of the system? (c) Assuming that the rifle is held loosely, with what speed does it recoil as the bullet leaves the barrel?

79. If two electrons approach each other with a relative velocity of 4.5×10^2 m/s, what is the total momentum of the two-particle system? (Hint: find the mass of an electron in Table 1 of Appendix I.)

80. A 2.0-kg sphere with a velocity of 6.0 m/s collides head-on and elastically with a stationary 10-kg sphere. Describe the motions of the spheres after the collision.

81. Four particles, m_1, m_2, m_3, and m_4, are located at $(0, 0)$, $(0, 4.0\text{ m})$, $(4.0\text{ m}, 4.0\text{ m})$, and $(4.0\text{ m}, 0)$, respectively. Locate the center of mass of the system (a) if all the masses are equal; (b) if $m_2 = m_4 = 2m_1 = 2m_3$; and (c) if $m_1 = 100$ g, $m_2 = 200$ g, $m_3 = 300$ g, and $m_4 = 400$ g. (Hint: make a sketch of the system.)

82. In an elastic, head-on collision with a stationary target particle, a moving particle recoils at $\frac{1}{3}$ of its incident speed. (a) What is the ratio of the particle masses (m_1/m_2)? (b) What is the speed of the target particle after the collision in terms of the initial speed of the incoming particle?

83. Water is ejected from a hose at a rate of 30 kg/s and leaves the nozzle with a speed of 50 m/s. How much horizontal force must a firefighter exert to hold the nozzle in a stationary horizontal position?

84. An isolated 3.0-kg object moves along the y axis between the third and fourth quadrants at a speed of 2.0 m/s. When it reaches the origin, an internal explosion sends a 1.0-kg fragment in the positive x direction at a speed of 3.7 m/s. What is the velocity of the remainder of the object after the explosion? (Hint: use components.)

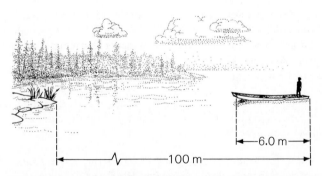

Figure 6.27 **Walking to the other end of the boat to get closer to shore—by how much?** See Problem 77.

Circular Motion and Gravitation

7

hapter 3 covered one type of circular motion, uniform circular motion. But not all circular motion is uniform—there may be an acceleration. In fact, this is so common that circular motion deserves another look. In addition to cars going around race tracks and curves and planets moving in nearly circular orbits about the Sun, there are also circular motions within rotating bodies. For example, the particles that make up a car's wheel rotate, or travel in circles, about the wheel's axle, and the Earth rotates on its axis as it circles the Sun. In fact, as a "particle" on the Earth, *you* are continually in circular motion.

In Chapter 3, uniform circular motion was described in terms of the distance traveled along the circular path. Circular motion can be described equally well in terms of angular quantities. This kind of description is particularly useful for rotational motion, which will be considered in Chapter 8. If you get familiar with the basic principles of circular motion, your study of rotating objects will be much easier.

Gravity plays a large role in determining the motions of the planets, since it supplies the centripetal force necessary to maintain their nearly circular orbits. This chapter will consider Newton's law of gravitation, which describes this fundamental force, and will analyze planetary motion in terms of certain basic laws. As you will learn, the orbits of the planets are not perfectly circular, which does make a difference in their orbital speeds. Knowledge of circular motion will help you understand the motions of these solar satellites, as well as the motions of Earth satellites, of which there is one natural one (the moon) and many artificial ones.

7.1 Angular Measure

Motion is described as a time rate of change of position. As you might guess, angular speed and velocity also involve a time rate of change of position, expressed using an angle. Consider a particle traveling in a circular path, as shown in Fig. 7.1. At a particular instant, the particle's position (P) may be designated by the Cartesian coordinates x and y and its instantaneous speed by the vector components v_x and v_y. However, the position may also be designated by the polar coordinates r and θ. The distance r extends from the origin, and the angle θ is commonly measured counterclockwise from the x axis. The transformation equations that relate one set of coordinates to the other are

$$x = r \cos \theta \tag{7.1a}$$

$$y = r \sin \theta \tag{7.1b}$$

Note that r is the same for any point on a given circle. As a particle travels in a circle, the value of r is constant and only θ changes with time. Thus, circular motion can be described using only one polar coordinate (θ), which changes with time, instead of two Cartesian coordinates (x and y).

Analogous to linear distance is **angular distance**, which is given by

$$\Delta\theta = \theta - \theta_\text{o} \tag{7.2}$$

or simply $\Delta\theta = \theta$ when $\theta_\text{o} = 0°$. A unit commonly used to express angular distance (or angles) is the degree (°); there are 360° in one complete circle, or revolution. Each degree is divided into 60 minutes and each minute into 60 seconds (no relationship to time units).

It is important to be able to relate the angular description of circular motion to the orbital or tangential description, that is, to relate the angular distance θ to the arc length s [Fig. 7.2(a)]. The arc length is the distance traveled along the circular path, and the angle θ is said to subtend or underlie the arc length. A unit that is very convenient for relating the angular distance to the arc length is the **radian (rad)**, which is defined as the angle that subtends an arc length (s) that is equal to the radius (r). By this definition [see Fig. 7.2(b)],

$$1 \text{ rad} = 57.3°$$

and

$$2\pi \text{ rad} = 360°$$

An angle, *expressed in radians*, is related to the arc length it subtends by this simple equation:

$$s = r\theta \qquad (\theta \text{ in radians}) \tag{7.3}$$

The following examples illustrate the use of this equation. But you may be wondering why the radian is defined to be equivalent to 57.3°, since that seems an odd choice. If so, take a look at the Problem-Solving Hint first.

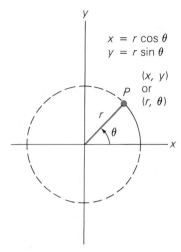

Figure 7.1 **Polar coordinates**
A point, or location, may be described using polar coordinates instead of Cartesian coordinates, that is, as (r, θ) instead of (x, y). For a circle, θ is called the angular distance and r is the radius. The two types of coordinates are related by the transformations $x = r \cos \theta$ and $y = r \sin \theta$.

PROBLEM-SOLVING HINT

Radian Measure The angle θ in degrees may be related to the subtended arc length s as follows: the ratio of the arc length s to the subtending angle θ is the same as the ratio of the circumference c of the circle to 360°. That is,

$$\frac{s}{\theta} = \frac{c}{360°} = \frac{2\pi r}{360°}$$

or $\quad s = \left(\dfrac{2\pi r}{360°}\right)\theta$

where θ is expressed in degrees and r is the radius of the circle.

This equation suggests a way of simplifying the relationship by defining a new unit of angular measure. Some angle θ subtends an arc length equal to the radius, $s = r$ [Fig. 7.2(b)]. This angle is defined to be 1 radian (rad). Then, solving the above equation for θ and substituting r for s gives

$$\theta = \left(\dfrac{360°}{2\pi r}\right)s = \left(\dfrac{360°}{2\pi r}\right)r = \dfrac{360°}{2\pi} \equiv 1 \text{ rad}$$

and

$$2\pi \text{ rad} = 360°$$

There are 2π radians in a complete circle, which can be used for convenient conversions of common angles (Table 7.1), and $1 \text{ rad} = 360°/2\pi = 57.3°$. Note in Fig. 7.2(b) that this sector of a circle is similar to an equilateral triangle with two sides pushed together so the third side bulges out.

The relationship between s and θ using radians is expressed by substituting 2π rad for $360°$:

$$\dfrac{s}{\theta} = \dfrac{c}{360°} = \dfrac{2\pi r}{2\pi} = r$$

and

$$s = r\theta$$

Since $\theta = s/r$, the angle in radians is the ratio of two lengths. This means that a radian measure is a pure number and is unitless.

In computing trigonometric quantities such as $\sin\theta$, the angle may be expressed in degrees or radians; that is, $\sin 30° = \sin(\pi/6)\,\text{rad} = 0.50$. In trigonometric tables, angles are often given in both degrees and radians. It is customary to express an angle in radians using a multiple of π for convenience in calculations. However, there is no simple π form for some angles; for example, $\sin 70° = \sin(1.22)\,\text{rad} = 0.940$. Also, you should note on your calculator that there is usually a way to change an angle entry between "deg" and "rad" if it computes trig functions.

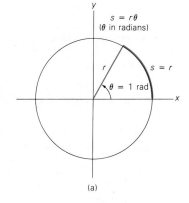

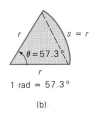

1 rad = 57.3°

(b)

Figure 7.2 **Radian measure** (a) Angular distance may be measured in either degrees or radians (rad). An angle θ subtends an arc of length s. When $s = r$, the angle subtending s is defined as 1 rad, and $s = r\theta$ with θ in radians. (b) A radian is equal to 57.3°, and the circular sector marked off by this angle is similar to an equilateral triangle with sides r.

Table 7.1

Equivalent Degree and Radian Measures

Degrees	Radians	Degrees	Radians
360°	2π	57.3°	1
180°	π	45°	$\pi/4$
90°	$\pi/2$	30°	$\pi/6$
60°	$\pi/3$		

Example 7.1 Arc Length

A spectator stands at the center of a circular running track, which has a radius of 256 m. She observes the start of a race at a point due east of her position and watches the runners run to the finish line, which is located due north of her position. (a) What is the distance run? (b) How long is one complete lap around the track?

Solution

Given: $r = 256$ m   *Find*: (a) s (for 90°)
       (a) $\theta = 90° = \pi/2$ rad   (b) $s = c$
       (b) $\theta = 360° = 2\pi$ rad

(a) Using Eq. 7.3,

$$s = r\theta = (256 \text{ m})\left(\frac{\pi}{2}\right) = 402 \text{ m}$$

Note that rad is omitted, and the equation is dimensionally correct.

(b) Similarly, with $\theta = 2\pi$ rad,

$$s = r\theta = (256 \text{ m})(2\pi) = 1610 \text{ m} = 1.61 \text{ km} \qquad \text{(about a mile)} \quad \blacksquare$$

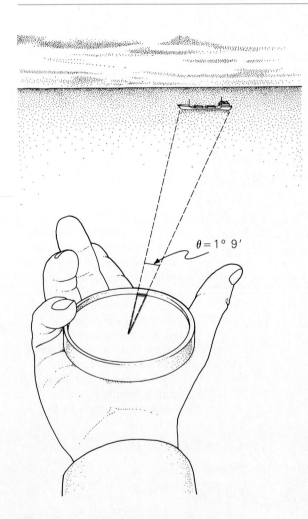

Figure 7.3 **Angular distance**
For small angles, the subtended arc length is approximately a straight line, or the chord length. Knowing the length of the tanker, the sailor can find how far away it is by measuring its angular size. See Example 7.2.

Example 7.2 Angular Distance
A sailor measures the length of a distant tanker as an angular distance of 1° 9′
(1 degree, 9 minutes) with a divided circle, as illustrated in Fig. 7.3. He knows
from the shipping charts that the tanker is 150 m in length. Approximately
how far away is the tanker?

Solution
Given: $\theta = 1°\, 9′$ *Find*: r
        $s = 150$ m

First, convert θ to radians by changing the minutes to degrees: $9′(1°/60′) = 0.15°$.
Then

$$\theta = 1.15°\left(\frac{2\pi\ \text{rad}}{360°}\right) = 0.020\ \text{rad}$$

and

$$r = \frac{s}{\theta} = \frac{150\ \text{m}}{0.020} = 7500\ \text{m} = 7.5\ \text{km}$$

Of course, s is a circular arc length. For small angles, however, the length of the
linear chord subtending an angle is approximately equal to the arc length. Note
that a conversion factor of the form 2π rad/360° was used. Any of the equivalent
relationships in Table 7.1 may be used, such as π rad/180° or even 1 rad/57.3°. ∎

7.2 Angular Speed and Velocity

The description of circular motion in angular form is analogous to the
description of linear motion. In fact, you'll notice that the equations are almost
identical. Different symbols are used to indicate that the quantities have
different meanings. The Greek letter omega is used to represent **average
angular speed** ($\bar{\omega}$), the angular distance divided by the total time to travel
the distance:

$$\bar{\omega} = \frac{\Delta\theta}{\Delta t} \tag{7.4}$$

The units of angular speed are rad/s.

Another common descriptive unit for angular speed is rpm (revolutions per
minute); for example, a turntable speed is $33\frac{1}{3}$ rpm. However, this unit is readily
converted to rad/s, since 1 revolution = 2π rad.

The instantaneous angular speed is given for an extremely small time
interval (Δt approaches zero). Also, $\bar{\omega} = \theta/t$ or $\theta = \bar{\omega}t$, where θ_0 and t_0 are taken
to be zero. If the angular speed is *constant*, $\bar{\omega} = \omega$, and

$$\theta = \omega t \tag{7.5}$$

The average and instantaneous angular velocities are analogous to their
linear counterparts. Being a vector, angular velocity has direction; however,
this is specified in a special way. The tangential velocity (**v**) of a particle in circular motion continually changes direction, as you learned in Chapter 3. But

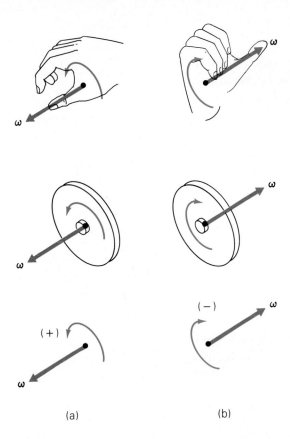

Figure 7.4 **Angular velocity**
The direction of the angular
velocity for an object in
rotational motion is given by
a right-hand rule: when the
fingers of the right hand are
curled in the direction of the
rotational motion, the
extended thumb points in the
direction of the vector.
Circular senses or directions
are commonly indicated by
plus and minus signs, as
shown in parts (a) and
(b), respectively.

in the angular sense, the particle goes in one angular direction or the other (analogous to one-dimensional linear motion). Thus, the vector of angular velocity of a particle in circular motion can have only two directions, which correspond to going around the circular path with either increasing or decreasing angular distance ($\Delta\theta$).

The direction of the angular velocity (a vector) is given by a right-hand rule, illustrated in Fig. 7.4(a). When the fingers of your right hand are curled in the direction of circular motion, your extended thumb points in the direction of ω. Note that circular motion can be in only one of two circular *senses*, clockwise and counterclockwise (not actually directions). Thus, plus and minus signs can be used to distinguish the directions of angular velocity. It is customary to take a counterclockwise rotation as positive (+) since the measurement of positive angular distance (and displacement) is conventionally done counterclockwise from the positive x axis.

A particle moving in a circle has an instantaneous velocity tangential to its circular path. For a constant angular velocity and speed, the particle's orbital speed v (the magnitude of the tangential velocity) is also constant (Fig. 7.5). How the angular and tangential speeds are related is revealed by starting with Eq. 7.3 ($s = r\theta$) and Eq. 7.5 ($\theta = \omega t$),

$$s = r\theta = r(\omega t)$$

The arc length, or distance, is also given by

$$s = vt$$

Combining the equations for s gives

$$v = r\omega \qquad (7.6)$$

where ω is in rad/s. Eq. 7.6 holds in general for instantaneous tangential and angular speeds.

Note that for an object rotating with a constant angular velocity, all the particles of the object have the same angular speed, but the tangential speeds are different at different distances from the axis of rotation (Fig. 7.5, Demonstration 6).

Example 7.3   Different Radii, Different (Tangential) Speeds

On a 45-rpm record, the beginning of the track is 8.0 cm from the center, and the end 5.0 cm from the center. What are (a) the angular speeds and (b) the tangential speeds at these distances when the record is spinning at 45 rpm?

Solution

Given: $\omega = 45$ rpm
$\qquad r_1 = 8.0$ cm $= 0.080$ m
$\qquad r_2 = 5.0$ cm $= 0.050$ m

Find: (a) ω_1 and ω_2
$\qquad$ (b) v_1 and v_2

(a) It should be apparent that $\omega_1 = \omega_2$; that is, the record as a whole rotates with the same angular speed. So every point on the record has the same angular speed, 45 rpm (rev/min), which can be converted to rad/s:

$$\omega = \left(\frac{45 \text{ rev}}{\text{min}}\right)\left(\frac{2\pi \text{ rad}}{1 \text{ rev}}\right)\left(\frac{1 \text{ min}}{60 \text{ s}}\right) = 4.7 \text{ rad/s}$$

(Note that a convenient conversion factor is 1 rev/min $= \pi/30$ rad/s.)

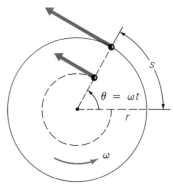

Figure 7.5   Tangential and angular speeds
The tangential and angular speeds are related by $v = r\omega$, with ω in rad/s. Note that all of the particles of a rotating object travel in circles. For uniform angular motion, all the particles have the same angular speed (ω), but particles at different distances from the central axis of rotation have different tangential speeds.

Demonstration 6

Constant Angular Velocity, but not Tangential Velocity

A demonstration of uniform circular motion that shows tangential velocities are different at different radii. Neon bulbs are placed at different radii. The bulbs fire (gas discharge begins) when the ac voltage reaches about 85 V and extinguishes at about 55 V, and the bulbs are on once every 1/120 second. The bulbs at the greater radii have greater tangential speeds, as indicated by longer path lengths (for the same time periods).

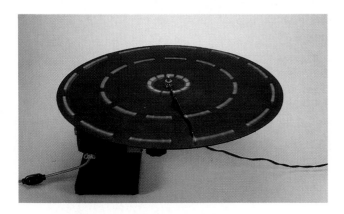

(b) The tangential speeds at different locations on the record are different. All of the particles that make up the record go through one revolution in the same time. Therefore, the farther a point is from the center of the record, the longer its orbital path will be, and its tangential speed must be greater (see Fig. 7.5), as Eq. 7.6 indicates. Thus,

$$v_1 = r_1\omega = (0.080 \text{ m})(4.7 \text{ rad/s}) = 0.38 \text{ m/s}$$

$$v_2 = r_2\omega = (0.050 \text{ m})(4.7 \text{ rad/s}) = 0.24 \text{ m/s}$$

where the unitless rad is included in the angular quantities for clarity.

The outer part of the record has a greater tangential speed than the inner part, as you might expect. The outer part must travel faster than the inner part to make one revolution in the same time. ∎

Some other quantities commonly used to describe circular motion are period and frequency. The **period** (T) is the time it takes for an object in circular motion to make one complete revolution, or cycle. For example, the period of revolution of the Earth about the Sun is 1 year, and the period of the Earth's axial rotation is 24 hours. The standard unit for the period is the second (s). Descriptively, the period is sometimes given in s/rev or s/cycle.

Closely related to the period is the **frequency** (f), which is the number of revolutions, or cycles, made in a given time, generally a second. For example, if a particle traveling uniformly in a circular orbit makes 5.0 revolutions in 1.0 s, the frequency (of revolution) is $f = 5.0 \text{ rev}/1.0 \text{ s} = 5.0 \text{ rev/s}$, or 5.0 cycle/s (cps). Revolution and cycle are merely descriptive terms and not really units. The SI unit of frequency is 1/s, or s^{-1}, which is called the **hertz** (Hz).

The unit is named for Heinrich Hertz (1857–1894), a German physicist and pioneering investigator of electromagnetic waves, which are also characterized by frequency.

Since the descriptive units for frequency and period are inverses of one another (cycles/s and s/cycle) it follows that the two quantities are related by

$$f = \frac{1}{T} \tag{7.7}$$

The frequency can also be related to the angular speed. For uniform circular motion, the orbital speed may be written as $v = 2\pi r/T$, which gives the distance traveled in one revolution per the time for one revolution (1 period). Similarly, for the angular case, when a distance of 2π rad is traveled in 1 period,

$$\omega = \frac{2\pi}{T} = 2\pi f \tag{7.8}$$

7.3 Angular Acceleration

Another important quantity for describing angular motion is the **average angular acceleration** ($\bar{\alpha}$). (The bar over the alpha indicates that it is an average.) This is the time rate of change of the angular velocity:

$$\bar{\alpha} = \frac{\Delta\omega}{\Delta t}$$

With $t_o = 0$ and the angular acceleration constant,

$$\alpha = \frac{\omega - \omega_o}{t}$$

or $\omega = \omega_0 + \alpha t$ (7.9)

The standard units for angular acceleration are radians per second squared (rad/s^2).

No boldface symbols are used in Eq. 7.9, since in general plus and minus signs will indicate angular directions. As with linear motion, if the angular acceleration increases the angular velocity, both quantities have the same sign, meaning that their vector directions are the same ($\boldsymbol{\alpha}$ is in the same direction as ω as given by the right-hand rule). If the angular acceleration decreases the angular velocity, the two quantities have opposite signs, meaning that their vector directions oppose one another ($\boldsymbol{\alpha}$ is in the direction opposite to ω as given by the right-hand rule, or is an angular deceleration, so to speak).

Example 7.4   An Accelerating Record

A $33\frac{1}{3}$-rpm record drops onto a turntable, and accelerates uniformly to its operating speed in 0.42 s. (a) What is the angular acceleration of the record during this time? (b) What is the angular acceleration after this time?

Solution

Given: $\omega_0 = 0$ *Find*: (a) α when $t < 0.42$ s
          $\omega = 33.3$ rpm $= 3.49$ rad/s   (b) α when $t > 0.42$ s
          $t = 0.42$ s

(a) Using Eq. 7.9,

$$\alpha = \frac{\omega - \omega_0}{t} = \frac{3.49 \text{ rad/s} - 0}{0.42 \text{ s}} = 8.3 \text{ rad/s}^2$$

in the direction of the angular velocity. [Record turntables rotate clockwise when viewed from above. Can you specify the vector directions of ω and α?]
(b) After the turntable reaches its operating speed of $33\frac{1}{3}$ rpm, the angular speed remains constant at that value, so $\alpha = 0$.   ∎

Similar to arc length and angle ($s = r\theta$) and tangential and angular speeds ($v = r\omega$), the magnitudes of the tangential and angular accelerations are related by a factor of r:

$$a_t = r\alpha$$ (7.10)

The **tangential acceleration** (a_t) is written with a subscript t to distinguish it from the radial, or *centripetal, acceleration* (a_c), which is necessary for circular motion.

Centripetal, or center-seeking, acceleration can also be written in angular form. For uniform circular motion (as described in Chapter 3),

$$a_c = \frac{v^2}{r} = r\omega^2$$ (7.11)

where v is the tangential, or orbital, speed and $v = rw$ [Fig. 7.6(a)]. For uniform circular motion, there is no angular acceleration ($\alpha = 0$) or tangential acceleration, as can be seen from Eq. 7.10.

However, when there is an angular acceleration α (and therefore a tangential acceleration, $a_t = r\alpha$), there is a change in *both* the angular and the tangential velocities ($v = r\omega$). As a result, the centripetal acceleration must increase or decrease if the object is to maintain the same circular orbit (if r is to

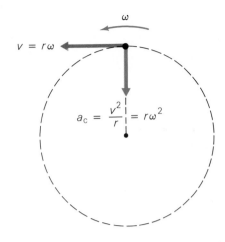

$$v = r\omega$$

$$a_c = \frac{v^2}{r} = r\omega^2$$

(a) Uniform circular motion
$$(a_t = r\alpha = 0)$$

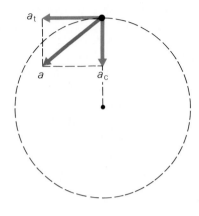

(b) Nonuniform circular motion
$$(\mathbf{a} = \mathbf{a}_t + \mathbf{a}_c)$$

Figure 7.6 **Acceleration and circular motion**
(a) With uniform circular motion, there is centripetal acceleration, but no angular acceleration ($\alpha = 0$) or tangential acceleration ($a_t = r\alpha = 0$). (b) With nonuniform circular motion, there are angular and tangential accelerations, and the total acceleration is the vector sum of the tangential and centripetal accelerations.

stay the same). When there are both tangential and centripetal accelerations, the total instantaneous acceleration is their vector sum [Fig. 7.6(b)]. The tangential acceleration vector and the centripetal acceleration vector are perpendicular to each other at any instant.

Example 7.5 Tangential and Centripetal Accelerations

An object travels with uniform circular motion at an orbital speed of 3.0 m/s and a radius of 1.5 m. The object then experiences an angular acceleration of 0.40 rad/s^2 for 2.0 s and remains in the same circular orbit. (a) By what factor does the centripetal acceleration increase during the 2.0-s interval? (b) What is the total acceleration at $t = 1.0$ s?

Solution

Given: $v_o = 3.0$ m/s   *Find:* (a) a_c
$\quad\quad\quad r = 1.5$ m   (b) $\mathbf{a}$
$\quad\quad\quad \alpha = 0.40$ rad/s^2
$\quad\quad\quad$ (a) $t = 2.0$ s
$\quad\quad\quad$ (b) $t = 1.0$ s

(a) The magnitude of the initial centripetal acceleration is

$$a_{c_o} = \frac{v_o^2}{r} = \frac{(3.0 \text{ m/s})^2}{1.5 \text{ m}} = 6.0 \text{ m/s}^2$$

and the initial angular speed is

$$\omega_o = \frac{v_o}{r} = \frac{3.0 \text{ m/s}}{1.5 \text{ m}} = 2.0 \text{ rad/s}$$

Thus, by Eq. 7.9, in 2.0 s, the angular speed increases to

$$\omega = \omega_o + \alpha t = 2.0 \text{ rad/s} + (0.40 \text{ rad/s}^2)(2.0 \text{ s}) = 2.8 \text{ rad/s}$$

and the centripetal acceleration increases to

$$a_c = r\omega^2 = (1.5 \text{ m})(2.8 \text{ rad/s}^2)^2 = 12 \text{ m/s}^2$$

The centripetal acceleration has doubled, which means that the centripetal force acting on the object must have doubled; otherwise, the object would spiral outward. [The factor of change could also have been found from the ratio $a_c/a_{c_o} = (v/v_o)^2 = (\omega/\omega_o)^2$.]

(b) At $t = 1.0$ s, by Eq. 7.9, $\omega = 2.4$ rad/s, so

$$a_c = r\omega^2 = (1.5 \text{ m})(2.4 \text{ rad/s})^2 = 8.6 \text{ m/s}^2$$

The tangential acceleration is

$$a_t = r\alpha = (1.5 \text{ m})(0.40 \text{ rad/s}^2) = 0.60 \text{ m/s}^2$$

Then, at that instant ($t = 1.0$ s), $\mathbf{a} = a_t\mathbf{t} + a_c\mathbf{r}$, where $\mathbf{t}$ and $\mathbf{r}$ are unit vectors directed tangentially and radially inward, respectively. You should be able to find the magnitude of $\mathbf{a}$ and the angle it makes relative to $\mathbf{a}_t$ [see Fig. 7.6(b)]. However, to find the actual direction of the vector, you would need to know or be given where the object was at $t_o = 0$. ∎

By now, the direct correspondence between the linear and angular kinematic equations should be apparent to you. The other angular equations can be derived, as was done for the linear ones in Chapter 2. That development will not be shown; the set of angular equations with their linear counterparts for constant accelerations are listed in Table 7.2. A quick review of Chapter 2 (with a change of symbols) will show you how the angular equations are derived.

Example 7.6   Rotational Kinematics

A child on a merry-go-round selects a horse located 5.5 m from the central axis. When the ride begins, the merry-go-round accelerates at a rate of 0.069 rad/s² for 12 s and then maintains a constant angular speed. (a) How many revolutions does the child make before the constant operating speed is reached? (b) What is the constant angular speed of the ride and the magnitude of the child's tangential velocity? (c) Are the angular speed and the tangential speed the same for all the riders on the merry-go-round?

Table 7.2
**Equations for Linear and Angular
Motion with Constant Acceleration***

Linear	Angular
$x = \bar{v}t$	$\theta = \bar{\omega}t$
$\bar{v} = \dfrac{v + v_o}{2}$	$\bar{\omega} = \dfrac{\omega + \omega_o}{2}$
$v = v_o + at$	$\omega = \omega_o + \alpha t$
$x = v_o t + \frac{1}{2}at^2$	$\theta = \omega_o t + \frac{1}{2}\alpha t^2$
$v^2 = v_o^2 + 2ax$	$\omega^2 = \omega_o^2 + 2\alpha\theta$

* For these equations, $x_o = 0$, $\theta_o = 0$, and $t_o = 0$. The first equation in each column is general, that is, is not limited to situations where the acceleration is constant.

Solution

Given: $r = 5.5$ m $\qquad\qquad$ *Find*: (a) θ (in revolutions)
$\qquad\quad\;\alpha = 0.069$ rad/s^2 $\qquad\qquad\qquad$ (b) ω and v
$\qquad\quad\;t = 12$ s $\qquad\qquad\qquad\qquad\;\;$ (c) whether ω and v are the
$\qquad\qquad\qquad\qquad\qquad\qquad\qquad\qquad\qquad$ same for all values of r

(a) Using Eq. 7.11 (Table 7.2) gives

$$\theta = \omega_o t + \tfrac{1}{2}\alpha t^2 = 0 + \tfrac{1}{2}(0.069 \text{ rad/s}^2)(12 \text{ s})^2 = 5.0 \text{ rad}$$

Since 2π rad $= 1$ rev (or $360°$),

$$(5.0 \text{ rad})\left(\frac{1 \text{ rev}}{2\pi \text{ rad}}\right) = 0.80 \text{ rev}$$

(b) Using Eq. 7.9 gives

$$\omega = \omega_o + \alpha t = 0 + (0.069 \text{ rad/s}^2)(12 \text{ s}) = 0.83 \text{ rad/s}$$

Then, by Eq. 7.6,

$$v = r\omega = (5.5 \text{ m})(0.83 \text{ rad/s}) = 4.6 \text{ m/s}$$

(c) The angular speed (ω) is the same for each rider on the merry-go-round. The tangential speed (v) varies, depending on the rider's radial distance from the center of the merry-go-round ($v = r\omega$). All riders at the same distance from the central axis (on a circle with the same radius) have the same tangential speed. ∎

7.4 Newton's Law of Gravitation

Another of Isaac Newton's many accomplishments was the formulation of what is called the **universal law of gravitation**. This law is very useful in analyzing the nearly circular orbital motions of planets. It gives a simple mathematical relationship for the gravitational interaction between two particles, or point masses, m_1 and m_2, separated by a distance r (Fig. 7.7). Basically, every particle in the universe has an attractive interaction with every other particle. The gravitational attraction or force (F) decreases as the distance (r) between two point masses increases; that is, the gravitational force and the distance separating the two particles are related as follows:

$$F \propto \frac{1}{r^2} \qquad (7.12)$$

(This type of relationship is called an inverse square law.)

The magnitude of the mutually attractive gravitational forces between two masses is given by

$$F = \frac{Gm_1m_2}{r^2} \qquad (7.13)$$

where G is a constant called the **universal gravitational constant**:

$$G = 6.67 \times 10^{-11} \text{ N-m}^2/\text{kg}^2$$

This constant is often referred to as "big G" to distinguish it from "little g," the

(a) Point masses

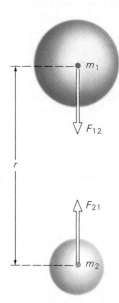

(b) Homogeneous spheres

$$F = \frac{Gm_1 m_2}{r^2}$$

Figure 7.7 Universal law of gravitation
(a) Any two particles or point masses are gravitationally attracted to each other with a force that has a magnitude given by Newton's universal law of gravitation. (b) For the case of homogeneous spheres, the masses may be considered to be concentrated at their centers.

acceleration due to gravity. Note that the forces of the mutual interaction are equal and opposite, a force pair as described by Newton's third law. Note also that F approaches zero only when r is infinitely large. Thus, the gravitational force has or acts over an infinite range.

You might wonder how Newton came to the conclusion that the force of gravity varied by a factor of $1/r^2$. Legend has it that his insight came after he observed an apple fall from a tree to the ground. Newton had been wondering what supplied the centripetal force to keep the moon in orbit and might have had this thought: "If gravity attracts an apple toward the Earth, perhaps it also attracts the moon, and the moon is falling, or accelerating toward the Earth, under the influence of gravity." (See Fig. 7.8.)

Whether or not the legendary apple did the trick, Newton did assume that the moon and the Earth were attracted to each other and that they could be treated as point masses, with their total masses concentrated at their centers. The inverse square relationship had probably been speculated on by some of his contemporaries. Newton's achievement was demonstrating that that relationship could be deduced using one of Johannes Kepler's laws of planetary motion (see Section 7.5).

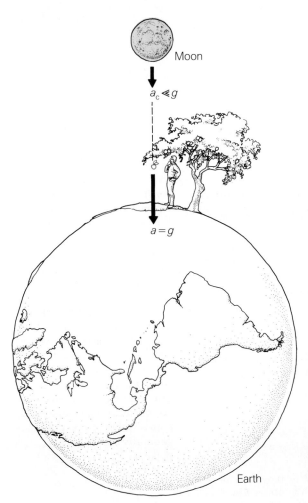

Figure 7.8 **Gravitational insight?**
Newton developed his law of gravitation while studying the orbital motion of the moon. A popular legend says that his thinking was spurred when he observed an apple falling from a tree. He supposedly wondered whether the force causing the apple to accelerate toward the ground could extend to the moon and cause it to fall toward Earth, that is, supply its orbital centripetal acceleration.

Newton expressed Eq. 7.13 as a proportion ($F \propto m_1 m_2 / r^2$) because he did not know the value of G. It was not until 1798 (71 years after Newton's death) that the value of the universal gravitational constant was experimentally determined by an English physicist, Henry Cavendish. He used a sensitive balance to measure the gravitational force between spherical masses.

As mentioned earlier, Newton considered the nearly spherical Earth and moon to be point masses located at their respective centers. It took him some years, using mathematical methods he developed, to prove that this is the case for spherical, *homogeneous* objects.* The general concept is illustrated in Fig. 7.9.

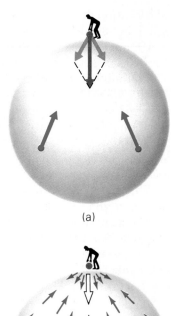

(a)

Example 7.7 Gravitational Attraction between the Earth and the Moon
Estimate the magnitude of the mutual gravitational force between the Earth and the moon. (You can assume that the Earth and the moon are homogeneous spheres.)

Solution
Given: (from Table 1 in Appendix I)   *Find*: F
$$M_e = 6.0 \times 10^{24} \text{ kg}$$
$$m_m = 7.4 \times 10^{22} \text{ kg}$$
$$r_{em} = 3.8 \times 10^8 \text{ m}$$

The average distance from the Earth to the moon (r_{em}) is taken to be the distance from the center of one to the center of the other, and

$$F = \frac{Gm_1 m_2}{r^2} = \frac{GM_e m_m}{r_{em}^2}$$

$$= \frac{(6.67 \times 10^{-11} \text{ N-m}^2/\text{kg}^2)(6.0 \times 10^{24} \text{ kg})(7.4 \times 10^{22} \text{ kg})}{(3.8 \times 10^8 \text{ m})^2}$$

$$= 2.1 \times 10^{20} \text{ N}$$

This is the magnitude of the centripetal force acting on the moon as it revolves around the Earth. It is a very large force, but the moon is a very large object to keep in orbit. ∎

An interesting relationship that is revealed when Newton's second law and gravitational attraction are used to measure mass is explained in the Insight feature. This feature shows how things are questioned in science and how implications arise.

The acceleration due to gravity can also be investigated using Newton's second law of motion and his law of gravitation. In general, the magnitude of g is found by setting mg, an object's weight, equal to the gravitational attraction on the surface of a spherical mass M with radius r:

$$mg = \frac{GmM}{r^2}$$

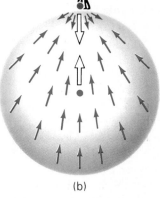

(b)

Figure 7.9 **Spherical masses** (a) Gravity acts between any two (or more) particles. The resultant gravitational force exerted on an object outside a homogeneous sphere by two particles at symmetric locations within the sphere is directed toward the center of the sphere. (b) Because of the sphere's symmetry and uniform mass distribution, the net effect is as though all the mass of the sphere were concentrated as a particle at the center (of symmetry). For this special case, the gravitational center of force and the center of mass coincide, but this is not generally true for other objects.

---

  * For a homogeneous sphere, the equivalent point mass is located at the center of mass. However, this is a special case. The center of gravitational force and the center of mass of a configuration of particles or an object do not generally coincide (see Problem 62).

Then

$$g(r) = \frac{GM}{r^2} \tag{7.14}$$

The expression $g(r)$ is read "g as a function of r." It is understood that $g(r)$ is the magnitude of a vector directed toward the center of the spherical mass M.

The above equation can be applied to the moon or any planet, but taking M_e as the mass of the Earth and R_e as its radius gives g at the Earth's surface:

$$g = \frac{GM_e}{R_e^2} \tag{7.15}$$

This equation has several interesting implications. First, it reveals that taking g to be constant everywhere on the surface of the Earth involves assuming that the Earth has a homogeneous mass distribution and that the distance from the center of the Earth to any location on its surface is the same. Since these two things are not true, taking g to be a constant is only an approximation, but one that works pretty well for most situations.

Also, you can see why the acceleration due to gravity is the same for all free-falling objects, that is, is independent of the mass of an object. This mass doesn't appear in Eq. 7.15, so all objects in free fall accelerate at the same rate.

Finally, if you're observant, you'll notice that Eq. 7.15 can be used to compute the mass of the Earth. All of the other quantities in the equation are measurable and their values are known, so M_e can readily be calculated. This is what Cavendish did after he determined the value of G experimentally. Then he also found the average density of the Earth (see Problem 60).

The value of g at the Earth's surface is referred to as the standard acceleration and is sometimes used as a unit. For example, when a spacecraft lifts off, astronauts are said to experience several g's. This means that their acceleration is several times the standard acceleration, g. On the surface of the Earth, you experience an acceleration of g, and the equivalent force of 1 g is your weight: $F = m(1\text{ g}) = mg = w$. Sometimes 1 g is used as a force standard, giving **g's of force**. If you experience a force of 2 g's, the force has a magnitude of twice your weight.

A jet pilot diving in a circular arc is said to "pull" some number of g's at the bottom of the arc. This is the magnitude of the reaction force on the pilot.

INSIGHT

Gravitational and Inertial Mass

To determine the mass of an object, you would find it easiest to measure its weight using a suitable scale. Then, in terms of weight at the Earth's surface, $w = mg$, or $m = w/g$, where the gravitational force is expressed using the acceleration due to gravity. But can the mass of an object be determined without reference to, or in the absence of, gravity? The answer is yes.

Newton's second law is $F = ma$. Therefore, by applying a force and measuring the resulting acceleration, you could determine the mass of an object. (This could be done in the presence of gravity by applying a horizontal force to an object on a nearly frictionless horizontal surface.) This type of measurement involves the property of inertia. Thus, you would be measuring *inertial* mass by this procedure and *gravitational* mass by weighing.

Are the masses determined by these two methods the same, or are there two different types or properties of mass? Comparisons have been made, and no significant difference has been found. Experimentally, the two types of measurements give results that differ by about only one part in a trillion (10^{12}).

Figure 7.10 **g's of force**
Lt. Col. John Stapp is shown before and during a high-speed rocket-propelled sled run.
He reached Mach 1.7 (1.7 times the speed of sound or about 1200 mi/h) in just seconds
and then quickly decelerated to rest.

A pilot might experience a blackout, or loss of consciousness, because this force causes a lack of blood flow to the brain. Similarly, an astronaut experiences as much as 8 g's of force on blast-off because of the upward acceleration (Fig. 7.10). This force (experienced as apparent weight) is exerted on the astronaut by the seat. If an astronaut were standing upright, such a force would cause the blood to go to the legs, where the blood vessels would be distended and the capillaries would rupture. Lack of blood circulation to the eyes and brain would cause temporary blindness and loss of consciousness. Consequently, astronauts are in a reclining position for blast-off.

The acceleration due to gravity does vary slightly with altitude. At a distance h above the Earth's surface, the acceleration is given by

$$g' = \frac{GM_e}{(R_e + h)^2} \tag{7.16}$$

The value of g' does not decrease very rapidly with increasing altitude. It has been determined that at an altitude of 16 km (10 mi), g' is still 99% of the value of g at the Earth's surface, and at an altitude of 160 km (100 mi), g' is 95% of g.

PROBLEM-SOLVING HINT

When comparing accelerations due to gravity or gravitational forces, you will often find it convenient to work with ratios. For example, comparing $g(r)$ to g (Eqs. 7.14 and 7.15) for the Earth gives

$$\frac{g(r)}{g} = \frac{GM_e/r^2}{GM_e/R_e^2} = \frac{R_e^2}{r^2} = \left(\frac{R_e}{r}\right)^2 \quad \text{or} \quad \frac{g(r)}{g} = \left(\frac{R_e}{r}\right)^2$$

Note how the constants cancel out. Taking $r = R_e + h$, you can easily compute g'/g, or the acceleration due to gravity at some altitude above the Earth compared to g on the Earth's surface (9.80 m/s^2).

Another way of looking at the gravitational interaction between bodies is by using a **gravitational field**. This is a mapping of the **gravitational intensity**

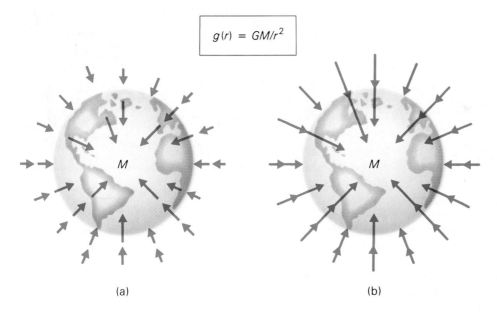

$$g(r) = GM/r^2$$

(a)

(b)

Figure 7.11 **Gravitational field**
The gravitational intensity is the force per unit mass, F/m. (a) Measured at different locations, the intensity may be represented as vectors. (b) Joined together, the vectors form the lines of force of a gravitational field.

(or gravitational field strength), a vector whose magnitude is $g(r)$, as given in Eq. 7.14,

$$g(r) = \frac{F}{m} = \frac{GM}{r^2}$$

As the term F/m indicates, gravitational intensity is the *gravitational force per unit of mass* (N/kg).

One method of mapping the gravitational field associated with a mass M, such as the Earth, is to put a test mass at different locations in space, measure the gravitational forces, and compute values of $g(r)$. Vector arrows may be used to depict these values [see Fig. 7.11(a)], and joining the arrows forms so-called lines of force. The closeness of the lines of force indicates the relative strength of the gravitational field, or the amount of gravitational force a mass would experience in that region. That is, the closer together the lines of force, the stronger the gravitational field is in that part of space.

The gravitational field approach essentially measures an effect rather than considering its cause. An object is seen as interacting with a gravitational field rather than with the Earth or some other body responsible for the field. Later chapters will extend the field concept to electric and magnetic forces. Einstein expanded the idea of force fields in the model of gravity included in his general theory of relativity, a model quite different from Newton's. In the theory of relativity, a gravitational field is considered to be a warping of four-dimensional space-time (see Chapter 26).

Another aspect of the change of g with altitude concerns potential energy. In Chapter 5, you learned that $U = mgh$ for an object at a height h above some zero reference point, since g is essentially constant near the Earth's surface. This potential energy is equal to the work done in raising the object a distance h in a *uniform* gravitational field. But what if there is a change in altitude and g is not constant while work is done in moving an object? In this case, the equation $U = mgh$ doesn't apply. In general, it can be shown (using mathematical methods that are beyond the scope of this book) that the **gravitational**

potential energy of two point masses separated by a distance r is given by

$$U = -\frac{Gm_1 m_2}{r} \tag{7.17}$$

The minus sign in Eq. 7.17 arises from the choice of the zero reference point (the point where $U = 0$), which is $r = \infty$. In terms of the Earth and some altitude h,

$$U = -\frac{Gm_1 m_2}{r} = -\frac{GmM_e}{R_e + h} \tag{7.18}$$

where r is the distance separating the Earth's center and the mass m which is at an altitude h. What this means is that on Earth we are in a negative gravitational potential energy well (Fig. 7.12) that extends to infinity because the force of gravity has an infinite range. [Recall from Chapter 5 that finite potential energy wells arise using the approximation $U = mgh$.] Thus, when work is done against gravity (an object moves higher in the well) or gravity does work (an object falls lower in the well), there is a *change* in potential energy. As with finite potential energy wells, this change in energy is usually what's important in analyzing situations.

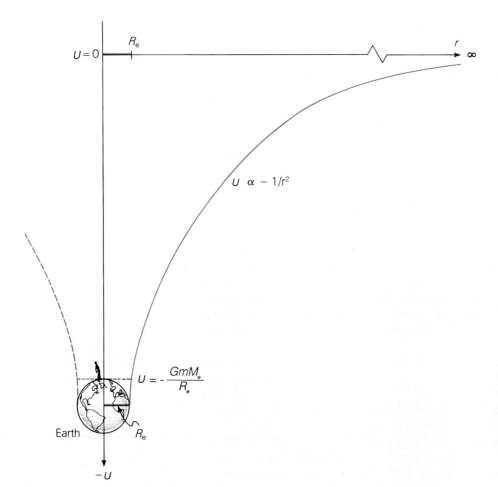

Figure 7.12 **Gravitational potential energy well** On Earth, we are in a negative gravitational potential energy well. As with an actual hole or well in the ground, work must be done against gravity to get higher in the well. The potential energy increases as an object moves higher in the well. This means that the value of U becomes less negative. The top of the Earth's gravitational well is at infinity, where the gravitational potential energy is zero.

Example 7.8 Gravitational Potential Energy

Two 50-kg weather satellites are in circular orbits about the Earth at altitudes of 1000 km (about 620 mi) and 35,000 km (about 22,000 mi). (The lower one monitors particles about to enter the atmosphere, and the higher one is synchronous with the Earth's rotation and takes weather pictures from its stationary position with respect to the Earth's surface.) How much more work was required to get the higher satellite into orbit?

Solution

Given: $m = 50$ kg

$h_1 = 1000$ km $= 1.0 \times 10^6$ m

$h_2 = 35,000$ km $= 35 \times 10^6$ m

$M_e = 6.0 \times 10^{24}$ kg

$R_e = 6.4 \times 10^6$ m

Find: W

Since the satellites have the same mass, the extra work required to get the higher one into orbit is simply the difference in their gravitational potential energies at the orbital altitudes. (This is the same as the work that would be required to raise the lower satellite to the higher orbit.) Air resistance is not a factor because both satellites are above the atmosphere. The path the upper satellite took to reach the higher altitude is not a factor either, since gravity is a conservative force (see Chapter 5). Thus,

$$W = \Delta U = U_2 - U_1 = -\frac{GmM_e}{R_e + h_2} - \left(-\frac{GmM_e}{R_e + h_1}\right)$$

$$= GmM_e\left(\frac{1}{R_e + h_1} - \frac{1}{R_e + h_2}\right)$$

$$= (6.67 \times 10^{-11} \text{ N-m}^2/\text{kg}^2)(50 \text{ kg})(6.0 \times 10^{24} \text{ kg})$$

$$\times \left[\frac{1}{(6.4 \times 10^6 \text{ m}) + (1.0 \times 10^6 \text{ m})} - \frac{1}{(6.4 \times 10^6 \text{ m}) + (35 \times 10^6 \text{ m})}\right]$$

$$= 2.2 \times 10^9 \text{ J}$$

This is the amount of work done in raising the satellite higher in the Earth's negative potential energy well. Note that U_2 has a smaller negative value than U_1 has. ∎

Using the gravitational potential energy (Eq. 7.17) gives the equation for the total mechanical energy a different form than it had in Chapter 5. For example, the total mechanical energy of a mass m_1 moving near a stationary mass m_2 is

$$E = K + U = \tfrac{1}{2}m_1v^2 - \frac{Gm_1m_2}{r} \tag{7.19}$$

This equation and the principle of the conservation of energy can be applied to the Earth moving about the Sun, by neglecting other gravitational forces. The Earth's orbit is not quite circular, but slightly elliptical. At perihelion (the point of the Earth's closest approach to the Sun), the mutual gravitational potential energy is less than it is at aphelion (the point farthest from the Sun). Therefore, by Eq. 7.19, the Earth's kinetic energy and orbital speed are greatest at perihelion (smallest value of r) and least at aphelion (greatest value of r). Or, in general, the Earth's orbital speed is greater when it is nearer the Sun than when it is farther away.

Mutual gravitational potential energy also applies to a group, or configuration, of more than two masses. That is, there is potential energy due to the masses being in a configuration. This is because work was done in bringing them together. Suppose that there is a single fixed mass m_1, and another mass m_2 is brought close to it, from an infinite distance (where $U = 0$). The work done is equal to the mutual potential energy of the masses, which are now separated by a distance r_{12}, that is, $U_{12} = -Gm_1m_2/r_{12}$. If a third mass m_3 is brought close to the other two fixed masses, there are then two forces of gravity acting on m_3, so $U_{13} = -Gm_1m_3/r_{13}$ and $U_{23} = -Gm_2m_3/r_{23}$. The total gravitational potential energy of the configuration is therefore

$$U = U_{12} + U_{13} + U_{23}$$

$$= -\frac{Gm_1m_2}{r_{12}} - \frac{Gm_1m_3}{r_{13}} - \frac{Gm_2m_3}{r_{23}} \tag{7.20}$$

A fourth mass could be brought in, but this development should be sufficient to show that the total gravitational potential energy of a configuration of particles is equal to the sum of the individual potential energies for all pairs of particles.

7.5 Kepler's Laws and Earth Satellites

The force of gravity determines the motions of the planets and of Earth satellites and holds the solar system (and galaxy) together. A general description of planetary motion had been set forth shortly before Newton's time by the German astronomer and mathematician Johannes Kepler (1571–1630). Kepler was able to formulate three empirical laws from observational data gathered during a 20-year period by the Danish astronomer Tycho Brahe (1546–1601). Brahe's extensive observations of stars and planets were done without the benefit of a telescope because it hadn't been invented yet. But the observations Brahe made were more accurate than those of most of his contemporaries because of better instrumentation he had developed. As a result, he is considered to have been one of the greatest practical astronomers.

Kepler went to Prague to assist Brahe, who was the official mathematician at the court of the Holy Roman Emperor. Brahe died the next year and Kepler succeeded him, inheriting his records of the positions of the planets. Analyzing this data, Kepler announced the first two of his three laws in 1609 (the year Galileo built his first telescope). These laws were applied initially only to Mars. Kepler's third law came 10 years later.

Interestingly enough, Kepler's laws of planetary motion, which took him about 15 years to deduce from observed data, can now be derived theoretically with a page or two of calculations. These three laws apply not only to planets, but to any system composed of a body revolving about a larger body, to which the inverse square law of gravitation applies (for example, the moon and solar-bound comets obey these laws).

Kepler's first law (the law of orbits) says that

Planets move in elliptical orbits with the Sun at one of the focal points.

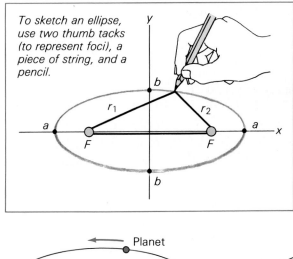

To sketch an ellipse, use two thumb tacks (to represent foci), a piece of string, and a pencil.

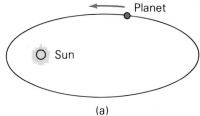

Planet

Sun

(a)

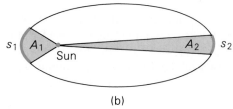

s_1 A_1

Sun

A_2 s_2

(b)

Figure 7.13 Kepler's first and second laws of planetary motion
(a) In general, an ellipse has an oval shape. The sum of the distances from the two focal points to any point on the ellipse is constant: $r_1 + r_2 = 2a$, where a is the length of the line joining the two points on the ellipse at the greatest distance from its center, called the semimajor axis. (The line joining the two points closest to the center is b, the semiminor axis.) Planets revolve about the Sun in elliptical orbits for which the Sun is one of the focal points. (b) A line joining the Sun and a planet sweeps out equal areas in equal times. Since $A_1 = A_2$, a planet travels faster along s_1 than along s_2.

An ellipse, shown in Fig. 7.13(a), has, in general, an oval shape, resembling a flattened circle. In fact, a circle is a special case of an ellipse in which the focal points, or foci (plural of focus), are at the same point (at the center of the circle). Although the orbits of the planets are elliptical, most do not deviate very much from a circle (Mercury and Pluto are notable exceptions; see Table 6 in Appendix I). For example, the difference between the perihelion and aphelion of the Earth (its closest and farthest distances from the Sun) is about 3 million miles. This may sound like a lot, but it is only 3% of 93 million miles, which is the average distance between the Earth and the Sun.

Kepler's second law (the law of areas) says that

A line from the Sun to a planet sweeps out equal areas in equal lengths of time.

This law is illustrated in Fig. 7.13(b). Since the time to travel the different orbital distances is the same, this law tells you that the orbital speed of a planet varies in different parts of its orbit. Because a planet's orbit is elliptical, its orbital speed is greater when it is closer to the Sun than when it is farther away. [This was deduced for the Earth using the conservation of energy in Section 7.4.]

Kepler's third law (the law of periods) says that

The square of the period of a planet is directly proportional to the cube of the average distance of the planet from the Sun; that is, $T^2 \propto r^3$.

Kepler's third law is easily derived for the special case of a circular orbit, using Newton's law of gravitation. Since the centripetal force is supplied by the force of gravity, the expressions for these forces can be set equal:

$$\frac{m_{\mathrm{p}}v^2}{r} = \frac{Gm_{\mathrm{p}}M_{\mathrm{s}}}{r^2}$$

centripetal   gravitational
  force   force

Where m_{p} and M_{s} are the masses of the planet and the Sun, respectively, and v is the orbital speed. But $v = 2\pi r/T$ (circumference/period), so

$$\left(\frac{m_{\mathrm{p}}}{r}\right)\left(\frac{2\pi r}{T}\right)^2 = \frac{Gm_{\mathrm{p}}M_{\mathrm{s}}}{r^2}$$

Solving for T^2 gives

$$T^2 = \left(\frac{4\pi^2}{GM_{\mathrm{s}}}\right)r^3$$

or $T^2 = Kr^3$ (7.21)

The constant K is easily evaluated from known data: $K = 2.97 \times 10^{-19}$ s²/m³. Use data from Table 6 of Appendix I to compare the values of the ratio of the period squared to the distance cubed ($K = T^2/r^3$) for the planets.

You might wonder how the mass of the Sun was determined. Take a look at Eq. 7.21. Could you compute M_{s} using known planetary data, for Earth, for example? (See Problem 45.)

Earth Satellites

We are just 30 years into the space age, taking its beginning as the first manned orbital flight in 1961, by the Soviet cosmonaut Yuri Gagarin. The first American to orbit the Earth was John Glenn in 1962. In 1969, Neil Armstrong stepped onto the surface of the moon. Before and since these events, numerous unmanned satellites have been put into orbit about the Earth, and now astronauts go to and from orbiting laboratories.

Putting a spacecraft into orbit about the Earth (or the moon) is an extremely complex task. However, you can get a basic understanding of the problem from fundamental principles. First, suppose that a projectile could be given the initial velocity required to take it just to the top of the Earth's potential energy well so that it could escape the Earth's gravitational attraction. (Such an initial thrust is not possible, even with rockets, but it gives the upper limit for orbital velocities.) At the exact top of the well (where $r = \infty$), the potential energy is zero. By the conservation of energy and Eq. 7.18,

                  initial   *final*

$$K_{\mathrm{o}} + U_{\mathrm{o}} = K + U$$

$$\tfrac{1}{2}mv_{\mathrm{e}}^2 - \frac{GmM_{\mathrm{e}}}{R_{\mathrm{e}}} = 0 + 0$$

where v_{e} is the **escape speed**. The final energy is zero since the projectile stops at the top of the well and $U = 0$ there. Solving for v_{e} gives

$$\tfrac{1}{2}mv_{\mathrm{e}}^2 = \frac{GmM_{\mathrm{e}}}{R_{\mathrm{e}}}$$

and

$$v_e = \sqrt{\frac{2GM_e}{R_e}} \qquad (7.22)$$

Since $g = GM_e/R_e^2$ (Eq. 7.15), it is convenient to write

$$v_e = \sqrt{2gR_e} \qquad (7.23)$$

Although derived for Earth, this equation may be used generally to find the escape speeds for the moon and the other planets.

Example 7.9   Escape Speed
What is the escape speed for the Earth?

Solution
Eq. 7.23 can be used with known data:

$$v_e = \sqrt{2gR_e} = \sqrt{2(9.80 \text{ m/s}^2)(6.4 \times 10^6 \text{ m})} = 11 \times 10^3 \text{ m}$$

$$= 11 \text{ km/s} \qquad \text{(about 7 mi/s)}$$

Thus, if a projectile or spacecraft could be given an initial speed of 11 km/s (about 40,000 km/h or 25,000 mi/h), it would leave the Earth and not fall back, or return. Note from the equation that the escape speed is independent of the mass of the spacecraft; this is the initial speed required to send any object to the top of the Earth's potential energy well. ∎

It is clear that a tangential speed smaller than the escape speed is required to orbit a satellite. Consider the centripetal force for a satellite in circular orbit about the Earth:

$$F = \frac{mv^2}{r} = \frac{GmM_e}{r^2}$$

Then

$$v = \sqrt{\frac{GM_e}{r}} \qquad (7.24)$$

where $r = R_e + h$. For example, suppose that a satellite is in a circular orbit at an altitude of 500 km (about 300 mi); its tangential speed is

$$v = \sqrt{\frac{GM_e}{r}} = \sqrt{\frac{GM_e}{R_e + h}}$$

$$= \sqrt{\frac{(6.67 \times 10^{-11} \text{ N-m}^2/\text{kg}^2)(6.0 \times 10^{24} \text{ kg})}{(6.4 \times 10^6 \text{ m}) + (0.50 \times 10^6 \text{ m})}} = 7.6 \times 10^3 \text{ m/s}$$

$$= 7.6 \text{ km/s} \qquad \text{(about 4.7 mi/s)}$$

This is about 27,000 km/h or 17,000 mi/h. As can be seen from Eq. 7.24, the orbital speed decreases with altitude.

In practice, a satellite is given a tangential speed by a component of the thrust from a rocket stage [Fig. 7.14(a)]. The inverse square relationship of Newton's law of gravitation means that the satellite orbits that are possible

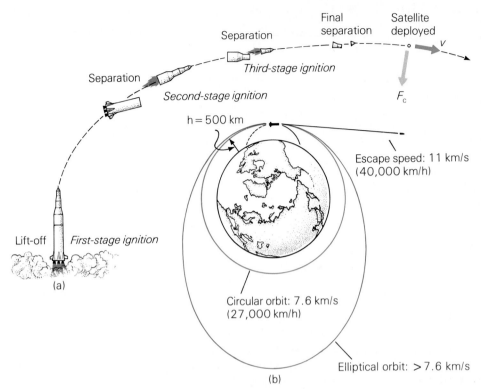

Final separation

Satellite deployed

Separation

Third-stage ignition

v

Separation

Second-stage ignition

Escape speed: 11 km/s (40,000 km/h)

F_c

h = 500 km

Lift-off   First-stage ignition

(a)

Circular orbit: 7.6 km/s (27,000 km/h)

Elliptical orbit: > 7.6 km/s

(b)

Figure 7.14   **Satellite orbits** (a) A satellite is put into orbit by giving it a tangential speed sufficient for maintaining a particular orbit. (b) A tangential speed of 7.6 km/s is required for a circular orbit at an altitude of 500 km. With a greater tangential speed, the satellite would move out of the circular orbit. Since it would not have the escape speed, however, it would "fall" around the Earth in an elliptical orbit with the Earth at one focal point. A tangential speed less than 7.6 km/s would also give an elliptical orbit, but below a certain speed, the satellite would spiral into the Earth.

about a large mass are ellipses, of which a circular orbit is a special case. This is illustrated in Fig. 7.14(b) for Earth, using the previously calculated values. If a satellite is not given a sufficient tangential speed, it will fall back to Earth (and possibly be burned up while falling through the atmosphere). If the tangential speed approaches the escape speed, the satellite may leave its orbit and go off into space.

Finally, the energy of an orbiting satellite is

$$E = K + U = \tfrac{1}{2}mv^2 - \frac{GmM_e}{r}$$

Substituting the expression for v from Eq. 7.24 in the kinetic energy term gives

$$E = \frac{GmM_e}{2r} - \frac{GmM_e}{r}$$

Thus,

$$E = -\frac{GmM_e}{2r} \tag{7.25}$$

That is, the total energy of an orbiting satellite is inversely proportional to the radius of its orbit (or its altitude, since $r = R_e + h$). Note, however, that this energy is negative. More work is required to get a satellite into a higher orbit, where it has more energy because it is higher in the Earth's potential energy well. The *numerical value* of its energy becomes smaller or less negative as the zero potential at the top of the well is approached. (You will learn in a later chapter that this result also holds for an electron orbiting the nucleus of a hydrogen atom.)

Also note from the development of Eq. 7.25 that the kinetic energy of an orbiting satellite is equal to the absolute value of its total energy:

$$K = \frac{GmM_e}{2r} = |E| \tag{7.26}$$

Example 7.10   Applying the Brakes to Speed Up
A spacecraft is in orbit about the Earth well outside the atmosphere. Describe what happens to the spacecraft if its retrorockets are fired.

Solution
The retrorockets apply a braking force to the spacecraft, work is done, and it loses some energy. That is, the total energy of the spacecraft,

$$E = -\frac{GmM_e}{2r}$$

is reduced. If E decreases (greater negative value), r must also decrease, and the spacecraft will spiral inward toward the Earth. From Eq. 7.26,

$$\tfrac{1}{2}mv^2 = \frac{GmM_e}{2r} \quad \text{or} \quad v^2 = \frac{GM_e}{r}$$

So, if r decreases, v increases, or the spacecraft speeds up.

The spacecraft would continue to spiral inward as long as the retrorockets were fired. When it reached the atmosphere, they could be shut off because air resistance would cause the spacecraft to lose further energy and continue its downward spiral. ■

7.6 Frames of Reference and Inertial Forces ◆

This is a good place to distinguish two types of reference frames and to discuss some of the effects commonly experienced or observed by observers in these different frames. It is easier to think of a reference frame as being fixed, or stationary, but observers in some systems are in circular motion (such as yourself on the rotating Earth).

A frame of reference is a set of mutually perpendicular axes relative to which positions in space are measured. The origin of the axes is arbitrarily fixed or attached to something, such as the Earth. This means that an object such as a car would be said to move with a speed of 60 km/h *relative* to the Earth. Where the origin is designated to be is arbitrary, but, as you will see, its position can make a difference.

An **inertial frame of reference** is defined as one for which Newton's laws are valid. Basically, an inertial frame is one for which the law of inertia (Newton's first law of motion) holds; that is, in a system referenced to this frame, an *isolated* object would be stationary or moving with a constant velocity. If this is the case, then Newton's second law (in the customary form of $F = ma$) also applies to the system as a whole or to any part of it.

Any reference frame moving with a constant velocity relative to an inertial frame is also an inertial frame, and $F = ma$ applies for it too. No acceleration effects are introduced by transferring from one frame to the other, so $F = ma$

Accelerated Reference Frame

A demonstration that shows an effect of a noninertial or accelerated reference frame.

(a) Two small holes are punched in a Styrofoam cup near its bottom on opposite sides. When the cup is filled with water (colored red here), two streams are observed coming from the cup.

(b) When the cup is dropped and in free fall, no streams are observed. In the accelerated reference frame falling with the cup, the water is weightless (that is, a layer of water does not feel the weight of the layer immediately above it). Hence there is no pressure and the water streams cease.

can be used by observers in either frame (with their own coordinate values) to analyze the same dynamic situation, and they will come up with similar descriptions.

A frame of reference that is accelerating relative to an inertial frame is called a **noninertial frame of reference** (see Demonstration 7). If an observer in a noninertial frame uses $F = ma$ to analyze a situation, the result will be different from what it would be for an observer in an inertial frame. For example, consider a person standing on a scale in an elevator that is accelerating upward (Fig. 7.15). From the point of view of an observer in an inertial reference frame (fixed to the Earth or the building), the motion is analyzed using $F = ma$, and isolating the forces on the person gives

$$F = ma$$

$$R - mg = ma$$

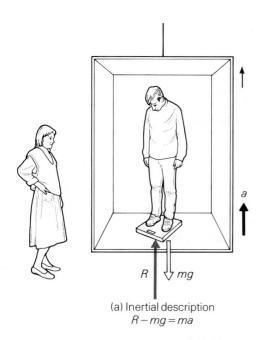
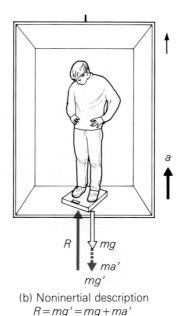

(a) Inertial description
$$R - mg = ma$$

(b) Noninertial description
$$R = mg' = mg + ma'$$

Figure 7.15 **Different frames of reference**
(a) The dynamic situation is described using Newton's laws by a stationary observer, or from an inertial frame of reference. (b) The noninertial frame of reference of a person in the elevator means that he must invent an inertial pseudo force to satisfy Newton's laws in the accelerating system.

where R is the magnitude of the upward reaction force of the surface of the scale on the person. This reaction force is

$$R = mg + ma = m(g + a)$$

By Newton's third law, the person exerts an equal and opposite force on the scale, so R is the magnitude of the scale reading.

Assume that the person does not know that the elevator is accelerating. The person's point of view (from a noninertial frame of reference) then is that he is motionless with respect to the elevator and two forces are acting on him. Applying Newton's second law, he finds

$$R - mg' = 0 \qquad \text{or} \qquad R = mg'$$

Considering himself not to be accelerating and to be in an inertial system, he suddenly notices that he weighs more than normal. He therefore concludes that there is another force acting downward on him, in addition to gravity. Because he knows his true weight, he can form this equation:

$$R = mg' = m(g + a) = mg + ma$$

By his analysis, there is a second force, $-ma$, acting downward.

The extra force is called an **inertial force**, or a pseudo (false) force. It results from Newton's law being incorrectly applied in a noninertial system. Essentially, the noninertial observer invents this inertial force to satisfy Newton's law in the accelerating reference frame. It appears to increase the force of gravity, but the noninertial observer cannot really distinguish it from an actual gravitational force.

The frame of reference of an object in circular motion is noninertial, since it is accelerating (centripetal acceleration). You have experienced an inertial force when riding as a passenger in a car rounding a curve at a fairly high speed.

Figure 7.16 **A common pseudo force**
A person in a car going around a curve feels a centrifugal force throwing him against the door, which supplies a reactive centripetal force. Then he is in equilibrium and travels around the curve with the car (his frame of reference). The stationary observer, who has an inertial frame of reference, can find no false centrifugal force and sees everything consistent with Newton's laws.

As the car turns (Fig. 7.16), you feel yourself being thrown or pushed outward. This effect is commonly said to be due to **centrifugal force** (centrifugal means "center-fleeing"). But this is a pseudo force and does not really exist.

The noninertial description of this situation goes something like this: you are thrown outward and hit the car door (if you're not wearing a seatbelt). The reaction force of the door on you balances the centrifugal force. Being in equilibrium, you are stationary with respect to the car (your frame of reference) and go around the curve with it. But the reaction force and the centrifugal force *act on the same body*, so they can't be a third law force pair. (What is the reaction force paired with the centrifugal force when you are being thrown outward?)

From an inertial frame such as the point of view of a spectator standing nearby, friction between the road and the car's tires supplies the centripetal force for the car to negotiate the circular arc of the curve. A person in the car continues to move in a straight line in accordance with Newton's first law, because friction between the car seat and the person's seat does not provide enough centripetal force. The car turns in front of the person, who hits up against the door, and the reaction force supplies the necessary centripetal force. Note that there is no need for a pseudo force in this inertial frame description.

Even though centrifugal force is nonexistent from an inertial point of view, the term is often used. Machines called **centrifuges** separate particles of different sizes and densities suspended in a liquid. For example, blood components and cells are separated in centrifuges in medical laboratories (Fig. 7.17). Such separation can also be done by gravitational sedimentation. Red blood cells settle in the bottom layer in the plasma because they reach their terminal sedimentation velocity after the white blood cells and platelets. [The viscous drag of the liquid plasma on the particles is analogous to the air resistance that determines the terminal velocity of falling objects; see Sections 4.6 and 10.5.] Each terminal sedimentation velocity is also proportional to g. The separation of blood components is greatly accelerated by centrifuging. Tubes are spun horizontally. The effective horizontal acceleration is equal to the centripetal acceleration, which for operational speeds of rotation is much greater than g.

Example 7.11 Acceleration in a Centrifuge

A laboratory centrifuge like that shown in Fig. 7.17 operates at a rotational speed of 12,000 rpm. (a) What is the centripetal acceleration on a blood cell at a radial distance of 8.00 cm from the centrifuge's axis of rotation? (b) How does this acceleration compare to g?

Solution

Given: $\omega = 1.20 \times 10^4$ rpm
$\qquad\quad = 1.26 \times 10^3$ rad/s
$\qquad\ r = 8.00$ cm $= 0.0800$ m

Find: (a) a_c
$\qquad$ (b) How a_c compares to g

(a) The centripetal acceleration is easily found:

$$a_c = r\omega^2 = (0.0800 \text{ m})(1.26 \times 10^3 \text{ rad/s})^2 = 1.27 \times 10^5 \text{ m/s}^2$$

(b) Dividing by 9.80 m/s^2 gives

$$a_c = (1.27 \times 10^5)\left(\frac{1 \text{ g}}{9.80 \text{ m/s}^2}\right) = 13,000 \text{ g}$$

Note that the separation rate is 13,000 times faster in the centrifuge than it would be with gravitational sedimentation. ∎

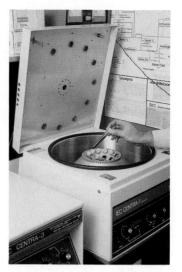

Figure 7.17 **Centrifuge**
Centrifuges are used to separate particles of different sizes and densities suspended a liquid. For example, red and white blood cells can be separated from each other as well as the plasma that makes up the liquid portion of the blood.

An astronaut in a spacecraft orbiting the Earth has a noninertial frame of reference. The astronaut observes that things, including himself, float inside the spacecraft. This is sometimes called a condition of weightlessness or zero g. Both of these are incorrect since gravity is acting; if it weren't, the spacecraft wouldn't be orbiting.

The astronaut's situation is similar to that of a person standing on a scale in

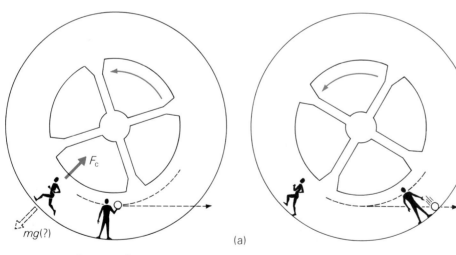

(a)

Figure 7.18 **Space colony**
(a) In the frame of reference of someone in a rotating spacecraft, objects are affected by an inward reactive centripetal force and an outward pseudo (centrifugal) force, which supplies artificial gravity. From the point of view of an inertial observer, the dropped ball follows a tangential straight-line path, but the colonist sees it as falling downward. (b) This rotating space colony appeared in *2001, A Space Odyssey*.

an elevator in free fall. Both the person and the scale are falling at the same rate, and the scale reads zero because there are action-reaction forces between it and the person. Similarly, an astronaut in orbit is in free fall and experiences no such reaction force. [See Problem 95 to find out how weightlessness is simulated on Earth.]

In the future, for extended stays in space colonies, artificial gravity may be supplied by rotation of the spacecraft (see Fig. 7.18). Rotation at the proper

---

INSIGHT

The Coriolis Force and the Foucault Pendulum

Probably one of the more famous inertial forces is the **Coriolis force**, which is named after the French engineer and mathematician Gaspard Coriolis (1792–1843), who first described it. Basically, what Coriolis showed was that Newton's laws of motion may be used in a rotating frame of reference *if* inertial forces are included in the kinematic equations.

For objects on the surface of the rotating Earth, these inertial forces are primarily the centrifugal force and the Coriolis force. The centrifugal force gives a slight correction to the acceleration due to gravity, and the Coriolis force acts or has a component horizontal to the Earth's surface. That is, objects moving along the Earth's surface experience an inertial deflection.

The mathematics of a rotating system are quite involved, but the example illustrated in Fig. 1 gives a general idea of the effect of the Coriolis force. Suppose that a projectile with a very long range is fired from the North Pole southward along a meridian. While the projectile is in flight, the Earth rotates beneath it, and the projectile lands to the west of the meridian, as seen from an inertial point of view. However, to the person firing the projectile, the projectile appears to be deflected and to curve to the right, with

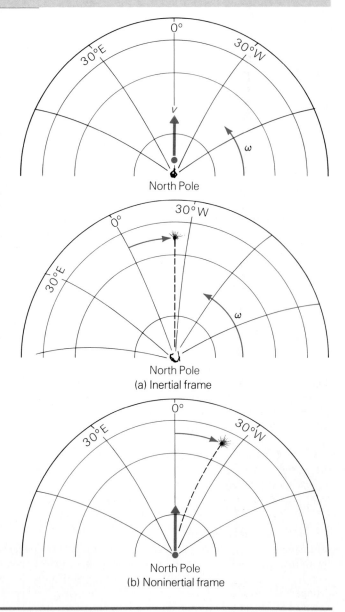

Figure 1 **The Coriolis force**
(a) If a projectile were shot southward from the North Pole along a meridian (the prime, or Greenwich, meridian here), an inertial observer would see it travel in a straight line while the Earth rotates beneath it. The projectile would land west of the 0° meridian. (b) To a noninertial observer on the Earth's surface, the projectile appears to be deflected, and it curves to the right. This requires a force according to Newton's laws, so the Coriolis force was invented.

speed would produce a simulation of gravity within the spacecraft. Dropped objects would fall down from the inhabitants' point of view, as illustrated in the figure, but the downward direction would be a continually changing one from an external perspective.

Another inertial force that is related to the rotating, noninertial Earth system is discussed in the Insight feature, which also describes a pendulum that proves experimentally that the Earth does rotate.

---

respect to the direction of its initial motion. Being a Newtonian thinker, the person in the rotating system (who does not think of it in that way) concludes that there must have been a horizontal force acting on the projectile to cause it to deflect. He calls this force invented to account for his observation the Coriolis force. In general, the Coriolis force deflects a moving object to the right as viewed in its direction of motion in the northern hemisphere and to the left in the southern hemisphere.*

The effect of the Coriolis force is very small and goes unnoticed for most common motions. But for long-range projectiles, such as rockets or shells from naval guns, the Coriolis effect is of real importance. During World War I, there was a naval battle between the British and the Germans near the Falkland Islands (at a latitude of about 50° S). The shots from the British long-range guns although well aimed, mysteriously landed as much as 100 m to the left of the targets. The gun sights had been accurately set, but for Great Britian's latitude (about 50° N). The opposite deflection due to the Coriolis force in the southern hemisphere had not been taken into account.

The Earth's rotation is now taken for granted, but it was not experimentally verified until 1851 by the French physicist Leon Foucault (1819–1868). He used a heavy iron ball suspended from a wire that was more than 60 m long. A simple pendulum arrangement used to demonstrate the Earth's rotation is called a Foucault pendulum (for obvious reasons) and may be seen in many science museums and exhibitions (see Fig. 2).

As the pendulum swings back and forth, the plane of this motion precesses, or turns (to the right or clockwise as viewed from above in the northern hemisphere). This is most easily understood for a Foucault pendulum at the North Pole. From an external observer's point of view (an inertial frame), the pendulum's plane of motion remains fixed while the

Figure 2 **A Foucault pendulum**
The pendulum in the United Nations Building.

Earth turns counterclockwise beneath it. However, to an observer on Earth, the rotation is not apparent and the plane precesses (because of a horizontal Coriolis force, no doubt).

The angular speed of the precession is ω_p, given by

$$\omega_p = \omega_e \sin\theta \tag{7.27}$$

where ω_e is the angular speed of the Earth's rotation and θ is the angle of latitude. At the North Pole ($\theta = 90°$), the pendulum precesses through a complete revolution in a period of 24 hours. The rate of precession decreases as the latitude decreases, and is zero at the Equator ($\theta = 0°$), which means there is no precession there.

---

* For motion along the Equator, the Coriolis force is vertical and, depending on the direction of the motion (east or west), contributes positively or negatively to the centrifugal force.

Important Formulas

Transformation equations for Cartesian and polar coordinates:

$$x = r \cos \theta$$

$$y = r \sin \theta$$

Arc length (angle in radians):

$$s = r\theta$$

Frequency and period:

$$f = \frac{1}{T}$$

Angular kinematic equations (see Table 7.2 for linear analogues):

$$\theta = \omega t$$

$$\bar{\omega} = \frac{\omega + \omega_o}{2}$$

$$\omega = \omega_o + \alpha t$$

$$\theta = \omega_o t + \tfrac{1}{2}\alpha t^2$$

$$\omega^2 = \omega_o^2 + 2\alpha\theta$$

Angular speed (with uniform circular motion):

$$\omega = \frac{2\pi}{T} = 2\pi f$$

Tangential and angular speeds:

$$v = r\omega$$

Tangential and angular accelerations:

$$a_t = r\alpha$$

Centripetal acceleration:

$$a_c = \frac{v^2}{r} = r\omega^2$$

Newton's law of gravitation:

$$F = \frac{Gm_1 m_2}{r^2}$$

$$G = 6.67 \times 10^{-11} \text{ N-m}^2/\text{kg}^2$$

Gravitational intensity:

$$g(r) = \frac{GM}{r^2}$$

Acceleration due to gravity at the Earth's surface:

$$g = \frac{GM_e}{R_e^2}$$

Acceleration due to gravity at an altitude h:

$$g' = \frac{GM_e}{(R_e + h)^2}$$

Gravitational potential energy:

$$U = -\frac{Gm_1 m_2}{r}$$

Total mechanical energy of moving mass, m_1 (where m_2 is stationary):

$$E = K + U = \tfrac{1}{2}m_1 v^2 - \frac{Gm_1 m_2}{r}$$

Kepler's law of periods:

$$T^2 = Kr^3$$

$$K = 2.97 \times 10^{-19} \text{ s}^2/\text{m}^3$$

Escape speed for Earth:

$$v_e = \sqrt{\frac{2GM_e}{R_e}} = \sqrt{2gR_e}$$

Total energy for satellite orbiting Earth:

$$E = -\frac{GmM_e}{2r}$$

Kinetic energy of Earth satellite:

$$K = \frac{GmM_e}{2r} = |E|$$

◆ *Angular speed of precession of Foucault pendulum (θ is angle of latitude):*

$$\omega_p = \omega_e \sin \theta$$

Questions

Circular Motion

1. What does it mean to say that a measurement in radians is a pure number? Explain why $s = r\theta$ only if θ is given in radians.

2. A phonograph turntable rotates clockwise as viewed from above. What is the direction of the angular velocity vector?

3. When clockwise (or counterclockwise) is used to describe rotational motion, why is a phrase such as "viewed from above" added?

4. Why are revolution and cycle not truly units? (Hint: consider circular motions with different radii.)

5. What is a hertz in standard SI units?

6. A rotating phonograph turntable is turned off.

What happens to the angular velocity and angular acceleration vectors during the time it takes the turntable to come to a stop?

7. Is it possible for a particle in circular motion to have a constant tangential acceleration and a constant centripetal acceleration at the same time? Explain.

Newton's Law of Gravitation

8. The gravitational force of attraction exerted by the Earth on the moon is quite large (about 10^{20} N—see Example 7.7), as is the equal and opposite force acting on the Earth. Why isn't the Earth attracted into the moon?

9. The backpacks used by astronauts on the moon weigh about 300 lb on Earth. On the moon, the astronauts walk around carrying the packs. Explain why they are so strong.

10. Is it possible to escape the Earth's gravitational pull, as spacecraft sent to the moon and other planets are sometimes said to do?

11. The ocean tides are caused primarily by the gravitational attraction of the moon and to a lesser extent by that of the Sun. Tides with a maximum difference between high and low, called spring tides (not associated with the season), occur when there is a particular alignment of the Earth, moon, and Sun. The tides with the minimum variation, called neap tides, occur with another alignment. What are these two alignments?

12. Ignoring the Sun's effect in producing ocean tides, explain how moon's gravitational attraction produces two tidal "bulges," resulting in two high tides each day. (Hint: consider the inverse square relation of the distance and the gravitational force acting on the water on the side of the Earth near the moon, on the Earth, and on the water on the far side. You may want to go to the library for a little help on this one.)

13. A gravitational field is a vector field. There are also scalar fields. Can you give an example?

14. Explain why the gravitational potential energy, $U = -Gm_1m_2/r$, is negative. Wouldn't it be more convenient to arbitrarily designate it as positive?

15. The negative value of the total energy of an object in a potential energy well is sometimes called its binding energy. Explain why this is a good descriptive term.

Planets and Satellites

16. When space probes are sent to the outer planets, they may be subjected to a so-called sling-shot effect caused by passing close to Jupiter. Describe how this would affect a probe's motion. What would its speed be at the distance of closest approach relative to the escape speed for Jupiter?

17. For the situation in Example 7.10, what would happen if the retrorockets were turned off before the spacecraft reached the Earth's atmosphere?

18. The discovery of Neptune was prompted by observed perturbations (irregularities) in the orbit of Uranus. Similarly, the search for Pluto was prompted by perturbations in the orbit of Neptune. Explain why such perturbations would suggest the existence of an unknown planet.

19. For a planet moving in an elliptical orbit about the Sun, relate work and mutual gravitational potential energy.

20. Are the orbital speeds the same for a large satellite and a small satellite maintaining the same constant altitude above the Earth? Explain. Does a satellite's orbital speed depend on the body being orbited?

21. A shuttlecraft transporting a replacement crew to a spacestation is in a circular orbit at an altitude lower than that of the spacestation. (a) Explain how the shuttlecraft can change its orbit in order to dock with the spacestation. What are the effects of this change of orbit on the shuttlecraft's energy and speed? (b) If the shuttlecraft moves to the altitude of the spacestation and establishes the same circular orbit, moving behind the spacestation, will it be able to overtake the spacestation and dock with it? Explain.

22. Must the Earth's rotation be taken into consideration when determining the direction in which a rocket is launched to put a satellite into orbit? Explain.

Frames of Reference and Inertial Forces ◆

23. As shown in Demonstration 7, when a Styrofoam cup with a hole from which water is escaping is dropped, no water comes out while the cup is falling. Explain this in terms of your and the cup's frames of reference. (Try dropping a cup yourself and observe this effect.)

24. Would you expect the Coriolis force to have an effect on the accuracy of long-range artillery or missiles? Explain. If corrections were made for this, would they differ north and south of the Equator?

25. Explain what happens when laundry is spinned "dry" in a washing machine from inertial and noninertial points of view.

26. (a) Cream is separated from milk by centrifuging. Explain in principle how this is done. (b) Would "centripuge" be a more correct name for a centrifuge?

27. What is the period of precession for a Foucault pendulum at the Equator? Explain your answer in terms of the speed of precession.

28. The apparatus shown in Fig. 7.19 on the next page is used to demonstrate forces in a rotating (accelerating) system. The floats are in jars of water. When the arm is rotated, which way will the floats move? Does it make a difference which way the arm is rotated?

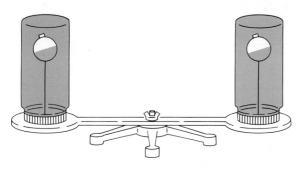

Figure 7.19 **A rotating system (when set into motion)**
See Question 28.

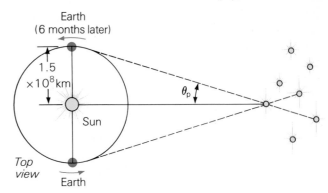

Figure 7.20 **Parallax and parsec**
See Problem 12.

Problems

7.1 Angular Measure

■**1.** Convert the following angles from degrees to radians: (a) 15°, (b) 270°, (c) 50°, and (d) 540°.

■**2.** Convert the following angles from radians to degrees: (a) $\pi/24$ rad, (b) $\pi/10$ rad, (c) 1.45 rad, and (d) 4π rad.

■**3.** What is the arc length subtended by an angle of $\pi/4$ rad on a circle with a radius of 6.0 cm?

■**4.** If an arc length is twice the radius of a circle, what angle subtends the arc?

■■**5.** A jogger on a circular track that has a radius of 0.25 km jogs a distance of 1.0 km. What angular distance does the jogger cover in (a) radians and (b) degrees?

■■**6.** A sector of a circle has two sides of length 0.50 m with an angle of 60° between them. What is the length of the third side?

■■**7.** The Cartesian coordinates of a point on a circle are ($\sqrt{2}$ m, $\sqrt{2}$ m). What are the polar coordinates (r, θ) of this point?

■■**8.** Assuming that the Earth's orbit around the Sun is circular, what is the approximate orbital distance the Earth travels in 3 months?

■■**9.** Two race cars start from rest and travel around a circular track whose diameter is 1.0 km. (a) One car develops engine trouble and comes to a stop after having traveled 220° around the track. What distance did this car travel? (b) The other car makes four complete laps. How far did this car travel?

■■**10.** The equation for a circle using Cartesian coordinates is $x^2 + y^2 = a^2$, where a is the radius of the circle.

What is the equation of a circle using polar coordinates?

■■■**11.** The Cartesian coordinates of a point on a circle are (0.40 m, 0.30 m). What is the arc length measured counterclockwise on the circle from the positive x axis to the point?

■■■**12.** An astronomical unit used to designate the distances of stars is the parsec. It is defined as the distance from the Sun to a hypothetical star for which the *para*llax angle (θ_p) is 1 *sec*ond, that is, the distance between the Earth and the Sun subtends 1 second of arc (Fig. 7.20). What is the equivalent length of 1 parsec in (a) kilometers and (b) light years? (Parallax is the apparent change in position of nearby stars relative to distant ones because of the Earth's orbital motion. Note that the parallax angle is smaller the more distant the star. The distance in parsecs is given by $1/\theta_p$, where θ_p is in seconds.)

■■■**13.** In 250 B.C., the Greek scientist Eratosthenes determined the circumference and from it the radius of the Earth. He knew that the Sun was directly overhead at noon at Syene (now Aswan), Egypt, on the first day of summer because deep wells there were lighted all the way to the bottom. On that day at Alexandria, 800 km due north of Syene, Eratosthenes measured the Sun's position to be $7\frac{1}{2}°$ south of the zenith (the point directly overhead), using a shadow cast by a vertical stick. (See Fig. 7.21.) Assuming the Earth to be a sphere and the Sun to be far enough away that the light rays reaching the Earth are parallel to each other, repeat Eratosthenes' calculation, and compare the result with the value for the Earth's radius given in Table 6 of Appendix I.

7.2 Angular Speed and Velocity

■**14.** A race car makes two laps around a circular track in 3.5 min. What is the car's average angular speed?

■**15.** What will be the angular speed of an object

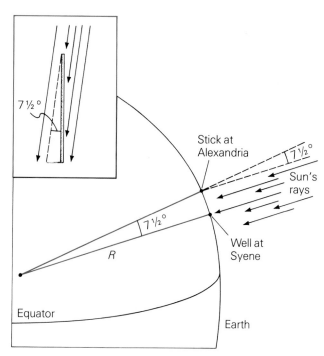

Figure 7.21 **A 2000-year-old calculation**
See Problem 13.

traveling with uniform motion at a tangential speed of 4.0 m/s in a circle with a diameter of 50 m?

■16. A satellite in a circular orbit has a period of 10 days. What is its frequency in revolutions per day?

■■17. A particle travels through 270° of a circular orbit in 4.0 s. What is its average angular speed?

■■18. Determine which has the greater angular speed: particle A, which travels 160° in 2.0 s, or particle B, which travels 3π rad in 7.0 s.

■■19. A runner running at a constant pace gets half-way around a circular track that has a diameter of 500 m in 2.0 min. What are the runner's (a) angular speed and (b) tangential speed?

■■20. A centrifuge used to separate blood samples rotates at a speed of 12,000 rpm (see Example 7.11). (a) What is the instantaneous tangential speed of a blood cell that is 8.00 cm from the rotational axis of the centrifuge? (b) How long does it take the centrifuge to make one revolution?

■■21. The driver of a motorboat sets the throttle and ties the wheel, making the boat travel at a uniform speed of 15.0 m/s in a circle with a diameter of 120 m. (a) Through what angular distance does the boat move in 4.00 min? (b) What arc distance does it travel in this time?

■■22. For the second hand and the hour hand of a (running) clock, find (a) the period, (b) the frequency, (c) the angular speed, and (d) the angular velocity.

■■■23. Assuming that the Earth revolves about the Sun in uniform circular motion, find (a) the frequency of revolution in hertz and (b) the angular speed of revolution. (Use data given in Appendix I.)

■■■24. What are (a) the frequency of the Earth's rotation and (b) its angular *velocity*? (c) Express the tangential and angular speeds of a person at a latitude of 45° N (or S) as percentages of those of a person at the Equator.

7.3 Angular Acceleration

■■25. A phonograph turntable set for 45 rpm is turned on and reaches its operating speed in 1.50 s. What is the average angular acceleration of the turntable? (Hint: turntables rotate clockwise as viewed from above.)

■■26. An automobile traveling at 60 km/h on a circular track with a diameter of 1.0 km speeds up to 90 km/h in 10 s. What is the magnitude of the average angular acceleration?

■■27. A particle initially at rest accelerates in a circular path at a rate of 2.0 rad/s² for 6.5 s. The circle has a radius of 15 cm. What are (a) the magnitude of the tangential acceleration during this time and (b) the tangential speed at the end of this time?

■■28. A merry-go-round accelerating uniformly from rest achieves its operating speed of 2.0 rpm in three revolutions. What is the magnitude of its angular acceleration?

■■29. Show that the magnitudes of the tangential and angular accelerations for uniform circular motion are related by $a_t = r\alpha$.

■■30. A 33⅓-rpm record on a turntable uniformly reaches its operating speed in 2.4 s once the record player is turned on. (a) What is the angular distance traveled during this time? (b) What is the corresponding arc length on the circumference of a 12-in. record?

■■31. The blades of a fan running at low speed turn at 250 rpm. When the fan is switched to high speed, the rotation rate increases uniformly to 350 rpm in 5.75 s. (a) What is the magnitude of the angular acceleration of the blades? (b) How many revolutions do the blades go through while the fan is accelerating?

■■32. A car moves into a curve with a radius of curvature of 250 m at a constant speed of 65 km/h. While still in the curve, the car accelerates at a rate of 0.48 rad/s² for

7.0 s. (a) What is the car's angular speed at the end of this time? (b) What is the magnitude of its tangential acceleration during this time?

■■**33.** A particle travels in a circular path with a radius of 25 cm at a uniform angular speed of 0.60 rad/s. What is the centripetal acceleration of the particle?

■■**34.** A 200-kg satellite in a circular orbit about the Earth at an altitude of 500 km makes one revolution in 15 min. (a) What is the centripetal force acting on the satellite? (b) What supplies this force?

■■**35.** Find the magnitude and direction (relative to the tangential acceleration) of the total acceleration of the object in Example 7.5.

■■■**36.** A phonograph record initially moving at $33\frac{1}{3}$ rpm accelerates uniformly to 45 rpm in 2.5 s. What is the angular distance traveled by the record during this time?

■■■**37.** A car on a circular track with a radius of 0.30 km accelerates from rest with a constant angular acceleration whose magnitude is 4.5×10^{-3} rad/s². (a) How long does it take the car to make one lap around the track? (b) What is the total acceleration of the car when it has completed half of a lap?

■■■**38.** An object initially at rest undergoes an angular acceleration of 3.4 rad/s² and travels in a circular orbit that has a radius of 0.15 m. (a) What is the centripetal acceleration of the object when it has been moving for 3.0 s? (b) What is the instantaneous total acceleration at this time?

■■■**39.** A student investigating circular motion places a dime on a $33\frac{1}{3}$-rpm record on a turntable at a distance of 10 cm from the center. Switching on the turntable, the student notes that the dime slides outward when the record is turning at 90% of its operating speed. (a) Why does the dime slide outward? (b) What is the coefficient of friction between the dime and the record?

■■■**40.** Suppose that the object described in Example 7.5 is just crossing the positive x axis, traveling counterclockwise, when the angular acceleration is initially experienced. Give the exact direction of the tangential acceleration vector when the acceleration ends.

■■■**41.** A 60-kg boy prepares to swing out over a pond on a rope, as shown in Fig. 7.22. What tension in the rope is required if he is to maintain a circular arc (a) when he starts his swing by stepping off the platform, (b) when he is 5.0 m above the pond surface, and (c) at the lowest point of the swinging motion? (Hint: treat the boy and the rope as a simple pendulum.)

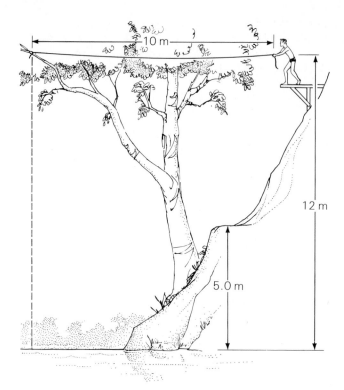

Figure 7.22 **A swinging time**
See Problems 41 and 42.

■■■**42.** Suppose that the boy swinging on the rope in Fig. 7.22 (see Problem 41) starts his swing at a lower point and has a tangential speed of 8.5 m/s when the rope is at an angle of 30° to the vertical. (a) What is his tangential acceleration at that time? (b) What is the vector sum of the tangential and centripetal accelerations at that time? (c) From what height above the pond surface did he start his swing?

7.4 Newton's Law of Gravitation

■**43.** Two homogeneous spheres of equal mass, 3.0 kg, are separated by a center-to-center distance of 1.2 m. What is the gravitational force experienced by each sphere?

■**44.** Using Eq. 7.15, compute the value of g at the surface of the Earth.

■**45.** Compute the mass of the Sun using Eq. 7.21.

■**46.** What is the gravitational intensity at a point 4.5 m from the center of a homogeneous sphere whose mass is 100 kg?

■■**47.** (a) Two objects, each with a mass of 1 metric ton, are in contact with each other. How far apart are the centers of gravitational force, if there is a mutual gravitational attraction of 0.00010 N between them? (b) If the objects are homogeneous spheres of the same size, what is their minimum uniform density? (Hint: remember that they are in contact.)

■■48. Two point masses, $m_1 = 0.50$ kg and $m_2 = 0.75$ kg, are located at the positions $(0,0)$ and $(0.60\,m, 0)$, respectively. (a) What is the gravitational force on a third point mass, $m_3 = 0.25$ kg, located midway between the first two? (b) What is the gravitational intensity midway between m_1 and m_2? (c) At what point on the x axis would the net force on m_3 be zero? Are there any other zero points?

■■49. Four identical masses of 2.0 kg each are located at the corners of a square with 1.0-m sides. (a) What is the net force on one of the masses? (b) What is the gravitational intensity at the center of the square?

■■50. For a spacecraft going directly from the Earth to the moon, after what point will lunar gravity begin to dominate; that is, where will the lunar gravitational force be greater than the Earth's gravitational force? Are the astronauts on board truly weightless at this point? Explain.

■■51. Find an approximate value of the acceleration due to gravity on the surface of Mars. (See Table 6 in Appendix I for data.)

■■52. What is the acceleration due to gravity on the top of Mt. Everest? (The summit is about 8.8 km above sea level.)

■■53. An astronaut has a mass of 80 kg on Earth. What would be the astronaut's weight be in a spacecraft in a circular orbit at an altitude of 400 km?

■■54. How far above the Earth's surface would a 75-kg person have to go to "reduce" his weight by 10%?

■■55. A plane catapulted from an aircraft carrier goes from rest to a speed of 54 m/s in a distance of 50 m. How many g's of force does the pilot experience because of the catapult action?

■■56. An elevator traveling downward with a constant speed of 2.54 m/s comes to an emergency stop in 1.25 m. How many g's of force are experienced by the passengers in the elevator because of the deceleration?

■■57. (a) What is the mutual gravitational potential of the configuration shown in Fig. 7.23 if all the masses are 1.0 kg? (b) What is the gravitational intensity at the center of the configuration?

■■■58. Using Kepler's law of periods, show that the centripetal force acting on the moon as it moves in its nearly circular orbit is an inverse square force. (Hint: use the general form of the force and recall that $v = 2\pi r/T$.)

■■■59. Compare the centripetal acceleration of the moon with its acceleration due to the Earth's gravity (at the distance of the moon). (Hint: compute a_c from lunar orbital data, and compute g' using a ratio with g at the Earth's surface.)

■■■60. (a) Find the average density of the Earth. (b) The average density of the Earth's outer crust is about 2.5 times that of water. What does the answer from part (a) tell you about the Earth's interior?

■■■61. Show that the value of the acceleration due to gravity on the moon's surface is about one-sixth of that on the Earth's surface. (Hint: use a ratio.)

■■■62. For the configuration of 1.0-kg masses in Fig. 7.23, suppose that another 1.0-kg mass (m_4) is placed on the y axis the same distance below the origin as m_2 is above. (a) Find the net gravitational force on m_4. (b) Show that the result from part (a) is *not* the same as the gravitational force on m_4 calculated by considering all of the mass of the initial triangular configuration to be located at its center of mass. (Hint: recall that a homogeneous sphere is

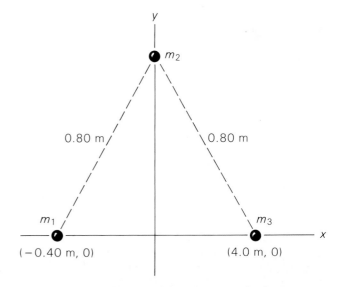

Figure 7.23 **Gravitational potential, gravitational force, and center of mass**
See Problems 57 and 62.

a special case; in general, the center of mass of a body is not the same as the center of gravitational force.)

7.5 Kepler's Laws and Earth Satellites

■**63.** Compute the constant K of Kepler's third law for (a) Earth and (b) Venus. (See Table 6 in Appendix I for data.)

■■**64.** The value of the Kepler constant is $K = 2.97 \times 10^{-19}$ s^2/m^3, and $T^2 = (2.97 \times 10^{-19}$ s^2/m$^3)r^3$. With nonstandard units, K can be made equal to 1, and the equation may be written $T^2 = r^3$. Find the units of K that make it equal to 1, using Earth data. [Hint: the average, or mean, distance of the Earth from the Sun is used as a unit in astronomy and is called an astronomical unit (AU).]

■■**65.** The mean (average) distance from the Sun to Neptune is about 30 times the distance from the Sun to Earth. What is Neptune's period of revolution? (Hint: you might want to use astronomical units for distance. See Problem 64.)

■■**66.** The period of the highly eccentric orbit of Halley's comet is about 76 years. (It was near the Sun and visible from Earth in late 1986 and early 1987.) (a) What is the comet's average distance from the Sun? (b) What is the outermost planet whose orbit would be crossed at this distance? (Halley's comet in fact goes beyond Neptune's orbit. Does this agree with your result? If not, why not?)

■■**67.** An instrument package is projected vertically upward to collect data at the top of the Earth's atmosphere (at an altitude of about 800 km). (a) What initial speed is required at the Earth's surface for the package to reach this height? (b) What percentage of the escape speed is this?

■■**68.** A 50-kg instrument package is put into a circular orbit about the Earth at an altitude of 800 km. What is the orbital kinetic energy of the satellite?

■■**69.** Two satellites are in circular orbits around the Earth at altitudes of 450 km and 600 km. Which satellite has the (a) greater kinetic energy and (b) greater potential energy? (c) How many times greater is each of these values than that for the other satellite?

■■■**70.** In 1610, Galileo discovered four of the sixteen moons of Jupiter, the largest of which is Ganymede. This Jovian moon revolves around the planet in a nearly circular orbit whose diameter is about 1.07×10^6 km in 7.16 days. Using this data, find the mass of Jupiter.

■■■**71.** Syncom satellites are communications satellites whose orbits are synchronized with the Earth's rotation. That is, they remain relatively stationary above some

location on the Earth's surface because the period of satellite revolution equals the period of Earth's rotation. What is the altitude of syncom satellites?

■■■**72.** Show that the escape speed for Earth is about 41% greater than the orbital speed of a satellite in a circular orbit near the surface of the Earth ($h \cong 0$), that is, that $v_e = 1.41v$.

■■■**73.** What is the escape speed for a satellite in a circular orbit about the Earth at an altitude of 650 km?

7.6 Frames of Reference and Inertial Forces ◆

■■**74.** Analyze the situation of a person standing on a scale in an elevator (as in Fig. 7.15) when the elevator is accelerating downward.

■■**75.** For a person on the rotating Earth, the outward centrifugal effect produces a small reduction in the force or acceleration due to gravity. What is the effective acceleration due to gravity at the Equator? (Hint: the inertial centrifugal force is equal in magnitude to the centripetal force relative to an inertial frame of reference.)

■■**76.** A pilot flies a jet plane in a vertical circular loop that has a diameter of 1.60 km. (a) At what speed would the plane have to go around the loop to make the pilot feel completely weightless at the top of the loop? (b) How many g's of force would the pilot experience at the bottom of the loop if the plane were flying at that speed? (c) Suppose that on the second time around the loop the pilot increases the speed. What would be experienced at the top and the bottom of the loop in this case?

■■**77.** What is the rate of precession of a Foucault pendulum (a) at a latitude of 45° N (or S) and (b) at your latitude?

■■**78.** Helena, Montana (46.5° N, 112° W), lies almost directly north of Phoenix, Arizona (33.5° N, 112° W), with a separation distance of about 1400 km (875 mi). If a projectile were shot northward from Phoenix with a speed of 1000 km/h, would it pass exactly through Helena? Justify your answer mathematically. (Hint: different latitudes have different tangential speeds. Latitude is the angle between the equatorial plane and a line from the center of the Earth to a location on the surface.)

■■**79.** At what latitude would it take a Foucault pendulum two days to go around in a circle?

■■**80.** Suppose that a wheellike space colony such as the one shown in Fig. 7.18 has a diameter of 1.6 km (about 1 mi). (a) With what angular speed (in rpm) would the colony have to rotate to give the effect of gravity on Earth?

(b) Would objects dropped anywhere on the inside circumference (the colony surface) have the same acceleration? Explain.

Additional Problems

81. Eventually, manned spacecraft may be sent to Mars. How would the weight of an astronaut on the surface of the red planet compare to his or her weight on Earth? (Hint: use a ratio.)

82. Compare the escape speeds for (a) Earth and Mercury and (b) Earth and Jupiter. (Hint: use a ratio.)

83. An asteroid (a minor planet) travels in a nearly circular orbit between Mars and Jupiter. The orbit has a radius of 4.1×10^6 km, and the period of the asteroid is 4.6 years. (a) What is the angular speed of the asteroid? (b) What is the orbital distance it travels in 1.0 year?

84. A particle accelerates from rest along a circular path with a radius of 0.80 m at a rate of 0.26 rad/s^2. (a) What is the centripetal acceleration of the particle after 5.0 s? (b) How many revolutions does the particle make during this time?

85. (a) What is the angular width of a full moon in degrees and radians as viewed from Earth? (b) What is the angular width of a full Earth in degrees and radians as viewed from the moon? (Hint: use data from Appendix I.)

86. Is the centripetal acceleration required to keep a person at the Equator rotating with the Earth the same as g, the acceleration due to gravity? Explain. (Justify your answer with actual values.)

87. At sunset, the Sun has an angular width of about 0.50°. From the time the lower edge of the Sun just touches the horizon, about how long does it take it to disappear?

88. A 500-kg spacecraft is in a circular orbit about the Earth at an altitude of 250 km. What are its (a) total mechanical energy, (b) kinetic energy, and (c) potential energy?

89. For small angles, a good approximation is to take the arc length subtended by the angle as being equal to the chord length, as was done in Example 7.2. The formula for the length of a chord subtended by an angle θ is $L = 2r \sin \theta / 2$, where r is the radius. For what angle will the error of this approximation be on the order of 1%, that is, will L be 1% less than s? (Hint: $\sin x \cong x - x^3/6$.)

90. Shortly after blast-off, the upward acceleration of a Saturn V rocket is about 80 m/s^2. (a) How many g's of force are experienced by an astronaut on board as a result of this acceleration? (b) What is a 75-kg astronaut's apparent weight because of this acceleration?

91. When a motorized potter's wheel is turned on, it accelerates uniformly at a rate of 2.6 rad/s^2 for 5.0 s to achieve its operating speed. What is that speed in rpm?

92. Many astronomers believe that there may be a tenth planet in our solar system. The existence of Planet X was first theorized because Pluto's orbit showed variations from its computed value. Suppose that the tenth planet were observed with newly developed, more powerful telescopes, and its orbital change of position showed it to have a period of revolution of 254 years. How far would Planet X be on the average from the Sun?

93. The centripetal acceleration of an object is 24 cm/s^2 as it travels in a circular orbit with a radius of 6.0 cm. (a) What is the angular speed of the object? (b) What is the angular distance in degrees traveled by the object in 1.5 s?

94. The same side of the moon always faces the Earth, because the period of the moon's revolution around the Earth is the same as that of the Earth's rotation. Prove that these periods being equal means that the same hemisphere of the moon is always visible from Earth.

95. A procedure used to study apparent weightlessness and to allow astronauts to experience this condition is done with an airplane (see Fig. 7.24). A weightless feeling is produced for a brief interval of 30 to 40 s by putting the plane into a flight path called a ballistic trajectory (a parabolic arc). The pilot first noses down to increase speed, then turns the nose upward and flies in the arc. During the short interval of flying through the arc, no lift is produced on the wings, and there is no reaction force on the passengers, so they float around the cabin. (Essentially, this is like traveling inside a projectile with the same velocity as the projectile. You may have experienced a similar feeling when in a car that goes over the crest of hill at a high speed or when riding on a roller coaster.) Assume that the arc is circular and subtended by an angle of 120°. If the desired conditions could be achieved for 30 s, what would the radius of the arc be?

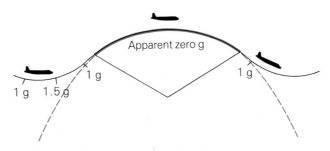

Figure 7.24 **Apparent weightlessness**
See Problem 95.

Rotational Motion

8

Rotational motion is very important in physics because rotating objects are all around us: wheels in vehicles, gears and pulleys in machinery, planets in our solar system, and even some bones in the human body. (Can you think of a few bones that rotate in sockets?)

In Chapter 7, the circular motions of particles in rotating bodies were considered. But the description of the rotational motion of a body as a whole must take into account the motions of all its particles. As you may have noticed in previous chapters, the study of physics proceeds in a stepwise manner, going from simple, ideal descriptions to more difficult, more realistic ones, which better describe actual physical events. This chapter takes another step into the real world of things, going beyond the circular motions of particles to the rotations of configurations of particles, or extended bodies.

Fortunately, the dynamic equations describing rotational motion can be written as almost direct analogues of those for translational (linear) motion. In Chapter 7, this similarity was pointed out for the kinematic equations. With the addition of equations describing rotational dynamics, you will be able to analyze the general motions of real, extended objects.

8.1 Rigid Bodies, Translations, and Rotations

It was convenient initially to consider motions of objects or particles with the understanding that the motion of an object may be represented by a particle located at its center of mass. Rotation, or spinning, was not a consideration then because a point mass has no physical dimensions. Rotational motion, on the other hand, is generally applied to solid extended objects or rigid bodies, which this chapter will focus on.

> A **rigid body** is an object or system of particles in which the interparticle distances (distances between particles) are fixed and remain constant.

A quantity of liquid water is obviously not a rigid body, but the ice that would be formed if the water were frozen would be. The discussion of rigid body rotation

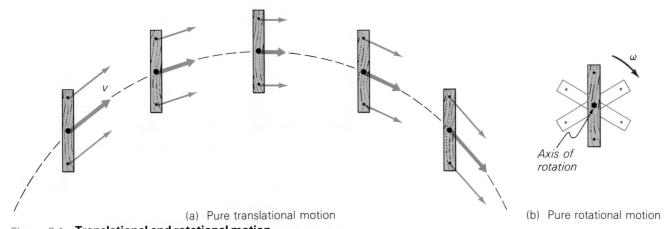

(a) Pure translational motion   (b) Pure rotational motion

Figure 8.1 **Translational and rotational motion**
(a) For a body in translational motion only, every particle has the same instantaneous velocity. (b) For a body in rotational motion only, all the particles have the same instantaneous angular velocity about an axis of rotation.

is thus conveniently restricted to solids. Actually, the concept of a rigid body is an idealization. In reality, the particles (atoms and molecules) of a solid vibrate constantly. Also, solids can undergo elastic (and inelastic) deformations (Chapter 6). Even so, most solids can be considered to be rigid bodies for purposes of analyzing rotational motion.

A rigid body may be subject to either or both of two different types of motion. If an object has only **translational motion** [Fig. 8.1(a)], every particle of the object has the same instantaneous velocity, which means that there is no rotation. (Why?) Any particle in the object can be used to describe the general motion of the object as a whole. Usually, the center of mass is chosen to be that particle because Newton's second law can be applied to this point ($F = Ma_{cm}$; see Chapter 6).

An object may also have only **rotational motion** about a fixed axis [Fig. 8.1(b)]. In this case, all the particles of the object have the same instantaneous angular velocity and travel in circles about the axis of rotation. Although the axis of rotation is commonly taken to be through the center of mass (think of the axle of a wheel), this is not always the case. For example, you could make an axis of rotation through one end of a meterstick by pivoting it there. In analyzing rotational motion about this axis using a particle representation, you might consider revolutions of the center of mass of the meterstick. But for extended bodies the mass distribution relative to the axis of rotation affects the motion, as you will learn shortly.

General rigid body motion is a combination of translational and rotational motions (Fig. 8.2). When you throw a ball, the translational motion is described by the motion of the center of mass. The ball may also spin, or rotate, usually about an axis through its center of mass.

Another common example of rigid body motion that combines translational and rotational motions is rolling, illustrated in Fig. 8.3. The axis of rotation is through the point or line of contact with the surface and is called the **instantaneous axis of rotation**. The successive points of contact of the body with the surface are instantaneously at rest (zero velocity), as you can see from the vector addition of the combined motions.

The words "rotation" and "revolution" are commonly used synonymously. In general, this book uses rotation when the axis of rotation goes through the body (the Earth's rotation on its axis) and revolution when the axis is outside the body (the revolution of the Earth about the Sun).

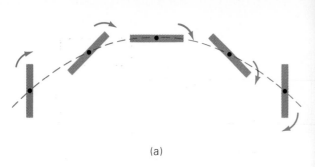

(a)

(b)

Figure 8.2 **General rigid body motion—a combination of translational and rotational motions**

(a) After being projected, a stick translates as it rotates. Note that the center of mass follows a parabolic arc as though it were a particle.
(b) The center of mass of the rotating diver also follows a parabolic arc.

Translational   +   Rotational   =   Rolling

v

$v = r\omega$

$2v$

v

ω

+

r

v

=

ω

v

$v = 0$

Point of contact

v

$v = r\omega$

Instantaneous axis of rotation

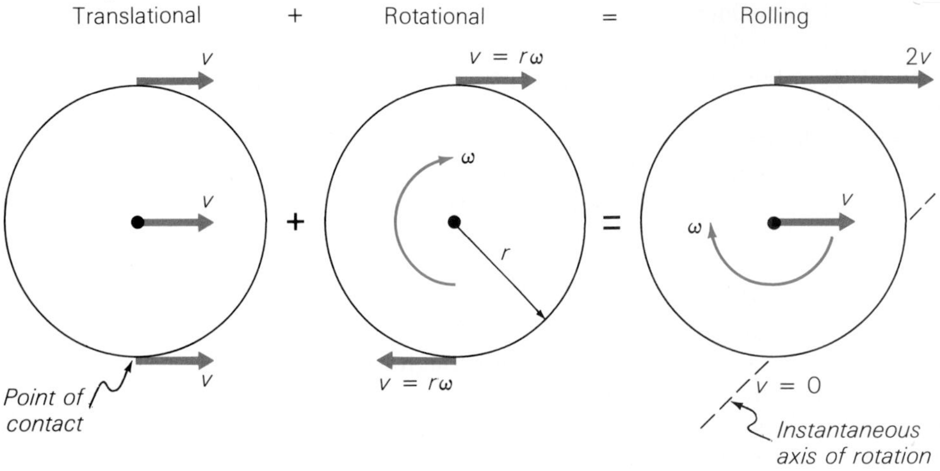

Figure 8.3 **Rolling—a combination of translational and rotational motions**

(a) As shown here with vectors, the center of mass of a rolling object translates at a speed v while rotating at an angular speed ω. The point of contact is instantaneously at rest and is on a line called the instantaneous axis of rotation. Note that the center of the object moves linearly and remains over the point of contact.

When an object rolls without slipping, the translational and rotational motions are related simply. For example, when a uniform ball (or cylinder) rolls in a straight line on a flat surface (Fig. 8.4), it turns through an angle θ, and a point on the object which was initially in contact with the surface moves through an arc distance s. From Chapter 7, you know that $s = r\theta$. The center of mass of the ball is directly over the point of contact and moves a linear distance s. Then

$$v = \frac{\Delta s}{\Delta t} = \frac{r\,\Delta\theta}{\Delta t} = r\omega$$

In terms of the speed of the center of mass and the angular speed, the **condition for rolling without slipping** is

$$v = r\omega \qquad (8.1)$$

The condition is also expressed by $s = r\theta$, where s is the distance an object rolls, or the distance the center of mass moves. Carrying Eq. 8.1 one step further gives $a = r\alpha$ for accelerated rolling without slipping.

Example 8.1 Rolling with Slipping and then Without

A good example of rolling with slipping occurs when a bowling ball is thrown with no initial spin, or rotation (Fig. 8.5). As the ball slides down the lane, a force of sliding friction opposes its motion. The frictional force causes the ball to rotate *and* reduces the speed at which the center of mass moves. (Looking at only the horizontal translational motion of the center of mass reveals that there is a force in one direction and a velocity in the other.)

Thus, the angular speed of the ball increases and the linear speed decreases until the condition $v = r\omega$ is met, and then the ball rolls down the alley without slipping (ideally on its way to a strike). Once the ball is rolling without slipping, there is no longer any sliding friction (there are rolling and static frictions, which will be discussed later in this chapter). ∎

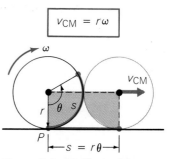

Figure 8.4 Rolling without slipping
As an object rolls without slipping, the length of the arc between two points of contact on the circumference is equal to the linear distance traveled and $s = r\theta$. The speed of the center of mass is $v = r\omega$.

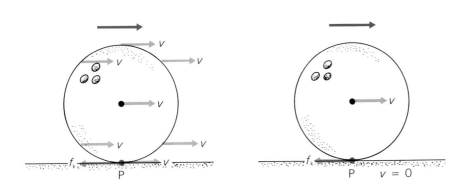

(a) Sliding (slipping)

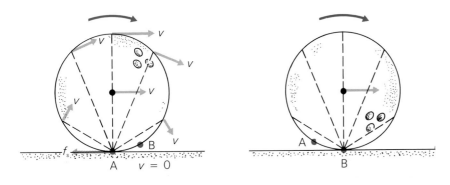

(b) Rolling (without slipping)

Figure 8.5 Rolling with slipping
(a) When a bowling ball is thrown without spin, it slides down the alley. The point of contact is in motion and the motion is opposed by the force of sliding friction. This decreases the linear speed and increases the angular speed until the condition $v = r\omega$ is reached. (b) Then the ball rolls without slipping and $v = 0$ for the point of contact.

8.2 Torque and Moment of Inertia

In Example 8.1, a force gave rise to a rotation. As with translational motion, a force is necessary to produce rotational motion. But rotational motion is not always produced when a force acts on a rigid body. The motion or angular acceleration depends on *where* the force is applied. If a force acts through the axis of rotation, no rotation is produced [Fig. 8.6(a)]. But when the line of action of the force does not go through the axis of rotation, the body rotates [Fig. 8.6(b) and (c)].

As a practical example, think of applying a force to a heavy glass door that swings in and out. Where you apply the force makes a great difference in how easily the door opens, or rotates on its axis (through the hinges). Have you ever tried to open such a door and inadvertently pushed on the side near the hinges?

As shown in Fig. 8.6, the rate of rotation depends not only on the magnitude of the force, but also on the distance of its line of action from the axis of rotation. These factors of force and distance are expressed in terms of their product, which is called **torque** (τ) (from the Latin *torquere*, meaning "to twist"):

$$\tau = r_\perp F = rF \sin \theta \qquad (8.2)$$

where $r_\perp = r \sin \theta$ is the perpendicular distance from the axis of rotation to the line of action of the force. This perpendicular distance is called the **moment arm**, or the **lever arm**. [Torque is sometimes referred to as the moment of force, and, equivalently, $\tau = rF_\perp = r(F \sin \theta)$.] The SI units for torque are meter-newton (m-N). Note that these are the same as the units for work, $W = Fd$ (N-m, or J). However, the units for torque are commonly written backwards so to speak, for distinction.

Torque in rotational motion may be thought of as the analogue of force in translational motion. An unbalanced force changes translational motion, and an unbalanced torque changes rotational motion. As the product of a force and a moment arm (vectors), torque is a vector. Its direction is always perpendicular to the plane of the force and moment arm vectors and is given by a right-hand rule. If the fingers of the right hand are curled around the axis of rotation in the direction of the rotational motion, the extended thumb points in the direction of the torque producing the motion (see Fig. 8.6).

For a constant force, the magnitude of the torque on a particle is

$$\tau = rF_\perp = rma_\perp = mr^2\alpha \qquad (8.3)$$

where $a_\perp = r\alpha$ is the tangential acceleration (a_t in Chapter 7.) For the rotation of a system of fixed particles (a rigid body) about a fixed axis, this equation can be applied to each particle and the results summed over the entire body to find the total torque. Since, as you recall, all the particles of rotating body have the same angular acceleration,

$$\tau = \tau_1 + \tau_2 + \tau_3 + \cdots + \tau_n$$
$$= m_1 r_1^2 \alpha + m_2 r_2^2 \alpha + m_3 r_3^2 \alpha + \cdots + m_n r_n^2 \alpha$$
$$= (m_1 r_1^2 + m_2 r_2^2 + m_3 r_3^2 + \cdots + m_n r_n^2)\alpha$$

or $\quad \tau = \left(\sum_{i=1}^{n} m_i r_i^2 \right) \alpha \qquad (8.4)$

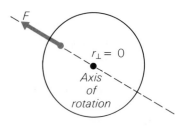

(a) Zero torque

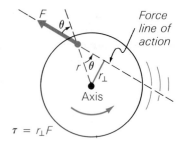

(b) Counterclockwise torque

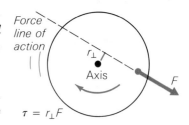

(c) Smaller clockwise torque

Figure 8.6 Torque and moment arm
(a) When a force acts through the axis of rotation, $r_\perp = 0$ and $\tau = 0$. (b) The perpendicular distance $r_\perp$ from the axis of rotation to the line of action of a force is called the moment arm (or lever arm) and is equal to $r \sin \theta$. The torque, or twisting force, that produces rotational motion is given by $\tau = r_\perp F$. (c) The same force in the opposite direction with a smaller moment arm produces a smaller torque in the opposite direction.

But for a rigid body, the masses (m_i's) and the distances from the axis of rotation (r_i's) are constant. Therefore, the quantity in the parentheses is constant and can be designated by the letter I:

$$I = \Sigma m_i r_i^2 \qquad (8.5)$$

The torque is then

$$\tau = I\alpha \qquad (8.6)$$

This is the rotational form of Newton's second law ($\mathbf{F} = m\mathbf{a}$ and $\boldsymbol{\tau} = I\boldsymbol{\alpha}$ in vector form). The constant I is called the **moment of inertia** and is a measure of rotational inertia, or a body's resistance to any change in its rotational motion.

Although I is said to be constant for a rigid body and is the rotational analogue of mass, you must keep in mind that, unlike the mass of a particle, the moment of inertia is referenced to a particular axis and can have different values for different axes. The moment of inertia also depends on the mass distribution *relative* to the axis of rotation, as the following example illustrates.

Example 8.2   Changes in Rotational Inertia
Find the moment of inertia for each of the simple one-dimensional dumbbell configurations in Fig. 8.7. (Consider the mass of the connecting bar to be negligible.)

Solution
Given: Values of m and r from figure   *Find*: $I = \Sigma m_i r_i^2$

With $I = m_1 r_1^2 + m_2 r_2^2$,

(a) $I = (30 \text{ kg})(0.50 \text{ m})^2 + (30 \text{ kg})(0.50 \text{ m})^2 = 15 \text{ kg-m}^2$

(b) $I = (30 \text{ kg})(0.50 \text{ m})^2 + (10 \text{ kg})(0.50 \text{ m})^2 = 10 \text{ kg-m}^2$

(c) $I = (30 \text{ kg})(1.5 \text{ m})^2 + (30 \text{ kg})(1.5 \text{ m})^2 = 135 \text{ kg-m}^2$

(d) $I = (30 \text{ kg})(0 \text{ m})^2 + (30 \text{ kg})(3.0 \text{ m})^2 = 270 \text{ kg-m}^2$ ∎

Example 8.2 clearly shows how the moment of inertia depends on the mass distribution relative to a particular axis of rotation. In general, the moment of inertia is larger the farther the mass is from the axis of rotation. This principle is important in the design of flywheels, which are used in automobiles to keep the engine running smoothly between cylinder firings. The mass of a flywheel is concentrated near the rim, giving a large moment of inertia. Once the flywheel is rotating, its large rotational inertia resists changes in motion.

The calculations for the moments of inertia of extended rigid bodies require math that is beyond the scope of this book. The results for some common shapes are given in Figure 8.8. The rotational axes are generally taken along axes of symmetry, that is, running through the center of mass so as to give a symmetrical mass distribution. One exception is the rod with an axis of rotation through one end [Fig. 8.8(f)]. This axis is parallel to an axis of rotation through the center of mass of the rod [Fig. 8.8(c)]. The moment of inertia about such a parallel axis is given by a useful theorem called the **parallel axis theorem**:

$$I = I_{\text{CM}} + Md^2 \qquad (8.7)$$

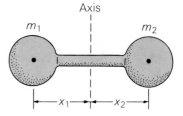

(a) $m_1 = m_2 = 30$ kg
    $x_1 = x_2 = 0.50$ m

(b) $m_1 = 30$ kg, $m_2 = 10$ kg
    $x_1 = x_2 = 0.50$ m

(c) $m_1 = m_2 = 30$ kg
    $x_1 = x_2 = 1.5$ m

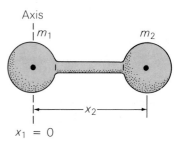

$x_1 = 0$

(d) $m_1 = m_2 = 30$ kg
    $x_1 = 0, x_2 = 3.0$ m

Figure 8.7   **Moment of inertia**
The moment of inertia depends on the distribution of mass relative to a particular axis of rotation. See Example 8.2.

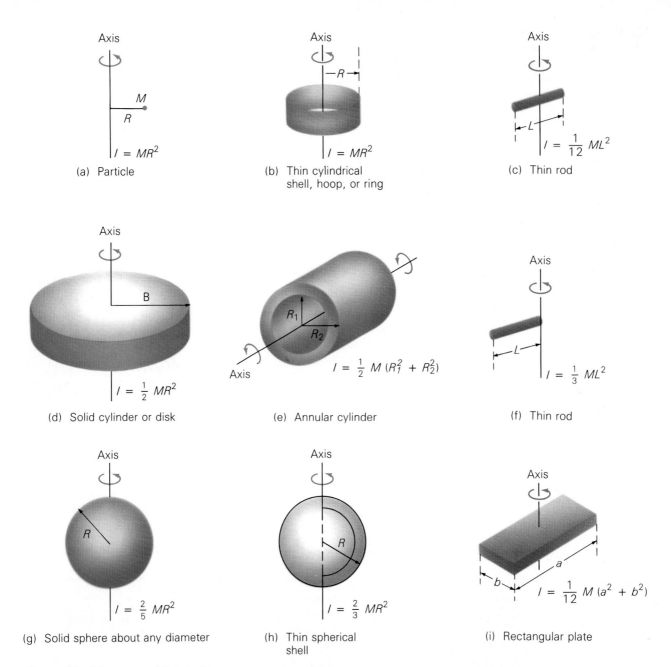

(a) Particle

$$I = MR^2$$

(b) Thin cylindrical shell, hoop, or ring

$$I = MR^2$$

(c) Thin rod

$$I = \tfrac{1}{12}ML^2$$

(d) Solid cylinder or disk

$$I = \tfrac{1}{2}MR^2$$

(e) Annular cylinder

$$I = \tfrac{1}{2}M(R_1^2 + R_2^2)$$

(f) Thin rod

$$I = \tfrac{1}{3}ML^2$$

(g) Solid sphere about any diameter

$$I = \tfrac{2}{5}MR^2$$

(h) Thin spherical shell

$$I = \tfrac{2}{3}MR^2$$

(i) Rectangular plate

$$I = \tfrac{1}{12}M(a^2 + b^2)$$

Figure 8.8 **Moments of inertia for some common shapes**

Where I is the moment of inertia about an axis that is parallel to one through the center of mass and at a distance d from it (Fig. 8.9), and M is the total mass of the body. I_{CM} is the moment of inertia about an axis through the center of mass, like the axes shown for parts (b)–(e) and (g)–(i) in Fig. 8.8. For the axis through the end of the rod [Fig. 8.8(f)], the moment of inertia is

$$I = I_{CM} + Md^2 = \tfrac{1}{12}ML^2 + M\left(\frac{L}{2}\right)^2 = \tfrac{1}{12}ML^2 + \tfrac{1}{4}ML^2 = \tfrac{1}{3}ML^2$$

8.3 Applications of Rotational Dynamics

The rotational form of Newton's second law allows you to analyze dynamic rotational situations. The examples in this section will show you how.

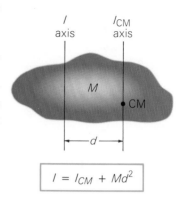

Example 8.3   Unwrapping a Cylinder

A uniform cylinder with a mass of 2.0 kg and a radius of 0.15 m is suspended by two strings wrapped around it (Fig. 8.10). As the cylinder descends, the strings unwind from it. What is the acceleration of the center of the cylinder? (Neglect the mass of the string.)

Solution

Given: $M = 2.0$ kg   Find: a
$\qquad R = 0.15$ m
$\qquad I = \frac{1}{2}MR^2$ [from Fig. 8.8(d)]

The string tensions produce torques on the cylinder about an axis through its center, and since $\tau = I\alpha$,

$$\Sigma\tau = RT + RT = I\alpha = (\tfrac{1}{2}MR^2)\left(\frac{a}{R}\right)$$

where $a = R\alpha$ is used. Solving for a,

$$a = \frac{4T}{M}$$

Then, applying Newton's second law to the translational motion and taking downward as positive,

$$Mg - 2T = Ma$$

or $\quad T = \dfrac{Mg - Ma}{2}$ (2)

Using Eq. 2 to eliminate T from Eq. 1 gives

$$a = \tfrac{2}{3}g = \tfrac{2}{3}(9.80 \text{ m/s}^2) = 6.53 \text{ m/s}^2$$

Note that the M and R canceled out and their values were not needed. This means that the center of any uniform cylinder unwinding in this way will have this same acceleration. ■

In problems involving pulleys in Chapter 4, the mass (and inertia) of the pulley was always neglected to simplify things. Now you know how to include this and can treat pulleys more realistically.

Example 8.4   Pulley with Inertia

A block of mass m hangs from a string wrapped around a frictionless pulley of mass M and radius R, as shown in Fig. 8.11. If the block descends from rest under the influence of gravity, what is the angular acceleration of the pulley? (Neglect the mass of the string.)

Figure 8.9   Parallel axis theorem
The moment of inertia about an axis parallel to one through the center of mass of a body is given by $I = I_{\text{CM}} + Md^2$, where M is the total mass of the body and d is the distance separating the two axes.

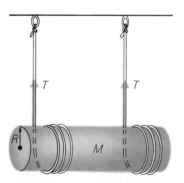

Figure 8.10   Unwrapping a cylinder
The tensions in the strings produce torques that cause the cylinder to rotate as it descends. See Example 8.3.

Solution

The pulley is a disk and thus has a moment of inertia of $I = \frac{1}{2}MR^2$ [Fig. 8.8(d)]. It experiences a torque due to the tension force in the string (T). With $\tau = I\alpha$ (considering only the upper dashed box in Fig. 8.11),

$$\tau = r_\perp F = RT = I\alpha = \tfrac{1}{2}MR^2\alpha$$

so

$$\alpha = \frac{2T}{MR} \qquad (1)$$

But T is unknown. Looking at the descending mass (the lower dashed box) and summing the forces gives

$$mg - T = ma$$

The linear acceleration of the block and the angular acceleration of the pulley are related by $a = R\alpha$, so

$$T = mg - mR\alpha \qquad (2)$$

Using Eq. 2 to eliminate T from Eq. 1 gives

$$\alpha = \frac{2T}{MR} = \frac{2(mg - mR\alpha)}{MR}$$

and

$$\alpha = \frac{2mg}{R(2m + M)} \qquad (3)$$

Note that you could use the same equations to find the linear acceleration of the block, in case you want to know how long it takes to fall a certain distance. ∎

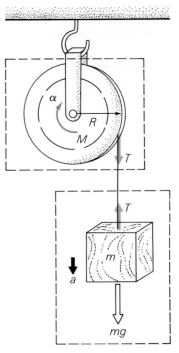

Figure 8.11 **Pulley with inertia**
Taking the mass, or rotational inertia, of a pulley into account allows a more realistic description of its motion. See Example 8.4.

PROBLEM-SOLVING HINT

For problems like those of Examples 8.3 and 8.4, dealing with coupled rotational and translational motions, keep in mind that the accelerations are usually related by $a = r\alpha$ (and $v = r\omega$ relates the velocities if the angular velocity is constant). Applying Newton's second law (in rotational or linear form) to different parts of the system gives equations that can be combined using such a relationship. Also, for rolling without slipping, $a = r\alpha$ and $v = r\omega$ relate the angular quantities to the linear motion of the center of mass.

Pulleys can be analyzed even more accurately. In Example 8.4, friction was neglected, but the next example includes it. Neglecting the mass of the string will still give a good approximation because the string is relatively light. Taking the mass of the string into account would give a continuously varying mass, and such a problem is beyond the scope of this study.

Example 8.5   One Up, One Down

For a pulley with two suspended masses, as shown in Fig. 8.12, the pulley itself has a mass of 0.20 kg, a radius of 0.15 m, *and* a constant torque of 1.2 m-N due to friction in the bearings. What is the acceleration of the suspended masses if $m_1 = 0.40$ kg and $m_2 = 0.80$ kg? (Neglect the mass of the string.)

Solution

Given: $M = 0.20$ kg   *Find*: a
$R = 0.15$ m
$\tau_f = 1.2$ m-N
$m_1 = 0.40$ kg
$m_2 = 0.80$ kg

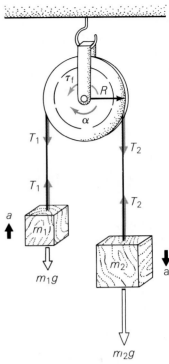

As before, Newton's second law is applied: the rotational form with the net torque acting on the pulley and the translational form with the net forces acting on the masses. Summing the torques on the pulley gives

$$\Sigma \tau = \tau_2 - \tau_1 - \tau_f = I\alpha$$

or $RT_2 - RT_1 - \tau_f = (\tfrac{1}{2}MR^2)\left(\dfrac{a}{R}\right)$ (1)

where the frictional torque opposes the rotation. Summing the forces acting on the masses gives

$$m_2g - T_2 = m_2a \qquad (2)$$

and

$$T_1 - m_1g = m_1a \qquad (3)$$

In terms of the physics, the problem is solved. There are three equations (Eqs. 1, 2, and 3) and three unknowns (a, T_1, and T_2). The rest of the problem (computing the value of a) is simple mathematics. ∎

Figure 8.12 The Atwood machine revisited
For this pulley, rotational inertia and frictional torque are taken into account. See Example 8.5.

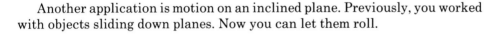

Another application is motion on an inclined plane. Previously, you worked with objects sliding down planes. Now you can let them roll.

Example 8.6 Rolling Down

A rigid spherical ball of mass M and radius R is released at the top of a hard-surfaced inclined plane. It rolls without slipping, with only static friction between it and the plane (Fig. 8.13). What is the acceleration of the ball's center of mass?

Solution

Since the ball rolls without slipping, there is no relative motion between it and the plane at the point of contact, and the force of static friction (f_s) acts on the ball. Without this force, the ball would slide down the plane. (Note that f_s is directed up the plane.)

You can analyze the motion using two different axes of rotation. First, consider an axis through the center of mass of the ball so you can see the effect of a frictional torque. The acceleration of the ball's center of mass is down the plane, and the ball's angular acceleration about an axis through that center has the clockwise sense. Since the reaction force (the normal force) and the weight force act through this axis of rotation, the angular acceleration is produced by the frictional torque, and f_s must be directed up the plane. Then, applying $\tau = I\alpha$ and using the moment of inertia about an axis through the ball's center of mass from Fig. 8.8(g), and the condition for rolling without slipping ($a = r\alpha$) gives

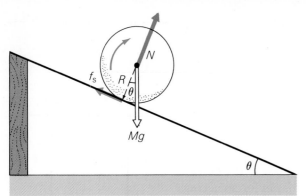

Figure 8.13 **Rolling down an inclined plane**
For a rigid ball rolling down a hard-surfaced inclined plane, the force of static friction acts on the ball in response to the parallel component of the weight force. See Example 8.6.

$$\tau = Rf_\text{s} = I\alpha = (\tfrac{2}{5}MR^2)\left(\frac{a}{R}\right)$$

and

$$f_\text{s} = \tfrac{2}{5}Ma$$

Looking at the translational motion of the center of mass and summing the forces acting parallel to the plane gives

$$Mg\sin\theta - f_\text{s} = Ma$$

or $\quad Mg\sin\theta - \tfrac{2}{5}Ma = Ma$

Thus,

$$a = \tfrac{5}{7}g\sin\theta$$

Here is another situation where the acceleration does not depend on M or R. Therefore, uniform spheres of any size or density will roll down the same plane with the same acceleration. But the acceleration will be *different* for objects with different shapes (for example, cylinders or hoops) because they have different moments of inertia. This reflects the dependence of rotational motion on mass distribution.

Now consider the rotational motion of the ball around the instantaneous axis of rotation, which runs through the contact point of the sphere with the plane's surface. This point is instantaneously at rest, as shown earlier. Note from Fig. 8.13 that the normal force and the force of static friction both act through this point and therefore produce no torque. The torque on the ball from this perspective comes from the weight force. Prove to yourself that the moment arm for this force is $r_\perp = R\sin\theta$. Then

$$\tau = r_\perp F = (R\sin\theta)(Mg) = I_\text{i}\alpha \tag{1}$$

But I_i, the moment of inertia about the instantaneous axis of rotation, is *not* equal to $\tfrac{2}{5}MR^2$, the expression used above. This is the moment of inertia about an axis *through the center of mass* of the ball (that is, $I_\text{CM} = \tfrac{2}{5}MR^2$). You find the moment of inertia about the instantaneous axis using the parallel axis theorem:

$$I_\text{i} = I_\text{CM} + Md^2 = \tfrac{2}{5}MR^2 + MR^2 = \tfrac{7}{5}MR^2 \tag{2}$$

Substituting for I_i from Eq. 2 into Eq. 1 gives

$$(R \sin \theta)(Mg) = I_i \alpha = (\tfrac{7}{5} M R^2)\left(\frac{a}{R}\right)$$

and

$$a = \tfrac{5}{7} g \sin \theta$$

This is the same as the acceleration found using the other axis of rotation (as it should be). ∎

8.4 Rotational Work and Kinetic Energy

This section gives the rotational analogues associated with work and kinetic energy for constant torques. Because their development is very similar to that given for their linear counterparts, detailed discussion is not needed.

Rotational Work   For rotational motion, $W = Fs$ for a force acting along an arc length s:

$$W = Fs = F(r\theta) = \tau\theta$$

Thus,

$$W = \tau\theta \tag{8.8}$$

In this book, both torque (τ) and angular displacement (θ) vectors are almost always along the axis of rotation, so you will not need to be concerned about parallel components as you were for translational work. The torque and angular displacement may be in opposite directions, in which case the applied torque decreases, or slows down, the rotation of the body. Then the rotational work is negative, which is analogous to F and d being in opposite directions for translational motion.

Rotational Power   An expression for power, or the time rate of doing work, is easily obtained from Eq. 8.8:

$$P = \frac{W}{t} = \tau\left(\frac{\theta}{t}\right) = \tau\omega \tag{8.9}$$

The Work-Energy Theorem and Kinetic Energy

The relationship between rotational work and rotational kinetic energy may be derived as follows, starting with the equation for rotational work:

$$W = \tau\theta = I\alpha\theta$$

For a constant angular acceleration, $\omega^2 = \omega_o^2 + 2\alpha\theta$, and therefore

$$W = I\left(\frac{\omega^2 - \omega_o^2}{2}\right) = \tfrac{1}{2}I\omega^2 - \tfrac{1}{2}I\omega_o^2$$

Thus,

$$W = \tfrac{1}{2}I\omega^2 - \tfrac{1}{2}I\omega_o^2 = K - K_o = \Delta K \tag{8.10}$$

where

$$\Delta K \equiv \tfrac{1}{2}I\omega^2. \tag{8.11}$$

and K is the **rotational kinetic energy**. Thus, the angular work is equal to the change in the rotational kinetic energy. Consequently, in order to change the rotational energy of an object, a torque must be applied, since Eq. 8.8 shows that rotational work involves a torque.

It is possible to derive the kinetic energy of a rotating rigid body directly. Summing the instantaneous tangential velocities and kinetic energies of the individual particles of the body gives

$$K = \tfrac{1}{2}\Sigma\, m_i v_i^2 = \tfrac{1}{2}(\Sigma\, m_i r_i^2)\omega^2 = \tfrac{1}{2}I\omega^2$$

Thus, Eq. 8.11 doesn't represent a new form of energy. It is simply another expression for kinetic energy, in a form that is more convenient for rigid body rotation.

A summary of the analogous equations for translational and rotational motion is given in Table 8.1.

When an object has both translational and rotational motions, its total kinetic energy may be divided into parts to reflect this.

For example, for a cylinder rolling without slipping on a level surface, the motion is purely rotational relative to the instantaneous axis of rotation, which is instantaneously at rest (see Fig. 8.3). The kinetic energy is

$$K = \tfrac{1}{2}I_i\omega^2$$

where I_i is the moment of inertia about the instantaneous axis. However, there is a translation and a rotation about an axis through the center of mass. By the parallel axis theorem, the moment of inertia about the instantaneous axis of rotation is related to the moment of inertia about a parallel axis through the center of mass. That is, $I_i = I_{CM} + MR^2$, where R is the radius of the cylinder. Then

$$K = \tfrac{1}{2}I_i\omega^2 = \tfrac{1}{2}(I_{CM} + MR^2)\omega^2 = \tfrac{1}{2}I_{CM}\omega^2 + \tfrac{1}{2}MR^2\omega^2$$

But since there is no slipping, $R\omega = v_{cm}$, and

$$K = \tfrac{1}{2}I_{CM}\omega^2 + \tfrac{1}{2}Mv_{CM}^2 \tag{8.12}$$

Table 8.1
Translational and Rotational Quantities and Equations

Translational		Rotational	
Force:	F	Torque (moment of force):	$\tau = rF\sin\theta$
Mass (inertia):	m	Moment of inertia:	$I = \Sigma m_i r_i^2$
Newton's second law:	$F = ma$	Newton's second law:	$\tau = I\alpha$
Work:	$W = Fd$	Work:	$W = \tau\theta$
Power:	$P = Fv$	Power:	$P = \tau\omega$
Kinetic energy:	$K = \tfrac{1}{2}mv^2$	Kinetic energy:	$K = \tfrac{1}{2}I\omega^2$
Work-energy theorem:	$W = \tfrac{1}{2}mv^2 - \tfrac{1}{2}mv_0^2 = \Delta K$	Work-energy theorem:	$W = \tfrac{1}{2}I\omega^2 - \tfrac{1}{2}I\omega_0^2 = \Delta K$

Note that although a cylinder was used as an example here, this is a general result and applies to any object that is rolling without slipping. Thus, the total kinetic energy of such an object is the sum of two contributions: the translational kinetic energy of the center of mass and the rotational kinetic energy relative to a horizontal axis through the center of mass.

Example 8.7   A Division of Energy

A uniform 1.00-kg cylinder rolls without slipping on a flat surface at a speed of 2.40 m/s. (a) What is the total kinetic energy of the cylinder? (b) What percentage of this is rotational kinetic energy?

Solution

Given: $M = 1.00 \text{ kg}$ *Find*: (a) K
$\qquad v_{\text{CM}} = 2.40 \text{ m/s}$ (b) $K_{\text{r}}/K \, (\times \, 100\%)$
$\qquad I_{\text{CM}} = \frac{1}{2}MR^2$

(a) The cylinder rolls without slipping, so the condition $v_{\text{CM}} = R\omega$ applies. Then

$$K = \tfrac{1}{2}I_{\text{CM}}\omega^2 + \tfrac{1}{2}Mv_{\text{CM}}^2 = \tfrac{1}{2}(\tfrac{1}{2}MR^2)\left(\frac{v_{\text{CM}}}{R}\right)^2 + \tfrac{1}{2}Mv_{\text{CM}}^2$$

$$= \tfrac{3}{4}Mv_{\text{CM}}^2 = \tfrac{3}{4}(1.00 \text{ kg})(2.40 \text{ m/s})^2 = 4.32 \text{ J}$$

(b) The rotational kinetic energy of the cylinder is

$$K_{\text{r}} = \tfrac{1}{2}I_{\text{CM}}\omega^2 = \tfrac{1}{2}(\tfrac{1}{2}MR^2)\left(\frac{v_{\text{CM}}}{R}\right)^2 = \tfrac{1}{4}Mv_{\text{CM}}^2$$

Then

$$\frac{K_{\text{r}}}{K} = \frac{\tfrac{1}{4}Mv_{\text{CM}}^2}{\tfrac{3}{4}Mv_{\text{CM}}^2} = 0.33 \, (\times \, 100\%) = 33\%$$

Thus, the total kinetic energy of the cylinder is split between rotational and translational components, with 33% being rotational.

Don't think that this division of energy is a general result. It is easy to show that the percentage is different for objects with different moments of inertia. For example, you should expect a rolling sphere to have a smaller percentage of rotational kinetic energy than a cylinder has since it has a smaller moment of inertia. Note that the radius of the cylinder was not needed at all and that the mass was not needed in part (b). ∎

Potential energy can be brought into the act by applying the conservation of energy to a cylinder rolling down an inclined plane.

Example 8.8   Rolling Down versus Sliding Down

A uniform cylinder is released from rest at a height of 0.45 m near the top of an inclined plane (Fig. 8.14). If the cylinder rolls down the plane without slipping and there is no energy loss due to friction, what is the linear speed of the cylinder's center of mass at the bottom of the incline?

Solution

Given: $h = 0.45 \text{ m}$ *Find*: v_{CM}
$\qquad I_{\text{CM}} = \frac{1}{2}MR^2$ [from Fig. 8.8(d)]

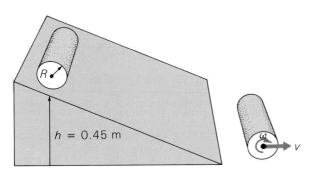

Figure 8.14 **Rolling motion and energy**
When an object rolls down an inclined plane, potential energy is converted to translational *and* rotational kinetic energy. This makes rolling slower than frictionless sliding. See Example 8.8.

Since the total mechanical energy of the cylinder is conserved, you can write

$$E_\text{o} = E$$

or $\quad Mgh = \tfrac{1}{2}I_\text{CM}\omega^2 + \tfrac{1}{2}Mv_\text{CM}^2$

$\quad\quad$ at rest $\quad$ at bottom of incline

Using the condition $v_\text{CM} = R\omega$ gives

$$Mgh = \tfrac{1}{2}(\tfrac{1}{2}MR^2)\left(\frac{v_\text{CM}}{R}\right)^2 + \tfrac{1}{2}Mv_\text{CM}^2$$

which, solved for v_CM, is

$$v_\text{CM} = \sqrt{\tfrac{4}{3}gh} = \sqrt{\tfrac{4}{3}(9.80 \text{ m/s}^2)(0.45 \text{ m})} = 2.4 \text{ m/s}$$

Again, not much data were needed here. If the surface of the inclined plane were frictionless and the cylinder slid down, the speed of the center of mass at the bottom would be $v_\text{CM} = \sqrt{2gh}$, since $Mgh = \tfrac{1}{2}Mv_\text{CM}^2$ in that case. Thus, v_CM would be greater if the cylinder slid down without friction. Do you know why? (Think about rotation. If an object rotates, some of the energy is rotational.) ∎

As Example 8.8 shows, v_CM is independent of M and R. The masses and radii cancel out, so all objects of a particular shape (with the same formula for the moment of inertia) roll with the same speed, regardless of size or density. But the rolling speed does vary with the moment of inertia, which varies with the shape. Therefore, rigid bodies with different shapes roll with different speeds. For example, if you released a cylindrical hoop, a solid cylinder, and a uniform sphere at the same time from the top of an inclined plane, the sphere would win the race to the bottom, followed by the cylinder, with the hoop coming in last—every time! (See Problem 39.)

You can try this experiment using a couple of food cans or other cylindrical containers—one full of some solid material (a rigid body) and one empty with the ends cut out—and a smooth ball of some sort. Remember that the masses and the radii make no difference. You might think that an annular cylinder [a hollow cylinder with inner and outer radii that vary appreciably—Fig. 8.8(e)] would be a possible front-runner, or front-roller, in such a race, but it wouldn't win. The rolling race down an incline is fixed even when you vary the masses and the radii. (See Problem 42.) Thus, you could save yourself some time by noting in Figure 8.8 that the shape with the greatest moment of inertia is a cylindrical hoop or ring.

Friction Associated with Rolling Motion

There are three types of friction that can come into play for rolling objects: static (adhesive) friction, sliding (kinetic) friction, and rolling friction. The static adhesive friction associated with rolling is analogous to the static friction for objects at rest that you studied earlier (Chapter 4). It exists only in response to an applied force and does no work and dissipates no energy; that is, it doesn't transform mechanical energy into heat. Oddly enough, static friction can accelerate or decelerate rolling motion, depending on the directions of the static force and rotation. (See Question 18.)

Sliding friction, on the other hand, is a force that dissipates energy and performs work. It too can accelerate or decelerate motion. Example 8.1 showed how sliding friction decelerated the linear motion of a bowling ball (and accelerated the rotational motion). Conversely, if a rotating cylinder is placed on a surface, sliding friction will accelerate the linear motion (and decelerate the rotational motion).

Rolling friction occurs when one surface rotates but does not slide on another at the area of contact. This type of friction is primarily due to the deformation of materials. A rolling object or the surface it rolls on or both are

---

INSIGHT

Slide or Roll to a Stop?

In an emergency situation, you may instinctively jam on the brakes of a car, trying to come to a quick stop, that is, to stop in the shortest distance. But with the wheels locked, the car skids, or slides, to a stop, often out of control. In this case, the force of sliding friction is acting on the wheels.

In order to prevent this, you learn to pump the brakes in order to roll rather than slide to a stop, particularly on wet or icy roads. Some newer automobiles have computerized braking systems that do this automatically. When the brakes are applied firmly and the car begins to slide, sensors in the wheels note this and a computer takes over control of the braking system. It momentarily releases the brakes and then varies the brake-fluid pressure with a rapid pumping action so that the wheels will continue to roll without slipping.

In the absence of sliding, both rolling friction and static friction act. In many cases, however, the force of rolling friction is very small and only static friction must be taken into account.

Does sliding instead of rolling make a big difference in the stopping distance for an automobile? It is easy to calculate the difference by assuming that rolling friction is negligible. Although the external force of static friction does no work to dissipate energy in slowing a car (this is done internally by friction on the brake pads), it does determine the rolling condition of the wheels.

In Example 2.5, a vehicle stopping distance was given by

$$x = \frac{v_o^2}{2a}$$

By Newton's second law,

$$F = f = \mu N = \mu mg = ma$$

and the stopping acceleration is

$$a = \mu g$$

Thus,

$$x = \frac{v_o^2}{2\mu g} \tag{1}$$

But, as was noted in Chapter 4, the coefficient of sliding (kinetic) friction is generally less than that of static friction; that is, $\mu_k < \mu_s$. The general difference between rolling and sliding stops can be seen by using the same initial velocity (v_o for both cases. Then, using Eq. 1 to form a ratio,

$$\frac{x_{roll}}{x_{slide}} = \frac{\mu_k}{\mu_s} \quad \text{or} \quad x_{roll} = \left(\frac{\mu_k}{\mu_s}\right) x_{slide}$$

The value of μ_k for rubber on wet concrete from Table 4.1 is 0.60, and the value of μ_s for these surfaces is 0.80. Using these values for a comparison of the stopping distances gives

$$x_{roll} = \left(\frac{0.60}{0.80}\right) x_{slide} = (0.75) x_{slide}$$

The car comes to a rolling stop in 75% of the distance required for a sliding stop, for example, 15 m instead of 20 m. Although this may vary for different conditions, it could be an important, perhaps life-saving, difference.

deformed under a load (see Fig. 8.15). At a surface depression, the displaced material builds up to form a mound that the rolling object must climb to continue in the direction of motion. Thus, rolling friction always resists and decelerates motion. For very hard objects and surfaces, rolling friction may be so small as to be negligible (as in Example 8.6).

This important component of friction is often overlooked. You know that a wheel will roll freely to a stop on a level surface (Fig. 8.16). What causes this? There is no rotational static friction, since there is no external applied force in the horizontal direction. [In Example 8.6, there was a static force in reaction to a component of the gravitational force.] Even if there were a static force, it would do no work in dissipating energy to bring the wheel to a stop; so it must be rolling friction that does the job. This type of friction is quite complex to analyze, but basically there is a rolling frictional torque (τ_f in the figure) that decelerates the wheel and brings it to rest.

People discovered long ago that it is much easier to move a load by putting it on rollers (wheels) than by sliding it. Another way of saying this is that the coefficient of rolling friction is much less than that of sliding friction. Ball bearings or roller bearings are used in machinery to replace sliding motion with rolling motion to take advantage of this. However, in some cases, such as for the driving (powered) wheels of an automobile, soft materials such as rubber are used to provide greater friction between the tires and the road surface for traction.

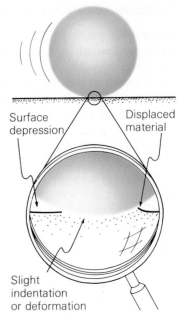

Figure 8.15 Rolling friction
Rolling friction is due to the deformation of materials. The displaced material builds up in the direction of motion, causing resistance to the motion.

8.5 Angular Momentum

Another important quantity of rotational motion is angular momentum. Like torque, **angular momentum (L)** involves a moment arm. For a particle of mass m, we have $p = mv$ and the magnitude of the angular momentum is

$$L = r_\perp p = mr_\perp v = mr_\perp^2 \omega \tag{8.13}$$

where v is the tangential speed of the particle, $r_\perp$ the moment arm, and ω the angular speed. For circular motion, $r_\perp = r$, since **v** is perpendicular to **r**.

For a system of particles or a rigid body, the total angular momentum is

$$L = (\Sigma \, m_i r_i^2)\omega = I\omega \tag{8.14}$$

which, in vector notation, is

$$\mathbf{L} = I\boldsymbol{\omega} \tag{8.15}$$

Thus, **L** is the direction of the angular velocity (ω) as given by the right-hand rule.

For linear motion, momentum is related to force by $\mathbf{F} = \Delta\mathbf{p}/\Delta t$ (Chapter 6). Angular momentum is analogously related to torque, as can be easily shown:

$$\tau = I\boldsymbol{\alpha} = \frac{I\Delta\omega}{\Delta t} = \frac{\Delta(I\omega)}{\Delta t} = \frac{\Delta\mathbf{L}}{\Delta t}$$

That is,

$$\tau = \frac{\Delta\mathbf{L}}{\Delta t} \tag{8.16}$$

Thus, torque is equal to the time rate of change of angular momentum.

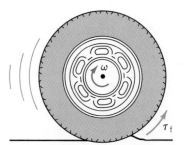

Figure 8.16 Rolling to a stop
When a wheel rolls freely to a stop on a level surface, the work is done by rolling friction, or a frictional torque.

Conservation of Angular Momentum

Eq. 8.16 was derived using $\tau = I\alpha$, which applies to a rigid system of particles or a rigid body having a constant moment of inertia. However, Eq. 8.16 is a general one that also applies to a nonrigid system of particles. In such a system, there may be a change in the mass distribution and a change in the moment of inertia. As a result, there may be an angular acceleration even in the absence of a torque. How can this be?

If the net torque on a system is zero, then, by Eq. 8.16, $\tau = \Delta L/\Delta t = 0$, and

$$\Delta \mathbf{L} = \mathbf{L} - \mathbf{L}_o = I\omega - I_o\omega_o = 0$$

or $\quad I\omega = I_o\omega_o \qquad\qquad\qquad\qquad\qquad\qquad\qquad (8.17)$

Thus, the condition for the **conservation of angular momentum** is as follows:

> In the absence of an external, unbalanced torque, the total angular momentum of a system is conserved (remains constant).

As with total linear momentum, the internal torques arising from internal forces cancel out.

For a rigid body with a constant moment of inertia (that is, $I = I_o$), the angular speed remains constant ($\omega = \omega_o$) in the absence of a net torque. But it is possible for the moment of inertia to change in some systems, giving rise to a change in the angular velocity, as the following example illustrates.

Example 8.9   Conservation of Angular Momentum

A small ball at the end of a string that passes through a tube is swung in a circle, as illustrated in Fig. 8.17. When the string is pulled downward through the tube, the angular speed of the ball increases. (a) Is this caused by a torque due to the pulling force? (b) If the ball is initially swung in a circle with a radius of 0.30 m

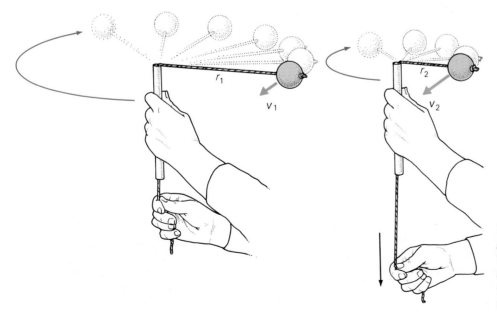

Figure 8.17   **Conservation of angular momentum**
When the string is pulled downward through the tube, the revolving ball speeds up. See Example 8.9.

at a speed of 2.8 m/s, what will be its tangential speed if the string is pulled down far enough to reduce the radius of the circle to 0.15 m?

Solution

Given: $r_1 = 0.30$ m

$\quad\quad\; r_2 = 0.15$ m

$\quad\quad\; v_1 = 2.8$ m/s

Find: (a) Cause of the increase in angular speed

$\quad\quad$ (b) v_2

(a) The change in the angular velocity, or the angular acceleration, is not caused by a torque due to the pulling force. The force on the ball as transmitted by the string acts through the axis of rotation, and therefore the torque is zero. As the rotating portion of the string is shortened, the moment of inertia of the ball [$I = mr^2$, from Fig. 8.8(a)] decreases. This causes the ball to speed up, since in the absence of an external torque, the angular momentum of the ball is conserved.

(b) Since the angular momentum is conserved,

$$I_o \omega_o = I\omega$$

Then, using $I = mr^2$ and $\omega = v/r$ gives

$$mr_1v_1 = mr_2v_2$$

and

$$v_2 = \left(\frac{r_1}{r_2}\right)v_2 = \left(\frac{0.30 \text{ m}}{0.15 \text{ m}}\right)2.8 \text{ m/s} = 5.6 \text{ m/s}$$

When the radial distance is shortened, the ball speeds up. ∎

(a)

(b)

Figure 8.18 **Moment of inertia change**
(a) When spinning slowly with arms outstretched, the person's moment of inertia is relatively large (r is large). Note that when on the stool, the person is isolated and there are no external torques (neglecting friction in the stool bearings), and the angular momentum, $L = I\omega$, is conserved. (b) Tucking his arms inward decreases the moment of inertia, and he goes into a dizzy spin since ω must increase.

Example 8.9 should help you understand Kepler's law of areas and why an orbiting satellite speeds up or slows down when the altitude of the orbit is changed (Chapter 7) in terms of the conservation of angular momentum. [Note that the person holding the string in Fig. 8.17 would have to supply a greater force to hold the ball in the smaller circular path. Check the centripetal forces.]

You can also look at this situation in terms of work and energy. The pulling force produces no torque, but it does do work, and the kinetic energy of the ball increases. Similarly, the conservation of energy may be used to analyze the motions of planets. Recall from Chapter 7 that the total mechanical energy of a body moving under the influence of a gravitational attraction is $E = K + U = \frac{1}{2}m_1v^2 - Gm_1m_2/r$ (Eq. 7.18). Since gravity is a conservative force, the total energy is conserved. If m_1 is a planet and m_2 the Sun, you can see that when the planet is closest to the Sun (smallest value of r), its kinetic energy and speed are at a maximum ($K = E + Gm_1m_2/r$). Similarly, when the planet is farthest from the Sun (largest value of r), its kinetic energy and speed are at a minimum.

A popular lecture demonstration of the conservation of angular momentum is shown in Fig. 8.18. A person sitting on a stool that rotates holds weights with his arms outstretched and is started slowly rotating. An external torque to start this rotation must be supplied by someone else, because the person on the stool cannot initiate the motion by himself. (Why?) Once rotating, if the person brings his arms inward, the angular speed increases and he spins much faster. Extending his arms again slows him down. Can you explain this? Refer to

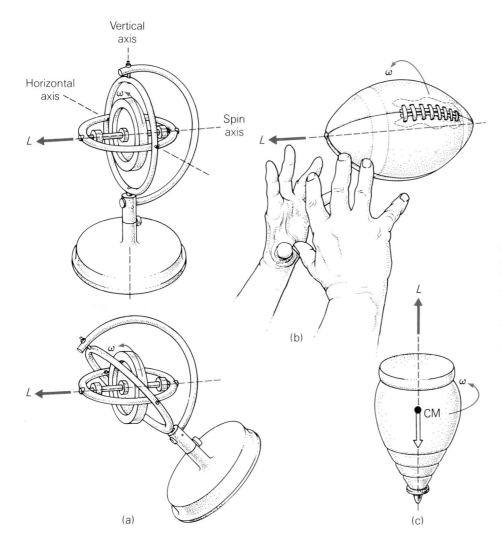

Vertical axis

Horizontal axis

Spin axis

L

ω

L

ω

L

(a)

(b)

L

ω

CM

(c)

Figure 8.19 **Constant direction of angular momentum**
When the angular momentum is conserved, its direction is constant in space. (a) This may be demonstrated with a gyroscope, a rotating wheel that is universally mounted on gimbals (rings) so it is free to turn about any axis. This is the principle of operation for gyrocompasses. (b) Gyroscopic action also occurs when a football is passed. (c) The axis of rotation of a sleeping top passes through its center of gravity. Thus, there is no torque due to the top's weight.

Eq. 8.15. If L is constant, what happens to ω when I is made smaller by reducing r? Similarly, a diver spins during a high dive by tucking in and changing his or her moment of inertia [see Fig. 8.2(b)].

When angular momentum is conserved, the direction of the vector **L** remains fixed in space. This principle is applied in gyrocompasses, as well as in passing a football accurately (Fig. 8.19). A football is thrown with a spiraling rotation. This spin, or gyroscopic action, stabilizes the ball's motion in its most streamlined orientation. Giving the football a sufficient spin ensures that it will travel end-first and not tumble because of torque produced by air resistance. The gyroscopic action will maintain the alignment of the spin axis. Similarly, a spinning top or gyroscope will stand straight up with its angular momentum vector fixed in space for some time. A top in this condition is called a sleeping top. Note in Fig. 8.19(c) that the axis of rotation passes through the top's center of gravity, and thus there is no torque due to the top's weight.

However, you know that a gyroscope or a sleeping top eventually slows down because of frictional torques. When playing with a gyroscope or top, you

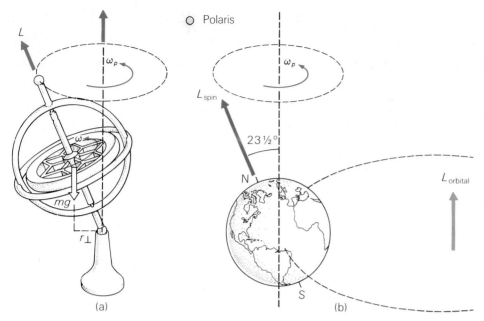

Figure 8.20 **Precession**
An external torque causes a change in angular momentum. (a) For a spinning gyroscope, this change is directional, and the axis of rotation precesses about a vertical line. (The torque due to the weight force would point out of the page as drawn here, as would **ΔL**.) (b) Similarly, the Earth's axis precesses because of gravitational torques caused by the Sun and the moon. We don't notice this since the period of precession is about 26,000 years.

have probably noticed how its spin axis revolves, or precesses, about a vertical axis as its rotation slows down [Fig. 8.20(a)]. Since the gyroscope precesses, **L** is no longer constant in direction, which means there is a torque acting to give a change (**ΔL**) with time. As can be seen from the figure, the torque arises from the vertical component of the weight force, since the center of gravity no longer lies directly above the point of support. The instantaneous torque is such that the gyroscope's axis moves about the vertical axis. [Taking the angular momentum vector at the beginning and the end of a short interval (Δt), by vector addition **ΔL** is in a direction nearly tangent to the circular path of precession. The instantaneous change in angular momentum and the torque are tangent to the circle. Prove this to yourself.]

It is interesting to note that the Earth's rotational axis precesses. The axis is tilted $23\frac{1}{2}°$ with respect to a line normal to the plane of revolution about the Sun. This precession is due to slight gravitational torques exerted on the Earth by the Sun and the moon. These torques arise because the Earth is not perfectly spherical—it has an equatorial bulge. [Note in Fig. 8.20(b) that the Earth's total angular momentum has orbital and spin components.]

The period of the precession of the Earth's axis is about 26,000 years, so it has little day-to-day effect. However, it does have an interesting long-term effect. Polaris will not always be (nor has it always been) the North Star, that is, the star "pointed at" by the Earth's axis of rotation. About 5000 years ago, Alpha Draconis was the North Star, and 5000 years from now, Alpha Cephei, which is at an angular distance of about 68° away from Polaris on the circle described by the precession of the Earth's axis, will be.

It is interesting to note that the Earth's rate of rotation (and therefore its spin angular momentum) is changing. On the average, the Earth's spin has been slowing down. This is primarily due to tidal friction, which give rise to a torque. Among other things, the slowing rate of rotation causes the average day to be longer; this century will be about 25 seconds longer than the last. But this is an average deceleration. At times, the Earth's rotation speeds up for a relatively short period. This is thought to be associated with the rotational inertia of the liquid outer layer of the Earth's core (see the Insight feature in Chapter 14).

The tidal torque on the Earth indirectly results from the moon's gravitational attraction, which is the main cause of ocean tides. This torque is *internal* to the Earth-moon system, so the total angular momentum of that system is conserved. Therefore, since the Earth is losing angular momentum, the moon must be gaining angular momentum to keep the total angular momentum constant. The Earth loses rotational, or spin, angular momentum, and the moon gains orbital angular momentum. As a result, the moon drifts slightly farther from Earth and its orbital speed decreases (see Chapter 7). The moon moves away from the Earth at about 1 cm per lunar revolution, or lunar month (calculated from the rate of the slowing down of the Earth's rotation). Thus, the moon moves in a slowly widening spiral.

Important Formulas

Condition for rolling without slipping:

$v = r\omega$

(or $s = r\theta$ or $a = r\alpha$)

Torque:

$\tau = rF\sin\theta$

Torque on a particle:

$\tau = mr^2\alpha$

Moment of inertia:

$I = \Sigma\, m_i r_i^2$

Rotational form of Newton's second law:

$\tau = I\alpha$

Parallel axis theorem:

$I = I_{CM} + Md^2$

Rotational work:

$W = \tau\theta$

Rotational power:

$P = \tau\omega$

Rotational kinetic energy:

$K = \frac{1}{2}I\omega^2$

Work-energy theorem:

$W = \Delta K = \frac{1}{2}I\omega^2 - \frac{1}{2}I\omega_o^2$

Kinetic energy of a rolling object:

$K = \frac{1}{2}I_{CM}\omega^2 + \frac{1}{2}Mv_{CM}^2$

Angular momentum of a particle in circular motion:

$L = mr^2\omega$

Angular momentum of a rigid body:

$L = I\omega$

Torque in terms of angular momentum:

$\tau = \dfrac{\Delta L}{\Delta t}$

Conservation of angular momentum (with $\tau = 0$):

$I\omega = I_o\omega_o$

Questions

Rigid Bodies, Translations, and Rotations

1. Is it possible for a rigid body to have pure translational motion and pure rotational motion at the same time? Explain.

2. If a hard-boiled egg is given the proper torque by starting it spinning on a level surface with a quick twist of the wrist, it will rise up and spin on one end like a top. A raw egg will not spin like this. Explain why.

3. What happens if v is less than $R\omega$ for a rolling cylinder? Is it possible for v to be greater than $R\omega$?

4. Could a moment of inertia be negative? If so, what would that mean?

Torque and Moment of Inertia

5. Is the moment of inertia of a rigid body related in any way to the center of mass? Explain.

6. Write rotational forms of Newton's first and third laws. Are they valid? Explain.

7. Why is it easier to do sit-ups with your arms extended in front of you than with your hands behind your head?

8. The string of a stationary yo-yo is pulled as shown in Fig. 8.21. (a) Which way does the yo-yo roll and why? (Hint: it may not be as you think. Check the center of mass. If you want to try this and don't have a yo-yo, a spool of thread also works.) (b) Explain what happens if the string is pulled at a greater and greater angle to the horizontal.

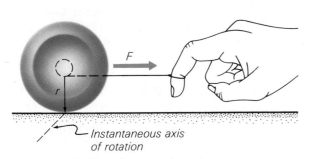

Figure 8.21 **Pulling the yo-yo's string**
See Question 8.

9. How can the moment of inertia be reduced? (Give more than one way.)

10. Why does the moment of inertia for a rigid body have different values for different axes of rotation? What does this mean physically?

11. Explain what happens and why when strings of equal length wrapped around a cylinder as in Fig. 8.10 unwrap completely. The ends of the strings are fixed to the cylinder. (Hint: think of a yo-yo.)

12. Why is it easier to balance a meterstick vertically on your finger than to balance a pencil? The pencil has a lower center of mass.

13. Tightrope walkers often carry long poles. How does the pole help a performer maintain his or her balance? [Hint: when walking on a thin bar or rail (for example, railroad rail), you would find it helpful to hold your arms outstretched.]

14. The depletion of the atmospheric ozone layer due to the release of chlorofluorocarbon gases and the release of vast amounts of carbon dioxide are both implicated in the so-called greenhouse effect, which may result in an increase in the average temperature and melting of the polar ice caps. If this occurred and the ocean level rose substantially, what effect would it have on the Earth's rotation and the length of the day?

Rotational Dynamics and Kinetic Energy

15. A softball, a volleyball, and a basketball are released at the same time from the top of an inclined plane. Give the results of the race down the plane.

16. If the ball in Fig. 8.13 were rolling up the plane, what would be the direction of the force of static friction?

17. If a ball rolls without slipping on a level surface at a constant angular speed, is a force of friction acting on the ball? Explain. (Hint: consider the motion of the center of mass.)

18. (a) For the car and trailer in Fig. 8.22, what will be the direction of the force of sliding friction on the rear (powered) wheels of the car if they are locked and the car is sliding to a halt? (b) What will be the direction of the force if the car is moving forward but the rear wheels are slipping in mud? (c) What will be the direction of the force of static friction on each wheel (car and trailer) when the wheels roll without slipping? (Hint: consider the motions of the centers of mass.) (d) What is the circular sense of the rolling frictional torque in each of the preceding cases?

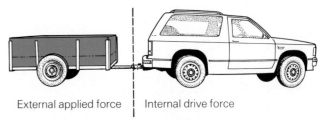

External applied force | Internal drive force

Figure 8.22 **Frictional directions**
See Question 18.

19. Is there a rotational analogue of impulse? Explain.

20. Can you add some more shapes and equations to Fig. 8.8? If so, what are they?

21. For linear motion, the kinetic energy can be expressed in terms of the momentum: $K = p^2/2m$. Can this be done similarly for rotational motion?

22. For a classroom demonstration, illustrated in Fig. 8.23, a person on a stool that rotates holds a rotating bicycle wheel (by the handles on it). When the wheel is held horizontally, she rotates one way (clockwise viewed from above). When the wheel is turned over, she

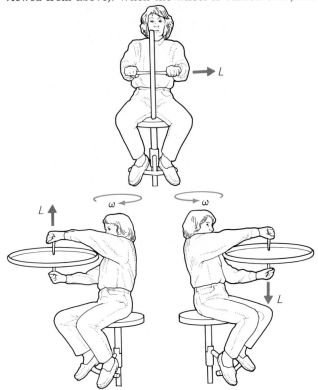

Figure 8.23 **A double rotation**
See Question 22 and Problems 51 and 54.

rotates in the opposite direction. Explain why this occurs. (Hint: consider angular momentum vectors.)

23. Why doesn't a sleeping top fall over? If it weren't spinning, it would fall. Explain why? Why can you balance a spinning basketball on your finger?

24. (a) Large helicopters have two rotors that rotate in opposite directions on the same axis. Explain why. (b) Small helicopters have a large top rotor and a small tail rotor. What is the purpose of the tail rotor? (Hint: what would happen without it or if it failed?)

25. Unlike helicopters (Question 24), single-engine airplanes have only one propeller, or set of rotor blades. What keeps the body of this type of aircraft from rotating?

26. An ice skater goes into a fast spin by tucking her arms in, and a ballerina does a pirouette with her arms over her head. Why are the arms held that way in each case?

27. What is the purpose of the spiraling grooves (called rifling) inside the barrel of a rifle?

28. If you lean to one side when riding a bicycle, the bicycle leans in the opposite direction. Explain why.

29. Cats always seem to land on their feet when they fall, even if held upside down and dropped. While a cat is falling, there is no external torque and the center of mass falls as a particle. How are cats able to turn themselves over while falling?

30. If a gyroscope such as that in Fig. 8.19(a), with its spin axis horizontal, were kept spinning at the North Pole, what would be observed and why?

31. Tidal friction slows the Earth's rate of rotation slightly. Could this have an effect on the length of the lunar month?

Problems

8.1 Rigid Bodies, Translations, and Rotations

■**1.** A circular disk with a radius of 0.10 m rolls without slipping on a level surface with an angular speed of 2.0 rad/s. (a) What is the linear speed of the center of mass of the disk? (b) What is the instantaneous tangential speed of the top of the disk?

■**2.** Show that $a = r\alpha$ is a condition for rolling without slipping.

■■**3.** A cylinder with a radius of 0.250 m rolls with an angular speed of 1.66 rad/s on a horizontal surface without slipping. What is the instantaneous tangential speed of a particle 36.0 cm above the instantaneous axis of rotation?

■■**4.** A disk with a radius of 0.10 m rotates through 270° as it travels 0.47 m. Does the disk roll without slipping?

■■**5.** A cylinder with a radius of 20 cm rolls without slipping on a level surface. The center of the cylinder moves 80 cm in 4.0 s. Through what angular distance does the cylinder rotate during this time?

■■**6.** A cylinder with a radius of 20 cm rolls with an

angular speed of 2.5 rad/s. If the center of the cylinder moves 96 cm in 2.0 s, does the cylinder roll without slipping?

■■7. A friend tells you that he measured the translational speed of the center of mass of a sphere whose diameter is 0.182 m to be 0.465 m/s when the sphere was rolling with an angular speed of 4.98 rad/s. Is his measurement accurate?

8.2 and 8.3 Torque, Moment of Inertia, and Rotational Dynamics

■8. A force of 12 N is applied to a particle located 0.20 m from its axis of rotation. What is the magnitude of the torque about this axis if the angle between the direction of the applied force and the moment arm is (a) 60°, (b) 90°, and (c) 120°?

■9. If the mass of the particle in Problem 8 is 0.10 kg, what is the angular acceleration in each case?

■10. A light meterstick is loaded with masses of 2.0 kg and 4.0 kg at the 30-cm and 80-cm positions, respectively. (a) What is the moment of inertia about an axis through the 0-cm end of the meterstick? (b) What is the moment of inertia about an axis through the center of mass of the system? (c) Use the parallel axis theorem to find the moment of inertia about the axis through the 0-cm end of the stick, and compare the result with the result of part (a).

■11. Show that $r_\perp F = rF_\perp$ for a torque.

■12. A fixed 0.15-kg pulley with a radius of 0.050 m is acted on by a net torque of 4.6 m-N. What is the angular acceleration of the pulley?

■13. A piece of machinery with a moment of inertia of 2.6 kg-m² rotates with a constant angular speed of 4.0 rad/s when experiencing a frictional torque of 0.56 m-N. What is the net external torque acting on the piece?

■■14. Four 0.25-kg point masses connected by rods of negligible mass form a square rigid body with 0.10-m sides. Find the moment of inertia of the body about each of the following axes: (a) an axis perpendicular to the plane of the square and through its center, (b) an axis perpendicular to the plane and through one of the masses, (c) an axis in the plane of the square and along one side through two of the masses, and (d) an axis in the plane running diagonally through two of the masses.

■■15. Show that if two rigid bodies with the same axis of rotation are stuck together, the moment of inertia of the combined rigid body about the common axis is equal to the sum of the individual moments, or $I = I_1 + I_2$.

■■16. The *radius of gyration (k)* is defined as the distance from an axis of rotation at which all the mass (M) of an object would have to be concentrated to give the same moment of inertia as would be calculated with the actual mass distribution, that is, $I = Mk^2$. This represents an object as a rotating particle of mass M. For the axes shown in Fig. 8.8, determine the radius of gyration for (a) a uniform sphere, (b) a uniform cylinder, and (c) a particle.

■■17. What is the magnitude of the tangential force that must be applied to the rim of a 2.0-kg wheel that is disk-shaped and has a radius of 0.55 m in order to give it an angular acceleration of 4.0 rad/s²? (Neglect friction.)

■■18. A man starts his lawn mower by applying a constant tangential force of 100 N to the flywheel using a starter rope. The flywheel is a cylindrical ring with a mass of 0.30 kg and a diameter of 20 cm. What is the angular speed of the wheel after it has turned through 1 revolution? (Neglect friction and motor compression.)

■■19. In Fig. 8.11, let $m = 2.0$ kg, $M = 0.25$ kg, and $R = 0.15$ m. (a) What is the linear acceleration of the descending block (neglecting friction and the mass of the string)? (b) If the pulley experiences a frictional torque of 0.13 m-N, what is the acceleration?

■■20. For the system shown in Fig. 8.24, $m_1 = 3.0$ kg and $m_2 = 4.0$ kg, and the mass and radius of the pulley are 0.20 kg and 0.10 m, respectively. (a) With what linear acceleration does m_2 descend? (Neglect friction and the string's mass.) (b) What would be the minimum frictional torque on the pulley that would keep the system stationary? (c) If the kinetic frictional torque is $\frac{1}{2}$ of the static

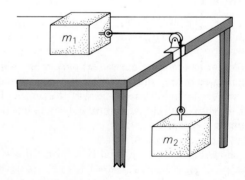

Figure 8.24 **Linear and rotational dynamics**
See Problems 20 and 25.

frictional torque, what is the acceleration of the masses after they are given a slight tap to set them in motion?

■■**21.** If the Earth rotated about a north-south axis tangential to the Equator, how much more rotational inertia would it have?

■■**22.** A disk-shaped 0.25-kg machine part with a radius of 6.0 cm rotates eccentrically about an axis that is normal to its flat surface and located $\frac{2}{3}$ of the distance from the center to the circumference. (a) What torque is required to rotate the part about this off-center axis with an angular acceleration of 2.0 rad/s^2? (b) How much greater is this than the torque required to rotate the part with the same angular acceleration about a parallel axis through the

■■**23.** For the system shown in Fig. 8.25, $m_1 = 8.0$ kg, $m_2 = 3.0$ kg, $\theta = 30°$, and the radius and mass of the pulley are 0.10 m and 0.10 kg, respectively. (a) What is the acceleration of the masses? (Neglect friction and the string's mass.) (b) If the pulley has a constant frictional torque of 0.050 m-N when the system is in motion, what is the acceleration? (Hint: isolate the forces. The tensions in the strings are different. Why?)

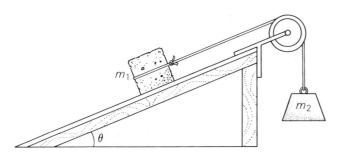

Figure 8.25 **Inclined plane and pulley**
See Problem 23.

■■■**24.** For the system shown in Fig. 7.23 the masses form a rigid body (connecting rods of negligible mass) and $m_1 = m_2 = 0.50$ kg and $m_3 = 0.75$ kg. Find the moment of inertia about the following axes: (a) axes along each side of the triangle and passing through the two equal masses, (b) an axis perpendicular to the plane of the triangle and passing through the center, (c) an axis perpendicular to the plane and passing through m_1, and (d) an axis perpendicular to the plane and passing through m_3.

■■■**25.** For the system of Fig. 8.24 with m_2 greater than m_1 and a static frictional torque on the pulley, find (a) the minimum value of the coefficient of static friction between m_1 and the table that would keep the system at rest, and (b) the value of the acceleration of the masses with ki-

netic friction between m_1 and the plane and a kinetic frictional torque on the pulley.

■■■**26.** For the ball rolling down the incline in Fig. 8.13, what would be the maximum angle of incline in terms of the coefficient of static friction for it to roll without slipping?

8.4 Rotational Work and Kinetic Energy

■**27.** A constant retarding torque of 20 m-N stops a rolling wheel whose diameter is 1.0 m in a distance of 25 m. How much work is done by the torque?

■**28.** A person opens a door by applying a force of 12 N perpendicular to it at a distance of 0.80 m from the hinges. The door is pushed wide open (to 90°) in 2.0 s. (a) How much work was done on the average? (b) What was the average power delivered?

■**29.** Assume that the Earth is a uniform sphere, and compute its rotational kinetic energy.

■**30.** Consider the Earth to be a particle in a circular orbit, and compute its kinetic energy due to revolution about the Sun.

■■**31.** A 20-kg flywheel with its mass concentrated on the outer rim can be considered to be an annular cylinder with an inner diameter of 0.70 m and an outer diameter of 0.80 m. (a) If the flywheel is driven by a motor at a constant speed of 1800 rpm, what is its kinetic energy? (b) After the power to the motor is shut off, the flywheel comes to rest in 30 s as a result of a constant frictional torque. What is the magnitude of this torque?

■■**32.** A 1.5-hp motor drives a machine disk having a mass of 2.5 kg and a diameter of 0.40 m from rest to an operating speed of 600 rpm. (a) What is the average work done by the motor? (b) How long does it take to reach the operating speed?

■■**33.** A 5.0-kg sphere with a diameter of 0.60 m starts from rest at a height of 1.2 m above the base of an inclined plane and rolls down under the influence of gravity. What is the linear speed of the sphere's center of mass just as it leaves the incline and rolls onto a horizontal surface? (Neglect friction.)

■■**34.** A 0.05-kg phonograph record with a radius of 0.15 m drops onto a turntable and is soon rotating at $33\frac{1}{3}$ rpm. How much work must be supplied to get the record to rotate at this speed, and what supplies it?

■■**35.** A spring is twisted 60° about its linear axis (axis of symmetry) by a torque of 100 m-N. How much torsional, or rotational, energy is stored in the spring? [Hint: let the restoring torque of the spring be $\tau = k\theta$, the rotational analogue of $F = kx$ (Hooke's law—see Chapter 5) with the minus sign omitted.]

■■**36.** Use the conservation of mechanical energy to find the linear speed of the descending mass ($m = 1.0$ kg) of Fig. 8.11 after it has descended a vertical distance of 2.0 m from rest. (For the pulley, $M = 0.30$ kg and $R = 0.15$ m. Neglect friction and mass of the string.)

■■**37.** A thin 1.0-m rod pivoted at one end falls (rotates) frictionlessly from a vertical position, starting from rest. What is the angular speed of the rod when it is horizontal? (Hint: consider the center of mass and use the conservation of mechanical energy.)

■■**38.** A sphere with a radius of 25 cm rolls on a level surface with a constant angular speed of 7.5 rad/s. How high on a $30°$ inclined plane will the sphere roll before coming to rest? (Neglect frictional losses.)

■■**39.** A cylindrical hoop, a cylinder, and a sphere are released at the same time from the top of an inclined plane. Show that the sphere always gets to the bottom of the incline first and the hoop last.

■■**40.** For the following objects, which are rolling without slipping, determine the rotational kinetic energy about the center of mass as a percentage of the total instantaneous kinetic energy: (a) solid sphere, (b) a thin spherical shell, (c) a thin cylindrical shell.

■■**41.** A car is traveling at 90 km/h on a straight, level road when the brakes are applied. The coefficients of friction between the car's tires and the dry pavement are $\mu_s = 1.20$ and $\mu_k = 0.85$. Determine (a) the minimum stopping distance and (b) the stopping distance with the wheels locked, considering only static and kinetic frictional forces.

■■■**42.** Prove that a thin hoop or ring starting from rest will roll more slowly down an inclined plane than an annular cylinder will (see Fig. 8.8). (Remember that masses and radii do not have to be taken into account.)

■■■**43.** The initial setup for a classroom demonstration is shown in Fig. 8.26. A hinged board is propped up with a stick. In a depression on the upper end of the board is a steel ball, and a plastic cup is fixed to the board nearer the hinge. When the stick is removed, the board and the ball fall and the ball lands in the cup. (a) Show that this happens only when the stick is propped at an angle less than or equal to a critical angle of $\theta_0 = 35°$. (b) If the hinged board

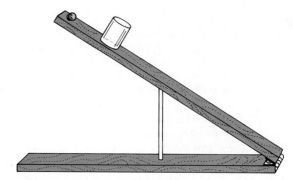

Figure 8.26 A ball-and-cup trick
See Problem 43.

is 1.0 m long and the stick props it up at an angle of $35°$, where should the cup be located on the stick?

■■■**44.** A steel ball rolls down an incline into a loop-the-loop of radius R [Fig. 8.27(a)]. (a) What is the minimum speed the ball must have at the top of the loop in order to stay on the track? (b) At what vertical height (h) on the incline, in terms of the radius of the loop, must the ball be released in order for it to have the required minimum speed at the top of the loop? (Neglect frictional losses.) (c) Fig. 8.27(b) shows a loop-the-loop for a roller coaster.

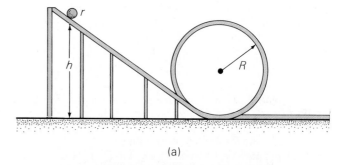

(a)

(b)

Figure 8.27 Loop-the-loop and rotational speed
See Problem 44.

What are the sensations of the riders if the roller coaster has the minimum speed or a greater speed at the top of the loop? (Hint: in case the speed is below the minimum, seat and shoulder straps hold the riders in.)

8.5 Angular Momentum

■45. What is the angular momentum of a 2.0-g particle moving in a horizontal circle of 20-cm radius counterclockwise as viewed from above and with an angular speed of 6π rad/s? (Give magnitude and direction.)

■46. A bicycle wheel has a diameter of 0.60 m and a mass of 2.0 kg. (a) If the bicycle travels with a linear speed of 4.5 m/s, what is the angular speed of the wheel? (b) What is the angular momentum of the wheel? (Consider the wheel to be a thin ring.)

■47. A 0.010-kg ball in an arrangement like that shown in Fig. 8.17 swings initially in a circle having a 10-cm radius at an angular speed of 3π rad/s. If the force on the string is reduced enough that the ball swings in a circle with a radius of 25 cm, what is its angular speed?

■■48. The Earth's orbit about the Sun is slightly elliptical. The points on the orbit at maximum and minimum distances from the Sun are called the aphelion and the perihelion, respectively. (a) Show that the Earth's orbital speed is greater at the perihelion than at the aphelion. (b) If $r_a = 1.52 \times 10^8$ km and $r_p = 1.47 \times 10^8$ km, how many times greater is the Earth's speed at the perihelion than its speed at the aphelion? (c) What is the difference between r_a and r_p in miles?

■■49. Compute the Earth's orbital angular momentum and its spin angular momentum. Are these momenta in the same direction?

■■50. A skater has a moment of inertia of 100 kg-m^2 when his arms are outstretched and a moment of inertia of 75 kg-m^2 when his arms are tucked in close to his chest. If he starts to spin at an angular speed of 2.0 rps with his arms outstretched, what will his angular speed be when they are tucked in?

■■51. If the bicycle wheel of Fig. 8.23 has an angular speed of 8.0 rad/s, what is the approximate angular speed of the stool and the person holding the wheel if the angular momentum is conserved? (Hint: estimate the masses and radii of the rotating bodies. Treat the person as a solid cylinder and the stool as a disk, and assume that they have a combined moment of inertia of $I = I_1 + I_2$. See Problem 15.)

■■52. Suppose that the person holding the string in Fig. 8.17 gets tired and hangs a weight on the string to supply the force when the angular speed of the 0.010-kg ball is 37 rad/s and the radius of the circular path is 0.15 m. (a) What mass should the weight have to keep the ball in this circular path? (b) If the weight had 10% less mass than this, what would happen?

■■53. A 75-kg person stands at the outer edge of a merry-go-round platform whose radius is 8.0 m and whose mass is 450 kg. The platform is rotating at a speed of $\pi/2$ rad/s. Assuming the platform to be a freely rotating disk (no axle friction), what would be its angular speed if the person walked to its center? (Hint: see Problem 15.)

■■■54. A man sitting on a stool that rotates is handed a rotating bicycle wheel with an angular momentum $\mathbf{L}_o$ (see Fig. 8.23). In one instance, the axis of rotation of the bicycle wheel is horizontal when it is handed to him, and he turns it so it is vertical with $\mathbf{L}_o$ pointing downward. In another instance, the bicycle wheel is handed to him with the axis of rotation vertical and $\mathbf{L}_o$ pointing downward, and he turns the wheel over so $\mathbf{L}_o$ is pointing upward. Is the final angular speed of the stool and the man the same in each case? Explain.

■■■55. A 0.50-kg kitten stands on the edge of a lazy susan (a kind of turntable) that has a mass of 1.5 kg and a radius of 0.30 m. Assume that the lazy susan has frictionless bearings and is initially at rest. (a) What will happen if the kitten starts walking around the edge of the lazy susan? (b) If the kitten walks at a speed of 0.25 m/s relative to the ground, what will the angular speed of the lazy susan be? (c) When the kitten has walked completely around the edge and is back at its starting point, will that point be above the same point on the ground as it was at the start? If not, where is the kitten relative to the starting ground point? (Speculate on what might happen if everyone on Earth suddenly started to run eastward. What effect might this have on the length of a day?)

■■■56. A disk with a moment of inertia I about its axis of symmetry is dropped onto another disk with a moment of inertia twice as large as I that is spinning freely at a rate of $33\frac{1}{3}$ rpm. (Think of a record dropping onto a turntable.) (a) What is the angular speed of the combination if there are no external torques acting on the system? (b) Show that this is an inelastic collision in the sense that the rotational kinetic energy is not conserved. Where did the lost energy go?

Additional Problems

57. A 1.0-kg ball rolls without slipping on a level surface. If the ball rolls in a straight line at a speed of 0.50 m/s, what is its kinetic energy?

58. A mouse running inside a rotating cylindrical cage with a radius of 30 cm turns the cage at an average rate of 20 rpm. If the mouse were running on a level surface, (a) what would be its average speed, and (b) how far would it run in 1 min?

59. An ice skater doing a toe spin with outstretched arms has an angular speed of 6.0 rad/s. The skater then tucks in her arms, decreasing her moment of inertia by 7.5%. (a) What is the resulting change in the angular speed? (b) By what factor does the skater's kinetic energy change? (Neglect any frictional effects.)

60. A 15-kg uniform sphere with a radius of 25 cm rotates about an axis tangent to its surface at a rate of 3.0 rad/s. A constant torque of 10 N-m then increases the rotational speed to 6.0 rad/s. Through what angle does the sphere rotate while accelerating?

61. A 0.25-kg pulley has a diameter of 12 cm. Two masses, 0.10 kg and 0.15 kg, are suspended from the ends of a light string passing over the pulley. (a) Neglecting friction, what will the acceleration be when the masses are moving under the influence of gravity? (b) What is the magnitude of the combined kinetic frictional torque (due to both friction of the pulley bearings and friction between the string and the pulley) that would keep the masses moving at a constant speed? (Hint: isolate the forces. The tensions in the strings are different. Why?)

62. A thin rectangular plate has dimensions of 25 cm by 40 cm and a mass of 0.50 kg. For which axis of rotation does the plate have the greater rotational inertia, an axis tangent to the midpoint at one end or an axis tangent to the midpoint of a side? How many times more? (Both axes are perpendicular to the plate surface.)

63. A pencil 20 cm long is initially held so that it stands vertically on its sharpened point on a table by a person with a finger on the eraser end. When the finger is removed, the pencil falls over onto the table. Assuming that the point does not slip, at what speed is the eraser end of the pencil moving when it strikes the horizontal surface?

64. A uniform hoop rolls without slipping down a plane with an incline of 20° above the horizontal. What is the tangential acceleration of the hoop?

65. Two 0.015-kg bullets travel at a speed of 200 m/s in opposite directions along straight-line parallel paths separated by a distance of 1.0 m. What is the magnitude of the angular momentum of the bullet system? (Does this depend on the origin of the chosen reference frame? Explain.)

66. A ball with a mass of 0.25 kg is suspended from the ceiling with a 1.5-m string of negligible mass. The ball is pulled to the side so the taut string makes an angle of 37° with the vertical. If the ball is then released from rest, what is its angular speed as it passes through the lowest point on its path?

67. An ultracentrifuge (a high-speed centrifuge) can be brought from rest to a rotational speed of 500,000 rpm uniformly in 3.0 min. If the rotating part of the centrifuge has a moment of inertia of 2.8 kg-m^2, how much power is delivered by the driving motor to do this? (Neglect frictional losses.)

68. In an ice show, two 70-kg skaters, both traveling at 8.0 m/s, approach each other along straight-line parallel paths that are separated by a distance of 4.0 m. When directly opposite each other, the skaters grab the ends of a light rod 4.0 m in length. What is the initial rotational speed of the joined skaters? (Hint: use the conservation of angular momentum.)

69. A uniform sphere and a uniform cylinder with the same mass and radius roll side by side on a level surface without slipping and at the same velocity. If the sphere and the cylinder approach an inclined plane and roll up it without slipping, will they be at the same height on the plane when they come to a stop? If not, what will be the percent height difference?

70. A comet approaches the Sun as illustrated in Fig. 8.28 and is deflected by the Sun's gravitational attraction. This may be considered a collision, and b is called the impact parameter. Find the distance of closest approach (d) in terms of the impact parameter and the velocities. Consider the radius of the Sun to be negligible compared to d. (As the figure shows, the tail of a comet always "points" away from the Sun.)

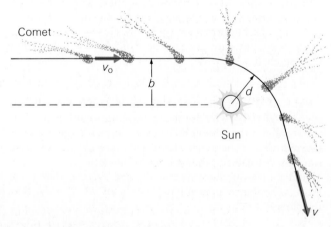

Figure 8.28 **A comet collision**
See Problem 70.

Mechanical Properties

9

A major focus of this course of study has been unbalanced forces or torques and the resulting changes in the motion of an object that is subjected to such a force. A very different situation results, though, when the net force or net torque on an object is zero. The body is then said to be in **equilibrium**. You may think of an object in equilibrium as being static, or at rest, but it could just as well have unchanging motion (a constant velocity).

This chapter is concerned with bodies that are in static equilibrium, that is, at rest. It is particularly comforting to know, for instance, that a bridge you are crossing is in static equilibrium and not subject to major unbalanced forces. And you'd feel the same at the top of a tall building. Seemingly rigid bodies such as bridges and buildings may, and often do, sway or vibrate in response to minor external forces without falling down. Such motions indicate internal restoring forces, which are generally characterized by the elasticity of materials.

Elasticity usually brings to mind an elastic rubber band, a spring, or something that will resume its original dimensions after being deformed. However, all materials are elastic to some degree, even very hard steel. (Recall elastic collisions in Chapter 6.)

A simplistic description of solids is that they are made up of small indivisible particles called atoms that are held rigidly together by interatomic forces. This concept of the *ideal* rigid body was used in Chapter 8 to describe rotational motion. Real solid bodies, though, are not absolutely rigid and can be elastically (and inelastically) deformed by external forces. But you will learn that there is a limit—an elastic limit.

9.1 Mechanical Equilibrium

According to Newton's first law, when the sum of the forces on a body is zero, it remains either at rest or in motion with a constant velocity. In either case, the body is in **translational equilibrium**. Stated another way, the **condition for translational equilibrium** is $\Sigma_i \mathbf{F}_i = 0$. That this condition is satisfied is readily apparent for the situations illustrated in Fig. 9.1(a) and (b). Forces with lines of action through the same point are called **concurrent forces**.

But what about the situation pictured in Fig. 9.1(c)? Here $\Sigma_i \mathbf{F}_i = 0$, but the opposing forces will cause the object to rotate, and it will clearly not be in a state of static equilibrium. Such a pair of equal and opposite forces not having the same line of action is called a **couple**. Thus, the condition $\Sigma_i \mathbf{F}_i = 0$ is a necessary but not sufficient condition for static equilibrium. For an extended body, the point of application of a force, in addition to its magnitude and direction, is important.

Since $\Sigma_i \mathbf{F}_i = 0$ is the condition for translational equilibrium, you might predict (and correctly so) that $\Sigma_i \tau_i = 0$ is the **condition for rotational equilibrium**. That is, if the sum of the torques acting on an object is zero, then the object is rotationally at rest or rotates with a constant angular velocity.

> A body is said to be in mechanical equilibrium when the conditions for both translational and rotational equilibrium are satisfied:
>
> $$\sum_i \mathbf{F}_i = 0 \qquad \textbf{(for translational equilibrium)}$$
>
> $$\sum_i \tau_i = 0 \qquad \textbf{(for rotational equilibrium)}$$
>
> (9.1)

A rigid body in mechanical equilibrium may be either at rest or moving with a constant linear and/or angular velocity. An example is an object rolling without slipping on a level surface with the center of mass having a constant velocity. Of course, this is an ideal condition because there is always some friction in reality. Of greater practicality is **static equilibrium**, when a rigid body is at rest. There are many instances in which we do not want things to move, and this absence of motion can only occur if the equilibrium conditions are satisfied, as the following examples show.

Example 9.1 Translational Static Equilibrium

A picture hangs on a wall as shown in Fig. 9.2. If it has a mass of 5.0 kg, what are the tension forces in the wires?

Solution

Since the forces are concurrent, that is, acting through one point, and the point is on what would be the axis of rotation (along the nail), the condition for rotational equilibrium ($\Sigma_i \tau_i = 0$) is automatically satisfied (zero moment arms). Thus, you only have to consider translational equilibrium.

You will find it helpful to isolate the forces in a free-body diagram, as shown in the figure. In this case, the diagram shows the forces acting on the nail. The reaction force of the wall on the nail is equal in magnitude to the weight of the picture. Note that R and mg are not a third law force pair because they act on the same object—the nail. Since the nail doesn't move, the forces are equal and opposite by Newton's second law.

With the picture in static equilibrium, the net force is zero; that is, $\Sigma_i \mathbf{F}_i = 0$. Thus, the sums of the rectangular components are also zero: $\Sigma_i \mathbf{F}_{x_i} = 0$ and $\Sigma_i \mathbf{F}_{y_i} = 0$. Then

$$\sum_i \mathbf{F}_{x_i}: \qquad T \cos 45° - T \cos 45° = 0$$

This really doesn't tell you anything, except that the x components of the forces are equal and opposite by symmetry.

Note that Newton's first law for translational motion applies to a particle or the center of mass of an extended object.

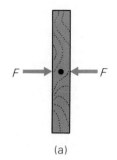

(a)

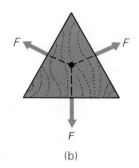

(b)

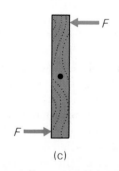

(c)

Figure 9.1 **Equilibrium and forces**
Forces with lines of action through the same point are said to be concurrent. The net force of the concurrent forces acting on the objects in parts (a) and (b) is zero, and the objects are in (static) equilibrium. In part (c), the object is in translational equilibrium, but it will rotate.

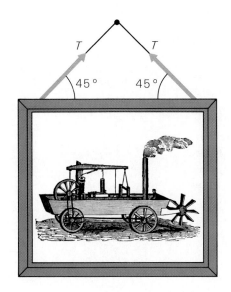

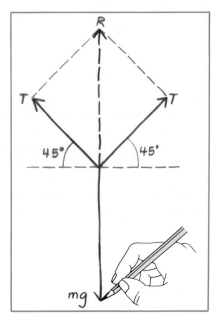

Figure 9.2 **Translational static equilibrium**
Since the picture hangs motionless on the wall, the sum of the forces acting on the nail must be zero. The forces are illustrated in a free-body diagram. See Example 9.1.

Free–body diagram

For the y components,

$$\sum_i \mathbf{F}_{y_i}: \qquad R - T\sin 45^\circ - T\sin 45^\circ = 0$$

or $\quad mg - 2T\sin 45^\circ = 0$

Thus,

$$T = \frac{mg}{2\sin 45^\circ} = \frac{(5.0\ \text{kg})(9.8\ \text{m/s}^2)}{2(0.707)} = 35\ \text{N} \quad \blacksquare$$

Example 9.2   Rotational Static Equilibrium

Three masses are suspended from a meterstick as shown in Fig. 9.3. How much mass must be suspended on the right side to have the system be in static equilibrium? (Neglect the mass of the meterstick.)

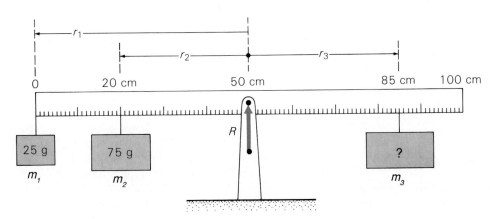

Figure 9.3 **Rotational static equilibrium**
For the meterstick to be in rotational equilibrium, the sum of the torques acting on it must be zero. See Example 9.2.

Solution

Given: $m_1 = 25$ g   *Find:* m_3
$\quad\quad\quad r_1 = 50$ cm
$\quad\quad\quad m_2 = 75$ g
$\quad\quad\quad r_2 = 30$ cm
$\quad\quad\quad r_3 = 35$ cm

The condition for translational equilibrium is automatically satisfied by the upward vertical reaction force of the stand on the stick. This reaction force balances the stick's weight at the center of mass; therefore, $R = Mg$, or $R - Mg = 0$, where M is the total mass. However, the weight forces of the suspended masses produce torques that tend to rotate the stick about an axis passing through the support point. So you need to apply the condition for rotational equilibrium. Summing the torques about that axis gives

$$\sum_i \tau_i: \quad\quad m_1 g r_1 + m_2 g r_2 - m_3 g r_3 = 0$$

or $\quad m_1 g r_1 + m_2 g r_2 = m_3 g r_3$

The counterclockwise torques equal the clockwise torques. [Note that the moment arms are measured from the support (pivot) point. The torques tending to produce a counterclockwise rotation, or an increasing angular distance as measured from the positive horizontal axis, are taken to be positive. This is arbitrary. The key thing to remember is that the magnitude of the sum of the counterclockwise torques must equal that of the clockwise torques for rotational equilibrium.]

The g's cancel out, so solving for m_3 gives

$$m_3 = \frac{m_1 r_1 + m_2 r_2}{r_3}$$

$$= \frac{(25 \text{ g})(50 \text{ cm}) + (75 \text{ g})(30 \text{ cm})}{35 \text{ cm}} = 100 \text{ g}$$

[The mass of the stick was neglected. If the stick is uniform, however, its mass will not affect the equilibrium. Why?] ∎

Sometimes the conditions for both translational and rotational equilibrium need to be written explicitly to solve the problem. The next example is one such case.

Example 9.3 Static Equilibrium

A ladder with a mass of 15 kg rests against a smooth wall (Fig. 9.4). A painter, who has a mass of 78 kg, stands on the ladder as shown in the figure. What frictional force must act on the bottom of the ladder to keep it from slipping?

Solution

Given: $m_l = 15$ kg   *Find:* f_s
$\quad\quad\quad m_m = 78$ kg
$\quad\quad\quad$ Distances given in figure

Since the wall is smooth, there is negligible friction between it and the ladder, and only the normal reaction force of the wall acts on the ladder at this point. In applying the conditions for static equilibrium, you are free to choose any axis of

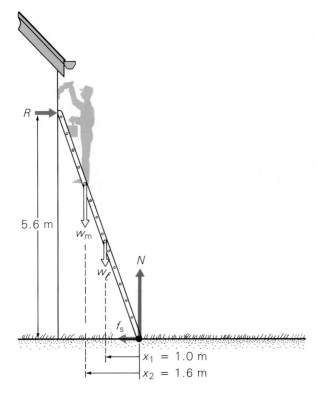

R

5.6 m

w_m

w_ℓ

N

f_s

$x_1 = 1.0$ m

$x_2 = 1.6$ m

Figure 9.4 **Static equilibrium**
For the painter's sake, the ladder has to be in static equilibrium; that is, both the sum of the forces and the sum of the torques must be zero. See Example 9.3.

rotation for the rotational condition. (The conditions must hold for any part of a system for static equilibrium; that is, there can't be motion in any part of the system.) Note that choosing an axis at the end of the ladder where it touches the ground eliminates the torques due to f_s and N since the moment arms are zero. Then you can write three equations (using mg for w):

$$\sum_i \mathbf{F}_{x_i}: \qquad R - f_s = 0$$

$$\sum_i \mathbf{F}_{y_i}: \qquad N - m_m g - m_l g = 0$$

and

$$\sum_i \tau_i: \qquad Ry - (m_l g)x_1 - (m_m g)x_2 = 0$$

The weight of the ladder is considered to be concentrated at its center of gravity. Solving the third equation for R and substituting the given values for the masses and distances gives

$$R = \frac{(m_l g)x_1 + (m_m g)x_2}{y}$$

$$= \frac{(15 \text{ kg})(9.8 \text{ m/s}^2)(1.0 \text{ m}) + (78 \text{ kg})(9.8 \text{ m/s}^2)(1.6 \text{ m})}{5.6 \text{ m}}$$

$$= 250 \text{ N}$$

From the first equation, then,

$$f_s = R = 250 \text{ N} \quad \blacksquare$$

When working problems involving static equilibrium, you should do two things:

1. Always draw free-body diagrams for situations involving concurrent forces (static translational equilibrium).
2. Select a convenient axis of rotation about which to sum torques. An axis through the line of action of one or more forces eliminates torques due to those forces from the sum because it establishes zero moment arms. Make certain that the moment arms are measured *from* the axis of rotation.

9.2 Stability and the Center of Gravity

The equilibrium of a particle or a rigid body can also be described as being stable, unstable, or neutral in a gravitational field. For rigid bodies, these categories of equilibria are conveniently analyzed in terms of the center of gravity. Recall from Chapter 6 that the center of gravity is the point at which all the weight of an object may be considered to be concentrated in representing it as a particle. And, in a uniform gravitational field, the center of gravity and the center of mass coincide. Basically, stable equilibrium occurs when the gravitational potential energy is at a minimum, and unstable equilibrium when it is at a maximum. For neutral equilibrium, the potential energy is constant. How is potential energy related to stability and the center of gravity?

If an object is in **stable equilibrium**, any small displacement results in a restoring force or torque, which tends to return the object to its original equilibrium position. As illustrated in Fig. 9.5, a ball in a bowl is in stable equilibrium. Analogously, the center of gravity of an extended body in stable equilibrium is essentially in a potential energy bowl. Any slight displacement raises its center of gravity, and a restoring force tends to return it to the position of minimum potential energy. This force is actually a torque that is due to a component of the weight force and that tends to rotate the object about a pivot point back to its original position.

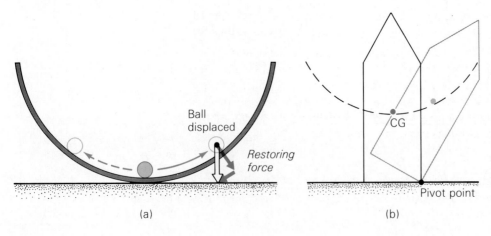

(a)

(b)

Figure 9.5 **Stable equilibrium**
Any small displacement from an equilibrium position results in a force or torque that tends to return an object to that position. (a) A ball in a bowl returns to the bottom after being displaced. (b) Note how the center of gravity of an extended object can be depicted as being in a potential energy bowl.

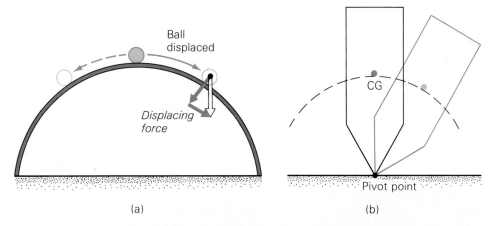

Ball
displaced

Displacing
force

(a)

CG

Pivot point

(b)

Figure 9.6 **Unstable equilibrium**

For an object in unstable equilibrium any small displacement from its equilibrium position results in a force or torque that tends to take it further away from that position. (a) This ball is on top of an overturned bowl. (b) For this extended object, note how the center of gravity is on top of an inverted potential energy bowl.

For an object in **unstable equilibrium**, any small displacement from equilibrium results in a force or torque that tends to take the object further away from its equilibrium position. This situation is illustrated in Fig. 9.6. Note that the center of gravity is at the top of an overturned, or inverted, potential energy bowl, that is, the potential energy is at a maximum in this case.

For an object **neutral equilibrium**, a slight displacement produces no restoring force or torque (Fig. 9.7). Here the potential energy of the object remains constant. The center of gravity is neither raised nor lowered.

Small displacements or slight disturbances have profound effects on objects in unstable equilibrium—it doesn't take much to cause such an object to change its position [Fig. 9.8(a)]. On the other hand, the angular displacement of an object in stable equilibrium can be quite substantial, and the object will still be restored to its equilibrium position. Such a case is also illustrated in Fig. 9.8(a). An object lying on a long side can be rotated through quite a distance and will still fall back to that original position.

This is another way of expressing the condition of stable equilibrium:

> An object is in stable equilibrium as long as its center of gravity lies above and inside its original base of support.

When this is the case, there will always be a restoring torque. However, when the center of gravity or center of mass falls outside the base of support, over goes the object—because of a gravitational torque that rotates it away from its equilibrium position.

Rigid bodies with wide bases and low centers of gravity are therefore more stable and less likely to tip over. This relationship is evident in the design of high-speed race cars, which have wide wheel bases and centers of gravity close to the ground [Fig. 9.8(b)]. Also, the location of the center of gravity of the human body has an effect on certain physical abilities. For example, women can generally bend over and touch their toes or touch their palms to the floor more easily than can men, who often fall over trying. On the average, men have higher centers of gravity (larger shoulders) than do women (larger pelvises), so it is more likely that a man's center of gravity will be outside his base of support when he bends over. (See Demonstration 8 on page 254.)

Example 9.4 Stack Them Up!

Uniform, identical bricks 20 cm long are stacked so that 4.0 cm of a brick extends beyond the brick beneath, as shown in Fig. 9.9. How many bricks can be stacked in this way before the stack falls over?

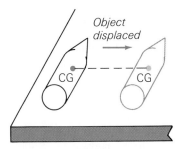

Object
displaced

CG   CG

Figure 9.7 **Neutral equilibrium**
A slight displacement of an object in neutral equilibrium produces no restoring force or torque, and the potential energy remains constant. (The center of gravity is neither raised nor lowered.)

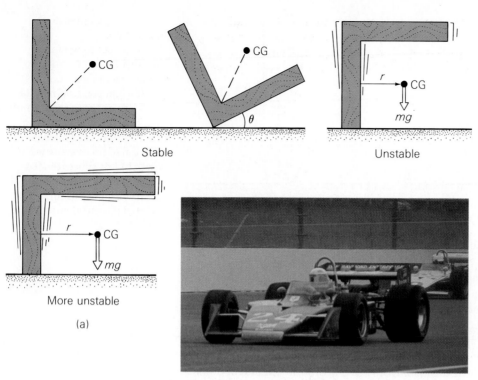

Stable   Unstable

More unstable

(a)

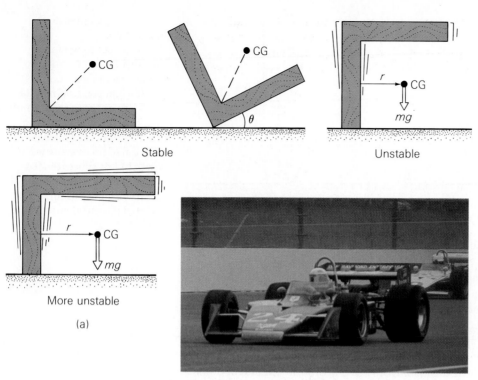

(b)

Figure 9.8 **Stable and unstable equilibrium** (a) When the center of gravity is outside an object's base of support, the object is in unstable equilibrium (there is a displacing torque). When the center of gravity is above and inside the base of support, the object is in stable equilibrium (there is a restoring torque). (b) Race cars have high stabilities because of their wide wheel bases and low centers of gravity.

Solution

The stack will fall over when its center of mass is no longer above its base of support, which is the bottom brick. All of the bricks have the same mass, and the center of mass of each is located at its midpoint. The horizontal coordinate of the center of mass (or center of gravity) for the first two bricks of the stack ($m_1 = m_2 = m$) is

$$X_{CM2} = \frac{mx_1 + mx_2}{m + m}$$

$$= \frac{m(x_1 + x_2)}{2m} = \frac{x_1 + x_2}{2} = \frac{0 + 4.0 \text{ cm}}{2} = 2.0 \text{ cm}$$

Note that the masses of the bricks cancel out (since they are all the same) and that the origin is taken to be at the center of mass of the bottom brick. For three bricks,

$$X_{CM3} = \frac{m(x_1 + x_2 + x_3)}{3m} = \frac{0 + 4.0 \text{ cm} + 8.0 \text{ cm}}{3} = 4.0 \text{ cm}$$

For four bricks,

$$X_{CM4} = \frac{m(x_1 + x_2 + x_3 + x_4)}{4m} = \frac{0 + 4.0 \text{ cm} + 8.0 \text{ cm} + 12.0 \text{ cm}}{4} = 6.0 \text{ cm}$$

and so on.

This series of results shows that the center of mass of the stack moves horizontally 2.0 cm for each brick added to the bottom one. For a stack of six bricks, the center of mass is 10 cm from the origin and directly over the edge of

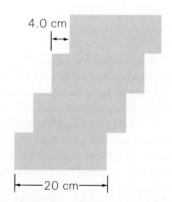

4.0 cm

20 cm

Figure 9.9 **Stack them up!** How many bricks can be stacked like this before the stack falls over? See Example 9.4.

the bottom brick (2.0 cm × 5 *added* bricks = 10 cm, which is $\frac{1}{2}$ of the length of the bottom brick), so the stack is in unstable equilibrium. The stack may not topple if the sixth brick is positioned very carefully, but it is doubtful that this could be done in practice. In any case, the seventh brick would definitely cause the stack to fall. ■

Example 9.5 Leaning into a Curve
When riding a bicycle on a level surface and going around a curve or making a circular turn, the rider usually leans inward, or into the curve (Fig. 9.10). Why is this? (The gentleman on the bicycle knew.)

Solution
When a vehicle goes around a level circular curve, a centripetal force is needed to maintain the circular path. As you learned in Chapter 7, this force is supplied by the force of static friction between the tires and the road [Fig. 9.10(b)]. The reaction force **R** of the ground on the bicycle is the sum of the friction force $\mathbf{f}_s$ and the normal force **N**. Note that **R** does not pass through the system's center of gravity when the rider does not lean inward when going around the curve [Fig. 9.10(a)]. If a rider did not lean inward, the wheels would slide inward underneath him, since there would be a counterclockwise torque that would rotate them inward. However, if the rider leans inward at the proper angle, **R** goes through the center of gravity and there is no instability.

The need to lean into a curve is readily apparent in bicycle or motorcycle races on level tracks. Things can be made easier for the riders if tracks or roadways are banked to provide a natural lean. ■

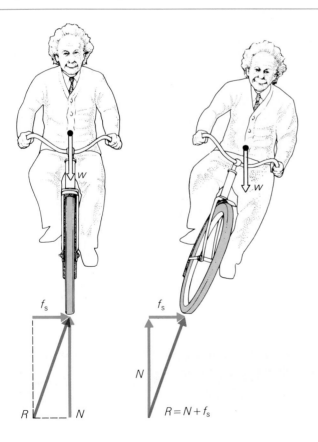

Figure 9.10 **Leaning into a curve**
When rounding a curve or making a turn, a bicycle rider must lean into the curve. See Example 9.5.

Stability and Center of Gravity

A demonstration to show that women generally have more stability than men—because of mass distribution and location of the center of gravity.

(a) An object is placed just beyond the fingertips of a man kneeling with elbows against his knees. Maintaining his balance, he can topple the object with his nose.

(b) With hands now clasped behind his back, the man's center of gravity shifts beyond his base of support (knees) as he bends over to topple the object, and he topples instead.

(c) A woman can generally do it both ways.

(d) Because her center of gravity is positioned lower in the body, it does not move outside the base of support (past her knees) as she bends with hands clasped behind her back.

9.3 Mechanical Advantage

In general, people use machines to make it easier to do work. A machine gives a mechanical advantage but does not save mechanical work or energy. In essence,

> **A machine is any device used to change the magnitude or direction of a force.**

Recall from Chapter 5 that work is done when an applied force acts through a parallel distance ($W = Fd$). Machines make tasks easier by multiplying or increasing the applied force (or by increasing the speed with which the task is performed). But *work or energy is never multiplied by a machine.* You can't get something for nothing.

A simple machine is usually used to exert an output force or torque greater than the input, or applied, force. The magnitude of the force multiplication is called the **mechanical advantage**. This is simply the ratio of the output force to the input force.

In real situations, frictional losses always occur. Thus, the output force is always less in practice than it would be for an ideal frictionless case. So the **actual mechanical advantage (AMA)** is

$$\text{AMA} = \frac{F_o}{F_i} \tag{9.2}$$

Note that $F_o = (\text{AMA})F_i$, and the AMA is the force multiplication factor for a machine. For example, if a machine has an AMA of 2, this means that if an input force of 100 N is applied, the output force will be 200 N. Does this sound too good to be true? What has to be given up to achieve this gain can be seen by considering the work done.

In order for the input and output forces to do work, they must travel through the parallel distances d_i and d_o, respectively. Then, by the conservation of work and energy,

work in = work out

or $\qquad F_i d_i = F_o d_o + W_f$

where $F_o d_o$ is the useful work output and W_f is the work done against friction, or the energy lost. Note that if the output force (F_o) is greater than the input force (F_i), then d_i must be greater than d_o. This means that an increase in force (a force multiplication) is gained at the expense of moving through a greater distance. That is, the input force moves a greater distance than the output force or the load does.

This is easy to see for the theoretical case of an ideal or frictionless machine. If $W_f = 0$, then $F_o d_o = F_i d_i$. For this ideal case, the **theoretical mechanical advantage (TMA)** is

$$\text{TMA} = \frac{F_o}{F_i} = \frac{d_i}{d_o}$$

That is, in general,

$$\text{TMA} = \frac{d_i}{d_o} \tag{9.3}$$

The TMA is the mechanical advantage, or force multiplication factor, for a machine if frictional losses could be eliminated. It is a maximum theoretical limit that can be conveniently computed solely from geometrical measurements (d_i and d_o) without considering forces. The theoretical mechanical advantage is sometimes called the *ideal mechanical advantage* (IMA).

It's useful to apply these observations to some simple machines. In general, there are six simple mechanical machines: the lever, the pulley, the wheel and axle, the inclined plane, the screw, and the wedge. By similarity of operating principles, these can be grouped into two basic classes: levers and inclined planes. The pulley and the wheel and axle are types of levers and the screw and the wedge are types of inclined planes.

The lever has many practical applications, but is basically used to lift loads, as illustrated in Fig. 9.11. A very large output force may be achieved by exerting a relatively small input force. These forces produce torques about the fulcrum, that is, the pivot point or axis of rotation. Thus, the forces act through arc lengths. The lever arms on each side of the fulcrum travel through the same angular distance θ, so

$$d_i = s_i = L_i\theta \quad \text{and} \quad d_o = s_o = L_o\theta$$

Substituting from these equations into Eq. 9.3 gives

$$\text{TMA} = \frac{L_i}{L_o} \quad \text{(for a lever)} \tag{9.4}$$

The mechanical advantage for a lever can be made quite large by increasing L_i and/or shortening L_o. The Greek scientist Archimedes no doubt understood this when he said, "Give me a lever long enough and a fulcrum on which to rest it, and I will move the Earth." Simple levers are quite common. You use one when you pull out a nail with a claw hammer, pry open a lid with a screw driver, or nod your head. (The fulcrum in the last case is at the top of the spinal column, a pivot joint. The human body has quite a few levers.) Keep in mind that the force advantage gained by doing something with a lever (and with all machines) is gained at the expense of acting through a greater distance.

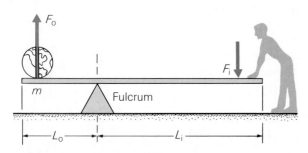

(a) Static equilibrium: $\Sigma\tau = 0 = F_oL_o - F_iL_i$

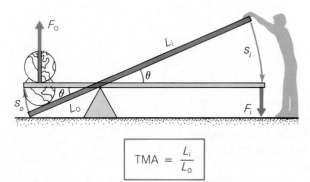

$$\boxed{\text{TMA} = \frac{L_i}{L_o}}$$

(b) Work input = work output

$$F_i s_i = F_o s_o$$

Figure 9.11 **The lever**
(a) When the lever is in static equilibrium, the sum of the torques about the fulcrum is zero. (b) When a lever is used to lift an object, s_i and s_o are the input and output distances, respectively, and, in general, $s = L\theta$ and $\text{TMA} = L_i/L_o$.

The lever is very efficient, having little frictional loss at the small contact area of the fulcrum. The efficiency of a machine is an important quantity. It tells you what you get out for what you put in, so to speak. More precisely, **mechanical efficiency (ε)** is the ratio of the *useful* work output to the work input:

$$\varepsilon = \frac{\text{useful work output}}{\text{work input}}$$

$$\varepsilon = \frac{F_o d_o}{F_i d_i} = \frac{F_o/F_i}{d_i/d_o} = \frac{\text{AMA}}{\text{TMA}} \tag{9.5}$$

This ratio is a fraction, but is commonly expressed as a percentage.

Note that to have an efficiency of 1.0, or 100%, the AMA would have to equal the TMA, an ideal frictionless condition. If 100% efficiency could be attained, a so-called perpetual motion machine could be devised. Using the machine's output as input would cause it to run forever. Even better, if the efficiency were greater than 100%, some energy could be siphoned off somehow and the machine would still continue to run. What you're probably thinking is correct. An efficiency greater than 100% would mean that energy is being created—that more is coming out than was put in. This has never been done, and if it were, the basic law of conservation of energy would be violated and would have to be reformulated.

Example 9.6 Mechanical Advantage and Efficiency

A worker in a warehouse uses a lever to lift a 330-N crate by applying a force of 145-N to the opposite end of the lever. If the lever arms are 1.50 m and 0.60 m, what are the (a) AMA, (b) TMA, and (c) efficiency of the lever? (Neglect the mass of the lever.)

Solution

Given: $F_o = 330$ N   *Find*: (a) AMA
           $d_o = 0.60$ m   (b) TMA
           $F_i = 145$ N   (c) ε
           $d_i = 1.50$ m

(a) By Eq. 9.2, $\text{AMA} = \dfrac{F_o}{F_i} = \dfrac{330 \text{ N}}{145 \text{ N}} = 2.28$

(b) By Eq. 9.4, $\text{TMA} = \dfrac{L_i}{L_o} = \dfrac{1.50 \text{ m}}{0.60 \text{ m}} = 2.5$

(c) By Eq. 9.5, $\varepsilon = \dfrac{\text{AMA}}{\text{TMA}} = \dfrac{2.28}{2.5} = 0.91 \ (\text{x } 100\%) = 91\%$

This is a relatively high efficiency compared to those for some complex machines. ∎

Machines are not always used to increase force. In some instances, a distance multiplication is desired, and it can be obtained at the expense of force. A familiar example is shoveling something, such as dirt, from one place to another. The fulcrum in this case is at the end of the shovel's handle, and the input force (applied to the handle between the fulcrum and the load) is greater than the output force. The TMA is less than 1 in this case (Why?)

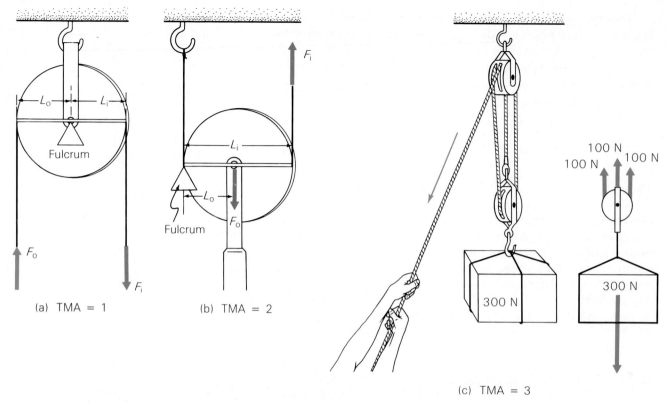

Figure 9.12 **The pulley**
A pulley is a continuous lever with the lever arms as shown here. (a) For a fixed
pulley, TMA = L_i/L_o = 1, and the pulley is simply a direction changer. (b) For a single
movable pulley, TMA = L_i/L_o = 2. (c) A group of pulleys (or sheaves) is called a block,
and the ropes (or supports) are referred to as the tackle; the combination is therefore
known as a block and tackle. The TMA of a block and tackle is equal to the number
of strands of rope supporting the movable block.

Fig. 9.12 illustrates why pulleys are classified as levers. Pulleys are said to be
continuous levers. The lever arms in Fig. 9.12(a) show that a fixed pulley has a
TMA of 1, which means it simply acts as a direction changer and not a force
multiplier. A single movable pulley has a TMA of 2 [Fig. 9.12(b)]. This requires
the input force to move twice as far as the load itself moves. For example, to
raise the load a distance of 1.0 m, you would have to pull up a 2.0-m length of
rope. This principle applies to the system of pulleys called a block and tackle
[Fig. 9.12(c)]; in general, *the number of support strands of the movable block is
equal to the TMA.*

The other basic type of simple machine is the inclined plane (Fig. 9.13). It
may be difficult to think of an inclined plane as a machine, but it does supply a
mechanical advantage: applying a relatively small input force through a long
distance will raise a heavier load (a larger force) a shorter distance. Work is
done against only a component of the weight force of a load in moving it up an
inclined plane, so less force is required to do this than to lift the load directly to
the top of the plane. The TMA is

$$\text{TMA} = \frac{d_i}{d_o} = \frac{L}{h} = \frac{1}{\sin\theta} \qquad \text{(for an inclined plane)} \qquad (9.6)$$

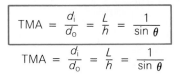

$$\boxed{\text{TMA} = \frac{d_i}{d_o} = \frac{L}{h} = \frac{1}{\sin\theta}}$$

$$\text{TMA} = \frac{d_i}{d_o} = \frac{L}{h} = \frac{1}{\sin\theta}$$

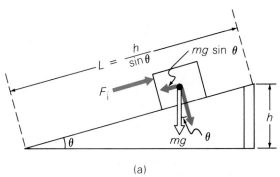

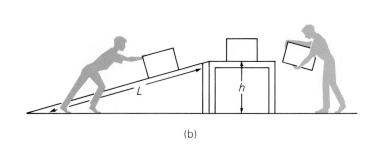

(a)

(b)

Figure 9.13 **The inclined plane**
(a) An inclined plane makes it easier to raise an object because work is done against only a component of the weight force. (b) The input force must act through a greater distance than when raising the object directly upward; L versus h here.

where L is the length of the plane's inclined surface and h is its vertical height. You can increase the TMA by decreasing the angle of incline, but in doing so you make the length of the incline longer for the same vertical height.

9.4 Elastic Moduli

All materials are elastic to some degree. That is, a body that is slightly deformed by an applied force will return to its original dimensions when the force is removed. The deformation may not be noticeable for many materials, but it's there. You may be able to visualize why materials are elastic if you think in terms of the simplistic model of a solid in Fig. 9.14. The atoms of the solid

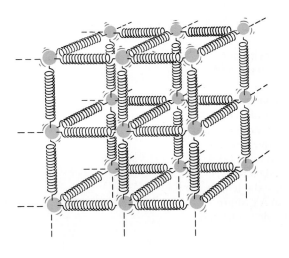

Figure 9.14 **A springy solid**
The elastic nature of interatomic forces is indicated by simplistically representing them as springs, which similarly resist deformation.

substance are imagined to be held together by springs. The elasticity of the springs represents the resilient nature of the interatomic forces. The springs resist permanent deformation, as do the short-range forces between atoms.

The elastic properties of solids are commonly discussed in terms of stress and strain. **Stress** is the quantity that describes the force causing a deformation. **Strain** is a relative measure of the deformation a stress causes. Quantitatively, *stress is the applied force per cross-sectional area*:

$$\text{stress} = \frac{F}{A} \tag{9.7}$$

Here F is the magnitude of the applied force normal (perpendicular) to the cross-sectional area. The SI units for stress are newtons per square meter (N/m^2).

As illustrated in Fig. 9.15, a force applied to the ends of a rod gives rise to either an elongating **tension**, or **tensile stress**, or a **compressional stress**, depending on the direction of the force. In these cases, *the strain is the ratio of the change in length to the original length*:

$$\text{strain} = \frac{\text{change in length}}{\text{original length}} = \frac{\Delta L}{L_o} \tag{9.8}$$

where $\Delta L = L - L_o$. Note that strain is a unitless quantity (length/length). It is the *fractional change* in length. For example, if the strain is 0.05, the material has changed in length by 5% of its original length.

As might be expected, the strain is proportional to the stress; that is, strain $\propto$ stress. For relatively small stresses, this is a direct or linear proportion. The constant of proportionality depends on the nature of the material and is called the **elastic modulus**. Thus,

$$\text{stress} = \text{elastic modulus} \times \text{strain}$$

or

$$\text{elastic modulus} = \frac{\text{stress}}{\text{strain}} \tag{9.9}$$

That is, the elastic modulus is the stress divided by the strain, or the ratio of stress to strain.

Three general types of elastic moduli (plural of modulus) are associated with stresses that produce changes in length, shape, or volume. These are called Young's modulus, shear modulus, and bulk modulus, respectively.

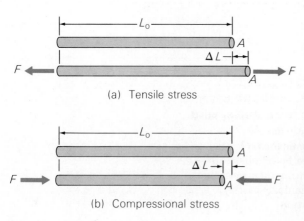

(a) Tensile stress

(b) Compressional stress

Figure 9.15 **Tensile and compressional stress and strain** Tensile and compressional stresses are due to forces applied normally to the surface area of the ends of bodies, as shown here. (a) A tension, or tensile stress, tends to increase the length of an object. (b) A compressional stress tends to shorten the length.

Feat of Strength or Knowledge of Materials?

The breaking of wooden boards or concrete blocks with a bare hand or foot is an impressive demonstration often done by karate experts (Fig. 1). The physics of this feat can be analyzed in terms of properties of the materials involved. In delivering the blow, the expert imparts a large impulse force (Chapter 6) to the board or block, which bends under the pressure. Note that the board or block is struck midway between the end supports. At the same time, the bones of the hand are being compressed as well. Fortunately, human bone can withstand more compressive force than can wood or concrete, which is why the expert's bones aren't damaged. (The ultimate compressive strength of bone is four times or more than that of concrete.*)

When the board or block is hit, the upper surface is compressed and the lower surface is elongated, or subjected to a tension force. Both wood and concrete are weaker under tension than under compression. (The ultimate tensile strength of concrete is only about a twentieth of its ultimate compressive strength.) Thus, the board or block begins to crack at the bottom surface first. As the descending hand follows through, the crack is widened and becomes a complete break.

The amount of force required to break a board or block in this way depends on several factors. For a wooden board, these are the type of wood, the width and thickness, and the distance between the end supports. Also, the edge of the hand must strike the board parallel to the grain of the wood. Similar considerations apply for a concrete block. Because concrete is more rigid, more force is generally required to break such a block. Some karate experts are able to break through a stack of boards and more than one concrete block. The force required does not increase by a factor equal to the number of boards or blocks. High-speed photography shows that the hand makes contact with only one or two boards at

* Ultimate strength is the maximum strength a material can stand before it breaks or fractures.

the top of the stack. Then, as each board breaks, it collides with and breaks through the board below it.

Even though the properties of the materials seem to guarantee the success of this demonstration, you should not attempt the feat unless you are an expert and know what you're doing. The board or block might win.

Figure 1 **Feat of strength or knowledge of materials?** Breaking wooden boards or concrete blocks using karate blows depends on both the physical strength of the expert *and* material strengths. Wood and concrete have different maximum tensile and compressional stresses, but the maximum compressional strength of bone is greater than either of them.

Change in Length: Young's Modulus

Fig. 9.16 is a graph of the tensile stress versus the strain for a typical metal rod. The curve is a straight line up to a point called the **proportional limit**. Beyond this point, the strain begins to increase more rapidly to another critical point called the **elastic limit**. If the tension is discontinued at this point, the material will return to its original length. If the tension is applied beyond the elastic limit and then removed, the material will recover somewhat but will retain some permanent deformation.

The straight-line part of the graph shows a direct proportionality between stress and strain. This relationship was first formalized by Robert Hooke in

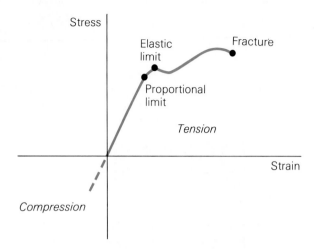

Figure 9.16 **Stress versus strain**
The plot of stress versus strain for a typical metal rod is a straight line up to the proportional limit. Then elastic deformation continues until the elastic limit is reached. Beyond that, the rod will be permanently deformed and will eventually fracture, or break.

1678 and is now known as *Hooke's law*. (It is the same general relationship as that given for a spring in Chapter 5.) The elastic modulus for a tension or a compression is called **Young's modulus** (Y):

$$\frac{F}{A} = Y\!\left(\frac{\Delta L}{L_\text{o}}\right)$$

or $\quad Y = \dfrac{F/A}{\Delta L/L_\text{o}}$ (9.10)

The units for Young's modulus are the same as for stress (N/m^2). Some typical values of Young's modulus are given in Table 9.1. (See Demonstration 9.)

Thomas Young (1773–1829) was a British physicist who investigated the mechanical properties of materials and optical phenomena.

Example 9.7 Tensile Stress and Young's Modulus

What mass would have to be suspended from a steel wire with a diameter of 0.20 cm to increase its length by 0.10%?

Solution
Given: $d = 0.20$ cm, so $r = 0.10$ cm   *Find*: m
$\qquad\qquad = 1.0 \times 10^{-3}$ m
$\qquad$ strain $= \Delta L/L_\text{o} = 0.10\%$
$\qquad\qquad = 0.0010 = 1.0 \times 10^{-3}$
$\qquad Y_\text{steel} = 20 \times 10^{10}$ N/m^2
$\qquad\qquad$ (from Table 9.1)

The force required to produce the elongation is, by Eq. 9.10,

$$F = Y\!\left(\frac{\Delta L}{L_\text{o}}\right)A = Y\!\left(\frac{\Delta L}{L_\text{o}}\right)\pi r^2$$

$$= (20 \times 10^{10} \text{ N/m}^2)(1.0 \times 10^{-3})\pi(1.0 \times 10^{-3} \text{ m})^2 = 6.3 \times 10^2 \text{ N}$$

This force is produced by the weight of the suspended mass, or $F = mg$. Thus,

$$m = \frac{F}{g} = \frac{6.3 \times 10^2 \text{ N}}{9.8 \text{ m/s}^2} = 64 \text{ kg} \quad \blacksquare$$

Table 9.1
Elastic Moduli for Various Materials (in N/m²)

Substance	Young's Modulus (Y)	Shear Modulus (S)	Bulk Modulus (B)
Solids			
Aluminum	7.0×10^{10}	2.5×10^{10}	7.0×10^{10}
Brass	9.0×10^{10}	3.5×10^{10}	7.5×10^{10}
Copper	11×10^{10}	3.8×10^{10}	12×10^{10}
Glass	5.7×10^{10}	2.4×10^{10}	4.0×10^{10}
Iron	15×10^{10}	6.0×10^{10}	12×10^{10}
Steel	20×10^{10}	8.2×10^{10}	15×10^{10}
Liquids			
Alcohol, ethyl			1.0×10^{9}
Glycerin			4.5×10^{9}
Mercury			26×10^{9}
Water			2.3×10^{9}

Demonstration 9

Stress, Strain, and Elasticity

A demonstration to show a material property—the elasticity of glass.

(a) A glass bottle with colored water is fitted with a capillary tube in a rubber stopper.

(b) Squeezing the bottle on the flat sides raises the water level in the tube because the bottle's cross section is reduced.

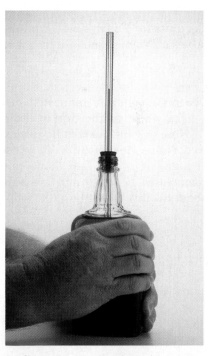

(c) Squeezing the bottle on its narrow sides lowers the water level because the bottle's cross section is increased.

Change in Shape: Shear Modulus

Another way an elastic body can be deformed is by a **shear stress**. In this case, the deformation is due to an applied force that is tangential to the surface area (Fig. 9.17). A change in shape results without a change in volume. [A torsional shear stress is due to a twisting action, for example, in shearing off the head of a bolt when tightening it.] The **shear strain** is given by x/h, where x is the relative displacement of the faces and h is the distance between them.

The shear strain is sometimes defined in terms of the **shear angle** (ϕ). As can be seen from the figure, $\tan\phi = x/h$. But the shear angle is usually quite small, so a good approximation is $\phi \cong \tan\phi = x/h$.

The **shear modulus** (sometimes called the modulus of rigidity) is then

$$S = \frac{F/A}{x/h} \cong \frac{F/A}{\phi} \tag{9.11}$$

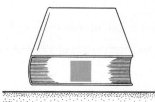

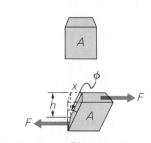

As you may note in Table 9.1, the shear modulus is generally less than Young's modulus; S is approximately $Y/3$ for many materials, which indicates that there is a greater response to a shear stress than to a tensile stress.

Change in Volume: Bulk Modulus

Suppose that a force directed inward acts over the entire surface of a body (Fig. 9.18). Such a **volume stress** is often applied by pressure transmitted by a fluid. An elastic material will be compressed by a volume stress, that is, will show a change in volume but not in general shape. The (change in) pressure is equal to the volume stress, or $\Delta p = F/A$. The **volume strain** is the ratio of the volume change (ΔV) to the original volume (V_o). The **bulk modulus** (B) is then

$$B = \frac{F/A}{-\Delta V/V_o} = -\frac{\Delta p}{\Delta V/V_o} \tag{9.12}$$

The minus sign is introduced to make B a positive quantity, since $\Delta V = V - V_o$ is negative for an increase in external pressure.

Note in Table 9.1 that liquids as well as solids have bulk moduli. But liquids do not have shear moduli. A shear stress cannot be effectively applied to a liquid or a gas; it is said that fluids cannot support a shear. (Why?) Gases can be compressed, however, and thus have bulk moduli. For a gas, it is common to talk about the inverse of the bulk modulus, which is called the **compressibility** (k):

Figure 9.17 **Shear stress and strain** A shear stress is produced when a force is applied tangentially to a surface area. The strain is measured in terms of the relative displacement of the object's faces or the shear angle (ϕ).

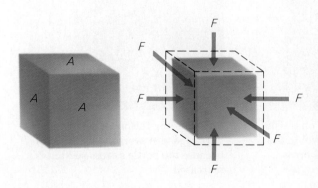

Figure 9.18 **Volume stress and strain** A volume stress is applied when a normal force acts over the entire surface area of a body. This most commonly occurs in fluids. The resulting strain is a change in volume.

$$k = \frac{1}{B} \qquad\qquad (9.13)$$

Solids and liquids are relatively incompressible and have small values of compressibility. Gases, on the other hand, are easily compressed and have large compressibilities, which vary with pressure and temperature.

Example 9.8 Volume Stress and Bulk Modulus

How much pressure must be exerted on a liter of water to compress it by 0.10%?

Solution

Given: $-\Delta V/V_0 = 0.0010$ Find: $p = F/A$

           $V_0 = 1.0\ \text{L} = 1000\ \text{cm}^3$

           $B_{H_2O} = 2.3 \times 10^9\ \text{N/m}^2$

                  (from Table 9.1)

Note that $-\Delta V/V_0$ is the *fractional* change in the volume, and for a compression of 0.10%, this is 0.0010 (with no units since it expresses a ratio of volumes). Since $V_0 = 1000\ \text{cm}^3$, the volume reduction is

$$-\Delta V = (0.0010)\,V_0 = (0.0010)(1000\ \text{cm}^3) = 1.0\ \text{cm}^3$$

Then, using Eq. 9.12 to find the pressure (or change or increase in pressure) gives

$$\Delta p = B\left(\frac{-\Delta V}{V_0}\right) = (2.3 \times 10^9\ \text{N/m}^2)(0.0010) = 2.3 \times 10^6\ \text{N/m}^2$$

(This is about 23 times normal atmospheric pressure.) ∎

Important Formulas

Conditions for translational and rotational mechanical equilibrium:

$$\Sigma_i \mathbf{F}_i = 0 \qquad \text{and} \qquad \Sigma_i \tau_i = 0$$

Actual mechanical advantage:

$$\text{AMA} = \frac{F_0}{F_i}$$

Theoretical (or ideal) mechanical advantage:

$$\text{TMA} = \frac{d_i}{d_0}$$

Theoretical mechanical advantage for a lever:

$$\text{TMA} = \frac{L_i}{L_0}$$

Theoretical mechanical advantage for an inclined plane:

$$\text{TMA} = \frac{L}{h} = \frac{1}{\sin\theta}$$

Mechanical efficiency:

$$\varepsilon = \frac{F_0 d_0}{F_i d_i} = \frac{\text{AMA}}{\text{TMA}}$$

Stress:

$$\text{stress} = \frac{F}{A}$$

Strain:

$$\text{strain} = \frac{\Delta L}{L_0} = \frac{L - L_0}{L_0}$$

Young's modulus:

$$Y = \frac{F/A}{\Delta L/L_0}$$

Shear modulus:

$$S = \frac{F/A}{x/h} \simeq \frac{F/A}{\phi}$$

Bulk modulus:

$$B = \frac{F/A}{-\Delta V/V_0} = -\frac{\Delta p}{\Delta V/V_0}$$

Compressibility:

$$k = \frac{1}{B}$$

Questions

Mechanical Equilibrium, Stability, and the Center of Gravity

1. Can a body be in both translational equilibrium and rotational equilibrium and still be moving? Explain.

2. When an object in stable equilibrium is slightly displaced, what happens to its center of gravity? What happens if an object in unstable equilibrium is slightly displaced?

3. A meterstick is pivoted at one end and hangs down like a pendulum. What are its positions of equilibrium?

4. What is the most stable position you can assume? What is your most stable configuration when standing on a moving bus?

5. Why does a lower center of gravity give an object greater stability?

6. A rod with a movable ball, like that shown in Fig. 9.19, is more easily balanced on a finger if the ball is in a higher position. If a lower center of gravity gives an object greater stability, why do the rod and ball seemingly have more stability when the ball (and the center of gravity) is raised?

(a)   (b)

Figure 9.20   **A test of strength or instability**
See Question 13.

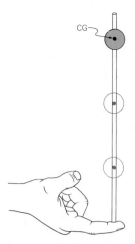

CG

Figure 9.19   **Greater stability with a higher center of gravity?**
See Question 6.

7. Does it make any difference how the trailer of a tractor-trailer rig is loaded? Explain. (Hint: think about an eighteen-wheeler going around a banked curve.)

8. (a) Why do people commonly extend the free arm outward when carrying a heavy suitcase? (b) Why can't everyone touch their toes with their legs straight? (Ignore possible problems with back muscles.)

9. A popular children's toy is an inflated shape with a rounded, heavy base, which always rights itself when knocked over. Explain the principle of this punching toy.

10. How many different positions of stable equilibrium and unstable equilibrium are there for a cube? (Consider each surface, edge, and corner to be a different position.)

11. A wooden letter L is 3 cm thick. How many different positions of stability and instability does the L have? (Consider each surface, edge, and corner to be a different position.)

12. Metastable equilibrium occurs when an object is in stable equilibrium in one dimension and in unstable equilibrium in another dimension. Can you give some examples of metastable equilibrium?

13. A situation that makes an interesting party challenge is as follows: Stand with your feet together two foot lengths (the length of your own foot) out from a wall. A chair is placed against the wall. Leaning forward with your head against the wall, pick up the chair and try to stand up [Fig. 9.20(a)]. (a) In general, women can do this but men can't. Explain why. (You might try this yourself with a friend.) (b) The young man in Fig. 9.20(b) shifted the chair behind his back and then stood up. Explain why this worked.

14. Locate the center of gravity of the arrangement shown in Fig. 9.21.

Mechanical Advantage

15. Explain why a screw can be classified as a type of inclined plane.

16. Because it is a wedge, a knife is a type of inclined plane. Explain how the TMA for a knife is changed when it is sharpened.

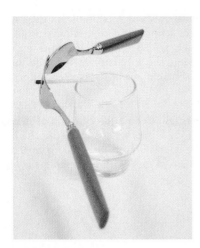

Figure 9.21 **Balancing act**
See Question 14.

17. Explain how the following qualify as levers: (a) a person doing pushups, (b) a set of nutcrackers, (c) a wheelbarrow, (d) the lower jaw, (e) a pair of hedge cutters, and (f) an oar.

18. A wheel and axle (one is shown in Fig. 9.31) is said to be a continuous lever with unequal lever arms. Explain this description. Could a wheel and axle be used for distance multiplication instead of force multiplication? Explain.

19. An inventor designs a machine and says it has a TMA greater than 100%. What would you say about the AMA of the machine and why?

Elastic Moduli

20. In what kind of situation might the strain be greater than the stress?

21. One material has a greater Young's modulus than another. What does this tell you?

22. Of two rods, one made of copper and the other of brass, which would suffer a greater deformation under the same tensile stress?

23. In general, which is greater for metals, (a) Young's modulus or the shear modulus or (b) the shear modulus or the compressibility?

24. Why are scissors sometimes called shears? Is this a descriptive name?

25. One material has a compressibility that is 2.5 times that of another material. What does this tell you about their bulk moduli?

26. Is it possible for the bulk modulus to be negative? Explain and give an example if possible.

Problems

9.1 Mechanical Equilibrium

■**1.** An applied force of 20 N moves an object with a constant velocity. (a) What is the net force on the object? (b) If there are two forces acting on the object, what is the other force?

■**2.** A 10-kg box is pulled along a level surface by a horizontal force. The box has a constant velocity of 2.5 m/s, and the coefficient of kinetic friction between it and the surface is 0.55. What is the magnitude of the applied force?

■**3.** A uniform meterstick pivoted at its center, as in Example 9.2, has a 100-g mass suspended at the 25.0-cm position. (a) At what position should a 75.0-g mass be suspended to put the system in equilibrium? (b) What mass would have to be suspended at the 80.0-cm position for the system to be in equilibrium?

■**4.** A 35-kg child sits on a uniform seesaw 2.0 m from the pivot point (or fulcrum). How far from the pivot point on the other side will her 30-kg playmate have to sit for the seesaw to be in equilibrium?

■■**5.** Three horizontal forces act on a stationary but movable object. One force is 50 N directed northward, and another is 60 N directed toward the southeast. Find the third force.

■■**6.** A 3.0-kg picture is hung as shown in Fig. 9.22. What is the tension in each of the wires?

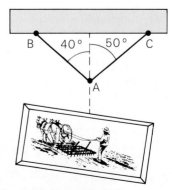

Figure 9.22 **Lop-sided equilibrium**
See Problems 6 and 7.

■■**7.** What would the tensions in the wires in Fig. 9.22 be if the angles they made with the vertical line were equal?

■■**8.** A flat bridge 20 m long is supported only by a column at each end. If the bridge has a mass of 100 metric tons and is slightly nonuniform, so its center of gravity is 1.0 m away from its geometrical center, what forces do the supporting columns exert on the bridge?

■■**9.** A mass is suspended by two cords as shown in Fig. 9.23. What are the tensions in the cords?

■■**10.** If the cord attached to the vertical wall in Fig. 9.23 were horizontal (instead of at a 30° angle), what would the tensions in the cords be?

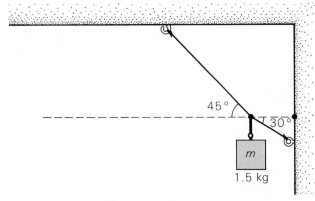

Figure 9.23 **Strung-out equilibrium**
See Problems 9 and 10.

■■**11.** A 80-kg man doing pushups holds his body up with his arms perpendicular to the floor. The distance from his hands to his toes is 1.8 m, and his center of gravity is 1.0 m from his toes. (a) How much force is exerted on the floor by his hands? (b) How much force is exerted on the floor by his feet?

■■**12.** A truck is on a flat 40-m bridge supported by end piers. The truck weighs 2.0×10^5 N, and the bridge weighs 8.5×10^7 N. If the center of mass of the truck is 15 m from one end of the bridge, what are the forces exerted on the bridge by the end piers?

■■**13.** A uniform meterstick weighing 4.0 N is pivoted so it can rotate about a horizontal axis through one end. If a 0.15-kg mass is suspended 70 cm from the pivoted end, what force must be applied to the free end of the meterstick to maintain it in a horizontal position?

■■**14.** Telephone and electrical lines are intentionally allowed to sag between poles so that the tension will not be too great when something hits or sits on the line. Suppose that a line were stretched perfectly horizontally between two poles, which are 30 m apart. If a bird with a mass of 0.25 kg perched on the wire midway between the poles and the wire sagged 1.0 cm, what would the tension in the wire be?

■■■**15.** A 75-kg painter paints the side of a house while standing on a long board resting on a scaffold, as shown in Fig. 9.24. If the board has a mass of 20 kg, how close to the end can the painter stand without tipping the board over?

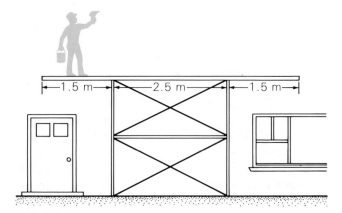

Figure 9.24 **Not too far!**
See Problems 15 and 16.

■■■**16.** Suppose that the board in Fig. 9.24 were suspended from vertical ropes attached to each end instead of resting on scaffolding. If the painter stood 1.5 m from one end of the board, what would the tensions in the ropes be? (See Problem 15 for additional data.)

■■■**17.** A boom system is pictured in Fig. 9.25. If the uniform boom has a mass of 50 kg, what is the tension in the cable?

■■■**18.** If the boom in Fig. 9.25 were lowered to the horizontal, what would be the tension in the cable? (The distance from the pivot to the top of the pulley is 9.0 m.)

■■■**19.** A uniform meterstick has a linear mass density (mass per unit length) of 0.50 g/cm. If a 15-g mass is suspended from one end, at what position should the stick be pivoted (see Example 9.2) to put it in stable equilibrium?

■■■**20.** A variation of Russell traction (Fig. 9.26) helps support the lower leg when in a cast. Suppose that the patient's leg and cast have a combined mass of 15 kg and m_1 is 4.0 kg. (a) What is the reaction force of the leg muscles to the traction? (b) What must m_2 be to keep the leg horizontal?

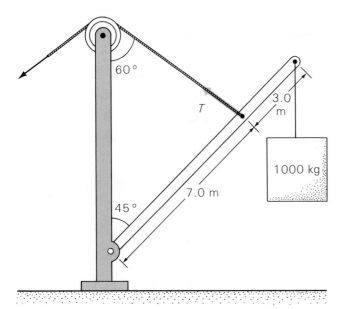

Figure 9.25 **Crane and boom**
See Problems 17 and 18.

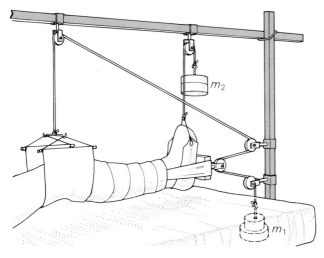

Figure 9.26 **Static traction**
See Problem 20.

9.2 Stability and the Center of Gravity

■■21. A 10-kg solid uniform cube with 0.50-m sides rests on a level surface. What is the minimum amount of work necessary to put the cube in unstable equilibrium?

■■22. A uniform L-shaped object has a leg that is 0.50 m long and a shorter leg that is 0.30 m long (both measured along the outside edge). The L has a width of 6.0 cm and a thickness of 2.0 cm. Locate the center of gravity. (Hint: divide the L into two rectangles and locate the center of gravity of each. Express mass in terms of density.)

■■23. (a) How many uniform, identical textbooks with a width of 24 cm can be stacked on top of each other on a level surface without the stack falling over if each successive book is displaced 3.0 cm in the width direction relative to the next lower book? (b) If the books are 4.0 cm thick, what will be the height of the center of mass of the stack above the level surface?

■■24. The location of a person's center of gravity relative to his or her height can be found using the arrangement shown in Fig. 9.27. The scales are initially adjusted to read zero with the board alone. Locate the center of gravity of the person relative to the horizontal dimension. Would you expect the location of the center of gravity in other dimensions to be exactly at the midway points? Explain.

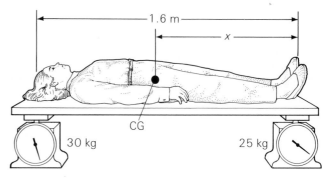

Figure 9.27 **Where's the center of gravity?**
See Problem 24.

■■25. Using bricks identical to those in Example 9.4, you are asked to create a stack of nine bricks (maximum) without its toppling, with each added brick displaced an equal distance horizontally. (a) What is the maximum displacement that will allow this? (b) What is the height of the center of mass of the stack if the uniform bricks are 8.0 cm thick?

■■■26. A tractor-trailer rig has an outer wheel base of 3.66 m. When the trailer is loaded, its center of gravity is equidistant from its sides and 3.58 m above the road surface. What banking angle of the road will put the trailer in unstable equilibrium?

■■■27. A uniform rectangular bar has end dimensions of 6.0 cm by 6.0 cm and a height of 40 cm. The bar is sitting on one end on a level surface. (a) What is the range of the horizontal displacement of the bar's center of gravity if it is to remain in stable equilibrium? (b) What is the range of the vertical displacement?

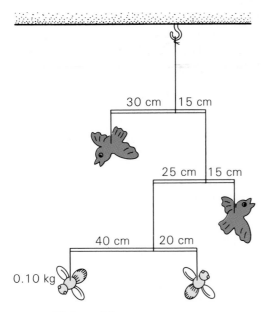

Figure 9.28 Birds and bees
See Problem 28.

■■■28. An artist wishes to construct a birds-and-bees mobile, as shown in Fig. 9.28. If the bee on the lower left has a mass of 0.10 kg and each vertical support string has a length of 30 cm, where is the center of gravity of the mobile relative to the support hook? (Neglect the masses of the bars and strings.)

9.3 Mechanical Advantage

■29. The input and output forces for a particular machine are 60 N and 80 N, respectively. (a) What is the AMA of the machine? (b) If the input force moves through a parallel distance of 1.5 m and 30 J of energy is lost to friction, how far does the load move?

■30. The AMA of a machine is 2.5. If the input force is 100 N, what is the output force?

■31. If the machine in Problem 30 has a TMA of 4, what is its efficiency?

■32. A worker uses a crowbar to pry up a steel plate. If 3.0 cm of the 1.0-m bar is on one side of the pivot point, what mechanical advantage does the bar give the worker?

■33. A 20° inclined plane has a base length of 10 m. What is the TMA of the plane?

■■34. An input force of 40 N acts through a parallel distance of 30 cm in a simple machine. (a) If the AMA of the machine is 2.2, what is the output force? (b) What is the machine's total work output? (c) Is it possible to find the TMA? Explain.

■■35. A force of 15 N applied perpendicularly downward at one end of a straight rod 2.4 m long balances a load of 50 N on the opposite end. (a) Where is the fulcrum located? (b) What is the TMA of the lever?

■■36. In a machine with an efficiency of 60%, an input force of 200 N moves through a parallel distance of 0.55 m. How much energy is lost by the machine?

■■37. A machine is used to move a load of 400 N through a distance of 2.5 m. If 300 J of energy is lost to friction in the process, what is the efficiency of the machine?

■■38. A machine with an AMA of 5.0 loses 10% of the work input to friction. (a) What is the efficiency of the machine? (b) What is its TMA?

■■39. A machine has an AMA of 2.0 and a TMA of 3.0. If the machine is used to raise a 25-kg load to a height of 1.4 m, what is the work input?

■■40. A wide board with a length of 4.0 m is used as an inclined plane. What is the difference between the TMA of the board at a 10° incline and that at a 20° incline?

■■41. A 20-kg box is pushed up a 15° incline by a force of 75 N parallel to the inclined surface. What are the inclined plane's (a) AMA and (b) efficiency?

■■42. A 100-kg crate slides a distance of 4.0 m down an 11° inclined plane under the influence of gravity. The coefficient of kinetic friction between the crate and the plane's surface is 0.090. (a) What is the efficiency of the inclined plane? (b) What is its AMA?

■■43. For a particular pair of pliers, the force is applied (by gripping them) 12 cm from the pivot axis, and an object in the jaws is 3.0 cm from the axis. (a) What is the TMA of the pliers? (b) If a force of 25 N is applied to the grips, what is the force exerted on the object in the jaws? (Neglect friction.)

■■44. A wheelbarrow contains a 50-kg load. The center of gravity of the load is 24 cm from the wheel, and the handles are 1.3 m from the wheel. (a) What is the TMA of the wheelbarrow? (b) What total force is applied normal to the handles when someone is wheeling the load? (Neglect friction and the weight of the wheelbarrow.)

■■45. A man holds a shovel full of dirt that weighs 50 N. The man's right hand is on the handle 1.6 m from the center of the load and acts as a fulcrum. His left hand exerts an upward force 0.80 m from the load. (a) What is the force exerted by the man's right hand? (b) What is the TMA? Comment on its value.

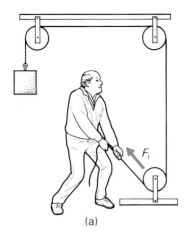

(a)

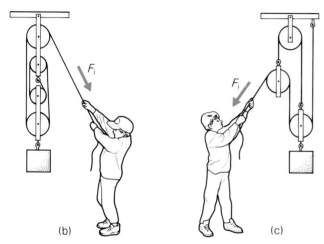

(b)　　　　　　　　　　(c)

Figure 9.29　**Pulling an advantage**
See Problem 46.

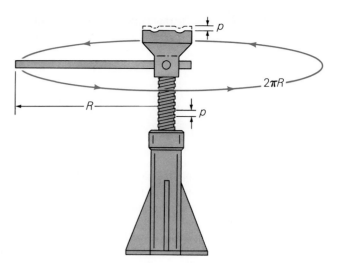

Figure 9.30　**A screw jack**
See Problem 48.

■■46. (a) For the pulley arrangements in Fig. 9.29, what's the TMA in each case? (b) If the block and tackle in Fig. 9.29(b) were used to hold a 50-kg load, what would the required input force be? (Neglect friction for the pulleys.)

■■47. A block and tackle with six supporting strands of rope on the movable block is used to pull a car that is stuck in some mud. The car must be moved 1.5 m. (a) What is the displacement of the free end of the rope? (b) If a 1200-N force is required just to move the car, and if the block and tackle has an efficiency of 60%, with what force must the rope be pulled?

■■■48. Show that for a screw jack such as that shown in Fig. 9.30, TMA = $2\pi R/p$, where p is the pitch of the screw, that is, the distance between the threads or the vertical distance the screw moves in one revolution.

■■■49. A vise has a screw pitch of 0.65 cm and a handle length of 9.0 cm. An object is held between the jaws with a force of 750 N. (a) Ideally, how much force must be applied to the handle? (b) If a force of 10 N is applied, what are the AMA and the efficiency of the vise? (Hint: see Problem 48.)

■■■50. Screw sizes are commonly expressed as two numbers, for example, 10–32. The first number refers to the diameter of the shaft (size 10 has a 0.190-in. diameter), and the second number gives the threads per inch (for the example, 32 threads/in.). What is the TMA of a 10–32 screw? (Hint: see Problem 48.)

■■■51. Show that for a wheel and axle like that shown in Fig. 9.31, TMA = R/r, where R is the radius of the wheel and r is the radius of the axle. (Hint: see Question 18.)

■■■52. A door knob has a diameter of 5.0 cm, and its connecting rod has a diameter of 5.0 mm. (a) What kind of

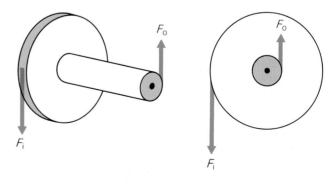

Figure 9.31　**A wheel and axle**
See Problem 51.

machine is this, and what is the TMA? (b) If a tangential force of 10 N is applied to the knob, what is the magnitude of the tangential force on the rod (neglecting friction)? (Hint: see Problem 51.)

9.4 Elastic Moduli

■**53.** A 100-kg mass is suspended using a cable with a diameter of 2.0 cm. What is the stress in the cable?

■**54.** A push rod in a machine receives a force of 750 N. If the rectangular cross section of the rod measures 1.0 cm by 0.50 cm, what is the compressional stress on the rod?

■**55.** A carpenter applies a tangential force of 125 N to the upper surface of a block of wood. If the dimensions of the surface are 20 cm by 30 cm, what is the shear stress on the block?

■**56.** A force of 800 N is applied at an angle of 30° to the end of a bar. That surface is 4.0 cm on a side. What are (a) the compressional stress and (b) the shear stress on the bar?

■■**57.** A metal wire 1.0 mm in diameter and 2.0 m long hangs vertically with a 6.0-kg mass suspended from it. If the wire stretches 1.4 mm under the tension, what is the value of Young's modulus for the metal?

■■**58.** An aluminum rod with a diameter of 1.0 cm and a length of 0.50 m is used to support a load of 100 N. By how much is the length of the rod compressed?

■■**59.** A support cable is initially 130.00 cm long and 2.00 mm in diameter and is stretched to a length of 130.26 cm by a force of 600 N. What is Young's modulus for the material of the cable?

■■**60.** A rectangular steel beam with a cross-sectional area of 24 cm^2 is used to support a sagging floor in a building. If the beam supports a load of 10,000 N, by what percentage is the beam compressed?

■■**61.** A rectangular block of Jell-O (any flavor you like) with length, width, and height of 10 cm, 8.0 cm, and 4.0 cm, respectively, is subjected to a shear force of 0.40 N on its upper surface. If the top surface is displaced 0.30 mm relative to the bottom surface, what is the shear modulus of the gelatin?

■■**62.** A shear force of 500 N is applied to one face of a cube of aluminum measuring 10 cm on a side. What is the relative displacement of the opposite face?

■■**63.** Two metal plates are held together by two rivets with diameters of 0.40 cm. If the maximum shear stress a single rivet can withstand is 5.0×10^8 N/m^2, how much force must be applied parallel to the plates to shear off both rivets?

■■■**64.** The maximum shear stress for aluminum is 1.5×10^8 N/m^2. How much force has to be applied to a punch with a diameter of 2.0 cm to punch a hole in a sheet of aluminum that is 0.50 cm thick? (Hint: draw a sketch and show that this requires a shear stress.)

■■**65.** What pressure difference is required to compress a volume of water by 0.10%?

■■**66.** A brass cube 6.0 cm on a side is placed in a pressure chamber and subjected to a normal force of 1.2×10^7 N on all its surfaces. What is the volume of the cube under this pressure?

■■■**67.** A 45-kg traffic light is suspended with two steel cables of equal length and radii of 0.50 cm. If the cables make an angle of 15° with the horizontal, what is the fractional increase in their length due to the weight of the light?

■■■**68.** A length of steel wire with a diameter of 0.10 cm is lengthened by 0.40% by a tension force. If a piece of copper wire with the same initial length is lengthened by the same percentage by a tension force of the same magnitude, what is the diameter of the copper wire?

■■■**69.** A copper wire 100.00 cm long is stretched to a length of 100.02 cm when it supports a certain load. If an aluminum wire of the same diameter is used to support the same load, what should its initial length be if its stretched length is to be 100.02 cm also?

Additional Problems

70. How many times greater is the TMA of a 30° inclined plane than the TMA of a 15° inclined plane?

71. A cube of aluminum 10 cm on a side receives equal pressure on all its faces. What would be the magnitude of the force required to compress the cube's volume by 0.01%?

72. In a laboratory experiment, a centrally pivoted meterstick and suspended weights are used to investigate torques and rotational equilibrium. A student is instructed to suspend a 0.15-kg mass at the 10-cm mark on the stick, a 0.10-kg mass at the 70-cm mark, and another 0.10-kg mass at the position that will put the stick in equilibrium. Where should the third mass be suspended, and how can the student do this?

73. An inclined plane has a TMA of 3.6. What is the angle of the incline?

74. In a circus act, a 70-kg tightrope walker stands on the high wire $\frac{2}{3}$ of the way between the end supports which are 30 m apart. If the wire sags 4.0 cm from the horizontal at that point, what is the tension in the wire? (Neglect the mass of the wire and assume any number of significant figures.)

75. A wheel with a circumference of 15.7 cm is attached to an axle with a diameter of 1.0 cm. (a) What is the TMA of this wheel and axle? (b) Suppose that the input force is applied tangentially to the axle rather than the wheel. What is the TMA in this case, and is there any practical advantage to doing this?

76. A 5.0-m diving board is fixed to two supports, one at one end and the other 0.75 m from that end. If a 65-kg diver stands on the other end of the board (over the pool), what forces do the supports exert on the board? (Neglect the mass of the board.)

77. Write the general form of Hooke's law and find the units of the "spring constant" for elastic deformation.

78. Students investigating stability stack their textbooks in a pile on a level surface. All the books measure 20 cm by 25 cm by 5.0 cm. The second book is displaced 1.0 cm relative to the bound edge of the bottom one, and the third book is displaced 2.0 cm relative to the bound edge of the second. This displacement pattern is repeated until the stack topples. (a) What is the maximum number of books that can be stacked in this way before the stack falls over? (b) What is the height of the center of mass of the stack?

79. (a) Sketch the rope arrangement for a block and tackle consisting of a single movable pulley and a fixed block of two sheaves (pulleys) to give the block and tackle a TMA of 3. (b) Sketch a block and tackle that has two and three sheaves in the blocks and a TMA of 5.

80. Two masses are suspended from the ends of a string running over a fixed pulley with a diameter of 10 cm. If the 2.0-kg mass descends and the 1.9-kg mass rises while the pulley rotates with a constant angular velocity, what is the frictional torque of the pulley?

81. The input force to a particular machine is constrained to act through a distance of 8.0 cm, and the output force acts through 3.0 cm. (a) What is the TMA of the machine? (b) What can you say about the ratio of the output to the input force?

82. A uniform meterstick is pivoted at its center (see Example 9.2). A student suspends a 100-g mass at the 20-cm position and then suspends a 25-g mass at the 70-cm position. A 50-g mass is also available. Where should it be suspended to put the system in equilibrium?

83. A nonuniform 0.40-kg meterstick pivoted at one end can rotate about a horizontal axis. With a 0.50-kg mass suspended 60 cm from the pivot point, an upward force of 4.8 N applied to the free end holds the stick horizontally in static equilibrium. What is the location of the meterstick's center of gravity?

84. (a) Which of the liquids in Table 9.1 has the greatest compressibility? (b) For equal volumes of ethyl alcohol and water, which would require more pressure to be compressed by 0.10%, and how many times more?

Fluids: Liquids and Gases

10

On the basis of general physical distinctions, matter is said to have three phases: solid, liquid, and gas. A solid has a definite shape and volume (ideally a rigid body). A liquid has a definite volume but assumes the shape of its container. A gas takes on the shape and volume of its container.

Solids and liquids are sometimes called condensed matter. Liquids and gases are referred to collectively as fluids. A **fluid** is a substance that can flow, so liquids and gases qualify, but true solids do not. Fluids are important in everyday life. You are surrounded by a fluid and, for the most part, are composed of fluids. Mechanically, *a fluid is a substance that cannot support a shear stress* (Chapter 9). That is, there is little or no elastic response to a shear stress (the molecules of a fluid go with the flow, so to speak).

Because of their fluidity, liquids and gases have many properties in common, and it is convenient to study them together. Of course, liquids and gases also have some important differences—a major one is their relative compressibility. Liquids are not very compressible, and gases are usually easily compressed. You should keep this distinction in mind.

10.1 Pressure and Pascal's Principle

A force can be applied to a solid at a point of contact, but this won't work with a fluid. With fluids, a force must be applied over an area. Such an application of force is expressed in terms of **pressure**, or the force per unit area:

$$p = \frac{F}{A} \qquad (10.1)$$

The force in this equation is understood to be acting normally (perpendicularly) to the surface area. It may be only a component of a force that is perpendicular to the surface (Fig. 10.1). Pressure is a scalar quantity (with magnitude only), but the force producing it or the force produced within a fluid by a pressure does have direction. Also, depending on whether a force is constant or not, constant pressure or average pressure is measured.

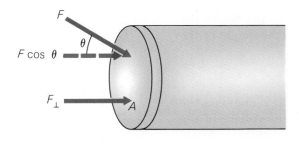

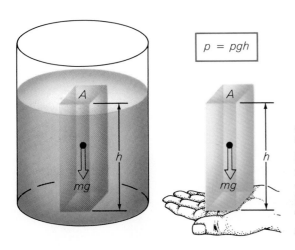

$$p = \frac{F_\perp}{A} = \frac{F \cos \theta}{A}$$

Figure 10.1 **Pressure**
Pressure is usually written $p = F/A$, where it is understood that F is normal to the surface. This may be a component of a force and $p = (F \cos \theta)/A$.

Pressure has SI units of newtons per square meter (N/m^2). This combined unit is given the special name **pascal (Pa)** in honor of the French scientist Blaise Pascal (1623–1662):

$$1 \text{ Pa} \equiv 1 \text{ N/m}^2$$

(One of Pascal's major contributions has to do with fluids and will be discussed shortly.) In the British system, a common unit of pressure is $lb/in.^2$ (pounds per square inch, or psi). Other units, some of which will be introduced later, are used in special applications.

Pressure and Depth

If you have done any diving, you know that pressure increases with depth, because you have felt increased pressure on your eardrums. An opposite effect is commonly felt when flying in a plane or riding in a car going up a mountain. With increasing altitude, your ears may "pop" because of reduced air pressure.

How the pressure in a fluid varies with depth can be demonstrated by considering a container of liquid at rest, like the one in Fig. 10.2. With the liquid at rest (in static equilibrium) all points at the same depth must have the same pressure; otherwise, there would be a pressure difference and horizontal

$$p = \rho g h$$

Figure 10.2 **Pressure and depth**
The pressure at a depth h in a liquid is due to the weight of the liquid above: $p = \rho g h$, where ρ is the density of the liquid (assumed constant). This is shown here for an arbitrary column of liquid.

movement of the liquid. For the rectangular column shown in the figure, the force on the surface at the bottom of the container is equal to the weight of the liquid making up the column:

$$F = mg = \rho V g = \rho g A h$$

where ρ (the Greek letter rho) is the density of the liquid ($\rho = m/V$, and since the liquid is assumed to be incompressible, ρ is constant), V is the volume of the column, A is the cross-sectional area of the column, and h is its height, or depth. Thus, the pressure at a depth h due to the weight of the column is

$$p = \frac{F}{A} = \rho g h \qquad (10.2)$$

The pressure is the same everywhere on a horizontal plane at a depth h. Note that the total cross-sectional area of the bottom of the container can be taken as the base of a circular column with the same result.

The derivation of Eq. 10.2 did not take into account pressure being applied to the open surface of the liquid. This adds to the pressure at a depth h to give a *total* pressure of

$$p = p_o + \rho g h \qquad (10.3)$$

pressure-depth equation

where p_o is the pressure applied to the liquid surface (that is, at $h = 0$).

For an open container, p_o is atmospheric pressure, or the weight (force) per area due to the gases in the atmosphere above the liquid's surface. The average atmospheric pressure at sea level is sometimes used as a unit, called an **atmosphere (atm)**:

$$1 \text{ atm} \equiv 101.325 \text{ kPa} = 1.01325 \times 10^5 \text{ N/m}^2 = 14.7 \text{ lb/in.}^2$$

How atmospheric pressure is measured will be described shortly.

Example 10.1 Pressure and Force

(a) What is the total pressure on the back of a scuba diver in a lake at a depth of 8.00 m? (b) What is the force on the diver's back due to the water alone, taking the surface of the back to be a rectangle 60.0 cm by 50.0 cm?

Solution

Given: $h = 8.00$ m   Find: (a) p
         $A = 0.600$ m $\times$ 0.500 m   (b) F
         $= 0.300$ m^2
         $\rho_{H_2O} = 1.00 \times 10^3$ kg/m^3
               (from Table 10.1)
         $p_a = 1.01 \times 10^5$ N/m^2

(a) The total pressure is the sum of the pressure due to the water and atmospheric presssure (p_a), by Eq. 10.3:

$$p = p_a + \rho g h$$

$$= (1.01 \times 10^5 \text{ N/m}^2) + (1.00 \times 10^3 \text{ kg/m}^3)(9.80 \text{ m/s}^2)(8.00 \text{ m})$$

$$= (1.01 \times 10^5 \text{ N/m}^2) + (0.78 \times 10^5 \text{ N/m}^2) = 1.79 \times 10^5 \text{ N/m}^2$$

(b) Since $p = F/A$,

$$F = pA = (1.79 \times 10^5 \text{ N/m}^2)(0.300 \text{ m}^2)$$

$$= 5.37 \times 10^4 \text{ N} \qquad (\text{about } 1.21 \times 10^4 \text{ lb or 6 tons!})$$

You might think this answer is wrong—how could the diver support such a force? Look at the force on the diver's back from just atmospheric pressure:

$$F_a = p_a A = (1.01 \times 10^5 \text{ N/m}^2)(0.300 \text{ m}^2)$$

$$= 3.03 \times 10^4 \text{ N} \qquad (\text{about } 6.82 \times 10^3 \text{ lb or 3.4 ton!})$$

This is roughly the force on your back right now. Our bodies don't collapse under atmospheric pressure because cells are filled with fluids that react with an equal outward pressure (equal and opposite forces). As with forces, it is a pressure *difference* that gives rise to dynamic effects. ∎

Pascal's Principle

When the pressure (for example, the air pressure) is increased on the open surface of an incompressible liquid at rest, the pressure at any point in the liquid or on the boundary surfaces increases by the same amount. The effect is the same if pressure is applied by means of a piston to any surface of an enclosed fluid (Fig. 10.3). The transmission of pressure in fluids was studied by Blaise Pascal (after whom the SI pressure unit is named), and the observed effect is called **Pascal's principle**:

> **Pressure applied to an enclosed fluid is transmitted without being reduced to every point in the fluid and to the walls of the container.**

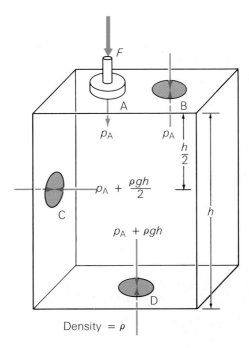

Figure 10.3 **Pascal's principle**
The pressure applied at A is fully transmitted to all parts of the fluid and to the walls of the container. There is also pressure due to the weight of the fluid at different depths.

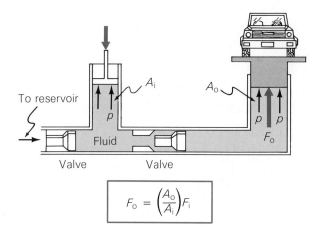

$$F_o = \left(\frac{A_o}{A_i}\right)F_i$$

Figure 10.4 **The hydraulic lift**
Because the input and output pressures are equal (Pascal's principle), a small input force gives a large output force, proportional to the ratio of the piston areas.

For an incompressible liquid, the pressure change is transmitted instantaneously. For a gas, the pressure change is transmitted throughout the fluid, and after equilibrium has been reestablished (following changes in volume and/or temperature), Pascal's principle is valid.

Common practical applications of Pascal's principle include the hydraulic braking systems used on automobiles. A relatively small force on the brake pedal transmits a large force to the wheel brake cylinder. Also, hydraulic lifts and jacks are used to raise automobiles and other heavy objects (see Fig. 10.4). The input pressure p_i, supplied by compressed air for a garage lift, gives an input force F_i on a small piston area A_i. The pressure is transmitted at full magnitude to the output piston, which has an area A_o. Since $p_i = p_o$,

$$\frac{F_i}{A_i} = \frac{F_o}{A_o}$$

and

$$F_o = \left(\frac{A_o}{A_i}\right)F_i \qquad\qquad (10.4)$$

And since A_o is larger than A_i, F_o will be larger than F_i. The input force is greatly multiplied. [Note that this is a mechanical advantage (Chapter 9), and the hydraulic lift or jack is a simple machine.]

Example 10.2 The Hydraulic Lift

A garage lift has input and lift pistons with diameters of 10 cm and 30 cm, respectively. The lift is used to hold up a car with a weight of 1.4×10^4 N. (a) What is the force on the input piston? (b) What pressure is applied to the input piston?

Solution

Given: $d_i = 10$ cm   *Find*: (a) F_i
$\ d_o = 30$ cm   (b) p_i
$\ F_o = 1.4 \times 10^4$ N

(a) Rearranging Eq. 10.4 and using $A = \pi r^2 = \pi d^2/4$ for the circular piston $(r = d/2)$ gives

$$F_i = \left(\frac{A_i}{A_o}\right)F_o = \left(\frac{\pi d_i^2/4}{\pi d_o^2/4}\right)F_o = \left(\frac{d_i}{d_o}\right)^2 F_o$$

$$F_i = \left(\frac{10\ \text{cm}}{30\ \text{cm}}\right)^2 F_o = \frac{F_o}{9} = \frac{1.4 \times 10^4\ \text{N}}{9} = 1.6 \times 10^3\ \text{N}$$

The input force is $\frac{1}{9}$ of the output force.

(b) Then

$$p_i = \frac{F_i}{A_i} = \frac{F_i}{\pi r_i^2} = \frac{1.6 \times 10^3\ \text{N}}{\pi (0.050\ \text{m})^2}$$

$$= 2.0 \times 10^5\ \text{N/m}^2\ (= 200\ \text{kPa})$$

This pressure is about 30 lb/in.^{2}, a common pressure used in automobile tires, and about twice atmospheric pressure (about 100 kPa or 15 lb/in.^{2}). ∎

Pressure Measurement

Pressure can be measured with a variety of mechanical devices that are often spring-loaded. Another type of instrument uses a liquid, usually mercury, and is called a manometer. An **open-tube manometer** is illustrated in Fig. 10.5(a). One end of the U-shaped tube is open to the atmosphere, and the other is connected to the container of gas whose pressure is to be measured. The liquid in the U-tube acts as a reservoir through which pressure is transmitted according to Pascal's principle.

The pressure of the gas (p) is balanced by the weight of the column of liquid

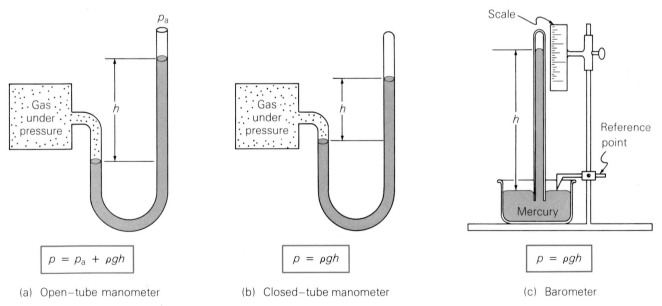

(a) Open–tube manometer (b) Closed–tube manometer (c) Barometer

Figure 10.5 **Manometers**
(a) For an open-tube manometer, the pressure of the gas in the container is balanced by the pressure due to the liquid column and atmospheric pressure acting on the open surface of the liquid. The absolute pressure of the gas equals atmospheric pressure plus ρgh, or the gauge pressure. (b) Since atmospheric pressure is not a consideration, a closed-tube manometer in a closed system shows gauge pressure only. (c) A barometer is a closed-tube manometer exposed to the atmosphere and thus reads atmospheric pressure.

(of height h measured from the lower surface level) and the atmospheric pressure (p_a) on the open liquid surface:

$$p = p_a + \rho g h \qquad (10.5)$$

The pressure p is called the **absolute pressure**, and $p - p_a$ is called the **gauge pressure**. The gauge pressure is the pressure that registers on a gauge where atmospheric pressure is not a consideration, for example, on a tire-pressure gauge.

For the manometer illustrated in Fig. 10.5(a), the absolute pressure of the gas is greater than the atmospheric pressure. Thus, the gauge pressure and the height (h) are positive ($p - p_a = \rho g h$). But if the absolute pressure of the gas in the container were equal to the atmospheric pressure, the gauge pressure would be zero. In this case, the column height would be zero ($h = 0$), or the liquid levels in the tubes would be at the same height. If the absolute pressure of the gas were less than atmospheric pressure (a partial vacuum), the gauge pressure and the column height would be negative.

If the open end of the U-shaped tube is sealed, the device is a **closed-tube manometer** [Fig. 10.5(b)]. Atmospheric pressure then has no effect on the liquid in the closed system, and the manometer reads only gauge pressure. A practical application of a closed-tube manometer is shown in Fig. 10.5(c). This is a **barometer**, which is used to measure atmospheric pressure. Essentially, the container of gas whose pressure the barometer measures is the atmosphere.

The mercury barometer was invented by Evangelista Torricelli (1608–1647), who was Galileo's successor as professor of mathematics at the university in Florence. A tube filled with mercury is inverted into a reservoir. Some mercury runs out, but as much as is supported by the air pressure on the surface of the reservoir pool remains in the tube. (You may be wondering about the space in the tube above the liquid column. Since the tube was initially full, there is no air above the liquid surface and therefore only vapor pressure on this surface, which is taken to be negligibly small.) The atmospheric pressure is equal to the pressure due to the weight of the column of mercury, or

$$p_a = \rho g h \qquad (10.6)$$

A **standard atmosphere** is defined as the pressure supporting a column of mercury exactly 76 cm in height at sea level and 0 °C.

Example 10.3   Standard Atmospheric Pressure
If a standard atmosphere supports a column height of exactly 76 cm of mercury (chemical symbol Hg), what is the standard atmospheric pressure in pascals? (The density of mercury is 13.5951×10^3 kg/m^3 at 0 °C, and $g = 9.80665$ m/s^2.)

Solution
Given: $h = 76$ cm $= 0.76$ m (exact)   *Find*: p_a
$\qquad\quad \rho_{Hg} = 13.5951 \times 10^3$ kg/m^3
$\qquad\quad g = 9.80665$ m/s^2

Using Eq. 10.6,

$$p_a = \rho g h = (13.5951 \times 10^3 \text{ kg/m}^3)(9.80665 \text{ m/s}^2)(0.760000 \text{ m})$$

$$= 101.325 \times 10^3 \text{ N/m}^2 = 101{,}325 \text{ Pa} \qquad (\text{or } 101.325 \text{ kPa}) \quad \blacksquare$$

Changes in atmospheric pressure are observed as changes in the height of the mercury column. Atmospheric pressure is commonly reported in terms of the height of the barometer column, and weather forecasters say that the barometer is rising or falling. As noted previously, 1 standard atmosphere is a unit for air pressure expressed in height units as

$$1 \text{ atm} = 76 \text{ cm Hg} = 760 \text{ mm Hg} = 29.92 \text{ in. Hg} \quad \text{(about 30 in. Hg)}$$

In honor of Torricelli, a pressure supporting 1 mm of mercury is given the name **torr**:

$$1 \text{ mm Hg} \equiv 1 \text{ torr} \quad \text{and} \quad 1 \text{ atm} = 760 \text{ torr}$$

Since mercury is very toxic, it is sealed inside a barometer. A safer and less expensive device widely used to measure atmospheric pressure is the aneroid (without fluid) barometer. In an aneroid barometer, a sensitive metal diaphragm on an evacuated container (something like a drum head) responds to pressure changes, which are indicated on a dial. This is the kind of barometer you frequently find in homes in decorative wall mountings.

Since air is compressible, the density and pressure of the atmosphere increase with depth as measured from the top of the atmosphere.* Atmospheric pressure is greatest at the Earth's surface and decreases with altitude. The rate of decrease is fairly uniform near the Earth. The pressure is approximately halved for each 5 km increase in altitude up to about 20 km. Above this, the atmospheric pressure decreases rapidly to about 6 Pa at an altitude of 70 km. The result is that half of the entire atmosphere is below 11 km (about 7 mi), and 99% is below 30 km (about 19 mi). The air is quite thin in the upper part of the atmosphere, which extends for several hundred kilometers.

We live at the bottom of the atmosphere but don't notice its pressure very much in our ordinary daily activities. Remember that our bodies are largely fluids, which exert a matching outward pressure.

Another unit sometimes used in weather reports is the millibar (mb). By definition, $1 \text{ atm} = 1.01325 \times 10^5 \text{ N/m}^2 = 1.01325 \text{ bar} = 1,013.25 \text{ mb}$. Normal atmospheric pressures are around 1000 mb.

10.2 Buoyancy and Archimedes' Principle

When an object is placed in a fluid, it will either sink or float. This is most commonly observed with liquids, for example, objects floating or sinking in water. But the same effect occurs in gases. A falling object is sinking in the atmosphere, and other bodies float (Fig. 10.6).

Things float because they are buoyant, or are buoyed up. For example, if you immerse a cork in water and release it, the cork will be buoyed up to the surface and float there. From your knowledge of forces, you know that such action requires an upward net force on the object. That is, there must be an upward force acting on the object that is greater than its downward weight force. The forces are equal when the object floats or stops sinking and is stationary. The upward force resulting from an object being wholly or partially immersed in a fluid is called the **buoyant force**.

How the buoyant force arises can be seen by considering a buoyant object being held under water, as shown in Fig. 10.7(a). The pressures on the upper and lower surfaces of the block are $p_1 = \rho g h_1$ and $p_2 = \rho g h_2$. Thus, there is a

Figure 10.6 **Fluid buoyancy** The air is a fluid in which objects float as a result of buoyant forces.

---

* The pressure and volume (density) of a quantity of gas are temperature-dependent; that is, $pV \propto T$, where T is temperature.

pressure difference, $\Delta p = p_2 - p_1 = \rho g(h_2 - h_1)$, between the top and bottom of the block, which gives rise to an upward force (the buoyant force).

The buoyant force (F_b) is equal in magnitude to the weight of the fluid ($m_f g$) displaced by a body. This is known as **Archimedes' principle**:

> A body immersed wholly or partially in a fluid is buoyed up by a force equal in magnitude to the weight of the volume of fluid it displaces.

$$F_b = m_f g = \rho_f g V_f \qquad\qquad (10.7)$$

Archimedes (287–212 B.C.) was given the task of determining whether a crown made for a certain king was pure gold or contained some other, cheaper metal. Legend has it that the solution to the problem came to him when he was bathing, perhaps from seeing the water level rise when he got into the tub and experiencing the buoyant force on his limbs. In any case, it is said that he was so excited that he ran through the streets of the city shouting "Eureka!" (Greek for "I have found it"). Although Archimedes' solution to the problem involved density and volume (see Problem 15), it presumably got him thinking about buoyancy.

To visualize Archimedes' principle, imagine that the block in Fig. 10.7(a) is able to float while completely immersed, that is, is in static equilibrium without a finger holding it under the water. (Submarines can do this.) The magnitudes of the weight of the block and the buoyant force are then equal. If you replaced the block with the "block" of water it was displacing, the water too would be in equilibrium with its surroundings. The magnitude of the buoyant force and the weight of the volume of water displaced by the block are therefore equal. Archimedes' principle can be demonstrated and proven directly by immersing a heavy object using an arrangement like that in Fig. 10.7(b).

Example 10.4 Lighter Than Air

What is the buoyant force on a helium balloon with a radius of 30 cm in air if $\rho_{air} = 1.3 \text{ kg/m}^3$?

Solution

Given: $r = 30 \text{ cm} = 0.30 \text{ m}$ *Find*: F_b
     $\rho_{air} = 1.3 \text{ kg/m}^3$

The volume of the balloon is

$$V = \tfrac{4}{3}\pi r^3 = \tfrac{4}{3}\pi(0.30 \text{ m})^3 = 0.11 \text{ m}^3$$

Then, by Eq. 10.7, the weight of the air displaced by the balloon's volume, or the magnitude of the upward buoyant force, is

$$F_b = m_{air}g = (\rho_{air} V)g = (1.3 \text{ kg/m}^3)(0.11 \text{ m}^3)(9.8 \text{ m/s}^2) = 1.4 \text{ N}$$

Note that the buoyant force depends on the *density of the fluid and the volume of the body*. Shape makes no difference. ∎

We commonly say that helium and hot-air balloons float because they are lighter than air. However, technically, they are *less dense than air*. An object's density will tell you whether it will sink or float in a fluid, as long as you also

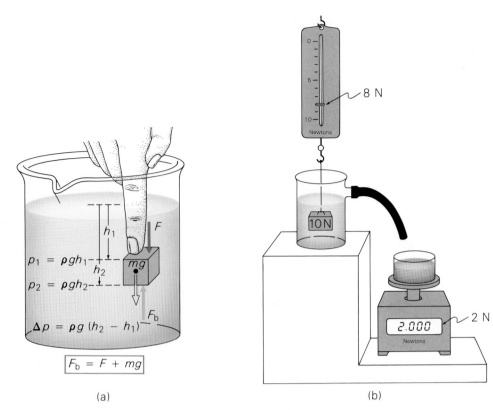

$p_1 = \rho g h_1$

$p_2 = \rho g h_2$

$\Delta p = \rho g (h_2 - h_1)$

$$\boxed{F_b = F + mg}$$

(a)

(b)

Figure 10.7 **Buoyancy and Archimedes' principle**
(a) A buoyant force arises from the pressure difference between different depths. The pressure on the bottom of the block is greater than that on the top, so there is a (buoyant) force directed upward. (b) Archimedes' principle can be demonstrated as shown. The buoyant force on the object is equal to the weight of the volume of fluid it displaces. (The scale is set to read zero with an empty container.)

know the density of the fluid. Consider an object totally immersed in a fluid. The weight of the object is

$$w_o = m_o g = \rho_o V_o g$$

The weight of the volume of fluid the object displaces, or the magnitude of the buoyant force, is

$$F_b = w_f = m_f g = \rho_f V_f g$$

But if the object is completely submerged, $V_o = V_f$, and dividing the second equation above by the first gives

$$\frac{F_b}{w_o} = \frac{\rho_f}{\rho_o} \quad \text{or} \quad F_b = \left(\frac{\rho_f}{\rho_o}\right) w_o \tag{10.8}$$

Thus, if ρ_o is less than ρ_f, then F_b will be greater than w_o, and the object will be buoyed to the surface and float. If ρ_o is greater than ρ_f, then F_b will be less than w_o, and the object will sink. Also, if ρ_o equals ρ_f, then F_b will be equal to w_o, and

Table 10.1
Densities of Some Common Substances (in kg/m³)

Substance	Density (ρ)	Substance	Density (ρ)	Substance	Density (ρ)
Solids		*Liquids*		*Gases**	
Aluminum	2.7×10^3	Alcohol, ethyl	0.79×10^3	Air	1.29
Brass	8.7×10^3	Alcohol, methyl	0.82×10^3	Helium	0.18
Copper	8.9×10^3	Blood, whole	1.05×10^3	Oxygen	1.43
Glass	2.6×10^3	Blood plasma	1.03×10^3	Water vapor (100°C)	0.63
Gold	19.3×10^3	Gasoline	0.68×10^3		
Ice	0.92×10^3	Kerosene	0.82×10^3		
Iron	7.9×10^3	Mercury	13.6×10^3		
Lead	11.4×10^3	Sea water (4°)	1.03×10^3		
Silver	10.5×10^3	Water, fresh (4°C)	1.00×10^3		
Steel	7.8×10^3				
Wood, oak	0.81×10^3				

* At 0°C and 1 atm unless otherwise specified.

the object will remain in equilibrium at any submerged depth (as long as the density of the fluid is constant).

These three conditions can be expressed in words:

An object will float in a fluid if the density of the object is less than the density of the fluid.

An object will sink in a fluid if the density of the object is greater than the density of the fluid.

An object will be in equilibrium at any submerged depth in a fluid if the densities of the object and the fluid are equal.
(See Demonstration 10.)

The densities of some solids and fluids are given in Table 10.1. A quick look will tell you if an object will float in a fluid, regardless of the shape or volume of the object. The conditions stated above also apply to a fluid in a fluid, provided the two are immiscible (do not mix).

In general, the densities of objects or fluids will be assumed to be uniform and constant in this book. Of course, the density of the atmosphere varies with altitude but is relatively constant near the surface of the Earth. In some instances, the overall density of an object may be purposefully varied. For example, a submarine submerges by flooding its tanks with sea water (called taking on ballast), which increases its overall density. When the sub is to surface, the water is pumped out of the tanks, and the density of the sub becomes less than that of the surrounding sea water.

Example 10.5   Float or Sink?
A cube of material 10 cm on a side has a mass of 700 g. (a) Will the cube float in water? (b) If so, how much of its volume will be submerged?

Solution

Given: $m = 700$ g
$L = 10$ cm
$\rho_w = 1.00$ g/cm^3 (density of water from Table 10.1)

Find: (a) Whether the cube will float in water
(b) The percentage of the volume submerged

It is sometimes convenient to work in cgs units in comparing quantities, particularly when working with ratios. For densities in g/cm^3, drop the "x 10^3" from the values given in Table 10.1 for solids and liquids and add "x 10^{-3}" for gases.

(a) The density of the cube material is

$$\rho_c = \frac{m}{V_c} = \frac{m}{L^3} = \frac{700 \text{ g}}{(10 \text{ cm})^3} = 0.70 \text{ g/cm}^3$$

Since ρ_c is less than ρ_w, the cube will float.

Demonstration 10

Buoyancy and Density

A demonstration of buoyancy that shows that the overall density of a can of Diet Coke is less than that of water, while the density of a can of Classic Coke is greater.

(a) Unopened cans of Coke are dropped into a container of water.

(b) The can of Classic Coke sinks and the can of Diet Coke floats.

Consider the following questions: Does one can have a greater volume of metal? higher gas pressure inside? more fluid volume? Do calories make a difference? You can investigate the possibilities yourself to determine the reason(s) for the different densities.

(b) The weight of the cube is $w_c = \rho_c g V_c$. When the cube is floating (in equilibrium), its weight is balanced by the buoyant force. That is, $F_b = \rho_w g V_w$, where V_w is the volume of water the submerged part of the cube displaces. Equating the expressions for weight and buoyant force gives

$$\rho_w g V_w = \rho_c g V_c$$

or $\quad \dfrac{V_w}{V_c} = \dfrac{\rho_c}{\rho_w} = \dfrac{0.70 \text{ g/cm}^3}{1.00 \text{ g/cm}^3} = 0.70$

Thus, $V_w = (0.70)V_c$, and 70% of the cube is submerged. (Similarly, most of an iceberg floating in the ocean is submerged; all that is visible is the proverbial tip of the iceberg.) ■

You may have heard of a quantity called specific gravity, which is related to density. It is commonly used for liquids but also applies to solids. Basically, it is a comparison of the weight of a volume of a substance with the weight of an equal volume of water. The **specific gravity** of a substance is equal to the ratio of the density of the substance (ρ_s) to the density of water (ρ_w, at 4 °C):

$$\text{sp. gr.} = \frac{\rho_s}{\rho_w}$$

Because it is a ratio of densities, specific gravity has no units. Since $\rho_w = 1.00 \text{ g/cm}^3$,

$$\text{sp. gr.} = \frac{\rho_s}{1.00} = \rho_s$$

That is, the specific gravity of a substance is equal to the numerical value of its density in cgs units. For example, if a liquid has a density of 1.5 g/cm^3, its specific gravity is 1.5, which tells you that it is 1.5 times denser than water.

Measuring the specific gravity of a liquid involves an application of Archimedes' principle and a device called a **hydrometer**. A common type of hydrometer consists of a sealed glass tube with a weighted bulb (to make the tube float upright) and an enclosed calibrated scale in the upper stem (Fig. 10.8). The hydrometer is calibrated so that its scale reads 1.000 at the surface level when it is floating in water. When the hydrometer is placed in another liquid with a different density, it floats higher or lower, depending on the liquid's density (why?), and the specific gravity of the liquid is read from the scale.

Hydrometers are used in numerous applications. They are used medically to measure the specific gravities of body fluids and commercially to measure the specific gravities of acids, alcohols, gasoline, and various solutions. A common use of a hydrometer is as a battery-acid tester. The liquid in a lead storage battery, such as those used in automobiles, is a solution of sulfuric acid and water. When the battery is charging, the concentration of sulfuric acid, which is denser than water, increases. When the battery is discharging, water is formed and sulfuric acid is depleted. Checking the specific gravity of the liquid in the battery is an easy way to determine how charged a battery is in terms of the sulfuric acid concentration. Since sealed, maintenance-free batteries have become more common recently, perhaps a more frequent application of the hydrometer is for determining the concentration of antifreeze in an automobile's cooling system.

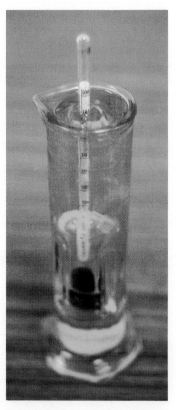

Figure 10.8 **A hydrometer** The specific gravity of a liquid in which the hydrometer is immersed is read from the calibrated scale on its upper stem. This hydrometer is constructed for measuring specific gravities greater than that of water, or 1.0000.

10.3 Surface Tension and Capillary Action

A fluid will not support a shear stress, yet you have probably seen water spiders walking on the surface of a pond (Fig. 10.9). Also, a razor blade or a needle placed carefully on the surface of water will float, even though steel is denser than water (about eight times denser according to Table 10.1). These things are possible because of an interesting property of liquids. A free liquid surface acts like a thin membrane that can be placed under a slight tension.

The molecules of a liquid exert small attractive forces on each other. Even though molecules are electrically neutral overall, there is often some slight asymmetry of charge that gives rise to attractive forces between molecules (called van der Waals forces). Within a liquid, where any molecule is completely surrounded by other molecules, the net force is zero [Fig. 10.10(a)]. However, for molecules at the surface of the liquid, there is no attractive force acting from above the surface. (The effect of air molecules is small and considered negligible.) As a result, the molecules of the surface layer experience net forces due to the attraction of neighboring molecules just below the surface. This inward pull on the surface molecules causes the surface of the liquid to contract, giving rise to a property called **surface tension**. The contraction is balanced by internal repulsive forces.

If something is placed on a liquid surface, work must be done in bringing more molecules from the interior of the liquid to the surface in order to increase the liquid surface area [Fig. 10.10(b)]. This work increases the potential energy of the surface molecules, and the surface acts like stretched elastic membrane [Fig. 10.10(c)].

The net effect of surface tension is to make the surface area of a liquid as small as possible. That is, a given volume of a liquid tends to assume the shape that has the least surface area so that the potential (surface) energy is a

Figure 10.9 **Surface tension**
Insects like this water strider can walk on water because of surface tension. This is analogous to your walking on a large trampoline.

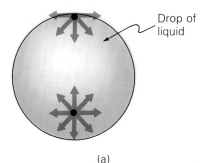

(a)

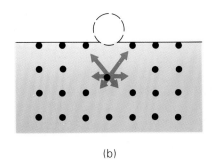

(b)

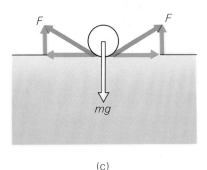

(c)

Figure 10.10 **Surface tension**
(a) The net force on a molecule in the interior of a liquid is zero, since it is surrounded by other molecules. However, a molecule at the surface experiences a nonzero net force due to the attractive forces of the neighboring molecules just below the surface. (b) To form a surface depression, work is done in bringing more interior molecules to the surface to increase the area. (c) As a result, the surface acts like a stretched elastic membrane and can support objects, such as a needle.

minimum. A sphere has the minimum surface area for a given volume, so soap bubbles and drops of liquid (for example, dew drops and rain drops) have spherical shapes. As a drop is formed, surface tension pulls the molecules together to minimize the surface area.

Quantitatively, the surface tension (γ) in a liquid film is defined as the force per unit length acting along a line (for example, on a length of wire) when stretching the surface:

$$\gamma = \frac{F}{L} \tag{10.9}$$

The SI units for surface tension are newtons per meter (N/m), as can be seen from the equation. Some surface tensions for liquids are given in Table 10.2. As might be expected, surface tension is highly dependent on temperature.

An apparatus used to measure surface tension is shown in Fig. 10.11. Basically, the device measures the force required to overcome the surface tension. For a circular wire loop, L is the length of the circumference, and $\gamma = F/2L$, since there are two film surfaces (one on each side of the wire).

Another way of looking at surface tension is in terms of the work or energy needed to stretch the surface area. If a straight piece of wire of length L is used to stretch a surface by a parallel distance Δx, the work done against the surface tension is

$$\Delta W = F \Delta x = \gamma L \Delta x = \gamma \Delta A$$

since $F = \gamma L$ and $\Delta A = L \Delta x$ (the change in surface area). Thus,

$$\gamma = \frac{\Delta W}{\Delta A}$$

The surface tension, or the force per unit length, is equivalently the work per unit of change in the surface area, with units of J/m^2. (Is J/m^2 equivalent to N/m?)

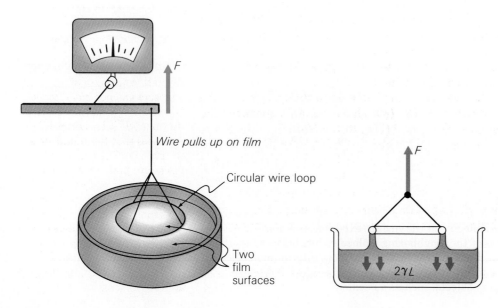

Wire pulls up on film

Circular wire loop

Two film surfaces

$2\gamma L$

Figure 10.11 **Measuring surface tension**
The force required to just overcome the surface tension is measured, and it is equal to $2\gamma L$ since there are two film surfaces.

Table 10.2
Surface Tensions for Some Liquids (in N/m)

Liquid		Surface tension (γ)
Alcohol, ethyl	(20 °C)	0.022
Blood, whole	(37 °C)	0.058
Blood plasma	(37 °C)	0.072
Mercury	(20 °C)	0.45
Soapy water	(20 °C)	0.025
Water	(0 °C)	0.076
Water	(20 °C)	0.073
Water	(100 °C)	0.059

Adhesion, Cohesion, and Capillary Action

Note the relatively small surface tension given in Table 10.2 for soapy water. Soaps and detergents have the effect of lowering the surface tension of water.* Such substances are called surfactants. The relatively high surface tension of plain water tends to prevent it from getting into small places, such as between the fibers of clothing. (You can also see from the table why warm water is generally used for cleaning.)

Soaps and detergents also act as **wetting agents**. Whether or not a liquid wets a surface depends on the relative strengths of the adhesive and cohesive forces between molecules. **Adhesive forces** (or adhesion) are attractive forces between unlike molecules. **Cohesive forces** (or cohesion) are attractive forces between like molecules. Cohesive forces hold a substance together, and adhesive forces hold different substances together. (Adhesives such as glue hold things together.)

If the adhesive forces between the molecules of the liquid and those of the surface are greater than the cohesive forces among the molecules of the liquid, the liquid wets the surface. On the other hand, if the cohesive forces are greater than the adhesive forces, the liquid does not wet the surface. Water beading up on a recently waxed car is a good example of the latter situation. Water does not adhere well to waxes or oils on a surface. (Wetting, as well as surface tension, is a factor enabling water spiders to walk on water. The spiders' feet are covered with a waxlike substance that prevents wetting.)

Although adhesive and cohesive forces are difficult to analyze, a relative measure of their effects is the **contact angle (ϕ)**. This is the angle between the surface and a line drawn tangent to the liquid (Fig. 10.12). Note from the figure that ϕ is less than 90° if the liquid wets the surface and greater than 90° if it does not. Water on clean glass has a contact angle of 0°, and water on paraffin has a contact angle of 107° (Table 10.3). A little detergent on the paraffin will cause the water to spread out, or wet the surface more, and the contact angle will decrease. The cleansing action of soaps and detergents is due in large part to their enhancing of the capability of water to wet dirt particles so they can be washed away.

* The general distinction is that a detergent is a soaplike cleaning compound that is not made from fats and lye (soap is).

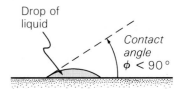

(a) Wetting condition

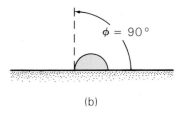

(b)

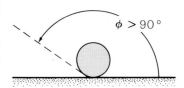

(c) Nonwetting condition

Figure 10.12 Contact angle
A relative measure of adhesive and cohesive forces is given by the contact angle. (a) An angle ϕ less than 90° indicates that the liquid wets the surface. The adhesive forces are greater than the cohesive forces. (b) When ϕ is equal to 90°, the drop forms a hemisphere. (c) An angle ϕ greater than 90° indicates that the liquid does not wet the surface. The cohesive forces are greater than the adhesive forces.

Table 10.3
Contact Angles for Some Liquids on Solids

Liquid-solid	Contact angle (ϕ)
Alcohol-glass	0°
Kerosene-glass	26°
Mercury-glass	140°
Water-glass	0°
Water-silver	90°
Water-paraffin	107°

In a container, the free liquid surface curves upward (is concave) if the liquid wets the container wall and curves downward (is convex) if it does not (Fig. 10.13). The curved shape of the liquid surface is referred to as the meniscus (from a Greek word meaning "crescent moon"). If a tube with a small diameter is positioned vertically with one end submerged in a liquid that wets the walls of the tube, the liquid in the tube will rise some distance above the surface of the surrounding liquid. This is called **capillary action** (or capillarity) and is a consequence of both surface tension and adhesion.

As might be expected, the height to which a liquid rises in a capillary tube depends on the diameter. (Capillary comes from a Latin word meaning "hairlike." Your smallest blood vessels are called capillaries and are so narrow that blood cells must pass through them in single file.) At equilibrium, the upward component of the surface tension force and the downward weight force of the liquid column must be equal in magnitude. The surface tension force is

$$F = \gamma L = \gamma(2\pi r)$$

where $L = 2\pi r$ since the liquid is in contact with the tube at all points around its circumference. The vertical component of this force has a magnitude of

$$F \cos \phi = \gamma(2\pi r)(\cos \phi)$$

The weight of the liquid column is given by

$$w = mg = \rho Vg = \rho(\pi r^2 h)g$$

where mass in terms of density is $m = \rho V$ and the volume of the cylinder of

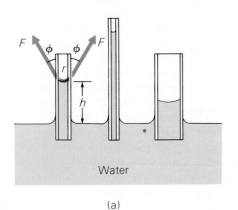

Water

(a)

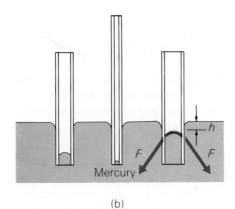

Mercury

(b)

Figure 10.13 **Capillary action**
(a) Liquids rise in small (capillary) tubes because of adhesion (wetting) and surface tension. The upward force (balanced by the weight of the liquid column) and the contact angle are shown on one tube. (b) If a liquid does not wet a capillary tube, there is a depression.

liquid is $V = \pi r^2 h$. (Atmospheric pressure is not a consideration because it is the same on both surfaces.) Equating these force magnitudes ($F \cos \phi = w$) and solving for h gives

$$h = \frac{2\gamma \cos \phi}{\rho g r} \qquad\qquad (10.10)$$

A similar analysis for capillary depression, which occurs when the liquid does not wet the tube surface, gives the same equation. In this case, ϕ is greater than $90°$, and h is negative.

Example 10.6   Soap Makes a Difference
How much higher will plain water rise in a 1.0-mm-diameter glass capillary tube than soapy water? (Assume that the densities are the same.)

Solution
Given: $d = 1.0$ mm $= 1.0 \times 10^{-3}$ m, $\qquad$ *Find*: Δh
$\qquad\qquad$ so $r = 5.0 \times 10^{-4}$ m
$\qquad \rho_w = \rho_{sw} = 1.0 \times 10^3$ kg/m^3
$\qquad\qquad$ (from Table 10.1)
$\qquad \gamma_w = 0.073$ N/m
$\qquad\qquad$ (from Table 10.2)
$\qquad \gamma_{sw} = 0.025$ N/m
$\qquad\qquad$ (from Table 10.2)

From Table 10.3, the contact angle for water and glass is $\phi = 0°$. This also applies to soapy water, which wets the glass more. With $\cos \phi = \cos 0° = 1$, Eq. 10.10 becomes $h = 2\gamma/\rho g r$. For water,

$$h_w = \frac{2\gamma_w}{\rho_w g r}$$

$$= \frac{2(0.073 \text{ N/m})}{(1.0 \times 10^3 \text{ kg/m}^3)(9.8 \text{ m/s}^2)(5.0 \times 10^{-4})}$$

$$= 0.030 \text{ m} \qquad \text{(or 3.0 cm)}$$

The height for the soapy water could be calculated using the same equation, but using a proportion saves some time:

$$h_{sw} = \left(\frac{\gamma_{sw}}{\gamma_w}\right) h_w = \left(\frac{0.025 \text{ N/m}}{0.073 \text{ N/m}}\right)(0.030 \text{ m})$$

$$= 0.010 \text{ m} \qquad \text{(or 1.0 cm)}$$

Thus,

$$\Delta h = h_w - h_{sw} = 0.030 \text{ m} - 0.010 \text{ m} = 0.020 \text{ m} \qquad \text{(or 2.0 cm)} \qquad \blacksquare$$

Capillary action is important in liquid transportation. Common examples are the absorbing of water by paper towels, the wicking of oil in oil lamps, the distribution of water and nutrients throughout plants, and the holding of water in the soil. If it were not for the latter, the ground would be pretty dry down to the water table. Less desirable results of capillary action are the wetting of concrete blocks in building walls and the rise of water in cracks and joints.

10.4 Fluid Dynamics and Bernoulli's Equation

In general, fluid motion is difficult to analyze. For example, think of trying to describe the motion of a particle (a molecule) of water in a rushing stream. The overall motion of the stream may be apparent, but a mathematical description of the motion of some particle of it may be virtually impossible because of eddy currents (small whirlpool motions), the gushing of water over rocks, frictional drag on the stream bottom, and so on. A basic description of fluid flow is conveniently obtained by ignoring such complications and considering an ideal fluid. Actual fluid flow can then be approximated in reference to this more simple theoretical model.

In this simple approach to fluid dynamics, it is customary to consider four characteristics of ideal fluids: steady, irrotational, nonviscous, and incompressible flow.

Steady flow means that all the particles of the fluid have the same velocity as they pass a given point. Steady flow might be called smooth or regular flow. The path of steady flow can be depicted in the form of **streamlines** (Fig. 10.14). Every particle that passes a particular point moves along a streamline. That is, at any point in the fluid, a streamline coincides with the tangential direction of the velocity at that point. Streamlines never cross. If they did, a particle would have alternate paths and changes in its velocity, in which case the flow would not be steady. Steady flow requires low velocities. For example, it is approximated by the flow relative to a canoe that is gliding slowly through still water.

Streamlines also indicate the relative magnitude of the fluid velocity. The number of streamlines passing through a small area is proportional to the magnitude of the velocity. That is, the velocity is greater where the streamlines are closer together.

Irrotational flow means that a fluid element (a small volume of the fluid) has no net angular velocity, which eliminates the possibility of whirlpools and eddy currents (the flow is nonturbulent). Consider the small paddle wheel in Fig. 10.14(a). With a zero net torque on it, its angular velocity is zero, and it does not rotate. Thus, the flow is irrotational. (Fig. 10.14b).

In real fluids, the flow becomes turbulent when the velocity is above a certain value or because a boundary condition causes streamlines to cross, giving rise to abrupt and erratic changes in velocity.

Nonviscous flow means that viscosity is neglected. Viscosity refers to a fluid's internal friction, or resistance to flow. A truly nonviscous fluid would flow freely with no energy lost within it. Also, there would be no frictional drag between the fluid and the walls containing it. In reality, when a liquid flows through a pipe, the velocity is less near the walls because of frictional drag and is greater toward the center of the pipe. (Viscosity is discussed in more detail in Section 10.5.)

Incompressible flow means that the fluid's density is constant. Liquids can usually be considered incompressible. Gases, on the other hand, are quite compressible. Sometimes, however, gases approximate incompressible flow, for example, in air flowing relative to the wings of an airplane traveling at low speeds.

Theoretical or ideal fluid flow is not characteristic of most real situations. But the analysis of ideal flow provides results that approximate, or generally describe, a variety of applications. This analysis is not done in terms of

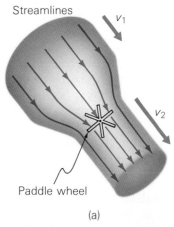

Streamlines

v_1

v_2

Paddle wheel

(a)

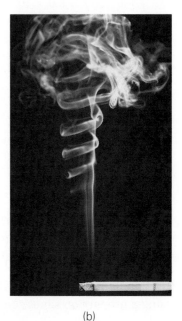

(b)

Figure 10.14 **Streamline flow**
(a) Streamlines never cross and are closer together in regions of greater fluid velocity. The stationary paddle wheel indicates that the flow is irrotational, or without whirlpools and eddy currents. (b) The smoke begins to rise in nearly streamline flow, but quickly becomes rotational and turbulent.

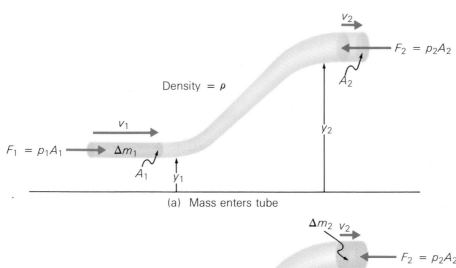

(a) Mass enters tube

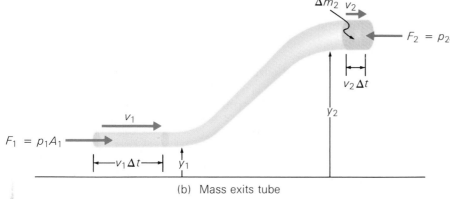

(b) Mass exits tube

Figure 10.15 **Bernoulli flow**
Ideal fluid flow can be
described by the equation of
continuity and Bernoulli's
equation. See text for
description.

Newton's laws as is usually the case, but in terms of two basic principles: the
conservation of mass and the conservation of energy.

Equation of Continuity

If there are no losses of fluid within a uniform tube, the mass of fluid flowing
into the tube in a given time must be equal to the mass flowing out of the tube in
the same time (by the conservation of mass). For example, in Fig. 10.15, the mass
(Δm_1) entering the tube during a short time (Δt) is

$$\Delta m_1 = \rho_1 \Delta V_1 = \rho_1 (A_1 v_1 \Delta t)$$

where A_1 is the cross-sectional area of the tube at the entrance, and in a time Δt,
a fluid particle moves a distance equal to $v_1 \Delta t$. Similarly, the mass leaving the
tube in the same time interval is

$$\Delta m_2 = \rho_2 \Delta V_2 = \rho_2 (A_2 v_2 \Delta t)$$

Since the mass is conserved, $\Delta m_1 = \Delta m_2$, and

$$\rho_1 A_1 v_1 = \rho_2 A_2 v_2$$

or $\quad \rho A v = \text{constant}$ (10.11)

This general result is called the **equation of continuity**.

For an incompressible fluid, the density ρ is constant, so

$$A_1 v_1 = A_2 v_2$$

or $\quad A v = \text{constant}$ (for an incompressible fluid)   (10.12)

This is sometimes called the **flow rate equation**, since Av has the standard units of cubic meters per second (m^3/s, or volume/time). (In the British system, gal/min. is often used.) Note that the flow rate equation shows that the fluid velocity is greater where the cross-sectional area of the tube is smaller. That is,

$$v_2 = \left(\frac{A_1}{A_2}\right)v_1$$

and v_2 is greater than v_1 if A_2 is less than A_1. (The streamlines will be closer together in a smaller tube.) This effect is evident in the common experience that the velocity of water is greater from a hose fitted with a nozzle than from the same hose without a nozzle.

Bernoulli's Equation

The conservation of energy or the general work-energy theorem leads to another relationship that has great generality for fluid flow. This relationship was first derived in 1738 by the Swiss mathematician Daniel Bernoulli (1700–1782) and is named for him.

Consider an ideal fluid flowing in the pipe in Fig. 10.15. The total work (W) done in moving the fluid through the pipe (to a higher elevation) is equal to the change in kinetic energy (ΔK) plus the change in gravitational potential energy (ΔU):

$$W = \Delta K + \Delta U$$

At the entrance and exit of the pipe section shown in the figure, work is done by two forces: F_1 is the driving force in the direction of the flow, and F_2 is a force in the opposite direction against which work is done. Since a fluid element travels a distance $\Delta x = v \Delta t$ in a time interval Δt, the work done by each force is

$$W_1 = F_1 \Delta x_1 = (p_1 A_1)(v_1 \Delta t)$$
$$W_2 = -F_2 \Delta x_2 = -(p_2 A_2)(v_2 \Delta t)$$

where force is expressed in terms of pressure ($F = pA$).

In the derivation of the equation of continuity, you saw that

$$\Delta m = \rho \Delta V = \rho(Av \Delta t) \qquad \text{or} \qquad \frac{\Delta m}{\rho} = Av \Delta t$$

Combining this with the general work equation gives

$$W = \frac{(\Delta m)p}{\rho}$$

The total work may then be written (recall that work is a scalar quantity)

$$W = W_1 + W_2 = \frac{\Delta m(p_1 - p_2)}{\rho}$$

The changes in kinetic and potential energies involve only the change in the energy of the mass Δm:

$$\Delta K = \tfrac{1}{2}\Delta m(v_2^2 - v_1^2)$$

and $\Delta U = (\Delta m)g(y_2 - y_1)$

Then, by the work-energy theorem ($W = \Delta K + \Delta U$),

$$\frac{\Delta m(p_1 - p_2)}{\rho} = \tfrac{1}{2}\Delta m(v_2^2 - v_1^2) + (\Delta m)g(y_2 - y_1)$$

Canceling the masses and rearranging gives

$$p_1 + \tfrac{1}{2}\rho v_1^2 + \rho g y_1 = p_2 + \tfrac{1}{2}\rho v_2^2 + \rho g y_2$$

or $\quad p + \tfrac{1}{2}\rho v^2 + \rho g y = \text{constant}$ $\hfill$ (10.13)

which is called **Bernoulli's equation**. Note that the kinetic and potential energy terms are expressed as energy per volume, that is, in terms of the density.

Bernoulli's equation can be applied to many situations (see Demonstration 11). If there is horizontal flow ($y_1 = y_2$), then $p + \tfrac{1}{2}\rho v^2 = \text{constant}$, which indicates that the pressure decreases if the fluid velocity increases (and vice versa). Chimneys and smokestacks are tall in order to take advantage of the more consistent and higher wind speeds at greater heights. The faster the wind blows over the top of a chimney, the lower the pressure, and the greater the pressure difference between the bottom and top of the chimney. Thus, the chimney draws better. Bernoulli's equation and the continuity equation ($Av = \text{constant}$) also tell you that if the cross-sectional area of a pipe is reduced (velocity greater), the pressure is reduced.

Suppose that a fluid is at rest ($v_2 = v_1 = 0$). Bernoulli's equation then becomes

$$p_2 - p_1 = \rho g(y_1 - y_2)$$

This should look familiar; it is the pressure-depth relationship ($p = \rho g h$) derived earlier.

Example 10.7   Flow Rate from a Tank

Water escapes through a small hole in a tank as depicted in Fig. 10.16. What is the flow rate of the water from the tank?

Solution

By Bernoulli's equation,

$$p_1 + \tfrac{1}{2}\rho v_1^2 + \rho g h = p_2 + \tfrac{1}{2}\rho v_2^2 + 0$$

where h is the height of the liquid surface above the hole. The atmospheric pressures acting on the open surface and at the hole, p_1 and p_2, respectively, are essentially equal and cancel from the equation, as does the density, giving

$$v_2^2 - v_1^2 = 2gh$$

The equation of continuity says that $A_1 v_1 = A_2 v_2$, where A_1 is the cross-sectional area of the tank and A_2 is that of the hole. Since A_1 is much greater than A_2, then v_2 is much greater than v_1. So a good approximation is

$$v_2^2 = 2gh \qquad \text{or} \qquad v_2 = \sqrt{2gh}$$

The flow rate (volume/time) is then

$$\text{flow rate} = A_2 v_2 = A_2\sqrt{2gh}$$

Given the area of the hole and the height of the liquid above it, you can find the instantaneous speed of the water coming from the hole and the instantaneous flow rate. ∎

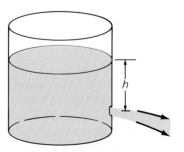

Figure 10.16 **Fluid flow from a tank**
The flow rate is given by Bernoulli's equation. See Example 10.7.

Bernoulli-trapped Ball

How does blowing downward through a funnel keep a Ping Pong ball from falling? In this demonstration, seeing is believing and Bernoulli.

(a) The demonstrator prepares to place the ball in an inverted funnel and to blow downward through the funnel.

(b) With air blowing through, the ball remains "trapped" in the funnel.

(c) Once the air is all blown out, the ball falls.

The air stream above and around the ball (along the wall of the funnel) is moving faster than the air below the ball. The faster air moves, the less pressure it exerts, therefore the greater pressure of the slower air below the ball holds it up. Without an air stream, the pressures are equalized (both atmospheric), and the ball falls.

Example 10.8   Throwing a Curve

A pitcher can cause a baseball to curve as it moves toward the catcher by giving it an appropriate spin. Explain why a curve ball curves in terms of Bernoulli's equation and the viscous properties of air.

Solution

If air were an ideal fluid, the spin of a pitched ball would have no effect in changing the direction of its motion. [See Fig. 10.17(a); the streamlines are those that would be seen by the ball or an observer moving with the ball.] However, since air is viscous, friction between it and the ball causes a thin layer

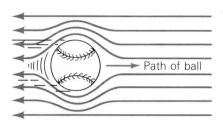

(a) With no spin, air flow is the same on both sides.

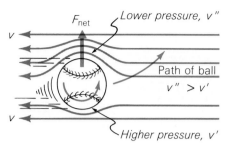

(b) With spin, air drag causes a difference in velocities.

Figure 10.17 **Curve ball** Throwing a curve ball requires viscosity. See Example 10.8.

of air to be dragged around by the spinning ball. The relative air speed is increased on one side of the ball and decreased on the other [Fig. 10.17(b)]. By Bernoulli's equation, the low-velocity side (v') has a greater pressure than the high-velocity side (v''). Thus, the ball experiences a net force toward the low-pressure side and is deflected (curves). ■

As applied to rotating objects, the effect described in Example 10.7 is sometimes called the Magnus effect, after the nineteenth-century German scientist who explained it. Ships have been powered using tall rotating cylinders in place of sails. The pressure difference between the opposite sides of a vertical rotating cylinder produces a propelling force lateral (perpendicular) to the wind direction, depending on which way the cylinder is rotated. However, if you don't want to go in either of these directions, you have to tack into the wind, which is slow. In 1985, Jacques Cousteau, the famed environmentalist and underwater explorer, launched a wind-powered boat called the *Alcyone* (after the daughter of the Greek god of wind), which is a variation of this application. The *Alcyone* is equipped with two nonrotating cylinders that use fans to provide the air flow and a flap system to deflect the air to give a force in the desired direction.

Fluid dynamics is an important practical consideration in the transporting of liquids and gases. Fluids move as a result of pressure differences, which are often mechanically generated by pumps. Water, oil, gasoline, natural gas, and many other fluids are moved by pumps. A vital natural pumping system is discussed in the Insight feature.

10.5 Viscosity, Poiseuille's Law, and Reynolds Number ◆

All real fluids have an internal resistance to flow, which is described as **viscosity**. Viscosity may be considered to be friction between the molecules of a fluid. In liquids, it is caused by short-range cohesive forces, and in gases, by collisions between molecules (see discussion of air resistance in Chapter 4). The viscous drag for both liquids and gases depends on velocity and may be directly proportional to it in some cases. However, the relationship varies depending on conditions; for example, the drag may be proportional to v^2 or v^3 for turbulent flow.

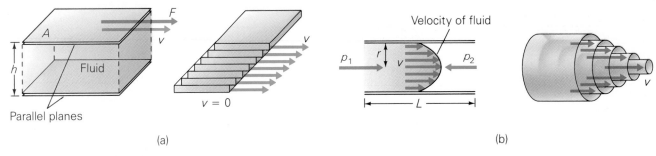

Figure 10.18 Laminar flow
(a) A shear stress causes fluid layers to move over each other in laminar flow. The shear force and the flow rate depend on the viscosity of the fluid. (b) For laminar flow through a pipe, the fluid velocity is less nearer the walls of the pipe because of frictional drag between the walls and the fluid.

Internal friction causes fluid layers to move relative to each other in response to a shear stress [Fig. 10.18(a)]. This layered motion, called laminar flow, is characteristic of steady flow for viscous liquids at low velocities. At greater velocities, the flow becomes rotational, or turbulent. As Fig. 10.18(a) indicates, the magnitude of the shear stress can be used as a measure of viscosity. Viscosity is characterized by a **coefficient of viscosity** (η):

$$\eta = \left(\frac{F}{A}\right)\left(\frac{h}{v}\right) = \frac{Fh}{Av} \qquad (10.14)$$

where F/A is the shear stress needed to maintain laminar flow between two parallel planes separated by a distance h at a relative velocity of magnitude v. Actually, η (commonly referred to as simply the viscosity) is the ratio of the shear stress to the rate of change of the shear strain: $(\Delta x/h)\,\Delta t = v/h$. Viscosities of some fluids are given in Table 10.4.

The SI units for viscosity are pascal-seconds:

$$(N/m^2)(m/m/s) = (N/m^2)(s) = Pa\text{-}s$$

This combined unit is called the **poiseuille (Pl)**, in honor of the French scientist Jean Poiseuille (1799–1869), who studied the flow of liquids, particularly blood. The cgs unit for viscosity is the **poise (P)**. A smaller multiple, the centipoise (cP), is widely used because of its convenient size.

As might be expected, viscosity and fluid flow vary with temperature, which is evident from the old saying "slow as molasses in January." A familiar application is the viscosity grade given to motor oil used in automobiles. In winter, a low-viscosity, or relatively thin, oil should be used (such as SAE grade 10W or 20W) because it will flow more readily, particularly when the engine is cold at start-up. In summer, a higher-viscosity, or thicker, oil is used (SAE 30W, 40W, or even 50W).

Of course, seasonal changes in the grade of motor oil are not necessary if you use the relatively new multigrade year-round oils. These contain additives called viscosity improvers, which are polymers whose molecules are long, coiled chains. A temperature increase causes the molecules to uncoil and intertwine with each other. Thus, the normal decrease in viscosity is counteracted. The action is reversed on cooling, and the oil maintains a relative constant viscosity over a temperature range. Such motor oils are graded as, for example, SAE 10W-20W-40 (or 10W-40 for short).

SAE stands for Society of Automotive Engineers, an organization that designates the grades of motor oils based on flow rate or viscosity.

Table 10.4
Viscosities of Various Fluids*

Fluid	Viscosity (η) Pl	cP
Liquids		
Alcohol, ethyl	1.2×10^{-3}	1.2
Blood, whole (37 °C)	1.7×10^{-3}	1.7
Blood plasma (37 °C)	2.5×10^{-3}	2.5
Glycerin	1.5	1.5×10^3
Mercury	1.55×10^{-3}	1.55
Oil, light machine	0.11	1.1×10^3
Water	1.00×10^{-3}	1.00
Gases		
Air	1.9×10^{-5}	1.9×10^{-2}
Oxygen	2.2×10^{-5}	2.2×10^{-2}

* At 20 °C unless otherwise indicated. 1 poiseuille (Pl) = 10^3 centipoise (cP).

Poiseuille's Law

When a fluid flows through a pipe, there is frictional drag between the liquid and the walls, and the fluid velocity is greater toward the center of the pipe [Fig. 10.18(b)]. The average **flow rate** ($Q = A\bar{v} = \Delta V/\Delta t$) depends on properties of the fluid and dimensions of the pipe, as well as on the pressure difference between the ends of the pipe. Jean Poiseuille studied flow in pipes and tubes, assuming constant viscosity and steady or laminar flow. He derived a relationship for the flow rate, which is known as **Poiseuille's law**:

$$Q = \frac{\pi r^4 \, \Delta p}{8 \eta L} \qquad (10.15)$$

where r is the radius of the pipe and L is its length.

As should be expected, the flow rate (m³/s) is inversely proportional to the viscosity and the length of the pipe. Also as expected, the flow rate is directly proportional to the pressure difference. Somewhat suprisingly, however, the flow rate is proportional to r^4, which makes it highly dependent on the radius of the tube, more so than might have been thought.

Example 10.9 Fourth-Power Dependence
The radius of a length of pipe carrying a liquid is decreased by 5.0% because of deposits on the inner surface. By how much would the pressure difference between the ends of the constricted pipe have to be increased to maintain a constant flow rate?

Solution
Given: $r_2 = 95\% \times r_1$, or   *Find:* Δp
$r_2 = (0.95)r_1$
$Q_2 = Q_1$

Blood Pressure

Basically, a pump is a machine that transfers mechanical energy to a fluid, causing it to flow. There are a wide variety of pumps, but one of interest to everyone is the human heart. The heart is a muscular pump that drives blood through the body's circulatory network of arteries, capillaries, and veins.

At the start of each pumping cycle, the heart's interior chambers enlarge and fill with freshly oxygenated blood arriving from the lungs (see Fig. 1). On contraction, the blood is forced out through the aorta into the arterial network. Smaller and smaller arteries branch off from the main ones, until the very small capillaries are reached. There food and oxygen being carried by the blood are exchanged with surrounding tissues, and wastes are picked up. The blood then flows into the veins to complete its circuit back to the heart.

The walls of the arteries are elastic and expand and contract with each pumping cycle. When the heart contracts, the blood pressure in the arteries increases. When the heart relaxes, the pressure decreases. The maximum pressure is called the systolic pressure, and the minimum pressure is called the diastolic pressure (after the two parts of the pumping cycle, the systole and diastole).

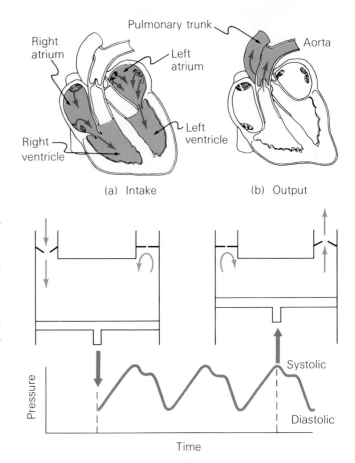

(a) Intake   (b) Output

Figure 1 **The heart as a force pump**
The human heart is analogous to a mechanical force pump. Its pumping action, consisting of (a) intake and (b) output gives rise to variations in blood pressure.

With $Q_2 = Q_1$, by Eq. 10.15,

$$\frac{Q_2}{Q_1} = \frac{r_2^4 \, \Delta p_2}{r_1^4 \, \Delta p_1}$$

and

$$\Delta p_2 = \left(\frac{r_1}{r_2}\right)^4 \Delta p_1 = \left(\frac{1}{0.95}\right)^4 \Delta p_1 = (1.23)\,\Delta p_1$$

The pressure difference must be increased by 23% to maintain the same flow rate.

Thus, the blood pressure has to increase to maintain the normal flow of blood in constricted arteries. As pointed out in the Insight feature, this makes the heart work harder. A similar application of Poiseuille's law concerns administering gases in respiratory therapy for conditions that reduce air flow to the lungs because of obstructions in the air passages. ∎

Taking a person's blood pressure is a measurement of the pressure of the blood on the arterial walls. This is done with a sphygmomanometer (the Greek word *sphygmo* means "pulse"). An inflatable cuff is used to shut off the blood flow temporarily. The cuff pressure is slowly released, and the artery is monitored with a stethoscope (Fig. 2). When the cuff pressure is slightly below the maximum blood pressure, blood will begin to flow through the artery with each beat of the heart. This produces a distinct beating sound that can be heard with the stethoscope. As soon as the sound is heard, the pressure is noted, normally about 120 mm Hg. This is the systolic pressure. As more air is released from the cuff, the blood flows more steadily, until at some pressure the distinct beating sound is no longer heard. The pressure at this point, normally about 80 mm Hg, corresponds to the diastolic pressure. Blood pressure is commonly reported as a ratio, for example, 120/80 (which is read as "120 over 80"). Normal blood pressure ranges between 100 and 140 for the systolic pressure and between 70 and 90 for the diastolic pressure.

High blood pressure is a common health problem. The elastic walls of the arteries expand under the hydraulic force of the blood pumped from the heart. Their elasticity may diminish with age, however. Fatty deposits (of cholesterols) can narrow and roughen the arterial passageways, impeding the blood flow and giving rise to a form of arteriosclerosis, or hardening of the arteries. Because of these defects, the driving pressure must increase to maintain a normal blood flow. The heart must work harder, which places a greater demand on its muscles. A relatively slight decrease in the

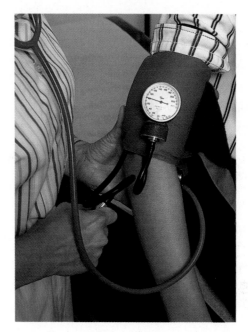

Figure 2 **Measuring blood pressure**
The pressure is indicated on the gauge in mm Hg or torr.

effective cross-sectional area of a blood vessel has a rather large effect on the flow rate. (See discussion of Poiseuille's law in Section 10.5.)

Reynolds Number

When the flow rate of a fluid exceeds a certain velocity, the flow ceases to be laminar and becomes turbulent. Poiseuille's law no longer applies. Analyzing turbulent flow is a difficult task, but a value that tells when the onset of turbulence will occur has been determined experimentally. This is expressed in terms of a dimensionless number called the **Reynolds number** (R_n):

$$R_n = \frac{\rho \bar{v} d}{\eta} \qquad (10.16)$$

Where ρ is the density of the fluid, $\bar{v}$ the average flow speed, d the diameter of a cylindrical pipe or tube, and η the viscosity.

In a pipe with smooth walls, the flow is laminar if R_n is below 2000. Turbulence generally sets in when R_n is about 2000 or more ($R_n \geq 2000$). It is possible to have laminar flow if R_n is above 2000, but that flow will be very unstable. Any slight disturbance will cause it to become turbulent.

Important Formulas

Pressure:

$$p = \frac{F}{A}$$

Pressure-depth equation:

$$p = p_o + \rho g h$$

Barometric (atmospheric) pressure:

$$p_a = \rho g h$$

Archimedes' principle:

$$F_b = m_f g = \rho_f g V_f$$

Surface tension:

$$\gamma = F/L$$

Capillary height:

$$h = \frac{2\gamma \cos \phi}{\rho g r}$$

Equation of continuity:

$$\rho_1 A_1 v_1 = \rho_2 A_2 v_2 \qquad \text{or} \qquad \rho A v = \text{constant}$$

Flow rate equation (for an incompressible fluid):

$$A_1 v_1 = A_2 v_2 \qquad \text{or} \qquad Av = \text{constant}$$

Bernoulli's equation:

$$p_1 + \tfrac{1}{2}\rho v_1^2 + \rho g y_1 = p_2 + \tfrac{1}{2}\rho v_2^2 + \rho g y_2$$

$$\text{or} \qquad p + \tfrac{1}{2}\rho v^2 + \rho g h = \text{constant}$$

◆ *Coefficient of viscosity*:

$$\eta = \frac{Fh}{Av}$$

◆ *Flow rate*:

$$Q = A\bar{v} = \frac{\Delta V}{\Delta t}$$

◆ *Poiseuille's law*:

$$Q = \frac{\pi r^4 \Delta p}{8 \eta L}$$

◆ *Reynolds number*:

$$R_n = \frac{\rho \bar{v} d}{\eta}$$

Questions

Pressure and Pascal's Principle

1. (a) How many ways could a uniform solid rectangular brick be placed on a table to yield different pressures on the table? (Consider positions of stable and unstable equilibrium.) (b) What position would produce (1) the greatest weight and (2) the greatest pressure on the table?

2. Two dams form artificial lakes with equal water depths. However, one lake backs up 5 km behind the dam, and the other backs up 10 km. (a) What effect does the difference in length have on the pressures on the dams? (b) What effect does it have on the areas of the lakes? (c) What structural considerations apply in dam construction?

3. The water in each of the containers in Fig. 10.19 exerts the same pressure *and* the same total force on the bottom of the container (equal base areas). But it is obvious that the weight of the water in each container is different. (a) What is the explanation of this so-called hydrostatic paradox? (b) Suppose that water could be added to the containers until the depth in each was doubled. How would this affect the weight of the contained water and the pressure on the bottom of each container?

4. A glass is filled with water, and an index or playing card is placed on top. If you hold the card in place and turn the glass over, you can remove your hand and the card and water will stay in place. Explain why. Could the glass be turned sideways (horizontally) with the same result?

5. A hypodermic needle and syringe are used to give shots. The relatively small force on the syringe alone would not be enough to do this. How does the needle help (even though it hurts)? What is the principle behind the new air-injected shots that are given without a needle?

6. An altimeter, a device that measures altitude (height above the Earth's surface), is really an aneroid barometer. Explain the principle of its operation.

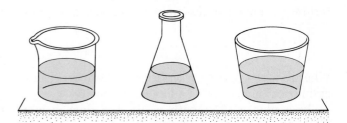

Figure 10.19 **Same pressure and same force?** See Question 3.

7. Automobile tires are inflated to pressures of about 200 kPa (30 psi), and bicycle tires are inflated to pressures twice as high. Explain why.

8. (a) Liquid storage cans, such as gasoline cans, generally have capped vent holes. What is the purpose of the vent, and what happens if you forget to remove the cap before pouring the liquid? (b) Explain how a medicine dropper works. (c) Explain how we breathe, i.e., inhalation and exhalation.

9. (a) A water dispenser for pets has an inverted plastic bottle, as shown in Fig. 10.20(a). (The water is dyed blue for contrast.) When a certain amount of water is drunk from the bowl, more water flows automatically from the bottle into the bowl. The bowl never overflows. Explain the operation of the dispenser. Does the height of the water in the bottle depend on the surface area of the water in the bowl? (b) A liquid tornado is shown in Fig. 10.20(b). When the bottle is swirled, the liquid flows into the lower bottle as shown. What would happen if the bottle were simply overturned?

10. (a) If you place a finger over the open end of a straw that is in a glass of liquid and remove it from the glass, the liquid in the straw does not run out. Why? (Pipettes are used in chemistry labs to transfer liquids in this manner.) (b) A siphon can be used to transport a liquid through a height change. Explain how a siphon works in terms of the pressure difference. (Hint: draw a picture.)

Buoyancy and Archimedes' Principle

11. Will a metal ball float in mercury? Explain.

(a)   (b)

Figure 10.20 **Water bottles**
See Question 9.

12. (a) What is the criterion for constructing a life jacket that will keep a 180-lb person afloat? (b) Why is it so easy to float in the Great Salt Lake in Utah?

13. A container filled with water just to the level of an exit tube sits on a scale, which reads 40 N. A 5.0-N object that is less dense than water is placed in the container, and the water it displaces runs out the exit tube into another container that is not on the scale. (a) Will the scale read 45 N? (b) Would it if the object had a density greater than that of water?

14. Describe a way to determine the density of an irregularly shaped object (a) if it sinks in water (explain how Archimedes used this principle in solving the crown problem), and (b) it floats in water.

15. An ice cube floats in a glass of water. As the ice melts, how does the level of water in the glass change? Would it make a difference if the ice cube were hollow? Explain.

16. When a helium balloon is released and rises in the atmosphere, it gets larger. (Why?) Does the buoyant force increase as the volume does, so the balloon will rise indefinitely? Explain.

17. Describe how you might use a wooden pencil and a thumb tack to construct a crude hydrometer that would indicate whether the specific gravity of a liquid is greater or less than 1.

18. The center of buoyancy (CB) is the point in a boat corresponding to the location of the center of gravity (CG) of the volume of water it has displaced. The buoyant force acts vertically upward through the center of buoyancy. Show that the stability of a boat (resistance to capsizing) depends on the relative positions of these two centers. (Hint: make sketches depicting different amounts of evenly distributed cargo.)

Surface Tension and Capillary Action

19. Why is the sand wet well above the water line on a beach? (Assume that there are no waves.) Why is wet sand firmer and easier to walk on than dry sand?

20. Explain why the bristles of a watercolor paintbrush spread out when the brush is immersed in water, but cling together to form a fine point when it is removed from the water.

21. Why are small drops of mercury on a surface spherical and larger drops noticeably flattened? (Drops of water on a surface they do not wet exhibit similar behavior.)

22. Spherical lead shot is manufactured by pouring molten lead through a sieve and letting the small drops solidify as they fall toward a container of water, in which they are cooled. Why do the lead droplets come out spherical?

23. A demonstration of surface tension is illustrated in Fig. 10.21. A wire loop with a thread loop inside it is immersed in a soap solution to form a film in the loops. When the film in the thread loop is punctured, a circle is formed. Explain why.

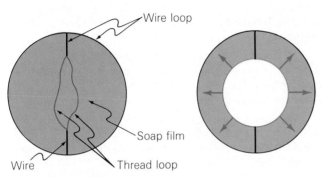

Figure 10.21 **Surface tension in action**
See Question 23.

24. The velocity of blood flow is greater in arteries than in capillaries. However, the flow rate equation ($Av =$ constant) seems to predict that the velocity should be greater in the smaller capillaries. Can you resolve this apparent inconsistency?

25. Why is it easier to blow bubbles with soapy water than with plain water?

26. (a) Explain why hot water is commonly used with soap or detergent for cleaning. (b) If detergent is applied to a duck's feathers, the bird will not be able to float. Why?

27. A liquid will rise to a certain maximum height in a capillary tube of a given diameter. An inventor proposes to make a liquid perpetual motion machine using a capillary tube shorter than the maximum length so that the liquid will continually rise to the top of the tube and spill over. Do you think this will work? Explain.

28. Explain how the depression distance (negative height) would be computed for a liquid such as mercury in a capillary tube.

Fluid Dynamics and Bernoulli's Equation

29. An ideal fluid flows in a pipe whose cross section varies. Explain where and why the flow velocity changes.

30. When a car is overtaken and passed by a fast-moving semitrailer truck, the driver of the car notices a force that tends to push the car toward the truck. What causes this? (This also occurs with ships, which is why they do not pass each other too closely.)

31. (a) The shape of an airplane wing causes the air to move faster over its upper surface than across its lower surface. Explain how this supplies lift to the plane, in terms of Bernoulli's equation. (b) What supplies lift for a helicopter?

32. Why must an airplane reach a certain take-off speed before leaving the runway? (Hint: see Question 31.) (b) Why do large commercial planes take off with their flaps down? (A flap is a movable curved section on the trailing edge of the wing that may be extended to change the curvature of the wing.)

33. If you hold a narrow strip of paper in front of your mouth and blow over the top surface, the strip will rise. (Try it.) Explain why.

34. Two Bernoulli effects are shown in Fig. 10.22. One photo shows a faucet aspirator commonly used to provide suction in chemistry labs, and the other shows a plastic egg trapped by a Bernoulli effect. Explain these effects.

(a)   (b)

Figure 10.22 **Bernoulli effects**
See Question 34.

Viscosity, Poiseuille's Law, and Reynolds Number ◆

35. (a) In general, how does the viscosity of a liquid vary with temperature? Is the same true for gases? (b) What is the temperature-dependent quantity in Eq. 10.14?

36. Where would you steer a boat in a river to take advantage of the current (a) when drifting downstream and (b) when rowing upstream?

37. Which has a greater viscosity, milk or cream? Which has the greater density?

38. Explain why Poiseuille's law is important in treating respiratory conditions.

39. How does viscosity affect the flow rate and the Reynolds number? Is this what you would expect?

Problems

■1. Which has the greater volume and how many times greater, 10 kg of aluminum or 2.4 kg of lead?

■2. A 100-mL graduated cylinder is filled halfway with ethyl alcohol, and then the other half is filled with gasoline. What is the total mass of the combined liquids?

■3. A 60-kg woman balances herself on the heel of one of her high-heeled shoes. If the round tip has a diameter of 1.5 cm, what is the pressure exerted on the floor? (Express your answer in lb/in.^{2} and atm.)

■4. What is the pressure due to the water on a diver who is 10 m below the surface of a lake?

■■5. A pure gold nugget with a volume of 15 cm^3 is placed on a pan balance. What would be the volume of the brass weights needed to balance the nugget?

■■6. What is the pressure exerted on a 15° ski slope by a 90-kg skier going down it? (Assume that the area of contact of the skis with the snow is 0.40 m^2.)

■■7. A scuba diver dives to a depth of 12 m in a lake. If the circular glass plate on the diver's face mask has a diameter of 18 cm, what is the force on it?

■■8. A force of 100 N is applied to a piston that has a circular cross-sectional area of 0.20 m^2. The piston is in a vertical cylinder that is 0.50 m long and filled with a hydraulic fluid whose density is 1.4×10^3 kg/m^3. What is the resulting pressure change due to the force?

■■9. In 1960, the U.S. Navy's bathscaphe *Trieste* descended to a depth of 10,912 m (about 35,000 ft) in the Mariana Trench in the Pacific Ocean. (a) What was the pressure at that depth? (Assume that sea water is incompressible.) (b) What was the force on a circular observation window with a diameter of 15 cm?

■■10. What is force exerted on a person's body by the atmosphere if the surface area is 1.30 m^2. (Express your answer in both newtons and pounds.)

■■11. The pressure exerted by a person's lungs can be measured by having the person blow as hard as possible into one side of a manometer. If a person blowing into one side of an open-tube manometer produces a 70-cm difference in the heights of the columns of water in the manometer arms, what is the lung pressure?

■■12. A vacuum pump is capable of producing a 50-cm difference in the column heights of a mercury barometer. What is the pressure produced by the pump?

■■13. Water is poured into one arm of an open U-tube containing mercury. What column height of water will cause the mercury level in the other arm to rise 5.0 cm?

■■14. The output piston of a hydraulic garage lift has a cross-sectional area of 0.20 m^2. (a) How much pressure on the input piston is required to support a car with a mass of 1.4 metric tons? (b) What force is applied to the input piston if it has a cross-sectional area of 5.0 cm^2?

■■15. Suppose that Archimedes found that the king's crown had a mass of 0.750 kg and a volume of 3.980×10^{-5} m^3. (a) How did he determine the volume? (b) Was the crown pure gold?

■■16. Show that the theoretical mechanical advantage (see Chapter 9) of a hydraulic jack or press is TMA = d_i/d_o, where d_i and d_o are the distances the input and output forces move, respectively. (Hint: consider displacements of liquid volumes.)

■■17. A hydraulic jack has an AMA of 5.0. (a) If a force of 80 N is applied to the driving piston, which has a diameter of 3.0 cm, what is the output force of the jack? (b) What is the area of the output piston? (See Problem 16.)

■■18. A hydraulic press has a TMA of 10. If an input piston supplying the driving force moves 4.0 cm on each stroke, how many strokes are required to move the output piston 50 cm? (See Problem 16.)

■■■19. Oil is poured into one side of an open tube manometer containing mercury. What is the density of the oil if a 10-cm column of mercury supports a 40-cm column of oil?

■■■20. A syringe with a plunger diameter of 2.0 cm is attached to a hypodermic needle with a diameter of 1.5 mm. What minimum force must be applied to the plunger to inject a fluid into a vein, where the blood pressure is 12 torr?

■■■21. What is the fractional decrease in pressure when a barometer is raised 30 m to the top of a three-story building? (Assume that the density of air remains constant.)

■■■22. The pressure of a gas in a container is measured with both an open-tube mercury manometer and a closed-tube mercury manometer. If the gauge pressure is 100 Pa, what is the height difference between the columns for each manometer? Is an open-tube manometer practical for measuring pressures this high?

■■■23. An open-tube mercury manometer has a height difference between the columns of −10 cm. What are (a) the gauge pressure and (b) the absolute pressure?

■■■24. A hydraulic balance used to detect small mass changes is shown in Fig. 10.23. If a mass of 0.25 g is placed on the balance platform, by how much will the height of the water in the cylinder have changed when the balance comes to equilibrium?

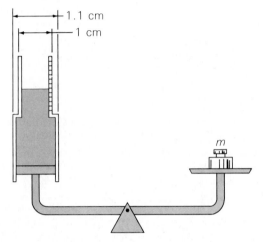

Figure 10.23 **A hydraulic balance**
See Problem 24.

10.2   Buoyancy and Archimedes' Principle

■■25. A cube 8.5 cm on a side has a mass of 0.65 kg. Will the cube float in water?

■■26. A cylindrical piece of plastic material with a radius of 8.0 cm and a length of 20 cm has a mass of 0.78 kg. Will the cylinder float if it is placed in a container of water?

■■27. A helium-filled balloon supports a load of 1 metric ton. (a) If the load includes the balloon's mass, what is its volume? (b) If the balloon were spherical, what would be its radius?

■■28. A submarine has a mass of 10,000 metric tons. How much water must be displaced for the sub to be in equilibrium just below the ocean surface?

■■29. A crane lifts a rectangular iron bar whose dimensions are 0.25 m by 0.20 m by 10 m from the bottom of a lake. What is the minimum upward force the crane must supply when the bar is (a) in the water and (b) out of the water?

■■30. A girl floats in a lake with 90% of her body beneath the water. What are (a) her mass density and (b) her weight density?

■■31. An iceberg floats in the Arctic Ocean. What percentage of its total volume is the exposed tip?

■■32. A flat-bottomed rectangular boat has a length of 4.0 m and a width of 1.5 m. If the load is 2000 kg (including the mass of the boat), how much of the boat will be submerged when it floats in a lake?

■■33. Show that specific gravity is the ratio of densities, given that by definition it is the ratio of the weight of a given volume of a substance and the weight of an equal volume of water.

■■■34. An oceangoing barge is 50.0 m long and 20.0 m wide and has a mass of 145 metric tons. Will the barge clear a reef 1.50 m below the surface of the water?

■■■35. A block of iron quickly sinks in water, but ships constructed of iron float. A solid cube of iron 1.0 m on a side is made into sheets. To make an open cube that will not sink from these sheets, what should the minimum length of the sides be?

■■■36. A bar of 24-karat (pure) gold is supposed to be completely solid. To check this, a scientist takes the bar into the lab, and using a scale finds that it has a mass of 1.00 kg in air and a measured mass of 0.93 when completely immersed in water. Is the bar solid?

■■■37. A hydrometer (Fig. 10.8) floats in a liquid with its 12.00-cm³ bulb completely immersed. If the hydrometer stem is 8.00 cm long with a cross-sectional area of 0.250 cm², what portion of the stem's length will be above the surface when floating in a liquid with a specific gravity of 1.25? (The mass of the hydrometer is 16.5 g.)

■■■38. A student in a rubber lifeboat containing some rocks paddles to the center of a large swimming pool. He notes the water level on the side of the pool and then throws the rocks into the water. Will the water level change? If so, how?

10.3   Surface Tension and Capillary Action

■■39. If a circular wire loop with a diameter of 3.0 cm is used to measure the surface tension of (a) soapy water and (b) mercury at room temperature (20 °C), what would be the vertical force required to lift the ring from the liquid in each case?

■■40. To what height will water rise in a glass capillary tube with a diameter of 2.0 mm?

■■41. If water rises to a height of 4.5 cm in a glass capillary tube, what is the diameter of the tube?

■■42. (a) Would water or blood rise higher in a glass capillary tube having a diameter of 0.80 mm? (b) How many times greater would the height difference be? (Assume a contact angle of 0° for each.)

■■43. In order to have water rise a distance of 5.0 cm in a glass capillary tube, what would the diameter of the tube have to be?

■■44. An eyedropper has an opening that is 2.5 mm in diameter. What is the minimum amount of water in a drop from the dropper?

■■45. Water rises in plants through tiny capillaries (called xylem), which have diameters that range from about 0.01 mm to 0.30 mm. What is the maximum height to which water can rise in a xylem system under capillary action? (Obviously, this does not account for moisture's getting to the tops of tall trees. Atmospheric pressure will support a column of water only about 10 m in height. It is believed that water is raised to tall heights in trees by a negative pressure created by cohesive forces between water molecules. Water evaporates from leaves, and the cohesive forces bring in more molecules from below.)

■■■46. A measurement instrument like that shown in Fig. 10.11 is used to apply a vertical force to a 4.0-cm-diameter wire loop to determine the surface tension of water. If the same force were used to determine the surface tension of mercury, what would be the diameter of the loop?

■■■47. A device like that shown in Fig. 10.11 is used to measure the surface tension of a liquid. The wire ring has an inside diameter of 3.9 cm. If the vertical force needed to just pull the ring free of the liquid is 0.013 N, what is the surface tension of the liquid?

■■■48. A glass capillary tube with a diameter of 1.0 mm is placed vertically in a dish of mercury. How far is the mercury in the tube depressed below the surface level in the dish?

10.4 Fluid Dynamics and Bernoulli's Equation

■49. Show that the pressure-depth equation can be derived from Bernoulli's equation.

■50. A container 100 cm deep is filled with water. If a small hole is punched in the side of the container 25 cm from the top, with what initial velocity will the water flow from the hole?

■51. What is the flow rate of water that moves at an average speed of 2.8 m/s through a pipe with a diameter of 15 cm?

■■52. The spout heights for the container shown in Fig. 10.24 are 10 cm, 20 cm, and 30 cm. The water level is maintained at a height of 35 cm. (a) What is the velocity of the water coming from each hole? (b) Which stream of water has the greatest range relative to the base of the container? Justify your answer. (You may be surprised.)

Figure 10.24 **Streams as projectiles**
See Problem 52.

■■53. Water flows from a 2.0-cm-diameter pipe at a speed of 0.35 m/s. How long will it take to fill a 10-L container?

■■54. Water as an ideal liquid flows through a horizontal pipe with a cross-sectional area of 10 cm² at a speed of 20 cm/s. At a constriction in the pipe, the cross-sectional area narrows to 2.5 cm². (a) What is the flow speed at the constriction? (b) What is the difference between the pressure at the constriction and the pressure in the nonconstricted portion of the pipe?

■■55. The flow rate of water through a garden hose is 66 cm³/s, and the hose and nozzle have cross-sectional areas of 6.0 cm² and 1.0 cm², respectively. (a) If the nozzle is held 10 cm above the spigot, what are the flow speeds through the spigot and the nozzle? (b) What is the pressure difference between these points? (Consider the water to be an ideal fluid.)

■■56. The speed of blood in a main artery, which has a diameter of 1.0 cm, is 4.5 cm/s. (a) What is the flow rate in the artery? (b) If the capillary system has a total cross-sectional area of 2500 cm², the average speed of blood through the capillaries is what percentage of that through the main artery? (c) Why is there a need for a low blood speed through the capillaries?

■■**57.** The water level in a 15-cm-diameter cylinder like that in Fig. 10.25 is maintained at a constant height of 0.45 m. If the diameter of the spout pipe is 0.50 cm, how high is the vertical stream of water? (Assume the water to be an ideal fluid.)

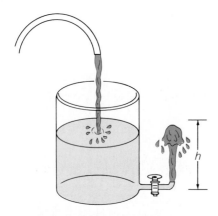

Figure 10.25 **How high a fountain?**
See Problem 57.

■■■**58.** Water flows through a horizontal 7.0-cm-diameter pipe under a pressure of 6.0 Pa with a flow rate of 25 L/min. At one point, calcium deposits reduce the cross-sectional area of the pipe to 30 cm². What is the pressure at this point? (Consider the water to be an ideal fluid.)

10.5 Viscosity, Poiseuille's Law, and
       Reynolds Number ◆

■**59.** Show that the units for viscosity are N-s/m² or Pa-s.

■**60.** Show that the Reynolds number is a dimensionless quantity.

■■**61.** A horizontal pipe with an inside diameter of 3.0 cm and a length of 6.0 m carries water, which has a flow rate of 40 L/min. What is the required pressure difference between the ends of the pipe?

■■**62.** A hypodermic needle with an inner diameter of 0.10 cm and a length of 3.0 cm is attached to a syringe whose plunger has cross-sectional area of 5.0 cm². The syringe is filled with saline (salt) solution. If a force of 100 N is applied to the plunger, what is the flow rate from the needle when the solution is squirted into the air? (Assume that the

solution has the same viscosity as water.)

■■**63.** A hospital patient receives a 500-cc blood transfusion through a needle with a length of 5.0 cm and an inner diameter of 1.0 mm. If the blood bag is suspended 0.85 m above the needle, how long does the transfusion take? (Neglect the viscosity of the blood flowing in the plastic tube between the bag and the needle.)

■■**64.** Glycerin is poured into a 1.0-L bottle through a funnel with a cylindrical neck that is 2.0 cm long and has an inner diameter of 1.0 cm. If the depth of the glycerin in the funnel is maintained at a constant 10 cm, how long will it take for the bottle to become half full? (The density of glycerine is 1.26×10^3 kg/m³.)

■■**65.** What is the maximum flow rate of water for laminar flow in a 5.0-cm-diameter pipe?

■■■**66.** The resistive force on a small spherical object falling through a fluid with a viscosity η is given by an expression known as Stokes' law:

$$F = 6\pi\eta r\upsilon$$

where η is the viscosity, r the radius of the object, and υ its velocity. Use Stokes' law to show that the terminal velocity of the spherical object is given by

$$\upsilon_t = \frac{\frac{2}{9}r^2 g(\rho_o - \rho_f)}{\eta}$$

where ρ_o and ρ_f are the densities of the object and the fluid, respectively. (Hint: the net force on the falling object is zero when the viscous force and the buoyant force balance the weight force.)

■■■**67.** (a) What is the terminal velocity of a 2.0-mm-diameter raindrop? (b) How does this compare with a hailstone of the same size? (See Problem 66.)

Additional Problems

68. A steel machine part has a volume of 0.36 m³. If it is suspended from a scale and immersed in gasoline, what will the scale read?

69. A room measures 3.0 m by 4.5 m by 6.0 m. If the heating and air conditioning ducts to and from the room have a diameter of 30 cm, what is the average flow rate of air necessary for all the air in the room to be exchanged every 10 min? (Assume that the density of the air is constant.)

70. If the atmosphere had a constant density equal to its density at the Earth's surface, how high would it extend?

71. Fluid A flows three times as fast as fluid B in a pipe with a particular cross-sectional area. Which has the greater density, and how many times greater is it?

72. A 75-kg athlete does a single handstand. If the area of the hand in contact with the floor is 125 cm², what pressure is exerted on the floor?

73. Which will rise higher in a 1.4-mm-diameter glass capillary tube, water or ethyl alcohol? Find the ratio of the heights.

74. Show that a water barometer would be impractical.

75. How high will water rise in a 0.75-mm-diameter capillary tube made of silver?

76. A vertical force of 0.014 N is required to lift a wire ring with a radius of 1.5 cm away from the surface of a liquid. What is the surface tension of the liquid? (Neglect the diameter of the wire.)

77. (a) Show that the average flow rate for turbulent flow is given by

$$Q = \frac{\pi d R_n \eta}{4\rho}$$

where d is the diameter of the pipe. (b) What is the flow rate of water through a 3.0-cm-diameter pipe?

78. In movies, people sometimes hide from pursuers by submerging themselves under water and breathing through straws or tubes. Estimate how deeply you could submerge yourself before it became impractical to breathe

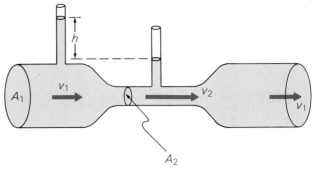

Figure 10.26 **A flow speed meter**
See Problem 80.

this way. (Hint: estimate the chest area and assume a force or weight on it that would make breathing difficult.)

79. A hydraulic lift is operated by a pump that supplies a gauge pressure of 600 kPa through a 4.0-cm-diameter pipe. What is the smallest diameter the lift piston may have if it is to lift a load of 24 metric tons?

80. A Venturi meter can be used to measure the flow speed of a liquid. A simple one is shown in Fig. 10.26. Show that the flow speed for an ideal fluid is given by

$$v_1 = \sqrt{\frac{2gh}{(A_1^2/A_2^2) - 1}}$$

81. A irregularly shaped piece of metal has a mass of 90 g in air. It is suspended from a scale and the scale reads 75 g when the piece is submerged in water. What is the volume and density of the piece of metal?

Temperature

11

T emperature and heat are frequent subjects of conversation, but if you had to state what the words really mean, you might find yourself at a loss. Certainly temperature and heat are related. You know from experience about how hot or cold something will be if you know its temperature. You know, too, that a temperature change generally results from the application or removal of heat, and that temperature is therefore some sort of indication, or measure, of heat. But what is heat?

An early theory of heat considered it to be a fluidlike substance called caloric (from the Latin word *calor*, meaning "heat") that could be made to flow in and out of a body. Even though this theory has been abandoned, scientists still speak of heat as flowing from one object to another. Heat is now characterized as energy, and temperature and thermal properties are explained by considering the atomic and molecular behavior of substances. This and the next two chapters examine the nature of temperature and heat in terms of microscopic (molecular) theory and macroscopic observations.

11.1 Temperature and Heat

A good way to begin studying thermal physics is with definitions of temperature and heat. **Temperature** is a relative measure, or indication, of hotness or coldness. A hot stove is said to have a high temperature, and an ice cube to have a low temperature. An object that has a higher temperature than another object does is said to be hotter. Note that hot and cold are relative terms, like tall and short.

We can perceive temperature by touch. However, this temperature sense is somewhat unreliable, and its range is too limited to be useful for scientific purposes. Methods and scales have been devised to measure temperature and temperature differences quantitatively. Temperature measurements tell *how much* hotter or colder one thing is than another (as length measurements tell how much taller or shorter something is than something else). Temperature is measured with thermometers, devices that will be discussed in the next section.

Heat is related to temperature and describes the process of energy transfer from one object to another. **Heat** describes energy that is transferred from one object to another because of a temperature difference. Thus, heat is energy in motion, so to speak. Once transferred, the energy becomes part of the total energy of the molecules of the object or system, its **internal energy**.

However, a higher temperature does not necessarily mean that one system has a greater internal energy than another. For example, in a classroom on a cold day, the air temperature is relatively high compared to that of the outdoor air. But the warm air in your classroom has far less internal energy as a whole than does all of the cold outdoor air. If this were not the case, heat pumps would not be practical (see Chapter 13). The internal energy of a system depends on its mass, or the number of molecules in the system.

Kinetic theory (to be covered in Section 11.5) shows that temperature is a measure of the average random kinetic energy of the molecules. Besides kinetic energy due to translational motion, molecules may also have kinetic energy due to linear and rotational vibrations (Fig. 11.1) as well as potential energy due to attractive forces between molecules. The sum of all these energies is the internal energy. Therefore, *heat is internal energy that is added to or removed from a body because of a temperature difference compared to another body.*

Average random kinetic energy is sometimes referred to as thermal energy, but that term is used to mean different things. When this energy is associated with random translational motion, it is the "temperature energy." An object moving as a whole, such as a thrown ball, is in translational motion, but this is ordered motion not random motion.

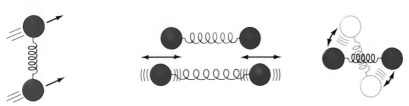

(a) Translational motion   (b) Linear and rotational vibrations

Figure 11.1 **Molecular motions**
A molecule may (a) move as a whole and/or (b) have linear and rotational vibrations. Temperature is associated with random translational motion, and the internal energy of a system is the total energy.

When heat is transferred between two objects, whether or not they are touching, they are said to be in **thermal contact**. Heat will be transferred between them as long as there is a temperature difference. When there is no longer a net heat transfer between objects in thermal contact, they are at the same temperature and are said to be in **thermal equilibrium**.

A statement concerning thermal equilibrium is known as the zeroth law of thermodynamics: two objects or systems that are separately in thermal equilibrium with a third are in thermal equilibrium with each other.

11.2 The Celsius and Fahrenheit Temperature Scales

A measure of temperature is obtained using a **thermometer**, a device constructed to make evident some property of a substance that changes with temperature. Fortunately, many physical properties of materials change sufficiently with temperature to be used as the bases for thermometers. By far the most obvious and commonly used property is **thermal expansion**, a change in the dimensions or volume of a substance that occurs when the temperature changes. Galileo is said to have constructed one of the first thermometers, a "thermoscope" that used the expansion of air in a bulb to measure temperature.

Other temperature-dependent properties on which thermometers are based include the change in electrical resistance (these devices are called resistance thermometers), the variation of electric voltage at the junction of two wires made of different metals (thermocouples), and the wavelength of emitted light or electromagnetic waves (pyrometers).

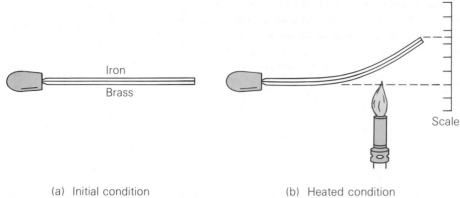

Iron

Brass

(a) Initial condition

Scale

(b) Heated condition

Figure 11.2 **Thermal expansion**
(a) A bimetallic strip is made of strips of two metals bonded together. (b) When such a strip is heated, it bends because of unequal expansions of the two metals. Here brass expands more than iron. The deflection of the end of such a strip can be used to measure temperature.

Figure 11.3 **Bimetallic coils**
Helical bimetallic coils are used in dial thermometers.

Almost all substances expand with increasing temperature, but to different extents. Most substances also contract with decreasing temperature. (Thermal expansion is used to mean both expansion and contraction; contraction is considered to be a negative expansion.) Because some metals expand more than others, a bimetallic strip (strips of two different metals bonded together) can be used to measure temperature changes. As heat is applied, the composite strip will bend away from the side made of the metal that expands more (Fig. 11.2). Helical coils formed from such strips are used in dial thermometers and in common household thermostats (Fig. 11.3).

A common thermometer is the liquid-in-glass type, which is based on the thermal expansion of a liquid. A liquid in a glass bulb expands into a glass stem, rising in a capillary bore. Mercury and alcohol (usually colored with a red dye to make it more visible) are the liquids used in most liquid-in-glass thermometers. These substances were chosen because of their relatively large thermal expansion and because they remain liquids over normal temperature ranges.

Thermometers are calibrated so that a numerical value may be assigned to a given temperature. For the definition of any standard scale or unit, two fixed reference points are needed. Since all substances change dimensions with temperature, an absolute reference for expansion is unavailable. However, the necessary fixed points may be correlated to physical phenomena which always occur at the same temperatures.

The ice point and the steam point of water are two convenient fixed points. Also commonly known as the freezing and boiling points, these two points are the temperatures at which pure water freezes and boils under a pressure of 1 atm (standard pressure).

The two familiar temperature scales are the **Fahrenheit temperature scale** and the **Celsius temperature scale**, named for their originators. As shown in Fig. 11.4, the ice and steam points have values of 32 °F and 212 °F, respectively, on the Fahrenheit scale and 0 °C and 100 °C on the Celsius scale. (See the Insight feature for some historical background.) On the Fahrenheit scale, there are 180 equal intervals, or degrees, between the two reference points, and on the Celsius scale, there are 100. Therefore, a Celsius degree is larger than a Fahrenheit degree; since $180/100 = 9/5 = 1.8$, it is almost twice as large.

A relationship for converting between the two scales may be obtained from a graph of Fahrenheit temperature (T_F) versus Celsius temperature (T_C), like the one in Fig. 11.5. The equation of the straight line (in general slope-intercept form, $y = mx + b$) is

$$T_F = \left(\frac{180}{100}\right)T_C + 32$$

or $\quad T_F = \frac{9}{5}T_C + 32$ \hfill (11.1)

where $\frac{9}{5}$ is the slope of the line and 32 is the intercept on the vertical axis. Thus, to change from a Celsius temperature (T_C) to its equivalent Fahrenheit temperature (T_F), you simply multiply the Celsius reading by $\frac{9}{5}$ and add 32.

The equation may be solved for T_C to convert from Fahrenheit to Celsius:

$$T_C = \frac{5}{9}(T_F - 32)$$ \hfill (11.2)

Example 11.1 Changing Temperatures

What are (a) $100\,°C$ on the Fahrenheit scale, (b) $68\,°F$ on the Celsius scale, and (c) normal body temperature, $98.6\,°F$, on the Celsius scale?

Solution

Given: (a) $T_C = 100\,°C$ *Find*: (a) T_F
       (b) $T_F = 68\,°F$ (b) T_C
       (c) $T_F = 98.6\,°F$ (c) T_C

(a) Eq. 11.1 is for changing Celsius readings to Fahrenheit:

$$T_F = \tfrac{9}{5}T_C + 32 = \tfrac{9}{5}(100) + 32 = 180 + 32 = 212\,°F$$

(b) Eq. 11.2 is for changing Fahrenheit readings to Celsius:

$$T_C = \tfrac{5}{9}(T_F - 32) = \tfrac{5}{9}(68 - 32) = \tfrac{5}{9}(36) = 20\,°C$$

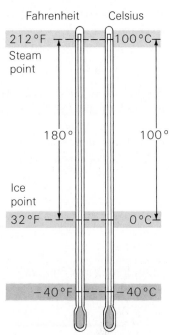

Fahrenheit   Celsius

212°F — Steam point — 100°C

180°

100°

Ice point

32°F — 0°C

−40°F — −40°C

Figure 11.4 Celsius and Fahrenheit temperature scales
Between the ice and steam fixed points, there are 100 degrees on the Celsius scale and 180 degrees on the Fahrenheit scale. Thus, a Celsius degree is 1.8 times larger than a Fahrenheit degree.

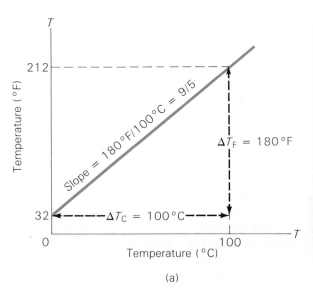

(a)

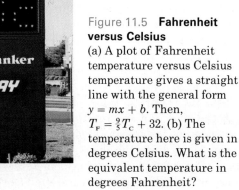

(b)

Figure 11.5 Fahrenheit versus Celsius
(a) A plot of Fahrenheit temperature versus Celsius temperature gives a straight line with the general form $y = mx + b$. Then, $T_F = \frac{9}{5}T_C + 32$. (b) The temperature here is given in degrees Celsius. What is the equivalent temperature in degrees Fahrenheit?

This temperature is commonly taken to be room temperature, and is a good one to remember.

(c) Again, Eq. 11.2 can be used:

$$T_{\rm C} = \tfrac{5}{9}(T_{\rm F} - 32) = \tfrac{5}{9}(98.6 - 32) = \tfrac{5}{9}(66.6) = 37.0\,^{\circ}{\rm C}$$

On the Celsius scale, normal body temperature has a whole-number value. Keep in mind that a Celsius degree is 1.8 (almost twice) as large as a Fahrenheit degree, so a temperature elevation of several degrees on the Celsius scale makes a big difference. For example, a temperature of $39.5\,^{\circ}{\rm C}$ represents an elevation of $2.5\,^{\circ}{\rm C}$ over normal body temperature. This is an elevation of $2.5 \times 1.8 = 4.5^{\circ}$ on the Fahrenheit scale, or a temperature of $98.6 + 4.5 = 103.1\,^{\circ}{\rm F}$. ∎

PROBLEM SOLVING HINT

Using a known temperature like the boiling point of water as in part (a) of Example 11.1 is a good way to make sure that you have written the conversion equation correctly.

Example 11.2 Equal Celsius and Fahrenheit Readings?

Weather reports commonly give temperatures as both Celsius and Fahrenheit readings. Could the Celsius and Fahrenheit temperatures reported ever be the same, that is, have equal numerical values?

Solution

To find out if the Celsius temperature can equal the Fahrenheit temperature, $T_{\rm C} = T_{\rm F}$ (or vice versa), you essentially "ask" the conversion equation if this is possible by letting $T_{\rm C} = T_{\rm F}$. You are assuming that this can be the case. When using an equation gives you a ridiculous answer, this generally means that it or the physical situation doesn't apply. If $T_{\rm C} = T_{\rm F}$, then $T_{\rm F}$ can be substituted for $T_{\rm C}$ in the conversion equation:

$$T_{\rm F} = \tfrac{9}{5}T_{\rm C} + 32 = \tfrac{9}{5}T_{\rm F} + 32$$

Solving for $T_{\rm F}$ gives

$$\tfrac{9}{5}T_{\rm F} - T_{\rm F} = \tfrac{4}{5}T_{\rm F} = -32 \qquad \text{and} \qquad T_{\rm F} = -40\,^{\circ}{\rm F}$$

So the temperature readings can be equal at -40°; that is, $-40\,^{\circ}{\rm C} = -40\,^{\circ}{\rm F}$. Whether or not you will ever hear this on a weather report depends on where you live. ∎

Liquid-in-glass thermometers are adequate for many temperature measurements, but problems arise when very accurate determinations are needed. A material may not expand uniformly over a wide temperature range. When calibrated to the ice and steam points, an alcohol thermometer and a mercury thermometer have the same readings at these points. But because of different expansion properties, the thermometers will not have exactly the same reading at an intermediate temperature, such as room temperature. For very sensitive

Daniel Fahrenheit and Anders Celsius

The names Fahrenheit and Celsius are probably mentioned more often than those of any other scientists. When talking about degrees Fahrenheit or degrees Celsius, you may not have realized that you were referring to the originators of the common temperature scales (see Fig. 1).

One of the achievements of Daniel Gabriel Fahrenheit (1686–1736), a German scientist and instrument maker, was the construction of one of the first practical mercury-in-glass thermometers. At that time, producing a glass tube with a uniform bore (open inside column) was a problem. Another problem was that impurities in the mercury would block the narrow bore. Fahrenheit managed to make glass tubes with reasonably uniform bores and developed a method of filtering mercury through leather to remove its impurities.

You may wonder why Fahrenheit chose 32° and 212° for the temperatures of the ice and steam points of water. Actually, he didn't. Initially, he chose for his zero point the lowest temperature that he could obtain in the laboratory, that of an ice-salt mixture. Normal human body temperature was used as the upper fixed point and was (incorrectly) assigned a value of 96°. The ice and steam points of water corresponded to 32° and 212° on this scale.

Anders Celsius (1701–1744) was a Swedish astronomer. While making astronomical and weather observations, he became aware of the need for more accurate temperature measurements. In 1741, he produced a thermometer with a 100-degree scale. However, this original version of the scale was a bit different from the one that now carries his name. Celsius had chosen the ice point as 100° and the steam point as 0°. The values of these ice and steam fixed points were reversed shortly thereafter.

Figure 1 **A. Celsius**

For a long time, the 100-degree scale was not associated with Celsius or called by his name. The symbol °C was used, but the scale was commonly called the centigrade scale. This name referred to the 100-degree interval between the fixed points (*centi* meaning "a hundred" and *grade* being German for "degree"). In 1948, the Ninth General Conference on Weights and Measures made Celsius the official name. This was done to eliminate confusion with the French angular unit *centigrade* (a *grade* is $\frac{1}{100}$ of a right angle, or 100 grade = 90° = π/rad).

---

temperature measurements and to define intermediate temperatures precisely, some other type of thermometer must be used. One such thermometer is discussed in the next section.

11.3 Gas Laws and Absolute Temperature

Different liquid-in-glass thermometers show slightly different readings for temperatures other than the fixed points because of differing expansion properties. A thermometer whose readings are independent of the substance used is one that uses a gas. Experiments show that all gases at very low densities have the same expansion behavior.

The variables that describe the behavior of a given quantity of gas are pressure, volume, and temperature (p, V, and T). When temperature is held constant, the pressure and volume of a quantity of gas are related as follows:

$$pV = \text{constant} \quad \text{or} \quad p_1 V_1 = p_2 V_2 \qquad (11.3)$$

(at constant temperature)

That is, the product of the pressure and volume is a constant. This relationship is known as **Boyle's law**, after Robert Boyle (1627–1691), the English chemist who discovered it. (See Demonstration 12.)

When the pressure is held constant, the volume of an amount of gas is related to the *absolute* temperature (to be defined shortly):

$$\frac{V}{T} = \text{constant} \quad \text{or} \quad \frac{V_1}{T_1} = \frac{V_2}{T_2} \qquad (11.4)$$

(a)

(b)

That is, the ratio of the volume and the temperature is a constant. This relationship is known as **Charles' law**, after the French scientist Jacques Charles (1747–1823), who made early hot-air balloon flights and was therefore quite interested in the relationship between the volume and temperature of a gas. A popular demonstration of Charles' law is shown in Fig. 11.6.

Low-density gases obey these laws, and for a constant mass of gas, the combined relationship is known as the **perfect gas law**, or the **ideal gas law**:

$$\frac{pV}{T} = Nk \quad \text{or} \quad \frac{p_1 V_1}{T_1} = \frac{p_2 V_2}{T_2} \qquad (11.5)$$

The product Nk is a constant because N, the number of molecules in the quantity of gas, reflects the constant mass of the gas, and k is a constant called Boltzmann's constant ($k = 1.38 \times 10^{-23}$ J/K). The perfect gas law applies to all gases at low densities and describes fairly accurately the behavior of gases at normal densities.

The perfect gas law is often written as $pV = nRT$, where n is the number of moles of the gas and R is the universal gas constant ($R = 8.31$ J/mole-K). The number of moles (n) is related to the number of molecules (N) by **Avogadro's number**, $N_A = 6.02 \times 10^{23}$. A mole of any pure substance contains Avogadro's number of atoms or molecules and $N = nN_A$. The number of moles gives the amount of a substance and is related to the mass by $n = m/M$, where M is the formula (atomic) weight, or the mass expressed in grams. (See Problems 19–22.)

The important point for the purpose of this section is that the pressure and volume are directly proportional to temperature: $pV \propto T$. This relationship allows a gas to be used to measure temperature in a **constant-volume gas thermometer**. Holding the volume of a gas constant, which can be done easily in a rigid container (Fig. 11.7), gives $p \propto T$. Thus, with a constant-volume gas thermometer, temperature is read in terms of pressure. A plot of pressure versus temperature gives a straight line in this case, as shown in Fig. 11.8(a).

As can be seen in Fig. 11.8(b), measurements of real gases (plotted data points) deviate from the values predicted by the perfect gas law at low temperatures. This is because the gases liquefy at low temperatures. However, the relationship is linear over a large temperature range, and it looks like the pressure might reach zero with decreasing temperature if the gas continued to be a gas (perfect or ideal).

The absolute minimum temperature for an ideal gas may therefore be inferred by extrapolating, or extending the straight line to the axis, as in Fig. 11.8. This temperature is found to be $-273.16\,°\text{C}$ and is designated as **absolute zero**. Absolute zero is believed to be the lower limit of temperature, but it

(c)

Figure 11.6 Charles' law in action
(a) When inflated balloons are put into liquid nitrogen (in the white pan), their temperature decreases, since the liquid nitrogen has a temperature of $-196\,°\text{C}$. With $V \propto T$, their volume decreases. Now the air (mostly nitrogen) inside the balloons has condensed into liquid and the balloons are flat. (b) As the temperature of the balloons increases, the liquid air returns to a gas and (c) the balloons blow themselves up, so to speak ($V \propto T$).

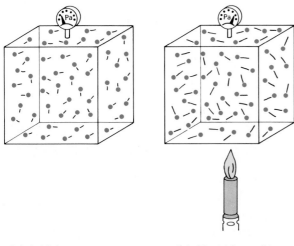

(a) Initial temperature   (b) Heated condition

Figure 11.7 **Constant-volume gas thermometer**
Such a thermometer indicates temperature as a function of pressure, since for a low-density gas, $p \propto T$. (a) At some initial temperature, the pressure reading has a certain value. (b) When the gas thermometer is heated, the pressure reading is higher.

has never been attained. In fact, there is a physical law that says it never can be (Chapter 13).

As you will learn shortly, there is an absolute temperature scale that takes absolute zero ($-273.16\,°C$) as its zero point. With the perfect gas law, $pV/T = $ constant, a temperature of zero gives an undefined relationship (because dividing by zero is not defined mathematically). Even so, this absolute temperature scale is the one that must be used with the perfect gas law. Temperatures on the Celsius or Fahrenheit scale do not apply. For a gas at a temperature $0\,°C$, the perfect gas law relationship would be undefined, but gases

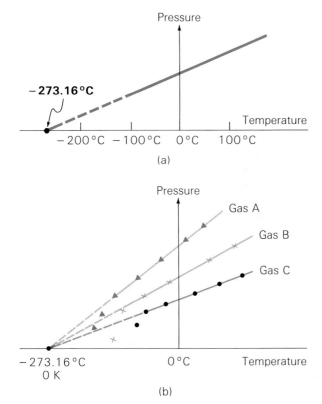

Figure 11.8 **Pressure versus temperature**
(a) A low-density gas kept at a constant volume gives a straight line on a p versus T graph [$pV = NkT$, or $p = (Nk/V)T$]. When the line is continued toward the zero-pressure value, a temperature of $-273.16\,°C$ is obtained, which is taken to be absolute zero. (b) Extrapolation of lines for all low-density gases indicates the same absolute zero temperature. The behavior of the gases deviates from the straight-line relationship at low temperatures because they start to liquefy.

Demonstration **12**

Boyle's Shaving Cream

A demonstration that shows the inverse relationship of pressure (p) and volume (V) of Boyle's law, $p \propto 1/V$.

(a) A swirl of shaving cream is placed in a transparent vacuum chamber.

(b) A vacuum pump begins to evacuate the chamber.

(c) As the pressure is reduced, the swirl of cream grows in volume from the expansion of the air bubbles trapped in the cream.

(d) In a dramatic volume reduction, the shaving cream cannot stand up to the sudden increase in pressure when the chamber is vented to the atmosphere.

Note: a similar effect occurs for a Boyle's marshmallow.

have definite pressures and volumes at this temperature. Since absolute zero is believed to be unattainable, this problem does not arise if an absolute scale is used.

There is no known upper limit to temperature. The temperatures at the center of stars are estimated to be greater than 100 million degrees.

Absolute zero is the foundation of the **Kelvin temperature scale**, named after the British scientist Lord Kelvin. On this scale, $-273.16\,°C$ is taken as the zero point, that is, as 0 K (Fig. 11.9). The size of the unit for Kelvin temperature is the same as the Celsius degree, so temperatures on these scales are related by

$$T_{\text{K}} = T_{\text{C}} + 273.16 \qquad (11.6)$$

where T_{K} is the temperature in **kelvins** (not degrees Kelvin). The kelvin is abbreviated as K (not °K).

The absolute Kelvin scale is the official SI temperature scale; however, the Celsius scale is usually used for everyday temperature readings. The absolute temperature in kelvins is used primarily in scientific applications. Keep in mind that Kelvin temperatures must be used with the perfect gas law. It is a common mistake to use Celsius or Fahrenheit temperatures in that equation. Also note that there can be no negative temperatures on the Kelvin scale.

Lord Kelvin, born William Thomson (1824–1907), developed devices to improve telegraphy and the compass and was involved in the laying of the first transatlantic cable. When he became a lord, it is said that he considered choosing Lord Cable or Lord Compass as his title, but decided on Lord Kelvin after a river that runs near the University of Glasgow in Scotland where he was a professor of physics for 50 years.

Example 11.3   Absolute Temperature

What is absolute zero on the Fahrenheit scale?

Solution

Given: $T_K = 0$ K   *Find*: T_F

Temperatures on the Kelvin scale are related directly to Celsius temperatures by $T_K = T_C + 273.16$, so first you convert 0 K to a Celsius value:

$$T_C = T_K - 273.16 = 0 - 273.16 = -273.16\,°C$$

Then converting to Fahrenheit gives

$$T_F = \tfrac{9}{5}T_C + 32 = \tfrac{9}{5}(-273.16) + 32 = -459.69\,°F$$

Thus, absolute zero is about $-460\,°F$.   ∎

Example 11.4   Absolute Temperature and the Perfect Gas Law

A quantity of low-density gas in a rigid container is initially at room temperature (20 °C) and a particular pressure (p_1). If the gas is heated to a temperature of 60 °C, by what factor does the pressure change?

Solution

Given: $T_1 = 20\,°C + 273 = 293$ K   *Find*: $\dfrac{p_2}{p_1}$
      $T_2 = 60\,°C + 273 = 333$ K
      $V_1 = V_2$ (because the container is rigid)

Note that the temperatures were changed to kelvins, with a rounded value of 273 in Eq. 11.6, which is appropriate in most cases. Using the perfect gas law in the form $p_2 V_2 / T_2 = p_1 V_1 / T_1$, and since $V_1 = V_2$,

$$p_2 = \left(\frac{T_2}{T_1}\right)p_1 = \left(\frac{333\text{ K}}{293\text{ K}}\right)p_1 = (1.14)p_1$$

Thus, the pressure increases by a factor of 1.14, or p_2 is 1.14 times p_1. (What would the factor be if the Celsius temperatures were *incorrectly* used? It would be much larger, as you can readily show.)   ∎

Initially, gas thermometers were calibrated using the ice and steam points. The Kelvin scale uses absolute zero and a fixed point adopted in 1954 by the International Committee on Weights and Measures. The second fixed point is the **triple point of water**, which represents a unique set of conditions where water coexists simultaneously in equilibrium as a solid, a liquid, and a gas. The conditions for the triple point are a pressure of 4.58 mm Hg, or about 610 Pa (760 mm Hg is normal atmospheric pressure), and a temperature taken to be 273.16 K. The kelvin is then defined as 1/273.16 of the temperature at the triple point of water.

11.4 Thermal Expansion

Changes in the dimensions and volumes of materials are common thermal effects. As you learned earlier, thermal expansion provides a means of temperature measurement. The thermal expansion of gases is generally described

There is also an absolute temperature scale whose basic unit is the same size as the Fahrenheit degree. This is called the Rankine scale, after William Rankine (1820–1872), a Scottish engineer, and $T_R = T_F + 459.69$, where T_R is in degrees Rankine (°R).

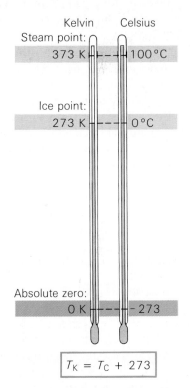

Figure 11.9   The Kelvin temperature scale
The lowest temperature on the Kelvin scale (corresponding to $-273.16\,°C$) is taken as absolute zero. An interval on the Kelvin scale, called a kelvin and abbreviated K, is equivalent to a Celsius degree, and $T_K = T_C + 273.16$.

by the perfect gas law, and is very obvious. Less dramatic, but by no means less important, is the thermal expansion of solids and liquids.

Thermal expansion results from a change in the average distance separating the atoms of a substance. The atoms are held together by bonding forces, which can be simplistically represented as springs in a simple model of a solid (Chapter 9). The atoms vibrate back and forth, and with increased temperature (more internal energy) become increasingly active and vibrate over greater distances. The solid as a whole expands as a result.

The change in one dimension of a solid (length, width, or thickness) is called linear expansion. For small temperature changes, linear expansion is approximately proportional to ΔT, or $T - T_o$ [Fig. 11.10(a)]. The fractional change in length is $(L - L_o)/L_o$, or $\Delta L/L_o$, where L_o is the original length (at the initial temperature). This is related to the change in temperature by

$$\frac{\Delta L}{L_o} = \alpha \Delta T \qquad \text{or} \qquad \Delta L = \alpha L_o \Delta T \qquad (11.7)$$

where α is the **coefficient of linear expansion**. Note that the unit of α is inverse temperature: $1/°C$, or $°C^{-1}$. Values of α for some materials are given in Table 11.1.

A solid may have different coefficients of linear expansion for different directions, but for simplicity this book will assume that the same coefficient applies to all directions (that solids show isotopic expansion). Also, the coefficient of expansion may vary slightly for different temperature ranges. Since this variation is negligible for most common applications, α will be considered to be constant and independent of temperature.

Eq. 11.7 may be rewritten to give the final length (L) after a temperature change:

$$\Delta L = \alpha L_o \Delta T$$
$$L - L_o = \alpha L_o \Delta T$$
$$L = L_o + \alpha L_o \Delta T$$
$$\text{or} \qquad L = L_o(1 + \alpha \Delta T) \qquad (11.8)$$

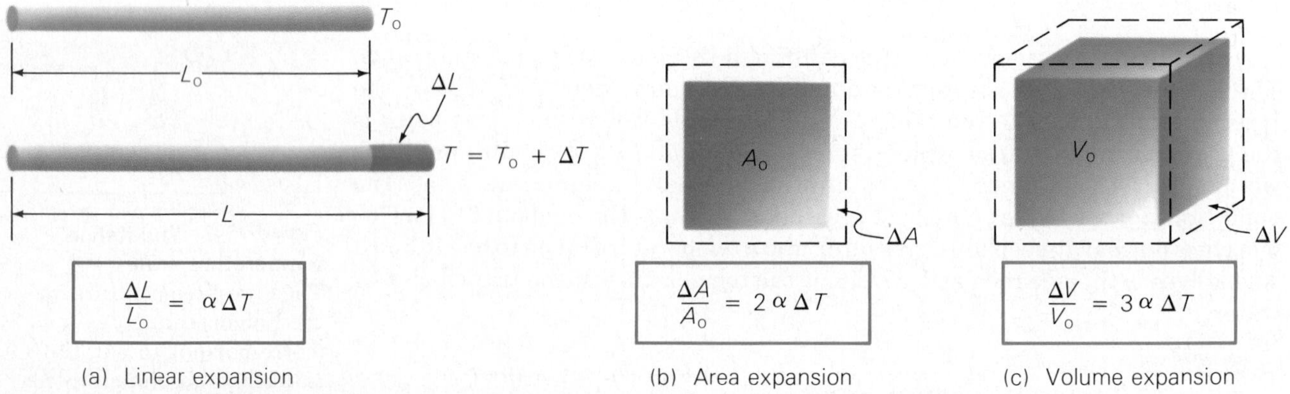

$$\frac{\Delta L}{L_o} = \alpha \Delta T$$

$$\frac{\Delta A}{A_o} = 2\alpha \Delta T$$

$$\frac{\Delta V}{V_o} = 3\alpha \Delta T$$

(a) Linear expansion (b) Area expansion (c) Volume expansion

Figure 11.10 **Thermal expansion**
(a) Linear expansion is proportional to the temperature change; that is, ΔL is proportional to ΔT, and $\Delta L = \alpha L_o \Delta T$, where α is the linear coefficient of expansion.
(b) For isotropic expansion, the coefficient of area expansion is approximately 2α.
(c) The coefficient of volume expansion for solids is about 3α.

Table 11.1
Values of Expansion Coefficients (in $°C^{-1}$) for Some Materials at 20 °C

Material	Coefficient of linear expansion (α)	Material	Coefficient of volume expansion (β)
Aluminum	24×10^{-6}	Alcohol, ethyl	1.1×10^{-4}
Brass	19×10^{-6}	Gasoline	9.5×10^{-4}
Brick or concrete	12×10^{-6}	Glycerin	4.9×10^{-4}
Copper	17×10^{-6}	Mercury	1.8×10^{-4}
Glass, window	9.0×10^{-6}	Water	2.1×10^{-4}
Glass, Pyrex	3.3×10^{-6}		
Ice	52×10^{-6}	Air (and most other gases at 1 atm)	3.5×10^{-3}
Iron and steel	12×10^{-6}		

Since area (A) is length squared (L^2),

$$A = L^2 = L_0^2(1 + \alpha \Delta T)^2 = A_0(1 + 2\alpha \Delta T + \alpha^2 \Delta T^2)$$

Since the values of α for solids are much less than 1 ($\sim 10^{-6}$, as shown in Table 11.1), the second-order term (containing α^2) may be dropped with negligible error. As a first-order approximation, then,

$$A = A_0(1 + 2\alpha \Delta T) \qquad \text{or} \qquad \frac{\Delta A}{A_0} = 2\alpha \Delta T \qquad (11.9)$$

Thus, the **coefficient of area expansion** [Fig. 11.10(b)] is twice as large as the coefficient of linear expansion (that is, is equal to 2α).
   Similarly, a first-order expression for volume expansion is

$$V = V_0(1 + 3\alpha \Delta T) \qquad \text{or} \qquad \frac{\Delta V}{V_0} = 3\alpha \Delta T \qquad (11.10)$$

The **coefficient of volume expansion** [Fig. 11.10(c)] is equal to 3α.
   The following example illustrates that the equations for thermal expansions are approximations. Even though an equation is a description of a physical relationship, you should always keep in mind that it may be only an approximation of physical reality and/or may only apply in certain situations. Because calculations of thermal expansion involve such small numbers, you may assume that all quantities are exact, that is, that they have any number of significant figures you wish.

Example 11.5   A Paradox or an Approximation?
An aluminum rod has a length of 1.0 m at 30 °C. The temperature is decreased to 10 °C and then raised back to 30 °C again. Is the length of the rod still 1.0 m?

Solution
Given: $L_0 = 1.0 \text{ m} = 100 \text{ cm}$ *Find*: Final length of the rod
   $T_0 = 30 °C$
   $T = 10 °C$
   $\Delta T_1 = T - T_0 = 10 °C - 30 °C = -20 °C$
   $\Delta T_2 = T_0 - T = 30 °C - 10 °C = 20 °C$
   $\alpha = 24 \times 10^{-6} °C^{-1}$ (from Table 11.1)

Intuitively, you know the rod should have the same length after its temperature is lowered and then raised by the same number of degrees. But what result does Eq. 11.8 give? With the first temperature change, the rod will contract to a length L_1:

$$L_1 = L_o(1 + \alpha \Delta T_1) = (100 \text{ cm})[1 + (24 \times 10^{-6} \, ^{\circ}\text{C}^{-1})(-20 \, ^{\circ}\text{C})]$$

$$= (100 \text{ cm})(1 - 0.00048) = 99.952 \text{ cm}$$

The minus from the negative temperature difference indicates a thermal contraction. Using the smaller units of centimeters better indicates the small effect. Any units of length can be used in the equation.

With the temperature rise, the rod will expand to a length L_2:

$$L_2 = L_1(1 + \alpha \Delta T_2) = (99.952 \text{ cm})[1 + (24 \times 10^{-6} \, ^{\circ}\text{C}^{-1})(20 \, ^{\circ}\text{C})]$$

$$= (99.952 \text{ cm})(1 + 0.00048) = 99.999 \text{ cm}$$

Thus, the result obtained by applying the equation for linear expansion to this situation does *not* reflect reality—the rod would not be shorter when it came back to its starting temperature. If this were true, putting the rod through numerous cycles would make it shorter and shorter, and it would eventually disappear! ∎

The thermal expansion of materials is an important consideration in construction. Seams are put in concrete highways and sidewalks to allow room for expansion and prevent cracking. Expansion gaps in large bridge structures and between railroad rails are necessary to prevent damage (Fig. 11.11). The thermal expansion of steel beams and girders can produce tremendous pressures, as the following example shows.

Example 11.6 Thermal Expansion and Stress

A steel I-beam is 5.0 m long at a temperature of 20 °C (68 °F). On a hot day, the temperature rises to 40 °C (104 °F). (a) What is the change in the beam's length due to thermal expansion? (b) Suppose that the ends of the beam are initially in contact with rigid vertical supports. How much force will the expanded beam exert on the supports?

Solution

Given: $L_o = 5.0$ m   *Find*: (a) ΔL
$\quad\quad\quad$ $T_o = 20 \, ^{\circ}\text{C}$ (b) F
$\quad\quad\quad$ $T = 40 \, ^{\circ}\text{C}$
$\quad\quad\quad$ $\alpha = 12 \times 10^{-6} \, ^{\circ}\text{C}^{-1}$
$\quad\quad\quad\quad$ (from Table 11.1)

(a) Using Eq. 11.7 with $\Delta T = T - T_o = 40 \, ^{\circ}\text{C} - 20 \, ^{\circ}\text{C} = 20 \, ^{\circ}\text{C}$,

$$\Delta L = L_o \alpha \, \Delta T = (5.0 \text{ m})(12 \times 10^{-6} \, ^{\circ}\text{C}^{-1})(20 \, ^{\circ}\text{C})$$

$$= 1.2 \times 10^{-3} \text{ m} = 1.2 \text{ mm}$$

This may not seem like much of an expansion, but it can give rise to a great deal of force if the beam is constrained and kept from expanding.

(b) The force exerted by the beam as it tries to expand by a length ΔL is the same as the force that would be required to compress the beam by that length. Using

Figure 11.11 **Expansion gaps** Expansion gaps can be found built into bridge roadways. They are designed to prevent contact stresses produced by thermal expansion.

the Young's modulus form of Hooke's law (Chapter 9) with $Y = 20 \times 10^{10}$ N/m^2, the stress on the beam is

$$\frac{F}{A} = \frac{Y\Delta L}{L_{\text{o}}} = \frac{(20 \times 10^{10} \text{ N/m}^2)(1.2 \times 10^{-3} \text{ m})}{5.0 \text{ m}}$$

$$= 4.8 \times 10^7 \text{ N/m}^2$$

The force is given by $F = (Y\Delta L/L_{\text{o}})A$, and estimating the end of the beam to have an area (A) of 60 cm^2 or 6.0×10^{-3} m^2 gives

$$F = (4.8 \times 10^7 \text{ N/m}^2)A = (4.8 \times 10^7)(6.0 \times 10^{-3} \text{ m}^2)$$

$$= 2.9 \times 10^5 \text{ N} \quad \text{(about 65,000 lb, or } 32\tfrac{1}{2} \text{ tons)} \quad \blacksquare$$

Example 11.7   A Bigger or Smaller Hole?

A circular piece with a diameter of 8.00 cm is cut from an aluminum sheet at room temperature [Fig. 11.12(a)]. If the sheet is then placed in an oven and heated to 150 °C, what will the area of the hole be?

Solution

Given: $d_{\text{o}} = 8.00$ cm   *Find*: A
  $T_{\text{o}} = 20\,°\text{C}$
  $T = 150\,°\text{C}$
  $\alpha = 24 \times 10^{-6}\,°\text{C}^{-1}$
    (from Table 11.1)

It is a common misconception to think that the area of the hole will shrink because of the expansion of the metal around it. Actually, the hole gets bigger. Think of the circular piece of metal removed from the hole, rather than of the hole itself: the circular piece will expand with increasing temperature. The metal in the sheet reacts as if the removed piece were still part of it, so the hole expands. [Think of putting the piece of metal back into the hole and heating, Fig. 11.12(b).]

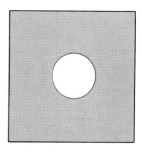

(a) Metal plate with hole

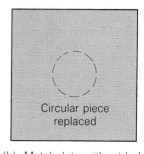

Circular piece replaced

(b) Metal plate without hole

Figure 11.12 **Area expansion** Does a hole in a metal plate get larger or smaller when the plate is heated? See Example 11.7.

Thus,

$$\Delta T = 150\,°C - 20\,°C = 130\,°C$$

and

$$A_{\mathrm{o}} = \frac{\pi d_{\mathrm{o}}^2}{4} = \frac{\pi(8.00\ \mathrm{cm})^2}{4} = 50.3\ \mathrm{cm}^2$$

The area of the hole (or the metal piece) after heating will therefore be

$$A = A_{\mathrm{o}}(1 + 2\alpha\,\Delta T) = (50.3\ \mathrm{cm}^2)[1 + 2(24 \times 10^{-6}\,°C^{-1})(130\,°C)]$$

$$= (50.3\ \mathrm{cm}^2)[1 + 0.0062] = 50.6\ \mathrm{cm}^2\quad\blacksquare$$

Liquids, like solids and gases, normally expand with increasing temperature. Because fluids have no definite shape, only volume expansion can be analyzed. The expression is

$$\frac{\Delta V}{V_{\mathrm{o}}} = \beta\,\Delta T \tag{11.11}$$

where β is the coefficient of volume expansion for fluids. Note in Table 11.1 that the values for β for fluids are typically larger than the values of 3α for solids.

Water exhibits an anomalous volume expansion near its freezing point. The volume of a given amount of water decreases as it is cooled from room temperature, until its temperature reaches $4\,°C$ (Fig. 11.13). Below $4\,°C$, the volume increases, and therefore the density decreases. This means that water has a maximum density ($\rho = m/V$) at $4\,°C$ (actually $3.98\,°C$). When water freezes, the molecules form a hexagonal (six-sided) lattice pattern. (This is why snowflakes have hexagonal shapes.) It is the open structure of this lattice that gives water its almost unique property of being less dense as a solid than as a liquid. (This is why ice floats in water.) The variation of the density of water over the temperature range of $4\,°C$ to $0\,°C$ indicates that the open lattice structure is beginning to form at about $4\,°C$, rather than arising instantaneously at the freezing point.

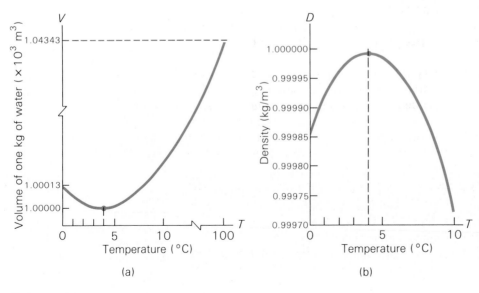

Figure 11.13 **Thermal expansion of water** Water exhibits nonlinear expansion behavior near its freezing point. (a) Above $4\,°C$ (actually $3.98\,°C$) water expands with increasing temperature, but from $4\,°C$ down to $0\,°C$, it expands with decreasing temperature. (b) As a result, water has its maximum density near $4\,°C$.

This property has an important environmental effect: bodies of water, such as lakes and ponds, freeze at the top first. As a lake cools toward 4 °C, water near the surface loses energy to the atmosphere, becomes denser, and sinks. The warmer, less dense water near the bottom rises. However, once the colder water on top reaches temperatures below 4 °C, it becomes less dense and remains at the surface, where it freezes. If water did not have this property, lakes and ponds would freeze from the bottom up, which would destroy much of their animal and plant life (and would make ice skating a lot less popular).

11.5 The Kinetic Theory of Gases

The notion that matter consists of tiny particles dates back to the Greek philosopher Democritus, who lived around 450 B.C. For centuries, this concept of the nature of matter was purely speculative. There was no experimental evidence that matter was made up of particles (no application of the scientific method).

In the early nineteenth century, the English chemist John Dalton developed explanations of several chemical phenomena using atomic theory. Not all of the explanations were satisfactory because Dalton considered atoms to be indivisible particles, and mistakenly thought gas molecules were atoms. (Molecules often divide in chemical reactions.) Although there were some errors in his approach, Dalton's work helped give atomic theory a firm foundation. Today, atoms are no longer considered to be indivisible; many subatomic particles have been observed (Chapter 30). However, in common reactions and interactions, atoms maintain their identity as separate particles.

More direct evidence for the atomic theory was accidentally discovered in 1827 by Robert Brown, a Scottish botanist. Brown observed under a microscope that pollen grains suspended in water move about in a jerky fashion. This so-called Brownian motion is readily explained by atomic theory. The tiny pollen grains are being pushed here and there because of collisions with randomly moving water molecules. Oddly enough, this simple explanation was not suggested until 1905—by Albert Einstein.

If atoms and molecules are viewed as colliding particles, the laws of mechanics can be applied to a quantity (system) of gas in describing its characteristics in terms of molecular motion, pressure (force/area), energy, and so on. Because of the large number of particles involved, statistical analysis is necessary.

One of the major accomplishments of theoretical physics was to derive the perfect gas law from mechanical principles by interpreting the temperature in terms of the kinetic energy of the gas molecules. Theoretically, the molecules of the perfect or ideal gas are viewed as essentially point masses in random motion with relatively large distances separating them. As point masses, the molecules have no internal structure, so vibrational and rotational motions are not a consideration.

The molecules make perfectly elastic collisions with each other and with the walls of the container. The forces between molecules are considered to act only over a short range, so the molecules interact with each other only during collisions. Since the molecules of a perfect gas interact only during collisions, such a gas would remain a gas at low temperatures and would not liquefy as real gases do. This gives the theoretical perfect gas its idealness. Colliding molecules lose and gain kinetic energy, with corresponding changes in their potential energy. This potential energy is ignored in summing the total energy of the system because molecules spend a negligible fraction of the time in collisions.

From Newton's laws of motion, the force on the walls of the container can be calculated from the change in momentum of the gas molecules when they collide with the walls (Fig. 11.14). If this force is expressed in terms of pressure (force/area), the following equation is obtained (see Appendix II for derivation):

$$pV = \tfrac{1}{3}Nm\bar{v}^2 \qquad\qquad (11.12)$$

where V is the volume of the container or gas, N the number of gas molecules in the closed container, m the mass of a gas molecule, and $\bar{v}$ a special average speed of the molecules.

Solving Eq. 11.5 for pV and equating that expression with Eq. 11.12 shows how the temperature came to be interpreted as a measure of the kinetic energy:

$$pV = NkT = \tfrac{1}{3}Nm\bar{v}^2 \qquad \text{or} \qquad \tfrac{1}{2}m\bar{v}^2 = \tfrac{3}{2}kT \qquad (11.13)$$

Thus, the temperature of a gas (and that of the walls of the container or a thermometer bulb in thermal equilibrium with the gas) is directly proportional to its average random kinetic energy (per molecule): $K = \tfrac{1}{2}m\bar{v}^2 = \tfrac{3}{2}kT$.

There are a couple of interesting points concerning Eq. 11.13. First, it predicts that at absolute zero ($T = 0$ K) all molecular motion of a gas would cease. According to classical theory, this would correspond to absolute zero energy. However, modern theory says that there would still be some zero-point motion. The zero-point energy associated with this motion is the energy minimum for the gas.

Also note that the total average kinetic energy of a gas is its internal energy. Thus, the internal energy of a perfect gas is directly proportional to its absolute temperature. This means that if the absolute temperature of a gas is doubled (by heat transfer), for example, from 200 K to 400 K, then its internal energy is also doubled. This does not apply to the Celsius and Fahrenheit temperatures, since their zero points are not referenced to zero energy. If a gas is at a temperature of 0 °C (or 0 °F) and its internal energy is doubled, the Celsius (or Fahrenheit) temperature is not doubled—doubling zero gives zero.

Diffusion

We depend on our sense of smell to detect odors, such as the smell of smoke from something burning. That you can smell something from a distance implies that molecules get from one place to another in the air, from the source to your nose. This process of random molecular mixing in which particular molecules move from a region where they are present in higher concentration to one where they are in lower concentration is called **diffusion**. Diffusion also occurs readily in liquids; think about what happens to a drop of ink in a glass of water (Fig. 11.15). It even occurs to some degree in solids.

The random molecular motion that produces gaseous diffusion can be understood in terms of kinetic theory. Consider the hypothetical situation of a container of gas in which most of the gas molecules are concentrated on one side (Fig. 11.16). Although the motions of the molecules are entirely random, there will be a net movement of molecules across the central section (of length Δx) of the container, from the region of higher concentration to the region of lower concentration. Molecules from both regions enter that central section,

The value $\bar{v}$ is a special kind of average speed. It is obtained by averaging the squares of the speeds and taking the square root of the average

$$\sqrt{\overline{v^2}} = \bar{v},$$

and $\bar{v}$ is called the root-mean-square (rms) speed.

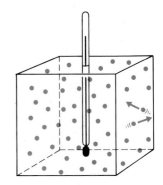

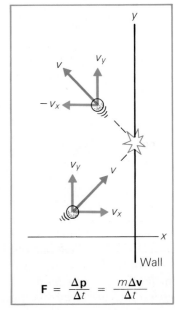

$$\mathbf{F} = \frac{\Delta \mathbf{p}}{\Delta t} = \frac{m\Delta \mathbf{v}}{\Delta t}$$

Figure 11.14 **Kinetic theory of gases**
The pressure a gas exerts on the walls of a container is due to the force resulting from the change in momentum of each gas molecule that collides with a wall.

Figure 11.15 **Diffusion**
Diffusion occurs in liquids. Random molecular motion would eventually distribute the dye of the ink throughout the water—here some of the distribution is due to mixing. The distribution would take some time with diffusion only.

but more enter on the average from the region of higher concentration because there are more molecules there. Since only one gas is present, this is called self-diffusion.

In general, the diffusion rate of gases and liquids is found to be proportional to $D(C_2 - C_1)/\Delta x$, where D is a diffusion coefficient and depends on the substance that is diffusing, and $(C_2 - C_1)/\Delta x$ is called the concentration gradient [concentration, e.g., kg/m^3, per length]. Note that C_1 and C_2 are the concentrations in adjoining regions, and when $C_2 = C_1$, the concentrations are equal, or the mixture is uniform, and the diffusion ceases.

Gases can also diffuse through porous materials or permeable membranes. Energetic molecules enter the material through the pores (openings), and, colliding with the pore walls, they slowly meander through the material. Such gaseous diffusion can be used to physically separate different gases from a mixture.

The kinetic theory of gases says that the average kinetic energy (per molecule) is proportional to the absolute temperature of a gas, $\frac{1}{2}m\bar{v}^2 = \frac{3}{2}kT$. So, on the average, the molecules of different gases (having different masses) move at different speeds at a given temperature. For example, at a particular temperature, molecules of oxygen (O_2) move faster on the average than the more massive molecules of carbon dioxide (CO_2). Because of this difference in molecular speed, oxygen can diffuse through a barrier faster than carbon dioxide can. Suppose that a mixture of equal volumes of oxygen and carbon dioxide is contained on one side of a porous barrier, as illustrated in Fig. 11.17. After a while, some O_2 molecules and CO_2 molecules will have diffused through

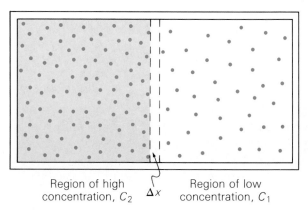

Region of high concentration, C_2 $\quad \Delta x \quad$ Region of low concentration, C_1

Figure 11.16 **Diffusion**
With regions of different concentrations, there is a net movement from the region of higher concentration to the region of lower concentration. The rate of this diffusion is proportional to $(C_2 - C_1)/\Delta x$. Once the concentrations are equal ($C_2 = C_1$), there is a uniform distribution.

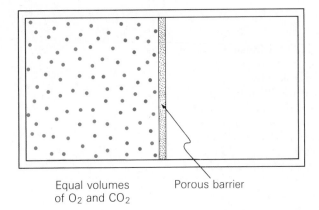

Equal volumes   Porous barrier
of O_2 and CO_2

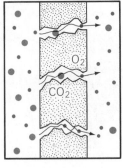

Diffusion begins

Figure 11.17 Separation by gaseous diffusion
The molecules of both gases diffuse through the porous barrier. But, since oxygen molecules have the greater average speed, more of them pass through. Thus, there is a greater concentration of oxygen molecules on the other side of the barrier.

the barrier. Of course, the diffused gases would still be mixed. After another diffusion stage, however, the oxygen concentration would become even greater. Almost pure oxygen can be obtained by repeating the process many times.

Separation by gaseous diffusion is a key process in obtaining enriched uranium, which was used in the atomic bomb and is currently used in nuclear reactors that generate electricity (Chapter 30).

Fluid diffusion is very important to plants and animals. In plant photosynthesis, carbon dioxide from the air diffuses into leaves, and oxygen and water (vapor) diffuse out. The diffusion of liquid water across a permeable membrane because of a concentration gradient is called **osmosis** and is vital in living cells. Oxygen in the lungs diffuses into the blood and is distributed to cells throughout the body, as carbon dioxide and other wastes are carried off. Osmotic diffusion is also important to kidney functioning: tubules in the kidneys concentrate waste matter and toxins from the blood in much the same way oxygen is removed from mixtures.

Important Formulas

Celsius-Fahrenheit conversion:

$T_F = \frac{9}{5}T_C + 32$

$T_C = \frac{5}{9}(T_F - 32)$

Boyle's law:

$pV = $ constant   or $p_1 V_1 = p_2 V_2$

Charles' law:

$\frac{V}{T} = $ constant   or $\frac{V_1}{T_1} = \frac{V_2}{T_2}$

Perfect (or ideal) gas law

$pV = NkT$ or $\frac{p_1 V_1}{T_1} = \frac{p_2 V_2}{T_2}$

or $pV = nRT$ (always absolute temperature):

Kelvin-Celsius conversion:

$T_K = T_C + 273.16$ or

$T_K = T_C + 273$ (general calculations)

Thermal expansion of solids

linear: $\frac{\Delta L}{L_o} = \alpha \Delta T$ or $L = L_o(1 + \alpha \Delta T)$

area: $\frac{\Delta A}{A_o} = 2\alpha \Delta T$ or $A = A_o(1 + 2\alpha \Delta T)$

volume: $\frac{\Delta V}{V_o} = 3\alpha \Delta T$ or $V = V_o(1 + 3\alpha \Delta T)$

Thermal expansion of fluids—volume:

$\frac{\Delta V}{V_o} = \beta \Delta T$

Kinetic theory of gases:

$pV = \frac{1}{3}Nm\bar{v}^2$

$\frac{1}{2}m\bar{v}^2 = \frac{3}{2}kT$

Questions

Temperature and Heat

1. What property of matter is most commonly used to measure temperature? Is there an absolute reference for this property?

2. Explain why temperature is a relative measurement.

3. Heat always flows from a body with a higher temperature that is in thermal contact with a body at a lower temperature. Does it always flow from one with more internal energy to one with less internal energy? Explain.

4. Given equal masses of helium gas and oxygen gas at the same temperature, which would have more internal energy? (Hint: the chemical symbols for these gases are He and O_2.)

5. (a) What effect does the expansion of glass have on a liquid-in-glass thermometer? (b) What would happen in a liquid-in-glass thermometer if the glass expanded more than the liquid?

6. If a hot piece of metal is dropped into a container of water, when will the metal and the water be in thermal equilibrium? When will thermal equilibrium be reached when an ice cube is dropped in a large container of water?

7. Can two objects be in thermal equilibrium and not in thermal contact? Explain.

The Celsius and Fahrenheit Temperature Scales

8. Are the Celsius and Fahrenheit degrees equal? If not, which scale has the larger 10-degree interval?

9. What is the difference in degrees between room temperature and normal body temperature on (a) the Celsius scale and (b) the Fahrenheit scale?

10. Devise a temperature scale of your own, different from the ones discussed in this chapter. (Give it a name, too.)

11. Celsius and Fahrenheit readings are equal at a temperature of $-40°$. How do they compare (a) just above and (b) just below this temperature?

12. A liquid-in-glass thermometer usually has an expansion chamber, an enlargement of the capillary bore at the top. What is the purpose of this chamber?

Gas Laws and Absolute Temperature

13. The temperature of a quantity of gas is increased. (a) How is its density affected if the pressure is held constant? (b) How is its density affected if the volume is held constant?

14. In general, what happens when the temperature of a gas is decreased?

15. If all gases at low densities are described by the

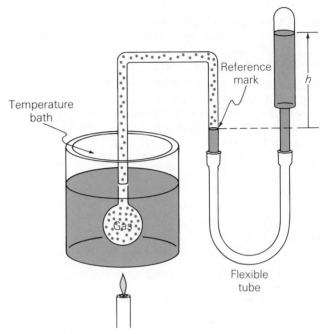

Figure 11.18 **A type of constant-volume gas thermometer**
See Question 17.

perfect gas law, why are there several different lines in the graph in Fig. 11.8(b)?

16. What are the advantages of a constant-volume gas thermometer?

17. A type of constant-volume gas thermometer is shown in Fig. 11.18. Describe how it operates.

18. Describe how a constant-pressure gas thermometer might be constructed.

19. What are the fixed points used for a constant-volume gas thermometer, and how is the kelvin defined?

20. How does a kelvin compare to a Fahrenheit degree?

Thermal Expansion

21. (a) Do all substances expand with increasing temperature? (b) If the linear expansion (L) of a rod were plotted against its temperature (T), would the result be a straight line? Explain.

22. Why are coffee pots and baking dishes made out of Pyrex glass instead of ordinary window glass, which is cheaper?

23. When a liquid-in-glass thermometer is placed in hot water, the liquid inside falls slightly before rising. Account for this observation.

24. Discuss the effect of thermal expansion on two pieces of material that are glued together.

25. Drinking glasses sometimes stick together when stacked one inside the other. A way of getting them apart is to put hot water on one and cold water on the other. Does it make a difference which is put on which glass?

26. Consider a cube sitting on a bimetallic strip as depicted in Fig. 11.19. What will happen if the cube is ice and (a) the upper strip aluminum and the lower strip brass, or (b) the upper strip iron and the lower strip copper? (c) If the two strips are brass and copper, which should be on top to keep a hot metal cube from falling off?

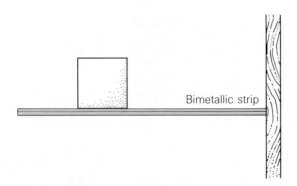

Figure 11.19 **Which way will the cube go?** See Question 26.

27. The graduated marks are engraved on a metal ruler when the ruler is at 20 °C. How is the accuracy of the rule affected when the temperature is 30 °C?

28. When a shower is turned on and left running to let the water get hot, a noticeable (sometimes audible) decrease in flow is observed after a while. Why?

29. A demonstration of thermal expansion is shown in Fig. 11.20. Initially, the ball goes through the ring. (a) When the ball is heated, it does not go through the ring. (b) If both the ball and the ring are heated, the ball again goes through the ring. Explain what is being demonstrated.

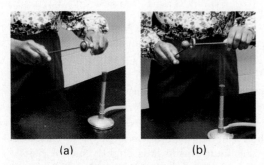

Figure 11.20 **Ball-and-ring expansion** See Question 29.

30. A solid metal disk rotates freely, so the conservation of angular momentum applies (Chapter 8). If the disk is heated while it is rotating, will there be any effect on the rate of rotation (the angular speed)?

31. Use the coefficients of volume expansion to show why mercury and alcohol are used in thermometers rather than other liquids. Why isn't water used in liquid-in-glass thermometers? Explain what would happen if it were. Do mercury thermometers have any advantages over alcohol thermometers?

32. (a) Is it more economical (on a really penny-pinching scale!) to buy gasoline in the summer or winter? (Assume that the underground tanks show a seasonal temperature variation.) (b) On a hot day, gasoline is sometimes seen dripping from an automobile tank. Explain why.

33. Will a block of ice float higher in water at 4 °C or in water at 40 °C, or is there no difference? (Neglect any melting and remember that ice also expands.)

The Kinetic Theory of Gases

34. According to kinetic theory, what causes the pressure that a gas exerts on the walls of its container?

35. If the temperature of an ideal gas is varied, what are the effects at the molecular level?

36. What does kinetic theory predict would happen to a gas at absolute zero?

37. If the temperature of an ideal gas is doubled from 10 °C to 20 °C, is the internal energy of the gas doubled? Explain.

38. Give several common examples of diffusion.

39. Equal volumes of helium gas (He) and neon (Ne) gas at the same temperature (and pressure) are on opposite sides of a porous membrane, as shown in Fig. 11.21. Describe what happens after a period of time, and why.

40. An aqueous saline (salt) solution with a concentration of 0.85% is said to be isotonic. There is no osmosis through the cell membranes of red blood cells placed in an isotonic solution. Explain what would happen and why if red blood cells were placed in hypertonic and hypotonic solutions. (Hint: think in terms of *water* concentrations.)

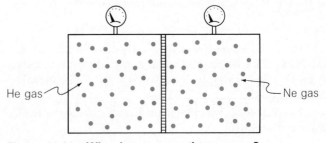

Figure 11.21 **What happens as time passes?** See Question 39.

Problems

■**1.** If the temperature drops by $10\,°C$, what is the corresponding temperature change on the Fahrenheit scale? If the temperature rises by $10\,°F$, what is the corresponding change on the Celsius scale?

■**2.** Convert the following to Celsius readings: (a) $0\,°F$, (b) $1800\,°F$, (c) $-20\,°F$, (d) $-40°C$.

■**3.** Convert the following to Fahrenheit readings: (a) $32\,°C$, (b) $110\,°C$, (c) $-10\,°C$, (d) $-273.16\,°C$.

■**4.** To conserve energy, thermostats in an office building are set at $78\,°F$ in the summer and $65\,°F$ in the winter. What would the settings be if the thermostats had a Celsius scale?

■■**5.** The highest and lowest recorded air temperatures in the world are $58\,°C$ (Libya, 1922) and $-89\,°C$ (Antarctica, 1983). What are these temperatures on the Fahrenheit scale?

■■**6.** The highest and lowest recorded air temperatures in the United States are $134\,°F$ (Death Valley, California, 1913) and $-80\,°F$ (Prospect Creek, Alaska, 1971), respectively. What are these temperatures on the Celsius scale?

■■**7.** The largest temperature drop recorded in the United States is $100\,°F$ in 1 day (from $44\,°F$ to $-56\,°F$, in Browning, Montana, 1916). The largest rise is $49\,°F$ in 2 min (from $-4\,°F$ to $45\,°F$, in Spearfish, South Dakota, 1943). (a) What are the corresponding temperatures and temperature ranges on the Celsius scale? (b) What are the rates of both temperature changes on both scales?

■■**8.** In the troposphere (the lowest part of the atmosphere), the temperature decreases rather uniformly with altitude at a lapse rate of about $6.5\,°C/km$. What are the temperatures (a) near the top of the troposphere (which has an average thickness of 11 km) and (b) outside a commercial aircraft flying at a cruising altitude of 34,000 ft? (Assume that the ground temperature is normal room temperature.)

■**9.** A quantity of an ideal gas initially at atmospheric pressure is maintained at a constant temperature while it is compressed to half of its volume. What is the final pressure of the gas?

■**10.** The pressure on a low-density gas in a cylinder is kept constant as its temperature is increased from $10\,°C$ to $40\,°C$. (a) Does the piston in the cylinder advance or recede? (b) What is the fractional change in the volume of the gas?

■**11.** Convert the following temperatures to kelvin: (a) $0\,°C$, (b) $100\,°C$, (c) $20\,°C$, and (d) $-25\,°C$.

■**12.** Convert the following temperatures to degrees Celsius: (a) 0 K, (b) 310 K, (c) 94 K, and (d) 273.16 K.

■■**13.** The surface temperature of the Sun is about 6000 K. (a) What is this temperature on the Fahrenheit and Celsius scales? (b) The surface temperature is sometimes reported to be $6000\,°C$. Assuming that 6000 K is correct, what is the percentage error of this Celsius value?

■■**14.** Assuming that the interior temperature of the Sun decreases uniformly from the central core to the surface, what is the Kelvin temperature gradient (K/km) if the core temperature is 15 million kelvins? (Use data given in Problem 13 and in Appendix I.)

■■**15.** A constant-volume gas thermometer has a pressure of 1000 Pa at $20\,°C$. If the pressure increases to 2000 Pa, what is the new Celsius temperature?

■■**16.** A constant-pressure gas thermometer is initially at a temperature of $20\,°C$. If the volume of gas in the thermometer increases by 5.0%, what is the final Celsius temperature?

■■**17.** On a warm day ($95\,°F$), the air in a balloon occupies a volume of $0.20\,m^3$ and has a pressure of $20\,lb/in.^2$ If the balloon is placed in a refrigerator and cooled to $41\,°F$, the pressure decreases to $14.7\,lb/in.^2$ What is the volume of the balloon? (Assume that the air behaves as a perfect gas.)

■■**18.** An automobile tire is inflated to 200 kPa when the air temperature is at the freezing point. It later warms up to $12.0\,°C$ and the air pressure in the tire is found to be 204 kPa. Does the volume of the tire change? If so, by what percentage? (Assume that the air is a perfect gas and that atmospheric pressure is constant.)

■■**19.** A mole of a substance is its atomic or molecular mass expressed in grams. For example, from the table in Appendix IV, the atomic mass of helium (He) is 4.0, so a mole of helium has a mass of 4.0 g. (a) How many grams are there in a mole of (1) CO_2, (2) H_2SO_4, and (3) HCl? (b) How many moles are there in (1) 36 g of H_2O, (2) 147 g of H_2SO_4, and (3) 117 g of NO_2?

■■**20.** A mole of any substance contains Avogadro's number of molecules ($N_A = 6.02 \times 10^{23}$ molecules/mole).

Find the number of molecules in each of the amounts of the substances given in Problem 19(b).

■■21. A perfect gas occupies a container with a volume of 0.75 L at STP (standard temperature and pressure, 0 °C and 1 atm). Find (a) the number of moles and (b) the number of molecules of the gas. (c) If the gas is carbon monoxide (CO), what is its mass?

■■22. Show that 1 mole of any dilute gas at STP (see Problem 21) occupies a volume of 22.4 L. (What can you say about a 22.4-L container of any gas at STP?)

■■■23. The geothermal gradient, the rate at which temperature increases with depth in the Earth's crust, is determined in deep mines and wells to be about 1 °F/150 ft. Assuming that this rate is uniform to any depth, what is the absolute temperature at the center of the Earth? Is this a reasonable assumption? (See Problem 14.)

■■■24. Derive a single formula that may be used to directly convert Kelvin temperatures to Fahrenheit (and vice versa).

11.4 Thermal Expansion*

■25. A copper wire has a length of 0.500 m at 20 °C. If the temperature is increased to 80 °C, what is the change in the wire's length?

■26. A rectangular steel plate whose area is 0.060 m² is cooled from 300 °C to room temperature. By what percentage does the area decrease?

■27. Suppose that a copper cube with sides 5.0 cm long undergoes a 150 °C temperature change. What is the cube's new volume?

■28. What temperature change would cause a 0.10% increase in the volume of a quantity of water that was initially at room temperature?

■■29. A piece of copper tubing used in plumbing has a length of 60 cm and an inner diameter of 1.50 cm at room temperature. When hot water at 80 °C flows through the tube, what are (a) its new length and (b) the change in its cross-sectional area? Does the latter affect the flow rate?

■■30. Steel train rails are 12.0 m long when the track is laid, on a day when the temperature is 20 °C. To prevent contact stress enough space must be left between the ends of adjacent rails so they will not touch up to a temperature of 45 °C. What is the width of the required gap between the rails?

* Assume that quantities given are exact numbers.

■■31. Concrete highway slabs are poured in lengths of 10.0 m. How wide should the expansion gaps between the slabs be to ensure that there will be no contact stress over a temperature range of −25 °C to 45 °C?

■■32. Show that the coefficient of volume expansion for solids is approximately equal to 3α.

■■33. Compute the final length of the rod in Example 11.5 with the thermal cycle reversed, that is, heating first and then cooling.

■■34. What temperature increase would produce a stress of 7.5×10^7 N/m² on a rigidly held steel beam?

■■35. A new metal alloy is made into a 50-cm rod to determine its coefficient of linear expansion. After being heated from room temperature to 250 °C, the rod is found to have a length of 50.044 cm. Is this a potentially valuable alloy? Justify your answer.

■■36. A square aluminum sheet 20 cm on a side is heated from room temperature to 300 °C. What is the change in the length of a side?

■■37. A brass rod has a circular cross section with a radius of 0.500 cm. It fits into a circular hole in a copper sheet with a clearance of 0.010 mm completely around it when both it and sheet are at 20 °C. (a) At what temperature will the clearance be zero? (b) Would such a tight fit be possible if the sheet were brass and the rod were copper?

■■38. A solid aluminum sphere has a diameter of 8.00 cm at room temperature. If the sphere is heated to 360 °C, what is the change in its diameter?

■■39. A mercury thermometer has a uniform capillary bore whose cross-sectional area is 0.012 mm². The volume of the mercury in the thermometer bulb at 10 °C is 0.130 cm³. If the temperature is increased to 50 °C, how much will the height of the mercury column in the capillary bore change? (Neglect the expansion of the mercury in the bore and of the glass of the thermometer.)

■■40. One morning when the temperature is 15 °C, an employee at a rent-a-car company fills the 20-gal gas tank of an automobile to the top and then parks the car a short distance away. That afternoon, when the temperature is 25 °C, gasoline drips from the tank onto the pavement. How much gas will be lost? (Neglect the expansion of the tank.)

■■■41. Use algebraic formulas (not numbers) to demonstrate the discrepancy shown in Example 11.5.

■■■42. A Pyrex beaker that has a capacity of 1000 cm³ at

20 °C contains 990 cm^3 of mercury at that temperature. Is there some temperature at which the mercury will completely fill the beaker? Justify your answer.

■■■43. A mercury thermometer has a bulb volume of 0.200 cm^3 and a capillary bore diameter of 1.25 mm. How far up the bore will the column of mercury move if the overall temperature of the thermometer is increased by 30 °C?

11.5   The Kinetic Theory of Gases

■■44. What is the average kinetic energy per molecule of a low-density gas at (a) 20 °C and (b) 100 °C?

■■45. If the temperature of a perfect gas is raised from 30 °C to 90 °C, what will the percentage change in the average (rms) velocity of the gas molecules be?

■■46. What is the average speed of the molecules in low-density oxygen gas at room temperature? (The mass of an oxygen molecule, 0_2, is 5.31×10^{-26} kg.)

■■47. Heat is added to a quantity of an ideal gas that is initially at 20 °C until its internal energy is doubled. What is the final Celsius temperature of the gas?

■■48. The temperature of a perfect gas is doubled from 20 °C to 40 °C. What is the percentage increase in the internal energy of the gas?

■■■49. Equal quantities of two ideal gases are at temperatures of 10 °C and 10 °F. (a) If the temperatures of the gases are raised to 60 °C and 60 °F, respectively, which one has a greater increase in internal energy? (b) How much greater is that increase than the other one?

■■■50. A quantity of an ideal gas has a temperature of 0 °C. An equal quantity of another ideal gas is twice as hot. What is its temperature?

Additional Problems

51. Is there a temperature that has the same numerical value on the Kelvin and Fahrenheit scales? Justify your answer.

52. The core of the Sun is estimated to have a temperature of about 15 million kelvins. What is this temperature on the Fahrenheit and Celsius scales?

53. Using the perfect gas law, express the coefficient of volume expansion in terms of temperature and pressure.

54. A quantity of a perfect gas at 10 °C occupies 3.0 L and has a pressure of 150 kPa. (a) What will the volume be if the temperature is kept constant and the pressure is decreased to 120 kPa? (b) What will the pressure be if the temperature is kept constant and the volume is compressed to 2.5 L? (c) What will the Celsius temperature be at a pressure of 120 kPa and 2.5 L?

55. Mercury has a density of 13.59 g/m^3 at room temperature. What is its density at 100 °C?

56. Which is the lower temperature, -45 °C or -45 °F, and how much lower?

57. (a) What is the average kinetic energy of molecules in a volume of gas at a temperature of 27 °C? (b) What is the average (rms) speed of the molecules if the gas is helium? (A helium molecule is a single atom, which has a mass of 6.65×10^{-27} kg.)

58. Steel rails 7.50 m long are laid end to end on a cold day when the temperature is 0 °C. The engineer on the project knows that the highest recorded summer temperature for that region is 40 °C. If the engineer adds 20% to this value as a safety factor, what minimum expansion space should be left between the rails to avoid contact stress?

59. A constant-volume gas thermometer has a pressure of 2.5×10^5 Pa at 10 °C. (a) What is the pressure for a temperature of -10 °C? (b) What is the temperature for a pressure of 1 atm?

Heat

12

Think for a moment how important heat is in our lives. Our bodies must work incessantly to exchange heat with the environment to maintain the temperature required for life processes. Most of us must heat our homes in winter and cool (remove heat from) them in summer. Heat processes are important in industry. Heating is required in the refinement of oil and metals and in many other industrial processes. The seasonal heating and cooling of the environment affect what we wear and our actions. Some of us wait for snow to fall and lakes to freeze so we can enjoy skiing and ice skating. Others look forward to the melting of ice and snow and enjoy basking in the warm rays of the summer sun. (Many lucky people enjoy both.)

Thus, heat, heat transfer, and phase changes not only affect the quality of our lives but are some of the most important topics in the study of physics. An understanding of them will allow you to explain many everyday things, as well as providing a basis for understanding the conversion of thermal energy into useful mechanical work.

12.1 Units of Heat

Heat is energy that is transferred from one object or system to another because of a temperature difference. Like work, heat represents a transfer of energy. It is commonly said that heat is a form of energy, but this is not true in the strictest sense. References to "heat energy" in this book refer to energy that is added or removed from a body or system.

Even though heat is energy in transit, it is measurable as energy losses or gains and is described by standard energy units like any other quantity of energy. Recall that the SI standard unit of energy is the joule (J), or newton-meter (N-m). Thus, it is correct to say, for example, that 20 J of heat energy is transferred from one body to another. However, there are other commonly used units of heat. A chief one is the **calorie (cal)** [Fig. 12.1(a)]:

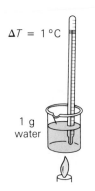

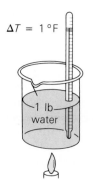

(a) Gram calorie (cal)

(b) Kilocalorie (kcal) or Calorie (Cal)

(c) British thermal unit (Btu)

Figure 12.1 **Heat units** (a) A calorie raises the temperature of 1 g of water by 1 °C. (b) A kilocalorie raises the temperature of 1 kg of water by 1 °C. (c) A Btu raises the temperature of 1 lb of water by 1 °F.

One calorie is defined as the amount of heat needed to raise the temperature of 1 g of water 1 °C (from 14.5 °C to 15.5 °C).

Because the calorie is a relatively small unit, its larger multiple, the **kilocalorie (kcal)**, is often used (1 kcal = 1000 cal). *One kilocalorie is the amount of heat needed to raise the temperature of 1 kg of water by 1 °C (from 14.5 °C to 15.5 °C).* See Fig. 12.1(b).

A familiar use of the kilocalorie is for the specification of the energy values of foods. However, in this context the word is usually shortened to calorie (Cal). That is, people on diets really count kilocalories, and a piece of cake may contain 400 Cal. A capital C is used by scientists to distinguish the larger kilogram-Calorie, or kilocalorie, from the smaller gram-calorie. They are sometimes referred to as "big Calorie" and "little calorie." The Insight feature tells how the caloric values are determined (for foods and other substances).

A unit of heat that is commonly used in industry is the **British thermal unit (Btu)**. *One Btu is the amount of heat needed to raise the temperature of 1 lb of water by 1 °F (from 63 °F to 64 °F).* See Fig. 12.1(c). A Btu is somewhat larger than a calorie, but only about a fourth of a kilocalorie (1 Btu = 252 cal = 0.252 kcal). If you buy an air conditioner or an electric heater, you will find it rated in Btu's; for example, window air conditioners range from 5000 to 25,000 Btu's. This number is actually Btu's *per hour* and specifies how much heat the unit will transfer (or deliver in the case of a heater) during that time.

The calorie is technically known as the 15° calorie. The temperature range is specified because the energy needed to raise 1 g of water by 1 °C varies slightly with temperature. It is a minimum in the temperature interval between 14.5 °C and 15.5 °C at a pressure of 1 atm. Over the temperature range in which water is a liquid, the variation is small and can be ignored for most purposes.

The Mechanical Equivalent of Heat

The current idea that heat is a transfer of energy is the result of work by many scientists on the relationship of heat and energy. Some early observations were made by Count Rumford while he was supervising the boring of cannon barrels in Germany. (See Fig. 12.2.) Rumford noticed that water put into the bore of the cannon to prevent overheating during drilling boiled away and had to be frequently replenished. He did several experiments, including ones in which he tried to detect "caloric fluid" by changes in the weights of heated substances. He eventually concluded that mechanical work was responsible for the heating of the water.

This conclusion was later proven quantitatively by James Joule, the English scientist after whom the SI unit for work and energy is named. Using an

Benjamin Thompson (1753–1814) was born in New England, but went to England at the time of the American Revolution. He later worked in Bavaria, where he was made a count. He took the name Count Rumford after his birthplace, now known as Concord, New Hampshire.

Heat of Combustion

Fossil fuels are burned to obtain energy for heating homes and for a variety of industrial applications. In a similar sense, our bodies use foods as fuels. The intrinsic energy values of foods and fuels are expressed in terms of the **heat of combustion** (H), which is the heat produced per unit mass of substance when it is burned in oxygen. That is,

$$H = \frac{Q}{m} \quad \text{or} \quad Q = mH \qquad (12.1)$$

where Q is the amount of heat released and m is the mass. The common units for heat of combustion are cal/g, kcal/kg, and Btu/lb. (See Table 1 for some typical values of the heat of combustion.)

For example, the heat of combustion of soft (bituminous) coal is approximately 7500 kcal/kg. This means that when 1 kg of coal burns completely (undergoes complete combustion), 7500 kcal of heat energy are liberated. Compared to fuel oil, which has a heat of combustion of 10,300 kcal/kg, coal has less intrinsic heat value per kilogram. The heat of combustion of ice cream is 2100 kcal/kg. In a food chart, this would be listed in terms of the "calories" (really Calories) in an average-size serving, for example, 1300 Cal per 8-oz serving.

Heats of combustion are ordinarily measured using a device called a bomb calorimeter. The bomb is a heavy steel cylinder that is leakproof. A known mass of a substance is placed in a cup inside the bomb in an atmosphere of pure oxygen. Electric current is used to ignite the sample, and combustion takes place in the form of an explosion. The heat of combustion is transmitted to a quantity of water surrounding the bomb. Each kilocalorie of heat given off in the explosion raises the temperature of the water by 1 °C/kg.

Table 1
Heats of Combustion for Some Substances

Substance	J/kg	kcal/kg
Fuels		
Alcohol	2.7×10^7	6,400
Coal, hard	3.4×10^7	8,000
Coal, soft	3.2×10^7	7,500
Coke	2.5×10^7	6,000
Gasoline	4.8×10^7	11,400
Natural gas*	4.4×10^7	(10,500 kcal/m^3)
Oil, diesel	4.4×10^7	10,500
Oil, fuel	4.3×10^7	10,300
Wood (average)	2.1×10^7	5,000
Foods		
Bread, white	0.84×10^7	2,000
Butter	3.4×10^7	8,000
Eggs, boiled	0.67×10^7	1,600
Eggs, scrambled	0.88×10^7	2,100
Ice cream	0.88×10^7	2,100
Meat, lean	0.50×10^7	1,200
Potatoes, boiled	0.38×10^7	970
Sugar, white	1.7×10^7	4,000

* At STP (standard temperature and pressure, 0 °C and 1 atm)

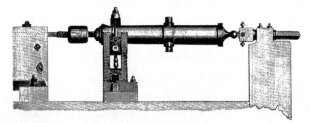

Figure 12.2 **Count Rumford (1753–1814)** Engravings of the statesman and physicist Count Rumford, and a cannon borer. Rumford's experience with the boring of cannon barrels led to his experiments on the nature of heat.

(a)

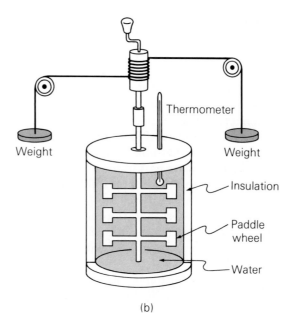
(b)

Thermometer

Insulation

Paddle wheel

Water

Weight

Weight

Figure 12.3 **Mechanical equivalent of heat** (a) Joule's apparatus for determining the mechanical equivalent of heat. (b) As the weights descend, the paddle wheel churns the water, and the water temperature rises.

apparatus like the one illustrated in Fig. 12.3, Joule showed that when a given amount of mechanical work was done, the water was heated, as indicated by an increase in temperature. He found that for every 4.19 J of work done the temperature of the water rose 1 °C per gram, or that 4.19 J was equivalent to 1 cal:

      1 cal = 4.19 J   (actually 4.186 J)

or   1 kcal = 4.19 × 10³ J

This relationship is called the **mechanical equivalent of heat** and provides a conversion factor between calories and joules.

Example 12.1   Work and Temperature
How many joules of mechanical work are required to raise the temperature of 10 kg of water from 15 °C to 25 °C?

Solution
Given: $m = 10$ kg   *Find*: W (in joules)
        $\Delta T = 25\,°C - 15\,°C = 10\,°C$

From the definition of the kilocalorie, you know that it takes 1.0 kcal to raise the temperature of 1.0 kg of water by 1 °C. This heat-mass-temperature relationship can be represented as $c = 1.0$ kcal/kg-°C. Then the amount of heat (Q) needed to raise the temperature of 10 kg of water by 10 °C is

$$Q = mc\,\Delta T = (10 \text{ kg})\left(1.0\,\frac{\text{kcal}}{\text{kg-°C}}\right)(10\,°C) = 100 \text{ kcal}$$

(Note that the units cancel appropriately.) Using the mechanical equivalent of heat gives the number of joules of work:

    $(100 \text{ kcal})(4.19 \times 10^3 \text{ J/kcal}) = 4.19 \times 10^5 \text{ J}$ ■

12.2 Specific Heat

If equal amounts of heat are added to equal masses of different substances, in general, the resulting temperature changes will not be the same. For example, suppose that you had 1 kg of water and 1 kg of aluminum and added 1 kcal of heat to each. From the definition of the kilocalorie, you know that the temperature of the water would rise 1 °C. You would find that the temperature of the aluminum would increase by 4.5 °C. The reason for this difference is that more of the added energy goes into the nonkinetic part of the internal energy of the water than of the aluminum. Different substances have different molecular configurations and bonding, which is why temperature changes vary. (Recall from Chapter 11 that when heat is added to a substance, the energy may go into increasing the random molecular motion, which results in a temperature change, and into increasing the potential energy associated with the molecular bonds.)

The amount of heat (Q) required to change the temperature of a substance is proportional to the mass (m) of the substance and to the change in the temperature (ΔT). That is, $Q \propto m \, \Delta T$, or, in equation form:

$$Q = mc \, \Delta T \qquad\qquad (12.2)$$

Here $\Delta T = T_f - T_i$ is the temperature change, or the difference between the initial temperature (T_i) and the final (T_f) temperature, and c is called the specific heat capacity. The constant c is commonly referred to as simply the **specific heat**. It is characteristic of, or *specific* for, a given substance and gives an indication of its internal molecular configuration and bonding. The specific heats for some common substances are given in Table 12.1.

Writing Eq. 12.2 as $c = Q/m \, \Delta T$ shows us that the units of specific heat are J/kg-K or kcal/kg-°C (or cal/g-°C in cgs units). Note the value of the specific

Table 12.1
Specific Heats of Various Substances
(20 °C and 1 atm)

Substance	Specific heat (c)	
	J/kg-°C	kcal/kg-°C (or cal/g-°C)
Air (50 °C)	1050	0.25
Alcohol, ethyl	2430	0.58
Aluminum	920	0.22
Copper	390	0.093
Glass	840	0.20
Ice (−5 °C)	2100	0.50
Iron or steel	460	0.11
Lead	130	0.031
Mercury	140	0.033
Soil (average)	1050	0.25
Steam (110 °C)	2010	0.48
Water (15 °C)	4190	1.00
Wood (average)	1680	0.40

heat of water used in Example 12.1. The standard SI unit for specific heat is the one with the temperature in kelvins. However, the use of Celsius temperature in this unit (giving J/kg-°C) is commonly accepted because the size of a Celsius degree is the same as the size of a kelvin. Note that *the specific heat is the amount of energy required to raise the temperature of 1 kg of a substance by 1 °C.* The specific heat depends somewhat on temperature (and pressure), but you can consider this effect to be negligible.

· The greater the specific heat (capacity) of a substance, the more energy must be transferred to it to change the temperature of a given mass. That is, a substance with a greater specific heat has a greater heat capacity, or accepts or yields more heat for a given temperature change (and mass).

Water has relatively large specific heat of 1.00 kcal/kg-°C. The value is exactly 1.00 because the definition of the kilocalorie states that 1 kcal raises the temperature of 1 kg of water by 1 °C.

You have been the victim of the large specific heat of water if you have ever burned your mouth on a baked potato or on the cheese on pizza. These foods have a high water content and a large heat capacity, so they don't cool off as quickly as some other foods do.

Example 12.2   Specific Heat

How much heat is required to raise the temperature of 0.20 kg of water from 15 °C to 45 °C?

Solution

Given: $m = 0.20$ kg   *Find*: Q
$\Delta T = T_f - T_i = 45 \, °C - 15 \, °C$
$\quad\quad = 30 \, °C$
$c = 4190 \text{ J/kg-}°C$
$\quad\quad$ (from Table 12.1)

Eq. 12.2 can be used directly with the quantities given:

$$Q = mc \, \Delta T = (0.20 \text{ kg})(4190 \text{ J/kg-}°C)(30 \, °C) = 2.5 \times 10^4 \text{ J} \quad \blacksquare$$

Note that when there is temperature increase, ΔT and Q are positive. This corresponds to energy being *added to* a system. Conversely, ΔT and Q are negative when energy is *removed from* a system.

Example 12.3   Finding Temperature

A half-liter of water at 30 °C is cooled, with the removal of 15 kcal of heat. What is the final temperature of the water?

Solution

Given: $m = 0.50$ kg   *Find*: T_f
$\quad\quad$ (1 L of water = 1 kg)
$T_i = 30 \, °C$
$Q = -15$ kcal (negative
$\quad\quad$ because energy is removed)
$c = 1.00 \text{ kcal/kg-}°C$
$\quad\quad$ (from Table 12.1)

Writing out the ΔT term of Eq. 12.2 gives

$$Q = mc\,\Delta T = mc(T_f - T_i)$$

Solving for T_f gives

$$T_f = \frac{Q}{mc} + T_i = \frac{-15\text{ kcal}}{(0.50\text{ kg})(1.00\text{ kcal/kg-}°\text{C})} + 30\,°\text{C}$$

$$= 0\,°\text{C}$$

The (liquid) water is at its freezing point; the removal of more heat would cause the water to freeze. However, Eq. 12.2 *does not apply* when the temperature change (ΔT) is an interval that includes a change of phase, as you will learn in the next section. ∎

Example 12.4   Greater Capacity
Equal masses of aluminum (Al) and copper (Cu) are at the same temperature. Which will require the greater heat to raise its temperature by a given amount, and how many times greater is this than the heat that would have to be added to the other metal?

Solution

Given: $m_{Al} = m_{Cu}$
$\quad\quad\quad \Delta T_{Al} = \Delta T_{Cu}$
$\quad\quad\quad c_{Al} = 0.22\text{ kcal/kg-}°\text{C}$
$\quad\quad\quad c_{Cu} = 0.093\text{ kcal/kg-}°\text{C}$
$\quad\quad\quad$ (from Table 12.1)

Find: The greater value of Q and how many times greater it is

Since aluminum has a larger specific heat, it has a larger heat capacity, and more heat is required to raise its temperature by a given amount. How many times more heat is required than for an equal mass of copper is given by a ratio:

$$\frac{Q_{Al}}{Q_{Cu}} = \left(\frac{m_{Al}}{m_{Cu}}\right)\left(\frac{c_{Al}}{c_{Cu}}\right)\left(\frac{\Delta T_{Al}}{\Delta T_{Cu}}\right) = \frac{c_{Al}}{c_{Cu}}$$

and

$$Q_{Al} = \left(\frac{c_{Al}}{c_{Cu}}\right)Q_{Cu} = \left(\frac{0.22\text{ kcal/kg-}°\text{C}}{0.093\text{ kcal/kg-}°\text{C}}\right)Q_{Cu}$$

$$Q_{Al} = (2.4)Q_{Cu}$$

That is, 2.4 times more heat is required for aluminum than for copper. ∎

PROBLEM-SOLVING HINT

When the quantities given in a problem are equal, like the masses and temperature intervals in Example 12.4, it is a good indication that a ratio can be used to cancel out the equal quantities. Also, phrases like "how many times more" imply a factor that is a ratio.

Calorimetry

The specific heat of a substance is determined by measuring the quantities in Eq. 12.2 ($Q = mc\,\Delta T$) other than c. A simple laboratory apparatus for measuring specific heats is shown in Fig. 12.4. A substance of known mass and temperature is put into a quantity of water in a calorimeter. The water is at a different temperature, usually a lower one. The calorimeter is an insulated container that allows little (ideally, no) heat loss. The principle of the conservation of energy is then applied to determine c. This procedure is sometimes called the **method of mixtures**.

Example 12.5 Calorimetry

Students in a physics lab are to determine the specific heat of copper experimentally. They heat 0.150 kg of copper shot to 100 °C and then carefully pour the hot shot into a calorimeter cup (Fig. 12.4) containing 0.200 kg of water at 20 °C. The final temperature of the mixture in the cup is measured to be 25 °C. If the aluminum cup has a mass of 0.037 kg, what is the specific heat of copper? (Assume that there is no heat loss.)

Solution

Given: $m_m = 0.150$ kg   *Find*: c_m
$m_w = 0.200$ kg
$c_w = 1.00$ kcal/kg-°C
    (from Table 12.2)
$m_c = 0.037$ kg
$c_c = 0.22$ kcal/kg-°C
    (from Table 12.2)
$T_h = 100$ °C, $T_f = 25$ °C
    and $T_i = 20$ °C

The subscripts m, w, and c refer to the metal, water, and calorimeter cup, respectively, and the subscripts h, f, and i refer to hot, final, and initial temperatures, respectively. The water and the cup are taken to be in thermal equilibrium, so they have the same initial temperatures.

Figure 12.4 **Calorimetry apparatus**
The calorimeter cup (center with black insulating ring) goes into the larger container and the cover with thermometer and stirrer is seen at the right. Metal shot or pieces of metal are heated in the cup that is inserted into the hole at the top of the steam generator on the tripod.

Assuming that no heat is lost from the system, its energy is conserved and

$$\Delta Q_{sys} = 0 = Q_{w+c} - Q_m$$

That is, when the system is in thermal equilibrium, the heat lost by the metal ($-Q_m$) is equal to the heat gained by the water and the calorimeter cup ($+Q_{w+c}$):

heat loss = heat gained

$$Q_m = Q_{w+c}$$

Substituting for these heats from Eq. 12.2 gives

$$m_m c_m (T_h - T_f) = m_w c_w (T_f - T_i) + m_c c_c (T_f - T_i)$$

Solving for c_m gives

$$c_m = \frac{(m_w c_w + m_c c_c)(T_f - T_i)}{m_m (T_h - T_f)}$$

$$= \frac{[(0.200\ \text{kg})(1.00\ \text{kcal/kg-}^\circ\text{C}) + (0.037\ \text{kg})(0.22\ \text{kcal/kg-}^\circ\text{C})] \times (25\,^\circ\text{C} - 20\,^\circ\text{C})}{0.150\ \text{kg}(100\,^\circ\text{C} - 25\,^\circ\text{C})}$$

$$= 0.093\ \text{kcal/kg-}^\circ\text{C}$$

The experimentally determined value would actually be slightly less than this calculated value, because some heat is lost in transferring the shot to the cup and from the calorimeter while the mixture comes to thermal equilibrium. ∎

12.3 Phase Changes and Latent Heat

Matter normally exists in three phases: solid, liquid, and gas (see Fig. 12.5). The phase that a substance is in depends on its internal energy (as monitored by its temperature) and the pressure on it. You probably think more readily of adding or removing heat to change the phase of a substance because most of your experience with phase changes has been at normal atmospheric pressure, which is relatively constant.

In the **solid phase**, molecules are held together by attractive forces, or bonds. (Simplistically, these bonds can be represented as springs, as was done in Chapter 9.) Adding heat causes increased motion about the molecular equilibrium positions. If enough heat is added to a solid to provide sufficient energy to break the intermolecular bonds, the solid undergoes a phase change and becomes a liquid. The temperature at which this occurs is called the **melting point**. The temperature at which a liquid becomes a solid is called the **freezing point**. In general, these temperatures are the same, but they can differ slightly.

Ice, table salt, and most metals are crystalline solids. That is, they have orderly molecular or atomic arrangements. Other substances, however, such as glass, are noncrystalline, or amorphous. Instead of melting at a particular temperature, these substances melt over a temperature range. This discussion will be concerned primarily with substances that have definite melting points.

In the **liquid phase**, molecules of a substance are relatively free to move, and a liquid assumes the shape of its container. In certain liquids, there may be some

Solid, liquid, and gas are sometimes referred to as states of matter, but the state of a system has a different meaning in physics, as you will learn in Chapter 13.

ordered structure, giving rise to so-called liquid crystals, such as are used in LCD's (liquid crystal displays) of calculators and clocks (Chapter 24). Adding heat increases the motion of the molecules of a liquid, and when they have enough energy to become separated by large distances (compared to their diameters), the liquid changes to the **gaseous phase**, or **vapor phase**. (The distinction between a gas and a vapor will be made shortly.) This change may occur slowly by the process of evaporation or rapidly at a particular temperature called the **boiling point**. The temperature at which a gas condenses and becomes a liquid is the **condensation point**.

Some solids, such as dry ice (solid carbon dioxide), moth balls, and certain air fresheners, change directly from the solid to the gaseous phase. This is called **sublimation**. Like the rate of evaporation, the rate of sublimation increases with temperature. A phase change from a gas to a solid is called **deposition**. Frost, for example, is solidified water vapor deposited directly on grass, car windows, and other objects. Frost is not frozen dew, as is sometimes mistakenly assumed.

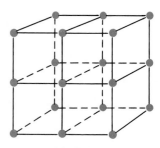

(a) Solid

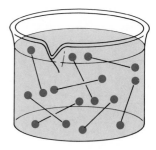

(b) Liquid

Latent Heat

In general, when heat energy is transferred to a substance, its temperature increases. However, when added (or removed) heat causes only a phase change, the temperature of the substance does *not* change. For example, if heat is added to a quantity of ice ($c = 0.50$ kcal/kg-°C) at -10 °C, the temperature of the ice increases until it reaches its melting point of 0 °C. At this point, the addition of more heat does not increase the temperature but causes the ice to melt, or change phase. (Of course, the heat must be added slowly so that the ice and melted water remain in thermal equilibrium.) Once the ice is melted, adding more heat will cause the temperature of the water to rise. A similar situation occurs during the liquid-gas phase change at the boiling point. Adding more heat to boiling water only causes more vaporization, not a temperature increase.

From the earlier description of the molecular nature of the different phases of matter, you can see that during a phase change the heat energy goes into the work of breaking bonds and separating molecules, rather than into increasing the temperature. The heat involved in a phase change is called the **latent heat** (L), and

$$Q = mL \qquad (12.3)$$

where m is the mass of the substance. The latent heat has units of joules per kilogram (J/kg) (or kcal/kg). The latent heat for a solid-liquid phase change is called the **latent heat of fusion** (L_f), and that for a liquid-gas phase change is called the **latent heat of vaporization** (L_v). These are often referred to as simply the heat of fusion and the heat of vaporization. The latent heats of some substances, along with their freezing and boiling points, are given in Table 12.2. (The latent heat for the less common solid-gas phase change is called the latent heat of sublimation and symbolized by L_s.)

It is helpful to focus on the fusion and vaporization of water. A plot of temperature versus heat energy for a quantity of water is shown in Fig. 12.6. Note that when heat is added (or removed) at the temperature of a phase change, 0 °C or 100 °C, the temperature remains constant. Once the phase

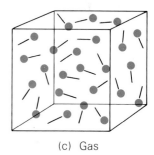

(c) Gas

Figure 12.5 **The three phases of matter** (a) The molecules of a solid are held together by bonds, and a solid has a definite shape and volume. (b) The molecules of a liquid are relatively free to move, and a liquid has a definite volume but assumes the shape of its container. (c) The molecules of a gas are separated by relatively large distances, and a gas has no definite shape or volume.

Table 12.2
Temperatures of Phase Changes and Latent Heats for Various Substances (1 atm)

Substance	Melting point	L_f		Boiling point	L_v	
		J/kg	kcal/kg		J/kg	kcal/kg
Alcohol, ethyl	−114°C	1.0×10^5	25	78°C	8.5×10^5	204
Gold	1063°C	0.645×10^5	15.4	2660°C	15.8×10^5	377
Helium*	—	—		−269°C	0.21×10^5	5
Lead	328°C	0.25×10^5	5.9	1744°C	8.67×10^5	207
Mercury	−39°C	0.12×10^5	2.8	357°C	2.7×10^5	65
Nitrogen	−210°C	0.26×10^5	6.1	−196°C	2.0×10^5	48
Oxygen	−219°C	0.14×10^5	3.3	−183°C	2.1×10^5	51
Tungsten	3410°C	1.8×10^5	44	5900°C	48.2×10^5	1150
Water	0°C	3.3×10^5	80	100°C	22.6×10^5	540

* Not a solid at 1 atm pressure; melting point −272°C at 26 atm.

change is complete, adding more heat causes the temperature to increase. For water, the latent heats of fusion and vaporization are

latent heats for water

$L_f = 3.3 \times 10^5 \text{ J/kg}$ (or 80 kcal/kg)

$L_v = 22.6 \times 10^5 \text{ J/kg}$ (or 540 kcal/kg)

That is, 3.3×10^5 J (or 80 kcal) of heat are needed to melt 1 kg of ice at 0 °C, and 22.6×10^5 J (or 540 kcal) of heat are needed to convert 1 kg of water to steam at 100 °C. Note that the latent heat of vaporization is almost 7 times the latent heat of fusion. This indicates that more energy is needed to separate the molecules in going from water to steam than to break up the lattice structure in going from ice to water.

The word "latent" means "hidden," and its use in this context may be

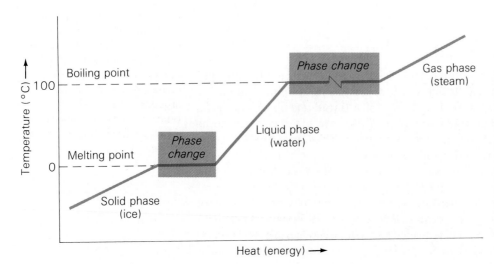

Figure 12.6 **Temperature versus heat for water**
As heat is added to the various phases of water, the temperature increases. During a phase change, however, the heat energy does the work of separating the molecules and the temperature remains constant. Note the different slopes of the phase lines, which indicate different values for the specific heat.

understood by considering a situation involving human skin. Since 540 kcal of heat energy are required to convert 1 kg of water into steam, the conservation of energy tells you that when 1 kg of steam condenses into water, 540 kcal of energy must be given up. As a result, burns from steam are usually more serious than those from boiling water. The condensing of the steam on the skin provides an additional 540 kcal/kg of heat that was seemingly hidden.

Example 12.6 Latent Heat

Heat is added to 0.50 kg of ice at $-10\,°C$. How many kilocalories are required to change the ice to steam at $110\,°C$?

Solution

Given: $m = 0.50$ kg   Find: Q (in kcal)
     $T_i = -10\,°C$, $T_f = 110\,°C$
     $L_f = 80$ kcal/kg,
     $L_v = 540$ kcal/kg (from Table 12.2)
     (specific heats from Table 12.1)

The temperature intervals and the heats required in the process are shown below:

$-10\,°C$

$\quad\quad \Delta T_1 \quad\quad Q_1 = mc_i\,\Delta T_1 = (0.50\text{ kg})(0.50\text{ kcal/kg-°C})(10\,°C) = 2.5\text{ kcal}$

$0\,°C \quad\quad\quad Q_L = mL_f = (0.50\text{ kg})(80\text{ kcal/kg}) = 40\text{ kcal}$

$\quad\quad \Delta T_2 \quad\quad Q_2 = mc_w\,\Delta T_2 = (0.50\text{ kg})(1.0\text{ kcal/kg-°C})(100\,°C) = 50\text{ kcal}$

$100\,°C \quad\quad Q_L = mL_v = (0.50\text{ kg})(540\text{ kcal/kg}) = 270\text{ kcal}$

$\quad\quad \Delta T_3 \quad\quad Q_3 = mc_s\,\Delta T_3 = (0.50\text{ kg})(0.48\text{ kcal/kg-°C})(10\,°C) = 2.4\text{ kcal}$

$110\,°C$

Then

$$Q_{total} = \sum_i Q_i = 2.5 + 40 + 50 + 270 + 2.4 = 364.9\text{ kcal} \quad ∎$$

PROBLEM-SOLVING HINT

Note that you must compute the latent heat at each phase change. It is a common error to use the specific heat equation with a temperature interval that includes a phase change.

Example 12.7 Thermal Equilibrium

A 0.30-kg piece of ice at $0\,°C$ is placed in a liter of water at room temperature $(20\,°C)$ in an insulated container. Assuming that no heat is lost to the container, what is the final temperature of the water?

Solution

Given: $m_i = 0.30$ kg   *Find*: T
$\quad\quad\quad T_i = 0\,°C$
$\quad\quad\quad V_w = 1.0$ L, so $m_w = 1.0$ kg
$\quad\quad\quad T_w = 20\,°C$

The subscripts i and w refer to ice and water, respectively. You know that 1.0 L of water has a mass (m_w) of 1.0 kg and that the water supplies the heat to melt the ice. If all of the ice melts, you could view the system as being two masses of water at different temperatures, which come to equilibrium at some intermediate temperature. Since there is no heat loss, it is tempting to write an equation equating the amount of heat lost by the water to the amounts of heat used to melt the ice and to warm up the ice water.

But does all the ice melt? To melt 0.30 kg of ice requires

$$Q_i = m_i L_f = (0.30\text{ kg})(80\text{ kcal/kg}) = 24\text{ kcal}$$

Then, looking at the water, you must ask how much heat it can supply to the ice. The maximum amount would be that given up in lowering the temperature of the water to $0\,°C$, which would be a temperature decrease of $\Delta T = -20\,°C$. This maximum amount of heat would be

$$Q_w = m_w c_w \Delta T = (1.0\text{ kg})(1.0\text{ kcal/kg-}°C)(-20\,°C) = -20\text{ kcal}$$

Thus, all of the ice does not melt since the water does not have enough energy. The final temperature of the water is therefore $0\,°C$.

The heat given up by the water (20 kcal) will melt a mass of ice equal to

$$m_i = \frac{Q_i}{L_f} = \frac{20\text{ kcal}}{80\text{ kcal/kg}} = 0.25\text{ kg}$$

Thus, the final mixture would be 0.30 kg − 0.25 kg = 0.05 kg of ice in thermal equilibrium with 1.25 kg (or 1.25 L) of water at $0\,°C$.

Information about phase changes is represented on graphs called **phase diagrams**. An example is the p-T (pressure-temperature) diagram for water shown in Fig. 12.7. The curves are formed of the points (p, T), or the pressure-temperature combinations at which different phases are in equilibrium. For example, the point at 1 atm and $100\,°C$ corresponds to the normal boiling point, at which liquid water and steam are in equilibrium.

The triple point is the point at which all three phases coexist. This is the unique point used as a reference for the Kelvin scale (Chapter 11). The three curves branching out from this point separate the phase regions. The fusion curve slopes upward to the left, reflecting the fact that the freezing point of water decreases slightly with increasing pressure. (You will learn in Chapter 13 that this type of slope for a fusion curve occurs only for a few substances that, like water, expand on freezing.) Note that if you started heating a quantity of ice that was below $0\,°C$ at 1 atm, the plotted state of the system would first cross the fusion curve (along the horizontal dashed line extending from 1.0 atm in the graph) into the liquid phase region and, with continued heating, would cross the vaporization curve into the vapor phase region at $100\,°C$.

"Vapor" is another term commonly used for "gas." Water vapor is water in the gas phase. (Vapor is also used in a nontechnical sense to mean visible

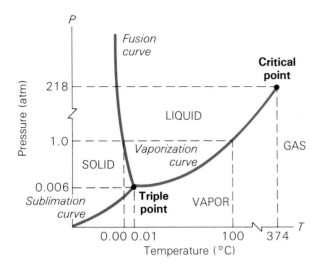

Figure 12.7 **Phase diagram for water**
The curves branch out from the triple point where water exists in all three phases. On the transition curves, it exists in two phases.

droplets of water, such as condensed steam or clouds.) The distinction between vapor and gas is often made relative to the critical point at the end of the vaporization curve. At temperatures less than the critical temperature (374 °C for water), a gas will change to a liquid if sufficient pressure is applied. If a gas is above its critical temperature, no amount of pressure will cause it to become a liquid. It becomes denser and denser with increasing pressure, but never quite becomes a liquid. A substance that is gaseous and has a temperature above its critical temperature is called a gas, and a substance that is gaseous but with a temperature below its critical temperature is known as a vapor.

12.4 Heat Transfer

Since heat is defined as energy in transit, how the transfer takes place is important. Heat moves from place to place (from a higher-temperature region to a lower-temperature region) by three mechanisms: conduction, convection, and radiation.

Conduction

You can keep a pot of coffee hot on a stove because heat is conducted through the coffee pot from the hot burner. The process of **conduction** is visualized as resulting from molecular interactions. Molecules in one part of a body at a higher temperature vibrate faster and collide with less energetic molecules located toward the cooler part of the body. In this way, energy is conductively transferred from a higher-temperature region to a lower-temperature region.

Solids can be divided into two general categories: metals and nonmetals. Metals are good conductors of heat, or **thermal conductors**. Modern theory views metals as having a large number of electrons that are free to move around (not permanently bound to a particular molecule or atom). These free electrons are believed to be primarily responsible for the heat conduction of metals. Nonmetals, such as wood or cloth, have relatively few free electrons and are poor heat conductors. A poor heat conductor is called a **thermal insulator**.

You will learn in Chapter 17 that the free electrons of a metal are also responsible for its electrical conductivity.

In general, the ability of a substance to conduct heat depends on its phase. Gases are poor thermal conductors because their molecules are relatively far apart, and collisions are therefore infrequent. Liquids are better thermal conductors than gases are because their molecules are closer together and can interact more readily.

Heat conduction may be described quantitatively as the time rate of heat flow ($\Delta Q/\Delta t$) in a material for a given temperature difference (ΔT), as illustrated in Fig. 12.8. It has been established through experiments that the rate of heat flow through a substance depends on the temperature difference between its boundaries. Heat conduction also depends on the size and shape of an object. Thus, the analysis of heat flow is generally done using a uniform slab of the material.

A moment's thought should convince you that the heat flow through a slab of material is directly proportional to its surface area (A) and inversely proportional to its thickness (d). That is,

$$\frac{\Delta Q}{\Delta t} \propto \frac{A\,\Delta T}{d}$$

The term $\Delta T/d$ is called **thermal gradient** (the change in temperature per unit of length). Using a constant of proportionality allows the relation to be written as an equation:

$$\frac{\Delta Q}{\Delta t} = \frac{kA\,\Delta T}{d} \tag{12.4}$$

The constant k is called the **thermal conductivity** and characterizes the heat-conducting ability of a material. The greater the value of k for a material, the more heat it will conduct.

The units of k are J/m-s-°C (or kcal/m-s-°C). The thermal conductivities of various substances are listed in Table 12.3. These values actually vary slightly over different temperature ranges, but can be considered to be constant over normal temperature ranges and differences. Compare the relatively large thermal conductivities of the good thermal conductors, the metals, with the relatively small thermal conductivities of some good thermal insulators, such as Styrofoam and wood. Plastic foams are good insulators mainly because they contain pockets of air. Recall that gases are poor conductors, and note the low thermal conductivity of air in the table.

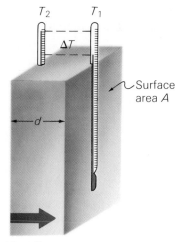

Figure 12.8 **Thermal conduction**
Heat conduction is characterized by the time rate of heat flow ($\Delta Q/\Delta t$) in a material with a temperature difference (ΔT). For a slab of material, $\Delta Q/\Delta t$ is directly proportional to the cross-sectional area (A) and the thermal conductivity of the material, and inversely proportional to the thickness (d).

Example 12.8   Thermal Conductivity

A room has a pine ceiling that measures 3.0 m × 5.0 m × 2.0 cm thick. On a cold day, the temperature inside the room is 20 °C, and the temperature in the attic above is 5 °C. (a) Assuming that the temperatures remain constant, how much heat is lost by conduction through the ceiling in 1.0 h? (b) If 6.0 cm of glass wool insulation were put in above the ceiling, how much energy would be saved?

Solution

Given: $A = 3.0 \text{ m} \times 5.0 \text{ m} = 15 \text{ m}^2$ *Find*: (a) ΔQ
$\qquad d_1 = 2.0 \text{ cm} = 0.020 \text{ m}$ (b) Amount of energy saved
$\qquad d_2 = 6.0 \text{ cm} = 0.060 \text{ m}$
$\qquad \Delta T = 20\,°\text{C} - 5\,°\text{C} = 15\,°\text{C}$
$\qquad \Delta t = 1.0 \text{ h}$
$\qquad k_1 = 0.12 \text{ J/m-s-°C}$
$\qquad k_2 = 0.042 \text{ J/m-s-°C}$ (from Table 12.4)

Table 12.3
Thermal Conductivities of Some Substances

Substance	Thermal conductivity (k)	
	J/m-s-°C	kcal/m-s-°C
Metals		
Aluminum	240	5.7×10^{-2}
Copper	390	9.4×10^{-2}
Iron and steel	46	1.1×10^{-2}
Silver	420	10×10^{-2}
Liquids		
Transformer oil	0.18	4.2×10^{-5}
Water	0.57	14×10^{-5}
Gases		
Air	0.024	0.57×10^{-5}
Hydrogen	0.17	4.0×10^{-5}
Oxygen	0.024	0.58×10^{-5}
Other materials		
Brick	0.71	17×10^{-5}
Concrete	1.3	31×10^{-5}
Cotton	0.075	1.8×10^{-5}
Fiberboard	0.059	1.4×10^{-5}
Floor tile	0.67	16×10^{-5}
Glass (typical)	0.84	20×10^{-5}
Glass wool	0.042	1.0×10^{-5}
Ice	2.2	53×10^{-5}
Styrofoam	0.042	1.0×10^{-5}
Wood		
Wood, oak	0.15	3.5×10^{-5}
Wood, pine	0.12	2.8×10^{-5}
Vacuum	0	0

(a) Using Eq. 12.4,

$$\frac{\Delta Q_1}{\Delta t} = \frac{k_1 A \, \Delta T}{d_1}$$

$$= \frac{(0.12 \text{ J/m-s-°C})(15 \text{ m}^2)(15 \text{ °C})}{0.020\text{m}} = 1.4 \times 10^3 \text{ J/s}$$

With $\Delta t = 1.0 \text{ h} = 3600 \text{ s}$,

$$\Delta Q = (1.4 \times 10^3 \text{ J/s}) \, \Delta t = (1.4 \times 10^3 \text{ J/s})(3.6 \times 10^3 \text{ s}) = 5.0 \times 10^6 \text{ J}$$

(b) Let T_i be the temperature at the interface of the two layers of material and T_h and T_c refer to the hotter and colder temperatures, respectively. Then

$$\frac{\Delta Q_1}{\Delta t} = \frac{k_1 A (T_h - T_i)}{d_1} \quad \text{and} \quad \frac{\Delta Q_2}{\Delta t} = \frac{k_2 A (T_i - T_c)}{d_2}$$

When the heat conduction is steady, $\Delta Q_1/\Delta t = \Delta Q_2/\Delta t$, so T_i may be found by equating the above two equations. Then T_i may be substituted into either equation to give the general expression for $\Delta Q/\Delta t$ for the combined layers:

$$\frac{\Delta Q}{\Delta t} = \frac{A(T_h - T_c)}{(d_1/k_1) + (d_2/k_2)}$$

$$= \frac{(15 \text{ m}^2)(15\,°\text{C})}{(0.020 \text{ m})/(0.12 \text{ J/m-s-}°\text{C}) + (0.060 \text{ m})/(0.042 \text{ J/m-s-}°\text{C})} = 1.4 \times 10^2 \text{ J/s}$$

In 1 h or 3600 s,

$$\Delta Q = (1.4 \times 10^2 \text{ J/s})(3.6 \times 10^3 \text{ s}) = 5.0 \times 10^5 \text{ J}$$

Thus, the heat loss is decreased by $(50 \times 10^5 \text{ J}) - (5.0 \times 10^5 \text{ J}) = 45 \times 10^5$ J. Ideally, this represents a savings of

$$\frac{45 \times 10^5 \text{ J}}{50 \times 10^5 \text{ J}} \times 100\% = 90\% \quad \blacksquare$$

Convection

In general, compared to solids, liquids and gases are not good thermal conductors. The mobility of molecules in fluids permits heat transfer by another process—**convection**. Heat transfer by convection involves mass transfer. For example, when cold water is run over a hot object, the object transfers heat to the water by conduction, and the water carries the heat away with it by convection.

Natural convection cycles occur in liquids and gases. Such cycles are important in atmospheric processes, as illustrated in Fig. 12.9. During the day, air in contact with the warm ground is heated by conduction. That air expands, becoming less dense than in surrounding cooler air. As a result, the warm air rises (air currents) and other air moves horizontally (winds) to fill the space—creating a sea breeze near a large body of water. Cooler air descends, and a thermal convection cycle is set up, which transfers heat away from the Earth. At night, the water surface is warmer than the land (because of the large specific heat of water), and the cycle is reversed.

You can see convection currents in the air above a hot road surface in the summer and in transparent liquids, such as heated water in a glass container. This is because regions of different temperatures have different densities, which

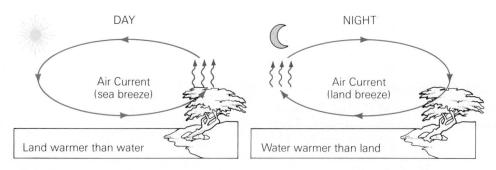

Figure 12.9 Convection cycles
During the day, natural convection cycles give rise to sea breezes near large bodies of water. At night, the situation is reversed and land breezes blow.

cause a bending, or refraction, of light (see Chapter 22).

Convection can also be forced, which means that the medium of heat transfer is moved mechanically. Common examples of forced convection systems are forced-air heating systems in homes (Fig. 12.10), the human circulatory system, and the cooling system of an automobile engine. The human body does not use all of the energy obtained from food; a great deal is lost. [Keep in mind that there's usually a temperature difference between your body and the surroundings.] So that body temperature will stay normal, the internally generated heat energy is transferred close to the surface by blood circulation. From the skin, it is conducted to the air or lost by radiation (the other heat-transfer mechanism, to be discussed shortly).

Water or some other coolant is circulated (pumped) through most automobile cooling systems (some engines are air-cooled). The fluid medium carries heat to the radiator (a heat exchanger), where forced air flow produced by the fan carries it away. The radiator of an automobile is actually misnamed—most of the heat is transferred from it by convection rather than radiation.

Radiation

Conduction and convection require some material as a transport medium. The third mechanism for heat transfer needs no medium and is called **radiation**, which refers to energy transfer by electromagnetic waves (see Chapter 20). This is how heat is transferred to the Earth from the Sun through empty space. Visible light and other forms of electromagnetic radiation are commonly referred to as radiant energy.

You have experienced heat transfer by radiation if you've ever stood near an open fire (Fig. 12.11). You can feel the heat on your exposed hands and face. This heat transfer is not due to convection or conduction, since heated air rises and air is a poor conductor. Visible radiation is emitted from the burning material, but most of the heating effect comes from the invisible **infrared radiation** emitted by the glowing embers or coals. You feel this radiation because it is absorbed by water molecules in your skin. (Body tissue is about 85% water.) The water molecule has an internal vibration whose frequency coincides with that of infrared radiation, which is therefore absorbed readily. [This is called resonance absorption. The electromagnetic wave drives the molecular vibration, and energy is transferred to the molecule, somewhat like pushing a swing; see Chapter 14.]

Infrared radiation is sometimes referred to as "heat rays." You may have noticed the red infrared lamps used to keep food warm in cafeterias. A common example of heating by radiation absorption is described in the Insight feature.

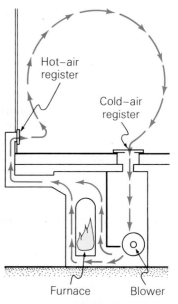

Figure 12.10 **Forced convection**
Houses are commonly heated by forced convection. In older homes, natural convection was (and still is) used. Registers in the floors allow heated air to enter and cooler air to return.

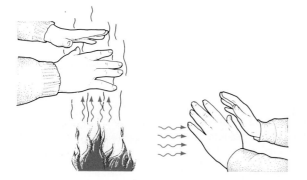

Figure 12.11 **Convection and radiation heating**
The person whose hands are over the fire warms them through the convection of rising hot air and some radiation. The hands of the person by the fire are warmed by radiation.

Although infrared radiation is invisible to the human eye, it can be detected by other means. For example, you can buy special infrared film for some cameras. A picture taken with this film will be an image consisting of contrasting light and dark areas corresponding to regions of higher and lower temperatures. Special instruments that apply such thermography are used in industry and medicine (Fig. 12.12). The frequency of the infrared radiation is proportional to the temperature of its source. This fact is the basis of infrared thermometers, which can measure temperature remotely.

The rate at which an object radiates energy has been found to be proportional to the fourth power of the absolute temperature (T). This is expressed in an equation known as **Stefan's law**:

$$P = \sigma A e T^4 \tag{12.5}$$

where P is the power radiated in watts (W) or joules/s (J/s). [The calorie is not generally used to measure radiation. Radiated power can be written $\Delta Q/\Delta t$ to indicate a rate of heat loss if desired.] The symbol σ (the Greek letter sigma) is called the **Stefan-Boltzmann constant**, and $\sigma = 5.67 \times 10^{-8}$ W/m²-K⁴. The radiated power is also proportional to the surface area (A) of the object. The **emissivity** (e) is a number between 0 and 1 that is characteristic of the material (e is unitless). Dark surfaces have emissivities close to 1, and shiny surfaces have emissivities close to 0. The emissivity of human skin is about 0.70.

Dark surfaces are not only better emitters of radiation, they are also good absorbers. In general, *a good emitter is also a good absorber*. An ideal, or perfect, absorber (and emitter) is referred to as a **black body** ($e = 1.0$). Shiny surfaces are poor absorbers, since most of the incident radiation is reflected. This fact may be demonstrated easily as shown in Fig. 12.13. (You should see why it is better to wear light-colored clothes in the summer and dark-colored clothes in the winter.)

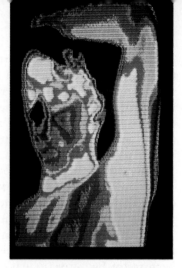

Figure 12.12 **Thermogram** A thermogram of a man holding his arm above his head. The skin temperature varies by about 1 °C for each color, from white (hottest) to black (coldest). Note the high temperature of the armpit due to the proximity of blood vessels to the skin and the colder nose and hair.

INSIGHT

Heat Transfer by Radiation: the Microwave Oven

The microwave oven has quickly become a common kitchen appliance. It is both time-saving and energy-saving, since the oven doesn't warm up and the heating is direct. The principle of operation of the microwave oven is heat transfer by radiation.

Microwaves are a form of electromagnetic radiation; they have a frequency range just below that of infrared radiation (see Chapter 22). Like infrared radiation, microwaves are absorbed by water molecules (in a molecular resonance). In a microwave oven, the microwaves are generated electronically and distributed by reflection from a metal stirrer or fan and the metal walls. Because the walls reflect the radiant energy, they do not get hot.

Microwaves pass through plastic wrap, glass, or dishes made of other "microwave-safe" materials and are absorbed by water molecules in the food, which causes it to be heated. (Metal utensils or objects can become electrically charged, causing sparking and possibly damage.) The microwaves do not penetrate the food completely but are absorbed near the surface. Heat is then conducted to the interior of the food, just as it is in conventional oven heating. This is why it is advisable to let large items or portions sit for a time after the microwave oven has shut off—so they will be warmed or cooked throughout.

Since microwaves could be absorbed by water molecules in the skin, causing burns, microwave ovens have several important safety features. The door is tight-fitting so that microwaves cannot leak out. You will note that the glass in the door is fitted with a metal shield that has small holes through which food can be viewed without opening the door. Microwaves are reflected by this shield and prevented from coming through the glass (essentially, the waves are larger than the holes). Also, the door has a mechanism that automatically shuts off the oven when the door is opened, so a person cannot get into the oven while it is running. In fact, the oven cannot be turned on when the door is open.

When an object is in thermal equilibrium with its surroundings, its temperature is constant. Thus, it must be emitting and absorbing radiation at the same rate. If the temperatures of the object and its surroundings are different, there must be a net flow of radiant energy. If an object is at a temperature T and its surroundings are at a temperature T_s, the net rate of energy loss or gain is given by

$$P_{net} = \sigma Ae(T_s^4 - T^4) \qquad (12.6)$$

Note that if T_s is less than T, then P (or $\Delta Q/\Delta t$) will be negative, indicating a net energy loss. *Keep in mind that the temperatures used in calculating radiated power are the absolute temperatures.*

It is sometimes convenient to know the **intensity** (I) of radiation. This is simply the power per area ($I = P/A$) or the energy per area per time ($I = E/At$). In terms of Eq. 12.6,

$$I = \frac{P_{net}}{A} = \sigma e(T_s^4 - T^4) \qquad (12.7)$$

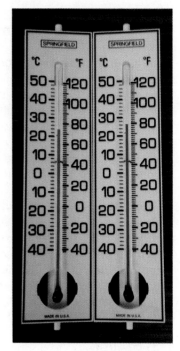

Figure 12.13 Good absorber
Black objects are good absorbers. The bulb of the thermometer on the right has been painted black. Note the difference in the temperatures.

Example 12.9   Radiant Heat Transfer

Suppose that your skin has an emissivity of 0.70 and its exposed area is 0.27 m². (a) How much net energy will be radiated per second from this area if the air temperature is 20 °C? (Assume that your body temperature is normal.) (b) What is the intensity of this radiation?

Solution

Given: $T_s = 20\,°\text{C} + 273 = 293$ K   *Find*: (a) P_{net}
$T = 37\,°\text{C} + 273 = 310$ K   (b) I
$e = 0.70$
$A = 0.27$ m²
$\sigma = 5.67 \times 10^{-8}$ W/m²-K

(a) Using Eq. 12.6,

$$P_{net} = \sigma Ae(T_s^4 - T^4)$$

$$= (5.67 \times 10^{-8} \text{ W/m}^2\text{-K})(0.27 \text{ m}^2)(0.70)[(293 \text{ K})^4 - (310 \text{ K})^4]$$

$$= -20 \text{ W} \qquad (\text{or } -20 \text{ J/s})$$

Since $P_{net} = \Delta Q/\Delta t$, 20 J of energy is radiated or *lost* (as indicated by the minus sign) each second.

(b) Since $I = P/A$,

$$I = \frac{P_{net}}{A} = \frac{-20 \text{ W}}{0.27 \text{ m}^2} = -74 \text{ W/m}^2 \quad \blacksquare$$

PROBLEM-SOLVING HINT

Note that in part (a) of Example 12.9 the fourth powers of the temperatures were found first and then their difference. It is *not* correct to find the temperature difference and then raise it to the fourth power: $T_s^4 - T^4 \neq (T_s - T)^4$.

12.5 Evaporation and Relative Humidity ◆

The evaporation of water from an open container becomes evident only after a relatively long period of time. This phenomenon can be explained in terms of kinetic theory. The molecules in a liquid are in motion, at different speeds. A faster-moving molecule that is near the surface may momentarily leave the liquid. If its velocity is not too large, it will return to the liquid because of the attractive forces exerted by the other molecules. Occasionally, however, a molecule has a large enough velocity that it leaves the liquid entirely and becomes part of the air. The higher the temperature of the liquid, the more likely this is to occur. (Why?)

Since the escaping molecules take their energy with them, the energy and temperature of the remaining liquid will be reduced. Thus, *evaporation is a cooling process.* You have probably noticed this when drying off after taking a bath or shower. About 600 kcal are required to evaporate 1 kg of water from the skin. This energy requirement can be estimated by considering a different process. The amount of heat per kilogram (Q/m) required to raise the temperature of water from $35\,°C$ (approximately skin temperature) to $100\,°C$ ($\Delta T = 100\,°C - 35\,°C = 65\,°C$) is $c\,\Delta T$ (since $Q = mc\,\Delta T$). Adding the latent heat of vaporization, which is also the heat per unit mass ($Q = mL_v$ and $L_v = Q/m$), gives

$$\frac{Q}{m} = c\,\Delta T + L_v = (1\ \text{kcal/kg-}°C)(65\,°C) + 540\ \text{kcal/kg} = 605\ \text{kcal/kg}$$

Of course, evaporation doesn't involve heating to the boiling point, but this estimate gives a limiting approximation.

Evaporation is a relatively slow process, but it is often important in preventing our bodies from overheating. Usually, radiation and conduction are sufficient to maintain a rate of heat loss to the air that keeps us comfortable, given the temperature difference between our bodies and the surroundings. However, when the air gets really hot and the temperature difference narrows (or disappears), we start to perspire. The evaporation of perspiration helps to cool our bodies. On a hot summer day, a person may stand in front of a fan and remark how cool the blowing air feels. But the fan is merely blowing hot air from one place to another. The air feels cool because the flow promotes evaporation, which removes heat energy. Of course, evaporation depends on the humidity (the amount of moisture already in the air). We feel less comfortable on hot, humid days because evaporation is reduced.

Air normally contains water vapor (water in the gaseous phase), mainly from evaporation. Consider an evacuated container partially filled with water or some other liquid (Fig. 12.14). Energetic molecules escape into the space above the liquid. In bouncing around, some strike the liquid surface and again become part of the liquid phase. Eventually, equilibrium will be reached, when the average number of molecules in the space is constant, with the same number of molecules entering the liquid as leaving it. The region above the liquid is then said to be saturated, and the pressure of the vapor is called the **saturated vapor pressure**. As you might expect, the saturated vapor pressure of any liquid depends on temperature. At higher temperatures, more molecules have enough kinetic energy to get into the vapor phase.

The same general description applies for a liquid evaporating into air. Evaporation takes place until an equilibrium is reached between escaping and

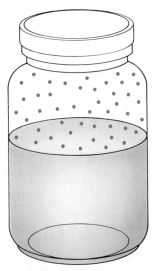

Figure 12.14 **Vapor pressure**
In a closed container, vapor fills the space above a liquid until equilibrium is reached, when the same number of molecules are reentering the liquid as leaving it. The region is then said to be saturated, and the pressure of the vapor is called the saturated vapor pressure.

returning molecules. The effect of the air molecules is that collisions with them may lengthen the time to reach equilibrium. The "container" is not closed in this case, so the air does not generally become saturated. If it does hold all of the water it can at *a given temperature*, it usually gives up the moisture as rain or some other form of precipitation. Filling a bucket with water provides a crude analogy for saturation of the air. When the bucket is full ("saturated"), water can easily overflow or be lost from it.

Heating a liquid speeds up the evaporation process. As mentioned above, the saturated vapor pressure increases with temperature. When the saturated vapor pressure equals the external pressure (generally atmospheric pressure), the liquid boils. Tiny bubbles form in the heated region as the boiling point is approached. When the saturated vapor pressure inside these bubbles equals or exceeds the external pressure, they increase in size and rise to the surface, giving evidence that boiling has begun. With continued heating, the liquid will boil vigorously.

As you can see from Fig. 12.7 (in Section 12.3), the boiling point of water decreases with decreasing pressure. In fact, a container of water in a vacuum chamber will boil at room temperature. The cooling effect of the boiling (the removal of latent heat) will eventually cause the remaining water to freeze if the vacuum is maintained. At high altitudes, where there is lower atmospheric pressure, the boiling point of water is lowered. For example, at Pike's Peak, Colorado, at an elevation of about 4300 m, the atmospheric pressure is about 600 torr, and water boils at about 94 °C rather than at 100 °C. This lengthens the cooking time of food. A pressure cooker may be used to reduce the cooking time (at Pike's Peak or at sea level)—by increasing the pressure, a pressure cooker raises the boiling point.

The vapor pressure of water in the air is commonly expressed in terms of **relative humidity**. In a mixture of gases such as air, the pressure of the individual gases (called the partial pressures) make up the total pressure. The relative humidity is defined as the ratio of the partial pressure of the water vapor to the saturated water vapor pressure at a given temperature. Relative humidity is usually expressed as a percentage. For example, a relative humidity of 60% means that a volume of air is 60% full of water, so to speak, at a particular temperature.

If the air temperature falls, the saturated vapor pressure, or maximum moisture capacity, decreases, and the relative humidity increases. When the air is completely saturated, the relative humidity is 100%. This occurs at a temperature called the **dew point**. If the temperature drops further, the air is said to be supersaturated. When air is at the dew point or below, condensation occurs, and the result is precipitation such as dew, rain, or snow. Since particles, such as dust or ice crystals, are needed for the formation of raindrops, air may be supersaturated and no precipitation will occur. The principle involved in rain-making is that the clouds are seeded with crystals of silver iodide, which have a structure similar to ice, or with dry ice pellets that sublime, and cause the freezing of ice crystals. Ideally, the moisture in supersaturated air in the clouds will condense on these particles and produce rain.

The optimum range of relative humidity for human comfort and health is 40–50%. High humidity causes us to feel hot and uncomfortable because perspiration does not evaporate, and cooling is therefore lessened. Low humidity can cause dry skin, eyes, and nasal membranes. Many people use dehumidifiers in the summer to remove moisture from the air and humidifiers in the winter.

Important Formulas

Heat of combustion:

$Q = mH$

Mechanical equivalent of heat:

$1 \text{ kcal} = 4.19 \times 10^3 \text{ J}$

Specific heat (capacity):

$Q = mc\,\Delta T$

Latent heat:

$Q = mL$

Thermal conduction:

(*for water*)

$L_f = 3.3 \times 10^5 \text{ J/kg (or 80 kcal/kg)}$
$L_v = 22.6 \times 10^5 \text{ J/kg (or 540 kcal/kg)}$

$$\frac{\Delta Q}{\Delta t} = \frac{kA\,\Delta T}{d}$$

Stefan's law:

$P = \sigma A e T^4$

Radiant power loss or gain:

$P_{net} = \sigma A e (T_s^4 - T^4)$

Intensity:

$$I = \frac{P_{net}}{A} = e(T_s^4 - T^4)$$

Questions

Units of Heat

1. Which unit of heat energy is the biggest, the joule, the calorie, the kilocalorie, or the Btu?

2. Why are both furnaces and air conditioners rated in Btu's?

3. A banana contains about 90 Cal of food energy. If this were converted into heat energy, would it raise the temperature of 90 g of water $1\,^\circ$C? Explain.

4. What is the source of the heat of combustion?

5. Gasohol is gasoline that contains about 10% ethyl alcohol by volume. Does gasoline or gasohol have a greater instrinsic heat value?

Specific Heat

6. Why is specific heat "specific"?

7. Heat is supplied to equal masses of mercury and lead. Which will require more heat to raise its temperature by $10\,^\circ$C?

8. Swimming in the ocean at night may feel pleasantly warm even when the air is quite cool. Why?

9. What is the conversion factor between kcal/kg-K and cal/g-$^\circ$C?

10. Is it possible to have a negative specific heat? Explain.

11. Does it make any sense to talk about the value of the specific heat for a phase change? Explain.

Phase Changes and Latent Heat

12. Why is there no temperature change during a phase change?

13. Why do substances have different freezing and boiling points? Would you expect the latent heats to be different for different substances? Explain.

14. What is the principle of operation of frost-free refrigerators? That is, why doesn't frost form, or how is it removed?

15. Why does water condense on the outside of a glass containing an iced beverage?

16. Freeze-dried coffee is made from frozen brewed coffee. How is this done?

17. Why is the melting point of helium not often listed in tables?

18. An ice skater glides primarily because friction between the skate blades and the ice causes the ice to melt, providing a thin lubricating film of water. Does it ever become too cold to ice skate? Explain.

19. When liquid nitrogen is transferred from an insulated container through a hose, moist areas are noted on the hose. What is this moisture? (Hint: note the boiling points of the atmospheric gases in Table 12.2.)

20. Automobile cooling systems operate under pressure. What is the purpose of this? What would happen if you removed the radiator cap shortly after turning off a hot engine? (*Don't* do this—it is quite dangerous!)

Heat Transfer

21. Thermal underwear and thermal blankets are loosely knitted with lots of small holes. Wouldn't it be better if the material were denser?

22. Underground water pipes sometimes freeze during extended cold spells. Why don't they freeze in a day or two?

23. Two large blocks of aluminum and copper are at

room temperature. Would they feel the same if you touched them? Explain.

24. How do clothes keep us warm? Why do drawn drapes in front of a picture window help prevent heat loss?

25. Why are thermopane (double-pane) windows more efficient than single-pane windows? What do storm windows and doors do?

26. In some homes and office buildings, the heating system uses hot water. Does this type of heating system apply forced or natural convection?

27. Could someone carrying a bucket of hot water be referred to as a convection process? Explain.

28. Foam insulation is sometimes blown between the inner and outer walls of a house. Air is a good thermal insulator, so why is more insulation needed?

Figure 12.15 **A cold warning**
See Question 29.

29. Why is the warning shown on the highway road sign in Fig. 12.15 so often necessary?

30. In Table 12.3, the thermal conductivity of a vacuum is listed as zero. Why?

31. What is the purpose of the fins on the radiator shown in Fig. 12.16?

32. Why don't microwave ovens brown foods?

Figure 12.16 **Radiator with fins**
See Question 31.

33. Newton's law of cooling (Sir Isaac was a busy man) states that, in general, the heat loss from an object is proportional to the temperature difference between it and the surroundings: $\Delta Q / \Delta t = K \Delta T$, where K is a constant that includes losses by conduction, convection, and radiation. Apply this law to the following question: would a cup of hot coffee stay hotter longer if you put cream in it right away or waited until you were ready to drink it, at a later time?

34. It is sometimes said that a pan of hot water will freeze faster than a pan of cold water of equal mass if both are placed in a freezer compartment or deep freeze. This could be possible under certain conditions. What might these be? (Hint: consider evaporation and thermal contact with the freezer shelf, and see Question 33.)

35. A Thermos bottle, shown in Fig. 12.17, keeps cold beverages cold and hot ones hot. It consists of a double-walled, partially evacuated container with silvered walls (mirrored interior). The bottle is constructed to counteract all three mechanisms of heat transfer. Explain how.

36. Discuss what is meant by the wind-chill factor that is sometimes discussed by weather forecasters. That is, why does it feel chillier when the wind is blowing?

Evaporation and Relative Humidity ◆

37. In desert areas, drinking water is sometimes carried on the bumper of a vehicle in a porous canvas bag that stays wet. What is the principle of the bag and the purpose of its location?

38. Fogs are basically low-lying clouds. Why do fogs sometimes form in valleys overnight? What do people mean when they say that the Sun burns off the fog?

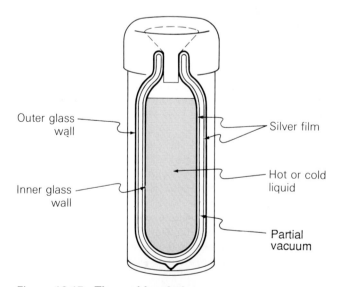

Figure 12.17 **Thermal insulation**
The Thermos bottle is an application involving all three methods of heat transfer. See Question 35.

39. Does the saturated vapor pressure of a liquid in a closed container depend on the volume of space above the liquid? Explain.

40. Even when the temperature and humidity are the same for miles inland, people will travel to the beach, not necessarily to swim but to be more comfortable. How is this possible when the conditions are the same?

41. It is raining, but a local radio station reports that the relative humidity is less than 100%. How is this possible?

Problems*

12.1 Units of Heat

■**1.** An apple is listed on a food chart as having 90 Cal. What is its energy value in joules?

■**2.** A person goes on an 1800-Cal diet to lose weight. What is the equivalent daily allowance in joules?

■**3.** A window air conditioner has a rating of 20,000 Btu. What would this rating be in (a) joules and (b) watts?

■■**4.** Suppose that an 80-kg person wants to "work off" a 600-Cal hot fudge sundae (including nuts, whipped cream, and a cherry) by climbing up a rope. How high would the person have to climb to do an equivalent amount of work?

■■**5.** A waterfall is 100 m high. If all of the gravitational potential energy of the water were converted into heat energy, by how much would the temperature of the water increase in going from the top to the bottom of the falls? (Hint: consider a gram or kilogram of water going over the falls.)

■■**6.** How much energy is released by the complete combustion of 1 gal of gasoline? (Hint: see Table 10.1.)

■■**7.** On the average, how much wood would have to be burned to give the amount of heat that is obtained from the burning of 100 m³ of natural gas?

■■**8.** The label on a container of ice cream says that there are 240 Cal per serving and 4 servings per container. How much ice cream does the container hold?

■■■**9.** Gasohol is about 90% gasoline and 10% ethyl alcohol by volume. How does the energy content of a gallon of gasohol compare to that of a gallon of gasoline? That is, what is the energy difference? (Hint: see Table 10.1.)

■■■**10.** Estimate the average caloric intake (in kilocalories) of a person who eats buttered toast and two eggs for breakfast and a meat-and-potatoes lunch and supper, with a portion of ice cream for dessert at supper.

12.2 Specific Heat

■**11.** If 200 J of energy is added to 10 g of water at 25 °C, what is the final temperature of the water?

■**12.** A 4-g pellet of aluminum at 30 °C gains 88 cal of heat. What is its final temperature?

■**13.** When 50 cal of heat is removed from 10 g of a substance, its temperature is observed to decrease from 40 °C to 15 °C. What is the specific heat of the substance?

■■**14.** A 100-g piece of aluminum at 90 °C is immersed in 200 mL of water at 20 °C. Assuming that no heat is lost to the surroundings or the container, what is the temperature of the aluminum and water when they reach thermal equilibrium?

■■**15.** A student mixes 1 L of water at 40 °C with 1 L of ethyl alcohol at 20 °C. Assuming that there is no heat loss to the container or the surroundings, what is the final temperature of the mixture? (Hint: see Table 10.1.)

■■**16.** A 0.25-kg coffee cup at room temperature is filled with 250 cc of boiling coffee. The cup and the coffee come to thermal equilibrium at 80 °C. If no heat is lost, what is the specific heat of the cup? (Hint: consider the coffee to be essentially boiling water.)

■■**17.** Initially at room temperature, 0.50 kg of aluminum and 0.50 kg of iron are heated to 100 °C. Which metal gains more heat and how much more?

■■**18.** Equal amounts of heat are added to different quantities of copper and lead. The temperature of the copper increases by 10 °C, and the temperature of the lead by 5 °C. Which piece of metal has the greater mass and how much greater?

■■**19.** In a calorimetry experiment, 0.35-kg of aluminum shot at 100 °C is carefully poured into 50 mL of water that has been chilled to 10 °C. Neglecting any losses to the container (or otherwise), what is the final equilibrium temperature of the mixture?

■■**20.** In a calorimetry experiment, 0.50 kg of a metal at 100 °C is added to 0.50 L of water at room temperature in an aluminum calorimeter cup. The cup has a mass of 250 g. If the final temperature of the mixture is 25 °C, what is the specific heat of the metal? What would be the effect if some water splashed out of the cup when the metal was added?

■■21. An electric immersion heater has a power rating of 1200 W. If the heater is placed in 0.5 L of water at room temperature, how long will it take to bring the water to a boil? (Assume that there is no heat loss.)

■■22. At what average rate would heat have to be removed from 1.5 L of (a) water and (b) mercury to reduce the liquid's temperature from room temperature to its freezing point in 5.0 min?

■■■23. A student doing an experiment pours 150 g of heated copper shot into a 300-g aluminum calorimeter cup containing 200 mL of water at 25 °C. The mixture (and the cup) comes to thermal equilibrium at 28 °C. What was the initial temperature of the shot?

■■■24. An aluminum rod has a diameter of 5.0 cm and a length of 20 cm. The rod, initially at room temperature, absorbs 24 kcal of heat. What is the change in the rod's length?

■■■25. A 0.030-kg lead bullet hits a steel plate and melts and splatters on impact. (This action has been photographed.) Assuming that the bullet receives 80% of the collision energy, at what minimum speed must it be traveling to melt on impact? (Assume that the bullet and the steel plate are initially at room temperature.)

12.3 Phase Changes and Latent Heat

■26. The mass of an average ice cube at 0 °C is about 25 g. How much heat must be added to an ice cube to melt it?

■27. How much heat is required to boil away 1.0 L of water that is at 100 °C?

■28. How much more heat is needed to convert 1.0 kg of ice at 0 °C to steam at 100 °C than to raise the temperature of 1.0 kg of water from 0 °C to 100 °C?

■■29. How much heat must be removed from 1.5 L of water at room temperature to form ice whose temperature is −5 °C?

■■30. If 20 g of steam at 120 °C is converted to water at room temperature (20 °C), how much heat is given up in the process?

■■31. How much heat is required to convert 0.45 kg of ice at −10 °C to water at 50 °C?

■■32. How much heat must be added to 0.75 kg of lead at room temperature to cause it to melt?

■■33. (a) A 0.25-kg piece of ice at −20 °C is converted to steam at 115 °C. How much heat must be supplied to do this? (b) To convert the steam back to ice at −5 °C, how much heat would have to be removed?

■■34. Heat is added to 40 mL of water and 40 mL of ethyl alcohol, both initially at room temperature, at a rate of 0.27 kcal/s. How long will it take for each liquid to change completely to a gas?

■■35. Ice is added to 0.50 L of tea at room temperature to make iced tea. If enough ice is added to bring ice and liquid to thermal equilibrium, how much liquid is in the pitcher when this occurs?

■■36. After a barrel jump, a 70-kg ice skater traveling at 25 km/h glides to a stop. If 40% of the frictional heat generated by the skate blades goes into melting ice at 0 °C, how much ice is momentarily melted? Where does the other 60% of the energy go?

■■■37. A 0.20-kg piece of ice at −10 °C is placed in 0.10 L of water at 50 °C. How much liquid is there when the system reaches thermal equilibrium?

■■■38. Steam at 100 °C is added to 200 cm³ of water at room temperature in an aluminum calorimeter cup. How much steam will have been added when the water in the cup is at 60 °C?

■■■39. In an experiment, a 150-g piece of ceramic super-conductor material is placed in liquid nitrogen (N_2) to cool. Assuming that this is done in a perfectly insulated flask, how many liters of liquid nitrogen will be boiled away in doing this? (Take the specific heat of the ceramic material to be the same as that of glass, and the density of liquid nitrogen to be 0.80×10^3 kg/m³.)

12.4 Heat Transfer

■40. How many times faster would heat be conducted from your bare feet by a tile floor than by an oak floor? (Assume that both floors have the same temperature and thickness.)

■41. The temperatures at the ends of a 25-cm metal bar are 20 °C and 70 °C. What is the thermal gradient in the bar?

■42. What is the rate of heat conduction through a slab of steel that is 0.50 m by 0.50 m by 0.010 m if there is a temperature difference of 40 °C between its sides?

■43. If an object has an emissivity of 0.75 and an area of 0.20 m², how much energy does it radiate at room temperature?

■■44. Of the metals listed in Table 12.3, for which two do the thermal conductivities form the greatest ratio? What does this tell you?

■■45. A wall is made up of equal areas of wood and glass. The glass is 0.25 cm thick, and the wood is 3.0 cm thick. For the same temperature gradient, which conducts more heat and relatively how much more?

■■46. A 0.75 m by 1.5 m window has a single pane of glass 0.32 cm thick. On a cold day, it is 5 °C outside and a comfortable 22 °C inside. Assuming that there is continuous air circulation on both sides of the window, how much heat is lost by conduction in 1 h? (Without air circulation, inside heat may not be transferred to the window by convection, and a layer of air could form outside that would reduce the temperature difference and the heat conduction through the window.)

■■47. A copper teakettle with a circular bottom that is 30 cm in diameter and has a uniform thickness of 2.5 mm sits on a burner whose temperature is 180 °C. (a) If the teakettle is full of boiling water, what is the rate of heat conduction through its bottom? (b) Assuming that the heat from the burner is the only heat input, how much water is boiled away in 5.0 min? Is your answer reasonable? If not, explain why.

■■48. An aluminum bar and a copper bar have the same temperature difference between their ends and conduct heat at the same rate. Which bar is longer and how much longer?

■■49. A frozen pond has a layer of ice 3.0 cm thick. If the air temperature is 0 °C, what is the rate of heat conduction through the ice?

■■50. The emissivity of an object is 0.60. How many times greater would the radiation be from a similar black body at the same temperature?

■■51. If the temperature of an object at room temperature is doubled to 40 °C, how are its (a) emissivity and (b) radiation rate affected? (c) What are the effects if the object's temperature (room temperature) in kelvins is doubled?

■■52. A metal sphere with a diameter of 10 cm has an emissivity of 0.58. If it is heated to a temperature of 300 °C when the temperature of the surroundings is 20 °C, approximately how much radiant energy does the sphere lose in 3.0 s? (Why is this value approximate without further heating?)

■■53. What is the Celsius temperature of a black body that radiates with an intensity of 462 W/m²?

■■54. Determine the ratio of the intensities of radiation from the exposed skin of a person when the air temperature is 55 °F and 95 °F?

■■■55. A Styrofoam ice chest is filled with a block of ice at 0 °C. The dimensions of the chest are 0.75 m by 0.40 m by 0.40 m, with walls 2.0 cm thick. (a) How much heat is conducted through the walls (including top and bottom) in 30 min if the outside temperature is 25 °C? (Assume that the temperature difference remains constant.) (b) How much ice will melt in this time? (Assume that ice and water are in thermal equilibrium.) (c) Will water overflow from the chest?

■■■56. A large picture window measures 2.0 m by 3.0 m. At what rate will heat be conducted through the window when the room temperature is 20 °C and the outside temperature is 0 °C (a) if the window has a single pane of glass 4.0 mm thick and (b) if the window has a double pane of glass (a thermopane), each pane 2.0 mm thick with an intervening air space of 1.0 mm? Do thermopanes save energy? (Assume that there is a constant temperature difference.)

■■■57. A steel cylinder with a radius of 5.0 cm and a length of 4.0 cm is placed in end-to-end thermal contact with a copper cylinder of the same dimensions. If the free ends of the two cylinders are maintained at constant temperatures of 95 °C (steel) and 15 °C (copper), how much heat will flow through the cylinders in 10 min?

■■■58. For the metal cylinders in Problem 57, what is the temperature at the interface of the cylinders?

■■■59. A lamp filament radiates 100 W of power when the temperature of the surroundings is 20 °C and 99.5 W of power when the temperature of the surroundings is 30 °C. If the temperature of the filament is the same in each case, what is that temperature on the Celsius scale?

12.5   Evaporation and Relative Humidity ◆

■■60. How much heat would be removed from a body by the evaporation of 10 cm³ of perspiration (water)? (Assume that $Q/m = 600$ kcal/kg.)

■■61. The body generates heat at a rate of about 60 kcal/h. (a) How much water would have to evaporate each hour to remove all of this heat? (b) Assuming that 20% of this heat goes into body functions and 10% is lost by other heat transfer mechanisms, how much water evaporation would be required in one day?

Additional Problems

62. If 50 g of ice at 0 °C is added to 300 cm³ of water at 25 °C in a 100-g aluminum calorimeter cup, what is the final temperature of the water?

63. How much heat must be added to 4.0 kg of copper to raise its temperature by 10 °C?

64. In calorimetry, usually a solid substance at a higher temperature is placed in a liquid. The mixture then comes to equilibrium at some final temperature (T_f). Neglecting the calorimeter cup, show that the final temperature is given by

$$T_f = \frac{m_s c_s T_i + m_l c_l T_i}{m_s c_s + m_l c_l}$$

where the subscripts s and l refer to solid and liquid, respectively.

65. A 0.025-kg lead bullet traveling with a speed of 250 m/s hits a 4.0-kg wooden block, penetrates, and stops inside the block. Assuming no heat loss, what is the temperature increase of the bullet and the block when they come to thermal equilibrium? (Assume that both had the same initial temperature.)

66. A mercury diffusion vacuum pump contains 0.015 kg of mercury vapor at a temperature of 630 K. Suppose that you wanted to condense the vapor. How much heat would have to be removed?

67. What is the Calorie content of a slice of bread (15 g) spread with 4.0 g of butter?

68. The thermal insulation used in building and construction is commonly rated in what is called the R-value, defined as L/k, where L is the thickness of the insulation in inches and k is the thermal conductivity.

For example, 3.0 in. of foam plastic would have an R-value of 3.0 in./0.30 = 10, where, in British units, $k = 0.30$ Btu-in./ft²-h-°F. This value is expressed as R-10. (a) What does the R-value tell you, that is, how does thermal insulation vary with R-value? (b) What thicknesses of (1) fiberboard and (2) brick would give an R-value of R-10?

69. Pine wood 14 in. thick has an R-value of 19. What thickness of (a) glass wool and (b) foam plastic would have the same R-value? (See Problem 68.)

70. The average intensity of solar radiation received at the top of the atmosphere is called the solar constant and is about 0.33 kcal/m²-s. How much radiant energy is emitted by the Sun each second? (Hint: the Earth intersects a cross-sectional area of the radiation at the Earth's average distance from the Sun. Think of the Sun's emitted energy as being spread evenly over a sphere with a radius equal to the distance of the Earth from the Sun.)

71. When exercising, a 75-kg person generates energy at a rate of 180 kcal/h. Assuming that a fourth of this goes into doing work and the rest goes into increasing body temperature and that no heat is lost to the environment, what would the person's body temperature be after 15 min of exercising? (Take the average specific heat of the human body to be 0.82 kcal/kg-°C.)

72. How many joules of energy are required to boil away 1.0 L of liquid nitrogen? (See Problem 39.)

73. If the ignition temperature of a type of wood is 300 °C, how much more heat is gained from the combustion of 1.0 kg of this wood than is needed to raise its temperature from 20 °C to the ignition temperature?

74. On a hot day, 0.025 kg of perspiration (water) is evaporated from a worker's body in 1.0 h. What is the average heat loss rate due to evaporation in kcal/s?

75. Suppose that a 1-kg piece of each of three metals, gold, lead, and tungsten, is at its melting point but still a solid. Which would require the most heat for melting, and how many times greater is this value than that for the other metals?

Thermodynamics

As the word implies, **thermodynamics** deals with the transfer or the actions (dynamics) of heat (the Greek word for "heat" is *therme*). The development of thermodynamics started about 200 years ago and grew out of efforts to develop heat engines. The steam engine was one of the first of such devices, which convert heat energy to mechanical work. Although this course of study is primarily concerned with heat and work, thermodynamics is a broad and comprehensive science that includes a great deal more than heat engine theory.

Classical thermodynamics was not based on hypotheses about the structure of matter, but on experimental observations. However, it is possible to gain deeper insight into the principles of thermodyamics by applying modern molecular and kinetic theory and statistical mechanics, which deals with large numbers of particles. In this chapter, you will learn about the two general laws on which thermodynamics is based, as well as the concept of entropy.

13.1 Thermodynamic Systems, States, and Processes

You need to be familiar with some of the terms used in describing thermodynamic systems. As you know, a **system** (thermodynamic or other) is a definite quantity of matter enclosed by boundaries, either real or imaginary. A system may be open or closed. An **open system** is one that mass may be transferred into or out of. A **closed system** is one for which there is no transfer of mass across its boundaries, or a system with constant mass.

An important thermodynamic occurrence is the interchange of energy between a system and its surroundings through a transfer of heat and/or the performance of mechanical work. For example, an expanding system does work on its surroundings; if the air in a balloon is heated, the balloon will expand. But if the transfer of heat is excluded, the system is said to be a **thermally isolated system** (no heat flows into or out of the system). Work may be done by or on a thermally isolated system. For example, a thermally isolated balloon can be compressed by an external force or pressure (work is done on the system). If there is no interaction at all between a system and its surroundings, the system is said to be a **completely isolated system** (or simply an isolated system).

When heat does enter or leave a system, it is absorbed from or given up to the surroundings or heat reservoirs, which have constant temperatures. A **heat reservoir** is a system with an unlimited heat capacity. This implies that any quantity of heat may be withdrawn from or added to the reservoir without appreciably changing its temperature. (Think of the change in the volume of an ocean that occurs when a bucket of water is added or removed.)

State of a System

Just as there are kinematic equations to describe aspects of the motion of an object, there are **equations of state** to describe the conditions of a thermodynamic system. Such an equation expresses a mathematical relationship of the thermodynamic variables of a system. The perfect gas law, $pV = NkT$, is a simple equation of state. This expression establishes a relationship involving the pressure (p), volume (V), absolute temperature (T), and mass (N, the number of molecules) of a gas, which are state variables. Different states have different sets of values for these variables. A set of these variables that satisfies the perfect gas law specifies a state of a system of ideal gas completely. Of course, the system must be in thermal equilibrium and have a uniform temperature.

The state of a given mass of gas in a closed system can be specified by the variables p, V, and T. It is convenient to plot the states using the thermodynamic coordinates (p, V, T), similar to plotting graphs with Cartesian coordinates (x, y, z). The three-dimensional pVT plot for carbon dioxide is shown in Fig. 13.1. Each point (p, V, T) on the surface represents an equilibrium state.

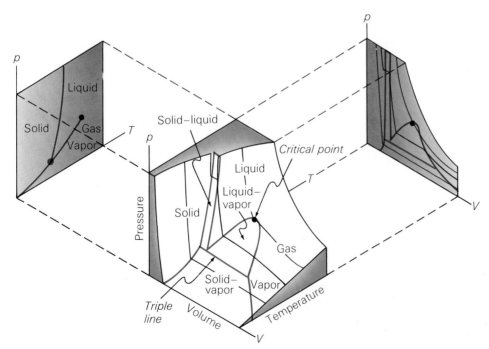

Figure 13.1 **The *pVT* surface for carbon dioxide** The thermodynamic states are points (p, V, T) on a three-dimensional pVT surface. The two-dimensional projections on the pT and pV planes illustrate the behavior of the substance at constant volume and constant temperature, respectively.

Note that the two-dimensional projections in the figure are phase diagrams, like those introduced in Chapter 12. The fusion curve between the solid and liquid phase regions on the p-T diagram for carbon dioxide slopes slightly upward and to the right. [Recall that the slope of the fusion curve for water was upward and to the left (Fig. 12.7).] As you can see from the pVT surface, when a quantity of carbon dioxide goes from liquid to solid, its volume is reduced; that is, it contracts on freezing (unlike water, which expands on freezing). The pVT plot of a substance is very helpful for understanding and predicting its behavior when the state variables change.

Processes

A **process** is a change in the state, or the thermodynamic coordinates, of a system. That is, when a system undergoes a process, the set of coordinates on the pVT plot describing it changes. Processes are said to be either reversible or irreversible.

Suppose that a system of gas in equilibrium is allowed to expand quickly. The state of the system will change rapidly and unpredictably, but will eventually return to equilibrium with another set of thermodynamic coordinates, in another state. A graph, such as a p-V diagram, could show the initial equilibrium point, or state, and the final state, but not what happened in between. Since the intermediate states changed so fast, there would be no data describing them. This is called an **irreversible process**, that is, one for which the intermediate states are unknown (the path is not known). "Irreversible" does not mean that the system can't be taken back to the initial state; it only means the process path can't be retraced, because it isn't known.

If, however, the gas expands very, very slowly, passing from one known equilibrium state to the neighboring one and eventually arriving at the final state, then the process path between the initial and final states would be known. This is called a **reversible process**, that is, one whose path is known. In fact, a perfectly reversible process cannot be achieved. All real thermodynamic processes are irreversible to some degree because they follow complicated paths with many intermediate states. However, the concept of an ideal reversible process is useful.

These thermodynamic terms and ideas are applied in the laws of thermodynamics.

Note that a change of state (of a system) does not necessarily mean a change of phase, such as from liquid to solid. You may avoid confusion if you refer to phases of matter rather than states of matter.

13.2 The First Law of Thermodynamics

The conservation of energy is considered to be valid for any system, and the **first law of thermodynamics** is simply a statement of the conservation of energy for thermodynamic systems. Heat, internal energy, and work are the quantities involved in a thermodynamic system. Suppose that some heat (Q) is added to a system. Where does it go? Into increasing the system's internal energy (ΔU) is certainly a possibility. Another possibility is that it could result in work (W) being done by the system. For example, when a heated gas expands, it does work on its surroundings (think of expanding air in a balloon).

Thus, it is possible for the added heat to go into internal energy or work, or both. Writing this as an equation gives a mathematical expression for the first law:

$$Q = \Delta U + W \qquad\qquad (13.1)$$

For this equation, a positive value of heat $(+Q)$ means that heat is *added to* the system and a positive value of work $(+W)$ means that work is *done by* the system. Negative quantities mean that heat *is removed* $(-Q)$ from the system and work is *done on* $(-W)$ the system.

The first law can be applied to several processes for an ideal gas in which one of the thermodynamic variables is kept constant. These processes have names beginning with iso- (from the Greek *isos*, meaning "equal").

Isobaric Process. A constant-pressure process is called an **isobaric process**. One for an ideal gas is illustrated in Fig. 13.2. On a p-V diagram, the path of an isobaric process is along a horizontal line called an **isobar**. When heat is added to the gas in the cylinder, the ratio V/T must remain constant ($pV = NkT$, so $V/T = Nk/p =$ a constant when p is constant). The heated gas expands; there is an increase in its volume. The temperature must also increase, which means the internal energy of the gas increases. [Recall from Chapter 11 that kinetic theory says that the internal energy of an ideal gas is directly proportional to its (absolute) temperature.]

Work is done by the gas as it expands, in moving the piston, and

$$W = F \Delta x$$

In terms of pressure ($p = F/A$), the force may be written $F = pA$, where A is the area of the piston. Then,

$$W = pA\,\Delta x$$

But $A\,\Delta x$ is simply the change in the volume of the gas, $A\,\Delta x = \Delta V = V_2 - V_1$. Thus,

$$W = p\,\Delta V = p(V_2 - V_1) \qquad \text{(for an isobaric process)} \qquad (13.2)$$

In Fig 13.2, you can see that $p\,\Delta V$ is the area under the isobar on the p-V diagram. For a nonisobaric process (one in which the pressure changes), the

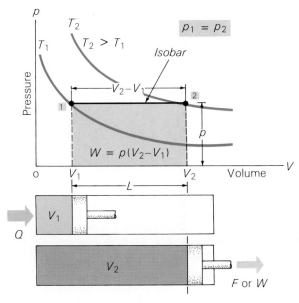

Figure 13.2 **Isobaric (constant-pressure) process** The heat added to the gas goes into increasing the internal energy and doing work (the expanding gas moves the piston): $Q = \Delta U + W$. The work is equal to the area under the process path (from state 1 to state 2 here) on the p-V diagram.

work is also equal to the area under the line showing the process path. Thus, the work depends on the process path as well as on the initial and final states (there will be different areas under different paths).

On the other hand, since the internal energy of a quantity of an ideal gas depends only on its (absolute) temperature, a change in this internal energy is *independent* of the process path and depends only on the initial and final states, or the temperatures of these states ($\Delta U = U_2 - U_1 \propto T_2 - T_1$).

Since V_2 is greater than V_1 for an expanding gas, work is done by the system ($+W$). In terms of the first law, then,

$$Q = \Delta U + W = \Delta U + p\,\Delta V \qquad\qquad (13.3)$$

That is, the heat added to the system goes into both increasing the internal energy ($+\Delta U$, since $\Delta U = Q - W$, which is greater than zero) and into work done by the system. If the process were reversed and the gas were being compressed by an external force doing work *on* the system, all the quantities would be negative. Heat would go out of the system ($-Q$), and the internal energy, or the temperature, of the gas would decrease ($-\Delta U$).

Isometric Process.   An **isometric process** (short for isovolumetric process) is a constant-volume process, sometimes called an isochoric process. As illustrated in Fig. 13.3, the process path on a p-V diagram is along a vertical line, commonly called an **isomet**. No work is done ($W = p\,\Delta V = 0$, since $\Delta V = 0$), so all the added heat goes into increasing the internal energy, and therefore the temperature, of the gas. By the first law,

$$Q = \Delta U + W = \Delta U + 0$$

and

$$Q = \Delta U \qquad \text{(for an isometric process with an ideal gas)}$$

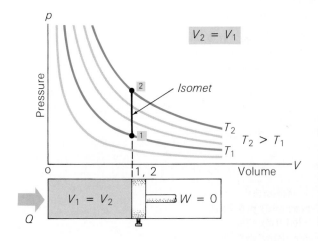

Figure 13.3 **Isometric (constant-volume) process** All of the heat added to the gas goes into increasing the internal energy when the volume is held constant: $Q = \Delta U$. This causes an increase in temperature.

Isothermal Process.   An **isothermal process** is a constant-temperature process (Fig. 13.4). In this case, the process path is along an **isotherm**, or a line of constant temperature. Since $p = (NkT)/V = (\text{constant})/V$ for an isothermal

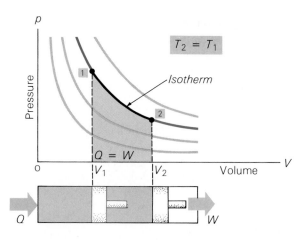

Figure 13.4 **Isothermal (constant-temperature) process**
All of the heat added to the gas goes into doing work (the expanding gas moves piston): $Q = W$. The work is equal to the area under the process path on the p-V diagram.

process, an isotherm is a hyperbola on a p-V diagram (the general form of the equation for a hyperbola is $y = a/x$).

In going from state 1 to state 2 in Fig. 13.4, heat is added to the system, and both the pressure and volume change in order to keep the temperature constant (pressure decreases and volume increases). The work done by the expanding system ($+ W$) is again equal to the area under the process path.

For an isothermal process, the internal energy of the ideal gas remains constant ($\Delta U = 0$), since the temperature is constant. By the first law,

$$Q = \Delta U + W = 0 + W$$

and

$$Q = W \qquad \text{(for an isothermal process)}$$

The constant level of internal energy applies in the case of an ideal gas, not generally.

Thus, for an ideal gas an isothermal process is one in which heat energy is converted to mechanical work (or vice versa for the reverse path).

Adiabatic Process. There's one more type of process in which a thermodynamic condition remains constant (Fig. 13.5). An **adiabatic process** is one in which no heat is transferred into or out of the system; that is, $Q = 0$ (the Greek

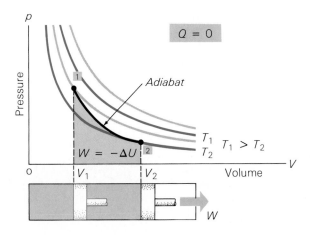

Figure 13.5 **Adiabatic (no heat transfer) process**
In an adiabatic process, no heat is added to or removed from the system: $Q = 0$. Work is done at the expense of the internal energy: $W = -\Delta U$. The pressure, volume, and temperature all change in the process.

word *adiabatos* means "impassable"). The condition $Q = 0$ is satisfied for a thermally isolated system. Processes that are nearly adiabatic can take place in systems that are not thermally isolated if they occur rapidly enough that there isn't time for energy to be transferred into or out of the system.

As the system follows the process path, a curve called an **adiabat**, all three thermodynamic coordinates change. From the first law,

$$Q = 0 = \Delta U + W$$

and

$$W = -\Delta U \qquad \text{(for an adiabatic expansion)}$$

Thus, in an adiabatic expansion, work (the area under the process path) is done by the system with a corresponding decrease in its internal energy. The decrease in internal energy is evidenced by a decrease in temperature in going from state 1 to state 2.

Example 13.1 Joule's Free Expansion Experiment

In 1843, Joule performed a simple experiment that produced some important information about the internal energy of gases. His experimental apparatus is shown in Fig. 13.6. Two containers connected by a valve are submerged in a water bath. Container A is filled with a gas at high pressure, and container B is empty (evacuated). If the valve is opened quickly, the gas in A undergoes *free expansion* into the evacuated container.

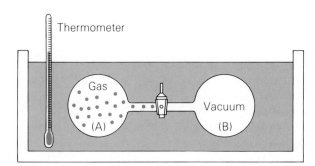

Figure 13.6 **Joule's free expansion experiment** Joule used this type of apparatus to show that the internal energy of a gas is not a function of pressure or volume. See text for more detail.

With free expansion, no work is done by the gas on the surroundings, so $W = 0$. During the process, the gas remaining in A does do work on the gas that has already flowed into B. This is work done on one part of the system by another, however, not by the system as a whole on the surroundings.

The temperature of the water bath is the same after the gas has come to equilibrium as it was before the valve was opened. Thus, there is no heat transfer between the system and its surroundings (the water bath), so $Q = 0$. By the first law,

$$Q = \Delta U + W$$

or

$$\Delta U = Q - W$$

Since $Q = W = 0$,

$$\Delta U = 0$$

The internal energy of the gas remains constant, which shows that this energy is not a function of pressure or volume, since these quantities changed during the process. The temperature of the gas remains unchanged, which indicates that temperature is proportional to internal energy. ∎

Example 13.2 Isobaric Expansion

A quantity of an ideal gas has a volume of 22.4 L at STP. While absorbing 331 cal of heat from the surroundings, the gas expands isobarically to 32.4 L (a) What is the change in the internal energy of the gas? (b) What is the equilibrium temperature of the gas after the expansion?

Solution

$Given$: $p_1 = p_2 = 1$ atm $Find$: (a) ΔU
$\quad\quad\quad = 1.01 \times 10^5$ N/m^2 (Pa)   (b) T_2
$\quad V_1 = 22.4$ L $= 22.4 \times 10^{-3}$ m^3
$\quad V_2 = 32.4$ L $= 32.4 \times 10^{-3}$ m^3
$\quad T_1 = 0\,°C = 273$ K
$\quad Q = 315$ cal $= 1.32 \times 10^3$ J

(a) The work done by the gas in expanding is

$$W = p\,\Delta V = (1.01 \times 10^5 \text{ N/m}^2)[(32.4 \times 10^{-3} \text{ m}^3) - (22.4 \times 10^{-3} \text{ m}^3)]$$

$$= 1.01 \times 10^3 \text{ J}$$

Then, by the first law,

$$\Delta U = Q - W = (1.32 \times 10^3 \text{ J}) - (1.01 \times 10^3 \text{ J}) = 310 \text{ J}$$

(b) Since $p_1 V_1/T_1 = p_2 V_2/T_2$ and $p_1 = p_2$,

$$T_2 = \left(\frac{V_2}{V_1}\right)T_1 = \left(\frac{32.4 \times 10^{-3} \text{ m}^3}{22.4 \times 10^{-3} \text{ m}^3}\right)(273 \text{ K}) = 395 \text{ K}$$

or

$$T_2 = 395 \text{ K} - 273 = 122\,°C \quad \blacksquare$$

13.3 The Second Law of Thermodynamics and Entropy

Suppose that a piece of hot metal is placed in an insulated container of cool water. Heat will be transferred from the metal to the water, and the two will come to thermal equilibrium at some intermediate temperature. For a thermally isolated system, the total energy remains constant, in accordance with the first law of thermodynamics. Could there be a reverse process in which heat would be transferred from the cooler water to the hot metal? Everyone realizes that this would not happen naturally. But if it did, the total energy of the system would remain constant, and the energy transfer would not violate the first law.

Clearly, there must be another principle, not expressed in the first law of thermodynamics, that specifies the *direction* in which a process can take place. This principle is embodied in the **second law of thermodynamics**, which says that certain processes do not take place, or have never been observed to take place, even though they are consistent with the first law.

There are many equivalent statements of the second law, which are worded differently according to their application. One of these, which is applicable to the above situation, is

Heat will not flow spontaneously from a colder body to a warmer body.

Another common statement of the second law is this:

Heat energy cannot be transformed completely into mechanical work (or vice versa).

This statement applies to heat engines, which are discussed in more detail in the next section.

In general, the second law applies to all forms of energy. It is considered to be true because no one has ever found an exception to it (regardless of the form of its statement). If it were not valid, a perpetual motion machine could be built. Such a machine could transform heat from a reservoir completely into work and motion (mechanical energy), with no energy loss. The mechanical energy could then be transformed back into heat and be used to reheat the reservoir (again with no loss). Since the processes could be repeated indefinitely, the machine would run perpetually. All of the energy is accounted for, so this stuation does not violate the first law. However, it is obvious that real machines are always less than 100% efficient, that the work output is always less than the energy input. Another statement of the second law is therefore:

It is impossible to construct an operational perpetual motion machine.

Many attempts have been made to construct perpetual motion machines, but there have been no known successes.

It would be convenient to have some way of expressing the direction of a process in terms of the thermodynamic properties of a system. One such property is temperature. In analyzing a conductive heat transfer process, you need to know the temperatures of the system and its surroundings. Knowing the temperature difference allows you to state the direction in which the heat transfer will take place, into or out of the system (indicated by $+Q$ or $-Q$, respectively).

A more general property that indicates the direction of a process was first described by Rudolf Clausius (1822–1888), a German physicist. This property is called **entropy**, a name coined by Clausius. The physical meaning of entropy is discussed in the Insight feature.

Entropy is related to temperature and heat. The change in entropy (ΔS) when an amount of heat (Q) is added to (or removed from) a system by a reversible process at a constant temperature is given by

$$\Delta S = \frac{Q}{T} \tag{13.4}$$

where T is the Kelvin temperature. [Note what would happen if you used the Celsius temperature in this equation for a phase change from ice to water at $0\,°C$.] Thus, the units for entropy are joules per kelvin (J/K).

If the temperature changes during the process, the change in entropy can be calculated using advanced mathematics. This discussion will be limited to

Perpetual motion machines are distinguished as being of the first kind or the second kind. A perpetual motion machine of the first kind would violate the first law of thermodynamics, the law of the conservation of energy. That is, energy would be created, or the machine would have an efficiency *greater* than 100%. A perpetual motion machine of the second kind would not create energy but would violate the second law, which specifies the direction of spontaneous heat flow. In this case, 100% efficiency would be enough.

isothermal processes or ones with relatively small temperature changes. For the latter, reasonable approximations of entropy changes may be obtained using average temperatures.

Example 13.3   Change in Entropy for an Isothermal Process

What is the change in entropy when 0.25 kg of ethyl alcohol ($L_v = 1.0 \times 10^5$ J/kg) vaporizes at its boiling point of 78 °C?

Solution

Given: $m = 0.250$ kg   *Find*: ΔS
$\qquad T = 78\,°C + 273 = 351$ K
$\qquad L_v = 1.0 \times 10^5$ J/kg

The amount of heat associated with the phase change is

$$Q = mL_v = (0.25 \text{ kg})(1.0 \times 10^5 \text{ J/kg}) = 2.5 \times 10^4 \text{ J}$$

Then

$$\Delta S = \frac{Q}{T} = \frac{2.5 \times 10^4 \text{ J}}{351 \text{ K}} = 71 \text{ J/K}$$

Note that Q is positive because heat is added to the system. The change in entropy, then, is also positive, and the entropy of the alcohol increases. ∎

Example 13.4   An Increase and Decrease in Entropy

A piece of metal at 24 °C is placed in 1.00 L of water at 18 °C. The thermally isolated system comes to equilibrium at a temperature of 20 °C. Find the approximate change in the entropy of the system.

Solution

Given: $T_{m_i} = 24\,°C$ *Find*: ΔS (of the system)
$\qquad T_{w_i} = 18\,°C$
$\qquad m_w = 1.00$ kg
$\qquad T_f = 20\,°C$
$\qquad c_w = 4190$ J/kg-°C
$\qquad\quad$ (from Table 12.2)

With $\Delta T = T_f - T_{w_i} = 20\,°C - 18\,°C = 2\,°C$, the heat gained by the water is

$$Q_w = m_w c_w \Delta T = (1.00 \text{ kg})(4190 \text{ J/kg})(2\,°C) = 8.38 \times 10^3 \text{ J}$$

This is also the amount of heat *lost* by the metal (heat gained must equal heat lost by the conservation of energy). Therefore, $Q_m = -8.38 \times 10^3$ J, where the minus sign indicates a loss of heat.

The average temperatures are

$$\bar{T}_w = \frac{T_f + T_{w_i}}{2} = \frac{20\,°C + 18\,°C}{2} = 19\,°C = 292 \text{ K}$$

$$\bar{T}_m = \frac{T_{m_i} + T_f}{2} = \frac{24\,°C + 20\,°C}{2} = 22\,°C = 295 \text{ K}$$

Using the average temperatures to compute the approximate entropy changes gives

Time's Arrow and the Heat Death of the Universe

In developing the concept of entropy, Clausius realized that when heat flows from a hot body to a colder one, there is something more involved than energy transfer. There must also be a temperature difference, or different levels of energy. When the bodies reach thermal equilibrium, their temperatures are the same, as are the energy levels, and the heat flow ceases. This is analogous to the different levels of gravitational potential energy at different heights. In order to get a ball to fall, for example, a height difference must be involved. Similarly, to get heat to flow, a temperature difference is necessary.

Even though there is a transfer of heat, the total energy of the isolated two-body system is the same before and after their temperatures come to equilibrium. But something is lost or reduced through the heat transfer—*the ability to do useful work*. With a temperature difference, there can be energy transfer with useful work output. With no temperature difference, there is no thermal ability to do work. As a system loses its ability to do work, its entropy increases.

According to the second law of thermodynamics, the entropy of the universe increases in every natural process. What would a decrease in the entropy of the universe mean? Suppose that you filmed a match being lit or a car rolling down a hill. Both of these processes have a natural direction. If the developed film were run backward, the match would ''unburn,'' or the car would roll uphill. Such impossible processes would correspond to negative changes in the entropy of the universe. The increase in entropy provides direction not only for processes but also for time. Thus, entropy is sometimes referred to as **time's arrow**. It points out the forward direction of the flow of events. Basically, entropy distinguishes the past from the future. There will be more entropy in the future, and there was less entropy in the past. However, the entropy form of the second law says nothing about the rate at which time passes.

As time goes on and natural processes occur, the total entropy of the universe increases, and differences in energy levels or temperatures are reduced. For example, if a system is at a high temperature, heat will naturally be transferred to its surroundings until thermal equilibrium is reached. In a universe of ongoing natural processes, there is a continual net entropy increase, and heat is continually being transferred to systems with lower temperatures. Over a long period of time, this will lead to the so-called **heat death of the universe**. When the entropy of the universe has reached a maximum, everything will be at the same temperature. This will pressumably be near absolute zero—not too warm a future, but not too imminent either.

Another interpretation of entropy is that it is a measure of disorder. Systems naturally move toward states of greater disorder or disarray. For example, a raked (ordered) pile of leaves does not naturally stay that way for long, nor does a clear (ordered) dorm room. (These things can only regain their ordered states if someone expends energy in raking or cleaning up, which causes net increases in entropy.)

Thus, an increase in disorder seems to correspond directly to an increase in entropy. For example, when ice cubes melt in a tray, the orderly crystalline structure of the ice is replaced by a more disorderly arrangement of the molecules in the liquid. Another example is the system for Joule's free expansion experiment (see Example 13.1 and Fig. 13.6), which is more ordered when all of the gas is in one container. When the expansion occurs, the disorder and the entropy increase. You know that the expansion will not spontaneously reverse, with all the gas returning to the original container, thereby increasing the order. Statistically, the probability of this occurring is very small. However, statistics does not say that it is impossible, only that the likelihood is virtually zero. The validity of the second law is based not on the impossibility of certain processes (although they tend to be viewed as such), but on the fact that they have never been observed.

Such observations can be summed up by stating the second law in this form: *entropy can be created, but not destroyed*. In contrast, the first law effectively states that *energy can neither be created nor destroyed*.

$$\Delta S_{\mathrm{w}} = \frac{Q_{\mathrm{w}}}{\overline{T}_{\mathrm{w}}} = \frac{8.38 \times 10^3 \text{ J}}{292 \text{ K}} = 28.7 \text{ J/K}$$

$$\Delta S_{\mathrm{m}} = \frac{Q_{\mathrm{m}}}{\overline{T}_{\mathrm{m}}} = \frac{-8.38 \times 10^3 \text{ J}}{295 \text{ K}} = -28.4 \text{ J/K}$$

The change in the entropy of the whole system is the sum of these individual changes:

$$\Delta S = \Delta S_{\mathrm{w}} + \Delta S_{\mathrm{m}} = 28.7 \text{ J/K} - 28.4 \text{ J/K} = +0.3 \text{ J/K}$$

The entropy of the metal decreased because heat flowed out of it. The entropy of the water increased by a greater amount, so the total entropy change of the system was positive. ∎

Note that the total entropy change for the system in Example 13.4 is positive. In general, the direction of any process is toward an increase in entropy. That is, *the entropy of an isolated system never decreases*. It can only increase or remain constant ($\Delta S \geq 0$). An adiabatic process ($Q = 0$) is a constant-entropy process, or an **isentropic process**, since $\Delta S = Q/T = 0$.

Another way to state this observation about entropy is to say that *the entropy of an isolated system increases for every natural process ($\Delta S > 0$)*. The water and metal in coming to an intermediate temperature in Example 13.4 are undergoing a natural process, which is termed that because such a situation is always observed to go that way. If the metal in coming to thermal equilibrium with the water somehow spontaneously became hotter while the water became cooler, this would certainly be an *unnatural* process ($\Delta S < 0$). Similarly, water at room temperature in an isolated ice cube tray will not naturally turn into ice.

If a system is not isolated, there may be a decrease in the entropy *of the system*. For example, if an ice cube tray filled with water is put into a freezer compartment, the water will freeze, with a decrease in entropy. But there will be a larger increase in entropy somewhere else in the environment or the universe. Work has to be done or energy expended to bring about the transfer of heat involved in the change from water to ice. Thus, this is a general statement of the second law of thermodynamics in terms of entropy:

> **The total entropy of the universe increases in every natural process.**

Some other interpretations of entropy are given in the Insight feature.

The entropy of a system is a function of its state. Each state of a system has a particular value of entropy, and a change in entropy depends only on the initial and final states for a process, or $\Delta S = S_f - S_i$ (similar to the change in internal energy for an ideal gas, $\Delta U = U_f - U_i$). An entropy change can be represented on a T-S diagram, as shown in Fig. 13.7. An isentropic process is a straight horizontal line. Just as the area under a curve on a p-V diagram for a gas is equal

An adiabatic process for which $\Delta S = 0$ is ideal and is therefore excluded from this general statement.

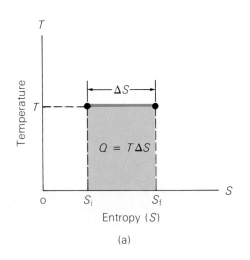

(a)

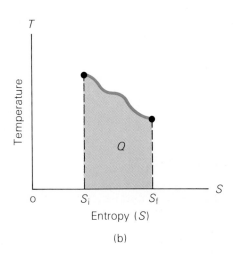

(b)

Figure 13.7 ***T-S* diagrams** When temperature (T) is plotted versus entropy (S), the area under the process path is equal to the heat (Q) transferred in the process. (This is similar to work being equal to the area under a process path on a p-V diagram.) (a) This is a T-S diagram for an isothermal process. Note that the area under the path is that of a rectangle. (b) This T-S diagram is for a general process.

to the work ($W = p \Delta V$), the area under a T-S curve is equal to the change in the heat energy ($Q = T \Delta S$), which is easily seen for the isothermal process in the figure. Like work for a process with variable pressure, heat transferred in a process with variable temperature is also equal to the area under the T-S curve (which generally requires advanced mathematics to calculate).

13.4 Heat Engines and Heat Pumps

A **heat engine** is any device that converts heat energy to work. Since the second law says that perpetual motion machines are impossible, some of the heat supplied to a heat engine will necessarily be lost. In studying thermodynamics, there is no need to be concerned with the usual mechanical components of an engine, such as pistons, cylinders, and gears. A generalized heat engine is usually represented as shown in Fig. 13.8(a). For theoretical purposes, a heat engine is simply a device that takes heat from a high-temperature source (a hot reservoir), converts some of it to useful work, and transfers the rest to its surroundings (a cold, or low-temperature, reservoir).

In general, practical heat engines operate in a cycle, or a series of processes, which brings the engine or system back to its original condition. A thermodynamic cycle with a net work output is shown in Fig. 13.8(b). Steam engines and internal combustion engines, such as automobile engines, are cyclic heat engines. Simply adding heat to a gas in a piston-cylinder arrangement will produce work in a single process. But since a continual output is usually wanted, an engine is run in a cycle.

Example 13.5 The Drinking Bird—A Heat Engine
An unusual example of a cyclic heat engine is a novelty item called a drinking bird, shown in Fig. 13.9. The glass bird contains a volatile liquid (usually ether). When the absorbent material on the head and beak is wet, evaporation occurs. Heat is removed from the head, which lowers its temperature and internal vapor

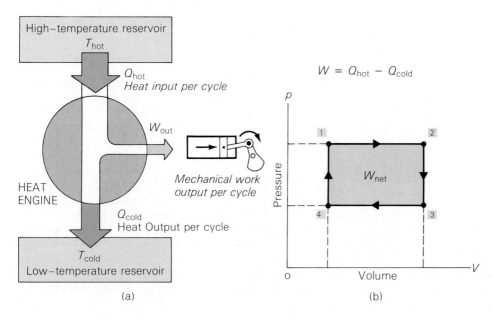

(a) (b)

Figure 13.8 **Heat engine** (a) In this diagram of energy flow for a generalized cyclic heat engine, note that the width of the arrow representing Q_{hot} is equal to the combined widths of the arrows representing W and Q_{cold}. (b) A cyclic process with work output is shown here. The work output is the area of the rectangle formed by the two isobars and the two isomets.

pressure. Because of the resulting pressure difference between the head and body, liquid rises to the head. This raises the bird's center of gravity above the pivot point, and a torque causes a forward rotation. The bird then rewets its beak in the glass of water.

The liquid from the body runs into the head warming it, and the pressures are equalized when the liquid does not fill the tube. The liquid then drains back into the body. With the lowering of its center of gravity below the pivot point, the bird swings back to the vertical position, and the cycle begins again. The net effect is that heat is transferred from a high-temperature reservoir (the body) to a low-temperature reservoir (the head), with work being done (evidenced by the motion). Of course, gravity and the atmosphere also play roles in the operation of this specialized heat engine. ■

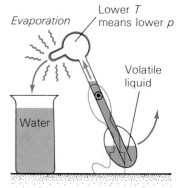

The most common cyclic heat engine is the internal combustion engine used in automobiles and for a variety of other applications. Two different cycles of operation are in general use: a two-stroke cycle and a four-stroke cycle. (The number of strokes tells how many up and down motions a piston makes in a cycle; for example, a two-stroke cycle has one upstroke and one downstroke per cycle.) Engines with a two-stroke cycle are used in motorcycles, outboard motors, chain saws, and some lawn mowers.

Most automobile engines have a four-stroke cycle. The steps in this cycle are shown in Fig. 13.10, along with a p-V diagram of the theoretical thermodynamic processes that the cycle approximates. The theoretical cycle is called the **Otto cycle**, for the German engineer Nikolas Otto (1832–1891), who built one of the first successful gasoline engines.

During the intake stroke (1–2), an isobaric expansion, the air-fuel mixture is admitted through the open valve. This mixture is adiabatically compressed on the compression stroke (2–3). On ignition, there is a quick isometric heating (3–4), which is followed by an adiabatic expansion during the power stroke (4–5). Next is an isometric cooling of the system when the piston is at its lowest position (5–2). The final, exhaust stroke is along the isobaric leg of the Otto cycle (2–1).

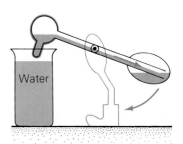

Figure 13.9 **A heat engine?** See Example 13.5.

Automobile engines typically have four, six, or eight cylinders operating in alternating cycles, so there is a continual work output. There are four strokes (two up and two down) as the cylinder goes through a cycle, which corresponds to two revolutions of the crankshaft. Since there is one power stroke per cycle, a four-cylinder engine is timed to have a cylinder fire every $\frac{2}{4}$, or $\frac{1}{2}$, revolution of the crankshaft; a six-cylinder engine has a cylinder firing every $\frac{2}{6}$, or $\frac{1}{3}$, revolution; and so on.

A two-stroke engine does not have an isobaric leg on its theoretical cycle. The air-fuel mixture is added to the cylinder at the same time as the previously combusted gas is expelled, during the isometric cooling process (5–2). This results in some loss of fuel (with the exhaust gas), which is the major disadvantage of the two-stroke engine. The advantage is that there is one power stroke for every crankshaft revolution for a two-stroke engine, compared to only one power stroke for two crankshaft revolutions for a four-stroke engine.

The diesel engine has a two-stroke cycle. This cycle is similar to the Otto cycle, but there are other differences between diesel and gasoline engines. One is the type of fuel—the diesel engine runs on oil, not gasoline. Another is that the diesel engine lacks an ignition system or sparkplugs. During the compres-

The diesel engine was developed by Rudolf Diesel (1858–1913), a German engineer.

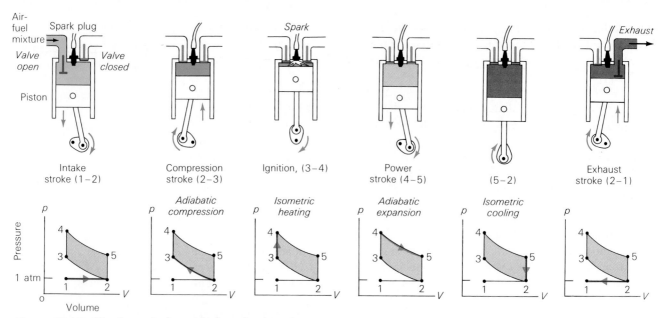

Figure 13.10 The four-stroke cycle for a heat engine
The process steps for the four-stroke Otto cycle. The piston moves up and down two times each cycle, or makes four strokes.

sion stroke, oil is injected into the cylinder. Combustion occurs spontaneously when the temperature of the air-fuel mixture gets high enough (because it is being compressed).

The higher compression and pressure present in a diesel engine give it a higher efficiency than a gasoline engine, about 40% as opposed to 30%. Thermal efficiency is used to rate heat engines (it is analogous to the mechanical efficiency used to rate simple machines, as described in Chapter 9). The **thermal efficiency** (ε_{th}) is

$$\varepsilon_{\text{th}} = \frac{\text{work out}}{\text{heat in}} = \frac{W}{Q_{\text{in}}} \tag{13.5}$$

For one cycle of a cyclic heat engine, W is the work output per cycle and $Q_{\text{in}} = Q_{\text{hot}} - Q_{\text{cold}}$ [the net heat input per cycle, see Fig. 13.8(a)].

By the first law,

$$Q = \Delta U + W \qquad \text{or} \qquad Q_{\text{hot}} - Q_{\text{cold}} = 0 + W$$

since the system returns to its original state after completing a cycle, ΔU is zero. Thus, the thermal efficiency (per cycle) of a cyclic heat engine is

$$\varepsilon_{\text{th}} = \frac{W}{Q_{\text{in}}} = \frac{Q_{\text{hot}} - Q_{\text{cold}}}{Q_{\text{hot}}} = 1 - \frac{Q_{\text{cold}}}{Q_{\text{hot}}} \quad (\times 100\%) \tag{13.6}$$

In general, an engine is expected to have the same efficiency each cycle.

Like mechanical efficiency, thermal efficiency is a fraction and is commonly expressed as a percentage. From Eq. 13.6, you can see that a heat engine could have 100% efficiency if Q_{cold} were zero. This would mean that no heat energy would be lost and all the heat input would be converted to useful work, which is impossible, according to the second law of thermodynamics. In 1851, Lord Kelvin stated this observation, another form of the second law:

> **No heat engine operating in a cycle can convert its heat input completely to work.**

Example 13.6 Thermal Efficiency

With each cycle a heat engine removes 1500 J of heat energy from a high-temperature reservoir and exhausts 1000 J to a low-temperature reservoir. (a) What is the engine's thermal efficiency? (b) What is its work output?

Solution

Given: $Q_{\text{hot}} = 1500$ J $\qquad\qquad$ *Find*: (a) ε_{th}
$\qquad\quad Q_{\text{cold}} = 1000$ J $\qquad\qquad\qquad$ (b) W

(a) $\quad \varepsilon_{\text{th}} = 1 - \dfrac{Q_{\text{cold}}}{Q_{\text{hot}}}$

$\qquad\quad = 1 - \dfrac{1000 \text{ J}}{1500 \text{ J}} = 0.33 \,(\times\, 100\%) = 33\%$

(b) The work is the difference between the heat input and output:

$$W = Q_{\text{hot}} - Q_{\text{cold}} = 1500 \text{ J} - 1000 \text{ J} = 500 \text{ J}$$

There are 500 J of work output each cycle. ∎

Heat Pumps

A heat pump performs the opposite function to a heat engine. That is, a **heat pump** is a device that transfers heat energy from a low-temperature reservoir to a high-temperature reservoir (Fig. 13.11). To do this, there must be work input, since the second law says that heat will not spontaneously flow from a cold body

As used here, the term "heat pump" is general and does not refer specifically to the devices that are commonly used for heating and cooling of buildings. These will be discussed shortly.

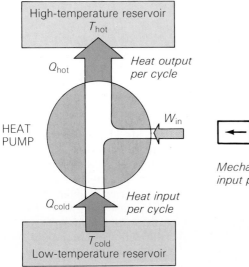

Figure 13.11 **Heat pump**
An energy flow diagram for a generalized cyclic heat pump.

to a hot body. Another statement of the second law, applied to heat pumps, was given by Clausius in 1850:

> No heat pump operating in a cycle can transfer heat energy from a low-temperature reservoir to a high-temperature reservoir without the application of work.

A familiar example of a heat pump is a refrigerator. With work input (from electrical energy), heat is transferred from inside the refrigerator (low-temperature reservoir) to the surroundings (high-temperature reservoir). Have you ever wondered how a refrigerator (or an air conditioner, another heat pump) operates? The knowledge you have gained about thermodynamic processes will allow you to understand the basic operation (see Fig. 13.12).

The refrigerant, or heat-transferring medium, is a substance with a relatively low boiling point. Ammonia (boiling point of $-33.3\,°C$ at 1 atm), sulfur dioxide ($-10.1\,°C$), and Freon ($-29.8\,°C$) can be used as refrigerants. Freon is used in most domestic refrigerators. The other compounds were once used, and their smells were quite evident if the system happened to leak.

It is convenient to think of the system diagrammed in Fig. 13.12 as having high and low (temperature and pressure) sides. On the low side, heat is transferred to the evaporator coils from inside the refrigerator. The heat causes the refrigerant to boil and is carried away as latent heat of vaporization. The gaseous vapor is drawn into the compressor chamber on the downstroke of the

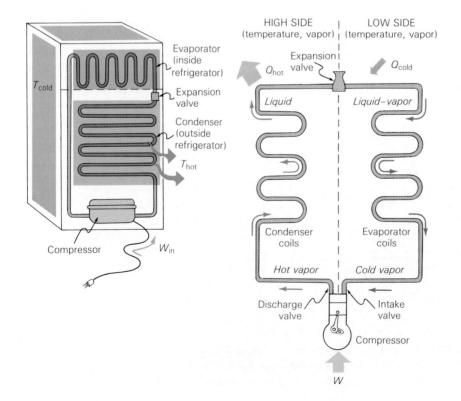

Figure 13.12 **Refrigerator operation**
Heat (Q_{cold}) is carried away from the interior by the refrigerant as latent heat. This heat energy and that of the work input (Q_{hot}) are discharged from the condenser to the surroundings.

piston and is compressed on the upstroke (work input by compressor motor). The compression increases the temperature of the gas, which is discharged from the compressor as a superheated vapor (on the high side). This vapor condenses in the cooler condenser unit, where circulating air (or water in larger units) carries away the latent heat of condensation and the heat of compression. The condensed liquid collects in a receiver.

An expansion valve maintains a balance between the high and low sides, thereby controlling the rate at which the refrigerant passes back to the low side. On being admitted to the low side, the liquid immediately boils and vaporizes because of the low pressure. This is a cooling process because the heat of vaporization is supplied by the internal energy of the liquid. The cooled refrigerant is drawn into the evaporator, and the cycle begins again.

Heat pumps are commonly used to cool homes and offices in the summer and to heat them in the winter. In the summer, the interior of the building is the low-temperature reservoir. The heat pump cools the air inside by transferring heat to the outside (the high-temperature reservoir), thus operating somewhat like a refrigerator. In the winter, these conditions are reversed—the interior is the high-temperature reservoir. Operating in reverse, the heat pump removes heat from the air outside and transfers it into the building. This may seem hard to believe since it is cold outside. However, keep in mind that air has internal energy regardless of its temperature.

If the outside temperature is very low, a heat pump may not be efficient enough to heat a building adequately. Then it must be supplemented by some conventional heating system, such as an electrical one. Some heat pumps use water from underground reservoirs or wells as a heat source and sink. These are more efficient than the ones that use the outside air because water has a larger specific heat than air, and the average temperature difference between the water and the inside air is smaller.

13.5 The Carnot Cycle and Ideal Heat Engines

Lord Kelvin's statement of the second law of thermodynamics says that any *cyclic* heat engine, regardless of its design, will always lose some heat energy. But how much heat must be lost? In other words, what is the maximum possible efficiency of a heat engine? In designing heat engines, engineers strive to make them as efficient as possible, but there must be some theoretical limit, and, according to the second law, it must be less than 100%.

Sadi Carnot (1796–1832), a French engineer, solved this problem. The first thing he considered was the thermodynamic cycle an ideal heat engine would use, that is, the most efficient cycle. A heat engine absorbs heat from a *constant* high-temperature reservoir and exhausts it to a *constant* low-temperature reservoir. These are ideally reversible isothermal processes and may be represented as two isotherms on a *p*-*V* diagram. But what are the processes that complete the cycle and make it the most efficient cycle? Carnot showed that these are reversible adiabatic processes, called adiabats when represented on a graph [(Fig. 13.13(a)].

Thus, the ideal **Carnot cycle** consists of two isotherms and two adiabats and is conveniently represented on a *T*-*S* diagram, where it forms a rectangle [(Fig. 13.13(b)]. The area under the upper isotherm (1–2) is the heat added to the system from the high-temperature reservoir: $Q_{hot} = T_{hot} \Delta S$. Similarly, the area

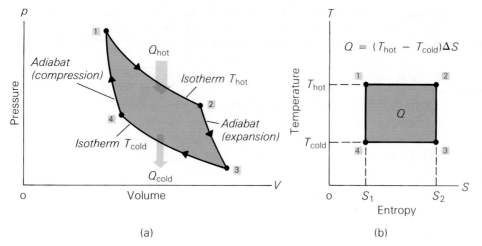

(a)

(b)

Figure 13.13 **Carnot cycle**
(a) The Carnot cycle consists of two isotherms and two adiabats. Heat is absorbed during the isothermal expansion and exhausted during the isothermal compression. (b) On a T-S diagram, the Carnot cycle forms a rectangle, and Q is equal to the area of the rectangle.

under the lower isotherm (3–4) is the heat exhausted: $Q_{cold} = T_{cold} \Delta S$. The difference between these is the work output, which is equal to the area of the rectangle enclosed by the processes (the shaded area on the diagram):

$$W = Q_{hot} - Q_{cold} = (T_{hot} - T_{cold}) \Delta S$$

Since ΔS is the same for the areas under both isotherms, these expressions can be used to relate the temperatures and heats. That is, since

$$Q_{hot} = T_{hot} \Delta S \quad \text{and} \quad Q_{cold} = T_{cold} \Delta S$$

then

$$\frac{Q_{hot}}{T_{hot}} = \frac{Q_{cold}}{T_{cold}} \tag{13.7}$$

This equation can be used to express the efficiency of an ideal heat engine in terms of temperature. From Eq. 13.6, this ideal **Carnot efficiency** (ε_c) is

$$\varepsilon_c = 1 - \frac{Q_{cold}}{Q_{hot}} = 1 - \frac{T_{cold}}{T_{hot}}$$

or $\quad \varepsilon_c = 1 - \dfrac{T_{cold}}{T_{hot}} = \dfrac{T_{hot} - T_{cold}}{T_{hot}}$ $\tag{13.8}$

Note that T_{cold} and T_{hot} are Kelvin temperatures.

The Carnot efficiency expresses the theoretical upper limit on thermodynamic efficiency for a cyclic heat engine, which can never be achieved. It corresponds to the theoretical mechanical advantage for a simple machine. A true Carnot engine cannot be built because the necessary reversible processes can only be approximated. Although reversible adiabatic processes can be approached, reversible isothermal processes are virtually impossible to approximate during the heat transfer processes in a real engine.

However, the Carnot efficiency does show that the greater the difference in the temperatures of the heat reservoirs, the greater the efficiency. For example, if T_{hot} is two times T_{cold}, or $T_{cold}/T_{hot} = \frac{1}{2} = 0.50$, the Carnot efficiency will be

$$\varepsilon_c = 1 - \frac{T_{cold}}{T_{hot}} = 1 - 0.50 \, (\times \, 100\%) = 50\%$$

But if T_{hot} is four times T_{cold}, or $T_{cold}/T_{hot} = \frac{1}{4} = 0.25$, then

$$\varepsilon_c = 1 - \frac{T_{cold}}{T_{hot}} = 1 - 0.25 \ (\times \ 100\%) = 75\%$$

Example 13.7 Carnot Efficiency

An engineer is designing a cyclic heat engine to operate between temperatures of $150\,°C$ and $27\,°C$. What is the maximum theoretical efficiency that can be achieved?

Solution

Given: $T_{hot} = 150\,°C = 423$ K   *Find*: ε_c
         $T_{cold} = 27\,°C = 300$ K

Using Eq. 13.8,

$$\varepsilon_c = 1 - \frac{T_{cold}}{T_{hot}} = 1 - \frac{300 \text{ K}}{423 \text{ K}} = 0.291 \ (\times \ 100\%) = 29.1\% \quad \blacksquare$$

The Third Law of Thermodynamics

Another interesting conclusion can be drawn from the expression for the Carnot efficiency (Eq. 13.8). To have ε_c equal to 100%, T_{cold} would have to be (absolute) zero. Since a heat engine with 100% efficiency is impossible by the second law, the **third law of thermodynamics** is

> It is impossible to reach a temperature of absolute zero.

Absolute zero has never been observed experimentally. If it were, this would in effect violate the second law. In cryogenic (low-temperature) experiments, scientists have come close to absolute zero—to about 0.000001 (one-millionth) of a kelvin—but have never reached it. Near absolute zero, reducing the temperature by an order of magnitude becomes more difficult at each step.

Important Formulas

First law of thermodynamics:

$$Q = \Delta U + W$$

Work:

$$W = p\Delta V = p(V_2 - V_1)$$

Change in entropy:

$$\Delta S = \frac{Q}{T}$$

Thermal efficiency:

$$\varepsilon_{th} = \frac{W}{Q_{in}} = \frac{Q_{hot} - Q_{cold}}{Q_{hot}} = 1 - \frac{Q_{cold}}{Q_{hot}}$$

Carnot efficiency:

$$\varepsilon_c = \frac{T_{hot} - T_{cold}}{T_{hot}} = 1 - \frac{T_{cold}}{T_{hot}}$$

Questions

Thermodynamic Systems, States, and Processes

1. Must a closed system be completely isolated? Does a thermally isolated system have to be completely isolated? Explain.

2. The pVT diagram for water is shown in Fig. 13.14 (compare this with Fig. 12.7). (a) Explain the processes on the liquid-vapor and solid-vapor surfaces. (b) Why is there a triple-point line?

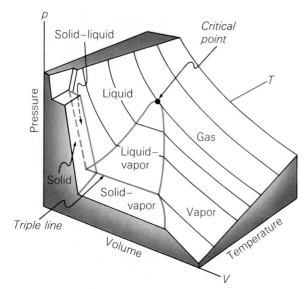

Figure 13.14 The *pVT* surface for water
See Questions 2 and 3 and Problem 3.

Labels on figure: p, Solid–liquid, Critical point, Pressure, Liquid, T, Gas, Liquid–vapor, Solid, Solid–vapor, Vapor, Triple line, Volume, Temperature, V

3. Explain how you can show from Fig. 13.14 that a quantity of water has a greater volume when it is frozen than when it is melted.

4. What is the difference between a reversible process and an irreversible process? Is this also the difference between a natural process and an unnatural process?

The First Law of Thermodynamics

5. Is it possible for a closed system to do work and have its internal energy increased? Explain.

6. (a) Why does the body of a hand pump become hot when a tire is pumped up? (Neglect friction. The plunger is usually lubricated to prevent air leakage.) (b) Why does the valve of a tire become cold when air is released from the tire?

7. Apply the first law to the stretching of a rubber band to determine the effect on its internal energy. (Hint: stretch a wide rubber band and immediately hold it to your forehead; note its temperature.)

8. Why does work depend on the process path? Is it possible for two different process paths to yield the same work? Explain.

9. What form does an isotherm have on a *p*-*V* diagram for a perfect gas?

10. Can an isobaric process and an isometric process occur simultaneously in a closed system of ideal gas?

11. Is an adiabatic process always isothermal for an ideal gas? Explain.

12. Use the first law to explain why the temperature of an ideal gas decreases in an adiabatic expansion.

The Second Law of Thermodynamics and Entropy

13. Does a perpetual motion machine violate the first law of thermodynamics? (Hint: consider efficiencies of 100% and greater.)

14. (a) What is the basic principle of the second law of thermodynamics? (b) How is the change of entropy defined for a reversible process?

15. Give three or more equivalent statements of the second law of thermodynamics.

16. Does a change in entropy depend on the path of a reversible process? Does anything else depend on the path of such a process?

17. Tell what entropy is in terms of (a) the incapability of a system to do work or transfer heat, (b) natural processes, and (c) the disorder of a system.

18. When a quantity of hot water is mixed with a quantity of cold water, the combined system comes to thermal equilibrium at some intermediate temperature. In the same terms as in Question 17, how does the entropy change?

19. You often hear people say "You can't get something for nothing," "You can't get more out than you put in," "You can't even break even," or "I'll never sink that low." Are these expressions descriptive of laws of thermodynamics? Explain.

Heat Engines and Heat Pumps

20. Why must a heat engine exhaust heat energy? Draw a generalized diagram of a heat engine with (a) 100% efficiency and (b) over 100% efficiency.

21. Could the human body be considered a heat engine? Explain. Discuss the body's metabolism in terms of the first law.

22. Is any external work done on the drinking bird in Example 13.5? How would the bird be represented in a generalized heat engine diagram?

23. Lord Kelvin's statement of the second law as applied to heat engines refers to their operating *in a cycle*. Why is this phrase included?

24. In atmospheric convection cycles, colder air from a higher altitude is transferred to a lower, warmer level. Does this violate the second law? Explain.

25. Is leaving a refrigerator door open a practical way to air condition a room? Is operating a window air conditioner in the middle of a room a practical way to cool it?

The Carnot Cycle and Ideal Heat Engines

26. Automobile engines can be either air-cooled or water-cooled. Which type of engine would you expect to be more efficient and why?

27. What are the differences between the Otto cycle and the Carnot cycle?

28. How is the Carnot efficiency similar to the theoretical mechanical advantage of a simple machine?

29. How may the Carnot efficiency of a heat engine be increased?

30. Will doubling the temperature difference between the high-temperature and low-temperature reservoirs of a heat engine double the Carnot efficiency? Explain.

31. Is the third law of thermodynamics really an independent law? Explain.

Problems

13.1 Thermodynamic Systems, States, and Processes

■■**1.** On a p-T diagram, sketch the general paths for the following reversible processes for an ideal gas: (a) isothermal. (b) isobaric, (c) isometric, and (d) adiabatic.

■■**2.** On a V-T diagram, sketch the general paths for the following reversible processes for an ideal gas: (a) isothermal, (b) isobaric, (c) isometric, and (d) adiabatic.

■■**3.** Sketch the projections of the pVT surface shown in Fig. 13.14 on the (a) pT plane and (b) pV plane.

13.2 The First Law of Thermodynamics

■**4.** An ideal gas in a cylinder expands adiabatically while doing 5.0×10^3 J of work in moving the piston. What is the change in the internal energy of the gas? Does the temperature of the gas increase or decrease?

■**5.** A quantity of ideal gas goes through a cyclic process and does 200 J of work. (a) Is heat added or removed from the system, and how much heat is involved? (b) Does the temperature of the gas increase or decrease?
(a) $W = 200$ J   (b) $T_f = T_i$

■**6.** During a thermodynamic process, a system absorbs 150 cal of heat, and its internal energy is decreased by 380 J. (a) Is work done on or by the system? (b) How much work is done?

■■**7.** A system of gas at low density has an initial pressure of 1.28×10^4 Pa and occupies a volume of 0.25 m³. The slow addition of 200 cal of heat to the system causes it to expand isobarically to a volume of 0.30 m³. (a) How much work is done by the system in the process? (b) Did the internal energy of the system change? If so, by how much?

■■**8.** A rigid container contains 1 mole of nitrogen gas that slowly receives 1.5 kcal of heat. What is the change in the internal energy of the gas?

■■**9.** In a piston-cylinder arrangement 2.0 L of water at room temperature slowly receive 3.0 kcal of heat with the piston locked in place. (a) What is the change in the

internal energy of the water? (b) What is the final temperature of the water? (Neglect pressure effects.)

■■**10.** A gram of water (1.00 cm³) at 100 °C is converted to 1671 cm³ of steam at atmospheric pressure. What is the change in the internal energy of the system?

■■**11.** During an isobaric process at 1 atm of pressure, a quantity of gas absorbs 150 cal of heat while expanding its volume from 1.0 L to 1.2 L. What is the change in the internal energy of the system?

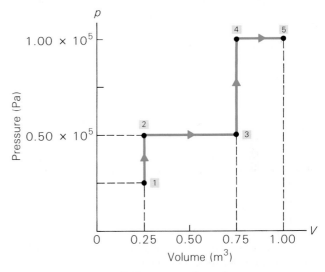

Figure 13.15 **A p-V diagram and work**
See Problems 12 and 13.

■■**12.** A quantity of gas undergoes the reversible changes illustrated in the p-V diagram in Fig. 13.15. How much work is done in each process?

■■**13.** Suppose that after the final process shown in Fig. 13.15 (see Problem 12) the pressure of the gas is decreased isometrically from 1.0×10^5 Pa to 0.70×10^5 Pa, and then the gas is compressed isobarically from 1.0 m³ to 0.80 m³. What is the total work done in all of the processes?

■■**14.** A quantity of gas is compressed as shown on the

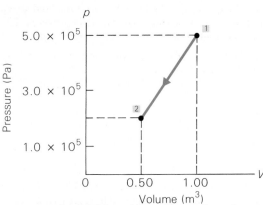

Figure 13.16 **A variable *p-V* process and work**
See Problems 14, 16, and 17.

p-V diagram in Fig. 13.16. How much work is done on the system?

■■■**15.** A gas is enclosed in a piston-cylinder arrangement with a diameter of 20 cm. Heat is slowly added to the system while the pressure is maintained at 1.0 atm. During the process, the piston moves 5.0 cm. (a) What type of process is this? (b) If the quantity of heat transferred to the system during the expansion is 0.10 kcal, what is the change in the internal energy of the gas?

■■■**16.** Suppose that after the process shown in Fig. 13.16 (see Problem 14) the gas undergoes an isobaric expansion to 1.0 m³ and then the pressure is increased isometrically to 5.0×10^5 Pa. (a) How much work is done in each of these processes? (b) What is the total work done in taking the gas from its initial state [(1) in the figure] to its final state? (c) Is heat added or removed from the system? How much heat?

■■■**17.** After the process shown in Fig. 13.16 (see Problem 14), the pressure of the gas is increased isometrically by 1.0 atm and is then allowed to expand isobarically to 1.0 m³. How much work is done in taking the gas from its initial state (state 1 in the figure) to the final state?

■■■**18.** The temperature of a cube of aluminum 10 cm on a side is slowly raised from 20 °C to 150 °C. (a) How much work is done by the system? (Hint: consider thermal expansion.) (b) What is the change in the internal energy of the aluminum?

13.3   The Second Law of Thermodynamics and Entropy

■**19.** What change in entropy is associated with the reversible phase change of 1.0 kg of ice to water at 0 °C?

■**20.** What is the change in entropy when 0.50 kg of steam condenses to water at 100 °C?

■**21.** Which process has the greater change in entropy—0.75 kg of water changing to ice at 0 °C or 0.25 kg of water changing to steam at 100 °C? Comment on the entropy changes in terms of order and disorder.

■■**22.** A quantity of an ideal gas initially at STP undergoes a reversible isothermal expansion and does 2000 J of work on its surroundings in the process. What is the change in the entropy of the gas?

■■**23.** A certain system undergoes an isothermal process at 27 °C, and there is an entropy change of 20 J/K. If 1.5×10^3 J of work is done on the system in the process, what is the change in the system's internal energy?

■■**24.** An ice tray is filled with 250 mL of water at 10 °C from a cold-water tap and placed in a freezer, where the temperature is −6.0 °C. Estimate the change in entropy that will have occurred when the water freezes and the ice comes to thermal equilibrium with its surroundings.

■■**25.** Suppose that two automobiles with equal masses of 3.0×10^3 kg are both traveling at 40 km/h when they have a head-on collision. Estimate the change in entropy for this process.

■■**26.** (a) How much heat is transferred in the processes shown on the *T-S* diagram in Fig. 13.17? (b) What type of process takes place as the system goes from state b to state c?

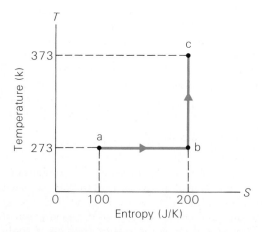

Figure 13.17 **Entropy and heat**
See Problems 26 and 27.

■■**27.** Suppose that the system described by the *T-S* diagram in Fig. 13.17 is returned to its original state (state a) by a reversible process that is depicted by a straight line from c to a. What is the change in entropy for the total cycle? How much heat is transferred in the cyclic process?

■■28. A 50-g ice cube at $0\,^\circ$C is placed in 500 mL of water at room temperature. Estimate the change in entropy when all the ice has melted (a) for the ice, (b) for the water, and (c) for the universe.

■■■29. A half-liter of ethyl alcohol at $18.0\,^\circ$C is mixed with a half-liter of water at $25.0\,^\circ$C. What is the change in entropy once the mixture has reached thermal equilibrium? (Consider the system to be thermally isolated.)

■■■30. If a 12-g piece of ice at $-2.0\,^\circ$C is added to 48 mL of water at $5.0\,^\circ$C, estimate the change in entropy once the system has reached thermal equilibrium. (Consider the system to be thermally isolated.)

■■■31. For the two-step reversible process illustrated in Fig. 13.18, the same amount of heat is added to and removed from the system. Prove that the total entropy (of the universe) increases in the process.

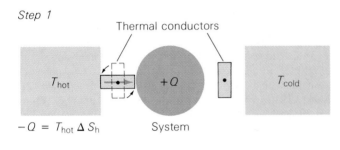

Step 1

Thermal conductors

T_{hot} $+Q$ T_{cold}

$-Q = T_{\text{hot}}\,\Delta S_{\text{h}}$ System

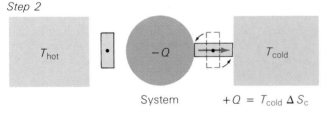

Step 2

T_{hot} $-Q$ T_{cold}

System $+Q = T_{\text{cold}}\,\Delta S_{\text{c}}$

Figure 13.18 **Does the entropy increase?** See Problem 31.

13.4 Heat Engines and Heat Pumps

■32. In each cycle, a heat engine absorbs 150 kcal of heat from a high-temperature reservoir and exhausts 100 kcal. (a) What is the engine's thermal efficiency? (b) What is its work output?

■33. A heat engine has a thermal efficiency of 30%. It absorbs 600 J from a high-temperature reservoir each cycle. (a) What is the work output of the engine? (b) How much of the heat input is not converted to work?

■34. A heat engine with an efficiency of 40% does 800 J of work each cycle. How much heat is rejected to the surroundings (the low-temperature reservoir)?

■35. The work output of a heat engine is 20,930 J per cycle, and its thermal efficiency is 25%. (a) What is the heat input per cycle? (b) How much heat is exhausted per cycle?

■■36. A steam engine does 8000 J of useful work each cycle but loses 520 J to friction and exhausts 2.0 kcal of heat energy. What is the engine's efficiency?

■■37. A theoretical cyclic heat engine absorbs 1800 kcal of heat from a high-temperature reservoir and exhausts 600 kcal to a low-temperature reservoir. When run in reverse as a heat pump with the same reservoirs, 5870 J of work must be input to deliver 1800 kcal of heat energy to the high-temperature reservoir. (a) What is the engine's thermal efficiency? (b) Is the heat engine reversible in the sense that the work input and output are the same when the engine is run between the same reservoirs?

■■38. A gasoline engine consumes 10 L of fuel per hour. (a) What is the energy input during a period of 2.0 h? (b) If the engine delivers 1900 kW of power during this time, what is its efficiency?

■■■39. Use the general definition of efficiency, $\varepsilon = W/Q_{\text{in}}$, to derive the Carnot efficiency. [Hint: consider the cycle in terms of entropy, and see Fig. 13.13(b).]

■■■40. Heat pumps are rated by a *coefficient of performance* (cop), which is defined as cop $= Q_{\text{cold}}/W$, or the ratio of what you get to what you pay for. (a) Show that

$$\text{cop} = \frac{1}{[(Q_{\text{hot}}/Q_{\text{cold}}) - 1]}$$

(b) Can the coefficient of performance of a heat pump be greater than 1? Justify your answer.

■■■41. (a) The coefficient of performance of a refrigerator (see Problem 40) is around 2.5. What does this tell you about the amount of heat exhausted per cycle? (b) On moderately cold days, a heat pump used to heat a house has a cop of 3.0. If the work input is 4000 J per cycle, how much heat is delivered per cycle? In terms of energy, how much of this is cost-free?

■■■42. The energy or useful work lost as a result of a change in the entropy of the universe is given by $W = T_{\text{cold}}\,\Delta S_{\text{u}}$, where T_{cold} is the temperature of the cold reservoir for the process. (a) Show that for heat (Q) transferred from a high-temperature reservoir to a low-temperature reservoir $W = Q[1 - (T_{\text{cold}}/T_{\text{hot}})]$. (b) What is the quantity within the brackets?

■43. A heat engine operates between a reservoir at 27 °C and one at 200 °C. What is the maximum theoretical efficiency of the engine? Is this ever achieved?

■44. An ideal heat engine takes in heat from a high-temperature reservoir at 150 °C and exhausts heat to a low-temperature reservoir at 0 °C. What is the Carnot efficiency of this engine?

■■45. An ideal heat engine with a Carnot efficiency of 35% takes in heat from a high-temperature reservoir at 147 °C. What is the temperature of the low-temperature reservoir?

■■46. An ideal heat engine takes 6.5 kcal of heat from a high-temperature reservoir at 320 °C and exhausts some of it to a low-temperature reservoir at 120 °C. How much work is done by the engine?

■■47. A Carnot engine with an efficiency of 35% operates with a low-temperature reservoir at 50 °C and exhausts 1200 J of heat each cycle. What is the heat input and the temperature of the high-temperature reservoir?

■■48. The maximum temperature for the superheated steam used in a turbine for electrical generation is about 540 °C because of material limitations. (a) If the steam condenser operated at room temperature, what would the ideal efficiency be? (b) The actual efficiency is about 35–40%. What does this tell you?

■■49. Which has the greatest theoretical efficiency—a heat engine operating between reservoirs at 300 °C and 100 °C or one operating between reservoirs at 300 K and 100 K? What are the efficiencies in each case?

■■50. An engineer wants to run a heat engine with an efficiency of 40% between a high-temperature reservoir at 200 °C and a low-temperature reservoir. Below what temperature must the low-temperature reservoir be for practical operation of the engine?

■■51. An ideal heat engine takes in heat from a reservoir at 327 °C and has an efficiency of 30%. If the exhaust temperature did not vary and the efficiency increased to 40%, what would be the increase in the temperature of the hot reservoir?

■■52. An inventor claims to have developed a heat engine that, on each cycle, takes in 120 kcal of heat from a high-temperature reservoir at 400 °C and exhausts 48 kcal to the surroundings at 125 °C. Would you invest your money in the production of this engine? Explain.

■■53. It has been proposed that the temperature difference in the ocean could be utilized to run a heat engine to generate electricity. In tropical regions, the water temperature is about 25 °C at the surface and about 5 °C at a low depth. (a) What would the maximum theoretical efficiency of such an engine be? (b) Do you think a heat engine with such a relatively low efficiency would be practical? Explain.

■■■54. An ideal heat pump is equivalent to a Carnot engine running in reverse. Show that the work required for an ideal heat pump is given by

$$W = Q_{cold}\left[\left(\frac{T_{hot}}{T_{cold}}\right) - 1\right]$$

■■■55. Show that the coefficient of performance for a Carnot engine is given by

$$cop_C = \frac{T_{cold}}{T_{hot} - T_{cold}}$$

(See Problem 40.)

Additional Problems

56. The working substance of a cyclic heat engine is 0.75 kg of an ideal gas. The cycle consists of two isobaric processes and two isometric processes as shown in Fig. 13.19. (a) What is the engine's work output? (b) What would be the efficiency of a Carnot engine operating with the same high-temperature and low-temperature reservoirs?

57. In a thermodynamic process, 2000 J of heat is transferred to a system, and 500 J of work is done on the system. What is the change in the internal energy of the system?

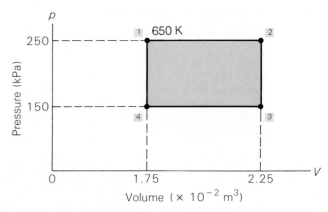

Figure 13.19   **Thermal efficiency**
See Problem 56.

58. A coal-fired power plant produces 850 MW of electricity and operates at an overall thermal efficiency of 35%. (a) What is the thermal input to the plant? (b) What is the rate of heat discharge from the plant? (c) Why is the water heated by the discharged heat cooled in a cooling tower before being discharged into a nearby river?

59. It can be shown that the work done in an isothermal process with an ideal gas is

$$W = nRT\left[\ln\left(\frac{V_2}{V_1}\right)\right]$$

If 2.0 moles of an ideal gas expand isothermally from 3.8×10^{-2} m^3 to 7.6×10^{-2} m^3 at room temperature, (a) how much heat is added to the system? (b) What is the change in the internal energy of the gas? (c) What is the change in the entropy of the system?

60. How much heat is added to a system if its internal energy is increased by 2.75×10^3 J and 1.50×10^3 J of work is done (a) on the system and (b) by the system?

61. A reservoir at 30 °C supplies heat to melt 1.5 kg of ice at 0 °C. What is the total change in entropy once all the ice is melted?

62. A thermally isolated quantity of gas has an initial volume of 10 L and is compressed isobarically at a pressure of 300 kPa. If 6.0×10^2 J of work is done on the system, what is the final volume of the gas?

63. In each cycle, a Carnot heat engine takes 774 J of heat from a high-temperature reservoir and discharges 258 J to a low-temperature reservoir. (a) What is the Carnot efficiency of the engine? (b) How many times greater is the temperature of the high-temperature reservoir than that of the low-temperature reservoir?

64. What is the change in entropy when 0.50 kg of mercury vapor ($L_v = 2.7 \times 10^5$ J/kg) condenses to a liquid at its boiling point of 357 °C?

65. The exhaust temperature of a Carnot engine is 95 °C, and the engine has a heat output of 100 kcal per cycle at this temperature. If the engine has an efficiency of 30%, what is its heat input?

66. A quantity of ideal gas is taken through processes that are depicted as straight-line paths on a p-V diagram: from (1 atm, 1 m^3) to (4 atm, 1 m^3), then to (4 atm, 3 m^3), and back to (1 atm, 1 m^3). What is the net work done?

67. Show that the change in entropy for a cycle of a heat engine is

$$\Delta S = \frac{Q_{\text{cold}}}{T_{\text{cold}}} - \frac{Q_{\text{hot}}}{T_{\text{hot}}}$$

68. A Carnot engine operating between reservoirs at 27 °C and 227 °C does 1500 J of work each cycle. (a) What is the efficiency of the engine? (b) What is the change in entropy each cycle? (Hint: see Problem 67.)

69. A heat engine has an efficiency that is half that of a Carnot engine that operates between temperatures of 120 °C and 350 °C. If the real engine absorbs heat at a rate of 50 kW per cycle, how much heat does it exhaust each cycle?

70. The change in entropy for a liquid that undergoes a temperature change can be shown to be

$$\Delta S = mc\left[\ln\left(\frac{T_{\text{f}}}{T_{\text{i}}}\right)\right]$$

where m is the mass, c the specific heat, and T_{f} and T_{i} the final and initial temperatures. If 1.0 kg of water at 0 °C and 1.0 kg of water at 100°C are mixed together in a thermally isolated container, will there be a net change in entropy? Justify your answer.

14

Vibrations and Waves

A vibration or an oscillation is a back-and-forth motion. There are many familiar examples. You may have felt a power tool or lawnmower vibrate. A car traveling on a rough road does some vibrating. A swinging pendulum has an oscillation, as does the dial on a bathroom scale as it comes to equilibrium to give a weight reading. Another example is a small ball in a bowl with a round bottom. If the ball is displaced from its equilibrium position at the bottom of the bowl, it will roll back and forth, or oscillate, and finally come to rest at that equilibrium position where its potential energy is lowest. Recall from Chapter 9 that the ball must be oscillating about a point of stable equilibrium, for which there is a restoring force or torque. Thus, a particle that undergoes vibrating or oscillatory motion must be in stable equilibrium.

A simplistic model of a solid pictures the intermolecular (bonding) forces as springs. The molecules (particles) joined by these elastic forces can vibrate back and forth when some form of energy, such as heat, is added. Energy may also be added to a material mechanically, in the form of a blow or (in the case of a gas) a compression. In any case, the added energy sets the molecules vibrating. Heat transfer by conduction is essentially the propagation (transmitting) of thermal energy through a medium by molecular vibrations. Similarly, mechanical energy may be carried through a medium, and this propagation of a disturbance is called a wave.

In this chapter, you will learn how to describe vibrations and wave motion. This description will be applied to sound waves in Chapter 15 and to light waves in Chapter 20.

14.1 Simple Harmonic Motion

The motion of an oscillating particle depends on the restoring force producing it. It is convenient to begin to study such motion by considering the simplest type of force, which is one that is directly proportional to the displacement. One such force is the spring force, described by **Hooke's law**,

$$F = -kx \tag{14.1}$$

where k is the spring constant. Recall from Chapter 5 that the minus sign indicates that the force is always in the opposite direction to the displacement; that is, it always tends to restore the spring to its equilibrium position.

Suppose that a mass on a horizontal frictionless surface is connected to a spring as shown in Fig. 14.1. When the mass is displaced to one side of its equilibrium position, it will move back and forth, will vibrate or oscillate. An **oscillation** or a **vibration** is by definition a *periodic motion*, that is, a motion that repeats itself again and again along the same path. For linear oscillations, like those of a mass attached to a spring, the path may be back and forth or up and down. For the angular oscillation of a pendulum, the path is back and forth along a circular arc.

Motion under the influence of the type of force described by Hooke's law is called **simple harmonic motion (SHM)**, because the force is the simplest force and because the motion can be described by harmonic functions (sines and cosines), as you will see later in the chapter. Descriptions of simple harmonic motion refer to the directional distance of the object from its equilibrium position as its **displacement**. Note in the figure that the displacement can be either positive or negative ($+x$ or $-x$). The maximum or greatest displacement from the equilibrium position is called the **amplitude** (A). Note there are two amplitude positions ($+A$ and $-A$).

Besides the amplitude, other important quantities for describing an oscillation are its period and frequency. The **period** (T) is the time required for one complete cycle of motion. A cycle is a complete round trip, a back-and-forth motion from some initial point and back to the same point. If an object starts at $x = A$, when it next returns to position A, it will have traveled one cycle in a time of one period.

The **frequency** (f) is the number of cycles per second. The frequency and the period are related by

$$f = \frac{1}{T} \tag{14.2}$$

The inverse relationship is reflected in the units: the period is the seconds per cycle, and the frequency is the cycles per second. For example, when $T = \frac{1}{2}$ s/cycle, $f = 2$ cycles/s.

The standard unit for frequency is the **hertz (Hz)**, which is one cycle per second (cps).* From Eq. 14.2, frequency has the unit $1/s$, or s^{-1}, since the period is a measure of time. Although cycle is not really a unit, you might find it convenient at times to express frequency in cycles/s to help with dimensional analysis. This is similar to the way the rad is used in the description of circular motion (see Chapter 7).

Energy and Speed with SHM

Recall from Chapter 5 that the potential energy stored in a stretched or compressed spring is given by

$$U = \frac{1}{2}kx^2 \tag{14.3}$$

---

*The unit is named for Heinrich Hertz (1857–1894), a German physicist and early investigator of electromagnetic waves.

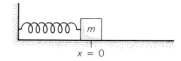

1 Equilibrium

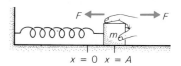

2 $t = 0$

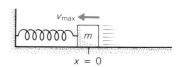

3 $t = \frac{1}{4}T$

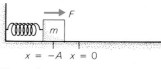

4 $t = \frac{1}{2}T$

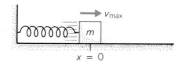

5 $t = \frac{3}{4}T$

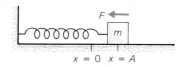

6 $t = T$

Figure 14.1 **Simple harmonic motion (SHM)** When displaced from its equilibrium position ($x = 0$) and released, the mass undergoes simple harmonic motion (no frictional losses). The time in which it completes one cycle is the period of oscillation (T).

This is equal to the work done on the spring, which is equal to the area under the curve on a graph of force versus displacement, such as the one in Fig. 14.2 (and see Chapter 5). A mass m oscillating on a spring has kinetic energy. Thus, the kinetic and potential energies together give the total mechanical energy of the system:

$$E = K + U = \tfrac{1}{2}mv^2 + \tfrac{1}{2}kx^2 \qquad (14.4)$$

When the mass is at one of the amplitude positions, $+A$ or $-A$, it is instantaneously at rest ($v = 0$). Thus, all the energy is potential energy at this time; that is,

$$E = \tfrac{1}{2}m(0)^2 + \tfrac{1}{2}kA^2 = \tfrac{1}{2}kA^2$$

or $\quad E = \tfrac{1}{2}kA^2 \qquad\qquad\qquad\qquad\qquad\qquad\quad (14.5)$

Neglecting any losses, the total energy is conserved. In this case,

The total energy of an object in simple harmonic motion is proportional to the square of the amplitude.

This fact allows the velocity to be expressed as a function of position:

$$E = K + U \qquad \text{or} \qquad \tfrac{1}{2}kA^2 = \tfrac{1}{2}mv^2 + \tfrac{1}{2}kx^2$$

Then

$$v^2 = \frac{k}{m}(A^2 - x^2)$$

and

$$v = \sqrt{\frac{k}{m}(A^2 - x^2)} \qquad\qquad\qquad\qquad (14.6)$$

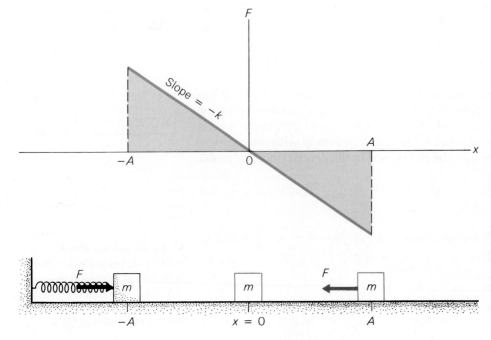

This discussion will be limited to *light* springs, for which the mass is considered negligible.

Figure 14.2 **Work and energy**
The work done by a force in stretching a spring is equal to the area under the curve on graph of force (F) versus displacement (x). Here the spring force is plotted ($F = -kx$), and the work is negative. Note at the amplitude positions: $W = U = \tfrac{1}{2}kA^2$.

Note that when the oscillating mass passes through the origin, or equilibrium position ($x = 0$), the potential energy is zero. At that instant, all the energy is kinetic and the mass is traveling at its maximum speed (v_{max}). The energy expression for this case is

$$E = \tfrac{1}{2}kA^2 = \tfrac{1}{2}mv_{max}^2$$

and

$$v_{max} = \sqrt{\frac{k}{m}}(A) \qquad\qquad (14.7)$$

Example 14.1   Motion—Simple and Harmonic

A block with a mass of 0.25 kg sitting on a frictionless surface is connected to a light spring that has a spring constant of 180 N/m (see Fig. 14.1). If the block is displaced 15 cm from its equilibrium position and released, what are (a) the total energy of the system, (b) the maximum speed of the block, and (c) the speed of the block when it is 10 cm from its equilibrium position?

Solution

Given: $m = 0.25$ kg   Find: (a) E
          $k = 180$ N/m   (b) v_{max}
          $A = 15$ cm $= 0.15$ m   (c) v
          $x = 10$ cm $= 0.10$ m

(a) The total energy is given directly by Eq. 14.5:

$$E = \tfrac{1}{2}kA^2 = \tfrac{1}{2}(180\text{ N/m})(0.15\text{ m})^2 = 2.0\text{ J}$$

(b) The block has its maximum speed when it passes through the equilibrium position ($x = 0$), where all of its energy is kinetic energy. From Eq. 14.7,

$$v_{max} = \sqrt{\frac{k}{m}}(A) = \sqrt{\frac{180\text{ N/m}}{0.25\text{ kg}}}(0.15\text{ m}) = 4.0\text{ m/s}$$

(c) The instantaneous speed of the block at a distance of 10 cm from the equilibrium position is given by Eq. 14.6:

$$v = \sqrt{\frac{k}{m}(A^2 - x^2)} = \sqrt{\frac{180\text{ N/m}}{0.25\text{ kg}}[(0.15\text{ m})^2 - (0.10\text{ m})^2]}$$

$$= \sqrt{9.0\text{ m}^2/\text{s}^2}$$

$$= 3.0\text{ m/s} \quad\blacksquare$$

Example 14.2   Determination of the Spring Constant

When a 0.50-kg mass is suspended from a spring, the spring stretches 10 cm (Fig. 14.3). The mass is then pulled down another 5.0 cm and released. (a) What is the total energy of the oscillating system? (b) What is the highest position of the oscillating mass?

Solution

Given: $m = 0.50$ kg   Find: (a) E
          $y_0 = 10$ cm $= 0.10$ m   (b) $+A$
          $-A = 5.0$ cm $= 0.050$ m

Since this oscillation is vertical, the displacement can be designated by y, the coordinate commonly used for this direction.

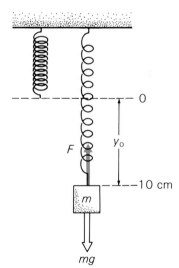

Figure 14.3 **Determination of the spring constant** When a mass suspended on a spring is in equilibrium, $F = ky_0 = mg$, and the spring constant (k) can be computed. See Example 14.2.

(a) You know that $E = \frac{1}{2}kA^2$, so the spring constant must be determined. When the suspended mass and the stretched spring are in equilibrium, the weight force and the spring force are equal and opposite, so

In the motion of a vertical spring, the mass oscillates through an equilibrium position equal to *mg/k*.

$$F = w$$

or $$ky_o = mg$$

Thus,

$$k = \frac{mg}{y_o} = \frac{(0.50 \text{ kg})(9.8 \text{ m/s}^2)}{0.10 \text{ m}} = 49 \text{ N/m}$$

The displacement of 5.0 cm from the equilibrium position is the amplitude of the oscillation, and

$$E = \frac{1}{2}kA^2 = \frac{1}{2}(49 \text{ N/m})(0.050 \text{ m})^2 = 1.2 \text{ J}$$

(b) Once set into motion, the mass oscillates up and down through the equilibrium position (y_o). Since the amplitude of the oscillation is 5.0 cm, the highest position of the mass will be 5.0 cm above y_o. ■

14.2 Equations of Motion

The equation of motion for an object or particle gives its position as a function of time. For example, the equation of motion with a constant linear acceleration is $x = v_o t + \frac{1}{2}at^2$, where v_o is the initial velocity (Chapter 2). However, the acceleration is not constant for simple harmonic motion, so the kinematic equations of Chapter 2 do not apply to this case.

The equation of motion for an object in simple harmonic motion can be derived by using a relationship between simple harmonic and uniform circular motion, illustrated in Fig. 14.4(a). Note that as the object moves in uniform circular motion, its shadow moves back and forth horizontally, following the same path as the mass on the spring, which is in simple harmonic motion. Since the shadow and the mass have the same position at any time, it follows that the equation of *horizontal* motion for the object in circular motion is the same as the equation of motion for horizontally oscillating mass on the spring.

From the reference circle in Fig. 14.4(b),

$$x = A \cos \theta$$

But the object moves with a constant angular velocity (ω), so $\theta = \omega t$ (Chapter 7):

$$x = A \cos \omega t \tag{14.8}$$

The angular velocity (in rad/s) is sometimes called the angular frequency, since $\omega = 2\pi f$, where f is the frequency of revolution or rotation. (Note in the figure that this is the same as the frequency of oscillation of the mass on the spring.) Thus,

$$x = A \cos 2\pi f t \tag{14.9}$$

Also, $f = 1/T$, so

$$x = A \cos \frac{2\pi t}{T} \tag{14.10}$$

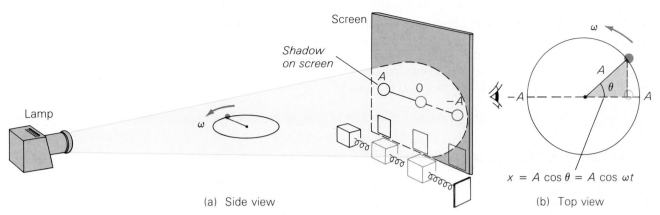

(a) Side view

(b) Top view

$x = A \cos \theta = A \cos \omega t$

Figure 14.4 Reference circle for horizontal motion
(a) The shadow of the object in uniform circular motion has the same horizontal motion as the mass on the spring in simple harmonic motion. (b) The motion can thus be described by $x = A \cos \theta = A \cos \omega t$.

Eqs. 14.8–14.10 are three forms of the equation of motion for a mass in simple harmonic motion. From Eq. 14.10, when $t = 0$, then $x = A$, and when $t = T$, then $x = A$ again. In other words, these three equations describe periodic motion, for which the period is T.

The period of a mass oscillating on a spring can also be determined in terms of the system parameters using the reference circle in Fig. 14.4(b). Note that in one revolution the object in circular motion travels a distance $d = vt$, so

$$2\pi A = v_{\text{o}} T \qquad \text{and} \qquad T = \frac{2\pi A}{v_{\text{o}}} \tag{14.11}$$

The mass oscillating on the spring has a speed of v_{o} when it passes through the origin, or equilibrium position, and by the conservation of energy

$$E = \tfrac{1}{2}kA^2 = K + U = \tfrac{1}{2}mv_{\text{o}}^2 + 0$$

Thus,

$$\frac{A}{v_{\text{o}}} = \sqrt{\frac{m}{k}}$$

Substituting $\sqrt{m/k}$ for A/v_{o} in Eq. 14.11 gives

$$T = 2\pi \sqrt{\frac{m}{k}} \tag{14.12}$$

Thus, the greater the mass, the greater the period; and the greater the spring constant (the stiffer the spring), the smaller the period.

Since $f = 1/T$,

$$f = \frac{1}{2\pi\sqrt{m/k}} = \frac{1}{2\pi}\sqrt{\frac{k}{m}} \tag{14.13}$$

Thus, the greater the spring constant (the stiffer the spring), the faster or more frequently the spring vibrates, as you might expect. This may help you remember that the frequency has k over m (rather than m over k, as for the period).

A simple pendulum (with no energy losses) undergoes simple harmonic motion for small angles of oscillation. The equation for the period of a pendulum in simple harmonic motion is similar in form to that for a mass oscillating on a spring. It can be shown that the period of a simple pendulum oscillating through a small angle ($\theta \leq 20°$) is given by

$$T = 2\pi \sqrt{\frac{L}{g}} \qquad\qquad (14.14)$$

where L is the length of the pendulum and g is the acceleration due to gravity.

Example 14.3   Applying the Equation of Motion
A mass on a spring oscillates horizontally on a frictionless surface with an amplitude of 15 cm and a frequency of 0.20 Hz. (a) If the equation of motion for the mass is $x = A \cos \omega t$, what is its displacement at $t = 3.1$ s? (b) How many oscillations have been made during this time?

Solution
Given: $A = 15$ cm $= 0.15$ m   *Find*: (a) x
           $f = 0.20$ Hz   (b) n
           $t = 3.1$ s
           $x = A \cos \omega t$

(a) First, the equation of motion can be written in the form $x = A \cos 2\pi f t$, since $\omega = 2\pi f$ and the frequency is given. At $t = 0$, $x = A$, so initially the mass was released from the amplitude position. Then, at $t = 3.1$ s,

$$x = A \cos 2\pi f t = (0.15 \text{ m})[\cos 2\pi(0.20 \text{ s}^{-1})(3.1 \text{ s})] = -0.11 \text{ m}$$

(b) The number of oscillations (cycles) is equal to the product of the frequency (cycles/s) and the elapsed time (s), which are both given:

$$n = f t = (0.20 \text{ cycle/s})(3.1 \text{ s}) = 0.62 \text{ cycle}$$

Thus, at $t = 3.1$ s, the mass has traveled to its negative amplitude position (in 0.50 cycle) and is back at $x = -0.11$ m approaching its equilibrium position ($x = 0$). ∎

PROBLEM SOLVING HINT

Note in part (a) of Example 14.3 that $\cos 2\pi(0.20 \text{ s}^{-1})(3.1 \text{ s}) = \cos 3.9$, where $\theta = 3.9$ rad. Don't forget to set your calculator to radians (instead of degrees) when finding the value of the cosine in equations for simple harmonic or circular motion.

Example 14.4   Period and Length of a Pendulum
A Foucault pendulum (see the Insight feature in Chapter 7) has a period of 6.35 s. What is the length of the pendulum?

Solution
Given: $T = 6.35$ s   *Find*: L

You can consider the Foucault pendulum to be a simple pendulum and find its length from Eq. 14.14. Squaring both sides of that equation and solving for L gives

$$L = \frac{T^2 g}{4\pi^2} = \frac{(6.35 \text{ s})^2 (9.80 \text{ m/s}^2)}{4\pi^2} = 10.0 \text{ m} \quad \blacksquare$$

A vertical reference circle can also be used to describe simple harmonic motion, as illustrated in Fig. 14.5 for a suspended mass oscillating on a spring. In this case, the vertical displacement is given by $y = A \sin \theta$. A development similar to that for the horizontal case leads to this equation of motion:

$$y = A \sin \omega t$$

Since $y = 0$ at $t = 0$, the mass initially started its upward motion from that point.

However, if the mass starts from a different position, as shown in Fig. 14.6, the resulting curve is a cosine, and

$$y = A \cos \omega t$$

Note that in this case the mass is released from $x = +A$ at $t = 0$.

Thus, the equation of motion for an oscillating mass may be either a sine or a cosine function, and both of these functions are referred to as being sinusoidal. That is, simple harmonic motion is described by a sinusoidal function of time.

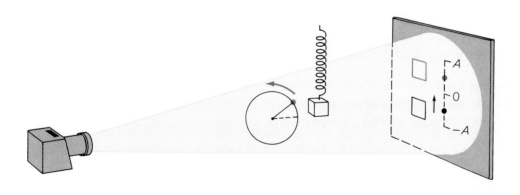

(a)

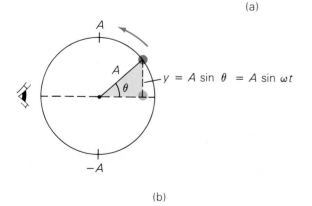

$y = A \sin \theta = A \sin \omega t$

(b)

Figure 14.5 **Reference circle for vertical motion**
(a) The shadow of the object in uniform circular motion has the same vertical motion as the mass on the spring in simple harmonic motion.
(b) The motion can thus be described by
$y = A \sin \theta = A \sin \omega t$.

Initial Conditions   You may be wondering how to decide whether to use a sine or cosine function to describe a particular case of simple harmonic motion. In general, the form of the function is determined by the initial displacement of the oscillating mass, which is one of the initial conditions for the system. The others are the system parameters, or the properties of the system that determine ω, f, or T. For a mass oscillating on a spring, these quantities depend on the system parameters k and m, which describe the particular spring and mass. Given k and m for such a system, the numerical value of ω, f, or T can be computed:

$$\omega = 2\pi f = \frac{2\pi}{T} = \frac{\sqrt{k}}{m}$$

The initial displacement determines whether a sine or cosine function is used as the equation of motion. This initial condition is the value of the displacement at $t = 0$ and tells how the system was initially set into motion. If an object in simple harmonic motion has an initial displacement of $y = 0$ at $t = 0$, then $y = A \sin \omega t$. That is, a sine curve satisfies this initial condition, since $y = A \sin \omega(0) = 0$. If $y = A \cos \omega t$ were used, at $t = 0$, $y = A \cos 0 = A$, which is not the case ($\cos 0 = 1$). The cosine function does not satisfy the initial condition. The situation for which a cosine function is appropriate in the equation of motion is shown in Fig. 14.6. Note that $y = A$ at $t = 0$, so $y = A \cos \omega t$ satisfies this initial condition.

For the general case,

$$y = A \sin(\omega t + \delta) \tag{14.15}$$

where $(\omega t + \delta)$ is the phase angle and δ is the **phase constant**. The phase constant essentially matches the appropriate sinusoidal function to the motion (Fig. 14.7, Demonstration 13). For $\delta = 0$, $y = A \sin \omega t$, and the object is set into motion at the equilibrium position. For $\delta = 90°$ (or $\pi/2$ rad), the equation of motion is $y = A \cos \omega t$, and a mass oscillating up and down on a spring would be initially released from the $+A$ position.* The curves for $\delta = 0$ and $\delta = 90°$ (or $\pi/2$ rad) are said to be $90°$ out of phase, or shifted by a quarter cycle, with respect to one another. The cases where $\delta = 180°$ (or π rad) and $\delta = 270°$ (or $3\pi/2$ rad) are also out of phase by $90°$, as illustrated in Fig. 14.7.

Two masses oscillating in phase (having the same δ) will oscillate together. Two that are completely out of phase ($180°$ difference in δ) will always be going in opposite directions or be at opposite amplitudes.

If the initial displacement is not zero or $+A$, then δ is not a multiple of $90°$, and things get a bit more complicated. The equation of motion will have both a sine term and a cosine term (this can be shown using the trigonometric identity given in the footnote).

Keep in mind that the phase angle is usually expressed in radians. For example, if $y = \sin 0.50t$, then at $t = 20$ s, $y = \sin(0.50)(20) = \sin 10$, or the sine of 10 rad. If you're using a calculator for such computations, make sure it is in the "rad" mode and not the "deg" mode.

---

* In general, $\sin(a + b) = \sin a \cos b + \cos a \sin b$, so $y = A \sin(\omega t + 90°) = A [\sin \omega t \cos 90° + \cos \omega t \sin 90°] = A \cos \omega t$, since $\cos 90° = 0$ and $\sin 90° = 1$.

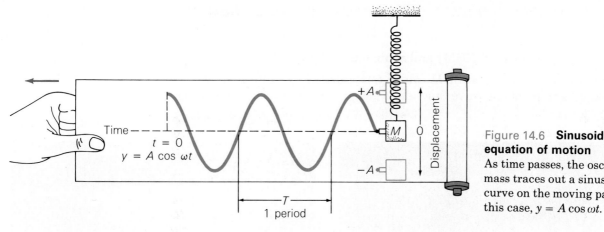

Figure 14.6 **Sinusoidal equation of motion**
As time passes, the oscillating mass traces out a sinusoidal curve on the moving paper. In this case, $y = A \cos \omega t$.

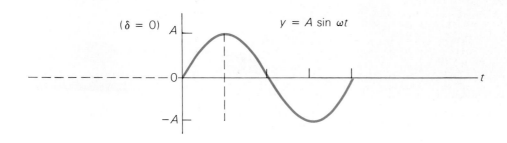

$(\delta = 0)$ $y = A \sin \omega t$

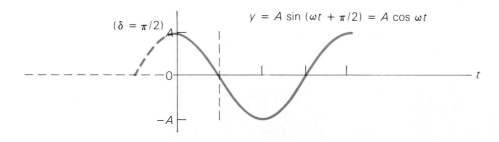

$(\delta = \pi/2)$ $y = A \sin (\omega t + \pi/2) = A \cos \omega t$

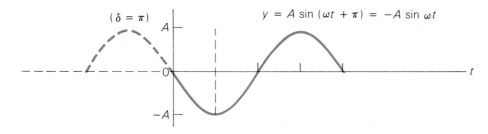

$(\delta = \pi)$ $y = A \sin (\omega t + \pi) = -A \sin \omega t$

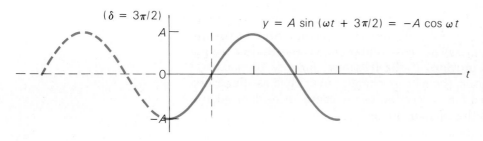

$(\delta = 3\pi/2)$ $y = A \sin (\omega t + 3\pi/2) = -A \cos \omega t$

Figure 14.7 **Phase differences**
With the general equation $y = A \sin(kx + \delta)$, motions are described by sine or cosine terms for the phase constants shown. The initial displacement determines δ. Each curve is 90° (or $\pi/2$ rad) out of phase with the preceding one; note that this is equivalent to shifting the wave a quarter of a cycle.

Demonstration 13

Simple Harmonic Motion (SHM) and Sinusoidal Wave

A demonstration to show that SHM can be represented by a sinusoidal wave function. A "graph" of the wave function is generated with an analogue of a strip chart recorder.

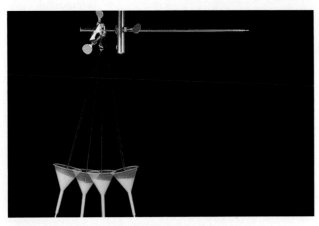

(a) A salt-filled funnel oscillates from a bi-filar suspension.

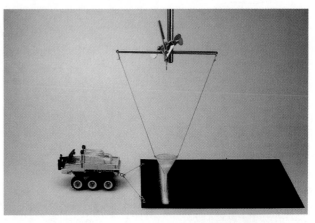

(b) The salt falls on a black-painted poster board that will be pulled in a direction perpendicular to the plane of the funnel's oscillation.

(c) Away we go.

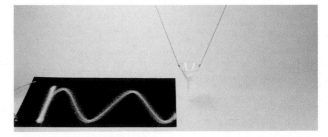

(d) The salt trail traces out a plot of displacement versus time, or $y = A \sin(\omega t + \delta)$. Note that in this case the phase constant is about $\delta = 90°$, and $y = A \cos \omega t$. (Why?)

Simple harmonic motion with a constant amplitude implies that there are no losses of energy, but in practical applications there are always some frictional losses. Therefore, to maintain a constant-amplitude motion, energy must be added to the system by some driving force (Fig. 14.8). Without a driving force, the amplitude and the energy of an oscillator decrease with time, giving rise to **damped harmonic motion**. If the frictional force is proportional to the velocity of the oscillator, the motion changes over time as illustrated in Fig. 14.8. The time required for the oscillations to cease, or be damped out, depends on the magnitude of the damping force.

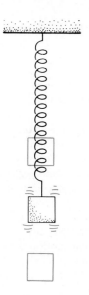

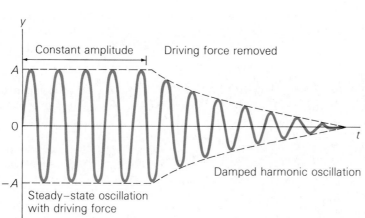

Constant amplitude   Driving force removed

Steady–state oscillation with driving force

Damped harmonic oscillation

Figure 14.8 Damped harmonic motion
When a driving force adds energy to a system equal to its losses, the oscillation is steady with a constant amplitude. When the driving force is removed, the oscillation decays, and its amplitude decreases exponentially with time (is damped).

In many applications involving continuous periodic motion, damping is unwanted and necessitates energy input. However, in some instances, damping is desirable. For example, the dial in a spring-operated bathroom scale oscillates briefly before stopping at the weight. If not properly damped, these oscillations would continue for some time (you would have a wait before you could read your weight). The necessity for damping also applies to shock absorbers on automobiles and needle indicators on instruments measuring electrical quantities.

Velocity and Acceleration of SHM

Expressions for velocity and acceleration with simple harmonic motion may also be obtained from reference circles (see Fig. 14.9). For vertical simple harmonic motion,

$$y = A \sin \omega t$$

The speed for the object in uniform circular motion is given by $v_o = A\omega = A2\pi f$, since $v = r\omega$. But the velocity of the object is continually changing (direction), and its y component, corresponding to the velocity for the vertical simple harmonic motion, is

$$v = v_o \cos \omega t$$

or

$$v = A\omega \cos \omega t \tag{14.16}$$

The acceleration of the object in uniform circular motion is the centripetal acceleration. It is directed toward the center of the circle and has a magnitude of

$$a_c = \frac{v_o^2}{A} = \frac{(\omega A)^2}{A} = A\omega^2$$

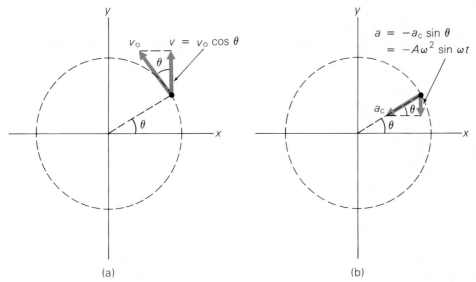

The *y* component of this acceleration is equal to the acceleration for the vertical simple harmonic motion:

$$a = -A\omega^2 \sin \omega t = -\omega^2 y \qquad (14.17)$$

Note that the functions for the velocity and acceleration are out of phase with that for the displacement (see Fig. 14.7). Since the velocity is 90° out of phase with the displacement, the velocity is greatest when the oscillating mass is at (passing through) its equilibrium position. The acceleration is 180° out of phase with the displacement (as indicated by the minus sign on the right-hand side of Eq. 14.17). Therefore, the acceleration is a maximum when the displacement is a maximum, or when the mass is at an amplitude position. At any position except the equilibrium position, the directional sign of the acceleration is the opposite of that of the displacement, as it should be for an acceleration resulting from a restoring force. At the equilibrium position, the displacement and acceleration are both zero. (Why?)

14.3 Wave Motion

The world is full of waves of various types—some examples are water waves, shock waves, sound waves, waves generated by earthquakes, and light waves. Any type of wave results from a disturbance. Thus, **wave motion** is the propagation of a disturbance, a process of energy transfer (Fig. 14.10). This

Figure 14.10 **Energy transfer**
The propagation of a disturbance or a transfer of energy is seen in a row of falling dominos.

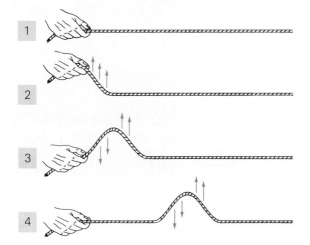

Figure 14.11 **Wave pulse**
The hand disturbs the stretched rope, and a wave pulse propagates along it. Note that the rope "particles" move up and down as the pulse passes.

chapter will consider mechanical waves, or those that are propagated in some material. (Light waves do not require a propagating medium and will be considered in more detail in later chapters.)

The transfer of energy from one place to another by mechanical wave motion involves the movement of particles in the propagating medium. For example, if the end of a stretched rope is given a quick shake, the disturbance transfers energy from the hand to the rope, as illustrated in Fig. 14.11. The forces acting between the rope "particles" cause them to move in response to the motion of the hand, and a **wave pulse** travels down the rope. Each "particle" goes up and then back down as the pulse passes by.

A continuous wave motion, or **periodic wave**, requires a constant disturbance from an oscillating source (Fig. 14.12). In this case, the particles move up and down continuously. If the driving source is such that a constant amplitude is maintained (and the restoring force has the form of Hooke's law), the particle motion can be described as simple harmonic motion. Also, the wave itself will have a sinusoidal form in both space and time. Being sinusoidal in space means that if you took a photograph of the wave at any instant (freezing it in time), you

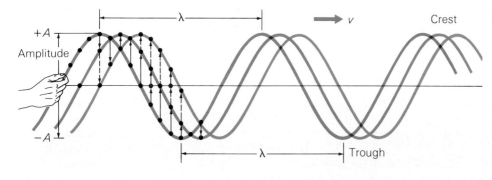

Figure 14.12 **Periodic wave**
A continuous disturbance can set up a sinusoidal wave in a stretched rope or string, and the wave travels down the rope with a wave speed v. Note that the rope "particles" oscillate in simple harmonic motion. The distance between two adjacent identical points on the waveform is the wavelength.

would see a sinusoidal waveform. Also, if you looked through a thin slit at a wave passing by, you would see a particle oscillating up and down in the slit, sinusoidally with time. This would be like looking through a slit at the moving paper in Fig. 14.6.

Specific characteristics of sinusoidal waves are used to describe them. As for a particle in simple harmonic motion, the **amplitude** of a wave is the maximum displacement and corresponds to the height of a crest or the depth of a trough (see Fig. 14.12). The distance between two successive crests (or troughs) is called the **wavelength** (λ). Actually, it is the distance between any two successive particles that are in phase (at identical points on the wave form). The crest and trough positions are usually used for convenience. Note that a wavelength corresponds spatially to one cycle. The **frequency** of a wave is the number of cycles per second, that is, the number of complete waveforms, or wavelengths, that pass by a given point during a second. Keep in mind that the wave is moving. The period $T = 1/f$ is the time for one complete waveform (a wavelength) to pass by a given point.

Since a wave moves, it has a **wave speed** (or velocity if direction is specified). You should be able to convince yourself that the wave (or a particular point, such as a crest) travels a distance of one wavelength in a time of one period. Then, since $v = d/t$,

$$v = \frac{\lambda}{T} = \lambda f \qquad (14.18)$$

Note that the dimensions are correct (length/time). In general, the wave speed depends on the nature of the medium.

Example 14.5 Wave Speed

A person on a pier observes incoming waves that have a sinusoidal form with a distance of 1.6 m between the crests. If a wave laps against the pier every 4.0 s, what are (a) the frequency and (b) the speed of the waves?

Solution

Given: $\lambda = 1.6$ m   *Find:* (a) f
$\qquad\quad T = 4.0$ s   (b) v

(a) The lapping indicates the arrival of a wave crest, so 4.0 s is the period of the wave (time it takes to travel one wavelength, or the crest-to-crest distance). Then

$$f = \frac{1}{T} = \frac{1}{4.0 \text{ s}} = 0.25 \text{ s}^{-1} = 0.25 \text{ Hz}$$

(b) The frequency or the period can be used in Eq. 14.18 to find the wave speed:

$$v = \lambda f = (1.6 \text{ m})(0.25 \text{ s}^{-1}) = 0.40 \text{ m/s}$$

or

$$v = \frac{\lambda}{T} = \frac{1.6 \text{ m}}{4.0 \text{ s}} = 0.40 \text{ m/s} \quad \blacksquare$$

Waves may be divided into two types based on the direction of the particles' oscillations relative to the wave velocity. A **transverse wave** is one for which

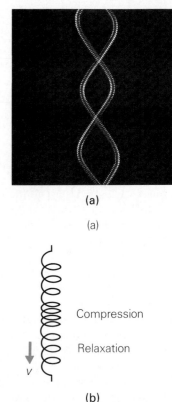

(a)

(a)

Compression

Relaxation

v

(b)

Figure 14.13 **Transverse and longitudinal waves** (a) The particle motion is perpendicular to the direction of the wave velocity in a transverse wave, as shown here in a vibrating guitar string. (b) The particle motion is parallel to (or *along*) the direction of the wave velocity in a longitudinal wave.

the particle motion is perpendicular to the direction of the wave velocity. The wave produced in a stretched string (Fig. 14.12) is an example of a transverse wave, as is the wave shown in Fig. 14.13(a). A transverse wave is sometimes called a shear wave because the disturbance supplies a force that tends to shear the medium. Shear waves can propagate only in solids, since a liquid or a gas cannot support a shear. That is, a liquid or a gas does not have sufficient restoring forces between its particles to propagate a transverse wave.

In a **longitudinal wave**, the particle oscillation is parallel to the direction of the wave velocity. (A memory aid is that *long*itudinal means particle motion a*long* the direction of the velocity.) A longitudinal wave may be produced in a stretched spring by moving the coils back and forth along the spring axis [Fig. 14.13(b)]. Alternating compression and relaxation pulses move along the spring. A longitudinal wave is sometimes called a compressional wave. Sound waves are longitudinal waves. A periodic disturbance produces compressions in the air. The intervening relaxations are called rarefactions because the density of the air in these regions is reduced, or rarefied.

Longitudinal waves can propagate in solids, liquids, and gases. All phases of matter can be compressed to some extent. The propagations of transverse and longitudinal waves in different media give information about the Earth's interior structure, as is discussed in the Insight feature.

The sinusoidal profile of water waves might make you think that these are transverse waves. Actually, they reflect a combination of longitudinal and transverse motions (Fig. 14.14). The particle motion may be nearly circular at

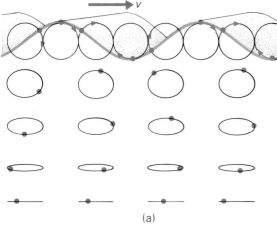

(a)

Figure 14.14 **Water waves**
Water waves are a combination of longitudinal and transverse motions. (a) At the surface, the water particles move in circles, but their motion becomes more longitudinal with depth. (b) When a wave approaches shore, the lower particles are forced into steeper paths until finally the wave breaks and forms a surf.

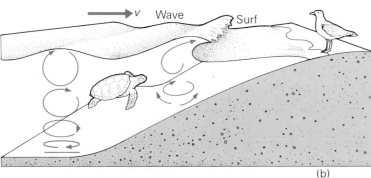

(b)

Earthquakes and Seismology

The Earth's interior structure is still something of a mystery. The deepest mine shafts and drillings extend only a few kilometers into the Earth. Using waves to probe the Earth's structure is one way to investigate it further. Appropriately, waves generated by earthquakes have proved to be especially useful for this purpose.

Earthquakes are caused by the sudden release of built-up stress along cracks and faults, such as the famous San Andreas Fault in California. The geological theory of plate tectonics views the outer layer of the Earth as being a series of rigid plates, or huge slabs of rock, that are in (very slow) motion relative to one another. Stresses are continually being built up, particularly along boundaries between plates.

The energy from a stress-relieving disturbance propagates outwardly in the form of seismic waves. These are of two general types: surface waves and body waves. The surface waves, which move along the Earth's surface, account for most earthquake damage. Body waves, as the name implies, travel through the Earth. There are both longitudinal and transverse vibrations. The compressional (longitudinal) waves are called P waves, and shear (transverse) waves are called S waves (see Fig. 1). The P and S stand for primary and secondary and indicate the waves' relative speeds or actually their arrival times at monitoring stations. Primary waves travel through materials faster than do secondary waves and are detected first. An earthquake's rating on the Richter scale is related to the amplitude or energy of the seismic waves.

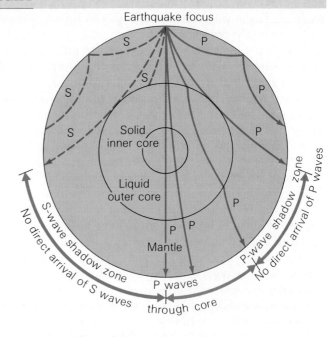

Figure 1 **Earth waves**
Earthquakes produce waves that travel through the Earth. Because the transverse S waves are not detected on the opposite side of the Earth, scientists believe that part of the Earth's interior is liquid.

the surface but becomes more elliptical with depth, eventually becoming longitudinal. A hundred meters or so below the surface of a large body of water, the wave disturbances have little effect. For example, a submarine at these depths is undisturbed by large waves on the ocean's surface. As a wave approaches shallower water near shore, the water particles have difficulty completing their elliptical paths. Finally, when the water becomes too shallow, the particles can no longer move through the bottom parts of their paths and the wave breaks. Its crest falls forward to form surf.

14.4 Wave Phenomena

Among the phenomena common to all waves are interference, superposition, reflection, refraction, and diffraction.

Seismic stations around the world monitor these waves with sensitive detecting instruments called seismographs (Fig. 2). With this data, the paths of the waves through the Earth can be mapped, giving knowledge of the interior structure. The Earth's interior seems to be divided into three general regions: the crust, the mantle, and the core, which has a solid inner part and a liquid outer part.* The locations of the boundaries of these regions are determined by the refraction, or bending, of the waves (see Section 14.4).

Longitudinal waves can travel through solids or liquids, but transverse waves can travel only through solids. When an earthquake occurs at a particular location, P waves are detected on the other side of the Earth and S waves are not (see Fig. 1). The absence of S waves in a shadow zone leads to the conclusion that the Earth must have a region near its center that is in the liquid phase. This region is a highly viscous metallic liquid, but definitely a liquid since it does not support a shear (transverse waves are not propagated). When the transmitted P waves enter and leave the liquid region, they are refracted (bent). This gives rise to a P wave shadow zone, which indicates that only the outer part of the core is liquid.

As you will learn in Chapter 19, the combination of a liquid outer core and rotation may be responsible for the Earth's magnetic field.

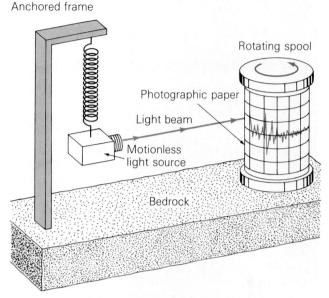

Figure 2 A simple seismograph
The recorded amplitude of the vibrations is proportional to the energy of the waves and is expressed on the Richter scale, which is used to report the relative magnitudes of earthquakes.

* The crust is about 24–30 km (15–20 mi) thick; the mantle is 2900 km (1800 mi) thick; and the core has a radius of 3450 km (2150 mi). The solid inner core has a radius of about 1200 km (750 mi).

Interference and Superposition

Strange as it may seem, when two or more waves meet, or pass through the same region of a medium, they pass through each other and proceed without being altered. While they are in the same region, the waves are said to be interfering.

What happens during interference, or what does the combined waveform look like? The relatively simple answer to this is given by the **principle of superposition**:

> At any time, the combined waveform of two or more interfering waves is given by the sum of the displacements of the individual waves at each point in the medium.

This principle is illustrated in Fig. 14.15. The displacement of the combined waveform at any point is given by $y = y_1 + y_2$, where y_1 and y_2 are the displacements of the individual pulses at that point (directions are indicated by plus and minus signs). Interference, then, is the physical addition of waves.

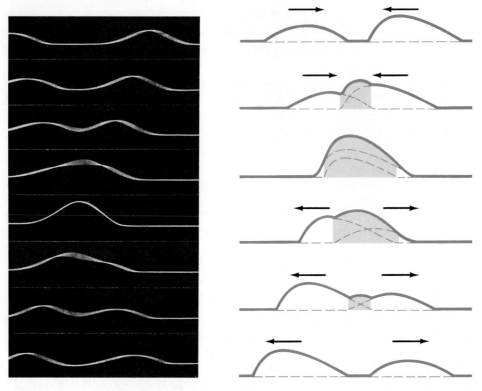

Figure 14.15 **Principle of superposition**
When two waves meet, they interfere. The displacement at any point on the combined wave is equal to the sum of the displacements of the individual waves.

In Fig. 14.15, the vertical displacements of the two pulses are in the same direction, and the amplitude of the combined waveform is greater than that of either pulse. This is called **constructive interference**. On the other hand, if two pulses tend to cancel each other when they overlap (one pulse has a negative displacement), the amplitude of the combined waveform is smaller than that of either pulse. This is called **destructive interference**.

Special cases of constructive and destructive interference are shown in Fig. 14.16. These occur when waves with the same frequency and amplitude meet. When these interfering waves are exactly in phase (crest coincides with

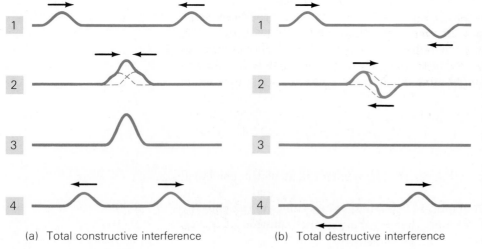

(a) Total constructive interference

(b) Total destructive interference

Figure 14.16 **Interference**
(a) When two waves of the same frequency and amplitude meet and are in phase, they interfere constructively (shown here with wave pulses). When the waves are exactly superimposed, this is referred to as total constructive interference. (b) When the interfering waves are completely out of phase (180°) and exactly superimposed, total destructive interference occurs.

crest), the amplitude of the combined waveform is twice that of either individual wave. This is sometimes referred to as **total constructive interference**. When these interfering waves are completely out of phase (180° difference, or crest coinciding with trough), the waveforms disappear; that is, the amplitude of the combined wave is zero. This is called **total destructive interference**.

The word "destructive" unfortunately tends to imply that the energy as well as the form of the waves is destroyed. This is not the case. At the point of total destructive interference, the wave energy is stored in the medium. Also, keep in mind that the total destructive interference of waves is an instantaneous condition. As time progresses, the waveforms reappear.

QUESTION: Since the energy of a wave is proportional to the square of its amplitude ($E \propto A^2$), wouldn't the exact superposition of two waves of equal amplitude give a quadrupling of the energy rather than a doubling? That is, $y = y_1 + y_2 = A + A = 2A$, and $E \propto (2A)^2 = 4A^2$.

ANSWER: This would be a violation of the law of the conservation of energy. Consider two identical pulses traveling toward each other in a stretched string. The potential energy is associated with the stretching of the string, and the kinetic energy with its vertical motion. When the pulses start to overlap, the addition of vertical displacements in the overlapping region gives a horizontal segment that is no longer stretched but is moving vertically with twice the original velocity. Thus, the potential energy in this region is zero, but the kinetic energy increases fourfold. As the pulses overlap further, there will be regions near the edges of the horizontal segment where the string is vertical and the velocity is zero. There the potential energy is quadrupled, while the kinetic energy is zero. For identical pulses, then, either the kinetic energy or the potential energy is zero in each overlapping region, so the total energy doubles.*

Reflection, Refraction, and Diffraction

Besides meeting other waves, waves can (and do) meet objects or a boundary with another medium. In such cases, several things may occur. One of these is reflection. **Reflection** occurs when a wave strikes an object or comes to a boundary of another medium and is at least partly diverted backward. An echo is an example of reflection (of sound waves), and mirrors reflect light waves.

Two cases of reflection are illustrated in Fig. 14.17. If the end of the string is fixed, the reflected pulse is inverted, or undergoes a 180° phase shift [Fig. 14.17(a)]. This is because the pulse causes the string to exert an upward force on the wall, and the wall exerts an equal and opposite downward force on the string (by Newton's third law). Thus, the reflected pulse is downward, or inverted. If the end of the string is free to move, then the reflected pulse is not inverted. This is illustrated in Fig. 14.17(b), where the string is attached to a light ring that can move freely on a smooth pole. The ring is accelerated upward by the incoming pulse and is then brought downward.

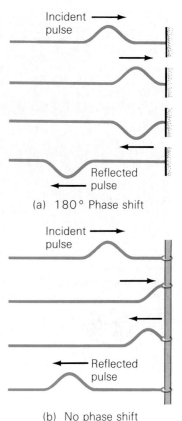

(a) 180° Phase shift

(b) No phase shift

Figure 14.17 **Reflection** (a) When a wave in a string is reflected from a fixed boundary, the reflected wave is inverted, or undergoes a 180° phase shift. (b) If the string is free to move at the boundary, there is no phase shift of the reflected wave.

---

* This explanation was provided by Peter Palff-Muhoray of the Liquid Crystal Institute, Kent State University, Kent, Ohio (in personal correspondence). He suggests sketching triangular wave pulses to see the horizontal and vertical effects. For his explanation of electromagnetic pulses, see *American Journal of Physics*, vol. 57, no. 3 (March, 1989), p. 201.

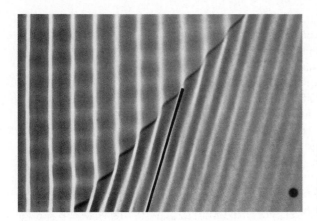

Figure 14.18 **Refraction**
Water waves are refracted in a ripple tank at the boundary between shallow and deep sections of the tank.

In general, when a wave strikes a boundary, some of its energy is reflected and some is transmitted or absorbed. When a wave crosses a boundary into another medium, its velocity changes because the new material has different characteristics. Thus, the transmitted wave may move in a direction different from that of the incident wave. This phenomenon is called **refraction**. An example of refraction for water waves is shown in Fig. 14.18. Here the velocity change is due to a change in depth rather than a change in medium.

Diffraction refers to the bending of waves around an edge of an object. For example, if you stand along an outside wall of a building near the corner, you can hear people talking around the corner. Assuming there are no reflections or air motion (wind), this would not be possible if the sound waves traveled in a straight line [Fig. 14.19(a)].

In general, the effects of diffraction are greater when the size of the diffracting object or opening is about the same or smaller than the wavelength of the waves. For example, water waves in a pond or lake pass by blades of grass or reeds with little noticeable effect because the widths of these objects are much smaller than the wavelength of the waves. The dependence of diffraction on wavelength and size of the object is illustrated in Fig. 14.19(b) using a ripple tank.

Reflection, refraction, and diffraction will be considered for light waves in Chapter 22.

14.5 Standing Waves and Resonance

If you shake one end of a stretched rope, waves travel down it to the fixed end and are reflected back. The waves going down and back interfere. In most cases, the combined waveforms will have a changing, jumbled appearance. But if the rope is shaken at just the right frequency, a waveform appears to stand in place along the rope. Appropriately, this phenomenon is called a **standing wave** (Fig. 14.20, Demonstration 14).

Some points on the rope are stationary and are called **nodes**. By the principle of superposition, the interfering waves must cancel each other completely at these points; that is, a crest exactly coincides with a trough, and the rope does not undergo a displacement. At all other points, the rope

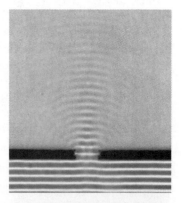

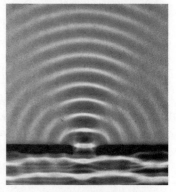

Figure 14.19 **Diffraction**
Diffraction effects are greater when the opening (or object) is about the same size as or smaller than the wavelength of the waves. This is shown here for water waves.

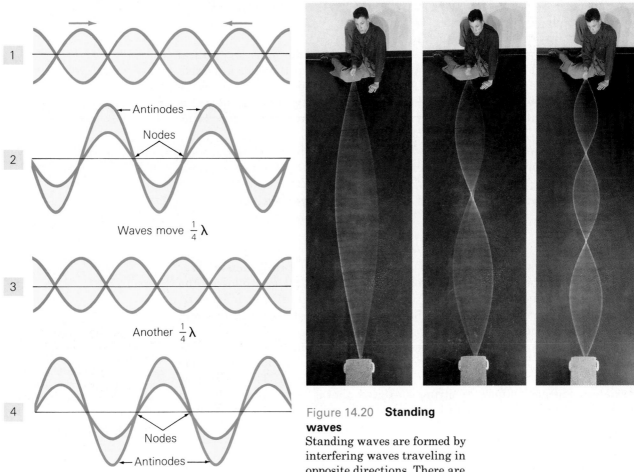

Figure 14.20 Standing waves
Standing waves are formed by interfering waves traveling in opposite directions. There are conditions of destructive and constructive interference as each wave travels a distance of $\lambda/4$. This gives rise to a standing wave with stationary nodes and maximum amplitude at antinodes.

oscillates back and forth with the same frequency. The points of maximum amplitude, where constructive interference is greatest, are called **antinodes**. As you can see in Fig. 14.20, adjacent antinodes are separated by half of a wavelength ($\lambda/2$), or one loop; adjacent nodes are also separated by half of a wavelength. The wavelength is that of the interfering waves that produce the standing wave.

Standing waves can be generated in a rope by more than one driving frequency; the higher the frequency, the greater the number of oscillating half-wavelength loops in the rope. The frequencies at which large-amplitude standing waves are produced are called **natural frequencies**, or **resonant frequencies**. The standing wave patterns are called normal, or resonant, modes of vibration. In general, all things have one or more natural frequencies, which depend on such factors as mass, elasticity or restoring force, and geometry (boundary conditions). The natural frequencies of a system are sometimes called its characteristic frequencies.

A stretched string or rope can be analyzed to determine its natural frequencies (Fig. 14.21). The boundary conditions are that the ends are fixed; thus, there must be a node at each end. The number of loops of a standing wave that will fit between the nodes at the ends (along the length of the string) is equal to an integral number of half-wavelengths. Note that $L = \lambda_1/2$, $L = 2\lambda_2/2$, $L = 3\lambda_3/2$, $L = 4\lambda_4/2$, and so on. In general,

$$L = n\frac{\lambda_n}{2} \quad \text{or} \quad \lambda_n = \frac{2L}{n} \quad \text{(for } n = 1, 2, 3, \ldots\text{)}$$

The natural frequencies of oscillation, where v is the wave speed, are

$$f_n = \frac{v}{\lambda_n} = \frac{nv}{2L} \quad \text{(for } n = 1, 2, 3, \ldots\text{)} \tag{14.19}$$

The lowest natural frequency ($f_1 = v/2L$) is called the **fundamental frequency**. All of the other natural frequencies are integral multiples of the fundamental frequency: $f_n = nf_1$ ($n = 1, 2, 3, \ldots$). The set of frequencies f_1, $f_2 = 2f_1$, $f_3 = 3f_1$, and so on, is called a harmonic series; f_1 is the **first harmonic**, f_2 the **second harmonic**, and so on.

Strings fixed at each end are found in musical instruments such as violins and guitars. When such a string is excited, the resulting vibration will include several harmonics in addition to the fundamental frequency. The number of harmonics depends on how the string is excited, that is, plucked or bowed. In any case, it is the combination of harmonic frequencies that gives a particular

A musical note or tone is referenced to the fundamental vibrational frequency. In musical terms, the first overtone is the second harmonic, the second overtone is the third harmonic, and so on.

Demonstration 14

Flame Standing Wave Pattern

A demonstration of a different type of standing wave form, where nodes and antinodes correspond to pressure minima and maxima, respectively, in the gas column.

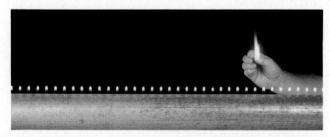

(a) Regularly spaced holes in a piece of downspout allow a line of uniformly sized flames when (natural) gas is supplied to the pipe. One end of the pipe is closed by a metal plate and the other end is fitted with a rubber diaphragm and loudspeaker.

(b) When an audio oscillator driving the speaker is tuned to give standing waves in the pipe, the presence of pressure maxima (antinodes) and minima (nodes) are indicated by higher and lower flames, respectively. [See *The Physics Teacher*, *17*, 307 (1979).]
Note: some fast, contemporary music with good bass instead of an oscillator output produces a rapidly varying, dancing flame pattern.

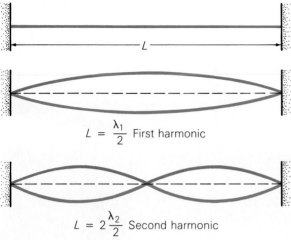

$L = \dfrac{\lambda_1}{2}$ First harmonic

$L = 2\dfrac{\lambda_2}{2}$ Second harmonic

$L = 3\dfrac{\lambda_3}{3}$ Third harmonic

Figure 14.21
(resonant) frequencies
A stretched string can have standing waves only at certain frequencies. These correspond to the numbers of half-wavelength loops that will fit along the length of string between the nodes at the fixed ends.

instrument its characteristic sound quality (Chapter 15). As Eq. 14.19 shows, the fundamental frequency of a stretched string, as well as the other harmonics, depends on the length of the string. (Think of how different notes are obtained on a particular string on a violin or a guitar.)

Natural frequencies also depend on other parameters, such as mass and force, which affect the wave speed. For a stretched string, the wave speed (v) can be shown to be

$$v = \sqrt{\frac{F}{\mu}} \tag{14.20}$$

where F is the tension force in the string and μ is the linear mass density (mass per unit length, $\mu = m/L$). Thus, Eq. 14.19 can be written

$$f_n = \frac{nv}{2L} = \frac{n}{2L}\sqrt{\frac{F}{\mu}} \qquad \text{(for } n = 1, 2, 3, \ldots) \tag{14.21}$$

Note that the greater the linear mass density of a string, the lower its natural frequencies are. As you may know, the low-note strings on a violin or guitar are thicker, or more massive, than the high-note strings.

Example 14.6   Fundamental Frequency
A particular string in a piano has an effective length of 1.15 m and a mass of 20.0 g and is under a tension of 6.30×10^3 N. (a) What is the fundamental frequency of the tone produced when the string is struck? (b) What are the frequencies of the next two harmonics?

Solution
Given: $L = 1.15$ m   *Find*: (a) f_1
$\quad\quad\quad m = 20.0$ g $= 0.0200$ kg   (b) f_2 and f_3
$\quad\quad\quad F = 6.30 \times 10^3$ N

Unwanted Resonance

When a large number of soldiers march over a small bridge, they are often ordered to break step, because it is possible that the marching frequency will correspond to one of the natural frequencies of the bridge and set it into resonant vibration. One famous incidence of bridge vibration wasn't due to marching soldiers, however, but to the driving force of the wind.

On the morning of November 7, 1940, winds gusting to 40–45 mi/h started the main span of the Tacoma Narrows Bridge (in Washington state) vibrating. The bridge had been opened to traffic only 4 months earlier. It was 2800 ft (855 m) long and 39 ft (12 m) wide. The steel girders used in its construction were 8 ft (2.4 m) long.

During the first month of use, small transverse modes of vibration had been observed. But on November 7, the wind drove the bridge in near resonance, and the main span vibrated with a frequency of 36 Hz and an amplitude of 1.5 ft. At 10 a.m., the main span began to vibrate in a torsional (twisting) mode in two segments with a frequency of 14 Hz. This was apparently due to a loosening of the cables by which the roadway was suspended. The wind continued to drive the bridge in resonance and the vibrational amplitude increased. Shortly after 11 a.m., the main span collapsed (Fig. 1).

"Galloping Gertie" (the nickname given to the bridge)

Figure 1 Galloping Gertie
The collapse of the Tacoma Narrows Bridge on November 7, 1940.

was rebuilt using the same tower foundations. However, the new design made the structure stiffer to increase its resonant frequency so that high winds could not produce unwanted resonance.

(a) The linear mass density of the string is

$$\mu = \frac{m}{L} = \frac{0.0200 \text{ kg}}{1.15 \text{ m}} = 0.0174 \text{ kg/m}$$

Then, using Eq. 14.21,

$$f_1 = \frac{1}{2L} \sqrt{\frac{F}{\mu}} = \frac{1}{2(1.15 \text{ m})} \sqrt{\frac{6.3 \times 10^3 \text{ N}}{0.0174 \text{ kg/m}}} = 262 \text{ Hz}$$

This is approximately the frequency of middle C (C_4) on a piano (261.6 Hz).
(b) Since $f_2 = 2f_1$ and $f_3 = 3f_1$,

$$f_2 = 2f_1 = 2(262 \text{ Hz}) = 524 \text{ Hz}$$

and

$$f_3 = 3f_1 = 3(262 \text{ Hz}) = 786 \text{ Hz}$$

The second harmonic corresponds approximately to C_5 on a piano, since the frequency doubles with each octave (or every eighth white key). ∎

When an oscillating system is driven at one of its natural, or resonant, frequencies, maximum energy transfer to the system occurs. The system is physically suited to a natural frequency, or wants to vibrate at it, so to speak. The condition of vibrating at a natural frequency is referred to as **resonance**.

A common example of a system in mechanical resonance is someone being pushed on a swing. Basically, a swing is a simple pendulum and has only one resonant frequency for a given length. $[f = (1/2\pi)\sqrt{g/L}]$. If you push the swing with this frequency, the amplitude and energy increase. (What happens if you push at a slightly different frequency? The energy transfer is no longer a maximum.)

Unlike a simple pendulum, a stretched string has many natural frequencies. A standing wave will be set up in such a string by a disturbance at any frequency. However, if the frequency of the driving force is not equal to one of the natural frequencies, the amplitude of the standing wave will be relatively small. When the frequency of the driving force is at one of the natural frequencies, more energy is transferred to the string, and the amplitude of the antinodes is relatively large. Although the driven end of the string is not a node, it is close to a node when the string is in resonance.

Mechanical resonance is not the only type. When you tune a radio, you are changing the resonance frequency of an electrical circuit (Chapter 21) so that it will be driven by, or pick up, the frequency signal of the station you want. A classic example of undesirable mechanical resonance is described in the Insight feature.

Important Formulas

Hooke's law:

$$F = -kx$$

Frequency and period for SHM:

$$f = \frac{1}{T}$$

Total energy of a spring and mass in SHM:

$$E = \tfrac{1}{2}kA^2 = \tfrac{1}{2}mv^2 + \tfrac{1}{2}kx^2$$

Velocity of oscillating mass:

$$v = \sqrt{\frac{k}{m}(A^2 - x^2)}$$

Maximum velocity of oscillating mass:

$$v_{\max} = \sqrt{\frac{k}{m}}(A)$$

Period of a mass oscillating on a spring:

$$T = 2\pi\sqrt{\frac{m}{k}}$$

Frequency of a mass oscillating on a spring:

$$f = \frac{1}{T} = \frac{1}{2\pi}\sqrt{\frac{k}{m}}$$

Displacement of a mass in SHM:

$$y = A\sin(\omega t + \delta)$$

Velocity of a mass in SHM ($\delta = 0$):

$$v = A\omega\cos\omega t$$

Acceleration of a mass in SHM ($\delta = 0$):

$$a = -A\omega^2\sin\omega t = -\omega^2 y$$

Angular frequency:

$$\omega = 2\pi f = \sqrt{\frac{k}{m}}$$

Wave speed:

$$v = \frac{\lambda}{T} = \lambda f$$

Period of a simple pendulum (small-angle approximation):

$$T = 2\pi\sqrt{\frac{L}{g}}$$

Natural frequencies (in a stretched string):

$$f_n = \frac{nv}{2L} = \frac{n}{2L}\sqrt{\frac{F}{\mu}} \quad (\text{for } n = 1, 2, 3, \ldots)$$

Questions

Simple Harmonic Motion

1. What is the necessary condition for a particle to undergo simple harmonic motion?

2. If a mass oscillating on a spring is given additional energy by a driving force, how will this affect (a) the amplitude, (b) the frequency, and (c) the period of oscillation?

3. If the amplitude of a mass in simple harmonic motion is doubled, how are (a) the energy and (b) the maximum velocity affected?

4. How is the speed of a mass in simple harmonic motion affected as it gets farther from its equilibrium position? Explain.

Equations of Motion

5. The apparatus in Fig. 14.6 demonstrates that the motion of a mass oscillating on a spring can be described by a sinusoidal function of time. How could you demonstrate this for a pendulum?

6. What must you know to write the equation of motion for a particular object in simple harmonic motion?

7. Two identical masses oscillate with the same amplitude on springs that have the same spring constant. Describe the relative motions of the masses if they oscillate (a) in phase, (b) 90° out of phase, (c) 180° out of phase, (d) 270° out of phase, and (e) 360° out of phase.

8. Is it possible to have simple harmonic motion that is described by more than one sinusoidal function? Explain.

9. Could simple harmonic motion be described using a tangent function? Explain.

10. Determine the phase constant for each of the following equations of motion: (a) $y = -10 \sin 4t$ (where y is in centimeters) and (b) $y = -0.005 \cos t/2$ (where y is in meters).

11. How do the sinusoidal functions that describe damped harmonic motion differ from those for simple harmonic motion?

Wave Motion

12. What is the net effect of wave motion?

13. What type(s) of waves will propagate in (a) solids, (b) fluids, and (c) stretched strings?

14. A motorboat moves on a lake, and an observer on shore notices that waves continually lap the shore for some time after the boat has passed. Does the speed of these waves depend on the speed of the boat? Explain.

15. Explain why the Earth's outer core is believed to be liquid.

Wave Phenomena

16. Two waves of equal amplitude and frequency are traveling in opposite directions and interfering. Describe the combined waveform when the waves are (a) in phase, (b) 90° out of phase, (c) 180° out of phase, and (d) 270° out of phase.

17. What is destroyed when destructive interference occurs?

18. Distinguish between refraction and diffraction.

Standing Waves and Resonance

19. When a stretched string is vibrating at its fifth harmonic, how many wavelengths are there along the string?

20. Is it possible to have a standing wave in a stretched string for $n = 2.5$? Explain what this would mean.

21. How many modes of oscillation are possible for (a) a stretched string and (b) a mass on a spring?

22. If a heavy rubber hose and a light rubber hose are stretched with the same tension, in which one would a wave travel faster and why?

23. What determines the wave speed in a medium? In general, how would you expect wave speeds to vary in the three phases of matter?

24. A person swinging on a swing often puts energy into the system by doing what is known as pumping. At what frequency should this action be performed? Explain.

25. A swing has only one natural frequency, f_o. Yet it can be driven or pushed smoothly at frequencies of $f_o/2$, $f_o/3$, $2f_o$, and $3f_o$. How is this possible?

Problems

14.1 Simple Harmonic Motion

■1. A particle in simple harmonic motion has a frequency of 20 Hz. What is the period of oscillation?

■2. A mass oscillating on a spring completes a cycle every 0.050 s. What is the frequency of the oscillation?

■**3.** A pendulum makes 6.0 complete cycles in a time of 10 s. What are the frequency and period of the pendulum's oscillation?

■**4.** The frequency of a simple harmonic oscillation is doubled from 40 Hz to 80 Hz. What is the change in the period of oscillation?

■**5.** Show that the total energy of a system in simple harmonic motion is given by $\frac{1}{2}m\omega^2 A^2$.

■■**6.** A 0.25-kg object suspended on a light spring is released from a position 15 cm above the equilibrium position. The spring has a spring constant of 75 N/m. (a) What is the total energy of the system? (b) Does this energy depend on the mass of the object? Explain.

■■**7.** What is the speed of the object in Problem 6 when it is (a) 5.0 cm above its equilibrium position and (b) 5.0 cm below its equilibrium position? (c) What is the object's maximum speed, and where does this occur?

■■**8.** A 0.20-kg mass is oscillating on a spring that has a spring constant of 40 N/m. The instantaneous speed of the mass is 0.95 m/s as it passes through its equilibrium position. What is the total energy of the system?

■■**9.** What is the spring constant of a spring that stretches 4.0 cm when a 0.25-kg mass is suspended from it?

■■**10.** A 0.350-kg mass resting on a horizontal friction-less surface is attached to a spring with a spring constant of 150 N/m. If the mass is pulled 0.100 m from its equilibrium position and released, what is the force on the mass and its acceleration at (a) $t = 0$, (b) $x = 0.050$ m, and (c) $x = 0$?

■■**11.** A spring-loaded toy gun is cocked by applying a force of 30 N acting through a distance of 7.0 cm. With what speed is a 0.065-kg ball propelled by the spring? (Neglect friction.)

■■**12.** Atoms in a solid are in continual vibration due to thermal energy. At room temperature, the amplitude of the atomic vibrations is about 10^{-9} cm, and the frequency of oscillation is about 10^{12} Hz. (a) What is the approximate period of oscillation of an atom? (b) What is the magnitude of its velocity?

■■■**13.** Two objects oscillate on springs in simple har-monic motion. One has twice the mass of the other and oscillates with half the amplitude. The more massive object's spring has a spring constant that is $\frac{2}{3}$ of that of the other spring. How do the total energies of the two systems compare?

■■■**14.** A 75-kg circus performer jumps from a height of 5.0 m onto a trampoline and stretches it a distance of 0.30 m. Assume that the trampoline obeys Hooke's law. (a) How far will it stretch if the performer jumps from a height of 8.0 m? (b) How far will it be stretched when the performer stands still on it while taking a bow?

■■■**15.** A 0.10-kg mass suspended on a spring is pulled to 8.0 cm below its equilibrium position and released. When the mass passes through the equilibrium position, it has a speed of 0.40 m/s. What is the speed of the mass when it is 3.0 cm from the equilibrium position?

14.2 Equations of Motion

■**16.** A mass of 0.25 kg oscillates in simple harmonic motion on a spring with a spring constant of 1.5×10^2 N/m. What are (a) the period and (b) the frequency of the oscillation?

■**17.** Write the general equation of motion for a mass that is on a horizontal frictionless surface and connected to a spring (a) if the mass is initially given a quick push away from the spring and (b) if the mass is pulled away from the spring and released.

■**18.** Make sketches showing two masses oscillating on springs (a) in phase, (b) 90° out of phase, (c) 180° out of phase, (d) 270° out of phase, and (e) 30° out of phase.

■**19.** For an object in simple harmonic motion with a frequency of 3.0 Hz and an amplitude of 10 cm, what are the magnitudes of (a) the maximum displacement, (b) the maximum velocity, and (c) the maximum acceleration? (d) Where is the object when each of these conditions occurs?

■**20.** Compute the percentage between the angle θ in rad and the sine of θ for (a) $\theta = 10°$, (b) $\theta = 20°$, and (c) $\theta = 25°$.

■■**21.** The equation of motion for a 10-g mass in simple harmonic motion is given by $y = 7.0 \cos 10t$, where y is in centimeters. What are (a) the period and (b) the phase constant for this motion? (c) What is the total energy of the system?

■■**22.** The bob of a simple pendulum has a mass of 100 g. (a) If the pendulum is to be a one-second pendulum (have a period of 1.0 s) how long should it be? (b) Thomas Jefferson suggested that such a pendulum could be used as a time standard. Comment on the precision of such a standard.

■■**23.** The equation of motion of a particle in simple harmonic motion is given by $y = 10 \sin 0.50t$ (where y is in centimeters). What are the particle's (a) displacement, (b) velocity, and (c) acceleration at $t = 1.0$ s?

■■24. A 0.50-kg mass oscillates in simple harmonic motion on a spring with a spring constant of 2.4×10^3 N/m. The spring was initially compressed 12 cm. (a) Write the equation of motion. (b) Determine the phase constant. (c) What is the displacement of the mass at $t = 0.25$ s?

■■25. The simple harmonic motion of a 0.20-kg mass on a spring is described by $y = 20 \sin 2\pi t$ (where y is in centimeters). At $t = \frac{1}{8}$ s, what are the particle's (a) displacement and (b) velocity? (c) What is the total energy of the system?

■■26. The motion of a particle is described by the curve in Fig. 14.22. Write the equation of motion in three equivalent forms (see Eqs. 14.8–14.10).

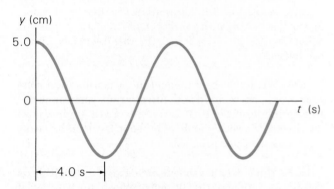

y (cm)

5.0

0

t (s)

—4.0 s—

Figure 14.22 **A traveling wave**
See Problem 26.

■■27. Use Newton's second law to show that the acceleration of a mass oscillating on a spring is generally given by $a = -\omega^2 A \sin(\omega t + \delta)$.

■■28. If the phase constant for the motion of the particle in Problem 26 is $\delta = 90°$, what are the magnitudes (in general terms), the times, and the locations of (a) the maximum velocity and (b) the maximum acceleration?

■■29. A 0.20-kg mass is suspended on a spring, which stretches a distance of 5.0 cm. The mass is then pulled down an additional distance of 10 cm and released. What are (a) the period of oscillation, (b) the equation of motion of the mass, (c) the total energy of the system, and (d) the displacement of the mass at $t = T/6$ s?

■■30. The motion of an object is described by $x = A \cos \omega t$. What are the general equations for the object's velocity and acceleration?

■■31. The equation of motion for a particle in simple harmonic motion is $x = 15 \sin 2\pi t$ (where x is in centimeters). At $t = 2.5$ s, what are the particle's (a) displacement, (b) velocity, and (c) acceleration? (Hint: see Problem 30.)

■■■32. A mass resting on a horizontal frictionless surface

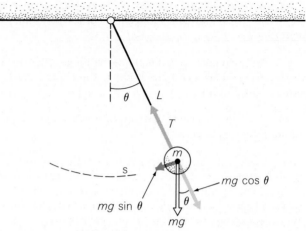

Figure 14.23 **SHM of a pendulum**
See Problem 36.

is connected to a fixed spring. The mass is displaced 16 cm from its equilibrium position and released. At $t = 0.50$ s, the mass is 8.0 cm from its equilibrium position (and has not passed through it yet). What is the period of oscillation of the mass?

■■■33. A grandfather clock has a pendulum that is 75 cm long. It is accidentally broken, and when repaired the length is shorter by 2.0 mm. (a) Will the repaired clock gain or lose time? (b) By how much will the repaired clock differ from the correct time (taken to be the time determined by the original pendulum in 24 h? (c) If the pendulum were metal, would temperature make a difference in the time-keeping of the clock? Explain.

■■■34. For a reference circle for a vertically oscillating shadow, as shown in Fig. 14.5, show that $a_y/y = -a_c/A =$ constant (where a_c is the centripetal acceleration of the object in uniform circular motion).

■■■35. A simple pendulum has a length of 50 cm and a bob whose mass is 0.10 kg. If the pendulum is initially released from an angle of 15° with the vertical, what is its position at (a) $t = 0.45$ s and (b) $t = 10$ s? (Hint: see Question 5.)

■■■36. The forces acting on a simple pendulum are shown in Fig. 14.23. (a) Show that for the small-angle approximation ($\sin \theta \cong \theta$) the force producing the motion has the same form as Hooke's law. (b) Show by analogy with a mass on a spring that the period of a simple pendulum is given by $T = 2\pi\sqrt{L/g}$.

14.3 Wave Motion

■37. A transverse wave has a wavelength of 1.5 m and a frequency of 20 Hz. What is the wave speed?

■38. A longitudinal sound wave has a speed of 340 m/s in air. If this produces a tone having a frequency of 5000 Hz, what is the wavelength?

■39. Radio broadcasting frequencies (of radio waves) have magnitudes in the kHz and MHz ranges. What are the corresponding wavelength and period ranges for radio waves? (The speed of radio waves is about 10^8 m/s.)

■■40. Light waves travel in a vacuum at a speed of 300,000 km/s. The frequency of visible light is about 10^{14} Hz. What is the approximate wavelength of visible light?

■■41. Red light has a frequency of 3.8×10^{14} Hz. What is the wavelength of this light in a vacuum?

■■42. A wave with a frequency of 60 Hz has a velocity of 12 m/s in a particular medium. (a) What is the wavelength? (b) If the wave is transmitted into another medium, in which it is propagated at a speed of 20 m/s, by how much will the wavelength change? (The frequency remains the same.)

■■43. The speed of longitudinal waves traveling in a long solid rod is given by $v = \sqrt{Y/\rho}$, where Y is Young's modulus and ρ is the density of the solid. If a disturbance has a frequency of 40 Hz, what is the wavelength of the waves it produces in (a) an aluminum rod and (b) a copper rod? (Hint: see Tables 9.1 and 10.1.)

■■44. As noted in Problem 43, the speed of longitudinal waves in a solid rod is given by $v = \sqrt{Y/\rho}$. Someone strikes a steel train rail with a hammer at a frequency of 0.50 Hz, and someone else puts his or her ear to the rail 1.0 km away. (a) How long after the first strike does the observer hear the sound? (b) What is the time interval between successive sound pulses heard by the observer? (Hint: see Tables 9.1 and 10.1.)

■■45. The speed of longitudinal waves in liquids is given by $v = \sqrt{B/\rho}$, where B is the bulk modulus and ρ is the density of the liquid. (a) What is the speed of longitudinal waves in water? (b) What is the speed of transverse waves in water? (Hint: see Tables 9.1 and 10.1.)

■■46. As noted in Problem 45, the speed of longitudinal waves in a liquid is given by $v = \sqrt{B/\rho}$. (a) Determine which is greater and by what factor: the speed of longitudinal waves in water or the speed of longitudinal waves in mercury. (b) If the frequency of a periodic disturbance is 200 Hz, by how much would the wavelengths differ in the two media? (Hint: see Tables 9.1 and 10.1.)

■■47. A sonar generator on a submarine produces periodic ultrasonic waves at a frequency of 2.50 MHz. The wavelength of the waves in sea water is 4.80×10^{-4} m. When the generator is directed downward, an echo reflected from the ocean floor is received 16.7 s later. How deep is the ocean at that point?

■■■48. The general form of the equation of motion for a traveling wave, expressing both spatial and temporal dependence, is

$$y = A \sin(kx - \omega t)$$

where k is the wave number (*not* the spring constant) and is equal to $2\pi/\lambda$. (a) Show mathematically or graphically that this equation represents a wave traveling in the positive x direction. (b) Show that

$$y = A \sin(kx + \omega t)$$

represents a wave traveling in the negative x direction.

■■■49. Assume that P and S waves from an earthquake with a focus near the Earth's surface travel through the Earth at average speeds of 8.0 km/s and 6.0 km/s, respectively. Assume that there is no deflection or refraction of the waves. (a) How long is the delay between the arrivals of successive waves at a seismic monitoring station located an angular distance of 90° from the epicenter of the quake? (b) Do the waves cross the boundary of the mantle? (c) How long does it take for the waves to arrive at a monitoring station on the opposite side of the Earth?

14.5 Standing Waves and Resonance

■50. The fundamental frequency of a stretched string is 220 Hz. What is the frequency of (a) the third harmonic and (b) the fourth harmonic?

■51. A standing wave is formed in a stretched string that is 4.0 m long. What is the wavelength of (a) the first harmonic and (b) the third harmonic?

■■52. A piece of rubber tubing with a linear mass density of 0.125 kg/m is stretched by a force of 8.0 N. (a) What will be the wave speed in the tubing? (b) If the stretched tubing has a length of 10 m, what are its natural frequencies?

■■53. Find the first four harmonics for a string that is 1.5 m long, has a linear mass density of 2.3×10^{-3} kg/m, and is under a tension of 30 N.

■■54. Will a standing wave be formed in a 10-m length of stretched string that transmits waves with a speed of 12 m/s if it is driven at a frequency of (a) 15 Hz or (b) 20 Hz?

■■55. A stretched string is observed to have four equal loops in a standing wave when driven at a frequency of

420 Hz. What driving frequency will set up a wave with two equal segments?

■■**56.** Two stretched strings have the same tension and linear mass density. Are any of the first six harmonics of the strings equal if the string lengths are (a) 1.0 m and 3.0 m or (b) 1.5 m and 2.0 m?

■■■**57.** In a common laboratory experiment on standing waves, the waves are produced in a stretched string by an electrical vibrator that oscillates at 60 Hz. The string runs over a pulley, and a hanger is suspended from the end. The tension in the string is varied by adding weights to the hanger. If the active length of string (the part that vibrates) is 1.5 m and this length of string has a mass of 10 g, what weights must be suspended to produce the first four harmonics in that length?

■■■**58.** A thin, flexible metal rod is 1.0 m long. It is clamped at one end to a table, and the other end can vibrate freely. What are the natural frequencies of the rod if the wave speed in the material is 4.0×10^3 m/s?

Additional Problems

59. Show that for a pendulum to oscillate with the same frequency as a mass on a spring, the pendulum's length must be $L = mg/k$.

60. A father pushes a 20-kg child on a swing. If the length of the swing is 4.0 m, how often does the father have to push to drive the system in resonance?

61. A 0.15-kg mass oscillates on a spring that has a spring constant of 500 N/m. (a) What is the energy of the system if the mass oscillates with an amplitude of 0.10 m? (b) What is the energy of the system if the 0.15-kg mass is replaced with a 0.30-kg mass that oscillates with the same amplitude?

62. The amplitude of a damped harmonic oscillation is given as a function of time by $A = A_0 e^{-t/\tau}$, where e is the base of the natural logarithm, A_0 is the maximum amplitude at $t = 0$, and τ is a constant that depends on the damping forces. (a) How long does it take for the amplitude to be halved if $\tau = 0.12$ s? (b) How long does it take for the amplitude to go to zero?

63. The range of sound audible to the human ear has frequencies from about 20 Hz to 20 kHz. The speed of sound in air is 345 m/s. What are the limits of this audible range in wavelengths?

64. A steel piano wire is 60 cm long and has a mass of 3.0 g. If the tension in the wire in 550 N, what are the fundamental frequency and wavelength?

65. A system consisting of a mass connected to a spring is to oscillate with a frequency of 0.64 Hz. How should the magnitude of the spring constant compare to that of the mass for this to occur?

66. A 0.25-kg mass on a spring oscillates in simple harmonic motion with an amplitude of 0.10 m. If the spring constant is 640 N/m, what is the maximum speed of the mass and where does this occur in a cycle?

67. A standing wave has nodes at $x = 0$ cm, $x = 6$ cm, $x = 12$ cm, and $x = 18$ cm. (a) What is the wavelength of the waves that are interfering to produce this standing wave? (b) At what positions are the antinodes?

68. On a violin, a correctly tuned A string has a frequency of 440 Hz. If an A string is found to produce a sharp note of 450 Hz when under a tension of 500 N, to what should the tension be adjusted to produce the correct frequency?

69. An object in simple harmonic motion is described by $y = 0.20 \sin 1.8\pi t$ (where y is in centimeters). What is the speed of the object at $t = 10$ s?

70. A mass on a spring is set into motion with the initial conditions $A = 0.15$ m and $\delta = 30°$. Sketch plots of y versus t for the motion of the mass (m) over the first period of oscillation for (a) $m = 0.75$ kg and (b) $m = 0.25$ kg.

71. A physical pendulum is a rigid body that is suspended so it swings freely about a horizontal axis. The body could be a meterstick, this book, or any object suspended so it swings back and forth in a vertical arc. The period of a physical pendulum can be shown to be $T = 2\pi\sqrt{I/mgL}$, where I is the moment of inertia of the body, m is its mass, and L is the distance between the axis of rotation and the center of mass of the body (see Chapter 8). Show that this expression reduces to that for the period of a simple pendulum if the physical pendulum is a small mass m suspended from a cord of length L.

Sound

15

Sound waves provide us with our major form of communication (speech), a favorite source of enjoyment (music), and a common kind of distraction (noise). Sound waves become speech, music, or noise only when our ears perceive them as disturbances (usually in the air). Physically, sound waves are longitudinal waves that are propagated in solids, liquids, and gases. Without a medium, there can be no sound. That is, in a vacuum, there would be no sound.

This distinction between the sensory and physical meanings of sound gives you a way to answer this old question: if a tree falls in the forest and there is no one there to hear it, will there be a sound? The answers are no in terms of sensory hearing and yes in terms of physical disturbances or longitudinal waves. That is, the answer depends on how sound is defined. The definition of sound covers three aspects: its *source*, the *medium* of its propagation (in the form of longitudinal sound waves), and its *detector*, which may be human ears.

Since sound waves are all around us most of the time, we are exposed to many interesting and important sound phenomena. A few of these, which often have distinct effects on the sounds we perceive, are covered in this chapter.

15.1 Sound Waves

For there to be sound waves, there must be a disturbance or vibrations in some medium. This disturbance may be the clapping of hands or the squeal of tires as a car comes to a sudden stop. Underwater you can hear the click of rocks against one another. If you put your ear to a thin wall, you can hear sounds from the other side through it. It is clear that **sound waves** are longitudinal waves that are propagated in solids, liquids, and gases (compressible media).

The characteristics of sound waves can be visualized by considering those produced by a tuning fork (see Fig. 15.1). A tuning fork is essentially a metal bar bent into a U shape. The prongs, or tines, vibrate when struck. The fork vibrates at its fundamental frequency (with an antinode at the end of each tine), so a

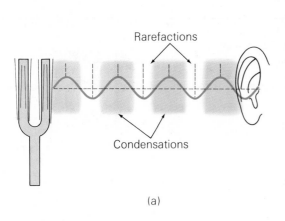

Rarefactions

Condensations

(a)

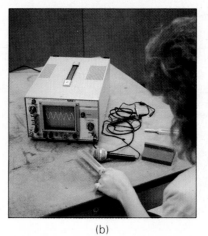

(b)

single tone is heard. The vibrations disturb the air, producing alternating higher-density (compressed) regions called **condensations** and lower-density regions called **rarefactions**. As the fork vibrates, these disturbances propagate outward, and a series of them can be described by a sinusoidal wave.

As the disturbances traveling through the air reach the ear, the eardrum (a thin membrane) is set into vibration by the pressure variations. On the other side of the eardrum, tiny bone structures (the hammer, anvil, and stirrup) carry the vibrations to the inner ear, where they are picked up by the auditory nerve. Characteristics of the human ear limit the perception of sound. Only sound waves with frequencies between about 20 Hz and 20 kHz (kilohertz) initiate nerve impulses that are interpreted by the brain as sound. This frequency range is called the **audible region** of the sound frequency spectrum (Fig. 15.2). Frequencies lower than 20 Hz are in the **infrasonic region**. The longitudinal waves generated by earthquakes have infrasonic frequencies. Above 20 kHz is the **ultrasonic region**. Ultrasonic waves can be generated by high-frequency vibrations in crystals.

Ultrasonic waves, or ultrasound, cannot be detected by humans but can be by other animals. The audible region for dogs extends beyond that of humans, so ultrasonic whistles can be used to call dogs without disturbing people. A more recent application of this idea is a cone-shaped whistle that is attached to a car or truck to generate ultrasonic waves when the vehicle travels above a certain (air) speed. Since deer can hear these sounds, the intention is to scare them away from the road so they won't run in front of a car or truck. According to road statistics, these deer whistles are effective.

There are many other practical applications of ultrasound. Since ultrasound can travel for kilometers in water, it is used in sonar. Also, underwater communications are made possible by combining audible sound waves with ultrasound. Ultrasonic baths are used to clean metal machine parts and jewelry. The high-frequency (short-wavelength) ultrasound vibrations loosen particles in otherwise inaccessible places. In medicine, ultrasound is used to examine internal tissues and organs that are nearly invisible to X-rays. Perhaps the best known medical application of ultrasound is its use to view a fetus without exposing it to the dangerous effects of X-rays (Fig. 15.3). Ultrasonic generators (transducers) made of quartz crystals produce high-frequency waves that are

The vibrations are transmitted to a liquid in the inner ear, where the disturbances stimulate hairlike cells that initiate nerve signals.

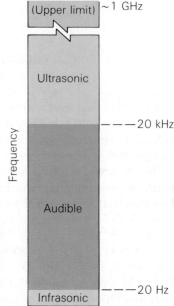

Figure 15.2 **Sound frequency spectrum**
The audible region of sound for humans lies between about 20 Hz and 20 kHz. Below this is the infrasonic region, and above it is the ultrasonic region.

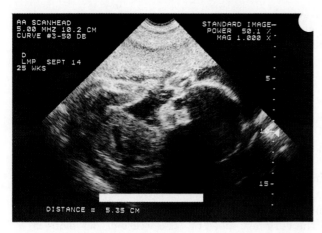

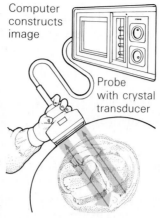

Computer constructs image

Probe with crystal transducer

Figure 15.3 **Ultrasound and echograms**
Ultrasound generated by crystal transducers is transmitted through tissue and is reflected from internal structures. The reflected waves are detected by the transducers and are used to construct an image, or echogram, as shown here for a 25-week-old fetus.

used to scan a designated region of the body from several angles. Reflections from the scanned areas are monitored, and images are recorded several times each second. The series of images provides a "moving picture" of an internal structure, such as the heart or a fetus. A still shot, or echogram, is shown in the figure.

Ultrasonic frequencies extend into the megahertz (MHz) range, but the sound frequency spectrum does not continue indefinitely. There is an upper limit of about 10^9 Hz, or 1 GHz (gigahertz), which is determined by the upper limit of the elasticity of materials.

15.2 The Speed of Sound

In general, the speed at which a disturbance moves through a medium depends on the elasticity of the medium and the density of its particles. For example, as you learned in Chapter 14, the wave speed in a stretched string is given by $v = \sqrt{F/\mu}$, where F is the tension in the string and μ is the linear mass density.

Solids are generally more elastic than liquids, which in turn are more elastic than gases. In a highly elastic material, the restoring forces between the atoms or molecules cause a disturbance to be propagated faster. Thus, the speed of sound is, in general, greater in solids than in liquids, and greater in liquids than in gases (Table 15.1). Sound travels about four times faster in solids than in liquids and about fifteen times faster in solids than in gases (such as air).

In air, the speed of sound is 331.5 m/s (about 740 mi/h) at a temperature of $0\,°C$. Temperature affects the speed of sound in gases. As the temperature increases, so do the speeds of the molecules. As a result, the molecules collide more frequently, and a disturbance is transmitted more quickly. For normal environmental temperatures, the speed of sound in air increases by about 0.6 m/s for each degree Celsius above $0\,°C$. Thus, a good approximation of the speed of sound in air for a particular temperature is given by

$$v = 331.5 + 0.6T_c \tag{15.1}$$

where v is in meters per second (m/s) and T_c is the air temperature in degrees Celsius. (Although they are not generally written explicitly, the units associated with the factor 0.6 are m/s-°C.)

The wave speeds in solids and liquids depend on Young's modulus and the bulk modulus, respectively (see Chapter 9), as well as the mass densities. That is, $v_s = \sqrt{Y/\rho}$ and $v_l = \sqrt{B/\rho}$.

Example 15.1 Speed of Sound and Temperature
What is the speed of sound in air at room temperature (20 °C)?

Solution
Given: $T_c = 20\,°C$ Find: v

Directly from Eq. 15.1,

$$v = 331.5 + 0.6T_c = 331.5 + (0.6)(20) = 343.5 \text{ m/s} \quad \blacksquare$$

A generally useful figure for the speed of sound in air is $\frac{1}{3}$ km/s (or $\frac{1}{5}$ mi/s). Using this figure, you can, for example, estimate how far away lightning has struck by counting the number of seconds between the time the flash is observed and the time the associated thunder is heard. Because of the very fast speed of light, the lightning flash is seen almost instantaneously. The sound waves of the thunder travel relatively slowly, at about $\frac{1}{3}$ km/s. For example, if the interval between the two is measured to be 6 s (often by counting "one thousand one, one thousand two, . . ."), the lightning stroke was approximately 2 km away (that is, $\frac{1}{3}$ km/s × 6 s).

You may also have noticed the delay of sound behind light at a baseball game. If you're sitting in the outfield stands, you see the batter hit the ball and hear the crack of the bat later.

15.3 Sound Intensity

Wave motion involves the propagation of energy. The rate of the energy transfer is expressed in terms of **intensity**, which is the energy transported per unit time across a unit area. Since energy/time is power, intensity is power/area:

$$\text{intensity} = \frac{\text{energy/time}}{\text{area}} = \frac{\text{power}}{\text{area}}$$

The standard units for intensity (power/area) are watts per meter squared (W/m^2).

Just as the component of a force is perpendicular to an area in the definition of pressure (Chapter 10), the surface area is perpendicular to the direction of wave propagation for the definition of sound intensity. For example, consider a point source that sends out spherical sound waves, as shown in Fig. 15.4. If there are no losses, the sound intensity at a distance R from the source is

$$I = \frac{P}{A} = \frac{P}{4\pi R^2} \tag{15.2}$$

where P is the power of the source and $4\pi R^2$ is the area of a sphere with a radius R, through which the sound energy passes perpendicularly.

The intensity for a point source is therefore inversely proportional to the square of the distance from the source (an inverse square relationship). Two intensities at different distances from a source of constant power may be compared using a ratio:

$$\frac{I_2}{I_1} = \frac{P/4\pi R_2^2}{P/4\pi R_1^2} = \frac{R_1^2}{R_2^2}$$

Table 15.1
The Speed of Sound in Various Media

Medium	Speed (m/s)
Solids	
Aluminum	5000
Copper	3750
Iron	4480
Glass	5170
Polystyrene	1840
Liquids	
Alcohol, ethyl	1160
Mercury	1450
Water	1480
Gases	
Air (0 °C)	331
Air (100 °C)	367
Helium (0 °C)	965
Hydrogen (0 °C)	1284
Oxygen (0 °C)	316

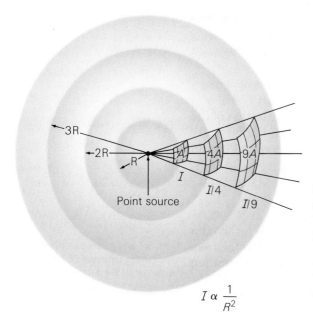

$$I \propto \frac{1}{R^2}$$

Point source

Figure 15.4 Intensity of sound
The energy emitted from a point source spreads out equally in all directions. Since $I = P/A = P/4\pi R^2$, the intensity is related to the distance from the source by $1/R^2$.

or
$$\frac{I_2}{I_1} = \left(\frac{R_1}{R_2}\right)^2 \qquad\qquad (15.3)$$

Suppose that the distance from a point source is doubled; that is, $R_2 = 2R_1$, or $R_1/R_2 = \frac{1}{2}$. Then

$$\frac{I_2}{I_1} = \left(\frac{R_1}{R_2}\right)^2 = \left(\frac{1}{2}\right)^2 = \frac{1}{4}$$

and

$$I_2 = \frac{I_1}{4}$$

Since the intensity decreases by a factor of $1/R^2$, doubling the distance decreases the intensity to a quarter of its original value. (The same energy is spread over a larger surface.)

Sound intensity is perceived by the ear as loudness. On the average, the human ear can detect sound waves (at 1 kHz) with an intensity as low as 10^{-12} W/m^2. This intensity is referred to as the **threshold of hearing**. Thus, for us to hear a sound, it must not only have a frequency in the audible range, but also be of sufficient intensity. As the intensity is increased, the perceived sound becomes louder. At an intensity of 1.0 W/m^2, the sound is uncomfortably loud and may be painful to the ear. Thus, this intensity is called the **threshold of pain**.

Note that the thresholds of pain and hearing differ by a factor of 10^{12}:

$$\frac{I_p}{I_h} = \frac{1.0 \text{ W/m}^2}{10^{-12} \text{ W/m}^2} = 10^{12}$$

That is, the intensity at the threshold of pain is a *trillion* times greater than that at the threshold of hearing. Within this enormous range, the perceived loudness is not directly proportional to the intensity. That is, if the intensity is doubled, the perceived loudness does not double. A doubling of perceived loudness corresponds approximately to an increase in intensity by a factor of 10. For example, a sound with an intensity of 10^{-5} W/m^2 would be perceived to be twice as loud as one with an intensity of 10^{-6} W/m^2 (the smaller the negative exponent, the larger the number).

It is convenient to compress the large range of sound intensities by using a logarithmic scale (base 10) to express intensity levels. The intensity level of a sound must be referenced to a standard intensity, which is taken to be that of the threshold of hearing, $I_o = 10^{-12}$ W/m^2. Then, for any intensity I, the intensity level is the log of the ratio of I to I_o: $\log I/I_o$. For example, if a sound has an intensity of $I = 10^{-6}$ W/m^2,

$$\log \frac{I}{I_o} = \log \frac{10^{-6} \text{ W/m}^2}{10^{-12} \text{ W/m}^2} = \log 10^6 = 6$$

(Recall that $\log_{10} 10^x = x$.) Thus, a sound with an intensity of 10^{-6} W/m^2 has an intensity level of 6 bel (B) on this scale. In this way, the intensity range from 10^{-12} W/m^2 to 1.0 W/m^2 is compressed into a scale of intensity levels running from 0 B to 12 B.

The bel was named in honor of Alexander Graham Bell, the inventor of the telephone.

A finer intensity scale is obtained by using a smaller unit, the **decibel (dB)**, which is a tenth of a bel. The 0–12 B range corresponds to 0–120 dB. In this case, the equation for the relative **sound intensity level**, or **decibel level** (β), is

$$\beta = 10 \log \frac{I}{I_o} \qquad \text{where } I_o = 10^{-12} \text{ W/m}^2 \tag{15.4}$$

The decibel intensity scale and familiar sounds at some intensity levels are shown in Fig. 15.5.

Example 15.2 Sound Intensity Levels
What are the intensity levels for sounds with intensities of (a) 10^{-12} W/m^2 and (b) 5.0×10^{-6} W/m^2?

Solution
Given: (a) $I = 10^{-12}$ W/m^2 *Find*: (a) β
       (b) $I = 5.0 \times 10^{-6}$ W/m^2 (b) β

Using Eq. 15.4,

(a) $\beta = 10 \log \dfrac{I}{I_o} = 10 \log \dfrac{10^{-12} \text{ W/m}^2}{10^{-12} \text{ W/m}^2}$

$= 10 \log 1 = 0 \text{ dB}$

The intensity is the same as that at the threshold of hearing. (Recall that $\log 1 = 0$, since $1 = 10^0$ and $\log 10^0 = 0$.)

(b) $\beta = 10 \log \dfrac{I}{I_o} = 10 \log \dfrac{5.0 \times 10^{-6} \text{ W/m}^2}{10^{-12} \text{ W/m}^2}$

$= 10 \log (5.0 \times 10^6) = 10(\log 5.0 + \log 10^6)$

$= 10(0.70 + 6.0) = 67 \text{ dB}$

Note that a half an order of magnitude for the intensity (5.0 is half of 10) does not correspond to half an order on the dB scale (65 dB) since the scale is logarithmic, not linear. ∎

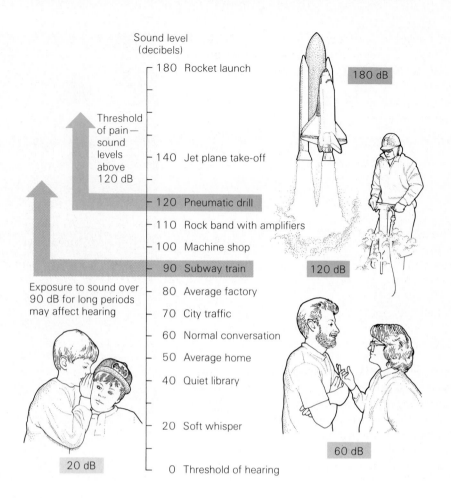

Sound level
(decibels)

180 Rocket launch   180 dB

Threshold
of pain—
sound
levels
above
120 dB

140 Jet plane take-off

120 Pneumatic drill

110 Rock band with amplifiers

100 Machine shop

90 Subway train   120 dB

Exposure to sound over
90 dB for long periods
may affect hearing

80 Average factory

70 City traffic

60 Normal conversation

50 Average home

40 Quiet library

20 Soft whisper   60 dB

20 dB

0 Threshold of hearing

Figure 15.5 Sound intensity levels
This shows the levels of some common sounds on the decibel (dB) scale.

Example 15.3 Intensity Level Factors

(a) What is the difference in the intensity level if the intensity of a sound is doubled? (b) By what factors does the intensity increase for intensity level differences of 10 dB and 20 dB?

Solution

Given: (a) $I_2 = 2I_1$ *Find*: (a) $\Delta\beta$
     (b) $\Delta\beta = 10$ dB   (b) I_2/I_1
          $\Delta\beta = 20$ dB

(a) The intensity is doubled, so $I_2/I_1 = 2$. This ratio can be used directly to find the intensity level *difference* because of the properties of logarithms (see Problem 39):

$$\Delta\beta = 10 \log \frac{I_2}{I_1} = 10 \log 2 = 3 \text{ dB}$$

Thus, doubling the intensity increases the intensity level by 3 dB (for example, an increase from 60 dB to 63 dB).

(b) For a 10-dB difference,

$$\Delta\beta = 10 \text{ dB} = 10 \log \frac{I_2}{I_1}$$

and

$$\log \frac{I_2}{I_1} = 1.0$$

Since $\log 10^1 = 1$,

$$\frac{I_2}{I_1} = 10^1 \quad \text{and} \quad I_2 = 10 I_1$$

Similarly, for a 20-dB difference,

$$\Delta \beta = 20 \text{ dB} = 10 \log \frac{I_2}{I_1}$$

and

$$\log \frac{I_2}{I_1} = 2.0$$

Since $\log 10^2 = 2$,

$$\frac{I_2}{I_1} = 10^2 \quad \text{and} \quad I_2 = 100 I_1$$

Thus, an intensity level difference of 10 dB corresponds to increasing (or decreasing) the intensity by a factor of 10. An intensity level difference of 20 dB corresponds to increasing (or decreasing) the intensity by a factor of 100. You should be able to guess the factor that corresponds to an intensity level difference of 30 dB. In general, the factor of the intensity change is $10^{\Delta B}$, where ΔB is the level difference in bels. Since 30 dB = 3 B and $10^3 = 1000$, the intensity changes by a factor of 1000 for an intensity level difference of 30 dB. ∎

Figure 15.6 **Sound intensity protectors**
Over a period of time, large sound intensities can cause ear damage, so ear protectors or valves are worn by workers in a variety of jobs.

Sound intensities can have detrimental effects on hearing, and because of this the government has set occupational noise-exposure limits. Also, people working at some jobs must wear ear protectors (see Fig. 15.6).

15.4 Sound Phenomena

There are many interesting sound phenomena. Several you may have experienced are discussed in this section.

Reflection, Refraction, and Diffraction

A sound wave can be reflected. An echo is a familiar example of sound reflection.

Sound refraction is less common, but you may have experienced this effect on a calm summer evening. Then it is sometimes possible to hear distant voices or other sounds that ordinarily would not be audible. This effect is due to the refraction, or bending, of the sound waves as they pass into a region where the air density is different. The effect is similar to what would happen if the sound passed into another medium.

The required conditions are a layer of cooler air near the ground with a layer of warmer air above it. These conditions occur frequently over bodies of water after sunset because of cooling. The sound waves spread out from their source and would ordinarily be too faint to be heard by someone at a distance. But, as the waves pass through the overlying layer of warm air, they may be bent, or refracted, toward the distant person. This increases the intensity of the sound received by that person. If the intensity is above the threshold of hearing, the distant sound can be heard.

Sound is also diffracted or bent around corners, as you know from experience.

Reflection, refraction, and diffraction of light can be seen, and these phenomena will be considered in Chapters 22 and 24.

Interference

Like waves of any kind, sound waves interfere when they meet. Suppose that two loudspeakers separated by some distance emit sound waves in phase at the same frequency [see Fig. 15.7(a)]. Consider the speakers to be point sources—the waves spread out spherically and interfere. The lines in the figure represent wave crests (or condensations), and the troughs (or rarefactions) lie between the lines.

At particular points in space, there will be constructive and destructive interferences. For example, if the waves are exactly in phase as they arrive at a point, there will be total constructive interference. This is illustrated for point C in Fig. 15.7(b). It is convenient to describe the path lengths traveled by the waves in terms of wavelengths (λ) to determine if they arrive in phase. The path lengths in this case are AC $= 4\lambda$ and BC $= 3\lambda$. The phase difference ($\Delta\theta$) is related to the path difference (PD) by this simple relationship:

$$\Delta\theta = \frac{2\pi}{\lambda}(\text{PD}) \tag{15.5}$$

Since there are 2π rad per wavelength ($2\pi/\lambda$), multiplying that ratio by the path difference *in wavelength units* gives the phase difference.

For the example illustrated in Fig. 15.7,

$$\Delta\theta = \frac{2\pi}{\lambda}(\text{AC} - \text{BC}) = \frac{2\pi}{\lambda}(4\lambda - 3\lambda) = 2\pi \text{ rad}$$

This is the same as $\Delta\theta = 0°$, or the waves being in phase. Thus, the waves interfere constructively at point C, increasing the intensity, or loudness, of the sound detected there.

From Eq. 15.5, it should be clear that the sound waves are in phase at any point where the path difference is zero or an integral multiple of the wavelength. That is,

$$\text{PD} = n\lambda \qquad (\text{for } n = 0, 1, 2, 3, \ldots)$$

$$\tag{15.6}$$

condition for constructive interference

A similar analysis for point D in Fig. 15.7(b) gives

$$\Delta\theta = \frac{2\pi}{\lambda}(3\lambda - 2.5\lambda) = \pi \text{ rad}$$

or $\Delta\theta = 180°$. At point D, the waves are completely out of phase, and destructive interference occurs.

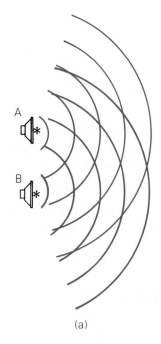

(a)

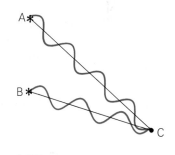

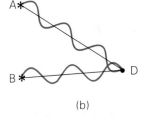

(b)

Figure 15.7 **Interference** (a) Sound waves from two (point) sources spread out and interfere. (b) At points where the waves arrive in phase, such as point C, constructive interference occurs. Where the waves are completely out of phase, such as at point D, destructive interference occurs. The phase difference depends on the path lengths the waves travel to reach a point.

Sound waves will be out of phase at any point where the path difference is an odd number of half-wavelengths ($\lambda/2$), or

$$PD = m\left(\frac{\lambda}{2}\right) \qquad (\text{for } m = 1, 3, 5, \ldots)$$

(15.7)

condition for destructive interference

At these points, a softer or less intense sound will be heard or detected. If the amplitudes of the waves are equal, the destructive interference is total, and no sound is heard!

Example 15.4 Interference and Path Difference

At an open-air concert on a hot day (air temperature of 25°C), a person sits at a location that is 7.0 m and 9.1 m, respectively, from speakers at each side of the stage. A musician, warming up, plays a single 494-Hz tone. What does the spectator hear? (Consider the speakers to be point sources.)

Solution

Given: $d_1 = 7.0$ m and $d_2 = 9.1$ m   *Find:* Whether there is inter-
$\qquad\quad f = 494$ Hz   ference and what type
$\qquad\quad T_c = 25\,°C$

The phase difference between the waves arriving at the spectator's location is found by expressing the path lengths in wavelengths. The speed of sound depends on the temperature, and by Eq. 15.1,

$$v = 331.5 + 0.6T_c = 331.5 + 0.6(25) = 346.5 \text{ m/s}$$

The wavelength of the sound waves is

$$\lambda = \frac{v}{f} = \frac{346.5 \text{ m/s}}{494 \text{ Hz}} = 0.701 \text{ m}$$

Thus,

$$d_1 = (7.0 \text{ m})\left(\frac{\lambda}{0.701 \text{ m}}\right) = 10\lambda$$

and

$$d_2 = (9.1 \text{ m})\left(\frac{\lambda}{0.701 \text{ m}}\right) = 13\lambda$$

The path difference is

$$PD = d_2 - d_1 = 13\lambda - 10\lambda = 3\lambda$$

This is an integral number of wavelengths ($n = 3$), so constructive interference occurs, and a loud sound is heard by the spectator.

If the path difference had been 2.5λ, for example, $2.5\lambda = 5(\lambda/2)$, which is a condition for destructive interference ($m = 5$), no sound would be heard if the waves from the speakers had equal amplitudes. Of course, during a concert the sound would not be a single-frequency tone, but would have a variety of frequencies, wavelengths, and amplitudes. Spectators at certain locations might not hear a small portion of the sound but probably wouldn't notice. ∎

Another interesting interference effect occurs when two tones of nearly the same frequency ($f_1 \cong f_2$) are sounded simultaneously. The ear senses pulsations in loudness known as **beats**. [The human ear can detect as many as 7 beats per second.]

For example, suppose that two sinusoidal waves with the same amplitude have similar frequencies. As these waves interfere, the total displacement at some location is (by the principle of superposition)

$$y = y_1 + y_2 = A \sin 2\pi f_1 t + A \sin 2\pi f_2 t$$

The trigonometric identity

$$\sin a + \sin b = 2 \cos \left(\frac{a - b}{2} \right) \sin \left(\frac{a + b}{2} \right)$$

allows the equation for y to be rewritten as

$$y = \left[2A \cos 2\pi t \left(\frac{f_1 - f_2}{2} \right) \right] \left[\sin 2\pi t \left(\frac{f_1 + f_2}{2} \right) \right] \tag{15.8}$$

As illustrated in Fig. 15.8, the sine term of Eq. 15.8 represents the sinusoidal oscillation having an average frequency of $(f_2 + f_1)/2$. The amplitude of this wave is modulated by an envelope represented by the bracketed cosine term. Note that the amplitude is not constant, but varies with time. The frequency of $(f_1 - f_2)/2$ means that two beats, or two maximum amplitudes of $2A$, pass by each cycle. Thus, the **beat frequency** (f_b) that is perceived is twice the wave frequency, or

$$f_b = |f_1 - f_2| \tag{15.9}$$

(The absolute value is taken because there cannot be a negative frequency.)

Beats may be produced when tuning forks of nearly the same frequency are vibrating at the same time. For example, using forks with frequencies of 516 Hz and 513 Hz, the beat frequency is f_b = 516 Hz − 513 Hz = 3 Hz, and 3 pulsations, or beats, are heard each second. Musicians tune two stringed instruments to the same note by adjusting the strings until the beats disappear ($f_1 = f_2$).

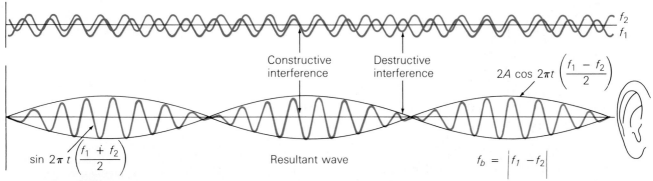

Figure 15.8 **Beats**
Two traveling waves of equal amplitude and slightly different frequencies interfere and give rise to pulsating tones called beats. The beat frequency is given by $f_b = |f_1 - f_2|$.

The Doppler Effect

If you stand along a highway and a car or truck approaches you with the horn blowing, the pitch (the frequency) of the sound is higher as the vehicle approaches and lower as it recedes. You can also hear variations in the frequency of the motor noise when watching a racing car going around a track. A variation in the perceived sound frequency due to the motion of the sound source is an example of the **Doppler effect**.

As Fig. 15.9 shows, the sound waves emitted by a moving source tend to bunch up in front of the source and spread out in back. The Doppler shift in frequency can be found by assuming that the air is at rest in a reference frame such as that depicted in Fig. 15.10. The speed of sound in air is v and the speed of the moving source is v_s. The frequency of the sound produced by the source is f_s. In one period, $T = 1/f_s$, a wave crest moves a distance $d = vT = \lambda$. (The sound wave would travel this distance in still air in any case, whether or not the source were moving.) But, in one period, the source travels a distance $d_s = v_s T$ before emitting another wave crest. The distance between the successive wave crests is thus shortened to a wavelength λ':

$$\lambda' = d - d_s = vT - v_s T$$

$$= (v - v_s)T = \frac{v - v_s}{f_s}$$

The frequency heard by the observer (f_o) is related to the shortened wavelength by $f_o = v/\lambda'$, and substituting for λ' gives

$$f_o = \frac{v}{\lambda'} = \left(\frac{v}{v - v_s}\right)f_s$$

or

$$f_o = \left(\frac{1}{1 - v_s/v}\right)f_s \qquad (15.10)$$

source moving toward a stationary observer

Since $(1 - v_s/v)$ is less than 1, f_o is greater than f_s in this situation. For example, suppose that the speed of the source is a tenth of the speed of sound: $v_s = v/10$, or $v_s/v = \frac{1}{10}$. Then, by Eq. 15.10, $f_o = \frac{10}{9}f_s$.

Similarly, when the source is moving away from the observer ($\lambda' = d + d_s$), the observed frequency is given by

$$f_o = \left(\frac{v}{v + v_s}\right)f_s = \left(\frac{1}{1 + v_s/v}\right)f_s \qquad (15.11)$$

source moving away from a stationary observer

Here f_o is less than f_s. (Why?)

Combining Eqs. 15.10 and 15.11 yields a general equation for perceived frequency with a moving source and a stationary observer:

$$f_o = \left(\frac{v}{v \mp v_s}\right)f_s = \left(\frac{1}{1 \mp v_s/v}\right)f_s \qquad (15.12)$$

$\begin{cases} - \text{ for source approaching stationary observer} \\ + \text{ for source moving away from stationary observer} \end{cases}$

As you might expect, the Doppler effect also occurs with a moving observer

The Austrian physicist Christian Doppler (1803–1853) first described what we now call the Doppler effect.

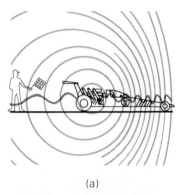

(a)

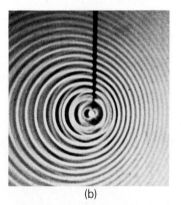

(b)

Figure 15.9 **The Doppler effect**
(a) The sound waves bunch up in front of a moving source, giving a higher frequency. They stretch out behind the source, giving a lower frequency. (b) The Doppler effect occurs in water, as shown by a moving disturbance in a ripple tank.

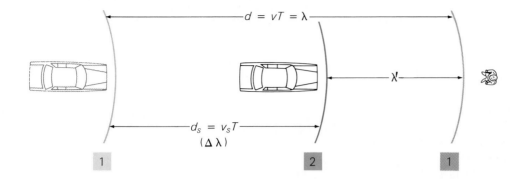

Figure 15.10 **Doppler effect and wavelength** Sound from a moving car's horn travels a distance d in a time T. During this time, the car (the source) travels a distance d_s, thereby shortening the observed wavelength of the sound.

and a stationary source. This situation is a bit different. As the observer moves toward the source, the distance between successive wave crests is the normal wavelength (or $\lambda = v/f_s$), but the measured wave speed is different. Relative to the approaching observer the sound from the stationary source has a wave speed of $v' = v + v_o$, where v_o is the speed of the observer and v is the speed of sound in still air. (The observer moving toward the source is moving in the opposite direction to the propagating waves and thus meets more wave crests in a given time.) The observed frequency (f_o) is then (with $\lambda = v/f_s$):

$$f_o = \frac{v'}{\lambda} = \left(\frac{v + v_o}{v}\right) f_s$$

or

$$f_o = \left(1 + \frac{v_o}{v}\right) f_s \qquad (15.13)$$

observer moving toward stationary source

Similarly, for an observer moving away from a stationary source, the perceived wave speed is $v' = v - v_o$, and

$$f_o = \frac{v'}{\lambda} = \left(\frac{v - v_o}{v}\right) f_s$$

or

$$f_o = \left(1 - \frac{v_o}{v}\right) f_s \qquad (15.14)$$

observer moving away from a stationary source

Eqs. 15.13 and 15.14 can be combined into another general equation:

$$f_o = \left(\frac{v \pm v_o}{v}\right) f_s = \left(1 \pm \frac{v_o}{v}\right) f_s \qquad (15.15)$$

$\begin{cases} + \text{ for observer approaching stationary source} \\ - \text{ for observer moving away from stationary source} \end{cases}$

There are also the cases when both the source and the observer are moving toward one another or both are moving away from one another. However, these will not be considered here.

Example 15.5 The Doppler Effect

As a truck traveling at 96 km/h approaches and passes a person standing along the highway, the driver sounds the horn. If the horn has a frequency of 400 Hz, what are the frequencies of the sound waves heard by the person (a) as the truck approaches and (b) after it has passed? (Assume that the speed of sound is 346 m/s.)

Solution

Given: $v_s = 96$ km/h $= 27$ m/s   *Find*: (a) f_o
$\qquad\quad f_s = 400$ Hz   (b) f_o
$\qquad\quad v = 346$ m/s

(a) Using Eq. 15.12 with a minus sign (source approaching stationary observer),

$$f_o = \left(\frac{v}{v - v_s}\right)f_s = \left(\frac{343 \text{ m/s}}{343 \text{ m/s} - 27 \text{ m/s}}\right)(400 \text{ Hz}) = 434 \text{ Hz}$$

(b) A plus sign is used when the source is moving away:

$$f_o = \left(\frac{v}{v + v_s}\right)f_s = \left(\frac{343 \text{ m/s}}{343 \text{ m/s} + 27 \text{ m/s}}\right)(400 \text{ Hz}) = 371 \text{ Hz}$$

PROBLEM-SOLVING HINT

You may find it difficult to remember whether a plus or minus sign is used in the general equations for the Doppler effect. Let your experience help you. For the commonly experienced case of being a stationary observer, the sound frequency increases when the source approaches, so the denominator in Eq. 15.12 must be smaller than the numerator. For this case, you use the minus sign. When the source is receding, the frequency is lower. The denominator in Eq. 15.12 must then be larger than the numerator, and you use the plus sign for this case. Similar thinking will help you choose a plus or minus sign for the numerator in Eq. 15.15.

The Doppler effect also occurs for light waves, although the formulas describing it are different from those given above. It has been found that the wavelengths of light from distant galaxies have increased when the light reaches Earth (shifted toward the red end of the visible spectrum). This indicates that the galaxies are moving away from the Earth. This so-called Doppler red shift is evidence for the theory of an expanding universe. The Doppler shift of light from stars in our galaxy, the Milky Way, indicates that the galaxy is rotating.

You have been subjected to a practical application of the Doppler effect if you have ever been caught speeding in your car by police radar, which uses reflected radio waves. [Radar stands for *radio detecting and ranging* and is similar to underwater sonar, which uses ultrasound.] If the radio waves are reflected from a parked car, the reflected waves return to the source with the same frequency. But for a car that is moving toward a patrol car, the reflected waves have a higher frequency, or are Doppler-shifted. Actually, there is a double Doppler shift. The moving car acts like a moving observer in receiving the wave (first Doppler shift), and in reflecting it, the car acts like a moving source emitting a wave (second Doppler shift). The magnitudes of the shifts depend on the speed of the car. A computer quickly calculates this speed and displays it for the police officer.

Sonic Booms

Consider a jet plane that can travel at supersonic speeds. As the speed of a moving source of sound approaches the speed of sound, the waves ahead of the source come closer together (Fig. 15.11). When a plane is traveling at the speed of sound, the waves can't outrun it, and they pile up in front. At supersonic speeds, the waves overlap, even behind the plane. This overlapping of a large number of waves produces many points of constructive interference, forming a large pressure ridge, or shock wave. [This is sometimes called a bow wave because it is analogous to the wave produced by the bow of a boat moving through water at a speed greater than the speed of the water waves.]

From an aircraft traveling at supersonic speed, the shock wave trails out to the sides and downward. When this pressure ridge passes over an observer on the ground, the large concentration of energy produces what is known as a **sonic boom**. There is really a double boom because shock waves are formed at both ends of the aircraft. Under certain conditions, the shock waves can break windows and cause other damage. [Sonic booms are no longer heard as frequently as in the past. Pilots are now instructed to fly supersonically only at high altitudes and to avoid populated areas.]

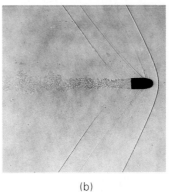

(b)

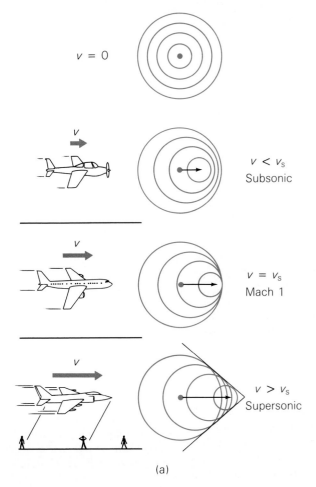

(a)

Figure 15.11 **Sonic boom!** (a) When an aircraft exceeds the speed of sound, the sound waves form a pressure ridge, or shock wave. As the trailing shock wave passes over the ground, observers hear a sonic boom—actually two booms because shock waves are formed at the front and the tail of the plane. (b) A .22-caliber rifle bullet traveling faster than the speed of sound is photographed in silhoutte to show its shock waves. Note the turbulence behind the bullet.

A common misconception is that a sonic boom occurs only when the plane breaks the sound barrier. As an aircraft approaches the speed of sound, the pressure ridge in front of it is essentially a barrier that must be overcome with extra power. However, once supersonic speed is reached, this barrier is no longer there, and the trailing shock wave produces booms along its ground path.

Ideally, the sound waves produced by a supersonic aircraft form a cone-shaped shock wave (Fig. 15.12). The waves travel outward with a speed v, and the speed of the plane (the source) is v_s. Note from the figure that the angle between a line tangent to the spherical waves and the line along which the plane is moving is given by

$$\sin \theta = \frac{vt}{v_s t} = \frac{v}{v_s} \qquad (15.16)$$

The inverse ratio is called the **Mach number** (M):

$$M = \frac{1}{\sin \theta} = \frac{v_s}{v} \qquad (15.17)$$

If v equals v_s, the plane is flying at the speed of sound, and the Mach number is 1 (that is, $v_s/v = 1$). Therefore, a Mach number less than 1 indicates a subsonic speed, and a Mach number greater than 1 indicates a supersonic speed. In the latter case, the Mach number tells the speed of the aircraft in terms of a multiple of the speed of sound.

Mach number, not surprisingly, was named for a physicist, Ernst Mach (1838–1916), also an Austrian.

15.5 Sound Characteristics

Perceived sounds are described by terms whose meanings are similar to those used to describe the physical properties of sound waves. Physically, a wave is generally characterized by intensity, frequency, and waveform (harmonics). The corresponding terms used to describe the sensations of the ear are loudness, pitch, and quality (or timbre). These general correlations are shown in Table 15.2. However, the correspondence is not perfect. The physical properties are objective and can be measured directly. The sensory effects are subjective and vary from person to person. (As an analogy, think of temperature as measured by a thermometer and by the sense of touch.)

Sound intensity and its measurement on the decibel scale were covered in Section 15.3. **Loudness** is related to intensity, but the human ear responds differently to sounds of different frequencies. For example, two tones with different frequencies but the *same* intensities (in W/m^2) may be judged by the ear to have different loudnesses.

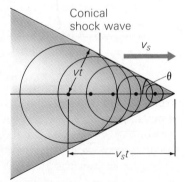

Figure 15.12 **Shock wave cone**
When the speed of the source (v_s) is greater than the speed of sound in air (v), the interfering spherical sound waves form a V-shaped pressure ridge (side view), or a conical shock wave. The angle θ is given by $\sin \theta = v/v_s$, and the inverse ratio v_s/v is called the Mach number.

Table 15.2
General Correlation between Perception and Physical Characteristics of Sound

Sensory Effect	Physical Wave Property
Loudness	Intensity
Pitch	Frequency
Quality (timbre)	Waveform (harmonics)

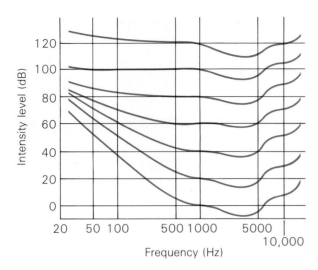

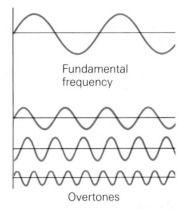

Figure 15.13 **Equal-loudness contours**
Curves indicate tones that are judged to be equally loud, though they have different frequencies and intensity levels. For example, on the next-to-lowest contour, a 1000-Hz tone at 20 dB sounds equally as loud as a 100-Hz tone at 50 dB. Note that the frequency scale is logarithmic to compress the large frequency range.

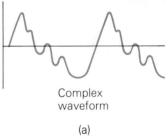

Fundamental frequency

Overtones

Complex waveform

(a)

Frequency and **pitch** are often used synonymously, but again there is an objective-subjective difference. If the same low-frequency tone is sounded at two intensity levels, most people will say that the more intense sound has a lower pitch, or perceived frequency.

The curves in the graph of intensity level versus frequency shown in Fig. 15.13 are called equal loudness contours. They join points representing intensity-frequency combinations that the average person judges to be equally loud. The top curve shows that the decibel level of the threshold of pain does not vary much from 120 dB, regardless of the frequency of the sound. However, the threshold of hearing (0 dB at 1000 Hz) varies widely with frequency, as indicated by the lowest curve in the graph. A tone with a frequency of 100 Hz must have an intensity of almost 40 dB to be heard. The dips (or minima) in the curves indicate that the human ear is most sensitive to sounds whose frequencies are between 1000 Hz and 5000 Hz. Note that a tone with a frequency around 3300 Hz can be heard at intensity levels *below* 0 dB.

The **quality** of a tone is the characteristic that enables it to be distinguished from another of basically the same intensity and frequency. Tone quality depends on the waveform, specifically, on the number of harmonics (overtones) present (Fig. 15.14). The tone of a voice depends in large part on the vocal resonance cavities. One person can sing a tone with the same basic frequency and intensity as another, but different combinations of overtones give the two voices different qualities.

Sound quality is highly subjective. Some sounds are pleasing to some people and cultures, but not to others. Certain music may seem discordant to some people, but others find it quite pleasing.

The notes of a musical scale correspond to certain frequencies; for example, middle C (C_4) has a frequency of 262 Hz. When a note is played on an instrument, its assigned frequency is that of the first harmonic, the fundamental frequency. This frequency is dominant over the accompanying overtones that determine the sound quality of the instrument. Recall from Chapter 14 that the overtones produced depend on how an instrument is played in some cases. Whether a violin string is plucked or bowed can be discerned from the quality of identical notes.

(b)

Figure 15.14 **Waveform and quality**
(a) The superposition of sounds of different frequencies and amplitudes gives a complex waveform. The overtones determine the quality of the sound. (b) A voice print. The waveform of a student's voice tone is electronically displayed on an oscilloscope.

Musical instruments provide good examples of standing waves and boundary conditions. On some stringed instruments, different notes are produced by varying the lengths of the strings using finger pressure (Fig. 15.15). As you learned in Chapter 14, the natural frequencies of a stretched string (fixed at each end, as is the case for the strings on an instrument) are $f_n = nv/2L$, where $v = \sqrt{F/\mu}$. Initially adjusting the tension in a string tunes it to a particular (fundamental) frequency. Then the effective length of the string is varied by finger pressure.

Standing waves are also set up in wind instruments. For example, consider a pipe organ with fixed pipe lengths, which may be open or closed (Fig. 15.16). An open pipe is open at both ends, and a closed pipe is closed at one end and open at the other (the antinode end). Analysis similar to that done in Chapter 14 for a stretched string with the proper boundary conditions shows that the natural frequencies for the pipes are (where v is the speed of sound in air):

$$f_n = \frac{v}{\lambda_n} = \frac{nv}{2L} \qquad \text{for } n = 1, 2, 3, \ldots \tag{15.18}$$

natural frequencies for an open pipe

and

$$f_m = \frac{v}{\lambda_m} = \frac{mv}{4L} \qquad \text{for } m = 1, 3, 5, \ldots \tag{15.19}$$

natural frequencies for a closed pipe

Figure 15.15 A shorter vibrating string, a higher frequency
Different notes are produced with a stringed instrument by placing a finger on a string to change its effective, or vibrating, length.

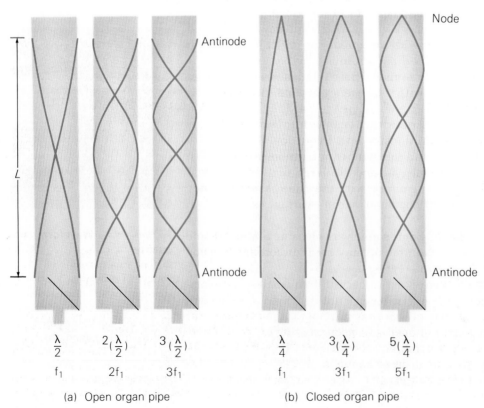

| $\frac{\lambda}{2}$ | $2(\frac{\lambda}{2})$ | $3(\frac{\lambda}{2})$ | $\frac{\lambda}{4}$ | $3(\frac{\lambda}{4})$ | $5(\frac{\lambda}{4})$ |
| f_1 | $2f_1$ | $3f_1$ | f_1 | $3f_1$ | $5f_1$ |

(a) Open organ pipe   (b) Closed organ pipe

Figure 15.16 Organ pipes
Longitudinal standing waves are formed in vibrating air columns in pipes (illustrated here with sinusoidal curves). (a) An open pipe has antinodes at both ends. (b) A closed pipe has a closed (node) end and an open (antinode) end.

Important Formulas

Speed of sound (in m/s):

$v = 331.5 + 0.6T_c$

Intensity of a point source:

$$I = \frac{P}{4\pi R^2} \quad \text{and} \quad \frac{I_2}{I_1} = \left(\frac{R_1}{R_2}\right)^2$$

Intensity level (in dB):

$$\beta = 10 \log \frac{I}{I_o} \quad \text{where } I_o = 10^{-12} \text{ W/m}^2$$

Phase difference (where PD is path difference):

$$\theta = \frac{2\pi}{\lambda} \text{ (PD)}$$

Condition for constructive interference:

$PD = n\lambda \quad$ where $n = 0, 1, 2, 3, \ldots$

Condition for destructive interference:

$$PD = m\left(\frac{\lambda}{2}\right) \quad \text{where } m = 1, 3, 5, \ldots$$

Beat frequency:

$f_b = |f_1 - f_2|$

Doppler effect:

$$f_o = \left(\frac{v}{v \mp v_s}\right)f_s = \left(\frac{1}{1 \mp v_s/v}\right)f_s$$

$$\begin{cases} - \text{ for source approaching stationary observer} \\ + \text{ for source moving away from stationary observer} \end{cases}$$

$$f_o = \left(\frac{v \pm v_o}{v}\right)f_s = \left(1 \pm \frac{v_o}{v}\right)f_s$$

$$\begin{cases} + \text{ for observer approaching stationary source} \\ - \text{ for observer moving away from stationary source} \end{cases}$$

Mach number:

$$M = \frac{1}{\sin \theta} = \frac{v_s}{v}$$

Natural frequencies of open organ pipe:

$$f_n = \frac{nv}{2L} \quad \text{for } n = 1, 2, 3, \ldots$$

Natural frequencies of closed organ pipe:

$$f_m = \frac{mv}{4L} \quad \text{for } m = 1, 3, 5, \ldots$$

Questions

Sound Waves

1. An alarm clock is hanging in a sealed glass jar. When the alarm goes off, the ringing can be heard. Why? What would happen if the air were evacuated from the jar?

2. What is necessary for there to be (a) physical sound and (b) perceived sound?

3. How do the wavelengths of infrasonic, audible, and ultrasonic waves compare?

4. Why do some flying insects produce buzzing sounds and some do not?

The Speed of Sound

5. How many times faster is the speed of light in air than the speed of sound in air?

6. Is the speed of sound the same for all frequencies? How could this be proven experimentally?

7. The speed of sound in air is affected by humidity. Why? (Hint: compare the molecular masses of the gases.)

8. (a) A rifle flash is seen in the distance, and the report of the shot is heard afterwards. Why? (b) With large explosions, an observer may feel the ground tremor before hearing the explosion. Why? (c) When you hear the sound of a jet plane flying over and look for the plane, you may have trouble spotting it even though it is within view. Why?

9. In a popular lecture demonstration, the instructor breathes helium gas. This gives the instructor's voice a high-pitched sound, like Donald Duck. Why? (Hint: the wavelengths in the vocal cavities remain the same.)

10. There are some noticeable differences in the speeds of sound in different media in Table 15.1. Can you offer some explanations for these?

11. Does wind affect the speed of sound? Explain.

Sound Intensity

12. The Richter scale used to measure the intensity levels of earthquakes is a logarithmic scale, as is the decibel scale. Why are logarithmic scales used?

13. Can there be negative decibel levels, such as -10 dB? What would these mean?

14. The intensity level of a sound doubles, from 20 dB to 40 dB, and then doubles again, from 40 dB to 80 dB. Are the effects the same? Explain. What is the total effect of an increase from 20 dB to 80 dB?

15. A sound is 10,000 times more intense than the threshold of hearing. What is the intensity level of the sound? Give an example of a familiar sound that is at this level.

Sound Phenomena

16. Give examples of (a) reflection, (b) refraction, and (c) diffraction of sound.

17. Explain how the interference of waves at a point is determined and how frequency and amplitude affect it.

18. Standing waves are said to be due to spatial interference (interference in space) and beats to be due to temporal interference (interference in time). Explain what is meant by this.

19. Do interference beats have anything to do with the beat of music?

20. Is there a Doppler effect if a sound source and an observer are moving with (a) the same velocity or (b) at right angles? (c) What would be the effect if a moving source accelerated toward a stationary observer?

21. Binary (double) stars revolve around each other. The Sun rotates. How were these facts discovered?

22. A sonic boom sounds like the noise of a nearby explosion. Discuss the similarities and differences.

23. How fast would a "jet fish" have to swim to create an aquatic boom?

24. Why is the Mach number not defined as $\sin \theta$ instead of $1/\sin \theta$, where θ is the half-angle of the shock wave cone?

Sound Characteristics

25. If two tones of different frequencies are judged to be equally loud, is it safe to assume that their intensities are equal?

26. A 1000-Hz tone has an intensity level of 20 dB. What would be the intensity level of a 200-Hz tone that would sound equally loud?

27. (a) Why do people's voices sound richer or fuller when they are singing in the shower? (b) Normal voice sounds range from 500 to 5000 Hz, but high-fidelity music requires an extended frequency range. Why?

28. (a) How does air temperature affect the frequencies of pipe organs? (b) When the base of vibrating tuning fork is placed against a tabletop, an amplified sound is heard. Why? Is there a similar effect for stringed instruments? Why do stringed instruments have holes in the tops of their bodies? (c) Why are some strings on stringed instruments wrapped with fine wire? Which ones are these? If you broke a string on a guitar and had to substitute a lighter string, how would you compensate for the change in tone? (d) Guitars have raised ridges or frets on their necks to show where to place the fingers. Why are the frets closer together near the bridge (the wooden piece at the node end on the body of the guitar)? Violins do not have frets. Why? (e) How are different notes or fundamental frequencies obtained on instruments such as trombones, trumpets, and clarinets?

29. (a) Why are curtains as well as sound-reflecting backdrops hung at the back of some auditorium stages? (b) Why do sounds in an empty room seem hollow? (c) After a snowfall, it seems particularly quiet. Why?

30. Loud music played in another room can often be heard through the walls (or ceiling or floor), particularly the bass parts. What does this tell you about the natural frequency of a wall?

Problems

15.1 and 15.2 Sound Waves and the Speed of Sound

■1. What is the speed of sound in air at (a) 10 °C and (b) 30 °C?

■2. Give the approximate divisions and upper limit of the sound spectrum in wavelengths. ($v = 340$ m/s)

■3. The thunder from a lightning flash is heard by an observer 8.0 s after she sees the flash. What is the approximate distance to the lightning strike in (a) kilometers and (b) miles?

■4. What is the air temperature if the speed of sound is 0.340 km/s?

■■5. Particles that are about 2.9×10^{-2} cm in diameter are to be scrubbed loose in an aqueous ultrasonic cleaning bath. Above what frequency should the bath be operated to produce wavelengths of this size and smaller?

■■6. How long does it take sound to travel 3.5 km in air if the temperature is 24 °C?

■■7. A tuning fork vibrates at a frequency of 512 Hz. What is the wavelength of the sound coming from the fork when the air temperature is (a) 0 °C and (b) 20 °C?

■■8. If the room temperature of the air increases by 10 °C, what will be the percentage change in the speed of sound?

■■9. According to Eq. 15.1, at what temperature will the speed of sound be twice what it is at 0 °C?

10. An engineer who uses units of feet and degrees Fahrenheit wants an equation for the speed of sound in air. Rewrite Eq. 15.1 to meet this need.

11. At a baseball game on a cool day (air temperature of 18 °C), a fan hears the crack of the bat 0.50 s after observing a batter hit a ball. How far is the fan from home plate?

12. A person fires a rifle at a target. The bullet has a muzzle velocity of 460 m/s, and the person hears the bullet strike the target 2.00 s after firing it. The air temperature is 72 °F. What is the distance to the target?

13. One hunter sees another who is 1.0 km away fire his rifle (sees smoke come from the barrel). If the air temperature is 5 °C, how long will it be until the first hunter hears the report of the shot?

14. The speed of sound in steel is about 4500 m/s. A steel rail is struck with a hammer, and there is an observer 0.30 km away with one ear to the rail. (a) How much time will elapse from the time the sound is heard through the rail to the time it is heard through the air? Assume that the air temperature is 20 °C and there is no wind blowing. (b) How much time would elapse if the wind were blowing toward the observer at 36 km/h from where the rail was struck?

15. The note A (440 Hz) is sounded on a stringed musical instrument. The air temperature is 20 °C. Approximately how many vibrations does the string make before the sound reaches a person 30 m away?

16. A hiker gives a shout and hears the echo reflected from a rock wall 5.0 s later. If the air temperature is 10 °C, how far is the hiker from the wall?

17. (Here's an old one.) A person drops a stone into a deep well and hears the splash from its hitting the water 3.16 s later. How deep is the well? (Assume the air temperature in the well to be 10 °C.)

18. Sound propagating through air at 20 °C passes through a vertical cold front into air that is 5.0 °C. If the sound has a frequency of 2400 Hz, by what percentage does its wavelength change as it crosses the boundary?

19. A resonance tube is an apparatus used to determine the speed of sound in air (see Fig. 15.17). The length of the empty part of the tube (L) is adjusted by raising and lowering the water level (by moving the reservoir can). If, when a vibrating tuning fork is held over the tube, the length of the empty part is such that there is a node at the water boundary and an antinode at the top of the tube, a

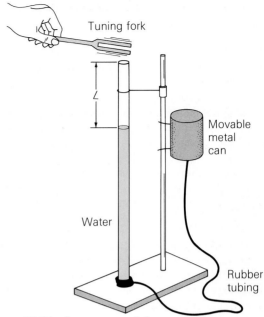

Figure 15.17 A resonance tube apparatus
See Problems 19 and 65.

loud resonance sound is heard. The wavelength can be determined by measuring the length of the empty tube. When a fork with a frequency of 512 Hz is used, the first resonance is heard when the water drops 16.5 cm below the top of the tube. (a) What is the speed of sound? (b) What is the air temperature? (c) If the overall tube length is 1.00 m, how many different resonances could be produced?

20. In a lab experiment using a resonance tube apparatus (see Problem 19), the second resonance is heard when the water level is 36 cm below the top of the tube. If the air is at room temperature, what is the frequency of the tuning fork?

15.3 Sound Intensity

21. What is the decrease in intensity if the distance from a point source is tripled?

22. By how many times must the distance from a point source be increased to *reduce* the sound intensity by $\frac{1}{3}$?

23. Find the intensity levels in decibels for sounds with intensities of (a) 10^{-2} W/m^2, (b) 10^{-4} W/m^2, and (c) 10^{-13} W/m^2.

24. What is the intensity of a sound that has an intensity level of (a) 60 dB and (b) 100 dB?

25. An observer 9.0 m from a stationary point source moves 5.0 m closer to the source. (a) By what factor does the intensity of the sound change? (b) How much farther away from the source would the observer then have to move to reduce the intensity by $\frac{1}{2}$?

■■26. If the intensity of one sound is 10^{-4} W/m² and the intensity of another is 10^{-2} W/m², what is the difference in their intensity levels?

■■27. A tape player has a signal-to-noise ratio of 58 dB. How many times larger is the intensity of the signal than that of the backgound noise?

■■28. A point source emits energy at a rate of 6.0×10^{-3} W. (a) What is the intensity at a distance of 1.55 m from the source? (b) What is the intensity level at that location?

■■29. A person standing 4.0 m from a wall shouts so that the sound strikes the wall with an intensity of 2.5×10^{-5} W/m². Assuming that the wall absorbs 10% of the incident energy and reflects the rest, what is the sound intensity level of the reflected sound?

■■30. Two identical balloons burst simultaneously, producing a sound with an intensity level of 90 dB. What would the intensity level have been if only one balloon had burst?

■■31. A particular machine produces a sound with an intensity of 2.0×10^{-5} W/m². (a) What is the intensity level of this sound? (b) What would the intensity level be if (1) two such machines were operating and then (2) ten more were added? (c) If only five of the machines were running at once, by how much would the intensity level of part (b2) be reduced?

■■32. Two people speak at the same time with intensity levels of 60 dB and 65 dB. What is the intensity level of the combined sounds heard by another person?

■■■33. A person has lost 30 dB of the normal range of hearing response. How many times must the intensity of a sound with an intensity level of 70 dB be amplified by a hearing aid to allow the person to hear it at a normal intensity?

■■34. A factory whistle is heard by one worker at an intensity level of 60 dB and by another at 80 dB. How many times farther away from the whistle is the first worker?

■■35. The average sound intensities due to continuously operating machinery in two parts of a factory are 10^{-3} W/m² and 10^{-1} W/m², respectively. What is the difference between these intensities in decibels?

■■36. Outdoor power equipment is rated as shown in

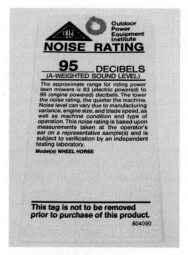

Figure 15.18 **Noise rating**
The sound level of an engine-powered riding mower is rated at 95 dB, as compared to 83 dB for an electric-powered mower. See Problem 36.

Fig. 15.18 for a riding lawn mower. (a) What is the intensity for this particular mower? (b) How many times more intense is an engine-powered mower than an electric-powered mower?

■■37. What would be the intensity level of a 33-dB sound after being amplified a million times?

■■38. Show that $10 \log I/I_0$ is $\frac{1}{10}$ of a bel.

■■39. Show that the difference in the intensity levels for intensities I_2 and I_1 is given by $10 \log I_2/I_1$.

■■■40. A 1000-Hz tone issuing from a loudspeaker has an intensity level of 100 dB at a distance of 2.5 m. If the speaker is assumed to be a point source, how far from the speaker will the sound have intensity levels of (a) 60 dB and (b) just barely enough to be heard?

■■■41. A flying bee produces a buzzing sound that is just barely audible to a person 5.0 m away. How many bees would have to be buzzing at that distance to produce a sound with an intensity level of 40 dB?

■■■42. A stereo speaker is rated at 60 W of output at 1000 Hz. At the lower limit of the audible sound range, the output decreases by 1.5 dB. What is the power output of the speaker at the lower frequency?

15.4 Sound Phenomena

■■43. Two sound waves with the same wavelength, 1.5 m, arrive at a point after having traveled (a) 13.50 m and

16.50 m and (b) 9.00 m and 11.25 m. What type of interference occurs in each case?

■■**44.** Two adjacent point sources, source 1 and source 2, are in front of an observer and emit identical 600-Hz tones. How far directly behind source 2 would source 1 have to be moved for the observer to hear no sound? (Assume that the air temperature is 20 °C.)

■■**45.** What is the beat frequency of two tones that have equal amplitudes and frequencies of 256 Hz and 260 Hz? How could the beat frequency be made zero?

■■**46.** A violinist tuning an instrument to a piano note of 256 Hz detects three beats per second. What are the possible frequencies of the violin tone?

■■**47.** While standing near a railroad crossing, a person hears a train horn. The frequency emitted by the horn is 400 Hz. If the train is traveling at 100 km/h and the air temperature is 25 °C, what is the frequency heard by the bystander when (a) the train is approaching and (b) when it has passed by?

■■**48.** What is the frequency heard by a person driving 50 km/h toward a blowing factory whistle ($f = 800$ Hz) if the air temperature is 0 °C?

■■**49.** How fast must a sound source be moving toward you to make the observed frequency 5.0% greater than the true frequency? (Assume that the speed of sound is 340 m/s.)

■■**50.** The driver of a truck traveling along a straight road sounds the horn, which has a frequency of 500 Hz. If an observer beside the road ahead of the truck measures a frequency of 520 Hz, how fast is the truck going? (Assume that the speed of sound is 340 m/s.)

■■**51.** A jet flies at a speed of Mach 1.5. What is the half-angle of the conical shock wave formed by the aircraft?

■■**52.** The half-angle of the conical shock wave formed by a supersonic jet is 30°. What are (a) the Mach number of the aircraft and (b) the actual speed of the aircraft if the air temperature is 0 °C?

■■■**53.** Two point-source loudspeakers are a certain distance apart and a person stands 12.00 m in front of one of the speakers on a line perpendicular to the base line of the speakers. If the speakers emit identical 1000-Hz tones, what is the minimum nonzero separation distance of the speakers so the observer hears no sound? (Let the speed of sound be 340 m/s.)

■■■**54.** Two identical strings on different cellos are tuned to the 440-Hz A note. The peg holding one of the strings slips, so its tensions is decreased by 1.5% What is the beat frequency heard when the strings are then played together?

■■■**55.** A stationary submerged submarine tracks an approaching submarine with sonar. A short pulse of ultrasound with a frequency of 400 kHz is sent toward the sub and returns 10 s later with an observed frequency of 412 kHz. (a) What is the speed of the approaching sub? Take the speed of sound in seawater to be 1200 m/s. (Hint: the ultrasound received by the moving sub is Doppler-shifted, and the moving sub acts as a moving source of sound with this shifted frequency. That is, the stationary sub receives a reflected pulse that is doubly Doppler-shifted.) (b) What is the approximate range of the approaching sub at the time of reflection?

■■■**56.** Show that the general equation for the Doppler effect for a moving source and a moving observer is given by

$$f_o = f_s \left(\frac{v \pm v_o}{v \mp v_s} \right)$$

where the sign convention is that used for Eqs. 15.12 and 15.15.

■■■**57.** A fire engine travels down a road at a speed of 90 km/h with its siren emitting sound at a frequency of 500 Hz. What is the frequency heard by a passenger in a car traveling at 65 km/h in the opposite direction to the fire engine and (a) approaching it and (b) moving away from it? (Hint: see Problem 56 and take the speed of sound to be 354 m/s.)

■■■**58.** The drivers of two trucks, one stationary and the other approaching it along a straight road, sound their horns at the same time. Both of the horns produce a sound whose frequency is 400 Hz. If an observer near the stationary truck hears a beat frequency of 4.0 Hz, how fast is the moving truck approaching the observer? Assume the speed of sound to be 340 m/s.

15.5 Sound Characteristics

■■**59.** An open organ pipe has a fundamental frequency of 440 Hz at room temperature. (a) What is the length of the pipe? (b) If the air temperature rose to 25 °C, what would the change in the fundamental frequency be?

■■**60.** A closed organ pipe has a length of 0.60 m. At room temperature, what are the frequencies of (a) the second harmonic and (b) the third harmonic?

■■**61.** Is it possible for an open and a closed organ pipe of the same length to produce notes of the same frequency? Justify your answer.

■■■**62.** A closed organ pipe has a fundamental frequency of 528 Hz (a C note) at room temperature. What is the fundamental frequency of the pipe when the temperature is 0 °C?

■■■**63.** A closed organ pipe is filled with helium. The pipe has a fundamental frequency of 660 Hz in air at 0 °C. What is the fundamental frequency with the helium?

Additional Problems

64. A violinist and a pianist simultaneously sound notes with frequencies of 352 Hz and 355 Hz, respectively. What beat frequencies will be heard by the musicians?

65. A tuning fork with a frequency of 440 Hz is held above a resonance tube partially filled with water (see Fig. 15.17). Assuming that the speed of sound in air is 342 m/s, for what heights of the air column will resonances occur?

66. A bystander hears a siren vary in frequency from 476 Hz to 404 Hz as a fire truck approaches, passes by, and moves away on a straight street. What is the speed of the truck? (Take the speed of sound in air to be 343 m/s.)

67. The sound intensity levels for a machine shop and a quiet office are 100 dB and 40 dB, respectively. (a) How many times greater is the intensity of the sound in the machine shop than that in the office? (b) What is each intensity?

68. A truck is traveling on a mountain road with a speed of 90 km/h. The driver gives the air horn a quick blast as the truck is moving directly toward a distant cliff. The frequency of the horn is 440 Hz, and the driver hears the echo 4.0 s later. (a) Approximately how far is the truck from the cliff if the speed of sound is 335 m/s? (b) What is the frequency of the echo heard by the driver?

69. Two sources of 440-Hz tones are located 6.97 m and 9.66 m from an observation point. If the air temperature that day is 15 °C, how do the waves interfere at the point?

70. If a person standing 25 m from a 550-Hz point source walks 5.0 m toward the source, by what factor does the sound intensity change?

71. The intensity level of a sound is 90 dB at a certain distance from its source. How much energy falls on a 1.5-m^2 area in 5.0 s?

72. At a distance of 4.0 m from a point source, a person measures the sound intensity level to be 70 dB. How far from the source would the person have to move to measure the intensity level as 50 dB?

73. Is the speed of sound greater in water or in mercury? How many times greater? (Hint: see Problem 46, Chapter 14.)

74. The first three natural frequencies of an organ pipe are 136 Hz, 408 Hz, and 680 Hz. (a) Is the pipe an open or closed one? (b) Taking the speed of sound in air to be 340 m/s, find the length of the pipe.

75. At a rock concert, the sound intensity level for a person in a front-row seat is 110 dB for a single band. If all the bands scheduled to play produce sound of that same intensity, how many of them would have to play simultaneously for the sound level to be at or above the threshold of pain?

Electric Charge, Force, and Energy

16

Our dependence on electricity becomes dramatically evident when the power goes off. Electricity is something that we generally take for granted. Yet less than a century ago there were no power lines crossing the land, no electric lights or appliances—none of the seemingly endless electrical applications surrounding us today.

The word "electricity" is derived from *elecktron*, the Greek word for amber. The early Greeks were familiar with the attraction a piece of amber rubbed with cloth or fur has for certain other materials. For centuries, such electrical properties were not understood. Only during the last 400 years have theories of electricity been developed.

Less than 200 years ago, the discovery was made that electricity and magnetism are in fact related phenomena. Today, the electromagnetic force is recognized as being one of the four fundamental forces (along with gravity and the strong and weak nuclear forces). It is customary in introductory physics courses to consider the electrical part of this force before the magnetic part and only then to combine electricity and magnetism into electromagnetism. This will be the approach followed in this book.

16.1 Electric Charge

What is electricity? This question has a variety of possible responses. One is that it has to do with electric charge. That is, electricity is a collective term describing phenomena associated with the interaction (or force) between electric charges. Like mass, **electric charge** is a fundamental property. Although no one knows what electric charge really is, particles of matter exhibiting certain interactions are said to possess this property.

Electric charge is associated with atomic particles, the electron and the proton. The simplistic solar system model of the atom likens its structure to planets orbiting the Sun. (Fig. 16.1). The electrons are viewed as orbiting a nucleus, a core containing protons and other particles. The centripetal force for the planets is supplied by gravity; the electrical force supplies it in the case of electrons. However, there is an important difference between the gravitational and electrical forces.

Gravitational "charges" (mass particles) are apparently of one type and give rise to only attractive gravitational forces. Electric charges, on the other hand, are of two types, distinguished as being positive ($+$) or negative ($-$). A positive charge is associated with the proton and a negative charge with the electron. Different combinations of the two kinds of charges produce *both* attractive and repulsive forces.

The directions of the electrical forces when charges interact with one another are given by the **law of charges**:

Like charges repel, and unlike charges attract.

That is, two negatively charged particles or two positively charged particles experience mutually repulsive forces, and particles with opposite charges are mutually attracted (Fig. 16.2).

The charge of an electron and the charge of a proton are equal in magnitude. The electronic charge (e) is taken as the fundamental unit of charge since it is the smallest charge that has been observed in nature.* An electric charge q is a charge with an integral multiple of fundamental charges (either positive or negative). That is,

$$q = ne \qquad (16.1)$$

where n is an integer. It is sometimes said that charge is quantized, which means that it only occurs in integral amounts of the fundamental electronic charge. Mass, on the other hand, is not quantized.

The standard unit of charge is the **coulomb (C)**, named for the French physicist Charles A. de Coulomb (1736–1806), who discovered a relationship between electrical force and charge (to be considered later in this chapter). The charges of the electron and the proton in terms of the coulomb are given in Table 16.1.

There are some other terms frequently used in discussing electrical properties. Saying that an object has a **net charge** means that it has an excess

---

*According to a recent theory, protons may be made up of particles called quarks that carry charges of $\frac{1}{3}$ and $\frac{2}{3}$ of the electronic charge. There is experimental evidence for the existence of quarks within the nucleus, but not outside the nucleus in common electrical interactions. See Chapter 30.

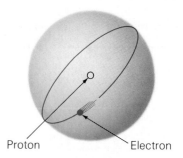
Proton   Electron
(a) Hydrogen atom

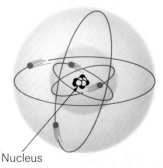
Nucleus
(b) Beryllium atom

Figure 16.1 **Simplistic model for atoms**
The so-called solar system model of (a) the hydrogen atom and (b) a beryllium atom views the electrons as orbiting the nucleus, analogous to the planets orbiting the Sun. (The electronic structure of atoms is much more complicated than this.)

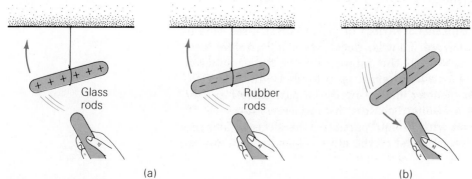
Glass rods

Rubber rods

(a)   (b)

Figure 16.2 **The law of charges**
(a) Like charges repel and (b) unlike charges attract.

Table 16.1
Particles and Electric Charge

Particle	Electric Charge	Mass
Electron	-1.6×10^{-19} C	$m_e = 9.11 \times 10^{-31}$ kg
Proton	$+1.6 \times 10^{-19}$ C	$m_p = 1.672 \times 10^{-27}$ kg
Neutron*	0	$m_n = 1.674 \times 10^{-27}$ kg

* The neutron is an electrically neutral particle found in the nucleus of an atom.

of either positive or negative charges. As you will learn in the next section, excess charges are obtained by a transfer of electrons. For example, if an object has a net charge of $+1.6 \times 10^{-19}$ C, it could be a charged atom (an ion) that has had one electron removed. This means that it has one proton whose charge is not offset by that of an electron, so it is not electrically neutral. Also, a point charge is similar to a point mass. But in this case, only the charge and not the particle is of interest.

Example 16.1 Quantized Charge
An object has a net charge of -1.0 C. How many excess electrons does this represent?

Solution
Given: $q = -1.0$ C   *Find*: n

The net charge is made up of an integral number of electronic charges. Using Eq. 16.1,

$$n = \frac{q}{e} = \frac{-1.0 \text{ C}}{-1.6 \times 10^{-19} \text{ C/electron}} = 6.3 \times 10^{18} \text{ electrons.} \quad \blacksquare$$

An important principle is the **conservation of charge**:

The net charge of an isolated system remains constant.

For example, if the system consists initially of two electrically neutral objects and electrons are transferred from one to the other, the former will have a net positive charge and the latter will have a net negative charge of equal magnitude. If an electrically neutral universe is assumed, the conservation of charge requires that positively and negatively charged particles always occur in pairs.

This does not mean that charged particles cannot be created or destroyed. Physicists have done this in this century, as you will learn in the chapters on modern physics. Mass-energy conversion involves a particle and its antiparticle, such as an electron (e^-) and a positron (e^+). (A positron is a particle with the same mass as an electron but having a positive charge; it and other antiparticles will be discussed in Chapter 28). However, by the conservation of charge, charged particles are created or destroyed only in such pairs. Like the other conservation laws of physics, the concept of the conservation of charge is called a law because no violation of it has ever been observed.

16.2 Electrostatic Charging

That there are two types of electrical charges and attractive and repulsive forces can be demonstrated easily. Before learning how this is done, you need to be able to distinguish between electrical conductors and insulators. What distinguishes these broad groups of substances is their ability to conduct, or transmit, electrical charge. Some materials, particularly metals, are good **conductors** of electrical charge. Others, such as glass, rubber, and most plastics, are **insulators**, or poor electrical conductors. A comparison of the relative magnitudes of the conductivities of some materials is given in Fig. 16.3.

A general picture is that in conductors, valence electrons (the ones in the outermost orbits) of atoms are relatively free, that is, are not permanently bound to a particular atom. (This electron mobility also plays a major role in thermal conduction.) In insulators, on the other hand, the valence electrons are more tightly bound. (The conduction of electrical charge can create an electrical current, which will be considered in more detail in Chapter 17.)

As Fig. 16.3 shows, there is an intermediate class of materials called

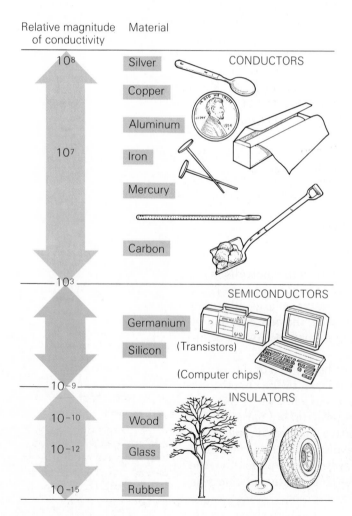

Figure 16.3 **Conductors, semiconductors, and insulators**
This is a comparison of the relative magnitudes of the electrical conductivities of various materials.

semiconductors. These do conduct electricity, but their resistance to charge flow is much greater than that of metals and much less than that of insulators. The conductivity of a semiconductor can be adjusted by adding certain types of atomic impurities in varying concentrations. Semiconductors form the basis of transistors and solid-state circuits, which have almost completely replaced vacuum tubes in electronic applications.

The electroscope is a simple device that demonstrates characteristics of electrical charge (Fig. 16.4). When charged objects are brought close to the insulated metal bulb, electrons in it are either attracted or repelled, according to the law of charges, and the metallic foil leaves separate. In general, **electrostatic charging** is the process by which an insulator or an isolated conductor receives a net charge and quickly comes to a static condition.

It was discovered in the early eighteenth century that when certain insulator materials were rubbed with cloth or fur, they became electrically charged. For example, if a hard rubber rod is rubbed with fur, the rod will have a net negative charge; and rubbing a glass rod with silk gives the rod a net positive charge. This is called **charging by friction**, although the transfer of charge is due to the close contact of the materials and is not associated with friction.

You have almost certainly experienced a result of electrostatic charging when after walking across a carpet on a dry day, you get "zapped" by a spark when you reach for a metal object, such as a doorknob. This happens because you have picked up a sufficient net charge to make the electric force great enough to ionize air molecules when your hand comes close to the metal object, giving rise to a spark discharge. This doesn't occur on humid days. With adequate humidity, a thin film of moisture on objects prevents the build-up of charge by conducting it away. (This should give you an idea of how sprays that prevent static cling work.)

As with heat, when a charge moves in a conductor, it is common to talk about a flow. Like heat flow, the flow of electricity was once thought to be due to some type of fluid transfer. Ben Franklin proposed a single-fluid theory of electricity. He assumed that all bodies had some normal amount of "electrical fluid." When some of this was transferred, for example, by rubbing two bodies together, one body would then have an excess of it and the other a deficiency. Franklin

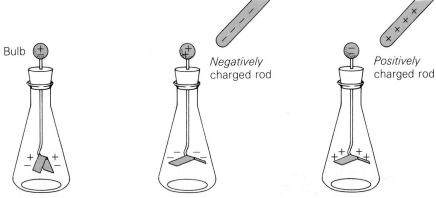

Bulb

Negatively charged rod

Positively charged rod

(a) Neutral electroscope has charges evenly distributed; leaves are close together.

(b) Electrostatic forces cause leaves to diverge.

Figure 16.4 **The electroscope**
An electroscope can be used to determine if an object is electrically charged. When a charged object is brought near the bulb, the leaves diverge.

(a) Electroscope is touched with negatively charged rod; charges transferred to bulb.

(b) Electroscope is negatively charged (net negative charge).

(c) Positively charged rod attracts electrons; leaves collapse.

Figure 16.5 **Charging by contact**
Charge is transferred to the electroscope when the charged rod touches the bulb. Then, when an oppositely charged rod is brought near the bulb, the leaves collapse, or come closer together.

indicated these conditions by plus and minus signs, respectively, which is the origin of the sign convention for charge.

Bringing a charged rod close to an electroscope bulb will reveal that the rod is charged, but won't tell you *how* the rod is charged (positively or negatively). This distinction can be made, however, if the electroscope is first given a known type of charge. For example, electrons can be transferred from one object to another if the two objects touch, as illustrated in Fig. 16.5(a) for an electroscope bulb and negatively charged rod. This is called **charging by contact**. If a negatively charged rod is brought close to the now negatively charged bulb, the leaves will diverge further [Fig. 16.5(b)]. An oppositely (positively) charged rod will cause the leaves to collapse, or come closer together [see Fig. 16.5(c)].

Since it is electrons that are transferred, you might wonder how an electroscope can be positively charged. This can be done by **charging by induction** (Fig. 16.6). Touching the bulb with a finger grounds the electroscope, that is, provides a path by which electrons can escape from the bulb. Then when a negatively charged rod is brought close to the bulb, it repels electrons

An electrical ground refers to earth (hence "ground") or some other object that can receive or supply electrons without significantly changing its own electrical condition.

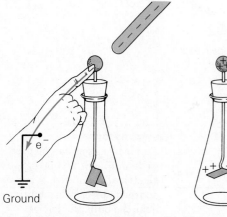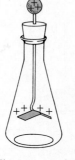

Ground

(a) Electrons are transferred to ground.

(b) Electroscope is left positively charged.

Figure 16.6 **Charging by induction**
(a) Touching the bulb provides a path to ground for charge transfer. (b) When the finger is removed, the electroscope has a net charge.

from the bulb. Removing the finger leaves the electroscope with a net positive charge.

Charging by induction does not have to involve a removal of charge from an object. Charge can be moved *within* an object to give different regions of charge. In this case, induction brings about **polarization**, or separation of charge (Fig. 16.7). Now you can understand why a balloon will stick to the wall or ceiling after being rubbed on someone's hair or sweater. The balloon is charged by friction, and the charged balloon induces an opposite charge on the surface, creating an attractive electrical force. Polarization of molecules is an important consideration in the storing of electrical energy.

Electrostatic charging can be annoying (even dangerous), but it can also be beneficial in a variety of practical applications. Clothes often stick together because of static cling, and an electrostatic spark discharge can start a fire or cause an explosion in the presence of a flammable gas. On the other hand, the air we breathe is cleaner because of electrostatic precipitators used in smoke-stacks. In these devices, electrical discharges cause the particles that are a byproduct of fuel combustion to be charged; then this particulate matter is removed from the flue gases through the use of electrical force. Another, almost indispensable, application is the electrostatic copier. (See the Insight feature on page 450.)

Nonpolar molecule

Induced molecular dipole

(a)

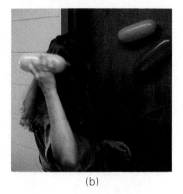

(b)

Figure 16.7 Polarization
(a) Some molecules can be electrically polarized so they have a separation of charge. The electric forces induce a molecular dipole, resulting in regions of charge. (b) Charging by friction and the polarization of the molecules in the door material by the charge on the balloons cause them to "stick" to the door because of attractive electrical forces.

16.3 Electric Force and Electric Field

The magnitude of the force between two electrical charges (q_1 and q_2) is described by an equation developed by Coulomb and (naturally) called **Coulomb's law**:

$$F = \frac{kq_1q_2}{r^2} \tag{16.2}$$

where k is a constant that is equal to 9.0×10^9 N-m^2/C^2 and r is the distance between the charges (Fig. 16.8).

Note that Coulomb's law is an inverse square relationship and that there are equal and opposite forces on the charges, by Newton's third law. The direction of the forces is given by the law of charges. Also note the similarity between this expression for the electrical force and the one for the force of gravity, $F = Gm_1m_2/r^2$. The relative strength of these two forces is the subject of Example 16.3.

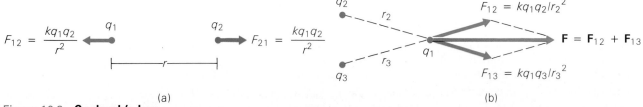

(a)

(b)

Figure 16.8 **Coulomb's law**
(a) The mutual electrostatic forces on two point charges are equal and opposite.
(b) For a configuration of more than two charges, the force on a particular charge is the vector sum of the forces on it due to all the other charges.

Xerography and Electrostatic Copiers

Xerography, coined from the Greek words *xeros* (meaning "dry") and *graphein* (meaning "to write"), refers to a dry process by which almost any printed material can be copied. This process makes use of a photoconductor, which is a light-sensitive semiconductor, such as selenium. When kept in darkness, a photoconductor is a good insulator that can be electrostatically charged. However, when light strikes the material, it becomes conductive and the electrical charge can be removed.

In transfer xerography, a photoconductor-coated plate, drum, or belt is electrostatically charged and then receives a projected image of the page to be copied. The illuminated portions of the photoconductor coating become conducting and are discharged, but the areas corresponding to the dark print remain charged. Essentially, this creates an electric charge copy of the original sheet. Then the photoconductor copy comes into contact with a negatively charged powder called toner, or dry ink. The toner is attracted to and adheres on the charged regions. Paper is placed over the inked photoconductor and given a positive charge. The toner is then attracted to the paper, and heating causes it to be permanently fused to the paper. All of this takes place very quickly—out comes your copy!

Example 16.2   Coulomb's Law

Two point charges of $-1.0 \ \mu C$ and $2.0 \ \mu C$ [$1 \ \mu C$ (1 microcoulomb) $= 10^{-6} \ C$] are separated by a distance of 0.30 m. What is the electrostatic force on each particle?

Solution

Given: $q_1 = -1.0 \ \mu C$ *Find*: F_{12} and F_{21}
$\quad\quad = -1.0 \times 10^{-6} \ C$
$\quad q_2 = 2.0 \ \mu C$
$\quad\quad = 2.0 \times 10^{-6} \ C$
$\quad r = 0.30 \ m$
$\quad k = 9.0 \times 10^9 \ \text{N-m}^2/\text{C}^2$

Eq. 16.2 gives the magnitude of the force acting on each particle:

$$F_{12} = F_{21} = F = \frac{kq_1q_2}{r^2} = \frac{(9.0 \times 10^9 \ \text{N-m}^2/\text{C}^2)(1.0 \times 10^{-6} \ C)(2.0 \times 10^{-6} \ C)}{(0.30 \ m)^2}$$

$$= 0.20 \ N$$

Since the charges are unlike, the force is attractive. By Newton's third law, F_{21} is equal to F_{12} but in the opposite direction.   ∎

PROBLEM-SOLVING HINT

The signs of the charges may be used explicitly in Eq. 16.2; if they are, a positive value for F indicates a repulsive force and a negative value an attractive force. However, it is usually less confusing to calculate the magnitude of the force without using the charge signs (as was done in Example 16.2) and apply the law of charges to determine whether the force is attractive or repulsive.

Example 16.3   Electrical Force versus Gravitational Force

How do the strengths, or magnitudes, of the electrical and gravitational forces between a proton and an electron compare?

Solution

Given: $q_1 = e = -1.6 \times 10^{-19}$ C   Find: $\dfrac{F_e}{F_g}$
$\quad\quad\quad q_2 = p = 1.6 \times 10^{-19}$ C
$\quad\quad\quad m_e = 9.11 \times 10^{-31}$ kg
$\quad\quad\quad m_p = 1.67 \times 10^{-27}$ kg
$\quad\quad\quad k = 9.0 \times 10^9$ N-m^2/C^2
$\quad\quad\quad G = 6.67 \times 10^{-11}$ N-m^2/kg^2

The expressions for the forces are

$$F_e = \frac{kq_1 q_2}{r^2} \quad \text{and} \quad F_g = \frac{Gm_1 m_2}{r^2}$$

Forming a ratio for comparison (and to eliminate r) gives

$$\frac{F_e}{F_g} = \frac{kq_1 q_2}{Gm_1 m_2}$$

$$= \frac{(9.0 \times 10^9 \text{ N-m}^2/\text{C}^2)(1.6 \times 10^{-19} \text{ C})^2}{(6.67 \times 10^{-11} \text{ N-m}^2/\text{kg}^2)(9.11 \times 10^{-31} \text{ kg})(1.67 \times 10^{-27} \text{ kg})}$$

$$= 0.23 \times 10^{40}$$

or $\quad F_e = (0.23 \times 10^{40}) F_g$

The magnitude of the electrostatic force between a proton and an electron is almost 10^{40} times greater than the gravitational force. For this reason, the gravitational force between charged particles is usually neglected.   ■

Example 16.4   Repulsive Force in the Nucleus

What is the magnitude of the repulsive electrostatic force between two protons in a nucleus? The distance between nuclear protons is about 10^{-13} cm.

Solution

Given: $r \cong 10^{-13}$ cm $= 10^{-15}$ m   Find: F
$\quad\quad\quad q_1 = q_2 = p = 1.6 \times 10^{-19}$ C
$\quad\quad\quad k = 9.0 \times 10^9$ N-m^2/C^2

Using Coulomb's law,

$$F = \frac{kq_1 q_2}{r^2} \cong \frac{(9.0 \times 10^9 \text{ N-m}^2/\text{C}^2)(1.6 \times 10^{-19} \text{ C})^2}{(10^{-15} \text{ m})^2}$$

$$\cong 230 \text{ N}$$

This is more than 50 lb of force. Large atoms contain many protons in their nuclei, so the repulsive force on any of these protons would be even larger, which would tend to cause the nucleus to fly apart. Since this doesn't generally happen, there must be a stronger attractive force holding the nucleus together (the nuclear force, which will be discussed in Chapter 29).   ■

It is convenient to represent electrical force in terms of an electric field. A particular arrangement, or configuration, of charges will have an effect on an additional charge placed anywhere nearby (or anywhere in space). Placing a test charge at various locations allows the force per unit charge at these points to be found. With these values, the electric field can be mapped. Knowing the *effect* of a charge configuration throughout space, you can ignore the configuration itself and talk in terms of the electric field it produces. The electric field tells you what force a charge experiences at a particular position. You can think of the force per unit charge (F/q_o) as being analogous to the price per pound used for food items. Knowing how much you want of an item, you can compute how much it will cost. Given the magnitude of a charge placed in an electric field, you can compute the force on it.

The electric field (intensity) is similar to the gravitational intensity (gravitational force per unit mass), discussed in Chapter 7: $F/m_o = g(r) = Gm/r^2$. In an electric field, the direction of the force on a charge depends on whether the charge is positive or negative, so this must be specified in a field representation. By convention, a *positive* test charge (q_o) is used for plotting the field. The **electric field** ($\mathbf{E} = \mathbf{F}/q_o$) is a vector field, and a vector at any point in it has the direction of the force that would be experienced by a positive charge placed there. The magnitude of the electric field (E), or the force per unit charge, at a distance r from a charge q is given by

$$E = \frac{F}{q_o} = \frac{kq_oq}{q_or^2} = \frac{kq}{r^2}$$

That is,

$$E = \frac{kq}{r^2} \tag{16.3}$$

The test charge is used simply to specify or determine the direction. Note in the equation that the q_o's cancel. Thus, the magnitude of the electric field is independent of the magnitude of the test charge. A *unit* test charge could be used ($q_o = 1.0$ C) or any other size positive charge.

The direction of the field is given by the law of charges, with q_o taken as positive. The units for the electric field strength can be seen to be newtons per coulomb (N/C).

Some electric field vectors in the vicinity of a positive charge are illustrated in Fig. 16.9. Note that the vectors point away from the positive charge, which is the direction of the force the charge experiences, and that the magnitude of the vectors decreases with distance from the charge (an inverse square relationship). Like a gravitational field, an electric field is represented by lines of force. For a configuration of charges, the total electric field at any point is the vector sum of the electric fields due to the individual charges.

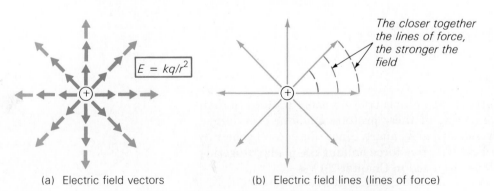

$E = kq/r^2$

(a) Electric field vectors

The closer together the lines of force, the stronger the field

(b) Electric field lines (lines of force)

Figure 16.9 **Electric field**
(a) Electric field vectors are determined using a positive test charge. Note that the magnitude of the field (lengths of vectors) becomes smaller as the distance from the charge increases (an inverse square relationship). (b) The vectors are connected to give electric field lines, or lines of force.

Example 16.5 Electric Field

What is the electric field at the point $(0, 3.0 \text{ m})$ due to two charges, $q_1 = 1.0 \ \mu\text{C}$ and $q_2 = -1.0 \ \mu\text{C}$, located at $(4.0 \text{ m}, 0)$ and $(-4.0 \text{ m}, 0)$, respectively? (See Fig. 16.10.)

Solution

Given: $q_1 = 1.0 \ \mu\text{C} = 1.0 \times 10^{-6} \text{ C}$ *Find:* E_t
$\qquad\quad q_2 = -1.0 \ \mu\text{C}$
$\qquad\qquad = -1.0 \times 10^{-6} \text{ C}$

Values of r from triangles
in figure

The distances of the point from the charges are both 5.0 m, from the 3-4-5 right triangles in the figure:

$$r = \sqrt{(3.0 \text{ m})^2 + (4.0 \text{ m})^2} = \sqrt{25 \text{ m}^2} = 5.0 \text{ m}$$

The electric field at the point due to each charge has the same magnitude, which is

$$E = \frac{kq}{r^2} = \frac{(9.0 \times 10^9 \text{ N-m}^2/\text{C}^2)(1.0 \times 10^{-6} \text{ C})}{(5.0 \text{ m})^2}$$

$$= 3.6 \times 10^2 \text{ N/C}$$

The directions of the electric fields for a positive test charge are indicated in the figure. The y components of the electric field vectors cancel, and the magnitude of the total electric field (E_t) is

$$E_t = 2E_x = 2(E \cos \theta)$$

Since $\theta = \tan^{-1}(3.0/4.0) = 37°$,

$$E_t = 2(E \cos 37°) = 2(3.6 \times 10^2 \text{ N/C})(0.80) = 5.8 \times 10^2 \text{ N/C}$$

in the $-x$ direction. ∎

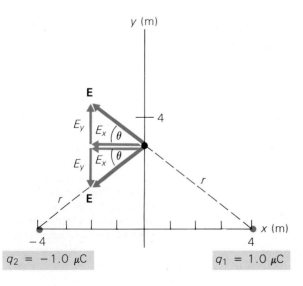

Figure 16.10 **Adding electric field vectors**
See Example 16.5.

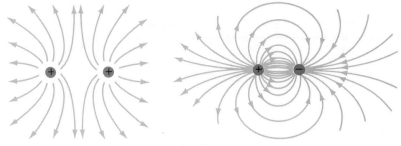

(a) Like and unlike point charges

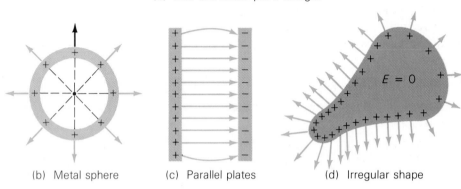

(b) Metal sphere

(c) Parallel plates

$E = 0$

(d) Irregular shape

Figure 16.11 **Electric fields**
Electric fields for various
charge configurations: (a)
Like and unlike point
charges, (b) charged metal
sphere, (c) differently charged
parallel plates, and (d) an
irregular shape.

The electric fields for some charge configurations and charged conductors
are depicted in Fig. 16.11. Note that the electric field lines begin on positive
charges and end on negative charges (or at infinity where there is no nearby
negative charge). Also note that the lines do not cross.

The electric fields associated with charged conductors (that are isolated or
insulated) have several interesting properties. Suppose that a metal sphere is
given a charge $+Q$, as shown in Fig. 16.11(b). Because of the mutually repelling
forces between them, the like charges will be distributed evenly over the sphere
and will eventually come to rest. At this point, the conductor is in electrostatic
equilibrium. Because of symmetry, the electric field outside a charged sphere is
as though all the excess charge on the sphere were concentrated at its center;
that is, $E = kQ/r^2$, where r is greater than the radius of the sphere.

As you might expect, the charges try to get as far away from each other as
possible. Thus,

> Any excess charge on an isolated conductor resides entirely on the
> surface of the conductor.

Also, it is easy to show that

> The electric field is zero everywhere inside a charged conductor.

If this were not the case, free charges inside the conductor would experience a
force, and the conductor wouldn't be in electrostatic equilibrium.

Another property of the electric field of a charged conductor is this:

> The electric field at the outer surface of a charged conductor is
> perpendicular to the surface.

Again, if this were not the case, there would be a component of the force
tangential to the surface of the conductor, which would cause charges to move
and would contradict the condition of electrostatic equilibrium.

These general characteristics apply to all charged conductors, not just the spherical one used for illustration. For irregularly shaped, or asymmetric, charged conductors, however, the electric field is not symmetric. As Fig. 16.11(d) shows, the field lines are not distributed evenly around an irregular shape. There is an accumulation of charge in certain places:

Charge tends to accumulate at sharp points, or locations of greatest curvature, on asymmetric charged conductors.

To understand why, consider the forces acting *between* charges on the surface of the conductor. Where the surface is slightly curved, these forces will be directed nearly parallel to the surface (they would be parallel for a completely flat surface). At a sharp end, the intercharge forces will be directed more nearly perpendicular to the surface, and there will be less tendency for the charges to move or be forced apart. Thus, a concentration of charge results. (Draw some surfaces and force vectors to prove this to yourself.)

If there is a large concentration of charge on a charged conductor with a sharp point, the molecules of air near the point may be ionized and a spark discharge may occur. More charge can be placed on a gently curved conductor, such as a sphere, before a spark discharge will occur. The concentration of charge at the sharp point of a conductor may be one reason for the effectiveness of lightning rods.

A general theory of how lightning develops is as follows. During the formation of a storm cloud, there is a separation of charge, presumably due to the separation of water droplets. This gives rise to different regions of charge in the cloud and induces a charge at the Earth's surface (Fig. 16.12). Air is a poor

(a)

(b)

Figure 16.12 **Lightning and lightning rods**
(a) Cloud polarization induces a charge on the Earth's surface. When the field becomes large enough, an electrical discharge results, which we call lightning.
(b) A lightning rod or arrestor provides a path to ground so as to prevent damage.

conductor, but when the charge build-up is great enough, air molecules become ionized. These charged ions provide a means by which the excess charge on the cloud can be discharged. The discharge is the lightning, which may be from cloud to cloud or from cloud to Earth (via the shortest distance to a tree or a building). The downward leader of the discharge is nearly invisible. When it makes contact with positive charges near the Earth, there is a large flow of charge. The bright flash that is most prominent is actually the return stroke, due to the charge flow propagating back along the original leader path.

The idea of using a rod to prevent a lightning stroke was based on the fact that charge tends to accumulate at the sharp edges or points of a conductor. Thus, it was thought that charges could leak off, or be attracted to these regions. If the cloud-induced charge leaked off, this might prevent a leader from coming into contact with the rod. However, it is now generally thought that the effectiveness of lightning rods results from providing a contact for the downward ionized leader and path for the flow of charge between the cloud and ground.

16.4 Electrical Energy and Electric Potential

As with any force, work and energy are associated with an electrical force. When two or more charges are brought closer together or further apart, work is done and energy is expended or stored. Recall how intermolecular (electrical) forces in matter can be seen as analogous to springs.

The expressions for the electrical force and the gravitational force are mathematically identical, and so are those for the potential energies, except for the sign. The **electrostatic potential energy** for two charges separated by a distance r is mutual and is given by

$$U = \frac{kq_1q_2}{r} \tag{16.4}$$

The electrical potential energy can be either positive or negative, since the force between two charges can be either attractive or repulsive, depending on the signs of the charges. For *unlike* charges, the electrical force is attractive. In that case, by direct analogy with an attractive gravitational force, the electrostatic potential energy is taken to be negative (Chapter 7). For *like* charges, the electrical force is repulsive, and the potential energy is positive. The signs of the charges can be used to keep things straight mathematically, as shown in the following example. Remember that energy is a scalar quantity.

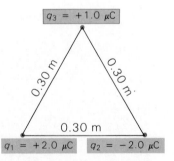

Example 16.6   Electrostatic Potential Energy
What is the total electrostatic potential energy of the charge configuration shown in Fig. 16.13?

Solution
Given: $q_1 = +2.0 \ \mu C$ *Find:* U_t
             $= +2.0 \times 10^{-6}$ C
      $q_2 = -2.0 \ \mu C = -2.0 \times 10^{-6}$ C
      $q_3 = +1.0 \ \mu C = +1.0 \times 10^{-6}$ C
      $r = 0.30$ m (same for all three cases)
      $k = 9.0 \times 10^9$ N-m^2/C^2

Figure 16.13 **Electrostatic potential energy**
A charged configuration has electrical potential energy due to the work done in bringing the charges together. This energy can be easily computed. See Example 16.6.

The total potential energy is the algebraic sum of the mutual potential energies of all pairs of charges:

$$U_t = U_{12} + U_{23} + U_{13} = k\left(\frac{q_1 q_2}{r} + \frac{q_2 q_3}{r} + \frac{q_1 q_3}{r}\right)$$

$$= (9.0 \times 10^9 \text{ N-m}^2/\text{C}^2)\left[\frac{(+2.0 \times 10^{-6}\text{ C})(-2.0 \times 10^{-6}\text{ C})}{0.30\text{ m}}\right.$$

$$+ \frac{(-2.0 \times 10^{-6}\text{ C})(+1.0 \times 10^{-6}\text{ C})}{0.30\text{ m}}$$

$$\left.+ \frac{(+2.0 \times 10^{-6}\text{ C})(+1.0 \times 10^{-6}\text{ C})}{0.30\text{ m}}\right]$$

$$= -0.12\text{ J}$$

This configuration of charge has potential energy because work had to be done in bringing the charges together. ■

It is important to consider the *change* in potential energy when charges are moved. For example, consider two positive charges separated by a distance r_A (see Fig. 16.14). Then mutual potential energy is

$$U_A = \frac{k(+q_1)(+q_2)}{r_A} = \frac{+kq_1q_2}{r_A}$$

If the charges are brought closer together (by work that is done by an external force—compressing the electrical spring, so to speak) so their separation is r_B (where $r_B < r_A$), then the mutual potential energy is

$$U_B = \frac{+kq_1q_2}{r_B}$$

The change in potential energy is

$$U = U_B - U_A = \frac{kq_1q_2}{r_B} - \frac{kq_1q_2}{r_A}$$

$U_A = kq_1q_2/r_A$

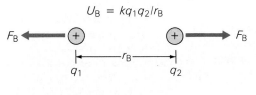

$U_B = kq_1q_2/r_B$

$\Delta U = U_B - U_A > 0$

Figure 16.14 **Change in potential energy** If two charges are brought closer together (or moved farther apart), a change in potential energy occurs as a result of work being done. In the case illustrated here, there is an increase in potential energy because work is done against the mutually repulsive force between the charges.

Since r_B is less than r_A $(r_B < r_A)$, ΔU is greater than zero, or the change is positive, which indicates an increase in potential energy. As you can easily show, if two charges are brought further apart, the change in their mutual potential is negative, indicating a decrease in the potential energy of the system.

Suppose that after the two charges are brought closer together, they are released. They will move apart because of the mutually repulsive force and there will be a decrease (a negative change) in their mutual potential energy. When the distance separating them is again r_A, the change in their potential energy is manifested as the kinetic energy of the particles. By conservation of energy, for the system in this case,

initial energy   final energy

$$U_B = K_A + U_A$$

$$\frac{kq_1q_2}{r_B} = \tfrac{1}{2}m_1v_1^2 + \tfrac{1}{2}m_2v_2^2 + \frac{kq_1q_2}{r_A}$$

The energy per unit charge is more commonly useful for electrical applications than the mutual potential energy. This quantity is determined by using a positive test charge, as for mapping an electric field. For such a test charge (q_o) located at a distance r from a charge q, the mutual electrostatic potential energy is

$$U = \frac{kq_oq}{r}$$

The **electric potential** (V) is the energy per unit charge, or

$$V = \frac{U}{q_o} = \frac{kq}{r} \tag{16.5}$$

Equivalently,

$$V = \frac{W}{q_o} \tag{16.6}$$

where W is the work done in bringing the test charge in from infinity, which is the zero reference point for the electrostatic potential energy and the electrical potential. As you can see, the units for electrical potential are joules per coulomb (J/C). This combination of units has been given the special name **volt** (**V**). That is, $1V \equiv 1$ J/C. Also, the electrical potential is commonly referred to as *voltage*.

The unit was named for Alessandro Volta (1745–1827), an Italian scientist who invented the electroscope and developed one of the first practical batteries.

What is usually of practical interest is the voltage *difference* between two points. This difference is equal to the work that must be done to move a test charge between the points (W_{AB})

$$\Delta V = V_B - V_A = \frac{W_{AB}}{q_o} \tag{16.7}$$

The work may be positive, negative, or zero. For these three possibilities, the potential at point B will be lower than, higher than, or the same as the potential at A, respectively. However, keep in mind that V is not potential energy but electric *potential* (energy/charge). The potential energy for a charge with potential V is $U = qV$.

Note that if $V_A = 0$ (or r_A taken to be infinite), Eq. 16.7 reduces to Eq. 16.6.

The selection of infinity as the zero reference point for measuring electrical potential difference is arbitrary. Later, when you study electrical circuit applications, you will learn that the potential of the Earth, or ground, is taken as a reference and assigned a value of zero (this is similar to taking $h = 0$ for the reference position of the gravitational potential).

The Relationship of Electric Potential to the Electric Field

The relationship between the electric potential and the electric field can be explored by moving a positive test charge in a uniform electric field, as illustrated in Fig. 16.15. This is done by an external force **F**, which acts against an opposing electrical force $q_o\mathbf{E}$. The work done in moving the charge through a distance d (from A to B) is

$$W_{AB} = Fd = q_o Ed$$

By Eq. 16.7, the potential difference is

$$\Delta V = V_B - V_A = \frac{W_{AB}}{q_o} = Ed$$

The potential at B is higher than that at A because work was done against the field. This is analogous to lifting an object to a greater height in the Earth's gravitational field.

The potential difference, or voltage, ΔV, is commonly written as simply V (not to be confused with the electric potential *at* a point). Recall that this is also done with Δx and Δt for convenience. From the equation $V = Ed$, you can see that units for an electric field are volts per meter (V/m). Since E also has units of newtons per coulomb (N/C), as can be seen from $E = F/q_o$, then N/C must be equivalent to V/m.

When the displacement of the test charge is not parallel to the field, the potential difference is found using the *component* of the electric field parallel to the displacement. For a uniform electric field, this is expressed as

$$V = -Ed \cos \theta \tag{16.8}$$

where θ is the angle between the electric field and the displacement. In Fig. 16.15, **E** and **d** are in opposite directions ($\theta = 180°$), so

$$V = -Ed \cos 180° = -Ed(-1) = Ed$$

As mentioned above, the potential difference, rather than the electric field, is usually specified for practical applications. For example, chemical energy produces a potential difference between the terminals of a battery. A D-cell flashlight battery has a terminal voltage of 1.5 V, that is, a potential difference of 1.5 V between its terminals.

You should keep in mind that the electrical force is conservative and that the work done in moving a charged particle between two points (and the potential difference between those points) is independent of path. That is, the potential difference depends only on the end points, not on how the charge is moved between them. Some points in space may have the same (equal) potentials. Lines connecting such points are called **equipotentials**. Some equipotentials for a point charge are shown in Fig. 16.16. The lines are at right angles to the field lines. Thus, no work is done in moving a charge along an

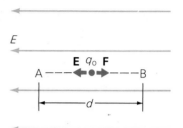

Figure 16.15 Electric field and electric potential
An external force (**F**) does work against the electric field, or electric force ($q_o\mathbf{E}$), in moving the charge a distance d. The change in potential is $\Delta V = Ed$. (Note that q_o is a positive charge, since it experiences an electrical force in the direction of the field.)

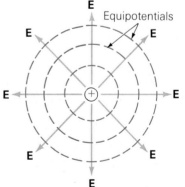

Figure 16.16 Equipotential
Lines connecting positions at which there are equal electrical potentials are called equipotentials. These are at right angles to the field lines, and no work is done in moving a charge along an equipotential.

equipotential since the displacement is at an angle of $90°$ to the electric field ($W = -Ed \cos \theta$ and $\cos 90° = 0$).

The concept of electric potential provides a convenient unit of energy that is used in calculating the kinetic energy of charged particles that have been accelerated through potential differences. An **electron volt (eV)** is the energy acquired by an electron in moving through a potential difference of 1 V, and

$$\Delta U = e \, \Delta V = (1.6 \times 10^{-19} \text{ C})(1.0 \text{ V})$$
$$= 1.6 \times 10^{-19} \text{ J}$$

That is,

$$1 \text{ eV} = 1.6 \times 10^{-19} \text{ J}$$

The energy of any charged particle accelerated through any potential difference can be expressed in electron volts. For example, if an electron is accelerated through a potential difference of 1000 V, its kinetic energy (K) is

$$K = eV = (1 \text{ e})(1000 \text{ V}) = 1000 \text{ eV} = 1 \text{ keV}$$

This is a kiloelectron volt. Although an electron volt is defined for an electron, any integral number of units of electronic charge can be considered.

Similarly, if a particle with a $+2$ charge were accelerated through this voltage, it would have $K = (2e)(1000 \text{ V}) = 2000 \text{ eV} = 2 \text{ keV}$. In this case, a charge of $+2$ is the equivalent of two units of electronic charge ($2e$).

Larger units for energy are sometimes needed, and mega–electron volts (MeV) and giga–electron volts (GeV) are used (1 MeV $= 10^6$ eV, and 1 GeV $= 10^9$ eV). At one time, a billion electron volts was referred to as BeV, but this was abandoned because confusion arose. In some countries, such as Great Britain and Germany, a billion means 10^{12} (which Americans call a trillion).

The Van de Graaff Generator

A method of generating large electrostatic potentials was developed by the American physicist Robert Van de Graaff around 1930. A diagram illustrating the operation of a simple Van de Graaff generator is given in Fig. 16.17. Such generators are commonly used for classroom demonstrations and can develop potential differences of more than 50,000 V.

A continuous, motor-driven, insulating (rubber) belt runs vertically around two pulleys. When the generator is turned on, frictional contact with the lower pulley transfers electrons from the belt to the pulley. The positively charged belt moves upward to the upper pulley, where electrons flow from the pulley to the belt. With continuous running, charges build up on both pulleys.

After a short time, the built-up charge is sufficient to ionize the air in the vicinity of the metal electrodes, or combs (C_1 and C_2 in the figure). When this occurs at the upper electrode (C_2), electrons are drawn from the metal sphere and transferred to the belt, which takes them to the lower pulley. When the charge on the lower pulley is sufficient to cause an ionizing discharge, the excess electrons are transferred to the lower electrode (C_1), which is grounded. The net effect is that electrons are transferred from the metal sphere to a ground, leaving the sphere positively charged. Since the electric field is nearly zero inside the sphere, electrons are continuously removed from it and a build-up of positive charge results, giving a high electric potential. (Another method of

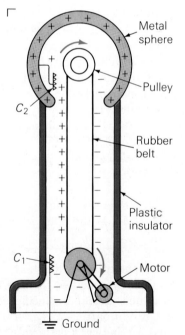

Figure 16.17 **The Van de Graaff generator**
This diagram shows how a charged, continuously moving belt transports charge from the metal sphere, giving rise to large potential differences between the sphere and ground.

charging the belt is through electric arcing, or corona discharge, from a high-voltage source.)

An electric potential between 50,000 V and 100,000 V is great enough to cause a so-called corona discharge through the ionization of air molecules around the sphere, and small electric arcs can be seen (and heard) between the sphere and a nearby object. In a demonstration of this, an instructor may arc the discharge to his or her knuckles. A voltage of 50,000–100,000 V may sound impressive, but there is no shock hazard for a healthy person because of the small amount of charge involved. The corona discharge can be surpressed and higher voltages achieved by surrounding the sphere with a gas that has a greater ionization potential than air. Because of this possibility of producing large voltages, Van de Graaff generators are used to accelerate charged particles in particle accelerators (Chapter 30).

16.5 Capacitance and Dielectrics

The electric field that exists between charged conductors provides a means for storing electrical energy. Consider the parallel metal plates shown in Fig. 16.18(a). Such an arrangement of conductors is called a **capacitor**. Work must be done to put charge on the plates. Suppose that one electron was moved to a plate. Putting the next electron on would be more difficult since it would be repelled by the original charge. The transfer of more charge requires more and more work as the charge on the plate accumulates. (This is analogous to

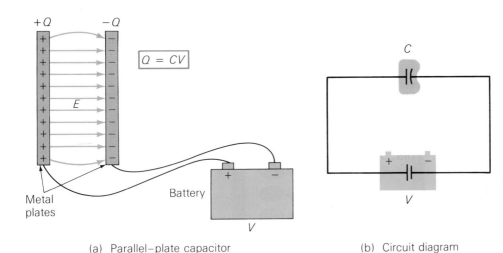

(a) Parallel–plate capacitor   (b) Circuit diagram

(c) Capacitors

Figure 16.18 **Capacitor and circuit diagram**
(a) Two parallel plates are charged by a battery to a charge Q, and $Q = CV$, where C is the capacitance. Work is done in charging the capacitor, and energy is stored in the electric field. (b) This circuit diagram shows the symbols used for a battery and a capacitor. The longer line of the battery symbol is taken as the positive terminal and the short line as the negative terminal. The symbol for a capacitor is similar, but with a curved line for the negative terminal. (At one time, it was common to represent a capacitor with two equal parallel lines, but the curved-line symbol is preferred to prevent confusion with the symbol for a battery.) (c) Various types of commercial capacitors.

potential energy and electric potential and (b) between electric potential and voltage?

23. When is a charge at a location of higher potential relative to some other position in an electric field?

24. Sketch the equipotentials in an electric field between two oppositely charged parallel plates, including those near the ends of the plates.

25. Explain why an electron volt is a unit of energy. Which is larger, a GeV or an MeV?

26. When an insulated person touches the metal sphere of a charged Van de Graaff generator, his or her hair may stand on end, as shown in Fig. 16.25(a). Explain why. Why must the person be insulated?

Capacitance and Dielectrics

27. In terms of the definition of capacitance, what implication does a large capacitance have for (a) the charge on the plates and (b) the energy of a capacitor?

28. Why must work be done in charging a capacitor even when it does not have a dielectric? Why is more work required when there is a dielectric?

29. What is an electric dipole? Do all materials have electric dipoles?

30. Why must a dielectric be an insulating material? The dielectric constant of a conductor is said to be infinite. Explain.

31. Explain why a stream of water is deflected, or bent, when a charged rubber rod is brought close to it, as shown in Fig. 16.25(b). (Try this yourself with a hard rubber comb or a balloon.) Will it work on a humid day? Explain.

(a)   (b)

Figure 16.25 Electrostatic effects
See Questions 26 and 31.

32. As you can see from the formulas given in this chapter, the capacitance and energy-storing capability of a parallel-plate capacitor can be increased by reducing the plate separation. This fact seems to imply that any amount of energy could be stored. Why is this not the case?

33. Which would result in a greater increase of the energy of a capacitor, doubling the capacitance or doubling the voltage across the plates?

34. Tell how the insertion of a dielectric affects each of the following properties of a capacitor: (a) the capacitance, (b) the electric field, and (c) the charge at a given voltage.

Problems

16.1   Electric Charge

■**1.** Two charges of -1.0 C are placed at opposite ends of a meterstick. Where could (a) a free electron and (b) a free proton be placed on the meterstick and be in electrostatic equilibrium?

■**2.** Charges of -1.0 C and $+1.0$ C are placed at opposite ends of a meterstick. Where could (a) a free electron and (b) a free proton be placed on the meter and be in electrostatic equilibrium?

■**3.** How many electrons are required to make up a net charge of $-0.10~\mu$C?

■**4.** What net electric charge would 10,000 protons have?

■**5.** A glass rod rubbed with silk acquires a charge of $+8.0 \times 10^{-10}$ C. (a) What is the charge on the silk? (b) How many electrons have been transferred to the silk?

16.3   Electric Force and Electric Field

■**6.** An electron and a proton are separated by 10 cm. (a) What is the force on the electron? (b) What is the net force on the system?

■**7.** Two charges originally separated by 50 cm are moved farther apart, until the force between them has decreased by a factor of 10. How far apart are the charges then?

■**8.** An isolated electron experiences an electrical force of 3.2×10^{-4} N. What is the magnitude of the electric field at the electron's location?

■**9.** What is the electric field at a point 0.25 cm away from a charge of 4.0 μC?

10. Two charges are attracted by a force of 20 N when separated by 25 cm. What is the force between the charges when the distance between them is 50 cm?

11. A charge of 6.0 μC at the origin of a set of coordinate axes experiences a force of 1.8 N in the y direction due to another charge located 0.30 m away. What are the magnitude and the position of the other charge?

12. Two charges, q_1 and q_2, are located at the origin and at $(0.50 \text{ m}, 0)$. Where on the x axis must a third charge, q_3, of arbitrary sign be placed to be in electrostatic equilibrium if (a) q_1 and q_2 are like charges of equal magnitude, (b) q_1 and q_2 are unlike charges of equal magnitude, and (c) $q_1 = +2.0 \mu$C and $q_2 = -8.0 \mu$C?

13. Two charges, -2.0μC and -5.0μC, are 0.40 m apart. (a) Where should a charge of -1.0μC be placed to put it in electrostatic equilibrium? (b) Where should a charge of $+1.0 \mu$C be placed to be in electrostatic equilibrium with the two original charges?

14. Three charges are located at the corners of an equilateral triangle, as depicted in Fig. 16.26. What are the magnitude and the direction of the force on q_1?

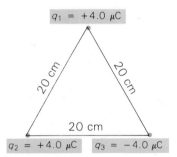

Figure 16.26 **A charge triangle**
See Problems 14, 17, 18, 26, 32, 34, 35, and 45.

15. Four charges are located at the corners of a square as illustrated in Fig. 16.27. What are the magnitude and the direction of the force on (a) q_2 and (b) q_4?

16. Two charges of $+2.0 \mu$C and $+6.0 \mu$C are 30 cm apart. Where on a line joining the charges is the electric field zero?

17. What is the electric field at the center of the triangle in Fig. 16.26?

18. Compute the electric field at a point midway between charges q_1 and q_2 in Fig. 16.26.

19. What is the electric field at the center of the square in Fig. 16.27?

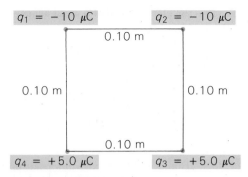

Figure 16.27 **A charge rectangle**
See Problems 15, 19, 20, 27, 33, 36, 37, and 46.

20. Compute the electric field at a point midway between charges q_3 and q_4 in Fig. 16.27.

21. A thin, hollow spherical conductor with a radius of 0.20 m has a uniform charge density of -10 C/m^2 on its surface. What is the electric field (a) inside the sphere and (b) at the surface of the sphere?

22. What are the magnitude and the direction of a vertically oriented electric field that would just support the weight of an electron?

23. An electron is released from rest in a uniform electric field that has a magnitude of 300 N/C in the $+x$ direction. (a) How long will it take the electron to travel 1.0 m? (b) What will its position be (in xy coordinates) at half that time?

24. Two 0.10-g pith balls are suspended from the same point by threads 30 cm long. (Pith is a light insulating material once used to make helmets worn in tropical climates.) When the balls are given equal charges, they come to rest 18 cm apart. What is the magnitude of the charge on each ball? (Neglect the mass of the thread.)

25. Two charged parallel plates separated by a distance of 5.0 cm have an electric field of 2.0×10^3 V/m between them (Fig. 16.28). (a) What is the potential difference between the plates? (b) What would be the change in the potential energy of a charge of 1.0 μC if it were moved from A to B? From A to C to B?

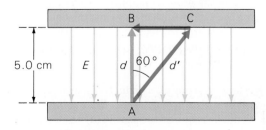

Figure 16.28 **Potential and energy**
See Problem 25.

■■■26. What is the electric field at a point 5.0 cm from q_1 and 15.0 cm from q_3 in Fig. 16.26?

■■■27. Compute the electric field at a point 4.0 cm from q_2 along a line running toward q_3 in Fig. 16.27.

16.4 Electrical Energy and Electric Potential

■28. Show that N/C is equivalent to V/m.

■■29. A charge of $-4.0\ \mu C$ is initially 40 cm from a fixed charge of $-6.0\ \mu C$ and is then moved to a position 90 cm from the fixed charge. (a) What is the change in the mutual potential energy of the charges? (b) Does this change depend on the path through which the one charge is moved?

■■30. How much work is required to bring two charges of $+3.0\ \mu C$ and $+5.0\ \mu C$ that are separated by an infinite distance to a distance of 0.50 m?

■■31. (a) What is the electrical potential energy of an electron located 10 cm from a charge of $+6.0\ \mu C$? (b) How much work is required to move the electron to an infinite distance from the charge?

■■32. Compute the energy necessary to bring together the charges in the configuration shown in Fig. 16.26.

■■33. Compute the energy necessary to bring together the charges in the configuration shown in Fig. 16.27.

■■34. What is the electric potential at the center of the triangle in Fig. 16.26?

■■35. Compute the electric potential at a point midway between q_2 and q_3 in Fig. 16.26.

■■36. What is the electric potential at the center of the square in Fig. 16.27?

■■37. Compute the electric potential at a point midway between q_1 and q_4 in Fig. 16.27.

■■38. A proton is moved 10 cm on a path parallel to the field lines of a uniform electric field of 10^5 V/m. (a) What is the change in the proton's potential? Consider both cases, of moving the proton with and against the field. (b) What is the change in energy in electron volts? (c) How much work would be done if the proton were moved perpendicularly to the electric field?

■■39. Two horizontal, conductive parallel plates are separated by a distance of 1.5 cm. What voltage across the plates would be required to suspend (a) an electron or (b) a

proton in midair between them? (c) What would be the signs of the charges on the plates in each case?

■■40. Two electrons that are initially at rest 10 cm apart are released. What will be their speed when they are 20 cm apart?

■■41. A proton moving directly toward another fixed proton has a speed of 0.50 m/s when the two are 1.0 m apart. How close to the stationary proton will the moving proton be when it stops and reverses course?

■■42. The electric field between two parallel plates separated by 6.0 cm has a magnitude of 3.0×10^4 V/m. How much energy is gained by an electron in going from one plate to the other?

■■43. In a television picture tube, electrons are accelerated from rest through a potential difference of 10 kV in an electron gun. (a) What is the "muzzle velocity" of the electrons emerging from the gun? (b) If the gun is directed at a screen 35 cm away, how long does it take the electrons to reach the screen?

■■■44. At STP (standard temperature and pressure, 0°C and 1 atm), the electric field required for the ionization of air is about 3 million volts per meter. (a) A particular Van de Graaff generator has a sphere with a diameter of 0.80 m. How much charge must be placed on this sphere to ionize air under these conditions? (b) What is the voltage of the sphere relative to a ground?

■■■45. What is the electric potential at a point 12 cm from q_2 and 8 cm from q_1 in Fig. 16.26?

■■■46. Compute the electric potential at a point 4.0 cm from q_4 along the line joining it to q_1 in Fig. 16.27.

16.5 Capacitance and Dielectrics

■47. What is the capacitance of a capacitor that acquires a charge of 0.40 μC when connected to a 200-V source?

■48. A capacitor has a capacitance of 50 pF, which increases to 175 pF with a dielectric material between its plates. What is the dielectric constant of the material?

■49. A parallel-plate capacitor without a dielectric has a capacitance of 0.20 μF for a certain voltage. When a film of one of the polymers in Table 16.2 is placed between the plates and the capacitor is charged to the same voltage, the capacitance is 0.46 μF. What material was used?

■■50. A 12-V battery is connected to a parallel-plate air capacitor with plate areas of $0.20\,\text{m}^2$ and a plate separation of $1.0\,\text{cm}$. (a) What is the resulting charge on the capacitor? (b) How much energy is stored in the capacitor?

■■**51.** If the capacitor in Problem 50 is discharged, immersed in silicone oil, and again connected to the battery, what will be the charge on it and the amount of stored energy?

■■**52.** A $1.0\text{-}\mu\text{F}$ parallel-plate capacitor with a distance of $0.50\,\text{mm}$ between the plates is to be built. (a) What is the required area of the plates without a dielectric between them? (b) What is the required area of the plates if a polystyrene film $0.50\,\text{mm}$ thick is used between them?

■■**53.** How much more energy is stored at a given voltage in a capacitor with a paper dielectric than in a capacitor with the same dimensions containing a polyethylene dielectric?

■■**54.** What are the maximum and minimum equivalent capacitances that can be obtained by combinations of three capacitors of $1.5\,\mu\text{F}$, $2.0\,\mu\text{F}$, and $2.5\,\mu\text{F}$?

■■**55.** Three capacitors with values $C_1 = 0.15\,\mu\text{F}$, $C_2 = 0.25\,\mu\text{F}$, and $C_3 = 0.30\,\mu\text{F}$ are connected in a circuit as shown in Fig. 16.22. (a) What is the equivalent capacitance of the arrangement? (b) If the circuit is connected to a 12-V battery, how much charge will be drawn from the battery? (c) What is the voltage difference across each capacitor?

■■**56.** For the arrangement of three capacitors in Fig. 16.22, with $C_1 = 20\,\text{pF}$ and $C_2 = 40\,\text{pF}$, what value of C_3 will give a total equivalent capacitance of $50\,\text{pF}$?

■■**57.** For the arrangement of three capacitors in Fig. 16.22, with $C_2 = 0.20\,\mu\text{F}$ and $C_3 = 0.30\,\mu\text{F}$, what value of C_1 will give a total equivalent capacitance of $0.17\,\mu\text{F}$?

■■**58.** Three capacitors of $0.20\,\mu\text{F}$ each are connected in parallel to a 12-V battery. (a) What is the charge on each capacitor? (b) How much charge is drawn from the battery?

■■**59.** Two capacitors, C_1 and C_2, are connected in parallel, and this combination is connected in series with another capacitor, C_3. If $C_1 = C_3 = 1.0\,\mu\text{F}$ and the total equivalent capacitance is $\frac{3}{4}\,\mu\text{F}$, what is the capacitance of C_2?

■■**60.** What is the charge on each of the capacitors in the circuit in Fig. 16.29?

■■■**61.** (a) A dielectric is inserted into a charged capac-

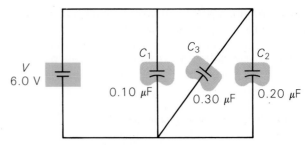

Figure 16.29 **A capacitor triad**
See Problem 60.

itor with a constant charge (Q) on the plates, as shown in Fig. 16.20(b). Show that the energy of the capacitor after the dielectric is introduced is less by a factor of $1/K$. Where does the "missing" energy go? (Hint: the dielectric will be attracted into the capacitor. Note the surface charges on the plates and the dielectric in the figure.) (b) Show that if a dielectric is inserted into a capacitor and the voltage across the plates is kept constant, the energy of the capacitor is increased by a factor of K. Where does the "added" energy come from?

■■■**62.** Four capacitors are connected in a circuit as illustrated in Fig. 16.30. Find (a) the charge on and (b) the voltage difference across each of the capacitors.

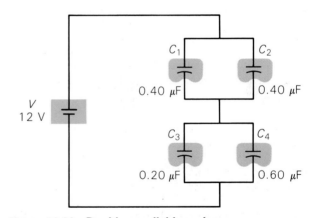

Figure 16.30 **Double parallel in series**

Additional Problems

63. Moving electrons may be deflected from a straight-line path by passing them between charged parallel plates. Consider an electron initially moving with a constant velocity and then passing between charged vertical parallel plates separated by a distance of $1.0\,\text{cm}$ and having a potential difference of 0.50 mV. (a) What is the lateral (horizontal) acceleration of the electron? (b) Show how you could use a small-angle approximation to calculate the angle of deflection of the electron during its transit between the plates if the electron's large initial velocity and the plate dimensions are known.

64. A particle with a mass of 1.0×10^{-5} kg and a charge of $+2.0$ μC is released in a uniform electric field of 12 N/C. How far does it travel in 0.50 s?

65. A rubber rod rubbed with fur acquires a charge of -2.4×10^{-9} C. (a) What is the charge on the fur? (b) How much mass is transferred to the rod?

66. In an unexcited hydrogen atom, which consists of a proton and an electron, the electron moves in a circular orbit of radius 0.053 nm about the proton. (a) Compute the centripetal force acting on the electron. (b) What is its orbital speed?

67. What is the equivalent capacitance of two capacitors of 0.04 μF and 0.80 μF when they are connected (a) in series and (b) in parallel?

68. Three charges, $+2.5$ μC, -4.8 μC, and -6.3 μC, are located at $(-0.20$ m, 0.15 m$)$, $(0.50$ m, -0.35 m$)$, and $(-0.42$ m, -0.32 m$)$, respectively. What is the electric field at the origin?

69. What speed will an electron reach if it is accelerated from rest through a potential difference of 50 V?

70. What is the mutual electrical potential energy of an electron and a proton in a hydrogen atom, as described in Problem 66?

71. Two charges, -4.0 μC and -5.0 μC, are separated by a distance of 20 cm. What is the electric field halfway between the charges?

72. How much charge flows through a 12-V battery when a 1.0-μF capacitor is connected across its terminals?

73. Compute the capacitance of a parallel-plate capacitor that has a plate separation of 1.0 mm and a plate area the size of a football field (300 yd by 600 yd, end zones not included).

74. Two charges, -3.0 μC and -4.0 μC, are located at $(-0.50$ m, $0)$ and $(0.75$ m, $0)$, respectively. (a) Where on the x axis is the electric field zero? (b) Is there a position (or positions) where the electric field has only a y component? Explain.

75. How much work is done in moving an electron along two legs of an equilateral triangle with sides 0.25 m long in an electric field of 15 V/m parallel to the other side of the triangle? (Take the electron as being initially at one end of the side of the triangle parallel to the electric field.)

76. When a series combination of two uncharged capacitors is connected to a 12-V battery, 173 μJ of energy is drawn from the battery. If one of the capacitors has a capacitance of 4.0 μC, what is the capacitance of the other?

77. The Earth has an electric field of about 150 V/m near its surface, directed toward its center. Assume that this field is due to a uniform distribution of charge on the Earth's surface. What is the surface charge density expressed in (a) coulombs/m^2 and (b) electrons/m^2?

Electric Current and Resistance

17

Most of the practical applications of electricity involve electric current. When you turn on the headlights of your car, a current of electrons flows through the circuit of which the filaments of the headlights are a part. Because of the resistance of the filaments, electrical energy is converted into light, or radiant energy. In the home, electrons flow in the wiring of electrical appliances when they are plugged in and turned on. In this case, the electrons surge back and forth 60 times a second, in an alternating current.

For sustained currents, voltage sources are required. These are devices such as batteries and generators, which produce electrical energy. That is, potential differences are produced by the conversion of other forms of energy into electrical energy. This chapter considers direct current, which is produced by batteries. The source of alternating current will be discussed in Chapter 21. However, the general relationships for voltage, current, and resistance developed here apply to both types of current and are useful in analyzing a variety of everyday practical applications.

17.1 Batteries and Direct Current

After studying electric force and energy in Chapter 16, you can probably guess what is required to produce an electric current, or a flow of charge. A flow of fluid (mass flow) may occur when there is a gravitational potential energy difference. Spontaneous heat flow occurs only when there is a temperature difference. Similarly, a flow of electric charge is dependent on a *difference* in electric potential, or voltage.

In solid conductors, particularly metals, outer atomic electrons of atoms are relatively free to move. The positively charged protons of the nuclei are fixed in the lattice structure of the material. (In liquid conductors, both positive and negative ions can move.) Energy is required to do the work of moving electric charge. Electrical energy is produced, or generated, through the conversion of other forms of energy into electrical energy, giving rise to a potential, or voltage, difference.

A **battery** is a device that converts chemical energy into electrical energy. The Italian scientist Alessandro Volta (1745–1827) is credited with constructing one of the first practical batteries. Basically, a battery consists of two electrodes in an electrolyte, which is a solution that conducts electricity (Fig. 17.1). With the appropriate electrodes and electrolyte (Volta used zinc and copper electrodes in dilute sulfuric acid), a potential difference develops across the electrodes as a result of chemical action. One electrode becomes negatively charged and is the **cathode**, or negative ($-$) terminal of the battery. The other electrode develops a deficiency of electrons, becoming positively charged; this is the **anode**, or positive ($+$) terminal of the battery.

The potential difference across the terminals of a battery is called the **electromotive force (emf)**. This is a somewhat misleading name because the electromotive force is not a force—it is the energy per unit charge given by the battery to each charge as it passes through. Because the emf is a potential difference, it is measured in volts. The emf is determined when the battery is not connected to an external circuit. When a battery is connected to a circuit and charge is flowing, the voltage across the battery is slightly less than the emf because of the battery's internal resistance. This is shown in the circuit diagram in Fig. 17.2. In this chapter, the emf will be taken to be the operating voltage unless otherwise specified, and the connecting wires in a circuit will be considered to have negligible resistance.

Current is supplied by a battery as long as the internal chemical action is maintained. The battery is said to deliver current to a circuit, or the circuit or its components draw current from the battery. Since the electrons can flow in only one direction in such a circuit, from the negative ($-$) terminal to the positive ($+$) terminal, this is called **direct current (dc)**.

There are a wide variety of batteries, constructed with different electrode materials and electrolytes. Two common types are the 1.5-V D-cell flashlight battery and the 12-V lead storage battery used in automobiles. A battery is called a (chemical) cell. The flashlight battery is referred to as a dry cell, but it is really only relatively so compared to a storage battery. The electrolyte

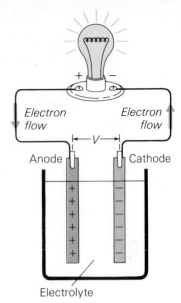

Figure 17.1 **The chemical battery or cell**
Chemical processes involving the electrolyte and two unlike metal electrodes cause one to become positively charged (the anode) and the other to become negatively charged (the cathode). The electric potential, or voltage, developed across the electrodes can cause a current, or a net flow of electrons, in a circuit.

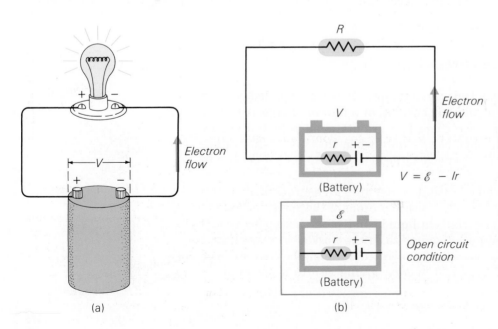

Figure 17.2 **Electromotive force (emf)**
(a) The potential difference across the terminals of a battery is called the electromotive force (emf), symbolized by $\mathscr{E}$. This is the voltage when the battery is not connected to or delivering current to an external circuit. (b) When the battery is operating in a circuit as diagrammed here, the voltage is slightly less than the emf because of the battery's internal resistance (r).

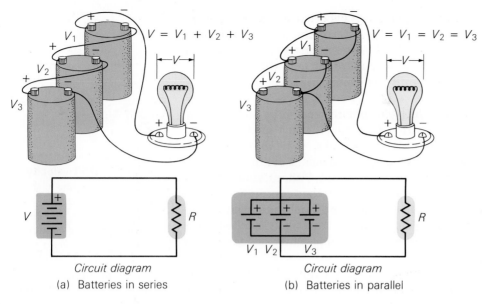

$$V = V_1 + V_2 + V_3$$

$$V = V_1 = V_2 = V_3$$

Circuit diagram

(a) Batteries in series

Circuit diagram

(b) Batteries in parallel

Figure 17.3 Batteries in series and in parallel
(a) When batteries are connected in series, their voltages add. (b) When similar batteries are connected in parallel, the voltage is the same, and each battery supplies a fraction of the total current.

(ammonium chloride and other additives) in a dry cell is in the form of a paste. When such a battery is dead, it cannot be regenerated because the active materials have been depleted.

A storage battery, which has many plate electrodes, may be recharged.* If a current is passed through a storage battery in the direction opposite to that of the normal current flow, the chemical reactions are reversed, and the battery is recharged. A 12-V automobile battery actually consists of six 2-V cells connected in series. That is, the positive terminal of each cell is connected to the negative terminal of the next cell in the circuit [Fig. 17.3(a)]. When the cells are connected in this fashion, their voltages add.

If cells are connected in parallel, all their positive terminals have a common connection, as do their negative terminals [Fig. 17.3(b)]. If identical batteries are connected in this way, the potential difference is the same across all of them, and each one supplies a fraction of the current to the circuit (for three batteries with equal voltages, each one supplies $\frac{1}{3}$ of the current).

Other sources of voltage, such as generators and photocells, will be considered in later chapters.

17.2 Current and Drift Velocity

A battery or some other voltage source connected to a continuous conducting path forms a complete circuit. *A sustained electric current requires a voltage source and a complete circuit.* As shown in Fig. 17.4, a circuit may have a switch (symbolized by S), which is used to open or close the circuit. When the switch is open, the current or circuit component is turned off; when the switch is closed, the current flows.

Since it is electrons that move in the wires of an external circuit, the net flow

---

* The anode of a lead storage battery is lead oxide (PbO_2) and the cathode is spongy, metallic lead (Pb). The electrolyte is a water solution of sulfuric acid (H_2SO_4). When the battery discharges (delivers current to an external circuit), both electrodes change to lead sulfate ($PbSO_4$), and the electrolyte becomes more dilute as sulfuric acid is converted to water.

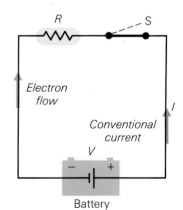

Figure 17.4 Conventional current
For historical reasons, circuit analysis is usually done using conventional current, which is in the direction in which positive charges would flow, or opposite to the actual electron flow.

of charge is in the direction in which an electron experiences a force—away from the negative terminal of the battery and toward the positive terminal (recall the law of charges). However, historically, circuit analysis has generally been done in terms of the **conventional current**, which is in the direction in which positive charges would flow, or the direction opposite to the electron flow (see Fig. 17.4).

Early investigators thought that electrical phenomena were the result of positive and negative fluids. (Recall from Chapter 12 that heat was also considered to be a fluid.) Benjamin Franklin advanced a single-fluid theory of electricity in which he postulated the existence of a special fluid in all matter. Objects could contain a certain quantity of this fluid without showing any electrical properties, but did show such properties when they contained either a surplus or a deficit of it. An object with an excess of the fluid was considered to be positively excited, and one with a deficit to be negatively excited. The designation of conventional (positive) current flow is apparently a consequence of Franklin's single-fluid theory.

An electric current is then a flow of charge. Quantitatively, the **electric current** (I) in a wire is defined as the net amount of charge (q) passing through a cross-sectional area of the wire per unit time:

$$I = \frac{q}{t} \tag{17.1}$$

As you can see, the units of current are coulombs per second (C/s). This combination is called an **ampere (A)** in honor of the French physicist André Ampère (1775–1836), an early investigator of electrical and magnetic phenomena. A value such as 10 A is usually read as "ten amps." Small currents are expressed in milliamperes (mA, 10^{-3} A) or microamperes (μA, or 10^{-6} A), which are shortened to milliamps and microamps.

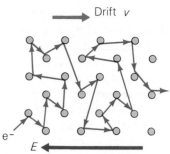

Figure 17.5 **Drift velocity** Because of collisions with the atoms of the conductor, electron motion is random. However, when the conductor is connected in a circuit, there is a relatively small net motion in the direction opposite the electric field (or toward the positive terminal of the battery or other voltage source). This net motion is characterized by a drift velocity.

Example 17.1 Coulombs of Charge Passing By
A current of 0.50 A flows in a circuit for 2.0 min. How much charge passes through a cross-sectional area of one of the connecting wires during this time?

Solution
Given: $I = 0.50$ A   *Find*: q
     $t = 2.0$ min $= 120$ s

Eq. 17.1 is $I = q/t$, so $q = It$:

    $q = It = (0.50$ A$)(120$ s$) = 60$ C

This is quite a few electrons passing by (about 10^{20}). ∎

Although current flow is frequently mentioned, you should be aware that electric charge does not flow in a conductor exactly like mass flow (such as of water through a pipe). In the absence of a potential difference, the free electrons in a metal wire move about randomly at very high speeds. As a result, there is no net flow of charge one way or the other in the wire. (On the average, the same amounts of charge pass through a cross-sectional area in opposite directions in a given time.) When a potential difference is applied across the ends of the wire (for example, by a battery), this creates an electrical bias in one direction and thus a net flow of charge in that direction. This does *not* mean electrons are moving directly from one end of the wire to the other. They still move about in all directions because they collide with the atoms of the conductor (Fig. 17.5). But their motions are less random, and more of them move toward the positive terminal of the battery than away from it.

The net electron flow is characterized by an average velocity called the **drift velocity**, which is much smaller than the random velocities of the electrons themselves. The magnitude of the drift velocity is on the order of 0.10 cm/s. Motion at this speed is relatively slow. A quick calculation using this drift velocity would show that on the average it would take an electron about 17 minutes to travel 1 m along a wire. Yet a lamp comes on almost instantaneously when you turn the switch (close the circuit), and the electronic signals carrying telephone conversations travel almost instantaneously over miles of wire.

Evidently, something must be moving faster than the net electron motion. This is the electric field. When a potential difference is applied, the associated electric field that drives the electrons in their net motion travels through the conductor at a speed close to the speed of light (on the order of 10^8 m/s). The electric field therefore influences the motion of electrons throughout the conductor almost instantaneously.

17.3 Ohm's Law and Resistance

Given that an applied voltage causes a current to flow in a conductor, how much current flows for a given voltage? As you might expect, the current is directly proportional to the voltage:

$$I \propto V$$

That is, the greater the voltage, the greater the current. This is analogous to more water flowing through a pipe when there is a greater gravitational potential.

However, another factor besides voltage affects current flow. Just as internal friction (viscosity) affects fluid flow, the internal resistance of materials affects the flow of electric charge. The **resistance** (R) is inversely proportional to the current. For many materials, the relationship between current, voltage, and resistance is given by

$$I = \frac{V}{R} \quad \text{or} \quad V = IR \tag{17.2}$$

This equation is called **Ohm's law**, after Georg Ohm (1789–1851), a German physicist who investigated the relationship between current and voltage. The units for resistance can be seen to be volts per ampere (V/A). This combined unit is called an **ohm (Ω)**.* A plot of voltage versus current for an ohmic conductor (one for which Ohm's law holds) gives a straight line with a slope equal to R (Fig. 17.6).

---

* The Greek letter omega is used as the symbol for ohm because O might be taken for a zero.

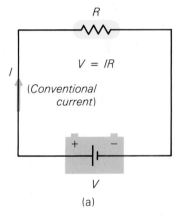

$$R$$

$$V = IR$$

I (Conventional current)

V

(a)

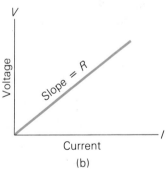

(b)

Figure 17.6 **Ohm's law**
(a) Resistive materials that show a voltage-current relationship of $V = IR$ are said to obey Ohm's law, or to be ohmic. (b) A plot of voltage versus current gives a straight line with a slope equal to R.

Although commonly referred to as Ohm's law, Eq. 17.2 is not a fundamental law in the sense that Newton's law of gravitation or the first and second laws of thermodynamics are. Eq. 17.2 is simply a *definition* of resistance as the constant of proportionality between voltage and current. The voltage-current relationship expressed by the equation applies to a wide range of materials, particularly metals. However, keep in mind that the relationship is not linear (that is, Ohm's law does not apply) for other materials, such as semiconductors.

Example 17.2 Electrical Resistance

When a 12-V battery is connected across an unknown resistor, 8.0 mA of current flows in the circuit. What is the value of the resistor?

Solution

Given: $V = 12$ V   *Find*: R
$\quad\quad\quad I = 8.0$ mA $= 8.0 \times 10^{-3}$ A

Resistors are used in circuits to control voltage and current and are manufactured in a wide variety of resistance values (see Fig. 17.7). In this case, using Ohm's law,

$$R = \frac{V}{I} = \frac{12 \text{ V}}{8.0 \times 10^{-3} \text{ A}} = 1.5 \times 10^3 \ \Omega$$

That is, the resistor is a 1.5-kΩ (kilo-ohm) resistor. Resistors in the kΩ and MΩ (megaohm) ranges are quite common. ∎

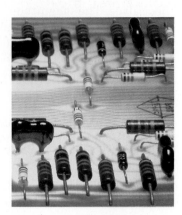

Figure 17.7 Resistors Various types of resistors of different values being used on a circuit board. See Example 17.2.

On the atomic level, resistance arises from collisions of electrons with the atoms or ions making up a material. Thus, resistance is a material property; that is, it depends on the type of material. The resistance of a particular conductor of uniform cross section, such as a length of wire, depends on several factors (Fig. 17.8). These factors directly affecting resistance are

1. Type of material
2. Length
3. Cross-sectional area
4. Temperature

As you might expect, the resistance of a conductor is directly proportional to its length (L) and inversely proportional to its cross-sectional area (A):

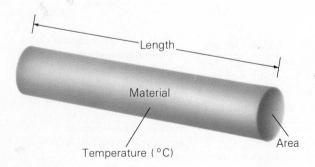

Figure 17.8 Resistance factors Factors directly affecting the electrical resistance of a uniform conductor are the type of material, the length (L), the cross-sectional area (A), and the temperature.

$$R \propto \frac{L}{A}$$

For example, a uniform metal wire 4 m long offers twice as much resistance as a similar wire 2 m long. Also, the larger the cross-sectional area of a conductor, the more current flow (less resistance) there is in it. These geometrical conditions are analogous to those for liquid flow in a pipe. The longer the pipe, the more resistance (drag) there is; but the larger the diameter (or cross-sectional area) of the pipe, the greater the amount of liquid it can carry.

The resistance of a material also depends on intrinsic properties, which are characterized by the **resistivity** (ρ). This acts as a constant of proportionality for the relationship of resistance to length and area:

$$R = \frac{\rho L}{A} \quad \text{or} \quad \rho = \frac{RA}{L} \tag{17.3}$$

You should be careful not to confuse resistivity with mass density, which has the same symbol (rho).

You can see that ρ has units of ohm-meters (Ω-m).

The resistivities of some common conductors (and insulators) are given in Table 17.1. The values given are for a particular temperature (20 °C, or room temperature), because resistivity is somewhat temperature-dependent. *For most metallic conductors, the resistivity increases as the temperature rises.* Higher temperatures cause increased atomic or molecular vibrations in a con-

Table 17.1
Resistivities (at 20°C) and Temperature Coefficients of Reistivity for Various Materials*

	$\rho(\Omega\text{-}m)$	$\alpha(°C^{-1})$
Conductors		
Aluminum	2.82×10^{-8}	4.29×10^{-3}
Copper	1.70×10^{-8}	6.80×10^{-3}
Iron	10×10^{-8}	6.51×10^{-3}
Mercury	98.4×10^{-8}	0.89×10^{-3}
Nichrome alloy of Nickel and Chromium	100×10^{-8}	0.40×10^{-3}
Nickel	7.8×10^{-8}	6.0×10^{-3}
Platinum	10×10^{-8}	3.93×10^{-3}
Silver	1.59×10^{-8}	6.1×10^{-3}
Tungsten	5.6×10^{-8}	4.5×10^{-3}
Semiconductors		
Carbon	3.6×10^{-5}	-0.0005
Germanium	4.6×10^{-1}	-0.05
Silicon	2.5×10^{2}	-0.07
Insulators		
Glass	10^{12}	
Rubber	10^{15}	
Wood	10^{10}	

* Values for semiconductors are general values, and resistivities for insulators are typical orders of magnitude.

ductor. Thus, more collisions with electrons take place, which hinder the net flow of charge.

A quantity that is sometimes mentioned is the electrical conductivity of a material (Chapter 16). As you might guess, the **conductivity** (σ) and the resistivity (ρ) are inversely proportional:

$$\sigma = \frac{1}{\rho} \tag{17.4}$$

The unit of conductivity is $(\Omega\text{-m})^{-1}$.

Example 17.3   Resistance of a Wire
A total length of 1.5 m of insulated copper wire is used to connect the components of a circuit. If the wire has a diameter of 2.3 mm (equivalent to American Wire Gauge, or AWG, No. 12), what is its resistance at room temperature?

Solution

Given: $L = 1.5$ m   *Find*: R
$\quad\quad\quad d = 2.3$ mm $= 2.3 \times 10^{-3}$ m
$\quad\quad\quad \rho_{\text{copper}} = 1.70 \times 10^{-8}\ \Omega\text{-m}$
$\quad\quad\quad\quad$ (from Table 17.1)

The cross-sectional area of the wire is

$$A = \pi r^2 = \frac{\pi d^2}{4} = \frac{\pi (2.3 \times 10^{-3}\ \text{m})^2}{4} = 4.2 \times 10^{-6}\ \text{m}^2$$

Then, using Eq. 17.3,

$$R = \frac{\rho L}{A} = \frac{(1.70 \times 10^{-8}\ \Omega\text{-m})(1.5\ \text{m})}{4.2 \times 10^{-6}\ \text{m}^2}$$

$$= 0.0061\ \Omega$$

This result demonstrates why the resistances of the connecting wires in an electrical circuit are usually considered to be negligible. The resistances of the circuit components are generally much larger than this. ∎

The temperature-dependence of resistivity is nearly linear if the temperature change is not too great. An expression similar to the one for thermal linear expansion (Chapter 11) may be written for this relationship. That is, the resistivity (ρ) at a temperature T after a temperature change $\Delta T = T - T_o$ is given by

$$\rho = \rho_o(1 + \alpha\,\Delta T) \tag{17.5}$$

Here α is a constant (over a small temperature range) called the **temperature coefficient of resistivity**, and ρ_o is a reference resistivity at T_o (usually $20\,°\text{C}$ or $0\,°\text{C}$). Eq. 17.5 may also be written like this:

$$\Delta\rho = \rho_o \alpha\,\Delta T \tag{17.6}$$

where $\Delta\rho = \rho - \rho_o$ is the change in resistivity for a given change in temperature (ΔT). Since the ratio $\Delta\rho/\rho_o$ is dimensionless, α must have the unit of $°\text{C}^{-1}$ (or

Over a large temperature range, higher-order terms may be needed to accurately describe the resistivity, for example, $\rho = \rho_o[1 + \alpha\,\Delta T + \beta\,(\Delta T)^2 + \gamma\,(\Delta T)^3]$. The coefficients β and γ are very small, but the higher-order terms become significant for large ΔT's.

$1/^\circ C$). The temperature coefficients of resistivity for some materials are listed in Table 17.1. This discussion will consider these coefficients to be constant over the temperature ranges used in the examples and problems.

Resistance is directly proportional to the resistivity, and Eqs. 17.3, 17.5, and 17.6 can be used to derive an expression for the resistance of a conductor of uniform cross section, where R_o is the resistance of the conductor at the reference temperature:

$$R = R_o(1 + \alpha \Delta T) \qquad \text{or} \qquad \Delta R = R_o \alpha \Delta T \qquad (17.7)$$

The variation of resistance with temperature provides a means of measuring temperature in the form of an electrical resistance thermometer.

Example 17.4   Variation in Resistance with Temperature

What is the variation (as a percentage) in the resistance of platinum wire over the range from $0\,^\circ C$ to $100\,^\circ C$? (Assume that α is constant over this temperature range.)

Solution

Given: $T_o = 0\,^\circ C$ *Find*: $\Delta R/R_o$
       $T = 100\,^\circ C$ (expressed as a percentage)
       $\alpha = 3.93 \times 10^{-3}\,^\circ C^{-1}$
          (from Table 17.1)

The ratio $\Delta R/R_o$ is the fractional change in the initial resistance R_o (at $0\,^\circ C$). Using Eq. 17.7 with the values given for α, T, and T_o,

$$\frac{\Delta R}{R_o} = \alpha(T - T_o) = (3.93 \times 10^{-3}\,^\circ C^{-1})(100\,^\circ C - 0\,^\circ C)$$

$$= 0.393\ (\times 100\%) = 39.3\% \quad \blacksquare$$

Example 17.5   Electrical Resistance Thermometer

The resistance of a coil of platinum wire is measured as $250\ m\Omega$ at room temperature ($20\,^\circ C$). When the coil is placed in a hot oven, its resistance is measured as $496\ m\Omega$. What is the temperature of the oven?

Solution

Given: $R = 496\ m\Omega$ *Find*: T
        $R_o = 250\ m\Omega$
        $T_o = 20\,^\circ C$
        $\alpha = 3.93 \times 10^{-3}\,^\circ C^{-1})$
          (from Table 17.1)

First, find the change in temperature ($\Delta T = T - T_o$) from the change in resistance using Eq. 17.7:

$$\Delta T = \frac{R - R_o}{R_o \alpha}$$

$$= \frac{496\ m\Omega - 250\ m\Omega}{(250\ m\Omega)(3.93 \times 10^{-3}\,^\circ C^{-1})} = 250\,^\circ C$$

Then

$$T = \Delta T + T_o = 250\,^\circ C + 20\,^\circ C = 270\,^\circ C$$

Note that it was dimensionally correct and convenient to work directly with the quantities in milliohms rather than converting them to ohms. ∎

Carbon and other semiconductors have negative temperature coefficients of resistivity (see Table 17.1). This means that the resistance of a semiconductor decreases with increasing temperature or increases with decreasing temperature. Most materials have positive temperature coefficients of resistivity, and their resistances decrease with decreasing temperature. You might be wondering how far electrical resistance can be reduced by lowering the temperature. In certain cases, the resistance can be reduced to zero! This condition of superconductivity is discussed in the Insight feature.

17.4 Electric Power

When a sustained current flows in a circuit, the electrons are given energy by the voltage source, such as a battery. As these charge carriers pass through a circuit component, they collide with the atoms of the material (experience resistance) and lose energy because of these collisions (Chapter 6). The energy transfer results in a temperature increase of the component, and electrical energy is transformed into thermal energy. Overall, all of the energy given to the charge carriers by the battery must be lost in the circuit. That is, a charge carrier traversing the circuit has the same electric potential when it returns to the battery terminal as it had when it left the other terminal. (Otherwise, a charge carrier could gain energy with each round trip, which would violate the law of the conservation of energy.)

The energy gained by a charge q from a voltage source that has an electric potential V is qV (this is the work done by the voltage source on the charge, $W = qV$). Over a period of time, the rate at which energy is obtained or expended as thermal energy in a circuit is given in terms of **electric power** (P):

$$P = \frac{W}{t} = \frac{qV}{t}$$

Since $I = q/t$ by Eq. 17.1,

$$P = IV \tag{17.8}$$

As you learned in Chapter 5, the SI unit of power is the watt (W).

Using Ohm's law ($V = IR$), power can be expressed in two other convenient forms:

$$P = IV = \left(\frac{V}{R}\right)R = \frac{V^2}{R}$$

and

$$P = IV = I(IR) = I^2R$$

That is,

$$P = \frac{V^2}{R} = I^2R \tag{17.9}$$

Superconductivity

The electrical resistance of metals and alloys generally decreases with decreasing temperature. At relatively low temperatures, some materials exhibit what is called **superconductivity**; that is, the electrical resistance vanishes, or goes to zero.

Superconductivity was discovered in 1911 by Heike Kamerlingh Onnes, a Dutch physicist. The phenomenon was first observed in solid mercury at a temperature of about 4 K (−269 °C). The mercury was cooled to this temperature using liquid helium. The boiling point of helium (the temperature at which it condenses to a liquid) is about −267 °C at 1 atm. Lead also exhibits superconductivity when cooled to such a temperature.

An electrical current established in a superconducting loop should persist indefinitely with no resistive losses. Currents introduced in superconducting loops have been observed to remain constant for several years. In 1957, a theory was presented by the American physicists John Bardeen, Leon Cooper, and Robert Schrieffer in an attempt to explain various aspects of metallic superconductors. This BCS theory is a quantum-mechanical model in which electrons are viewed as waves traveling through a material. According to the theory, lattice vibrations and imperfections, which scatter electrons and give rise to resistance and energy loss in metals under normal conditions, have no effect on the electrons in superconductors. In the absence of this electron scattering, the resistance is zero, and a current can persist as long as the metal is in a superconducting state.

Keeping a superconductor at the low temperatures of liquid helium is somewhat difficult and costly. Thus, there has been an ongoing search for materials that become superconducting at higher temperatures. Other superconducting metals and alloys were found for which the critical temperature was about 18 K (−255 °C). In 1973, a material with a critical temperature of 23 K (−250 °C) was discovered.

In 1986, there was a major breakthrough—a new class of superconductors was discovered. These were the so-called ceramic alloys of rare earth elements, such as lanthanum and yttrium. These superconductors were prepared by grinding the mixture of metallic elements together and heating it to a high temperature to produce a ceramic material. For example, one such mixture consisted of barium, yttrium, and copper oxide. The critical temperature for these ceramic mixtures was about 57 K (−216 °C).

In 1987, a ceramic material with a critical temperature of 98 K (−175 °C) was discovered. This was a major advance because it meant that superconductivity could be obtained using liquid nitrogen, which has a boiling point of 77 K (−196 °C). (See Fig. 1.) Liquid nitrogen is relatively plentiful (nitrogen is the chief constituent of air) and inexpensive compared to liquid helium (liquid nitrogen costs cents per liter compared to dollars per liter for liquid helium).

There have been more recent reports of higher critical temperatures, and even suggestions that certain copper oxide compounds may lose their resistance to electrical current at room temperature and above. As scientists study the new superconductors, better understanding will be gained, and no doubt the critical temperature will rise. One problem is the difficulty of confirming a superconducting state in small samples or regions of samples. Another is that the BCS theory does not appear to apply to the new high-temperature superconductors. The mechanism for their superconductivity is not well understood.

There are many possible applications for superconductors. One is superconducting magnets. The strength of an electromagnet depends on the magnitude of the current in the windings (Chapter 19). If there were no resistance, there would be greater current and no losses. Used in motors or engines, superconducting electromagnets would provide more power. (Superconducting magnets cooled by liquid helium have been used in ships' engines for some time.) Such magnets could be used to levitate and propel trains and electric cars. Another application of superconductors might be underground transmission cables with no resistive losses. However, among the many technological problems still to overcome is how to form wires out of ceramic materials (which are generally brittle). Probably the most likely application to be realized soon will be in making faster computer circuits.

Imagine what it would mean if someone discovered a material that was superconducting at refrigerator or room temperature. The absence of electrical resistance opens many possibilities. You're likely to hear more about superconductor applications in the near future.

Figure 1 Magnetic levitation
A magnetic cube levitates above a piece of "high-temperature" superconductor material that is cooled with liquid nitrogen. When the material becomes cold enough to superconduct, the magnet is repelled and levitates.

The thermal energy expended in a current-carrying conductor is sometimes referred to as **joule heat**, or ***I* squared *R* (*I²R*) loss**. In many instances, joule heat loss is undesirable, for example, in electrical transmission lines. However, in other instances, the conversion of electrical energy to thermal energy is of practical use. These applications include the heating elements (burners) of electric stoves, hair dryers, and toasters.

Electric light bulbs are rated directly in watts (power), for example, 100-W or 60-W (Fig. 17.9). The greater the wattage of a bulb, the greater the energy loss per unit time (joules per second). Incandescent lamps are relatively inefficient light sources. Only about 5% of the electrical energy is converted to visible light (most of the radiant energy produced is invisible infrared radiation; see Section 20.4).

Electrical appliances are tagged or stamped with their power ratings. Either the voltage and power requirements or the voltage and current requirements are given. In either case, the current, power, and effective resistance may be found using Eqs. 17.8 and 17.9. The power requirements of some household appliances are given in Table 17.2. You should keep in mind that lamps and appliances generally operate on alternating current, which will be discussed in Chapters 20 and 21. However, the current-voltage equations for alternating current are of the same form as those for direct current.

Figure 17.9 **Power rating** Light bulbs are rated directly in watts. This 60-W bulb uses 60 J of energy each second.

Example 17.6 Power and Current
A 60-W light bulb operates on 120-V household voltage. (Household voltage varies between 110 V and 120 V and is commonly given as 110 V, 115 V, or 120 V.)

Table 17.2
Typical Power and Current Requirements for
Household Appliances (120 V)

Appliance	Power	Current
Air conditioner, room	1500 W	12.5 A
Air conditioning, central	5000 W	41.7 A
Blender	800 W	6.7 A
Coffee maker	1625 W	13.5 A
Dishwasher	1200 W	10.0 A
Electric blanket	180 W	1.5 A
Hair dryer	1200 W	10.0 A
Heater, portable	1500 W	12.5 A
Microwave oven	625 W	5.2 A
Radio–cassette player	14 W	0.12 A
Refrigerator, regular	400 W	3.3 A
Refrigerator, frost-free	500 W	4.2 A
Stove, top burners	6000 W	50.0 A
Stove, oven	4500 W	37.5 A
Television, black-and-white	50 W	0.42 A
Television, color	100 W	0.83 A
Toaster	950 W	7.9 A
Water heater	4500 W	37.5 A

(a) How much current does the light bulb draw? (b) What is the resistance of the bulb?

Solution

Given: $P = 60$ W   Find: (a) I
            $V = 120$ V   (b) R

(a) The current may be found using $P = IV$:

$$I = \frac{P}{V} = \frac{60 \text{ W}}{120 \text{ V}} = 0.50 \text{ A}$$

(b) Using Ohm's law ($V = IR$), the resistance is

$$R = \frac{V}{I} = \frac{120 \text{ V}}{0.50 \text{ A}} = 240 \text{ } \Omega$$

(R could also be found using the relationship $P = V^2/R$.) ■

Example 17.7 Joule Heat

A radiant heater is rated at 1500 W for 120 V. The uniform wire filament breaks near one end, and the owner repairs it by unlooping and reconnecting it. The filament then has a length 10% shorter than its original length. What effect will this have on the heater's power output?

Solution

Given: $P_o = 1500$ W   Find: P
            $V = 120$ V
            $R = 0.90 R_o$

Since 10% of the filament's length is gone, it will have 90% of its original resistance (R_o) after being reconnected. (Why?) This means that the current will increase. With the same applied voltage before and after,

$$IR = I_o R_o$$

and

$$I = \left(\frac{R_o}{R}\right) I_o = \left(\frac{1}{0.90}\right) I_o = (1.11) I_o$$

Under the initial conditions, the current was

$$I_o = \frac{P_o}{V} = \frac{1500 \text{ W}}{120 \text{ V}} = 12.5 \text{ A}$$

Then

$$I = (1.11) I_o = (1.11)(12.5 \text{ A}) = 13.9 \text{ A}$$

The power output could be computed directly from $P = IV$ or by computing the resistance and then using Eq. 17.9. By the latter method, using Ohm's law,

$$R_o = \frac{V}{I_o} = \frac{120 \text{ V}}{12.5 \text{ A}} = 9.6 \text{ } \Omega$$

and

$$R = 0.90 R_o = (0.90)(9.6 \text{ } \Omega) = 8.6 \text{ } \Omega$$

For the repaired heater, then,

$$P = I^2 R = (13.9 \text{ A})^2 (8.6 \ \Omega) = 1660 \text{ W}$$

The power output of the heater has been increased.

You should not attempt such a repair job. With reduced resistance and increased current, the filament could overheat, melt, and possibly start a fire. ∎

Note that by Eq. 17.9 the power increases with the *square* of the current—double the current and the power output increases by a factor of 4 (for constant resistance). As shown in Example 17.7, heating elements have relatively low resistances to allow large currents to flow in order to take advantage of the I^2 term. Conversely, joule heat losses can be reduced significantly by reducing the current. If the current is halved, the power expenditure is reduced to a fourth of its original value (for a constant resistance).

Example 17.8   Electrical Cost

If a frost-free refrigerator runs 15% of the time, how much does it cost to operate (a) per day and (b) per month if the power company charges 11¢ per kilowatt-hour?

Solution

(a) Since the refrigerator operates 15% of the time, in one day it runs

$$t = (0.15)(24 \text{ h}) = 3.6 \text{ h}$$

Taking the power requirement of the refrigerator to be 500 W (see Table 17.2), the electrical work done, or the energy expended, per day is, since $P = W/t$,

$$W = Pt = (500 \text{ W})(3.6 \text{ h}) = 1800 \text{ Wh} = 1.8 \text{ kWh}$$

Then

$$(1.8 \text{ kWh})(\$0.11/\text{kWh}) = \$0.20$$

It costs 20¢ a day to operate the refrigerator. (Note that power companies charge for electricity in units of energy.) (b) For a 30-day month,

$$(\$0.20/\text{day})(30 \text{ day}/\text{month}) = \$6.00/\text{month}$$

[The cost of electricity varies around the country. On the average, it ranges from 5¢ to 18¢ per kilowatt-hour. Do you know the price of electricity in your locality? Check an electric bill to find out.] ∎

Electrical Efficiency

In our electrical society, use of and demand for electricity are continually increasing. About 25% of the electricity generated in the United States goes into lighting (Fig. 17.10). This is roughly equivalent to the output of 100 electrical generating (power) plants. Refrigerators consume about 7% of the electricity produced in the United States (the output of about 25 power plants).

This huge consumption of electricity has prompted the federal government to set minimum efficiency limits for refrigerators, freezers, air conditioners, water heaters, dishwashers, heat pumps, and so forth (Fig. 17.11). It is estimated

Figure 17.10 **All lit up**
A satellite mosaic of the United States at night. Can you identify the major population centers and some cities in your area?

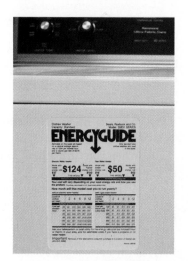

Figure 17.11 **Energy Guide**
Consumers are made aware of appliance efficiencies in terms of the average yearly cost of operation. The yearly cost is also given for different kilowatt-hour (kWh) rates, which vary around the country.

that these new standards will save over 30 terawatt-hours (10^9 kilowatt-hours) of electricity by 1995. Also, more efficient fluorescent lighting is being developed. The most efficient fluorescent lamp now in use consumes about 25–30% less energy than the average fluorescent lamp and roughly 75% less energy than incandescent lamps with an equivalent illumination output. Researchers are using new techniques in an effort to develop fluorescent lamps with a 40–50% increase in efficiency.

Important Formulas

Current:

$$I = \frac{q}{t}$$

Ohm's law:
$$V = IR$$

Resistivity:

$$\rho = \frac{RA}{L}$$

Temperature-dependence of resistivity:
$$\rho = \rho_o(1 + \alpha \Delta T)] \quad \text{or} \quad \Delta\rho = \rho_o \alpha \Delta T$$

Conductivity:

$$\sigma = \frac{1}{\rho}$$

Temperature-dependence of resistance: (for a conductor with a uniform cross section):
$$R = R_o(1 + \alpha \Delta T) \quad \text{or} \quad \Delta R = R_o \alpha \Delta T$$

Electrical power:

$$P = IV \quad \text{and} \quad P = \frac{V^2}{R} = I^2 R$$

Questions

Batteries and Direct Current

1. What is the function of a battery, and what are its basic components?

2. Is the electromotive force really a force? Explain.

3. Distinguish between a dry cell and a storage battery. Why does an automobile storage battery need to have a large number of plate electrodes?

4. On a very cold day, a car engine may be sluggish in starting if the car has been parked outside for some time. Why is this?

5. When a car with a dead battery is jump-started using the battery of another car, are the two batteries connected in series or in parallel? Explain why they are connected in this way.

6. What makes the current supplied by a battery direct current? If you switched the ends of wires connected in a circuit rapidly back and forth between the terminals of a battery, would there be an alternating current in the circuit? Explain.

Current and Drift Velocity

7. What is necessary in order to have a sustained electric current?

8. When you turn a light switch on and off, what are you doing to the circuit?

9. A circuit has a net flow of 1.0 C of charge across a cross-sectional area of a wire conductor in a time of 4.0 s. Describe the conventional current for the circuit.

10. The drift velocity of electrons is about 0.10 cm/s. Why do an automobile's headlights come on almost immediately after being switched on? It would take an electron much longer than an instant to get from the battery to the headlights at that rate.

Ohm's Law and Resistance

11. Is the voltage-current relationship for all materials described by Ohm's law? Explain.

12. If voltage (V) were plotted against current (I) for two conductors with different resistances on the same graph, what would be the general results and why?

13. If the voltage across a conductor is increased and its temperature remains constant, how are (a) the resistance and (b) the current affected?

14. One type of switch for dimming lights uses a wire coil for resistance. What causes the lights to dim when the switch knob is rotated?

15. (a) If a connecting wire in a circuit is replaced with one of the same material that is twice as long and has twice the cross-sectional area, how will the current through the wire be affected? (Assume that the potential difference remains the same.) (b) How will the current be affected if the new wire is twice as long as the old and has half the diameter?

16. A closed switch has very little resistance. What is the resistance of an open switch?

17. Distinguish between resistance and resistivity.

18. (a) Speculate why different materials have different resistivities. (b) What does the negative temperature coefficient of resistivity for a semiconductor imply about the conducting electrons in such a material?

19. If resistance versus temperature were plotted for a conductor, would the graph be a straight line? If so, what would the slope and intercept be?

Electric Power

20. At a given voltage, does a good conductor or a good insulator dissipate more energy (per unit time)?

21. How can you tell how many amps of current an appliance draws if this information is not given on a sticker or tag?

22. Which has the greater resistance, the filament of a 60-W light bulb or the filament of a 100-W light bulb? If the tungsten filament in each bulb is the same length, how do the bulbs otherwise differ?

23. A lamp at room temperature is switched on. (a) Comment on the power output of the lamp between the time it is turned on and reaching normal operation. What are the general changes in current and resistance? (b) Discuss the normal operating condition of the lamp in terms of thermal equilibrium. (c) Considering the time interval between the cold start and the normal operating condition, comment on the rate of conversion of electrical to thermal energy.

Problems

17.1　Batteries and Current

■1. (a) Two 1.5-V dry cells are connected in series with the same polarity direction. What is the total voltage of the combination? (b) What would be the total voltage if the cells were connected in parallel?

■2. What is the voltage across six 1.5-V batteries when they are conneced (a) in series and (b) in parallel? (Assume same polarity direction.)

■3. A net charge of 20 C passes through a cross-sectional area of a wire in 1.0 min. What is the current carried by the wire?

■■4. Two 6.0-V batteries and one 12-V battery are connected in series with the same polarity direction. (a) What is the voltage across the whole arrangement? (b) What arrangement of these three batteries would give a total voltage of 12 V?

■■5. A battery is rated at 60 A-h at 3.0 A. This means that the life of the battery is 20 h (60 A-h/3.0 A) when fully charged. How much charge will the battery deliver in this time?

■■6. What is the net number of electrons that move past a point in a wire carrying 500 mA of current in 10 min?

■■7. An electrical appliance draws 0.50 A of current. What is the net number of charge carriers through the appliance in 1 h?

8. A wire carries a current of 100 mA. How many electrons have to pass through a cross-sectional area of the wire in 2.0 s to produce this current?

9. Fig. 17.12 shows charge carriers having a charge q and moving with a speed v_d (drift speed) in a conductor of cross-sectional area A. Let n be the number of free charge carriers per unit volume. (a) Show that the total charge (ΔQ) free to move in the volume element shown is given by $\Delta Q = (nAx)q$. (b) Show that the current in the conductor is given by $I = nqv_d A$.

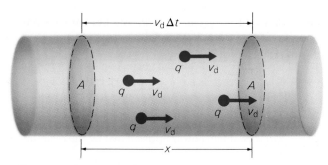

Figure 17.12 **Total charge and current**
See Problem 9.

10. A copper wire with a cross-sectional area of 13.3 mm² (AWG No. 6) carries a current of 1.2 A. If there are 8.5×10^{22} free electrons per cubic centimeter, what is the drift velocity of the electrons? (Hint: see Problem 9.)

17.3 Ohm's Law and Resistance*

11. What potential difference is required to cause 2.0 A of current to flow through a 6.0-Ω resistor?

12. What voltage must a battery have to produce 3.0 A of current through a 2.0-Ω resistor?

13. How much current is drawn from a 12-V battery when a 100-Ω resistor is connected across its terminals?

14. A circuit component with a resistance of 14 Ω draws 1.5 A of current when connected to a dc power supply. What is the voltage of the power supply?

15. A copper wire is 1.5 m long and has a diameter of 0.10 cm. What is the resistance of the wire?

16. A 100-Ω resistor is placed in a circuit with two 9.0-V batteries connected in series. How much current flows through the resistor?

---

* Assume that the temperature coefficients of resistivity given in Table 17.1 apply over large temperature ranges.

17. An automobile starter is connected to a 12-V battery. If the starter has an effective resistance of 3.0 Ω, what is the net rate at which electrons move in the circuit?

18. Find the resistance of a piece of copper wire that is 50 cm long and has a diameter of 2.0 mm.

19. Two copper wires have equal cross-sectional areas and lengths of 2.0 m and 0.50 m. What is the ratio of the resistance of the shorter wire to the resistance of the longer wire?

20. Two copper wires have equal lengths, but the diameter of one is twice that of the other. What is the ratio of the resistance of the thicker wire to the resistance of the thinner wire?

21. When a 1.0-m length of wire is connected to a 6.0-V source, 0.25 A of current flows in it. Calculate the current in a 2.0-m length of the same type of wire connected to the same voltage source.

22. A certain material is formed into a long rod with a square cross section 0.50 cm on a side. When a voltage of 100 V is applied across a length of the rod 2.0 m long, 5.0 A of current flows. What are (a) the resistivity and (b) the conductivity of the material? Is it a good conductor?

23. What is the resistance of a 4.0-km length of aluminum cable with an effective diameter of 1.0 cm?

24. Find the ratio of the resistance of a copper wire to the resistance of an aluminum wire (a) if the wires have the same length and diameter and (b) if the copper wire is twice as long as the aluminum wire and has a diameter that is half that of the aluminum wire.

25. What length of wire is required to make a 20-Ω resistor by winding AWG No. 10 nichrome wire in a coil? (AWG No. 10 wire has a diameter of 2.588 mm.)

26. A bus bar used to conduct large amounts of current at a power station is made of copper and is 1.5 m long and 8.0 cm by 4.0 cm in cross section. What voltage difference across the ends of the bar will cause a current of 3000 A to flow through it?

27. When a wire 1.0 mm in diameter and 2.0 m in length is connected to a 3.00-V source, a current of 11.8 A flows. (a) What is the resistivity of the wire? (b) What material is the wire made of?

28. What is the percentage variation of the resistivity of copper over the range from room temperature to 100 °C?

29. A copper wire has a resistance of 25 mΩ at 20 °C. When the wire is carrying a current, joule heat causes its temperature to increase to 30 °C. What is the change in the wire's resistance?

■■**30.** What is the fractional change in the resistance of an aluminum wire 1.0 m long over a temperature range from the freezing point to the boiling point of water?

■■■**31.** An aluminum wire has a mass of 10 g and a length of 50 cm. Find the current in the wire when it is connected to a 10-V source.

■■■**32.** Suppose that the bus bar described in Problem 26 is drawn into a wire with a diameter of 5.0 mm. (a) Will the resistance of the wire be the same as that of the bus bar? (b) Find the resistance of the wire.

■■■**33.** Pieces of carbon and copper have uniform cross sections and the same resistance at room temperature. (a) If the temperature of each piece is increased by 10 °C, which will have the greater resistance? (b) Find the ratio of the resistances, and how many times greater one is than the other.

■■■**34.** An electrical resistance thermometer made of platinum has a resistance of 5.0 Ω at 20 °C and is in a circuit with a 1.5-V battery. What is the change in the current in the circuit when the thermometer is heated to 2020 °C?

■■■**35.** Pieces of aluminum and copper wire are identical in length and diameter. At some temperature, one of the wires will have the same resistance as the other has at room temperature. What is the temperature? (Is there more than one temperature?)

17.4 Electric Power

■**36.** An appliance is rated for 1.5 A at 120 V. How much electric power does it require?

■■**37.** How much power will be expended in a 10-kΩ resistor when it is connected across a 100-V voltage source?

■■**38.** A hair dryer is rated for 1200 W at 120 V. (a) How much current does the dryer use? (b) What is its resistance?

■■**39.** A 100-Ω resistor is rated at $\frac{1}{4}$ W. (a) What is the maximum current the resistor can carry? (b) What is the maximum operating voltage that should be applied across the resistor?

■■**40.** An electric heater is designed to produce 50 kW of heat when connected to a 240-V source. What must the resistance of the heater be?

■■**41.** An electric iron with a 14-Ω heating element

operates on 120 V. How much heat does the iron produce in 30 min?

■■**42.** A 1500-W hair dryer is used for 10 min every day. If the cost of electricity is 8.5¢/kWh, what is the cost of using the dryer for a month (30 days)?

■■**43.** A wire 5.0 m long and 3.0 mm in diameter has a resistance of 100 Ω. A potential difference of 15 V is applied across the wire. Find (a) the current in the wire, (b) the resistivity of its material, and (c) the rate at which heat is being produced in it.

■■**44.** A coil of tungsten wire initially dissipates 500 W of power when connected to a voltage source. In a short time, the temperature of the coil increases by 150 °C because of joule heat. What is the corresponding change in the power?

■■**45.** A current of 0.10 A flows in a resistor in a circuit with a 40-V voltage source. (a) What is the resistance of the resistor? (b) How much power is dissipated by the resistor? (c) How much energy is dissipated in 2.0 min?

■■**46.** An electric heater that operates on 120 V is rated at 850 W. (a) How much current does the heater draw? (b) What is the resistance of its heating element?

■■**47.** A 150-W light bulb operates on 120-V household voltage. What is the resistance of the bulb?

■■**48.** A 5500-W water heater operates on 240 V. (a) Should the heater circuit have a 20-A or a 30-A circuit breaker? (A circuit breaker is a safety device that opens the circuit at its rated current.) (b) Assuming that there is no energy loss, how long will the heater take to heat the water in a 55-gal tank from room temperature to 80 °C?

■■**49.** By what factor is the power expended in a 100-Ω resistor increased when the voltage across it is increased from 15 V to 45 V?

■■**50.** A 0.50-kΩ resistor is in a circuit with three 12-V batteries. What is the joule heat loss of the resistor (a) if the batteries are connected in series and (b) if the batteries are connected in parallel.

■■**51.** If the price of electricity is 8¢/kWh, how much would it cost to operate ten 75-W light bulbs for 6 hours?

■■**52.** A 20-Ω, 5-W resistor is used in a circuit. (a) What is the maximum voltage that should be applied across the resistor? (b) Could the resistor safely carry a current of 0.60 A?

■■**53.** What size resistor (amount of resistance) should

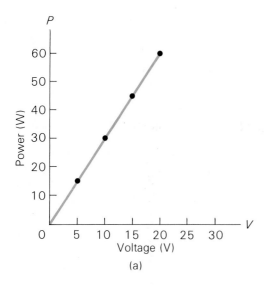

(a)

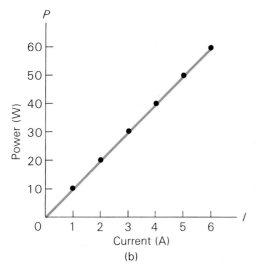

(b)

Figure 17.13 **Power versus voltage and power versus current**
See Problem 54.

be used to generate 10 kJ of heat per minute when connected to a 120-V source?

■■**54.** (a) Fig. 17.13(a) is a graph of power versus voltage. What is the value of the slope of the line, and what does it represent physically? How was the straight line obtained? (b) Answer the same questions for the graph of power versus current in Fig. 17.13(b).

■■■**55.** What is the efficiency of a 350-W motor that draws 4.0 A from a 120-V line?

■■■**56.** An immersion heater with a resistance of 9.0 Ω

operates on 120 V. The heater is placed in a cup of water (0.25 L) at room temperature (20 °C) to heat it up to make coffee. Assuming that no heat loss occurs, how long will the heater have to operate to bring the water to a boil?

■■■**57.** How long will the heater in Problem 56 have to operate to boil all the water away?

■■■**58.** A 500-W heating element made of nichrome wire whose diameter is 4.0 mm operates on 120 V. (a) What is the length of the wire? (b) What power would be dissipated if the voltage were reduced to 105 V?

■■■**59.** Find the total charged on a monthly (30-day) electric bill for the following appliance usage if the utility rate is 9.5¢/kWh: central air conditioning runs 30% of the time; a blender is used 0.50 h/month; a dishwasher is used 8.0 h/month; a microwave oven is used 15 min/day; a frost-free refrigerator runs 15% of the time; a stove (range and oven) is used a total of 10 h/month; and a color television is operated 120 h/month. (Use the information given in Table 17.2.)

Additional Problems

60. A resistance thermometer made of platinum shows a 25% change in resistance over a certain temperature range. What is the change in temperature for this range?

61. A large motor draws 45 A of current when starting up. If the start-up time is 0.75 s, what is the net charge passing through a cross-sectional area of the circuit in this time?

62. When a resistor is connected to a 12-V source, it draws 185 mA of current. When the same resistor is connected to a 90-V source, it draws 1.38 A of current. Is the resistor ohmic? Justify your answer mathematically.

63. A carbon resistor has a resistance of 200 Ω at room temperature. In a circuit with a voltage of 12 V across it, its operational temperature is 90 °C. How much more current flows through the resistor at 90 °C than when the circuit is initially closed?

64. An electric water heater automatically operates for 2.0 h each day. (a) If the cost of electricity is 12¢/kWh, what is the cost of operating the heater during a 30-day month? (b) What is the effective resistance of a typical water heater? (Hint: see Table 17.2.)

65. A particular application requires a 20-m length of aluminum wire to have a resistance of 1.0 mΩ. What diameter should the wire have?

66. A net charge of 25 C passes through a wire with a

cross-sectional area of 0.30 cm^2 in 1.25 min; a net charge of 35 C passes through another wire with a cross-sectional area of 0.450 cm^2 in 1.52 min. Which wire carries more current and how much more?

67. An electric toy rated at 2.5 W is operated by four 1.5-V batteries connected in series. (a) How much current does the toy draw? (b) What is its effective resistance?

68. A portable radio draws 150 mA of current from a battery. If the radio is played for a half-hour, what is the net number of electrons that move through the circuit?

69. A 1.0-m length of copper wire with a diameter of 2.0 mm is stretched out; its length increases by 25% while its cross-sectional area decreases but remains uniform. Is the resistance of the wire the same before and after? Justify your answer mathematically.

70. A welding machine draws 16 A of current at 240 V. (a) How much energy does the machine use per second? (b) What is its effective resistance?

71. An iron conductor with a resistance of 2.50 Ω at 20 °C is connected to a 12-V source. If the conductor is placed in an oven and heated to 300 °C, what is the change in the current flowing through it?

72. A 10-m length of 1.0-mm-diameter nichrome wire is wound in a coil and connected to a 1.5-V flashlight battery. How much current will initially flow in the coil?

73. Show that V^2/R has the same units as power.

74. The tungsten filament of an incandescent lamp has a resistance of 50 Ω at room temperature. What will the filament's resistance be at an operating temperature of 1600 °C?

75. A toaster oven is rated for 1500 W at 120 V. When the filament ribbon of the oven burns out, the owner makes a repair that reduces the length of the filament by 10%. How does this affect the power output of the oven, and what is the new value?

Electric Circuits

18

Electric circuits are designed for many specific purposes. Some contain many components and are quite complex. Armed with the principles learned in Chapters 16 and 17, you are now ready to analyze some relatively simple circuits. This will give you an appreciation of how electricity actually works.

Circuit analysis most often deals with voltage, current, and power requirements. For example, a circuit component with a given resistance may be designed to operate at a particular voltage. Supplying that voltage will depend on specific current and power requirements. How two or more components are connected in a circuit affects the voltage and/or current. A circuit may be analyzed theoretically before actually being hooked up. The analysis might show that the circuit will not function properly as designed or that there is a safety problem (such as overheating due to joule heat).

Circuit diagrams are used to visualize circuits in order to understand their functioning. A few of these diagrams were included in figures in Chapter 17. In actuality, the wires of a circuit are not placed in the rectangular pattern of a circuit diagram. The rectangular form is simply a convention that provides a neater presentation and easier visualization of the circuit arrangement.

18.1 Resistances in Series, Parallel, and Series-Parallel Combinations

The resistance symbols in a circuit diagram can represent any resistive element—a commercial resistor, a light bulb, and appliance, and so on. This discussion will consider all of these to be ohmic and to have a constant operating resistance, unless otherwise stated. Also, the resistance of the connecting wires in a circuit will be considered to be negligible.

In analyzing a circuit, you need to keep in mind that the sum of the voltages around a circuit loop is zero. For example, for the circuit in Fig. 18.1(a) the sum of the voltage drops (decreases) across the circuit components, or resis-

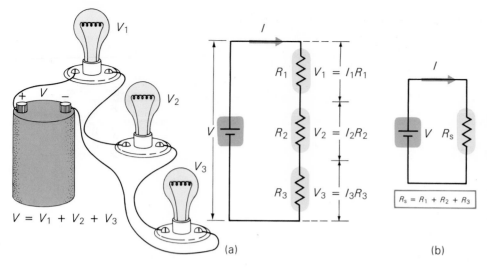

Figure 18.1 Resistors in series
(a) When resistors are connected in series, the current through each of them is the same. Also, the sum of the voltage drops across the resistors is equal to the voltage (rise) of the battery. (b) The equivalent resistance R_s of the resistors in series.

tances (V_r), equals the voltage rise by the battery (the voltage supplied by the battery, V_b):

$$V_b = \sum V_r = \sum_i IR_i$$

or $V_b - \sum_i IR_i = 0$ 　　　　　　　　　　　　　　　　　　　(18.1)

The sum of the voltages around the circuit loop is zero. This is essentially a statement of the conservation of energy.

Recall from Chapter 17 that for the charge carriers traversing the circuit, the energy given to them by a voltage source must be lost in the circuit. That is, they must return to the voltage source with the same electric potential they had when they left it. Otherwise, the charge carriers would gain energy with each round trip of the circuit, which would violate the law of the conservation of energy.

The resistances or resistors in Fig. 18.1 are in series, or connected end to end, so to speak. *When resistors are connected in series, the current is the same through all of them*, as required by the conservation of charge. As an analogy, think of an incompressible fluid flowing through a series of pipe sections. The amount of fluid leaving a section must be equal to the amount going into it by the conservation of mass. Thus, the volume of fluid flow must be the same through all the pipe sections. Otherwise, there would be build-up of mass somewhere along the way. Similarly, the current, or charge flow, is the same through each resistor connected in series. Otherwise, there would be a build-up of charge somewhere along the way.

The total voltage drop around a circuit is equal to the sum of the individual voltage drops across the resistors. With a current I, Eq. 18.1 can be written explicitly (see Fig. 18.1):

$$V = V_1 + V_2 + V_3$$
$$= IR_1 + IR_2 + IR_3$$
$$= I(R_1 + R_2 + R_3)$$

The operating voltage (V) of a battery is related to its emf ($\mathscr{E}$) by $V = \mathscr{E} - Ir$, where r is the internal resistance of the battery (see Fig. 17.2). Note that V varies with the amount of current being drawn from the battery and that $V = \mathscr{E}$ for an open circuit ($I = 0$). In general, this book uses the operating voltages of batteries in circuits.

Thus, since $V = IR$, three resistors in series have an equivalent resistance R_s:

$$R_s = R_1 + R_2 + R_3$$

That is, the equivalent resistance of resistors in series is simply the sum of the individual resistances. The three resistors of Fig. 18.1 could be replaced with a single resistor with a resistance value of R_s without having any effect on the voltage source. For example, if each of the resistors in Fig. 18.1 had a value of 10 Ω, then R_s would be 30 Ω.

This result may be extended to any number of resistors in series:

$$R_s = R_1 + R_2 + R_3 + \cdots \tag{18.2}$$

equivalent resistance for resistors in series

QUESTION: Suppose that one of the light bulbs in the circuit in Fig. 18.1(a) blew out. What would happen?

ANSWER: All of the bulbs would go out because the circuit would no longer be complete. If the filament of one of the bulbs is broken or blows out, the circuit is open and no current will flow.

Another basic way of connecting resistors in a circuit is in parallel [Fig. 18.2(a)]. In this case, all the resistors have a common connection (one side of all of them connected together). *When resistors are connected in parallel, the voltage drop across each resistor is the same and equal to the voltage rise of the battery.* However, the current from the battery divides among the different paths, as shown in Fig. 18.2(a),

$$I = I_1 + I_2 + I_3$$

$$\tag{18.3}$$

If the individual resistances are equal, the current divides equally. But if the resistances are not equal, the current divides among the resistors proportionately; that is, the greatest current flows through the path of least resistance. (It may be helpful to think of how a liquid flowing in a pipe would divide at a junction into pipes with different cross-sectional areas.)

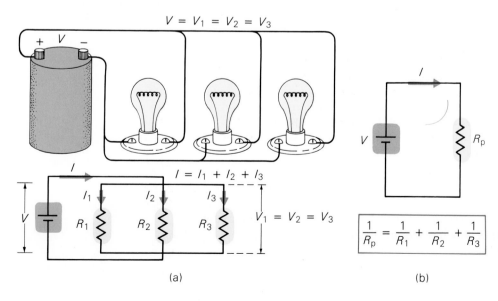

(a)

(b)

Figure 18.2 **Resistors in parallel**
(a) When resistors are connected in parallel, the voltage drop across each one is the same. The current from the battery divides proportionally among the resistors. (b) The equivalent resistance R_p of the resistors in parallel.

Since the voltage drop (V) across each resistor is the same, by Ohm's law,

$$I = I_1 + I_2 + I_3$$

$$= \frac{V}{R_1} + \frac{V}{R_2} + \frac{V}{R_3}$$

$$= V\left(\frac{1}{R_1} + \frac{1}{R_2} + \frac{1}{R_3}\right) = \frac{V}{R_p}$$

Thus, the equivalent resistance R_p of three resistors in parallel is given (in reciprocal form) by

$$\frac{1}{R_p} = \frac{1}{R_1} + \frac{1}{R_2} + \frac{1}{R_3}$$

That is, the reciprocal of the equivalent resistance is equal to the sum of the reciprocals of individual resistors connected in parallel. The three resistors could be replaced with a single resistor with a resistance value of R_p (see Fig. 18.2(b).

This result may also be generalized to include any number of resistors in parallel:

$$\frac{1}{R_p} = \frac{1}{R_1} + \frac{1}{R_2} + \frac{1}{R_3} + \cdots \qquad (18.4)$$

equivalent resistance for resistors in parallel

PROBLEM-SOLVING HINT

Eq. 18.4 gives $1/R_p$. Don't forget to take the reciprocal of this value to get R_p.

If there are only two resistors in parallel, the formula for the equivalent resistance may be written in a nonreciprocal form, which is sometimes more convenient:

$$\frac{1}{R_p} = \frac{1}{R_1} + \frac{1}{R_2} = \frac{R_1 + R_2}{R_1 R_2}$$

or
$$R_p = \frac{R_1 R_2}{R_1 + R_2} \qquad (18.5)$$

equivalent resistance for two resistors in parallel

The equivalent resistance of resistors in parallel is always less than the smallest individual resistance. You can easily demonstrate this for two resistors in parallel by putting some values in Eq. 18.5. More generally, you can see that

$$\frac{1}{R_p} = \sum_i \frac{1}{R_i}$$

where $\Sigma_i 1/R_i$ is the sum of the reciprocals of the individual resistances. This sum is greater than the reciprocal of any individual resistance:

$$\frac{1}{R_p} = \sum_i \frac{1}{R_i} > \frac{1}{R_i}$$

Therefore, R_p must be less than any R_i, where R_i may be the smallest individual resistance.

Thus, series connections provide a way to increase resistance, and parallel connections a way to reduce resistance.

QUESTION: Suppose that one of the light bulbs in the circuit in Fig. 18.2(a) blew out. What would happen?

ANSWER: The other two bulbs would continue to burn because there is still a complete circuit for current flow through each of them. (Trace these circuits in the figure with your finger.)

Strings of Christmas tree lights used to be wired in series. When one bulb blew out, all the others on that string also went out and you had to hunt for the faulty bulb (and you had a real problem if more than one bulb went out at the same time). With newer lights, one or more bulbs may burn out, but the others remain lit. Are the bulbs now wired in parallel? No, this would make the total equivalent resistance very small, and a large current would flow through the circuit, which is not desirable. Also, more costly wire would be required. What is done is to put a resistor (called a shunt resistor) in parallel with each bulb filament (Fig. 18.3). Then, if the filament breaks, there is still a path for the current to follow (it is shunted through the resistor), and the other bulbs continue to burn.

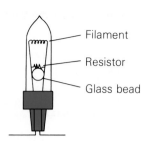

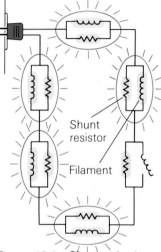

Figure 18.3 **Shunt-wired Christmas tree lights**
Shunt resistors in parallel with the bulb filaments maintain a complete circuit when one (or more) of the lights blows out (the filament breaks) by shunting the current around the broken filament. Without the shunt resistors, if one bulb blew out, all the lights would go out.

Example 18.1 Resistors in Series and in Parallel

What is the equivalent resistance of three resistors ($1.0\,\Omega$, $2.0\,\Omega$, and $3.0\,\Omega$) when they are connected (a) in series [Fig. 18.1(a)], and (b) in parallel [Fig. 18.2(a)]? (c) What current will be delivered from a 12-V battery for each of these arrangements?

Solution

Given: $R_1 = 1.0\,\Omega$ *Find:* (a) R_s
$\quad\quad\quad R_2 = 2.0\,\Omega$ (b) R_p
$\quad\quad\quad R_3 = 3.0\,\Omega$ (c) I
$\quad\quad\quad V = 12\text{ V}$

(a) The equivalent series resistance is simply the sum of the resistors:

$$R_s = R_1 + R_2 + R_3 = 1.0\,\Omega + 2.0\,\Omega + 3.0\,\Omega = 6.0\,\Omega$$

(b) For resistors in parallel, Eq. 18.4 applies:

$$\frac{1}{R_p} = \frac{1}{R_1} + \frac{1}{R_2} + \frac{1}{R_3} = \frac{1}{1.0\,\Omega} + \frac{1}{2.0\,\Omega} + \frac{1}{3.0\,\Omega}$$

With a common denominator,

$$\frac{1}{R_p} = \frac{6.0}{6.0\,\Omega} + \frac{3.0}{6.0\,\Omega} + \frac{2.0}{6.0\,\Omega} = \frac{11}{6.0\,\Omega}$$

and

$$R_p = \frac{6.0\,\Omega}{11} = 0.55\,\Omega$$

Note that R_p is found by inverting the value of $1/R_p$ and it is less than the smallest resistance. Another way to find R_p would be to apply Eq. 18.5 twice. That is, you could combine two of the resistances to get R_{p_1} and then combine that value with the third resistance to get R_p. This avoids working with a reciprocal.

(c) Using Ohm's law with the equivalent resistance for the series arrangement gives the current:

$$I = \frac{V}{R_s} = \frac{12 \text{ V}}{6.0 \text{ }\Omega} = 2.0 \text{ A}$$

Note that the voltage drop across each resistor is

$$V_1 = IR_1 = (2.0 \text{ A})(1.0 \text{ }\Omega) = 2.0 \text{ V}$$

$$V_2 = IR_2 = (2.0 \text{ A})(2.0 \text{ }\Omega) = 4.0 \text{ V}$$

$$V_3 = IR_3 = (2.0 \text{ A})(3.0 \text{ }\Omega) = 6.0 \text{ V}$$

and the sum of the voltage drops equals the battery voltage.

Similarly, for the parallel arrangement, the current is

$$I = \frac{V}{R_p} = \frac{12 \text{ V}}{0.55 \text{ }\Omega} = 22 \text{ A}$$

Note that the current for the parallel combination is large relative to that for the series combination. The current through each of the resistors in parallel is

$$I_1 = \frac{V}{R_1} = \frac{12 \text{ V}}{1.0 \text{ }\Omega} = 12 \text{ A}$$

$$I_2 = \frac{V}{R_2} = \frac{12 \text{ V}}{2.0 \text{ }\Omega} = 6.0 \text{ A}$$

$$I_3 = \frac{V}{R_3} = \frac{12 \text{ V}}{3.0 \text{ }\Omega} = 4.0 \text{ A}$$

The current divides proportionally according to the individual resistances.

The power dissipated by each resistor ($P = IV = I^2 R$) could easily be found for either arrangement. ■

Series-Parallel Combinations

Resistors may be connected in a circuit in a variety of series-parallel combinations. As shown in Fig. 18.4, circuits with only one voltage source may be reduced or collapsed, theoretically into a single loop containing the voltage source and one equivalent resistance by applying the equations given above.

A general procedure for analyzing circuits with different series-parallel combinations of resistors is to find the voltage drops across and the currents through the various resistors as follows:

1. Starting with the resistor combination farthest from the voltage source, find the equivalent series and parallel resistances.
2. Reduce the circuit until there is a single loop with one total equivalent resistance.
3. Find the current delivered to the reduced circuit using Ohm's law.
4. Expand the circuit by reversing the reduction steps, and use the current for

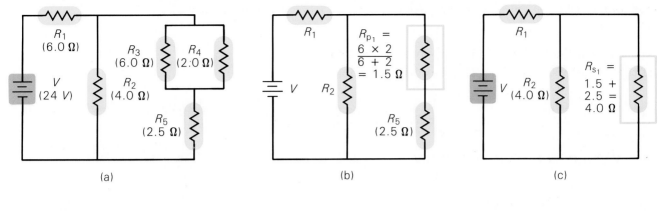

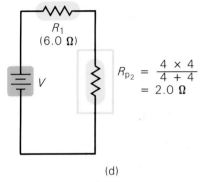

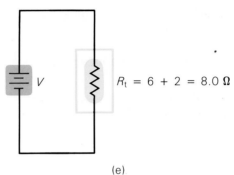

Figure 18.4 **Series-parallel combinations and circuit reduction** Reducing series combinations and parallel combinations to equivalent resistances reduces a circuit with one voltage source to a single loop with a single equivalent resistance.

the reduced circuit to find the currents and voltage drops for the resistors in each step.

The following example illustrates this procedure.

Example 18.2   Series-Parallel Combination of Resistors

(a) What is the voltage drop and current for each of the resistors R_1 through R_5 in Fig. 18.4(a)? (b) How much power is dissipated in R_4?

Solution
Given: Values in figure (any number   *Find*: V's, I's, and P_4
of significant figures may
be assumed)

(a) The parallel combination at the right side of the circuit diagram (farthest from the battery) is first reduced to the equivalent resistance R_{p_1} [see Fig. 18.4(b)]. This leaves a series combination along that side, which is reduced to R_{s_1} [Fig. 18.4(c)]. The final equivalent series resistance is the *total* equivalent resistance (R_t) of the circuit [Fig. 18.4(e)].

Then, by Ohm's law, the battery delivers to the reduced circuit a current of

$$I = \frac{V}{R_t} = \frac{24 \text{ V}}{8.0 \text{ }\Omega} = 3.0 \text{ A}$$

This is the current through R_1 and R_{p_2} since they are in series [Fig. 18.4(d)]. Their voltage drops are

$$V_1 = IR_1 = (3.0 \text{ A})(6.0 \text{ }\Omega) = 18 \text{ V}$$

and

$$V_{p_2} = IR_{p_2} = (3.0 \text{ A})(2.0 \text{ } \Omega) = 6.0 \text{ V}$$

Since R_{p_2} is made up of R_2 and R_{s_1} [Fig. 18.4(c)], there is a 6.0-V drop across each of these resistors. Since $R_2 = R_{s_1}$, the current divides equally:

$$I_2 = 1.5 \text{ A} \qquad \text{and} \qquad I_{s_1} = 1.5 \text{ A}$$

Then I_{s_1} is the current through R_{p_1} and R_5 in series [Fig. 18.4(b)]. Their voltage drops are therefore

$$V_{p_1} = I_s R_{p_1} = (1.5 \text{ A})(1.5 \text{ } \Omega) = 2.25 \text{ V}$$

$$V_5 = I_s R_5 = (1.5 \text{ A})(2.5 \text{ } \Omega) = 3.75 \text{ V}$$

Note that these add up to 6.0 V.

Finally, the voltage drop across R_3 and R_4 is the same as V_{p_1}:

$$V_3 = V_4 = 2.25 \text{ V}$$

The current of 1.5 A (I_{s_1}) divides at the R_3-R_4 junction, and

$$I_3 = \frac{V_3}{R_3} = \frac{2.25 \text{ V}}{6.0 \text{ } \Omega} = 0.375 \text{ A}$$

$$I_4 = \frac{V_4}{R_4} = \frac{2.25 \text{ V}}{2.0 \text{ } \Omega} = 1.125 \text{ A}$$

Note that $I_3 + I_4 = I_{s_1}$.

To see how the current divides proportionally between R_3 and R_4, consider these equations:

$$V_3 = V_4 \qquad \text{and} \qquad I_3 R_3 = I_4 R_4$$

or $\qquad I_3 = \left(\dfrac{R_4}{R_3} \right) I_4 = \left(\dfrac{2.0 \text{ } \Omega}{6.0 \text{ } \Omega} \right) I_4 = \dfrac{I_4}{3}$

That is, I_3 is $\frac{1}{3}$ of I_4. This means that when the current I_{s_1} divides at the junction, $\frac{1}{4}$ of it goes through R_3 and $\frac{3}{4}$ goes through R_4, according to their relative resistances. For R_3, the relative resistance expressed as a fraction of the total resistance is $R_3/(R_3 + R_4) = \frac{6}{8} = \frac{3}{4}$, and for R_4 it is $R_4/(R_3 + R_4) = \frac{2}{8} = \frac{1}{4}$.

(b) The power expended in R_4 is

$$P_4 = I_4 V_4 = (1.125 \text{ A})(2.25 \text{ V}) = 2.53 \text{ W} \quad \blacksquare$$

Circuit analysis like that done in Example 18.2 may seem involved, but it is simply repeated applications of equations for equivalents of series and parallel combinations and Ohm's law, both of which are quite simple.

18.2 Multiloop Circuits and Kirchhoff's Rules

Simple circuits with a single voltage source that can be reduced to a single loop can be analyzed using Ohm's law, as you learned in the preceding section. However, in general, circuits may contain several loops, with each one having voltage sources and/or resistances (collectively referred to as circuit elements

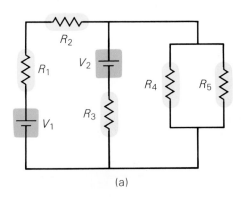

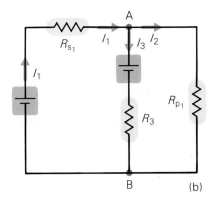

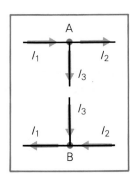

(a)

B   (b)

Figure 18.5 Multiloop circuit
In general, a circuit that contains more than one loop may not be able to be further reduced by series and parallel resistance reductions, but some reductions within each loop may be possible, such as from (a) to (b) here. At a circuit junction, where three or more wires come together, the current divides or comes together [point A or B in part (b)]. The path between two junctions is called a branch. There are three branches in part (b).

or components). A simple multiloop circuit is shown in Fig. 18.5(a). Some combinations of resistors may be replaced by equivalent resistances, but this circuit can be reduced only so far.

A **junction** is a point in a circuit at which three or more connecting wires are joined together, for example, at point A in Fig. 18.5(b). The current either divides or comes together at a junction. A path connecting two junctions is called a **branch** and may contain one or more elements.

A general method for analyzing multiloop circuits is by applying Kirchhoff's rules. These rules embody the conservation of charge and the conservation of energy. (Although they were not stated specifically, Kirchhoff's rules were applied to the simple circuits analyzed in Section 18.1.)

Kirchhoff's first rule, or **junction theorem**, is that the algebraic sum of the currents at any junction is zero:

Kirchhoff's rules were developed by the German physicist Gustav Kirchhoff (1824–1887).

$$\sum_i I_i = 0 \qquad (18.6)$$

This simply means that the sum of the currents going into a junction (taken as positive) is equal to the sum of the currents leaving the junction (taken as negative), or that charge is conserved. For the junction at A in Fig. 18.5(b), the algebraic sum of the currents is $I_1 - I_2 - I_3 = 0$, or

$$I_1 = I_2 + I_3$$

current in = current out

(This rule was applied in analyzing parallel resistances in Section 18.1.)

Of course, you cannot tell whether a particular current flows into or out of a junction simply by looking at a multiloop circuit diagram. You have to *assume* current directions at a particular junction. If these assumptions are wrong, you will find out later from the mathematical results. (You will see this in Example 18.3.)

Kirchhoff's second rule, or **loop theorem**, is that the algebraic sum of the voltage differences across all of the elements of any closed loop is zero:

$$\sum_i V_i = 0 \qquad\qquad (18.7)$$

This means that the sum of the voltage rises equals the sum of the voltage drops around a closed loop, which must be true if energy is conserved. (This rule was used in analyzing series resistances in Section 18.1.)

Since traversing a circuit loop in either a clockwise or a counterclockwise sense will give rise to either a voltage rise or a voltage drop across each circuit element, respectively, it is important to establish a sign convention for voltage changes. This book uses the one illustrated in Fig. 18.6. The voltage change across a battery is taken to be positive (a voltage rise) if the loop is traversed toward the positive terminal (in the direction of conventional current flow from the battery), as shown in Fig. 18.6(a). The voltage change across a battery is taken to be negative if the loop is traversed in the opposite direction, that is, toward the negative terminal. (Note that the assigned branch currents have nothing to do with determining the sign of the voltage change across a battery. The sign of the change depends only on the direction in which the loop is traversed.)

The voltage change across a resistor is taken to be negative (a voltage drop) if the loop is traversed in the direction of the assigned current in that branch [Fig. 18.6(b)], and positive if the loop is traversed in the opposite direction.

This sign convention allows you to go around a loop either clockwise or counterclockwise. Either way, Eq. 18.7 remains mathematically the same.

The general steps in applying Kirchhoff's rules are as follows:

1. Assign a current and current direction for each branch in the circuit. This is done most conveniently at junctions.
2. Indicate the loops and the arbitrarily chosen directions in which they are to be traversed (see Fig. 18.7). Every branch *must* be in at least one loop.
3. Apply Kirchhoff's first rule and write the equations for the currents, one for each junction that gives a different equation. (In general, this gives a set of equations that includes all branch currents.)
4. Traverse the loops applying Kirchhoff's second rule and write the equations using the adopted sign convention.

Steps 3 and 4 give a set of N equations with N unknowns (the currents), which may be solved for the unknowns. If more loops are traversed than necessary, you will have redundant equations. Only the number of loops that includes all the branches is needed.

This procedure may seem complicated, but it's really straightforward, as the following example shows.

Example 18.3 Kirchhoff's Rules

For the circuit diagrammed in Fig. 18.7, find the branch currents.

Solution

The branch currents and their directions as well as the loops have been arbitrarily assigned in the diagram. Note that there is a current in every branch and that every branch is in at least one loop (some branches are in more than one loop but that is acceptable).

Applying Kirchhoff's first rule at the upper junction gives

$$I_1 = I_2 + I_3 \qquad\qquad (1)$$

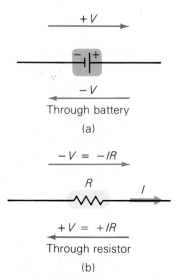

Figure 18.6 **Sign convention**
(a) When Kirchhoff's rules are applied in going around a circuit loop, the voltage change is taken to be positive if a loop is traversed going toward the positive terminal of the battery (or in the direction conventional current would normally flow from the battery) and negative if the loop is traversed going toward the negative terminal. (b) The voltage change across a resistance is taken to be negative if the resistance is traversed in the direction of the assigned branch current and positive if it is traversed in the opposite direction.

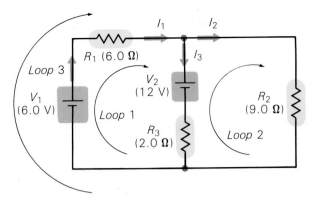

Figure 18.7 **Application of Kirchhoff's rules**
A current and current direction are assigned for each branch in the circuit (most conveniently done at junctions), and then loops are traversed to write equations. Every branch must be in at least one loop; sign conventions must be used. See Example 18.3.

For the lower junction, you could write $I_2 + I_3 = I_1$ (currents in = currents out), but this is the same equation. (If there were another branch, that would have a different current equation for its junctions.)

Going around loop 1 as indicated in the figure and applying Kirchhoff's second rule with the sign convention gives

$$V_1 - I_1 R_1 - V_2 - I_3 R_3 = 0$$

Then, putting in the numerical values from the figure gives

$$6 - I_1(6) - 12 - I_3(2) = 0$$

and

$$6I_1 + 2I_3 = -6 \quad \text{or} \quad 3I_1 + I_3 = -3 \quad\quad\quad (2)$$

Units are omitted for simplicity, and any number of significant figures can be assumed.

Similarly, for loop 2,

$$V_2 - I_2 R_2 + I_3 R_3 = 0$$

and

$$12 - I_2(9) + I_3(2) = 0$$

Thus,

$$9I_2 - 2I_3 = 12 \quad\quad\quad (3)$$

Equations 1, 2, and 3 form a set of three equations with three unknowns. Physically, the problem is solved. The rest is math. You can solve for the I's in several ways. First substitute from Eq. 1 into Eq. 2 to eliminate I_1:

$$3(I_2 + I_3) + I_3 = -3$$

This simplifies to

$$3I_2 + 4I_3 = -3 \quad\quad\quad (4)$$

Then, substituting from Eq. 4 into Eq. 3 eliminates I_2:

$$9(-1 - \tfrac{4}{3}I_3) - 2I_3 = 12$$

That is,

$$-14I_3 = 21 \quad \text{or} \quad I_3 = -1.5 \text{ A}$$

The minus sign on the result tells you that the wrong direction was assumed for I_3.

Putting the value of I_3 into Eq. 4 gives I_2:

$$3I_2 + 4(-1.5) = -3$$

$$I_2 = 1.0 \text{ A}$$

Then, by Eq. 1,

$$I_1 = I_2 + I_3 = 1.0 \text{ A} - 1.5 \text{ A} = -0.5 \text{ A}$$

The minus sign here indicates that I_1 was also assigned the wrong direction. So, at the upper junction in Fig. 18.7, I_3 really flows into the junction and I_1 and I_2 flow out. (You might have suspected this to be the case because of the larger 12-V battery in the I_3 branch.)

Note that loop 3 was not used in this analysis. The equation for this loop would be redundant, giving four equations and three unknowns. However, loop 3 could have been used with either loop 1 or loop 2 in solving the problem. ■

This application of Kirchhoff's rules is called the branch current method. A similar loop current method, which some consider to be simpler mathematically, is described in Problem 69.

18.3 RC Circuits

The previous sections dealt with circuits having constant currents. In some dc circuits, the current may vary with time. A good example are series RC circuits, which have a resistor (R) (or other resistive component) and a capacitor (C) in series (Fig. 18.8). You may be quick to notice that even when the switch is closed, the circuit is still open, or incomplete, because of the plate separation of the capacitor. This is true, but charge does flow when the switch is closed while the capacitor is charging.

The maximum amount of charge built up (Q_o) on the capacitor depends on the capacitance (C) and the voltage of the battery (V_o). Recall from Chapter 16 that this charge is $Q_o = CV_o$. Immediately after the switch is closed, at $t = 0$, there is an initial current in the circuit, which by Ohm's law is $I_o = V_o/R$ through the resistor. Also recall that as charge accumulates on the capacitor's plates, it takes more work to add charge because of the repulsion between like charges. Eventually, the capacitor is charged to the maximum, and the current goes to zero.

The resistance helps determine how fast the capacitor is charged, since the larger its value, the greater the resistance to charge flow. The voltage across the capacitor varies with time according to this equation:

$$V = V_o(1 - e^{-t/RC})$$

(18.8)

charging voltage for an RC circuit

where $e = 2.718$ is the base of the natural logarithm. A graph of V versus t and a graph of I versus t are presented in Fig. 18.9. The current varies with time according to this equation:

$$I = I_o e^{-t/RC}$$

(18.9)

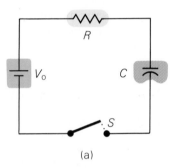

(a)

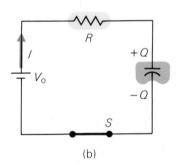

(b)

Figure 18.8 **A series RC circuit**
(a) Even though the circuit is open because of the space between the plates of the capacitor, (b) a current flows in the circuit when the switch is closed until the capacitor is charged to its maximum value. The rate of charging (and discharging) depends on the product of the values of the resistance and capacitance, which is called the time constant of the circuit: $\tau = RC$.

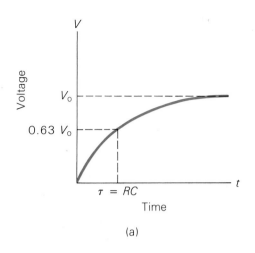

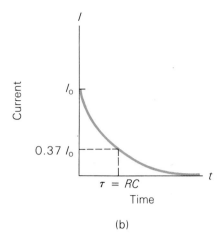

Figure 18.9 **Capacitor charging**
(a) In a series RC circuit, the voltage across the capacitor increases exponentially with time, and rises to 63% of its maximum voltage (V_o) in one time constant, or $t = \tau = RC$.
(b) The current in the circuit is initially a maximum ($I_o = V_o/R$) and decays exponentially with time, falling to 37% of its initial value in one time constant.

The current is said to decay exponentially.

According to Eq. 18.8, it would theoretically take an infinite amount of time for the capacitor in an RC circuit to become fully charged (to reach the battery voltage, V_o). However, in practice, such a capacitor charges and discharges in relatively short times. It is customary to use a special value to express the charging and discharging rates. This **time constant** (τ) for an RC circuit is

$$\tau = RC \tag{18.10}$$

At a time equal to the time constant, $t = \tau = RC$, the voltage across the charging capacitor is

$$V = V_o(1 - e^{-t/\tau}) = V_o(1 - e^{-\tau/\tau}) = V_o(1 - e^{-1})$$

That is,

$$V = 0.63\,V_o$$

After one time constant, the voltage across the capacitor is at 63% of its maximum (or the capacitor is 63% charged since $Q = CV$). Note that at that time the current has decayed to 37% of its initial maximum value (I_o).

At the end of two time constants, $t = 2\tau = 2RC$, the capacitor is charged to more than 86% of its maximum value, and so on. Practically, the capacitor is considered to be fully charged after only a few time constants.

When a fully charged capacitor is discharged through a resistance, the voltage across the capacitor decays exponentially with time (as does the current):

$$V = V_o e^{-t/RC} \tag{18.11}$$

discharging voltage for an RC circuit

In one time constant, the voltage across the capacitor falls to 37% of its original value (see Fig. 18.10).

Example 18.4 Time Constant
The capacitance and resistance in the RC circuit in Fig. 18.8 are 6.0 μF and 0.25 MΩ, respectively, and the battery is a 12-V one. (a) What is the voltage

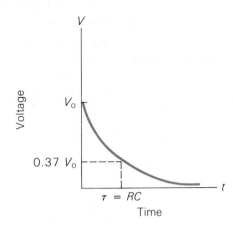

Figure 18.10 **Capacitor discharging**
When a charged capacitor is discharged through a resistance, the voltage across the capacitor (and the current in the circuit) decays exponentially with time, falling to 37% of its initial value in one time constant, $t = \tau = RC$.

across the capacitor at one time constant after the switch is closed if it was initially uncharged? (b) What is the voltage across the capacitor and its charge at $t = 5.0$ s?

Solution

Given: $C = 6.0\ \mu\text{F} = 6.0 \times 10^{-6}$ F   *Find:* (a) V
$\qquad\quad R = 0.25\ \text{M}\Omega = 2.5 \times 10^{5}\ \Omega$ $\qquad\qquad$ (b) V and Q
$\qquad$ (a) $t = \tau$
$\qquad$ (b) $t = 5.0$ s

(a) In one time constant, the capacitor is charged to 63% of its maximum charge, and

$$V = 0.63\,V_{\text{o}} = (0.63)(12\ \text{V}) = 7.6\ \text{V}$$

(b) When a specific time is given, you also need to know the time constant for the circuit, which in this case is

$$\tau = RC = (2.5 \times 10^{5}\ \Omega)(6.0 \times 10^{-6}\ \text{F}) = 1.5\ \text{s}$$

Then, for $t = 5.0$ s, using Eq. 18.8,

$$V = V_{\text{o}}(1 - e^{-t/\tau}) = (12\ \text{V})(1 - e^{-5.0\,\text{s}/1.5\,\text{s}})$$

$$= (12\ \text{V})(1 - e^{-3.3}) = (12\ \text{V})(1 - 0.036)$$

$$= (12\ \text{V})(0.964) = 11.6\ \text{V}$$

Note that after $t = 3.3\tau$, the voltage across the capacitor is more than 96% of its maximum value.

Expressing τ as RC and multiplying both sides of Eq. 18.8 by C gives

$$CV = CV_{\text{o}}(1 - e^{-t/RC})$$

Since $Q = CV$,

$$Q = Q_{\text{o}}(1 - e^{-t/RC})$$

Thus, the charge varies with time just as the voltage does.

The maximum charge (Q_{o}) for the capacitor is

$$Q_{\text{o}} = CV_{\text{o}} = (6.0 \times 10^{-6}\ \text{F})(12\ \text{V}) = 7.2 \times 10^{-5}\ \text{C}$$

508 CH. 18 Electric Circuits

At $t = 3.3\tau = 5.0$ s, the charge on the capacitor is 96.4% of the maximum charge (the same percentage as for the voltage):

$$Q = 0.964Q_o = (0.964)(7.2 \times 10^{-5}\ \text{C}) = 6.9 \times 10^{-5}\ \text{C} \quad \blacksquare$$

A practical application of an RC circuit is diagrammed in Fig. 18.11(a). This is called a blinker circuit (or, more impressively, a neon-tube relaxation oscillator). The resistor and capacitor are in series, and a small neon tube is connected across (in parallel with) the capacitor. (These neon tubes are about the size of miniature Christmas tree lights.)

When the circuit is closed, the voltage across the capacitor (and the neon tube) rises from 0 to V_b, which is the breakdown voltage of the neon gas in the tube (about 80 V). At that voltage, the gas is ionized and begins to conduct electricity, and the tube lights up. When the tube is in a conducting state, the capacitor discharges through it. But when the voltage across the discharging capacitor can no longer sustain the tube discharge, the tube stops conducting. During this short time, the voltage across the capacitor and tube drops from V_b to V_m, which is called the maintaining voltage. The voltage then rises from V_m to V_b, and the cycle begins again. The continual repetition of this cycle causes the tube to blink on and off. The period of the oscillator, or the time between flashes, depends on the RC time constant (see Problem 66).

18.4 Ammeters and Voltmeters

As the names imply, an ammeter measures current (amps) and a voltmeter measures voltage. A basic component of both of these meters is a galvanometer (Fig. 18.12). The galvanometer operates on magnetic principles that will be covered in Chapter 19. In this chapter, it will simply be considered to be a circuit element having an internal resistance (r).

The deflection of the needle on the galvanometer's dial is directly proportional to the current through its wire coil. The resistance of the coil (r) is relatively small. In effect, a galvanometer measures current, but because of the

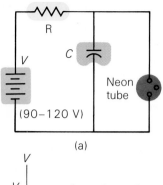

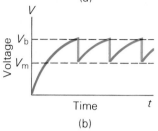

(a)

(b)

Figure 18.11 **Neon-tube relaxation oscillator circuit** (a) When a neon tube is connected across the capacitor in a series RC circuit having the proper voltage source, the voltage across the tube (and capacitor) will relax, or oscillate, with time. As a result, the tube periodically flashes or blinks. (b) A graph of voltage versus time shows the oscillating effect.

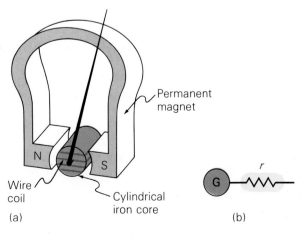

Figure 18.12 **Galvanometer** (a) A galvanometer is a current-sensitive device whose needle deflection is proportional to the current through its coil. (b) The circuit symbol is a circle containing a G. The internal resistance (r) of the meter is indicated explicitly.

small coil resistance, only currents in the microamp range can be measured without burning out the coil. An ammeter that can be used to measure larger currents has a small shunt resistor in parallel with a galvanometer (Fig. 18.13). This provides an alternate path by which part of a large current (I) can bypass the galvanometer ($I = I_g + I_s$, at the left-hand junction in the figure).

The voltages across the galvanometer and the shunt resistor in parallel are equal,

$$V_g = V_s$$

By Ohm's law, then,

$$I_g r = I_s R_s$$

Using $I = I_g + I_s$,

$$I_g r = (I - I_g) R_s$$

and

$$I_g = \frac{IR_s}{r + R_s} \tag{18.12}$$

This equation allows you to select the proper shunt resistance for a given current range and galvanometer. The following example illustrates how to do this.

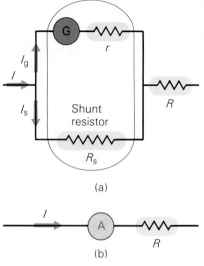

(a)

(b)

Figure 18.13 **dc ammeter**
(a) A shunt resistor (R_s) in parallel with a galvanometer is an ammeter capable of measuring various current ranges, depending on the value of R_s. (b) The circuit symbol for an ammeter is a circle with an A inside it.

Example 18.5   A dc Ammeter

A galvanometer that can safely carry a maximum coil current of 200 μA (called the full-scale sensitivity) has a coil resistance of 50 Ω. It is to be used in an ammeter designed to read currents up to 3.0 A (at full scale). What is the required shunt resistance?

Solution
Given: $I_g = 200 \ \mu A = 2.0 \times 10^{-4}$ A   *Find*: R_s
$\quad\quad\quad r = 50 \ \Omega$
$\quad\quad\quad I_{max} = 3.0$ A

A full-scale reading of 3.0 A means that when a current of 3.0 A enters the ammeter, I_g should be 200 μA. Solving Eq. 18.12 for the shunt resistance and plugging in the values given (assumed to be exact) gives

$$R_s = \frac{I_g r}{I_{max} - I_g}$$

$$= \frac{(2.0 \times 10^{-4} \text{ A})(50 \ \Omega)}{3.000 \text{ A} - 0.0002 \text{ A}} = 0.0033 \ \Omega$$

Note the small size of the shunt resistance as compared to r (50 Ω). This allows most of the current (2.9998 A at full scale) to pass through the shunt resistor branch. This resistor is made of a material that does not burn out as readily as the thin wire of the galvanometer coil. The ammeter will read currents linearly up to 3.0 A. For example, if a current of 1.5 A flowed into the ammeter, there would be a current of 100 μA in the coil of the galvanometer, which would give a half-scale reading. ∎

QUESTION: When an ammeter is used to measure the current through a circuit element, it is connected in series so that the current through the element flows through the ammeter. Since the ammeter has resistance, wouldn't the instrument itself affect the current?

ANSWER: Yes, but the resistance of the ammeter is the equivalent parallel resistance for r and R_s, which is very small. In fact, it may usually be considered negligible compared to the circuit or component resistance. Recall that the equivalent resistance for a parallel combination is less than any of the individual resistances. (What would it be for the resistances in Example 18.5?)

A voltmeter that is capable of reading voltages higher than the microvolt range is constructed by connecting a large multiplier resistor in series with a galvanometer (Fig. 18.14). When connected across a circuit element, the voltmeter (or galvanometer branch) experiences a voltage drop of $V = V_g + V_m$, and most of this drop is across the multiplier resistor rather than the galvanometer coil. Because the voltmeter has a large internal resistance due to the multiplier resistor, it draws little current from the main circuit. By Ohm's law,

$$V = V_g + V_m$$
$$= I_g r + I_g R_m = I_g (r + R_m)$$

and

$$I_g = \frac{V}{r + R_m} \tag{18.13}$$

This voltage is also the potential difference across the circuit element having a resistance R because of the parallel connection (see Fig. 18.14). If the galvanometer scale is calibrated in volts (instead of amps), the voltage drop across that circuit element can be read from the meter.

Example 18.6 A dc Voltmeter
Suppose that the galvanometer described in Example 18.5 is to be used instead in a voltmeter with a full-scale reading of 3.0 V. What is the required value of the multiplier resistor?

Solution
Given: $I_g = 2.0 \times 10^{-4}$ A   *Find*: R_m
         (from Example 18.5)
         $r = 50\ \Omega$ (from Example 18.5)
         $V_{max} = 3.0$ V

Eq. 18.13 can be used to find R_m:

$$R_m = \frac{V_{max} - I_g r}{I_g}$$

$$= \frac{3.0\ \text{V} - (2.0 \times 10^{-4}\ \text{A})(50\ \Omega)}{2.0 \times 10^{-4}\ \text{A}}$$

$$= 1.5 \times 10^4\ \Omega = 15\ \text{k}\Omega \quad \blacksquare$$

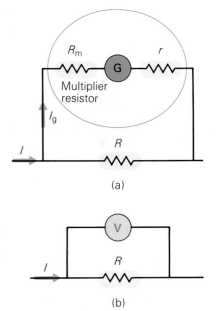

Figure 18.14 **dc voltmeter** (a) A multiplier resistor (R_m) in series with a galvanometer is a voltmeter capable of measuring various voltage ranges depending on the value of R_m. (b) The circuit symbol for a voltmeter is a circle with a V inside it.

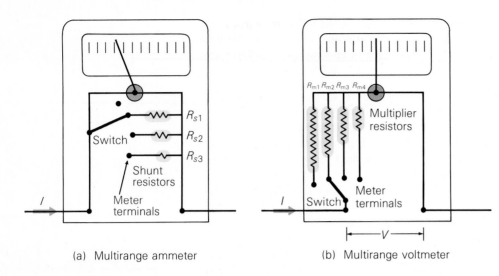

Figure 18.15 **Multirange meters**
(a) An ammeter or (b) a voltmeter can be made to measure several ranges of current and voltage by providing for switching among different shunt or multiplier resistors, respectively. (Instead of a switch, there may be an exterior terminal for each resistance; see Figs. 18.13 and 18.14.)

(a) Multirange ammeter

(b) Multirange voltmeter

For versatility, ammeters and voltmeters may be multiranged. This is accomplished by having several shunt or multiplier resistors (Fig. 18.15).

18.5 Household Circuits and Electrical Safety

Although household circuits generally use alternating current and that has not yet been discussed, they include practical applications of some of the principles already studied.

For example, would you expect the elements in a household circuit (lamps, appliances, and so on) to be connected in series or in parallel? From the discussion of Christmas tree lights connected in series, it should be apparent that the elements must be connected in parallel. When the bulb in a lamp blows out, other elements in the circuit continue to work. This would not be the case for a series circuit. Moreover, household appliances and lamps are generally rated for 120 V. If these elements were connected in series, the voltage drops across them would add up to a *total* of 120 V.

Power is supplied to a house by a three-wire system (Fig. 18.16). There is a potential difference of 240 V between the two hot, or high-potential, wires, and each of these has a 120-V potential difference with the ground. The third wire is grounded at the point where the wires enter the house, usually by a metal rod driven into the ground. This wire has a zero potential and is the ground, or neutral wire.

The potential difference of 120 V needed for most household appliances is obtained by connecting them between the grounded wire and either of the high-potential wires: $\Delta V = 120 \text{ V} - 0 \text{ V} = 120 \text{ V}$ or $\Delta V = 0 \text{ V} - (-120 \text{ V}) = 120 \text{ V}$. (See Fig. 18.16.) Even though the grounded wire has a zero potential, it is a *current-carrying* wire because it is part of a circuit. Large appliances such as central air conditioners, ovens, and hot water heaters need 240 V for operation, and this is obtained by connecting them between the two hot wires: $\Delta V = 120 \text{ V} - (-120 \text{ V}) = 240 \text{ V}$.

As you learned in Chapter 17, the amount of current an appliance draws (uses) may be given on a rating tag but can also be determined from the power

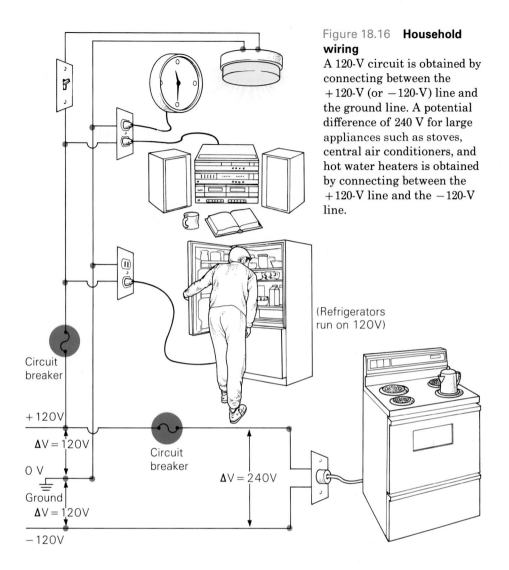

Figure 18.16 **Household wiring**
A 120-V circuit is obtained by connecting between the +120-V (or −120-V) line and the ground line. A potential difference of 240 V for large appliances such as stoves, central air conditioners, and hot water heaters is obtained by connecting between the +120-V line and the −120-V line.

Circuit breaker

+120V

ΔV = 120V

0 V

Ground

ΔV = 120V

−120V

Circuit breaker

ΔV = 240V

(Refrigerators run on 120V)

rating (using $P = IV$). For example, a stereo rated at 180 W at 120 V would draw 1.5 A ($I = P/V$). There are limitations to the number of elements that can be put in a circuit. In particular, the joule heat (or I^2R loss) must be considered. Generally, the more elements (resistances) connected in parallel, the smaller the equivalent resistance. In any case, the equivalent resistance is smaller than the resistance of the smallest element. Thus, it is possible to overload a household circuit and draw too much current; the joule heat from it could then start a fire.

Overloading is prevented by limiting the current in a circuit by means of two types of devices: fuses and circuit breakers. Fuses are common in older homes. An Edison-base fuse has threads like those on the base of a light bulb (see Fig. 18.17). Inside the fuse is a metal strip that melts because of joule heat when the current is larger than the rated value (which might be 15 A). The melting of the strip opens the circuit, and electrically everything comes to a halt.

One problem with Edison-base fuses is that they are interchangeable. That is, a 30-A fuse can be put in a circuit that is rated for 15 A. This problem is avoided with another type of fuse, called a Type-S fuse. A nonremovable adapter is installed in the socket for a fuse; the adapter is specific for a particular rated value (the fuses and adapters for different ratings have different threads). That is, a 30-A Type-S fuse will not screw into a 15-A Type-S adapter, for example, so the wrong fuse can't be used.

Circuit breakers are now used exclusively in wiring new homes. One type of circuit breaker, shown in Fig. 18.18, uses a bimetallic strip (see Chapter 11). As the current through the strip increases, it becomes warmer and bends. At the rated current value, the strip will be bent sufficiently to cause the circuit to open mechanically. The strip quickly cools, so the breaker may be reset. However, when a fuse blows or a circuit breaker trips, this is an indication that the circuit is drawing or attempting to draw too much current. You should investigate to find the cause before replacing the fuse or resetting the circuit breaker.

Switches, fuses, and circuit breakers are placed in the hot (high-potential) side of the line. They would work in the grounded side, but the circuit and the elements would still be connected to a high potential, which could be dangerous if a person made electrical contact. Even with fuses or circuit breakers in the hot side of the line, there is still a possibility of getting an electrical shock from a defective appliance that has a metal casing, such as a hand drill. A wire may come loose inside and make contact with the casing, which would then be hot, or at a high potential. As illustrated in Fig. 18.19, a person could provide a path to the ground and become part of the circuit, thus getting a shock.

To prevent this from happening, a third dedicated grounding wire is added to the circuit that grounds the metal casing [Fig. 18.20(a)]. Note that this wire is not normally a current-carrying wire. If a hot wire comes into contact with the casing, the circuit is completed to this grounded wire, and the fuse is blown or the circuit breaker tripped.

On three-prong plugs, the large round prong connects with the dedicated grounding wire. Adapters can be used between a three-prong plug and a two-prong socket. Such an adapter has a grounding lug or grounding wire. This should be fastened to the receptacle box by the plate-fastening screw or some other means. The receptacle box is grounded by means of the grounding wire. If the adapter lug or wire is not connected, the system is left unprotected, which defeats the purpose of the dedicated grounding safety feature.

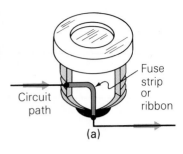

(b)

Figure 18.17 Fuses
(a) A fuse contains a metallic strip, or ribbon, that melts when the current exceeds a rated value. This opens the circuit and prevents overheating. (b) Edison-base fuses (left) have threads similar to light bulbs; fuses with different ampere ratings can be interchanged. Type-S fuses (right) have different threads for different ratings and cannot be interchanged with specific adapters in the fuse socket.

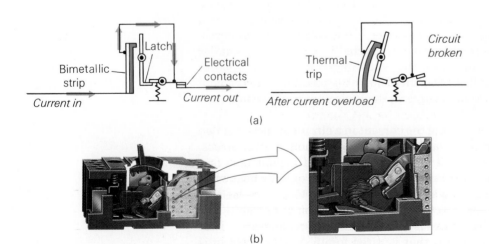

(a)

(b)

Figure 18.18 Circuit breaker
(a) A diagram of a thermal trip element. With increased current and joule heating, the element bends until it opens the circuit at some preset current value. Magnetic trip elements are also used. (b) An exposed view of a circuit breaker shows the trip element at the lower right.

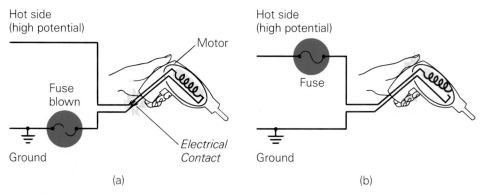

Hot side
(high potential)

Motor

Fuse
blown

Ground

*Electrical
Contact*

(a)

Hot side
(high potential)

Fuse

Ground

(b)

Figure 18.19 **Electrical safety**
(a) Switches, fuses, or circuit breakers should always be wired in the hot side of the
line, *not* in the grounded side as shown here. When these elements are wired in the
grounded side, the line is at a high potential even when the fuse is blown or a switch
is open. (b) Even when the fuse or circuit breaker is in the hot side of the line, a
potentially dangerous situation exists. If an internal wire comes into contact with the
metal casing of an appliance or power tool, a person who touches the casing can get a
shock.

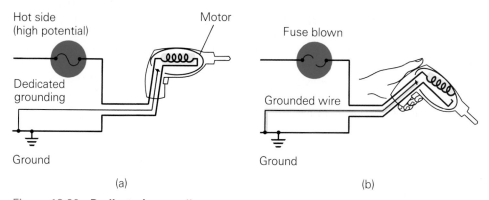

Hot side
(high potential)

Motor

Dedicated
grounding

Ground

(a)

Fuse blown

Grounded wire

Ground

(b)

Figure 18.20 **Dedicated grounding**
(a) For electrical safety, a third wire is connected from an appliance to a ground. This
dedicated grounding wire normally carries no current (as opposed to the grounded
wire). (b) If a loose wire comes into contact with the grounded casing, the shorting to
ground through the dedicated grounding wire blows the protective fuse, and the
appliance will not be at a high potential.

You have probably noticed that there is another type of plug, a two-prong
plug that only fits in the socket one way because one prong is larger than the
other and one of the slits of the receptacle is also larger. This is called a
polarized plug. Polarizing in the electrical sense is a method of identifying the
sides of the line (or the voltages) so that particular connections may be made.

Polarized plugs and sockets are also a safety feature. Wall receptacles are
wired so that the small slit connects to the hot side and the large slit connects to
the neutral, or ground, side. The casing of an appliance can always be connected
to the ground side by means of a polarized plug. If a hot wire inside the
appliance came loose and made contact with the metal casing, the effect would
be similar to that with a dedicated grounding system. The hot side of the line

Electricity and Personal Safety

Safety precautions are necessary to prevent injuries when people work with and use electricity. Electricity conductors (such as wires) are coated with insulating materials so that they can be handled safely. However, when a person comes into contact with a charged conductor, a potential difference may exist across part of the body. A bird can sit on a high-voltage line without any problem because the bird and the line are at the same potential, and there is no potential difference or circuit path. But if a person carrying an aluminum (conducting) ladder touches it to an electrical line, a potential difference exists from the line to the ground, and the ladder and the person are part of the circuit.

The extent of personal injury in such a case depends on the amount of electric current that flows through the body and on the circuit path. The current is given by

$$I = \frac{V}{R_{body}}$$

where R_{body} is the resistance of the body. Thus, for a given voltage, the current depends on the body resistance.

Body resistance varies. If the skin is dry, the resistance may be 0.50 MΩ ($0.50 \times 10^6\,\Omega$) or more. For a potential difference of 120 V, there would be a current of

$$I = \frac{V}{R_{body}} = \frac{120\text{ V}}{0.50 \times 10^6\,\Omega} = 0.24 \times 10^{-3}\text{ A} = 0.24\text{ mA}$$

This current is almost too weak to be felt.

Suppose, however, that the skin is wet with perspiration.

Then R_{body} is about 5.0 kΩ ($5.0 \times 10^3\,\Omega$). In this case, the current is

$$I = \frac{V}{R_{body}} = \frac{120\text{ V}}{5.0 \times 10^3\,\Omega} = 24 \times 10^{-3}\text{ A} = 24\text{ mA}$$

which could be very dangerous.

A basic precaution to take is to avoid coming into contact with an electrical conductor that might cause a potential difference to exist across the body or part of it. The effect of such contact depends on the path of the current. If that path is from the little finger to the thumb on one hand, probably only a burn would result from a large current. However, if the path is from hand to hand through the chest, the effect can be much worse. Some of the possible effects of this circuit path are given in Table 1.

Injury results because the current interferes with muscle functions and/or causes burns. Muscle functions are regulated by electrical impulses through the nerves, and these can be influenced by currents from external voltages. Muscle reaction and pain can occur from a current of a few milliamps. At about 10 mA, muscle paralysis can prevent a person from releasing the conductor. At about 20 mA, contraction of the chest muscles occurs, which may cause breathing to be impaired or to stop. Death can occur in a few minutes. At 100 mA, there are rapid uncoordinated movements of the heart muscles (called ventricular fibrillation), which prevent the proper pumping action.

Working safely with electricity requires a knowledge of fundamental electrical principles *and* common sense. Electricity must be treated with respect.

Table 1
Effects of Electric Current on the Human Body*

Current (*approximate*)	Effect
1.0 mA (0.001 A)	Mild shock or heating
10 mA (0.01 A)	Paralysis of motor muscles
20 mA (0.02 A)	Paralysis of chest muscles, causing respiratory arrest; fatal in a few minutes
100 mA (0.1 A)	Ventricular fibrillation, preventing coordination of the heart's beating; fatal in a few seconds
1000 mA (1 A)	Serious burns; fatal almost instantly

* The effect on the human body of a given amount of current depends on a variety of conditions. This table gives only general and relative descriptions.

would be shorted to the ground, which would blow a fuse or trip a circuit breaker.

The three-wire dedicated grounding system is preferred over polarized plugs and sockets. There could be a wiring error in an appliance or receptacle, which might cause a problem even with a polarized plug. (What would the problem be?)

Important Formulas

Equivalent series resistance:

$R_s = R_1 + R_2 + R_3 + \cdots$

Equivalent parallel resistance:

$$\frac{1}{R_p} = \frac{1}{R_1} + \frac{1}{R_2} + \frac{1}{R_3} + \cdots$$

Equivalent parallel resistance for two resistors:

$$R_p = \frac{R_1 R_2}{R_1 + R_2}$$

Kirchhoff's rules:

(1) $\quad \sum_i I_i = 0$

(2) $\quad \sum_i V_i = 0$

Time constant:

$\tau = RC$

Charging voltage for an RC circuit:

$V = V_o(1 - e^{-t/RC}) = V_o(1 - e^{-t/\tau})$

Discharging voltage for an RC circuit:

$V = V_o e^{-t/RC} = V_o e^{-t/\tau}$

Current (charging and discharging) in an RC circuit:

$I = I_o e^{-t/RC} = I_o e^{-t/\tau}$

Ammeter current:

$$I_g = \frac{IR_s}{r + R_s}$$

Voltmeter current:

$$I_g = \frac{V}{r + R_m}$$

Questions

Resistances in Series, Parallel, and Series-Parallel Combinations

1. Are the voltage drops across resistors in series generally the same? If not, could they be so in certain cases?

2. Are the headlights (or taillights) of an automobile wired in series or in parallel?

3. Are the currents in resistors in parallel generally the same? If not, could they be in certain cases?

4. As more resistors are added to a parallel combination, how is the equivalent resistance affected? (Give some numerical examples with individual resistances greater than and less than $1\ \Omega$.)

5. How many different values of resistance could be obtained using one or more of three different resistors (R_1, R_2, and R_3)?

Multiloop Circuits and Kirchhoff's Rules

6. What would be the implications if Kirchhoff's rules were not satisfied for a circuit loop?

7. Why is it necessary to use a sign convention when applying Kirchhoff's rules in going around a loop?

8. What would happen if a branch were not included in one of the loops in applying Kirchhoff's rules?

9. Traverse loop 3 in Fig. 18.7, and show that the equation for this loop is not needed or could be used with the equations for one of the other loops.

RC Circuits

10. What must the unit of the time constant (τ) be?

11. How long does it take to charge a capacitor in an RC circuit (a) theoretically and (b) practically? What determines the time needed to change the plates?

12. Why doesn't the voltage across the neon tube in an oscillator circuit fall below the maintaining voltage and go to zero?

Ammeters and Voltmeters

13. Ammeters and voltmeters are termed low-resistance and high-resistance instruments, respectively. Explain why.

14. What will happen if the full-scale reading on an ammeter or voltmeter is exceeded?

15. An ammeter should always be connected in series in a circuit. What would happen if it were not?

16. A voltmeter should always be connected in parallel across a circuit element. What would happen if it were connected in series?

Household Circuits and Electrical Safety

17. Would there be any advantage to having series-parallel combinations in household circuits? Explain.

18. Why are switches and fuses placed in the high-potential side of a household circuit?

19. If a fuse were wired in parallel with circuit elements, what would be the effect?

20. (a) If a live power line dropped on a parked car in which you were sitting, what should you do? (b) Suppose that you found a person lying on the ground in contact with a fallen live power line. What should you do?

21. Tell why each of the following actions is unsafe (perhaps dangerous): (a) putting a penny behind a blown Edison-base fuse instead of replacing the fuse, (b) cutting off the grounding prong of a three-prong plug so that it can be used in two-prong receptacles, (c) not using the ground-ing lug or wire on a three-to-two-prong plug adapter, and (d) continually resetting a circuit breaker to get things to work again.

22. On the handle of a hair dryer is this warning: "Danger. Electrocution is possible if used or dropped in a tub. Unplug after each use." Explain why this warning is necessary.

23. Bodily injury depends on the magnitude of the current and its path, yet you usually see signs that warn "Danger. High Voltage." Shouldn't the warning refer to high current? Explain.

Problems

18.1   Resistances in Series, Parallel, and Series-Parallel Combinations

■1. Three 30-Ω resistors are connected in series. What is the total equivalent resistance?

■2. Three resistors that have values of 20 Ω, 30 Ω, and 40 Ω are to be connected together. (a) What is the maximum equivalent resistance? (b) What is the minimum equivalent resistance?

■3. Five 50-Ω resistors are connected in parallel. What is the total equivalent resistance?

■4. A combination of two 20-Ω resistors in parallel is connected in series to a 40-Ω resistor. What is the total equivalent resistance?

■■5. Three 2.0-Ω resistors can be connected in different ways to produce different equivalent resistances. Draw a circuit diagram showing each way, and find the total resistance in each case.

■■6. Three resistors with values of 5.0 Ω, 10 Ω, and 15 Ω are connected in series in a circuit with a 9.0-V battery. (a) What is the total equivalent resistance? (b) What is the current through each resistor? (c) What is the rate at which energy is dissipated in the 10-Ω resistor?

■■7. Find the equivalent resistances for all possible combinations of one or more of the three resistors in Problem 2.

■■8. Three resistors with values of 2.0 Ω, 4.0 Ω, and 6.0 Ω are connected in series and put in a circuit with a 12-V battery. (a) How much current is delivered to the circuit by the battery? (b) What is the current through each resistor? (c) How much power is dissipated by each resistor? (d) How does this compare with the power dissipated by the total equivalent resistance?

■■9. Suppose that the resistors in Problem 8 are connected in parallel. (a) How much current is drawn from the battery? (b) How much current flows through each resistor? (c) How much power is dissipated by each resistor? (d) How does this compare with the power dissipated by the total equivalent resistance?

■■10. Resistors of 10 Ω and 15 Ω are connected in parallel in a circuit with a 9.0-V battery. How much energy is dissipated each second by the resistors?

■■11. Four identical light bulbs are connected across a 120-V line ("across" means in parallel). The resistance for each bulb is 240 Ω. (a) What is the current through each bulb? (b) What is the voltage drop across each bulb?

■■12. How many 60-W light bulbs can be connected in parallel with a 120-V source without blowing a 15-A fuse in the circuit? (The fuse is in series with the parallel bulb arrangement. Why?)

■■13. Two 75-W light bulbs are connected to a 110-V power source. What are the current in each bulb and the total power dissipated if the bulbs are connected (a) in series and (b) in parallel?

■■14. Two 40-W and one 60-W light bulb are connected across (in parallel with) a 120-V source. (a) What is the current through each bulb? (b) How much power is dissipated by each bulb?

■■15. What is the equivalent resistance for the resistors in Fig. 18.21?

■■16. What is the equivalent resistance between points A and B in Fig. 18.22?

■■17. Find the equivalent resistance for the arrangement of resistors shown in Fig. 18.23.

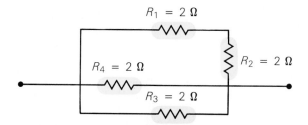

Figure 18.21 **Series-parallel combination**
See Problems 15 and 23.

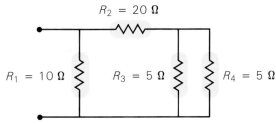

Figure 18.22 **Series-parallel combination**
See Problems 16 and 25.

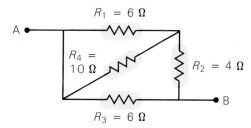

Figure 18.23 **Series-parallel combination**
See Problem 17.

■■18. Arrange three 50-Ω resistors and a 120-V source in a circuit in any arrangement. (a) What is the maximum power for the circuit? (b) What is the minimum power?

■■19. Find the current and the voltage drop for the 2.0-Ω resistor in Fig. 18.24.

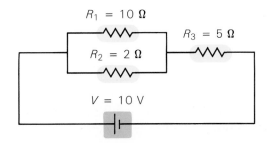

Figure 18.24 **Current and voltage drop for a resistor**
See Problem 19.

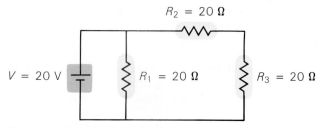

Figure 18.25 **Circuit reduction**
See Problems 20 and 38.

■■20. For the circuit shown in Fig. 18.25, find (a) the current through each resistor, (b) the voltage across each resistor, and (c) the total power dissipated.

■■21. A three-way light bulb can produce 50 W, 100 W, or 150 W of power at 120 V. (a) Find the current for each power setting. (b) Find the resistance for each power setting.

■■22. A 150-W light bulb is connected in series with a 75-W bulb at 120 V. (a) What is the total resistance of the two bulbs? (b) What is the voltage drop across each bulb?

■■23. Suppose that the resistor arrangement in Fig. 18.21 is connected to a 12-V battery. (a) What will the current through each resistor be? (b) What will the voltage drop across each resistor be? (c) What will the total power dissipated be?

■■24. Three 100-Ω immersible resistors and a 12-V source are used to heat 0.50 kg of water initially at 10 °C as rapidly as possible. (Assume that there is no heat loss.) (a) How much time is required to heat the water to 100 °C? (b) How much additional time is required to boil away all of the water?

■■25. The terminals of a 6.0-V battery are connected to points A and B in Fig. 18.22. (a) How much current flows through each resistor? (b) How much power is dissipated by each resistor? (c) Compare the sum of the individual power dissipations to the power dissipation of the equivalent resistance for the circuit.

■■26. Two 1.0-kΩ resistors are connected in parallel. That arrangement is connected in series to another 1.0-kΩ resistor in a circuit containing a battery. If each resistor is rated at 0.25 W, find the maximum voltage of the battery that can be safely used in the circuit.

■■27. A battery has three cells, each with an internal resistance of 0.02 Ω and an emf of 1.5 V. The battery is connected in parallel with a 10-Ω resistor. (a) Determine the voltage across the resistor. (b) How much current passes through each cell? (The cells in a battery are connected in series.)

■■28. A 10-Ω resistor and a 15-Ω resistor are connected in series. This combination is connected in parallel with a 5.0-Ω resistor, and the resulting arrangement is connected in series with a 12-V battery. What is the current through each resistor?

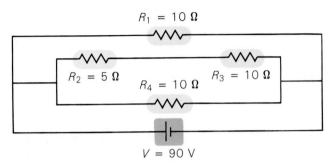

Figure 18.26 **How much power dissipated?**
See Problem 29.

■■29. Four resistors are connected to a 90-V source as shown in Fig. 18.26. (a) Which resistor dissipates the most power and how much? (b) What is the total power dissipated in the circuit?

■■■30. How many different values of resistance can be obtained by connecting four 10-Ω resistors in any arrangement, using any number of them?

■■■31. What is the equivalent resistance of the arrangement in Fig. 18.27?

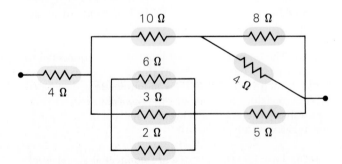

Figure 18.27 **Equivalent resistance replacement**
See Problem 31.

■■■32. If a 120-V source were connected across the leads (open ends) of the circuit shown in Fig. 18.28, how much current would be drawn from it?

■■■33. Four resistors are connected in a circuit with a

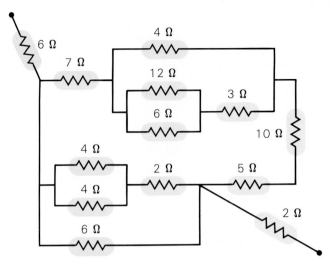

Figure 18.28 **Connect it to a voltage source**
See Problem 32.

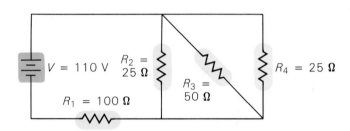

Figure 18.29 **Joule heat losses**
See Problem 33

110-V source, as shown in Fig. 18.29. (a) What is the current through each resistor? (b) How much power is dissipated by each resistor?

■■■34. What is the total power dissipated in the circuit shown in Fig. 18.30?

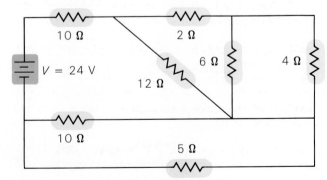

Figure 18.30 **Power dissipation**
See Problem 34.

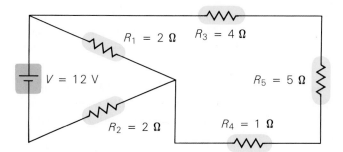

Figure 18.31 **How much current?**
See Problem 35.

■■■35. Find the current through each of the resistors in the circuit in Fig. 18.31.

18.2 Multiloop Circuits and Kirchhoff's Rules

■36. Find the current in each resistor in Fig. 18.32 using Kirchhoff's rules.

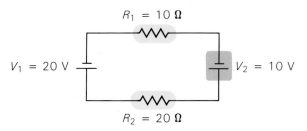

Figure 18.32 **Single-loop circuit**
See Problem 36.

■37. (a) For the circuit in Fig. 18.7, reverse the directions of the interior loops and show that equations equivalent to those in Example are obtained. (b) Using the exterior loop (loop 3) and one of the interior loops (1 or 2), show that the equations resulting from applying Kirchhoff's rules give a solution to the problem.

■38. Apply Kirchhoff's rules to the circuit in Fig. 18.25 to find the current through each resistor.

■■39. Apply Kirchhoff's rules to the circuit in Fig. 18.33, find (a) the current in each resistor and (b) the rate at which energy is being dissipated in the 8-Ω resistor.

■■40. Find the current through each resistor in the circuit in Fig. 18.34.

■■41. Find the currents in the circuit branches in Fig. 18.35.

■■42. Find the current through each resistor in the circuit in Fig. 18.36.

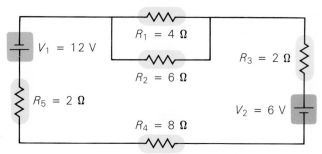

Figure 18.33 **A loop in a loop**
See Problem 39.

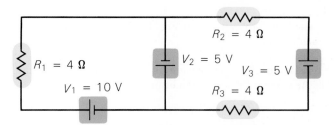

Figure 18.34 **Double-loop circuit**
See Problem 40.

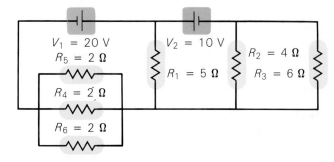

Figure 18.35 **How many loops?**
See Problem 41.

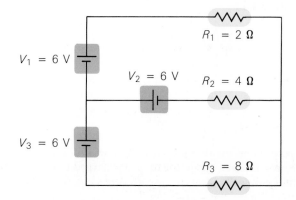

Figure 18.36 **Kirchhoffs rules**
See Problem 42.

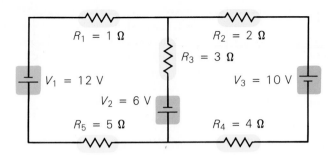

Figure 18.37 **Three voltage sources**
See Problem 43.

■■**43.** For the circuit in Fig. 18.37, find the current in each branch.

■■■**44.** For the multiloop circuit shown in Fig. 18.38, what is the current through each branch?

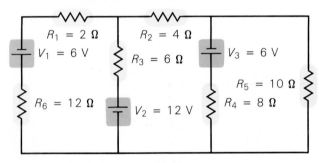

Figure 18.38 **Triple-loop circuit**
See Problem 44.

18.3 RC Circuits

■**45.** A capacitor in a single-loop RC circuit is charged to 63% of its final voltage in 5.0 s. Find the time constant for the circuit.

■**46.** For an RC circuit with a 2.5 MΩ resistance, what capacitance will give a time constant of 1.0 s?

■**47.** An RC circuit has a time constant of 2.25 s. The capacitor is in the microfarad range and has the same numerical prefix as the resistor. What are the values of the capacitor and the resistor?

■**48.** How many time constants will it take for an initially charged capacitor to be discharged to half of its initial voltage?

■■**49.** A 4.0-μF capacitor in series with a resistor is connected to a 12-V source and charged. (a) What re-

sistance is necessary to cause the capacitor to have only 37% of its initial charge 1.5 s after starting to discharge? (b) What is the voltage across the capacitor at $t = 3\tau$ when it is charging?

■■**50.** An RC circuit with $C = 40$ μF and $R = 6.0$ Ω has a 12-V source. With the capacitor initially uncharged, an open switch in the circuit is closed. (a) What is the potential difference across the 6-Ω resistor immediately afterward? (b) What is the potential difference across the capacitor at that time? (c) What is the current in the resistor at that time?

■■**51.** (a) In the circuit in Problem 50 after the switch has been closed for $t = 4\tau$, what is the charge on the capacitor? (b) After a long time, what are the voltages across the capacitor and the resistor?

■■**52.** An RC circuit with a resistance of 5.0 MΩ and a capacitance of 0.40 μF is connected to a 12-V source. If the capacitor is initially uncharged, what is the voltage change across it between $t = 2\tau$ and $t = 4\tau$?

18.4 Ammeters and Voltmeters

■■**53.** A galvanometer with a full-scale sensitivity of 600 μA and a coil resistance of 50 Ω is to be used in an ammeter designed to read 5 A at full scale. What is the required value of the shunt resistor?

■■**54.** A galvanometer has a coil resistance of 20 Ω. A current of 200 μA deflects the needle through 10 divisions full-scale. What is needed to convert the galvanometer to a full-scale 10-V voltmeter?

■■**55.** An ammeter has a resistance of 1.0 mΩ. Find the current in the ammeter when it is properly connected to a 10-Ω resistor and a 6-V source.

■■**56.** An ammeter with a resistance of 2.0 mΩ is advertised to have an accuracy of $\pm 5\%$. What will the low reading be if the ammeter is connected properly to a 5.0 Ω resistor and a 10-V battery?

■■■**57.** An ammeter and a voltmeter can be used to measure the value of a resistor in a circuit. Suppose that the ammeter is connected in series with the resistor, and the voltmeter is placed across the resistor *only*. Show that the value of the resistance when measured like this is given by

$$R = \frac{V}{I - (V/R_v)}$$

where I is the current measured by the ammeter, V is the voltage measured by the voltmeter, and R_v is the resistance of the voltmeter. (Hint: draw a circuit diagram to help analyze the problem.)

■■■**58.** An ammeter and a voltmeter are used to measure the value of a resistor in a circuit. The ammeter is connected in series with the resistor, and the voltmeter is placed across *both* the ammeter and the resistor. Show that the value of the resistance when measured like this is given by

$$R = (V/I) - R_a$$

where V is the voltage measured by the voltmeter, I is the current measured by the ammeter, and R_a is the resistance of the ammeter. (Hint: draw a circuit diagram to help analyze the problem.)

Additional Problems

59. For the circuit in Fig. 18.39, find (a) the current through each resistor and (b) the voltage drop across each resistor.

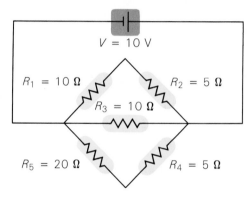

Figure 18.39 **Resistors and currents**
See Problem 59.

60. How many different integral (whole-number) values of resistance can be obtained by using one or more of three resistors with values R, $2R$, and $3R$?

61. A galvanometer with a full-scale sensitivity of 2000 μA has a coil resistance of 100 Ω. If it is to be used in an ammeter with a full-scale reading of 3.0 A, what is the necessary shunt resistance?

62. If the galvanometer in Problem 61 were to be used in a voltmeter with a full-scale reading of 1.5 V, what would the required value of the multiplier resistor be?

63. A length of wire with a resistance of 15 $\mu\Omega$ is cut into three equal segments. The segments are then joined together to form a conductor a third as long as the original wire. What is the resistance of the shortened conductor?

64. A 4.0-Ω resistor and a 6.0-Ω resistor are connected in series. A third resistor is connected in parallel with the 6.0-Ω resistor. This gives a total equivalent resistance of 7.0 Ω for the whole arrangement. What is the value of the third resistor?

65. Two 8.0-Ω resistors are connected in parallel, and two 4.0-Ω resistors are connected in parallel. These combinations are then connected in series and put in a circuit with a 9.0-V battery. What are the current through and the voltage across each resistor?

66. The time between flashes of the neon tube, or the period of the oscillator circuit, in Fig. 18.11 is the time it takes for the voltage to rise from V_m to V_b. Show that the period of such an oscillator is given by

$$T = t_b - t_m = RC \ln\left(\frac{V_o - V_m}{V_o - V_b}\right)$$

where V_o is the maximum voltage, or the voltage of the battery in the circuit.

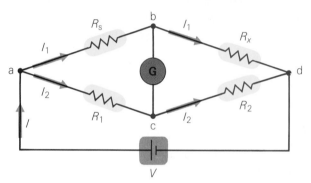

Figure 18.40 **Wheatstone bridge**
See Problem 67.

67. The circuit shown in Fig. 18.40 is called a Wheatstone bridge after Sir Charles Wheatstone (1802–1875) and is used to make accurate measurements of resistance without the errors of ammeter-voltmeter measurements (see Problems 57 and 58). The resistances R_1, R_2, and R_s are all known, and R_x is the unknown resistance to be measured; R_s is variable and is adjusted until the bridge circuit is balanced, when the galvanometer (G) shows a zero reading (no current through the galvanometer branch). Show that when the bridge is balanced R_x is given by

$$R_x = \left(\frac{R_2}{R_1}\right) R_s$$

68. Show that the unit of the time constant is the second.

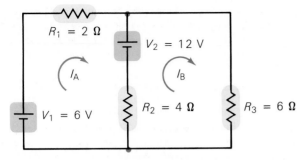

Figure 18.41 **The loop-current method**
See Problem 69.

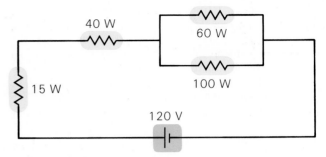

Figure 18.42 **Watt's up?**
See Problem 70.

69. A procedure called the loop current method is considered by some to be mathematically simpler than Kirchhoff's branch current method. In the loop current method, a particular current is assigned to each loop (in Fig. 18.41, I_A and I_B). Kirchhoff's second rule is then used to write an equation for each loop, incorporating the sign convention. The equations are solved for the loop currents, and then Kirchhoff's first rule is applied to each junction in the circuit to find the branch currents by comparison with the loop currents. Show that this loop current method gives the same currents for the circuit in Fig. 18.41 as the branch current method. (Remember that all the loop currents in a branch must be taken into account when applying the loop and junction theorems.)

70. Light bulbs with the wattage ratings shown in Fig. 18.42 are connected in a circuit. (a) What current does the voltage source deliver to the circuit? (b) Find the power dissipated by each bulb. (Assume normal operating resistances of the bulbs to be constant at any voltage.)

Magnetism

19

Children, and many adults, are fascinated by the behavior of magnets. Experiencing the attraction and repulsion between magnets or between a magnet and a piece of metal gives a hands-on demonstration of a magnetic force. Magnets are readily available today and are produced commercially, but they were once quite scarce and existed only as natural magnets (magnetized stones found in nature). It was known as early as about 600 B.C. that a certain type of rock, called lodestone, could attract pieces of iron and other pieces of the same kind of rock. Today, we know that lodestone is a type of iron ore called magnetite (Fe_3O_4, iron oxide).

The early history of of human use of magnetism is largely undocumented. It is generally believed that magnetic rocks were first found in a region called Magnesia (in what is now Turkey), from which the name "magnet" is derived. For centuries, the attractive properties of magnets were attributed to supernatural forces. Early Greek philosophers believed that a magnet had a soul that caused it to attract a piece of iron. We now associate magnetism with electricity (electromagnetism) in the context of a fundamental force or interaction (the electromagnetic force). We have put electromagnetism to use in motors, generators, radios, telephones, and so many other familiar applications that ours might be called an electromagnetic society.

Although electricity and magnetism are manifestations of the same fundamental force, it is instructive to consider them individually and then put them together, so to speak. This and the next chapter will investigate magnetism and its intimate relationship to electricity.

19.1 Magnets and Magnetic Poles

One of the first things anyone notices in examining a common bar magnet is that it has two poles, or "centers" of force, one at or near each end. Rather than being distinguished as positive and negative, these poles are called north and south. This terminology comes from the early use of the magnetic compass. The north pole of a compass magnet is the north-seeking end.

The attraction and repulsion between poles of magnets are similar to the behavior of like and unlike electric charges. That is, analogous to the law of charges is the **law of poles**:

Like magnetic poles repel, and unlike magnetic poles attract.

That is, north and south poles (N-S) attract each other, and north and north poles (N-N) or south and south (S-S) poles repel each other.

Magnetic poles always occur in pairs, as a so-called magnetic dipole. Although an isolated, single magnetic pole, a magnetic monopole, has been postulated to exist, it has yet to be verified experimentally. You might think you could break a bar magnet in half and get two isolated monopoles. However, you would find that the pieces are two shorter magnets, both with north and south poles. This fact may make you wonder about the nature of magnetism. As will be discussed shortly, magnetism is due to electric charges in motion, which occurs with electric currents, orbiting atomic electrons, and so on. This understanding of the source of magnetism and some knowledge of magnetic materials (to be covered) in a later section will allow you to see why pieces of a magnet are (dipole) magnets themselves.

From your knowledge of electrostatics, you may be tempted to think of a pole as a magnetic "charge" and to wonder if the magnetic force can be expressed in a form similar to Coulomb's law for electric charges. In fact, such a law was developed by Coulomb using the magnetic pole strengths in place of the electric charges. However, this law is now used rarely because it doesn't match present understanding of magnetism and because it is more convenient to work with magnetic fields.

In Chapter 16, you learned how convenient it is to describe the interaction of electrically charged objects in terms of electric fields. There is an electric field surrounding any electric charge, and this is represented by electric field lines, or lines of force ($E = F/q$). Similarly, it is convenient to describe magnetic interactions in terms of magnetic fields, and there is a magnetic field surrounding any magnet. Like an electric field, a **magnetic field** is a vector quantity and is represented by the symbol $\mathbf{B}$. The pattern of the magnetic field lines surrounding a magnet can be visually demonstrated by sprinkling iron filings over a magnet covered with a piece of paper or glass (Fig. 19.1). As you will learn, the iron filings become induced magnets and line up with the field.

To describe any vector field, you must specify both its magnitude, or

The technical term is magnetic induction, but it is common to talk in terms of magnetic field, or B field.

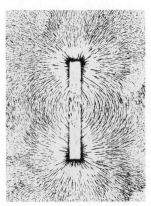 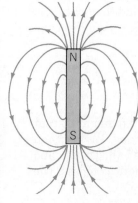

Figure 19.1 **Magnetic field** The magnetic field lines may be outlined using iron filings. The filings become tiny magnets and line up with the field.

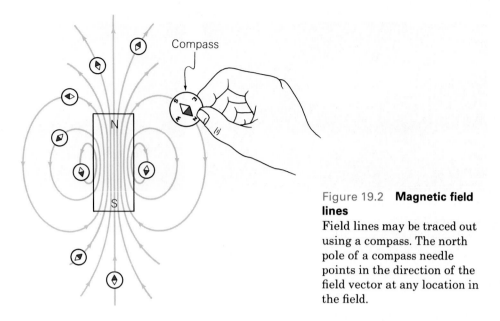

strength, and its direction. The direction of a magnetic field, or B field, can be defined using another north magnetic pole:

> The direction of a B field at any location is in the direction that the north pole of a compass will point with the compass at that location.

This definition provides another method of mapping a B field by placing a small compass near a magnet. The torque on the compass needle (a small magnet) due to the magnetic force causes the needle to line up with the field. If the compass is moved in the direction in which the needle points, the path of the needle traces out a field line, as illustrated in Fig. 19.2. Thus, *the direction of the magnetic field at any point is in the direction of the force on a magnetic north pole.* The field, then, by the law of poles, is away from the north pole of a magnet and toward the south pole. As with an electric field (Chapter 16), the closer together the field lines, the stronger the field.

Since a magnetic north pole can be used to map out a magnetic field, you might think that B would be the magnetic force per unit pole (as E is the electric force per unit charge). The magnetic field was once thought of in this manner. However, the magnetic field is now defined in terms of the force on a moving electric charge, as will be discussed in the next section.

19.2 Electromagnetism and the Source of Magnetic Fields

An example of an *electromagnetic* interaction occurs when a particle with a positive charge (q) moving with a constant velocity enters a region with a uniform magnetic field in a path such that the velocity and the magnetic field are at right angles [Fig. 19.3(a)]. When the charge enters the field, it is deflected into a curved circular path.

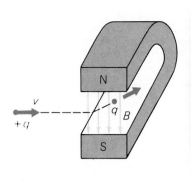

(a)

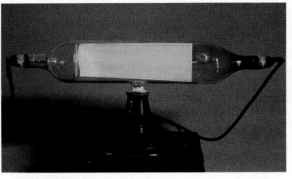

(b)

(c)

Figure 19.3 **Force on a moving charged particle**
(a) When a charged particle enters a magnetic field, it experiences a force, which is evident because it is deflected from its original path. (b) A charged particle beam in a discharge tube is made evident by a fluorescent strip in the tube. (c) The magnetic field of a magnet deflects the beam.

From the study of dynamics, you know that this deflection must be caused by a force perpendicular to the particle's velocity. But what gives rise to this force? No electric field is present (other than that of the charge itself). And the force of gravity is very weak (because the particle is so small), but it would deflect the particle downward rather than sideways. Evidently, the force is due to the interaction of the moving charge and the magnetic field. *A charged particle moving in a magnetic field experiences a force.*

Varying the magnitude of the charge, its velocity, and the magnetic field shows that the magnitude of the deflecting force is directly proportional to each of these quantities. That is,

$$F \propto qvB$$

In equation form,*

$$F = qvB \qquad (19.1)$$

This gives an expression for the strength (magnitude) of the magnetic field in terms of familiar quantities:

$$B = \frac{F}{qv} \qquad (19.2)$$

That is, B is the magnetic force per moving charge. From Eq. 19.2, you can see that a magnetic field has SI units of N/C-(m/s). This combination of units is given the name **tesla (T)**. The magnetic field is sometimes given in webers per square meter (Wb/m²), and 1 T = 1 Wb/m².

If the direction of the charged particle's velocity is not perpendicular to the magnetic field, the magnitude of the magnetic force on the particle is not given by Eq. 19.1. It is known that the magnitude of this force depends on the angle (θ) between the velocity and field vectors, in fact, on the sine of that angle ($\sin \theta$). That is, the force is zero when **v** and **B** are parallel and a maximum when those two vectors are perpendicular. In general,

$$F = qvB \sin \theta \qquad (19.3)$$

Another unit for the magnetic field is the gauss (G), and 1 Wb/m² = 10⁴ G. The weber, tesla, and gauss are named after early investigators of magnetic phenomena.

* The constant of proportionality has a value of 1 by the selection of the magnetic field unit.

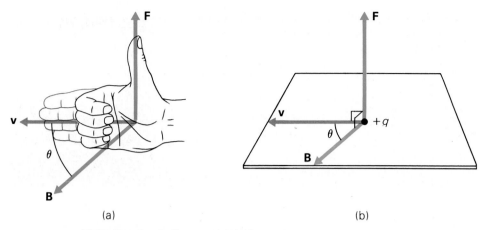

(a)   (b)

Figure 19.4 **Right-hand rule for magnetic force**
(a) When the fingers of the right hand are pointed in the direction of **v** and then turned or curled toward the vector **B**, the extended thumb points in the direction of the force **F** on a *positive* charge. (b) The force is always perpendicular to the plane of **B** and **v**, or always perpendicular to the direction of the particle's motion.

The direction of the magnetic force on a charged particle can also be determined by considering the velocity and field vectors. This method is stated in the form of a right-hand rule (Fig. 19.4):

> When the fingers of the right hand are pointed in the direction of **v** and then turned or curled toward the vector **B**, the extended thumb points in the direction of **F** for a *positive* charge.

The magnetic force is always perpendicular to the plane of the vectors **v** and **B**. If the moving charge is negative, the right-hand thumb points in the direction opposite to that of the force. (You can simply assume that the negative charge is positive, apply the right-hand rule, and then reverse the direction.)

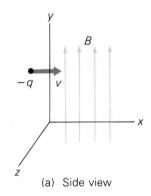

(a) Side view

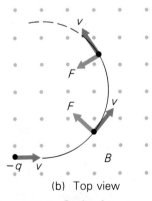

(b) Top view

Figure 19.5 **Path of a charged particle in a magnetic field**
(a) A charged particle entering a uniform magnetic field will be deflected.
(b) In the field, the particle moves in a circular path. See Example 19.1.

Example 19.1   Force on a Moving Charge

A negatively charged particle with a charge of -5.0×10^{-4} C moves at a speed of 1.0×10^2 m/s in the $+x$ direction toward a uniform magnetic field of 2.0 T in the $+y$ direction [Fig. 19.5(a)]. (a) What is the force on the particle when it enters the magnetic field? (b) Describe the path of the particle while it is in the field.

Solution

Given: $q = -5.0 \times 10^{-4}$ C   *Find*: (a) F
         $v = 1.0 \times 10^2$ m/s   (b) Path of the particle
         $B = 2.0$ T   in the field

(a) Eq. 19.3 can be used to find the magnitude of the force:

$$F = qvB \sin\theta = (5.0 \times 10^{-4}\ \text{C})(1.0 \times 10^2\ \text{m/s})(2.0\ \text{T}) = 0.10\ \text{N}$$

since $\sin\theta = \sin 90° = 1$.

The direction of the force just as the particle enters the magnetic field is in the $-z$ direction (into the page) in Fig. 19.5(a). By the right-hand rule, the force on a positive charge would be in the $+z$ direction (out of the page). But since the charge is negative, the force is in the opposite direction.

(b) Since the magnetic force is always perpendicular to the velocity of the particle, it is a centripetal force that causes the particle to move in a circular path [Fig. 19.5(b)]. The radius of the path (r) may be found using Newton's second law:

$$F = ma_c$$

This is, in this case,

$$qvB = \frac{mv^2}{r}$$

and

$$r = \frac{mv}{qB}$$

If the mass of the particle were known, the radius of its circular path could be computed easily. ■

Electric Currents and Magnetic Fields

Electric and magnetic phenomena are closely related. In fact, magnetic fields are produced by electric currents. This fact was discovered by the Danish physicist Hans Christian Oersted in 1820. He noted that an electric current could produce a deflection of a compass needle. This can be demonstrated using an arrangement like that in Fig. 19.6. When the circuit is open and there is no current flowing in it, the compass needle points in the northerly direction due to the Earth's magnetic field (to be discussed in Section 19.6). However, when the switch is closed and current flows in the circuit, the compass needle is deflected, indicating that another magnetic field (other than that of the Earth) is affecting the needle. When the switch is opened, the compass needle goes back to pointing north.

Developing expressions for the magnitude of the magnetic field around a current-carrying wire requires mathematics beyond the scope of this book. This section will simply present the resulting formulas for the magnitude of the B field for several arrangements that are used in many practical applications.

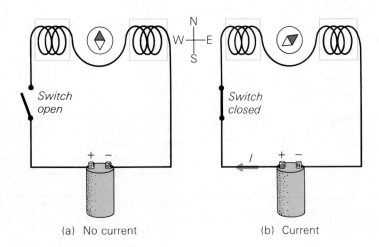

(a) No current   (b) Current

Figure 19.6 **Current and magnetic field**
(a) With no current in the wire, the compass needle points north. (b) With a current in the wire, the compass needle is deflected, indicating the presence of a magnetic field other than that of the Earth.

Magnetic Field Around a Long, Straight Wire. At a perpendicular distance d from a long, straight wire carrying a current I (Fig. 19.7), the magnitude of **B** is

$$B = \frac{\mu_{o}I}{2\pi d} \tag{19.4}$$

where

$$\mu_{o} = 4\pi \times 10^{-7} \text{ T-m/A} \quad \text{(or Wb/A-m)}$$

is a constant called the *permeability of free space*. The field lines are closed circles around the wire.

Note in Fig. 19.7 that the direction of **B** is given by a right-hand rule:

> If a current-carrying wire is grasped with the right hand with the extended thumb pointing in the direction of the conventional current, the curled fingers indicate the circular sense of the magnetic field. (The field vector is tangent to a circular field line at any point on the circle.)

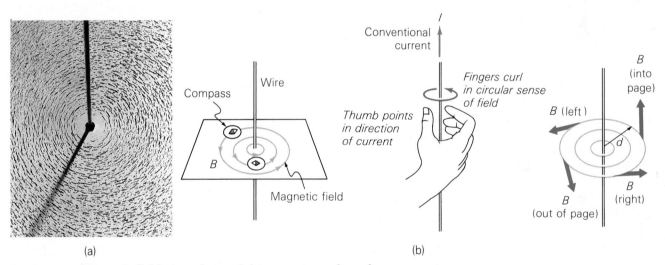

(a) (b)

Figure 19.7 **Magnetic field around a straight current-carrying wire**
(a) The field lines form concentric circles around the wire, as revealed by the pattern of iron filings. (b) The circular sense of the field lines is given by a right-hand rule, and the field vector is tangent to a circular field line at any point.

Magnetic Field at the Center of Circular Loop of Wire. At the *center* of a circular loop of wire of radius r carrying a current I (Fig. 19.8), the magnitude of **B** is

$$B = \frac{\mu_{o}I}{2r} \tag{19.5}$$

This equation is valid only at the center of the loop. The direction of **B** is given by the right-hand rule.

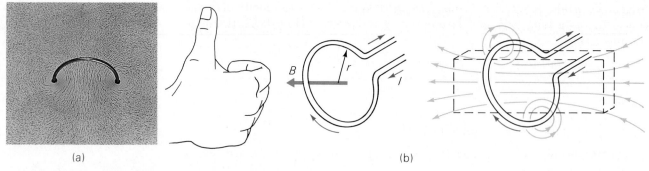

(a)

(b)

Figure 19.8 **Magnetic field at the center of a current-carrying loop**
(a) The magnetic field is perpendicular to the plane of the loop and similar in pattern
to that of a bar magnet. (b) The direction of the field is given by a right-hand rule.

Magnetic Field in a Solenoid. A solenoid is produced by winding a long wire
in a tight coil, or helix (many circular loops, as shown in Fig. 19.9). If the radius
of the loops is small compared to the length (L) of the coil, the magnitude of **B**
along a longitudinal axis through the center of a solenoid of N turns (loops)
carrying a current I is

$$B = \frac{\mu_\circ NI}{L} \tag{19.6}$$

The direction of **B** is given by the right-hand rule as applied to one of the loops
of the coil.

The expression $n = N/L$, or the number of turns per meter, is called the
linear turn density. Using it allows Eq. 19.6 to be written in simpler form:

$$B = \mu_\circ nI \tag{19.7}$$

The longer the solenoid, the more uniform the magnetic field across the
cross-sectional area within its coil. An ideal, or infinitely long, solenoid would
have a uniform internal field. Of course, this ideal condition can only be
approached in reality, but a long solenoid does produce a relatively uniform
magnetic field. Note how the pattern of the field lines for a loop (Fig. 19.8) and
especially that for a solenoid (Fig. 19.9) are similar to that for a bar magnet.

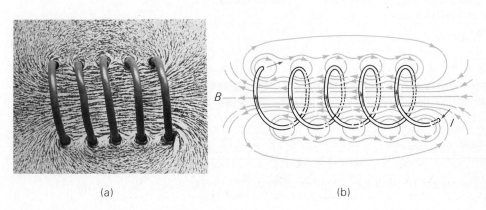

(a)

(b)

Figure 19.9 **Magnetic field
of a solenoid**
(a) The magnetic field is fairly
uniform near the central axis
of a solenoid. (b) The
direction of the field may be
determined by applying a
right-hand rule to one of the
loops.

Example 19.2 Magnetic Field Vector

A long, straight, horizontal power line carries a conventional current of 50 A. Considered as a conventional current, it flows in a south-to-north direction. What is the magnitude and direction of the magnetic field 1.0 m directly below the wire?

Solution

Given: $I = 50$ A   *Find*: **B** (magnitude and direction)
          $d = 1.0$ m

Use Eq. 19.4 to find the magnitude of the field at the point 1.0 m directly below the line:

$$B = \frac{\mu_o I}{2\pi d} = \frac{(4\pi \times 10^{-7} \text{ T-m/A})(50 \text{ A})}{2\pi(1.0 \text{ m})} = 1.0 \times 10^{-5} \text{ T}$$

By the right-hand rule, the vector points westward.

Example 19.3 Magnetic Field in a Solenoid

A solenoid 0.30 m long with 10^3 turns per meter carries a current of 5.0 A. What is the magnitude of the magnetic field through the center of the solenoid?

Solution

Given: $I = 5.0$ A   *Find*: B
          $n = 10^3$ turns/m

Using Eq. 19.7 for a solenoid,

$$B = \mu_o nI = (4\pi \times 10^{-7} \text{ T-m/A})(10^3 \text{ m}^{-1})(5.0 \text{ A}) = 2\pi \times 10^{-3} \text{ T}$$

PROBLEM-SOLVING HINT

In Example 19.2, with $\mu_0 = 4\pi \times 10^{-7}$ T-m/A, the 4π conveniently canceled. This is not the case in the second example. Note that π may be left in symbol form unless a numerical answer is needed.

19.3 Magnetic Materials

You may wonder why some materials are magnetic or easily magnetized and others are not. Knowing that a current produces a magnetic field and comparing the magnetic fields of a bar magnet and a current-carrying loop (see Figs. 19.1 and 19.8), you might be tempted to think that the magnetic field of the bar magnet is in some way due to internal currents.

Recall that, according to the simplistic solar system model of the atom, the electrons orbit the nucleus. They are thus charges in motion, and each orbiting electron is in effect a current loop that produces a magnetic field. This model pictures a material as having many internal atomic magnets. If some of these atomic magnets could be aligned, then a strong magnetic field could be produced and the material would be magnetic.

Although this is an attractive theory, it has a basic flaw. Physicists have discovered that, in general, the electron orbits in an atom are paired so that their magnetic effects cancel. That is, for each electron revolving clockwise in an orbit, there is another electron going counterclockwise in a similar orbit. The two resulting magnetic fields are in opposite directions. Therefore, for most materials, the magnetic effects due to the motions of atomic electrons are zero or very small—certainly not large enough to make a material magnetic.

What then is the source of the magnetism of magnetic materials? Modern theory says that magnetism is an atomic effect due to electron spin. The word "spin" is used because of an early idea that an additional contribution to the atomic magnetic field might come from an electron spinning on its axis. The classical picture likens a spinning electron to the Earth rotating on its axis. However, this is *not* the case. Electron spin is strictly a quantum mechanical effect with no direct classical analogue.

In atoms with two or more electrons, the electrons usually pair up, in pairs whose spins are opposite. The magnetic fields then cancel each other, and the material is not magnetic. Aluminum is an example. However, in certain strongly magnetic materials, electron spins do not pair, or cancel, completely. These are called **ferromagnetic materials**. In such materials there is a strong coupling or interaction between neighboring atoms, forming large groups of atoms, or **magnetic domains**, in which many of the electron spins are aligned. There aren't very many ferromagnetic materials. The most common are iron, nickel, and cobalt (gadolinium and certain alloys are also ferromagnetic).

In an unmagnetized ferromagnetic material, the domains are randomly oriented and there is no net magnetic effect [Fig. 19.10(a)]. However, when a ferromagnetic material is placed in an external magnetic field, two things may happen:

1. The domain boundaries change, and the domains with magnetic orientations parallel to the external field grow at the expense of the other domains.
2. The magnetic orientation of some domains may change slightly so as to be more aligned with the field.

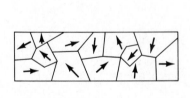
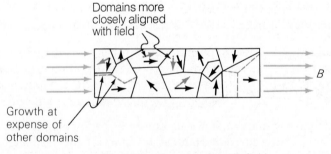

(a) No external magnetic field

(b) With external magnetic field

Figure 19.10 **Magnetic domains**
(a) With no external magnetic field, the magnetic domains of a ferromagnetic material are randomly oriented, and the material is unmagnetized. (b) In an external magnetic field, domains with orientations parallel to the field grow at the expense of other domains, and the orientations of some domains may become more aligned with the field. As a result, the material is magnetized, or exhibits magnetic properties.

You now can understand why an unmagnetized piece of iron is attracted to a magnet and why iron filings line up with a magnetic field. Essentially, the pieces of iron become induced magnets.

Ferromagnetic materials are used to make electromagnets, usually by wrapping a wire around an iron core (Fig. 19.11). Turning the current on and off, the magnet (or magnetic field) may be turned on and off at will. The iron used in an electromagnet is called soft iron. It is treated so that when the external field is removed, the magnetic domains quickly become unaligned and the iron is unmagnetized. ("Soft" does not refer to the metal's mechanical hardness but to its magnetic properties.)

In an electromagnet, the iron core becomes magnetized and adds to the field of the solenoid. The total field is expressed as

$$B = \mu nI$$

where μ is the **permeability** of the core material. The role permeability plays in magnetism is similar to that of permittivity ε in electricity: $\mu = K_m\mu_0$ and $\varepsilon = K\varepsilon_0$ (see Chapter 16). The magnetic analogue of the dielectric constant is then $K_m = \mu/\mu_0$, which is called the relative permeability. Both K and K_m are equal to one in a vacuum. Magnetic permeability is a useful material property. A core of a ferromagnetic material with a large permeability in an electromagnet can enhance its field thousands of times (compared to an air core).

Note that the strength of the field of an electromagnet depends on the current ($B = \mu nI$). Large currents produce large fields, but this field generation is accompanied by much greater joule heating (or I^2R losses). Large electromagnets need water-cooling coils to remove this heat. The problem may be alleviated by using a superconductor, which has zero resistance. Electromagnets made with the new high-temperature superconductors (Chapter 17) should find increasing applications.

Iron that retains some magnetism after being in an external magnetic field is called "hard" iron and is used to make so-called permanent magnets. You may have noticed that a paper clip or a screwdriver blade becomes slightly magnetized after being near a magnet. Permanent magnets are produced by heating pieces of some ferromagnetic material in an oven and then cooling them in a strong magnetic field to get the maximum effect. Above a certain critical temperature, called the **Curie temperature** (or Curie point), domain coupling disappears and a ferromagnetic material loses its ferromagnetism. The Curie temperature for iron is 770 °C.

A permanent magnet is not really permanent—its magnetism can be lost. Hitting such a magnet with a hard object or dropping it on the floor may cause it to lose its domain alignment. Also, heating can cause a loss of magnetism because the resulting increase in random motions of atoms tends to unalign the domains.

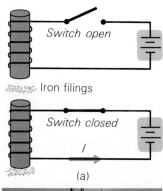

Iron filings

(b)

Figure 19.11 **Electromagnet** (a) With no current in the circuit, there is no magnetism. However, with a current in the coil, the iron core becomes magnetized and attracts other ferromagnetic materials. (b) An electromagnet in action picking up scrap metal.

Pierre Curie (1859–1906), a French physicist, discovered this effect and was the husband of Marie (Madame) Curie.

19.4 Magnetic Forces on Current-Carrying Wires

An electric charge moving in a magnetic field experiences a force (unless it is moving along or parallel to the field lines). Thus, a current-carrying wire in a magnetic field should experience such a force since a current is made up of moving charges. That is, the magnetic forces on the moving charges should give a resultant force on the wire conducting them, and this is the case.

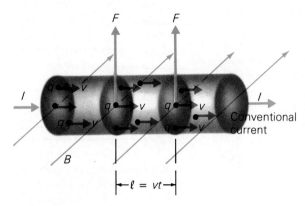

Figure 19.12 **Force on a wire segment** This diagram helps visualize and determine the force on a length of current-carrying wire in a magnetic field. Charge carriers are considered to be positive for convenience. See text for a description.

As you know, a conventional current can be thought of as positive charges moving in a straight wire, as depicted in Fig. 19.12. In a time t, a charge (q) would move on the average through a length given by $\ell = vt$, where v is the drift velocity and is perpendicular to the B field. Since all the moving charges $(\Sigma_i q_i)$ in this length of wire will experience a force due to the magnetic field, the magnitude of the total force on the length of wire is

Thinking of a current in terms of a positive charge is simply a convention. In reality, electrons are the (negative) charge carriers for electric current.

$$F = \left(\sum_i q_i\right)vB = \left(\sum_i q_i\right)\left(\frac{\ell}{t}\right)B = \left(\frac{\sum_i q_i}{t}\right)\ell B$$

But $\Sigma_i q_i/t$ is the current (I) since it represents the charge passing through a cross-sectional area of the wire per unit time. Therefore,

$$F_{\text{max}} = I\ell B \qquad (19.8)$$

Eq. 19.8 gives the maximum force on the wire (F_{max}) because the direction of the charges' motion (the current) is at an angle of $90°$ to the magnetic field. If the wire is not perpendicular to the field, but at some angle θ, then the force on the wire will be less. As for the force on a moving charge, for the general case,

$$F = I\ell B \sin \theta \qquad (19.9)$$

Note that if the wire is parallel to the field, the force on it is zero (as is the force on a charge moving parallel to a field).

The direction of the magnetic force on a current-carrying wire is also given by a right-hand rule. When the vector ℓ is crossed by the fingers of the right hand curled toward **B**, the extended thumb points in the direction of **F**. The vector ℓ for the length of wire is in the direction of the *conventional current*. The right-hand rule is often stated in terms of this direction:

> When the fingers of the right hand are pointed in the direction of the conventional current and then turned or curled toward the vector **B**, the extended thumb points in the direction of the magnetic force on the wire.

Two examples of magnetic forces on current-carrying wires are given Fig. 19.13. Apply the right-hand rule in each case to see how the direction of the force is found.

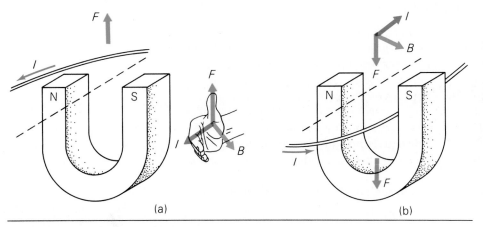

Example 19.4   Current, Magnetic Field, and Magnetic Force, and Definition of the Ampere

Two long parallel wires are separated by a distance d and have conventional currents I_1 and I_2 going in the same direction, as illustrated in Fig. 19.14. There is a force on each wire because of the magnetic field produced by the current in the other wire. By a right-hand rule, the magnetic field due to I_1 at the second wire is into the page, and its magnitude is (Eq. 19.4):

$$B_1 = \frac{\mu_0 I_1}{2\pi d}$$

The direction of B_1 is perpendicular to the wire carrying I_2, so the magnitude of the *force per unit length* (F_2/ℓ) on the wire is (Eq. 19.8):

$$\frac{F_2}{\ell} = I_2 B_1 = \frac{\mu_0 I_1 I_2}{2\pi d}$$

By a right-hand rule, with the fingers in the direction of I_2 and turning them into B_1, we see that the magnetic force on the wire carrying I_2 is *toward* the other wire.

The magnetic field due to I_2 also exerts a force per unit length on the wire carrying I_1, which is given by

$$\frac{F_1}{\ell} = \frac{\mu_0 I_1 I_2}{2\pi d}$$

Again, by a right-hand rule, this force is *toward* the other wire (equal and opposite forces of Newton's third law). Thus, two parallel wires carrying current in the same direction are attracted toward one another.

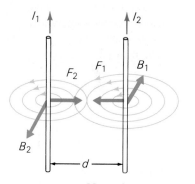

Figure 19.14 **Mutual interaction**
For two parallel current-carrying wires, the magnetic field due to each current interacts with the other current, and there is a force on each wire. If the currents are in the same direction, the forces are attractive. See text for a description and how this condition is used to define the ampere unit of current.

The magnetic force between parallel wires in an arrangement like that analyzed in Example 19.4 is used to define the ampere. According to the National Institute of Standards and Technology (formerly the National Bureau of Standards):

> The ampere is defined as that current which, if maintained in each of two long parallel wires separated by a distance of 1 m in free space, would produce a force between the wires (due to their magnetic fields) of 2×10^{-7} N for each meter of wire.

Force on a Current-Carrying Loop

Another important application of magnetic force on a current is that on a current-carrying loop, for example, the rectangular loop in Fig. 19.15(a). (The wires connecting the loop to a voltage source are not shown.) Suppose that the loop is free to rotate about an axis passing through opposite sides, as shown in the figure. There are no net forces or torques on the pivot sides of the loop (the sides through which the axis of rotation passes), because the forces on these sides are equal and opposite in the plane of the loop or zero when the sides are parallel to the field. (Why?) However, the equal and opposite forces on the other two sides of the loop (the sides parallel to the axis of rotation) produce torques that rotate the loop.

The magnitude of the magnetic force on each of the nonpivoted sides, whose length is L, is given by

$$F = ILB$$

since L and B are always at right angles to each other. The component of each of these forces perpendicular to the wire produces a torque, $rF_\perp$, or $\frac{1}{2}wF_\perp$, where the moment arm r is half the length of the pivot side of the loop ($r = \frac{1}{2}w$), as shown in Fig. 19.15(b). The magnitude of the total torque on the loop is then

$$\tau = \tfrac{1}{2}wF_\perp + \tfrac{1}{2}wF_\perp = wF\sin\theta = w(ILB)\sin\theta$$

$$\tau = IAB\sin\theta \tag{19.10}$$

Since wL is the area (A) of the loop, note that $F_\perp = F\cos\alpha = F\sin\theta$, where θ is the angle between $\mathbf{B}$ and a line normal to the plane of the loop. Even though it has been derived for a rectangular loop, Eq. 19.10 is valid for a flat loop of any shape.

The quantity IA is represented by m and referred to as the **magnetic moment** of the loop. Then $\tau = mB\sin\theta$. The magnetic moment is a vector perpendicular to the plane of the loop, so the torque tends to line up the magnetic moment with the field.

A loop in a magnetic field will rotate, or experience a torque, until $\sin\theta = 0$ (when the forces causing the torque are parallel to the plane of the loop). This occurs when the plane of the loop is perpendicular to the field. Then the loop stops rotating.

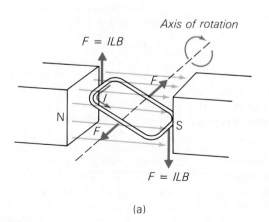

(a)

(b) Side view (pivot side)

Figure 19.15 **Force and torque on a current-carrying loop**
(a) A current-carrying rectangular loop oriented in a magnetic field as shown here experiences a force on each side. The forces on the sides parallel to the axis of rotation produce a torque that causes the loop to rotate. (b) A side view shows the geometry for determining the torque. See text for a description.

Current-carrying loops, even ones with such limited motion, have important applications. Since the current is directly related to the force, such a loop provides a means to measure current. Also, there are ways to keep a loop rotating, which allows continuous conversion of electrical energy to mechanical energy and is the operating principle of motors. These and other applications of electromagnetism are considered in the following section.

19.5 Applications of Electromagnetism

With the principles you have learned so far about electromagnetic interactions, you can understand the operation of several widely used scientific instruments.

The Galvanometer

In the discussion of ammeters and voltmeters in Chapter 18, the galvanometer was considered to be simply a circuit element that measured small currents. Now you are ready to see how it does this. As Fig. 19.16(a) shows, a galvanometer consists of a coil of wire loops on an iron core that is pivoted between the pole faces of a permanent magnet. When a current flows in the coil, the coil experiences a torque. A small spring supplies a counter (restoring) torque, and in equilibrium, a pointer indicates a deflection ϕ that is proportional to the current in the coil.

As you learned in Section 19.4, the torque on a current-carrying loop in a magnetic field is given by $\tau = IAB \sin \theta$. The dependence of the torque on the angle θ between the field lines and a line normal to the plane of the loop would cause a problem in measuring current with a galvanometer coil. As the coil rotates from its position of maximum torque ($\theta = 90°$, or the B field is perpendicular to the nonpivoted sides of the coil), the torque would become less, and the pointer deflection ϕ would not be proportional to the current alone. This problem is avoided by making the pole faces curved and wrapping the coil on a cylindrical iron core. The core tends to concentrate the field lines so that **B** is always perpendicular to the nonpivoted side of the coil [Fig. 19.16(b)]. The magnetic force is then always perpendicular to the plane of the coil, and the torque does not vary with the angle θ. Thus, the total torque on the coil is

$$\tau = NIAB \tag{19.11}$$

where N is the number of loops in the coil.

The restoring torque (τ_s) of the spring in a galvanometer is given by a rotational form of Hooke's law ($F = kx$ for linear restoring force):

$$\tau_s = k\phi \tag{19.12}$$

Here ϕ is the deflection angle as measured by the pointer. When a current flows in the coil, the coil and the pointer will rotate until the torque on the coil is balanced by that on the spring. In equilibrium, the magnitudes of the torques are equal, and

$$\phi = \frac{NIAB}{k} \tag{19.13}$$

Thus, the pointer deflection (ϕ) is directly proportional to the current (I).

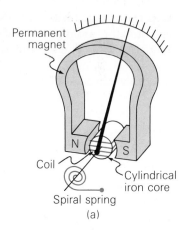

(a)

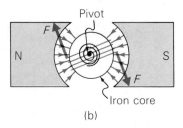

(b)

Figure 19.16 **The galvanometer**
(a) The deflection of the needle is proportional to the current in the coil, and therefore a galvanometer can be used to detect and measure currents. (b) A magnet with curved pole faces is used so the field lines are always perpendicular to the core surface and the torque does not vary with angle.

A properly calibrated galvanometer can be used to measure small currents (in the microamp range). With an appropriate shunt resistor, a galvanometer forms the basis of an ammeter for measuring larger currents. Also, with an appropriate series multiplier resistor, the galvanometer forms a voltmeter (Chapter 18).

The dc Motor

In general, *a motor is a device that converts electrical energy into mechanical energy*. An example of such a conversion occurs with a galvanometer: a current causes the coil to rotate mechanically. However, a practical motor must have continuous rotation for continuous energy output.

A pivoted, current-carrying loop in a magnetic field will rotate freely, but for only a half-cycle, or through a maximum angle of 180°. Recall that $\tau = IAB \sin \theta$, and when the magnetic field is perpendicular to the plane of the loop ($\sin \theta = 0$), the torque is zero and the loop is in equilibrium.

To provide for continuous rotation, the current is periodically reversed so that the torque forces are reversed. This is done by means of a split-ring commutator, an arrangement of two metal half-rings insulated from each other [Fig. 19.17(a)]. The ends of the wire that forms the loop are fixed to the half-rings, and the current is supplied to the loop through the commutator by means of contact brushes. Then, with one half-ring electrically positive (+) and the

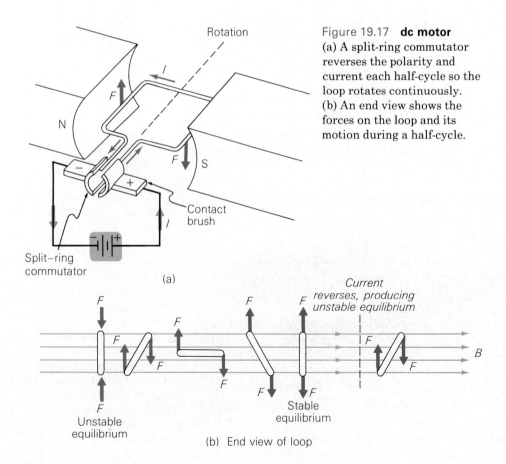

Figure 19.17 **dc motor**
(a) A split-ring commutator reverses the polarity and current each half-cycle so the loop rotates continuously.
(b) An end view shows the forces on the loop and its motion during a half-cycle.

(a)

(b) End view of loop

other negative (−), the loop and ring rotate. When they have gone through half a rotation, the half-rings come in contact with the opposite brushes. This reverses their polarity, and the current flows in the loop in the opposite direction. This changes the directions of the torque forces [Fig. 19.17(b)]. The torque is zero at the equilibrium position, but the loop is in unstable equilibrium. The loop has enough inertial motion to move through this point, and it rotates through another half-cycle. The process repeats, and the motor armature rotates continuously.

The Cathode Ray Tube (CRT)

The cathode ray tube (CRT) is a vacuum tube that is used in an oscilloscope [Fig. 19.18(a)]. Electrons are "boiled off" a hot filament in an electron gun and accelerated through a potential difference. Because they are emitted from the negative terminal, or cathode, the electrons are called cathode rays. (The positive terminal toward which the electrons are accelerated is the anode.) In 1877, the English physicist J. J. Thomson used a type of cathode ray tube to demonstrate the existence of the electron. Previously, it had not been known that the rays emitted from the hot filament of such a tube were electrons.

The electrons pass through a small hole in the anode, forming a beam. The beam passes through two sets of parallel plates, one for vertical deflections and the other for horizontal deflections. Input voltages (which usually vary with time) are placed across the plates, and the deflected beam makes patterns on a screen. (The fluorescent material coating the inside of the screen glows when struck by the electrons.) In normal operation, the horizontal input voltage from an internal voltage source sweeps the dot illuminated by the beam across the screen periodically. The vertical deflection is proportional to an external voltage input. Thus, the display on a CRT is a graph of the external voltage versus time [Fig. 19.18(b)].

The picture tube of a television set is a cathode ray tube (Fig. 19.19). In this case, magnetic coils are usually used instead of plates to deflect the electron beam. The beam scans the fluorescent screen in a 525-line pattern in a fraction of a second. For a black-and-white TV, the signals transmitted from a video camera reproduce an image on the screen as a mosaic of light and dark dots, forming a black-and-white picture.

Producing the images of color TV is somewhat more involved. A common

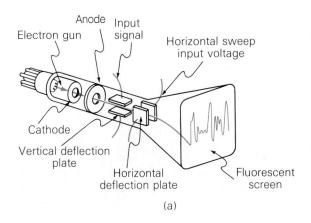

(a)

(b)

Figure 19.18 **Cathode ray tube (CRT)**
(a) A beam from an electron gun is controlled by deflection plates. (b) The motion of the deflected beam traces out a pattern on a fluorescent screen.

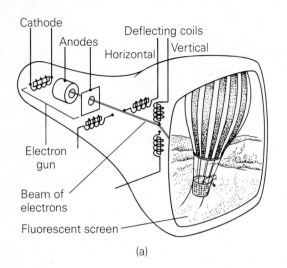 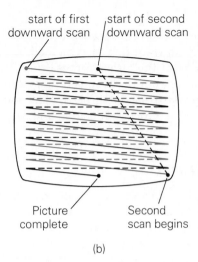

Cathode
Anodes
Horizontal
Deflecting coils
Vertical
Electron gun
Beam of electrons
Fluorescent screen

(a)

start of first downward scan
start of second downward scan
Picture complete
Second scan begins

(b)

Figure 19.19 Television tube
(a) A television picture tube is a CRT. (b) The beam scans every other line on the screen in one downward scan that takes $\frac{1}{60}$ s and then scans the lines in between in a second scan of $\frac{1}{60}$ s. This yields a complete picture of 525 lines in $\frac{1}{30}$ s.

color picture tube has three beams, one for each of the primary colors (Chapter 25). Phosphor dots on the screen are arranged in groups of three (triads) with one dot for each of the primary colors. The excitation of the appropriate dots and the resulting combination of colors produce a color picture. There are also single-beam color picture tubes in which color dots are excited sequentially in rapid succession to produce the image on the screen.

The fact that a picture tube is a CRT can be demonstrated by bringing a magnet near a black-and-white TV screen. The magnetic field will deflect the electrons and distort the picture. *Do not try this on a color TV screen!* Internal metallic parts may become magnetized, causing the picture to be permanently distorted.

Mass Spectrometer

Have you ever thought about how the mass of an atom or molecule is measured? Electric and magnetic fields provide a way in the form of a mass spectrometer (called a mass spec for short). Actually, the masses of ions are measured since electric and magnetic fields have motional effects only on charged particles.

Ions with a known charge ($+q$) are produced by heating a substance. A beam of ions introduced into the mass spec has a wide distribution of speeds. Ions with a particular velocity are selected by setting the values of the electric and magnetic fields between the plates of a *velocity selector* (Fig. 19.20). For a positively charged ion, the electric field produces a downward force ($F_e = qE$), and the magnetic field produces an upward force ($F_m = qvB_1$). [In the figure, a cross ($\times$) means into the page, and a dot ($\cdot$) means out of the page, similar to the feathered end and tip of an arrow.] If the beam is not deflected, the resultant force is zero, and

$$qE = qvB_1$$

or $$v = \frac{E}{B_1}$$

If E is written as V/d, since the voltage and the plate separation distance are the controllable or measured quantities,

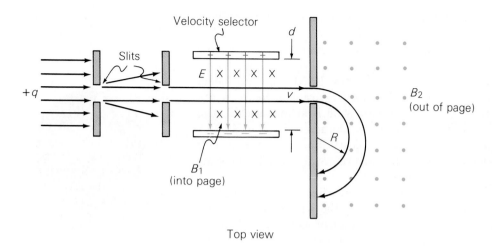

Velocity selector

Slits

d

E

$+q$

v

B_2 (out of page)

R

B_1 (into page)

Top view

$$v = \frac{V}{B_1 d} \tag{19.14}$$

Thus, a beam with a known speed can be selected.

The beam then passes through a slit into another magnetic field ($\mathbf{B}_2$), which is perpendicular to the direction of the beam. The force due to this magnetic field ($F = qvB_2$) is always perpendicular to the velocity of the ions, which are deflected in a circular arc. The magnetic force supplies the centripetal force for this motion, and

$$\frac{mv^2}{R} = qvB_2$$

Using the expression for the selected ion velocity (Eq. 19.14),

$$m = \left(\frac{qdB_1 B_2}{2V} \right) R \tag{19.15}$$

The greater the mass of an ion, the greater the radius of the circular path. Two circular paths of different radii are shown in Fig. 19.20, which indicates that the beam contains ions of two different masses.

19.6 The Earth's Magnetic Field ◆

The magnetic field of the Earth has been used for centuries. In ancient times, navigators used lodestones or magnetized needles to show them where north was. An early study of magnetism was done by the English scientist Sir William Gilbert around 1600. In investigating the magnetic field of a specially cut spherical lodestone that simulated the Earth, he came to believe that the magnetic field was associated with the entire Earth, or that the Earth as a whole acted as a magnet. Gilbert thought that the field might be produced by a large body of permanently magnetized material within the Earth.

In fact, the Earth's magnetic field has approximately the configuration that would be produced by a large interior bar magnet, as Fig. 19.21 shows. The magnitude of the horizontal component of the Earth's magnetic field at the magnetic equator is about 10^{-5} T, and the vertical component at the geomagnetic poles is about 10^{-4} T. It has been calculated that for a ferromagnetic material of maximum magnetization to produce this field, it would have to occupy about 0.01% of the Earth's volume.

The idea of a ferromagnet of this size within the Earth may not seem unreasonable at first, but this cannot be a true model. The interior temperature of the Earth is well above the Curie temperatures of iron and nickel, the ferromagnetic materials believed to be the most abundant in the Earth's interior. For iron, this temperature is 770 °C, which is attained at a depth of only 100 km. The temperature of the Earth is higher at greater depths. This means that such a permanent internal magnet is impossible.

The fact that an electric current produces a magnetic field has led scientists to speculate that the Earth's magnetic field is associated with motions in the liquid outer core. (Recall that the magnetic field of a loop of wire or a solenoid is similar to that of a bar magnet—compare Figs. 19.8 and 19.9 to Fig. 19.1.) These motions may be associated in some way with the Earth's rotation. Jupiter and Saturn have magnetic fields much larger than that of Earth. These planets are largely gaseous and rotate rapidly, with periods of 10–11 hours. Mercury and Venus, on the other hand, have very weak magnetic fields. These planets are more like Earth in density and rotate relatively slowly, with periods of 58 and 243 days, respectively.

Several theoretical models have been proposed to explain the Earth's magnetic field. For example, it has been suggested that it arises from currents associated with thermal convection cycles in the liquid outer core caused by heat from the inner core.

The Earth's magnetic field shows a variety of fluctuations. The axis of the magnetic field does not lie along the Earth's rotational axis, which defines the geographic poles. That is, the north magnetic pole and the geographic North Pole do not coincide. (We will refer to the north magnetic pole when talking about magnetic north, even though it is actually a south magnetic pole.) The magnetic pole is about 1500 km (930 mi) south of the geographic North Pole (true north), displaced by about 13° of latitude. The south magnetic pole is displaced even more from its geographic pole. Thus, the magnetic axis is not a straight line.

Another change in the Earth's magnetic field is that the magnetic poles wander. The north magnetic pole, on which there are more data, moves about 1° of latitude (roughly 110 km or 70 mi) in a decade. For some unknown reason, it has moved consistently northward from its 1904 latitude position of 69 °N and westward, crossing the 100 °W longitudinal meridian. Also, there are sometimes dramatic daily shifts of as much as 80 km (50 mi), with a return to the starting position. These are thought to be caused by charged particles from the Sun that reach the Earth's upper atmosphere and set up currents that disturb the magnetic field.

Charged particles from the Sun and cosmic rays entering the Earth's magnetic field give rise to other phenomena. A charged particle entering a uniform magnetic field not perpendicular to the field, spirals in a helix, [Fig. 19.22(a)]. This is because there is a component of the particle's velocity parallel to the field, and therefore a force perpendicular to the field (by a right-hand rule: v turned into $\mathbf{B}$). The motions of charged particles in a non-uniform field are quite complex. However, for a bulging field such as that

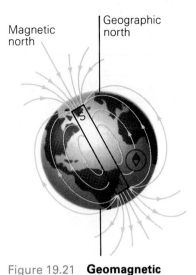

Magnetic north

Geographic north

Figure 19.21 **Geomagnetic field**
The Earth's magnetic field is similar to that of a bar magnet. However, a permanent magnet could not exist within the Earth because of high temperatures there. The field is believed to be associated with motions in the liquid outer core.

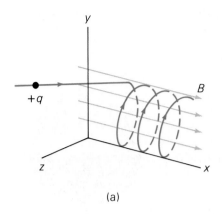

(a)

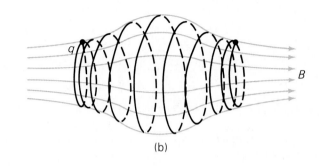

(b)

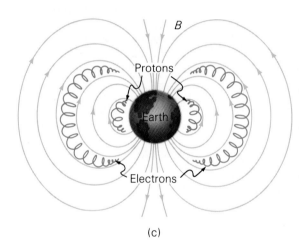

(c)

Figure 19.22 **Magnetic confinement** (a) A charged particle entering a uniform magnetic field at an angle moves in a spiraling path. (b) In a nonuniform, bulging magnetic field, particles spiral back and forth as though confined in a magnetic bottle. (c) Particles are trapped in the Earth's magnetic field, and the regions where they are concentrated are called Van Allen belts.

depicted in Fig. 19.22(b), analysis shows that the particles spiral back and forth as though in a magnetic container, a so-called bottle. This is an important consideration in confining a plasma (a gas of charged particles) in fusion research (Chapter 30).

An analogous phenomenon occurs in the Earth's magnetic field, giving rise to regions where there are concentrations of charged particles. These regions are called Van Allen belts [Fig. 19.22(c)]. Light emissions called the Aurora Borealis, or Northern Lights, in the Northern Hemisphere and the Aurora Australis, or Southern Lights, in the Southern Hemisphere are believed to be the result of the recombinations of electrons and ions of atmospheric gas molecules, which were previously ionized by particles from the lower Van Allen belt. Energy is needed to ionize a gas molecule, and energy is given up in the form of light when the electrons and ions recombine. Maximum aurora activity is noted to occur after a solar disturbance, which causes large numbers of charged particles to be incident on the Earth's magnetic field.

The belts are named after James Van Allen, an American scientist who discovered them in 1958 from data from instruments that were on board the first U.S. satellite, *Explorer I.*

Important Formulas

Magnitude of the magnetic force on a moving charge (q):

$F = qvB \sin \theta$

Directional right-hand rule: When the fingers of the right hand are pointed in the direction of **v** and then turned or curled toward the vector **B**, the extended thumb points in the direction of the force **F** on a *positive* charge. (**F** is in the opposite direction for a negative charge.)

Magnitude of the magnetic force around a long, straight current-carrying wire:

$B = \dfrac{\mu_0 I}{2\pi d}$

Magnitude of the magnetic force at the center of circular loop of current-carrying wire:

$B = \dfrac{\mu_0 I}{2r}$

Magnitude of the magnetic force in a solenoid (along the axis):

$B = \dfrac{\mu_0 NI}{L}$ or $B = \mu_0 nI$ (where $n = N/L$)

Permeability of free space:

$\mu_0 = 4\pi \times 10^{-7}$ T-m/A

Directional right-hand rule: When a current-carrying wire is grasped with the right hand with the extended thumb pointing in the direction of the conventional current, the curled fingers indicate the circular sense of the magnetic field. (The field vector is tangent to the circular field line at any point on the circle.)

Force on a straight current-carrying wire:

$F = I\ell B \sin \theta$

Directional right-hand rule: When the fingers of the right hand are placed in the direction of the conventional current and then turned or curled toward the **B** vector, the extended thumb points in the direction of the force on the wire.

Torque on a current-carrying loop:

$\tau = IAB \sin \theta$

Restoring spring torque in a galvanometer:

$\tau_s = k\phi$

Pointer deflection in a galvanometer:

$\phi = \dfrac{NIAB}{k}$

Velocity selector (mass spec):

$v = \dfrac{E}{B_1} = \dfrac{V}{B_1 d}$

Particle mass (mass spec):

$m = \left(\dfrac{qdB_1 B_2}{2V}\right) R$

Questions

Magnets and Magnetic Poles

1. How are the poles of a bar magnet distinguished and why?

2. Give a description of the poles of a bar magnet in terms of the magnetic field.

3. Electric fields have equipotentials, or paths along which no work is done in moving a charge. Are there analogous paths for magnetic fields? Explain. (Hint: see Question 7.)

4. Given two identical iron bars, one of which is a permanent magnet and the other unmagnetized, how could you tell the difference using only the two bars?

5. Suppose that a flat, current-carrying loop of wire placed on this page has a conventional current directed toward the top of the page at its right side. What is the direction of the magnetic field at the center of the loop?

Electromagnetism and the Source of Magnetic Fields

6. Will a moving electric charge entering a magnetic field always be deflected? Explain.

7. It is said that no work is done by the force acting on a charge moving in a magnetic field. Is this a true statement? Explain. (Hint: see Question 3.)

8. If a positively charged particle and a negatively charged particle were moving from left to right across the top of this page when a magnetic field directed into the page was turned on, which way would the particles initially be deflected?

9. If a negatively charged particle were moving downward along the right edge of this page, which way should a magnetic field be oriented so the particle would initially be deflected to the left?

10. Explain why a charged particle moving in a magnetic field generally follows a circular path.

11. A magnetic field can be used to determine the sign of charged carriers. Consider a wide conducting strip in a magnetic field oriented as shown in Fig. 19.23. The charge carriers are deflected by the magnetic force and accumulate on one side of the strip, giving rise to a measurable voltage across it. (This is known as the Hall effect.) If the sign of the charge carriers is unknown (they are either positive or negative), how does the measured voltage allow it to be determined?

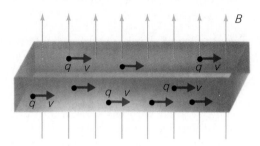

Figure 19.23 **The Hall effect**
See Question 11.

Magnetic Materials

12. What produces the magnetic properties of a material?

13. Can a magnetic monopole be obtained by breaking one end off a permanent bar magnet? Explain.

14. Why doesn't an electromagnet stay permanently magnetized?

15. Is it possible to make a permanent bar magnet with the same kind of pole at each end? Explain. (Hint: the answer is yes.)

Magnetic Forces on Current-Carrying Wires

16. A long, straight, horizontal wire has a conventional current directed toward the north. In what direction would the needle point if a compass were placed (a) above the wire, (b) below the wire, and (c) on each side of the wire? (Neglect the Earth's magnetic field.)

17. What would happen if the magnets in Fig. 19.13 were rotated around the wires so the ends pointed in the opposite directions and the currents were reversed?

18. Hold up your left hand, palm outward, with the index and middle fingers forming a V. Suppose that the middle finger points in the direction of a magnetic field and the index finger points in the direction of the conventional current in a straight wire. In what direction would the force on the wire be?

19. If a current-carrying solenoid is suspended horizontally from a string so it can rotate freely, what will happen?

20. Two straight wires that are parallel carry currents in opposite directions. What is the direction of the force on each wire? Explain how this is determined.

21. Two straight wires are at a right angle to each other (see Fig. 19.26, for example). If one wire carries a current toward the south and the other a current toward the east, what is the direction of the force on each wire?

Applications of Electromagnetism

22. Why are the pole faces of the magnet in a galvanometer curved?

23. Could a continuously operating dc motor be made without a split-ring commutator? (Hint: think about reversing the magnetic field.)

24. Could a velocity selector for charged particles be made with an electric field and a magnetic field parallel to each other? Explain.

25. Why does dust frequently cling to the screen of a television set?

26. Explain the operation of the door bell and door chimes illustrated in Fig. 19.24.

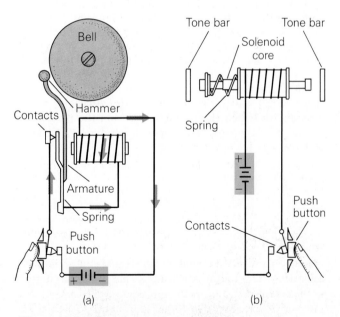

Figure 19.24 **Electromagnetic applications**
(a) A doorbell and (b) door chimes both have electromagnets. See Question 26.

27. How would a compass behave at one of the Earth's magnetic poles?

28. In general (except at the Equator), the Earth's magnetic field lines are not parallel to the surface and enter the Earth at an angle. What does this mean for the operation of a compass?

29. Charged particles from the Sun and cosmic rays are concentrated in the Earth's polar regions. Why is this?

Problems

19.2 Electromagnetism and the Source of Magnetic Fields

■1. A positive charge of 0.25 C moves horizontally at a speed of 2.0×10^2 m/s and enters a magnetic field of 0.01 T directed vertically downward. What is the force on the charge?

■2. A long, straight wire carries a current of 2.0 A. Find the magnitude of the magnetic field 50 cm from the wire.

■3. A loop of wire with a diameter of 40 cm carries a current of 0.50 A in a clockwise sense, as viewed from above. Find the magnetic field at the center of the loop.

■4. A horizontal beam of electrons travels at a speed of 10^3 m/s along a north-to-south line in a discharge tube. What is the force on the electrons due to the vertical component of the Earth's magnetic field if it has magnitude of 5.0×10^{-5} T at that location?

■■5. An electron traveling at a uniform speed of 100 m/s in the $+x$ direction enters a uniform magnetic field and experiences a force of 4.0×10^{-15} N in the $+y$ direction. What is the magnitude and direction of the magnetic field?

■■6. A particle with a charge of 4.0×10^{-8} C moves at a speed of 2.0×10^2 m/s through a magnetic field in the direction at which the magnetic force on the particle is maximum. If the force on the particle is 1.8×10^{-6} N, what is the magnitude of the magnetic field?

■■7. An electron travels at a speed of 2.0×10^4 m/s through a uniform magnetic field whose magnitude is 1000 T. What is the magnitude of the force on the electron if its velocity vector and the magnetic field vector (a) are perpendicular, (b) make an angle of 45° or (c) are parallel?

■■8. The magnetic field at the center of a coil of 250 turns whose radius is 20 cm is 6.0 mT. Find the current in the coil.

■■9. What current in a long, straight wire will produce a magnetic field with a magnitude of 4.0 μT at a perpendicular distance of 20 cm from the wire?

■■10. A solenoid 50.0 cm long with 2000 turns/m has a central magnetic field of 0.60 mT. Find the current in the coil.

■■11. A horizontal power line carries a current of 2500 A from west to east. What is the magnetic field 8.0 m below the line?

■■12. Two long, parallel wires separated by 50 cm carry currents of 10 A each in a horizontal direction. Find the magnetic field midway between the wires if the currents are (a) in the same direction and (b) in opposite directions.

■■13. A long, straight wire has a resistance of 2.0 Ω. What potential difference between the ends of the wire will produce a magnetic field of 10^{-4} T at a distance of 6.0 cm from the wire?

■■14. Two long, parallel wires separated by 20 cm carry equal currents of 1.5 A in the same direction. Find the magnetic fields 10 cm away from either side of each wire on a line running perpendicularly through both wires.

■■15. Two long, parallel wires carry currents of 8.0 A and 2.0 A, as illustrated in Fig. 19.25. (a) What is the magnetic field at the point midway between the wires? (b) Where on a line perpendicular to and joining the wires is the magnetic field zero?

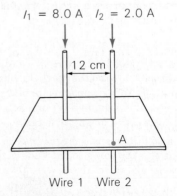

$I_1 = 8.0$ A $I_2 = 2.0$ A

12 cm

A

Wire 1   Wire 2

Figure 19.25 Parallel current-carrying wires
See Problems 15–17, 51, and 52.

■■16. Find the magnetic field (magnitude and direction) at point A in Fig. 19.25, which is located 9.0 cm away from wire 2 on a line perpendicular to the line joining the wires.

■■17. Suppose that the current in wire 1 in Fig. 19.25 were in the opposite direction. (a) What would be the magnetic field at a point midway between the wires? (b) Where on a line perpendicular to and joining the wires would the magnetic field be zero?

■■18. A circular loop of wire with a diameter of 12 cm is horizontal and carries a current of 0.50 A in a clockwise sense, as viewed from above. What is the magnetic field (magnitude and direction) at the center of the loop?

■■19. How much current must flow in a circular loop with a radius 15 cm to produce a magnetic field at its center that is approximately the same magnitude as the horizontal component of the Earth's magnetic field at the Equator?

■■20. A wire with four circular loops of radius 5.0 cm carries a current of 2.0 A in a clockwise sense, as viewed along a line perpendicular to the circles formed by the loops. What is the magnetic field at the center of the loops?

■■21. A circular loop of wire with a radius of 5.0 cm carries a current of 1.0 A. Another circular loop of wire is concentric with the first (has a common center with it) and carries a current of 2.0 A. The magnetic field at the center of the loops is measured to be zero. What is the radius of the second loop?

■■22. A solenoid 10 cm long is wound with 3000 turns of wire. How much current must flow through the windings to produce a magnetic field of 3.0×10^{-3} T at the solenoid's central axis?

■■23. A proton moves at a speed of 1.0×10^4 m/s perpendicularly to a uniform magnetic field of 80 mT. Find the radius of the resulting circular path.

■■24. An electron has a kinetic energy of 20 keV. It enters a uniform magnetic field of 40 T that is perpendicular to its line of motion. What is the radius of the resulting circular path?

■■25. A proton travels in a circular path with a radius of 60 mm in a uniform magnetic field of 12 μT. What is the kinetic energy of the proton?

■■■26. Two long, perpendicular wires carry currents of 15 A, as illustrated in Fig. 19.26. What is the magnetic field at the midpoint of the line joining the wires?

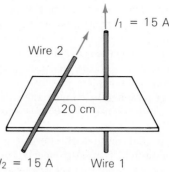

Figure 19.26 **Perpendicular current-carrying wires** See Problem 26.

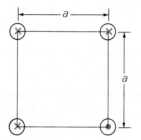

Figure 19.27 **Current-carrying wires in a square** See Problem 27.

■■■27. Four wires running through the corners of a square with sides of length a as shown in Fig. 19.27 carry equal currents I. Calculate the magnetic field at the center of the square in terms of these parameters.

■■■28. A solenoid is wound with 200 turns/cm. An outer layer of insulated wire with 180 turns/cm is wound over the first layer of wire. In operation, the inner coil carries a current of 5.0 A, and the outer coil carries a current of 7.5 A in the direction opposite to that of the current in the inner coil. What is the magnetic field at the central axis of this doubly wound solenoid?

■■■29. A solenoid 10 cm long has 3000 turns of wire and carries a current of 5.0 A. A 2000-turn coil of wire of the same length as the solenoid surrounds the solenoid and is concentric with it (shares a common center). If the outer coil carries a current of 10 A in the same direction as that of the current in the solenoid, find the magnetic field at the center of both coils.

■■■30. A proton is accelerated from rest through a potential difference of 1.0 kV. It enters a uniform magnetic field of 5.0 mT that is perpendicular to the direction of its motion. (a) Find the radius of the circular path of the proton. (b) Calculate the period of revolution of the proton.

■■■31. A fast-moving electron moves in a horizontal plane at a right angle to a uniform vertical magnetic field. (a) What is the frequency of the electron's revolution? (This frequency is called the cyclotron frequency.) (b) Show that the time required for the electron (or any charged particle) to make one complete revolution is independent of its speed and radius. (c) Compute the path radius and frequency for an electron if $v = 10^5$ m/s and $B = 10^{-4}$ T.

■■■32. The circular motion of charged particles in a uniform magnetic field is used to accelerate atomic particles in a machine called a cyclotron (see Fig. 19.28). Particles are injected into the center of the D-shaped sections, through which a uniform vertical magnetic field passes. Each time a particle crosses the gap between the D's, it receives energy because of a potential difference across the D's. The polarity of the voltage on the D's is reversed periodically by an oscillator circuit that has a frequency equal to that of the cyclotron frequency (see Problem 31). Since the particle frequency is independent of the radius, the particle and oscillator frequencies are in resonance. As a result, the particle accelerates in a spiral path, and the energy received by the charged particle depends on the radius of the D's. (a) Show that the kinetic energy (K) of a particle of charge q and mass m that is accelerated in a cyclotron with D's of radius R is

$$K = \frac{(qBR)^2}{2m}$$

(b) If the oscillator frequency of a cyclotron with D's of 0.50 m radius is 2.0×10^6 Hz, what is the magnitude of the magnetic field needed to accelerate protons in a spiral path?

19.3 Magnetic Materials

■■33. The magnitude of the magnetic moment (m) of a current loop is the product of the current (I) and the cross-sectional area of the loop (A): $m = IA$. Show that for an atomic electron in a circular orbit of radius r at an orbital speed v the magnetic moment is $m = evr/2$.

■■34. What is the magnetic moment of the electron in a hydrogen atom, which travels around the nuclear proton at a radius of 0.053 nm? (Hint: find the electron's speed by considering the centripetal force and see Problem 33.)

■■35. What is the magnitude of the magnetic moment of a rectangular loop of wire that measures 10 cm by 15 cm and carries a current of 2.0 A?

■■36. What is the magnetic field at the center of the circular orbit of the electron in a hydrogen atom (the orbital radius is 0.053 nm)? (Hint: find the electron's period by considering the centripetal force.) .

■■37. A solenoid with 100 turns/cm has an iron core with a relative permeability of 2000. The solenoid carries a current of 1.5 A. (a) What is the magnetic field at the central axis of the solenoid? (b) How much greater is the magnetic field with the iron core than it would be without?

19.4 Magnetic Forces on Current-Carrying Wires

■38. A straight, horizontal segment of wire is 1.0 m long and carries a current of 5.0 A in the $+x$ direction in a magnetic field of 0.60 T that is directed vertically upward. What is the force on the wire (magnitude and direction)?

■39. A 2.0-m length of straight wire carries a current of 20 A in a uniform magnetic field of 50 mT whose field lines make an angle of 40° with the direction of the conventional current. Find the force on the wire.

■40. Use a right-hand rule to find the direction of the conventional current in a wire in a magnetic field that results in the force on the wire shown for each case in Fig. 19.29.

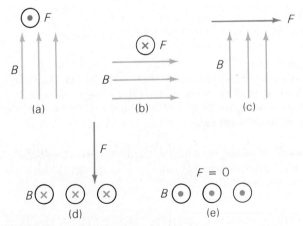

Figure 19.29 **The right-hand rule**
See Problem 40.

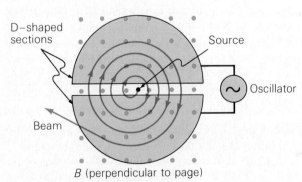

Figure 19.28 **The cyclotron**
See Problem 32.

■■41. A straight wire 50 cm long conducts a current of 10 A directed vertically upward. If the wire experiences a force of 2.0×10^{-2} N in the eastward direction due to a magnetic field at right angles to its length, what is the magnitude and direction of the magnetic field?

■■42. A horizontal magnetic field of 10^{-2} T is at an angle of 30° to the direction of the current in a straight, horizontal wire 60 cm long. If the wire carries a current of 20 A, what is the magnitude of the force on the wire?

■■43. A wire carries a current of 10 A in the $+x$ direction in a uniform magnetic field of 40 T. Find the magnitude of the force per unit length and direction of the force on the wire if the magnetic field is directed in (a) the $+x$ direction, (b) the $+y$ direction, (c) the $+z$ direction, (d) the $-y$ direction, and (e) the $-z$ direction.

■■44. How much current is flowing in a straight wire 4.0 m long if the force on it is 0.040 N when it is placed in a perpendicular magnetic field of 0.50 T?

■■45. A straight wire 25 cm long is oriented vertically in a uniform horizontal magnetic field of 3.0×10^2 T in the $-x$ direction. What current flowing in what direction in the wire will cause it to experience a force of 5.0 N in the $+y$ direction?

■■46. A wire carries a current of 10 A in the $+x$ direction. Find the force per unit length on the wire if the magnetic field has components $B_x = 0.020$ T and $B_y = 0.040$ T.

■■47. Two long, straight, parallel wires 10 cm apart carry equal currents of 3.0 A in opposite directions. What is the force per unit length on the wires?

■■48. Two parallel wires are 4.0 m long and 50 cm apart and carry equal currents of 2.5 A in opposite directions. (a) Find the force on each wire. (b) Find the force on each wire if their currents are in the same direction.

■■49. A set of jumper cables used to start one car from another's battery is connected to the terminals of the cars' batteries. If 15 A of current flows in the cables when the one car is started and the cables are parallel and 15 cm apart, what is the force per unit length on the cables?

■■50. Find the force per unit length on each of two long, straight, parallel wires that are 20 cm apart when one carries a current of 2.0 A and the other a current of 4.0 A in the same direction.

■■51. What is the force per unit length on wire 1 in Fig. 19.25?

■■52. What is the force per unit length on wire 2 in Fig. 19.25?

■■53. A current-carrying loop is in the form of a rectangle with dimensions of 30 cm by 40 cm. The loop carries a current of 10 A and is in a uniform magnetic field of 20 mT directed parallel to the direction of the current along the 40-cm sides. Find the torque on the loop.

■■■54. A 50.0-cm length of wire carries a current of 4.5 A in the $+x$ direction in a uniform magnetic field with components of $B_x = B_y = B_z = 100$ mT. Find the force on the wire.

■■■55. A rectangular wire loop of a cross-sectional area of 0.20 m^2 carries a current of 0.50 A. If the loop is free to rotate about an axis perpendicular to a uniform magnetic field of 3.0 T and the plane of the loop is initially at an angle of 30° to the direction of the magnetic field, what is the magnitude of the torque on the loop?

■■■56. A loop of wire with a magnetic moment of 1.6 A-m^2 is in a uniform magnetic field of 100 T. Find the magnitude of the torque on the loop if the magnetic field is (a) perpendicular to the plane of the loop and (b) at an angle of 30° to the plane of the loop.

■■■57. A current of 10 A maintained through a square loop creates a torque of 0.15 m-N about an axis through the loop's center and parallel to one of the sides in a magnetic field of 300 mT that is at an angle of 50° relative to the plane of the loop. Find the length of each side of the square.

19.5 Applications of Electromagnetism

■■58. A beam of protons traveling at a speed of 2.0×10^2 m/s passes through the space between two horizontal parallel plates, where a constant electric field and a constant magnetic field are at right angles to one another. If the electric field has a magnitude of 100 V/m and is directed from the bottom to the top plate, what must the magnitude and direction of the magnetic field be to allow the beam to pass undeflected? (Hint: make a sketch of the situation.)

■■59. In a velocity selector, a uniform magnetic field of 1.5 T from a large magnet and two parallel plates with a separation distance of 2.0 cm produce a perpendicular electric field. What voltage should be applied across the plates so that (a) a singly charged ion traveling at a speed of 8.0×10^4 m/s will pass through or (b) a doubly charged ion traveling at the same speed will pass through?

■■60. A charged particle travels undeflected through crossed electric and magnetic fields whose magnitudes are

6000 N/C and 300 mT, respectively. Find the speed of the particle if it is (a) a proton or (b) an alpha particle. (An alpha particle is a helium nucleus, a positive ion with a double positive charge.)

■■**61.** An alpha particle (see Problem 60) is accelerated in the $+x$ direction through a potential of 1000 V. The particle then moves in an undeflected path between two oppositely charged parallel plates in a uniform magnetic field of 50 mT in the $+y$ direction. (a) If the plates are horizontal in the xy plane, what is the magnitude of the electric field between them? (b) If the distance between the plates is 10 cm, what is the potential difference between them? (c) Which plate (top or bottom) is positively charged? Take the mass of the alpha particle to be 6.64×10^{-27} kg.

■■**62.** In a mass spectrometer, a singly charged ion having a particular velocity is selected using a magnetic field of 0.10 T perpendicular to an electric field of 1.0×10^3 V/m. The same magnetic field is used to deflect the ion, which moves in a circular path with a radius of 1.2 cm. What is the mass of the ion?

■■**63.** In a mass spectrometer, a doubly charged ion having a particular velocity is selected using a magnetic field of 100 mT perpendicular to an electric field of 1.0 kV/m. The same magnetic field is used to deflect the ion in a circular path with a radius of 15 mm. Find (a) the mass of the ion and (b) the kinetic energy of the ion. (c) Does the kinetic energy of the ion increase in the circular path? Explain.

■■**64.** A proton is accelerated through a potential difference of 2000 V. It then enters a region between two parallel plates that are separated by 20 cm and have a potential difference of 200 V. Find the magnitude of the magnetic field needed to allow the proton to pass between the plates undeflected.

Additional Problems

65. A rectangular loop of wire whose dimensions are 20 cm by 30 cm carries a current of 1.5 A. (a) What is the magnitude of the magnetic moment of the loop? (b) How should the loop be pivoted and oriented in a uniform magnetic field to obtain the maximum torque on it?

66. A proton with a speed of 3.5×10^6 m/s enters a uniform magnetic field of 0.80 T in such a way that the proton follows a circular path with a radius of 4.6 cm. What is the kinetic energy of the proton in circular motion?

67. How fast should an electron travel at right angles to a magnetic field of 10^{-4} T so that the magnetic force just balances the gravitional force on it?

68. How many turns should a solenoid 30 cm long have to produce a magnetic field of 2.5 mT at its axis when operating with a current of 6.0 A?

69. A nearly horizontal dc power line carries a current of 500 A directly eastward. If the only component of the Earth's magnetic field at that location is 5.0×10^{-5} T due north, what is the magnitude and direction of the magnetic force on a 75-m section of the line between two utility poles?

70. An electron is accelerated from rest through a potential difference of 200 V in a CRT. The electron then passes through a horizontal electric field of 0.10 V/m, moving a horizontal distance of 15 cm. Find the electron's deflection from its original straight-line path.

71. A wire is bent as shown in Fig. 19.30 and placed in a magnetic field with a magnitude of 10 T in the indicated direction. Find the force on each segment of the wire if $x = 50$ cm and the wire carries a current of 5.0 A.

72. An ionized deuteron (a particle with a $+1$ charge) passes through a velocity selector whose magnetic and electric fields have magnitudes of 40 mT and 4.0 kV/m, respectively. Find the speed of the ion.

73. A proton is accelerated from rest to a velocity of 3.0×10^6 m/s in a horizontal direction. It then enters a vertical magnetic field of 2.5 T. (a) Calculate the radius of the circular path of the particle. (b) Find the electric field that would cause the proton to move in a straight line.

74. A 4-keV proton passes in an undeflected straight line through a region of crossed electric and magnetic fields. If the magnetic field has a magnitude of 200 mT, what is the magnitude of the electric field?

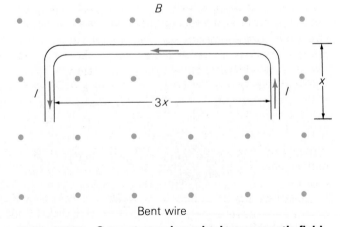

Bent wire

Figure 19.30 **Current-carrying wire in a magnetic field** See Problem 71.

Electromagnetic Induction

20

As you learned in Chapter 19, a current produces a magnetic field. After this was discovered, scientists wondered whether the reverse was possible. That is, can a magnetic field produce a current? The answer is yes, under certain conditions.

Chapter 19 considered only constant magnetic fields. No current is induced in a loop of wire that is stationary in a constant magnetic field. But if the magnetic field changes with time or if the wire moves within the field, a current can be induced in the wire.

These phenomena form the basis for the production of the electricity used in our homes and industry. Have you ever wondered how this electricity is produced? This chapter examines the underlying electromagnetic principles.

20.1 Induced Emf's: Faraday's Law and Lenz's Law

A current can be induced in a conductor in several ways. For example, if a magnet is moved toward a loop of wire, as shown in Fig. 20.1(b), the deflection of the galvanometer needle indicates that there is a current in the loop. If the magnet is moved away from the loop, as shown in Fig. 20.1(c), the galvanometer needle is deflected in the opposite direction, which indicates a reversal of the current's direction. Deflections of the galvanometer needle due to induced currents also occur if the loop is moved toward and away from a stationary magnet. The effect, then, depends on the *relative* motion of the loop and the magnet. Also, the extent of the needle's deflection, or the magnitude of the induced current, depends on the speed of the motion. However, if a loop is moved parallel to a uniform magnetic field (Fig. 20.2), there is no induced current.

Another way to induce a current in a stationary wire loop is to vary the current in another loop close to it. When the switch in the battery-powered circuit in Fig. 20.3(a) is closed, the current in its loop goes from zero to a constant value in a short time. During this time, the magnetic field caused by

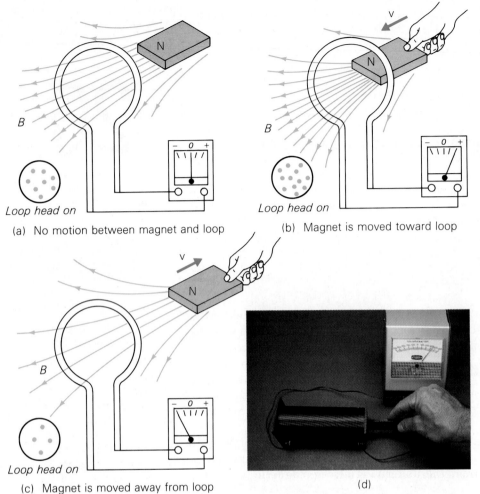

(a) No motion between magnet and loop

(b) Magnet is moved toward loop

(c) Magnet is moved away from loop

(d)

Figure 20.1 Electromagnetic induction

(a) When there is no relative motion between the magnet and the loop, the number of field lines through the loop is constant, and the galvanometer needle shows no deflection. (b) Moving the magnet toward the loop increases the number of field lines passing through the loop, and an induced current is detected. (c) Moving the magnet away from the loop decreases the number of field lines passing through the loop, and the needle deflection and induced current are in the opposite directions. (d) In an actual experiment of this type, a coil of wire, or solenoid, is used to magnify the effect.

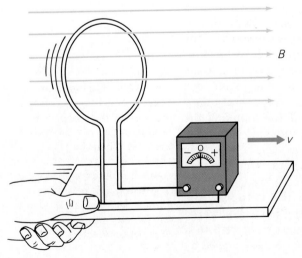

Figure 20.2 Relative motion and no induction

When a loop is moved parallel to a uniform magnetic field, no current is induced in the loop.

this current and passing through both loops builds up, and the galvanometer needle deflects. When the current in the loop in the powered circuit is at its maximum (constant) value, the resulting magnetic field is also constant, and the galvanometer reads zero. Similarly, when the switch is opened [Fig. 20.3(b)], the current and the field quickly decrease to zero, and the galvanometer deflects in the opposite direction. In both cases, the deflection and induced current occur only when the current and the magnetic field are *changing*.

The current induced in a loop is said to be set up by an induced electromotive force (emf) due to **electromagnetic induction**. Recall that an emf is energy capable of driving charges around a circuit (Chapter 17). For example, a battery is a chemical source of emf. In the case of a moving magnet and a stationary loop (Fig. 20.1), mechanical energy is converted to electrical energy. For the two stationary loops (Fig. 20.3), electrical energy is transferred by electromagnetic induction. The latter is an example of what is called *mutual induction*.

Experiments on electromagnetic induction were done independently around 1830 by Michael Faraday in England and Joseph Henry in the United States. Faraday realized that the important factor in electromagnetic induction was the time rate of change of the magnetic field through the loop. That is, *an induced emf is produced in a loop by a changing magnetic field, or more specifically, by a change in the number of field lines passing through the loop.* (Note that this is the case for the situations in Figs. 20.1 and 20.3, and think about the situation in Fig. 20.2.)

A measure of the number of field lines passing through a particular area (the area within the loop, for instance) is given by the **magnetic flux (Φ)**, which is defined as

$$\Phi = BA \cos \theta \qquad (20.1)$$

Here θ is the angle between the magnetic field lines or the vector **B** and a line normal to the plane of the loop (actually the area vector **A**, as shown in Fig. 20.4). The orientation of the loop to the magnetic field affects the number of field lines passing through it, and this is taken into account by including the cosine term in Eq. 20.1. If **B** and **A** are parallel ($\theta = 0$), then the magnetic flux is a maximum ($\Phi_{max} = BA \cos 0° = BA$), and the maximum number of field lines pass through the loop. If **B** and **A** are perpendicular, there are no field lines through the loop, and the flux is zero ($\Phi = BA \cos 90° = 0$).

A unit for magnetic flux is the weber (Wb). Recall that B has the units Wb/m^2, so $\Phi = BA = (Wb/m^2)(m^2) = Wb$, which is a T-m^2 in the SI system.

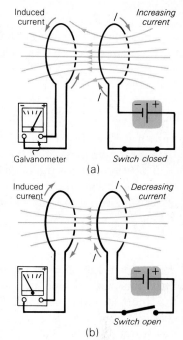

Figure 20.3 **Mutual induction** (a) When the switch is closed in the right-loop circuit, the current buildup produces a changing magnetic field through the other loop and a current is induced in it. (b) Similarly, when the switch is opened, a current is induced in the other loop, but in the opposite direction.

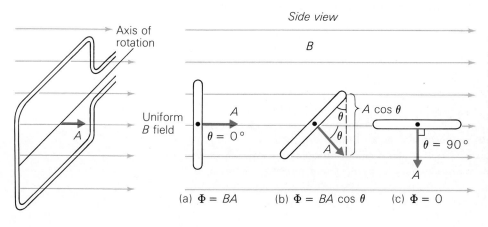

Figure 20.4 **Magnetic flux** Magnetic flux (Φ) is a relative measure of the number of field lines passing through an area. (a) When the plane of a rotating loop is perpendicular to the field and $\theta = 0°$, then $\Phi = BA$. (b) As the loop is rotated, less area is open and the flux decreases. Here $\Phi = BA \cos \theta$. (c) At $\theta = 90°$, $\Phi = 0$.

From his experiments, Faraday concluded that the emf induced in a loop depends on the time rate of change of the number of field lines through the loop, or the time rate of change of the magnetic flux:

$$\mathscr{E} = -N\frac{\Delta\Phi}{\Delta t} \tag{20.2}$$

That is, $\Delta\Phi$ is the change in flux through N loops of wire in a time Δt. Eq. 20.2 is known as **Faraday's law of induction**. Note that $\mathscr{E}$ is an *average* value over the time interval Δt.

The minus sign is included in Eq. 20.2 to give an indication of the polarity of the induced emf, which is found by considering the induced current and its effect, according to Lenz's law:

> An induced emf gives rise to a current whose magnetic field opposes the change in flux that produced it.

What this means is that the magnetic field due to the induced current is in a direction that tends to maintain the flux through the loop. For example, if the flux increases, the magnetic field due to the induced current will be in the direction opposite to the original field. This tends to cancel the increase in the flux, or oppose the change, since magnetic fields add vectorially. The current direction is given by a right-hand rule: with the fingers in the direction of the field, the extended thumb points in the direction of the conventional current. This is the reverse of the rule used to find the magnetic field caused by a current. (Try applying Lenz's law to the loops in Fig. 20.3. Also see Demonstration 15.)

Lenz's law incorporates the conservation of energy. Suppose that the magnetic field due to an induced current complemented the original field, that is, added to it, or increased the flux. This would give a greater induced emf and current, which would give a greater magnetic field, which would give a greater induced current, and so on. This would be a something-for-nothing situation that would contradict the law of the conservation of energy.

For the case of inducing an emf in a loop by moving a magnet (Fig. 20.1), recall that a current-carrying loop has a magnetic field similar to that of a bar magnet. The induced current sets up a magnetic field whose effect is such that the loop acts like a bar magnet with a polarity that opposes the motion of the real bar magnet. Thus, there is opposition to the motion; work must be done to move the magnet.

Substituting for the magnetic flux (Φ) in Eq. 20.2 from Eq. 20.1 shows that a change in flux can result from changes other than a change in magnetic field. In general,

$$\mathscr{E} = -N\frac{\Delta\Phi}{\Delta t}$$

$$= -N\left[\left(\frac{\Delta B}{\Delta t}\right)(A\cos\theta) + B\left(\frac{\Delta A}{\Delta t}\right)(\cos\theta) + BA\left(\frac{\Delta(\cos\theta)}{\Delta t}\right)\right] \tag{20.3}$$

The first term in the brackets represents a flux change due to a time-varying magnetic field (with the area and orientation of the loop constant). A time-varying magnetic field is easily obtained by varying the current producing it.

The second term represents a flux change with a constant magnetic field and a constant loop orientation (constant $\cos\theta$) but a time-varying loop area. This

The law was first stated by Heinrich Lenz (1804–1865), a Russian physicist.

Magnetic Levitation and Lenz's Law

A demonstration of magnetic levitation—not with a superconductor, but with electromagnetic induction and eddy currents.

The levitation height depends on the tangential speed of the disk. This can be shown by moving the magnet along a radial axis. Also, levitation because of air currents is ruled out by using a similar strip of paper without a magnet. The paper strip is necessary to prevent horizontal movement of the magnet.

(a) A strong, lightweight magnet is taped to a strip of Manila folder and held above an aluminum disk on a turntable.

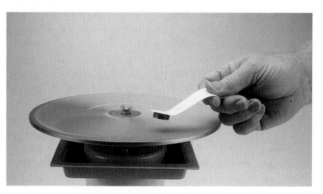

(b) When the disk is rotating, the magnet levitates above it because, by Lenz's law, there is a repulsive force due to induced eddy currents in the disk.

could occur if a circular loop was being stretched or flattened in a plane or if it had an adjustable circumference.

The third term represents a change in flux resulting from a change in the orientation of the loop over time with a constant magnetic field and a constant loop area. This occurs when a loop is rotated in a magnetic field. The change in the number of field lines passing through a loop is evident in the sequential views of the rotating loop in Fig. 20.4. Another way of looking at this is that there is a change in the *effective* area of the loop, or a change in the area that is open to the magnetic field lines.

The induced emf in a rotating loop will be considered in more detail in Section 20.2. The emf's resulting from the first two terms in Eq. 20.3 are analyzed in the following two examples.

Example 20.1   An Emf Due to a Time-Varying Magnetic Field

The magnitude of a magnetic field passing perpendicularly through a 20-loop wire coil with a radius of 6.0 cm goes from 5.3 T to zero in 0.010 s. (a) What emf is induced in the coil? (b) What is the direction and magnitude of the induced current if the resistance of the coil is 60 Ω?

Electromagnetic Induction in Action

You have probably used applications of electromagnetic induction many times without knowing it. One of these is magnetic tape—either audio or video. In an audio tape recorder, a plastic tape coated with a film of iron oxide or chromium oxide runs past a recording head, which consists of a coil wound around an iron core with a gap (Fig. 1). Current in the coil produces a magnetic field in the gap which magnetizes the tape's film. The strength and direction of the magnetization in the film are determined by the gap field, which is determined by the current pulses that are generated by sound from a microphone.

To reproduce the sound, the tape is run by the same head or one similar to it. The changing flux due to the magnetization of the film on the tape induces an emf in the coil, matching the original voltage or current fluctuations. These are amplified and converted to sound in a speaker (which also commonly uses electromagnetic induction; see Question 8 and Fig. 20.19). A tape may be erased by passing it over the gap in a head that has a high-frequency alternating current (ac) input. The resulting magnetic field diminishes to zero as the tape moves away from the gap, leaving its film in a demagnetized condition.

A computer diskette, or floppy disk, is a vinyl platter coated with a magnetic material (Fig. 2). When in use, the diskette spins inside its protective jacket. A read/write head comes into contact with the recording surface through a slot in the jacket. Information is written on or read from the magnetic surface of the diskette just as in a tape recorder. (The index hole in the diskette is used by the computer to sense the angular speed of the diskette. A light is directed on the index hole and there is a detector on the other side.)

Finally, the dark strip on the back of automatic teller cards and credit cards is of a magnetized material. Machines can read identifications from these magnetic strips for banking transactions and credit card authorizations. Don't let your cards (or a computer diskette) get near a magnet. You can probably guess why.

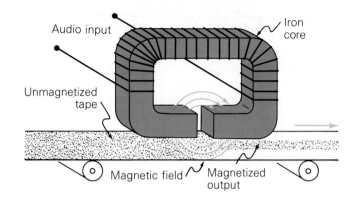

Figure 1 Tape recorder head
Current pulses in the coil on the iron core produce magnetic fields that magnetize the metallic coating on the tape as it passes by.

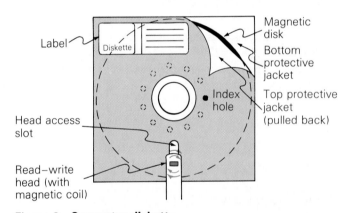

Figure 2 Computer diskette
Information is written on and read from the magnetic surface of a diskette in a manner similar to that of a tape recorder head.

Solution
Given: $\theta = 0°$
$N = 20$
$r = 6.0 \text{ cm} = 0.060 \text{ m}$
$\Delta B = B_2 - B_1$
$\quad = 0 \text{ T} - 5.3 \text{ T}$
$\quad = -5.3 \text{ T (or Wb/m}^2)$
$\Delta t = 0.010 \text{ s}$
$R = 60 \ \Omega$

Find: (a) $\mathscr{E}$
(b) I

(a) Only the first term in Eq. 20.3 contributes, since A and θ are constant (for example, $\Delta A/\Delta t = 0$), and the induced emf is

$$\mathcal{E} = -N\frac{\Delta\Phi}{\Delta t} = -N\left(\frac{\Delta B}{\Delta t}\right)(A\cos\theta) = -N\left(\frac{\Delta B}{\Delta t}\right)(\pi r^2)(\cos 0°)$$

$$= -(20)\left(\frac{-5.3\text{ T}}{0.010\text{ s}}\right)\pi(0.060\text{ m})^2 = 120\text{ V}$$

Note that this is an average value.

(b) Since the magnetic field is decreasing, the induced current is in the direction that sets up a magnetic field in the same direction as that field so as to add to it and oppose the decreasing change (by Lenz's law). The magnitude of the current is given by Ohm's law ($\mathcal{E} = IR$), and

$$I = \frac{\mathcal{E}}{R} = \frac{120\text{ V}}{60\ \Omega} = 2.0\text{ A}$$

Again, this is an average value. ∎

Example 20.2   An Emf Due to a Time-Varying Area

A metal rod is pulled along a metal frame at a constant velocity, as illustrated in Fig. 20.5. A uniform magnetic field is normal to the plane of the frame. What is the potential difference, or emf, across the rod?

Solution

In this situation, the magnetic field is constant and $\theta = 0°$, but the flux through the loop formed by the frame and the rod increases because its area increases. In a time Δt, the rod moves a distance $\Delta d = v\,\Delta t$. If ℓ is the length of the rod, the change in the area of the loop is

$$\Delta A = \ell\,\Delta d = \ell v\,\Delta t$$

and

$$\frac{\Delta A}{\Delta t} = \ell v$$

Thus, the emf that develops across the rod is (since $\cos 0° = 1$)

$$\mathcal{E} = -\frac{\Delta\Phi}{\Delta t} = -B\left(\frac{\Delta A}{\Delta t}\right) = -B\ell v$$

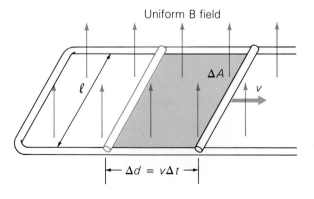

Uniform B field

ΔA

v

$\Delta d = v\Delta t$

ℓ

Figure 20.5 **Motional emf** As the metal rod moves on the metal frame, the area of the loop varies with time and a current is induced in the loop as a result of the changing flux. See Example 20.2.

This is called *motional emf*. It would set up an induced current in the loop that would produce a magnetic field to oppose the change in flux. Since the moving bar causes an increase in the area of the loop and thus an increase in the number of field lines through the loop, the induced field would be downward in the figure and the current would be directed away from the bar on the near side of the frame.

A potential difference would similarly develop across a bar moving on an insulated frame, but a sustained current would not flow in that case. Also, a very small emf develops across the wings of an airplane in flight as they "cut through" the Earth's magnetic field lines. ∎

Some common applications of electromagnetic induction are described in the Insight feature on page 558.

20.2 Generators and Back Emf

As was pointed out in the preceding section, one method to induce an emf or current in a loop is through a change in the loop's orientation, or a change in its effective area. Note again in Fig. 20.4 that the number of field lines through the loop changes as it *rotates*; the effective area of the loop (the area perpendicular to the field) is $A \cos \theta$. This way of producing a flux change is the principle of operation of a simple electrical generator.

A **generator** is device that converts mechanical energy into electrical energy. Basically, a generator performs the reverse function to that of a motor (Chapter 19). In some instances, a generator may be run "backward" as a motor, and vice versa; however, this is not usually the case.

A battery supplies direct current (dc). That is, the polarity of the voltage does not change. Generators can produce either direct current or **alternating current (ac)**, for which the polarity of the voltage and the direction of the current periodically change. The electricity used in homes and industry is primarily ac.

An **ac generator** is sometimes called an *alternator*. The basic elements of a simple ac generator are shown in Fig. 20.6. A wire loop is mechanically rotated in a magnetic field by some external means. The rotation of the loop (called an armature) causes the magnetic flux through it to change, and a current is induced in its wire. The ends of the wire loop are connected to an external circuit by means of slip rings and brushes. (In practice, generators have many loops, or windings, on their armatures.)

When the loop is rotated with a constant angular speed (ω), the angle (θ) between the magnetic field vector and the area vector of the loop (which is perpendicular to the plane of the loop) changes with time: $\theta = \omega t$. As a result, the cross-sectional area of the loop open to the magnetic field lines changes with time, and the flux at any time is

$$\Phi = BA \cos \theta = BA \cos \omega t$$

By Faraday's law, the induced emf is then

$$\mathscr{E} = -N \frac{\Delta \Phi}{\Delta t} = -NBA \left(\frac{\Delta (\cos \omega t)}{\Delta t} \right)$$

Although redundant, the expression ac current is often used, as is ac voltage (usually abbreviated VAC, for volts-ac).

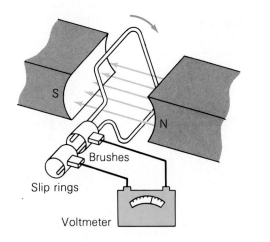

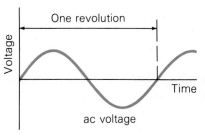

Figure 20.6 **A simple ac generator**
The rotation of a loop of wire in a magnetic field produces a voltage output whose polarity reverses with each half-cycle. This is an alternating voltage and gives rise to an alternating current.

It can be shown by mathematical methods beyond the scope of this book that as Δt approaches zero,

$$\frac{\Delta(\cos \omega t)}{\Delta t} = -\omega \sin \omega t$$

Thus, the instantaneous value of the emf is

$$\mathscr{E} = NBA\omega \sin \omega t$$

or $\quad \mathscr{E} = \mathscr{E}_o \sin \omega t$ $\qquad\qquad$ (20.4)

where $\mathscr{E}_o = NBA\omega$ is the maximum value of the emf ($\sin \omega t = \pm 1$).

Since the value of the sine function varies between $+1$ and -1, the sign, or polarity, of the emf changes with time (Fig. 20.7). Note from the figure that the emf has its maximum value when $\theta = 90°$ or $\theta = 270°$, that is, just when the area of the rotating loop becomes closed to field lines. This is because the *change* in flux is greatest at these points.

The direction of the current also periodically changes, and that is why it is called *alternating* current. Since $\omega = 2\pi f$, Eq. 20.4 can be written as

$$\mathscr{E} = \mathscr{E}_o \sin 2\pi f t$$ $\qquad\qquad$ (20.5)

where f is the rotational frequency of the generator's armature. The ac frequency in the United States is 60 Hz (or 60 cycles).

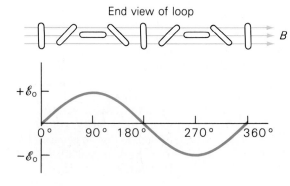

Figure 20.7 $\quad \varepsilon = \varepsilon_o \sin \omega t$
The graph of the sinusoidal generator output is shown with an end view of the corresponding orientations of the loop during the cycle.

Keep in mind that Eqs. 20.4 and 20.5 give the instantaneous value of the emf and that $\mathcal{E}$ varies between $+\mathcal{E}_o$ and $-\mathcal{E}_o$ over one period. You will learn how to determine more practical time-averaged values for ac voltage and current in Chapter 21.

Example 20.3   Generated Emf

A 60-cycle ac generator produces a maximum emf of 120 V. What is the value of the emf at $\frac{1}{120}$ s after it has a value of zero?

Solution

Given: $\mathcal{E}_o = 120$ V   *Find*: $\mathcal{E}$
$\qquad\quad f = 60$ Hz
$\qquad\quad t = \frac{1}{120}$ s

The emf at a given time is given by Eq. 20.5:

$$\mathcal{E} = \mathcal{E}_o \sin 2\pi ft = (120 \text{ V})[\sin 2\pi(60 \text{ Hz})(\tfrac{1}{120} \text{ s})] = 0$$

Thus, the instantaneous value of the emf at $\frac{1}{120}$ s after it is zero is again zero. Note that the period of rotation of the generator's armature is $T = 1/f = \frac{1}{60}$ s. The emf is zero at $t = 0$, and $\frac{1}{120}$ s is one half-cycle later. ■

Electromagnetic induction can also be used to generate direct current. A simple **dc generator** is illustrated in Fig. 20.8. Here a split-ring commutator is used to maintain the same polarity. One of the brushes is kept positively charged and the other negatively charged by an exchange of the ends of the armature just as the polarity changes. Note how this occurs from the figure. The output is a pulsating current, but because the polarity does not change, and the current is in one direction, or direct. In practice, an armature has many windings, and when the pulses from a large number of loops are superimposed, the resultant direct current is steady and almost free of fluctuations. However, since alternating current is easily rectified, or converted to direct current, direct current is not usually generated in this manner.

A generator that uses a permanent magnet is referred to as a magneto. Magnetos are common in single-cylinder engines, such as those used to power small lawn mowers. (In the gasoline combustion engine of automobiles, the

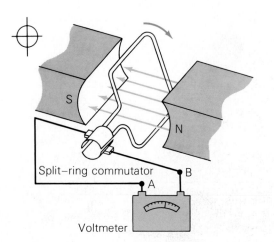

Split–ring commutator
A
B
Voltmeter

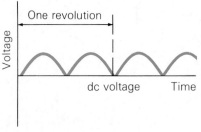

One revolution
Voltage
dc voltage   Time

Figure 20.8 **A simple dc generator**
The polarity of the induced emf reverses every half-cycle, but the split-ring commutator maintains the same polarity on terminals A and B and the output is a fluctuating dc (constant-polarity) voltage. (Compare this dc generator with the dc motor in Fig. 19.17.)

ignition voltage is usually supplied by an induction coil rather than a magneto.) The voltage output of a lawn mower's magneto is used to fire the spark plug that ignites the fuel mixture. In this application, permanent magnets mounted on a flywheel rotate about stationary armature coils. The moving and stationary parts of a generator (or a motor) are called the rotor and stator, respectively. The circuit is opened mechanically or by solid state device detection so that a built-up field collapses, producing a large induced voltage. About 15 kV is supplied to the spark plug at the proper time in the engine's cycle.

In most large-scale ac generators (power plants), the armature is stationary and magnets are revolved about it. The revolving magnetic field produces a time-varying flux through the coils of the armature and thus an ac output. The mechanical energy for the generator is supplied by a turbine that is powered by steam generated from the heat of combustion of fossil fuels, by nuclear reactions, or by water power (hydroelectricity). (See Fig. 20.9.)

Back Emf

Motors also generate emf's. Like a generator, a motor has a rotating armature in a magnetic field. The induced emf in this case is called a **back emf** (or sometimes a counter emf), because its polarity is opposite to that of the line voltage and tends to reduce the current in the armature coils. (What would happen if this were not the case?)

The back emf of a motor depends on the rotational speed of the armature and builds up from zero to some maximum value when the armature goes from rest to its normal operating speed. Ordinarily, a motor turns something; that is, it has a mechanical load. Without a load, the armature speed will increase until the back emf has built up to a point where it almost equals the line voltage, with just enough current in the coils to overcome friction and joule heat loss. Under normal load conditions, the back emf will be less than the line voltage. The larger the load, the slower the motor will rotate and the smaller the back emf

Figure 20.9 **Electrical generation**
A generator and turbine in an oil-fired power plant.

will be. If a motor is overloaded and turns very slowly, the back emf may be reduced to a point where a large enough current flows to burn out the coils. Thus, the back emf is involved in the regulation of a motor's operation.

In effect, a back emf in a dc motor circuit can be represented in a circuit diagram as a battery with opposite polarity to that of the driving voltage (see Fig. 20.10).

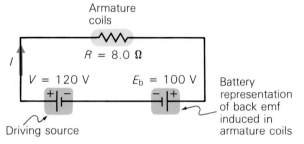

Armature coils

$R = 8.0\ \Omega$

$V = 120\ V$ $E_b = 100\ V$

I

Driving source

Battery representation of back emf induced in armature coils

Figure 20.10 **Back emf**
The back emf in the armature of a dc motor may be effectively represented as a battery with opposite polarity to that of the driving source. See Example 20.4.

Example 20.4   Back Emf

A dc motor with a resistance in its windings of 8.0 Ω operates on 120 V. With a normal load, there is a back emf of 100 V when the motor reaches full speed (see Fig. 20.10). What are (a) the starting current drawn by the motor and (b) the armature current at operating speed under a normal load?

Solution

Given: $R = 8.0\ \Omega$ *Find*: (a) I_s (starting)
        $V = 120\ V$ (b) I_a (operational)
        $\mathscr{E}_b = 100\ V$

(a) When the motor starts and its armature just begins to turn, the back emf is essentially zero. The current in the windings at that time is given by Ohm's law:

$$I = \frac{V}{R} = \frac{120\ V}{8.0\ \Omega} = 15\ A$$

(b) At operating speed, the back emf opposes the driving voltage. The effective voltage is the difference in these two voltages. Thus,

$$I = \frac{V - \mathscr{E}_b}{R} = \frac{120\ V - 100\ V}{8.0\ \Omega} = 2.5\ A$$

Note that the starting current of the motor is relatively large. In some instances, resistors are temporarily connected in series with a motor's coil to prevent the windings from burning out because of high starting currents. When a big motor, such as that of a central air conditioning unit for a whole building, starts up, you may notice the lights momentarily dim because of the large starting current drawn by the motor. ∎

Since motors and generators are opposites, so to speak, and a back emf develops in a motor, you may be wondering whether a back force develops in a generator. The answer is yes. When an operating generator is not connected to an external circuit, no current flows and there is no force on the coils of the armature due to the magnetic field. However, when the generator delivers power to a circuit and current flows in the coils (current-carrying wires in a magnetic field), there is a force that produces a **counter torque**, which opposes the rotation of the armature. As more current is drawn, the counter torque becomes greater, and a greater driving force is needed to turn the armature. Therefore, the higher the current output of a generator, the greater will be the energy expended (fuel consumed) in overcoming the counter torque.

20.3 Transformers and Power Transmission

Electricity is transmitted by power lines over long distances. Obviously, it is desirable to minimize the I^2R losses through these transmission lines. The resistance of a line is fixed, so reducing the I^2R losses means reducing the current. However, the power output of a generator is equal to current times voltage ($P = IV$), and for a fixed voltage (such as 120 V), a reduction in current would mean a reduced power output. It might appear that there is no way to reduce the current while maintaining the power supplied. Fortunately, however, electromagnetic induction can be applied to reduce power transmission losses by stepping up the voltage and reducing the current. This is done with a device called a transformer.

As Fig. 20.11 shows, a simple **transformer** consists of two coils of insulated wire wound on the same (closed) iron core. When ac voltage is applied to the input coil, or **primary coil**, the alternating current gives rise to an alternating magnetic flux that is concentrated in the iron core. The changing flux passes through the output coil, or **secondary coil**, inducing an alternating voltage and current in it.

The induced secondary voltage differs from the primary voltage depending on the ratio of the numbers of turns in the two coils. By Faraday's law, the secondary voltage is given by

For tranformers, it is customary to work with voltage rather than emf.

$$V_s = -N_s \frac{\Delta\Phi}{\Delta t}$$

where N_s is the number of turns in the secondary coil. The changing flux in the primary coil produces a back emf equal to

$$V_p = -N_p \frac{\Delta\Phi}{\Delta t}$$

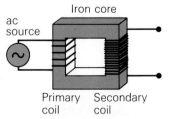

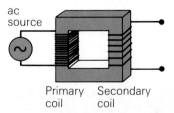

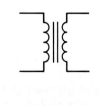

(a) Step–up transformer:
high–voltage
(low current) output

(b) Step–down transformer:
low–voltage
(high current) output

(c) Transformer
circuit symbol

Figure 20.11 **Transformers** (a) A step-up transformer has more turns in the secondary coil than in the primary. (b) A step-down transformer has more turns in the primary coil than in the secondary. (c) The circuit symbol for a transformer reflects its actual structure to some extent.

where N_p is the number of turns in the primary coil. If the resistance of the primary coil is neglected, this is equal to the external voltage applied to it. Then, forming a ratio gives

$$\frac{V_s}{V_p} = \frac{-N_s \Delta\Phi/\Delta t}{-N_p \Delta\Phi/\Delta t}$$

or

$$\frac{V_s}{V_p} = \frac{N_s}{N_p} \tag{20.6}$$

Here it is assumed that the core concentrates the field so the flux through each coil is the same.

If the transformer is assumed to be 100% efficient (no energy losses), the power input is equal to the power output, and since $P = IV$,

$$I_p V_p = I_s V_s \tag{20.7}$$

Then, using Eq. 20.6, the currents and voltages are related to the turn ratio by the relationship

$$\frac{I_p}{I_s} = \frac{V_s}{V_p} = \frac{N_s}{N_p} \tag{20.8}$$

Although some energy is always lost, this is good approximation, since a well-designed transformer may have an efficiency of more than 95%. The sources of energy losses will be discussed shortly.

With these equations, it is easy to see how a transformer affects the voltage and current. In terms of the output,

$$V_s = \left(\frac{N_s}{N_p}\right) V_p \qquad \text{and} \qquad I_s = \left(\frac{N_p}{N_s}\right) I_p$$

That is, if the secondary coil has a greater number of windings than the primary does ($N_s > N_p$ or $N_s/N_p > 1$) as in Fig. 20.11(a), the voltage is stepped up ($V_s > V_p$). However, less current flows in the secondary than in the primary ($N_p/N_s < 1$ and $I_s < I_p$). This type of arrangement is called a **step-up transformer**. For example, if the primary coil of a transformer has 50 turns and the secondary has 100 turns, $N_s/N_p = 2$ and $N_p/N_s = \frac{1}{2}$. Thus, a 220-V input at 10 A will be stepped up to a 440-V output at 5.0 A.

The opposite situation, where the secondary coil has fewer turns than the primary, characterizes a **step-down transformer** [Fig. 20.11(b)]. In this case, the voltage is stepped down, or reduced, and the current is increased. A step-up transformer may be used as a step-down transformer by simply reversing the output and input connections. Also, it should be apparent that transformers operate only on ac (not on dc).

The preceding equations apply to an ideal transformer, but actual transformers do have some energy losses. There is always some flux leakage; that is, not all of the flux passes through the secondary coil. In some transformers, one of the coils is wound over (on top of) the other rather than having the two on separate "legs" of a core. Also, when ac current flows in the primary coil, the changing magnetic flux through the loops gives rise to an induced emf in that coil. This is called *self-induction*. By Lenz's law, the self-induced emf will oppose the change in current and will thus limit the current (similar to a back emf in a motor). Self-induction may be thought of as a kind of electromagnetic inertia—like the inertia of material bodies, it opposes change.

Some energy is lost due to the resistances of the coil wires (I^2R losses), but this is generally small.

Another cause of energy loss is **eddy currents** in the transformer core. A highly permeable material is used for the core to increase the density of the magnetic flux, but such materials are usually good conductors. The changing magnetic flux sets up swirling movements of charge, or eddy currents, in the core material, and these dissipate energy. To reduce this effect, transformer cores are made of thin sheets of material (usually iron) laminated with an insulating glue between them. Because of the insulating layers between the sheets, the eddy currents are broken up, or confined to the sheets, which greatly reduces the loss due to them. Well-designed transformers generally have less than 5% internal energy loss.

An effect of eddy currents may be demonstrated by swinging a plate made of a nonmagnetic metal, such as aluminum, through a magnetic field, as illustrated in Fig. 20.12(a). Eddy currents are set up in the plate as a result of its motion in the field and the changing magnetic flux. By Lenz's law, the eddy currents set up opposing fluxes, or effectively opposite magnetic poles on the plate. This gives rise to a repulsive force that retards the swinging motion and brings the plate quickly to rest.

The breaking up of the eddy currents may be demonstrated using a plate with holes and slits [Fig. 20.12(b)]. When the end of the plate with the holes swings between the magnet's poles, the motion is retarded and damped, since eddy currents can have large paths around the holes. However, when the end with the slits swings between the pole faces, the plate swings relatively freely.

The damping effect of eddy currents is applied in the braking systems of rapid-transit cars that travel on rails. When an electromagnet on the car is turned on, it applies a magnetic field to a metal wheel or the rail. The repulsive force due to the induced eddy currents acts as a braking force. As the car slows, the eddy currents decrease, allowing a smooth braking action.

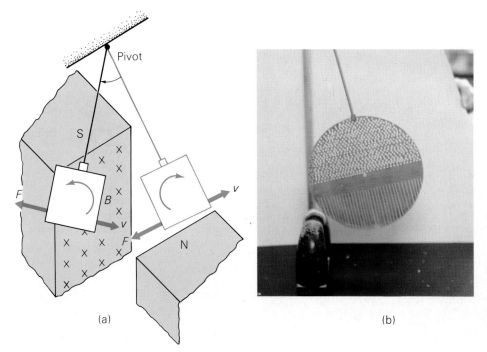

(a)   (b)

Figure 20.12 Eddy currents
(a) Eddy currents are induced in a nonmagnetic conductive plate moving in a magnetic field. The induced currents oppose the change in flux, and there is a retarding force that opposes the motion. Note that the currents reverse as the plate swings through the field. (b) If that half of the pendulum plate with the holes swings between the magnet, the motion is retarded. However, if the half with the slits is used, the plate swings relatively freely.

Power Transmission

For power transmission over long distances, transformers provide a means to increase (step up) the voltage and reduce the current of a generator's output in order to cut down the I^2R losses. A schematic diagram of an ac power distribution system is shown in Fig. 20.13. The voltage output of the generator is stepped up and transmitted over long distances to an area substation near the consumers. There the voltage is stepped down. There are further voltage step-downs at distributing substations and utility poles before 120–240 V are supplied to homes and businesses.

The following example illustrates the benefits of being able to step up the voltage (and step down the current) for electrical power transmission.

Example 20.5   Power Transmission

A generator produces 10 A at 440 V. The voltage is stepped up to 4400 V (by an ideal transformer) for transmission over 40 km of power line, which has a resistance of 0.50 Ω/km. [A power transmission line has two wires (for a complete circuit); however, for simplicity, you can take the length given to be the total wire length rather than doubling it.] (a) What percentage of the original energy would have been lost in transmission if the voltage had not been stepped up? (b) What percentage of the original energy is lost even with the voltage stepped up?

Solution

Given: $V_p = 440$ V
$\quad\quad\; I_p = 10$ A
$\quad\quad\; V_s = 4400$ V
$\quad\quad\; R_o/L = 0.50$ Ω/km
$\quad\quad\; L = 40$ km

Find: (a) Percentage energy loss without voltage step-up
(b) Percentage energy loss with voltage step-up

(a) The power output of the generator is

$$P = I_p V_p = (10 \text{ A})(440 \text{ V}) = 4400 \text{ W}$$

The resistance of 40 km of power line is

$$R = \left(\frac{R_o}{L}\right)L = \left(\frac{0.50 \text{ Ω}}{\text{km}}\right)(40 \text{ km}) = 20 \text{ Ω}$$

The energy loss (per unit time) in transmitting a current of 10 A is

$$P_{\text{loss}} = I_p^2 R = (10 \text{ A})^2(20 \text{ Ω}) = 2000 \text{ W}$$

Thus,

$$\% \text{ loss} = \frac{P_{\text{loss}}}{P}\,(\times 100\%) = \frac{2000 \text{ W}}{4400 \text{ W}}\,(\times 100\%) = 45\%$$

This is the percentage of energy loss because power is energy per time.

(b) When the voltage is stepped up to 4400 V, the transmitted current is

$$I_s = \left(\frac{V_p}{V_s}\right)I_p = \left(\frac{440 \text{ V}}{4400 \text{ V}}\right)(10 \text{ A}) = 1.0 \text{ A}$$

(Note that the voltage was stepped up by a factor of 10 and the current is stepped down by the same factor.) The power loss in this case is

$$P_{\text{loss}} = I_s^2 R = (1.0 \text{ A})^2(20 \text{ Ω}) = 20 \text{ W}$$

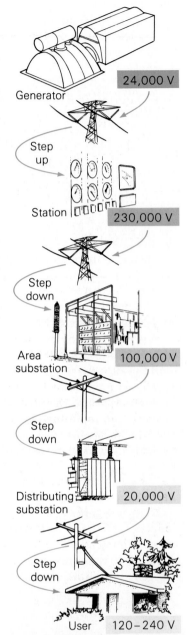

Figure 20.13 **Power transmission**
An illustration of an electrical power distribution system.

and

$$\% \text{ loss} = \frac{P_{\text{loss}}}{P} \left(\times 100\% \right) = \frac{20 \text{ W}}{4400 \text{ W}} \left(\times 100\% \right) = 0.45\% \quad \blacksquare$$

Example 20.5 shows the advantage of high-voltage, or high-tension, transmission lines. However, there is a practical limit to the degree of voltage step-up. At very high voltages, the molecules in the air surrounding a power line may be ionized, forming a conducting path to nearby trees, buildings, or the ground. This is referred to as a leakage loss. Crews from electric companies are continually clearing foliage from near power lines, and long insulators are used to hold the high-voltage wires away from the metal towers. Leakage losses are generally greater during wet weather because moist air is more easily ionized. Under certain conditions, you may see the arcing or corona discharge from a high-voltage line and/or hear the accompanying crackling noise.

20.4 Electromagnetic Waves

Electromagnetic waves (or radiation) as a means of heat transfer were considered in Chapter 12. Now you are ready to understand more fully the production and characteristics of electromagnetic radiation. As the name implies, these waves have both electric and magnetic properties, which may be described by quantities you have studied.

James Clerk Maxwell (1831–1879), a Scottish scientist, showed that four fundamental relationships could describe completely all observed electromagnetic phenomena. Maxwell also used this set of equations to predict the existence of waves of an electromagnetic nature. Because of his contributions, the set of equations is known as **Maxwell's equations**, although they were for the most part developed individually by other scientists (for example, Faraday discovered the law of induction).

Essentially, Maxwell's equations combine the electric force and the magnetic force into a single electromagnetic force; the separate forces or fields are shown to be symmetrically related. This symmetry is evident in the equations as presented in their advanced mathematical form. For the purposes of this course of study, a qualitative description is sufficient:

A time-varying magnetic field produces a time-varying electric field.
A time-varying electric field produces a time-varying magnetic field.

The first statement reflects the fact that a changing magnetic flux gives rise to an induced voltage in a wire, or to *an electric field in space*. The second statement implies that a changing electric field gives rise to a changing magnetic field. This symmetry is important in the analysis of electromagnetic waves.

Basically, electromagnetic waves are produced by accelerating electric charges, such as an electron oscillating in simple harmonic motion. This could be one of the many electrons in a radio transmitter, which oscillates with frequencies of about 10^6 Hz. As such an electron moves, it accelerates and radiates an electromagnetic wave (Fig. 20.14). The continually alternating motions of many such charges due to the alternating current in the transmitter rod produce time-varying electric and magnetic fields in the immediate vicinity

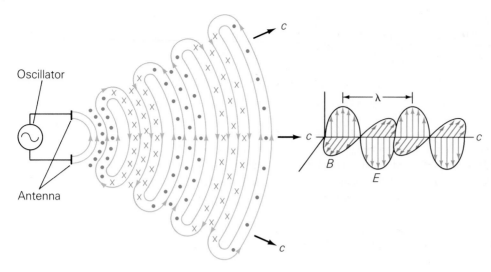

of the rod. The electric field (in the plane of the paper) continually changes direction, as does the magnetic field (into and out of the paper by a right-hand rule).

Both the electric and the magnetic fields store energy and propagate outward with the speed of light (c in a vacuum). Maxwell's equations show that, except in the immediate vicinity of the source, the electromagnetic waves at a fixed instant of time are of the form shown in Fig. 20.14. The electric vector of the E field (**E**) is perpendicular to the magnetic vector of the B field (**B**), and each varies sinusoidally with time. Both **E** and **B** are perpendicular to the direction of propagation. Thus, electromagnetic waves are transverse waves.

Radiation Pressure

An electromagnetic wave carries energy. Consequently, it can do work and can exert a force on a material it strikes. Consider light striking an electron at the surface of a material (Fig. 20.15). The electric field of the electromagnetic wave does work on the electron; assume that this gives the electron a velocity (v) as shown in the figure. As a result, there is a magnetic force on the electron due to the magnetic field component of the light wave. (Recall from Chapter 19 that a moving charged particle in a magnetic field experiences a force.) By the right-hand rule for the magnetic force on a moving charged particle, the force on the electron is in the direction shown in Fig. 20.15. That is, the electromagnetic wave produces a force on the electron in the direction in which the wave is propagating and therefore exerts a force on the material to which the electron is bound.

The radiation force per area is called the **radiation pressure** on an absorbing material. If the electromagnetic wave is re-emitted, the radiation pressure is doubled because of the recoil momentum and force. Radiation pressure is negligible for most common situations but is of importance in some atmospheric and astronomical phenomena. For example, radiation pressure plays a key role in the formation of the tail of a comet, which is made up of fine dust particles (recall that the tail always points away from the Sun).

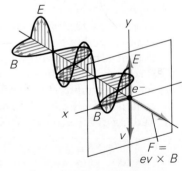

Figure 20.15 **Radiation pressure**
The electric field of an electromagnetic wave that is striking a surface acts on an electron, giving it a velocity (v). The magnetic field produces a force on the charge in the direction of the incident light. (Check this using the appropriate right-hand rule.)

Types of Electromagnetic Waves

Electromagnetic waves are classified by ranges of frequencies or wavelengths in a spectrum ($\lambda f = c$). The electromagnetic spectrum is continuous, so the limits of the various ranges are approximate. Table 20.1 lists these frequency and wavelength ranges for the general types of electromagnetic waves. See also Fig. 20.16.

Power Waves. Electromagnetic waves of 60-cycle frequency result from currents moving back and forth (alternating) in electrical circuits. As Table 20.1 indicates, these power waves have a wavelength of 5.0×10^6 m, or 5000 km (more than 3000 mi). Waves of this low frequency are of little practical importance. They may occasionally produce a so-called 60-cycle hum on your stereo.

Radio Waves. Electromagnetic waves with frequencies in the kilohertz (kHz) and megahertz (MHz) ranges are classified as radio waves. AM radio waves are in the kHz range, and FM radio waves in the MHz range. Frequencies slightly higher than those for FM are used for short-wave radio and television channels.

Microwaves. Microwaves, with frequencies in gigahertz (GHz) range, are produced by special vacuum tubes (called klystrons and megetrons). Microwaves are used in communications and radar applications. A very common use of microwaves today is in microwave ovens (Chapter 12).

Infrared Radiation. The infrared region of the electromagnetic spectrum lies adjacent to the low-frequency or long-wavelength end of the visible spectrum. The frequency at which a warm body emits radiation depends on its temperature. Actually, such a body emits electromagnetic waves of many different frequencies, but the frequency of the maximum intensity characterizes

Table 20.1
Classification of Electromagnetic Waves

Type of Wave	Approximate Frequency Range (Hz)	Approximate Wavelength Range (m)	Source
Power waves	60	5×10^6	Electric currents
Radio waves			Electric circuits
AM	0.53–1.6×10^6	570–186	
FM	88–108×10^6	3.4–2.8	
TV	54–890×10^6	5.6–0.34	
Microwaves	10^9–10^{11}	10^{-1}–10^{-3}	Special vacuum tubes
Infrared radiation	10^{11}–10^{14}	10^{-3}–10^{-7}	Warm and hot bodies
Visible light	4.0–7.0×10^{14}	10^{-7}	Sun and lamps
Ultraviolet radiation	10^{14}–10^{17}	10^{-7}–10^{-10}	Very hot bodies and special lamps
X-rays	10^{17}–10^{19}	10^{-10}–10^{-12}	Electron collisions
Gamma rays	above 10^{19}	below 10^{-12}	Nuclear reactions and processes in particle accelerators

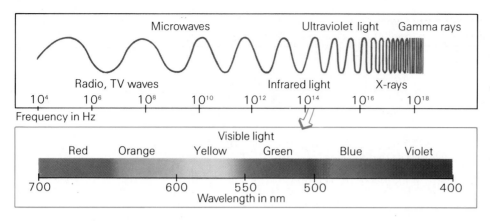

Figure 20.16 **The electromagnetic spectrum** The spectrum of frequencies or wavelengths is divided into regions, or ranges. Note that the visible region is a very small part of the total spectrum. The wavelengths are given in nanometers, nm (10^{-9} m).

the radiation. A body at about room temperature emits radiation in the far infrared region (meaning farthest from the visible region).

Recall from Chapter 12 that infrared radiation is sometimes referred to as heat rays. This is because this radiation is readily absorbed by some materials containing water molecules, producing an increase in the material's temperature because of increased molecular motion. Infrared lamps are used in therapeutic applications and to keep food warm in cafeterias.

Infrared radiation emitted by the Earth is absorbed in the atmosphere chiefly by water and carbon dioxide (CO_2) molecules in a process called the **greenhouse effect**. This process regulates the Earth's temperature and may be described in simple terms as follows: visible light from the Sun penetrates the atmosphere and warms the Earth's surface. The Earth reradiates waves in the infrared region, and this radiation is selectively absorbed by the so-called greenhouse gases (chiefly water vapor and carbon dioxide). Selective absorption means that only certain wavelengths (or frequencies) are absorbed (resonance absorption). When infrared radiation from the Earth is absorbed by the atmosphere, the atmosphere heats up, which causes the temperature of the Earth's surface to increase. As a result, the wavelengths of the emitted radiation are shifted. (The wavelength of infrared radiation depends on the temperature of the radiation body.) The different wavelengths then emitted are transmitted through the atmosphere into space. The Earth cools, the wavelengths of the emitted radiation shift back, and the process repeats. This is essentially a thermostatic cycle that helps regulate the average temperature of the Earth. Increased concentrations of the greenhouse gases, particularly carbon dioxide (from combustion) have led to concern about an overall increase in the Earth's average temperature.

Why is it called the greenhouse effect? Because ordinary glass used in greenhouses has absorption properties similar to those of the atmosphere—it transmits visible light and absorbs infrared radiation. This is one reason why the interiors of greenhouses stay warm on cool, sunny days. You have probably noticed the same effect in a closed car. However, the warmth of a greenhouse (and a closed car) is primarily due to the fact that the warm air heated within the glass enclosure is prevented from escaping.

Visible Light. The visible region is a very small portion of the total electromagnetic spectrum, from about 4×10^{14} Hz to about 7×10^{14} Hz. Only radiation of this frequency activates the sensors in our eyes, and light emitted or reflected from the objects around us provides us with much information about our world. Visible light is considered in some detail in Chapters 22–25.

Ultraviolet Radiation. Beyond the violet end of the visible region lies the ultraviolet frequency range. Ultraviolet (or uv) radiation is produced by special lamps and very hot bodies. The Sun emits large amounts of ultraviolet radiation, but fortunately most of it received by the Earth is absorbed in the ozone (O_3) layer in the atmosphere at an altitude of about 40–50 km. Because the ozone layer plays a protective role, there is concern about its depletion by chlorofluorocarbon gases (such as Freon, used in refrigerators) that drift upward and react with the ozone.

The small amount of ultraviolet radiation that reaches the Earth's surface from the Sun can cause human skin to burn or tan. Pigmentation in the skin acts as a protective mechanism against the penetration of ultraviolet radiation. The degree of penetration depends on the amount of a pigment called melanin in the skin and on the thickness of the skin's layers. Persons (except albinos) have varying amounts of melanin in their skin. Exposure to sunlight induces the production of more melanin, causing the skin to tan. Overexposure, especially initially, may cause the skin to burn and turn red (sunburn). Healing thickens the outer skin layer, which then offers greater protection. Creams and lotions containing so-called sunscreens are also available to keep skin from sunburning. The molecules of chemicals in these preparations absorb some of the ultraviolet radiation before it reaches the skin.

Exposure to sunlight is necessary for the natural production of vitamin D from compounds in the skin. In northern latitudes, where people's exposure to sunlight is relatively seasonal, diets need to be supplemented with synthetic vitamin D. You may have noticed that a vitamin D supplement is often provided in milk.

Ultraviolet radiation is absorbed by certain molecules in ordinary glass. Therefore, you cannot get a tan or a sunburn through glass windows. Welders wear special glass goggles or face masks to protect their eyes from the large amounts of ultraviolet radiation produced by the arcs of welding torches. Similarly, it is important to shield the eyes when using a sun lamp. The ultraviolet component of sunlight reflected from snow-covered surfaces can produce snowblindness in unprotected eyes.

X-rays. Beyond the ultraviolet region of the electromagnetic spectrum is the important X-ray region. We are all familiar with X-rays, primarily through medical applications. X-rays were discovered accidentally in 1895 by the German physicist Wilhelm Roentgen (1845–1923), when he noted the glow of a piece of fluorescent paper caused by some mysterious (unknown and therefore referred to as X) radiation coming from a cathode ray tube.

The basic elements of an X-ray tube are shown in Fig. 20.17. A potential difference of several thousand volts is applied across the electrodes in a sealed, evacuated tube. Electrons emitted from the negative electrode are accelerated toward the positive anode, which is called the target. When the electrons strike the target, they interact with the atomic electrons of its material, and the electrical repulsion abruptly decelerates them. This results in a loss of energy, which takes the form of high-frequency X-rays (called *Bremsstrahlung* in German, which means "braking rays"). Similar processes take place in color

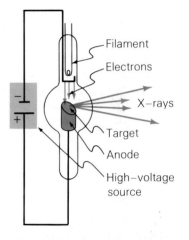

Filament
Electrons
X–rays
Target
Anode
High–voltage source

Figure 20.17 **An X-ray tube**
Electrons accelerated through a large potential difference strike a target electrode and interact with the atomic electrons of its material, causing the incoming electrons to be slowed down. Energy is emitted in the form of X-rays. The photo shows an early X-ray tube circa 1925.

television picture tubes, which use electron beams and high potential differences. Proper shielding is necessary to protect viewers from the resulting X-rays.

As you will learn in Chapter 27, the energy of electromagnetic radiation depends on its frequency. High-frequency X-rays have very high energies and can produce skin burns and other undesirable effects. However, at low intensities, X-rays can safely be used to view the internal structure of the human body and other opaque objects. X-rays can pass through materials that are opaque to other types of radiation. The denser the material, the greater its absorption of X-rays and the less intense the transmitted radiation will be. For example, as X-rays pass through the human body, many more of them are absorbed by bone than by tissue. If the transmitted radiation is directed onto a photographic plate or film, the exposed areas show variations in intensity that form a contrasted picture of internal structures.

The combination of a computer with modern X-ray machines permits the formation of three-dimensional images called CAT scans. (See the Insight feature.)

Gamma Rays.   The electromagnetic waves of the upper frequency range of the known electromagnetic spectrum are called gamma rays (γ-rays). This high-frequency radiation is produced in nuclear reactions and in particle accelerators. Gamma rays will be discussed in Chapter 29.

INSIGHT

CAT Scan

A relatively new but increasingly utilized X-ray technique is called a CAT scan, where CAT stands for computerized axial tomography. To make conventional X-ray images, the entire thickness of the body is projected on the film. Internal structures often overlap, and their images therefore mask one another and are difficult to distinguish. A tomographic image, on the other hand, pictures a slice through the body about a selected axis (the Greek word tomo means "slice" and graph means "picture").

For a CAT scan, an X-ray beam scans across a slice of the body, and the transmitted radiation is detected by a series of detectors and recorded in a computer. Measurements are made at a large number of points on each slice as the X-ray tube rotates about the body's longitudinal axis. With a narrow beam, it would take some time to produce a complete picture, so fan beams and multiple detectors are used to speed up the process. Using the measurements from all the slices, the computer constructs a three-dimensional image. In addition, any two-dimensional slice may be displayed for study. CAT scans provide doctors with much more information than can be obtained from a conventional X-ray (Fig. 1).

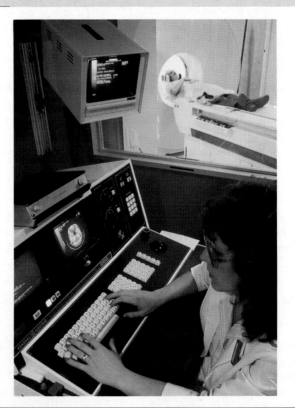

Figure 1 **CAT scan**
A CAT scan of the brain is taken.

Important Formulas

Magnetic flux:

$$\Phi = BA \cos \theta$$

Faraday's law of induction:

$$\mathscr{E} = -N \frac{\Delta \Phi}{\Delta t}$$

$$= -N \left[\left(\frac{\Delta B}{\Delta t} \right) (A \cos \theta) + B \left(\frac{\Delta A}{\Delta t} \right) (\cos \theta) + BA \left(\frac{\Delta (\cos \theta)}{\Delta t} \right) \right]$$

Generator emf (where $\mathscr{E}_o = NBA\omega$):

$$\mathscr{E} = \mathscr{E}_o \sin \omega t = \mathscr{E}_o \sin 2\pi f t$$

Back emf:

$$\mathscr{E}_b = V - IR$$

Currents, voltages, and turn ratio for a transformer:

$$\frac{I_p}{I_s} = \frac{V_s}{V_p} = \frac{N_s}{N_p}$$

Questions

Induced Emf's: Faraday's Law and Lenz's Law

1. Explain why a change in the magnetic flux through a loop is an indication of a change in the number of field lines that pass through the loop.

2. Sketch a graph of $\mathscr{E}$ versus t for the loop with the induced current in Fig. 20.3 for both cases: (a) when the switch is closed and (b) when the switch is opened.

3. A bar magnet is dropped through a coil of wire as shown in Fig. 20.18. (a) Describe what is observed on the galvanometer by sketching a graph of $\mathscr{E}$ versus t. (b) Does the magnet fall freely? Explain.

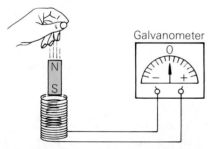

Figure 20.18 **A time-varying magnetic field**
Which way will the galvanometer's needle deflect?
See Question 3.

4. A bar magnet and a loop of wire move together with the same linear velocity. Is an emf induced in the loop? Explain.

5. A loop of wire moves in a region where there is a uniform magnetic field. If the plane of the loop moves perpendicularly to the magnetic field, is an emf induced in the loop? Explain.

6. Show that a tesla-meter squared per second (T-m²/s) is equivalent to a volt (V).

7. A magnetic field directed vertically upward through a horizontal loop of wire decreases with time. What is the direction of the current induced in the loop?

8. A diagram of a loudspeaker is presented in Fig. 20.19. Study the diagram and explain how such a speaker converts electrical pulses into sound.

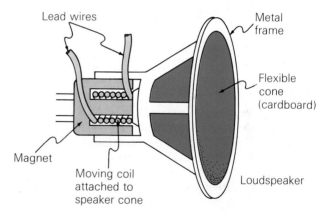

Figure 20.19 **An electromagnetic application**
This diagram shows a cross section of a loudspeaker.
See Question 8.

9. A telephone has both a speaker-transmitter and a receiver (Fig. 20.20). The transmitter has a diaphragm coupled to a carbon chamber (called the button), which contains loosely packed granules of carbon. As the diaphragm vibrates because of incident sound waves, the pressure on the granules varies, causing them to be more or less closely packed. As a result, the resistance of the button changes. The receiver is similar to the speaker shown in Fig. 20.19. Applying principles of electricity and magnetism you have learned, explain the basic operation of the telephone.

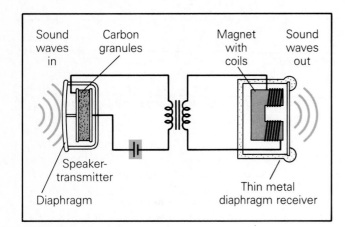

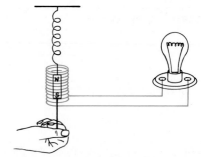

Figure 20.21 **Inventive genius?**
See Question 15.

Figure 20.20 **Telephone operation**
See Question 9.

Generators and Back Emf

10. What is the orientation of the armature loop in a simple ac generator when the value of the emf is a maximum? What is the orientation when the flux through the loop is a maximum?

11. Why does an ac generator have two slip rings?

12. For a dc generator, why is the armature always the rotor?

13. A simple dc generator contains two armatures whose loops are at right angles to each other. Sketch the emf or current output with time.

14. Explain why a back emf in a motor and a counter torque in a generator can be seen as opposite but symmetrical effects.

15. A student has a bright idea for a generator: for the arrangement shown in Fig. 20.21, the magnet is pulled down and released. With a highly elastic spring, the student thinks there should be a relatively continuous electrical output. What is wrong with this idea? (Discuss the output in terms of energy.)

Transformers and Power Transmission

16. Can a transformer be operated using a battery as the voltage source? Explain.

17. Does a transformer operate by mutual induction or self-induction?

18. What would happen with a transformer that had the same number of turns on the primary and secondary coils?

19. Why are the eddy currents in the plate in Fig. 20.12 in opposite directions at the different positions shown?

20. Distinguish between mutual induction and self-induction.

21. The voltage for ignition by a spark plug in an automobile engine is supplied by what is referred to as the coil but is actually a pair of induction coils (see Fig. 20.22). Explain how the voltage of a car's battery (usually 12 V) is raised to as high as 25 kV by this device. Explain the function of the distributor.

22. What limits the efficiency of high-voltage power transmission?

Electromagnetic Waves

23. Explain the significance of the symmetry expressed by Maxwell's equations in the production of electromagnetic waves; that is, explain how electromagnetic waves are produced.

24. Are electromagnetic waves longitudinal or transverse?

25. How are electromagnetic waves classified? Name five different types of electromagnetic waves.

26. Describe the wavelength and the frequency of each of the following relative to those of visible light: (a) infrared radiation, (b) ultraviolet radiation, and (c) microwaves.

27. Explain why some people tan and others burn easily. How does using a sunscreen help the latter?

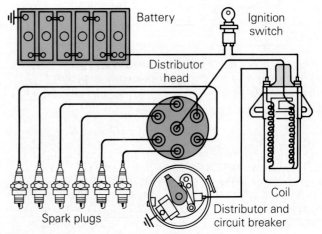

Figure 20.22 **Auto ignition with a coil**
See Question 21.

Problems

■1. A loop of wire in the form of a square 1.0 m on a side is perpendicular to a uniform magnetic field of 0.75 T. What is the flux through the loop?

■2. A circular loop of wire that encloses an area of 0.015 m² is in a uniform magnetic field of 0.25 T. What is the flux through the loop if its plane is (a) parallel to the field, (b) at an angle of 30° to the field, and (c) perpendicular to the field?

■3. A conductive loop enclosing an area of 2.0 × 10⁻³ m² is perpendicular to a uniform magnetic field of 2.5 T. If the field goes to zero in 0.0045 s, what is the average emf induced in the loop?

■■4. A circular loop with a radius of 10 cm is positioned at various locations in a uniform 0.6-T magnetic field. Find the magnetic flux if the normal to the plane of the loop is (a) perpendicular to the magnetic field, (b) parallel to the magnetic field, and (c) at an angle of 40° to the magnetic field.

■■5. A loop in the form of an equilateral triangle 40 cm on a side is oriented perpendicularly to a uniform magnetic field of 750 mT. What is the flux through the loop?

■■6. A square coil of wire with 10 turns is in a magnetic field of 0.025 T. The flux through the loop is 0.50 T-m². Find the effective area of the loop if the field is (a) perpendicular to the plane of the loop and (b) makes an angle of 60° with the plane of the loop.

■■7. What is the magnetic flux through an ideal solenoid whose windings have a radius of 3.0 cm if the turn density is 200 turns/m and a current of 1.5 A flows through the wire?

■■8. A circular wire loop is in a uniform magnetic field of 1.5 × 10⁻² T. The flux through the loop is 1.2 × 10⁻² T-m². What is the radius of the loop if the normal to the plane of the loop makes an angle of 45° with the field?

■■9. A circular wire loop with a diameter of 28 cm carries a current of 0.50 A. Assume that the magnitude of the magnetic field at the center of the loop is uniform across the whole area. What is the flux through the loop?

■■10. A magnetic field perpendicular to the plane of a wire loop with an area of 0.40 m² decreases by 0.10 T in 10⁻³ s. What is the average value of the emf induced in the loop?

■■11. A square loop of wire with 50-cm sides experiences a uniform, perpendicular magnetic field of 500 mT. If the field goes to zero in 0.01 s, what is the average emf induced in the loop?

■■12. The magnetic flux through a 20-turn coil of wire is reduced from 20 Wb to 5.0 Wb in 0.10 s. Find the average induced current in the coil, which has resistance of 12 Ω.

■■13. If the magnetic flux through a loop of wire increases by 30 T-m² and an average emf of 20 V is induced in the wire, over what period of time was the flux increased?

■■14. During a time of 0.20 s, a coil of wire with 50 loops has an average induced emf of +9.0 V due to a changing magnetic field perpendicular to the plane of the coil. If the radius of the coil is 10 cm and the initial value of the magnetic field is 1.5 T, what is the final value of the field?

■■15. A single strand of wire of adjustable length is wound around the circumference of a round balloon that has a diameter of 40 cm. A uniform magnetic field with a magnitude of 0.15 T is perpendicular to the plane of the loop. If the balloon is blown up so its diameter and the diameter of the wire loop increase to 50 cm in 0.040 s, what is the average value of the emf induced in the loop?

■■16. A metal rod is pulled at a uniform velocity of magnitude v perpendicularly to a uniform magnetic field of magnitude B. If the rod has a length L, what emf is induced across it?

■■17. A metal rod 20 cm long moves in a straight line at a speed of 4.0 m/s with its length perpendicular to a uniform magnetic field of 1.2 T. Find the potential difference between the ends of the rod.

■■■18. The flux through a fixed loop of wire changes uniformly from +40 Wb to −20 Wb in 3.0 ms. (a) What is the significance of the negative flux? (b) What is the average induced emf in the loop?

■■■19. A rectangular conductive loop measuring 20 cm by 30 cm has a resistance of 0.10 Ω. At what rate must a magnetic field perpendicular to the plane of the loop change with time to cause a current of 3.0 A to flow in the loop?

■■■20. The south pole of a bar magnet is pulled out of a 10-turn coil of wire that has a diameter of 10 cm in 0.50 ms. The magnitude of the magnetic field changes from 500 mT to zero. The wire is made of copper and has a diameter of

2.3 mm and a length of 1.5 m. Find (a) the magnitude and direction of the average current induced in the coil and (b) the average energy expended in the coil. (Make a sketch and indicate the direction of the induced current.)

■■■**21.** A length of 20-gauge copper wire is formed into a circular loop with a radius of 20 cm. (20-gauge wire has a diameter of 0.8118 mm). A magnetic field perpendicular to the plane of the loop increases from zero to 5.0 mT in 0.25 s. Find the average electrical energy dissipated in the process.

■■■**22.** A square conductive loop 50 cm on a side has a resistance of 200 mΩ. Find the rate at which a magnetic field perpendicular to the plane of the loop must change with time in order to cause an average of 5.0 J per unit time to be expended in the loop.

■■■**23.** A metal airplane with a wing span of 30 m flies horizontally at a constant speed of 320 km/h in a region where the vertical component of the Earth's magnetic field is 5.0×10^{-5} T. What is the induced motional emf across its wing tips?

20.2 Generators and Back Emf

■**24.** An ac generator is rated at 60 Hz. At what intervals of the period is the emf (a) a maximum, (b) zero, and (c) a minimum?

■**25.** A pivoted coil of wire is rotated 75 times per second in a uniform magnetic field. How often during each second does the induced emf in the coil have a value of zero?

■**26.** (a) What is the maximum value of the emf output from a simple ac generator having a single loop with an area of 90 cm^2 that rotates with a frequency of 60 Hz in a uniform magnetic field of 10^{-2} T? (b) What would the maximum value be if a coil of 10 such loops were used?

■■**27.** A simple ac generator consists of a coil having 10 turns of wire, with each loop having an area of 50 cm^2. The coil rotates in a uniform magnetic field of 0.25 T with a frequency of 60 Hz. (a) Write an equation showing how the emf of the generator varies as a function of time. (b) Compute the maximum emf.

■■**28.** A 60-Hz sinusoidal ac voltage has a maximum value of 120 V. What is the value of the voltage at $\frac{1}{180}$ s after it has a value of zero? (Is there more than one such value?)

■■**29.** A simple ac generator with a maximum emf value of at least 20 V is to be constructed, using loops of wire with

a radius of 0.10 m, a rotational frequency of 60 Hz, and a magnetic field of 10^{-2} T. How many loops of wire will be needed?

■■**30.** A simple ac generator has a rotational frequency of 60 Hz. What percentage of the maximum emf is the instantaneous emf (a) at 0.50 s after starting up and (b) at $\frac{1}{360}$ s after the emf passes through zero in changing to a negative polarity?

■■**31.** Which has the greater voltage output—a generator with a loop area of 100 cm^2 rotating in a magnetic field of 20 mT at 60 Hz or a generator with a loop area of 75 cm^2 rotating in a magnetic field of 200 mT at 120 Hz? Justify your answer mathematically.

■■**32.** A motor has a resistance of 4.0 Ω and draws a current of 2.5 A when operating at its normal speed on a 115-V line. What is the back emf of the motor?

■■**33.** A 120-V dc motor draws a current of 8.0 A and has a back emf of 96 V at its operating speed. (a) What starting current is required by the motor? (Assume that there is no additional resistance.) (b) What series resistance would be required to limit the starting current to 5.0 A?

■■■**34.** A coil with 200 concentric loops of wire with a radius of 16 cm is used as the armature of a generator with a magnetic field of 0.80 T. With what frequency should the armature be rotated to make the polarity change every 0.010 s?

■■■**35.** A 240-V dc motor with an armature whose resistance is 1.5 Ω draws a current of 16 A when running at its operating speed. (a) What is the back emf of the motor when it is operating normally? (b) What is the starting current? (Assume that there is no additional resistance.) (c) What series resistance would be required to limit the starting current to 20 A?

20.3 Transformers and Power Transmission

■■**36.** The number of turns on the secondary coil of an ideal transformer is 500 and on the primary it is 50. (a) What type of transformer is this? (b) What is the ratio of the current in the primary coil to the current in the secondary? (c) What is the ratio of the voltage in the primary coil to the voltage in the secondary?

■■**37.** A transformer has 800 turns on its primary coil and 600 turns on its secondary. (a) What type of transformer is this? (b) If the input to the primary coil is 4.0 A at 120 V, what is the output of the secondary?

■■**38.** An ideal transformer has 50 turns on its primary

coil and 250 turns on its secondary. If 3.0 A at 12 V is applied to the primary coil, what are (a) the current and (b) the voltage output of the secondary?

■■**39.** An ideal transformer steps up 8.0 V to 2000 V, and the 4000-turn secondary coil carries 2.0 A. (a) Find the number of turns on the primary coil. (b) Find the current in the primary.

■■**40.** The primary coil of an ideal transformer has 720 turns, and the secondary coil has 180 turns. If 20 A flows in the primary with a voltage of 110 V, what are (a) the voltage and (b) the current output of the secondary?

■■**41.** An ideal transformer has 120 turns on its primary coil and 840 turns on its secondary. If 2.5 A flows in the primary with a voltage of 120 V, what are (a) the voltage and (b) the current output of the secondary?

■■**42.** A transformer changes a 120-V input to a 6000-V output. Find the ratio of the number of turns on the primary coil to the number of turns on the secondary coil.

■■**43.** The primary coil of an ideal transformer is connected to a 120-V source and draws 10 A. The secondary coil has 800 turns and a current of 4.0 A. (a) What is the voltage on the secondary? (b) How many turns are on the primary?

■■**44.** An ideal transformer has 840 turns on its primary coil and 120 turns on its secondary. If the primary draws 1.2 A at 110 V, what are (a) the current and (b) the voltage output of the secondary?

■■**45.** A 600-V power line runs to a residential area, where a transformer with 100 turns on its primary coil steps down the voltage to 240 V. How many turns are on the secondary coil of this transformer?

■■**46.** A circuit component operates on 20 V at 0.50 A. A transformer with 300 turns on its primary coil is used to convert 120-V household electricity to the proper voltage. (a) How many turns must the secondary coil have? (b) How much current flows in the primary?

■■**47.** An electric arc-welding machine requires 200 A of current. The transformer of the machine has 1200 turns on the primary coil, which draws 2.5 A at 240 V. (a) How many turns are on the secondary coil? (b) What is the voltage output of the secondary?

■■**48.** An ac generator supplies 20 A at 440 V to a 10,000-V power line. If the step-up transformer has 132 turns on its primary coil, how many turns are on the secondary?

■■**49.** The electrical energy described in Problem 48 is

transmitted over an 80-km line ($R = 0.8\ \Omega$/km). How many kilowatt-hours are saved in 2.0 h by stepping up the voltage?

■■■**50.** At an area substation, the power-line voltage is stepped down from 100,000 V to 20,000 V. If 1.0 MW of power is delivered to the 20,000-V circuit, what are the primary and secondary currents in the transformer?

■■■**51.** The transformer on a utility pole steps down the voltage from 20,000 V to 220 V for home usage. If a household circuit uses 6.6 kW of power, what are the primary and secondary currents in the transformer?

■■■**52.** An electric-generating plant produces 50 A at 20 kV. The electricity is transmitted 25 km over transmission lines whose resistance is 1.2 Ω/km. (a) What would the power loss be if the energy is transmitted at 20 kV? (b) To what value should the output voltage of the generator be stepped up to decrease the energy loss by a factor of 15?

Additional Problems

53. The efficiency of a transformer is defined as the ratio of the power output to the power input: efficiency = $I_s V_s / I_p V_p$. Show that in terms of the ratios of currents and voltages given in Eq. 20.8, an efficiency of 100% is obtained. What does this imply?

54. A coil of wire with 10 turns and a cross-sectional area of 0.055 m^2 is placed in a magnetic field of 1.8 T and oriented so the area is perpendicular to the field. The coil is then flipped in 0.25 s and ends up with the area parallel to the field. What is the average emf induced in the coil?

55. The armature of a simple ac generator has 15 circular loops of wire with a radius of 10 cm. It is rotated with a frequency of 60 Hz in a uniform magnetic field of 800 mT. What is the maximum value of the emf induced in the loops, and how often is this value attained?

56. Suppose that the metal rod in Fig. 20.5 is 20 cm long and is moving at a speed of 10 m/s in a magnetic field of 1.5 T, but that the metal frame is covered with an insulating material. What would happen in this case? (Give a quantitative answer.)

57. A transformer in a door chime steps down the voltage from 120 V to 12 V and supplies a current of 0.50 A to the chime mechanism. (a) What is the turn ratio of the transformer? (b) What is the current input to the transformer?

58. The magnetic field perpendicular to the plane of a

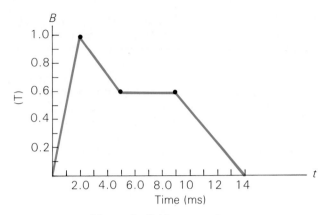

Figure 20.23 Magnetic field versus time
See Problem 58.

wire loop with an area of 0.10 m² changes with time as shown in Fig. 20.23. What is the average emf induced in the loop for each time interval on the graph (from 0 to 2.0 ms, for example)?

59. Electrical power is transmitted through a 250-km power line with a resistance of 1.2 Ω/km. The generator output is 50 A at its operating voltage. This voltage is stepped up for transmission. A step-down transformer at the power delivery point has a turn ratio of 500 and an output voltage of 220 V. (a) How much power is lost as joule heat? (b) What is the output voltage of the generator? (Neglect the voltage drop of the line.)

60. The armature of an ac generator has 100 turns, which are rectangular loops measuring 8.0 cm by 12 cm. The generator has a sinusoidal voltage output with an amplitude of 24 V. If the magnetic field of the generator is 100 mT, with what frequency does the armature turn?

61. The starter motor in an automobile has a resistance of 0.40 Ω in its armature windings. The motor operates on 12 V and has a back emf of 10 V when running at normal operating speed. How much current does the motor draw (a) when running at its operating speed and (b) when initially starting up?

62. A voltage of 230,000 V in a transmission line is reduced to 100,000 V at an area substation, to 7200 V at a distributing substation, and to 240 V at a utility pole outside a house. (a) What turn ratio is required for each reduction? (b) By what factor is the transmission line current stepped up in each voltage step-down and overall?

AC Circuits

21

Because alternating voltage and current are used in power transmission systems and in homes and industry, ac circuits are very common. In previous chapters, Ohm's law ($V = IR$) and expressions for power ($P = IV = I^2R$) were developed for direct current but also applied to situations in which there were alternating voltage and current. These forms of the relationships do apply to ac circuits, but the voltage and current must have special values.

Since ac voltage and current vary with time, ac circuits can have impeding effects in addition to ohmic resistance. You will learn that capacitors and coils (inductors) give rise to opposition to ac current in special ways. An example of the back emf of a motor was described in Chapter 20. Also, in certain ac circuits, there is an electrical analogue of mechanical resonance—the condition for maximum energy transfer.

21.1 Resistance in an AC Circuit

In general, an ac circuit is the one with an ac voltage source and one or more other circuit elements. The circuit diagram for an ac circuit with a single resistive element is shown in Fig. 21.1. If the source output is assumed to be sinusoidal, as is the case for a simple generator (Chapter 20), the voltage across the resistor varies with time according to this equation:

$$V = V_o \sin \omega t = V_o \sin 2\pi ft \qquad (21.1)$$

where ω is the angular frequency ($\omega = 2\pi f$). The voltage oscillates between maximum instantaneous values of $+ V_o$ and $- V_o$, where V_o is the **peak voltage.**

The current through the resistor also oscillates—its direction changes. From Ohm's lsw, the current can be expressed as a function of time:

$$I = \frac{V}{R} = \frac{V_o}{R} \sin 2\pi ft$$

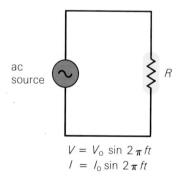

$$V = V_0 \sin 2\pi ft$$
$$I = I_0 \sin 2\pi ft$$

Figure 21.1 **A purely resistive circuit**
The ac source delivers a sinusoidal voltage to the circuit, and the voltage across and the current through the resistor are sinusoidal.

That is

$$I = I_o \sin 2\pi ft \qquad (21.2)$$

where $I_o = V_o/R$ is the **peak current.**

Both current and voltage for an ac circuit are plotted versus time in the graph in Fig. 21.2. Note that they are in step, or in phase, with each other, reaching zero values and maxima at the same times. The current oscillates back and forth and has equal positive and negative values during each cycle. Thus, the summation of all the instantaneous current values (positive and negative) over one complete cycle is zero, and the *average current is zero.* This does not mean that there is no joule heating (I^2R losses). The dissipation of electrical energy depends on the collisions of electrons with the atoms of the material in which the current is flowing; these occur when the electrons of an alternating current are going in either direction.

The instantaneous power is obtained using the instantaneous current (Eq. 21.2):

$$P = I^2R = I_o^2 R \sin^2 2\pi ft \qquad (21.3)$$

Even though the current changes sign each half-cycle, I^2 is always positive. Thus the average value of I^2R is *not* zero, even though the average value of I is zero. The average, or mean, value of I^2 is

$$\bar{I}^2 = \langle I_o^2 \sin^2 2\pi ft \rangle = I_o^2 \langle \sin^2 2\pi ft \rangle = \tfrac{1}{2}I_o^2 \qquad (21.4)$$

The average power is therefore

$$\bar{P} = \bar{I}^2 R = \tfrac{1}{2}I_o^2 R \qquad (21.5)$$

To write this in a form similar to the dc power ($P = I^2R$), we take the square root of $\bar{I}^2$, and with $1/\sqrt{2} = 0.707$ we have

$$I_{\text{rms}} = \sqrt{\bar{I}^2} = \sqrt{\tfrac{1}{2}I_o^2} = \frac{I_o}{\sqrt{2}} = 0.707 I_o \qquad (21.6)$$

This is called the rms current, or the *effective current* (rms stands for root-mean-square, in this case indicating the root of the mean value of the square of the current). The effective power is

$$\bar{P} = \tfrac{1}{2}I_o^2 R = I_{\text{rms}}^2 R \qquad (21.7)$$

By a similar development, the rms value of the ac voltage is

$$V_{\text{rms}} = \frac{V_o}{\sqrt{2}} = 0.707 V_o \qquad (21.8)$$

which is called the *effective voltage.*

Thus, for power dissipation through a resistance R, an alternating current and voltage with rms values equal to dc current and voltage values will have the same effect, or produce the same I^2R loss. For alternating current, Ohm's law is written as

$$V_{\text{rms}} = I_{\text{rms}}R \qquad (21.9)$$

It is customary to measure and specify rms values for ac quantities. For example, the household line voltage of 120 V is an rms value with a peak voltage of

The time *average value* of $\sin 2\pi ft$ over one or more complete cycles or periods is zero: $\langle \sin\theta \rangle = \langle \sin 2\pi ft \rangle = 0$ (where angular brackets are used to denote the average value instead of overbars). Similarly, $\langle \cos\theta \rangle = 0$.

The trigonometric identity $\sin^2\theta = \tfrac{1}{2}(1 - \cos 2\theta)$ gives $\langle \sin^2\theta \rangle = \tfrac{1}{2}$, since $\langle \cos 2\theta \rangle = 0$ just as $\langle \cos\theta \rangle = 0$.

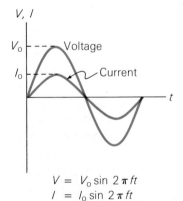

$$V = V_o \sin 2\pi ft$$
$$I = I_o \sin 2\pi ft$$

Figure 21.2 Voltage and current in phase
In a purely resistive circuit, the voltage and the current are in step, or in phase, with zero values and maxima occurring at the same times.

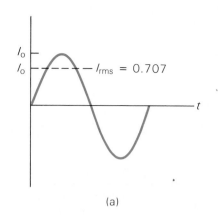

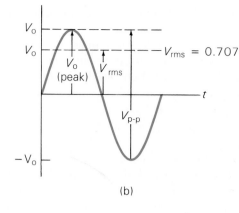

(a)

(b)

$$V_0 = \sqrt{2}\,V_{\text{rms}} = \frac{120\text{ V}}{0.707} = 170\text{ V}$$

The various mean and peak values of ac current and voltage are shown in the graphs in Fig. 21.3. Another voltage designation sometimes used is the peak-to-peak value.

Example 21.1   Rms Values

A lamp with a 100-W bulb is plugged into a 120-V outlet. (a) What are the rms and peak currents through the lamp? (b) What is the resistance of the bulb? (Neglect the resistance of the lamp's wire.) (c) What would happen if the lamp were plugged into a 240-V outlet?

Solution

Given: $\bar{P} = 100$ W   *Find*: (a) I_{rms} and I_0
$\quad\quad\quad V_{\text{rms}} = 120$ V   (b) R
$\quad\quad\quad\quad\quad\quad\quad\quad\quad\quad\quad$ (c) Describe the effect of 240 V

(a) The rms current is

$$I_{\text{rms}} = \frac{\bar{P}}{V_{\text{rms}}} = \frac{100\text{ W}}{120\text{ V}} = 0.833\text{ A}$$

and the peak current is

$$I_0 = \sqrt{2}\,I_{\text{rms}} = \frac{0.833\text{ A}}{0.707} = 1.18\text{ A}$$

(b) The resistance of the bulb is

$$R = \frac{V_{\text{rms}}}{I_{\text{rms}}} = \frac{120\text{ V}}{0.833\text{ A}} = 144\ \Omega$$

Note that the resistance could also be obtained from the peak values (V_0 for 120 V was computed above):

$$R = \frac{V_0}{I_0} = \frac{170\text{ V}}{1.18\text{ A}} = 144\ \Omega$$

(c) If the lamp were connected to a 240-V line, the voltage would be doubled, giving twice the current (since R is the same) and four times the power output

(400 W) because of the I^2 term in I^2R. This would melt the filament and blow out the bulb.

Fortunately, the mistake of plugging 120-V appliances into a 240-V line is not made because the plugs and sockets are different. In countries such as Great Britain where the normal line voltage is 240 V, you couldn't plug in a hair dryer or electric razor without using an adaptor, which contains a step-down transformer that converts 240 V to 120 V. ■

The formulas derived for dc circuits can be used for ac circuits if the rms, or effective values, of the quantities are inserted. For convenience and simplicity, the rms subscripts will be omitted from now on—but keep in mind that rms values are given for ac quantities unless otherwise specified.

Both alternating current and direct current are measured in amperes. But how is the ampere physically defined for alternating current? It cannot be derived from the mutual attraction between two parallel wires carrying ac current, as the dc ampere is defined (Chapter 19). An ac current changes direction with the source frequency (for example, 60 Hz), and the attractive force would average to zero. Thus, the ac ampere must be defined in terms of some property that is independent of the direction of the current. Joule heating is such a property, and *1 ampere of alternating current is said to flow in a circuit if the current produces the same average heating effect as 1 ampere of dc current would under the same conditions.* That is, the ampere is defined as an *effective* unit of ac current.

21.2 Capacitive Reactance

As you know, when a capacitor is connected to a voltage source in a dc circuit, current will flow for the short time required to charge the capacitor (Chapter 18). As charge accumulates on the capacitor's plates, the voltage across them increases, opposing the current. That is, a capacitor in a dc circuit will limit or oppose the current as it charges. When the capacitor is fully charged, the current in the circuit goes to zero.

When a capacitor is in a circuit with an ac source, as shown in Fig. 21.4(a), it limits, or regulates, the current, but does not completely prevent the flow. The capacitor is alternately charged and discharged as the current and voltage reverse each half-cycle.

Plots of ac current and voltage versus time for a circuit with a capacitor are shown in Fig. 21.4(b). Note that at $t = 0$, the voltage is zero. As the voltage starts to increase, there is a maximum current flow. With the plates initially uncharged, there is no opposition to current flow (other than the resistance of the connecting wires). However, when the voltage reaches its maximum value (in a quarter-cycle), the current goes to zero. During the next quarter-cycle, the voltage decreases and the current reverses the plates losing the previously accumulated charge. The process repeats during the next half-cycle with reversed polarities.

The current and voltage are not in step, or in phase, in this situation. The current reaches a maximum a quarter-cycle ahead of the voltage, which is stated in this way:

In a purely capacitive ac circuit, the current leads the voltage by 90°.

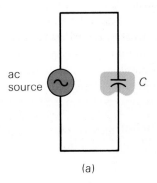

(a)

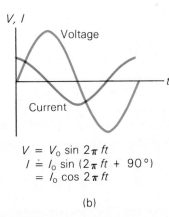

$$V = V_0 \sin 2\pi ft$$
$$I = I_0 \sin (2\pi ft + 90°)$$
$$= I_0 \cos 2\pi ft$$

(b)

Figure 21.4 A purely capacitive circuit
In a circuit with only capacitance, the current leads the voltage by 90°. (Notice a current zero is 90° or a quarter cycle ahead of a voltage zero.)

As in a dc circuit, there is opposition to the charging process in an ac circuit but it is not totally limiting (as in an open-circuit condition). The impeding effect of a capacitor on current flow in an ac circuit is expressed in terms of a quantity called the **capacitive reactance** (X_c):

$$X_c = \frac{1}{2\pi f C} \qquad (21.10)$$

Here C is the capacitance (in farads), and f is the frequency (in Hz). Like resistance, reactance is measured in ohms (Ω).

As you can see, the reactance is inversely proportional to the capacitance (C). Recall that capacitance is the charge per voltage ($C = Q/V$). For a particular voltage, therefore, the greater the capacitance, the more charge the capacitor can accommodate. Putting more charge on the plates requires a larger flow of charge, that is, a greater current, which means the opposition to the current, or the reactance, must be smaller. Also, the greater the frequency of the ac driving source, the shorter the charging time and the less the charge accumulation on the plates, and therefore the less the opposition to the current flow. Thus, the capacitive reactance is inversely proportional to the frequency. (Note that if $f = 0$, as is the case for dc, the capacitive reactance is infinite and there is no current flow, which corresponds to an open circuit.)

The capacitive reactance is related to the voltage across the capacitor and the current in the circuit by an equation that has the same form as Ohm's law:

$$V = IX_c \qquad (21.11)$$

(Remember that V and I are rms values.)

Example 21.2 Capacitive Reactance
A 15-μF capacitor is connected to a 120-V, 60-Hz source. What are (a) the capacitive reactance and (b) the current in the circuit?

Solution
Given: $C = 15\ \mu\text{F} = 15 \times 10^{-6}\ \text{F}$ *Find*: (a) X_c
         $V = 120\ \text{V}$ (b) I
         $f = 60\ \text{Hz}$

(a) The capacitive reactance is

$$X_c = \frac{1}{2\pi f C} = \frac{1}{2\pi (60\ \text{Hz})(15 \times 10^{-6}\ \text{F})} = 177\ \Omega$$

(b) Then, using Eq. 21.11, the current is

$$I = \frac{V}{X_c} = \frac{120\ \text{V}}{177\ \Omega} = 0.678\ \text{A} \quad \blacksquare$$

21.3 Inductive Reactance

Another important phenomenon in ac circuits is inductance. As you learned in Chapter 20, a coil (an inductor) in a circuit with a time-varying current has a reverse voltage, or back emf, set up in it as a result of a changing flux. This is known as self-induction. The induced emf opposes the change in flux (Lenz's law) and thus opposes the current flow in the circuit.

The self-induced emf is given by Faraday's law: $\mathscr{E} = -N\Delta\Phi/\Delta t$. But since the geometry of the coil is fixed, the time rate of change of the flux, or $\Delta\Phi/\Delta t$, is proportional to the time rate of change of the current in the coil, or $\Delta I/\Delta t$. Thus, the emf is

$$\mathscr{E} = -L\left(\frac{\Delta I}{\Delta t}\right) \qquad (21.12)$$

where L is the (self) **inductance** of the coil. As you can see from the equation, inductance has units of volt-second per ampere (V-s/A). This combination is called a **henry (H)**. The smaller unit of millihenry (mH) is commonly used (1 mH = 10^{-3} H).

The inductance of a coil is a constant that depends on the number of turns, the coil's diameter, its length, and the material of the core (if any). The greater the inductance, the greater the opposition to current flow. Also, the higher the frequency of the ac driving source, the greater the inductance opposition to current flow because of faster time variations.

The impeding effect of an inductor in an ac circuit is expressed in terms of **inductive reactance** (X_L):

$$X_L = 2\pi f L \qquad (21.13)$$

Here f is the frequency of the driving source and L is the inductance. Like capacitive reactance, inductive reactance is measured in ohms (Ω).

In terms of X_L, the equation for the voltage across an inductor has an Ohm's law form:

$$V = IX_L \qquad (21.14)$$

The circuit symbol for an inductor and the graphs of the voltage across the inductor and the current in the circuit are shown in Fig. 21.5. For this kind of circuit, when the voltage is a maximum, the current is zero; and when the voltage goes to zero, the current is a maximum. The current lags a quarter-cycle behind the voltage, which is also expressed like this:

In a purely inductive ac circuit, the voltage leads the current by 90°.

The phase relationships of current and voltage for purely inductive and purely capacitive circuits are opposite. A phrase that may help you remember the relationships is

ELI the ICE man

With E representing voltage, ELI indicates that with an inductance (L) the voltage leads the current (I). Similarly, ICE tells you that with a capacitance (C) the current leads the voltage.

Example 21.3   Inductive Reactance
A 125-mH inductor is connected to a 120-V, 60-Hz source. What are (a) the inductive reactance and (b) the current in the circuit?

Solution

Given: $L = 125$ mH = 125×10^{-3} H   Find: (a) X_L
$\qquad\quad V = 120$ V $\qquad\qquad$ (b) I
$\qquad\quad f = 60$ Hz

(a) The inductive reactance is

The henry was named for the American physicist Joseph Henry (1797–1879) in honor of his work on electromagnetic induction. Henry discovered this phenomenon independently at about the same time as Michael Faraday did, but Faraday published his findings first.

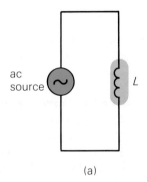

(a)

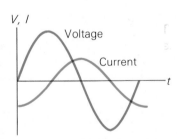

$V = V_0 \sin 2\pi ft$
$I = I_0 \sin(2\pi ft - 90°)$
$\quad = -I_0 \cos 2\pi ft$

(b)

Figure 21.5   A purely inductive circuit
In a circuit with only inductance, the voltage leads the current by 90°. (Notice a voltage zero is 90° or a quarter cycle ahead of a current zero.)

$$X_L = 2\pi f L = 2\pi (60 \text{ Hz})(125 \times 10^{-3} \text{ H}) = 47.1 \ \Omega$$

(b) Then, using Eq. 21.14 gives

$$I = \frac{V}{X_L} = \frac{120 \text{ V}}{47.1 \ \Omega} = 2.55 \text{ A} \quad \blacksquare$$

21.4 Impedance: RLC Circuits

The previous sections considered purely capacitive and inductive circuits, in which only *nonresistive* opposition to current flow was found. However, it is impossible to have purely reactive circuits, since there is always some resistance, at minimum that from the connecting wires or the wire in an inductor. Therefore, real ac circuits usually have appreciable resistance, and the current is impeded by both resistances and reactances (capacitive and/or inductive). Analysis of some simple circuits will illustrate these effects.

Series RC Circuit

Suppose that an ac circuit consists of a voltage source and a resistance and a capacitive element connected in series, as illustrated in Fig. 21.6(a). The phase difference between the current and the voltage is different for each of these circuit elements. As a result, a special method must be used to find the effective opposition to the current flow in the circuit. This is conveniently found by means of a **phase diagram**, such as the one shown in Fig. 21.6(b).

In a phase diagram, the resistance and reactance of the circuit are given vectorlike properties and their magnitudes are represented as arrows called **phasors**. On a set of xy coordinate axes, the resistance is plotted on the positive x axis, since the voltage-current phase difference for a resistor is zero ($\phi = 0$). The capacitive reactance is plotted along the negative y axis, to reflect a phase difference of $-90°$. (A negative phase angle implies that the voltage lags behind the current, as is the case for a capacitor.)

The phasor sum is the **impedance (Z)**, or the effective opposition to the current flow. Phasors are added in the same way vectors are, so for the series RC circuit,

$$Z = \sqrt{R^2 + X_c^2} \tag{21.15}$$

The unit for the impedance is the ohm.

The Ohm's law relationship for the RC circuit in terms of impedance is

$$V = IZ \tag{21.16}$$

Example 21.4 Capacitive Impedance
A series RC circuit has a 100-Ω resistor and a 15-μF capacitor. How much current flows in the circuit when driven by a 120-V, 60-Hz source?

Solution
Given: $R = 100 \ \Omega$ *Find*: I
$C = 15 \ \mu\text{F} = 15 \times 10^{-6} \text{ F}$
$V = 120 \text{ V}$
$f = 60 \text{ Hz}$

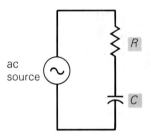

(a) Circuit diagram

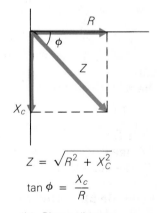

$$Z = \sqrt{R^2 + X_C^2}$$

$$\tan \phi = \frac{X_c}{R}$$

(b) Phase diagram

Figure 21.6 A series RC circuit
In a series RC circuit (a), the impedance Z is the phasor sum of the resistance R and the capacitive reactance X_c (b).

First, the current impedance is computed. From Example 21.2, the reactance for the capacitor is $X_c = 177\ \Omega$. Then

$$Z = \sqrt{R^2 + X_c^2} = \sqrt{(100\ \Omega)^2 + (177\ \Omega)^2} = 203\ \Omega$$

Since $V = IZ$, the current in the circuit is

$$I = \frac{V}{Z} = \frac{120\ \text{V}}{203\ \Omega} = 0.591\ \text{A} \quad \blacksquare$$

Series RL Circuit

The analysis of a series RL circuit (Fig. 21.7) is similar to that of a series RC circuit. However, the inductive reactance is plotted along the positive y axis in the phase diagram. (In this case, a positive phase angle implies that the voltage leads the current, as is the case for an inductor.)

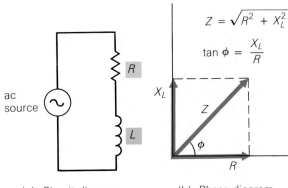

(a) Circuit diagram   (b) Phase diagram

$$Z = \sqrt{R^2 + X_L^2}$$

$$\tan \phi = \frac{X_L}{R}$$

Figure 21.7 **A series RL circuit**
In a series RL circuit (a), the impedance Z is the phasor sum of the resistance R and the inductive reactance X_L (b).

The impedance can be seen to be

$$Z = \sqrt{R^2 + X_L^2} \tag{21.17}$$

as for an RC circuit, $V = IZ$.

Series RLC Circuit

An ac circuit may contain all three circuit elements—resistor, inductor, and capacitor—as in Fig. 21.8. Again, phasor addition is used to determine the impedance by a component method. Summing the vertical components (the inductive and capacitive reactances) gives the total reactance ($X_L - X_c$, since X_c is defined as a negative phasor). The impedance is then the phasor sum of the resistance and the total reactance. From the phasor diagram, you can see that this is

$$Z = \sqrt{R^2 + (X_L - X_c)^2} \tag{21.18}$$

Again, $V = IZ$.

In this case, the **phase angle** (ϕ) between the voltage from the source and the current is given by

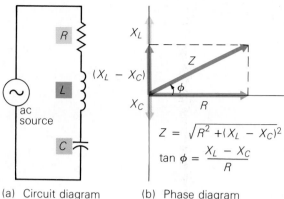

(a) Circuit diagram

(b) Phase diagram

$$Z = \sqrt{R^2 + (X_L - X_C)^2}$$

$$\tan \phi = \frac{X_L - X_C}{R}$$

Figure 21.8 **A series RLC circuit**
In a series RLC circuit (a), the impedance Z is the phasor sum of the resistance R and the total reactance $(X_L - X_C)$, as shown in (b).

$$\tan \phi = \frac{X_L - X_C}{R} \qquad\qquad (21.19)$$

Note that if X_L is greater than X_C (as is true in Fig. 21.8), the phase angle is positive $(+\phi)$, and the circuit is said to be an **inductive circuit**. If X_C is greater than X_L, the phase angle is negative $(-\phi)$, and the circuit is said to be a **capacitive circuit**.

A summary of impedances and phase angles for the three circuit elements and combinations considered is given in Table 21.1.

Example 21.5 Impedance in an RLC Circuit

A series RLC circuit has a resistance of 25 Ω, a capacitance of 50 μF, and an inductance of 0.30 H. If the circuit is driven by a 120-V, 60-Hz source, what are (a) the impedance of the circuit, (b) the current in the circuit, and (c) the phase angle between the current and the voltage supplied?

Solution

Given: $R = 25\ \Omega$ *Find:* (a) Z

$C = 50\ \mu\text{F} = 5.0 \times 10^{-5}\ \text{F}$ (b) I

$L = 0.30\ \text{H}$ (c) ϕ

$V = 120\ \text{V}$

$f = 60\ \text{Hz}$

Table 21.1
Impedances and Phase Angles

Circuit element(s)		Impedance, Z (in Ω)	Phase angle, ϕ
R		R	$0°$
C		X_C	$-90°$
L		X_L	$+90°$
RC		$\sqrt{R^2 + X_C^2}$	$-\phi$
RL		$\sqrt{R^2 + X_L^2}$	$+\phi$
RLC		$\sqrt{R^2 + (X_L - X_C)^2}$	$+\phi$, if $X_L > X_C$ $-\phi$, if $X_C > X_L$

(a) The reactances are

$$X_c = \frac{1}{2\pi fC} = \frac{1}{2\pi(60\text{ Hz})(5.0 \times 10^{-5}\text{ F})} = 53\ \Omega$$

and

$$X_L = 2\pi fL = 2\pi(60\text{ Hz})(0.30\text{ H}) = 113\ \Omega$$

Then

$$Z = \sqrt{R^2 + (X_L - X_c)^2}$$
$$= \sqrt{(25\ \Omega)^2 + (113\ \Omega - 53\ \Omega)^2} = 65\ \Omega$$

(b) Since $V = IZ$,

$$I = \frac{V}{Z} = \frac{120\text{ V}}{65\ \Omega} = 1.8\text{ A}$$

(c) Solving $\tan\phi = (X_L - X_c)/R$ for the phase angle gives

$$\phi = \tan^{-1}\left(\frac{X_L - X_c}{R}\right) = \tan^{-1}\left(\frac{113\ \Omega - 53\ \Omega}{25\ \Omega}\right) = 67° \quad\blacksquare$$

Power Factor

The power, or energy dissipated per unit time, is an interesting feature of an RLC circuit. *There are no power losses associated with pure capacitances and inductances in an ac circuit.* Capacitors and inductors simply store energy and then give it back. For example, during a half-cycle, a capacitor in an ac circuit is charged, and energy is stored in the electric field between the plates. During the next half-cycle, the capacitor discharges and returns the charge to the voltage source. Similarly, in an inductor, the energy is stored in the magnetic field associated with the induced current, or back emf, which builds up and reverses each cycle. (In Chapter 16, you saw that the energy stored in a capacitor was given by $U = \frac{1}{2}CV^2$. Similarly, it can be shown that the energy stored in an inductor is given by $U = \frac{1}{2}LI^2$, where L is the inductance and I is the current in the inductor.)

Thus, *the only element that dissipates energy in an RLC circuit is the resistive element.* As you learned earlier, the average power dissipated by a resistance R is $P = I^2R$ (where I is the rms value). The power can also be expressed in terms of the current and voltage, but the voltage in this case *must* be that across the resistive element (V_R), since this is the only dissipative element. That is, the power dissipated in an RLC circuit is

$$P = IV_R$$

The voltage across the resistive element is conveniently found from a voltage triangle that corresponds to the phasor triangle (see Fig. 21.9). As you can see from the figure, the voltage across the resistive element (V_R) depends on the phase angle:

$$V_R = V\cos\phi \qquad (21.20)$$

The term $\cos\phi$ is called the **power factor**, and from Fig. 21.9,

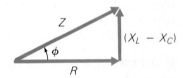

(a) Phasor triangle

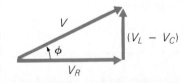

(b) Equivalent voltage triangle

Figure 21.9 **Phasor and voltage triangles**
Since the voltages across the components of a series RLC circuit are given by $V = IZ$, $V = IR$, and $V = IX$; and the current is the same through all three, the phasor triangle (a) may be converted to a voltage triangle (b). Note that $V_R = V\cos\phi$.

$$\cos\phi = \frac{R}{Z} \qquad\qquad (21.21)$$

Note that $\cos\phi$ varies from 1 to 0. When $\phi = 0$ and $\cos\phi = 1$, the circuit is said to be completely resistive. That is, there is maximum power dissipation (as though the circuit contained only a resistor). The power factor decreases as the phase angle increases [either $+\phi$ or $-\phi$, since $\cos(-\phi) = \cos\phi$], and the circuit becomes more inductive or capacitive. At $\phi = \pm 90°$, the circuit is completely inductive or capacitive. For these cases, it is as though the circuit contained only an inductor or a capacitor, respectively, and no power is dissipated ($P = IV\cos 90° = 0$).

Example 21.6 Power Factor
How much power is dissipated in the circuit described in Example 21.5?

Solution
In Example 21.5, the circuit was found to have an impedance of $Z = 65\,\Omega$, and its resistance is $R = 25\,\Omega$. The power factor of the circuit is therefore

$$\cos\phi = \frac{R}{Z} = \frac{25\,\Omega}{65\,\Omega} = 0.38$$

Using other data from Example 21.5 gives

$$P = IV\cos\phi = (1.8\text{ A})(120\text{ V})(0.38) = 82\text{ W} \quad \blacksquare$$

21.5 Circuit Resonance

When the power factor of an RLC circuit is equal to 1 ($\cos\phi = 1$), there is maximum power dissipation. For a particular voltage, the current in the circuit is then a maximum, since the impedance is a minimum. This occurs when the inductive and capacitive reactances effectively cancel each other.

The inductive and capacitive reactances are both frequency-dependent, so the impedance also depends on the frequency. From the expression for the impedance,

$$Z = \sqrt{R^2 + (X_L - X_c)^2}$$

the impedance is a minimum when

$$X_L = X_c$$

or $$2\pi f_o L = \frac{1}{2\pi f_o C}$$

Thus,

$$f_o = \frac{1}{2\pi\sqrt{LC}} \qquad\qquad (21.22)$$

Eq. 21.22 gives the frequency for the condition of minimum impedance, which is called the **resonance frequency**.

Thus, when a series RLC circuit is driven in resonance (at its resonance frequency), the impedance is a minimum, and the power transferred from the

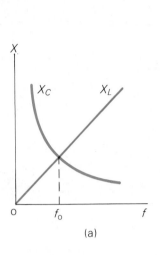

(a)

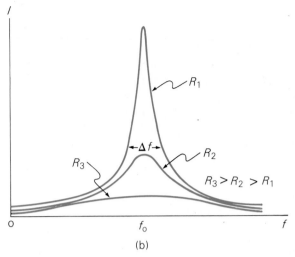

(b)

Figure 21.10 **Resonance frequency**
(a) At the resonance frequency (f_o), the reactances are equal ($X_L = X_c$). On a graph of X versus f, this is the frequency at which the X_c and X_L curves intersect. (b) On a graph of I versus f, the current is a maximum at f_o. The curve becomes sharper and narrower as the resistance decreases. For comparison, Δf, the full-width at half maximum, or the half-width of the resonance, is sometimes given.

source to the circuit is at a maximum. If $X_L = X_c$, then $Z = R$, or the impedance is completely resistive. A graph of the current versus frequency is shown in Fig. 21.10(b) for several different resistances in an RLC circuit with fixed capacitance and inductance. Note how the curve becomes sharper and narrower as the resistance decreases.

Resonant circuits have a wide variety of applications. Probably most common is in the tuning mechanism of a radio. Each radio station has an assigned broadcast frequency at which radio waves are transmitted. Signals of several different frequencies generally reach a radio, but the receiver circuit only picks up the signal whose frequency is at or near its resonance frequency (with a transfer of power to the circuit). A radio receiver circuit usually has a variable air capacitor so the resonance frequency of the circuit can be changed, allowing different stations to be tuned in (Fig. 21.11).

The broadcast frequency is the frequency of the carrier wave. The sound waves are electrically superimposed on this wave either by amplitude modulation (AM) or frequency modulation (FM).

Example 21.7   Electrical Resonance

A series RLC circuit has a 50-Ω resistor, a 0.0060-μF (or 6.0 pF) capacitor, and a 28-mH inductor. The circuit is connected to a wide-range, adjustable-frequency source with an output of 25 V. (a) What is the resonance frequency of the circuit? (b) How much current flows in the circuit when it is in resonance? (c) What is the voltage across each of the circuit elements for this condition?

Solution

Given: $R = 50\ \Omega$
$C = 0.0060\ \mu\text{F} = 6.0 \times 10^{-9}\ \text{F}$
$L = 28\ \text{mH} = 28 \times 10^{-3}\ \text{H}$
$V = 25\ \text{V}$

Find: (a) f_o
(b) I
(c) V_R, V_c, V_L

(a) By Eq. 21.22,

$$f_o = \frac{1}{2\pi\sqrt{LC}} = \frac{1}{2\pi\sqrt{(28 \times 10^{-3}\ \text{H})(6.0 \times 10^{-9}\ \text{F})}}$$

$$= 12.3 \times 10^3\ \text{Hz} = 12.3\ \text{kHz}$$

(b) At the resonance frequency, $X_L = X_c$, and $Z = R = 50\ \Omega$. The current in the circuit is then

$$I = \frac{V}{Z} = \frac{V}{R} = \frac{25\text{ V}}{50\ \Omega} = 0.50\text{ A}$$

(c) The reactances at f_o are

$$X_L = 2\pi f_o L = 2\pi(12.3 \times 10^3\text{ Hz})(28 \times 10^{-3}\text{ H}) = 2200\ \Omega$$

and $X_C = X_L = 2200\ \Omega$ at f_o. Then, using Ohm's law of relationships for voltage and current,

$$V_R = IR = (0.50\text{ A})(50\ \Omega) = 25\text{ V}$$

$$V_L = IX_L = (0.50\text{ A})(2200\ \Omega) = 1100\text{ V}$$

$$V_C = IX_C = (0.50\text{ A})(2200\ \Omega) = 1100\text{ V}$$

Note that the voltages across the capacitor and the inductor are significantly higher than the source voltage. Can you explain where the additional voltage comes from? ∎

Figure 21.11 **Variable air capacitor**
Rotating the movable plates between the fixed plates changes the overlap area and the capacitance. Such capacitors are commonly used in radio-tuning circuits.

When the resistance (R) in an RLC circuit is very small (there's always some), the circuit is an LC circuit. Energy in such a circuit oscillates back and forth between the inductor and capacitor at a frequency of f_o (and some energy is dissipated because of the small resistance). In an ideal LC circuit with no ohmic resistance, the oscillation would continue indefinitely.

Electrical resonance is analogous to mechanical resonance (Chapter 14). Maximum energy transfer to a mechanical system occurs when it is driven at its resonance frequency. Similarly, the oscillation of energy between a capacitor and an inductor is analogous to the exchange of kinetic and potential energy for a mass oscillating on a spring.

Oscillating resonance circuits are used in electronic devices where an output signal of a particular frequency is required. In operation, a small ac voltage of the proper frequency must be applied to the circuit to maintain strong, continuous oscillations. Otherwise, the electrical oscillator would run down as a result of I^2R and radiation losses (as its mechanical counterpart would be slowed down by friction).

Important Formulas

Instantaneous voltage in an ac circuit:

$$V = V_o \sin 2\pi f t$$

Instantaneous current in an ac circuit (where $I_o = V_o/R$):

$$I = I_o \sin 2\pi f t$$

Effective power in an ac circuit:

$$\bar{P} = I_{rms}^2 R$$

Ohm's law for ac:

$$V_{rms} = I_{rms} R$$

Effective (or rms) current:

$$I_{rms} = \frac{I_o}{\sqrt{2}} = 0.707 I_o$$

Effective (or rms) voltage:

$$V_{rms} = \frac{V_o}{\sqrt{2}} = 0.707 V_o$$

Capacitive reactance and Ohm's law:

$$X_c = \frac{1}{2\pi f C} \quad \text{and} \quad V = IX_c$$

Inductive reactance and Ohm's law:

$$X_L = 2\pi f L \quad \text{and} \quad V = IX_L$$

Impedance for an RLC circuit:

$$Z = \sqrt{R^2 + (X_L - X_c)^2}$$

Ohm's law for RC, RL, and RLC circuit:

$V = IZ$

Power factor:

$$\cos \phi = \frac{R}{Z}$$

Resonance frequency:

$$f_{\mathrm{o}} = \frac{1}{2\pi\sqrt{LC}}$$

Questions

Resistance in an AC Circuit

1. What is meant by peak current and peak voltage? What is the peak-to-peak voltage?

2. For the circuit in Fig. 21.1, why is the magnitude of V_{o} greater than I_{o}? Could it be the other way with the magnitude of I_{o} greater than V_{o}?

3. If the average current in an ac circuit is zero, why isn't the average power zero?

4. Is the 120-V line voltage commonly found in homes a peak or rms value? Explain.

5. Explain how the equation for the time-varying ac power, analogous to $P = I^2 R$ for dc power, is derived.

Capacitive and Inductive Reactances

6. Show explicitly that the unit for the capacitive and inductive reactances (X_c and X_L) is the ohm (Ω).

7. In an ac circuit, what can oppose current flow other than ohmic resistance and why?

8. Why is capacitive reactance inversely proportional to the capacitance and frequency?

9. Why is inductive reactance directly proportional to the inductance and frequency?

10. Give the phase difference between the current and voltage for (a) a purely resistive circuit, (b) a purely capacitive circuit, and (c) a purely inductive circuit.

11. Analyze the graph in Fig. 21.5(b) and explain why the current is zero when the voltage is a maximum, and vice versa. (Hint: $\Delta I/\Delta t$ is the slope of the curve for the current.)

Impedance: RLC Circuits

12. What is the impedance, and when is it a maximum?

13. When is a RLC circuit said to be (a) inductive and (b) capacitive? Are there any other possibilities?

14. What elements dissipate power in an RLC circuit?

15. Why can't the power in a RLC circuit be expressed generally as $P = IV$, where $V = IZ$?

16. In Fig. 21.9, why can the voltage be written as a phasor triangle?

17. What is the power factor and what does it tell you?

Circuit Resonance

18. How are the power factor and the resonance frequency related?

19. If the capacitance and the inductance of an RLC circuit are varied, how is the resonance frequency affected?

20. If the resistance in an RLC circuit were zero and the circuit were driven with a variable frequency effective voltage, what would a graph of I versus f look like? What would this mean physically?

Problems

21.1 Resistance in an ac Circuit

■**1.** What is the peak voltage of a 240-VAC line?

■**2.** An ac circuit has a rms current of 2.5 A. What is the peak current?

■**3.** The rms current in an ac circuit is 0.25 A. What is the maximum instantaneous current?

■**4.** The maximum potential difference across a resistor in an ac circuit is 156 V. Find the effective voltage.

■**5.** Based on the definition of the ac ampere, how much ac current must flow through a 6.0-Ω resistor to produce 15 J/s of joule heat?

■■**6.** An ac circuit with a resistance of 5.0 Ω has an effective current of 0.80 A. (a) Find the rms voltage and peak voltage. (b) Find the average power dissipated by the resistance.

■■**7.** A hair dryer rated at 1200 W is plugged into a 120-V outlet. (a) Find the effective current. (b) Find the peak current. (c) Find the resistance of the dryer.

8. The voltage across a resistor varies according to this equation:

$$V = (120 \text{ V})(\sin 120t)$$

What are the (a) effective voltage across the resistor and (b) the frequency with which the voltage changes polarity?

9. An ac voltage is applied to a 5.0-Ω resistor, and it dissipates 500 W of power. Find (a) the effective and peak currents and (b) the effective and peak voltages.

10. The current through a 60-Ω resistor is given by this equation:

$$I = 2.0 \sin 380t$$

where I is in amps and t is in seconds. (a) What is the frequency of the current? (b) What is the effective current? (c) How much power is dissipated by the resistor? (d) Write an equation expressing the voltage as a function of time.

11. Prove that $\langle \sin^2 \omega t \rangle = \frac{1}{2}$.

12. Find the effective and peak currents through a 40-W, 120-V light bulb.

13. A 50-kW heater is connected to a 220-V ac source. Find (a) the peak current and (b) the peak voltage.

14. The current and voltage outputs of an ac generator have peak values of 1.5 A and 12 V. What is the effective power output of the generator?

15. A coil of 50 loops of wire with a diameter of 30 cm rotates at 1000 rpm about an axis along a diameter and perpendicular to a uniform magnetic field of 0.60 T. What is the effective voltage produced in the coil?

16. A coil of 200 loops of wire with a radius of 10 cm is used as the armature of a simple generator with a magnetic field of 0.80 T. If the coil has a resistance of 2.0 Ω, how fast must it be rotated to have an effective induced current of 2.7 A?

17. Rotating coils are often used to measure magnetic fields. A coil of 50 loops with a radius of 7.5 cm rotates at 60 Hz about an axis along a diameter perpendicular to a uniform magnetic field. An *effective* voltage of 10 V, as read from an ac voltmeter, is induced in the coil. What is the magnitude of the magnetic field?

21.2 and 21.3 Capacitive and Inductive Reactances

18. A 1.0-μF capacitor is connected to a 120-V, 60-Hz source. (a) What is the capacitive reactance of the circuit? (b) How much current flows in the circuit? (c) What is the phase angle between the current and the supplied voltage?

19. What capacitance is needed in a 60-Hz ac circuit to have a reactance of 50 kΩ?

20. Find the frequency at which a 100-μF capacitor will have a reactance of 25 Ω.

21. A 4.0-μF capacitor is connected across a 60-Hz voltage source and a current of 2.0 mA is measured on an ammeter. What is the capacitive reactance of the circuit?

22. A 100-mH inductor is connected in a circuit with a 120-V, 60-Hz source. (a) What is the inductive reactance of the circuit? (b) How much current flows in the circuit? (c) What is the phase angle between the current and the supplied voltage?

23. An inductor has a reactance of 95 Ω in a 60-Hz ac circuit. What is the inductance of the inductor?

24. Find the frequency at which a 500-mH inductor will have a reactance of 200 Ω.

25. With a 150-mH inductor in a circuit with a 60-Hz voltage source, a current of 1.6 A is measured on an ammeter. (a) What is the voltage of the source? (b) What is the phase angle between the current and that voltage?

26. A 120-V, 60-Hz source is connected across a 0.70-H inductor. (a) Find the current through the inductor. (b) Find the phase angle between the current and the supplied voltage.

27. What inductance will have the same reactance in a 120-V, 60-Hz circuit as a 10-μF capacitor does?

21.4 and 21.5 Circuit Impedance and Resonance

28. A series RC circuit has a resistance of 200 Ω and a capacitance of 50 μF and is driven by a 120-V, 60-Hz source. (a) Find the capacitive reactance of the circuit. (b) How much current is drawn from the source?

29. To reduce the impedance in the RC circuit in Problem 28 by 2.0% while maintaining the 120-V source voltage, what would the required source frequency be?

30. In a series RC circuit with a 150-Ω resistor, what capacitance will give a current of 0.50 A if the circuit is driven by a 120-V, 60-Hz source?

31. What is the phase angle between the current and the source voltage in Problem 30?

32. A coil in a 60-Hz circuit has a resistance of 100 Ω and an inductance of 0.70 H. Calculate (a) the coil's reactance and (b) the circuit's impedance.

■■33. A circuit connected to a 110-V, 60-Hz source contains a single coil with an inductance of 100 mH and a resistance of 40 Ω. Find (a) the reactance of the coil, (b) the impedance of the circuit, (c) the current in the circuit, and (d) the power dissipated by the coil.

■■34. Calculate the phase angle between the current and the supplied voltage in Problem 33.

■■35. A series RL circuit has a resistance of 100 Ω and an inductance of 0.20 H. The circuit is driven by a 120-V, 60-Hz source. (a) What is the inductive reactance of the circuit? (b) How much power is dissipated by the circuit?

■■36. A coil with a resistance of 30 Ω and an inductance of 0.15 H is connected to a 120-V, 60-Hz source. (a) How much current flows in the circuit? (b) What is the phase angle between the current and the source voltage?

■■37. A series RLC circuit has a resistance of 200 Ω, an inductance of 0.30 H, and a capacitance of 5.0 μF. (a) What is the impedance of the circuit? (b) Find the resonance frequency of the circuit.

■■38. A circuit connected to a 220-V, 60-Hz power supply has the following components connected in series: a 10-Ω resistor, a coil with an inductive reactance of 120 Ω, and a capacitor with a reactance of 120 Ω. Compute the potential difference across (a) the resistor, (b) the inductor, and (c) the capacitor.

■■39. In a series RLC circuit with an inductance of 750 mH, what values of resistance and capacitance would give the circuit a resonance frequency of 60 Hz?

■■40. A series RLC circuit has a resistance of 30 Ω, an inductance of 0.25 H, and a capacitance of 8.0 μF. At what frequency should the circuit be driven to have the maximum power transferred from the driving source?

■■41. In a series RLC circuit, for a particular driving frequency, $R = X_c = X_L = 50$ Ω. If the driving frequency is doubled, what will the impedance of the circuit be?

■■42. A resistor, an inductor, and a capacitor have values of 500 Ω, 500 mH, and 3.5 μF, respectively, and are connected in series to a 240-V, 60-Hz power supply. Is the circuit driven in resonance? If not, what values of resistance and inductance would be required for this to occur?

■■43. How much power is dissipated in the circuit described in Problem 42?

■■■44. The circuit in Fig. 21.12(a) is called a low-pass

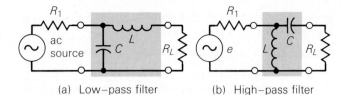

(a) Low-pass filter　　(b) High-pass filter

Figure 21.12 Low-pass and high-pass filters
See Problem 44.

filter because a large current is delivered to the load (R_L) only by a low-frequency source. The circuit in Fig. 21.12(b), on the other hand, is called a high-pass filter because a large current is delivered to the load only by a high-frequency source. Show mathematically why the circuits have these characteristics.

■■■45. A tuning circuit in a radio receiver has a fixed 2.5-mH inductance and a variable capacitor. If the circuit is tuned to a radio station broadcasting at 980 kHz on the AM dial, what is the capacitance of the capacitor?

■■■46. A series RLC circuit has components with these values: $R = 50$ Ω, $L = 0.10$ H, and $C = 20$ μF. The circuit is driven by a 120-V, 60-Hz source. What is the power loss of the circuit as a percentage of its power loss when in resonance?

■■■47. If the values of all three circuit components in Problem 45 were halved, what would be the change in the percentage power loss?

Additional Problems

48. An ac voltage has an amplitude of 85 V and a frequency of 60 Hz. (a) If $V = 0$ at $t = 0$, what is the instantaneous voltage at $t = 2.0$ s? (b) What is the effective voltage?

49. A series RLC circuit has a 25-Ω resistor, a 1.0-μF capacitor, and a 350-mH inductor. The circuit is connected to a variable-frequency source with a constant output of 12 V. If the supplied frequency is set at the circuit's resonance frequency, what is the voltage across each of the circuit elements?

50. Using the fact that the self-induced emf in a coil is given by $\mathscr{E} = -L\,\Delta I/\Delta t$, show that $L = N\,\Delta\Phi/\Delta I$, where Φ is the flux through the coil and N is the number of turns.

51. A radio receiver circuit containing an inductor of 150 mH is tuned to an FM station at 98.9 MHz by adjusting a variable capacitor. What is the capacitance of the capacitor when the circuit is tuned to the station?

52. An ac voltage has a rms value of 120 V. The voltage

goes from zero to its maximum value in 1.5 ms. Write an expression for the voltage as a function of time.

53. A machine operates on 240 V at 60 Hz. In operation, it uses 550 W of power, and the power factor is 0.75. Find the effective current in the circuit containing the machine.

54. A circuit has a current (in amps) given by $I = 8.0 \sin 4\pi t$ with an applied voltage (in volts) of $V = 60 \sin 40t$. What is the average power delivered to the circuit?

55. An RL circuit operates on 120 V at 60 Hz and contains a 60-Ω resistor. If the wattage rating of the resistor

is 5.0 W, what inductance should be used in the circuit so power is expended in the resistor at that rated limit?

56. A variable capacitor in a circuit with a 120-V, 60-Hz source is initially at a value of 0.25 μF and is then increased to 0.40 μF. What is the percentage change in the current in the circuit?

57. A series RLC circuit with a resistance of 400 Ω has capacitive and inductive reactances of 300 Ω and 500 Ω, respectively. (a) What is the power factor of the circuit? (b) If the circuit operates at 60 Hz, what additional capacitance should be connected with the original capacitance to give a power factor of zero, and how should the capacitances be connected?

Geometrical Optics: Reflection and Refraction of Light

22

Optics is the study of light and vision. Vision, of course, requires light, specifically what is called visible light. The broader meaning of light is any radiation in or near the visible region of the electromagnetic spectrum. Ultraviolet radiation, which is invisible to us, is still light. Properties shared by all the radiation in or near the visible region are a wave nature and the ability to reflect from mirror surfaces. For example, infrared radiation is reflected from a mirror surface.

The wave nature of light can be used to explain a variety of phenomena. In fact, wave properties are useful for analyzing many phenomena. In many instances, light and sound (also a wave) exhibit similar behaviors. However, a geometrical approach is used in this chapter to explain the reflection and refraction of light, which are the bases of many optical phenomena, for example, the formation of rainbows. Mirrors and lenses will be considered in Chapter 23.

22.1 Wave Fronts and Rays

Waves, electromagnetic or other, are conveniently described using wave fronts. A **wave front** is the line or surface defined by adjacent portions of a wave that are in phase. For example, if a line is drawn along one of the crests of a circular water wave moving out from a point source, all the particles on the line will be in phase [Fig. 22.1(a)]. A line along a wave trough would work equally well (and would also form a circle). For a three-dimensional spherical wave, such as sound or light emitted from a point source, the wave front is a spherical surface rather than a circle. At a distance far from the source, a short segment of a circular or spherical wave front may be approximated as a **plane wave front** [Fig. 22.1(b)]. A plane wave front may also be produced directly by a linear, elongated source. In a uniform medium, wave fronts propagate outward from the source at a particular wave speed. For light in a vacuum, this speed is $c = 3.0 \times 10^8$ m/s.

The geometrical description of a wave using wave fronts tends to neglect the actual characteristics of the wave. This simplification is carried a step further

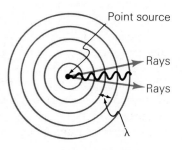

(a) Circular wave fronts

(b) Plane wave fronts

Figure 22.1 Wave fronts and rays
A wave front is defined by adjacent points on a wave that are in phase, such as those along wave crests or troughs. (a) Near a point source, the wave fronts are circular (two-dimensional) or spherical (three-dimensional). (b) Far from the source, the wave fronts are approximately linear or planar. A line perpendicular to a wave front in the direction of the wave's propagation is called a ray.

by the concept of a ray. As illustrated in Fig. 22.1, an arrow drawn perpendicularly to a series of wave fronts and pointing in the direction of propagation is called a **ray**. Note that a ray is in the direction of the energy flow of a wave. A plane wave front is assumed to travel in a straight line in a medium in the direction of its rays (Fig. 22.2). A beam of light may be represented by a group of parallel rays or simply as a single ray. The representation of light as rays is adequate and convenient for explaining some optical phenomena.

The use of the geometrical representations of wave fronts and rays to explain phenomena such as the reflection and refraction of light is called **geometrical optics**. However, certain other phenomena, such as the interference of light, cannot be treated in this manner and must be explained in terms of actual wave characteristics. These phenomena will be considered in Chapter 24.

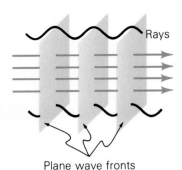

Figure 22.2 **Light rays** A plane wave front travels in a straight line in the direction of its rays. A beam of light may be represented by a group of parallel rays (or a single ray).

22.2 Reflection

The reflection of light is an optical phenomenon of enormous importance—if light did not reflect off objects around us, we couldn't see them at all. **Reflection** involves the absorption and emission of light by means of complex electronic vibrations in the atoms of the reflecting medium. However, the phenomenon is easily described using rays.

A light ray incident on a surface is described by an **angle of incidence** (θ_i). This is measured relative to a line perpendicular, or normal, to the reflecting surface, a line commonly referred to as a normal (see Fig. 22.3). Similarly, the reflected ray is described by an **angle of reflection** (θ_r). The relationship between these angles is given by the **law of reflection**:

The angle of incidence is equal to the angle of reflection; that is,

$$\theta_i = \theta_r \tag{22.1}$$

Another condition for reflection is that the incident ray, the reflected ray, and the normal all lie in the same plane.

When the reflecting surface is smooth, the reflected rays of a beam of light are parallel (Fig. 22.4). This is called **regular reflection**, or **specular reflection**. Reflection from a flat mirror is regular. If the reflecting surface is rough, on the other hand, the light rays are reflected in nonparallel directions because of the irregular nature of the surface. This is termed **irregular reflection**, or **diffuse reflection**. The reflection of light from this page is an example of diffuse reflection. Note in Fig. 22.4 on page 600 that the law of reflection applies to both types of reflection.

Experience with friction and direct investigations show that all surfaces are rough on a microscopic scale. What then determines whether reflection is specular or diffuse? In general, if the dimensions of the surface irregularities are greater than the wavelength of the light, the reflection is diffuse. Therefore, to make a good mirror, glass (with a metal coating) or metal must be polished until the surface irregularities are about the same size as the wavelength of light. Recall that the wavelengths of visible light are on the order of 10^{-5} cm.

It is because of diffuse reflection that we are able to see illuminated objects, for example, the moon. If the moon's spherical surface were smooth, only the reflected sunlight from a small region would come to an observer on Earth, and

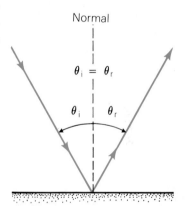

Figure 22.3 **The law of reflection** According to the law of reflection, the angle of incidence (θ_i) is equal to the angle of reflection (θ_r). Note that the angles are measured relative to a normal (a line perpendicular to the reflecting surface).

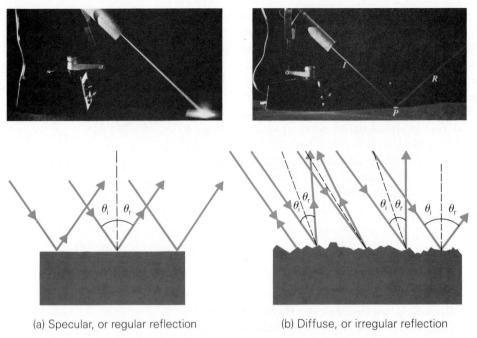

(a) Specular, or regular reflection   (b) Diffuse, or irregular reflection

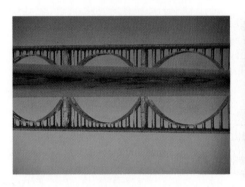

only a small illuminated area would be seen. Also, you can see the beam of light from a flashlight or spotlight because of diffuse reflection from dust and particles in the air (Fig. 22.5).

22.3 Refraction

Refraction refers to the change in direction of a wave at a boundary where it passes from one medium into another. In general, when a wave is incident on a boundary between media, some of its energy is reflected and some is transmitted. For example, when light traveling in air is incident on a transparent material such as glass (Fig. 22.6), it is partially reflected and partially transmitted. But the direction in which the transmitted light is propagated is different from the direction of the incident light, and the light is said to have been refracted, or bent.

This phenomenon can be analyzed conveniently using a geometrical method developed by the Dutch physicist Christian Huygens (1629–1696). **Huygens' principle** for wave propagation is

Every point on an advancing wave front can be considered to be a source of secondary waves, or wavelets, and the line or surface tangent to all these wavelets defines a new position of the wave front.

Huygens' principle is applied to incident and transmitted wave fronts at a media boundary as shown in Fig. 22.7(a). The wave speeds are different in the two media. In this case, $v_1 > v_2$. (In general, the speed of light is less in denser media. Intuitively, you might expect the passage of light to take longer through a medium with more atoms per volume. And, in fact, the speed of light in water is about 75% of that in air or a vacuum.)

The distances the wave fronts travel in a time t are $v_1 t$ in medium 1 and $v_2 t$ in medium 2 [Fig. 22.7(b)]. As a result of the smaller wave speed in the second medium (and depending on the angle of incidence), the direction of the transmitted wave front is different from that of the incident wave front. The particles of the second medium are driven or set in motion by the incident wave disturbance, and thus the waves' frequency is the same in both media. However, the wavelengths are different because of the different wave speeds ($\lambda f = v$).

The change in the direction of wave propagation is described using the **angle of refraction**. In Fig. 22.7(b), the angle of incidence is θ_1, and the angle of refraction is θ_2. From the geometry of two parallel rays, where d is the distance between the normals to the boundary at the points where the rays are incident,

$$\sin \theta_1 = \frac{v_1 t}{d} \qquad \text{and} \qquad \sin \theta_2 = \frac{v_2 t}{d}$$

Combining these two equations in ratio form gives

$$\frac{\sin \theta_1}{\sin \theta_2} = \frac{v_1}{v_2} \qquad\qquad (22.2)$$

This expression is known as **Snell's law**.

Figure 22.6 **Reflection and refraction**
A beam of light striking a piece of glass is partially reflected and refracted. Note the reflection and refraction at the second surface.

The Dutch physicist Willebord Snell (1591–1626) discovered it.

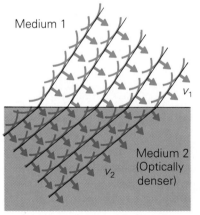

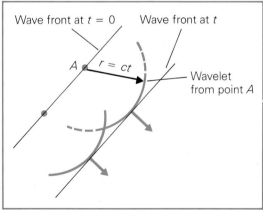

Wave front at $t = 0$ Wave front at t

A $r = ct$

Wavelet from point A

(a)

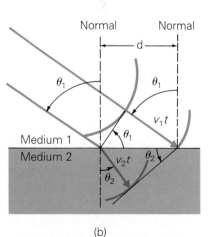

Normal   Normal

d

θ_1 θ_1

$v_1 t$

Medium 1 θ_1

Medium 2 $v_2 t$ θ_2

θ_2

(b)

Figure 22.7 **Huygens' principle**
(a) Each point on a wave front is considered to be the source of a wavelet. The line or surface tangent to all these wavelets defines a new position of the wave front. When a wave enters an optically denser medium, its speed is less, and the wave fronts (or rays) are bent, or refracted. (b) This diagram shows the geometry for the derivation of Snell's law. See the text for a description.

Refraction in Action

The refractive bending of light can produce some strange effects, but the principle is still the same as described by Snell's law.

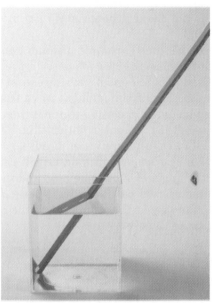

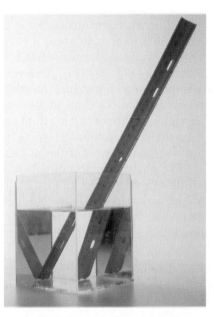

(a) Light rays entering a more optically dense medium are bent away from the normal, so the split ruler appears to be standing on legs.

(b) On looking down on the refractive surface, there is a difference in the perceived and actual depths of the immersed part of the ruler (see Example 22.3).

(c) A whole lot of refraction and reflection going on. Can you figure it out?

Thus, light is refracted when passing from one medium into another because the speed of light is different in the two media. (See Demonstration 16.) The speed of light is a maximum in a vacuum, and it is convenient to compare the speed of light in other media to this constant value (c). This is done by defining a ratio called the **index of refraction** (n):

$$n = \frac{c}{v} \frac{\text{(speed of light in a vacuum)}}{\text{(speed of light in a medium)}} \tag{22.3}$$

The indices of refraction of several substances are given in Table 22.1. Note that these values are for a specific wavelength of light. This is specified because n is slightly different for different wavelengths ($n = c/v = c/\lambda_{\mathrm{m}}f$, where λ_{m} is the wavelength of light in a particular material). The values of n given in Table 22.1 will be used in examples and problems in this chapter for all wavelengths of light in the visible region, unless noted otherwise. Remember that the frequency of light does not change when it enters another medium, but the wavelength of light in a material differs from the wavelength of that light in vacuum, as can be easily shown:

$$n = \frac{c}{v} = \frac{\lambda f}{\lambda_{\mathrm{m}} f}$$

Table 22.1

Indices of Refraction (at $\lambda = 590$ nm)*

Substance	n
Air	1.00029
Water	1.33
Ethyl alcohol	1.36
Fused quartz	1.46
Glycerine	1.47
Polystyrene	1.49
Oil (typical value)	1.50
Glass (depending on type)[†]	1.45–1.70
crown	1.52
flint	1.66
Zircon	1.92
Diamond	2.42

* A nanometer (nm) is 10^{-9} m.

[†] Crown glass is a soda-lime silicate glass and flint glass is a lead-alkali silicate glass. Flint glass is more dispersive than crown glass (see Section 22.5).

or $\quad n = \dfrac{\lambda}{\lambda_m}$ $\hspace{6cm}$ (22.4)

Note that n is always greater than 1 because the speed of light in a vacuum is greater than the speed of light in any material ($c > v$). Therefore, $\lambda > \lambda_m$.

The greater the index of refraction of a material, the more optically dense it is. (Optical density does not correlate directly with mass density. In some instances, a material having a greater optical density than another will have a lower mass density.) Also, the greater the index of refraction, the smaller the speed of light in the material. Since the speed of light in air is approximately equal to c,

$$n_{air} \cong \frac{c}{c} = 1$$

For practical purposes, the index of refraction is measured in air rather than in a vacuum.

Example 22.1 The Speed of Light in Water

What is the speed of light in water?

Solution

Given: $n = 1.33$ (from Table 22.1) $\hspace{2cm}$ *Find*: v
$\hspace{2cm} c = 3.0 \times 10^8$ m/s

Since $n = c/v$,

$$v = \frac{c}{n} = \frac{3.0 \times 10^8 \text{ m/s}}{1.33} = 2.3 \times 10^8 \text{ m/s}$$

Note that $1/n = v/c = 1/1.33 = 0.75$, or 75% of the speed of light in a vacuum. ∎

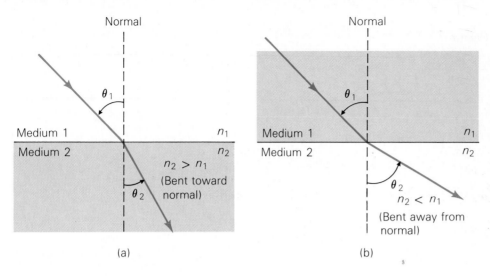

Figure 22.8 **Index of refraction and ray deviation** (a) When the second medium is more optically dense than the first $(n_2 > n_1)$, the refracted ray is bent toward the normal. (b) When the second medium is less optically dense than the first $(n_2 < n_1)$, the refracted ray is bent away from the normal. [Note that this is the case if the ray in part (a) is traced in reverse, going from medium 2 to medium 1.]

The index of refraction of a material may be determined experimentally using Snell's law:

$$\frac{\sin \theta_1}{\sin \theta_2} = \frac{v_1}{v_2} = \frac{c/n_1}{c/n_2} = \frac{n_2}{n_1}$$

or $\quad n_1 \sin \theta_1 = n_2 \sin \theta_2$ \hfill (22.5)

where n_1 and n_2 are the indices of refraction for the first and second media, respectively.

Eq. 22.5 is a very practical form of Snell's law. If the first medium is air, $n_1 \cong 1$, and $n_2 = \sin \theta_1 / \sin \theta_2$. Thus, only the angles of incidence and refraction need to be measured to determine the index of refraction of a material experimentally. If the index of refraction of a material is known, that value can be used in the practical form of Snell's law to find the angle of refraction for a given angle of incidence. In general, the following relationships can be easily deduced from Eq. 22.5:

■ If the second medium is more optically dense than the first medium $(n_2 > n_1)$, the refracted ray is bent toward the normal $(\theta_2 < \theta_1)$, as illustrated in Fig. 22.8(a).

■ If the second medium is less optically dense than the first medium $(n_2 < n_1)$, the refracted ray is bent away from the normal $(\theta_2 > \theta_1)$, as illustrated in Fig. 22.8(b).

Example 22.2   Snell's Law

A beam of light traveling in air strikes a glass plate an an angle of incidence of $45°$ (Fig. 22.9). The glass has an index of refraction of 1.5. (a) What is the angle of refraction for the light transmitted into the glass? (b) If the glass plate is 2.0 cm thick, how far is the beam displaced laterally from the normal in passing through the glass? (c) Prove that the emergent beam is parallel to the incident beam, that is, that $\theta_4 = \theta_1$.

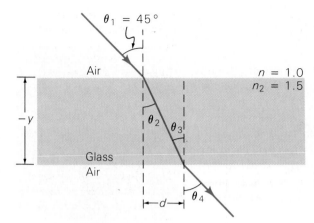

$\theta_1 = 45°$

Air

Glass

Air

$n = 1.0$
$n_2 = 1.5$

θ_2 θ_3

$-y$

θ_4

d

Figure 22.9 **Double refraction**
The refracted ray is displaced laterally (sideways) a distance d, and the emergent ray is parallel to the original ray. See Example 22.2.

Solution

Given: $\theta_1 = 45°$
$n_1 = 1.0$
$n_2 = 1.5$
$y = 2.0$ cm

Find: (a) θ_2
(b) d
(c) Show that $\theta_4 = \theta_1$

(a) Using the practical form of Snell's law, $n_1 \sin \theta_1 = n_2 \sin \theta_2$, with $n_1 = 1.0$ gives

$$\sin \theta_2 = \frac{\sin \theta_1}{n_2} = \frac{\sin 45°}{1.5} = \frac{0.707}{1.5} = 0.47$$

Thus,

$$\theta_2 = \sin^{-1}(0.47) = 28°$$

Note that the beam is bent toward the normal.

(b) From Fig. 22.9, you can see that the lateral displacement (d) is related to θ_2 by $\tan \theta_2 = d/y$.

$$d = y \tan \theta_2 = (2.0 \text{ cm})(\tan 28°) = (2.0 \text{ cm})(0.53) = 1.1 \text{ cm}$$

(c) If $\theta_1 = \theta_4$, then the emergent ray is parallel to the incident ray. Applying Snell's law to the beam at both surfaces gives

$$n_1 \sin \theta_1 = n_2 \sin \theta_2$$

and

$$n_2 \sin \theta_3 = n_1 \sin \theta_4$$

From the figure, it is evident that $\theta_2 = \theta_3$. Therefore,

$$n_1 \sin \theta_1 = n_1 \sin \theta_4$$

or $\theta_1 = \theta_4$

Thus, the emergent beam is parallel to the incident beam but displaced laterally a distance d. ∎

Two examples of refraction are shown in Fig. 22.10. In Fig. 22.10(a) the pencil appears to be almost bisected because the light by which you see it is refracted. If you traced a light ray back from your eyes, you would find that it is bent on entering the glass and on entering the water, and the ray would lead back to the lower part of the pencil.

Fig. 22.10(b) shows another example of refraction. The refraction of light from the sky by warm air near the surface of a road produces the "wet spot" mirage commonly seen on a road on a hot day. We also "see" hot air rising from a hot road surface. Of course, you cannot really see the air, but gases have a property that makes it possible for you to sense it visually. The index of refraction of a gas is proportional to its density, which is in turn inversely proportional to its temperature. Thus, you can perceive regions of air having different densities (temperatures) because of the refraction of light as it passes through the rising air. (Ordinarily, the air near a road's surface has a uniform density, and you do not see these effects.)

Refraction of light in air gives rise to other effects. At night, starlight travels to your eyes through air of varying densities, causing the light to be bent. This refraction and the motion of the air are responsible for stars' twinkling effect. Atmospheric refraction also accounts for the fact that the Sun can be seen for a short time before it rises above or after it sinks below the horizon. (Make a sketch and convince yourself of this.)

You may have experienced an effect of the refractive bending of light while trying to reach for something that is underwater, for example, a coin (Fig. 22.11). You are used to light traveling in straight lines from objects to your eyes, but the light reaching our eyes from the submerged object is bent at the water surface. (Note in the figure that the ray is bent away from the normal.) The object appears to be closer to the surface than it actually is, and therefore you tend to miss the object when reaching for it.

Figure 22.10 **Refraction in action**
(a) The refraction of light at the water–glass and glass–air boundaries makes part of the pencil look displaced. Note the magnification effect due to the lens action of the glass and water. (b) The refraction of sky's light by hot air near the road gives rise to a "wet spot" mirage. Note the image of the car on the road due to refracted light.

Example 22.3   Depth Perception
Find the apparent depth (d') of the coin in Fig. 22.11 for a small angle of incidence (θ_2) in terms of the index of refraction of the liquid.

Solution
From the figure,

$$\tan \theta_1 = \frac{a}{d'} \quad \text{and} \quad \tan \theta_2 = \frac{a}{d}$$

Thus,

$$\frac{d}{d'} = \frac{\tan \theta_1}{\tan \theta_2}$$

For small angles, since $\cos \theta \cong 1$,

$$\tan \theta = \frac{\sin \theta}{\cos \theta} \cong \sin \theta$$

Then, by Snell's law,

$$\frac{d}{d'} = \frac{\tan \theta_1}{\tan \theta_2} \cong \frac{\sin \theta_1}{\sin \theta_2} = n$$

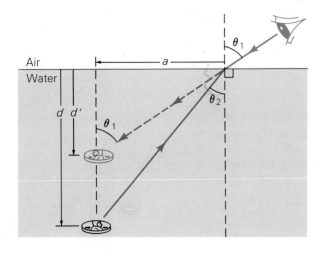

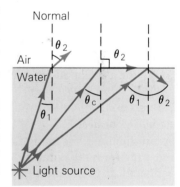

Figure 22.11 **Depth perception**
An object immersed in water appears closer to the surface than it actually is because of refraction. See Example 22.3.

Thus, a small-angle approximation is

$$d' \cong \frac{d}{n} \quad \blacksquare$$

22.4 Total Internal Reflection and Fiber Optics

An interesting phenomenon occurs when light traveling in one medium is incident on the boundary with another medium that is less optically dense, for example, when light goes *from* water *to* air. As you know, in such a case a transmitted beam will be bent away from the normal (Fig. 22.12), and Snell's law tells you that the greater the angle of incidence, the greater the angle of refraction will be. That is, as the angle of incidence increases, the farther the refracted ray diverges from the normal.

However, there is a limit. For a certain angle of incidence called the **critical angle** (θ_c), the angle of refraction is 90°, and the refracted ray is directed along the boundary. But what happens if the angle of incidence is even larger? If the angle of incidence is greater than the critical angle ($\theta_1 > \theta_c$), the light isn't transmitted but is internally reflected. This condition is called **total internal reflection** and the reflection process can be 100% efficient. See Fig. 22.13.

The critical angle for total internal reflection may be obtained using Snell's law. If $\theta_1 = \theta_c$ in the optically denser medium, $\theta_2 = 90°$, and

$$\frac{\sin \theta_1}{\sin \theta_2} = \frac{\sin \theta_c}{\sin 90°} = \frac{n_2}{n_1}$$

Since $\sin 90° = 1$,

$$\sin \theta_c = \frac{n_2}{n_1} \quad \text{(where } n_1 > n_2\text{)} \tag{22.6}$$

If the second medium is air, $n_2 \cong 1$, and the critical angle for total internal reflection at the boundary from a medium into air is given by

Figure 22.12 **Internal reflection**
When light enters a less optically dense medium, it is refracted away from the normal. At a critical angle (θ_c), the light is refracted along the interface (common boundary) of the media. At an angle greater than the critical angle ($\theta_1 > \theta_c$), there is total internal reflection.

$$\sin \theta_c = \frac{1}{n} \qquad \text{(where } n \text{ is the index of refraction of the material)} \qquad (22.7)$$

angle for total internal reflection at a material-air boundary

Example 22.4   Wide-Angle View

(a) What is the critical angle for light traveling in water and incident on a water-air boundary? (b) If a diver submerged in a pool looked up at the water surface at angles of $\theta < \theta_c$ and $\theta > \theta_c$, what would she see in each case? (Neglect any thermal or motional effects.)

Solution

Given: $n = 1.33$ (from Table 22.1)   *Find*: (a) θ_c
                                                      (b) Effects of changing θ

(a) The critical angle can be found directly from Eq. 22.7:

$$\theta_c = \sin^{-1}\left(\frac{1}{n}\right) = \sin^{-1}\left(\frac{1}{1.33}\right) = 48.8°$$

(b) For viewing angles of $\theta < \theta_c$, the diver would have a conical view of things above the surface (see Fig. 22.12 and mentally trace the rays in reverse for light coming from all angles in three dimensions). That is, light coming from the above-water 180° panorama could be viewed in a cone with a half-angle of 48.8°. As a result, objects would appear distorted. (Such panoramic views are seen in photographs made with "fish-eye" lenses.) For a viewing angle greater than θ_c, the diver would see the reflection of something on the bottom of the pool. ∎

(a)

Internal reflections enhance the brilliance of cut diamonds. The critical angle for a diamond-air surface is

$$\theta_c = \sin^{-1}\left(\frac{1}{n}\right) = \sin^{-1}\left(\frac{1}{2.42}\right) = 24°$$

A cut diamond has many facets, or faces (58 in the so-called brilliant cut), and most of the light entering the main and upper facets (or crown) is internally reflected. It then emerges from the upper facets, giving rise to the diamond's brilliance.

Figure 22.13 Internal reflection in a prism
(a) Because the critical angle of glass is less than 45°, prisms with 45° and 90° angles can be used to reflect light through 90° and 180°.
(b) The angles of incidence of the lower rays exceed the critical angle and the rays are internally reflected.

Fiber Optics

When a fountain is illuminated from below, the light is transmitted along the curved streams of water. This phenomenon was first demonstrated in 1870 by the British scientist John Tyndall (1820–1893), who showed how light was "conducted" along the curved path of a stream of water flowing from a hole in the side of a container. As you may have guessed, this phenomenon is observed because the light is internally reflected along the stream of water.

Internal reflection forms the basis of **fiber optics**, a relatively new and

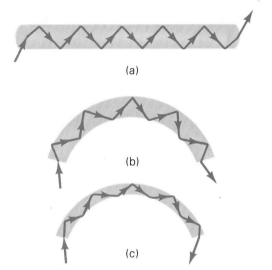

(a)

(b)

(c)

Figure 22.14 **Light pipes**
(a) When light is incident on the end of a cylindrical form of transparent material so that the internal angle of incidence is greater than the critical angle of the material, the light is reflected down the length of the light pipe.
(b) Light is also transmitted along curved light pipes by internal reflection. (c) As the diameter of the rod or fiber becomes smaller, the number of reflections per unit length becomes greater.

fascinating field centered on the use of transparent fibers to transmit light. Multiple internal reflections make it possible to "pipe" light along a transparent rod, even if the rod is curved (Fig. 22.14). Note from the figure that the smaller the diameter of the light pipe, the greater the number of internal reflections. In a small fiber, there can be as many as several hundred internal reflections per centimeter.

Internal reflection is an exceptionally efficient process. Optical fibers can be used to transmit light over very long distances with losses of about 25% per kilometer. These losses are primarily due to fiber impurities which scatter the light. Transparent materials have different degrees of transmission. Fibers are made out of certain plastics and special glass for maximum transmission, and the greatest efficiency is achieved with infrared radiation, for which there is less scattering.

The comparatively greater efficiency of multiple internal reflections over multiple external reflections can be illustrated by considering a good reflecting surface such as a plane mirror, which at best has a reflectivity of about 95%. Suppose that a number of plane mirrors were placed in two parallel rows, properly spaced so that light incident on the first mirror would be reflected back and forth down the rows. After each reflection, the beam intensity would be 95% of the intensity of incident beam (from the preceding reflection). The intensity of the reflected beam at any point is thus given by

$$I = (0.95)^n I_o$$

where n is the number of times the beam has been reflected and I_o is the initial light intensity. After 100 reflections,

$$I = (0.95)^{100} I_o = 0.006 I_o$$

Thus, with multiple reflections on plane mirrors there is only 0.6% transmission. Compare this to about 75% transmission with optical fibers over a kilometer in length even after thousands and thousands of reflections.

Fibers whose diameter is about 10 μm (10^{-5} m) are grouped together in flexible bundles that are 4–10 mm in diameter and up to several meters in length, depending on the application (Fig. 22.15). A fiber bundle with a cross-

Figure 22.15 **Fiber optic bundle**
Extremely thin fibers are grouped together to make a bundle.

sectional area of 1 cm² can contain as many as 50,000 individual fibers. To prevent light from being transmitted between fibers in contact with each other, they are coated with a film.

Fiber optics can be used for purely decorative purposes, such as lamps, but a much more important application involves the piping of light to and images from inaccessible or hard-to-reach places. To do this, the ends of a fiber bundle are cut and polished to form a flexible **fiberscope**. A beam of light can be transmitted along the bundle to illuminate an area, even if the bundle is bent and twisted. Equally important, an image or picture may be transmitted back by a fiberscope [Fig. 22.16]. Light travels through one set of fibers to illuminate an object and, after reflection, travels back in another set. The image has light and dark regions since each fiber has a circular cross section and transmits a dot. The overall image thus has a mosaic pattern, like a newspaper picture. The smaller the elements in the mosaic, the finer the detail. Thus, a fiberscope has a very large number of extremely fine fibers. A transmitted image can be magnified by a lens for viewing.

There are a number of interesting applications of fiber optics. For example, you're probably aware that many telephone transmissions are accomplished by means of fiber optics. Light signals, rather than electrical signals, are transmitted through optical telephone lines. Optical fibers have lower energy losses than current-carrying wires, particularly at higher frequencies. Also, these fibers are lighter than metal wires, have greater flexibility, and are not affected by electromagnetic disturbances (electric and magnetic fields) since they are made of materials that are electrical insulators.

Fiber optics has been widely applied in medicine. For example, endoscopes, or instruments used to view internal portions of the human body, previously consisted of lens systems in long narrow tubes. Some contained a dozen or more lenses and gave relatively poor images. Also, since the lenses had to be aligned in certain ways, the tubes had to have rigid sections, which limited the endoscope's maneuverability. Such an endoscope could be inserted down the throat into the stomach to observe the stomach lining. However, there would be blind spots due to the curvature of the stomach and the inflexibility of the instrument.

The flexibility of fiber bundles has eliminated this problem. A fiber endoscope is shown in Fig. 22.17. Lenses are used at the end of the fiber bundles to focus the light, and a prism is used to change the direction. The incident light is usually transmitted by an outer layer of fiber bundles, and the image is returned through a central core of fibers. Mechanical linkages allow maneuverability. The end of a fiber endoscope may be equipped with forceps to obtain specimens of the viewed tissues for biopsy (diagnostic examination).

A Fiber-optic cardioscope used for direct observation of heart valves typically has a fiber bundle about 4 mm in diameter and 30 cm long. Such a cardioscope passes easily to the heart through the jugular vein, which is about 15 mm in diameter. To displace the blood and provide a clear field of view for observations and photographing, a transparent balloon at the tip of the cardioscope is inflated with saline (salt water) solution.

Another application of fiber optics is the coding and decoding of information. To make a "coded" image of a classified picture, for example, the component fibers of a bundle are deliberately misaligned and randomly interwoven. As a result, a transmitted image is jumbled and unrecognizable unless it is viewed through an identically interwoven bundle that "decodes" it.

Figure 22.16 **Applications of fiber optics**
Images can be transmitted by fiber bundles.

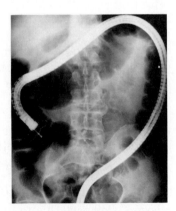

22.17 **Fiber endoscope**
An X-ray picture showing a fiber-optic endoscope in the colon. A small forceps may be seen extending from the end.

22.5 Dispersion

Light of a single frequency is called monochromatic light (from the Greek, *mono* meaning "one" and *chroma* meaning "color"). Visible light that contains all the component frequencies, or colors, is termed white light. Sunlight is white light. When a beam of white light passes through a glass prism as shown in Fig. 22.18, it is spread out, or dispersed, into a spectrum of colors. This phenomenon led Newton to believe that sunlight is a mixture of component colors. When the beam enters the prism, the component colors, corresponding to different wavelengths, must be refracted at slightly different angles so they spread out into a spectrum.

In a vacuum, the speed of light (c) is the same for all wavelengths. However, in a dispersive medium the wave speed is slightly different for different wavelengths. The frequency of a particular wave is constant, so the greater the wavelength, the greater its speed will be ($v = f\lambda$). Consequently, the index of refraction is slightly different for different wavelengths ($n = c/v = c/f\lambda$). In a transparent material that has different indices of refraction for different wavelengths of light, there is a separation of the wavelengths, and the material is said to exhibit **dispersion**. Dispersion may be neglected in many instances. Also, since the differences in the indices of refraction for different wavelengths are small, a representative value at some specified wavelength can be used for general purposes. (See Table 22.1.)

A good example of a dispersion material is diamond, which is about five times more dispersive than glass. In addition to the brilliance resulting from internal reflections off many facets, a cut diamond shows a play of colors, or "fire," resulting from the dispersion of the internally reflected light.

Another dramatic example of dispersion is the rainbow. A rainbow is produced by refraction, dispersion, and internal reflection of light within water droplets. When millions of water droplets remain suspended in the air after a rainstorm, a multicolored arc is seen, whose colors run from violet along the lower part up the spectrum (by increasing wavelength) to red along the upper

You can remember the sequence of the colors of the visible spectrum (from the long-wavelength end to the short-wavelength end) using the name ROY G. BIV, which is an acronym for Red, Orange, Yellow, Green, Blue, Indigo, and Violet.

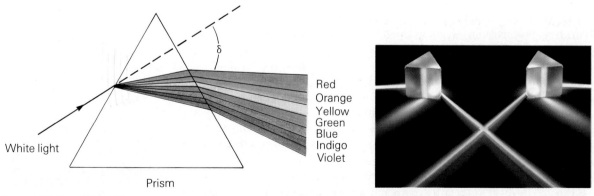

Red
Orange
Yellow
Green
Blue
Indigo
Violet

White light

Prism

Figure 22.18 **Dispersion**
White light is dispersed into a spectrum of colors by a glass prism. In a dispersive medium, the index of refraction differs slightly for different wavelengths. Red light, with the longest wavelength, has the smallest index of refraction and is bent least. The angle between the incident and refracted rays is called the angle of deviation (δ).

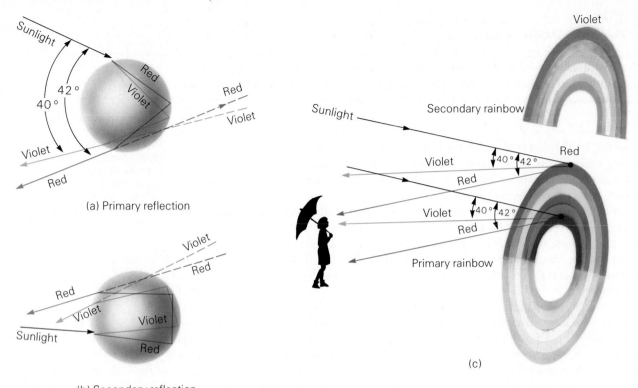

Figure 22.19 **The rainbow**
Rainbows are due to the refraction, dispersion, and internal reflection of sunlight in water droplets. (a) A single internal reflection gives rise to the primary rainbow. (b) A double internal reflection produces a secondary rainbow. (c) In the primary rainbow, an observer sees red light from the higher droplets because red light is deviated most. (The other components from those droplets pass above the observer's eyes.) Similarly, violet light is seen from the lower droplets.

[Fig. 22.19(a)]. Below the arc, the light from the droplets combines to form a region of brightness. Occasionally, more than one rainbow is seen; the main, or primary, rainbow is sometimes accompanied by a fainter and higher secondary rainbow.

The light that forms the primary rainbow is reflected once inside each water droplet [Fig. 22.19(b)]. Being also refracted and dispersed, the light is spread out into a spectrum of colors. However, because of the conditions for refraction and internal reflection in water, the angles of deviation (between incoming and outgoing rays) for violet to red light lie within a narrow range of 40°–42°. This means that you can see a rainbow only when the Sun is behind you, so the dispersed light is reflected to you through these angles.

The less frequently seen secondary rainbow is caused by a double internal reflection [Fig. 22.19(c)]. This results in an arc whose vertical color sequence is the inverse of the primary rainbow's. The angles of deviation in this case lie between 50.5° for red light and 54° for violet light.

The higher the Sun is in the sky, the less of a rainbow you will be able to see from the ground. In fact, you won't see a primary rainbow if the Sun's altitude is greater than 42°. (Altitude in this case is the angle between the plane of the surface and a line to an object above the horizon.) The primary rainbow can still be seen from a height, however. As an observer's elevation increases, more of the arc becomes visible. It is common to see a completely circular rainbow from an airplane. You may also have seen a circular rainbow in the spray from a garden hose.

QUESTION: Can there be a third-order, or tertiary, rainbow?

ANSWER: Yes, a third-order rainbow resulting from three internal reflections in a water droplet is possible. In a book entitled *Opticks*, Isaac Newton wrote, "The Light which passes through a drop of Rain after two Refractions, and three or more Reflexions, is scarcely strong enough to cause a sensible rainbow." (By "sensible," he meant intense enough to be seen.) However, his friend Edmund Halley (for whom the famous comet was named) showed that the reason the tertiary rainbow is not seen is not because the reflected light lacks intensity, but rather because the tertiary arc is formed in the general direction of the Sun and cannot be seen against the brightness of the sky.

Important Formulas

Law of reflection:

$\theta_i = \theta_r$

Snell's law:

$$\frac{\sin\theta_1}{\sin\theta_2} = \frac{v_1}{v_2} \quad \text{or} \quad n_1\sin\theta_1 = n_2\sin\theta_2$$

Index of refraction:

$$n = \frac{c}{v} = \frac{\lambda}{\lambda_m}$$

Critical angle for boundary between two materials (where $n_1 > n_2$):

$$\sin\theta_c = \frac{n_2}{n_1}$$

Critical angle for material-air boundary (where n is the index of refraction of the material):

$$\sin\theta_c = \frac{1}{n}$$

Questions

Wave Fronts and Rays

1. Define wave front and give an example.

2. What distinguishes spherical, circular, and plane wave fronts?

3. How are waves represented in geometrical optics?

Reflection

4. The reflections from a properly placed plane mirror allow you to see around a corner. Is a person around the corner able to see you? Explain.

5. Is the reflection from a movie screen specular or diffuse?

6. A metal surface is polished until the dimensions of its irregularities measure about 10^{-5} m. Will the surface make a good mirror?

Refraction

7. Is light in air perpendicularly incident on the boundary of a transparent material refracted in the material? Explain.

8. A glass container is filled with water. Describe how a light ray is refracted in going from air through the glass and into the water.

9. A marching band obliquely entering a muddy field is sometimes used as an analogy for refraction, since the direction of the band is changed. Explain this analogy in terms of cadence (stepping frequency) and slipping in the mud. (Hint: this is similar to how a marching band turns a corner while keeping straight rows.)

10. As shown in Fig. 20.10(b), mirages are formed by refraction when the temperature (density) of air decreases

with altitude. Sometimes there is a temperature inversion, and the air temperature increases with altitude. What kind of mirage might you expect to see in this case?

11. Explain the phenomena illustrated in Fig. 22.20. (a) Why does the rising or setting Sun sometimes appear flattened? (b) The pictures are taken with a camera from a fixed angle. Nothing is seen in the empty container. However, when water is added, a coin is seen. Why?

Figure 22.20 **Refraction effects**
See Question 11.

12. Why do designs painted on the bottoms of swimming pools appear distorted when viewed from poolside as shown in Fig. 22.21?

Figure 22.21 **Another refraction effect**
See Question 12.

Total Internal Reflection and Fiber Optics

13. Is it possible to have total internal reflection if the second medium is more optically dense than the first? Explain.

14. How could crown glass and flint glass (both are transparent) be used to make a mirror without any metal coating? (Hint: see Table 22.1.)

15. A refractometer is a device used to measure the index of refraction of a transparent material. With one type of refractometer, the index of refraction is measured by placing the material in contact with a piece of glass whose index of refraction is known and determining the critical angle for internal reflection in the glass. Describe the principle of this instrument and its limitations. (Hint: an oil having the same index of refraction as the glass is usually used between the material and the glass.)

Dispersion

16. If glass is dispersive, why don't we see a spectrum of colors when sunlight passes through a window pane?

17. Why are diamonds displayed under bright white lights in jewelry stores?

18. (a) Why do rainbows have the form of circular arcs? What is necessary to see a completely circular rainbow? (b) Why is a secondary rainbow fainter than the primary rainbow?

19. Explain why the red light is above the blue in a primary rainbow even though the red component of the sunlight is refracted less and actually comes out of a droplet below the blue.

20. Do two observers viewing a rainbow from different locations see exactly the same bow? Explain.

21. (a) A prism disperses white light into a spectrum. A second prism will not recombine the spectral components. Explain why. (b) It takes three additional prisms to recombine the spectral components into white light. Show that this is the case by tracing the paths of rays.

Problems

22.2 Reflection

■1. Laser light has been reflected from a mirror on the moon in order to measure how far the moon is from the Earth. If the measured distance is 3.8×10^5 km, how long did it take for light to travel to the moon and back?

■2. The angle of incidence of a light ray on a mirrored surface is 50°. What is the angle between the incident and reflected rays?

■■3. Light strikes a surface at an angle of 40° relative to the surface. What is the angle of reflection?

■■4. A beam of light is incident on a plane mirror at an angle of 30° to the normal. What is the angle between the reflected rays and the surface of the mirror?

■■5. Two people stand in a dark room. They are 3.0 m from a large plane mirror and 4.0 m apart. At what angle of incidence should one of the people shine a flashlight beam on the mirror so the reflected beam directly strikes the other person?

■■6. Two plane mirrors are placed 60 cm apart with their mirrored surfaces parallel and facing each other. If the mirrors are 25 cm wide, at what angle should a beam of light be incident at one end of one mirror so it just strikes the far end of the other?

■■■7. A person 1.8 m tall stands 3.0 m away from a plane mirror. His eyes are 20 cm from the top of his head. Find the minimum height of the mirror if the person is to see his complete image.

■■■8. A light beam is incident on a mirror at an angle θ_i. The mirror is rotated through an angle ϕ about an axis in the plane of the mirror and perpendicular to the point of incidence. The incident angle is then $\theta_i' = \theta_i - \phi$ (see Fig. 22.22). Show that the reflection ray rotates through an angle $\alpha = 2\phi$. (This is the principle of an optical lever, which is used to measure small angles of rotation.)

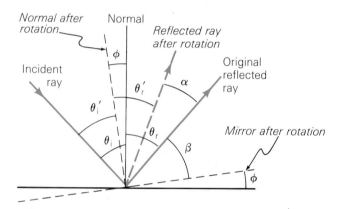

Figure 22.22 **Rotation of reflected ray**
See Problem 8.

22.3, 22.4 Refraction and Total Internal Reflection and Fiber Optics

■9. Find the speed of light in diamond.

■10. The speed of light in a material is determined to be 2.0×10^8 m/s. What is the index of refraction of the material?

■11. What is the speed of light in water?

■12. Light strikes water perpendicular to the surface. What is the angle of refraction?

■■13. A beam of light enters water at an angle of 40° with a normal to the water surface. Find the angle of refraction.

■■14. A beam of light in air is incident on the surface of a slab of fused quartz. Part of the beam is transmitted into the quartz at an angle of 30° with a normal to the surface, and part of the beam is reflected. What is the angle of reflection?

■■15. Light passes from air into water. If the angle of refraction is 15°, what is the angle of incidence?

■■16. A beam of light traveling in air strikes the flat surface of a transparent material at an angle of 45°. What is the index of refraction of the material if the angle of refraction is 40°?

■■17. A person shines a flashlight beam on the surface of a body of still water at an angle of 45°. (a) What is the angle of refraction in the water? (b) Compare that angle to the angle of refraction for crown glass for the same angle of incidence.

■■18. Light passes through a piece of flint glass in 20 ps (picoseconds). Find the thickness of the glass.

■■19. Light passes from material A, which has an index of refraction of $\frac{4}{3}$, into material B, which has an index of refraction of $\frac{5}{4}$. Find the ratio of the speed of light in material B to the speed of light in material A.

■■20. If the angle of incidence in Problem 19 is 30°, what is the angle of refraction?

■■21. The wavelength of red light in air is 680 nm. What are the wavelength and frequency of this light in water?

■■22. Monochromatic blue light having a frequency of 6.5×10^{14} Hz enters a piece of crown glass. What are the frequency and wavelength of the light in the glass?

■■23. A beam of light traveling in water strikes a glass surface at an angle of 40° with a normal to the surface. If the angle of refraction in the glass is 35°, what is the index of refraction of the glass?

■■24. A coin is placed at the bottom of an empty container and water is poured into the container to a depth of 50 cm. How far below the water surface would the coin appear to be to a person looking at it through the water at a small angle of incidence?

■■25. A layer of water 30.0 mm thick floats on a layer of another liquid, which is 20.0 mm thick. The other liquid has an index of refraction of 1.50. How far below the water surface will the bottom of the container appear to be for a small angle of incidence?

▪▪26. A light ray traveling in air is incident on a glass plate 10.0 cm thick at an angle of 40°. The glass has an index of refraction of 1.65. The emerging ray on the other side of the plate is parallel to the incident ray but laterally displaced. How far is the emerging ray displaced relative to the normal?

▪▪27. When looking into an opaque, empty container that is 10 cm deep at a viewing angle of 60° relative to the vertical side of the container, an observer sees nothing on the bottom of the container. When the container is filled with water, the observer fully sees the coin on the bottom of and just beyond the side of the container (from the same viewing angle). See Fig. 22.20(b). How far is the coin from the side of the container?

▪▪28. Prisms are sometimes used as optical components in place of mirrors. Find the minimum index of refraction for glass that is to be used in a prism that is meant to change the direction of light by 100° on reflections.

▪▪29. How far will a beam of light travel in flint glass in the time it takes light to travel 10 cm in air?

▪▪30. A submerged diver shines a light toward the surface of the water at angles of incidence of 40° and 50°. Will a person on the shore see a beam of light emerging from the surface in either case? Justify your answer mathematically.

▪▪31. A beam of light passes through a 45°-45°-90° prism as shown in Fig. 22.23. (a) What must the index of refraction of the prism be for this to occur? (b) What would the index of refraction have to be for the situation to occur under water?

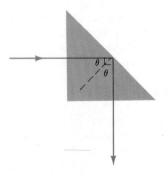

Figure 22.23 **Internal reflection in a prism**
See Problem 31.

▪▪▪32. A coin lies on the bottom of a pool under 1.5 m of water and 0.90 m from the side wall (Fig. 22.24). If a light

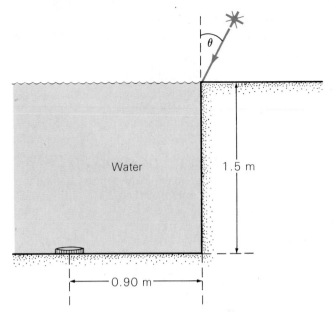

Figure 22.24 **Find the coin**
See Problem 32.

beam is incident on the water surface at the wall, at what angle (θ) relative to the wall must the beam be directed so it will illuminate the coin?

▪▪▪33. A crown-glass plate 4.0 cm thick is placed over a newspaper. How far beneath the top surface of the plate would the print appear to be to a person looking almost vertically downward through the plate?

▪▪▪34. Light passes from medium A into medium B at an angle of incidence of 30°. The index of refraction of A is 1.5 times that of B. (a) What is the angle of refraction? (b) What is the ratio of the speed of light in B to the speed of light in A? (c) What is the ratio of the frequency of the light in B to the frequency of light in A? (d) At what angle of incidence would the light be internally reflected?

▪▪▪35. Light travels from a material whose index of refraction is $n_1 = 2.0$ into another material whose index of refraction is $n_2 = \frac{5}{3}$. The opposite surface of the second material is exposed to the air. (a) Find the critical angle at which light will be reflected at the interface of the materials. (b) Find the critical angle at which the light will pass through the interface and then be reflected at the opposite surface of the second material.

▪▪▪36. The angle between incident and emergent rays of light for a prism is called the angle of deviation (Fig. 22.25). The angle of minimum deviation (D_m) occurs when a ray passes through the prism symmetrically, or parallel to the base of the prism. Using Snell's law, show that the angle D_m

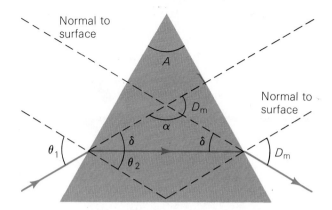

Figure 22.25 **Angle of minimum deviation**
See Problem 36.

and the prism angle A are related to the index of refraction (for a particular wavelength) by this equation:

$$n = \frac{\sin[(A + D_m)/2]}{\sin(A/2)}$$

22.5 Dispersion

■■**37.** The index of refraction of crown glass is 1.515 for red light and 1.523 for blue light. Find the angle separating rays of the two colors in a piece of crown glass if their angle of incidence is 30°.

■■**38.** White light passes through a prism made of crown glass and strikes an air interface at an angle of 41.15°. Using the indices of refraction given in Problem 37, describe what happens.

■■**39.** Fused quartz is a dispersive medium with an index of refraction of 1.470 for blue light (400 nm in air) and 1.445 for red light (680 nm in air). A beam of light composed of these two colors is incident on a plate of fused quartz at an angle of 45°. (a) What is the angle of separation between the two components of the beam in the quartz? (b) What is the ratio of the wavelengths of the two components in the quartz?

Additional Problems

40. A He-Ne (helium-neon) laser beam ($\lambda = 632.8$ nm in air) is directed through ethyl alcohol. What are the wavelength and frequency of the light in the alcohol?

41. A beam of light is incident on a flat piece of polystyrene at an angle of 60° relative to a surface normal. What angle does the refracted ray make with the plane of the surface?

42. To a submerged diver looking upward through the water, the altitude of the Sun appears to be 50° (the angle between the horizon and the Sun). What is the Sun's actual altitude?

43. At what angle must a diver submerged in a lake look toward the surface to see the setting Sun?

44. Is the speed of light greater in diamond or zircon? Express the difference as a percentage.

45. The glass of a prism has an index of refraction of $n_1 = 1.554$ for light with a wavelength of $\lambda_1 = 440$ nm and an index of refraction of $n_2 = 1.538$ for light with a wavelength of $\lambda_2 = 650$ nm. If a beam consisting of light of these two wavelengths is incident on one of the prism's surfaces at an angle of 70° with a normal to the surface, what is the angular separation of the emerging beams?

46. The critical angle for a certain type of glass is determined to be 40° (at which no light is seen to emerge from a sample). What is the index of refraction of the glass?

47. What percentage of the actual depth is the apparent depth of an object submerged in water? (Assume that the observer is looking almost straight downward.)

48. A 45°-90°-45° prism is made of a material with an index of refraction of 1.85. Can the prism be used to deflect a beam of light by 90° (a) in air or (b) in water?

49. Two upright plane mirrors touch along one edge, where their planes make an angle of 40°. If a beam of light is directed onto one of the mirrors at an angle of incidence of 60° and is reflected onto the other mirror, what will be the angle of reflection of the beam from the second mirror?

50. A beam of light traveling in air is incident on a transparent plastic material at an angle of 45° with a normal to the material's surface. The angle of refraction is measured to be 30°. What is the index of refraction of the material?

Mirrors and Lenses

23

You look into mirrors every day and may wear glasses or contact lenses. Mirrors and lenses are undoubtedly the most common optical components. The first mirror was probably the reflecting surface of a pool of water. Later people discovered that polished metals and glass have reflective properties. People also noticed that when they looked at things through pieces of glass, the objects looked different, depending on the shape of the glass. In some cases, the objects appeared to be enlarged, or magnified. Eventually, lenses were constructed to improve people's vision by purposefully shaping pieces of glass.

The optical properties of mirrors and lenses are based on the principles of the reflection and refraction of light. In this chapter, you will learn about these general properties by applying geometrical optics. The practical uses of mirrors and lenses in optical instruments such as microscopes and telescopes will be discussed in Chapter 25.

23.1 Plane Mirrors

Mirrors are smooth reflecting surfaces, usually made of polished metal or glass that has been coated with some metallic substance. As you know, even an uncoated piece of glass such as a window pane can act as a mirror (see Demonstration 17). However, when one side of a piece of glass is coated with a compound of tin, mercury, or silver, its reflectivity is increased, and light is not transmitted through the coating. A mirror may be front-coated or back-coated, depending on the application.

When you look directly into a mirror, you see the reflected images of yourself and objects around you. Usually, these images appear to be behind the mirror (on the other side of the surface). The geometry of a mirror's surface affects the size, orientation, and type of image.

A mirror with a flat surface is called a **plane mirror**. How images are formed by a plane mirror is illustrated by the ray diagram in Fig. 23.1. Diverging light

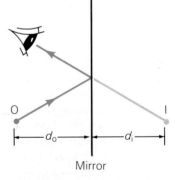

Figure 23.1 **Point image in a plane mirror**
The image of a point is seen as being in the mirror, with the image distance (d_i) equal to the object distance (d_o). The rays do not actually pass through the mirror and only appear to converge at the image position. The image in this case is said to be virtual.

A Candle Burning Underwater?

It would appear so, but you know this is not possible. It's really a matter of reflection and an image.

(a) The black frame holds a pane of glass, which acts as a plane mirror. The burning candle seen in the water is the image of the candle on the front stand. There is a container of water on a similar stand behind the glass, but no burning candle.

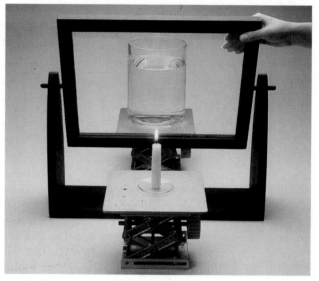

(b) The effect can be removed by tilting the glass—the image can no longer been seen from this viewing point. (Something to do with the law of reflection. What?)

rays from a point source in front of the mirror are reflected by the mirror, but to an observer they appear to come from another point source that is located behind the mirror surface. Images are formed where the reflected light rays intersect or converge or where they *appear* to do so to an observer. In the case of a plane mirror, the light only appears to converge at the image position. Such an image is referred to as a **virtual image**. With spherical mirrors (discussed in the next section), the reflected light rays may actually converge at and pass through the image point, and in this case a **real image** is formed.

The distance an object is from a mirror is called the **object distance** (d_o), and the distance its image appears to be behind the mirror is called the **image distance** (d_i). These distances are shown in Fig. 23.1 for a point and in Fig. 23.2 (on the next page) for an extended object. Rays leave the object from all points on it but only those from two points are shown in the figure. Each object point has a corresponding image point, but the two rays shown are enough to determine the size and orientation of the image. By similar triangles, $d_o = d_i$, which means that *the image formed by a plane mirror appears to be at a distance behind the mirror's surface that is equal to the distance from the object to the surface.*

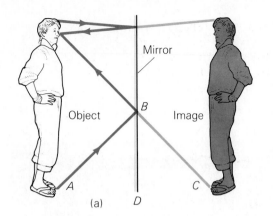

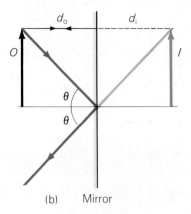

(a)

(b) Mirror

Figure 23.2 **Extended image in a plane mirror**
(a) Rays leave an object from all points, and each point has a point image. However, in most instances, two rays are sufficient to determine the size and orientation of the image. (b) An arrow is commonly used as the object to show this.

A difference between the size of an object and its image is expressed in terms of the **magnification** (M), which is defined as a ratio:

$$M = \frac{\text{image height}}{\text{object height}} = \frac{h_i}{h_o} \tag{23.1}$$

Again by similar triangles, $h_i = h_o$, so $M = 1$ for a plane mirror and there is no magnification. Note that M is a magnification *factor*, and $h_i = M h_o$.

Also, with a plane mirror, the image is upright, or erect. This means that the image is oriented in the same direction as the object. In ray diagrams, an arrow is commonly used as the object and a dashed arrow as the image so the orientation of the image relative to the object can be clearly seen from the directions of the arrowheads. Note in Fig. 23.2(b) that the object and image arrows are both pointing upward. (If an object and its image are both upside down, the image is still said to be upright to indicate it has the same orientation as the object.) With other types of mirrors, it is possible to have inverted images. In this case, the image arrow points in the direction opposite to that of the object arrow.

A characteristic of a plane mirror that you may have noticed is the right-left reversal of the images. If you stand in front of a plane mirror and raise your right hand, your image will raise its left hand. Also, if you part your hair, the part will appear to be on the wrong side (unless you part it down the middle).

Sometimes M is called the lateral magnification. For some nonplanar mirrors, M may be greater than or less than 1 and the image is enlarged or reduced in size relative to the object.

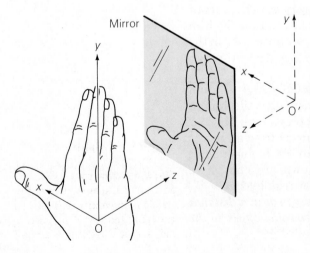

Figure 23.3 **Right-left reversal in a plane mirror**
An image in a plane mirror shows right-left reversal relative to the object. This is a matter of orientation, since the directions right and left depend on orientation. Note that the image x axis points in the same direction as the object x axis.

Figure 23.4 **Another right-left reversal**
The letters of the word AMBULANCE are printed backwards, or reversed, so that they appear in the proper order when seen in a rear-view mirror.

This right-left reversal of images in plane mirrors is illustrated in Figs. 23.3 and 23.4.

Characteristics of the image formed by a plane mirror are summarized in Table 23.1.

Table 23.1
Characteristics of Images Formed by Plane Mirrors

The image distance is equal to the object distance ($d_i = d_o$). That is, the image appears to be as far behind the mirror as the object is in front.

The image is virtual, upright, and unmagnified ($M = 1$).

The image shows right-left reversal.

23.2 Spherical Mirrors

As the name implies, a **spherical mirror** is a reflecting surface with spherical geometry. As Fig. 23.5 shows, if a portion of a sphere of radius R is sliced off along a plane, the severed section has the shape of a spherical mirror. Either the inside or outside surface of such a section can be reflective. If the inside surface reflects, the mirror is a **concave mirror** (think of looking into a cave to remember the indented surface). When the outside surface reflects, the mirror is a **convex mirror**.

The radial line through the center of the spherical mirror is called the optic axis, and it intersects the mirror surface at the vertex of the spherical section (Fig. 23.5). The point on the optic axis that corresponds to the center of the sphere of which the mirror forms a section is called the **center of curvature** (C). The distance between the vertex and the center of curvature is equal to the radius of the sphere and is called the **radius of curvature** (R).

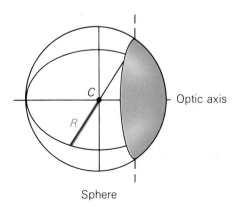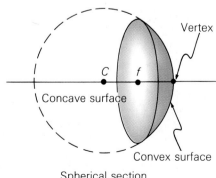

Figure 23.5 **Spherical mirrors**
A spherical mirror is a section of a sphere. Either the outside (convex) surface or the inner (concave) surface may be the reflecting surface.

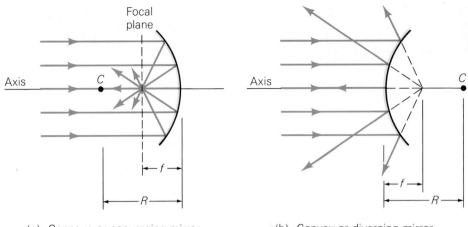

Focal plane

Axis — C — |←— f —→| ←— R —→|

(a) Concave or converging mirror

Axis — C — |←— f —→| ←— R —→|

(b) Convex or diverging mirror

Figure 23.6 **Focal point**
(a) Rays parallel and close to
the optic axis of a concave
mirror converge at a focal
point. Off-axis parallel rays
converge at a point in the
focal plane. (b) Rays parallel
to the optic axis of a convex
mirror are reflected along
paths as though they came
from a focal point behind the
mirror.

When rays parallel to the optic axis are incident on a concave mirror, the reflected rays intersect at a common point called the **focal point** (f).* As a result, a concave mirror is called a **converging mirror** [Fig. 23.6(a)]. Similarly, a beam parallel to the optic axis of a convex mirror diverges on reflection, as though the reflected rays came from a focal point behind the mirror's surface [Fig. 23.6(b)]. Thus, a convex mirror is called a **diverging mirror**. An example of a diverging mirror is shown in Fig. 23.7.

The distance from the vertex to the focal point of a spherical mirror is called the **focal length** (f). The focal length is related to the radius of curvature by this simple equation:

It is common practice to use the symbol *f* to represent both the focal point and the focal length.

$$f = \frac{R}{2} \tag{23.2}$$

focal length for a spherical mirror

In some instances, spherical mirrors form virtual images that appear to be behind the surface just as images with plane mirrors do. However, in other instances, spherical mirrors form images that can be seen on a screen placed at certain positions. These are real images formed by converging rays (a concentration of energy). The distinction between these two kinds of images is sometimes stated like this: *a real image is one that can be formed on a screen, and a virtual image is one that cannot.*

The characteristics of images formed by spherical mirrors can be determined using geometrical optics. The method involves drawing rays emanating from one or more points on an object. The law of reflection applies ($\theta_i = \theta_r$), and certain rays are defined with respect to the mirror's geometry as follows:

■ A **parallel ray** is a ray that is incident along a path parallel to the optic axis and is reflected through the focal point (as are all rays near and parallel to the axis).

■ A **chief ray**, or **radial ray**, is a ray that is incident through the center of curvature (C). Since it is incident normal to the mirror's surface, this ray is reflected back along its incident path, through C.

Figure 23.7 **Diverging mirror**
Note by reverse-ray tracing that a diverging mirror gives an expanded field of view, as can be seen in the store-monitoring mirror.

* This is the case for the approximation where the width of the mirror or the illuminated area is small compared to the radius of curvature, that is, for small angles of reflection.

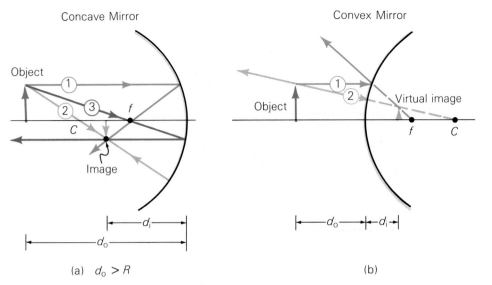

Figure 23.8 **Ray diagrams for spherical mirrors**
(a) A parallel ray (1) is reflected through the focal point (f). A chief ray (2) is incident through the center of curvature (C) and is reflected back along its path of incidence. A focal ray (3) is incident through the focal point and is reflected parallel to the optic axis. The image is real in this case and can be formed on a screen. (b) Only two rays are needed to determine the orientation of a mirror's image. As shown here, the image is virtual for a convex mirror.

These rays are illustrated in the ray diagrams in Fig. 23.8 for concave and convex mirrors. It is customary to use the head of the object arrow as the origin of the rays. This makes it easy to see whether the image is upright or inverted. The arrowhead of the image is at the point of intersection of the rays. Also, the object arrow is arbitrarily taken to be upright with its base on the optic axis.

Note in Fig. 23.8(a) that for a concave mirror with an object at a distance greater than the radius of curvature ($d_o > R$), a real image is formed. That is, the image may be seen on a screen (for example, a piece of white paper) positioned at a distance d_i from the mirror. The image has the characteristics of being real, inverted, and smaller than the object.

The rays reflected from a convex mirror diverge [Fig. 23.8(b)]. Projecting the rays behind the mirror to find where they intersect indicates that the image is virtual. (The image arrow is drawn with its arrowhead at the point of intersection.) This image is analogous to the virtual image formed by a plane mirror, and it cannot be projected on a screen. Since the reflected rays for an object at any distance from a convex mirror diverge, *a diverging mirror cannot form a real image.*

For a converging spherical mirror, three regions marked off by two points are of interest in the formation of images. These are shown in Fig. 23.9 on the next page. Along with ray diagrams for the cases where $R > d_o > f$, $d_o < f$, and $d_o = f$. [The case of $d_o > R$ was shown in Fig. 23.7(a); the case of $d_o = R$ is shown in Demonstration 18.] Note that an enlarged, inverted, real image is formed for $R > d_o > f$. When the object is at the focal point of the mirror, the reflected rays are parallel, and the image is said to be formed at infinity. By reverse ray tracing, rays from an object at a great (taken as "infinite") distance from the mirror can be shown to be essentially parallel when they are near the mirror and to form an image on a screen aligned in the focal plane. This fact provides a method for determining the focal length of a concave mirror. If the object is inside the focal point (between the focal point and the mirror's surface), a virtual image is formed.

The position and size of the image may be determined graphically by ray diagrams drawn to scale. However, these can be determined more quickly by

A focal ray is one that passes through (or appears to go through) the focal point and is reflected parallel to the optic axis. It is thus a mirror image, so to speak, of a parallel ray. A focal ray (3) is shown in Fig. 23.8(a) but is omitted in subsequent diagrams for clarity, since only two rays are needed to determine the image. You may prefer to use a focal ray in ray diagrams rather than one of the other rays.

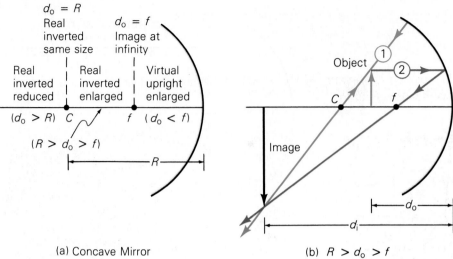

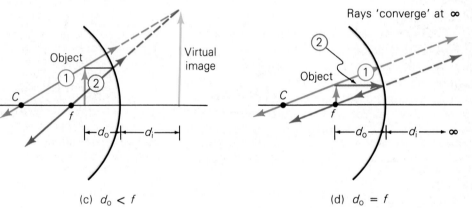

(a) Concave Mirror

(b) $R > d_o > f$

(c) $d_o < f$

(d) $d_o = f$

Figure 23.9 **Concave mirrors** (a) For a concave mirror, the location of the object may be within one of three regions defined by the center of curvature (C) and the focal point (f). For $d_o > R$, the image will be real, as shown by the ray diagram in Fig. 23.8(a). (b) For $R > d_o > f$, the image will also be real. (c) For an object at the focal point ($d_o = f$), the image is said to be formed at infinity. (d) For $d_o < f$, the image will be virtual.

analytical methods. For the ray diagram in Fig. 23.10, by the similarity of the triangles involving the object arrow and the image arrow,

$$\frac{y_o}{d_o} = \frac{y_i}{d_i} \qquad \text{or} \qquad \frac{y_i}{y_o} = \frac{d_i}{d_o} \tag{23.3}$$

Also, by the similarity of the triangles formed by the parallel and focal rays,

$$\frac{y_o}{f} = \frac{y_i}{d_i - f} \qquad \text{or} \qquad \frac{y_i}{y_o} = \frac{d_i - f}{f} \tag{23.4}$$

Equating the right-hand sides of Eqs. 23.3 and 23.4 and rearranging gives what is known as the **spherical mirror equation**:

$$\frac{1}{d_o} + \frac{1}{d_i} = \frac{1}{f} \tag{23.5}$$

Since the image distance is often the quantity to be found, a convenient alternative form of this equation is

$$d_i = \frac{d_o f}{d_o - f} \tag{23.6}$$

Demonstration 18

Candle Burning at Both Ends

Or is it? Notice that one flame is burning downward, which is rather strange. It's an illusion done with a spherical concave mirror.

(a) When an object is at the center of curvature of a spherical concave mirror, a real image is formed that is inverted and the same size as the object, and the image distance is the same as the object distance. What is seen here is a horizontal candle burning at one end (flame up) and its overlapping image (flame down). Viewed at the same level, the inverted flame image appears to be at the opposite end of the candle.

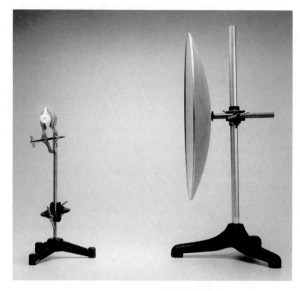

(b) A side view showing the burning end of the horizontal candle in front of the spherical mirror.

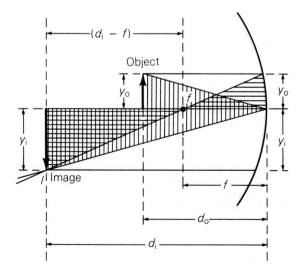

Figure 23.10 **The spherical mirror equation**
This diagram shows the geometry for deriving the spherical mirror equation. Note that the two triangles with vertical markings are similar, as are the two triangles with horizontal markings.

The ratio of the image size (y_i) to the object size (y_o) is called the (lateral) magnification:

$$M = y_i/y_o$$

Note M is a magnification factor that gives the height of the image relative to the object, $y_i = My_o$. From Eq. 23.3, the magnification can be expressed in terms of the image and object distances:

$$M = -\frac{d_i}{d_o} \tag{23.7}$$

magnification equation for a spherical mirror

The minus sign is added as a sign convention to indicate the orientation of the image. If $|M| > 1$, the image is magnified, or larger than the object. If $|M| < 1$, the image is reduced, or smaller than the object.

The signs on the various quantities are very important in the application of these equations. This book uses these sign conventions for spherical mirrors:

- The focal length f (or R) is positive for a concave mirror and negative for a convex mirror.

- The object distance d_o is always taken to be positive.

- Then the image distance d_i is positive for a real image (formed on the same side of the mirror as the object) and negative for a virtual image (formed behind the mirror).

- The magnification M is positive for an upright image and negative for an inverted image.

This sign convention is summarized in Table 23.2.

The use of the spherical mirror equations and sign convention can be illustrated using a plane mirror, which can be thought of as being a section of a sphere having an infinite radius of curvature ($R = \infty$). Then, since $f = R/2$,

$$\frac{1}{d_o} + \frac{1}{d_i} = \frac{1}{f} = \frac{2}{R} = \frac{2}{\infty} = 0$$

Thus,

$$\frac{1}{d_o} = -\frac{1}{d_i} \quad \text{and} \quad d_i = -d_o$$

This result indicates that the image distance is equal to the object distance and that the image is virtual (appears to be behind the mirror) because there is a minus sign. The magnification is

$$M = -\frac{d_i}{d_o} = -\frac{(-d_o)}{d_o} = 1$$

The image is the same size as the object ($y_i/y_o = 1$ or $y_i = y_o$). Since the magnification factor is positive, the image is upright. From experience, you know that these are the correct characteristics for an image that is formed by a plane mirror.

The following examples show how the spherical mirror equations are used.

A helpful hint for remembering that the magnification is d_i over d_o is that the ratio is in alphabetical order (i over o).

$|M|$ is the absolute value of M, that is, its magnitude without regard to sign. For example, $|+2| = |-2| = 2$.

Table 23.2
Sign Convention for Spherical Mirrors

Concave mirror: f positive
Convex mirror: f negative
d_o always positive

d_i	Image	M	Image
+	Real	+	Upright
−	Virtual	−	Inverted

Example 23.1 Images with a Concave Mirror

A concave mirror has a radius of curvature of 30 cm. If an object is placed (a) 45 cm, (b) 30 cm, (c) 20 cm, and (d) 10 cm from the mirror, where is the image formed and what are its characteristics? (Specify real or virtual, upright or inverted, and larger or smaller for each image.)

Solution

Given: $R = 30$ cm, so $f = R/2$ *Find*: d_i and image characteristics
$\qquad\qquad = 15$ cm
$\qquad\qquad$ (a) $d_o = 45$ cm
$\qquad\qquad$ (b) $d_o = 30$ cm
$\qquad\qquad$ (c) $d_o = 20$ cm
$\qquad\qquad$ (d) $d_o = 10$ cm

Note that these object distances correspond to the regions shown in Fig. 23.8.
(a) In this case, the object distance is greater than the radius of curvature $(d_o > R)$, and

$$\frac{1}{d_o} + \frac{1}{d_i} = \frac{1}{f}$$

or $\qquad \dfrac{1}{45} + \dfrac{1}{d_i} = \dfrac{1}{15}$

where the units (cm) have been omitted for simplicity. Then

$$\frac{1}{d_i} = \frac{2}{45}$$

or $\qquad d_i = \dfrac{45}{2} = 22.5$ cm

and

$$M = -\frac{d_i}{d_o} = -\frac{22.5 \text{ cm}}{45 \text{ cm}} = -\frac{1}{2}$$

Thus, the image is real (positive d_i), inverted (negative M), and $\frac{1}{2}$ as large as the object ($|M| = \frac{1}{2}$).
(b) The object distance is equal to the radius of curvature in this case ($d_o = R$). Since $f = R/2$, for convenience, the spherical mirror equation can be put in terms of R: that is, in this case,

$$\frac{1}{d_o} + \frac{1}{d_i} = \frac{1}{f}$$

becomes

$$\frac{1}{R} + \frac{1}{d_i} = \frac{2}{R}$$

Thus,

$$\frac{1}{d_i} = \frac{1}{R}$$

or $\qquad d_i = R = 30$ cm

Then

$$M = -\frac{d_i}{d_o} = -\frac{R}{R} = -1$$

The image is therefore real, inverted, and the same size as the object. (Draw a ray diagram to prove to yourself that the object and image arrows are both at the center of curvature pointing in opposite directions.)

(c) Here $R > d_o > f$. Using Eq. 23.6,

$$d_i = \frac{d_o f}{d_o - f} = \frac{(20\ \text{cm})(15\ \text{cm})}{20\ \text{cm} - 15\ \text{cm}} = 60\ \text{cm}$$

and

$$M = -\frac{d_i}{d_o} = -\frac{60\ \text{cm}}{12\ \text{cm}} = -5$$

In this case, the image is real, inverted, and five times the size of the object.

(d) For this case, $d_o < f$, and

$$d_i = \frac{d_o f}{d_o - f} = \frac{(10\ \text{cm})(15\ \text{cm})}{10\ \text{cm} - 15\ \text{cm}} = -30\ \text{cm}$$

Then

$$M = -\frac{d_i}{d_o} = -\frac{(-30\ \text{cm})}{10\ \text{cm}} = 3$$

In this case, the image is virtual (negative d_i), upright (positive M), and three times the size of the object.

From the denominator of the right-hand side of the equation for d_i (Eq. 23.6), you can see that d_i will always be negative when d_o is less than f. Therefore, a virtual image is always formed for an object inside the focal point of a converging mirror.

(Draw representative ray diagrams for each of these cases to see that the image characteristics are correct.) ∎

PROBLEM-SOLVING HINT

When using the spherical mirror equations to find image characteristics, it is helpful to first make a quick sketch of the ray diagram for the situation. Doing this shows you the image characteristics and helps you avoid making mistakes when applying the sign convention.

Example 23.2 Image with a Diverging Mirror

An object is 30 cm in front of a diverging mirror that has a focal length of 10 cm. Where is the image and what are its characteristics?

Solution

Given: $d_o = 30\ \text{cm}$ *Find*: d_i and image characteristics
$f = -10\ \text{cm}$

Note that the focal length is negative for a convex mirror (see Table 23.2). Using Eq. 23.6,

$$d_i = \frac{d_o f}{d_o - f} = \frac{(30 \text{ cm})(-10 \text{ cm})}{30 \text{ cm} - (-10 \text{ cm})} = -7.5 \text{ cm}$$

Then

$$M = -\frac{d_i}{d_o} = -\frac{(-7.5 \text{ cm})}{30 \text{ cm}} = +0.25$$

Thus, the image is virtual (negative d_o) and upright (positive M) and is reduced by a factor of 0.25 (or $\frac{1}{4}$). ∎

Example 23.3 Finding the Focal Length

Where is the image formed for a concave mirror if the object is at infinity? (An object at a great distance from a mirror, relative to the mirror's dimensions, may be considered to be at infinity.)

Solution
With $d_o = \infty$, we have

$$\frac{1}{d_o} + \frac{1}{d_i} = \frac{1}{\infty} + \frac{1}{d_i} = \frac{1}{f} \quad \text{or} \quad d_i = f$$

Thus, the image is real ($+d_i$) and is formed at the focal point (or in the focal plane for an extended object). This result provides an experimental means of determining the focal length of such a mirror. ∎

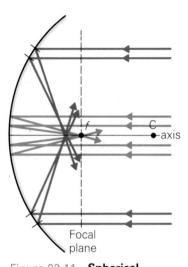

Figure 23.11 Spherical aberration for a mirror
According to a small-angle approximation, rays parallel to and near the mirror's axis converge at the focal point. However, when parallel rays not near the axis are reflected, they converge in front of the focal point. This effect is called spherical aberration and gives rise to blurred images.

The descriptions of image characteristics for spherical mirrors are true only for objects near the optic axis, that is, only for small angles of reflection. If these conditions do not hold, the images will be blurred (out of focus) or distorted, because not all of the rays will converge in the same plane. As illustrated in Fig. 23.11, incident parallel rays far from the optic axis do not converge at the focal point. The farther the incident ray is from the axis, the more distant is its reflected ray from the focal point. This effect is called **spherical aberration**. Spherical aberration does not occur with a parabolic mirror. All of the incident rays parallel to the optic axis of a parabolic mirror have a common focal point. However, parabolic mirrors are more difficult to make than spherical mirrors (and therefore more expensive).

23.3 Lenses

The word "lens" comes from the Latin word for lentil, which is a seed whose shape is similar to that of a common lens. An optical lens is made from some transparent material (most commonly glass but sometimes plastics or crystals). One or both surfaces usually have a spherical contour. Biconvex lenses (both surfaces convex) and biconcave lenses (both surfaces concave) are illustrated in Fig. 23.12.

The properties of lenses are due to the refraction of light passing through them. When light rays pass through a lens, they are bent, or deviated from their original paths, according to the law of refraction. You studied refraction for

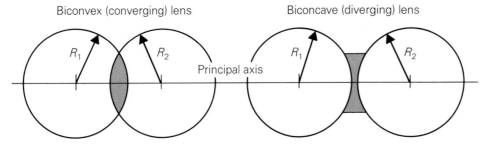

Figure 23.12 **Spherical lenses**

Biconvex (converging) lens

Biconcave (diverging) lens

R_1 R_2 Principal axis

R_1 R_2

Spherical lenses have surfaces defined by two spheres, and the surfaces may be either convex or concave. Biconvex and biconcave lenses are shown here. If $R_1 = R_2$, the lens is spherically symmetric.

prisms in Chapter 22. To analyze lens refraction, a biconvex lens may be approximated by two prisms placed base to base, as shown in Fig. 23.13(a). A biconvex lens is a **converging lens**: incident light rays parallel to the lens axis converge at a focal point on the opposite side of the lens. You may have focussed the Sun's rays with a magnifying glass (a biconvex, or converging, lens) and have witnessed the concentration of radiant energy that results. The parallel rays coming from the Sun or some other distant object (at infinity) converge at the focal point. This fact provides a way for experimentally determining the focal length of a converging lens.

Conversely, a biconcave lens can be approximated by two prisms placed point to point [Fig. 23.13(b)]. A biconcave lens is a **diverging lens**: incident parallel rays emerge from the lens as though they emanated from a focal point on the incident side of the lens.

There are several types of converging and diverging lenses (Fig. 23.14). Meniscus lenses are the type most commonly used for corrective eyeglasses. In general, a converging lens is thicker at its center than at its periphery, and a

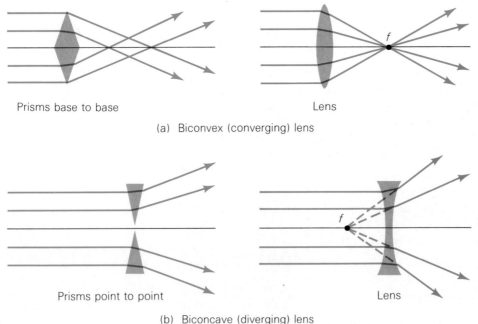

Prisms base to base

Lens

(a) Biconvex (converging) lens

Prisms point to point

Lens

(b) Biconcave (diverging) lens

Figure 23.13 **Converging and diverging lenses**
(a) A biconvex, or converging, lens may be approximated by two prisms placed base to base. For a thin biconvex lens, rays parallel to the axis converge at the focal point. (b) A biconcave, or diverging, lens may be approximated by two prisms placed tip to tip. Rays parallel to the axis of a biconcave lens appear to diverge from a focal point on the incident side of the lens.

diverging lens is thinner at its center than at its periphery. This discussion will be limited to spherically symmetric biconvex and biconcave lenses, that is, ones for which both surfaces have the same radius of curvature.

When light passes through a lens, it is refracted and displaced laterally, as was shown in the last chapter (see Example 22.2). If a lens is thick, this displacement may be fairly large and can complicate the analysis of the lens's characteristics. This problem does not arise with thin lenses, for which the refractive displacement of transmitted light is negligible. This discussion will be limited to thin lenses.

Like a spherical mirror, a lens with spherical geometry has *for each lens surface* a center of curvature, a radius of curvature, a focal point, and a focal length. For thin lenses, the focal length may be measured from either surface vertex with negligible error. If each surface has the same radius of curvature, the focal points are at equal distances on either side of the lens. Note, however, that for a spherical lens $f \neq R/2$, as it does for a spherical mirror (see Section 23.5). Usually, only the focal length of a lens is specified.

The general rules for drawing ray diagrams for lenses are similar to those for spherical mirrors, but obviously some modifications are necessary since light passes through a lens in either direction. Opposite sides of a lens are generally distinguished as the object and image sides. The object side is, of course, the side on which an object is positioned, and the image side is the opposite side of the lens (where a real image would be formed). The rays from a point on an object are drawn as follows:

- A **parallel ray** is a ray that is parallel to the lens axis on incidence and that after refraction either passes through the focal point on the image side of a converging lens, *or* appears to diverge from the focal point on the object side of a diverging lens.

- A **chief ray** is a ray that passes through the center of the lens and is undeviated.*

As with mirrors, only two rays are needed to determine the image. These rules are applied in Fig. 23.15 (on the next page) for several cases. A focal ray is shown only in the first diagram [ray 3 in part (a)]. This ray through the focal point is parallel to the lens axis after refraction.

For a lens, the image is real when formed on the side of the lens opposite the object (on the image side, Fig. 23.16 page 633) and virtual when formed on the same side of the lens as the object (on the object side). Another way of looking at this is that rays actually pass through a real image and only appear to pass through a virtual image. Note that for a diverging lens [Fig. 23.15(d)], a parallel ray diverges after being refracted and *appears* to come from a virtual image on the object side of the lens. Like diverging mirrors, diverging lenses can form only virtual images. Also, as with spherical mirrors, a real image from a lens can be formed on a screen but a virtual image cannot. (See Demonstration 19.)

Regions for the object distance with a converging lens could be similarly defined, as was done for a converging mirror in Fig. 23.9(a). The object distance $d_o = 2f$ for a converging lens has a similar significance as $d_o = R$ for a converging mirror. However, the center of curvature for a converging lens does not have the same distinction as it does for a mirror, since $R \neq 2f$ for a lens.

The image distances and characteristics for a lens can be found analytically.

* This incorporates the thin lens approximation (negligible lateral deviation).

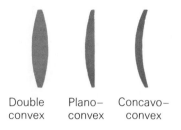

Double convex   Plano–convex   Concavo–convex

Converging lenses

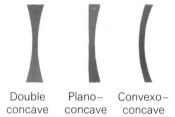

Double concave   Plano–concave   Convexo–concave

Diverging lenses

Figure 23.14 **Lens shapes** The wide variety of lens shapes are generally categorized as converging or diverging.

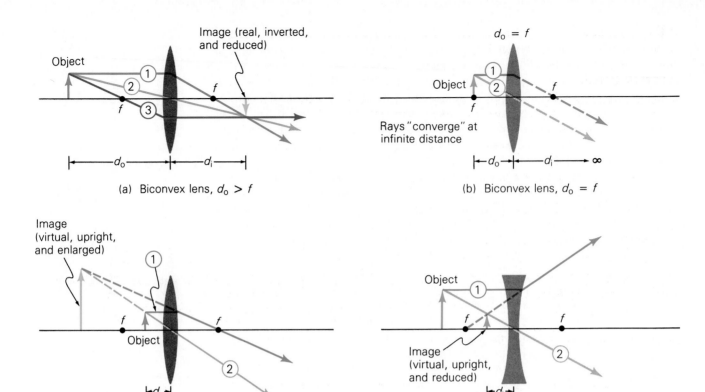

Figure 23.15 **Ray diagrams for lenses**
As in ray diagrams for mirrors, parallel rays (1) and chief rays (2) are usually drawn to analyze lenses. (a) Focal rays (3) can be drawn for lenses, too, as shown here for a biconvex lens with $d_o > f$. (b) This ray diagram illustrates the situation for a biconvex lens with $d_o = f$. (c) When $d_o < f$, a biconvex lens forms a virtual image. (d) This situation corresponds to that in part (a) but the lens is biconcave.

The equations for thin symmetrical biconvex and biconcave lenses are identical to those for spherical mirrors (see Problem 47). The **thin lens equation** is

$$\frac{1}{d_o} + \frac{1}{d_i} = \frac{1}{f} \tag{23.8}$$

thin lens equation

The (lateral) magnification factor is given by

$$M = -\frac{d_i}{d_o} \tag{23.9}$$

magnification equation for a thin lens

The sign conventions for these thin lens equations are similar to those for spherical mirrors:

Inverted Image

A demonstration of how a cylindrical convex lens produces an inverted image.

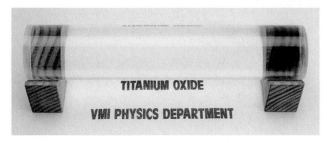

(a) Observe the words TITANIUM OXIDE as placed before a cylindrical plastic rod.

(b) Viewed through the rod, the letters in TITANIUM are inverted, but the letters in OXIDE appear not to be. What is wrong, or is there something wrong?

Note: a similar inversion occurs for spherical convex lens.

- The focal length (f) is positive for a converging lens (sometimes called a positive lens) and negative for a diverging lens (sometimes called a negative lens).

- The object distance (d_o) is always taken as positive.

- The image distance (d_i) is positive for a real image (which is located on the image side of the lens) and negative for a virtual image (which is located on the object side of the lens).

- The magnification (M) is positive for an upright image and negative for an inverted image.

Just as when working with mirrors, it is helpful to sketch a ray diagram before working a lens problem analytically.

Example 23.4 Images with a Converging Lens

A biconvex lens has a focal length of 12 cm. Where is the image formed and what are its characteristics for an object (a) 18 cm from the lens and (b) 4 cm from the lens?

Solution

Given: $f = 12$ cm   *Find*: d_i and image characteristics

      (a) $d_o = 18$ cm
      (b) $d_o = 4$ cm

(a) Using Eq. 23.8,

$$\frac{1}{d_o} + \frac{1}{d_i} = \frac{1}{f}$$

or $$\frac{1}{18} + \frac{1}{d_i} = \frac{1}{12}$$

Figure 23.16 Real image
A converging lens forms a real image of a candle flame on a posterboard screen. Note that the image is inverted.

With a common denominator,

$$\frac{2}{36} + \frac{1}{d_i} = \frac{3}{36}$$

Thus,

$$\frac{1}{d_i} = \frac{1}{36}$$

or $d_i = 36$ cm

Then

$$M = -\frac{d_i}{d_o} = -\frac{36}{18} = -2$$

The image is real (positive d_i), inverted ($-M$), and twice as tall as the object ($|M| = 2$).

(b) Solving Eq. 23.8 for d_i gives

$$d_i = \frac{d_o f}{d_o - f} = \frac{(4 \text{ cm})(12 \text{ cm})}{4 \text{ cm} - 12 \text{ cm}} = -6 \text{ cm}$$

Then

$$M = -\frac{d_i}{d_o} = -\frac{(-6 \text{ cm})}{4 \text{ cm}} = -1.5$$

In this case, the image is virtual (on the object side of the lens), upright, and magnified by a factor of 1.5.

   You should draw ray diagrams for these two cases.   ∎

Example 23.5   Image with a Diverging Lens

A biconcave lens has a focal length of 15 cm. If an object is 30 cm from the lens, where is the image and what are its characteristics?

Solution

Given: $f = -15$ cm   *Find*: d_i and image characteristics
         $d_o = 30$ cm

Note that the focal length is negative since the lens is diverging (biconcave). Then

$$d_i = \frac{d_o f}{d_o - f} = \frac{(30 \text{ cm})(-15 \text{ cm})}{30 \text{ cm} - (-15 \text{ cm})} = -10 \text{ cm}$$

and

$$M = -\frac{d_i}{d_o} = -\frac{(-10 \text{ cm})}{30 \text{ cm}} = \frac{1}{3}$$

Thus, the image is virtual, upright, and $\frac{1}{3}$ of the size of the object.   ∎

Example 23.6   Magnified and Reduced Images

As the object distance of a biconvex lens is varied, at what point does the real image go from being reduced to being magnified?

Solution

Recall that for a converging mirror, the real image of an object at $d_o = R$ is the same size as the object ($M = 1$). Also, for $d_o > R$, the image is smaller, and for

$R > d_o > f$, the image is larger. For a biconvex lens, R does not have the same significance, since $R \neq 2f$. However, an object distance $d_o = 2f$, or a distance $2f$ from the lens, does have this significance for the size of the real image. Substituting $2f$ for d_o in Eq. 23.8 gives

$$\frac{1}{d_o} + \frac{1}{d_i} = \frac{1}{2f} + \frac{1}{d_i} = \frac{1}{f}$$

which can be solved for d_i:

$$d_i = 2f$$

Thus,

$$M = -\frac{d_i}{d_o} = -\frac{2f}{2f} = -1$$

The real, inverted image formed equidistant from the lens is the same size as the object when the object is twice the focal length from the lens.

It can be easily shown that for $d_o > 2f$, the image is smaller than the object, and for $2f > d_o > f$, the image is larger. (You can do this analytically or by quickly sketching two ray diagrams.) ■

Combinations of Lenses

Many optical instruments such as microscopes and telescopes (Chapter 25) use a combination of lenses, or a compound lens system. When two or more lenses are used in combination, the overall image produced may be determined by considering the lenses individually in sequence. That is, the image formed by the first lens is the object for the second lens, and so on.

If the first lens produces an image in front of the second lens, that image is treated as a real object for the second lens [Fig. 23.17(a)]. If, however, the lenses

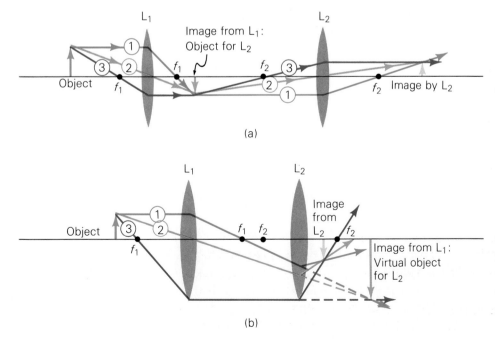

(a)

(b)

Figure 23.17 **Lens combinations** The final image produced by a compound lens system may be found by treating the image from one lens as the object for the adjacent lens. (a) If the image of the first lens (L_1) is formed in front of the second lens (L_2), the object for the second lens is said to be real. (b) If the rays pass through the second lens before the image is formed, the object for the second lens is said to be virtual, and the object distance is taken to be negative.

Fresnel Lenses

To focus or to produce a large beam of parallel light rays, a sizable converging lens is necessary. The large mass of glass necessary to form such a lens is bulky and heavy, and the lens is likely to show aberrations. A French physicist named Augustin Fresnel (1788–1827) developed a solution to this problem for lenses used in lighthouses. Fresnel recognized that the refraction of light takes place at the surfaces of a lens. Essentially, the interior of the lens could be deleted and its surface would still retain converging properties (Fig. 1).

Such a series of concentric curved surfaces that gives the lens segments shape is called a Fresnel lens. These lenses are widely used in overhead projectors and in beacons. A Fresnel lens is very thin and therefore much lighter than a conventional biconvex lens with the same optical properties. Also, Fresnel lenses are easily molded from plastic.

One disadvantage of Fresnel lenses is that concentric circles are visible when an observer is looking through such a lens or when an image magnified by one in an overhead projector is projected on a sceen.

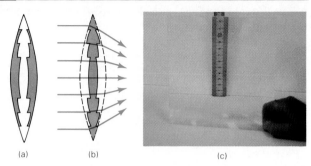

(a)   (b)   (c)

Figure 1 **Fresnel lens**
(a) The focusing action of a lens comes from refraction at its surfaces. Therefore, the interior of the lens may essentially be removed. (b) The resulting series of curved surfaces will have the same effect as the lens from which they are derived. (c) A flat Fresnel lens with concentric curved surfaces magnifies like a biconvex converging lens.

---

are close enough together that the image from the first lens is not formed before the rays pass through the second lens [Fig. 23.17(b)], then a modification must be made in the sign convention. In this case, the image from the first lens is treated as a *virtual* object for the second lens, and the object distance for it is taken to be *negative* in the lens equation.

The total magnification (M_t) of a compound lens system is the product of the magnification factors (absolute values) of all the component lenses. For example, for a two-lens combination, as in Fig. 23.17,

$$M_t = |M_1| \times |M_2| \qquad\qquad (23.10)$$

You should be aware that not all lenses are spherical. A different type of lens is discussed in the Insight feature.

23.4 Lens Aberrations

Lenses, like mirrors, also can have aberrations. Here are several lens aberrations.

Spherical Aberration

The discussion of lenses thus far has focused on thin lenses, for which the lateral deviation of light due to refraction is negligible. In general, with spherical mirrors and lenses, light rays parallel to and near the axis will be reflected or refracted to converge at the focal point. Converging lenses may, however, like

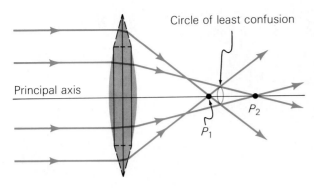

Circle of least confusion

Principal axis

P_1 P_2

Figure 23.18 **Spherical aberration for a lens**
In general, rays closer to the axis of a lens are refracted less and come together at a point farther from the lens than do rays passing through the periphery of the lens. The smallest cross section of the transmitted beam is called the circle of least confusion, and the best (least distorted) image will be formed in its plane.

spherical mirrors, show **spherical aberration**, the effect that occurs when parallel rays passing through different regions of a lens do not come together on a common focal plane. In general, rays close to the axis of a converging lens are refracted less and come together at a point farther away from the lens than do rays passing through the periphery of the lens (Fig. 23.18). The place where the transmitted light beam has the smallest cross section is called the *circle of least confusion*, and the best (least distorted) image will be formed at this location.

Spherical aberration can be minimized by using an aperture to reduce the effective area of the lens, so only light rays near the axis are transmitted. Also, combinations of converging and diverging lenses may be used. The aberration of one lens can be compensated for (nullified) by the optical properties of another lens.

Chromatic Aberration

Chromatic aberration is an effect that occurs because the index of refraction of the material making up a lens is not the same for all wavelengths (that is, the material is dispersive). When white light is incident on a lens, the transmitted rays of different wavelengths (colors) do not have a common focal point, and images of different colors are produced at different locations (Fig. 23.19). This

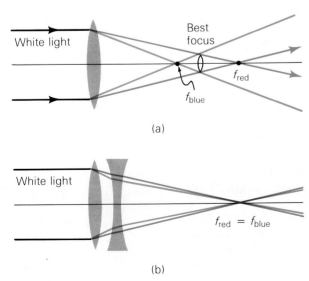

White light

Best focus

f_{red}

f_{blue}

(a)

White light

$f_{red} = f_{blue}$

(b)

Figure 23.19 **Chromatic aberration**
(a) Because of dispersion, different wavelengths (colors) of light are focused on different planes, which results in distortion of the overall image. (b) This aberration may be corrected by using a diverging lens that forms an achromatic doublet with the converging lens.

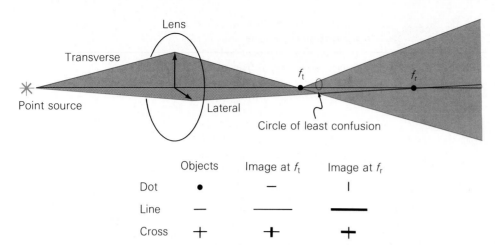

Figure 23.20 **Astigmatism** An off-axis beam forming an elliptical illuminated area gives rise to separate images in different planes.

aberration can be eliminated by using a compound lens system consisting of lenses of different materials, such as crown glass and flint glass. With a properly constructed two-component lens system, called an *achromatic doublet* (achromatic means without color), the images will coincide.

Astigmatism

When a cone of light (circular cross section) from an off-axis source falls on a surface of a spherical lens from some distance away, the light illuminates an elliptical area on the lens. The rays along the major and minor axes of the ellipse focus at different points after passing through the lens, as shown in Fig. 23.20. This condition is called **astigmatism**.

Note in the figure that the image of a point is no longer a point but two separate perpendicular lines. If the object is a horizontal line (a series of points), its image will be focused in one of the planes but not the other, as shown in the figure. (How would the image appear if the object were a vertical line?) Since extended objects usually have both horizontal and vertical dimensions, their images are blurred in both dimensions. The best image is formed somewhere between the horizontal and vertical images, at the location of the circle of least confusion.

23.5 The Lens Maker's Equation ◆

The biconvex and biconcave lenses considered so far in this chapter have been symmetric, with the same focal length for each side because of equal radii of curvature. However, there are converging and diverging lenses that have surfaces with different radii of curvature (see Fig. 23.14 and Fig. 23.21). The focal lengths of such lenses are also important in optical analyses.

In general, the focal length of a thin lens in air is given by the so-called **lens maker's equation**:

$$\frac{1}{f} = (n - 1)\left(\frac{1}{R_1} - \frac{1}{R_2}\right) \tag{23.11}$$

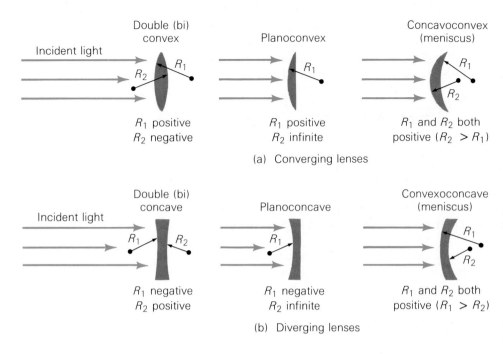

Double (bi) convex

Incident light

R_1 positive
R_2 negative

Planoconvex

R_1 positive
R_2 infinite

Concavoconvex (meniscus)

R_1 and R_2 both positive ($R_2 > R_1$)

(a) Converging lenses

Double (bi) concave

Incident light

R_1 negative
R_2 positive

Planoconcave

R_1 negative
R_2 infinite

Convexoconcave (meniscus)

R_1 and R_2 both positive ($R_1 > R_2$)

(b) Diverging lenses

Figure 23.21 Sign conventions for the lens maker's equation The radius of curvature is taken as positive if it is on the emergent side of the lens and negative if it is on the incident side of the lens. A plane surface has an infinite radius of curvature (with no sign).

For this equation, n is the index of refraction of the material, and R_1 and R_2 are the radii of curvature of the first and second lens surfaces, respectively (that is, the first surface is the one on which light from the object is first incident). The equation locates the focal point at which light passing through a lens will converge for a converging lens or the focal point from which transmitted light appears to diverge for a diverging lens.

The sign conventions for the radii of curvature in the lens maker's equation are included in Fig. 23.21 and summarized in Table 23.3. Note that the signs depend on identifying the location of the center of curvature (C) of a surface relative to either the side of the lens on which light is incident or the side from which it emerges. The lens maker's equation can be used to show why the focal length of a biconvex glass lens is not equal to half the radius of curvature ($f \neq R/2$), as it is for a spherical mirror. For a glass lens to have a focal length that was half of its radius of curvature would require an index of refraction of 2, which is greater than the values for known glasses (see Table 22.1).

Table 23.3

Sign Convention for Lens Maker's Equation

$+R$	when C is on side of lens from which light emerges
$-R$	when C is on side of lens on which light is incident
$R = \infty$	for a plane (flat) surface
$+f$	converging (positive) lens
$-f$	diverging (negative) lens

Example 23.7 Focal Length of a Plano-convex Lens
A thin plano-convex lens made of crown glass has an index of refraction of 1.5. If the radius of curvature for the convex surface is 20 cm, what is the focal length of the lens when light is incident on (a) the convex surface and (b) the plane surface?

Solution
Given: $n = 1.5$

    (a) $R_1 = +20$ cm and $R_2 = \infty$
    (b) $R_1 = \infty$ and R_2
        $= -20$ cm

Find: (a) f
      (b) f

(a) The first surface in this case is the convex surface, so its radius of curvature is R_1. The center of curvature of R_1 is on the emergent side of the lens, so R_1 is positive by the sign convention (see Fig. 23.19). The plane surface is the second surface, and $R_2 = \infty$ (with no positive or negative designation). Then

$$\frac{1}{f} = (n-1)\left(\frac{1}{R_1} - \frac{1}{R_2}\right) = (1.5 - 1)\left(\frac{1}{20 \text{ cm}} - \frac{1}{\infty}\right) = \frac{0.5}{20 \text{ cm}}$$

and

$$f = \frac{20 \text{ cm}}{0.5} = 40 \text{ cm}$$

Thus, the lens is converging (f is positive), and a beam parallel to the axis and incident on the convex surface will converge at a point 40 cm beyond the plane surface of the lens.

(b) In this case, the first surface is the plane surface, and the radius of curvature is negative for the convex surface since its center of curvature is on the incident side of the lens. Using the lens maker's equation with these conditions gives

$$\frac{1}{f} = (n-1)\left(\frac{1}{R_1} - \frac{1}{R_2}\right) = (1.5 - 1)\left[\frac{1}{\infty} - \frac{1}{(-20 \text{ cm})}\right] = \frac{0.5}{20 \text{ cm}}$$

and

$$f = \frac{20 \text{ cm}}{0.5} = 40 \text{ cm}$$

The lens is converging, which you may find unexpected. Draw an off-axis parallel ray and show that convergence is the case when the ray is refracted away from the normal in emerging from the glass lens. ∎

Optometrists express the power (P) of a lens in **diopters**, which is simply the reciprocal of the focal length of the lens expressed in meters:

$$P(\text{diopters}) = \frac{1}{f} \text{ (meters)} \tag{23.12}$$

Note that the lens maker's equation (Eq. 23.11) gives a value in diopters ($1/f$) if the radii of curvature are expressed in meters.

Converging and diverging lenses are referred to as positive ($+$) and negative ($-$) lenses, respectively. Thus, if an optometrist prescribes a corrective lens with a power of $+2$ diopters, this means a converging lens with a focal length of

$$f = 1/P = 1/2 = 0.50 \text{ m} = 50 \text{ cm}$$

The greater the power of a lens in diopters, the shorter its focal length and the more converging or diverging it is.

Example 23.8   Diopters and Lens Grinding

A lens maker must fill a prescription for a corrective convexo-concave lens with a power of -2.1 diopters. The lens is to be ground from a glass blank ($n = 1.7$) having a convex front surface whose radius of curvature is 50 cm. To what radius of curvature should the second (rear) surface of the lens be ground?

Solution

Given: $P = 1/f = -2.1$ diopters (m^{-1}) *Find*: R_2
  $n = 1.7$
  $R_1 = 50 \text{ cm} = 0.50 \text{ m}$

Since the center of curvature of the convex surface is on the emergent side of the lens, R_1 is positive (see Fig. 23.19).

$$P = \frac{1}{f} = (n - 1)\left(\frac{1}{R_1} - \frac{1}{R_2}\right)$$

or $-2.1 = (1.7 - 1)\left(\frac{1}{0.50 \text{ m}} - \frac{1}{R_2}\right)$

Solving for R_2 gives

$$\frac{1}{R_2} = \frac{1}{0.50 \text{ m}} + \frac{2.1}{0.7 \text{ m}} = 5.0 \text{ m}^{-1}$$

and

$$R_2 = \frac{1}{5.0 \text{ m}} = 0.20 \text{ m} = 20 \text{ cm}$$

A positive value for R_2 indicates that the center of curvature for the second surface is on the emergent (or back) side of the lens, as is that for R_1. With $R_1 > R_2$, the lens is diverging. ∎

Important Formulas

Focal length for spherical mirror:

$$f = \frac{R}{2}$$

Spherical mirror equation:

$$\frac{1}{d_o} + \frac{1}{d_i} = \frac{1}{f}$$

Thin lens equation (where $f \neq R/2$):

$$d_i = \frac{d_o f}{d_o - f}$$

Magnification:

$$M = -\frac{d_i}{d_o}$$

Total magnification with a two-lens system:

$$M_t = |M_1| \times |M_2|$$

Lens maker's equation:

$$\frac{1}{f} = (n - 1)\left(\frac{1}{R_1} - \frac{1}{R_2}\right)$$

Lens power in diopters (where f is in meters):

$$P = \frac{1}{f}$$

Sign conventions for spherical mirrors and lenses:

Concave mirror ⎫
Biconvex lens ⎬ f positive   Convex mirror ⎫
                                    Biconcave lens ⎬ f negative

d_o is always positive.

When d_i is positive, image is real; when d_i is negative, image is virtual.

When M is positive, image is upright; when M is negative, image is inverted.

◆ *Sign convention for lens maker's equation*:

  $+R$ when C is on side of lens from which light emerges
  $-R$ when C is on side of lens on which light is incident
  $R = \infty$ for a plane (flat) surface
  $+f$ converging (positive) lens
  $-f$ diverging (negative) lens

Questions

Plane Mirrors

1. If an object touches the surface of a plane mirror made of ordinary glass, do the object and its image touch? How about for a polished metal mirror?

2. If you walk toward a full-length plane mirror, what does your image do? Is your image in step with you?

3. Can a plane mirror produce a magnified image? Is the image of a plane mirror ever real? Explain.

4. (a) A transparent glass windowpane can serve as a mirror, but reflections in it are seen more clearly at night. Why? (Hint: the stars are still shining during the day even though you can't see them.) (b) When you look at a reflecting windowpane at night, you may see two similar images. Why are there two images?

5. One-way mirrors reflect on one side and can be seen through from the other. (This effect is also used on some sunglasses.) What principle is applied in making a one-way mirror? (Hint: at night a windowpane may be a one-way mirror.)

6. Plane mirrors show right-left reversal of images. Why is there no up-down reversal?

Spherical Mirrors

7. Explain why a chief ray is reflected back along its line of incidence.

8. (a) What is the reason for using a dual truck mirror such as the one shown in Fig. 23.22(a)? (b) Could the TV satellite dish shown in Fig. 23.22(b) be considered a converging mirror? Explain.

9. What are the image characteristics with a concave mirror if (a) $d_o > R$, (b) $d_o < f$, and (c) $d_o = f$?

10. What are the image characteristics with a convex mirror if (a) $R > d_o > f$, (b) $d_o = f$, and (c) $d_o < f$?

11. Describe the image changes in the image when an object is slowly moved from a distance greater than the radius of curvature ($d_o > R$) to a position near the surface ($d_o < f$) of (a) a concave mirror and (b) a convex mirror.

12. Can a concave mirror be used to burn a hole in a piece of paper the way a magnifying glass will? Explain.

13. When people stand in front of a type of mirror found in amusement parks, they see themselves with small heads and large lower torsos. How is this accomplished?

14. A popular novelty item consists of a concave mirror with a ball suspended at or slightly inside the center of curvature (see Fig. 23.23). When the ball swings toward the mirror, its image grows larger and suddenly fills the whole mirror. The effect is that the image appears to be jumping out of the mirror. Explain what is happening.

15. Some rear-view mirrors on the passenger side of automobiles have the written warning "OBJECTS IN MIRROR ARE CLOSER THAN THEY APPEAR." Explain why this is so. (Hint: such mirrors are not plane mirrors.)

Figure 23.23 **Spherical mirror toy**
See Question 14.

Lenses

16. Suppose that parallel light rays are incident on a biconvex lens. A biconcave lens having the same radii of curvature as the biconvex lens is placed on its other side midway between its surface and the focal point. In what direction(s) will the light rays emerge from the second lens?

17. How can the focal length be found experimentally for (a) a concave mirror and (b) a convex lens?

18. Can a biconcave lens be used to magnify images that are to be projected onto a screen? Explain. (Hint: consider a compound lens system.)

19. Show mathematically why application of the thin

Figure 23.22 **Mirror applications**
See Question 8.

lens equation (Eq. 23.8) to a diverging lens always gives a virtual image.

20. Does the thin lens equation apply to lenses whose surfaces have unequal radii of curvature? Explain.

21. What are the image characteristics with a biconvex lens if (a) $d_o > R$, (b) $d_o < f$, and (c) $d_o = f$?

22. What are the image characteristics with a biconcave lens if (a) $R > d_o > f$, (b) $d_o = f$, and (c) $d_o < f$?

23. What are the position and characteristics of the image produced by a biconvex lens when an object is at a distance that is twice the focal length of the lens? Is the object at the center of curvature of the lens? Explain.

24. Describe the changes in the image when an object is slowly moved from a distance greater than the radius of curvature ($d_o > R$) to a position near the surface ($d_o < f$) of (a) a convex lens and (b) a concave lens.

Lens Aberrations

25. Explain what is meant by (a) spherical aberration, (b) chromatic aberration, and (c) astigmatism. Do these effects occur with both mirrors and lens? Explain how each of them may be prevented.

26. What types of aberrations might affect the images formed by (a) front-coated convex mirrors and (b) back-coated convex mirrors.

The Lens Maker's Equation ◆

27. Why is the radius of curvature for a plane mirror not designated as positive or negative?

28. What does it mean to say that one lens has a greater power than another? Does lens power apply to both converging and diverging lenses? Explain.

Problems

23.1   Plane Mirrors

■1. A person stands 3.0 m away from the reflecting surface of a plane mirror. (a) What is the apparent distance between the person and his or her image? (b) What are the image characteristics?

■2. An object 3.0 cm tall is placed 100 cm from a plane mirror. Find (a) the distance from the object to the image, (b) the height of the image, and (c) the magnification of the image.

■■3. A small dog sits in front of a plane mirror at a distance of 2.0 m. (a) Where is the dog's image? (b) If the dog jumps at the mirror with a speed of 0.5 m/s, how fast does it approach its image?

■■4. A man 2.0 m tall stands in front of a plane mirror. (a) What is the minimum height the mirror must be to allow the man to view his complete image from head to foot? Assume that his eyes are 10 cm below the top of his head. (Hint: he will see his feet via a ray reflected from the bottom edge of the mirror and the top of his head via a ray reflected from the top edge of the mirror.) (b) Does the minimum height of the mirror depend on how far the man stands from the mirror? Explain.

■■5. Draw ray diagrams showing how three images of an object are formed in two plane mirrors at right angles as shown in Fig. 23.24(a). (Hint: consider rays from each end of the object arrow in the drawing for each image.) Fig. 23.24(b) shows a similar situation from a different point of view that gives four images. Can you explain the extra image in this case?

■■■6. A student who is 1.6 m tall stands in front of a plane mirror that is tilted at an angle of 20° away from her. If the student's eyes are 12 cm below the top of her head, find the minimum height the mirror must be to allow her to see her complete image (head to foot) in it.

■■■7. Find the minimum height of the mirror in Problem 6 if it is tilted at an angle of 20° toward the student.

23.2   Spherical Mirrors

■■8. An object 5.0 cm tall is located 100 cm from the concave surface of a mirror with a radius of curvature of 50 cm. Where is the image located and what are its characteristics?

■■9. A candle with a flame 2.5 cm tall is placed 5.0 cm from the front of a concave mirror. A virtual image is

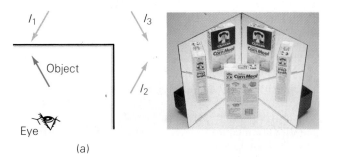

Figure 23.24   Two mirrors—multiple images See Problem 5.

produced that is 10 cm from the vertex of the mirror. (a) Find the focal length and radius of curvature of the mirror. (b) How tall is the image?

■■**10.** Show that the magnification for objects near the optic axis of a convex mirror is given by $M = d_i/d_o$. (Hint: use a ray diagram with rays reflected at the mirror's vertex.)

■■**11.** An object 3.0 cm tall is placed 20 cm from the front of a concave mirror with a radius of curvature of 30 cm. Where is the image formed and how tall is it?

■■**12.** If the object in Problem 11 is moved to a position 10 cm from the front of the mirror, what will the characteristics of the image be?

■■**13.** If an object is 15 cm in front of a convex mirror that has a focal length of 30 cm, how far behind the mirror will the image appear to an observer, and how tall will it be?

■■**14.** A concave mirror has a magnification of 4.0 for an object placed 50 cm in front of the mirror. (a) What type of image is produced? (b) Find the radius of curvature of the mirror.

■■**15.** A virtual image at a magnification of 0.50 is produced when an object is placed in front of a spherical mirror. (a) What type of mirror is it? (b) Find the radius of curvature of the mirror if the object is 5.0 cm from it.

■■**16.** A concave shaving mirror is constructed so that a man at a distance of 20 cm from the mirror sees his image magnified 1.5 times. What is the radius of curvature of the mirror?

■■**17.** An object 2.0 cm tall is placed 15 cm from the front of a convex mirror whose focal length is 30 cm. Find the location of the image and its height.

■■**18.** A dentist uses a spherical mirror that produces an upright image that is magnified four times. What is the focal length of the mirror in terms of the object distance?

■■**19.** A wooden cube 5.0 cm on a side is placed a distance of 30 cm in front of a converging mirror having focal length of 20 cm. Where is the image of the front surface of the block located and what are its characteristics?

■■**20.** Show that the spherical mirror equation and magnification equation give the correct image characteristics for a plane mirror.

■■**21.** A 15-cm pencil is placed with its eraser on the optic axis of a concave mirror and its point directed upward at a distance of 25 cm in front of the mirror. The radius of curvature of the mirror is 30 cm. (a) Where is the image of the pencil formed, and what are its characteristics? (b) Draw a ray diagram for the situation where the pencil point is directed downward from the optic axis.

■■**22.** The image of an object located 30 cm from a concave mirror is formed on a screen located 20 cm from the mirror. What is the mirror's radius of curvature?

■■**23.** A dentist holds a concave mirror whose radius of curvature is 40 mm at a distance of 15 mm from a tooth in order to view a small cavity. How large is the image of the cavity?

■■**24.** A converging mirror with a radius of curvature of 90 cm produces an upright, virtual image 45 cm tall and located 30 cm from the vertex of the mirror. What are the height and location of the object?

■■**25.** A child looks at a reflecting Christmas tree ball that has a diameter of 12 cm and sees an image of her face that is half the real size. How far is the child's face from the ornament?

■■**26.** An object 2.0 cm tall is placed 6.0 cm from the front of a mirror. An upright image 4.0 cm tall is formed. What kind of mirror is it, and what is its radius of curvature?

■■**27.** A mirror at an amusement park shows anyone who stands 2.5 m from it an upright image three times the person's height. What is the mirror's radius of curvature?

■■**28.** Use the spherical mirror equation and magnification equation to find the image distance and characteristics of an object when $d_o = f$. $d_i = \infty$, $M = \infty$

■■■**29.** (a) Sketch graphs of (1) d_i versus d_o and (2) M versus d_o for a concave mirror, for values of d_o from 0 to ∞. (b) Sketch similar graphs for a convex mirror.

■■■**30.** A convex mirror is used on the exterior of the passenger side of many cars. If the focal length of such a mirror is -40.0 cm, what will the location and height of the image of a car that is 2.0 m tall be if the car is (a) 100 m behind and (b) 10.0 m behind? See Question 15.

■■■**31.** Two students in a physics laboratory each have a concave mirror with the same radius of curvature, 40 cm. Each student places an object in front of a mirror. The image in both mirrors is three times the size of the object. However, when the students compare notes, they find that the object distances are not the same. Is this possible? If so, what are the object distances?

■■■32. Show that $f = R/2$ for a converging mirror. [Hint: draw a ray diagram for the situation where $R > d_o > f$ using all three types of rays, as in Fig. 23.8(a). Check for similar triangles.]

23.3 Lenses

■■33. An object 4.0 cm tall is placed in front of a converging lens whose focal length is 18 cm. Where is the image formed and what are its characteristics if the object distance is (a) 15 cm and (b) 36 cm?

■■34. A biconvex lens produces a real, inverted image of an object that is magnified 2.5 times when the object is 20 cm from the lens. What is the radius of curvature of the lens?

■■35. An object is placed in front of a biconcave lens whose focal length is 20 cm. Where is the image located and what are its characteristics if the object distance is (a) 10 cm and (b) 25 cm?

■■36. (a) Design a single-lens projector that will form a sharp image on a screen 4.0 m away with the transparent slides 6.0 cm from the lens. (b) If the object on a slide is 1.0 cm tall, how tall will the image on the screen be, and how should the slide be placed in the projector?

■■37. A single-lens camera (biconvex lens) is used to photograph a man 1.8 m tall who is standing 5.0 m from the camera. If the man's image fills the length of a frame of 35-mm film, what is the focal length of the lens?

■■38. A photographer uses a single-lens camera having a focal length of 60 mm to photograph a full moon. What will be the size of the moon's image on the film? (Hint: data on the moon may be found in Appendix I.)

■■39. An object is placed 20 cm from a screen. At what point between the object and the screen should a converging lens with a focal length of 5.0 cm be placed so it will produce a sharp image on the screen?

■■40. A lens with a power of $+5.0$ diopters is used to produce an image on a screen that is 2.0 m from the lens. How many times is the object magnified?

■■41. (a) If a book is held 30 cm from an eyeglass lens with a power of -2.5 diopters, where is the image of the print formed? (b) If an eyeglass lens with a power of $+2.5$ diopters is used, where is the image formed?

■■42. (a) For a biconvex lens, what is the minimum distance between an object and its image if the image is real? (b) What is the distance if the image is virtual? (Hint: use ray diagrams and the thin lens equation.)

■■■43. (a) Sketch graphs of (1) d_i versus d_o and (2) M versus d_o for a converging lens, for values of d_o from 0 to ∞. (b) Sketch similar graphs for a diverging lens.

■■■44. Two positive lenses, each having a power of 10 diopters, are placed 20 cm apart along the same axis. If an object is 60 cm from the first lens on the side opposite the second lens, where is the final image relative to the first lens, and what are its characteristics?

■■■45. An object is 15 cm from a converging lens whose focal length is 10 cm. On the opposite side of that lens at a distance of 60 cm from the object is another converging lens with a focal length of 20 cm. Where is the final image formed, and what are its characteristics?

■■■46. Two converging lenses, L_1 and L_2, have focal lengths of 30 cm and 20 cm, respectively. The lenses are placed 60 cm apart along the same axis, and an object is placed 50 cm from L_1 on the side opposite to L_2. Where is the image formed relative to L_2, and what are its characteristics?

■■■47. Using Fig. 23.25, derive (a) the thin lens equation and (b) the lens magnification equation.

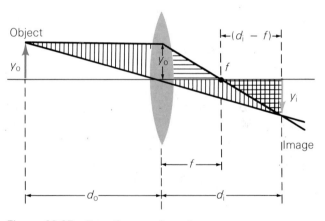

Figure 23.25 **Ray diagram for a lens**
See Problem 47.

■■■48. Show that for thin lenses with focal lengths f_1 and f_2 and in contact the effective focal length (f) is given by

$$\frac{1}{f} = \frac{1}{f_1} + \frac{1}{f_2}$$

■■■49. A method of determining the focal length of a diverging lens is called autocollimation. As Fig. 23.26 shows, first, a sharp image of a light source is projected on a screen by a converging lens. Second, the screen is replaced

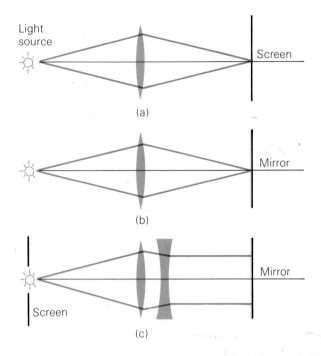

(a)

(b)

Screen

(c)

Figure 23.26 **Autocollimation**
See Problem 49.

with a plane mirror. Third, a diverging lens is placed between the converging lens and the mirror. Light will then be reflected by the mirror back through the compound lens system, and an image will be formed on a screen near the light source. This image is made sharp by adjusting the distance between the diverging lens and the mirror. The distance at which the image is clearest is equal to the focal length of the lens. Explain why this is true.

23.5 The Lens Maker's Equation ◆

■■**50.** A biconvex lens made of glass whose index of refraction is 1.5 has radii of curvature of 30 cm for one surface and 40 cm for the other. What is the focal length of the lens for light incident on each side?

■■**51.** Prove that each of the lenses shown in Fig. 23.21 is converging or diverging no matter which direction the incident light comes from. (Hint: turn the lenses around.)

■■**52.** An optometrist prescribes a corrective lens with a power of +1.5 diopters. The lens maker will start with glass blank with an index of refraction of 1.6 and a convex front surface whose radius of curvature is 20 cm. To what radius of curvature should the other surface be ground?

■■**53.** A plastic plano-concave lens has a radius of curvature of 50 cm for its concave surface. If the index of refraction of the plastic is 1.35, what is the power of the lens?

54. A concave mirror with a radius of curvature of 50 cm reflects sunlight. Find (a) the location and (b) the diameter of the Sun's image.

55. An object 3.0 cm tall is placed at different locations in front of a concave mirror whose radius of curvature is 30 cm. Determine the location of the image and its characteristics when the object distance is (a) 40 cm, (b) 30 cm, (c) 20 cm, (d) 15 cm, and (e) 5.0 cm.

56. Draw a ray diagram for each of the cases in Problem 55.

57. An object 5.0 cm tall is 10 cm from a concave lens. The resulting image is $\frac{1}{5}$ as large as the object. What are the focal length and power of the lens?

58. Find the location and magnification of the Sun's image with a convex lens whose focal length is 20 cm.

59. A boy runs toward a plane mirror with a speed of 6.0 m/s. What is the speed of the boy's image relative to him?

60. (a) Show that a ray parallel to the axis of a biconvex lens is refracted toward the axis at the incident surface and again at the exit surface. (b) Show that the refraction effect described in part (a) also holds for a biconcave lens?

61. The image of an object 18 cm from a convex mirror is half the size of the object. What is the focal length of the mirror?

62. A light source and a screen are separated by a distance d, and a converging lens with a focal length f can be placed anywhere between them. Determine whether sharp images will be formed on the screen (a) when $d < 4f$ and (b) when $d > 4f$.

63. A concave mirror has a radius of curvature of 20 cm. For what *two* object distances will the image have twice the height of the object?

64. A biconvex lens has a focal length of 5.0 cm. Where on the lens axis should an object be placed in order to get (a) a real, enlarged image with a magnification of 2.5 and (b) a virtual, enlarged image with a magnification of 2.5?

65. A symmetric biconvex lens has a focal length of 50 cm and is made of glass whose index of refraction is 1.5. What is the radius of curvature of both lens surfaces?

Physical Optics: The Wave Nature of Light

24

The phenomena of reflection and refraction are conveniently analyzed using geometrical optics. Ray diagrams show what happens when light is reflected from a mirror or passed through a lens. However, some other phenomena involving light cannot be adequately explained or described using rays, since this technique ignores the wave nature of light. These phenomena include interference, diffraction, and polarization.

Physical optics, or **wave optics**, takes into account wave motion and characteristics. The wave theory of light leads to satisfactory explanations of those phenomena that cannot be analyzed with rays. Thus, this chapter again considers waveforms.

24.1 Young's Double-Slit Experiment

It has been stated in this book that light is a wave, but no proof of this has been given. How would you go about demonstrating the wave nature of light? One method that involves the use of interference was first carried out in 1801 by the English scientist Thomas Young (1773–1829). Young's double-slit experiment not only demonstrated the wave nature of light but also allowed him to measure its wavelengths.

Recall from the discussion of wave interference in Chapter 14 that superimposed waves may interfere constructively or destructively. Total constructive interference occurs when two crests are superimposed, and total destructive interference occurs when a crest and a trough of two identical waves are superimposed. Interference can be observed with water waves (Fig. 24.1), for which constructive and destructive interference produce obvious interference patterns.

The interference of light waves is not as easily observed because of their relatively short wavelengths ($\cong 10^{-7}$ m). Also, a visually observable interference pattern can be produced only with *coherent sources*, that is, sources that produce light waves having a constant phase relationship to one another.

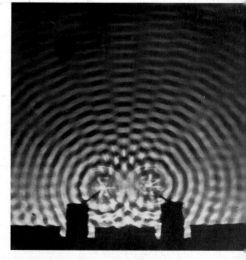

Figure 24.1 **Water wave interference**
The constructive and destructive interference of water waves from two sources in a ripple tank produce interference patterns.

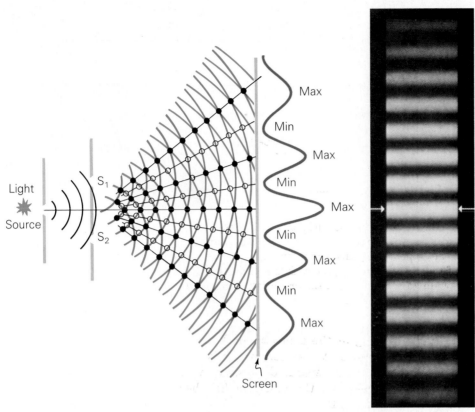

Figure 24.2 **Double-slit interference**
The coherent waves from two slits spread out and interfere, producing alternate maxima and minima, or bright and dark fringes, on a screen.

For example, for constructive interference to occur at some point, the waves meeting at that point must be in phase. As the waves meet, a crest must always overlap a crest and a trough must always overlap a trough. If a phase difference develops between the waves with time, the interference pattern changes, and a stable or stationary pattern will not be set up.

In an ordinary light source, the atoms are excited randomly, and the emitted light waves have amplitude and frequency fluctuations. Thus, light from two such sources is incoherent and does not produce a stationary interference pattern. Interference does occur, but the phase difference between the interfering waves changes so fast that the interference effects are not discernible.

To overcome this problem, Young used a single source to illuminate two narrow slits [Fig. 24.2(a)]. Light passing through the slits spreads out, so the slits essentially act as two sources. If monochromatic light is used (Young used sunlight), the light emerging from the slits is coherent, since the slits merely separate the original beam into two parts. Any random changes in the light from the source will also occur for the light passing through the slits, and the phase difference will be constant.

Coming from two coherent "sources," the interfering waves produce a stable interference pattern on a screen. The pattern consists of a series of bright and dark **interference fringes**, which mark the positions at which constructive and destructive interferences occur (Fig. 24.3). The pattern is symmetrical about a bright central maximum, and the bright fringes decrease in intensity on either side of this central maximum. This interference pattern demonstrates the wave nature of light.

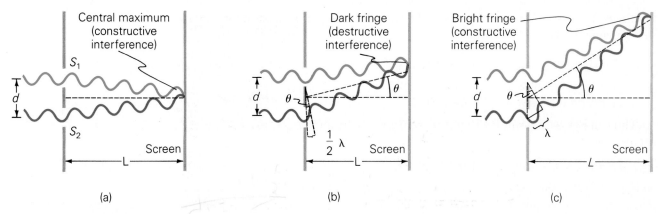

(a)   (b)   (c)

Figure 24.3 Interference
The interference that produces bright and dark fringes depends on the difference in path lengths for the light from the two slits. (a) The path difference at the position of the central maximum is zero, so the waves arrive in phase and interfere constructively. (b) At the position of the first dark fringe, the path difference is $\lambda/2$, and the waves interfere destructively. (c) At the position of the first bright fringe, the path difference is λ, and the interference is constructive.

Measuring the wavelength requires looking at the geometry of Young's experiment, shown in Fig. 24.4. Let P be a point on the screen, which is at a distance L from the double slits. The slits S_1 and S_2 are separated by a distance d. Note that the light path from one slit to P is longer than the path from the other slit to P. As the figure shows, the path difference (PD) is

   $\text{PD} = d \sin \theta$

The relationship of the phase difference of two waves to their path difference was discussed in Chapter 15 for the interference of sound waves. The conditions for interference given for sound waves hold for any sinusoidal waves, including light waves. Recall that constructive interference occurs at any point where the path difference between two coherent, in-phase waves is an integral number of wavelengths:

$$\text{PD} = n\lambda \qquad \text{for } n = 1, 2, 3, \ldots \tag{24.1}$$

condition for constructive interference

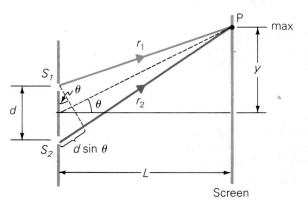

Screen

Figure 24.4 Geometry of Young's experiment
The difference in path length for light from the two slits traveling to a position P, where there is an arbitrary maximum, or bright fringe, is $r_2 - r_1 = d \sin \theta$. By the condition for constructive interference, $d \sin \theta = n\lambda$, where $n = 0, 1, 2, 3, \ldots$ is the order of interference. (Drawing is not to scale, $y \ll L$.)

Similarly, for destructive interference, the path difference is an odd number of half-wavelengths:

$$\text{PD} = \frac{m\lambda}{2} \qquad \text{for } m = 1, 3, 5, \ldots \tag{24.2}$$

condition for destructive interference

Thus, in Fig. 24.4, point P will be the location of a bright fringe (constructive interference) if

$$d \sin \theta = n\lambda \qquad \text{for } n = 0, 1, 2, 3, \ldots \tag{24.3}$$

where n is called the order number. The zeroth-order fringe ($n = 0$) corresponds to the central maximum, the first-order fringe ($n = 1$) is the first bright fringe on either side of the central maximum, and so on. Basically, varying path differences give rise to different phase differences at different points.

The wavelength can therefore be determined by measuring d and θ for a particular order bright fringe (other than the central maximum). The angle θ is related to the displacement of a fringe (y), which is conveniently measured from a photograph of the interference pattern. If P is a point at the center of a bright fringe, the distance from the center of the central maximum to that nth fringe is (see Fig. 24.4)

$$y = L \tan \theta = L \left(\frac{\sin \theta}{\cos \theta} \right) \tag{24.4}$$

Since θ is very small ($y << L$), $\cos \theta \cong 1$ and $\tan \theta$ may be approximated by $\sin \theta$. Therefore,

$$y \cong L \sin \theta \qquad \text{or} \qquad \sin \theta \cong \frac{y}{L} \tag{24.5}$$

Substituting this expression for $\sin \theta$ in Eq. 24.3 and then solving for y gives the distance of the nth bright fringe (y_n) from the central maximum:

$$y_n = \frac{nL\lambda}{d} \tag{24.6}$$

Thus, the wavelength of the light is

$$\lambda = \frac{y_n d}{nL} \qquad \text{for } n = 1, 2, 3, \ldots \tag{24.7}$$

A similar analysis gives the distances to the dark fringes (see Problem 10).

Example 24.1   Measuring the Wavelength of Light

Monochromatic light passes through two narrow slits that are 0.050 mm apart. A picture of the interference pattern is taken with a camera 1.0 m from the slits, and the second-order bright fringe is measured as being 2.4 cm from the center of the central maximum on the photograph. (a) What is the wavelength of the light? (b) What is the distance between the second-order and third-order bright fringes?

Solution

Given: $d = 0.050$ mm $L = 1.0$ m $= 10^3$ mm   Find: (a) λ
       $y_2 = 2.4$ cm $= 24$ mm   (b) $y_3 - y_2$
       $n = 2$

(a) Using Eq. 24.7 with distances in millimeters for convenience gives

$$\lambda = \frac{y_n d}{nL} = \frac{(24 \text{ mm})(0.050 \text{ mm})}{2(10^3 \text{ mm})} = 6.0 \times 10^{-4} \text{ mm} = 6.0 \times 10^{-7} \text{ m}$$

This is 600 nm, which is the wavelength of yellow-orange light (see Fig. 20.16). (b) With the wavelength, y_3 could be computed, and then the distance between the second-order and third-order fringes $(y_3 - y_2)$ could be found. However, the bright fringes for a given wavelength of light are evenly spaced; in general, the distance between adjacent bright fringes is constant:

$$\Delta = y_{n+1} - y_n = \frac{(n+1)L\lambda}{d} - \frac{nL\lambda}{d} = \frac{L\lambda}{d}$$

In this case,

$$\Delta = \frac{L\lambda}{d} = \frac{(1.0 \text{ m})(6.0 \times 10^{-7} \text{ m})}{0.050 \times 10^{-3} \text{ m}} = 1.2 \times 10^{-2} \text{ m} = 1.2 \text{ cm} \quad \blacksquare$$

24.2 Thin-Film Interference

Have you ever wondered what causes the rainbowlike colors when light is reflected from a thin film of oil or a soap bubble? This effect is due to interference of light reflected from opposite surfaces of the film and may be readily understood in terms of wave interference.

First, however, you need to see how the phase of a light wave is affected by reflection. Recall from Chapter 14 that a wave pulse undergoes a 180° phase shift (change) when reflected from a rigid support and no phase shift when reflected from a free support (Fig. 24.5). Similarly, as the figure shows, the phase change for the reflection of light waves at a boundary depends on the optical densities,

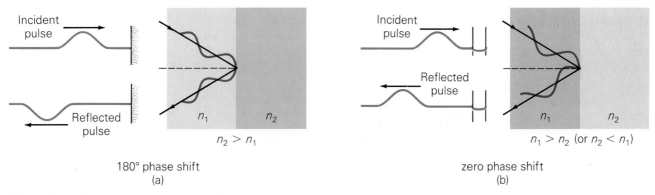

Figure 24.5 **Reflection and phase shifts**
The phase changes that light waves undergo on reflection are analogous to those for pulses in strings. (a) The phase of a pulse in a string is shifted 180° on reflection from a fixed end, and so is the phase of a light wave when it is reflected from a more optically dense medium. (b) A pulse in a string has phase shift of zero (is not shifted) when reflected from a free end. Analogously, a light wave is not phase-shifted when reflected from a less optically dense medium.

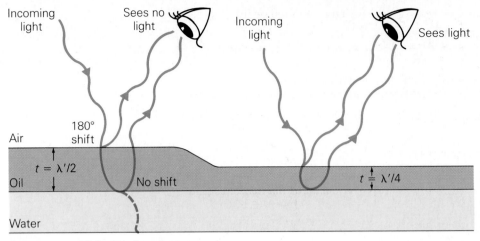

Figure 24.6 **Thin-film interference**
For an oil film on water, there is a 180° phase shift for light reflected from the air-oil interface and a zero phase shift at the oil-water interface. (a) Destructive interference occurs if the oil film has a thickness of $\lambda'/2$. (b) Constructive interference occurs with film thickness of $\lambda'/4$. (c) Thin-film interference in an oil slick.

or the indices of refraction, of the two materials:

■ A light wave traveling in one medium and reflected from the boundary of a second medium whose index of refraction is greater than that of the first medium ($n_2 > n_1$) undergoes a 180° phase change.

■ If the reflecting medium has the smaller index of refraction ($n_2 < n_1$), there is no phase change.

To understand the colors you see in an oil film (on water or a wet road), consider the reflection of monochromatic light from a thin film, as illustrated in Fig. 24.6. (The rays in the figure are drawn at an angle for purposes of illustration, but the discussion will assume they are incident normally.) The oil film has a greater index of refraction than air, and the light reflected from the air-oil interface undergoes a 180° phase shift. The transmitted waves pass through the oil film and are reflected at the oil-water interface. In general, the index of refraction of oil is greater than that of water (see Table 22.1); that is, $n_2 < n_1$, so a reflected wave in this instance is *not* phase shifted.

You might think that if the path length of the wave in the oil film (2t, twice the thickness, down and back) were an integral number of wavelengths, then the waves reflected from the two surfaces would interfere constructively. But keep in mind that the wave reflected from the top surface undergoes a 180° phase shift. The reflected waves from the two surfaces are therefore out of phase for this film thickness, and they interfere destructively. This means that the light of this wavelength is not reflected, but transmitted. Similarly, if the path length in the film were an odd number of half-wavelengths, the reflected waves would be in phase (as a result of the 180° phase shift of the incident wave), and they would interfere constructively. Light of this wavelength would be reflected from the film.

Because oil and soap films generally have different thicknesses in different regions, particular wavelengths (colors) of white light interfere constructively

The path length is in terms of the wavelength of the light *in the film*. Recall from Chapter 22 that the wavelength is different in different media. Here $\lambda' = \lambda/n$, where λ' is the wavelength in the oil and λ the wavelength in air. This may be seen from $n = c/v = f\lambda/f\lambda' = \lambda/\lambda'$.

in different regions and are reflected (see Demonstration 20). As a result, a vivid display of various colors appears, which may change if the film thickness changes with time. (A similar display of colors is seen if two glass slides are stuck together with an air film between them.)

From this analysis of thin-film interference, you can see that the word "destructive" in this context does not imply that energy is destroyed. Destructive interference is simply a description of a physical fact—that a light wave is not present at a particular location, but is somewhere else. When destructive interference occurs for light incident on the surfaces of a thin film, this tells you that the light is not reflected, but that the incident beam (energy) is transmitted. Similarly, the mathematical description of Young's double-slit experiment in the preceding section tells you that you *should not expect* to find light at the locations of the dark fringes. The light is at the bright fringes, and interference merely redistributes the light.

A practical application of thin-film interference is described in the Insight feature.

Nonreflecting Lenses

You have probably noticed the blue tint of the coated optical lenses used in cameras and binoculars. The coating makes the lenses "nonreflecting." If a lens is nonreflecting, the incident light is totally transmitted. Complete transmission of light is desirable for the exposing of photographic film and for viewing objects with binoculars.

A lens is made nonreflecting by coating it with a thin film of a material that has an index of refraction between those of air and glass (Fig. 1). If the coating is a quarter-wavelength thick, the difference in path length between the reflected rays is $\lambda'/2$, where λ' is the wavelength of light in the coating. In this case, both reflected waves undergo a 180° phase shift, and they are out of phase for a path difference of $\lambda'/2$ and interfere destructively. That is, the incident light is transmitted, and the coated lens is nonreflecting.

A quarter-wavelength thickness of film is, of course, specific for a particular wavelength of light. The thickness is usually chosen to be a quarter-wavelength of yellow-green light ($\lambda \cong 550$ nm), to which the human eye is most sensitive. The shorter wavelengths at the blue end of the visible region are reflected, giving the coated lens its bluish hue.

The coating on a nonreflecting lens actually serves a double purpose. It promotes nonreflection from the front of the lens and cuts down on back reflection. Some of the light transmitted through a lens is reflected from the back surface. This could be reflected again from the front surface of an uncoated lens, giving rise to poor images. However, the proper lens thickness allows back reflections to be transmitted through the coating.

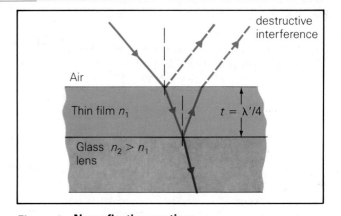

Figure 1 Nonreflective coating
For a thin film on a glass lens, there is a 180° phase shift at each interface when the index of refraction of the film is less than that of the glass. As a result, destructive interference occurs for a minimum film thickness of $\lambda'/4$, and the waves are transmitted rather than reflected, making the lens surface nonreflecting.

Nonreflective coatings are also applied to the surfaces of solar cells, which convert light into electrical energy (Chapter 27). Since the thickness of such a coating is wavelength-dependent, like that on a nonreflecting lens, not all of the light is transmitted. However, the reflective losses may be decreased from around 30% to 10%, making the cell more efficient.

Demonstration 20

Thin-Film Interference

A vivid display of colors from soap bubbles illuminated with white light from below.

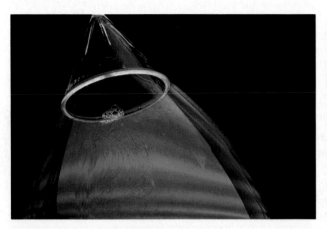

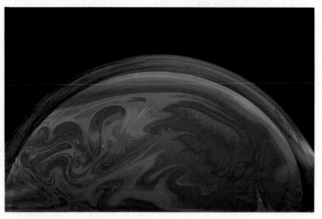

(a) Interference of light reflected from the inner and outer soap-bubble surfaces produces the colors. As the film thickness increases from top to bottom, the spectrum of colors is repeated several times.

(b) Swirling colors may be produced by blowing gently tangent to the film.

As you learned in the Insight feature on nonreflecting lenses, phase shifts of incident light take place at both the film and glass surfaces. In such a case, the condition for destructive interference is that the path difference for normally incident light is $\lambda'/2$, where λ' is the wavelength of the light in the film. Thus, $2t = \lambda'/2$, and the film thickness that will produce the least reflection is

$$t = \frac{\lambda'}{4}$$

The wavelength of the light in the film (λ') is related to that in air (λ) by

$$\lambda' = \frac{\lambda}{n}$$

(See the preceding marginal note on wavelengths.) Substituting for λ' in the equation for t gives

$$t = \frac{\lambda}{4n} \qquad (24.8)$$

Example 24.2   Nonreflecting Coatings

A glass lens ($n = 1.6$) is coated with a thin transparent film of magnesium fluoride (MgF_2, $n = 1.38$) to make it nonreflecting. What is the minimum thickness of the film for the lens to be nonreflecting for incident light whose wavelength is 550 nm?

Solution

Given: $n = 1.38$ *Find*: t

        $\lambda = 550$ nm

Using Eq. 24.8,

$$t = \frac{\lambda}{4n} = \frac{(550 \text{ nm})}{4(1.38)} = 99.6 \text{ nm} \quad \blacksquare$$

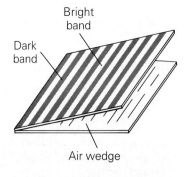

(a)

(b)

Figure 24.7 Optical flatness (a) An optical flat is used to check the smoothness of a reflecting surface by placing the flat so there is an air wedge between it and the surface. (b) If the surface is smooth, a regular interference pattern occurs.

Optical Flats and Newton's Rings

The phenomenon of thin-film interference is used to check the smoothness and uniformity of optical components such as mirrors and lenses. **Optical flats** are made by grinding and polishing glass plates until they are as flat and smooth as possible. The degree of flatness may be checked by putting two such plates together so there is a thin air wedge between them (Fig. 24.7). If the plates are smooth and flat, a regular interference pattern of bright and dark fringes, or bands, appears. This pattern is a result of the uniformly varying differences in path lengths between the plates. Any irregularity in the pattern indicates an irregularity in at least one of the plates. Once a good optical flat is verified, it can be used to check the flatness of a reflecting surface, such as that of a precision mirror.

A similar technique is used to check the smoothness and symmetry of lenses (Fig. 24.8). When a curved lens is placed on an optical flat, the air wedge is a ring (below the periphery of the circular lens). The regular interference pattern in this case is a set of concentric bright and dark circular fringes. They are called **Newton's rings**, after Isaac Newton, who first described this interference effect. Lens irregularities give rise to a distorted fringe pattern.

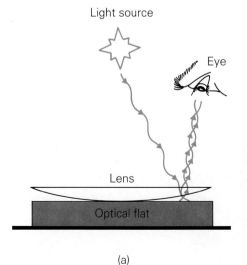

(a)

(b)

Figure 24.8 Newton's rings (a) A lens placed on an optical flat gives a ring-shaped air wedge, which gives rise to interference. (b) A regular interference pattern is a set of concentric fringes called Newton's rings. Lens irregularities produce a distorted pattern.

24.2 Thin-Film Interference   **655**

24.3 Diffraction

In geometrical optics, light is represented by rays and pictured as traveling in straight lines. If this model represented the real nature of light, however, there would be no interference effects in Young's double-slit experiment. Instead, there would be only two bright slit images on the screen with a well-defined shadow area [Fig. 24.9(a)]. In fact, there are interference patterns, which means that the light must deviate from a straight-line path and enter the regions that would otherwise be in shadow. Huygens' principle requires that the waves spread out from the slits [Fig. 24.9(b)], and this deviation, or bending, of light is called **diffraction**.

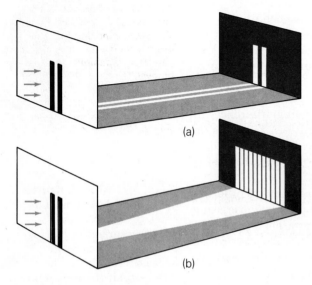

(a)

(b)

Figure 24.9 **Slit diffraction** (a) If there were no diffraction of light, two bright slit images would be observed on the screen in Young's double-slit experiment. (b) Because an interference pattern does occur, the light must be diffracted, or bent, on passing through the slits.

Diffraction generally occurs when waves pass through small openings or around sharp edges or corners. The diffraction of sound (Chapter 15) is quite evident. Someone can talk to you from another room or around the corner of a building, and even in the absence of reflections, you can easily hear them from your position in the acoustical shadow, so to speak. Diffraction phenomena for light waves often go unnoticed. However, careful observation will reveal that a shadow boundary is blurred or fuzzy. On close inspection, you can see that there

Straight edge

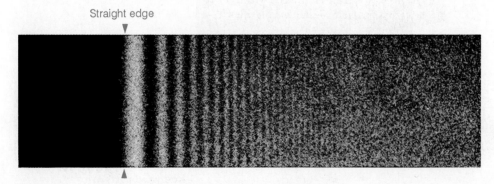

Figure 24.10 **Fringes at a shadow boundary**

is a pattern of bright and dark fringes (Fig. 24.10). These interference patterns are evidence of the diffraction of the light around the edge of the object.

An illustrative example of diffraction is that with a single slit (Fig. 24.11). Suppose that a single slit of width w is illuminated using monochromatic light whose wavelength is much smaller than the slit width. An interference pattern consisting of a bright central maximum and a symmetrical array of bright fringes on both sides is observed on a screen at a distance L from the slit (where $L \gg w$). Because the width of the slit is very much greater than the wavelength of the light, the slit with light passing through cannot be treated as a point source of Huygens' wavelets. However, various points on the wave front passing through the slit may be considered to be such point sources. Then, the interference of those wavelets can be analyzed similarly to the way double-slit interference was analyzed earlier. The result gives the displacement of the *dark* fringes from the center of the central maximum:

$$y_m = \frac{m\lambda L}{w} \qquad \text{for } m = 1, 2, 3, \ldots \tag{24.9}$$

Note that Eq. 24.9 has the same form as Eq. 24.6, the equation for the displacement of bright fringes for double-slit interference. Here w is the width of the slit, and there d is the distance between the slits.

The qualitative predictions from Eq. 24.9 are quite interesting and instructive:

- For a given slit width (w), the greater the wavelength (λ), the wider the diffraction pattern.

- For a given wavelength (λ), the narrower the slit width (w), the wider the diffraction pattern.

- The width of the central maximum is twice the combined widths of the other maxima.

The analysis of a single-slit interference is complicated but shows the condition for destructive interference, or interference minima (dark fringes), to be $d \sin\theta = m\lambda$ (where $m = 1, 2, 3, \ldots$). Then, as in the analysis of Young's experiment, for small angles, $\sin\theta \cong \theta \cong y/L$ (with $L \gg d$).

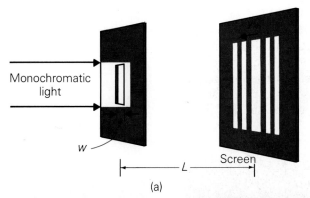

(a)

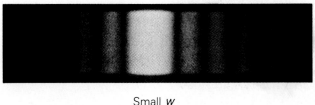

Small *w*

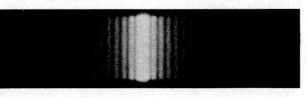

(b)

Large *w*

Figure 24.11 **Single-slit diffraction**
(a) The diffraction of light by a single slit gives rise to an interference pattern consisting of a large bright central maximum and a symmetric array of side fringes. (b) The widths of the fringes depend on the width of the slit.

Thus, as the slit is made narrower, the central maximum and the side fringes spread out and become larger [Fig. 24.11(b)]. Eq. 24.9 is not applicable to very small slit widths (because of the small-angle approximation). If the slit is decreased until it is the same order of magnitude as the wavelength of the light, the central maximum spreads out over the whole screen. That is, diffraction effects are greater when the slit width (or opening or object) is about the same size as the wavelength. Thus, diffraction effects are easily observed when $\lambda/w \geq 1$, or $\lambda \geq w$. For example, FM radio waves (88–108 MHz, or $\lambda = 2.8$–3.4 m) may be blocked by large objects such as hills or buildings and tend to travel in straight lines. However, AM radio waves (525–1610 kHz, or $\lambda = 186$–570 m) may be diffracted around objects and received in the shadow areas.

Conversely, if the slit is made wider when using a particular wavelength of light, the diffraction pattern becomes narrower. The fringes move closer together and eventually become difficult to distinguish ($\lambda \ll w$). The pattern then appears as a fuzzy shadow around the central maximum, which is the illuminated image of the slit. This type of pattern is observed for the image produced by sunlight entering a dark room through a hole in a curtain. Such an observation led early experimenters to investigate the wave nature of light, and the acceptance of this theory was, in large part, due to the explanation of diffraction offered by physical optics.

The width of the central maximum is the angular separation between the bounding minima on each side ($m = 1$), or a width of $2y_1$. From Eq. 24.9,

$$2y_1 = \frac{2\lambda L}{w} \tag{24.10}$$

Similarly, the width of the bright side fringes is given by

$$y_2 - y_1 = y_3 - y_2 = \frac{\lambda L}{w}$$

That is,

$$\Delta y_{m+1} - y_m = \frac{\lambda L}{w} \tag{24.11}$$

Thus, the width of the central maximum is twice that of the side fringes.

Example 24.3 Width of the Central Maximum

Monochromatic blue light ($\lambda = 425$ nm) passes through a slit whose width is 0.50 mm. What is the width of the central maximum on a screen located 1.0 m from the slit?

Solution

Given: $\lambda = 425$ nm $= 4.25 \times 10^{-5}$ cm   *Find*: $2y_1$
     $w = 0.50$ mm $= 0.050$ cm
     $L = 1.0$ m $= 10^2$ cm

Eq. 24.10 can be used directly:

$$2y_1 = \frac{2\lambda L}{w} = \frac{2(4.25 \times 10^{-5}\ \text{cm})(10^2\ \text{cm})}{0.050\ \text{cm}} = 0.17\ \text{cm} \quad \blacksquare$$

Note that diffraction effects would not be readily observed in this case since $\lambda \ll \omega$.

Diffraction Gratings

The angle of deviation of light diffracted by a slit is specific for a given wavelength. This fact provides a means of separating light and determining its spectral composition. A **diffraction grating**, which consists of a large number of parallel slits spaced at regular intervals (Fig. 24.12), is used instead of a single slit, however.

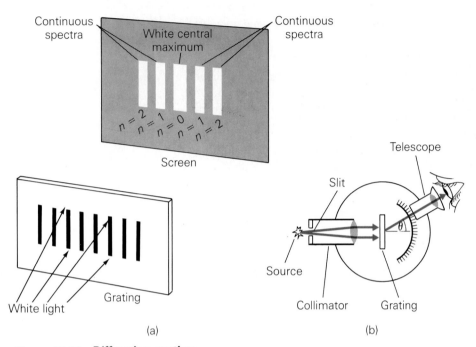

(a)   (b)

Figure 24.12   **Diffraction grating**
(a) A diffraction grating produces a sharply defined interference pattern in which the light components are separated in the side fringes since the deviation depends on the wavelength: $\theta = \sin^{-1}(n\lambda/d)$. (b) As a result, gratings are used in spectrometers to measure the wavelength of light by measuring the angle of diffraction of a particular component.

Diffraction gratings were first made of fine strands of wire. Their effects were similar to what may be seen by viewing a candle flame through a feather held close to the eye. Better gratings have a large number of fine lines or grooves on glass or metal surfaces. If light is transmitted through a grating, it is called a transmission grating. However, reflection gratings are more common. These are made by depositing a thin film of aluminum on an optically flat surface and then removing some of the reflecting metal by cutting regularly spaced, parallel lines. Precision diffraction gratings are made using two coherent laser beams, intersecting at an angle. The beams expose a layer of photosensitive material, which is then etched. The spacing of the grating lines is determined by the intersection angle of the beams. Precision gratings may have 10,000 lines per centimeter or more and are therefore expensive and difficult to fabricate. Most gratings used in laboratory instruments are replica gratings, which are plastic castings of high-precision master gratings.

Diffraction gratings are widely used in spectroscopy (the study of spectra), where they have almost replaced prisms. The forming of a spectrum and the measurement of wavelengths by means of a grating depend only on geometrical measurements such as lengths and/or angles. Wavelength determination using a prism, on the other hand, also depends on the dispersive characteristics of the glass or other material of which the prism is made.

The condition for interference maxima for a grating illuminated with monochromatic light is identical to that for a double slit:

$$d \sin \theta = n\lambda \qquad \text{for } n = 0, 1, 2, 3, \ldots \qquad (24.12)$$

interference maxima for a grating

Here n is the order of interference and θ is the angle of deviation of a particular wavelength. The zeroth-order maximum corresponds to the central maximum of the diffraction pattern. The spacing between adjacent slits (d) is obtained from the number of lines per unit length of the grating. For example, if $N = 5000$ lines/cm,

$$d = \frac{1}{N} = \frac{1}{5000} = 2.0 \times 10^{-4} \text{ cm}$$

In contrast to a prism, which deviates red light least and violet light most, a diffraction grating produces the least angle of deviation for violet light (short λ), and the greatest for red light (long λ). Also, a prism disperses white light into a single spectrum. A diffraction grating, however, produces a number of spectra, one for each order other than $n = 0$. There is no deviation of the components of the light for the zeroth order ($\sin \theta = 0$ for all wavelengths), so the central maximum is a white image. However, a spectrum is produced for each higher diffraction order, as illustrated in Fig. 24.12.

The number of spectral orders produced by a grating depends on the wavelength of the light (which may be infrared, visible, or ultraviolet) and on the grating's spacing (d). From Eq. 24.12, since $\sin \theta$ cannot exceed 1 ($\sin \theta \leq 1$),

$$\sin \theta = \frac{n\lambda}{d} \leq 1$$

The order number is therefore limited as follows:

$$n \leq \frac{d}{\lambda} \qquad (24.13)$$

Example 24.4   Grating Spacings and Spectral Orders

A particular diffraction grating produces an $n = 2$ spectral order at a deviation angle of 30° for light with a wavelength of 500 nm. (a) How many lines per centimeter does the grating have? (b) If the grating were illuminated with white light, how many orders of the *complete* visible spectrum would be produced?

Solution

Given: $n = 2$
$\lambda = 500 \text{ nm} = 5.0 \times 10^{-7} \text{ m}$
$\theta = 30°$

Find: (a) N
(b) Number of orders

(a) The grating spacing may be found using Eq. 24.12:

$$d = \frac{n\lambda}{\sin\theta} = \frac{2(5.0 \times 10^{-7} \text{ m})}{\sin 30°}$$

$$= 20 \times 10^{-7} \text{ m} = 2.0 \times 10^{-4} \text{ cm}$$

Then

$$N = \frac{1}{d} = \frac{1}{2.0 \times 10^{-4} \text{ cm}}$$

$$= 5000 \text{ lines/cm}$$

(b) The greatest angle of deviation for any order is for the longest wavelength, which is that of red light in the visible spectrum ($\lambda = 700$ nm $= 7.0 \times 10^{-7}$ m). With a spacing of $d = 2.0 \times 10^{-4}$ cm $= 2.0 \times 10^{-6}$ m, by Eq. 24.13, the order numbers for the visible spectrum for this grating are limited to

$$n \le \frac{d}{\lambda} = \frac{2.0 \times 10^{-6} \text{ m}}{7.0 \times 10^{-7} \text{ m}} = 2.9$$

Thus, for a grating with this spacing, only two orders of the complete visible spectrum are observed. A large portion of the spectrum is seen in the third order, but this is limited to wavelengths of

$$\lambda \le \frac{d}{n} = \frac{2.0 \times 10^{-6} \text{ m}}{3} = 6.7 \times 10^{-7} \text{ m} = 670 \text{ nm}$$

Thus, the red end of the visible spectrum is not seen in the third spectral order. ■

Note that it is possible for spectra produced by diffraction gratings to overlap at higher orders. That is, the angles of deviation for different orders may be the same for two different wavelengths. The spacing must be greater than or equal to the wavelength ($d \ge \lambda$) to have interference for the first spectral order. However, if the spacing is much greater than the wavelength ($d \gg \lambda$), the difference in the deviation angles for nearby wavelengths is quite small. This difference increases for higher orders, but then the overlapping of spectra becomes a problem. Thus, there is an optimal spacing of the lines on a diffraction grating for each spectral region of interest—infrared, visible, or ultraviolet. The spacing is typically chosen to be between 3λ to 6λ, where λ is the median (middle) wavelength of the spectral region.

The wavelength of any electromagnetic wave may be determined if a diffraction grating with the appropriate spacing is available. Diffraction was used to determine the wavelengths of X-rays early in this century. Experimental evidence indicated that the wavelengths of X-rays were probably around 10^{-8} cm, but it is impossible to construct a diffraction grating with this spacing. Around 1913, Max von Laue, a German physicist, suggested that the regular spacing of the atoms in a crystalline solid might make it act as a diffraction grating for X-rays, since the atomic spacing in crystals is on the order of 10^{-8} cm. X-rays were directed at crystals, and diffraction patterns were observed.

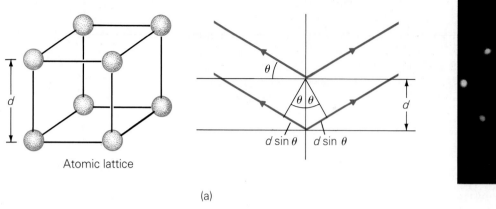

(a)

(b)

Atomic lattice

$d \sin \theta$ $\quad$ $d \sin \theta$

Figure 24.13 **Crystal diffraction**
(a) The array of atoms in a crystal lattice structure acts like a diffraction grating, and X-rays are diffracted from the planes of atoms. With a lattice spacing of d, the path difference for the X-rays diffracted from adjacent planes is $2d \sin \theta$. (b) X-ray diffraction pattern for a single crystal of sodium chloride.

Figure 24.13 illustrates diffraction by the planes of atoms in a crystal such as sodium chloride. You can see that the path difference is $2d \sin \theta$, where d is the distance between the crystal's internal planes. Thus, the condition for constructive interference is

$$2d \sin \theta = n\lambda \qquad \text{for } n = 1, 2, 3, \ldots \qquad (24.14)$$

This relationship is known as **Bragg's law**, after W. L. Bragg, the British physicist who first derived it.

The wavelengths of X-rays were experimentally determined by this means, and X-ray diffraction is now used to investigate the internal structure of crystals. Because of their short wavelengths, X-rays provide a diffraction "probe" for investigating the interatomic spacings of crystals.

24.4 Polarization

When you think of polarized light, you may visualize polarizing (or Polaroid) sunglasses since this is one of the more common applications of polarization. When something is polarized, this means that it has a preferential direction, or orientation (think of a polarized electrical plug, as described in Chapter 18). In terms of light waves, **polarization** refers to the orientation of the transverse oscillations.

Recall that light is an electromagnetic wave with vibrating electric and magnetic field vectors (**E** and **B**, respectively) perpendicular to the direction of propagation (Chapter 20). Light from most sources consists of a large number of electromagnetic waves emitted by the atoms of the source. Each atom produces a wave with a particular orientation, corresponding to the direction of the atomic vibration. However, since electromagnetic waves are produced by numerous atoms, all orientations of the E and B fields are possible in the composite light emitted. When the field vectors are randomly oriented, the light is said to be *unpolarized*. This is commonly represented schematically in terms of the electric field vector as shown in Fig. 24.14.

If there is some preferential orientation of the field vectors, the light is said to be *partially polarized*. Also, if the field vectors oscillate in only one plane, the light is *plane polarized*, or *linearly polarized*. Note that polarization is evidence that light is a transverse wave. Longitudinal waves, such as sound waves, cannot be polarized because there are no two-dimensional vibrations.

Light can be polarized in several ways. Polarization by reflection and double refraction will be discussed here. Polarization by scattering will be considered in Section 24.5.

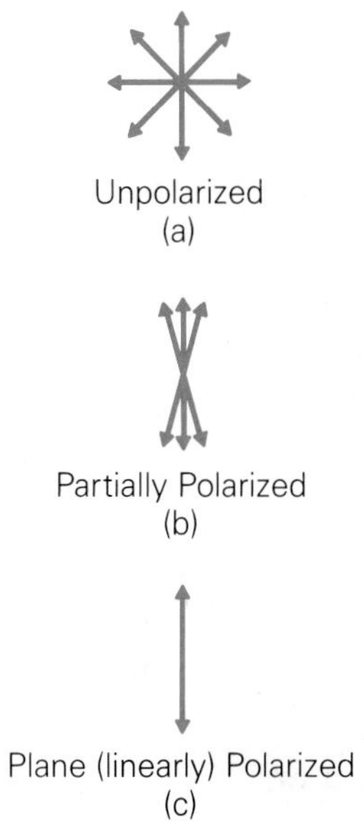

Polarization by Reflection

When a beam of unpolarized light strikes a smooth transparent medium such as glass, it is partially reflected and partially transmitted. Both the reflected and refracted beams are partially polarized. David Brewster (1781–1868), a Scottish physicist, found that the maximum polarization of the reflected beam occurs when the reflected and refracted beams are 90° apart (Fig. 24.15).

The incident angle for maximum polarization (θ_p), called the polarizing angle, or **Brewster angle**, is specific for a given material. As Fig. 24.15 shows, when $\theta_1 = \theta_p$, the reflected and refracted beams are 90° apart, and

$$\theta_1 + 90° + \theta_2 = 180°$$

Then

$$\theta_1 + \theta_2 = 90°$$

or $$\theta_2 = 90° - \theta_1$$

By Snell's law (Chapter 22), for incidence in air,

$$\frac{\sin \theta_1}{\sin \theta_2} = n$$

Unpolarized
(a)

Partially Polarized
(b)

Plane (linearly) Polarized
(c)

Figure 24.14 Polarization
Polarization refers to the orientation of the electromagnetic field vectors of light. (a) When the vectors are randomly oriented (as viewed along the direction of propagation), the light is unpolarized. (b) With preferential orientation of the vectors, the light is partially polarized. (c) When the vectors are in a plane, the light is plane polarized, or linearly polarized.

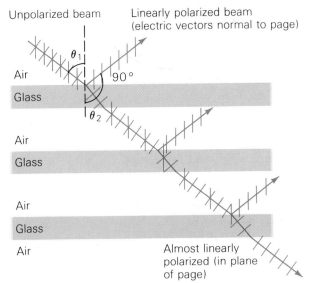

Figure 24.15 Polarization by reflection
When the reflected and refracted components of a beam of light are 90° apart, the reflected component is linearly polarized, and the refracted component is partially polarized. This occurs when $\theta_1 = \theta_p = \tan^{-1} n$. With a stack of glass plates (or layers of thin film), there are multiple reflections, which increase the intensity of the reflected polarized beam, and the refracted beam becomes more polarized.

In addition to linear polarization, there is another type of polarization, called circular or elliptical polarization. The field vectors describe circular or elliptical paths about the axis of propagation. The discussion in this book is limited to linear polarization.

In this case, $\sin \theta_2 = \sin(90° - \theta_1) = \cos \theta_1$. Thus,

$$\frac{\sin \theta_1}{\sin \theta_2} = \frac{\sin \theta_1}{\cos \theta_1} = \tan \theta_1 = n$$

With $\theta_1 = \theta_p$,

$$\tan \theta_p = n \tag{24.15}$$

(Since glass is dispersive, the Brewster angle also depends to some degree on the wavelength of the incident light.)

Example 24.5   Polarizing (Brewster) Angle

What is the incident angle for maximum polarization for reflection from crown glass having an index of refraction of 1.52?

Solution

Given: $n = 1.52$ *Find*: θ_p

From Eq. 24.15,

$$\theta_p = \tan^{-1} n = \tan^{-1}(1.52) = 56.7° \quad \blacksquare$$

As shown in Fig. 24.15, in a stack of glass plates, the reflections from successive surfaces increase the intensity of the reflected polarized beam. Note that the refracted beam becomes more linearly polarized with successive refractions. This effect is sometimes referred to as polarization by refraction. In practical applications, several thin films of a transparent material are used instead of glass plates.

Polarization by Double Refraction (Birefringence and Dichroism)

When monochromatic light travels in glass, its speed is the same in all directions and is characterized by a single index of refraction. Any material like this is said to be *isotropic*, meaning that it has the same characteristics in all directions. Some crystalline materials, such as quartz, calcite, and ice, are *anisotropic* with respect to the speed of light; that is, the speed of light is different in different directions within the material. Anisotropism gives rise to some unique optical properties, one of which is that of different indices of refraction in different directions. Such materials are said to be doubly refracting, or to exhibit **birefringence**.

For example, a beam of unpolarized light incident on a birefringent crystal of calcite ($CaCO_3$, calcium carbonate) is illustrated in Fig. 24.16. When the beam is at an angle to a particular crystal axis, it is doubly refracted and separated into two components, or rays. Also, the two rays are linearly polarized in mutually perpendicular directions. One ray, called the **ordinary (o) ray**, passes through the crystal in an undeflected path and is characterized by an index of refraction (n_o) that is the same in all directions. The second ray, called the **extraordinary (e) ray**, is refracted and is characterized by an index of refraction (n_e) that varies with direction. The particular axis direction indicated by dashed lines in Fig. 24.16 is called the optic axis. Along this direction, $n_o = n_e$, and nothing extraordinary is noted about the transmitted light.

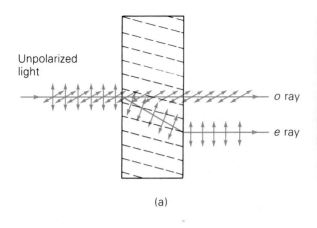

(a)

(b)

Figure 24.16 **Double refraction, or birefringence** (a) Unpolarized light incident normal to the surface of a birefringent crystal and at an angle to a particular crystal direction is separated into two components. The ordinary (o) ray and the extraordinary (e) ray are plane polarized in mutually perpendicular directions. (b) Birefringence seen through a calcite crystal.

Some birefringent crystals, such as tourmaline, exhibit the interesting property of absorbing one of the polarized components more than the other. This property is called **dichroism**. If a dichroic crystal is sufficiently thick, the more strongly absorbed component may be completely absorbed. In that case, the emerging beam is plane polarized (Fig. 24.17).

Another dichroic crystal is quinine sulfide periodide (commonly called herapathite, after W. Herapath, an English physician who discovered its polarizing properties in 1852). This crystal was of great practical importance in the development of modern polarizers. Around 1930, Edwin H. Land, an American scientist, found a way to align tiny, needle-shaped dichroic crystals in sheets of transparent celluloid. The result was a thin sheet of polarizing material that was given the commercial name Polaroid.

Better polarizing films have been developed using synthetic polymer materials instead of celluloid. During the manufacturing process, this kind of film is stretched in order to align the long molecular chains of the polymer. With proper treatment, the outer (valence) electrons of the molecules can move along the oriented chains. As a result, light with E vectors parallel to the oriented chains is readily absorbed, but light with E vectors perpendicular to the chains is transmitted. The direction perpendicular to the orientation of the molecular chains is called the **transmission axis**, or the **polarization direction**. Thus, when unpolarized light falls on a polarizing sheet, the sheet acts as a polarizer and polarized light is transmitted (Fig. 24.18).

A common analogy compares a polarizer to a picket fence. A randomly oriented wave in a rope will only pass through the fence parallel to the pickets and will be polarized in that direction. However, keep in mind that in a polarizing sheet the polarization direction is perpendicular to the oriented molecules.

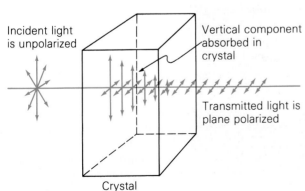

Figure 24.17 **Selective absorption** Dichroic birefringent crystals selectively absorb one polarized component more than the other. If the crystal is thick enough, the emerging beam is linearly polarized.

LCD's and Polarized Light

LCD's (Liquid Crystal Displays) are commonplace on watches, calculators, gas pumps, and even television screens. The name "liquid crystal" may seem self-contradictory. In general, when a crystalline solid melts, the resulting liquid no longer has an orderly atomic or molecular arrangement. Some organic compounds, however, pass through an intermediate state in which the molecules may rearrange somewhat but still maintain the overall order that is characteristic of a crystal.

A liquid crystal flows like a liquid, but its optical properties may depend on the order of its molecules. For example, certain liquid crystals with orderly arrangements are transparent. But the crystalline order can easily be disturbed by applied electrical forces. The disordered liquid then scatters light and is opaque.

A common type of LCD, called a twisted-nematic display, makes use of the effect of a liquid crystal on polarized light (Fig. 1). A display like those commonly used on wristwatches and calculators is made by putting a layer of liquid crystal material sandwiched between two glass plates that have fine parallel grooves, or channels, in their surfaces. One of the plates is then rotated or twisted 90°. The molecules in contact with the plates remain parallel to the grooves, and the result is a molecular orientation that rotates through 90° between the plates. In this configuration, the liquid crystal is optically active and will rotate the direction of polarization of linearly polarized light. The plates are then placed between crossed polarizing sheets and backed with a mirrored surface. Light entering and passing through the LCD is pola-

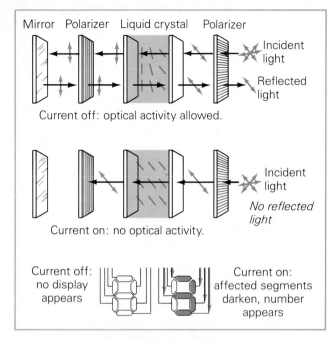

Figure 1 Liquid crystal display (LCD)
A twisted-nematic display is an application involving the optical activity of a liquid crystal and crossed Polaroids. When the crystalline order is disoriented by an electric field, the liquid crystal loses its optical activity in that region, and light is not transmitted. Numerals and letters are formed by applying voltages to segments of a block display.

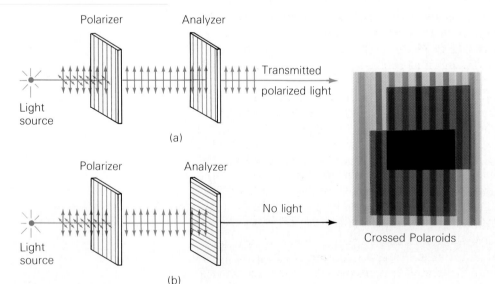

Figure 24.18 Polarizing sheets
(a) When polarizing sheets are oriented so their polarization directions are the same, polarized light is transmitted through them. (b) When one of the sheets is rotated 90° and the polarization directions of the sheets are perpendicular (crossed Polaroids), little light (ideally none) is transmitted. In this case, the second polarizer acts as an analyzer.

rized, rotated 90°, reflected, and again rotated 90° by its components. After the return trip through the liquid crystal, the polarization direction of the light is the same as that of the initial polarizer. Thus, the light is transmitted and leaves the display unit. Because of the reflection and transmission the display appears to be a light color (usually light gray) when illuminated with white, unpolarized light.

The light from the bright regions of an LCD can readily be shown to be polarized using an analyzer. The whole display appears dark when the analyzer is properly oriented. You may have noticed this effect if you have ever tried to see the time on the LCD of a wristwatch while wearing polarizing sunglasses (Fig. 2).

The dark numbers or letters on an LCD are formed by applying an electric field to parts of the liquid crystal layer. The molecules of the liquid crystal are polar, so an electric field can disorient them. Transparent, electrically conductive film coatings arranged in a seven-block pattern are put onto the glass plates. Each block, or display segment, has a separate electrical connection. When a voltage is applied across one or more of the segments, the electric field disorients the molecules of the liquid crystal in that area, and their optical activity is lost. (Note that the numerals 0 through 9 can all be formed by the segmented display.) The incident polarized light passing through the disoriented regions of the liquid crystal is absorbed by the second polarizer. Thus, these regions are opaque and appear dark. To produce images on small TV screens, the display segments of the LCD are coupled so that many of them may be energized with a single lead.

One of the major advantages of LCD's is their low power

Figure 2 **Polarized light**
The light from an LCD is polarized, as can be shown by using polarizing sunglasses as an analyzer.

consumption. Other similar displays, such as those using red light-emitting diodes (LED's), produce light themselves, using relatively large amounts of power. LCD's, on the other hand, produce no light, but instead use reflected light.

Some liquid crystals respond to temperature changes, and the orientation of the molecules affects the light-scattering properties of the crystal. Different wavelengths, or colors, of light are selectively scattered. The color of a liquid crystal can thus be an indication of the surrounding temperature, and liquid crystals are used to make thermometers.

The human eye cannot distinguish between polarized and unpolarized light. To tell whether or not light is polarized, we must use an analyzer, which may be a sheet of polarizing film. As shown in Fig. 24.18(b), if the transmission axis of an analyzer is perpendicular to the plane of polarization of polarized light, little light (ideally none) will be transmitted.

Now you can understand the principle of polarizing sunglasses. Light reflected from a smooth surface is partially polarized, as you learned earlier in this section. The direction of polarization is chiefly in the plane of the surface (see Fig. 24.15). Light reflected from water or snow may be so intense that it gives rise to visual glare (Fig. 24.19). To reduce this, the polarizing lenses of glasses are oriented with their transmission axes vertical so some of the partially polarized light from reflective surfaces is blocked or absorbed.

Polarizing glasses whose lenses show different polarization directions are used to view 3-D movies. The pictures are projected on the screen by two projectors that transmit slightly different images. The projected light is linearly polarized, but in directions that are mutually perpendicular. The lenses of the 3-D glasses also have polarization directions that are perpendicular, and one eye

Figure 24.19 **Glare reduction** Light reflected from a horizontal surface is partially polarized in the horizontal direction.

sees the image from one projector and the other eye sees the image from the other projector. The brain interprets the image difference as depth, or a third dimension.

Some transparent materials have the ability to rotate the direction of polarization of linearly polarized light. This property is called **optical activity** and is due to the molecular structure of the material (Fig. 24.20). The rotation may be clockwise or counterclockwise, depending on the molecular orientation. Optically active molecules include those of certain proteins, amino acids, and sugars.

Optical activity may be demonstrated by placing a transparent container of corn syrup between crossed Polaroids. Some light is transmitted, as a result of optical activity. Diluting the syrup with water will reduce the optical activity and the light transmission.

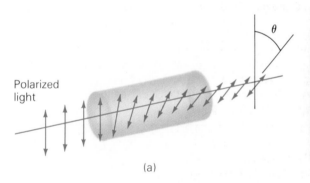

(a) (b)

Figure 24.20 **Optical activity** (a) Some substances have the property of rotating the polarization direction of linearly polarized light. This ability depends on molecular structure and is called optical activity. (b) Glasses and plastics become optically active under stress, and the points of greatest stress are apparent when the material is viewed through crossed Polaroids.

Glasses and plastics become optically active when under stress. The greatest rotation of the direction of polarization of the transmitted light occurs in the regions where the stress is the greatest. Viewing the stressed piece of material through crossed Polaroids allows the points of greatest stress to be identified. This determination is called optical stress analysis.

Another use of polarizing films is described in the Insight feature on page 666.

24.5 Atmospheric Scattering of Light

The scattering of sunlight in the atmosphere produces some interesting effects. Among these are the polarization of sky light (that is, sunlight that has been scattered by the atmosphere), the blueness of the sky, and the redness of sunsets and sunrises. When light is incident on a suspension of particles, such as the

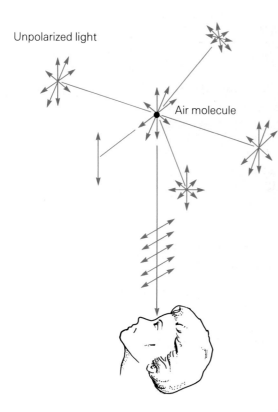

Unpolarized light

Air molecule

Figure 24.21 **Polarization by scattering**
When incident unpolarized sunlight is scattered by a molecule of a gas in the air, the light in the plane perpendicular to the direction of the incident ray is plane polarized. Light scattered at some arbitrary angle is partially polarized. An observer at a right angle to the direction of the sunlight receives plane-polarized light.

molecules of the air, some of the light may be absorbed and reradiated. This process is called **scattering**.

Atmospheric scattering causes the sky light to be polarized. When unpolarized sunlight is incident on air molecules, the electric field of the light wave sets electrons of the molecules into vibration (Fig. 24.21). The vibrations are complex, but the accelerating charges emit radiation, like the vibrating charges in the antenna of a radio broadcast station (see Section 20.4). The intensity of this emitted radiation is strongest along a line perpendicular to the oscillation. And, as illustrated in the figure, an observer viewing from an angle of 90° with respect to the direction of the sunlight will receive linearly polarized light because of the horizontal charge oscillations. The light also has a vertically polarized component. At other viewing angles, both components are present, and sky light seen through a polarizing filter appears partially polarized because of the stronger component.

Since the scattering of light with the greatest degree of polarization is at a right angle to the direction of the Sun, at sunrise and sunset the scattered light from directly overhead has the greatest degree of polarization. The polarization of sky light may be observed by viewing the sky through a polarizing filter (or a polarizing sunglass lens) and rotating the filter. Light from different regions of the sky will be transmitted in different degrees, depending on its degree of polarization. It is believed that some insects, such as bees, use polarized sky light to determine navigational directions relative to the Sun.

The scattering of sunlight by air molecules causes the sky to look blue. This is not a polarization effect but is caused by the selective absorption of light. As oscillators, air molecules have resonant frequencies in the ultraviolet region.

Consequently, light in the nearby blue end of the visible region is scattered more than that in the red end. That is, the sunlight is preferentially scattered, with the light of the blue end of the visible spectrum being scattered almost ten times more than the light of the red end.

For particles such as air molecules, which are much smaller than the wavelength of light, the scattering is found to be inversely proportional to the wavelength to the fourth power (that is, $1/\lambda^4$). This is called **Rayleigh scattering** after Lord Rayleigh (1842–1919), a British physicist who derived the wavelength relationship. This inverse relationship predicts that light of the shorter-wavelength, or blue, end of the spectrum will be scattered more than light of the longer-wavelength, or red, end. The scattered blue light is rescattered in the atmosphere and eventually is directed toward the ground. This is the sky light we see, so the sky appears blue.

Example 24.6   Rayleigh Scattering

How much more is light at the blue end of the visible spectrum scattered by air molecules than light at the red end?

Solution

As stated earlier, light at the blue end of the visible spectrum is scattered almost ten times more than light at the red end. This can be shown as follows. The Rayleigh scattering relationship is $S \propto 1/\lambda^4$, where S is the degree of scattering for a particular wavelength. Thus, you can form a ratio

$$\frac{S_b}{S_r} = \left(\frac{\lambda_r}{\lambda_b}\right)^4$$

where the subscripts stand for blue and red light. The blue end of the spectrum (violet light) has a wavelength of about $\lambda_b = 400$ nm and red light has a wavelength of about $\lambda_r = 700$ nm. Putting these values into the ratio gives

$$\frac{S_b}{S_r} = \left(\frac{\lambda_r}{\lambda_b}\right)^4 = \left(\frac{700 \text{ nm}}{400 \text{ nm}}\right)^4 = 9.4 \qquad \text{or} \qquad S_b = 9.4 S_r$$

Thus, blue light is scattered almost ten times as much as red light is.   ∎

You should keep in mind that all colors are present in sky light, but the dominant color is blue. (The sky doesn't appear violet because our eyes are more sensitive to blue and there is more blue than violet light in sunlight.) You may have noticed that the sky is more blue directly overhead than toward the horizon and appears white just above the horizon (look for this effect the next time you are outside on a clear day). This is because there are relatively fewer air molecules or scatterers directly overhead than toward the horizon. Multiple scatterings occur in the denser air near the horizon, giving rise to the white appearance. Analogously, if a drop or two of milk is added to a glass of water and the suspension is illuminated with intense white light, the scattered light has a bluish hue. But a glass of undiluted milk is white, because of multiple scatterings. Similarly, atmospheric pollution may give rise to a milky white appearance of most or all of the sky.

When the Sun is near the horizon, the sunlight travels a greater distance through the denser air near the Earth's surface. Since the light therefore undergoes a great deal of scattering, you might think that only the least scattered light, the red light, would reach observers on the Earth's surface. This would explain red sunsets. However, it has been shown that the dominant color of white light after only molecular scattering is orange.*

Thus, there must be other scattering that shifts the light from the setting (or rising) Sun toward the red. Red sunsets have been found to result from the scattering of sunlight by atmospheric gases *and* small foreign particles. These particles are not necessary for the blueness of the sky, but are necessary for deep red sunsets and sunrises. Red sunsets occur most often when there is a high-pressure air mass to the west, since the concentration of particles is generally greater in high-pressure air masses than in low-pressure air masses. Similarly, red sunrises occur most often when there is a high-pressure air mass to the east.

Red sunrises and sunsets are often made more spectacular by clouds. The clouds are colored pink by reflected red light. In general, clouds are white (when not shadowed) because the water droplets of which they are composed are not preferential scatterers and scatter visible light of all wavelengths (white light). Clouds do not affect the color of the incident light, but simply diffusely reflect the light from the Sun.

Important Formulas

Path difference for constructive interference:

$$PD = n\lambda \quad \text{for } n = 1, 2, 3, \ldots$$

Path difference for destructive interference:

$$PD = \frac{m\lambda}{2} \quad \text{for } m = 1, 3, 5, \ldots$$

Bright fringe condition (double-slit interference):

$$d \sin \theta = n\lambda \quad \text{for } n = 0, 1, 2, 3, \ldots$$

Width of bright fringes (double-slit interference):

$$\Delta = y_{n+1} - y_n = \frac{L\lambda}{d}$$

Wavelength measurement (double-slit interference):

$$\lambda = \frac{y_n d}{nL} \quad \text{for } n = 1, 2, 3, \ldots$$

Nonreflecting film thickness:

$$t = \frac{\lambda}{4n}$$

Location of dark fringes (single-slit diffraction):

$$y_m = \frac{m\lambda L}{w} \quad \text{for } m = 1, 2, 3, \ldots$$

Width of bright fringes (single-slit diffraction):

$$\Delta = y_{m+1} - y_m = \frac{\lambda L}{w}$$

Interference maxima for a grating:

$$d \sin \theta = n\lambda \quad \text{for } n = 0, 1, 2, 3, \ldots$$

Limit of the order number:

$$n \leq \frac{d}{\lambda}$$

Bragg's law:

$$2d \sin \theta = n\lambda \quad \text{for } n = 1, 2, 3, \ldots$$

Brewster (polarizing) angle:

$$\tan \theta_p = n$$

Questions

Young's Double-Slit Experiment

1. What would be observed with the set-up for Young's experiment if the light passing through the slits were not coherent?

2. What happens to the interference pattern if the distance between the slits is varied in Young's double-slit experiment?

* See "Colors of the Sky," by C. F. Bohren and A. B. Fraser, in *The Physics Teacher*, vol. 23, no. 5, p. 267.

3. Describe what would be observed if white light were used in Young's experiment.

Thin-Film Interference

4. Is light actually destroyed by thin-film destructive interference? Explain.

5. Two microscope slides are stuck together with an air film between them. Describe the phase changes for reflected light at the various interfaces.

6. Must the film coating on a nonreflecting lens have a thickness of only $\lambda/4$? Explain.

7. Suppose that you wanted to make a reflecting surface using thin-film interference. How would you go about doing this?

8. What would be the effect of nondirect incidence on thin-film interference?

9. A thin film shaped like a wedge is illuminated with monochromatic light and then white light. Describe what is observed in each case.

10. A glass surface is being ground and polished to make a plane mirror for a telescope. A glass known to be optically flat is placed over the surface to test its flatness. What will be observed if the test piece is not flat but has some uneven points?

Diffraction

11. Can diffraction phenomena be explained by geometrical optics? If not, why not?

12. Are the fringes of a single-slit diffraction pattern more easily observed with a relatively wide or a relatively narrow slit? Explain.

13. What does the equation for the width of the central maximum for single-slit diffraction predict when $d = \lambda$? Is the prediction correct? Explain.

14. Explain what happens in terms of diffraction effects where (a) $\lambda/d << 1$ and (b) $\lambda/d \geq 1$.

15. Do prisms and diffraction gratings separate the spectral components of light by the same means? Explain. What will be observed if beams of white light are incident on a prism and a diffraction grating?

16. Why is there no zeroth order ($n = 0$) in Bragg's law for crystal diffraction?

Polarization

17. Can sound waves be polarized? If so, how?

18. What is the Brewster angle, and how would you determine it experimentally?

19. If light reflected from a transparent surface is completely polarized, will the transmitted light be completely polarized? Explain.

20. Given two pairs of sunglasses, could you tell if they were polarizing or nonpolarizing? How?

21. Light that has passed through a birefringent crystal is viewed through a polarizing sheet. What is observed when the sheet is rotated?

22. Suppose that you held two polarizing sheets in front of you and looked through both of them. How many times would you see the sheets lighten and darken (a) if one of them were rotated through one complete rotation, (b) if both of them were rotated through one complete rotation at the same rate but in opposite directions, and (c) if both of them were rotated through one complete rotation at the same rate in the same direction?

23. What would you see if you viewed a 3-D movie without wearing polarizing glasses? What would you see if you wore polarizing sunglasses?

24. If corn syrup were poured into a transparent container between crossed Polaroids, what would be necessary to maintain no transmission of light through the Polaroids?

25. What would be observed if an LCD were illuminated with polarized light? Explain.

Atmospheric Scattering of Light

26. If the scattering of light by air molecules is proportional to $1/\lambda^4$, why doesn't the sky appear violet instead of blue?

27. Does the sky have a uniform blueness? Explain.

28. Why are spectacular red sunsets not seen every evening? Why is the setting Sun sometimes yellow-orange?

29. What would an astronaut on the moon see when looking at the sky?

Problems

24.1 Young's Double-Slit Experiment

■■**1.** What is the shortest path difference for two waves of equal frequency that arrive at a point with a phase difference of (a) $\pi/3$, (b) $\pi/2$, and (c) π? (Hint: see Eq. 15.5.)

■■**2.** Two point sources of coherent sound are located at (2 m, 5 m) and (4 m, 3 m) in the xy plane. What is the wavelength of the sound waves at the point (15 m, 20 m) if (a) they interfere destructively to the first order and (b) they have a phase difference of 45°?

■■**3.** Two parallel slits 0.50 mm apart are illuminated

with monochromatic light whose wavelength is 480 nm. Find the angle between the center of the central maximum and the center of an adjacent bright fringe formed on a screen 1.0 m away from the slits.

■■4. Monochromatic light whose wavelength is 500 nm falls on two slits separated by a distance of 40 μm. What is the distance between the first-order and third-order bright fringes formed on a screen 1.2 m away from the slits?

■■5. An interference pattern is formed on a screen 0.95 m away from two parallel slits 0.060 mm apart when they are illuminated with monochromatic light having a wavelength of 580 nm. What is the angular separation for the first-order fringes on either side of the central maximum?

■■6. In a double-slit experiment using monochromatic light, the angular separation between the central maximum and the second-order bright fringe is measured to be 0.16°. What is the wavelength of the light if the slit separation distance is 0.50 mm?

■■7. In a double-slit experiment with slits that are 0.25 mm apart, the interference pattern is formed on a screen 1.0 m away from the slits. The third-order bright fringe is 0.60 cm from the center of the central maximum. (a) What is the frequency of the monochromatic light? (b) What is the distance between the second-order and third-order bright fringes?

■■8. Monochromatic light illuminates two parallel slits that are 0.20 mm apart. The adjacent bright lines of the interference pattern on a screen 1.5 m away from the slits are 0.45 cm apart. What is the color of the light?

■■9. Monochromatic light passes through two narrow slits 0.25 mm apart and forms an interference pattern on a screen 2.0 m away. If light with a wavelength of 680 nm is used, what is the distance between the center of the central maximum and the center of the third-order bright fringe?

■■10. Derive a relationship giving the locations of the dark fringes in Young's double-slit experiment. What is the distance between the dark fringes?

■■■11. What would be the effect on the interference fringes if the apparatus for Young's double-slit experiment were completely immersed in still water?

■■■12. What would the distance in Problem 9 be if the entire system were immersed in still water?

■■■13. Light of two different wavelengths is used in a double-slit experiment. The location of the third-order bright fringe for yellow light ($\lambda = 600$ nm in air) coincides

with the location of the fourth-order bright fringe for the other light. What is the wavelength of the other light?

24.2   Thin-Film Interference

■■14. Light with a wavelength of 600 nm in air is normally incident on a glass plate ($n = 1.5$) whose thickness is 10^{-3} cm. (a) What is the total path length of the light in wavelengths in glass? (b) What is the phase difference in the beams reflected from each surface (that is, will they interfere constructively or destructively)?

■■15. A camera lens is coated with a thin layer of a material that has an index of refraction of 1.4. This makes the lens nonreflecting for light with a wavelength of 500 nm (in air) that is normally incident on the lens. What is the thickness of the thinnest film that will make the lens nonreflecting?

■■16. A film on a lens is 1.0×10^{-7} m thick and is illuminated with white light. The index of refraction of the film is 1.4. For what wavelength of light will the lens be nonreflecting?

■■17. A solar cell is to have a nonreflecting coating of a transparent material whose index of refraction is 1.2. (a) What is the minimum thickness of the film for light with a wavelength of 550 nm? (b) Does the index of refraction of the adjacent material in the solar cell make a difference?

■■18. A thin layer of oil ($n = 1.5$) floats on water. Destructive interference is observed for light with wavelengths of 480 nm and 600 nm. Find the thicknesses of the oil film.

■■■19. A thin air wedge between two flat glass plates forms bright and dark interference bands when illuminated with normally incident monochromatic light, as illustrated in Fig. 24.7. (a) Show that the thickness of the air wedge changes by $\lambda/2$ from one bright fringe to the next, where λ is the wavelength of the light. (b) What would the change in the wedge thickness between bright fringes be if the space were filled with a liquid with an index of refraction n?

24.3   Diffraction

■■20. A slit of width 0.20 mm is illuminated with monochromatic light having a wavelength of 480 nm, and a diffraction pattern is formed on a screen 1.5 m away from the slit. (a) What is the width of the central maximum? (b) What is the distance between the second-order and third-order bright fringes?

■■21. If the slit width in a single-slit experiment were

doubled, the distance to the screen reduced by a third, and the wavelength of the light changed from 600 nm to 450 nm, how would the width of the bright fringes be affected?

■■**22.** Red light ($\lambda = 700$ nm) passes through a slit with a width of 1.0 mm. (a) Find the angular width and the linear width of the central maximum of the diffraction pattern on a screen 2.0 m from the slit. (b) Find the angular and linear widths if the slit is illuminated with blue light ($\lambda = 440$ nm).

■■**23.** When illuminated with monochromatic light, a diffraction grating with 530 lines/mm produces a second-order maximum 1.5 cm from the central maximum on a screen 0.50 m away. What is the wavelength of the light?

■■**24.** Visible white light traveling in air contains components with wavelengths from about 400 nm to 700 nm. If the white light illuminates a diffraction grating with 5000 lines/cm, what is the angular width of the first-order spectrum produced?

■■**25.** How many orders of interference maxima occur when monochromatic light with a wavelength of 560 nm illuminates a diffraction grating with 5000 lines/cm?

■■**26.** White light whose components have wavelengths from 400 nm to 700 nm illuminates a diffraction grating with 4000 lines/cm. Do the first and second orders overlap? Justify your answer.

■■**27.** What is the angular width of the first-order spectrum in Problem 24 if the whole system is immersed in water?

■■■**28.** Show that for a diffraction grating the violet ($\lambda = 400$ nm) portion of the third-order spectrum overlaps the yellow-orange ($\lambda = 600$ nm) portion of the second-order spectrum regardless of the grating's spacing.

■■■**29.** A teacher standing inside a doorway 1.0 m wide blows a whistle with a frequency of 1000 Hz to summon children from the playground. All the children come, except two boys playing near the swings, about 100 m away from the school building. Later the teacher begins to reprimand them, but the boys maintain rather convincingly that they did not hear the whistle. The swings are at an angle of 19.6° from a line normal to the doorway, and the speed of sound is 335 m/s. Could they be telling the truth? Justify your answer mathematically.

24.4 Polarization

■**30.** A glass plate has an index of refraction of 1.5. At what angle of incidence will light be reflected from the plate with the maximum linear polarization?

■**31.** If the Brewster angle for a glass plate is 1.0 rad, what is the index of refraction of the glass?

■■**32.** The critical angle for internal reflection in a certain medium is 60°. What is the Brewster angle for light externally incident on the medium?

■■**33.** A light beam is incident on a glass plate ($n = 1.6$), and the reflected and refracted rays have maximum linear polarization. What is the angle of refraction for the beam?

■■**34.** Find the Brewster angle for a piece of glass ($n = 1.5$) that is submerged in water.

■■■**35.** Sketch a graph of the Brewster angle versus the index of refraction.

■■■**36.** A plate of crown glass is covered with a layer of water. A beam of light traveling in air is incident on the water and partially transmitted. Is there any angle of incidence for which the light reflected from the water-glass interface will have maximum linear polarization? Justify your answer mathematically.

Additional Problems

37. In a particular diffraction pattern, the red component ($\lambda = 700$ nm) in the second-order spectrum is deviated at an angle of 20°. (a) How many lines per centimeter does the grating have? (b) If the grating is illuminated with white light, how many orders of the complete visible spectrum are produced?

38. Blue light with a wavelength of 440 nm is used in a double-slit experiment where the slits are 0.35 mm apart. If the screen is 0.80 m from the slits, what is the angular separation of the first-order and third-order bright fringes?

39. Some types of glass have a range of indices of refraction of about 1.4 to 1.7. What is the range of the polarizing angle for these glasses?

40. A lens with an index of refraction of 1.6 is to be coated with a material ($n = 1.4$) that will make it nonreflecting for red light ($\lambda = 700$ nm). What is the minimum required thickness of the coating?

41. An interference pattern is formed when light whose wavelength is 550 nm is incident on two parallel slits 50 μm apart. The second-order bright fringe is 2.0 cm from the center of the central maximum. How far from the slits is the screen on which the pattern is formed?

42. What is the spectral order limit for a diffraction grating with 7500 lines/cm when illuminated by white light?

43. Two parallel slits 1.0 mm apart are illuminated with monochromatic light whose wavelength is 640 nm, and an interference pattern is observed on a screen 1.75 m away. (a) What is the separation between adjacent interference maxima? (b) If the distance between the slits were increased to 1.5 mm, how would this affect the separation of the maxima?

44. Show that the film thicknesses for a "nontransmissive" lens are given by

$$t = \frac{(m + 1)\lambda}{2n}$$

where $m = 0, 1, 2, 3, \ldots$, and n is the index of refraction of the film.

45. Magnesium fluoride ($n = 1.38$) is frequently used as a lens coating to make nonreflecting lenses. What is the difference in the minimum film thicknesses required for maximum transmission of blue light ($\lambda = 400$ nm) and red light ($\lambda = 700$ nm)?

46. A slit 0.20 mm wide is illuminated with red light ($\lambda = 680$ nm). How wide are (a) the central maximum and (b) the side maxima of the interference pattern formed on a screen 1.0 m from the slit?

47. The angle of incidence is adjusted so there is maximum linear polarization for the reflection of light from a transparent piece of plastic with $n = 1.25$. But some of the light is transmitted. What is the angle of refraction?

48. A diffraction grating is designed to have the red end of the visible spectrum an angular distance of 20° from the center of the central maximum for the second order. How many lines per centimeter does the grating have?

49. What angle of incidence is necessary for reflected light to have the maximum linear polarization at a water–crown glass interface?

50. A certain crystal gives a deflection angle of 25° for the first-order diffraction of monochromatic X-rays with a frequency of 5.0×10^{17} Hz. What is the lattice spacing of the crystal?

51. Find the locations of the blue ($\lambda = 420$ nm) and red ($\lambda = 680$ nm) components of the first-order and second-order spectra in a pattern formed on a screen 2.0 m from a diffraction grating with 6000 lines/cm.

Optical Instruments

25

Optical instruments are used for a variety of purposes, but their basic function is to improve and extend our powers of visual observation. For example, microscopes and telescopes are used to view objects and details that may not be visible to the unaided eye.

Although some optical instruments are extremely complex, they can generally be analyzed in terms of their basic components, usually mirrors and lenses, which you have already studied in Chapter 23. This chapter first considers the optical properties of the eye itself and then describes how optical instruments improve our view of the world.

25.1 The Human Eye

The human eye is the most fundamental optical instrument, since without it the field of optics would not exist. The eye is analogous to a simple camera in several respects (Fig. 25.1). A simple camera consists of a converging lens, which is used to focus images on light-sensitive film at the back of the camera's interior chamber. An adjustable diaphragm, or opening, and a shutter control the amount of light entering the camera. The eye also has a converging lens that *helps* focus images on the light-sensitive lining on the rear surface inside the eyeball. The iris is a circular diaphragm that opens and closes to adjust the amount of light entering the eye. The eyelid might be thought of as a shutter; however, the shutter of the camera, which controls the exposure time, is generally opened for only a fraction of a second, and the eyelid is normally open for continuous exposure. The human nervous system performs a function analogous to that of a shutter in analyzing image signals from the eye at a rate of about 30 times per second. The eye might therefore be likened to a movie or video camera, which exposes a similar number of frames (images) per second.

Although the optical functions of the eye are relatively simple, its physiological functions are quite complex. As Fig. 25.1(b) shows, the eyeball is a

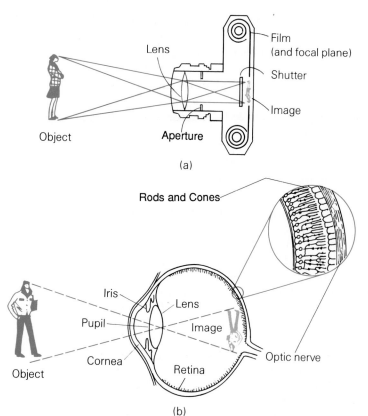

Object — Lens — Film (and focal plane) — Shutter — Image — Aperture

(a)

Rods and Cones

Object — Iris — Pupil — Cornea — Lens — Image — Retina — Optic nerve

(b)

Figure 25.1 **Camera and eye analogy**
In some aspects, a simple camera is similar to the eye. An image is formed on the film in a camera and on the retina in the eye.

nearly spherical chamber (with an internal diameter of about 1.5 cm) filled with a jellylike substance called the vitreous humor. The eyeball has a white outer covering called the sclera, part of which is visible as the white of the eye. Light enters the eye through a curved, transparent tissue called the cornea and passes into a clear fluid known as the aqueous humor. Behind the cornea is a circular diaphragm, the iris, whose central hole is called the pupil. The iris contains the pigment that determines eye color. Through muscle action, the iris can change the area of the pupil (2 to 8 mm in diameter), thereby controlling the amount of light entering the eye.

Behind the iris is a **crystalline lens**, a converging lens composed of microscopic glassy fibers. When tension is exerted on the lens by attached muscles, the glassy fibers slide over each other, causing the shape and curvature of the lens to change. On the back interior wall of the eyeball is a light-sensitive surface called the **retina**. From the retina, the optic nerve relays signals to the brain. The retina is composed of nerves and two types of light receptors, or photosensitive cells, called **rods** and **cones** because of their shapes. The rods are more sensitive to light and distinguish light from dark in low light intensities (twilight vision). The cones, on the other hand, are less sensitive to light but resolve sufficiently intense light into its component wavelengths, thereby enabling us to see colors. The cells of the retina are present in large numbers (estimated at $125,000/mm^2$). Most of the cones are clustered together around a center (called the fovea centralis). The more numerous rods are outside this region and distributed nonuniformly over the retina.

The optical adjustments of the eye are truly amazing. Most of the refraction of light occurs at the front surface of the cornea. The crystalline lens makes fine adjustments for focusing the images. Nerve signals cause the attached muscles to change the radius of curvature and thus the focal length of the lens. Thus, images of objects at different distances are focused sharply on the retina. When the eye is focused on distant objects, the muscles are relaxed, and the crystalline lens has its thinnest shape and a power of about 20 D (diopters). (Recall from Chapter 23 that the power of a lens in diopters is the reciprocal of its focal length in meters.) When the eye is focused on closer objects, the lens becomes thicker, and the radius of curvature and focal length are decreased. The lens power may increase to 30 D, or even more in young children. The adjustment of the focal length of the crystalline lens is called **accommodation**. (Look at a nearby object and then one in the distance and notice how fast accommodation takes place.)

The focusing adjustment of the eye is unlike that of a camera. A camera lens has a constant focal length, and the image distance is varied by moving the lens to produce sharp images on the film for different object distances. In the eye, the image distance is constant, and the focal length of the lens (or its radius of curvature) is varied to produce sharp images on the retina for different object distances.

The extremes of the range over which distinct vision (sharp focus) is possible are known as the far point and the near point. The **far point** is the greatest distance at which the normal eye can see objects clearly and is taken to be infinity. The **near point** is the position closest to the eye at which objects can be seen clearly and depends on the extent the lens can be deformed (thickened) by accommodation. The range of accommodation gradually diminishes with age as the crystalline lens loses its elasticity. That is, the near point gradually recedes with age. The approximate positions of the near point at various ages are listed in Table 25.1.

Children can see sharp images of objects that are within 10 cm (4 in.) of their eyes, and the crystalline lens of a normal young adult eye can be deformed to produce sharp images of objects as close as 12–15 cm (5–6 in.). However, at about the age of 40, the near point normally moves beyond 25 cm (10 in.). You may have noticed people over the age of 40 holding reading material at some distance from their eyes to bring it within the range of accommodation. When the print gets too small or the arm too short, corrective reading glasses are the solution. The recession of the near point with age is not considered an abnormal defect of vision, since it proceeds at about the same rate in all normal eyes.

Table 25.1

Approximate Near Points of the Normal Eye at Different Ages

Age (years)	Near point (centimeters)
10	10
20	12
30	15
40	25
50	40
60	100

Vision Defects

Speaking of the normal eye, which is the norm, or average eye, implies the existence of a large population with defective vision. That this is the case is apparent from the number of people who wear glasses or contact lenses. The eyes of many people cannot accommodate within the normal range of 25 cm to infinity. These people have one of the two most common visual defects: nearsightedness (myopia) and farsightedness (hyperopia). Both of these conditions can usually be corrected.

Nearsightedness is the condition of being able to see nearby objects clearly but not distant objects. That is, the far point is not infinity but some nearer

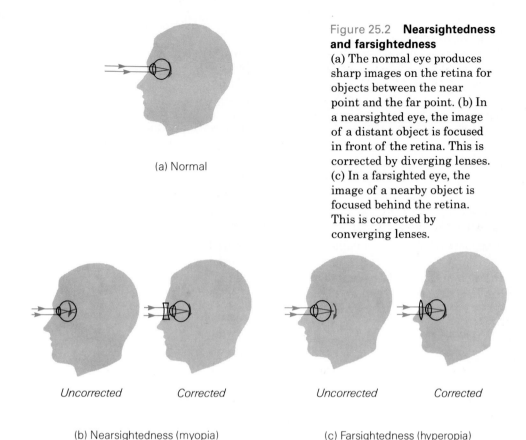

(a) Normal

Figure 25.2 **Nearsightedness and farsightedness**
(a) The normal eye produces sharp images on the retina for objects between the near point and the far point. (b) In a nearsighted eye, the image of a distant object is focused in front of the retina. This is corrected by diverging lenses. (c) In a farsighted eye, the image of a nearby object is focused behind the retina. This is corrected by converging lenses.

Uncorrected   *Corrected*   *Uncorrected*   *Corrected*

(b) Nearsightedness (myopia)   (c) Farsightedness (hyperopia)

point. When an object beyond the far point is viewed, the image is in focus in front of the retina. As a result, the image on the retina is blurred, or out of focus [Fig. 25.2(b)]. As the object is moved closer to the eye, its image moves back toward the retina. If the object is moved within the far point, a sharp image is seen.

Nearsightedness arises because the eyeball is too long or perhaps because the curvature of the cornea is too great. Whatever the reason, the images of distant objects are focused in front of the retina. Appropriate diverging lenses will correct this. Such a lens causes the rays to diverge, and the eye focuses the image farther back so it falls on the retina.

Farsightedness is the condition of being able to see distant objects clearly but not nearby objects. That is, the near point is not at the normal position but at some point farther from the eye. The image of an object that is closer to the eye than the near point is formed behind the retina [Fig. 25.2(c)]. Farsightedness arises because the eyeball is too short or perhaps because of insufficient curvature of the cornea.

Farsightedness is usually corrected by appropriate converging lenses. Such a lens causes the rays to converge, and the eye focuses the image on the retina. Converging lenses are also used to correct the farsightedness associated with the natural recession of the near point with age. Older people often must wear reading glasses, which have converging lenses.

Example 25.1 Correcting Nearsightedness

A certain nearsighted person cannot see objects clearly when they are more than 78 cm from either eye. (a) What power must the corrective lenses have if this person is to see distant objects clearly? (b) The person's near point is 23 cm. How is this affected when viewing through the corrective lenses? Assume that the lenses are in eyeglasses and are 3.0 cm in front of the eye.

Solution

Given: $d_f = 78$ cm   *Find*: (a) P (in diopters)
          $d_n = 23$ cm   (b) Near point (with glasses)
          $d = 3.0$ cm

(a) The lens must effectively put the image of a distant object ($d_o = \infty$) at the far point (d_f), which is 78 cm from the eye. The image, which acts as an object for the eye, is then within the range of accommodation. The image distance is measured *from* the lens, so $d_i = -75$ cm (78 cm $-$ 3 cm $=$ 75 cm), and a minus sign is used because the image is virtual on the object side of the lens. Then, assuming that the lens is described by the thin lens equation (Chapter 23),

$$\frac{1}{f} = \frac{1}{d_o} + \frac{1}{d_i} = \frac{1}{\infty} - \frac{1}{75\text{ cm}} = -\frac{1}{75\text{ cm}} \quad \text{or} \quad f = -75 \text{ cm}$$

The minus sign indicates a diverging lens. Recall that the lens power (P) in diopters (D) is the reciprocal of the focal length *in meters*. Thus,

$$P = \frac{1}{f} = \frac{1}{0.75\text{ m}} = -1.33 \text{ D}$$

A negative, or diverging, lens with a power of 1.33 D is needed.

(b) How wearing glasses affects near point vision can be seen by assuming that the image is formed by the corrective lens at the near point (d_n), which is 23 cm from the eye. The image is then 23 cm $-$ 3 cm $=$ 20 cm *from the lens*, so $d_i = -20$ cm (a minus sign is used since the image is virtual). Then, with $f = -75$ cm,

$$d_o = \frac{d_i f}{d_i - f} = \frac{(-20\text{ cm})(-75\text{ cm})}{-20\text{ cm} - (-75\text{ cm})} = 27 \text{ cm}$$

This means that the near point for this person is 27 cm in front of the lenses when the glasses are being worn. If an object is closer than this, its image (which acts as an object for the eye) will be inside the near point. ■

___

If the near point is changed by corrective lenses as in Example 25.1 and this causes a problem, bifocal lenses can be used. Bifocals were invented by Ben Franklin, who glued two lenses together. They are now made by grinding lenses with different curvatures in two different regions. Both nearsightedness and farsightedness can be treated at the same time with bifocals. (There are also trifocals, which are used to correct three defects.)

Example 25.2 Correcting Farsightedness

A farsighted person has a near point of 75 cm for one eye and a near point of 100 cm for the other. What powers should contact lenses have to allow the person to see clearly an object at a distance of 25 cm?

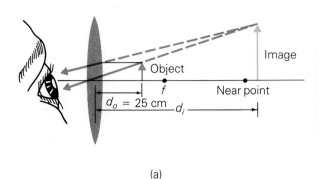

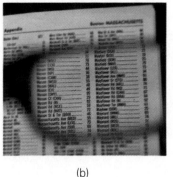

(a)

(b)

Figure 25.3 **Reading glasses** (a) When an object at the normal near point (25 cm) is viewed through reading glasses, the image is formed further away but within the eye's range of accommodation (beyond the receded near point). (b) Small print as viewed through the lens of reading glasses.

Solution

Given: $d_{i_1} = -75 \text{ cm} = -0.75 \text{ m}$ *Find*: P_1 and P_2
$d_{i_2} = -100 \text{ cm} = -1.0 \text{ m}$
$d_o = 25 \text{ cm} = 0.25 \text{ m}$

The eyes are distinguished as 1 and 2. Keep in mind that the optics of a person's eyes are usually different, and a different lens may be needed for each eye. In this case, each lens is to form an image at its eye's near point of an object that is at a distance (d_o) of 25 cm. This image will then act as an object within the eye's range of accommodation. This situation is illustrated in Fig. 25.3 and corresponds to a person wearing reading glasses.

The image distances are negative since the images are virtual. With contact lenses, the distance from the eye and the distance from the lens are taken to be the same. Then

$$P_1 = \frac{1}{f_1} = \frac{1}{d_{o_1}} + \frac{1}{d_{i_1}} = \frac{1}{0.25 \text{ m}} - \frac{1}{0.75 \text{ m}}$$

$$= \frac{2}{0.75 \text{ m}} = +2.67 \text{ D}$$

and

$$P_2 = \frac{1}{f_2} = \frac{1}{d_{o_2}} + \frac{1}{d_{i_2}} = \frac{1}{0.25 \text{ m}} - \frac{1}{1.0 \text{ m}}$$

$$= \frac{3}{1.0 \text{ m}} = +3.0 \text{ D}$$

Note that these are positive, or converging, lenses. ∎

Another common defect of vision is **astigmatism**, which is usually due to a refractive surface, most usually the cornea or crystalline lens, being out of round (nonspherical). As a result, the eye has different focal lengths in different planes (similar to the lens astigmatism for off-axis objects discussed in Section 23.4). Points may appear as lines, and the image of a line may be distinct in one direction and blurred in another or blurred in both directions in the circle of least confusion (the location of the least distorted image). A test for astigmatism is given in Fig. 25.4(a).

(a) Test for astigmatism

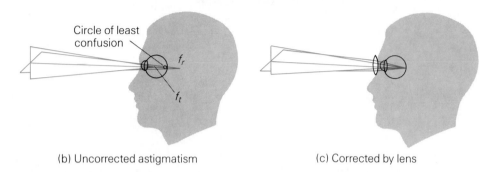

Circle of least confusion

f_r

f_t

(b) Uncorrected astigmatism

(c) Corrected by lens

Figure 25.4 **Astigmatism**
When one of the eye's refracting components is not spherical, the eye has different focal lengths in different planes. (a) If your eye is astigmatic, some or all of the lines in this diagram will appear blurred. (b) The effect occurs because rays along the transverse and radial axes are focused at different points, f_t and f_r, respectively. (c) Nonspherical lenses, such as plano-convex cylindrical lenses, are used to correct astigmatism.

Astigmatism may be corrected with lenses that have greater curvature in the plane in which the cornea or crystalline lens has deficient curvature. Generally, these corrective lenses are cylindrical-shaped, as shown in Fig. 25.4(c). Astigmatism is lessened in bright light because the pupil of the eye becomes smaller and the circle of least confusion is reduced.

QUESTION: What is meant by 20/20 vision?

ANSWER: Visual acuity is a measure of how vision is affected by object distance. This is commonly measured using a chart of letters placed at a given distance from the eyes. The result is usually expressed as a fraction: the numerator is the distance at which the tested eye sees a standard symbol, such as the letter E, clearly, and the denominator is the distance at which the letter is seen clearly by a normal eye. A 20/20 rating, which is sometimes called perfect vision, means that at a distance of 20 ft, the eye being tested can see standard-sized letters as clearly as can a normal eye.

A person's eyes may have differing visual acuity; for example, one eye may have 20/20 vision and the other 20/30. The latter means that the eye can read standard-sized letters from a chart at a distance of 20 ft and someone with normal vision could do so at 30 ft. That is, the eye in question has to be closer to the chart than a normal eye would to see the letters clearly.

25.2 Microscopes

Microscopes are used to magnify objects so that we can see more detail or see features that are normally indiscernible. Two basic types of microscope will be considered here.

The Magnifying Glass (Simple Microscope)

When we look at an object in the distance, it appears to be very small. As it is brought or comes closer to our eyes, it appears larger (Fig. 25.5). How large an

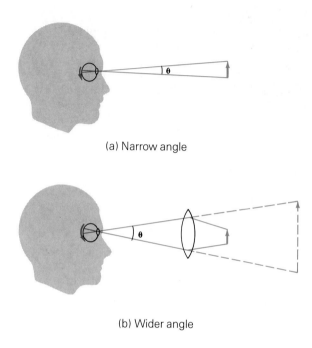

(a) Narrow angle

(b) Wider angle

Figure 25.5 **Angle and image size**
(a) The size of an image on the retina may be related to the angle θ the object subtends. (b) The greater the angle, the larger the image and the greater the amount of detail that can be seen. The angle may be made larger by bringing an object closer or by using a lens, as shown here.

object appears depends on the size of the image on the retina. This may be related to the angle subtended by the object; as shown in the figure, the greater the angle, the larger the image. When we want to examine detail, or look at something closely, we bring it close to our eyes so that it subtends a greater angle. For example, you may examine the detail of a figure in this book by bringing it closer to your eyes. The greatest amount of detail will be seen when the book is at your near point (assuming you are not wearing glasses). If your eyes were able to accommodate to shorter distances, an object brought very close to them would be further magnified. However, as you can easily prove by bringing this book very close to your eyes, images are blurred when objects are at distances inside the near point.

A **magnifying glass**, which is simply a single convex lens (sometimes called a simple microscope), allows a clear image to be formed of an object that is closer than the near point. In such a position, an object subtends a greater angle and therefore appears larger, or magnified. The lens produces a virtual image

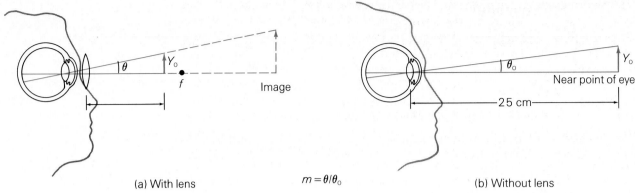

(a) With lens $m = \theta/\theta_o$ (b) Without lens

Figure 25.6 **Angular magnification**
The angular magnification of a lens is defined as the ratio of (a) the angular size of an object viewed through the lens to (b) the angular size of the object viewed without the lens.

beyond the near point on which the eye focuses (Fig. 25.6). If a hand-held magnifying glass is used, its position is usually adjusted until this image is seen clearly.

As illustrated in Fig. 25.6, the angle subtended by an object is much greater when a magnifying glass is used. The magnification of an object *viewed through a magnifying glass* is expressed in terms of this angle. This **angular magnification**, or magnifying power, is symbolized by m (not to be confused with M, the lateral magnification of a lens). The angular magnification is defined as the ratio of the angular size of the object viewed through the magnifying glass (θ) to the angular size of the object viewed without the magnifying glass (θ_o):

$$m = \frac{\theta}{\theta_o} \qquad (25.1)$$

angular magnification

The maximum angular magnification occurs when the image seen through the glass is at the eye's near point, that is, $d_i = -25$ cm. (A value of 25 cm will be assumed to be typical for near point in this discussion. The minus sign is used because the image is virtual.) The corresponding object distance may be calculated from the thin lens equation (Chapter 23):

$$d_o = \frac{d_i f}{d_i - f} = \frac{(-25 \text{ cm})f}{-25 \text{ cm} - f}$$

or $d_o = \dfrac{25f}{25 + f}$ (25.2)

where f is in centimeters.

The angular sizes of the object are related to its height by (see Fig. 25.6)

$$\tan \theta_o = \frac{y_o}{25} \qquad \text{and} \qquad \tan \theta = \frac{y_o}{d_o}$$

Assuming that a small-angle approximation ($\tan \theta \cong \theta$) is valid gives

$$\theta_o \cong \frac{y_o}{25} \qquad \text{and} \qquad \theta \cong \frac{y_o}{d_o}$$

Then the maximum angular magnification may be expressed as

$$m = \frac{\theta}{\theta_o} = \frac{y_o/d_o}{y_o/25} = \frac{25}{d_o}$$

Substituting for d_o from Eq. 25.2 gives

$$m = \frac{25}{25f/(25+f)}$$

This simplifies to

$$m = 1 + \frac{25 \text{ cm}}{f} \qquad (25.3)$$

angular magnification
for image at near point (25 cm)

Thus, lenses with shorter focal lengths give greater angular magnifications.

In the derivation of Eq. 25.3, the object for the unaided eye was taken to be at the near point, as was the image viewed through the lens. Actually, the eye can focus on an image located anywhere between the near point and infinity. At the extreme where the image is at infinity, the eye is more relaxed (the muscles attached to the crystalline lens are relaxed, and the lens is thin). For the image to be at infinity, the object must be at the focal point of the lens. In this case,

$$\theta \cong \frac{y_o}{f}$$

and the angular magnification is simply

$$m = \frac{25 \text{ cm}}{f} \qquad (25.4)$$

angular magnification
for image at infinity

Mathematically, it seems that the magnifying power may be increased to any desired value by using lenses with short enough focal lengths. Physically, however, lens aberrations limit the practical range of a single magnifying glass to about $3\times$ or $4\times$ (read as "three ex" and "four ex"), or a sharp image magnification of three or four times the size of the object.

Example 25.3   Magnifying Glass Magnification

A person uses a converging lens with a focal length of 12 cm to examine the fine detail on a painting. (a) What is the maximum magnification given by the lens? (b) What is the magnification for relaxed eye viewing?

Solution

Given: $f = 12$ cm

Find: (a) m (d_i = near point)
(b) m ($d_i = \infty$)

(a) The maximum magnification occurs when the image formed by the lens is at the near point of the eye, which for Eq. 25.3 was taken to be 25 cm. Using that equation,

$$m = 1 + \frac{25 \text{ cm}}{f} = 1 + \frac{25 \text{ cm}}{12 \text{ cm}} = 3.1$$

(b) The eye is most relaxed when viewing distant objects. Eq. 25.4 gives the magnification for the image formed by the lens at infinity:

$$m = \frac{25 \text{ cm}}{f} = \frac{25 \text{ cm}}{12 \text{ cm}} = 2.1 \quad \blacksquare$$

The Compound Microscope

A compound microscope provides greater magnification than is attained with a single lens, or simple microscope. A **compound microscope** consists of a pair of converging lenses, each of which contributes to the magnification [Fig. 25.7(a)]. A converging lens having a relatively short focal length ($f_o <$ 1 cm) is known as the **objective**. It produces a real, inverted, and enlarged image of an object positioned slightly beyond its focal point. The other lens, called the **eyepiece**, or **ocular**, has a longer focal length (f_e is a few centimeters) and is positioned so that the image formed by the objective falls just *inside* its focal point. This lens forms a magnified virtual image that is viewed by the observer. In essence, the objective acts as a projector, and the eyepiece is a simple microscope used to view the projected image.

The total magnification (M_t) of a lens combination is the product of the magnifications produced by the two lenses. The image formed by the objective is greater in size than its object by a factor M_o equal to the lateral magnification ($M_o = d_i/d_o$, with the minus sign omitted). In Fig. 25.7(a), note that the image distance for the objective lens is approximately equal to L, the distance between the lenses. That is, $d_i \cong L$. (The image I_o is formed by the objective just inside the focal point of the eyepiece, which has a very short focal length.) Also, since the object is very close to the focal point of the objective, $d_o \cong f_o$. With these approximations,

$$M_o \cong \frac{L}{f_o}$$

The angular magnification of the eyepiece for an object at the focal point is given by Eq. 25.4:

$$m_e = \frac{25 \text{ cm}}{f_e}$$

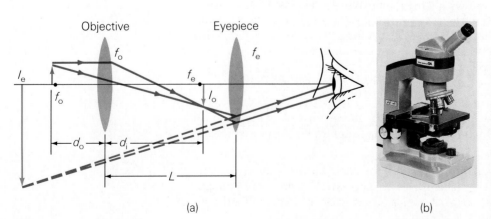

(a)

(b)

Figure 25.7 **The compound microscope**
(a) In the optical system of a compound microscope, the real image formed by the objective falls just within the focal point of the eyepiece and acts as the object for the eyepiece. An observer looking through the eyepiece sees an enlarged image. (b) A compound microscope.

Since the object for the eyepiece (the image formed by the objective) is very near the focal point of the eyepiece, a good approximation is

$$M_t = M_o m_e = \left(\frac{L}{f_o}\right)\left(\frac{25 \text{ cm}}{f_e}\right)$$

or $\quad M_t = \dfrac{25L}{f_o f_e}$ $\hspace{5cm}$ (25.5)

where L is in centimeters.

Example 25.4 Magnification with a Compound Microscope

A microscope has an objective with focal length of 10 mm and an eyepiece with a focal length of 4.0 cm. The lenses are fixed at 20 cm apart in the barrel. Determine the approximate total magnification of the microscope.

Solution

Given: $f_o = 10 \text{ mm} = 1.0 \text{ cm}$ $\hspace{2cm}$ *Find:* M_t
$\hspace{2.3cm} f_e = 4.0 \text{ cm}$
$\hspace{2.3cm} L = 20 \text{ cm}$

Using Eq. 25.5,

$$M_t = \frac{25L}{f_o f_e} = \frac{(25 \text{ cm})(20 \text{ cm})}{(1.0 \text{ cm})(4.0 \text{ cm})} = 125\times$$

(Note the relatively short focal length of the objective.) ∎

A modern compound microscope is shown in Fig. 25.7(b). Interchangeable eyepieces with magnifications from about $5\times$ to over $100\times$ are available. For standard microscopic work in biology or medical laboratories, $5\times$ and $10\times$ eyepieces are normally used. Microscopes are often equipped with rotating turrets, which usually contain three objectives for different magnifications, for example, $10\times$, $43\times$, and $97\times$. These objectives and the $5\times$ and $10\times$ eyepieces can be used in various combinations to provide magnifying powers from $50\times$ to $970\times$. The maximum magnification with a compound microscope is about $2000\times$.

Opaque objects are usually illuminated with a light source placed above them. In many instances, the specimens to be viewed are transparent, such as glass slides containing cells or thin sections of tissues. These are illuminated with a light source beneath the microscope stage so that light passes through the specimen. A modern microscope is usually equipped with a light condenser (converging lens) and diaphragm below the stage, which are used to concentrate the light and control its intensity. A microscope may have an internal light source. The light is reflected into the condenser from a mirror. Older microscopes have two reflecting surfaces: one a plane mirror for reflecting light from a high-intensity external source, and the other a concave mirror for converging low-intensity light such as sky light.

25.3 Telescopes

Telescopes apply the optical principles of mirrors and lenses to improve our ability to see distant objects. Used for both terrestrial and astronomical observations, telescopes allow some objects to be viewed in greater detail and

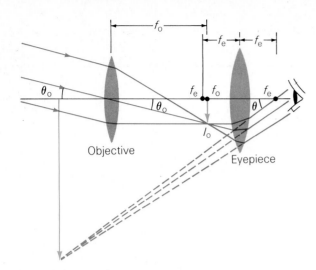

Figure 25.8 **The refracting astronomical telescope** In an astronomical telescope, rays from a distant object form an intermediate image at the focal point of the objective (f_o). The eyepiece is moved so the image is at or slightly inside its focal point (f_e). An observer sees an enlarged image at or near infinity (shown at a finite distance here for illustration).

others to be seen at all. Basically, there are two types of telescopes: refracting and reflecting. These depend on the gathering and converging of light by lenses and mirrors, respectively.

Refracting Telescope

The principle behind one type of refracting telescope is similar to that behind a compound microscope. The major components of this type of telescope are objective and eyepiece lenses, as illustrated in Fig. 25.8. The objective is a large converging lens with a long focal length, and the movable eyepiece has a relatively short focal length. Parallel rays from a distant object form an image (I_o) at the focal point (f_o) of the objective. This image acts as an object for the eyepiece, which is moved until the image lies just inside its focal point (f_e). A large, inverted image (I_e) is seen by an observer.

For relaxed viewing, the eyepiece is adjusted so that I_e is at infinity or I_o is at f_e, the focal point of the eyepiece. As you can see in the figure, the distance between the lenses is then the sum of the focal lengths ($f_o + f_e$), which is the length of the telescope tube. The magnifying power of a telescope focused for the final image at infinity can be shown to be (see Problem 35)

$$m = \frac{f_o}{f_e} \qquad (25.6)$$

Thus, to achieve the greatest magnification, the focal length of the objective should be made as great as possible and the focal length of the eyepiece as short as possible.

The telescope illustrated in Fig. 25.8 is called an **astronomical telescope**. Its final image is inverted, but this poses little problem to astronomers. (Why?) However, someone viewing an object on Earth through a telescope finds it more convenient to have an upright image. A telescope in which the final image is upright is called a **terrestrial telescope**. An upright final image can be obtained in several ways—two are illustrated in Fig. 25.9.

In the telescope diagrammed in Fig. 25.9(a), a diverging lens is used as an eyepiece. This type of terrestrial telescope is referred to as a Galilean telescope,

Galileo did not invent the telescope, but he built one after hearing about a Dutch instrument that could be used for distant viewing.

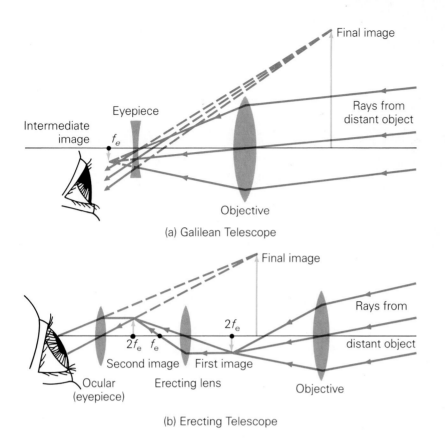

Intermediate image

Eyepiece

f_e

Final image

Rays from distant object

Objective

(a) Galilean Telescope

Final image

Ocular (eyepiece)

Second image

$2f_e$ f_e

Erecting lens

First image

$2f_e$

Rays from

distant object

Objective

(b) Erecting Telescope

(c)

Figure 25.9 **Terrestrial telescopes**
(a) A Galilean telescope uses a diverging lens as an eyepiece, which produces an upright image. (b) Another way to produce an upright image is to use an erecting lens between the objective and eyepiece in (c) an astronomical-type telescope.

because Galileo built one in 1609. The image of the objective is formed at the focal point of the eyepiece (f_e) on the side opposite the objective. Thus, the light rays pass through the eyepiece before the image is formed, and the image acts as a "virtual" object for the eyepiece (see Section 23.3). An observer sees a magnified, upright, virtual image.

The other type of terrestrial telescope, illustrated in Fig. 25.9(b), uses a third lens, called the erecting lens, between converging objective and eyepiece lenses. If the image is formed by the objective at a distance that is twice the focal length of the erecting lens ($2f_e$), then the lens merely inverts the image without magnification, and the telescope magnification is still given by Eq. 25.6. However, to achieve the upright image in this way requires a greater telescope length. Using the intermediate erecting lens to invert the image increases the length of the telescope by four times the focal length of the erectig lens ($2f_e$ on each side). An erecting lens is used in a spyglass, which is telescoped to a long length. The inconvenient length can be avoided by using reflecting prisms. This is the principle of prism binoculars, which are double telescopes—one for each eye (Fig. 25.10).

Reflecting Telescope

For viewing the Sun, moon, and nearby planets, magnification is necessary to see details. However, even when distant stars and galaxies are magnified, they appear only as points of light. In this case, it is more important to gather enough

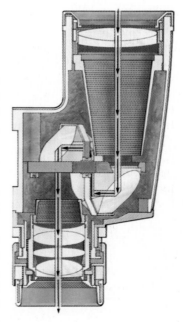

Figure 25.10 **Prism binoculars**
A schematic cutaway view of one ocular showing the internal reflection in the prisms, which reduce the overall length.

Figure 25.11 **Yerkes refracting telescope**

light so the object can be seen at all. The intensity of light from a distant source is very weak. In many instances, such a source may be detected only when the light is gathered and focused on a photographic plate over a long period of time.

Recall that intensity is energy per *area* per unit time. Thus, more light can be gathered if the size of the objective is increased. This increases the distance at which the telescope can detect objects. However, producing a large lens involves difficulties associated with glass quality, grinding, and polishing. Compound lens systems are required to reduce aberrations, and a very large lens may sag under its own weight, producing further aberrations. The largest objective lens in use has a diameter of 40 in. (102 cm, or about 1 m) and is part of the refracting telescope of the Yerkes Observatory at Williams Bay, Wisconsin (Fig. 25.11).

These problems are reduced with a reflecting telescope utilizing a large, concave, front-surface parabolic mirror (Fig. 25.12). A parabolic mirror does not

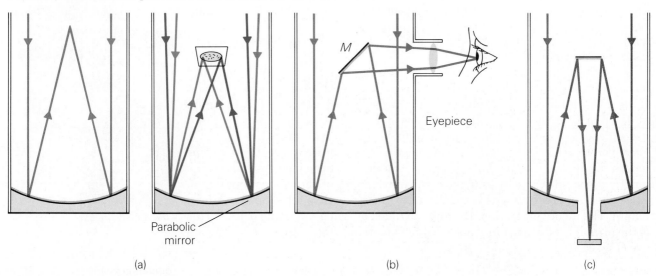

(a)   (b)   (c)

Figure 25.12 **Reflecting telescope**
A concave mirror may be used in a telescope to converge light to form an image of a distant object. The image may be at the prime focus as shown in part (a), or a small mirror may be used to focus the image at another location as in (b), called a Newtonian focus, and (c) a type of Cassegrain focus.

Figure 25.13 **Hale reflecting telescope**

exhibit spherical aberration (Chapter 23), and a mirror has no inherent chromatic aberration. High-quality glass is not needed, since the light is reflected by a mirrored front surface. And only one surface has to be ground, polished, and silvered.

As shown in Fig. 25.12, light from a distant source entering the telescope tube is focused by the concave mirror. A small plane mirror is sometimes used to direct the converging rays to a more convenient viewing location. Such a mirror blocks only a small percentage of the incoming light. The largest reflecting telescope in the United States has a mirror with a 200-in. (5.1-m) diameter and is located at Hale Observatories on Palomar Mountain in California. In the Hale design, the observer sits in a cage at the prime focus of the telescope (Fig. 25.13). The world's largest reflecting telescope has a reflector with a 6-m diameter and is located in the Soviet Union.

Even though reflecting telescopes have advantages over refracting telescopes, they also have their own problems. Like a large lens, a large mirror may sag under its own weight, and the weight necessarily increases with the size of the mirror. The weight factor is also a cost factor for construction of a telescope, since the supporting elements for a heavier mirror must be more massive.

These problems are being addressed by two new technologies. One of these reduces the weight of the mirror by making its back in the form of a honeycomb rather than a solid. Also, the casting of the mirror is done in a rotating furnace.

Telescopes Using Nonvisible Radiation

The word "telescope" usually brings to mind visual observations. However, the visible region is a very small part of the electromagnetic spectrum, and celestial objects emit radiation of many other types, including radio waves. This fact was discovered accidentally in 1931 by an electrical engineer named Carl Jansky while he was working on the problem of static interference with intercontinental radio communications. Jansky found an annoying static hiss that came from a fixed direction in space, apparently from a celestial source. It was immediately apparent that radio waves are another source of astronomical information, and radio telescopes were built.

A radio telescope operates similarly to a reflecting light telescope. A reflector with a large area collects and focuses the radio waves at a point where a detector picks up the signal (Fig. 1). The parabolic collector, called a disk, is covered with metal wire mesh. Since the wavelengths of radio waves range from a centimeter or so to several meters, the wire mesh is a good reflecting surface for such waves.

Radio telescopes supplement optical telescopes and provide some definite advantages. Radio waves pass freely through the huge clouds of dust in our galaxy that hide a large part of it from visual observation. Also, radio waves easily penetrate the Earth's atmosphere, which reflects and scatters a large percentage of the incoming visible light.

The Earth's atmosphere absorbs many wavelengths of electromagnetic radiation and completely blocks some from reaching the ground. Water vapor, a strong absorber of infrared radiation, is concentrated in the lower part of the atmosphere near the surface of the Earth. Thus, observations with infrared telescopes are made from high-flying aircraft or from orbiting spacecraft. The first orbiting infrared observatory was launched in 1983. Not only are atmospheric interferences eliminated in space, but a telescope may be cooled to a very low temperature without becoming coated with condensed water vapor. Cooling the telescope helps eliminate interfering infrared radiation generated by the telescope itself. The orbiting telescope launched in 1983 was cooled with liquid helium to about 10 K; it carried out an infrared survey of the entire sky.

The atmosphere is virtually opaque to ultraviolet radiation, X-rays, and gamma rays from distant sources. Orbiting satellites with telescopes sensitive to these types of radiation have mapped out portions of the sky, and other surveys are planned.

At the altitudes reached by orbiting satellites even observations in the visible region are not affected by air motion and temperature refraction. Perhaps in the not too distant future, a permanently staffed orbiting observatory carrying a variety of telescopes will help expand our knowledge of the universe.

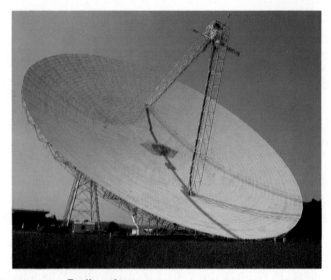

Figure 1 **Radio telescope**

The spinning gives the mirror a parabolic front surface, which reduces the amount of grinding needed and the amount of glass necessary for casting. Another way to solve the weight problem for a reflecting telescope is to use an array of small mirrors to do the job of a single large mirror. One of the first telescopes using this new technology was the Multiple Mirror Telescope (MMT) built in 1979 in Arizona. The MMT consists of six individual mirrors that are adjusted by a computer so that the light from all six focuses at the same point. Each mirror is 72 in. (1.8 m) in diameter, and when used together, the six provide the light-gathering power of a single 4.5-m mirror. A similar 10-m telescope is now under construction in Hawaii.

Not all telescopes are optical in nature. See the Insight feature for a look at a different type of reflecting telescope.

25.4 Diffraction and Resolution

The diffraction of light places a limitation on our ability to distinguish objects that are close together when using microscopes or telescopes. This effect can be understood by considering two point sources located far from a narrow slit of width d (Fig. 25.14). For example, the sources could be distant stars. In the absence of diffraction, two bright spots, or images, would be observed on a screen. As you know from Section 24.3, however, the slit diffracts the light, and each image consists of a central maximum with a pattern of weaker bright and dark fringes on either side. If the sources are close together, the two central maxima may overlap. Then the images cannot be distinguished, or are unresolved. For the images to be resolved, the central maxima must not overlap appreciably.

In general, images of two sources can be resolved if the central maximum of one falls at or beyond the first minimum (dark fringes) of the other. This generally accepted limiting condition for the resolution of two diffracted images was first proposed by Lord Rayleigh (1842–1919), a British physicist, and is known as the **Rayleigh criterion**:

> Two images are said to be just resolved when the central maximum of one image falls on the first minimum of the diffraction pattern of the other image.

The Rayleigh criterion may be expressed in terms of the angular separation (θ) of the sources (see Fig. 25.14). The first minimum ($m = 1$) for a single-slit diffraction pattern satisfies this relationship:

$$d \sin \theta = \lambda$$

or
$$\sin \theta = \frac{\lambda}{d}$$

This gives the minimum angular separation for two images to be just resolved according to the Rayleigh criterion. In general, for visible light, the wavelength is much smaller than the slit width ($\lambda \ll d$), so a good approximation is $\sin \theta \cong \theta$. The limiting, or minimum, angle of resolution (θ_{min}) for a slit of width d is then

$$\theta_{min} = \frac{\lambda}{d} \qquad (25.7)$$

resolution with a slit

(Note that θ_{min} is a pure number and is therefore expressed in radians.) Thus, the images of two sources will be *distinctly* resolved if the angular separation of the sources is greater than λ/d.

The apertures (openings) of microscopes and telescopes are generally circular. Thus, there is a circular diffraction pattern around the central maximum, which is a bright circular disk (Fig. 25.15). Detailed analysis for a circular aperture shows that the minimum angular separation for two objects

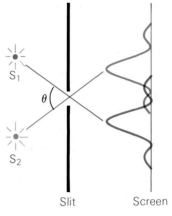

(a) Resolved

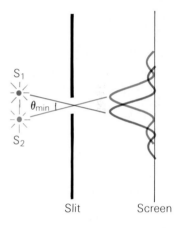

(b) Just resolved

Figure 25.14 Resolution
Two light sources that are from a slit produce diffraction patterns. (a) When the angle subtended by the sources at the slit is large enough such that the diffraction patterns are distinguishable, the images are said to be resolved. (b) At smaller angles, the central maxima are closer together, and at θ_{min}, when the central maximum of one pattern falls on the first dark fringe of the other, the images are said to be just resolved. For even smaller angles, they would be unresolved.

for their images to be just resolved is

$$\theta_{min} = \frac{1.22\lambda}{D} \qquad \text{(27.8)}$$

resolution with a circular aperture

where D is the diameter of the aperture.

Eq. 25.8 applies to the objective lens of a microscope or telescope, which may be considered to be a circular aperture for light. According to the equation, to make θ_{min} small so objects close together can be resolved, it is advantageous to make the aperture as large as possible. This is another reason for using large lenses (and mirrors) in telescopes, in addition to their greater light-gathering power.

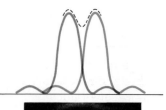

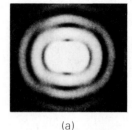

(a)

Example 25.5   Resolution

Determine the minimum angle of resolution by the Rayleigh criterion for (a) the pupil of the eye (diameter of about 4.0 mm), (b) the Lick Observatory 40-in. (102-cm) refracting telescope for visible light with a wavelength of 660 nm, and (c) a radio telescope of 25-m diameter for radiation with a wavelength of 21 cm.

Solution

Given: (a) $D = 4.0$ mm $= 4.0 \times 10^{-3}$ m $\qquad$ Find: (a) θ_{min}
$\qquad$ (b) $D = 102$ cm $= 1.02$ m $\qquad\qquad\qquad$ (b) θ_{min}
$\qquad\qquad$ $\lambda = 660$ nm $= 6.60 \times 10^{-7}$ m $\qquad\quad$ (c) θ_{min}
$\qquad$ (c) $D = 25$ m
$\qquad\qquad$ $\lambda = 21$ cm $= 0.21$ m

Eq. 25.8 applies to all three cases.
(a) For the eye,

$$\theta_{min} = \frac{(1.22)(6.60 \times 10^{-7}\text{ m})}{4.0 \times 10^{-3}\text{ m}} = 2.0 \times 10^{-4}\text{ rad}$$

(b) For the light telescope,

$$\theta_{min} = \frac{(1.22)(6.60 \times 10^{-7}\text{ m})}{1.02\text{ m}} = 7.9 \times 10^{-7}\text{ rad}$$

(c) For the radio telescope,

$$\theta_{min} = \frac{(1.22)(0.21\text{ m})}{25\text{ m}} = 0.01\text{ rad}$$

The smaller the angular separation, the better the resolution. What do the results tell you? ∎

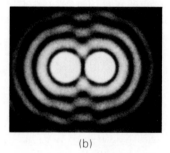

(b)

Figure 25.15   **Circular aperture resolution**
When the angular separation of two objects is very small, the images are not resolved. (a) The minimum angular separation for two images to be just resolved is given by the Rayleigh criterion: the central maximum of one image falls on the first minimum of the diffraction pattern of the other image. (b) When the objects are farther apart, the images are well resolved.

For a microscope, it is convenient to specify the actual separation (s) between two point sources. Since the objects are usually near the focal point of the objective,

$$\theta_{min} = \frac{s}{f} \qquad \text{or} \qquad s = f\theta_{min}$$

where f is the focal length of the lens. Then, using Eq. 25.8,

$$s = f\theta_{\min} = \frac{1.22\lambda f}{D} \qquad (25.9)$$

This minimum distance between two points whose images can be just resolved is called the **resolving power** of the lens. Practically, the resolving power of a microscope indicates the ability of the objective to distinguish fine detail in specimens' structures.

As with a telescope, the *useful* magnification of a microscope is not limited by the power (focal lengths) of its lenses but by the aperture, that is, the size of the objective lens. However, making the objective lens larger leads to another limiting factor on resolution by a lens—aberrations. As you learned in Chapter 23, spherical and other aberrations give rise to blurred images. Compound lens systems can be used to reduce these effects significantly, so the limit of resolution is determined mainly by diffraction.

Note from Eq. 25.8 that greater resolution may be gained by using radiation of a shorter wavelength. Thus, a telescope with a particular objective will show greater resolution with violet light than with red light. For microscopes, an overall reduction in wavelength may be accomplished. One of the objectives has an *oil immersion lens*. A film of transparent oil is used between the objective and the specimen. The wavelength of light in the oil is λ/n, where n is the index of refraction of the oil and λ the wavelength of the light in air. With values of n of about 1.50 or higher, the wavelength is significantly reduced and the resolving power increased.

The wavelength can be further reduced by using ultraviolet light and photographic film. However, ordinary glass absorbs ultraviolet light, so the lenses of ultraviolet microscopes must be made of other materials, such as quartz.

The fact that the resolving power of a microscope can be increased by decreasing the wavelength of the incident light opened up new possibilities for seeing fine details after it was discovered that a beam of electrons has wave properties (Chapter 28). These beams have wavelengths that can be adjusted to be very much smaller than those of light in or near the visible region. Using a short-wavelength electron beam, an electron microscope can magnify up to 1000 times more than a visible-light microscope. X-rays have even shorter wavelengths than electron beams and are used to attain even greater magnification and resolution.

Useful magnification refers to the magnification of resolvable details. Magnification beyond the limit of usefulness offers no further disclosure of the details of an object and is called empty magnification. For example, enlarging a newspaper picture is empty magnification.

25.5 Color and Optical Illusions ◆

In general, physical properties are fixed or absolute. For example, a particular radiation has a certain frequency or wavelength. However, visual perception of this radiation may vary from person to person. How we see radiation gives rise to the following interesting topics.

Color Vision

Color is perceived because of a physiological response to light excitation of cone receptors in the retina of the human eye. (Many animals have no cone cells and live in a black-and-white world.) The cones are sensitive to light with

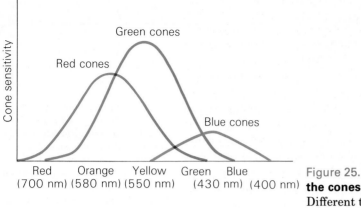

Figure 25.16 **Sensitivity of the cones**
Different types of cones in the eye respond to different frequencies of light to give three general color responses.

frequencies between 7.5×10^{14} Hz and 4.3×10^{14} Hz (wavelengths between 400 nm and 700 nm). Different frequencies of light are perceived by the brain as different colors. The association of a color with a particular frequency is subjective and may vary from person to person. As pitch is to sound and hearing (Chapter 15), color is to light and vision.

Color vision is not well understood. One of the most widely accepted theories is that there are three different types of cones in the retina, responding to different parts of the visible spectrum, particularly in the red, green, and blue regions (Fig. 25.16). Presumably, the types of cones absorb light of specific wavelengths and functionally overlap one another to form combinations that are interpreted by the brain as various colors of the spectrum. For example, when red and green cones are stimulated equally by light of a particular frequency, the brain interprets the color as yellow. But when the red cones are stimulated more strongly than the green cones, the brain senses orange.

Color-blindness results when one type of cone is missing. This condition is genetically inheritable, and most color-blind people are males (about 4% of the male population). For example, if a man lacks red cones, he can see green through orange-red by means of his green cones. However, he is unable to distinguish among these colors satisfactorily because there are no red cones to provide contrast. A similar condition occurs if the green cones are missing. In either case, this is termed red-green color-blindness because the person cannot effectively distinguish those colors.

The theory of color vision described above is based on the experimental fact that beams of varying intensities of red, green, and blue light will produce most colors, or hues. The red, blue, and green from which we interpret a full spectrum of colors are called the **additive primary colors** (or the additive primaries). When light beams of the additive primaries are projected on a white screen so they overlap, other colors will be produced, as illustrated in Fig. 25.17. This is called the **additive method of color production** or mixing.

Triad dots consisting of three phosphors that emit the additive primary colors when excited are used in television picture tubes to produce colored images.

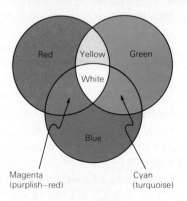

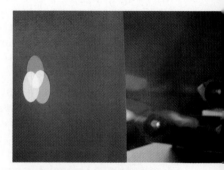

Figure 25.17 **Additive method of color production**
When light beams of the primary colors (red, blue, and green) are projected on a white screen, mixtures of them produce other colors. Varying the intensities of the beams allows most colors, or hues, to be produced.

Note in Fig. 25.17 that the proper combination of the primary colors appears white to the eye. Also, many *pairs* of colors appear white to the eye when combined. The colors of such pairs are said to be **complementary colors**. For example, the complement of blue is yellow, that of red is cyan, and that of green is magenta.

Edwin H. Land (the developer of Polaroid film—see Chapter 24) has shown that when the proper mixtures of only two wavelengths (colors) of light are passed through black and white transparencies (no color), they produce images of various colors. Land wrote, *"In this experiment we are forced to the astonishing conclusion that the rays are not in themselves color-making. Rather they are bearers of information that eye uses to assign appropriate colors to various objects in an image."* * From Land's experiments, it seems that information about colors, other than for the two wavelengths used, is developed in the brain via the cone cells. However, knowledge about color vision is far from complete.

Objects have color when they are illuminated with white light because they reflect (scatter) or transmit the light of the wavelength of the color they appear to be. The other wavelengths of the white light are absorbed. For example, when white light strikes a piece of transparent red glass, or a red filter, only the wavelengths in the red portion of the spectrum are reflected or transmitted. Otherwise, the glass would not appear red. This occurs because the color pigments in the glass are selective absorbers.

Pigments are mixed to form various colors, such as in the production of paints and dyes. You are probably aware that mixing yellow and blue paints produces green. This is because the yellow pigment absorbs most of the wavelengths except those in the yellow region of the visible spectrum, the blue pigment absorbs most of the wavelengths except those in the blue region. The wavelengths in the intermediate green region are not strongly absorbed by either pigment, and therefore the mixture appears green. The same effect may be accomplished by passing white light through stacked yellow and blue filters. The light coming through both filters appears green.

Mixing pigments results in the subtraction of colors, and the color that is perceived is the one not absorbed, or subtracted. This is an example of what is called the **subtractive method of color production** or mixing. Three particular pigments—cyan, magenta, and yellow—are called the **subtractive primary pigments** (or subtractive primaries). Various combinations of two of the three subtractive primaries produce the three additive primary colors (red, blue, and green), as illustrated in Fig. 25.18. When the subtractive primaries are mixed in the proper proportions, the mixture appears black (all wavelengths are absorbed). Painters often refer to the subtractive primaries as red, yellow, and blue. They are loosely referring to magenta (purplish-red), yellow, and cyan "true" blue). Mixing these paints in the proper proportions produces a broad spectrum of colors.

Note in Fig. 25.18 that the magenta pigment essentially subtracts the color green where it overlaps with cyan and yellow. As a result, magenta is sometimes referred to as "minus green." If a magenta filter were placed in front of a green light, no light would be transmitted. Similarly, cyan is called "minus red," and yellow is called "minus blue."

* From "Experiments in Color Vision," by Edwin H. Land, in *Scientific American*, May 1959, pp. 84–99.

Recall the colors of the visible spectrum may be remembered by ROY G. BIV (Chapter 22).

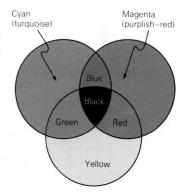

Figure 25.18 **Subtractive method of color production** When the primary pigments (cyan, magenta, and yellow) are mixed, different colors are produced by subtractive absorption. When all the wavelengths of visible light are absorbed, the mixture appears black.

Visual Images and Illusions

Everything you see and therefore most of what you know depends on the information conveyed through the tiny pupils of your eyes. Thus, the better you understand the function of your eyes, the better you can interpret your visual observations. Even though you often hear people say "seeing is believing," you can't take this statement too literally. For example, from the discussion on converging lenses, you know that the images of objects formed on the retinas of your eyes are inverted. But you don't perceive the world as upside down—by some means, the brain interprets the inverted image of the world as being right side up.

Knowledge of the eye will help you understand some of the things you see and do not see. For example, look intensely at the black cross in Fig. 25.19(a) with this book held at arm's length and your right eye closed. Then slowly bring the book toward you. At certain points, you will see the square and dot alternately disappear and reappear. Is there something wrong with your vision? No, the experiment demonstrates the eye's blind spot. This is the region where the optic nerve attaches to the retina and where there are no rods or cones. Consequently, there is no optical response. You don't ordinarily notice the blind spot because of movements of the eye and objects and because you have binocular vision.

The stimulation of one area of the retina may affect the sensations in an adjacent area, producing an illusion. For example, if you look at the pattern in Fig. 25.19(b), you will probably see fleeting patches of gray where the white lines intersect between the black diamonds. Also, stimulation of a particular area of the retina may be "remembered" after your gaze has shifted to another object. Look intensely at the white dot in the black Y in Fig. 25.19(c) for about 30 seconds. Then transfer your gaze to the white dot between the E's. You will see a ghostly white Y between the E's. This is an example of an afterimage. Afterimages are commonly seen when you shift your eyes from a bright object to a dark background, for example, looking at a light in a darkened room and then looking away from it.

Keep in mind that there are many ways people can be visually fooled. For example, your depth perception is tricked in 3-D movies when objects give the illusion of jumping out of the screen. Other optical illusions are caused by the geometry of the situation in which they occur. Several common illusions are shown in Demonstration 21. How does your eye (or brain) interpret them?

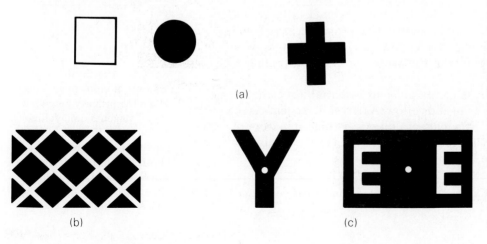

(a)

(b)   (c)

Figure 25.19 **Visual images** (a) To experience the blind spot, hold the book at arm's length and with the right eye closed, look intently at the black cross. Then bring the book slowly toward your face. The square and the dot will alternately disappear and reappear. (b) Induced stimulation causes fleeting patches of gray to appear on the white areas between the points of the black diamonds. (c) Similarly, an image of a Y can be induced between the E's by looking intently at the white dot in the black Y, and then shifting your gaze to the white dot between the E's.

Demonstration 21

Geometrical Optical Illusions

What do you see?

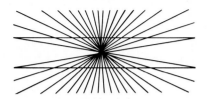
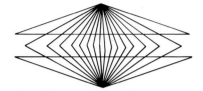

(a) Are the horizontal lines parallel?

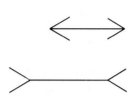

(b) Is the lower horizontal line longer?

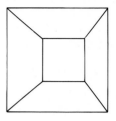

(c) Top view or recessed view?

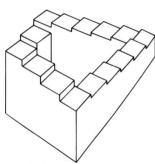
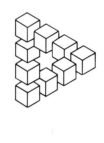

(d) How much work is done getting back to the starting point?

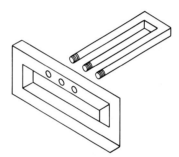

(e) Is something dimensionally wrong here?

(f) Which cube is complete?

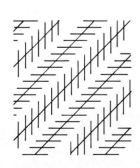

(g) Parallel diagonal lines?

(h) Big, bigger, biggest?

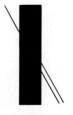

(i) Which lower line connects with the top one?

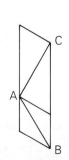

(j) Is line AB shorter than line AC?

Important Formulas

Angular magnification:

$$m = \frac{\theta}{\theta_{\text{o}}}$$

Magnification of a magnifying glass with image at near point (25 cm):

$$m = 1 + \frac{25 \text{ cm}}{f}$$

Magnification of a magnifying glass with image at infinity:

$$m = \frac{25 \text{ cm}}{f}$$

Magnification of a compound microscope (with L in centimeters):

$$M_{\text{t}} = M_{\text{o}} m_{\text{e}} = \left(\frac{L}{f_{\text{o}}}\right)\left(\frac{25}{f_{\text{e}}}\right) = \frac{25L}{f_{\text{o}} f_{\text{e}}}$$

Magnification of a refracting telescope:

$$m = \frac{f_{\text{o}}}{f_{\text{e}}}$$

Resolution for a slit:

$$\theta_{\min} = \frac{\lambda}{d}$$

Resolution for a circular aperture:

$$\theta_{\min} = \frac{1.22\lambda}{D}$$

Resolving power:

$$s = f\theta_{\min} = \frac{1.22\lambda f}{D}$$

Questions

The Human Eye

1. Name the parts of the eye whose functions are analogous to those of the following parts of a camera: (a) lens, (b) diaphragm, (c) shutter, and (d) film. (e) Is there an analogy between the lens of the eye and a zoom lens on a camera? Explain.

2. Explain the purpose of the eye's accommodation. How is it accomplished?

3. How do a camera and the eye differ with respect to focusing adjustments?

4. Although we can see objects in dim light such as moonlight, we do not perceive their colors. Why is this?

5. How would you determine the near points of your eyes experimentally?

6. (a) Distinguish between farsightedness and nearsightedness, and explain how these conditions may be corrected. (b) What characterizes the images seen by a person with astigmatism, and how is this vision defect corrected?

7. Would someone's eyeglasses and contact lenses have the same focal lengths? Explain.

8. Which ratio indicates better vision, 20/15 or 15/20? Explain.

Microscopes

9. What limits the angular magnification of a lens?

10. When an object is at the focal point of a magnifying glass, the magnification is given by $m = 25 \text{ cm}/f$. Can the magnification be increased indefinitely by using lenses with shorter focal lengths?

11. What are the disadvantages of using objectives and eyepieces with short focal lengths in compound microscopes?

12. What is empty magnification?

13. Explain why the total magnification for a compound microscope is the product of the *lateral* magnification of the objective and the *angular* magnification of the eyepiece.

Telescopes

14. On what does the length of a refracting telescope depend?

15. What is the difference between an astronomical telescope and a terrestrial telescope? Describe two types of terrestrial telescope.

16. Besides having only one surface to prepare and fewer aberrations, a reflecting telescope has some structural advantages over a refracting telescope. What are these? (Hint: consider the positions and mounting of the optical components.)

17. Compare the area of the 200-in. mirror of the Hale Observatories reflecting telescope with the floor area of a typical dormitory room.

18. What two major factors limit the resolution of a lens?

19. Explain the Rayleigh criterion for resolution.

20. Is it better to make θ_{min} larger or smaller? Explain.

21. What are the advantages of using large lenses and mirrors in telescopes?

22. What is the advantage of an oil immersion lens in a microscope? Does such a lens make any difference in the brightness of the image?

23. Astronomical observations are sometimes made through colored filters. What are the advantage and disadvantage of doing this?

Color and Optical Illusions ◆

24. Why do your eyes have a certain color, and why is the pupil black?

25. Is color television based on the additive or subtractive method of color production? Explain.

26. (a) Can white be obtained by the subtractive method of color production? Explain. (b) It is sometimes said that black is the absence of color or that a black object absorbs all of the incident light. If so, why do we see black objects?

27. Why does a white screen appear to be the color of whatever type of light is used to illuminate it?

28. Describe how the American flag would appear if it were illuminated with light of each of the primary colors.

29. What color would a green chalkboard appear to be when viewed through (a) amber, (b) blue, or (c) green sunglasses? What color would letters written on the board with white chalk appear to be?

Problems

25.1   The Human Eye*

■1. A certain farsighted person has a near point of 50 cm. What type and power of lens should an optometrist prescribe to enable the person to see clearly objects as close as 25 cm?

■■2. The far point of a certain nearsighted person is 200 cm. What type of lens with what focal length will allow this person to see distant objects clearly?

■■3. If the normal near point of the crystalline lens is taken to be 25 cm, what is the power of the lens? (Assume all refraction is done by the lens.)

■■4. A man cannot see objects clearly when they are closer than 1.5 m. (a) Is he nearsighted or farsighted? (b) What type of lens of what power (in diopters) will allow him to see objects clearly at 25 cm?

■■5. A woman cannot see objects clearly when they are farther than 5.0 m away. (a) Is she nearsighted or farsighted? (b) What type of lens of what power (in diopters) will allow her to see distant objects clearly?

■■6. A student can't see the print in a book clearly when holding the book at arm's length (0.80 m from the eyes). What type of lens with what focal length will allow the student to read the text at the normal near point?

* Assume corrective lenses are in contact with the eye (contact lenses) unless otherwise stated.

■■7. To correct a case of hyperopia, an optometrist prescribes positive contact lenses that effectively move the patient's near point from 100 cm to 25 cm. (a) What is the power of the lenses? (b) Will the person be able to see distant objects clearly when wearing the lenses or will she have to take them out?

■■8. A certain myopic person has a far point of 100 cm. (a) What power must a lens have to allow him to see distant objects clearly? (b) If he is able to read print at 25 cm while wearing his glasses, is his near point less than 25 cm? (c) Give an indication of his age.

■■9. An eyeglass lens with a power of $+2.5$ D allows a farsighted student to read a book held at a distance of 25 cm from her eyes. At what distance must the student hold the book to read it without glasses?

■■10. A farsighted person wears glasses whose lenses have a power of $+1.25$ D and is then able to see clearly objects as close as 25 cm. What is the person's near point?

■■11. A nearsighted person wears glasses whose lenses have a power of -0.15 D. What is the person's far point?

■■■12. A college professor can see objects clearly only if they are between 50 and 300 cm from her eyes. Her optometrist prescribes bifocals that enable her to see distant objects through the top half of the lenses and read students' papers at a distance of 25 cm through the lower half. What are the powers of the bifocal lenses?

■■■13. A middle-aged man starts to wear glasses with

lenses of $+2.0$ D that allow him to read a book held as closely as 25 cm. Several years later, he finds that he must hold a book no closer than 33 cm to read it clearly using the same glasses, so he gets new glasses. What is the power of the new lenses?

■■■**14.** The amount of light reaching the film in a camera depends on the lens aperture (the effective area) as controlled by the diaphragm. The f number is the ratio of the focal length of the lens to its effective diameter. For example, an f/8 setting means that the diameter of the aperture is $\frac{1}{8}$ of the focal length of the lens. The lens setting is commonly referred to as the f-stop. (a) Determine how much light each of the following lens settings will admit to the camera compared to f/8: (1) f/3.2, and (2) f/16. (b) The exposure time of a camera is controlled by the shutter speed. If a photographer correctly uses a lens setting of f/8 with a film exposure time of $\frac{1}{60}$ s, what exposure time should he use to get the same amount of light *exposure* if he sets the f-stop at f/5.6?

25.2 Microscopes*

■■**15.** A diamond cutter uses a jeweler's lens to examine a diamond. If the focal length of the lens is 15 cm, what is the maximum angular magnification of the diamond?

■■**16.** A person using a magnifying glass with a focal length of 10 cm views an object at the focal point of the lens. What is the magnification?

■■**17.** A student views the details of a dollar bill with a magnifying glass, achieving its maximum magnification of $3\times$. What is the focal length of the magnifying glass?

■■**18.** What is the maximum magnification of a magnifying glass with a power of $+2.5$ D for (a) a person with a near point of 25 cm and (b) a person with a near point of 10 cm?

■■**19.** When viewing an object with a magnifying glass whose focal length is 15 cm, a person positions the lens so there is minimum eyestrain. What is the observed magnification?

■■**20.** A detective looks at a fingerprint with a magnifying glass whose power is $+2.0$ D. What is the maximum magnification of the print?

■■**21.** A compound microscope has an objective with a focal length of 4.0 mm and an eyepiece with a magnification of $10\times$. If the objective and eyepiece are 15 cm apart, what is the total magnification of the microscope?

---

* The normal near point should be taken as 25 cm unless otherwise specified.

■■**22.** A $1500\times$ microscope has an objective whose focal length is 0.50 cm. If the tube of the microscope is 18 cm long, find the focal length of the eyepiece.

■■**23.** The tube of a compound microscope is 16 cm long. The focal length of the eyepiece is 3.0 cm, and the focal length of the objective is 0.45 cm. Find the total magnification of the microscope.

■■**24.** A microscope whose tube is 10 cm long has objective and eyepiece lenses with focal lengths of 8.0 mm and 12 mm, respectively. What is the total magnification of the microscope?

■■■**25.** A lens with a power of 10 D is used as a simple microscope. (a) How close can an object be brought to the eye to be examined through the lens? (b) What is the angular magnification at this point?

■■■**26.** A specimen is 5.0 mm from the objective of a compound microscope, which has a power of $+250$ D. What must the magnifying power of the eyepiece be if the total magnification of the specimen is $60\times$?

■■■**27.** Referring to Fig. 25.20, show that the magnifying power for a magnifying glass held at a distance d from the eye is given by

$$m = \left(\frac{25}{f}\right)\left(\frac{1-d}{D}\right) + \frac{25}{D}$$

when the actual object is located at the near point (25 cm). (Hint: use a small-angle approximation and note that $y_i/y_o = d_i/d_o$ by similar triangles.)

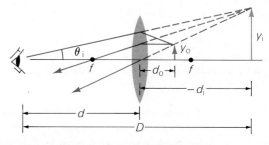

Figure 25.20 **Power of a magnifying glass**
See Problem 27.

■■■**28.** A magnifying glass with a focal length of 10 cm is held 4.0 cm from the eyes to view the small print of a book. What is the magnification if the magnifying glass is 5.0 cm from the book? (Hint: see Problem 27.)

■■■**29.** The objective of a compound microscope has a focal length of 6.0 mm and is located 6.5 mm from a specimen.

The specimen is viewed through an eyepiece ($f_e = 2.0$ cm) adjusted for minimum eye strain. (a) What is the magnification of the specimen? (b) How far apart are the lenses?

■■■30. A modern microscope is equipped with a turret having three objectives with focal lengths of 16 mm, 4.0 mm, and 1.6 mm and interchangeable eyepieces of 5× and 10×. A specimen is positioned so each objective produces an image 160 mm from the objective. What are the least and greatest magnifications possible?

25.3 Telescopes

■■31. An astronomical telescope has an objective and an eyepiece whose focal lengths are 60 cm and 15 cm, respectively. What are (a) the magnifying power and (b) the length of the telescope?

■■32. A telescope has an ocular with a focal length of 10 mm. If the length of the tube is 1.5 m, what is the angular magnification of the telescope when it is focused for an object at infinity?

■■33. The three lenses of a terrestrial telescope have focal lengths of 40 cm, 20 cm, and 15 cm for the objective, erecting lens, and eyepiece, respectively. (a) What is the magnification of the telescope for an object at infinity? (b) What is the length of the telescope?

■■■34. In a terrestrial telescope, an objective and eyepiece with focal lenths of 45 cm and 15 cm, respectively, are used. What should the focal length of the erecting lens be if the overall length of the telescope is to be 1.0 m?

■■■35. Referring to Fig. 25.21, show that the angular magnification of a refracting telescope focused for the final image at infinity is $m = f_o/f_e$. (Since telescopes are designed for viewing distant objects, the angular size of an object viewed with the unaided eye is the angular size of the object at its actual location rather than at the near point, as is true for a microscope.)

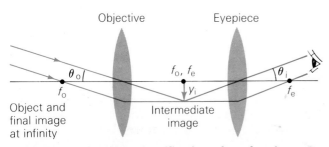

Objective   Eyepiece

f_o, f_e

θ_o θ_i

f_o y_i f_e

Object and   Intermediate
final image   image
at infinity

Figure 25.21 **Angular magnification of a refracting telescope**
See Problem 35.

25.4 Diffraction and Resolution

■■36. According to the Rayleigh criterion, what is the minimum angle of resolution for two point sources of red light ($\lambda = 680$ nm) in the diffraction pattern produced by a single slit with a width of 0.25 mm?

■■37. The minimum angular separation of the images of two monochromatic point sources in a single-slit diffraction pattern is 0.0055 rad. If a slit width of 0.10 mm is used, what is the wavelength of the sources?

■■38. What is the resolution limit due to diffraction for the 40-in. (102-cm) Yerkes Observatory refracting telescope for light with a wavelength of 550 nm?

■■39. The objective of a microscope is 2.0 cm in diameter and has a focal length of 40 mm. (a) If yellow light with a wavelength of 570 nm is used to illuminate a specimen, what is the minimum angular separation of two fine details of the specimen for them to be just resolved? (b) What is the resolving power of the lens?

■■40. For the microscope in Problem 39, how great a difference is there in (a) the minimum angular separation and (b) the resolving power for blue light ($\lambda = 400$ nm) and red light ($\lambda = 700$ nm)?

■■41. A refracting telescope with lens whose diameter is 30 cm is used to view a binary star that emits light in the visible region. (a) What is the minimum angular separation of the pair of stars for them to be barely resolved? (b) If the binary star is a distance of 6.0×10^{20} km from the Earth, what is the lateral separation between the stars of the pair?

■■42. The minimum angular separation of two stars emitting red light ($\lambda = 680$ nm) for a particular telescope is 5.6×10^{-6} rad. What is the diameter of the telescope's lens?

Additional Problems

43. A farsighted person with a near point of 100 cm gets contact lenses and can then read a newspaper held at a distance of 25 cm. What is the power of the lenses? (Assume that the lenses are the same for both eyes.)

44. A physics student uses a converging lens with a focal length of 10 cm to read a small measurement scale. (a) What is the maximum magnification that can be obtained? (b) What is the magnification for viewing with a relaxed eye?

45. A certain adult has a near point of 40 cm. What is the maximum magnification this person can obtain using a magnifying glass with a focal length of 15 cm?

46. An object is placed 10 cm in front of a converging

lens with a focal length of 20 cm. What are (a) the lateral magnification and (b) the angular magnification?

47. The focal length of the objective lens of a compound microscope is 4.5 mm. The eyepiece has a focal length of 2.0 cm. If the distance between the lenses is 20 cm, what is the magnification of a viewed image?

48. A refracting telescope has an objective with a focal length of 50 cm and an eyepiece with a focal length of 15 mm. The telescope is used to view an object that is 10 cm high and located 50 m away. What is the apparent angular height of the object as viewed through the telescope?

49. Light from two monochromatic 550-nm point sources is incident on a slit that is 0.050 mm wide. What is the minimum angle of resolution in the diffraction pattern formed by the slit?

50. A myopic person has glasses with a lens that corrects for a far point of 12.5 cm. What is the power of the lens?

51. A compound microscope has a 20-cm barrel and an ocular with a focal length of 8.0 mm. What power should the objective have to give a total magnification of $360\times$?

52. Find the magnification of a telescope whose objective has a focal length of 60 cm and whose eyepiece has a focal length of 3.1 cm.

Relativity

26

here was a bustle of scientific activity in so many areas at the beginning of the twentieth century that the study of modern physics could begin with any one of various topics. We begin here with the popular and intriguing subject of relativity, which was an outgrowth of nonclassical observations and thought. This introductory topic will illustrate the deficiencies in some classical concepts and how new theories and ideas were developed.

Relativity is concerned with physical phenomena when speeds approach that of light. This was not a subject of study in classical Newtonian kinematics and dynamics because objects with speeds of these magnitudes cannot ordinarily be observed. However, some particles do travel at speeds comparable to the speed of light in nature, and they can now be accelerated to such speeds in particle accelerators. In this chapter, you will learn how Einstein's theory of relativity explains such speed-dependent phenomena, ones that seem rather strange from the classical viewpoint.

26.1 Classical Relativity

The logic and main basis of physics come from observations that some aspects of nature occur in a consistent, invariant manner. That is, the ground rules by which nature plays are consistent, and our descriptions of nature or physical principles do not change from observation to observation. We emphasize this fact by referring to such principles as "laws"—for example, the laws of motion. Not only have physical laws proved valid over time, but they are the same for all observers.

This means that a physical principle or law is the same regardless of the observer's frame of reference. When we make a measurement or perform an experiment, we do so in reference to a particular frame or coordinate system, usually the laboratory, which is considered to be at rest. The experiment may also be observed by a person passing by (in motion). On comparing notes, the experimenter and the observer should find the results of the experiment or the physical principles involved to be the same in both frames of reference or coordinate systems. Basically, you can't change the laws of nature by the way you observe them. Descriptions may be different, but they describe the same thing. For example, the path of a particle moving in a circle of radius a may be described in Cartesian coordinates by $x^2 + y^2 = a^2$ and in polar coordinates by $r = a$.

In measuring relative velocity, however, there seems to be no "true" rest frame. Any reference frame may be considered to be at rest *relative* to another. In Chapter 7, it was emphasized that Newton's laws of motion were valid in an inertial reference frame. An **inertial reference frame** is one in which Newton's first law of motion holds. That is, in an inertial system, an *isolated* object is stationary or moves with a constant velocity. If it does, then Newton's second law in its customary $F = ma$ form applies to the system.

Any reference frame moving with a constant velocity relative to an inertial reference frame is also an inertial frame. Given a constant relative velocity, no acceleration effects are introduced in comparing one frame to another, and $F = ma$ can be used by observers in either frame (with their own coordinate values) to analyze a dynamic situation, and both observers will come up with similar results. Thus, there is no one inertial frame of reference that is preferred over another. This is sometimes called the **principle of classical, or Newtonian, relativity**:

The laws of mechanics are the same in all inertial reference frames.

Around the turn of the century, with the development of the theories of electricity and magnetism, some serious questions arose. Maxwell's equations (Chapter 20) predicted light to be an electromagnetic wave that traveled with a speed of 3.0×10^8 m/s in vacuum. But relative to *what reference frame* does light have this definite speed? Classically, we would expect the speed of light to differ, possibly to be even greater than 3.0×10^8 m/s in different frames of reference by simple vector addition.

Consider the situation illustrated in Fig. 26.1. If a person in a reference frame moving in relation to another with a constant velocity $\mathbf{v}'$ threw a ball with a velocity $\mathbf{v}_b$, then the "stationary" observer would say that the ball had a velocity of $\mathbf{v} = \mathbf{v}' + \mathbf{v}_b$ relative to the stationary reference frame. For example, suppose that a train were moving with a speed of 20 m/s relative to the ground and a ball were thrown with a speed of 10 m/s relative to the train in the direction of the train's motion. Then the ball would have a speed of 20 m/s + 10 m/s = 30 m/s relative to the ground.

Now suppose that a person on the train turned on a flashlight, projecting a beam of light. In this case we have $\mathbf{v} = \mathbf{v}' + \mathbf{c}$, and the speed of light measured by an observer on the ground would be greater than 3.0×10^8 m/s. Classically then, the measured speed of light could have almost any value, depending on the observer's frame of reference.

A particular speed of $c = 3.0 \times 10^8$ m/s must then be referenced to a particular frame. As a wave, light was thought to require some medium of transport. This was quite natural. All experience showed that physical disturbances must be transmitted through the stress or distortion of some medium, as when sound waves travel in air. Since the Earth receives light from

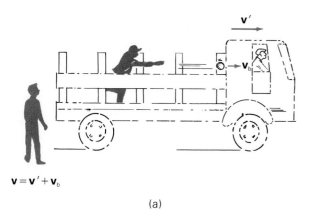

$$\mathbf{v} = \mathbf{v}' + \mathbf{v_b}$$

(a)

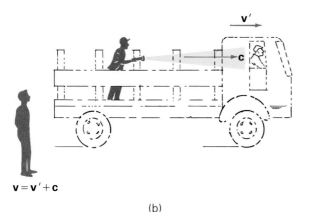

$$\mathbf{v} = \mathbf{v}' + \mathbf{c}$$

(b)

Figure 26.1 **Relative velocity**
(a) According to an observer on the ground, the velocity of the ball is $\mathbf{v} = \mathbf{v}' + \mathbf{v_b}$.
(b) Similarly, the velocity of light would classically be measured to be $\mathbf{v} = \mathbf{v}' + \mathbf{c}$, or greater than c.

the Sun and from distant stars and galaxies, it was thought that this special light-transporting medium in which $c = 3.0 \times 10^8$ m/s must permeate all space. The medium was given the name *luminiferous ether* and referred to simply as **ether**.

Don't confuse this ether with the anesthetic of the same name.

The idea of an undetected ether became popular in the latter part of the nineteenth century. Maxwell believed in the necessity of an etherlike substance:

> Whatever difficulties we may have in forming a consistent idea of the constitution of the ether, there can be no doubt that the interplanetary and interstellar spaces are not empty, but are occupied by a material substance or body which is certainly the largest, and probably the most uniform body of which we have any knowledge.

It would seem that Maxwell's equations, which describe the propagation of light, did not satisfy the Newtonian relativity principle as did other physical laws. On the basis of the preceding discussion, a preferential or special inertial reference frame would seem to exist, one that could be considered absolutely at rest.

As another example of the apparent violation of the relativity principle, consider two stationary charges separated by a distance r as illustrated in

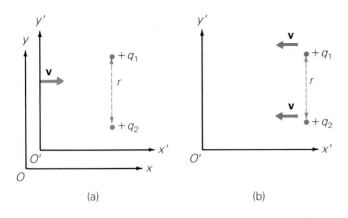

(a)   (b)

Figure 26.2 **Nonrelativity**
(a) The repulsive forces between the stationary charges in the O reference frame are given by Coulomb's law: $F = kq_1q_2/r^2$. (b) For an observer moving toward the charges with a velocity $\mathbf{v}$, the charges would be seen to be moving. In addition to the repulsive Coulomb forces, there would be attractive forces due to the production of an interaction with magnetic fields (similar to parallel current-carrying wires).

Fig. 26.2—stationary relative to the O reference frame. In this frame, an observer would measure a force of repulsion between the charges as given by Coulomb's law ($F = kq_1q_2/r^2$).

Now suppose the charges were investigated by an observer in the O' frame moving with a velocity $\mathbf{v}$ relative to the O frame and perpendicularly to a line separating the charges. This observer would see the two charges approaching with a velocity $-\mathbf{v}$. The repulsive Coulomb force would be evident as it was to the observer in the O frame. But, since a moving charge corresponds to a current, the observer in the O' system would also note attractive forces between the charges (similar to the forces for parallel current-carrying wires described in Section 19.4). In other words, observers in different inertial frames would measure *different* forces for the same situation, and the physical laws would not be the same in all inertial reference frames. Here, the laws of electricity and magnetism seem not to satisfy the classical relativity principle.

This was the state of affairs toward the end of the nineteenth century when scientists set out to investigate whether they had come upon a new dimension of physics or perhaps a flaw in what were considered to be established principles. One of the first attempts was to test the ether theory. If it could be proved that the ether existed, then presumably a true rest frame would be identified. This was the purpose of the famous Michelson-Morley experiment.

26.2 The Michelson-Morley Experiment

If the ether permeated all space, the Earth itself would be surrounded by the medium. In its orbit around the Sun, the Earth presumably moved through the ether, like an airplane moving through still air. Relative to the airplane, there is air motion or wind, and similarly the orbiting Earth would experience an "ether wind."

If the ether wind existed, this would provide a means to detect the ether experimentally. Observers could carefully measure the speed of light on Earth in several directions, for example, parallel and perpendicular to the orbital velocity of the Earth, and compare the results. Because of the relative motion, the speeds should be different in different directions.

As an analogy, consider two airplanes flying at the same air speed and the same distances back and forth as illustrated in Fig. 26.3(a). The plane flying perpendicular to the wind must be directed slightly into the wind on both legs to stay on a straight-line course, an action that slows the plane somewhat. The other plane flies into the wind on the first leg and with the wind on the return leg. The velocities at which the planes fly relative to the ground are given by vector addition. It is a simple matter to show that the plane flying perpendicular to the wind takes less time to complete the two-leg trip than does the other plane.

A similar effect should be seen if the airplanes are replaced by light beams and the atmospheric wind by the ether wind [Fig. 26.3(b)]. However, this is more easily said than done. The average speed of the Earth in its orbit is about 30 km/s, compared to 300,000 km/s for the speed of light. In the late nineteenth century, the American physicist Albert A. Michelson devised a technique capable of detecting the required differences using the interference of light and designed an extremely sensitive **interferometer**. The basic principle of Michelson's interferometer is shown in Fig. 26.4. A monochromatic light beam is incident on a partially silvered glass plate (P), which acts as a beam splitter, dividing the incident beam into transmitted and reflected components. The beams advance to the plane mirrors (M_1 and M_2), and are reflected back toward P. There, part of each beam is reflected and transmitted to the detector screen (D). Another glass plate (P′) is inserted in one beam, so both beams pass through an equal amount of glass.

Since the beams are from the same initial beam, they are coherent and, on combining, interfere according to their phase relationship. This is determined by the difference in the path lengths of the beams (see Chapter 15).

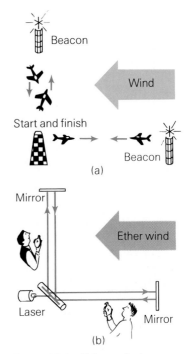

Figure 26.3 **Ether wind detection**
(a) When two airplanes fly with equal air speeds over equal distances, the plane flying perpendicular to the wind direction takes less time to fly a round trip.
(b) Classically, the ether wind should give rise to a similar difference in the times for light to be reflected back and forth.

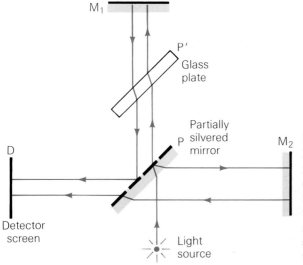

Figure 26.4 **Michelson interferometer**
If one of the arms of the interferometer were in the direction of the ether wind, the beams should arrive at the detector out of phase, and an interference pattern should be produced. See the text for detailed description.

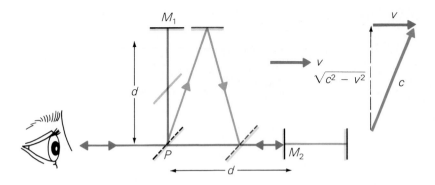

Figure 26.5 **Interferometer light paths**
An interferometer moving through the stationary ether would be observed to have light paths as shown. See the text for description.

Consider now the interferometer (and the Earth) traveling through the stationary ether with a velocity **v** with the interferometer arranged so one of its perpendicular arms is parallel to **v** (Fig. 26.5). The speed of the beam traveling to M_2 is $c - v$. On the return trip the relative velocity has a magnitude of $c + v$. The total time for the round trip is

$$t_2 = \frac{L_2}{c - v} + \frac{L_2}{c + v} = \frac{2L_2 c}{c^2 - v^2} \tag{26.1}$$

For the light beam traveling between P and M_1, the speed of light in the direction perpendicular to **v** is $\sqrt{c^2 - v^2}$ (as shown in the vector diagram in the figure). The speed is the same for the return path, so the time for the round trip on this arm is

$$t_1 = \frac{2L_1}{\sqrt{c^2 - v^2}} \tag{26.2}$$

Then, if $L_1 = L_2 = L$, we have with some rearranging of terms

$$\Delta t = t_2 - t_1 = \frac{2L}{c}\left[\frac{1}{1 - v^2/c^2} - \frac{1}{\sqrt{1 - v^2/c^2}}\right] \tag{26.3}$$

If there were a time difference, the beams would arrive at the detector out of phase, and an interference pattern would be observed. From the spacing of the fringes, one could determine the velocity of the light source relative to the ether.

However, it is difficult to ensure experimentally that $L_1 = L_2$ to the required degree of accuracy. This problem may be resolved by rotating the apparatus by 90°, which interchanges the arms. (With equal arms, the time difference, Δt, when the arms are interchanged becomes the negative of the value obtained using Eq. 26.3.) What would be observed then is a *shift* in the interference pattern or fringes.

Michelson performed this experiment with his colleague E. W. Morley in 1887. The interferometer was more than sensitive enough to detect the predicted fringe shift. But much to their surprise, Michelson and Morley found *no fringe shift at all.* Perhaps the experiment had been done just at a time when the Earth was nearly at rest relative to the ether. To check this possibility, many measurements were made at different times (day and night and at different seasons), with always the same null result.

Where then was the ether? Several hypotheses were suggested to explain the null result of the Michelson-Morley experiment. One suggested that the ether in the vicinity of the Earth was dragged along with the Earth, so the

The mirrors are arranged so that they are not quite perpendicular to the beams. As a result, the path difference of the two beams depends upon where the light strikes the mirrors, which gives rise to a pattern of fringes.

interferometer was at rest with respect to the ether (that is, $v = 0$). But because of the Earth's orbital motion, the viewing of a star perpendicular to the plane of the orbital plane throughout the year requires the telescope to be consistently tilted in the direction of the Earth's motion. This effect is called the *aberration of starlight* and was explained in 1725. The aberration requires that the Earth move in relation to the ether and *not* be stationary in relation to it (as would be the case if the ether were dragged along).

An Irish physicist, George F. FitzGerald, proposed that as result of the motion through the ether, perhaps the linear dimensions of objects in the direction of motion contracted or became smaller. If this occurred by a fractional amount of $\sqrt{1 - v^2/c^2}$, then the null result would be explained. Of course, this hypothesis cannot be experimentally investigated, since the measurement instrument would also contract by the same fractional amount.

Another scientist, H. A. Lorentz, who also suggested this possibility, considered it in terms of possible changes in the electromagnetic forces between the atoms of a material due to the motion. This suggested length contraction, which is known as the Lorentz-FitzGerald contraction, might be termed the right idea, but for the wrong reason. Oddly enough, such a length contraction was predicted by a theory proposed by Albert Einstein in 1905. His special theory of relativity is described in the next section.

26.3 The Special Theory of Relativity

The failure of the Michelson-Morley experiment to detect the ether left the scientific community in a quandary. The inconsistencies between Newtonian mechanics and electromagnetic theory remained unexplained. These problems were resolved, however, by a theory introduced in 1905 by Albert Einstein. Apparently, the Michelson-Morley experiment was not a motivation for the development of his theory of relativity. Einstein later could not recall whether he had even known about the experiment at the time when he was working as a clerk in the Swiss Patent Office (see Fig. 26.6 and the Insight feature).

Einstein's approach was that the inconsistencies in electromagnetic theory were due to the assumption that an absolute rest frame (the ether) existed. His theory, based on two postulates, did away with the need for an ether. The first postulate is an extension of the Newtonian principle of relativity that applies

Figure 26.6 **Einstein and Michelson**
A 1931 photo shows Michelson (front left) and Einstein during a meeting in Pasadena, California. On Einstein's left is Robert A. Millikan, the recipient of a Nobel Prize in 1923 for his experimental determination of the value of the electronic charge.

Albert Einstein (1879–1955)

Albert Einstein is one of the most celebrated scientific figures (Fig. 1). Over a life span of three quarters of a century, his name became a household word, and his theories changed scientific thought.

Born in the city of Ulm, Germany, Einstein grew up there with a younger sister, Maja. His parents noted that he was slow to talk and even slower to read. School seemed to bore him, but fortunately an uncle sparked his interest in mathematics. Einstein termed it ''a merry science.... When the animal that we hunt cannot be caught, we call it x temporarily and continue to hunt it until it is bagged.''

His parents moved, and Einstein continued his schooling at the Swiss Federal Polytechnic in Zurich. He received his diploma and took Swiss citizenship. Jobs were scarce, and Einstein was forced to take a low-paying job in the Swiss Patent Office in Berne as a probationary Technical Expert, Third Class. He needed the work to support his wife Mileva, who had been a classmate in Zurich, and their two sons.

Luckily, his job permitted spare time for his own scientific interests and work on his doctoral degree. Like Newton, Einstein was very productive during his twenties. In 1905, he received his Ph.D. degree from the University of Zurich, and in the same year he published three papers of significant scientific importance. One dealt with quantum theory and an explanation of the photoelectric effect (discussed in Section 27.2). This work formed the basis of his receipt of the Nobel Prize in physics in 1921. Another paper addressed molecular motions and sizes and the analysis of Brownian motion (Chapter 11). The third paper was on the special theory of relativity, which revolutionized ideas about space and time.

With these publications to his credit, Einstein had no difficulty in getting a job as a professor. He returned to Germany to become director of the newly created Kaiser Wilhelm Institute in Berlin. In 1915 he published his general theory of relativity, which provided a new theory for gravitation. Now divorced, Einstein married his cousin Elsa Einstein. She remained his devoted wife until her death in 1938. Her interests were in the home, not in science and mathematics. ''My interest in mathematics is mainly in household bills,'' she once said.

A pacifist during World War I, Einstein denounced the war as ''senseless violence.'' His theory of relativity was popularized, and so Einstein became a popular figure. He was modest. ''It is an irony of fate that I myself have been the recipient of excessive admiration and reverence from my fellow beings, through no fault and no merit of my own,'' he said. Money did not appear to interest him. ''I refuse to make money out of my science. My laurel is not for sale....'' However, he was irritated by the commonly held notion that only ten people in the world were capable of understanding his theory of relativity. ''Every earnest student of theoretical physics is in a position to comprehend it,'' he said.

Because of the rising anti-Semitism in Germany in the 1930s, Einstein went to Belgium. When his safety could not be guaranteed there, various countries offered asylum. Einstein chose the United States and went to the Institute for Advanced Study in Princeton, New Jersey. He settled there permanently and became a U.S. citizen in 1940.

Near the beginning of World War II, and following the discovery of nuclear fission, Einstein was asked by other scientists to write a letter to President Roosevelt. They, along with Einstein, recognized the tremendous military potential and the dire consequences of nuclear fission should Germany develop it first. Einstein's famous letter (see Appendix III) was instrumental in starting the Manhattan Project and in the development of the atomic bomb. After World War II, Einstein devoted much of his time to advocating agreements to end the threat of nuclear war.

Einstein was an outspoken supporter of human rights. He also supported the establishment of a Jewish state. When the state of Israel was formed, Einstein was offered the presidency, which he declined. Einstein died in 1955 at the age of 76, and eulogies from around the world poured in. A fitting epitaph for this warm and outgoing scientist would be some of his own words: ''I speak to everyone in the same way, whether he is the garbage man or the president of the university.''

Figure 1 **Young Einstein**

to the laws of mechanics. In Einstein's theory, the **principle of relativity** applies to *all* the laws of physics, including those of electricity and magnetism:

> **Postulate I (principle of relativity)**: The laws of physics are the same in all inertial reference frames.

The first postulate implies that all inertial reference frames are equivalent, with physical laws being the same in all of them. This means that there is no *absolute* reference frame with some unique physical property to distinguish it from other inertial reference frames. The first postulate seems quite reasonable, since one would expect the basic laws of nature to be the same for all inertial observers. One would not expect nature to play favorites.

Einstein's second postulate involves the speed of light. As he saw it, the problem came from the vector addition of velocities. With the appropriate relative velocity between inertial systems, the observed speed of light could have any value. With $(c + v)$, it could be greater than c; with $(c - v)$, it could be less than c; and with $(c - c)$, it could even reduce to zero! The last possibility was the source of Einstein's question "What would I see if I rode a beam of light?" In the same frame as the electromagnetic wave, electric and magnetic field vectors would be seen that varied in space but *not* with time. Wave motion is fundamental to waves, and such static fields were not consistent with electromagnetic theory.

Such variations in the speed of light were unacceptable to Einstein, so he made the speed constant in his second postulate:

> **Postulate II (constancy of the speed of light)**: The speed of light in a vacuum is constant in any inertial system.

These two postulates form the basis of Einstein's **special theory of relativity**. The "special" designation is to indicate that it deals only with the particular case of inertial reference frames. The *general* theory of relativity, which will be discussed later, deals with noninertial or accelerating reference frames.

The second postulate is perhaps more difficult to accept than the first. It implies that two observers in different inertial reference frames would measure the speed of light to be c independent of the speed of the source and/or the observer. For example, if a person moving toward you with a constant velocity turned on a light, both you and the other person would measure the speed of light to be c, irrespective of the magnitude of the relative velocity.

The second postulate is essential to the validity of the first, and the null result of the Michelson-Morley experiment is consistent with the constancy of the speed of light. By doing away with the ideas of the ether and an absolute reference frame, Einstein was able to reconcile the differences between mechanics and Maxwell's equations. Even so, the second postulate seemed to go against common sense. But then we have very little everyday experience in dealing with velocities at or near the speed of light. A theory may beautifully postulate a solution to a problem, but it must stand the test of the scientific method. What does Einstein's theory predict, and has it been experimentally verified?

The predictions of the special theory of relativity can be understood by imagining simple situations to see what the theory predicts. Einstein did this himself in what he called *gedanken*, or thought, experiments. We also look at some experimental evidence that supports the nonclassical predictions of the theory.

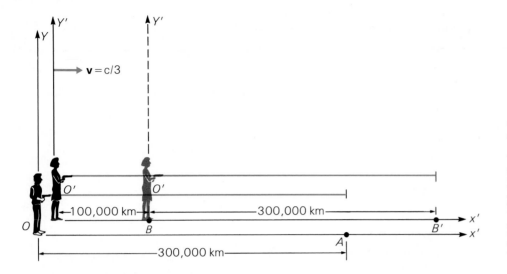

Figure 26.7 **Constancy of the speed of light**
Two inertial observers send out light beams when at the same location. In 1 s, the light beam from G travels to point A. During the 1 s, O' travels with a speed of $c/3$ to point B, and by the constancy of the speed of light, her beam has traveled to point B'.

Simultaneity and Time Dilation

The special theory of relativity, particularly the second postulate, offers new views of the concepts of space and time. To illustrate the nonclassical nature of the second postulate, consider the situation shown in Fig. 26.7. In this thought experiment, two observers, O and O', each with lasers, simultaneously turn the lasers on when the observers are next to each other. Now let O be stationary and O' be in motion to the right in the same direction as the light beams are traveling. Assume that O' is traveling with a constant speed of $\frac{1}{3}$ of the speed of light ($c/3$), or 100,000 km/s.

At the end of 1 second, O would say that the light from his laser had traveled 300,000 km to point A. According to the second postulate, the speed of light is independent of the speed of the source, so O' would say that in 1 second the light from her laser also traveled this distance relative to her. But at the end of 1 second, O' has traveled 100,000 km to point B, so relative to O', her light beam has reached point B'.

But how about O looking at the O' laser light? He too measures the speed of the light to be c (as required by the second postulate), and in 1 s relative to him the light from the other laser travels 300,000 km to point A. How can the light beam be *simultaneously* at two locations? It can also be shown that two events that are simultaneous to one observer are not necessarily simultaneous to a second inertial observer (see Problem 22).

Such radical results appear to be contradictory, but they lead to the conclusion that time and space are not absolute. The reason that the light beam appears to be at different points to the two observers in the preceding example is because the 1-second time interval is not absolute, or the same for both observers. What Einstein's theory predicts is that time is measured differently by different inertial observers.

To see this, try another thought experiment. For the comparison of time intervals in moving inertial reference frames, let's use a special "relativistic" light clock as illustrated in Fig. 26.8(a). A tick or unit time interval on the clock corresponds to the time for a light pulse to make a complete round trip between two mirrors. It is assumed that two observers, O and O' again, have light clocks

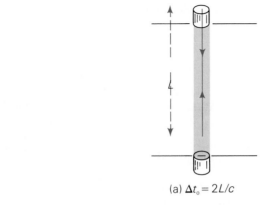

(a) $\Delta t_o = 2L/c$

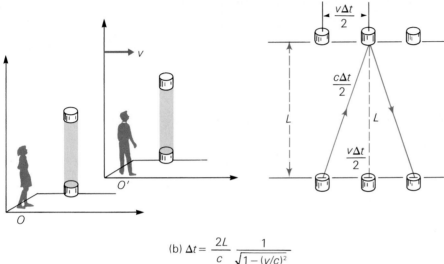

(b) $\Delta t = \dfrac{2L}{c}\,\dfrac{1}{\sqrt{1-(v/c)^2}}$

Figure 26.8 **Time dilation**
(a) A light clock that measures time in units of round-trip reflections of light. The time for light to travel up and back is $\Delta t_o = 2L/c$. (b) An observer O measures a time interval of

$$\Delta t = \frac{2L}{c}\left[\frac{1}{\sqrt{1-(v/c)^2}}\right]$$

on the moving clock, and it appears to run more slowly. See the text for further description.

and that the clocks have been synchronized and run at the same rate when both systems are at rest. In this case, the time interval for a light round trip is

$$\Delta t_o = \frac{2L}{c} \tag{26.4}$$

Now consider O' to be in motion with a velocity v relative to O. In the moving system, O' still measures the unit time interval as t_o. But, as seen by O, the clock in the O' system is moving, and the path of the light pulse is as shown in Fig. 26.8(b). That is, O sees a longer light path for the clock in the moving system. From O's point of view, the geometry of the path of the light pulse is, by the triangle in the figure,

$$\frac{c^2\,\Delta t^2}{2} = \frac{v^2\,\Delta t^2}{2} + L^2$$

Solving for Δt gives

$$\Delta t = \frac{2L}{c}\left[\frac{1}{\sqrt{1-(v/c)^2}}\right] \tag{26.5}$$

But the same time interval as measured by either observer in his or her own frame is given by Eq. 26.4. So the measured time intervals are different, and combining the equations, we have

$$\Delta t = \frac{\Delta t_0}{\sqrt{1 - (v/c)^2}} \tag{26.6}$$

Since $\sqrt{1 - (v/c)^2}$ is less than 1, O measures a longer time interval on the O' clock than does observer O'. This effect is called **time dilation**. With longer ticks, the O' clock appears to O to run more slowly than his own. Of course, the situation is relative, and O' would say that the clock in O's system ran slow. Thus, because of the relativistic time dilation, *moving* clocks are measured to run more slowly or not as fast as a clock in the observer's system.

To distinguish between the two time intervals, we refer to **proper time**. This is the time interval measured between two events in the reference frame where the events *occur at the same location or point in space*. As can be seen, this is Δt_0 or the time interval in the reference frame where the clock is at rest.

It is convenient to define

$$\gamma \equiv \frac{1}{\sqrt{1 - (v/c)^2}} \tag{26.7}$$

Gamma (γ) is always greater than or equal to 1 ($c > v$ and $v = 0$, respectively). The values of γ for several values of v expressed as fractions of c are given in Table 26.1. The values illustrate how the speed of an object must be an appreciable fraction of the speed of light for relativistic effects to be observed. The time dilation equation (Eq. 26.6) may then be written

$$t = \gamma t_0 \tag{26.8}$$

where the deltas have been eliminated and it is understood that the t's are intervals. In words, the time interval measured on a moving clock is γ times the proper time.

For example, suppose you were observing a clock that was in a system moving at a constant velocity of $0.6c$ relative to you. The gamma factor would then be 1.25 (Table 26.1). Because of the time dilation effect, when 20 minutes had elapsed on the moving clock, you would observe a time of $t = \gamma t_0 = (1.25)(20 \text{ min}) = 25 \text{ min}$ to have elapsed on your clock. The 20 minutes is the proper time, since the events defining the 20-minute interval took place at the same location (that of the clock, for example, for readings at 8:00 and 8:20). As was pointed out earlier, the moving clock runs more slowly (20 minutes elapsed as opposed to 25 minutes on your clock).

Table 26.1

Some Values of $\gamma = \dfrac{1}{\sqrt{1 - (v/c)^2}}$

v	γ
0	1.00
$0.1c$	1.01
$0.2c$	1.02
$0.3c$	1.05
$0.4c$	1.09
$0.5c$	1.15
$0.6c$	1.25
$0.7c$	1.40
$0.8c$	1.67
$0.9c$	2.29
$0.95c$	3.20
$0.99c$	7.09
$0.995c$	10.0
$0.999c$	22.4
c	∞

Example 26.1   Time Dilation

There are many subatomic elementary particles. One of these, called a muon, has the same charge as an electron but is about 200 times more massive. Muons are created in the Earth's atmosphere when cosmic rays (mostly protons) from outer space collide with the nuclei of the gas molecules of the air. The muons then approach the Earth with a speed near that of light (say about $0.998c$).

However, muons are unstable and quickly decay into other particles. The average lifetime of a muon has been measured in the laboratory as 2.2×10^{-6} s. In this time, a muon would travel a distance of

$$d = v_0 t = (0.998c)(2.2 \times 10^{-6} \text{ s})$$

$$= [(0.998)(3.0 \times 10^8 \text{ m/s})](2.2 \times 10^{-6} \text{ s}) = 660 \text{ m}$$

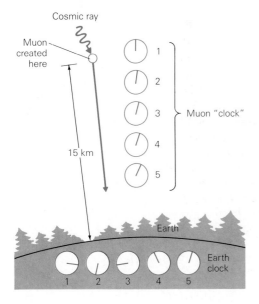

Figure 26.9 **Experimental evidence of time dilation** Muons are observed at the surface of the Earth as predicted by the theory of relativity.

This is only 0.66 km, or 0.41 mi. Since muons are created at altitudes of about 15–20 km, none of them should reach the Earth's surface.

However, an appreciable number of muons do reach the Earth. The paradox arises because the preceding calculation does not take time dilation into account. That is, a muon decays by its own clock, or in a time measured by a clock in its own reference frame (proper time, since in the rest frame of the muon, birth and death take place at the same coordinates). To an observer on Earth, a clock in the moving muon reference frame would appear to run more slowly than a clock on Earth (Fig. 26.9). So in a time $t_o = 2.2 \times 10^{-6}$ s on the muon clock, the Earth clock reads a time interval that is greater by a factor of

$$\gamma = \frac{1}{\sqrt{1 - (v/c)^2}} = \frac{1}{\sqrt{1 - (0.998c/c)^2}} = 16$$

and with $t = \gamma t_o$, we have the muon traveling a distance of

$$d = vt = \gamma v t_o = \gamma(0.66 \text{ km}) = (16)(0.66 \text{ km}) \cong 11 \text{ km}$$

Therefore, the theory of relativity predicts that the muons would travel a much greater distance than is predicted by classical physics. Since t_o is an *average* value, some muons do travel greater distances and are observed at the Earth's surface. ∎

Length Contraction

As was mentioned earlier, the null result of the Michelson-Morley experiment could be explained if it were assumed that the length of the interferometer arm in the direction of motion contracted by a factor of $\sqrt{1 - (v/c)^2}$, the Lorentz-FitzGerald contraction. The same length contraction also proves to be a consequence of the special theory of relativity. Its origin is not some ether interaction with interatomic forces but is explained by the two postulates of the relativity theory.

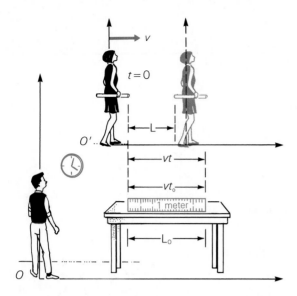

Figure 26.10 **Length contraction**
Observer O' measures the length of the rod by measuring the time interval between her passing the ends of the rod. Observer O similarly measures the length of the rod by using the same events. It is found the lengths are not the same. See the text for description.

To see how this contraction arises, consider a measuring rod of length L_o, as measured by observer O in his rest frame (Fig. 26.10). This **proper length**, L_o, is also to be used as a length standard by observer O', who is traveling with a velocity v in the direction parallel to the rod. To measure the length of the rod, O' simply measures the time interval between her passing of the ends of the rod. The length of the rod is then given by $L = vt$, where t is the measured time interval.

Observer O could also measure the length of the rod by the same means. If he noted the times on his clock when O' passed by the ends of the rod, he would measure a time interval t_o, and for him the length of the rod would be given by $L_o = vt_o$. Then, dividing one length by the other, we have

$$\frac{L}{L_o} = \frac{vt}{vt_o} = \frac{t}{t_o} \tag{26.9}$$

But in this case, the *proper time* is in the O' reference frame (time measurement done at same point), and the time t_o is greater than t or there is a time dilation, since moving clocks run more slowly. This changes the notation in Eq. 26.8, and $t_o = \gamma t$, or $t/t_o = 1/\gamma$. Eq. 26.9 then becomes

$$L = \frac{L_o}{\gamma} \tag{26.10}$$

Since γ is always greater than 1 with relative motion, we have $L < L_o$ or a **length contraction** by a factor of $1/\gamma$ or $\sqrt{1 - (v/c)^2}$, which is precisely the Lorentz-FitzGerald contraction.

Thus, not only time intervals but also space intervals, that is, lengths and distances, are affected by relative motion. It should be noted that the length contraction occurs only along the direction of motion. As a result, objects would appear to be contracted in the direction of their motion in a relativistic world.

Example 26.2 Length Contraction and Time Dilation
An observer sees a spaceship that is 100 m long when at rest (proper length) pass by in uniform motion with a speed of $0.5c$ (Fig. 26.11). While the observer is

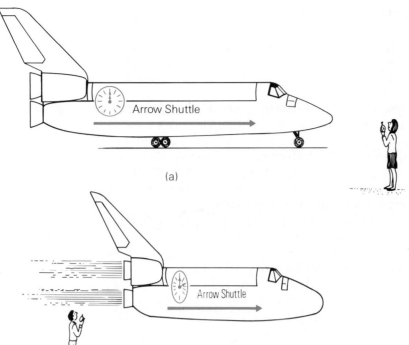

(a)

(b)

Figure 26.11 **Length contraction and time dilation** As a result of length contraction, moving objects appear to be shorter, or contracted in the direction of motion, and moving clocks appear to run more slowly because of time dilation. See Example 26.2.

watching the spaceship, a time of 2.0 s elapses on a clock on board the ship (proper time). (a) What would the observer measure the length of the moving spaceship to be? (b) What time interval elapses on the observer's clock for the 2.0 s interval measured on the ship's clock?

Solution

Given: $L_o = 100$ m   *Find*: (a) L
       $t_o = 2.0$ s   (b) t
       $v = 0.5c$

(a) By calculation or from Table 26.1, $\gamma = 1.15$ for $v = 0.5c$, and the length contraction is given by

$$L = \frac{L_o}{\gamma} = \frac{100 \text{ m}}{1.15} = 87 \text{ m}$$

Thus the observer measures the length of the spaceship to be considerably shorter than its proper length.
(b) The time dilation is given by

$$t = \gamma t_o = (1.15)(2.0 \text{ s}) = 2.3 \text{ s}$$

The Twin Paradox

Time dilation gave rise to another popular relativistic topic—the so-called twin, or clock, paradox. According to the special theory, a clock in a moving system runs more slowly than, or not as fast as, one viewed by an inertial observer in another reference frame. For example, with $\gamma = 4$, for a proper time interval of 15 minutes, an hour would elapse on the observer's clock. Similarly, 1 year of proper time in the moving system takes 4 years in the observer's time frame. Since our heartbeat and age are measured by proper time, the question arises: Does an observer in one system age more quickly than a person in a system that is in motion relative to the first system?

This situation is commonly applied to identical twins, one of whom goes on a high-speed space journey. The question then is: Will the space traveler come back younger than his Earth-bound twin? You might say that everything is relative and that the space twin sees the Earth clock run more slowly. He would claim that the Earth twin would age more slowly. They can't both be right.

The solution to the paradox lies in the fact that in leaving and returning to Earth, the space twin must experience accelerations and so is not in an inertial reference frame. However, if it is assumed that the acceleration periods occupy only a small part of the total trip time, the special theory can be applied to get an indication of what is likely to happen. (Einstein's general theory considers accelerating systems and confirms the result.) The prediction is that the space-traveling twin does return younger than the Earth-bound twin.

While traveling at a constant velocity on the outward and return trips, which occupy most of the total trip, both twins measure the same relative speed. However, the *proper length* of the trip is in the Earth twin's reference frame with fixed beginning and end points. The traveling twin thus measures a length contraction, travels a shorter time, and returns home younger than his twin.

In a sense, the twin paradox has been experimentally tested. In 1971, atomic clocks were flown around the world in jet planes. It was necessary to use these extremely accurate clocks to detect any effects because of the small γ factors, or low relative velocities. They were accurate enough to measure the predicted effect (with corrections made for accelerations, etc.), and a time dilation was experimentally verified. As was stated in the reporting article, "These results provide an unambiguous empirical resolution of the famous clock 'paradox' with macroscopic clocks."*

26.4 Relativistic Mass and Energy

The ramifications of the theory of relativity are particularly important in particle physics, in which the speeds of the particles may be appreciable fractions of the speed of light. Such speeds are possible in particle accelerators in which charged particles are accelerated through electrical potential differences. Of the three fundamental properties—length, time, and mass—the preceding discussion has shown that the first two are relativistic quantities. That is, their values depend on the measurement frame of reference. What about mass? Is it too a relativistic quantity?

Einstein showed that it is, and the mass of a particle or object increases with speed according to the relationship

$$m = \gamma m_\text{o} = \frac{m_\text{o}}{\sqrt{1 - (v/c)^2}} \tag{26.11}$$

Here m_o is the mass of the particle at rest, or its **rest mass** (proper mass), and m

* Hafele and Keating, *Science*, vol. 117, no. 4044 (July 14, 1972), pp. 166–170.

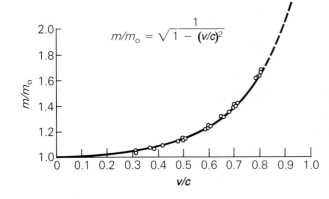

$$m/m_o = \sqrt{\dfrac{1}{1 - (v/c)^2}}$$

Figure 26.12 **Relativistic mass**
The data points on the curve are measured masses of accelerated particles. There is good agreement between the theoretical prediction for relativistic mass and experimental data.

is the **relativistic mass**, or the measured particle mass in an inertial reference frame moving with a speed v.

There is experimental evidence for this relativistic mass increase. Particle accelerators accelerate beams of charged particles to very high speeds using electric fields. Linear beams are deflected, or turned, by deflecting magnets. It is found that greater magnetic fields are needed to deflect high-speed particles than would be needed as predicted by using the particle rest mass. This is explained by a mass increase. Also, in circular accelerators called cyclotrons, accelerating particles are found to arrive at particular points in the orbital paths increasingly late. The lateness is explained by a relativistic increase in the masses of the particles.

According to Eq. 26.11, as v approaches c, the mass (m) of an object approaches infinity (Fig. 26.12). A basic result of the special theory of relativity is that no object can travel as fast as or faster than the speed of light. This may be seen from Eq. 26.11. As a particle approaches the speed of light, its mass approaches infinity. To accelerate an object until $v = c$ would require an infinite amount of energy, which makes c the maximum and unattainable speed limit.

Time dilation and length contraction (Eqs. 26.5 and 26.10) also imply this. If a reference frame could travel with a speed c, then time would be infinite, or would stand still, and the lengths of objects would contract to nothing, or the objects would vanish.

In the theory of relativity, the variation of mass with speed is a condition for the conservation of momentum. The basic conservation principle is valid if the **relativistic momentum** is defined as

$$p = mv = \gamma m_o v \qquad (26.12)$$

Einstein used the conservation of momentum to derive the relativistic mass equation.

Since the relativistic momentum has a different form from its classical counterpart, we might expect the expression for kinetic energy also to be different, and this is true. In the relativistic case, the kinetic energy K of a particle is no longer equal to $\frac{1}{2}mv^2$ (or $p^2/2m$), even if the relativistic mass and momentum are used. The mathematics is beyond the scope of this text, but Einstein showed, using the work-energy theory, that the **relativistic kinetic energy** is given by

$$K = mc^2 - m_o c^2 \qquad (26.13)$$

This expression may be written in other forms, for example,

$$K = \gamma m_o c^2 - m_o c^2 = m_o c^2 (\gamma - 1) \qquad (26.14)$$

The relativistic expression is more complicated than the classical $K = \frac{1}{2}mv^2$, but it must be used if the speed of a particle is comparable to c. Of course, if $v \ll c$, the relativistic expression must reduce to the classical form. This can be shown by expanding the $\gamma - 1$ term in the preceding equation in a series expansion:

$$K = m_o c^2 (\gamma - 1) = m_o c^2 (-1 + \gamma) = m_o c^2 \left[-1 + \left(1 - \frac{v^2}{c^2} \right)^{-\frac{1}{2}} \right]$$

$$= m_o c^2 \left[-1 + 1 + \frac{1}{2} \left(\frac{v^2}{c^2} \right) + \frac{3}{8} \left(\frac{v^4}{c^4} \right) + \cdots \right] \qquad (v < c)$$

and

$$K \cong m_o c^2 \left[\frac{1}{2} \left(\frac{v^2}{c^2} \right) \right] = \frac{1}{2} m_o v^2 \qquad \text{(to a first-order approximation, where } v \ll c)$$

Eq. 26.14 shows that an increase in kinetic energy is reflected as an increase in the relativistic mass m; and an increase in the kinetic energy increases the total energy by a corresponding amount. Since $m_o c^2$ is a constant, this suggests that the *total energy* can be written $E = mc^2$ (assuming no potential energy), and

$$E = K + E_o \qquad \text{or} \qquad mc^2 = K + m_o c^2 \qquad (26.15)$$

where

$$E_o = m_o c^2 \qquad (26.16)$$

is the **rest energy** of the particle. For a particle at rest in a given reference frame ($K = 0$), the total energy is the rest energy: $E = E_o = m_o c^2$.

Eq. 26.15 shows that the relativistic mass is a direct measure of the total energy of the particle. This is Einstein's famous **mass-energy equivalence formula**, which points out that *mass is a form of energy*. Classically, when work is done on an object, its speed and energy increase. But, relativistically, the mass of an object increases with increasing speed. Thus, the work not only goes into increasing the speed, but also contributes to increasing the mass, and we are led to a mass-energy equivalence, or to the conclusion that mass is a form of energy.

When a particle has potential energy U, the total energy is $E = mc^2 = K + U + m_o c^2$.

The mass-energy equivalence does not mean that we can convert rest mass into useful energy at will. If we could, our energy problems would be solved, since we have a lot of mass. Mass-energy conversion does take place on a limited scale in nuclear reactions as a later chapter will show. In the context of the present discussion, the important point is that any variation (increase or decrease) in the kinetic energy of a particle will give a variation in its relativistic mass.

The energy equivalent of a particle at rest depends on its rest mass, $E_o = m_o c^2$. For example, the rest mass of an electron is 9.11×10^{-31} kg, and the rest energy is

$$E_o = m_o c^2 = (9.11 \times 10^{-31} \text{ kg})(3.00 \times 10^8 \text{ m/s})^2 = 8.19 \times 10^{-14} \text{ J}$$

or $\quad E_o = (8.19 \times 10^{-14} \text{ J}) \left(\frac{1 \text{ eV}}{1.60 \times 10^{-19} \text{ J}} \right) = 5.11 \times 10^5 \text{ eV} = 0.511 \text{ MeV}$

In modern physics, particle energies are commonly expressed in units of electron volts because charged particles are often given energy by accelerating them through potential differences.

Example 26.3   Accelerating Energy

How much energy is required to give an electron initially at rest a speed of $0.9c$?

Solution

The energy needed is equal to the kinetic energy of the electron traveling at $0.9c$. However, we cannot use the classical expression $K = \frac{1}{2}mv^2$, since the increase in the energy of an electron traveling this fast significantly increases its mass, and the relativistic expression must be used. Since the rest mass of an electron has been shown to be $E_\text{o} = m_\text{o}c^2 = 0.511$ MeV, a convenient form of the relativistic kinetic energy is given by Eq. 26.14:

$$K = m_\text{o}c^2(\gamma - 1) = E_\text{o}(\gamma - 1)$$

With $v = 0.9c$, by calculation or from Table 26.1, $\gamma = 2.29$, and

$$K = E_\text{o}(\gamma - 1) = (0.511 \text{ MeV})(2.29 - 1) = 0.659 \text{ MeV}$$

Thus, 0.659 MeV would be required to accelerate an electron to a speed of $0.9c$. ∎

Example 26.4   Total Energy and Speed

An electron is accelerated from rest through a voltage difference of 500 kV. (a) What is the total energy of the electron? (b) What is its speed?

Solution

Given: $V = 500$ kV   *Find*: (a) E
$\qquad\quad E_\text{o} = 0.511$ MeV $\qquad\qquad\qquad$ (b) v

(a) The kinetic energy gained by the electron is

$$K = eV = (1e)(500 \text{ kV}) = 500 \text{ keV}$$

Notice why it is convenient to work in electron volts.

The rest energy of an electron is $E_\text{o} = 0.511$ MeV $= 511$ keV. Thus,

$$E = K + E_\text{o} = 500 \text{ keV} + 511 \text{ keV} = 1011 \text{ keV} \ (= 1.011 \text{ MeV})$$

(b) The speed v may be found from γ by

$$\gamma = \frac{m}{m_\text{o}} = \frac{mc^2}{m_\text{o}c^2} = \frac{E}{E_\text{o}} = \frac{1011 \text{ keV}}{511 \text{ keV}} = 1.98$$

(Notice from the development that $E = \gamma E_\text{o}$.)

Then, with $\gamma = 1/\sqrt{1 - (v/c)^2}$,

$$\frac{v}{c} = \left(1 - \frac{1}{\gamma^2}\right)^{\frac{1}{2}} = \left[1 - \frac{1}{(1.98)^2}\right]^{\frac{1}{2}} = 0.863$$

and

$$v = 0.863c = (0.863)(3.00 \times 10^8 \text{ m/s}) = 2.59 \times 10^8 \text{ m/s} ∎$$

Example 26.5 Classical or Relativistic?

Given the kinetic energy of a particle, how do we know whether to use the relativistic formula or the classical formula to find its speed?

Solution

Expressing the relativistic kinetic energy as Eq. 26.14,

$$K = m_\mathrm{o}c^2(\gamma - 1) = m_\mathrm{o}c^2 \left[\frac{1}{\sqrt{1 - (v/c)^2}} - 1 \right]$$

and solving for $(v/c)^2$, we have

$$\left(\frac{v}{c}\right)^2 = 1 - \frac{1}{[1 + K/m_\mathrm{o}c^2]^2} = 1 - \frac{1}{[1 + K/E_\mathrm{o}]^2}$$

Thus, if $K \ll E_\mathrm{o}$, or the kinetic energy is much less that the particle's rest energy, then $v \ll c$, and the classical formula $K = \frac{1}{2}m_\mathrm{o}v^2$ would be appropriate. The relativistic formula always applies, but more calculation is required to give essentially the same result when $v \ll c$. The relationship also shows the reverse result. That is, if $v \ll c$, then $K \ll E_\mathrm{o}$, or the kinetic energy is much less than the rest energy. ■

26.5 The General Theory of Relativity

Einstein's special theory of relativity applies to inertial systems and therefore not to accelerating systems. Accelerating systems require an extremely complex theory, which was described by Einstein in several papers published around 1915. Called the **general theory of relativity**, this is essentially a gravitational theory.

An important aspect of the general theory is the **principle of equivalence**:

> An inertial reference frame in a uniform gravitational field is equivalent to a reference frame that has a constant acceleration with respect to the inertial frame.

What this basically means is that:

> No experiment performed in a closed, accelerating system can distinguish between the effects of a gravitational field and the effects of the acceleration.

According to the principle of equivalence, an observer in an accelerating system would find the effects of a gravitational field and those of an acceleration to be equivalent or indistinguishable. Rotating frames are excluded, since rotational acceleration can be distinguished from gravitational acceleration by releasing an object on a smooth, almost frictionless table. (Why?)

To better understand the equivalence principle, consider the situations illustrated in Fig. 26.13. The observer in a closed spaceship has no way of knowing whether he is near a mass, that is, in a gravitational field, or whether his system is accelerating. If he released an object and it remained suspended in midair, either he could be in free space (negligible gravity) or his spaceship could be accelerating in free fall. The observed effects are equivalent, and he could not distinguish between the two situations.

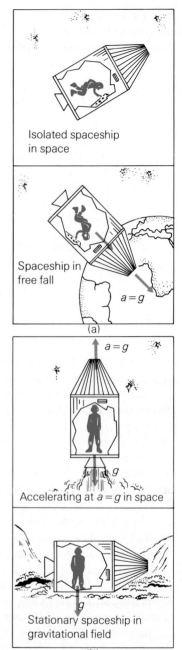

Isolated spaceship in space

Spaceship in free fall

$a = g$

(a)

$a = g$

Accelerating at $a = g$ in space

Stationary spaceship in gravitational field

g

(b)

Figure 26.13 **The principle of equivalence**
(a) The astronaut could perform no experiment in a closed spaceship that would determine whether he was in free space or in free fall.
(b) Similarly, he would not be able to distinguish between the effects of an acceleration and a gravitational field.

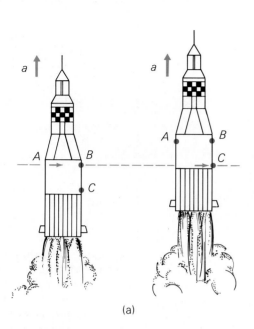

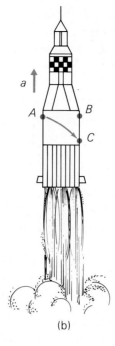

(a)

(b)

Figure 26.14 **Light bending**
(a) Light traversing an
accelerating, closed rocket
from point A arrives at
point C. (b) In the moving
system, the light would
appear to be bent. Since an
acceleration produces this
effect, by the principle of
equivalence, light should also
be bent in a gravitational
field.

Similarly, if the spaceship were in free space and accelerated with an acceleration a, or if it were stationary in a gravitational field with $g = -a$, the object would fall to the floor, and the observer in the closed system could not determine whether this was a gravitational or an acceleration effect. Thus, any effect that occurs on Earth with its acceleration due to gravity g will also occur on a spaceship traveling with an acceleration of g, and vice versa.

The principle of equivalence of the general theory of relativity leads to an important prediction—that light is bent in a gravitational field. To see how this prediction arises, let's use our imaginations in the thought experiment illustrated in Fig. 26.14. Suppose a beam of light traverses a spaceship that is accelerating at a very fast rate. If the spaceship were stationary, light emitted at point A would arrive at point B. However, because the spaceship is accelerating, the ship moves upward during the finite time it takes for the light to traverse the ship, and the light arrives at point C.

Now consider the situation as observed by a person on board the spaceship. From his point of view, he would observe the same effect of light emitted at A and arriving at C—*as though the light were bent*. The acceleration of the rocket produces the effect, so by the equivalence principle we conclude that light should also be bent in a gravitational field.

Of course, this effect must be very small, since we have observed no evidence of gravitational light bending in the Earth's gravitational field. However, this prediction of Einstein's theory was experimentally verified in 1919 during a solar eclipse (see Fig. 26.15). Distant stars appear to be motionless and are measured to have a constant angular distance apart at night when they can be clearly seen. Light from one of the stars may pass near the Sun, which has a relatively strong gravitational field. However, any bending would not be observed on Earth because most of the starlight would be masked by the glare of the Sun itself and by the sunlight scattered in the atmosphere (which is why we can't see stars during the day).

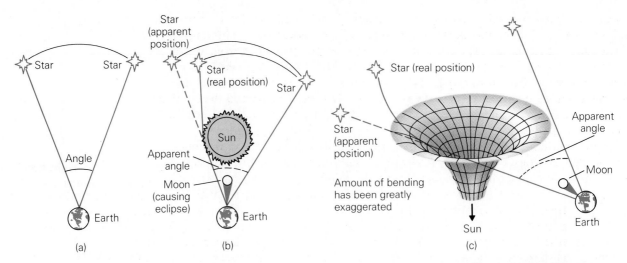

(a)
(b)
(c)

Figure 26.15 **Gravitational attraction of light** (a) Normally, two distant stars are observed to be a certain angular distance apart. (b) During a solar eclipse, the one star behind the Sun can still be seen because of the effect of solar gravity on the starlight, which is evidenced by a larger measured angular separation. (c) General relativity views a gravitational field as a warping of space and time. A simplified analogy of the surface of a warped rubber diaphragm, or sheet, is shown here.

But when a total solar eclipse occurs, the moon coming between the Earth and the Sun, an observer in the moon's shadow (the umbra) can see stars. If the light from a star passing near the Sun were bent, then the star would have an apparent location different from its actual position. As a result, the angular distance between stars would be measured to be slightly larger. Einstein's theory predicted the angular difference in the apparent and actual positions of the stars to be 1.75 seconds of arc, or an angle of about 0.00005°. The experimental angular distance was found to be 1.61 ± 0.30 seconds of arc.

General relativity views a gravitational field as a "warping" of space and time, as illustrated in Fig. 26.15(c). A light beam follows the curvature of space as would a ball rolling on a surface. The bending of light was verified during subsequent solar eclipses and also by signals from space probes. For example, signals from a Viking lander on the surface of Mars passed through the gravitational field of the Sun when Mars was on the far side of the Sun. The signals were observed to be delayed by about 100 μs. The delay was caused by the signals passing through the space-time warp, or "sink," of the Sun.

Another effect of the gravitational bending of light is known as *gravitational lensing*. In the late 1970s, a double quasar was discovered. (A quasar is an astronomical radio source.) The fact that it was a double quasar was not unusual, but everything about the two quasars seemed to be exactly the same, except that one was fainter than the other. It was suggested that perhaps there was only one quasar and that somewhere between it and the Earth, a massive object had deflected the light (radio waves), producing multiple images. The subsequent detection of a faint galaxy between the two quasar images confirmed this hypothesis, and other examples have since been discovered.

Gravitational lenses can be considered to be a test of general relativity, but this concept has also given general relativity a new role in astronomy. By examining multiple quasar images, their relative intensities, and so on, information can be gained about an intervening galaxy or cluster of galaxies that produce these effects.

The concept of gravity affecting light has become a popular notion with regard to black holes. A **black hole** is the gravitationally collapsed remnant of a star, which has a density so great and a gravitational field so intense that no light can escape (hence its blackness). In terms of a space-time warp analogy, a black hole would be a light sink or perhaps a bottomless hole from which nothing escapes. How then might we observe a black hole? The most likely possibility comes from binary star systems in which one star has collapsed to become a black hole. As matter is drawn into a spiraling disk around the black hole from the other star, the matter is accelerated and heated. Collision would give rise to X-rays, and the black hole would appear as an X-ray source.

Another result of the general theory is that gravity affects time by causing it to slow down—the greater the gravitational field, the greater the slowing of time. There is evidence of this effect in what is called the *gravitational red shift* of light from the Sun. The gravitational field of the Sun is quite large, and if the slowing of time in a gravitational field is correct, the electronic vibrations in the Sun's atoms should be slower. As a result, light from these atoms would have a lower frequency and be shifted toward the red end of the visible spectrum.

The measurement of a solar gravitational red shift was unsuccessful for many years. This was because of a lack of understanding of the conditions at the solar surface. For example, violent motions, with convection columns of rising hot gases and descending cooler gases, give rise to Doppler shifts. It wasn't until 20 to 30 years ago that the gravitational red shifts of solar lines were measured after a variety of effects were taken into account. The gravitational red shift was also verified in the laboratory in the 1960s by using complicated nuclear techniques.

Relativistic effects apply to all systems, no matter how complicated. Although these effects are most readily observed for atomic particles and light, they apply as well to real clocks, human beings, and so on. Although we don't commonly observe them, relativistic effects form the bases of some good science fiction and even some (not so good?) poetry, such as the following variation on a limerick that appeared in the British magazine *Punch* in 1923:

> A precocious student quite bright,
> Could travel much faster than light,
> He departed one day
> In an Einsteinian way
> And arrived the previous night.

Important Formulas

Time dilation:

$$t = \gamma t_o = \frac{t_o}{\sqrt{1 - (v/c)^2}}$$

Length contraction:

$$L = \frac{L_o}{\gamma} = L_o\sqrt{1 - (v/c)^2}$$

Relativistic mass:

$$m = \gamma m_o = \frac{m_o}{\sqrt{1 - (v/c)^2}}$$

Relativistic momentum:

$$p = mv = \gamma m_o v$$

Relativistic total energy:

$$E = K + E_o$$
$$mc^2 = K + m_o c^2$$
$$E = \gamma E_0$$

Relativistic kinetic energy:

$$K = E_o(\gamma - 1)$$

Rest energy:

$$E_o = m_o c^2$$

Questions

1. What is the principle of classical relativity? Does it apply to all situations? Explain.

2. Why was the existence of the ether postulated?

3. What would the existence of the ether imply?

The Michelson-Morley Experiment

4. How was the Michelson-Morley experiment intended to furnish evidence of the existence of the ether?

5. Why was it necessary to rotate the interferometer by 90° in the Michelson-Morley experiment?

6. How does the constancy of the speed of light explain the null result of the Michelson-Morley experiment?

The Special Theory of Relativity

7. Discuss the implications if the principle of relativity were not valid.

8. How is the simultaneity of events treated classically? According to the theory of relativity, do events occur simultaneously as measured in moving inertial systems?

9. (a) The driver of a moving truck blows his horn, which produces a sound whose frequency is 1000 Hz. A stationary observer in the path of the oncoming truck hears a sound of frequency 1010 Hz. (b) A woman on a moving train drops her purse, which falls vertically to the floor. To a stationary observer on the station platform, the purse appears to fall in a curved arc. In each of the two situations, which observer is correct? Explain.

10. Two observers in moving inertial reference frames watch half-hour programs on their TV sets. Which program runs during the proper time? What would one observer note if she compared the program in the other system to her own? (Assume that both are watching the same program.)

11. In a thought experiment, assume that you rode a beam of light coming from the face of a clock. How would time measured on the clock appear to you?

12. Does time dilation or length contraction depend on whether a system is moving away from or toward an observer?

13. Explain why there is no length contraction in the dimension perpendicular to the motion of an inertial reference frame.

14. An observer sees a friend passing by her in a rocketship with a uniform velocity that is an appreciable fraction of the speed of light. If the observer knows her friend to be 5 ft, 8 in. tall, does he still appear the same to the observer?

Relativistic Mass and Energy

15. How does the relative speed of a particle affect its momentum and energy as observed from another inertial reference frame?

16. Give some reasons why a particle cannot travel at the speed of light according to the theory of relativity.

17. Relativistically, what makes up the total energy of an object?

18. What would the theory of relativity predict if the speed of light were infinitely large ($c \cong \infty$)?

19. What is the relativistic energy of a stationary particle?

20. If an electron has a kinetic energy of 2 keV, could the classical expression for kinetic energy be used to compute its speed? Could it be used if the kinetic energy of the electron were 2 MeV?

The General Theory of Relativity

21. What is the distinction between the special and general theories of relativity?

22. What is the principle of equivalence and what are its implications for light in a gravitational field?

23. Would it be possible to detect a black hole using light from a distant star? Explain.

Problems

26.1–26.2 Classical Relativity and the Michelson-Morley Experiment

■■**1.** Car A is traveling eastward at 80 km/h. Car B is traveling with a speed of 70 km/h. Find the relative velocity of car B with respect to car A if car B is traveling (a) westward and (b) eastward.

■■**2.** A person 1.5 km from your location fires a gun; you observe the muzzle flash. If a 15 m/s wind is blowing, how long will it take the sound to reach you if the wind is

(a) toward you and (b) toward the other person? (The speed of sound is 345 m/s.)

■■**3.** A small airplane has an air speed of 200 km/h (speed with respect to air). Find the airplane's ground speed if there is (a) a head wind of 30 km/h and (b) a tail wind of 25 km/h.

■■**4.** A speedboat can travel 50 m/s in still water. If the boat is in a river that has a flow speed of 5 m/s, find the maximum and minimum values of the boat's speed relative to an observer on the river bank.

■■5. A small boat can travel 0.80 m/s in still water. The boat heads directly across a 50-m-wide river, which has a flow speed of 0.20 m/s. (a) How long will it take the boat to reach the opposite shore? (b) How far downstream will the boat travel? (c) How would the boat have to be steered if it were to arrive at a point across the river directly opposite the departure point?

■■■6. The apparatus used for measuring the speed of light by the French scientist Fizeau in 1849 is illustrated in Fig. 26.16. Teeth on a rotating disk periodically interrupt a beam of light. The light flashes travel to a plane mirror and are reflected back to an observer. Show that if the disk is rotated so that light passing through one tooth gap travels to the mirror and returns to be seen by the observer through the adjacent gap, the speed of light is given by $c = 2fNL$, where N is the number of gaps in the disk, f is the frequency of, or number of revolutions per second made by, the rotating disk, and L is the distance between the disk and the mirror.

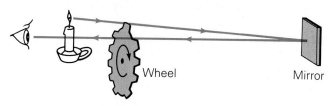

Figure 26.16 **Fizeau's apparatus**
See Problem 6.

■■■7. One proposed explanation of the null result of the Michelson-Morley experiment suggested that the interferometer arm in the direction of motion is shortened by a factor of $[1 - (v/c)^2]^{\frac{1}{2}}$ (the Lorentz-FitzGerald contraction). Show how this contraction would give a null result. (This hypothesis was disproved by making the lengths of the arms of the interferometer unequal.)

■■■8. The swimmer analogy of the Michelson-Morley experiment is illustrated in Fig. 26.17. Two swimmers start out from point A at the same time and swim with a constant speed c relative to the water. One swimmer swims across the river to point C and back to A, while the other swimmer swims to point B and back to A. Do the swimmers arrive back at A at the same time? Justify your answer mathematically.

■■■9. In the Michelson-Morley experiment, a difference in phase for the beams could be detected if the apparatus were rotated by 90°, since the interference pattern should change. Also, this would take care of the problem of

Figure 26.17 **The swimmer analogy**
See Problem 8.

assuming that the interferometer arm lengths are exactly equal, which should not be done arbitrarily. Show that the difference in the time differences for beam travel for the nonrotated and rotated conditions is

$$\Delta t - \Delta t' = \frac{2}{c}(l_1 + l_2)\left[\frac{1}{1 - v^2/c^2} - \frac{1}{\sqrt{1 - v^2/c^2}}\right]$$

where $\Delta t'$ is the time difference for the rotated condition and l_1 and l_2 are the lengths of the interferometer arms.

26.3 The Special Theory of Relativity

■■10. A person observes 5.0 min to elapse on her clock. How much time does she observe to elapse during this interval on a clock in another reference frame moving with a relative velocity of $0.9c$?

■■11. Imagine that you are moving with a velocity of $0.75c$ past a person who is reading this chapter in the textbook. If he takes 30 min to read the chapter by his clock, how much time would be observed to elapse on your clock?

■■12. A person has a pulse rate of 60 beats/min. What would the person's pulse rate be according to an observer moving with a velocity of $0.5c$ relative to the person?

■■13. A time of 15 min elapses on a clock, but an observer in an inertial reference frame measures 20 min to elapse. What is the relative velocity of the reference frames of the clocks?

■■14. A spaceship travels with a speed of 1.0×10^8 m/s. How much time does a clock on board the spaceship appear to lose in one day according to an observer in another inertial frame?

■■15. Alpha Centuri, the binary star closest to this solar system, is about 4.30 light years away. If a spaceship traveled a similar distance with a constant speed of $0.50c$ relative to Earth, how much time would elapse on a clock on the spaceship?

■■16. If a meterstick moved past you with a velocity $2c/3$ in a direction parallel to its length, what would you observe its length to be compared to a meterstick of your own?

■■17. A pole vaulter at the Relativistic Olympics sprints past you down the runway for a vault with a speed of $0.6c$. When he is at rest, his pole is 7.0 m long. What does the length of the pole appear to be to you as he passes you?

■■■18. What is the length contraction of an automobile 5.0 m long when it is traveling at 100 km/h? [Hint: $\sqrt{1 - x^2} \cong 1 - (x^2/2)$ for $x \ll 1$.]

■■■19. A spaceship has a length of 100 m in its rest frame. An observer in another inertial frame sees the length of the spaceship as 75 m. What is the relative velocity of the reference frames?

■■■20. An observer in a certain reference frame cannot understand why there is a problem in converting to the metric system because she notes the length of a meterstick in another moving reference frame to be the same length as her yardstick. Explain this situation quantitatively.

■■■21. Two reference frames move in relation to each other with a constant relative speed v in the x direction. Considering the O' frame to move in the $+x$ direction relative to the O frame, (a) show that the coordinate systems are classically related by the transformation equations: $x = x' + vt$, $y = y'$, and $z = z'$. These are the so-called Galilean transformations, together with $t = t'$, since time classically is a universal quantity. (b) Taking into account the relativistic length contraction, show that the transformation equations are given by: $x' = \gamma(x - vt)$, $y' = y$, and $z' = z$. These relativistic transformations are called the Lorentz transformations. (Time is also affected, but the derivation of this transformation is beyond the scope of this book.)

■■■22. Events that are simultaneous to an observer in one inertial frame (O) are not necessarily simultaneous when viewed by an observer in another inertial frame (O'). If the frames are at rest, as illustrated in Fig. 26.18, both observers would say that the lightning flash events were simultaneous, or occurred at the same time, because the signals are received at the same instant. However, let the O' frame be moving to the right with a constant velocity and the simultaneous events occur just as O' passes the O position. Show that the observers would disagree about the simultaneity of the events in this case.

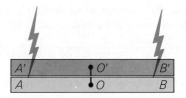

Figure 26.18 **Simultaneity or not**
See Problem 22.

26.4 Relativistic Mass and Energy

■■23. What is the rest mass of a proton in mega-electron volts?

■■24. If the mass of 1.0 g of coal (or any substance) could be completely converted into energy, how many kilowatt-hours of energy would be produced?

25. The United States uses over 3.0 trillion kWh of electricity annually. If 3.0 trillion kWh of electrical energy were supplied by nuclear generating plants, how much nuclear mass would have to be converted to energy, assuming a production efficiency of 35%?

■■26. If the kinetic energy of a particle is increased by 10 keV, what is the corresponding increase in its relativistic mass?

■■27. A person whose mass is 48 kg wishes to gain 12 kg relativistically with respect to another reference frame. (a) How fast would the person have to travel? (b) How would her height and width change? (c) How would her overall density change?

■■28. Sketch graphs of the (a) momentum, (b) energy, and (c) mass of an object as a function of speed.

■■29. An electron travels at a speed of 5.0×10^7 m/s. What is its total energy?

■■30. How much energy is required to accelerate an electron from rest to half the speed of light?

■■31. The kinetic energy of an electron is 75% of its total energy. Find the speed and momentum of the electron.

■■32. An electron has a total relativistic energy of 1.5 MeV. What is its relativistic momentum?

■■33. A proton moves with a speed of $0.2c$. What are its (a) total energy, (b) kinetic energy, and (c) relativistic momentum?

34. Show that the *classical* kinetic energy and momentum of a particle can be expressed in terms of the rest energy of the particle as

$$K = \tfrac{1}{2}E_o\left(\frac{v^2}{c^2}\right) \quad \text{and} \quad p = \frac{E_o v}{c^2}$$

35. The National Accelerator Laboratory at Batavia, Illinois, can accelerate particles to energies of giga-electron volts (1 GeV = 10^9 eV). (a) What is the speed of a proton accelerated from rest through a potential of 1 GeV? ($m_o = 1.67 \times 10^{-27}$ kg.) (b) The planned Superconducting Super Collider (SSC) accelerator will produce particles with energies of 40 GeV. What would be the speed of a proton in this case?

36. Each second the Sun gives off about 4.0×10^{26} J of radiant energy produced by converting mass to energy in nuclear reactions. (a) How many kilograms of mass are lost by the Sun each second? (b) If the Sun's mass is 2.000×10^{30} kg, at the rate calculated in part (a), how long in years will it be until the Sun's mass is reduced to 1.999×10^{30} kg?

26.5 The General Theory of Relativity

37. The escape velocity from an object with a mass M and radius R is given by $v_e = (2GM/R)^{\frac{1}{2}}$ (see Chapter 7). Show that an object of mass M and radius $R = 2GM/c^2$ or less can be classified as a black hole according to the theory of relativity. ($R = 2GM/c^2$ is called the Schwarzchild radius.)

38. Assume that a collapsing star has a mass of 5 solar masses. (a) What would the radius of the collapsed star have to be for it to qualify as a black hole? (See Problem 37.) (b) What would be the density of the black hole material?

Additional Problems

39. An electron travels at a speed of $c/2$. What is its total energy?

40. A spaceship travels with a speed of $c/3$. What time change does a clock on board the spaceship show in 1 hour according to an observer in another inertial frame?

41. An electron is accelerated from rest through a potential difference of 2.0 MV. Find the speed, kinetic energy, and momentum of the electron.

42. One 25-year-old twin decides to take a space trip while the other remains on Earth. The traveling twin travels at a speed of 0.9c for 39 years according to Earth time. Assuming that special relativity applies, what are their ages when the traveling twin returns to Earth?

43. A home uses approximately 15,000 kWh of electricity per year. How much matter would have to be converted directly to energy to supply this amount of energy?

44. How fast must an object travel for its mass to have an increase of (a) 10% and (b) 99% of its rest mass?

45. According to a navigator on a spaceship, the ship has traveled one light year at a speed of 0.9c. How long does it take for the spaceship to travel this distance according to an inertial observer in the frame to which this speed is referenced?

46. A relativistic rocket is measured to be 50 m long, 2.5 m high, and 2.0 m wide. What are its dimensions to an inertial observer when the rocket travels at 0.85c?

47. How fast must a meterstick travel for its rest length to contract by (a) 1%, (b) 10%, and (c) 99%?

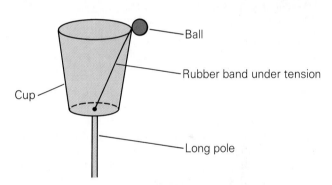

Figure 26.19 How to get the ball in the cup? See Problem 48.

48. An apparatus like the one in Fig. 26.19 was given to Albert Einstein on his 76th birthday by Eric M. Rogers, a physics professor at Princeton University. The goal is to get the ball into the cup without touching the ball. (Jiggling the pole up and down will not do it.) Einstein solved the puzzle immediately and then confirmed his answer with an experiment. How did Einstein get the ball into the cup? (Hint: he used a fundamental concept of general relativity.)*

* This problem is adapted from *Physics* by R. T. Weidner (Needham Heights, MA: Allyn and Bacon, 1985), p. 333.

Quantum Physics

27

instein's theory of special relativity (Chapter 26) helped to resolve the problems that classical physics had in describing particles moving at speeds comparable to that of the speed of light. However, there were other troublesome areas in which classical theory did not agree with experimental results. As scientists explored the nature of the atom in the early years of the twentieth century, attempts to apply classical physics to explain phenomena on the atomic level were unsuccessful in several major areas of study.

The period from about 1900 to 1930 was perhaps one of the most productive in the history of physics. Scientists looked at the problems of the day and came up with new hypotheses that explained them. These hypotheses were *new* in the respect that they required nonclassical approaches. It was during this period that the principles of modern physics were developed, and a so-called revolution in physics took place.

Chief among these new theories was the idea that light could be *quantized*, that it was made up of discrete amounts of energy. This concept lead to the formulation of *quantum mechanics*, which was highly successful in showing the correlations between waves and particles and in explaining the behavior of atoms. The idea of probability, rather than classically exact determination, was introduced. All this was, and is, quite mathematical and beyond the scope of this text. But a general overview of the important results is essential in the understanding of physics as we know it today. Toward this end, the important developments of "quantum" physics are presented in this chapter, and an introduction to quantum mechanics will be given in Chapter 28.

27.1 Quantization: Planck's Hypothesis

One of the problems scientists were having at the end of the nineteenth century was how to explain the spectra of radiation emitted by hot objects. This is sometimes called **thermal radiation**. You learned in Chapter 12 that the total intensity of the emitted radiation is proportional to the fourth power of the absolute Kelvin temperature (T^4). Since the temperatures of all objects are above absolute zero, all objects emit thermal radiation.

At normal temperatures, this radiation is in the infrared region and so is not detected visibly. At temperatures of about 1000 K, an object emits radiation in the long-wavelength end of the visible spectrum and has a reddish glow. A hot electric stove burner is a good example. Increased temperature causes the radiation and color to shift to a shorter wavelength, such as yellow-orange. Above a temperature of about 2000 K, an object glows with a yellowish-white color, like the glowing filament of a light bulb.

Analysis shows that, although we see only a particular color, there is a general continuous spectrum at all these temperatures as illustrated in the graph in Fig. 27.1(a). The curves shown here do not correspond to a real radiating body; they are for an idealized blackbody. A **blackbody** is an ideal system that absorbs (and emits) all radiation that is incident on it (see Section 12.4). An absorbing blackbody may be approximated by a small hole leading to an internal cavity inside a block of material [see Fig. 27.1(b)]. Radiation falling on the hole enters the cavity and is reflected back and forth by the cavity walls. If the hole is very small in comparison to the surface area of the cavity, only a small amount of radiation will be reflected through the hole. Since nearly all radiation incident on the hole is absorbed and very little comes back out, the hole, viewed from the outside, is an excellent approximation to the surface of a blackbody.

Notice in the figure that as the temperature increases, more radiation is emitted at every wavelength and that the wavelength of the maximum-intensity component becomes shorter. This shift is found to obey a relationship, named

(b)

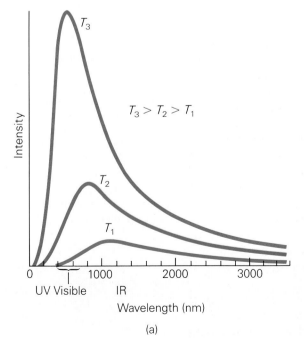

(a)

Figure 27.1 **Thermal radiation**
(a) Intensity versus wavelength curves for the thermal radiation from an idealized blackbody. Notice how the maximum component shifts to a lower wavelength with increasing temperature. (b) A blackbody may be approximated by a small hole leading to an interior cavity in a block of material.

after the scientist who developed it in 1893, known as **Wien's displacement law**:

$$\lambda_{\max} T = \text{a constant} = 2.90 \times 10^{-3}\ \text{m-K} \qquad (27.1)$$

where $\lambda_{\max}$ is the wavelength of the radiation component of maximum intensity in meters and T is the temperature of the body in kelvins.

Wien's law can be used to determine not only the wavelength of the maximum spectral component, but also the temperature of the emitter. Thus, it can be used to estimate the temperatures of stars from their radiations. For example, taking the wavelength of the maximum component of sunlight to be in the visible region with a $\lambda_{\max} = 500\ \text{nm} = 5.0 \times 10^{-7}\ \text{m}$, we have $T = 2.90 \times 10^{-3}\ \text{m-K}/5.0 \times 10^{-7}\ \text{m} = 5800\ \text{K}$. Thus, the surface of the Sun has a temperature of about 6000 K.

Classically, thermal radiation is pictured as resulting from the thermal agitation and acceleration of electric charges near the surface of an object. Electric charges in thermal agitation would undergo many different accelerations, so a continuous spectrum of emitted radiation would be expected. However, classical theoretical calculations based on the number of standing wave modes in a blackbody cavity predicted the intensity to be proportional to $1/\lambda^4$, that is, $I \propto 1/\lambda^4$.

At long wavelengths, the classical prediction agrees fairly well with experimental data. However, at shorter wavelengths, there is little correlation. Contrary to experimental observations, classical theory predicts that the radiation intensity should increase as the wavelength goes to zero (Fig. 27.2). This classical prediction is sometimes called the **ultraviolet catastrophe**—"ultraviolet" because the difficulty occurs for short wavelengths beyond the violet end of the visible spectrum and "catastrophe" because it predicts that the emitted intensity or energy will be very large, becoming infinitely large.

The failure of classical physics to explain the characteristics of thermal radiation led Max Planck, a German physicist (Fig. 27.3), to reexamine the phenomenon. In 1900, Planck formulated a theory that correctly predicted the observed distribution of the blackbody radiation spectrum, but only by introducing a radical new idea. Planck showed that if it were assumed that the thermal oscillators could have only *discrete*, or particular, amounts of energy, rather than a continuous distribution of energies, or any arbitrary energy, then theory would agree with experiment.

These discrete amounts of energy of thermal oscillators were related to the frequency f of oscillation by

$$E_n = nhf \qquad \text{for } n = 1, 2, 3, \ldots \qquad (27.2)$$

where h is a constant called **Planck's constant**:

$$h = 6.63 \times 10^{-34}\ \text{J-s}$$

This relationship is commonly called **Planck's hypothesis**. Rather than considering oscillator energy to be a continuous quantity as it is classically, by Planck's hypothesis the energy was *quantized*, that is, an oscillator could have only discrete amounts of energy. The smallest possible amount of oscillator energy is

$$E = hf \qquad (27.3)$$

which is called a **quantum** of energy (from the Latin word *quantus*, meaning

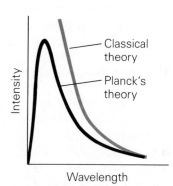

Figure 27.2 **The ultraviolet catastrophe**
Classical theory predicts the intensity of thermal radiation to be proportional to $1/\lambda^4$, which calls for an infinitely large amount of radiation energy as the wavelength approaches zero. Planck's quantum theory, on the other hand, predicts the observed radiation distribution.

"how much"). Thus, each oscillator could absorb or emit energy in quantum amounts, as Einstein would soon point out.

Although the theoretical predictions agreed with experiment, Planck himself was not completely convinced of the validity of his quantum hypothesis. However, the concept of quantization was quickly used to explain other phenomena that could not be explained classically. His quantum hypothesis earned Planck the Nobel Prize in 1918.

27.2 Quanta of Light: Photons and the Photoelectric Effect

The concept of the quantization of light was introduced in 1905 by Albert Einstein in a paper on light absorption and emission:

> In accordance with the assumption herein proposed, the radiant energy from a point source is not distributed continuously throughout an increasingly larger region, but, instead, this energy consists of a finite number of spatially localized energy quanta which, moving without subdividing, can only be absorbed and created in whole units.

If the energy of the molecular oscillators in a hot substance is quantized, then to conserve energy, the emitted radiation should also be quantized. A quantum or packet of light is referred to as a **photon**, and each photon has an energy

$$E = hf \tag{27.4}$$

This idea suggests that light is transmitted as discrete quanta (plural of quantum), or "particles" of energy, rather than as waves.

Einstein used this quantum concept to explain what is called the **photoelectric effect**, another area in which classical physical description was inadequate. Certain metallic materials are *photosensitive*. That is, when light strikes the surface of the material, electrons are emitted. The radiant energy supplies the work necessary to free the electrons from the material surface. Such materials are used to make photocells, as illustrated in Fig. 27.4(a) on the next page. The emitted photoelectrons complete the circuit, and a current is registered on the ammeter.

When a photocell is illuminated with monochromatic (single-frequency) light of different intensities, characteristic curves are obtained as shown in Fig. 27.4(b). The photocurrent increases as the voltage across the cell increases until a saturation current is reached. At this point, all of the emitted electrons are reaching the anode, and any further increase in voltage has no effect on the magnitude of the current.

As would be expected classically, the current is proportional to the intensity of the incident light—the greater the intensity ($I_2 > I_1$ in the figure), the more energy there is to free electrons.

An idea of the kinetic energy of the electrons being emitted from the photoelectric material can be gained by placing a negative voltage across the electrodes. As the voltage is decreased and made negative, the photocurrent decreases. Only the electrons with kinetic energy greater than $e|V|$ can make it to the negative plate and contribute to the current because of the retarding voltage. (The kinetic energy goes into the work of overcoming the negative potential difference, $W = U = -eV$.)

Figure 27.3 **Max Planck (1858–1947)**
Planck was the German physicist who introduced quantum theory to modern science. His work yielded an important constant that appears over and over again in physics and is known as Planck's constant, h. His research won him the Nobel Prize in physics in 1918.

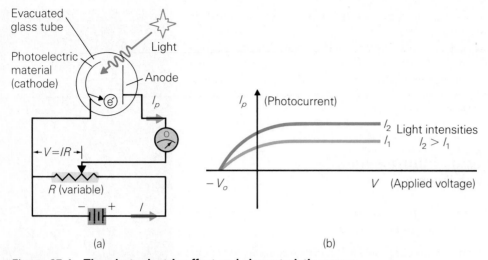

(a)   (b)

Figure 27.4 The photoelectric effect and characteristic curves
(a) Incident light on the photomaterial in a phototube causes the emission of
electrons, and a current flows in the circuit. The voltage applied to the tube is varied
by means of the variable resistor. (b) As the plots of current versus voltage for two
intensities of monochromatic light show, the current becomes saturated, or constant,
as the voltage is increased. For negative voltages, the current goes to zero at a
particular stopping potential (V_o).

At some voltage V_o, called the **stopping potential**, no current will flow
between the electrodes. The maximum kinetic energy (K_{max}) of the electrons
emitted from the photomaterial is then related to the stopping potential by

$$K_{max} = e V_o \qquad\qquad (27.5)$$

since $e V_o$ is the work needed to stop the most energetic electrons.

As can be seen from Eq. 27.5, the maximum kinetic energy of the
photoelectrons does not depend on the intensity of the light. However, it is
found that the stopping potential and the maximum kinetic energy *do* depend
on its frequency. The maximum kinetic energy of the electrons is linearly
dependent on the frequency of the incident light (Fig. 27.5). No photoemission is
observed to occur below a certain **cutoff frequency** f_o.

The electron emission begins instantaneously, or with no observable time
delay, when a photomaterial is illuminated with light $f > f_o$, even if the light
intensity is very low.

These four key characteristics of photoemission are summarized in Table 27.1 along with their correspondence with classical theoretical predictions.
Note that only the first observation is predicted correctly by classical wave
theory. Classically, the electric field of a light wave interacts with the electrons
at the surface of the material and sets them into oscillation. The intensity
(energy) of an oscillating electric field is proportional to the amplitude of the
field vector, so the average kinetic energy of the electrons is proportional to the
intensity of the light. Therefore, it is difficult to understand why the maximum
kinetic energy of the photoelectrons is independent of the intensity as
observed. Also, according to wave theory, photoemission should occur for any
frequency of light, provided that the intensity is sufficient.

The immediate emission of electrons raises an even more serious problem

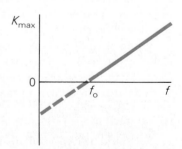

Figure 27.5 **Maximum
kinetic energy and frequency**
The maximum kinetic energy
(K_{max}) of photoelectrons is a
linear function of the
frequency of the incident
light. Below a certain cutoff
frequency f_o, no
photoemission occurs.

Table 27.1
Photoelectric Characteristics

Characteristic	Predicted by classical theory?
1. The photocurrent is proportional to the intensity of the light.	Yes
2. The kinetic energy of the emitted electrons is dependent on frequency but not on intensity.	No
3. No photoemission occurs for light with a frequency below a certain cutoff frequency f_o regardless of intensity.	No
4. A photocurrent is observed immediately when the light frequency is greater than f_o even if the intensity is low.	No

for classical theory. Classically, the time required for an electron to acquire enough energy to be freed and have a typically observed value of maximum kinetic energy is on the order of minutes for low light intensities.

An explanation of the photoelectric effect was put forth by Einstein in 1905 using the quantum idea suggested by Planck in his theory of thermal oscillators. As was stated previously, he considered the energy of light to be in quantum bundles or photons, each with an energy $E = hf$. Thus, light of frequency f could give energy to the electrons in a photomaterial only in discrete amounts equal to hf. An electron absorbs either the whole photon or nothing at all.

The electrons of a photomaterial are bound by attractive forces. Therefore, work must be done to free an electron, and some of the absorbed photon energy goes into doing this. Then, when a photon of energy hf is absorbed, by the conservation of energy we have

$$hf = K + \phi \qquad (27.6)$$

where ϕ is the work or energy needed to free a surface electron from the material. Thus, part of the energy of the absorbed photon goes into the work of freeing the electron, and the rest is carried off by the emitted electron as kinetic energy.

The least tightly bound electron will have the maximum kinetic energy (K_{max}), and in this case, the energy needed to free the electron is called the **work function** ϕ_o of the material. The energy conservation equation is then

$$hf = K_{max} + \phi_o \qquad (27.7)$$

Other electrons will require more energy than the ϕ_o minimum amount, and the kinetic energy of these electrons will be less than K_{max}.

Einstein's quantum theory was consistent with experimental observation. Increasing the light intensity merely increases the number of photons and thus increases the photocurrent. An increase in intensity does not change the energy of individual photons ($E = hf$), so K_{max} is independent of intensity. Also, K_{max} is linearly dependent on the frequency of light in Eq. 27.7, as has been observed experimentally.

The existence of a cutoff frequency is also established. Note that Einstein's formula has the same graphic form as the frequency dependence of K_{max} in Fig. 27.5. By comparison, when $K_{max} = 0$,

$$\phi_o = hf_o \qquad (27.8)$$

in which case the photon has just enough energy to free an electron from the material, but no extra energy to give it kinetic energy. Notice that the graph in Fig. 27.5 provides a means to determine the value of h, which is the slope of the line.

The frequency f_o is called the **threshold frequency** and corresponds to the cutoff frequency. If the incident light has a frequency less than f_o, then no matter how many protons are available (that is, no matter how intense the light, the photons will not have enough energy to cause photoemission). No time delay in the emission of electrons would be expected in the quantum theory, since the energy is "dumped on" an electron in a concentrated bundle instead of being continuously supplied and distributed over an area as in wave theory.

Einstein's theory of the photoelectric effect scored another success for the idea of quantization. It is interesting to note that Einstein won the Nobel Prize in physics in 1921 for his theory of the photoelectric effect rather than for his more famous theory of relativity.

Example 27.1   The Photoelectric Effect

The work function of a particular metal is 2.0 eV. (a) What is the threshold frequency for the material? (b) If the metal is illuminated with monochromatic light having a wavelength of 550 nm, what will be the maximum speed of the emitted electrons? (c) What is the stopping potential?

Solution

Given: $\phi_o = 2.0$ eV $\times$ (1.6×10^{-19} J/eV)   Find: (a) f_o
              $= 3.2 \times 10^{-19}$ J   (b) v_{max}
       $\lambda = 550$ nm $= 5.50 \times 10^{-7}$ m   (c) V_o

(a) The threshold frequency is related directly to the work function, $\phi_o = hf_o$, and

$$f_o = \frac{\phi_o}{h} = \frac{3.2 \times 10^{-19} \text{ J}}{6.63 \times 10^{-34} \text{ J} \cdot \text{s}} = 4.8 \times 10^{14} \text{ Hz}$$

which is in the visible region of the electromagnetic spectrum.

(b) The energy of a photon of light of this wavelength is

$$hf = \frac{hc}{\lambda} = \frac{(6.63 \times 10^{-34} \text{ J} \cdot \text{s})(3.0 \times 10^8 \text{ m/s})}{5.50 \times 10^{-7} \text{ m}}$$

$$= 3.6 \times 10^{-19} \text{ J}$$

Then

$$K_{max} = hf - \phi_o = 3.6 \times 10^{-19} \text{ J} - 3.2 \times 10^{-19} \text{ J}$$

$$= 0.4 \times 10^{-19} \text{ J} = \tfrac{1}{2}mv_{max}^2$$

where m is the mass of an electron. Solving for v_{max}, we get

$$v_{max} = \sqrt{\frac{2K_{max}}{m}} = \left[\frac{2(0.4 \times 10^{-19})}{9.11 \times 10^{-31} \text{ kg}} \right]^{\frac{1}{2}}$$

$$= 3.0 \times 10^5 \text{ m/s}$$

(c) The stopping potential is related to K_{max}, that is, $K_{max} = eV_o$, and

$$V_o = \frac{K_{max}}{e} = \frac{0.4 \times 10^{-19}}{1.6 \times 10^{-19} \text{ C}} = 0.25 \text{ V} \quad \blacksquare$$

27.3 Quantum "Particles": The Compton Effect

In 1923, the American physicist Arthur H. Compton (1892–1962) explained a phenomenon he observed in the scattering of X-rays by considering the radiation as being composed of quanta. This explanation of the observed effect provided convincing evidence that light, and electromagnetic radiation in general, is composed of quanta, or "particles" of energy.

Compton had observed that when a beam of monochromatic X-rays was scattered by a material, the scattered radiation had a wavelength longer than the wavelength of the incident beam. Also, the change in the wavelength was dependent on the angle through which the X-rays were scattered but not on the scattering material. See Fig. 27.6.

Classically, scattered radiation should have the same frequency or wavelength as the incident radiation. The electrons in the atoms of the scattering material absorb the radiation and oscillate at the same frequency as the

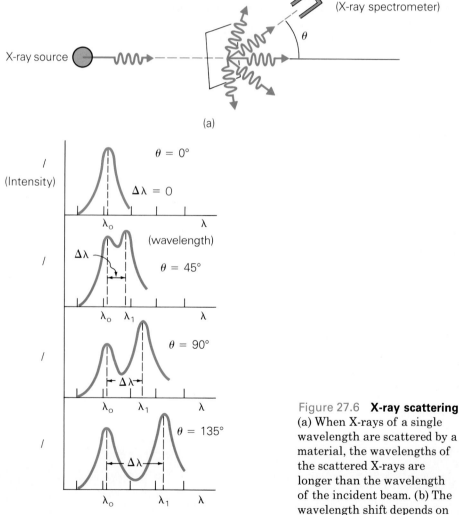

Figure 27.6 **X-ray scattering** (a) When X-rays of a single wavelength are scattered by a material, the wavelengths of the scattered X-rays are longer than the wavelength of the incident beam. (b) The wavelength shift depends on the scattering angle.

incident wave. On emission, the scattered radiation should have the same frequency in all directions.

Compton noted that the wavelength of the shifted component was always longer than the wavelength of the incident radiation and that it increased with a larger scattering angle. The frequency of the scattered radiation, on the other hand, was less ($f = c/\lambda$) and decreased as the scattering angle increased. According to Einstein's photon theory, the frequency of a quantum is directly proportional to its energy, so a change in frequency would correspond to a change in energy. The effect of the scattering angle on the energy of the scattered radiation reminded Compton of a similar effect in the elastic collision of particles. Could the same principles apply in the scattering of quanta? Pursuing this idea, Compton developed a simple theory that correctly predicted the wavelength shift of the scattered radiation.

In elastic collisions (Chapter 6), the kinetic energy and momentum are conserved. The energy of a quantum of radiation is given by $E = hf$, but what is its momentum? Since a quantum travels at the speed of light and is never at rest, it must have no rest energy or be a "massless" particle. Otherwise, it would relativistically have infinite energy. Recall from Chapter 26 that the relativistic energy is $E = mc^2 = \gamma m_o c^2 = m_o c^2 / \sqrt{1 - (v/c)^2}$. Thus, the total relativistic energy of a quantum is entirely kinetic energy.

The relationship between the total relativistic energy and the relativistic momentum of a particle is given by

$$E^2 = p^2 c^2 + (m_o c^2)^2 \qquad (27.9a)$$

where m_o is the rest mass of the particle. Since a quantum is treated as a massless particle ($m_o = 0$), its momentum is

$$p = \frac{E}{c} = \frac{hf}{c} = \frac{h}{\lambda}$$

Compton assumed that the collisions took place between quanta and individual electrons that were free and initially stationary. He then applied the principles of conservation of momentum and conservation of total relativistic energy to a general collision as illustrated in Fig. 27.7. By the conservation of momentum,

	before	*after*
x component:	$p_o = p_1 \cos \theta + p_e \cos \phi$	
y component:	$0 = p_1 \sin \theta + p_e \sin \phi$	

If we square and add these equations, the trigonometric identity $\cos^2 \phi + \sin^2 \phi = 1$ can be used to obtain

$$p_o^2 + p_1^2 - 2p_o p_1 \cos \theta = p_e^2 \qquad (27.9b)$$

By the conservation of total relativistic energy,

	before	*after*
	$E_o + m_o c^2 = E_1 + K + m_o c^2$	

where m_o is the rest mass of the electron. Thus,

$$E_o - E_1 = c(p_o - p_1) = K \qquad (27.9c)$$

where $E = pc$ for a quantum. The total relativistic energy of the moving

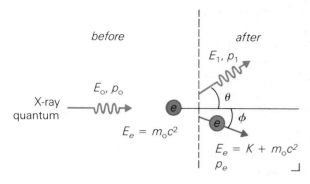

before

after

E_1, p_1

E_0, p_0

X-ray
quantum

θ

$E_e = m_o c^2$

ϕ

$E_e = K + m_o c^2$

p_e

Figure 27.7 **Compton scattering**
By considering an X-ray quantum to be a particle and analyzing the scattering in a particle collision with an electron, Compton was able to explain the X-ray wavelength shift.

electron can also be written as follows, using the relationship given in Eq. 27.9a:

$$(K + m_o c^2)^2 = p_e^2 c^2 + (m_o c^2)^2 \qquad (27.9d)$$

Then, substituting from Eqs. 27.9c and 27.9d into Eq. 27.9b to eliminate p_e, after some algebraic manipulation, we obtain

$$\frac{1}{p_1} - \frac{1}{p_0} = 1 - \frac{\cos \theta}{m_o c^2}$$

Multiplying through by h and recalling that for a quantum $p = h/\lambda$, we have Compton's result:

$$\lambda_1 - \lambda_0 = \lambda_c (1 - \cos \theta) \qquad (27.10)$$

The constant $\lambda_c = h/m_o c^2 = 2.4 \times 10^{-12}$ m is called the *Compton wavelength*. The equation correctly predicted the observed wavelength shift ($\Delta \lambda = \lambda_1 - \lambda_0$), and Compton was awarded a Nobel Prize in 1927 for his work.

Example 27.2   The Compton Effect

A monochromatic beam of X-rays with a wavelength of 1.3×10^{-10} m is scattered by a metal foil. By what percentage is the wavelength shifted for the scattered component observed at an angle of 90°?

Solution

Given: $\lambda_0 = 1.3 \times 10^{-10}$ m   *Find*: % shift
$\qquad \theta = 90°$

The change or shift in the wavelength, $\Delta \lambda$, is given by Eq. 27.10, and the fractional change is $\Delta \lambda / \lambda_0$. Then

$$\frac{\Delta \lambda}{\lambda_0} = \frac{\lambda_c}{\lambda_0} (1 - \cos \theta)$$

$$= \frac{2.4 \times 10^{-12} \text{ m}}{1.3 \times 10^{-10} \text{ m}} (1 - \cos 90°) = 1.8 \times 10^{-2}$$

and

$$\frac{\Delta \lambda}{\lambda_0} \times 100\% = 1.8\% \quad \blacksquare$$

Einstein's and Compton's successes in explaining electromagnetic phenomena in terms of quanta left scientists with two theories of electromagnetic radiation. Classically, the radiation is pictured as a continuous wave, and classical theory explains satisfactorily such things as interference and diffraction. On the other hand, quantum theory was necessary to satisfactorily explain the photoelectric and Compton effects.

The two theories gave rise to a description that is called the **dual nature of light**. That is, light apparently behaves sometimes as a wave and at other times as photons or "particles." The reconciliation of the two theories will be considered in the next chapter. Basically, light energy is carried as photons, and a wave acts as a guide for the quanta of energy and other particles.

27.4 The Bohr Theory of the Hydrogen Atom

In the 1800s, a great deal of experimental work was done with gas discharge tubes—for example, those containing hydrogen, neon, and mercury (vapor). (The common fluorescent lamp is a gas discharge tube, in which an electrically

Continuous spectrum (tungston lamp)

Helium

Neon

Lithium

(a)

Emission spectrum

Absorption spectrum

(b)

Figure 27.8 **Line spectra**
(a) A continuous spectrum and the line spectra of various elements. Each line corresponds to a particular wavelength. (b) The emission spectrum for a hot gas has bright lines on a dark background. An absorption spectrum, made by passing white light through a relatively cool gas, has dark lines on a light background. The emission and absorption lines coincide for the same gas.

excited gas emits radiation.) When light emissions from these tubes were analyzed, discrete, or line, spectra were observed instead of continuous spectra as from incandescent sources (Fig. 27.8).

Light coming directly from a tube gives an emission or bright-line spectrum. If white light is passed through a relatively cool gas, certain frequencies or wavelengths are found to be missing in the resulting absorption spectrum, which shows dark lines. Atoms absorb light as well as emitting it; and when the absorption and emission lines of a particular gas were compared, they were found to occur at the same frequencies. The reason for line spectra was not understood at the time, but they provided a clue to the electronic structure of the atoms.

Hydrogen, with a relatively simple visible spectrum, received much attention. Hydrogen is also the simplest atom, with only one electron and one proton. In the late nineteenth century, a Swiss physicist, J. J. Balmer, found an empirical formula that gives the wavelengths of the spectral lines of hydrogen in the visible region:

$$\frac{1}{\lambda} = R\left(\frac{1}{2^2} - \frac{1}{n^2}\right) \qquad \text{for } n = 3, 4, 5, \ldots \tag{27.11}$$

where R is called the *Rydberg constant* and has a value of 1.097×10^{-2} nm^{-1}. The spectral lines of hydrogen in the visible region, called the **Balmer series**, were found to fit the formula, but it was not understood why. Similar formulas were found to fit the spectral lines in the ultraviolet and infrared regions.

An explanation of the spectral lines was given in a theory of the hydrogen atom put forth in 1913 by the Danish physicist Niels Bohr (Fig. 27.9). Bohr assumed that the hydrogen electron orbited the nuclear proton in a circular orbit (analogous to a planet orbiting the Sun). The electrical Coulomb force supplies the necessary centripetal force for the circular motion, and we may write

$$F = \frac{ke^2}{r^2} = \frac{mv^2}{r} \tag{27.12}$$

where e is the charge of the proton and the electron, v the electron's orbital speed, and r the radius of the orbit (Fig. 27.10).

The total energy of the electron is the sum of its kinetic and potential energies:

$$E = K + U = \tfrac{1}{2}mv^2 - \frac{ke^2}{r}$$

From the first equation, the kinetic energy may be written $\tfrac{1}{2}mv^2 = ke^2/2r$. Using this, the total energy becomes

$$E = \frac{ke^2}{2r} - \frac{ke^2}{r} = -\frac{ke^2}{2r} \tag{27.13}$$

Notice that as the radius approaches infinity, E approaches zero. With $E = 0$, the electron is no longer bound to the proton, and the atom, having lost its electron, would be ionized.

Thus far, only classical principles have been applied. But at this point in the theory, Bohr made a radical assumption—radical in the sense that he introduced a quantum concept. His critical assumption was that the angular momentum of the electron was quantized and could have only discrete values

Figure 27.9 **Niels Bohr (1885–1962)**
Bohr, one of the foremost scientists of the twentieth century, contributed greatly to the development of quantum theory. His initial application of quantum theory to the hydrogen atom, for which he was awarded the Nobel Prize in physics in 1922, produced much of the present–day insight into the atom. Bohr's subsequent work in nuclear theory played an important part in the understanding of nuclear fission.

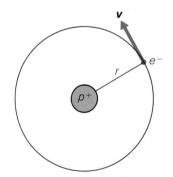

Figure 27.10 **The Bohr model of the hydrogen atom**
The electron of the hydrogen atom is pictured as going around the nuclear proton in a circular orbit, similar to the way a planet orbits the Sun.

that were integral multiples of $h/2\pi$, where h is Planck's constant. In equation form,

$$mvr = \frac{nh}{2\pi} \qquad \text{for } n = 1, 2, 3, 4, \ldots \qquad (27.14)$$

The integer n is called a *quantum number*, specifically the **principal quantum number**. According to this assumption, the orbital speed of the electron (v) may have the values

$$v = \frac{nh}{2\pi mr}$$

Putting this expression for v into Eq. 27.12 and solving for r, we find that

$$\frac{ke^2}{r^2} = \frac{mv^2}{r} = \frac{m(nh/2\pi mr)^2}{r}$$

and

$$r = \left(\frac{h^2}{4\pi^2 ke^2 m}\right) n^2 \qquad (27.15)$$

Thus, electron orbits of only certain radii are possible as determined by the principle quantum number n. The energy for a particular orbit can be found by substituting this expression for r into Eq. 27.13:

$$E = -\frac{ke^2}{2r} = -\frac{ke^2}{2}\left(\frac{4\pi^2 ke^2 m}{h^2 n^2}\right)$$

and

$$E = -\left(\frac{2\pi^2 k^2 e^4 m}{h^2}\right)\frac{1}{n^2} \qquad (27.16)$$

The quantities in the parentheses on the right-hand sides of Eqs. 27.15 and 27.16 are constants and are easily evaluated. The important results are

$$r_n = 0.53\, n^2 \text{ Å} = 0.053\, n^2 \text{ nm}$$

$$E_n = \frac{-13.6}{n^2} \text{ eV} \qquad \text{for } n = 1, 2, 3, 4, \ldots \qquad (27.17)$$

where the radius is commonly expressed in angstroms or nanometers and the energy in electron volts. The n subscripts indicate that r and E are different for different values of n.

Example 27.3 Radius and Energy of a Bohr Orbit

Find the radius and energy of an electron in a hydrogen atom characterized by the principal quantum number $n = 2$.

Solution
For $n = 2$,

$$r_2 = 0.53 n^2 \text{ Å} = 0.53(2)^2 \text{ Å} = 2.12 \text{ Å} = 0.212 \text{ nm}$$

and

$$E_2 = \frac{-13.6}{n^2} \text{ eV} = \frac{-13.6}{(2)^2} \text{ eV} = -3.40 \text{ eV} \quad \blacksquare$$

However, there is still a classical problem with Bohr's theory. Classically, an accelerating electron radiates electromagnetic energy, and even in discrete circular orbits the electron would be centripetally accelerating. Thus, an orbiting electron would lose energy and spiral into the nucleus, like an Earth satellite in an orbit decaying because of frictional losses. This doesn't happen in the hydrogen atom, so Bohr had to make another nonclassical assumption. He postulated that the hydrogen electron does not radiate energy when it is in a bound, discrete orbit but does so only when it makes a transition to another orbit.

The possible allowed orbits of the hydrogen electron are commonly expressed in terms of energy levels as illustrated in Fig. 27.11. In this context, we tend to ignore the orbital motion and simply refer to the electron as being in a particular energy level or state. The electron, being bound to the nuclear proton, is in a potential energy well, much like a gravitational potential well (Chapter 7). The principal quantum number designates the energy level. The lowest energy level ($n = 1$) is called the **ground state**. The energy levels above the ground state are called **excited states**. For example, $n = 2$ is the first excited state and so on.

The electron is normally in the ground state and must be given energy to excite it, or to raise it up in the well to an excited state. The hydrogen electron can be excited only by discrete amounts. The energy levels are somewhat analogous to the rungs or steps on an energy ladder. Analogous to a person going up and down a real ladder, this must be done in discrete steps on the ladder rungs. Notice, however, that the energy rungs of the hydrogen atom are not evenly spaced. If enough energy is absorbed to raise the electron to the top of the energy well so that the electron is no longer bound, the atom is ionized. For example, to raise an electron from the ground state to $n = \infty$ requires 13.6 eV of

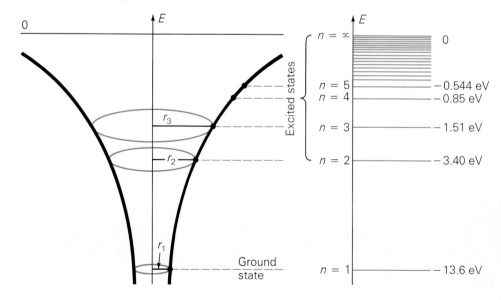

Figure 27.11 Orbits and energy levels
The Bohr theory predicts that the hydrogen electron can be in only certain orbits having discrete radii (drawn here in a $1/r$ potential well for illustration). Each allowed orbit has a corresponding energy. These are conveniently displayed as energy levels. The lowest energy level ($n = 1$) is called the ground state, and those above ($n > 1$) are called excited states.

energy. This is called the **binding energy** of the electron. If the electron is initially in an excited state, its binding energy is less (that is, equal to $-E_n$).

An electron does not remain in an excited state for long and decays or makes a transition to a lower energy level in a short time. The time an electron spends in an excited state is called the **lifetime** of the excited state. For many states, the lifetime is about 10^{-8} s. In making a transition to a lower state, energy is emitted as a photon of light (Fig. 27.12). The energy of the photon is equal to the energy difference of the levels, that is,

$$\Delta E = E_{n_i} - E_{n_f} = \frac{-13.6}{n_i^2} \text{ eV} - \left(\frac{-13.6}{n_f^2} \text{ eV}\right)$$

or $\quad \Delta E = 13.6\left(\frac{1}{n_f^2} - \frac{1}{n_i^2}\right) \text{ eV}$ \hfill (27.18)

where the subscripts of n_i and n_f refer to the initial and final states, respectively. Since $\Delta E = hf = hc/\lambda$, only photons of particular frequencies and wavelengths may be emitted. These correspond to the various discrete transitions. Solving for the wavelength λ, we get

$$\lambda = \frac{hc}{\Delta E} = \frac{12{,}400}{\Delta E} \text{ Å}$$ \hfill (29.19)

$$= \frac{1{,}240}{\Delta E} \text{ nm}$$

where ΔE is expressed in electron volts and the constant hc has been evaluated to give the wavelength in either angstroms or nanometers.

Example 27.4 Discrete Spectral Lines

What is the wavelength of the emitted light when excited electrons in hydrogen atoms make transitions from the $n = 3$ energy level to the $n = 2$ energy level?

Solution

Given: $n_i = 3$ \hfill *Find*: λ
$\qquad\quad\, n_f = 2$

The energy of the emitted photon is

$$E = 13.6\left(\frac{1}{n_i^2} - \frac{1}{n_f^2}\right) \Delta E = 13.6\left(\frac{1}{4} - \frac{1}{9}\right) \Delta E = 1.89 \text{ eV}$$

and, using Eq. 27.19,

$$\lambda = \frac{12{,}400}{\Delta E} \text{ Å} = \frac{12{,}400}{1.89} \text{ Å} = 6561 \text{ Å} \quad (= 656.1 \text{ nm})$$

which is light in the red end of the visible spectrum. ∎

Therefore, since the hydrogen electron can make transitions only between discrete energy levels, discrete wavelengths, producing spectral lines, are emitted (Fig. 27.13). Transitions may occur between two or more energy levels as the electron returns to the ground state. The wavelengths of the various transitions were computed and they were found to correspond to the experimentally observed spectral lines.

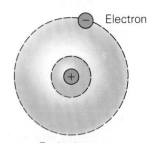
Electron

Excited atom

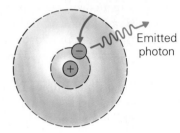
Emitted photon

De-excitation

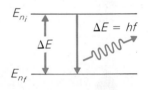

Figure 27.12 Electron transitions and photon emission When the electron in an excited atom decays or makes a transition to a lower orbit or energy level, a photon is emitted. The energy of the photon is equal to the energy difference in the energy levels.

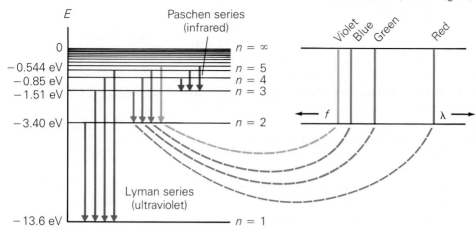

Figure 27.13 **Hydrogen spectrum**
Transitions may occur between two or more energy levels as the excited electron returns to the ground state. Transitions to the $n = 2$ state give spectral lines that are in the visible region of the spectrum and are called the Balmer series. Transitions to other levels give other series as shown.

Although Bohr's results gave exact agreement with the experiment for hydrogen, it could not successfully handle multielectron atoms. Bohr's theory was incomplete in the sense that it patched new quantum ideas into a basically classical framework. The theory contains many correct ideas, but a complete description of the atom did not come until quantum mechanics was developed (Chapter 28).

In general, a discrete line spectrum is characteristic of the atoms or molecules of a particular material, by which it may be identified spectroscopically.

27.5 A Quantum Success: The Laser

The development of the laser was a major technological success derived from theoretical atomic physics. Many scientific discoveries have been made accidentally, and many applications or inventions have come about by tri· l and error. Roentgen's discovery of X-rays and Edison's electric lamp are examples. However, the laser was developed on purpose. By using quantum physics, the principle of the laser was first predicted theoretically, then practically applied. (The word "laser" is an acronym and stands for *l*ight *a*mplification by *s*timulated *e*mission of *r*adiation. Stimulated emission will be discussed shortly.)

This relatively new invention has found widespread usefulness. Laser light beams are used in medicine to control bleeding, to weld torn retinas, and to treat skin cancer. In industry, lasers are used to drill holes in materials, to weld, and to cut cloth. They are used in surveying, and laser light is used to carry telephone and television signals over fiber optic cables. There are laser printers, laser pickups in video and compact disc players, and lasers in supermarket checkouts.

The concept of atomic energy levels allows us to understand this important technological instrument. Bohr's model of the atom explains spectral absorption and emission lines in terms of quantum jumps between energy levels. However, other questions arise. Given several lower energy levels, what determines the energy level to which an excited electron will decay? Also, how long does an electron stay in an excited state before making a transition? These questions were answered by the use of quantum mechanics, which is described in Chapter 28. The general results are that different transitions have different probabilities and, crucial to laser application, the time an electron stays in an excited state varies, depending on the atom.

Usually, an excited electron makes a transition to a lower energy level almost immediately, remaining in an excited state for only about 10^{-8} s. However, the lifetimes of an excited electron in some energy levels are appreciable. An energy level in which an excited electron remains for some time is called a **metastable state**. *Phosphorescent* materials are examples of substances with metastable states. These materials are used on luminous watch dials, toys, and items that "glow in the dark." When a phosphorescent material is exposed to light, atomic electrons are excited. Many of the excited electrons return to their normal state fairly soon, but there are metastable states in which electrons remain for seconds, minutes, even more than an hour. Consequently, the material, being made of billions of atoms, emits light or glows for some time.

A major consideration in laser operation is the emission process. As Fig. 27.14 shows, spontaneous absorption and emission of radiation occur between two energy levels. That is, a photon is absorbed and a photon is emitted. However, there is another possible emission process called **stimulated emission**. Einstein proposed this process in 1919. If a photon with an energy equal to an allowed transition strikes an atom in an excited state, it may stimulate the atom to make a transition and emit a photon. Two photons with the same frequency and phase then go off in the same direction.

The monochromatic (single-frequency), coherent (same-phase) and directional properties of laser light give it unique properties (Fig. 27.15). Light from sources like incandescent lamps is incoherent. The atoms emit randomly and at different frequencies (many different transitions). As a result, the light is out of phase or incoherent. Such beams spread out and become less intense. The properties of laser light allow the formation of a very tight beam, which with amplification can be very intense.

Stimulated emission was first applied to microwave radiation, and the instrument was called a "maser" (*microwave amplification by stimulated emission of radiation*).

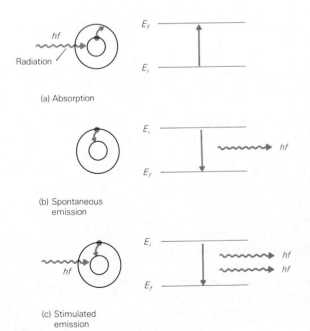

(a) Absorption

(b) Spontaneous emission

(c) Stimulated emission

Figure 27.14 **Photon absorption and emission** (a) In absorption, a photon is absorbed, and the electron is excited to a higher energy level. (b) After a short time, the electron spontaneously decays to a lower energy level with the emission of a photon. (c) If another photon with an energy equal to an allowed transition strikes an excited atom, stimulated emission occurs, and two photons with the same frequency and phase go off in the same direction as the incident photon.

Notice that stimulated emission is an amplification process—one photon in, two out. Of course, this is not a case of getting something for nothing, since the atom must initially be excited. However, it does provide a means to amplify light, analogous to the way in which we can amplify much lower frequency electronic signals. Ordinarily, when light passes through a material, photons are more likely to be absorbed than to give rise to stimulated emission. This is because there are normally many more atoms in the ground state than in excited states.* However, under special circumstances, it is possible to have more atoms in an excited state than in the ground state. This condition is known as a **population inversion**. In this case, there may be more stimulated emission than absorption, and we have amplification. *Population inversion and stimulated emission are two basic conditions necessary for laser action.*

There are several types of lasers, but the helium-neon (He-Ne) gas laser is most commonly used for classroom demonstrations and laboratory experiments. The reddish-pink light produced by the He-Ne laser can be observed in an optical scanning system at supermarket checkouts. The gas mixture is about 85% helium and 15% neon. Essentially, the helium is used for energizing and the neon for amplification. The gas mixture is subjected to a high dc voltage discharge rectified from a radio-frequency power supply, and helium atoms are excited [Fig. 27.16(a)]. This process is referred to as *pumping*. Energy is pumped

* A multielectron atom is referred to as being in the ground state when its electrons occupy the lowest energy levels.

(a) Coherent

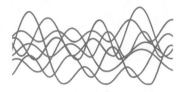

(b) Incoherent

Figure 27.15 **Coherent light** (a) Laser light is monochromatic (single frequency) and coherent, or in phase. (b) Light waves from sources in which atoms emit randomly are incoherent, or out of phase.

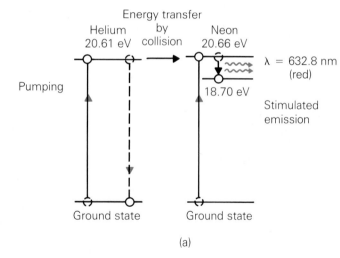

(a)

Figure 27.16 **A helium-neon laser**
(a) Helium atoms are excited or pumped into an excited state. Energy is transferred to neon atoms by collisions, and stimulated emission occurs between energy levels with the emission of red light. (b) Reflections from the end mirrors of the laser tube set up an intense beam parallel to the tube axis.

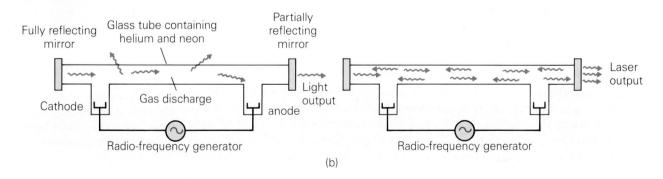

(b)

into the system, and the helium atoms are pumped into an excited state.

The excited state in the He atom has a relatively long lifetime (about 10^{-4} s) and has almost the same energy as an energy level in the Ne atom. There is a good chance that before an excited He atom spontaneously emits a photon, it will collide with a Ne atom. When this occurs, energy is transferred to the Ne atom. There is a possible transition to a nearby lower level, and this occurs spontaneously after a short time. However, the lifetime of the 20.66-eV neon state is relatively long, and the delay causes a piling up of neon atoms in this state, leading to a population inversion.

The stimulated emission and amplification of the light emitted by the neon atoms are caused by reflections from mirrors placed at the end of the laser tube [Fig. 27.16(b)]. Some excited Ne atoms spontaneously emit photons in all directions, and these photons induce stimulated emissions. In stimulated emission, the two photons leave the atom in the same direction as that of the incident photon. Photons traveling in the direction of the tube axis are reflected back through the tube by the end mirrors. The photons, in reflecting back and forth, cause more stimulated emissions, and an intense, highly directional, coherent, monochromatic beam develops along the tube axis. Part of the beam emerges through one of the end mirrors, which is only partially silvered.

We hear about the development of "Star Wars" laser weapons that would shoot down missiles, and lasers powerful enough to induce nuclear fusion (Chapter 30). Such lasers will require large power outputs. Carbon dioxide gas lasers can have a continuous 25,000-W output, which is capable of cutting steel. Pulsed ruby lasers may deliver as much as 50 MW, but only for a short fraction of a second. The He-Ne gas lasers used in college classrooms and laboratories have typical outputs of fractions of a milliwatt.

Some laser applications are shown in Fig. 27.17. It is important to remember that, although laser beams are used to weld detached retinas in the eye, viewing a laser beam directly can be quite hazardous. The beam is focused on the retina in a very small spot. If the beam intensity and the viewing time are sufficient, the photosensitive cells of the retina may be burned and destroyed.

You might own a laser yourself—see the Insight feature.

Another interesting application of laser light is the production of three-dimensional images in a process called **holography** (from the Greek word *holos*, meaning "whole"). The process does not use lenses as ordinary image-forming processes do, yet it re-creates the original scene in three dimensions. The key to holography is the coherent property of laser light, which gives the light waves a definite spatial relationship to each other.

In the photographic process of making a *hologram*, an arrangement such as that illustrated in Fig. 27.18 is used. Part of the light from the laser passes through a partial mirror to the object. The other part, or reference beam, is reflected to the film. The light incident on the object is reflected to the film, and it interferes with the reference beam. The film records the interference pattern of the two light beams, which essentially imprints on the film the information carried by the light wavefronts from the object.

When the film is developed, the interference pattern bears no resemblance to the object and appears as a meaningless pattern of light and dark areas. However, when the wavefront information is reconstructed by passing light through the film, a three-dimensional image is perceived. If part of the three-dimensional image is hidden from view, you can see it by moving your head to

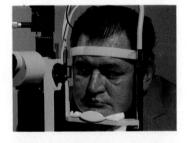

Figure 27.17 **Laser surgery** (a) An apparatus and laser beam used in detached retina operations, (b) A laser beam being used for glaucoma surgery.

The Compact Disc

The compact disc (CD) is a relatively new method of data storage, and most of us are familiar with it in terms of musical recordings. The CD system was introduced in 1980. There are now millions of CD players, and musical CD's outsell long-playing records. The disc itself is only 12 cm (5 in.) in diameter, but it can store more than 6 billion bits of information. This is the equivalent of more than 1000 floppy disks or over 275,000 pages of text, which could be stored on a CD for display on a television monitor. For audio use, a CD can store 74 minutes of music, and the sound reproduction is virtually unaffected by dust, scratches, or fingerprints on the disc.

The terms "digital" and "analog" are important to understanding how CD systems work. A digital format involves a discrete representation such as numbers. An analog format is a continuous one such as voltage or a sound wave (music). The CD has a digital format in that the recorded information is contained in pits impressed into its plastic surface. This is coated with a thin layer of aluminum to reflect the laser beam that "reads" the information.

The conversion of an analog signal, such as music, to a digital format is done by sampling and digital conversion. The continuous sound wave form is sampled 44,100 times per second. These samples (actually voltage pulses) are converted to digital numbers by an analog-to-digital converter.

The numbers are in the binary (base-2) system, which uses only 0 (zero) and 1 (one), or an off and on, to represent numbers. On the CD, each pit represents a binary 1, and the flat ("land") areas between the pits are read as binary 0's.

The pits are arranged in a spiral track like the grooves in a phonograph record, but the track is much narrower. The tracks on a CD are about 60 times closer together than the grooves on a long-playing phonograph record.

In the read-out system, a laser beam, which comes from a small semiconductor (solid state) laser, is applied from below the disc and focused on the aluminum coating of the track. The disc rotates at about 3.5 to 8 revolutions per second as the laser beam follows the spiral track. When the beam strikes a land area between two pits, it is reflected. When the beam spot overlaps a land area and a pit, the light reflected from the different areas interferes, and little or no light is reflected. The reflected light of varying intensity strikes a photodiode (solid state photocell), which reads the information. The signals are electronically converted back to analog form and into sound.*

Research is underway to produce an erasable disc so that CD's can be used to record sound as well as play it back.

---

* For a more complete and more technical discussion of the CD system, see an article entitled "The Compact Disc Digital Audio System," by T. Rossing, in *The Physics Teacher*, Vol. 25, No. 9, (December, 1987), pp. 556–562.

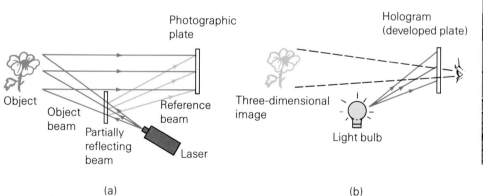

(a) (b)

Figure 27.18 Holography
(a) The coherent light from a laser is split into reference and object beams. The interference patterns of these beams are recorded on a photographic plate. (b) When the developed plate or hologram is illuminated, the viewer sees a three-dimensional image (c).

one side, just as you would to see a hidden part of a real object.

The technique of holography has many potential applications. Using coherent ultrasonic waves to form a hologram could provide a three-dimensional view of internal structures of the human body. Also, holograms made with X-ray lasers (yet to be developed) could provide better resolution of an indepth view of microscopic structures. Holographic three-dimensional television may be commonplace someday in the future.

Important Formulas

Wien's displacement law:

$$\lambda_{max} T = 2.90 \times 10^{-3} \text{ m-K}$$

Photon energy
(where Planck's constant is $h = 6.63 \times 10^{-34}$ J-s):

$$E = hf$$

K_{max} *and stopping potential in the photoelectric effect:*

$$K_{max} = e V_o$$

Energy conservation in the photoelectric effect:

$$hf = K + \phi$$

Energy conservation (with work function) in the photoelectric effect:

$$hf = K_{max} + \phi_o$$

Work function and threshold frequency in the photoelectric effect:

$$\phi_o = hf_o$$

Compton equation
(where $\lambda_c = h/m_o c^2 = 2.4 \times 10^{-12}$ m):

$$\lambda_1 - \lambda_o = \lambda_c (1 - \cos \theta)$$

(Quantum momentum: $p = h/\lambda$)

Bohr theory and orbit radius:

$$r_n = 0.53 \, n^2 \text{ Å} = 0.053 \, n^2 \text{ nm}$$

Bohr theory and electron energy:

$$E_n = \frac{-13.6}{n^2} \text{ eV} \qquad n = 1, 2, 3, \ldots$$

Bohr theory and transition energy (in eV):

$$\Delta E = 13.6 \left(\frac{1}{n_f^2} - \frac{1}{n_i^2} \right)$$

Bohr theory and photon wavelength:

$$\lambda = \frac{12,400}{\Delta E} \text{ Å} = \frac{1,240}{\Delta E} \text{ nm}$$

Questions

Quantization: Planck's Hypothesis

1. (a) What is thermal radiation? Why is a spectrum of frequencies emitted? (b) Do all bodies emit thermal radiation? Explain.

2. If a blackbody absorbs all the radiation incident on it, why doesn't its temperature increase without limit?

3. Does a blackbody at 200 K emit twice as much total radiation as when its temperature is 100 K? Explain.

4. Explain what is meant by the ultraviolet catastrophe.

5. Why was Planck's hypothesis a radical departure from classical theory?

Quanta of Light: Photons and the Photoelectric Effect

6. (a) Does a quantum of red light or a quantum of blue light have more energy? (b) Does a quantum of ultraviolet light have more energy than a quantum of visible light?

7. How is the maximum kinetic energy of photoelectrons determined? Does K_{max} depend on the light intensity? Explain.

8. What is meant by the cutoff frequency for photoemission?

9. What are the observed features of the photoelectric effect that classical theory fails to explain?

10. How did Einstein's explanation of the photoelectric effect resolve the discrepancies between experiment and theory?

11. Will increasing the intensity of light having frequencies below the cutoff frequency cause the emission of photoelectrons? Explain.

12. If electromagnetic radiation is made up of quanta, why don't we detect the discrete packets of energy, for example, when listening to a radio with the radio signal arriving in packets?

Quantum "Particles": The Compton Effect

13. Does classical theory predict a change in the wavelength of scattered radiation? Explain.

14. How is the momentum of a photon determined?

15. What is the maximum wavelength shift predicted by the Compton equation?

The Bohr Theory of the Hydrogen Atom

16. Why are continuous spectra of light emitted from incandescent objects and line spectra emitted from gas lischarge tubes?

17. What is the critical quantum assumption of the Bohr theory, and what are the resulting restrictions on the hydrogen electron?

18. How does the Bohr theory disagree with classical theory regarding the orbital acceleration of the hydrogen electron?

19. The binding energy is the amount of energy needed to ionize or remove an electron from an atom. What is the binding energy of a hydrogen electron in (a) the ground state and (b) the second excited state?

20. What happens if a hydrogen atom absorbs (a) more than 13.6 eV of energy and (b) less than 13.6 eV of energy?

21. How do we know that the Sun is made up of certain elements?

Problems

27.1 Quantization: Planck's Hypothesis

■■1. What is the wavelength and frequency of the most intense radiation component from a blackbody with a temperature of $0\,°C$?

■■2. Find the approximate temperature of a blue star that emits light with a wavelength of 450 nm.

■■3. Sketch a graph of temperature versus (a) frequency and (b) wavelength of the most intense radiation component of blackbody radiation.

■■4. What is the frequency of the most intense spectral component from a blackbody at room temperature $(20\,°C)$?

■■5. The walls of a blackbody cavity are at a temperature of $227\,°C$. What is the frequency of the radiation of maximum intensity?

■■6. What would be the approximate temperature of a blackbody if it appeared to be (a) red and (b) blue?

■■7. What would be the change in the frequency of the most intense spectral component of a blackbody that was initially at $200\,°C$ and was heated to $500\,°C$?

■■8. What is the energy of a thermal oscillator in a blackbody with a temperature of $212\,°F$?

■■■9. The temperature of a blackbody is 1000 K. If the total intensity of the emitted radiation of 1.0 W were due to the most intense frequency component, how many quanta would be emitted per second per square meter?

27.2 Quanta of Light: Photons and the Photoelectric Effect

■10. What is the energy of a quantum of light with a frequency of 6.0×10^{14} Hz?

■11. A photon has an energy of 3.3×10^{-15} J. Of what type of light is this photon?

■■12. Which has more energy and how many times more: a quantum of violet light or a quantum of red light?

■■13. How many quanta of red light $(\lambda = 700$ nm) would it take to have 1.0 J of energy?

■■14. A photon of ultraviolet light has a wavelength of 300 nm. How much energy does the photon have in (a) joules and (b) electron volts?

■■15. Light with an intensity I produces a photocurrent I_p. If the intensity of the light is doubled, how is the current affected?

■■16. Light with an intensity I produces a photoelectric emission, and the photoelectron has a kinetic energy of 5.0 eV. If the intensity of the light is tripled, what is the kinetic energy of emitted photoelectrons?

■■17. A beam of monochromatic light with a frequency of 6.0×10^{14} Hz strikes a photoelectric material that has a work function of 2.7×10^{-19} J. (a) What is the maximum kinetic energy of the emitted photoelectrons? (b) What is the threshold frequency for the photoemission of electrons for the material?

■■18. The work function of a particular metal is 5.0 eV. (a) What is the frequency of light that causes electrons to be emitted with a maximum kinetic energy of 2.0×10^{-19} J? (b) What is the stopping potential of the electrons?

■■19. A beam of monochromatic light with a wavelength of 450 nm strikes a photoelectric material for which the threshold wavelength is 650 nm. Will photoelectrons be emitted from the material? If so, what is the maximum kinetic energy of the electrons?

■■20. Light with a wavelength of 400 nm causes electrons to be emitted from a photoelectric material having kinetic energies up to 3.2×10^{-19} J. (a) What is the stopping potential of the electrons? (b) What is the threshold frequency of the material?

■■21. Blue light with a wavelength of 480 nm is observed to cause the emission of photoelectrons with a maximum kinetic energy of 1.0×10^{-19} J. Will red light ($\lambda = 700$ nm) cause the emission of photoelectrons from the material? Justify your answer mathematically.

■■22. The ionization energy of a hydrogen atom is 13.6 eV. Would the absorption of a photon having a frequency of 7.00×10^{15} Hz cause a hydrogen atom to be ionized? If so, what would be the kinetic energy of the emitted electron?

■■■23. When the surface of a particular material is illuminated with monochromatic light of various frequencies, the stopping potentials for the photoelectrons are determined to be:
Frequency (Hz)
7.0×10^{15} 5.5×10^{15} 4.0×10^{15} 3.5×10^{15}
Stopping potential (V)
12.6   10.3   4.10   1.97
Plot these data and determine the values of Planck's constant and the work function of the metal.

27.3   Quantum "Particles": The Compton Effect

■24. What is the momentum of a photon having a frequency of 6.5×10^{20} Hz?

■25. A red line in the hydrogen spectrum has a wavelength of 650 nm. Calculate the momentum of a photon of this light.

■26. Calculate the maximum wavelength shift for Compton scattering.

■27. At what scattering angle would a wavelength shift of 3.2×10^{-11} cm be observed for Compton scattering?

■28. What is the change in wavelength when monochromatic X-rays are scattered through an angle of 15°?

■■29. A photon with an energy of 10 keV is scattered by a free electron. If the electron recoils with a kinetic energy of 5.0 keV, what is the wavelength of the scattered photon?

■■30. A monochromatic beam of X-rays with a wavelength of 2.80×10^{-10} m is scattered through a metal foil. What is the wavelength of the scattered X-rays observed at an angle of 60° from the direction of the incident beam?

■■31. What is the incident energy of a beam of monochromatic X-rays that are observed to have a wavelength of 4.5×10^{-10} m at a Compton scattering angle of 90°?

■■32. For what scattering angle would the wavelength shift for Compton scattering be 30 percent of the Compton wavelength?

27.4   The Bohr Theory of the Hydrogen Atom

■■33. Use the Bohr theory to find the value of the Rydberg constant for the empirical formula (Eq. 27.11).

■■34. Give the radius of the electron orbit of the hydrogen atom for each of the following states: (a) $n = 3$, (b) $n = 6$, (c) $n = 10$.

■■35. Find the energy of the hydrogen electron for each of the following states: (a) $n = 3$, (b) $n = 6$, (c) $n = 10$.

■■36. Find the binding energy of the hydrogen electron for each of the following states: (a) $n = 1$, (b) $n = 5$, (c) $n = 20$.

■■37. Which of the following energies do not correspond to the energy levels of the electron in the hydrogen atom? (a) -0.278 eV, (b) -0.206 eV, (c) -0.190 eV, (d) -0.168 eV.

■■38. What is the energy of a photon emitted when a hydrogen electron makes a transition from the $n = 2$ state to the ground state?

■■39. What are the energies of the photons required to excite a hydrogen electron in the ground state to states (a) $n = 4$ and (b) $n = 7$?

■■40. What is the wavelength of the photon emitted for each of the following transitions of an electron in a hydrogen atom? (a) $n = 6$ to $n = 2$, (b) $n = 3$ to $n = 1$, (c) $n = 4$ to $n = 3$.

■■■41. A hydrogen electron in the ground state is excited to the $n = 5$ energy level. The electron makes a transition to the $n = 2$ level before returning to the ground state. (a) What are the wavelengths of the emitted photons? (b) Would the light emitted from a group of such hydrogen atoms be visible?

■■■42. For which one of the following transitions of a hydrogen electron is the photon of greatest energy emitted? (a) $n = 5$ to $n = 3$, (b) $n = 6$ to $n = 2$, (c) $n = 2$ to $n = 1$. Justify your answer mathematically.

■■■43. Determine the number of transitions of the electron in a hydrogen atom that will result in the emission of light in the visible region of the spectrum (400–700 nm).

■■■44. A hydrogen atom absorbs a photon of wavelength of 434.1 nm. (a) How much energy did the atom absorb? (b) What were the initial and final states of the hydrogen electron?

Additional Problems

45. A 100-W light bulb gives off 5.0% of its energy as visible light. How many photons of visible light are given off in 1 minute? (Use an average wavelength of 550 nm.)

46. A proton has a kinetic energy of 5.0 MeV. Calculate the energy of a photon that has the same momentum as the proton.

47. Find the energy needed to ionize a hydrogen atom whose electron is in the $n = 4$ state.

48. A metal surface has a work function of 1.5 eV. Find the maximum kinetic energy of an emitted photoelectron if the metal is illuminated by light having a wavelength of 550 nm.

49. The threshold wavelength for emission from a metallic surface is 500 nm. Calculate the maximum speed of emitted photoelectrons for light having a wavelength of (a) 400 nm and (b) 600 nm.

50. Find the energy required to excite a hydrogen electron from (a) the ground state to $n = 4$ and (b) $n = 2$ to $n = 3$.

51. The work function of a photoelectric material is

3.5 eV. If the material is illuminated with monochromatic light ($\lambda = 300$ nm), (a) find the stopping potential of emitted photoelectrons. (b) What is the cutoff frequency?

52. What is the longest wavelength of light that can cause the release of electrons from a metal that has a work function of 2.28 eV?

53. What is the frequency of a photon that would excite the electron of a hydrogen atom from (a) $n = 2$ to $n = 5$ and (b) $n = 2$ to $n = \infty$?

54. A metal with a work function of 2.4 eV is illuminated with a beam of monochromatic light. It is found that the stopping potential for the emitted electrons is 2.8 V. What is the wavelength of the light?

55. A muon or μ meson is a particle that has the same charge as an electron but a mass 207 times that of an electron. Using a muon to replace the electron in a "hydrogen" (muonium) atom is possible. For such an atom, (a) what would be the muon's energies and orbital radii for the $n = 1$ and $n = 2$ states? (b) What is the wavelength of the photon emitted for a muon transition from $n = 2$ to $n = 1$?

56. The Compton effect also occurs for protons. (a) What is the value of the Compton wavelength for a proton? (b) Compare the wavelength shifts for maximum electron and proton scatterings for incident photons of equal energies.

57. Fig. 27.19 shows a graph of stopping potential versus frequency for a photoelectric material. Determine (a) Planck's constant and (b) the work function of the material from the graph data.

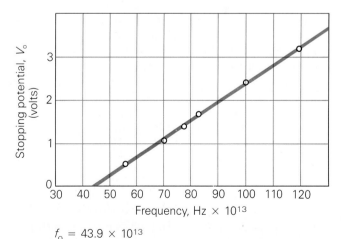

$f_0 = 43.9 \times 10^{13}$

Figure 27.19 **Stopping potential versus frequency** See Problem 57.

Quantum Mechanics

28

The initial successes of the quantum theory were impressive, but they also created a perplexing situation. They seemed to show that light (that is, electromagnetic radiation) has a dual nature. On one hand, in many experiments, light exhibits classical wave behavior. On the other hand, explanations of certain phenomena require light to have a quantum or particle nature. It was as though each theory worked well for its own experiments. Moreover, as you learned in Chapter 27, the classical and quantum theories of light are mutually contradictory in some instances. For example, classical wave theory predicts a time delay in the emission of photoelectrons, while the quantum theory does not.

Around 1925, a new kind of physics based on the synthesis of wave and quantum ideas was introduced. This new theory, called **quantum mechanics**, attempts to combine the wave-particle duality into a single consistent theory. It revolutionized scientific thought and provided the basis of our present understanding of microscopic (and submicroscopic) phenomena.

Quantum mechanics deals mainly with the minute aspects of the world of atoms. It has replaced the mechanistic view of the universe, in which all things moved according to exact natural laws, with a new concept of probability. All physical observations are now accepted as being to some degree uncertain. Even so, when quantum mechanics is applied to macroscopic phenomena, it must reproduce the results of classical physics in order to be a consistent theory. This means that there is no need to abandon the laws and principles of classical physics, since they provide an accurate description of everyday phenomena.

In a sense, the two theories complement each other. A more complete understanding of phenomena is obtained if the two theories are used in a complementary manner. It is more convenient to describe macroscopic phenomena by classical physics. But when dealing with phenomena in the atomic and subatomic worlds, we use quantum mechanics. One could make an analogy here. When objects are moving with speeds that are appreciable fractions of the speed of light, we must use the theory of relativity. But when things slow down, classical principles may be applied.

The application of quantum mechanics to specific phenomena is mathematically detailed and beyond the scope of the text. In this chapter we will generally discuss some of the basic ideas of quantum mechanics to see how they describe waves and particles. Also, some of the results of this new approach to describing phenomena will be presented.

28.1 Matter Waves: The de Broglie Hypothesis

Since a photon travels at the speed of light, we treat it relativistically as a massless particle with an energy $E = pc$ (see Section 27.3). The momentum of a photon is then $p = E/c$. The energy may also be written $E = hf = hc/\lambda$, so the momentum of a photon is related to its wavelength by

$$p = \frac{E}{c} = \frac{hf}{c} = \frac{h}{\lambda} \qquad (28.1)$$

Thus, the energy in electromagnetic waves of wavelength λ is carried by photon "particles" having momenta of h/λ.

Since nature exhibits a great deal of symmetry, the French physicist Louis de Broglie (Fig. 28.1) thought that there might be a wave-particle symmetry. That is, if light sometimes behaves as a wave and sometimes as a particle, perhaps material particles, such as electrons, also had wave properties. In 1924, de Broglie put forth a hypothesis that a moving particle has a wave associated with it. He proposed that the wavelength of a material particle was related to the particle's momentum by an equation similar to that for a photon (Eq. 28.1). More specifically, the **de Broglie hypothesis** states:

Whenever a particle has a momentum p, its motion is associated with (that is, "guided by") a wave whose wavelength is

$$\lambda = \frac{h}{p} = \frac{h}{mv} \qquad (28.2)$$

The waves associated with moving particles were called **matter waves** or, more commonly, **de Broglie waves**. One might say that an electromagnetic wave is the de Broglie wave for a photon, but the de Broglie waves associated with particles such as electrons and protons are *not* electromagnetic waves.

De Broglie's hypothesis met with a great deal of skepticism. The idea that the motions of photons were somehow governed by the electromagnetic wave properties of light did not seem unreasonable. But to extend this idea and say that the motion of a mass particle is governed by the wave properties of an associated *pilot wave* was difficult to accept. Moreover, there was no evidence that particles had wave properties.

In support of his hypothesis, de Broglie showed how it could give an interpretation of the quantization of the angular momentum in Bohr's theory of the hydrogen atom (Chapter 27). For a free particle, the associated wave would have to be a traveling wave. However, the bound electron of a hydrogen atom travels repeatedly in discrete circular orbits according to the Bohr theory. The associated matter wave would then be expected to be a standing wave. That is, only a certain number of wavelengths could fit into a given orbit (circular boundary condition), similar to the linear case of a standing wave in a stretched string fixed at both ends (see Fig. 28.2).

Figure 28.1 **Louis de Broglie (1892–1987)**
Prince Louis de Broglie, a French nobleman, studied medieval history at the Sorbonne. He enlisted in the French army and was a radioman in World War I. The experience gave him an interest in physics. His theory of matter waves won him a Nobel Prize in 1929.

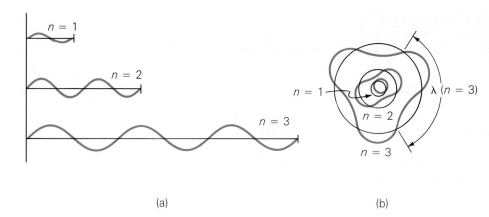

(a)   (b)

Figure 28.2 **De Broglie waves and Bohr orbits** (a) Similar to standing waves in a stretched string, de Broglie waves form circular standing waves on the circumferences of the Bohr orbits (b). Notice that the number of wavelengths in an orbit is equal to the principal quantum number of the orbit.

The circumference of a Bohr orbit of radius r_n is $2\pi r_n$, so this length would be equal to an integral number of wavelengths

$$2\pi r_n = n\lambda \qquad \text{for } n = 1, 2, 3, \ldots$$

Using the de Broglie equation $\lambda = h/mv$ for the wavelength, we have

$$2\pi r_n = \frac{nh}{mv} \qquad \text{or} \qquad mvr_n = \frac{nh}{2\pi}$$

Thus, the angular momentum is quantized just as Bohr proposed. For orbits other than those allowed by the Bohr theory, the de Broglie wave would not match or close on itself. This would give rise to destructive interference, and the wave would die out or have zero amplitude. The amplitude of the de Broglie wave must then be related in some way to the location of the electron.

Example 28.1   Short de Broglie Waves
A pitcher throws a fast ball with a speed of 40 m/s. If the mass of the ball is 0.15 kg, what is the wavelength of the de Broglie wave associated with the moving ball?

Solution
Given: $m = 0.15$ kg   *Find*: λ
         $v = 40$ m/s

Using the de Broglie hypothesis (Eq. 28.2),

$$\lambda = \frac{h}{mv} = \frac{6.63 \times 10^{-34} \text{ J} \cdot \text{s}}{(0.15 \text{ kg})(40 \text{ m/s})} = 1.1 \times 10^{-34} \text{ m} \quad \blacksquare$$

Notice the small wavelength for the baseball in the example. It is little wonder that we don't observe any effects of such matter waves. Recall that wave effects such as diffraction interference take place only when the size of the objects or slits is comparable to the wavelength, and $\lambda \cong 10^{-34}$ m is just too small. We cannot detect the wave properties of ordinary objects. For a 1000-kg jet plane flying near the speed of sound, say 350 m/s, the wavelength of the de Broglie wave is still on the order of 10^{-29} m.

However, particles with very small masses traveling at relatively low speeds are another matter, as the following example shows.

Example 28.2   Not-So-Short de Broglie Waves

What is the de Broglie wavelength of the wave associated with an electron that has been accelerated through a potential of 50.0 V?

Solution

Given: $V = 50.0$ V   *Find*: λ

The work done in accelerating the electron is $W = eV$ and is equal to its kinetic energy. Expressing this in terms of momentum, we have

$$\tfrac{1}{2}mv^2 = \frac{p^2}{2m} = eV \quad \text{or} \quad p = (2meV)^{\frac{1}{2}}$$

The de Broglie wavelength is then

$$\lambda = \frac{h}{p} = \frac{h}{(2meV)^{\frac{1}{2}}}$$

Putting in the values of h, e, and m leads to a convenient general formula for the de Broglie wavelength of an electron accelerated through a potential difference:

$$\lambda = \left(\frac{150}{V}\right)^{\frac{1}{2}} \times 10^{-10} \text{ m} = \left(\frac{150}{V}\right)^{\frac{1}{2}} \text{ Å} \qquad (28.3)$$

where V is in volts. For $V = 50.0$ V,

$$\lambda = \left(\frac{150}{50.0}\right)^{\frac{1}{2}} \times 10^{-10} \text{ m} = 1.73 \times 10^{-10} \text{ m} = 1.73 \text{ Å} \quad \blacksquare$$

Although wavelengths on the order of 10^{-10} m are extremely small, such waves can at least be detected. Slit widths on this order cannot be fabricated. However, recall from Chapter 24 that nature provides such slits in the form of crystal lattices.

In 1927, two physicists in the United States, C. J. Davisson and L. H. Germer, used a crystal to diffract a beam of electrons, thereby demonstrating a wavelike property of particles. A single crystal of nickel was used in the experiment. The crystal was cut to expose a spacing of $d = 2.15$ Å between the lattice planes. When a beam of electrons was directed normally on the crystal face, a maximum in the intensity of the scattered electrons was observed at an angle of 50° relative to the surface normal (Fig. 28.3). The scattering was most intense for an accelerating potential of 54 V.

According to wave theory, the first-order maximum should be observed at an angle given by (see Section 24.3):

$$d \sin \theta = \lambda$$

This condition requires a wavelength of

$$\lambda = d \sin \theta = (2.15 \text{ Å}) \sin 50° = 1.65 \text{ Å}$$

Using the general formula from the previous example (Eq. 28.3) to determine the de Broglie wavelength of the electrons, we find

$$\lambda = \left(\frac{150}{V}\right)^{\frac{1}{2}} = \left(\frac{150}{54}\right)^{\frac{1}{2}} = 1.67 \text{ Å}$$

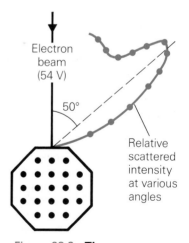

Figure 28.3   The Davisson-Germer experiment
When a beam of 54-eV electrons is incident on the face of a nickel crystal, a maximum in the scattering intensity is observed at an angle of 50°.

The agreement of the wavelengths was excellent within the limits of experimental accuracy, and the Davisson-Germer experiment gave convincing proof of the validity of de Broglie's hypothesis.

Another experiment carried out by G. P. Thomson in Great Britain in the same year added further proof. Thomson passed a beam of energetic electrons through a thin metal foil. The diffraction pattern of the electrons was the same as that of X-rays. A comparison of such patterns as shown in Fig. 28.4 leaves little doubt that particles exhibit wavelike properties. An example of a practical application of this is given in the Insight feature.

INSIGHT

The Electron Microscope

The de Broglie hypothesis helped to set the stage for the development of an important practical application of electron beams—the **electron microscope**. As you can see from Example 28.2, accelerated electrons have very short wavelengths. With such short wavelengths, greater resolution and magnification can be obtained (Chapter 25), so the resolving power of standard electron microscopes is on the order of a few nanometers. Technological developments made in the 1920s, in particular the focusing of electron beams by magnetic coils, permitted the construction of the first electron microscope in Germany in 1931.

In a *transmission electron microscope*, an electron beam is directed onto a very thin specimen. Different numbers of electrons pass through different parts of the specimen depending on its structure. The transmitted beam is then brought to focus by a magnetic objective lens as illustrated in Fig. 1. The general components of electron and light microscopes are analogous, but an electron microscope is housed in a high-vacuum chamber so that the electrons are not deflected by air molecules. As a result, an electron microscope looks nothing like a light microscope (Fig. 2). Magnifications up to 100,000X can be achieved with an electron microscope, whereas a light microscope is limited to a magnification of about 2000X.

Another difference is that the final lens in an electron microscope, called the projector lens, has to project a real image onto a fluorescent screen or photographic film, since the eye cannot perceive an electron image directly. Specimens for transmission electron microscopy must be very thin. Special techniques allow the preparation of specimen sections as thin as 100 Å to 200 Å (only about 100 atoms thick).

The surfaces of thicker objects may be examined by the reflection of the electron beam from the surface. This is done with the more recently developed *scanning electron microscope*. A beam spot is scanned across the specimen by means of deflecting coils, much as is done in a television tube. Surface irregularities cause directional variations in the

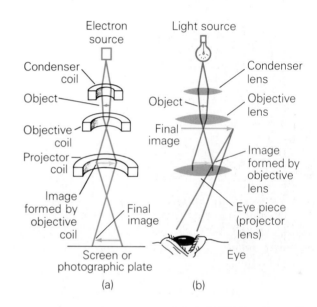

Figure 1   Electron and light microscopes
A comparison of the elements of (a) an electron microscope and (b) a compound light microscope. The light microscope is drawn upside down for a better comparison.

Figure 2   An electron microscope

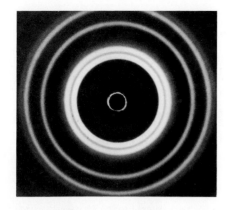 

Figure 28.4 **Diffraction patterns**
A comparison of an X-ray diffraction pattern (a) and an electron diffraction pattern (b) leaves little doubt that electrons exhibit wave like properties.

INSIGHT

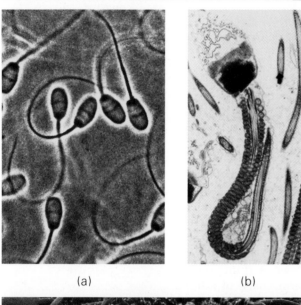

(a)   (b)

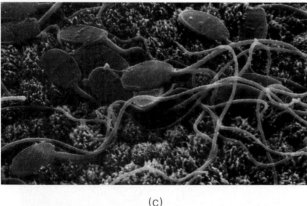

(c)

Figure 3 **A comparison of microscopic images**
Rabbit sperm cells as seen through (a) a light microscope, (b) a transmission electron microscope, and (c) a scanning electron microscope.

intensity of the reflected electrons, which gives contrast to the image. The specimens have to be coated with a thin layer of metal (such as gold or aluminum) to make them conducting. Otherwise, they would charge up nonuniformly from the electron beam and distort the image. Through such techniques, an electron microscope gives remarkable pictures such as those shown in Fig. 3(b,c).

A more recent microscopic instrument is the *scanning tunneling microscope*, which was developed in the 1970s and 1980s. It bears little resemblance to the scanning electron microscope. The scanning tunneling microscope uses a quantum phenomenon called *vacuum tunneling*. The tip of a metal needle probe is moved across the contours of a metalized specimen surface. Applying a small voltage between the probe and the surface then causes electrons to tunnel through the vacuum. The tunneling current is extremely sensitive to the separation of the needle tip and the surface. As a result, when the probe is scanned across the sample, surface features as small as atoms show up as variations in the tunneling current (Fig. 4).

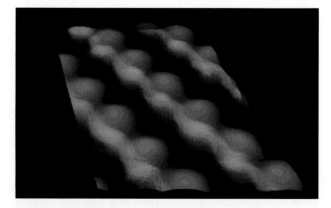

Figure 4 **Scanning tunneling microscope images**
Combined color images of the surface of a GaAs (gallium-arsenic) semiconductor compound. The Ga atoms are shown in blue, the As atoms in red.

28.2 The Schrödinger Wave Equation

De Broglie's hypothesis predicts that moving particles have associated waves that somehow govern or describe their behavior. However, it does not tell us the form of these waves. To have a useful theory, we need an equation that will give us the form of the matter waves, in particular the wave form of a particle moving under the influence of a force. Also, we must know how these waves govern particle motion. In 1926, Erwin Schrödinger, an Austrian physicist (Fig. 28.5), presented a general equation that describes the de Broglie matter waves.

Recall the conservation of energy that we discussed in Chapter 5. For a conservative mechanical system, we had

$$E = K + U = K_o + U_o = E_o$$

or $\quad \frac{1}{2}mv^2 + U = \frac{1}{2}mv_o^2 + U_o$

Now Schrödinger proposed an almost identical equation for de Broglie matter waves known as the **Schrödinger wave equation**

$$(K + U)\psi = E\psi \qquad (28.4)$$

Here K, U, and E are the kinetic, potential, and total energies, respectively, of the particle under consideration. The term denoted by ψ (Greek letter psi, pronounced "sigh") is called the **wave function** and describes the wave as functions of time and space.

In the early development of quantum mechanics (sometimes called wave mechanics), it was not clear how ψ should be interpreted. After much thought and investigation, it was concluded that ψ^2 (the wave function squared) represents the probability of finding the particle at a certain position and time (actually, the absolute value squared, $|\psi|^2$).

This interpretation involves the amplitude or displacement of ψ. Recall from Chapter 14 that the energy or intensity of a classical wave (function) is proportional to the square of its amplitude. Similarly, the intensity of a light wave is proportional to E^2, where E is the electric field amplitude. Looking at this in terms of "particle" photons, the *intensity* of a light beam is proportional to the number (n) of photons in the beam, so $n \propto E^2$. That is, the number of protons is proportional to the square of the electric field amplitude of the wave.

The wave function ψ is interpreted in an analogous manner. The wave function generally varies in magnitude in space and time. If ψ describes an electron beam, then ψ^2 will be proportional to the number of electrons that may be expected to be found in a small volume around a point at some time. However, when there are only a few electrons or a single particle, ψ^2 represents the probability of finding an electron at the point. Thus, in quantum mechanics the square of the wave function is proportional to the intensity of the probability of finding a particle in space and time, or the *probability density*.

Figure 28.5 **Erwin Schrödinger (1887–1961)** Schrödinger was one of the founders of quantum mechanics. Born in Austria, he served in World War I, afterward becoming a physics professor in Germany. In 1925 he wrote a paper that mathematically treated moving particles to be governed by waves or wave functions. He left Germany in 1933 when the Nazis came to power. Later he was a professor at the University of Dublin in Ireland.

Example 28.3 Wave Function for Particle in a Box

To help understand the idea of a wave function and probability density, consider a particle constrained to move in one dimension in a box, for example, a bead on a wire (Fig. 28.6). It is assumed that no forces act on the particle as long

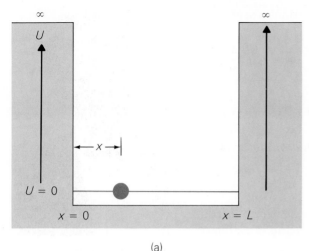

(a)

Figure 28.6 **Schrödinger wave functions**
(a) A particle confined to a one-dimensional rigid box may be considered to be in an infinitely deep potential well.
(b) The wave functions of the Schrödinger equation for the particle are sinusoidal with nodes at the boundaries.
(c) The ψ^2 probability densities for the first three quantum numbers.

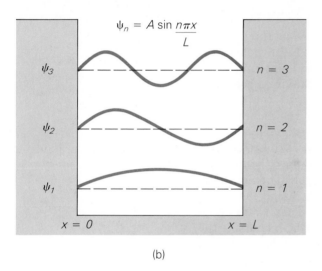

(b)

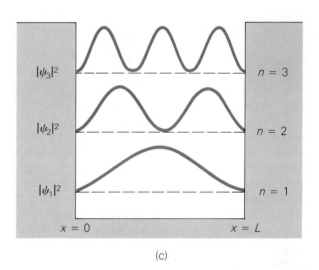

(c)

as it is in the interval $0 < x < L$ and that it hits the perfectly rigid walls if it tries to go outside the box. This situation is expressed by saying that the potential energy of a particle is zero inside the box and is infinite for $x < 0$ and $x > L$, or outside the box. That is, since the particle cannot get out of the box, it may be thought of as being at the bottom of an *infinite* potential well with $U = 0$.

With the particle never outside the box, $\psi = 0$ in this region. Inside the box, the particle can be anywhere on the X axis, but its de Broglie wave must have nodes ($\psi = 0$) at the boundaries. This situation is analogous to waves in a stretched string fixed at both ends (Chapter 14). Thus, the spatial wave function of the particle in the box can be expressed in sinusoidal form. The nodal conditions at $x = 0$ and $x = L$ require that only an integral number of half de Broglie wavelengths fit in the box (Fig. 28.6), that is,

$$\frac{n\lambda}{2} = L \quad \text{or} \quad \lambda = \frac{2L}{n} \quad \text{for } n = 1, 2, 3, \ldots \tag{28.5}$$

The wave function solutions satisfying these boundary conditions are given by

$$\psi_n = A \sin \frac{2\pi x}{\lambda} = A \sin \frac{n\pi x}{L} \qquad \text{for } n = 1, 2, 3, \ldots \qquad (28.6)$$

From the de Broglie equation and Eq. 28.5,

$$p = \frac{h}{\lambda} = \frac{nh}{2L}$$

The kinetic energy, which is the total energy since $U = 0$, has values given by

$$K = E = \frac{p^2}{2m} = \frac{n^2 h^2}{8mL^2} \qquad (28.7)$$

where m is the mass of the particle. The expression shows that only certain energy states are possible for the bound particle, and we have the same type of result as in the Bohr theory. ■

The Schrödinger equation is commonly used to find the wave functions of electrons in a bound state, for example, in atoms. As with the Bohr theory, it is convenient to view an electron bound to the nucleus to be in a potential well. However, in this case, the potential well is not infinite as is that for a bead on a wire, and the electron could be excited out of the well.

Some amazing predictions are given by solutions to the Schrödinger equation for particles in potential wells, even for a simple one-dimensional well as illustrated in Fig. 28.7. If the total energy E of a particle is less than the value of the potential energy at the top of the well, the particle is trapped (bound) and classically doesn't have enough energy to escape from the well. The Schrödinger wave equation predicts that the wave functions are sinusoidal inside the well *but* that there is an exponential tail outside the well. That is, quantum mechanics predicts that the particle has a probability (ψ^2) of being outside the energy well, where according to classical physics it does not have enough energy to be! This result is important in nuclear physics, as you will learn in the next chapter.

The idea of the *probability* of a particle's being in a location is quite different from classical determinism, which considered it possible to predict and determine the exact locations of particles. However, quantum mechanics did give a reasonable interpretation of the unsettling prediction that an electron could be outside a potential well in a classically forbidden region. The probability of this occurring (ψ^2 in this region) was so small that one would not expect to observe the electron outside its potential well. But keep in mind that quantum mechanics predicts that there is a *finite* probability of its being there.

The interpretation of ψ^2 as the probability of finding a particle at a particular place and of the wave function solutions to the Schrödinger equation for the hydrogen atom altered the idea that the hydrogen electron could be found only in orbits at discrete distances from the nucleus. The radial wave functions for each energy level were found to be finite for all distances from the nucleus. Then quantum mechanically, there is a finite probability of finding the electron in a given energy level at any distance from the nucleus. For example, the relative probability that an electron in the ground state is at a given radius is shown in Fig. 28.8.

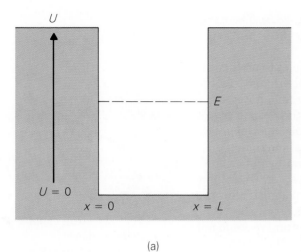

(a)

Figure 28.7 **Nonclassical prediction**
(a) A particle in a bound state, such as an electron in an atom, is in a finite potential well. With less total energy E than the value of the potential energy at the top of the well, the particle cannot classically be outside the well. But (b) the wave functions of the Schrödinger equation and (c) the probability densities have exponential tails *outside* the well, predicting that the particle might be found there.

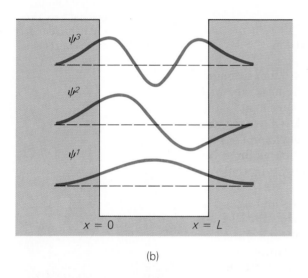

(b)

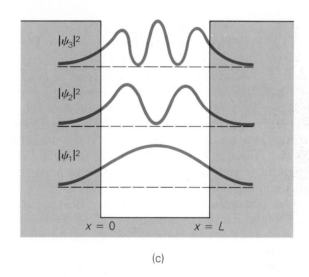

(c)

The maximum probability coincides with the Bohr radius of 0.53 Å, but it is possible (that is, there is a finite probability) that the electron could be at various distances from the nucleus. The wave function exists for distances beyond 2 Å, but there is little chance (probability) of finding an electron in the ground state beyond this distance. The probability density distribution gives

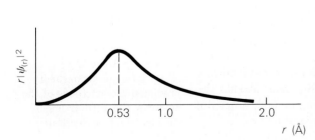

Figure 28.8 **Probability distribution for the hydrogen atom**
The radial possibility density $\psi^2(r)$ is shown as a function of r for an electron in the ground state of a hydrogen atom. The maximum probability is at 0.53 Å, which corresponds to the first Bohr radius.

rise to the idea of an electron cloud around the nucleus (Fig. 28.9). The cloud density reflects the probability density that the electron is in a particular region.

Thus, quantum mechanics uses the wave functions of particles to predict the probability of particle phenomena. This also applies to light and photon "particles." For example, the square of a classical electromagnetic wave function, which is the de Broglie wave for its photons, gives the probability of finding the photons at a point in space. In a single-slit diffraction experiment, we know that many photons will strike the region of the central maximum on a screen, but fewer photons will arrive in the regions of the first bright fringes, even fewer in the regions of the second bright fringes, and so on (see Chapter 24). The quantum mechanical probability distribution or density on the screen has the same relative distribution as the classical intensity.

However, if the experiment could be done with only one photon, we could not predict exactly to which region it would go. We could only give the probability that it would be found in a particular region.

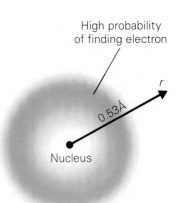

Figure 28.9 **Electron cloud** The probability density distribution gives rise to the idea of an electron cloud in the region of space where the electron is likely to be found. The cloud density reflects the probability density.

28.3 The Heisenberg Uncertainty Principle

Another important aspect of quantum mechanics has to do with measurement and accuracy. In classical mechanics, there is no limit to the accuracy of a measurement. Theoretically, by continual refinement of a measurement instrument and procedure, the accuracy could be improved to any degree so as to give *exact* values. This resulted in a *deterministic* view of nature. For example, if the position and velocity of an object are known *exactly* at a particular time, you can determine where it will be in the future and where it was in the past (assuming no future or past unknown forces).

However, quantum theory predicts otherwise and sets limits on the possibilities of measurement accuracy. This idea was introduced by the German physicist Werner Heisenberg in 1927 (Fig. 28.10), who had developed another approach to quantum mechanics that complemented Schrödinger's wave theory. The **Heisenberg uncertainty principle** as applied to position and momentum (or velocity) may be stated as follows:

> It is impossible to know simultaneously an object's exact position and momentum.

This concept is often illustrated with a simple thought experiment. Suppose that you wanted to measure the position and momentum (actually the velocity) of an electron. In order for you to "see," or locate, the electron, at least one photon must bounce off the electron and come to your eye, as illustrated in Fig. 28.11. However, in the collision process, some of the photon's energy and momentum are transferred to the electron (similar to the Compton effect, Chapter 27).

After the collision, the electron recoils. Thus, in the very process of trying to locate the position very accurately (trying to make the uncertainty of position Δx very small), you induce more uncertainty into your knowledge of the electron's velocity or momentum ($\Delta p = m \Delta v$), since the process of determining its position sent the electron flying off. In the macroscopic world, the uncertainty due to viewing an object would be negligible, since light does not appreciably alter the motion or position of an ordinary-sized object.

For our subatomic case, the position of an electron could be measured at best

Figure 28.10 **Werner Heisenberg (1901–1976)** Heisenberg (right) and P.A.M. Dirac at Cambridge University. Dirac's relativistic quantum theory predicted the existence of the positron, the antiparticle of the electron (see Section 28.4).

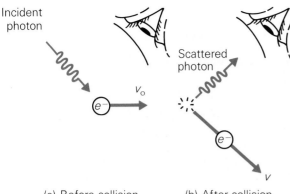

Incident photon

Scattered photon

v_0

e^-

e^-

v

(a) Before collision   (b) After collision

Figure 28.11 Measurement-induced uncertainty
(a) To measure the position and momentum (velocity) of an electron, at least one photon must collide with the electron and be scattered toward your eye. (b) But in the collision process, energy and momentum are transferred to the electron, which induces uncertainty in the velocity.

to an accuracy of about the wavelength λ of the incident light, that is, $\Delta x \cong \lambda$. The photon "particle" has a momentum of $p = h/\lambda$. Since we cannot tell how much of this momentum might be transferred during collision, the final momentum of the electron would have an uncertainty of $\Delta p \cong h/\lambda$.

The total uncertainty is given by the product of the individual uncertainties, and

$$(\Delta p)(\Delta x) \cong \left(\frac{h}{\lambda}\right)(\lambda) = h$$

This equation gives an estimate of the minimum uncertainties or maximum accuracies of simultaneous measurements of the momentum and position. In actuality, the uncertainties could be much worse depending on the amount of light (number of photons) used, apparatus, and technique. Through theoretical calculations, Heisenberg found that, at very best,

$$(\Delta p)(\Delta x) \geq \frac{h}{2\pi} \tag{28.8}$$

Thus, Heisenberg's uncertainty principle states that the product of the *minimum* uncertainties of position and momentum is on the order of Planck's constant ($\cong 10^{-34}$). These are the minimum uncertainties or the *best degree of accuracies* we can ever hope to achieve for *simultaneous* measurements. In the process of trying accurately to locate the position of the particle (that is, to make Δx small), the uncertainty in the momentum is made larger ($\Delta p \cong h/2\pi \Delta x$) and vice versa. Thus, the measurement procedure itself limits the accuracy to which we can simultaneously measure position and momentum. If we could measure the exact location of a particle ($\Delta x \to 0$), we would have no idea about its momentum ($\Delta p \to \infty$).

According to Heisenberg, "Since the measuring device has been constructed by the observer... we have to remember that what we observe is not nature in itself but nature exposed to our method of questioning."

Example 28.4   Uncertainty Principle

An electron and a 20-g bullet, both moving linearly, are measured to have equal speeds of 300 m/s to an accuracy of $\pm 0.010\%$. What is the minimum uncertainty in the position of each?

Solution

Given: $m_b = 20$ g $= 0.020$ kg

$(m_e = 9.11 \times 10^{-31}$ kg$)$

$v = 300$ m/s $\pm 0.010\%$

Find: Δx's

An uncertainty of 0.010% in velocity is

$$(300 \text{ m/s})(0.00010) = 0.030 \text{ m/s} \quad (= 3.0 \text{ cm/s})$$

The total uncertainty in velocity is twice this amount, since the measurements can be off by 0.010% above or below the actual values ($\pm 0.010\%$), that is,

$$\Delta v = 0.060 \text{ m/s} \quad (= 6.0 \text{ cm/s})$$

Then, for the electron,

$$\Delta x = \frac{h}{2\pi \, \Delta p} = \frac{h}{2\pi m_e \, \Delta v}$$

$$= \frac{6.63 \times 10^{-34} \text{ J} \cdot \text{s}}{2\pi (9.11 \times 10^{-31} \text{ kg})(0.060 \text{ m/s})} = 0.0019 \text{ m} \quad (= 0.19 \text{ cm})$$

and for the bullet,

$$\Delta x = \frac{h}{2\pi m_b \, \Delta v} = \frac{6.63 \times 10^{-34} \text{ J} \cdot \text{s}}{2\pi (0.020 \text{ kg})(0.060 \text{ m/s})} = 8.8 \times 10^{-32} \text{ m}$$

Notice that the uncertainty in the position of the bullet is considerably less than that of the electron. The uncertainty for relatively massive objects traveling at ordinary speeds is practically negligible. ∎

Another form of the uncertainty principle relates energy and time. To understand this, consider the position of the electron in the previous thought experiment to be known with an uncertainty of $\Delta x \cong \lambda$. The photon used to detect the particle travels with a speed c, and it takes a time of $\Delta x / c \cong \lambda / c$ for this photon to traverse a distance equal to the uncertainty in position. Thus, the time when the particle is at the measured position is uncertain by about

$$\Delta t \cong \lambda / c$$

Since we can't tell whether the photon transfers some or all of its energy ($E = hf = hc/\lambda$) to the particle, the uncertainty in the energy is

$$\Delta E = \frac{hc}{\lambda}$$

Then the total uncertainty is

$$(\Delta E)(\Delta t) \cong \left(\frac{hc}{\lambda}\right)\left(\frac{\lambda}{c}\right) = h$$

Similar to the position-momentum relationship, at very best we have

$$(\Delta E)(\Delta t) \geq \frac{h}{2\pi} \tag{28.9}$$

This form of the uncertainty principle indicates that the energy of an object may be uncertain by an amount ΔE for a time $\Delta t \cong h/2\pi \, \Delta E$. During this time, the energy is uncertain and might not even be conserved. This is an important consideration in particle interactions, as you will learn in a later chapter.

Also, we cannot measure the energy of a particle exactly unless we take an infinite amount of time to do so. If a measurement of energy is carried out in a time Δt, then it must be uncertain by an amount ΔE. For example, the measurement of the frequency of a photon emitted by an atomic electron is a measurement of the energy associated with the transition from an excited state to the ground state. The measurement must be carried out in a time comparable to the time the electron is in the excited state—that is, the lifetime of the excited state. As a result, the observed emission line in the frequency spectrum has a finite width, since $\Delta E = h\,\Delta f$ (Fig. 28.12). This so-called *natural broadening* was ignored in Chapter 27, where spectral lines were considered to have widths of single frequencies (that is, no width at all). This is the same as assuming that the excited states of the Bohr atom have infinite lifetimes.

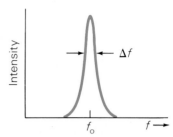

Figure 28.12 Natural line broadening
Because a measurement must be carried out in a time comparable to the lifetime (Δt) of an electron in an excited state, the energy is uncertain by an amount $\Delta E = h\,\Delta f$. The observed emission line then has a width Δf, rather than being a line of single frequency f_0.

Example 28.5   Natural Line Broadening

An electron in an excited state has a lifetime of 10^{-8} s. (a) What is the minimum uncertainty in the energy of the photons emitted on de-excitation? (b) What is the magnitude of the natural broadening of the spectral line?

Solution

Given: $\Delta t = 10^{-8}$ s   *Find*: (a) ΔE
                                                        (b) Δf

(a) The minimum uncertainty in the energy is

$$\Delta E = \frac{h}{2\pi\,\Delta t} = \frac{6.63 \times 10^{-34} \text{ J} \cdot \text{s}}{2\pi(10^{-8} \text{ s})} = 1.06 \times 10^{-26} \text{ J}$$

(b) The uncertainty or broadening of the frequency of the observed spectral line is then

$$\Delta f = \frac{\Delta E}{h} = \frac{1.06 \times 10^{-26} \text{ J}}{6.63 \times 10^{-34} \text{ J} \cdot \text{s}} = 1.60 \times 10^{7} \text{ Hz}$$

The uncertainty in the frequency of a spectral line is called the *natural line width*. The lifetimes of atomic electrons in their excited states are on the order of 10^{-8} s, and it might appear from the magnitude of the value of Δf that the spectral lines would be quite wide. However, recall that the frequency of visible light is on the order of 10^{14} Hz. The percent uncertainty is then about $10^7/10^{14}$ ($\times$ 100%) = 0.00001%. Thus, the spectral line has a frequency width on the order of $10^{14} \pm 0.00001\%$, which is quite narrow. ∎

28.4 Particles and Antiparticles

The quantum mechanics of Schrödinger and Heisenberg were successful in explaining observations and in predicting new atomic phenomena. When the quantum theory was extended to include relativistic considerations by the British physicist Paul A. M. Dirac (Fig. 28.10) in 1928, something new and very different was predicted—a particle called the **positron**. The positron should have the same mass as the electron but should carry a *positive* charge. The oppositely charged positron is said to be the **antiparticle** of the electron.

The positron was first observed experimentally in 1932 by the American physicist C. D. Anderson in cloud chamber experiments with cosmic rays. The curvature of the particle tracks in a magnetic field showed two types of

particles (Fig. 28.13). The tracks indicated that both particles had the same mass, but their spiral curvatures were opposite. From the magnetic force relationship, $F = qvB$, this observation requires the particles to be oppositely charged. Thus, Anderson discovered the positron, a particle with a mass equal to that of the electron and with the same magnitude of electrical charge, but opposite in sign.

By the conservation of charge, a positron can be created only with the simultaneous creation of an electron in a process called **pair production**. In Anderson's experiment, positrons were observed to be emitted from a thin lead plate exposed to cosmic rays from outer space, which contain highly energetic X-rays. Pair production occurs when an X-ray "photon" collides with a nucleus—the nuclei of the lead atoms of the plate in the Anderson experiment. In the collision process, the photon goes out of existence, and an "electron pair" (an electron and a positron) is created, as illustrated in Fig. 28.14, in a conversion of energy into mass. By the conservation of energy,

$$hf = 2m_e c^2 + K_{e^-} + K_{e^+} + K_{\text{nuc}}$$

where hf is energy of the photon, $2m_e$ is the rest mass of the electron pair, and the K's are the kinetic energies of the particles and the recoil nucleus. Because of its relatively large mass, the kinetic energy of the recoil nucleus can usually be considered negligible, and

$$hf \cong 2m_e c^2 + K_{e^-} + K_{e^+}$$

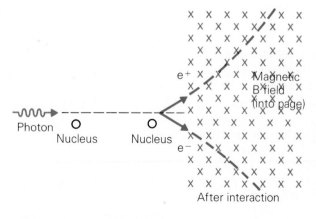

Figure 28.14 **Pair production**
An electron and a positron are created when an energetic photon interacts with a nucleus.

However, a photon cannot spontaneously decay into an electron pair, since a nucleus or some body must be present for the conservation of energy and momentum, even though we have considered its recoil energy negligible.

From the energy equation, we can see that a photon cannot create an electron pair unless

$$hf \geq 2m_e c^2 = 1.022 \text{ MeV} \qquad (28.10)$$

(Recall that the rest mass of an electron is $m_e c^2 = 0.511$ MeV.) This minimum energy is called the *threshold energy for pair production*.

But if positrons are created by cosmic rays, why are they not commonly found in nature? For example, why are they not evident in ordinary chemical processes? The answer is because positrons are taken out of existence by a process called **pair annihilation**. When an energetic positron appears in pair production, it loses kinetic energy in collisions as it passes through matter. Finally, moving at a low speed, it combines with an electron of the material and forms a hydrogenlike atom, called a *positronium atom*, with the positron substituting for a proton. The positronium atom is unstable and quickly decays ($\cong 10^{-10}$ s) into two photons (Fig. 28.15).

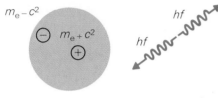

Before Annihilation (positronium atom) After Annihilation (photons)

Figure 28.15 Pair annihilation
The disappearance of a positronium atom is signaled by the appearance of two 0.511-MeV photons.

Before annihilation, the positronium atom is at rest, so by conservation of linear momentum, the photons are emitted in opposite directions and have equal magnitudes, that is,

$$\frac{hf_1}{c} = \frac{hf_2}{c}$$

By the conservation of energy,

$$2m_e c^2 = hf_1 + hf_2$$

Therefore, the energies of the photons are equal:

$$hf_1 = hf_2 = m_e c^2 = 0.511 \text{ MeV}$$

Pair annihilation is then a direct conversion of (rest) mass into electromagnetic energy, the inverse of pair production so to speak. These processes are striking examples of the mass-energy equivalence predicted by Einstein.

All subatomic particles have been found to have antiparticles, which are observed in cosmic rays from outer space and/or are produced in nuclear processes. For example, there is an antiproton with same mass as a proton but with a negative charge. There are also antineutrons. In our environment, there is a preponderance of electrons, protons, and neutrons. When antiparticles are created, they quickly combine with their respective particles in annihilation processes.

It is conceivable that antiparticles predominate in some parts of the universe. If so, the atoms of the **antimatter** in this region would consist of negatively charged nuclei composed of antiprotons and antineutrons, surrounded by positively charged positrons (antielectrons). It would be difficult to distinguish a region of antimatter visibly, since it would appear the same as ordinary matter. The physical behavior of antimatter atoms would presumably be the same as those of ordinary matter. (Recall that the assignment of plus and minus signs to electric charges is an arbitrary convention.) However, if antimatter and ordinary matter came into contact, they would annihilate each other with an explosive release of energy.

Important Formulas

Momentum of a photon:

$$p = \frac{E}{c} = \frac{hf}{c} = \frac{h}{\lambda}$$

de Broglie wavelength:

$$\lambda = \frac{h}{p} = \frac{h}{mv}$$

Electron wavelength when accelerated through potential V:

$$\lambda = \left(\frac{150}{V}\right)^{\frac{1}{2}} \text{Å} = \left(\frac{150}{V}\right)^{\frac{1}{2}} \times 10^{-10} \text{ m}$$

Wave function of a particle in a box:

$$\psi_n = A \sin \frac{n\pi x}{L}$$

Energy of a particle in a box:

$$E = \frac{n^2 h^2}{8mL^2}$$

Heisenberg uncertainty principle:

$$(\Delta p)(\Delta x) \geq \frac{h}{2\pi}$$

$$(\Delta E)(\Delta t) \geq \frac{h}{2\pi}$$

Condition for electron pair production:
$$hf \geq 2m_e c^2 = 1.022 \text{ MeV}$$

Questions

Matter Waves: The de Broglie Hypothesis

1. How are electromagnetic waves and de Broglie waves related?

2. Two electrons move with speeds of v_1 and $v_2 = 2v_1$. Which has the longer de Broglie wavelength? How much longer is it?

3. Explain how the de Broglie hypothesis gave an interpretation of the discrete Bohr orbits of the electron of the hydrogen atom.

4. Did the de Broglie hypothesis pass the test of the scientific method? If so, explain how.

5. Why are the effects of matter waves not observed for ordinary objects in motion, for example, a baseball?

6. Compare a light microscope to an electron microscope. What are the similarities and the dissimilarities?

The Schrödinger Wave Equation

7. The de Broglie equation gives only the wavelength of the wave. How can mathematical expressions of the de Broglie waves be obtained?

8. How are ψ and ψ^2 interpreted?

9. What was one of the surprising predictions from the wave function solutions of the Schrödinger equation for a particle bound in a potential well?

10. Given the radial wave function ψ of the hydrogen electron in the first excited state, at what distance from the nuclear proton would you expect ψ^2 to have a maximum? Is the electron at this radial distance?

The Heisenberg Uncertainty Principle

11. What is meant by a deterministic universe? Is this a classical or quantum mechanical concept?

12. Explain how a measurement by a microscope system inherently affects the accuracy of the measurement.

13. How accurately can the position and velocity of a particle be known simultaneously?

14. What is the relationship between the uncertainties in simultaneous energy and time measurements? Explain in terms of natural line broadening.

Particles and Antiparticles

15. Is charge created in pair production? Explain in terms of the conservation of charge.

16. What is the threshold energy for pair production?

17. Can a photon spontaneously decay into an electron pair?

18. (a) What is a positronium atom? (b) What role does this atom play in pair annihilation? What are the distinguishing features of the photons created in this process?

19. Would it be possible to distinguish a region of antimatter? Explain.

Problems

28.1 Matter Waves: The de Broglie Hypothesis

■■**1.** What is the de Broglie wavelength of (a) an electron and (b) a proton, both moving with a speed of 300 m/s?

■■**2.** Electrons are accelerated from rest through a potential of 250 kV. If the potential is increased to 500 kV, how is the de Broglie wavelength of the electrons affected?

■■**3.** The de Broglie wavelength of a proton is 2.0 pm. What is the speed of the particle?

■■**4.** Calculate the de Broglie wavelength of a 70-kg person running with a speed of 20 m/s.

■■**5.** A 0.20-kg mass has a kinetic energy of 500 J. What is the de Broglie wavelength of the mass?

■■**6.** Charged particles are accelerated through a potential difference V. By what factor would the de Broglie wavelength of the particles change if the voltage were tripled?

■■**7.** A photon has an energy of 5.0 MeV. What are (a) its momentum and (b) the wavelength of the associated de Broglie wave?

■■**8.** The frequency of an electromagnetic radiation is 2.0×10^{16} Hz. What is the de Broglie wavelength associated with a photon of the radiation?

■■**9.** What is the de Broglie wavelength for the matter wave associated with (a) a 250-g ball thrown at 30 m/s and (b) a 1800-kg automobile traveling at 80 km/h?

■■**10.** If the spacing between the lattice planes of a particular crystal is 1.90 Å, what voltage is needed to accelerate electrons (from rest) into a beam that exhibits a first-order diffraction maximum at an angle of 50°?

■■**11.** What is the energy of a beam of electrons that exhibits a first-order maximum at an angle of 30° when diffracted by a crystal grating with a spacing between the lattice planes of 0.15 nm?

■■■**12.** According to the Bohr theory of the hydrogen atom, the speed of the electron in the first Bohr orbit is 2.19×10^8 cm/s. (a) What is the wavelength of the matter wave associated with the electron? (b) How does this compare with the circumference of the first Bohr orbit?

■■■**13.** It is desired to observe details whose size is of the order of 10 Å with an electron microscope. Through what potential must the electrons be accelerated so that they have a de Broglie wavelength of this order?

28.2 The Schrödinger Wave Equation

■■**14.** Determine the positions at which there is a maximum probability of finding a particle in a box for the wave functions in Fig. 28.6.

■■**15.** Show that for a particle moving in one dimension between the fixed boundaries $-L < x < L$ the wave functions are given by $\psi = A \cos n\pi x/2L$, where $n = 1, 3, 5, \ldots$.

■■**16.** Where is the particle of Problem 15 most likely to be found? What is the probability of finding it there?

■■**17.** Sketch the form you would expect for the probability of the position of the hydrogen electron in the first three excited states as a function of the radial distance from the nucleus.

28.3 The Heisenberg Uncertainty Principle

■■**18.** What is the minimum uncertainty in the velocity of a 0.50-kg ball that is known to be at 1.0000 ± 0.0005 cm from the edge of a table?

■■**19.** With what accuracy would you have to measure the position of a moving electron so that its velocity is uncertain by 10^{-3} cm/s?

■■**20.** What is the minimum uncertainty in the velocity of an electron that is known to be somewhere between 0.50 Å and 1.0 Å from a proton?

■■**21.** An electron is accelerated from rest through a potential of 5.0 kV ± 10%. What is the minimum uncertainty in the position of the electron after the acceleration?

■■**22.** If an excited state of an atom is known to have a lifetime of 10^{-7} s, what is the minimum error within which the energy of the state can be measured?

■■**23.** How much greater is the width of a spectral line due to natural broadening for a transition from an excited state with a lifetime of 10^{-12} s than for one from a state with a lifetime of 10^{-8} s?

28.4 Particles and Antiparticles

■■**24.** A photon with a frequency of 1.5×10^{18} Hz collides with a nucleus. Will pair production occur? Justify your answer mathematically.

■■**25.** What is the frequency of the photons produced in electron pair annihilation?

■■**26.** What would be the required energy for a photon that could cause the production of a proton-antiproton pair?

Additional Problems

27. What is the de Broglie wavelength associated with photons of red light ($\lambda = 680$ nm)?

28. An electron travels with a speed of 3.60560 ± 0.00021 m/s. To what minimum uncertainty can its position be measured?

29. A proton and an electron are accelerated from rest through a potential difference V. What is the ratio of the de Broglie wavelength of an electron to that of a proton?

30. The energy of a 2.00-keV electron is known to 5.00%. How accurately can its position be measured?

31. An electron is accelerated from rest through a potential difference that gives it a matter wave with a wavelength of 2.5 cm. What is the potential difference?

32. A muon, or μ meson, has a negative charge like that of an electron, but the muon's mass is 207 times that of an electron. What would be the required energy for a photon that could cause the pair production of a meson and an antimeson?

33. A proton traveling with a speed of 5.0×10^4 m/s is accelerated through a potential difference of 37 eV. By what percentage does the de Broglie wavelength of the proton change?

The Nucleus

29

Great advances have been made in nuclear physics since the introduction of the nuclear model of the atom by Lord Rutherford in 1911. However, not everything is known about the nucleus. This small integral part of the atom presents a great challenge to the ingenuity of scientists, who would like to know its secrets and to apply its potential.

How does one study such a minute entity? Because of its unimaginably small size, all our knowledge of the nucleus hinges on indirect observations that are not fully explained by classical theory. Quantum mechanics, which was so helpful to our understanding of atomic physics, can be applied to the nucleus, but with limitations. The potential energy of a system must be known before we can attempt to solve the Schrödinger equation. Knowing the energy of the system requires an understanding of the force that holds the nucleus together. The exact form of this nuclear force is not known.

Even so, a great deal can be learned about the nucleus by looking at the properties of various nuclei. One of the most revealing phenomena of the nucleus is radioactivity. Some nuclei are unstable and decay into nuclei of other elements with the telltale emission of detectable particles. The study of radioactivity and nuclear stability gives us some general ideas about the nature of the nucleus, the energy it possesses, and how this energy can be released. The release of nuclear energy has become one of our major energy sources. This will be considered in the next chapter. First, let's take a look at the nucleus itself.

29.1 Nuclear Structure and the Nuclear Force

It is evident from the emission of electrons from heated filaments (thermionic emission) and the photoelectric effect that the atoms of certain materials, probably all materials, contain electrons. Since atoms are normally electrically neutral, they must also contain positive charge equal in magnitude to that of the electrons in an atom. Also, since the mass of an electron is small in comparison to the mass of even the lightest atoms, most of the mass appears to be associated with the positive charge.

J. J. Thomson, who experimentally proved the existence of the electron in 1897, based his model of atomic structure on these general observations. In the Thomson model, the negatively charged electrons were pictured as being uniformly positioned in a continuous distribution of positive charge (Fig. 29.1). The positive charge distribution was assumed to be spherical with a radius on the order of 10^{-8} cm, as calculated from the bulk properties of matter.

The modern concept of the atom is quite different. This model pictures all of the atom's positive charges and practically all of its mass in a nucleus concentrated in a small region of space, and the negatively charged electrons surrounding the nucleus. Such a model was proposed in 1910 by the British physicist (Lord) Ernest Rutherford (1871–1937). Rutherford's insight came from the results of alpha particle scattering experiments performed in his laboratory by Hans Geiger and Ernest Marsden in 1909. An alpha particle (α particle) is a doubly positively charged particle that is naturally emitted from some radioactive materials (to be discussed in more detail in Section 29.2). Scattering experiments with charged particles allow the study of the distribution of matter or electric charge in atoms.

Further alpha particle scattering experiments from thin metal foils, such as gold, in 1913 provided a critical test between the Thomson model and the Rutherford model of the atom. A beam of alpha particles from a radioactive source was directed toward a metal foil "target," and the scattering angles of the particles were observed (Fig. 29.2).

A critical result of the Geiger-Marsden experiments was the scattering of a small percentage of the incident alpha particles. Some alpha particles (only 1 in 8000) were found to be scattered through angles greater than 90°; that is, they were *back-scattered*. As described by Rutherford, this was "about as credible as if you fired a 15-inch shell at a piece of tissue paper and it came back and hit you."

Small deflections would be expected in the Thomson model due to multiple collisions with the light electrons as the alpha particle passed through a large gold atom. But because an alpha particle is over 7000 times more massive than an electron and the foil was so thin ($\cong 10^{-4}$ cm), calculations showed that scattering through the observed large angles had only a minuscule probability of taking place—certainly much less than 1 in 8000.

If all of the positive charge of an atom in the foil were concentrated in a very small region in the atom, then an alpha particle coming very close to this region would experience a large deflecting force. The mass of this nucleus of charge would be larger than that of the alpha particle, since most of the atomic mass is associated with the positive charge. Thus, back-scattering would be possible. Scattering for the Rutherford nuclear model of the atom is illustrated in

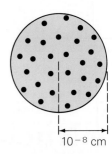

Figure 29.1 **Thomson model of the atom** The negatively charged electrons were pictured as being uniformly distributed in a continuous distribution of positive charge. The model was referred to as the plum pudding model because of its resemblance to the distribution of raisins in a plum pudding (a kind of cake).

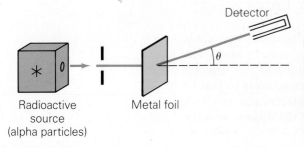

Figure 29.2 **Scattering experiment** A beam of alpha particles from a radioactive source is scattered by a thin foil, and the scattering is observed to be a function of the scattering angle θ.

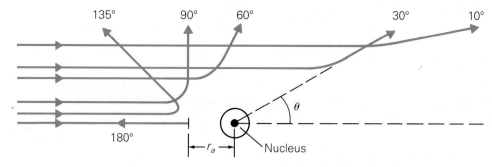

Fig. 29.3. The theoretical predictions of the so-called Rutherford scattering for this model were consistent with experimental results.

A simple calculation can give an idea of the size of the atomic nucleus. For a head-on collision, an alpha particle would attain its distance of closest approach to the nucleus (r_a in Fig. 29.3). That is, the alpha particle approaching the nucleus would be stopped at a distance r_a by the repulsive Coulomb force and would be accelerated back along its original path. Assuming a spherical or point charge distribution, the electric potential is $V = kZe/r$, where Z is the *atomic number* or the number of protons in the nucleus and Ze is the total charge. The work done by the Coulomb force in stopping the alpha particle is $W = qV = (2e)V$, since the alpha particle has a double positive electronic charge. This is equal to the initial kinetic energy of the alpha particle, that is,

$$\tfrac{1}{2}mv^2 = \frac{k(2e)Ze}{r_a}$$

and

$$r_a = \frac{4\,kZe^2}{mv^2} \tag{29.1}$$

By using the known values of the energy of the alpha particles from the particular source and other known values, r_a is found to be about 10^{-12} cm. This value is an *upper limit* for the nuclear radius, since the alpha particle does not reach the nucleus itself.

Although the nuclear model of the atom is useful, the nucleus is much more than simply a region of positive charge. For one thing, we now know that the atomic nucleus is composed of two types of particles—protons and neutrons—which are collectively referred to as **nucleons**. The nucleus of the common hydrogen atom is a proton. Rutherford suggested that the hydrogen nucleus be given the name *proton* (from a Greek word meaning "first") after he became convinced that there was no positively charged particle in an atom lighter than the hydrogen nucleus. As you know from an earlier chapter, a neutron is an electrically neutral particle with a mass slightly greater than that of a proton. The existence of the neutron was not experimentally verified until 1932.

The idea of electrons orbiting the nucleus comes from Bohr's theory of the hydrogen atom (Chapter 27), giving rise to the so-called Rutherford-Bohr model of the atom.

The Nuclear Force

Of the forces in the nucleus, we know there is an attractive gravitational force between nucleons (protons and neutrons). But as was explained in Chapter 16, the magnitude of the gravitational force is negligible in comparison with the mutually repulsive electrical force between positively charged protons. Taking only these repulsive forces into account, one might expect the nucleus to fly apart. Yet the nuclei of most atoms are stable, so there must be an additional force that holds the nucleus together. This strongly attractive force is called the **strong nuclear force**, usually referred to as simply the *nuclear force*.

The exact mathematical expression for the nuclear force is not known, and approximations indicate that it is extremely complicated. However, some general features of this force are as follows.

■ The nuclear force is strongly attractive and much larger in relative magnitude than the electrostatic and gravitational forces.

■ The nuclear force is very short-ranged; that is, a nucleon interacts only with its nearest neighbors.

■ The nuclear force acts between any two nucleons within a short range, that is, between two protons, a proton and a neutron, or two neutrons.

Thus, nuclear protons in close proximity repel each other by the electric force but attract each other (and nearby neutrons) by the strong nuclear force. Having no electric charge, neutrons are attracted to nearby protons and neutrons only by the nuclear force.

Nuclear Notation

To describe the nuclei of different atoms, it is convenient to use the notation illustrated in Fig. 29.4. The chemical symbol of the element is used with subscripts and a superscript. The left subscript is the **atomic number** (Z), which indicates the number of protons in the nucleus. For an atom, the atomic number is the number of electrons in an atom. In neutral atoms, the number of nuclear protons equals the number of electrons, so this term was also applied to the number of protons. A more descriptive term, which this book will generally use, is the **proton number**.

Recall that the number of protons in the nucleus of an atom defines the species of the atom—that is, the element to which the atom belongs. In the example in Fig. 29.4, the proton number $Z = 6$ indicates that this is a carbon nucleus. The proton number defines which chemical symbol is used. Electrons can be removed (or added) to an atom to form an ion, but this does not change its species. For example, a nitrogen atom with an electron removed, N^+, is still nitrogen—a nitrogen ion.

The left superscript is called the **mass number** (A) and is the total number of protons and neutrons in the nucleus. Since protons and neutrons have approximately equal masses, the mass numbers of nuclei give a relative comparison of nuclear masses. For the example in Fig. 29.4, the carbon nucleus, the mass number is $A = 12$, since there are 6 protons and 6 neutrons. The number of neutrons is indicated by the subscript on the right side, which is called the **neutron number** (N). The neutron number is often omitted, as in $^{12}_{6}C$, since it can easily be found from the other numbers: $N = A - Z$.

The atoms of an element, all of which have the same number of protons in their nuclei, may have different numbers of neutrons. For example, nuclei of different carbon atoms ($Z = 6$) may contain either 6, 7, or 8 neutrons and in nuclear notation would be written

There is also a weak nuclear force, which is much weaker than the strong nuclear force and is involved in certain types of radioactive decay (more on this force in Chapter 30).

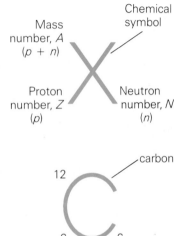

Figure 29.4 Nuclear notation
A nucleus is described by the chemical symbol of the particular element with the mass number A (number of protons and neutrons) as a superscript on the left and the proton (atomic) number Z and the neutron number N as subscripts on the left and right, respectively.

$$^{12}_{6}C_6 \qquad ^{13}_{6}C_7 \qquad ^{14}_{6}C_8$$

Atoms whose nuclei have the same number of protons but different numbers of neutrons are called **isotopes**, in this case isotopes of carbon.

Isotopes are like members of a family; they all belong to the same family (have the same Z number) and have the same surname, but the members of the family are distinct on the basis of the number of neutrons in their nuclei. However, it should be noted that different isotopes of the same family have the same electronic structure and thus the same chemical properties. Isotopes are referred to by their mass numbers, for example, the previous isotopes of carbon are called carbon-12, carbon-13, and carbon-14. There are also other isotopes of carbon, ^{11}C, ^{15}C, and ^{16}C. A particular nuclear species or isotope is called a **nuclide**. Thus, we have six nuclides or isotopes of carbon.

Another family of isotopes is that of hydrogen, which has three isotopes or nuclides: ^1H, ^2H, and ^3H. These isotopes are not generally referred to as hydrogen-1, and so on, but are given special names that refer to their atoms. ^1H is called ordinary hydrogen or, simply, hydrogen. ^2H is the nucleus of a *deuteron*, and a quantity of these atoms is called *deuterium*. Deuterium is sometimes referred to as heavy hydrogen and, when combined with oxygen as water, gives heavy water (commonly written D_2O). ^3H is the nucleus of a *triton* or an atom of *tritium*.

29.2 Radioactivity

Most elements have stable isotopes. It is the atoms of these stable nuclides with which we are most familiar in the environment. However, the nuclei of some isotopes are unstable and disintegrate spontaneously (decay) of their own accord, emitting energetic particles. Such isotopes are said to be *radioactive*. For example, the tritium isotope of hydrogen just mentioned is radioactive. The prefix "radio" refers to the emitted nuclear radiation, which in the modern context may be a particle or a wave, and a radioactive isotope is "active" in doing so. Of the nearly 1200 known unstable nuclides, only a small number occur naturally. The others are produced artificially (Chapter 30).

We know that radioactivity is completely unaffected by normal physical or chemical processes, such as heat and pressure or chemical reactions. Since chemical processes involve the outer electrons of atoms, the source of radioactivity must lie deeper in the atom, that is, in the nucleus. This instability cannot be explained directly by a simple imbalance of attractive and repulsive forces in the nucleus because the nuclear disintegrations of a given isotope occur as a function of time at a fixed rate. Classically, one would expect identical nuclei to do the same thing at the same time. Therefore, radioactive decay processes suggest quantum mechanical probability effects.

The discovery of radioactivity is credited to the French scientist Henri Becquerel. In 1896, while studying the fluorescence of a uranium compound, Becquerel discovered that a photographic plate in the vicinity of a sample had been darkened when the compound had not been activated by exposure to light and was not fluorescing. Apparently, the darkening was caused by some new type of radiation being emitted from the compound. In 1898, Pierre and Marie Curie announced the discovery of two radioactive elements, radium and polonium, which they had isolated from uranium pitchblende ore (Fig. 29.5).

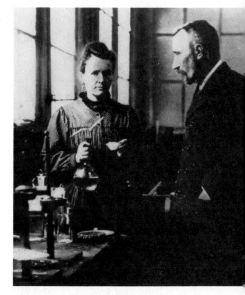

Figure 29.5 **The Curies** Marie Curie (1867–1934) was born in Poland and studied in France. There she met and married Pierre Curie (1859–1906), who was a physicist well known for his work on crystals and magnetism. In 1903, Madame Curie (as she is commonly known) and Pierre shared the Nobel Prize in physics with Henri Becquerel for their work on radioactivity. Mme. Curie was also awarded the Nobel Prize in chemistry in 1911 for the discovery of radium and the study of its properties.

Experiment easily shows the radiation emitted by radioactive isotopes to be of three different types. If a radioactive isotope is placed in a chamber, as illustrated in Fig. 29.6, so that the emitted radiation passes through a magnetic field to a photographic plate, one (and possibly two) of three distinct spots of radiation exposure is observed on the plate after it is developed. These spots are characteristic of what is known as alpha, beta, and gamma radiation. From the deflections of two of the types of radiation in the magnetic field, it is evident that positively charged particles are emitted from nuclei undergoing alpha decay and that negatively charged particles are emitted in beta decay. Also, the degree of deflection shows that alpha particles must be more massive than beta particles. The undeflected gamma radiation (gamma rays) is electrically neutral.

Investigations of the different radiations reveal that:

- **Alpha particles** are doubly charged (2+) ions containing two protons and two neutrons. They are identical to the nucleus of the helium atom ($^4_2$He).

- **Beta particles** are electrons.

- **Gamma rays** are particles or quanta of electromagnetic energy.

For a few radioactive elements, two spots are found on the film, indicating that these elements decay by two different modes.

Let's now look at what happens to a decaying nucleus in each of these three decay modes of radioactivity.

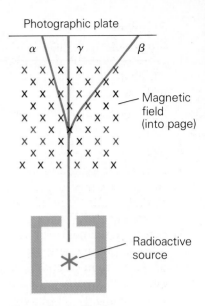

Photographic plate

Magnetic field (into page)

Radioactive source

Figure 29.6 **Nuclear radiations**
The radiations from radioactive sources can be distinguished by passing them through a magnetic field. Alpha and beta particles are deflected, and by $F = qvB$, alpha particles are positively charged, and beta particles are negatively charged. The radii of curvature (not to scale) allow the particles to be identified by mass. Gamma rays are not deflected and so are uncharged, being in fact "particles" or quanta of electromagnetic energy.

Alpha Decay

When an alpha particle is ejected from a radioactive nucleus, the nucleus loses two protons and two neutrons, so the mass number (A) is decreased by four, that is, $\Delta A = -4$, and the proton number (Z) is decreased by two, $\Delta Z = -2$. Since the parent nucleus loses two protons, the resulting daughter nucleus must then be the nucleus of another element as defined by the proton number. (The original and resulting nuclei are commonly referred to as the parent and daughter nuclei, respectively.) Thus, the decay process is one of nuclear transmutation in which the nuclei of one element change into the nuclei of a lighter element.

An example of an isotope or nuclide that undergoes **alpha decay** is polonium-214. The decay process is represented in the form of a nuclear equation, similar to a chemical equation:

$$^{214}_{84}\text{Po} \longrightarrow\ ^{210}_{82}\text{Pb}\ +\ ^4_2\text{He}$$

polonium   lead   alpha particle
(helium nucleus)

Note that the totals of the mass numbers and the proton numbers are equal on each side of the equation: (214 = 210 + 4) and (84 = 82 + 2), respectively. This reflects the fact that *two conservation laws apply to all nuclear processes.* The first is the **conservation of nucleons.** That is, the total number of nucleons (A) remains constant in any process. The second is the familiar **conservation of charge**, which applies to nuclear reactions.

Example 29.1 Alpha Decay
A $^{238}_{92}$U nucleus undergoes alpha decay. What is the resulting daughter nucleus?

Solution

Since $\Delta Z = -2$ for alpha decay, the uranium-238 (^{23}U) nucleus loses two protons, and the daughter nucleus has a proton number of $Z = 92 - 2 = 90$, which is the proton number of thorium (see Appendix IV). The equation for the process is

$$^{238}_{92}\text{U} \longrightarrow {}^{234}_{90}\text{Th} + {}^{4}_{2}\text{He} \quad \blacksquare$$

Alpha particles from a radioactive source were used in the scattering experiments that led to the Rutherford nuclear model (Section 29.1). The energies of the alpha particles from radioactive sources are on the order of a few megaelectron volts (MeV). For example, the energy of the alpha particle emitted from the decay of ^{214}Po is about 7.7 MeV, and that from ^{238}U decay is about 4.2 MeV. Scattering experiments give some idea about the size of the nucleus, as you learned in the preceding section. The Coulomb potential of the repulsive force of the nuclear charge falls off as $1/r$, where r is the distance from the nuclear charge. But within the nucleus, the potential is opposite in sign because of the strongly attractive nuclear force. These conditions are represented in the U versus r diagram in Fig. 29.7. This might represent the potential of a ^{238}U nucleus.

Consider alpha particles from a ^{214}Po source incident on a thin foil of ^{238}U. As illustrated in the figure, Rutherford scattering takes place. We say that this results from a potential barrier that the incident ^{214}Po alpha particles do not have enough energy to overcome. Instead, they are scattered away. If an incident particle did have enough energy to overcome or cross this Coulomb barrier, it would enter the nucleus. In this case, a nuclear reaction would result (to be discussed in Chapter 30).

However, the ^{238}U nucleus itself undergoes alpha decay, emitting an alpha particle with an energy of 4.4 MeV. This fact appears to contradict the scattering experiment. How can the lower-energy alpha particles cross a potential barrier that higher-energy incident alpha particles cannot? Classically, this is impossible, since it violates the conservation of energy. However, quantum mechanics offers an explanation.

Recall from Section 28.2 that quantum mechanics predicts a finite probability of finding a particle in a classically forbidden region as a result of a decaying wave function in this region. If the alpha particle exists as an entity in the nucleus, perhaps its wave function tails off in the barrier region and makes it through to the outside (Fig. 29.8). There would then be a finite probability (ψ^2) of finding the alpha particle outside the nucleus. This is called **tunneling**, or **barrier penetration**, since the alpha particle seems to tunnel through the barrier.

You might point out that there is still a violation of the conservation of energy. However, the uncertainty principle tells us that energy conservation can be violated (is uncertain) by an amount ΔE for a time Δt. Quantum mechanics then allows the conservation of energy to be violated for brief periods, which may be long enough for the alpha particle to tunnel through the barrier. The probability of finding an alpha particle outside the ^{238}U nucleus is extremely small, but this is reflected in the extremely slow decay rate of ^{238}U. The height and width of the potential barrier determine the time or probability an alpha particle has to escape from the nucleus. Thus, the barrier would also control the decay rate (which will be considered in the next section).

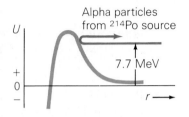

Figure 29.7 Nuclear potential barrier
Alpha particles from radioactive polonium with energies of 7.7 MeV do not have enough energy to overcome the nuclear potential barrier of ^{238}U and are scattered.

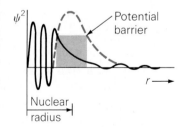

Figure 29.8 Tunneling effect, or barrier penetration
The probability (ψ^2) of finding an alpha particle tails off through the nuclear potential barrier, and there is a finite probability of finding the alpha particle outside the nucleus.

Beta Decay

The emission of an electron (a beta particle) in a nuclear decay process might seem contradictory to the proton-neutron model of the nucleus. It should be noted that the electron emitted in beta decay is not an orbital electron, but comes from the nucleus. In fact, the process of **beta decay** indicates that the electron is created in the nucleus itself. For example, the equation for carbon-14 beta decay is

$$\underset{\text{carbon}}{{}^{14}_{6}\text{C}_8} \longrightarrow \underset{\text{nitrogen}}{{}^{14}_{7}\text{N}_7} + \underset{\substack{\text{beta particle} \\ \text{(electron)}}}{{}^{0}_{-1}\text{e}}$$

In this and similar beta decay processes, the neutron number of the parent nucleus decreases by one ($\Delta N = -1$), and the proton number of the daughter nucleus increases by one ($\Delta Z = +1$). The mass number is unchanged. This indicates that a neutron within the nucleus decays into a proton and an electron:

$$\underset{\text{neutron}}{{}^{1}_{0}\text{n}} \longrightarrow \underset{\text{proton}}{{}^{1}_{1}\text{p}} + \underset{\text{electron}}{{}^{0}_{-1}\text{e}}$$

Notice that the neutron is written in nuclear notation as an n with a zero proton number and the electron (beta particle) as an e with a -1 proton number. The total proton numbers on each side of nuclear equations containing these particles are also equal.

There is another elementary particle, called a neutrino, emitted in beta decay, but for the sake of simplicity, it is not shown in the nuclear equation. The neutrino is a quantum of energy; its place in beta decay will be discussed in Chapter 30.

There are actually three modes of beta decay: β^-, β^+, and electron capture. The β^- decay involves the emission of an electron as above. Isotopes that decay by this means have more neutrons than protons. However, unstable isotopes with more protons than neutrons decay by β^+ **decay**, which involves the emission of a positron (${}^{0}_{+1}\text{e}$). An example of β^+ decay is

$$\underset{\text{oxygen}}{{}^{15}_{8}\text{O}_7} \longrightarrow \underset{\text{nitrogen}}{{}^{15}_{7}\text{N}_8} + \underset{\text{positron}}{{}^{0}_{+1}\text{e}}$$

As in β^- decay, the mass numbers of the parent and daughter nuclei do not change, but the proton number of the daughter nucleus in this case is one less than that of the parent nucleus. This implies that a proton disintegrates in the process into a neutron and a positron:

$$\underset{}{{}^{1}_{1}\text{p}} \longrightarrow {}^{1}_{0}\text{n} + {}^{0}_{+1}\text{e}$$

The third related process, **electron capture**, involves the absorption of one of the atomic electrons by a nucleus. An example of electron capture is

$$\underset{\text{electron}}{{}^{0}_{-1}\text{e}} + \underset{\text{beryllium}}{{}^{7}_{4}\text{Be}_3} \longrightarrow \underset{\text{lithium}}{{}^{7}_{3}\text{Li}_4}$$

As in β^+ decay, a proton changes into a neutron, but no particle is emitted in the electron capture process. An electron in the innermost K shell is usually captured, and this is referred to as K-capture. Since there is no charged particle emission, the process is detected by observing the emission of characteristic

X-rays produced in the filling of the K-shell vacancy by an electron from an outer shell and the filling of the subsequent resulting vacancies. Because X-ray emission must take place *after* the K-capture, the X-rays are characteristic of the daughter nucleus, not the parent.

Gamma Decay

In **gamma decay**, the nucleus emits a gamma (γ) ray, or a "particle" of electromagnetic energy. The emission of a gamma ray by a nucleus in an excited state is analogous to the emission of a photon by an excited atom. This may come about because of an energetic collision with another particle or, more commonly, because the nucleus is in an excited state after a previous radioactive decay.

Nuclei are thought of as having energy levels like atomic electrons. However, the nuclear energy levels are much farther apart than those of an atom. The nuclear energy levels are on the order of kilo-electron volts (keV) and mega-electron volts (MeV) apart, compared to a few electron volts of difference between atomic energy levels. As a result, gamma rays are very energetic, having frequencies similar to those of X-rays. The distinction is that gamma rays are produced in nuclear processes, while X-rays come from atomic electron interactions.

$$\underset{\text{nickel}}{^{61}_{28}\text{Ni*}} \longrightarrow \underset{\text{nickel}}{^{61}_{28}\text{Ni}} + \underset{\text{gamma ray}}{\gamma}$$

The asterisk indicates that the ^{61}Ni nucleus is in an excited state. The de-excitation process results in the emission of a gamma ray. Note that *in the gamma decay process, the mass and proton numbers do not change.* The daughter nucleus in this case is simply the parent nucleus with less energy.

Of the nearly 1200 known unstable nuclides, only a small number occur naturally. Most of the radioactive nuclides found in nature occur as products of the decay chains of heavy nuclei, that is, a chain of continual radioactive decay into lighter elements. The ^{238}U decay series is given in Fig. 29.9. It stops when the stable isotope ^{206}Pb is reached. Note that some nuclides in the series decay by two modes. As can be seen, radon (^{222}Rn) is part of this decay series. This radioactive gas has received a great deal of attention because it can accumulate in dangerous amounts in poorly ventilated buildings. The gas seeps upward from natural radioactive decay in the ground.

29.3 Decay Rate and Half-Life

The nuclei of a sample of a radioactive isotope do not decay all at once, but do so randomly at a characteristic rate. There is apparently some internal timing mechanism that is unaffected by any external stimulus. No one can tell exactly when particular nuclei will decay. All that can be determined is how many nuclei in a sample will decay over a period of time.

The **activity** of a sample of radioactive isotope is expressed in terms of the number (ΔN) of disintegrations, or decays, per time, that is, $\Delta N / \Delta t$. For a given amount of material, the activity decreases with time, since fewer nuclei are present. Also, each isotope has its own characteristic rate of decrease. Since the activity is proportional to the number of nuclei (N) present,

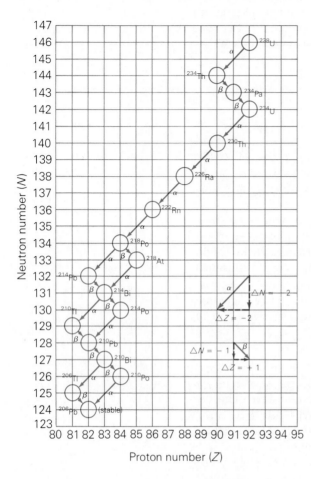

Figure 29.9 **Decay series for uranium-238**
On this plot of N versus Z, a diagonal transition from right to left is an alpha decay process, and a diagonal transition from left to right is a beta decay process. The decay series continues until a stable nucleus is reached.

$$\frac{\Delta N}{\Delta t} \propto N$$

This can be written in equation form with a constant of proportionality:

$$\frac{\Delta N}{\Delta t} = -\lambda N \qquad (29.2)$$

The constant λ is called the **decay constant** and is different for different isotopes. The greater the decay constant, the greater the rate of decay and the more radioactive the isotope. The minus sign in the equation indicates that N is decreasing.

It turns out that the number of parent nuclei N decreases with time as illustrated in Fig. 29.10. The curve and N are said to decay *exponentially* since the graph follows an exponential function $e^{-\lambda t}$ with time, where λ is the decay constant.

The number N of the undecayed parent nuclei at a particular time is given by

$$N = N_o e^{-\lambda t} \qquad (29.3)$$

where N_o is the initial number of nuclei ($t = 0$). The activity decreases proportionally.

The value $e = 2.718$ is the base of natural logarithms.

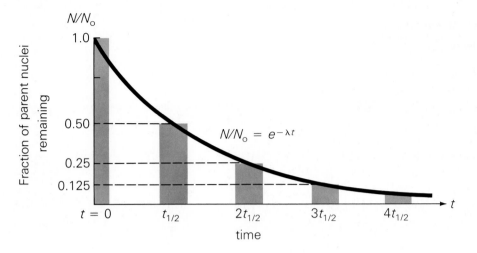

Figure 29.10 **Radioactive decay versus time**
When the fraction of remaining parent nuclei in a radioactive sample is plotted as a function of time, an exponential decay curve is obtained, the shape or steepness of which depends on the decay constant λ.

The decay rate of an isotope is commonly expressed in terms of its half-life rather than the decay constant. The **half-life** ($t_{\frac{1}{2}}$) is defined as the time it takes for half of the original nuclei in a radioactive sample to decay. This is the time corresponding to $N_0/2$ parent nuclei in Fig. 29.10, that is, $N/N_0 = \frac{1}{2}$. The activity also decreases to the fractional amount N/N_0 in a time t, since it is proportional to the number of nuclei present. That is, the number of decays per second decreases by half. This is what is actually measured to determine the half-life, since we cannot count the nuclei.

For example, as may be seen from the graph in Fig. 29.11, the half-life of strontium-90 (^{90}Sr) is 28 years. Another way of looking at what occurs in a half-life is to consider the amount or mass of the parent material. As illustrated in Fig. 29.11, if one had 100 micrograms (μg) of ^{90}Sr initially, only 50 μg would remain at the end of 28 years. The other 50 μg would have beta decayed by the reaction

$$^{90}_{38}\text{Sr} \longrightarrow ^{90}_{39}\text{Y} + ^{\;\;0}_{-1}\text{e}$$

strontium   yttrium   electron

After another 28 years, another half of the strontium nuclei will have decayed, leaving only 25 μg, and so on.

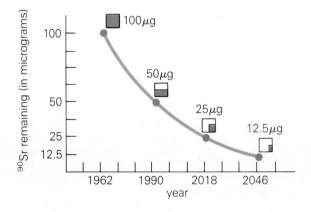

Figure 29.11 **Radioactive decay and amount of isotope**
As illustrated here for strontium-90 ($t_{\frac{1}{2}} = 28$ years), after each half-life, only half the amount of the original ^{90}Sr remains. The other half has decayed into ^{90}Y via beta decay.

Table 29.1
The Half-Lives of Some Radioactive Isotopes

Nuclide	Decay Mode	Half-Life
Beryllium-8 ($^8_4$Be)	Alpha	1×10^{-16} s
Polonium-213 ($^{213}_{84}$Po)	Alpha	4×10^{-6} s
Oxygen-19 ($^{19}_8$O)	β^-	27 s
Fluorine-17 ($^{17}_9$F)	β^+	66 s
Technetium-104 ($^{104}_{43}$Tc)	β^-	18 min
Krypton-76 ($^{76}_{36}$Kr)	Electron capture	14.8 h
Magnesium-28 ($^{28}_{12}$Mg)	β^-	21 h
Iodine-131 ($^{131}_{53}$I)	β^-	8 days
Cobalt-60 ($^{60}_{27}$Co)	β^-	5.3 years
Strontium-90 ($^{90}_{38}$Sr)	β^-	28 years
Radium-226 ($^{226}_{88}$Ra)	Alpha	1600 years
Carbon-14 ($^{14}_6$C)	β^-	5730 years
Plutonium-239 ($^{239}_{94}$Pu)	Alpha	2.4×10^4 years
Uranium-238 ($^{238}_{92}$U)	Alpha	4.5×10^9 years
Rubidium-87 ($^{87}_{37}$Rb)	β^-	4.7×10^{10} years

The half-lives of radioactive isotopes vary greatly, as may be seen from Table 29.1. Isotopes with very short half-lives are generally created in nuclear reactions, which will be considered in the next chapter. If these isotopes did exist when the Earth was formed, they would have long since decayed away. In fact, the elements technetium (Tc) and promethium (Pm) are not found on Earth (but can be produced in laboratories). On the other hand, the half-life of the naturally occurring ^{238}U isotope is 4.5 billion years, which is close to the estimated age of the Earth. Thus, we would expect about half of the original ^{238}U present when the Earth was formed to exist today.

The longer the half-life of an isotope, the more slowly it decays and the smaller the decay constant. The half-life and the decay constant have an inverse relationship ($t_{\frac{1}{2}} \propto 1/\lambda$). This can be seen from Eq. 29.3. The half-life is the time it takes for the number of nuclei present or the activity to decrease by half, that is, $N = N_0/2$. Thus,

$$\frac{N}{N_0} = e^{-\lambda t_{\frac{1}{2}}} = \frac{1}{2}$$

But

$$e^{-0.693} = \frac{1}{2}$$

so by comparison,

$$t_{\frac{1}{2}} = \frac{0.693}{\lambda} \tag{29.4}$$

Example 29.2 Half-Life and Activity
The half-life of iodine-131 used in medical thyroid uptake tests is 8.0 days. After a certain time, an amount of ^{131}I containing 4.0×10^{22} nuclei accumulates in a patient's thyroid gland. (a) What will be the observed activity in 24 hours? (b) How many of the ^{131}I nuclei remain at this time?

Solution

Given: $t_{\frac{1}{2}} = (8.0 \text{ days})(86{,}400 \text{ s/day})$ *Find*: (a) $\dfrac{\Delta N}{\Delta t}$
$\qquad\qquad = 6.9 \times 10^5 \text{ s}$
$\qquad N_o = 4.0 \times 10^{22} \text{ nuclei}$ (b) N
$\qquad \Delta t = 24 \text{ h} = 8.64 \times 10^4 \text{ s}$

(a) First, Eq. 29.4 is used to find the decay constant with $t_{\frac{1}{2}}$ in seconds:

$$\lambda = \frac{0.693}{t_{\frac{1}{2}}} = \frac{0.693}{6.9 \times 10^5 \text{ s}} = 1.0 \times 10^{-6} \text{ s}^{-1}$$

Then, using Eq. 29.2, we have

$$\Delta N / \Delta t = -\lambda N_o = -(1.0 \times 10^{-6} \text{ s}^{-1})(4.0 \times 10^{22})$$
$$= -4.0 \times 10^{16} \text{ decays/s}$$

where the minus indicates that the activity is decreasing.
(b) The actual number of parent nuclei present can be found from Eq. 29.3. With $t = 1$ day and $\lambda = 0.693/8.0$ day $= 0.087$ day^{-1}, we have

$$N = N_o e^{-\lambda t} = (4.0 \times 10^{22} \text{ nuclei})e^{-(0.087 \text{ day}^{-1})(1 \text{ day})}$$
$$= (4.0 \times 10^{22} \text{ nuclei})(0.917) = 3.7 \times 10^{22} \text{ nuclei} \quad \blacksquare$$

The strength of a radioactive sample or source may be specified at a given time by its activity. A common unit of radioactivity is named in honor of Pierre and Marie Curie. One **curie (Ci)** is defined as

$$1 \text{ Ci} \equiv 3.70 \times 10^{10} \text{ decays/s}$$

This is based on the activity of one gram of radium.

The curie is the traditional unit for expressing radioactivity. However, the proper SI unit is the **becquerel (Bq)**, which is simply defined as

$$1 \text{ Bq} \equiv 1 \text{ decay/s}$$

Therefore,

$$1 \text{ Ci} = 3.70 \times 10^{10} \text{ Bq}$$

Even with the emphasis on the SI system, radioactive sources are commonly rated in curies. The curie is a relatively large unit, however, so the millicurie (mCi), the microcurie (μCi), and sometimes even smaller multiples are used. As was mentioned earlier, there is concern about concentrations of radioactive radon gas in homes. The radon results from the alpha decay of uranium in the soil below the structure. The government-recommended limit for radon, a lung-cancer–causing agent, is 4 picocuries per liter (pCi/L) of air.

Example 29.3   Source Strength
A ^{90}Sr beta source has a strength of 10.0 mCi. How many decays/s will be observed to take place at the end of 84 years or 3 half-lives?

Solution

Given: strength $= 10.0$ mCi   *Find*: $\dfrac{\Delta N}{\Delta t}$
$\qquad \Delta t = 84 \text{ years}$
$\qquad t_{\frac{1}{2}} = 28 \text{ years (from Table 29.1)}$

After 3 half-lives, the activity will be only $\frac{1}{8}$ as great ($\frac{1}{2} \times \frac{1}{2} \times \frac{1}{2} = \frac{1}{8}$), and the strength of the source will then be

$$\frac{\Delta N}{\Delta t} = 10.0 \text{ mCi} \times \frac{1}{8} = 1.25 \text{ mCi} = 1.25 \times 10^{-3} \text{ Ci}$$

and

$$\frac{\Delta N}{\Delta t} = (1.25 \times 10^{-3} \text{ Ci})(3.70 \times 10^{10} \text{ decays/s/Ci})$$

$$= 4.63 \times 10^7 \text{ decays/s} \quad (= 4.63 \times 10^7 \text{ Bq}) \quad \blacksquare$$

29.4 Nuclear Stability and Binding Energy

Now that we have considered some of the properties of unstable isotopes, we turn our attention to the more common stable isotopes. Stable isotopes exist naturally for all elements having proton numbers from 1 to 83, except for those with $Z = 43$ (technetium) and $Z = 61$ (promethium), as was pointed out earlier. The nuclear interactions that give rise to nuclear stability are extremely complicated. However, by looking at some of the general properties of stable nuclei, it is possible to obtain some qualitative and quantitative criteria for nuclear stability. This will give us some general rules for determining whether or not a particular nuclide would be expected to be stable or unstable.

Nucleon Populations

One of the easiest things to consider is the relative number of protons and neutrons in stable nuclei. Nuclear stability must be related in some way to the dominance of either the repulsive Coulomb force between protons or the attractive nuclear force between nucleons, and this force dominance depends on the relative numbers, or ratio, of protons and neutrons.

For stable nuclei of low mass numbers for about $A < 40$, the nucleon ratio is approximately 1. That is, the number of protons and the number of neutrons are equal or nearly equal. For example, $_2^4\text{He}_2$, $_6^{12}\text{C}_6$, $_{11}^{23}\text{Na}_{12}$, and $_{13}^{27}\text{Al}_{14}$ have the same or about the same number of protons as neutrons. For stable nuclei of higher mass numbers ($A > 40$), the number of neutrons exceeds the number of protons. The heavier the nuclei, the more the neutrons outnumber the protons.

This trend is illustrated in Fig. 29.12, which shows a plot of the neutron number (N) versus proton number (Z) for stable nuclei. Notice how the heavier stable nuclei lie above the 45° or $N = Z$ line. Examples of heavy stable nuclei include $_{28}^{62}\text{Ni}_{38}$, $_{50}^{114}\text{Sn}_{64}$, $_{82}^{208}\text{Pb}_{126}$, and $_{83}^{209}\text{Bi}_{126}$. (Bismuth is the heaviest element that has a stable isotope.) The extra neutrons of such heavy stable nuclei presumably allow the attractive forces among the nucleons to be greater than the repulsive forces between the numerous protons.

Radioactive decay adjusts the proton and neutron numbers of an unstable isotope until a stable isotope on the neutron-proton stability curve in Fig. 29.12 is reached. Since alpha decay decreases the numbers of protons and neutrons by equal amounts, alpha decay alone would give nuclei with neutron populations that are larger than that of one of the stable isotopes on the curve. However,

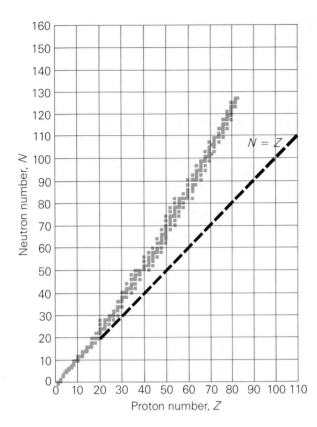

Figure 29.12 **A plot of *N* versus *Z* for stable nuclei** For nuclei with $A < 40$ ($Z < 20$ and $N < 20$), the number of protons and the number of neutrons are equal or nearly equal. For nuclei with $A > 40$, the number of neutrons exceeds the number of protons and the nucleus lies above the $N = Z$ line.

beta decay following alpha decay would lead to a stable combination, since the effect of beta decay is the loss of a neutron and the gain of a proton. Very heavy nuclei undergo a chain, or sequence, of alpha and beta decays until a stable nucleus is reached, as was illustrated in Fig. 29.9 for ^{238}U.

Pairing Effect

A fact that is not too evident from Fig. 29.12 is that many stable nuclei have even numbers of both protons and neutrons, and very few have odd numbers of both protons and neutrons. A survey of the stable isotopes (Table 29.2) shows that 168 stable nuclei have this even-even combination, 107 are even-odd or odd-even, and only four contain odd numbers of both protons and neutrons. These four are isotopes of the elements with the four lowest odd proton numbers: $^{2}_{1}\text{H}_1$, $^{6}_{3}\text{Li}_3$, $^{10}_{5}\text{B}_5$, and $^{14}_{7}\text{N}_7$.

The even and odd combinations seem to indicate that the protons and neutrons in stable nuclei above the very lightest elements tend to pair up. That is, two protons pair up and two neutrons pair up, but a proton and neutron do not pair up. (The converse seems to be the case for the light nuclei. For example, the deuterium nucleus is a p-n combination (proton-neutron), but there are no p-p or n-n particles.) When there are odd numbers of nucleons, there is generally an instability associated with the pairing up of the last proton and neutron. The instability is less if there is only one odd nucleon (odd-even or even-odd combination).

This so-called **pairing effect** gives a qualitative criterion for stability. For

Table 29.2
Pairing Effect of Stable Nuclei

Proton Number	Neutron Number	Number of Stable Nuclei
Even	Even	168
Even	Odd	107
Odd	Even	
Odd	Odd	4

example, you would expect the sodium isotope $^{23}_{11}Na_{12}$ to probably be stable, but not $^{22}_{11}Na_{11}$. This is actually the case.

The general criteria for nuclear stability can be summarized as follows:

1. All isotopes with proton numbers greater than 83 ($Z > 83$) are unstable.
2. (a) Most even-even nuclei are stable.
   (b) Many odd-even or even-odd nuclei are stable.
   (c) Only four odd-odd nuclei are stable (^2_1H, ^6_3Li, $^{10}_5B$, and $^{14}_7N$).
3. (a) Stable nuclei with mass numbers less than 40 ($A < 40$) have approximately the same number of protons and neutrons.
   (b) Stable nuclei with mass numbers greater than 40 ($A > 40$) have more neutrons than protons.

A single exception to these rules (an unsatisfied criterion) for a nucleus indicates that the isotope will be unstable.

Example 29.4   Nuclear Stability

Is the sulfur isotope $^{38}_{16}S$ likely to be stable?

Solution

Applying the general criteria:

1. *Satisfied.* Isotopes with $Z > 83$ are immediately known to be unstable. With $Z = 16$, this criterion is satisfied.

2. *Satisfied.* The isotope $^{38}_{16}S_{22}$ has an even-even nucleus and so has a good probability of being stable.

3. *Not satisfied.* $A < 40$, and $Z(= 16)$ and $N(= 22)$ are not approximately equal.

Therefore, the ^{38}S isotope is likely to be unstable. (The nucleus actually is unstable to beta decay.)   ∎

Binding Energy

An important quantitative aspect of nuclear stability is the binding energy of the nucleons. This can be calculated by considering the sums of nuclear and electron masses and of their corresponding energies.

Since the masses of nuclei are so small in relation to the standard kilogram mass unit, another standard, the *unified* **atomic mass unit (u)** is used to measure nuclear masses. A *neutral atom* of carbon-12 is defined as having an exact value of 12.000000 u, and

$$1\ u \equiv 1.6606 \times 10^{-27}\ kg$$

The various atomic particles then have masses in atomic mass units as shown in Table 29.3. The listed energy equivalents reflect Einstein's $E = m_o c^2$ relationship.

Note that the proton and the hydrogen atom (^1_1H) are listed separately, having different masses. The difference is the mass of the atomic electron. We deal almost exclusively with the masses of neutral atoms (nucleons plus Z electrons) rather than strictly with nuclei, since the atomic masses are what can be measured. Keep this in mind. We will soon be interested in very small mass differences, and the electronic mass is significant. We must balance out the masses of the electrons when we compare masses in nuclear processes.

Table 29.3
Particle Masses and Energy Equivalents

Particle	Mass		Energy (MeV)
	u	kg	
	1	1.6606×10^{-27}	931.50
Electron	0.000548	9.1095×10^{-31}	0.511
Proton	1.007276	1.67265×10^{-27}	938.28
$_1^1$H atom	1.007825	1.67356×10^{-27}	938.79
Neutron	1.008665	1.67500×10^{-27}	939.57

Given this information, we can look at the effect of mass-energy relationships on nuclear stability. For example, if you compare the mass of a helium nucleus to the total mass of nucleons that make it up, a significant inequity emerges: a neutral helium atom has a mass of 4.002603 u. (Atomic masses of some isotopes are given in Appendix V.) The total mass of two protons (with two electrons, or actually two ^{1}H atoms) and two neutrons is by addition,

$$2m(^1\text{H}) = 2.015650 \text{ u}$$

$$2m_n = \underline{2.017330 \text{ u}}$$

$$4.032980 \text{ u}$$

Notice that this is greater than the mass of the helium atom. The helium nucleus is then less massive than the sum of its parts by an amount

$$\Delta m = 4.032980 \text{ u} - 4.002603 \text{ u} = 0.030377 \text{ u}$$

where the electronic mass cancels out, since the mass of a hydrogen atom was used for the proton. This mass difference has an energy equivalent of

$$(0.030377 \text{ u})(931.5 \text{ MeV/u}) = 28.30 \text{ MeV}$$

which proves to be the same as the **total binding energy** (E_b) of the nucleus, or the amount of energy/mass given up when the separate nucleons combined to form a nucleus.

Another way of looking at the binding energy is that it is the energy that would be required to separate the constituent nucleons into free particles. This is illustrated in Fig. 29.13 for the helium nucleus. Exactly 28.30 MeV of energy would be necessary to do the work of separating the nucleons. If the mass of a nucleus and the mass of its nucleons were the same, then the nucleons would not be bound together, and the nucleus would come apart without energy input.

We can gain some insight into the nature of the nuclear force by considering the **average binding energy per nucleon** for stable nuclei. This is simply the total binding energy of a particular nucleus divided by its number of nucleons, or E_b/A, where A is the mass number (number of protons plus number of neutrons). For example, for the helium nucleus (^{4}He), the average binding energy per nucleon is

$$\frac{E_b}{A} = \frac{28.30 \text{ MeV}}{4} = 7.075 \text{ MeV}$$

Compared to the binding energy of atomic electrons, for example, 13.6 eV for a

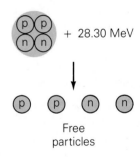

Figure 29.13 Binding energy
As illustrated here, 28.30 MeV of work is required to separate a helium nucleus into free protons and neutrons. Conversely, if two protons and two neutrons could be combined to form a helium nucleus, 28.30 MeV of energy would be given up, and this is the binding energy of the nucleus.

hydrogen electron in the ground state, the nuclear binding energy is about 10^6 times larger, which indicates a very strong binding force.

If E_b/A is calculated for various nuclei and plotted versus mass number, the values generally lie along a curve as shown in Fig. 29.14. The value of E_b/A rises rapidly with increasing A for light nuclei and levels off (around $A = 15$) at about 8.0 MeV, with a maximum value of about 8.7 MeV. The maximum of the curve occurs in the vicinity of iron, which has a very stable nucleus. For $A > 60$, the E_b/A values decrease slowly for heavier nuclei, a finding indicating that the nucleons are less tightly bound.

The maximum in the curve gives an important indication that will be considered in the next chapter. If a large nucleus could be split or fissioned into two lighter nuclei, the nucleons would be more tightly bound, and energy would be given up in the process—more than would be needed to induce the fission. (Recall how an atom gives up energy when one of its electrons becomes more tightly bound or makes a transition to a lower energy level.) Similarly, on the other side of the maximum, if we could fuse together two very light nuclei into a heavier nucleus, this would raise it up on the curve, and energy would be released. Such fission and fusion processes are the sources of nuclear energy (Chapter 30).

In general, the E_b/A curve shows us that, except for very light nuclei, the binding energy per nucleon does not change a great deal and is on the order of

$$E_b/A \cong 8 \text{ MeV}$$

Since E_b/A is relatively constant for most nuclei, an approximation is that

$$E_b \propto A$$

In other words, the total binding energy is proportional to the mass number.

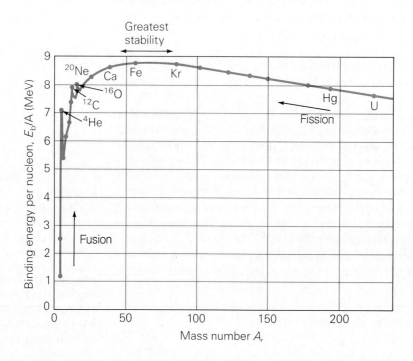

Figure 29.14 **A plot of binding energy per nucleon versus mass number** If the E_b/A of stable nuclei is plotted versus A, the curve has a maximum, which indicates that the nuclei in this region are on the average more tightly bound together with the greatest stability.

This proportionality indicates a characteristic of the nuclear force that was mentioned earlier, one that makes it quite different from the electrical Coulomb force. Imagine that the attractive nuclear force acts among *all* nucleons (protons and neutrons) in a nucleus. Each pair of nucleons would then contribute to the total binding energy. In a nucleus containing A nucleons, there are $A(A-1)/2$ pairs, so there would be $A(A-2)/2$ contributions to the total binding energy. For nuclei for which $A \gg 1$ (heavier nuclei), then, $A(A-1) \cong A^2$, and we would expect that

$$E_b \propto A^2$$

In fact, $E_b \propto A$ for most nuclei, which indicates that a given nucleon is not bound to all the other nucleons. This phenomenon, called *saturation*, implies that the nuclear forces act over a very short range.

29.5 Radiation Detection and Applications

Since the products of radioactive decay cannot be detected directly by our senses, detection must be done by some indirect means. We have already mentioned the detection of nuclear radiation by photographic film. People who work with radioactive materials and X-rays usually wear film badges, which indicate cumulative exposure to radiation by the degree of darkening of the film when it is developed. However, more immediate and quantitative methods of detection are desirable, and a variety of instruments have been developed for this purpose.

These detectors are based on the ionization or excitation of atoms by the passage of energetic particles through matter. The energy of charged particles, such as alpha and beta particles, is transferred to atoms along their paths by Coulomb interactions. Gamma rays are detected indirectly. The photons lose energy by Compton scattering or pair production. The energetic particles produced by these interactions cause the ionization in a detector.

One of the most common detectors is the **Geiger counter**, which was developed chiefly by Hans Gieger, a student and then colleague of Lord Rutherford. The principle of the Geiger counter is illustrated in Fig. 29.15. A voltage of 800–1000 V is applied across the wire electrode and metal tube of the Geiger tube. The tube contains a gas (such as argon) at low pressure. When an ionizing particle enters the tube through a thin window at one end, it ionizes a few atoms of the gas. The freed electrons are attracted and accelerated toward the positive wire anode. On their way, they strike and ionize other gas atoms.

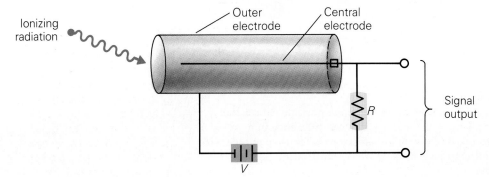

Figure 29.15 The Geiger counter
This diagram shows the basic elements of a Geiger tube. See text for description.

This process multiplies, and an "avalanche" discharge results that produces a voltage pulse. The pulse is amplified and sent to an electronic counter that counts the pulses or the number of particles detected. The pulses may also be used to drive a loudspeaker so that particle detection is heard as an audible click.

A major disadvantage of the Geiger counter is its relatively long dead time. This is the recovery time required between successive electrical discharges in the tube. The dead time of a Geiger tube is about 200 μs, which limits the counting rate to a few hundred counts per second. If particles are incident on the tube at a faster rate, not all of them will be counted.

Another method of detection, one of the oldest, has a much shorter dead time. This is used in the **scintillation counter** (Fig. 29.16). In the counter, atoms of a phosphor material (such as sodium iodide, NaI) are excited by an incident particle, and visible light is emitted when the atoms return to their ground state. The light pulse is converted to an electrical pulse by a photoelectric material. This is amplified in a *photomultiplier tube*, which consists of a series of successively higher potential electrodes. The photoelectrons are accelerated toward the first electrode and acquire sufficient energy to cause several secondary electrons to be emitted when they strike the electrode. This process continues, and relatively weak scintillations are converted into sizable electrical pulses, which are counted electronically.

The dead time of a scintillation counter is a few microseconds. Also, the magnitude of the photomultiplier current pulse is proportional to the number of photons generated by the incident particle, which in turn is proportional to its energy. Thus, the magnitude of the counted pulses gives a measure of the energy of the incident particle. Counters with this feature are called *proportional counters*.

Semiconductor materials provide a relatively new method of radiation detection. In a **solid state** or **semiconductor detector**, charged particles passing through a junction diode produce electron-hole pairs (solid state

Early detection was done with a scintillating phosphorescent screen that produced tiny flashes of light bright enough to be seen in the dark with the aid of a magnifying glass. Such a screen was used in the Rutherford scattering experiments with slow and tiring visual counting. It has been said that whenever Rutherford himself did the actual counting, he would "damn" vigorously for a few minutes and then have an assistant take over. More than a million scintillations or flashes were counted in the overall experiment.

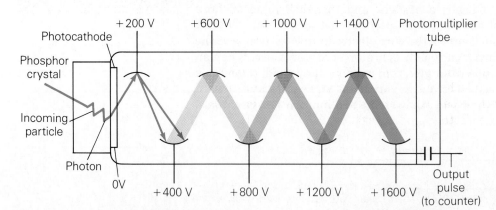

Figure 29.16 **The scintillation counter**
A photon emitted by a phosphor atom excited by an incoming particle causes the emission of a photoelectron. Accelerated through a potential difference in a photomultiplier tube, photoelectrons free secondary electrons when they collide with the electrode. Through successive steps, a relatively weak scintillation is converted into a measurable electrical pulse.

ionization, so to speak). With a voltage across the diode, the electron-hole pairs give rise to electric signals, which may be amplified and counted. Solid state detectors are very fast—that is, they are capable of very high counting rates.

Counting methods are used to determine the number of particles produced by decaying nuclei of a radioactive isotope. Other methods of detection allow the tracks of charged particles to be seen or recorded visually. One of the earliest used was the *photographic emulsion*. A particle passing through the emulsion ionizes atoms along its path. This exposes the emulsion at these points, and when the emulsion is developed, a record of the particle's path is observed.

Other types of detectors primarily used in research are the cloud chamber, the bubble chamber, and the spark chamber. In the first two, vapors and liquids are supercooled and superheated, respectively, by suddenly varying the volume and pressure. The **cloud chamber** was developed early in this century by C. T. R. Wilson, a British atmospheric physicist. In the chamber, supercooled vapor condenses into droplets on ionized molecules along the path of an energetic particle. When the chamber is illuminated, the droplets scatter the light, making the path visible.

The **bubble chamber**, which was invented by the American physicist D. A. Glazer in 1952, uses a similar principle. A reduction in pressure causes a liquid to be superheated and able to boil. Ions produced along the path of an energetic particle become nucleation sites where a trail of bubbles form (Fig. 29.17). When a magnetic field is applied across the chamber, charged particles are deflected, and the energy of a particle can be calculated from the radius of curvature of the particle's path measured from a photograph. Since the bubble chamber uses a liquid, commonly liquid hydrogen, the density of atoms in it is much greater than that in the vapor of a cloud chamber. As a result, the tracks are more readily observable, and bubble chambers have displaced cloud chambers.

Gamma rays do not leave visible tracks in a bubble chamber. However, their presence can be detected indirectly, for example, by electrons that have been Compton scattered and by the creation of electron-positron pairs in pair production.

The path of a charged particle is registered by a series of sparks in a **spark chamber**. Basically, a charged particle passes between a pair of electrodes with a high potential difference that are immersed in an inert (noble) gas. The charged particle is accelerated and causes the ionization of gas molecules, giving rise to a visible spark or flash between the electrodes. A spark chamber is merely an array of such electrode pairs in the form of parallel plates or wires. A series of sparks, which can be photographed, marks the particle's path.

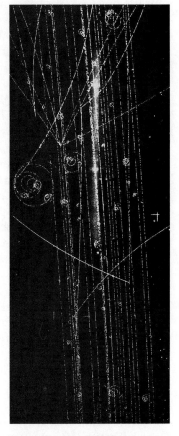

Figure 29.17 **Bubble-chamber photograph of particle tracks**

Radioactive Dating

Because of their constant decay rates, radioactive isotopes can be used as nuclear clocks. As we have seen, the half-life of a radioactive isotope may be used to project how much of a given amount of material will exist in the future. Similarly, by using the half-life to project backward in time, scientists can determine the ages of objects containing radioactive isotopes. As you might surmise, some idea of the initial composition or initial amount of an isotope must be known.

To illustrate the principle of radioactive dating, let's take a look at how it is done with ^{14}C. Carbon-14 is used to date the remnants of objects, such as bones or parchment, that were once alive. The process depends on the fact that living things—plants and animals (including yourself)—contain a known amount of radioactive ^{14}C. The concentration of carbon-14 is very small, about one ^{14}C atom for 7.2×10^{11} atoms of ordinary ^{12}C. Even so, this concentration cannot be due to an abundance of carbon-14 present when the Earth was formed, since a half-life of $t_{\frac{1}{2}} = 5730$ years for ^{14}C is brief in comparison to the estimated age of the Earth (over 4 billion years).

The observed concentration of ^{14}C is accounted for by its continuous production in the upper atmosphere. Cosmic rays from outer space cause reactions that produce neutrons (Fig. 29.18). The neutrons react with the nuclei of the nitrogen atoms of the air to produce ^{14}C by the reaction

$$^{14}_{7}N + ^{1}_{0}n \longrightarrow ^{14}_{6}C + ^{1}_{1}H$$

Recall that the carbon-14 then decays by beta decay ($^{14}_{6}C \longrightarrow ^{14}_{7}N + ^{0}_{-1}e$). Although the intensity of incident cosmic rays may not be constant, the concentration of ^{14}C in the atmosphere is relatively constant because of atmospheric mixing and the fixed decay rate.

The ^{14}C is oxidized into carbon dioxide (CO_2), so a small fraction of the CO_2 molecules of the air are radioactive. Plants take in this radioactive CO_2 by photosynthesis, and animals ingest the material produced. As a result, the concentration of carbon-14 in living organic matter is the same as the concentration in the atmosphere, 1 part in 7.2×10^{11}. However, once an organism dies, the ^{14}C is not replenished, and the concentration decreases with radioactive decay (with $t_{\frac{1}{2}} = 5730$ years). Measurement of the concentration of carbon-14 in dead matter relative to that in living things can then be used to establish when the organism died.

Since radioactivity is generally measured in terms of activity, we must first know the ^{14}C activity in a living organism. This is derived in the following example.

Example 29.5 Natural Carbon-14 Activity

Determine the ^{14}C activity, in decays per minute per gram of natural carbon, found in living organisms if the concentration of carbon-14 is the same as that in the atmosphere.

Solution

The relative concentration of carbon-14 in living organisms is one part in 7.2×10^{11} or

$$\frac{^{14}C}{^{12}C} = \frac{1}{7.2 \times 10^{11}} = 1.4 \times 10^{-12}$$

Carbon has an atomic mass of 12.0, so the number of nuclei (atoms) in 1 g of carbon may be found by using Avogadro's number N_A (6.02×10^{23}) and the number of moles ($n = N/N_A$, Chapter 11). With $n = 1\,g/(12\,g/mole) = 1/12$ mole,

$$N = nN_A = \left(\frac{1}{12\,\text{mole}}\right)(6.02 \times 10^{23}\ \text{nuclei/mole})$$

$$= 5.0 \times 10^{22} \quad \text{nuclei (per gram)}$$

Cosmic rays yield neutrons

which react with nitrogen atoms

to make carbon-14

which shows up as carbon dioxide throughout the atmosphere,

but when an organism dies, no fresh carbon-14 replaces the carbon-14 decaying in its tissues and the carbon-14 radioactivity decreases by half every 5730 years.

Figure 29.18 **Carbon-14 radioactive dating** This illustrates the formation of carbon-14 and its entry into the biosphere.

The number of ^{14}C nuclei per gram is

$$N\left(\frac{^{14}C}{^{12}C}\right) = (5.0 \times 10^{22} \text{ nuclei/g})(1.4 \times 10^{-12})$$

$$= 7.0 \times 10^{10} \quad ^{14}C \text{ nuclei/g}$$

With a half-life of $t_{\frac{1}{2}} = (5730 \text{ y})(5.26 \times 10^5 \text{ min/y}) = 3.01 \times 10^9 \text{ min}$, the decay constant is

$$\lambda = \frac{0.693}{t_{\frac{1}{2}}} = \frac{0.693}{3.01 \times 10^9 \text{ min}} = 2.30 \times 10^{-10} \text{ min}^{-1}$$

Then the activity, or the number of decays per minute per gram of carbon, is given by

$$\Delta N / \Delta t = \lambda N = (2.30 \times 10^{-10} \text{ min}^{-1})(7.0 \times 10^{10}) = 16 \text{ decays/g-min} \quad \blacksquare$$

If we determined that an artifact such as a bone or a piece of cloth had an activity of 8.0 counts/min per gram of carbon, the original living organism would have died one half-life or about 5700 years ago. This would put the date of the artifact near 3700 B.C.

The limit of radioactive carbon dating depends on the ability to measure the very low activity after an appreciable number of half-lives. Current techniques give an age dating limit of about 40,000–50,000 years, depending on the size of the sample. After about 9 half-lives, the radioactivity of a carbon sample has decreased so much that it is barely measurable (about 2 counts per gram per *hour*).

Another radioactive dating process uses lead-206 (^{206}Pb) and uranium-238 (^{238}U). This dating method is used extensively in geology because of the long half-life of ^{238}U. Lead-206 is the stable end isotope of the ^{238}U decay series (see Fig. 29.9). If a rock sample contains both these uranium and lead isotopes, the lead is assumed to be a result of the decay of the uranium that was there when the rock was first formed. Thus, the ratio of $^{206}Pb/^{238}U$ is a measure of the geologic age of the rock.

Medical Applications

In medicine, nuclear radiation has both advantages and disadvantages. It can be used beneficially in the diagnosis and treatment of some diseases but can also be potentially harmful if it is not properly handled and administered. Nuclear radiation and X-rays can penetrate human tissue without pain or any other sensation. However, early investigators quickly learned that large doses or repeated small doses led to red skin, lesions, and other conditions.

The chief hazard of radiation is damage to living cells, primarily due to ionization. Several effects can occur. Ions, particularly complex ions or radicals, produced by ionizing radiation may be highly reactive (for example, a hydroxyl ion OH^- from water). These interfere with the normal chemical operations of the cell. Also, radiation can alter the structure of a cell so that it cannot perform its normal functions. The cell may even die. This might not be serious, since similar cells would reproduce to make more cells. But if enough cells are damaged with large radiation doses, cell reproduction might not be fast enough, and the irradiated tissue could eventually die.

In other instances, there may be damage to a chromosome in the cell nucleus (genetic damage). Again, the cell might die, but it could instead live with damaged genetic code for cell reproduction. Damaged cells might reproduce, but the new cells might be different or defective. Also, such reproduction often occurs at an uncontrolled rate. This rapid, unregulated production of abnormal cells is called *cancer*.

Cancer cells may grow slowly in number with little effect on the surrounding normal cells, forming a benign tumor. They also may grow at the expense of the surrounding cells, producing a malignant tumor. Skin cancer and leukemia (cancer of the blood) are probable results of excessive radiation exposure. Leukemia is characterized by an abnormal increase in the number of white blood cells (leukocytes). The human cells most susceptible to radiation damage are those of the reproductive organs, bone marrow, and lymph nodes.

Radiation can cause cancer, but it can also be used to treat cancer. In this case, cell damage by radiation is used to control the growth of cancer cells. The radiation may be gamma rays from a radioactive ^{60}Co source or X-rays from a machine. Other particles produced in particle accelerators are also used to treat cancer. The electrical charge and energy of the particles determine the penetrating power of the radiation. X-rays and gamma rays are deeply penetrating, but generally, electrons (beta particles) penetrate only a few millimeters into biological tissue, and alpha particles penetrate only a fraction of a millimeter.

An important consideration in radiation therapy and radiation safety is the amount, or *dose*, of radiation. Several quantities are used to describe this in terms of *exposure*, *absorbed dose*, and *equivalent dose*. The earliest unit of dosage, the **roentgen** (R), was based on exposure and was defined in terms of the ionization produced in air. (One roentgen is the quantity of X-rays or gamma rays required to produce an ionization charge of 2.58×10^{-4} C/kg of air.)

The roentgen has been largely replaced by the rad (radiation *a*bsorbed *d*ose), which is an absorbed dose unit. One **rad** is an absorbed dose of radiation of 10^{-2} J of energy per kilogram in any absorbing material. The SI unit for absorbed dose is the gray (Gy):

$$1 \text{ Gy} \equiv 1 \text{ J/kg} \equiv 100 \text{ rad}$$

The gray was named in honor of Louis Harold Gray, a British radiobiologist whose studies laid the foundation for measuring absorbed dose.

These *physical* units give the energy absorbed per mass, but it is helpful to have some means for measuring the biological damage produced by radiation, since equal doses of different types of radiation produce different effects. For example, a relatively massive alpha particle with a charge of $2+$ moves through tissue rather slowly with a great deal of Coulomb interaction. The ionizing collisions thus occur close together along a short penetration path, and more localized damage is done than by a fast-moving electron or gamma ray.

This effect, or effective dose, is measured in terms of the **rem** unit (*r*oentgen or *r*ad *e*quivalent *m*an). The different degrees of effectiveness of different particles are characterized by the **relative biological effectiveness (RBE)** or *quality factor* (QF), which has been tabulated for the various particles in Table 29.4. The RBE is defined in terms of the number of rads of X-rays or gamma rays that produces the same biological damage as one rad of a given radiation. (Note in Table 29.4 that gamma rays have an RBE of 1.)

The effective dose is then given by the product of the dose in rads and the appropriate RBE:

$$\text{effective dose (in rem)} = \text{dose (in rad)} \times \text{RBE} \qquad (29.5)$$

Thus, one rem of any type of radiation does approximately the same amount of biological damage. For example, a 20-rem effective dose of alpha particles does the same damage as a 20-rem dose of X-rays, but 20 rad of X-rays are needed compared to 1 rad of alpha particles.

The SI unit of absorbed dose is the gray, and the effective dose with this unit is called the sievert (Sv):

$$\text{effective dose (in Sv)} = \text{dose (in Gy)} \times \text{RBE} \qquad (29.6)$$

Since 1 Gy $\equiv$ 100 rad, it follows that 1 Sv $\equiv$ 100 rem.

It is difficult to set a maximum permissible radiation dosage, but the general standard for humans is an average dose of 5 rem/year after the age of 18 with no more than 3 rem in any 3-month period. In the United States, the normal average annual dose per capita is about 200 mrem (millirem). About 125 mrem comes from the natural background of cosmic rays and naturally occurring radioactive isotopes (in the soil, building materials, and so on). The remainder is chiefly from diagnostic medical applications, mostly X-rays, and from miscellaneous sources such as television tubes.

Radioactive isotopes offer an important technique for diagnostic procedures. For example, a radioactive isotope behaves chemically like a stable isotope of the element, even though it is radioactive. That is, the radioactivity is independent of the chemical compound in which the isotope resides, and the atom of an isotope can participate in chemical reactions without affecting the radioactivity. This makes it possible to label or tag molecules with radioisotopes, which can then be used as tracers.

By using tracers, many body functions can be studied simply by monitoring the location and activity of the labeled molecules as they are absorbed in the body processes. For example, the activity of the thyroid gland in hormone production can be determined by monitoring its iodine uptake using radioactive ^{131}I. Similarly, radioactive solutions of iodine and gold are quickly absorbed by the liver. This can be done safely because the isotopes have short half-lives. One of the most commonly used diagnostic tracers is technetium-99 (^{99}Tc). It has a convenient half-life of 6 hours and combines with a large variety of compounds. Detectors outside the body can scan over a region and record the activity so that a complete activity image may be reconstructed.

It is possible to image gamma ray activity in a single plane or "slice" through the body. A gamma detector is moved around the patient to measure the emission intensity from many angles. A complete image can then be constructed by using computer-assisted tomography (from the Greek *tomo-*, meaning "slice") as in X-ray CAT scans. This is referred to as single-photon emission tomography (SPET). Another technique, positron emission tomography (PET), uses tracers that are positron emitters. These include ^{11}C and ^{15}O. When a positron is emitted, it is quickly annihilated, and two gamma rays are produced. Recall that the photons have equal energies and fly off in nearly opposite directions so as to conserve momentum (Section 28.4). The detection of the gamma rays is by a ring of detectors surrounding the patient.

Another medical tool is nuclear magnetic resonance (NMR). A compass

Table 29.4

Typical Relative Biological Effectiveness (RBE) of Radiations

Type	RBE (or QF)
X-rays and gamma rays	1
Beta particles and protons	1
Slow neutrons	4
Fast neutrons	10
Alpha particles	20

needle or magnet tends to line up in a magnetic field. The orbital motions of atomic electrons give rise to current loops resulting in magnetic moments (Chapter 19) to which the intrinsic electron spin also contributes. When atoms are placed in a magnetic field, the atomic energy levels are split into several closely spaced levels.

Many nuclei exhibit similar properties and have magnetic moments. For example, the simple hydrogen nucleus can have only two values of spin, similar to its electron. When a hydrogen atom is placed in a magnetic field, the energy levels split, and we have a "spin-up" state and a "spin-down" state, which correspond to the direction of the spin being parallel and antiparallel, respectively, to the magnetic field. It turns out that the spin-up state has the lower energy.

If varying radio-frequency (r-f) radiation is applied to a sample containing hydrogen nuclei, as all tissues do, when the frequency is in resonance or equal to that of the energy difference of the spin levels ($\Delta E = hf$), photons are absorbed. This is the basis of the name "nuclear magnetic resonance." The spin directions are reversed in the process. By measuring the emitted radiation when the nucleus returns to the lower state and using tomography techniques, computers construct an image based on the directions and intensities of the reradiated photons (Fig. 29.19).

Although a variety of atoms or nuclei exhibit nuclear magnetic resonance, most medical work is done with hydrogen atoms because of the varying water content of tissue. For example, muscle tissue has more water content than fatty tissue, so there is a distinct contrast in radiation intensity. Fatty deposits in blood vessels are distinct from the tissue of the vessel walls. (The water-rich blood is not seen because it has moved to a new tomography plane by the time it reradiates.) A tumor with a water content different from that of the surrounding tissue would show up in an NMR image.

Domestic and Industrial Applications

The major application of radioactivity in the home is the smoke detector. In the smoke detector, a weak radioactive source ionizes the air molecules, setting up a small current in the detector circuit. If smoke enters the detector, the ions there become attached to the smoke particles. This causes a reduction in the current because the heavier, charged smoke particles move more slowly. The current difference is electronically sensed, and an alarm is triggered.

Industry also makes good use of radioactive isotopes. Radioactive tracers are used to determine flow rates in pipes, to detect leaks, and to study corrosion and wear. It is possible to radioactivate certain compounds at a particular stage in a process by irradiating them with particles, generally neutrons. For example, iron in a corroding steel pipe can be activated to radioactive iron-59 (^{59}Fe) to indicate the presence and magnitude of the corrosion. This process is called *neutron activation analysis*, and it is a valuable method of identifying elements. In contrast to chemical analysis and spectroscopic methods, only a small amount of sample is needed.

Activation techniques are used to study the wear on piston rings and cylinders in motors and in transmission gears. Another method is to begin by placing a radioactive part in the machinery. The wear of the part may be monitored by measuring the radioactivity of the lubricating oil, which is due to the presence of fragments of the part, making periodic dismantling and inspection unnecessary.

It was originally thought that the splitting of the energy levels or spectral lines was due to the angular momentum associated with an electron spin. A spinning electron (analogous to the Earth spinning on its axis) would give rise to a current that would interact with the magnetic field due to the orbital electron motion and perhaps produce line splitting. However, this has been shown not to be the case, but the spin name remains. The two different spin states are due to some intrinsic atomic property other than electron spin, and this is a purely nonclassical quantum mechanical effect.

Figure 29.19 **NMR scan** A computer constructs an image of a person's head based on the directions and intensities of detected radiation.

Another industrial use of radiation is in *thickness gauging*, particularly in continuous manufacturing processes such as those producing sheet metal, paper, and plastic films. The amount of radiation passing through a material depends on the material's density and thus its thickness. When beta particles or X-rays are used, a variation in the intensity of the radiation transmitted through a material indicates a variation in its thickness The thickness gauge may be coupled to computer process controls so that adjustments can be made automatically to keep continuous sheets within thickness tolerances. Thickness gauges are extremely sensitive and have been used to measure the thicknesses of ink layers.

Another application of the same principle is the measuring of the levels of corrosive or dangerous chemicals and molten metals in closed containers. Gamma rays easily pass through the walls of a containing tank. When a source is placed on one side of the tank and a detector on the other side, the height of the liquid level can be determined. When the source is below the liquid level, gamma rays are absorbed by the liquid, and the intensity of the radiation received by the detector is less than when the source and detector are above the liquid level.

Important Formulas

Radius of nucleus (upper limit):

$$r_a = \frac{4kZe^2}{mv^2}$$

Activity of a radioisotope:

$$\frac{\Delta N}{\Delta t} = -\lambda N$$

$(1 \text{ Ci} = 3.70 \times 10^{10} \text{ decays/s} = 3.70 \times 10^{10} \text{ Bq})$

Number of undecayed nuclei:

$$N = N_0 e^{-\lambda t}$$

Half-life and decay constant:

$$t_{\frac{1}{2}} = \frac{0.693}{\lambda}$$

Questions

Nuclear Structure and the Nuclear Force

1. What are some differences in the gravitational force, the electromagnetic force, and the nuclear force.

2. If the protons in the nuclei repel each other, why don't nuclei generally fly apart?

3. What semiquantitative features of the nuclear force do we know about?

4. Write the nuclear notation for the following nuclei: (a) hydrogen-3, (b) oxygen-17, (c) iron-56, (d) gold-197, (e) mercury-200, and (f) plutonium-244.

5. Calcium ($_{20}$Ca) has 14 isotopes. Write the nuclear symbols for all the isotopes of calcium with mass numbers from 40 to 46. (The complete family of isotopes has mass numbers from 37 to 50.)

Radioactivity

6. What is the literal meaning of the word "radioactivity"?

7. What are the types of radioactive decay? What type of radiation is emitted in each process?

8. Could an electric field be used to distinguish the different radiations from radioactive decays? Explain.

9. What are the differences in the mass numbers, proton numbers, and neutron numbers of the parent and daughter nuclei in (a) alpha decay, (b) β^- decay, (c) β^+ decay, (d) electron capture, and (e) gamma decay?

10. Explain why alpha decay is believed to involve a quantum mechanical tunneling effect.

11. Write the nuclear equations for (a) the alpha decay

of neptunium-237, (b) the β^- decay of cobalt-60, (c) the β^+ decay of cobalt-56, (d) electron capture in cobalt-56, and (e) the gamma decay of potassium-42.

12. Tritium is radioactive. Would you expect it to β^+ or β^- decay? (Hint: write the nuclear equation for each.)

13. The difference in the nuclear energy levels is on the order of kilo-electron volts (keV) and mega-electron volts (MeV), compared to a few electron volts for the differences in electron energy levels. The existence of different nuclear energy levels allows the possibility of nuclear lasing action with the emission of gamma rays, or a "grazer." What would be the advantage of such a gamma ray or X-ray laser? (Hint: this would be a Star Wars type of laser.)

Decay Rate and Half-Life

14. Explain the relationships among activity, decay constant, number of parent nuclei present, and half-life.

15. If one isotope has a decay constant that is half that of another, will twice as many of its nuclei decay in a given time? Explain.

16. Would it be possible to determine the half-life of a radioactive isotope by weighing techniques? Explain.

17. What physical or chemical properties affect the decay rate or half-life of a radioactive isotope?

18. What are the units commonly used to measure radioactivity or a source strength, and how are they related?

Nuclear Stability and Binding Energy

19. Why are more neutrons than protons required for heavy nuclei?

20. Bismuth-209 is said to be the heaviest stable nucleus, yet it alpha decays with a half-life $\geq 2 \times 10^{18}$ years. How is it justified to say that this radioactive isotope is stable?

21. Explain what is meant by the pairing effect.

22. The binding energy of a nuclear proton is generally less than the binding energy of a neutron. Why is this?

23. Identify the regions of β^- and β^+ decays on a graph of N versus Z for stable nuclei (Fig. 29.12).

24. In general, how is the binding energy of a nucleus proportional to the mass number? How does this indicate that the nuclear force is short-ranged?

Radiation Detection and Applications

25. (a) How does a Geiger counter work and what are some of its advantages and disadvantages? (b) How are electrically neutral gamma rays and neutrons detected by a Geiger counter?

26. Explain the operations of (a) a scintillation counter, (b) a semiconductor detector, (c) a cloud chamber, and (d) a bubble chamber.

27. Why does every living thing contain ^{14}C? How is ^{14}C used in radioactive dating?

28. Why can't carbon-14 dating be used to determine the ages of objects over 50,000 years old?

29. A basic assumption of radiocarbon dating is that the cosmic ray intensity has been generally constant for the last 40,000 years or so. Suppose it were found that the intensity was much less 10,000 years ago than it is today. How would this affect carbon-14 dating?

30. Why are large doses of radioactivity dangerous? How are dosages expressed?

31. Explain the principles and uses of (a) nuclear magnetic resonance and (b) neutron activation analysis.

Problems

29.1 Nuclear Structure and the Nuclear Force

■■**1.** Two isotopes of cobalt are ^{60}Co and ^{59}Co. What is the number of protons, neutrons, and electrons in each if (a) the atom is electrically neutral, (b) the atom has a -2 charge, and (c) the atom has a $+1$ charge?

■■**2.** For the isotopes of uranium that have mass numbers from 232 through 239, what are the numbers of protons, neutrons, and electrons in a neutral atom of each isotope?

■■**3.** (a) Write the nuclear notations for the isotopes of hydrogen using the symbols H, D, and T. (b) Write the chemical symbols for six different forms of water using these symbols. Identify the radioactive forms.

■■**4.** A theoretical result for the approximate nuclear radius (R) is $R = R_o A^{1/3}$, where $R_o = 1.2 \times 10^{-15}$ m and A is the mass number of the nucleus. Find the nuclear radii of atoms of the noble gases: He, Ne, Ar, Kr, Xe, and Rn.

■■■**5.** Two isotopes of chlorine are ^{35}Cl (mass 34.968853 u) and ^{37}Cl (mass 36.965903 u). Natural chlorine contains 24.2% ^{37}Cl and 75.8% ^{35}Cl. Calculate the atomic weight of natural chlorine.

29.2 Radioactivity

■■**6.** Write the nuclear equations expressing (a) the beta decay of $^{60}_{27}$Co, and (b) the alpha decay of $^{226}_{88}$Ra.

■■7. Bismuth-214 ($^{214}_{83}$Bi) decays by alpha decay or beta decay. Identify the daughter nucleus in each case by writing the decay equations.

■■8. Thorium-229 ($^{229}_{90}$Th) undergoes alpha decay. (a) What is the daughter nucleus of this process? Write the decay equation. (b) This daughter nucleus undergoes beta decay. What is the resulting granddaughter nucleus? Write the decay equation.

■■9. Complete the following nuclear decay equations:

(a) $^{47}_{21}$Sc $\longrightarrow$ $^{47}_{22}$Ti + _____

(b) $^{226}_{88}$Ra $\longrightarrow$ $^{4}_{2}$He + _____

(c) $^{237}_{93}$Np $\longrightarrow$ $_{-1}^{0}$e + _____

(d) $^{210}_{84}$Po* $\longrightarrow$ $^{210}_{84}$Po + _____

(e) $^{11}_{6}$C $\longrightarrow$ $^{11}_{5}$B + _____

■■10. Complete the following nuclear decay equations:

(a) $^{238}_{92}$U $\longrightarrow$ $^{234}_{90}$Th + _____

(b) $^{40}_{19}$K $\longrightarrow$ $^{40}_{20}$Ca + _____

(c) $^{236}_{92}$U $\longrightarrow$ $^{131}_{53}$I + 3^{1}_{0}n + _____

(d) $^{23}_{11}$Na + γ $\longrightarrow$ _____

(e) $^{22}_{11}$Na + _____ $\longrightarrow$ $^{22}_{10}$Ne

■■■11. Actinium-227 ($^{227}_{89}$Ac) decays by alpha decay or by beta decay and is part of a decay sequence like the one in Fig. 29.9. Each daughter nucleus then decays to radium-223 ($^{223}_{88}$Ra), which subsequently decays to polonium-215 ($^{215}_{84}$Po). Write the nuclear equations for the decay process in the decay series from ^{227}Ac to ^{215}Po.

■■■12. Figure 29.20 shows the decay series for plutonium-239. (a) Identify each of the nuclei in this decay process. (b) Compare the totals of the half-lives before and after the ^{227}Ac nuclei.

29.3 Decay Rate and Half-Life

■13. A particular radioactive sample is observed to undergo 2.50×10^5 decays per second. What is the activity of the sample in (a) curies and (b) becquerels?

■14. A radioactive beta source has an activity of 20 mCi. How many beta particles per minute are emitted from the source?

■15. The half-life of a radioactive isotope is 1 h. What fraction of a sample of this isotope would be left after (a) 2 h and (b) 1 day?

■16. After 2 h, $\frac{1}{8}$ of the nuclei of a radioactive isotope remains. What is the half-life of the isotope?

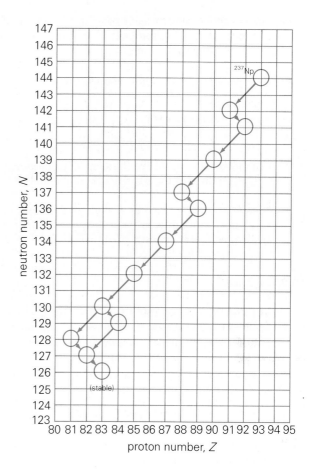

Figure 29.20 **Decay series for plutonium-239**
See Problem 12.

■■17. Which has a greater decay constant, $^{28}_{12}$Mg or $^{131}_{53}$I? By a factor of how much is it greater?

■■18. A sample of ^{215}Bi, which beta decays ($t_{\frac{1}{2}} = 2.4$ min), contains Avogadro's number of nuclei. How many bismuth nuclei are present after (a) 10 min and (b) 1 h? (c) What are the activities in curies and becquerels at these times, and is this a realistic radioactive source?

■■19. How long would it take for $\frac{2}{3}$ of a sample of strontium-90 to decay?

■■20. For a sample of cobalt-60, how long would it take for its activity to reduce to 10% of the original activity?

■■21. A soil sample contains 40 μg of ^{90}Sr. Approximately how much ^{90}Sr will be in the sample in the year 2077?

22. What period of time is required for a sample of radioactive tritium ($_1^3$H) to lose $\frac{4}{5}$ of its activity? Tritium has a half-life of 12.3 years.

23. The activity of an unknown beta source is observed to decrease by 87.5% in 54 min. What is the radioactive isotope?

24. If an amount of ^{131}I is introduced into a patient in a medical diagnostic procedure, what will be the approximate reduction in the activity of the sample in one month if all of ^{131}I is retained in the patient's thyroid gland?

25. The recoverable reserves of high-grade uranium-238 ore (high-grade ore contains about 10 kg of ^{238}U$_3$O$_8$ per ton) in the United States are estimated to be about 500,000 tons. Neglecting any geological changes, how much ^{238}U existed in this high-grade ore when the Earth was formed? (Hint: see Appendix V.)

26. In 1898, Pierre and Marie Curie isolated about 10 mg of radium-226 from 8 tons of uranium ore. If this sample had been placed in a museum how much of the radium would remain in the year 2898?

27. Gold-198 beta decays with a half-life of 2.7 days. (a) What is the activity of a sample of 1.0 g of pure $_{79}^{198}$Au? (b) What is the activity of the sample at the end of 1 week?

28. Nitrogen-13, with a half-life of 10 minutes, decays with the emission of positrons. (a) Write the nuclear equation for this decay process. (b) If a pure sample of ^{13}N has a mass of 0.0015 kg, what is the activity in 35 minutes? (c) What percentage of the sample is ^{13}N at this time?

29.4 Nuclear Stability and Binding Energy

29. The total binding energy of $_1^2$H is 2.224 MeV. What is the mass of a ^{2}H nucleus?

30. The binding energy per nucleon for $_2^4$He is 7.075 MeV. (a) What is the total binding energy? (b) What is the mass of a ^{4}He nucleus?

31. What are the equivalent rest energies of (a) a carbon-12 atom and (b) an alpha particle?

32. Determine whether the following isotopes are likely to be stable: (a) ^{20}C, (b) ^{100}Sn, (c) ^{6}Li, (d) ^{9}Be, (e) ^{22}Na.

33. Determine whether the following isotopes are likely to be unstable: (a) ^{23}Na, (b) ^{50}V, (c) ^{209}Bi, (d) ^{209}Po, (e) ^{44}Ca.

34. Only two isotopes of Sb (antimony, $Z = 51$) are stable. Pick the two stable isotopes from the following: (a) ^{120}Sb, (b) ^{121}Sb, (c) ^{122}Sb, (d) ^{123}Sb, (e) ^{124}Sb.

35. The mass of $_2^3$He is 3.016029 u. (a) What is the total binding energy of this nucleus? (b) What is the average binding energy per nucleon?

36. The mass of $_4^9$Be is 9.012183 u. What is E_b/A for this nucleus?

37. How much energy would be required to separate the nucleons of an oxygen-16 nucleus, the atom of which has a mass of 15.994915 u?

38. Determine the average binding energy per nucleon of sulfur-32 ($m = 31.972072$ u).

39. If an alpha particle could be removed intact from an aluminum-27 nucleus ($m = 26.981541$ u), a sodium-23 nucleus ($m = 22.989770$ u) would remain. How much energy would be required to do this?

40. On the average, are nucleons more tightly bound in an ^{27}Al nucleus or in a ^{23}Na nucleus? (See Problem 39).

41. The total binding energy of $_3^6$Li is 32.0 MeV. What is the mass of a ^{6}Li nucleus?

42. The value of E_b/A for $_{92}^{238}$U is 7.58 MeV. (a) What is the total binding energy of this nucleus? (b) What is its mass?

43. The mass of $_4^8$Be is 8.005305 u. (a) Which is less, the total mass of two alpha particles or the mass of the ^{8}Be nucleus? (b) Which is greater, the total binding energy of ^{8}Be nucleus or the total binding energy of an alpha particle? (c) Would you expect the ^{8}Be nucleus to decay spontaneously into two alpha particles?

29.5 Radiation Detection and Applications

44. An old bone yields, on the average, one beta emission per minute per gram of carbon. About how old is the bone?

45. A fossil specimen is believed to be about 18,000 years old. Explain how this could be confirmed.

46. An old artifact is found to contain 500 g of carbon and has an activity of 950 decays per minute. What is the approximate age of the artifact?

47. A patient receives a 2.0-rad dose of radiation from X-rays and a 2.0-rad dose from an alpha source. (a) Which has the greater biological effectiveness? (b) What are the doses in rems?

■■**48.** A technician working at a nuclear reactor facility is exposed to a slow neutron radiation and receives a dose of 1.25 rad. (a) How much energy is absorbed by 200 g of the worker's tissue? (b) Was the maximum permissible radiation dosage exceeded?

■■**49.** A person working with nuclear isotopes for a 2-month period receives a 0.5-rad dose from a gamma source, a 0.3-rad dose from a slow-neutron source, and a 0.1-rad dose from an alpha source. Was the maximum permissible radiation dosage exceeded?

■■**50.** What is the average annual dose (in rads) of each type of radiation listed in Table 29.4 that would equal the maximum permissible dosage for a person older than 18 years of age? (Neglect background radiation.)

■■**51.** In a diagnostic procedure, a patient in a hospital ingests 80 mCi of gold-198 ($t_{\frac{1}{2}} = 2.7$ days). What is the approximate activity at the end of two weeks if none is eliminated from the body by biological functions?

Additional Problems

52. Tin ($_{50}$Sn) has 23 isotopes that have mass numbers from 108 to 130. Write the complete nuclear notations for each of the isotopes of tin with mass numbers from 115 to 120.

53. (a) What is the decay constant of fluorine-17? (b) How long will it take for the activity of a sample of ^{17}F to decrease to 12.5 percent of its value at $t = 0$? ($t_{\frac{1}{2}} = 66$ s.)

54. Starting with uranium-234 ($^{234}_{92}$U), there is a decay sequence of four alpha decays and two beta decays. What is the resulting nucleus of the sequence?

55. The mass of $^{238}_{92}$U is 238.050786 u. Find the average binding energy per nucleon for this isotope.

56. Complete the following nuclear decay equations:

(a) $^{8}_{4}\text{Be} \longrightarrow {}^{4}_{2}\text{He} + \underline{\hspace{1.5cm}}$

(b) $^{240}_{94}\text{Po} \longrightarrow {}^{97}_{38}\text{Sr} + {}^{139}_{56}\text{Ba} + \underline{\hspace{1.5cm}}$

(c) $^{47}_{21}\text{Sc*} \longrightarrow {}^{47}_{21}\text{Sc} + \underline{\hspace{1.5cm}}$

(d) $^{29}_{11}\text{Mg} \longrightarrow {}_{-1}^{0}\text{e} + \underline{\hspace{1.5cm}}$

57. A sample of old parchment is found to have 4 beta decays/g-min. Approximately how old is the parchment?

58. Francium-283 ($^{283}_{87}$Fr) has a half-life of 21.8 min. (a) If there is initially a 25.0-mg sample of this isotope, how many nuclei are present? (b) How many nuclei will be present 1 h and 49 min later?

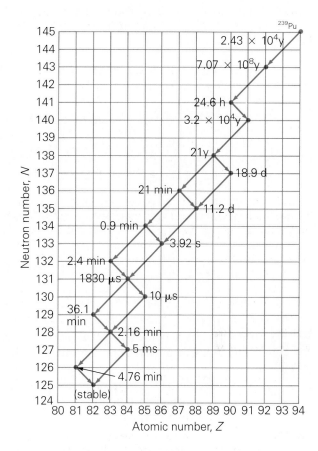

Figure 29.21 **Decay series for neptunium-237** See Problem 61.

59. Radon-222 ($^{222}_{86}$Rn), a radioactive inert gas, undergoes alpha decay. (a) What is the daughter nucleus? Write the nuclear equation for the decay process. (b) The daughter nucleus decays by either alpha decay or beta decay. What is the granddaughter nucleus for each of these processes?

60. A sample of technetium-104 has an activity of 10 mCi. What will be the activity of the sample in one hour? ($t_{\frac{1}{2}} = 18$ min.)

61. The decay series for neptunium-237 is shown in Fig. 29.21. (a) Identify the decay modes and each of the nuclei in the sequence. (b) Tell why each nucleus is likely to decay.

62. Determine whether each of the three isotopes of hydrogen is likely to be stable or unstable.

Nuclear Reactions and Elementary Particles

30

Thus far, we have considered radioactive processes in which the nucleus of one element decays into the nucleus of another with the emission of energetic particles. As soon as this process was understood, scientists wondered whether the reverse was possible. That is, could the nucleus of one element be bombarded with energetic particles and thus be converted into the nucleus of another element with the absorption of a particle (reverse radioactivity, so to speak)?

Indeed, such processes are possible, and physicists have learned how to manipulate nuclear particles to create one isotope from another by nuclear reactions. Such reactions have made available the many artificial radioactive isotopes that do not occur naturally.

Also, the knowledge of nuclear reactions has made it possible to release energy from the nucleus. The dropping of the so-called atomic bomb on the Japanese city of Hiroshima on August 6, 1945, was an announcement of the devastating power that could be released from the nucleus. On the more peaceful side, nuclear energy is now an important source of energy for generating electricity.

In this chapter, we will look at these topics and at elementary particles. Investigations have shown that a variety of particles other than the proton and neutron are associated with the nucleus. The discovery of the existence of these particles has given some insight into the still puzzling world of the nucleus and what holds it together.

30.1 Nuclear Reactions

In chemical reactions, substances react with each other to form compounds. Since an ordinary chemical reaction involves only the atomic electrons, the atoms of the reactants do not lose their nuclear identities in the process. On the other hand, in nuclear reactions, the nuclei of isotopes are converted into the nuclei of other isotopes, which in general are isotopes of completely different

(a)

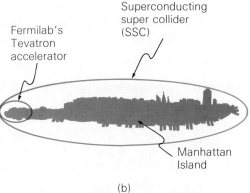

Fermilab's Tevatron accelerator

Superconducting super collider (SSC)

Manhattan Island

(b)

Figure 30.1 **Particle accelerators**
(a) Accelerator facilities at Fermilab in Batavia, Illinois. The circular ring of the main accelerator has a diameter of 2 km. (b) The Superconducting Super Collider (SSC) now being constructed will have a diameter of about 27 km and will use some 10,000 superconducting magnets to focus and guide particle beams that will approach the speed of light. Note the size of the SSC in comparison with that of the size of Manhattan Island.

elements. From the study of the nucleus, scientists have learned to initiate nuclear reactions by bombarding nuclei with energetic particles.

The first induced nuclear reaction was produced by Lord Rutherford in 1919. Nitrogen was bombarded with alpha particles from a natural source (bismuth-214). The occasional particles that came from the reactions were identified as protons. Rutherford reasoned that an alpha particle colliding with a nitrogen nucleus must sometimes induce a reaction that produces a proton. As a result, the nitrogen nucleus is *artificially* transmuted into an oxygen nucleus:

$$^{14}_{7}\text{N} \quad + \quad ^{4}_{2}\text{He} \quad \longrightarrow \quad ^{17}_{8}\text{O} \quad + \quad ^{1}_{1}\text{H}$$

nitrogen   alpha particle   oxygen   proton
(14.003074 u)   (4.002603 u)   (16.999133 u)   (1.007825 u)

where the atomic masses are given for a later purpose.

Nuclear reactions are sometimes written with intermediate compound nuclei. For example, the preceding reaction can be written,

$$^{14}_{7}\text{N} + ^{4}_{2}\text{He} \longrightarrow (^{18}_{9}\text{F*}) \longrightarrow ^{17}_{8}\text{O} + ^{1}_{1}\text{H}$$

The fluorine nucleus, $^{18}_{9}\text{F}$, is formed in an excited state, as is indicated by the asterisk. A compound nucleus rids itself of excess energy by ejecting a particle. This occurs in a very short time, so the compound nucleus is commonly omitted from the equation for the nuclear reaction.

Think of the implication of Rutherford's discovery. One element had been converted into another. This was the age-old dream of the alchemists, although their main concern was changing common metals, such as mercury and lead, into gold. Even this transmutation can be done today through nuclear reactions. Large machines called **particle accelerators** use electric and magnetic fields to accelerate charged particles to very high energies (Fig. 30.1). When the particles strike target nuclei, they initiate nuclear reactions. Different reactions require different particle energies. One nuclear reaction that is initiated when protons strike nuclei of mercury is

$$^{200}_{80}\text{Hg} \quad + \quad ^{1}_{1}\text{H} \quad \longrightarrow \quad ^{197}_{79}\text{Au} \quad + \quad ^{4}_{2}\text{He}$$

mercury   proton   gold   alpha particle
(199.968321 u)   (1.007825 u)   (196.96656 u)   (4.002603 u)

In this reaction, mercury is converted into gold, so it would seem that modern physics has fulfilled the alchemists' dream. However, making such small amounts of gold in an accelerator costs far more than the gold is worth.

Reactions like those written above have the general form

$$A + a \longrightarrow B + b$$

where the uppercase letters represent the nuclei and the lowercase letters represent the particles. Such reactions are often written in a shorthand notation like this:

$$A(a, b)B$$

For example, in this form, the two previous reactions are written as

$$^{14}\text{N}(\alpha, \text{p})^{17}\text{O} \quad \text{and} \quad ^{200}\text{Hg}(\text{p}, \alpha)^{197}\text{Au}$$

In general, the periodic table lists 105 elements, but only 90 elements occur naturally on Earth. Elements with proton numbers greater than uranium ($Z = 92$), as well as technetium ($Z = 43$) and promethium ($Z = 61$), are created artificially by nuclear reactions. The name technetium comes from the Greek word *technetos*, meaning "artificial," and technetium was the first unknown element to be created by artificial means. Elements up to $Z = 110$ have been reported, but those above $Z = 105$ are unconfirmed.

Conservation of Mass-Energy and the Q Value

In every nuclear reaction, the total relativistic energy is conserved ($E = K + m_0 c^2$, as shown in Chapter 26). Consider the reaction by which nitrogen is converted into oxygen, $^{14}\text{N}(\alpha, \text{p})^{17}\text{O}$. By the conservation of total relativistic energy,

$$(K_N + m_N c^2) + (K_\alpha + m_\alpha c^2) = (K_O + m_O c^2) + (K_p + m_p c^2)$$

where the subscripts refer to the energy and mass of a particular particle and the respective masses are rest masses. Rearranging the equation, we have

$$K_O + K_p - K_N - K_\alpha = (m_N + m_\alpha - m_O - m_p)c^2 \tag{30.1}$$

The **Q value** of the reaction is defined as

$$Q = (K_O + K_p) - (K_N + K_\alpha) \tag{30.2}$$

The Q value is a measure of the total energy released or absorbed in a reaction, that is, the kinetic energy of the products of the reaction minus the kinetic energy of the reactants of the reaction. In terms of final and initial kinetic energies, $Q = K_f - K_i$. The Q value is an important quantity in nuclear reactions. However, the kinetic energies of the reactants and products of the reaction do not have to be measured, since by Eqs. 30.1 and 30.2,

$$Q = (m_N + m_\alpha - m_O - m_p)c^2 \tag{30.3}$$

Similarly, in terms of a general reaction of the form $A + a \longrightarrow B + b$, or $A(a, b)B$,

$$Q = (m_A + m_a - m_B - m_b)c^2 = (\Delta m)c^2 \tag{30.4}$$

Thus, the Q value is given by the difference in the rest energies of the reactants and the products of a reaction. Note that this means that mass is generally converted into energy or vice versa.

Using the masses given in the parentheses under the reactants and products in the previous equation for $^{14}N(\alpha, p)^{17}O$ reaction, we obtain

$$Q = (m_N + m_\alpha - m_O - m_p)c^2$$

$$= [(14.003074 \text{ u} + 4.002603 \text{ u}) - (16.999133 \text{ u} + 1.007825 \text{ u})]c^2$$

$$= (-0.001281 \text{ u})c^2$$

or $\quad Q = (-0.001281 \text{ u})c^2[931.5 \text{ MeV}/(1 \text{ u})c^2] = -1.193 \text{ MeV}$

The negative Q value indicates that energy was absorbed in the reaction. When Q is less than zero (negative), the reaction is said to be **endoergic** (or endothermic). That is, energy (*ergic*) must be put into (*endo*) the reaction for it to proceed. In endoergic reactions, the kinetic energy of the reacting particles is at least partially converted into mass.

When the Q value of a reaction is positive ($Q > 0$), energy is released, and the reaction is said to be **exoergic** (or exothermic). That is, energy is produced by (*exo*) the reaction. In this case, mass is converted into energy and is carried away as kinetic energy of the products of the reaction. An example of an exoergic reaction is said to be **exoergic** (or exothermic). That is, energy is produced by

$$^2_1H \quad + \quad ^2_1H \quad \longrightarrow \quad ^3_2He \quad + \quad ^1_0n \quad (+ \text{ energy})$$

deuteron	deuteron	helium	neutron
(2.014102 u)	(2.014102 u)	(3.016029 u)	(1.008665 u)

In exoergic, or positive Q, reactions, the total rest mass of the reactants is greater than that of the products. In this example the mass difference is

$$\Delta m = 2m_H - m_{He} - m_n$$

$$= 2(2.014102 \text{ u}) - 3.016029 \text{ u} - 1.008665 \text{ u} = 0.00351 \text{ u}$$

Then

$$Q = (0.00351 \text{ u})(931.5 \text{ MeV}/\text{u}) = 3.27 \text{ MeV}$$

Computing the Q value with the mass difference and the energy-mass conversion factor in this manner eliminates the necessity of carrying along the c^2.

The Q value of radioactive decay is always positive ($Q > 0$), since energetic particles are emitted spontaneously. This is sometimes called the *disintegration energy*. The meanings of the positive and negative Q values are summarized in Table 30.1.

When the Q value of a reaction is negative or the mass of the products is greater than the mass of the reactants, then a certain amount of kinetic energy is converted into rest mass. Such reactions will not occur unless enough kinetic

Table 30.1
Interpretation of Q Values

Q value	Effect
Positive ($Q > 0$)	Exoergic, mass converted into energy
(mass of reactants greater than mass of products)	
Negative ($Q < 0$)	Endoergic, energy converted into mass
(mass of products greater than mass of reactants)	

energy is initially present. One might think that if a particle with a kinetic energy of $|-Q|$ is incident on a stationary nucleus, then a reaction would occur, since this energy is equivalent to the mass gain. However, momentum would not be conserved if this happened. For example, for the reaction $^{14}\text{N}(\alpha, \text{p})^{17}\text{O}$, which has a $-Q$, with $K_\alpha = |-Q| = Q$, Eq. 30.2 with the nitrogen nucleus stationary gives

$$-Q = K_\text{o} + K_p - K_\alpha = K_\text{o} + K_p - Q$$

or $\qquad K_\text{o} + K_p = 0$

Thus, the total momentum (P) of the products of the reaction would be zero. (Recall that $K = p^2/2m$.)

This violates the conservation of momentum, since the initial total momentum is *not* zero. Therefore, the kinetic energy of the incident particle must be greater than $K = |-Q|$ so the products of the reaction can have the same total momentum as the incident particle. The minimum kinetic energy (K_min) that an incident particle needs to have to initiate a reaction is called the **threshold energy**. For classical (nonrelativistic) collisions, the threshold energy can be shown to be

$$K_\text{min} = \left(1 + \frac{m_a}{M_A}\right)|Q| \qquad (30.5)$$

where m_a and M_A are the masses of the incident particle and the *stationary* target nucleus, respectively.

Example 30.1   Threshold Energy

What is the threshold energy for the reaction $^{14}\text{N}(\alpha, \text{p})^{17}\text{O}$?

Solution

Given: $m_a = m_\alpha = 4.002603$ u $\qquad$ *Find*: K_min
$\qquad\quad M_A = M_\text{N} = 14.003074$ u
$\qquad\quad Q \ = -1.193$ MeV

Using Eq. 30.5, we have

$$K_\text{min} = \left(1 + \frac{m_\alpha}{M_\text{N}}\right)|Q| = \left(1 + \frac{4.002603 \text{ u}}{14.003074 \text{ u}}\right)(1.193 \text{ MeV}) = 1.53 \text{ MeV} \quad \blacksquare$$

Reaction Cross Sections

More than one reaction is possible when a particle collides with a nucleus. As we have seen for an endoergic reaction, the incident particle must have a minimum kinetic energy to initiate a particular reaction. When a particle has more kinetic energy than the threshold energies of the several possible reactions, any of the reactions may occur. A measure of the probability that a particular reaction will occur is called the **cross section of the reaction**. If we knew the exact expression for the nuclear force and the form of the nuclear structure, we might be able to calculate the cross section for each possible reaction. However, at present we must determine the probabilities, or cross sections, of the possible reactions experimentally.

The cross section of a particular reaction is observed to vary with the kinetic energy of the incident particle. For positively charged incident particles, such

The term *cross section* arises from calculations of the effective reaction area of a thin foil, where the cross section is expressed explicitly as an area.

as protons and alpha particles, the Coulomb barrier of the nucleus influences the reaction cross section. Thus the probability of a given reaction increases as the kinetic energy of the incident particle increases. Since reactions do not occur for relatively low energy particles, there must be barrier penetration (from the outside in). The increase in the cross section with particle energy may be thought of as being due to more energetic particles having a greater probability of tunneling through the Coulomb barrier (see Figs. 29.7 and 29.8).

Being electrically neutral, neutrons are unaffected by the Coulomb barrier. As a result, the cross section for a given reaction may be quite large for low-energy neutrons, in particular for reactions such as $^{27}\text{Al}(n, \gamma)^{28}\text{Al}$. For low energies, the neutron reaction cross sections are often found to be proportional to $1/v$, where v is the speed of the neutron. That is, the probability of a reaction appears to be proportional to the time a neutron spends in the vicinity of a nucleus ($t \propto 1/v$). As the neutron energy increases, the cross section varies a great deal, as Fig. 30.2 shows. The peaks in the curve are referred to as resonances. The resonances are believed to be associated with nuclear energy levels.

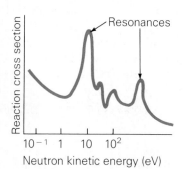

Figure 30.2 Reaction cross section
A typical graph of neutron reaction cross section versus energy. The peaks where the probabilities of reactions are greater are called resonances.

30.2 Nuclear Energy: Fission and Fusion

The two exoergic reactions used in the production of nuclear energy are fission and fusion reactions. These processes were mentioned in Chapter 29 with regard to the binding energy curve to more stable nuclei (see Fig. 29.14). One process involves the splitting, or fissioning, of a heavy nucleus, and the other involves the putting together, or fusion, of two light nuclei.

Fission

In early attempts to make heavier elements artificially, uranium, the heaviest element known at the time, was bombarded with particles, mainly neutrons. An unexpected result was that the uranium nuclei sometimes broke apart or "split" into fragments. In the late 1930s, three fragments were identified as the nuclei of lighter elements. The process was dubbed "fission" after biological fission (the dividing of living cells).

Thus, in a **fission reaction**, a heavy nucleus divides into two lighter nuclei with the emission of two or more neutrons. Energy is emitted in the process, being carried off primarily by the neutrons and fission fragments. Some heavy nuclei undergo spontaneous fission, but at very slow rates. However, fission may be *induced*, and this is the important process in energy production. For example, when a ^{235}U nucleus absorbs an incident neutron, it may fission according to the reaction

$$^{235}_{92}\text{U} + ^{1}_{0}\text{n} \longrightarrow (^{236}_{92}\text{U}^{*}) \longrightarrow ^{140}_{54}\text{Xe} + ^{94}_{38}\text{Sr} + 2(^{1}_{0}\text{n})$$

The capture of a neutron results in the formation of an excited uranium-236 nucleus (Fig. 30.3).

According to the so-called liquid drop model, the ^{236}U nucleus undergoes violent oscillations and becomes distorted like a liquid drop. The separation of the nucleons into different parts of the "drop" weakens the nuclear force, and the repulsive electrical force of the protons between the parts causes the

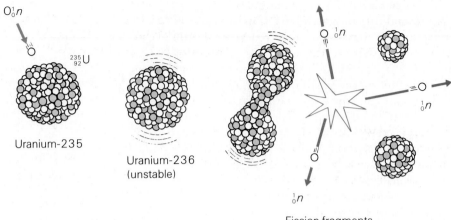

Figure 30.3 **Fission**
When an incident neutron is absorbed by a fissionable nucleus, such as ^{235}U, the unstable nucleus (^{236}U) undergoes violent oscillations and breaks apart like a liquid drop, emitting two or more neutrons.

nucleus to split or fission. Some isotopes, however, capture neutrons without fissioning.

The preceding reaction is only one of a variety of ways in which uranium-235, the fissionable isotope of uranium, can fission. Others include (compound nuclei omitted)

$$\,^1_0\text{n} + \,^{235}_{92}\text{U} \longrightarrow \,^{141}_{56}\text{Ba} + \,^{92}_{36}\text{Kr} + 3(\,^1_0\text{n})$$

$$\,^1_0\text{n} + \,^{235}_{92}\text{U} \longrightarrow \,^{150}_{60}\text{Nd} + \,^{81}_{32}\text{Ge} + 5(\,^1_0\text{n})$$

Only certain nuclei undergo fission, and the probability of a fission reaction for a fissionable isotope depends on the energy of the incident neutrons. For example, the greatest cross sections for fission reactions for ^{235}U and ^{239}Pu are for "slow" neutrons with energies less than 1 eV. However, for ^{232}Th, the greatest cross section is for "fast" neutrons with energies of 1 MeV or more.

Rather than calculating the energy released in an exoergic fission reaction, we can obtain an estimate of its magnitude from the E_b/A curve for stable nuclei (Fig. 29.14). When a nucleus with a large mass number A, such as uranium, splits into two nuclei, it is in effect moving up along the downward-sloping tail of the E_b/A curve to a more stable state. As a result, the average binding energy per nucleon increases from a value of about 7.5 MeV to a value of approximately 8.5 MeV. Thus, the energy liberated is about 1 MeV per nucleon. In the first of the preceding fission reactions, there are $140 + 94 = 234$ nucleons in the fission products, so the accompanying energy release is approximately

$$\frac{1 \text{ MeV}}{\text{nucleon}} \times 234 \text{ nucleons} \cong 234 \text{ MeV}$$

On a relative basis, this might seem to be a lot of energy. For example, the energy released *per atom* in a chemical exothermic process is on the order of a few hundred electron volts. On the other hand, this might not seem like much energy, since 200 MeV is only about 3×10^{-11} J. When you pick up your textbook from the desk, you do about 5 J of work or expend 5 J of energy. But keep in mind that there are billions of nuclei in a small sample of fissionable material. We simply need to have enough nuclei fissioning to produce practical amounts of energy.

This is accomplished by means of a **chain reaction**. For example, suppose a ^{235}U nucleus fissions with the release of two neutrons (Fig. 30.4). Ideally, the neutrons may then initiate two more fission reactions, a process that results in the availability of four neutrons. These neutrons may initiate more reactions, and the process multiplies, the number of neutrons doubling with each generation. When this occurs uniformly with time, we have exponential growth.

To have a sustained chain reaction, there must be an adequate quantity of fissionable material. The minimum mass required to produce a chain reaction is called the **critical mass**. In effect, this means that there is enough fissionable material that one neutron from each fission event, on the average, goes on to fission another nucleus. With more fissioning neutrons, the chain and energy output would grow.

Several factors determine the critical mass. Most evident is the amount of fissionable material. If the quantity of material is small, neutrons may escape from the sample before inducing a fission event, and the reaction would die out. Also, the nuclei of other isotopes in the sample may absorb neutrons, thereby limiting the chain reaction. As a result of both of these factors, the purity of the fissionable isotope affects the critical mass.

In general, natural uranium is made up of the isotopes of ^{238}U and ^{235}U. The concentration of the fissionable ^{235}U isotope is only 0.7%. The remaining 99.3% is ^{238}U, which may absorb neutrons for a nuclear reaction other than fission (and removes neutrons from contributing to a chain reaction). To have more fissionable ^{235}U nuclei in a sample and reduce the critical mass, the ^{235}U is concentrated or enriched. This enrichment varies from 3–5% ^{235}U for reactor-grade material used in electrical generation to over 99% for weapons-grade material. This is an important difference, since it is desirable that a nuclear reactor used for electrical generation not explode like an atomic bomb.

Chain reactions take place almost instantaneously. If such a reaction proceeds uncontrolled in a fissionable sample of critical mass, the quick and

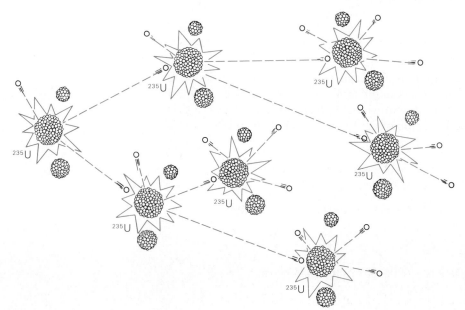

Figure 30.4 **Fission chain reaction**
The neutrons that result from one fission event initiate other fission reactions, and so on. When enough fissionable material is present, the sequence of reactions may form a chain of reactions.

Demonstration 22

A Simulated Chain Reaction (without fission)

A simulation of how neutrons from reactions induce reactions in other nuclei so that the process grows in a chain reaction.

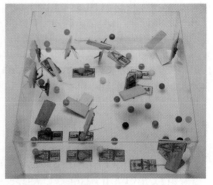

(a) Here, the "nuclei" are mousetraps loaded with a single Super Ball (neutron) that will be emitted when a reaction occurs. A plexiglas enclosure ensures a "critical mass," or that the balls (neutrons) are reflected instead of "escaping."

(b) The reaction begins.

(c) The reaction grows, with an increasing number of fissions and release of energy.

enormous release of energy from billions and billions of fissioning nuclei causes an explosion. This is the principle of the atomic bomb. For the practical production of nuclear energy, the chain reaction process must be controlled. We will discuss this shortly. (See Demonstration 22.)

Fusion

The other important type of reaction for the production of nuclear energy is fusion. In a **fusion reaction**, light nuclei fuse together to form a heavier nucleus, releasing energy ($Q > 0$). An example of a fusion reaction was given in the preceding section for the fusion of deuterium nuclei (sometimes called a D-D reaction). Another example is the fusion of deuterium and tritium nuclei (a D-T reaction):

$$\underset{(2.014102 \text{ u})}{{}_{1}^{2}\text{H}} \quad + \quad \underset{(3.016049 \text{ u})}{{}_{1}^{3}\text{H}} \quad \longrightarrow \quad \underset{(4.002603 \text{ u})}{{}_{2}^{4}\text{He}} \quad + \quad \underset{(1.008665 \text{ u})}{{}_{0}^{1}\text{n}}$$

Using the given atomic masses to find the mass difference between the reactants and products, we have

$$\Delta m = m_{{}^2\text{H}} + m_{{}^3\text{H}} - (m_{\text{He}} + m_n)$$

$$= 2.014102 \text{ u} + 3.016049 \text{ u} - (4.002603 \text{ u} + 1.008665 \text{ u})$$

$$= 0.018883 \text{ u}$$

and

$$Q = (0.018883 \text{ u})(931.5 \text{ MeV/u}) = 17.6 \text{ MeV}$$

Notice that the energy release from this and the previous fusion reaction is very small in comparison to the more than 200 MeV released from a fission reaction. However, think about equal masses of hydrogen and uranium. A given mass of hydrogen isotopes has many, many more nuclei than an equivalent mass of a heavy fissionable isotope. Also, notice that there is no critical mass for fusion, since there is no chain reaction to maintain.

Fusion is the source of energy for stars, including the Sun. In the initial stages of a star's life, there is hydrogen burning, or the fusion of hydrogen into helium. The process is somewhat complicated and consists of sets of reactions with the net result of

$$4(^1_1\text{H}) \longrightarrow {}^4_2\text{He} + 2(^{\,0}_{+1}\text{e})$$

Nuclear Reactors and Devices

Currently, the only type of practical nuclear reactor is based on the fission chain reaction. Research reactors designed primarily to produce artificial isotopes exist, but we will focus on the type of reactor that is commonly used to generate electricity. The technology for this is over 30 years old, the first nuclear reactor for the purpose of electrical generation having gone into operation in 1957 in Shippingport, Pennsylvania. A typical design of a nuclear reactor is shown in Fig. 30.5 on the following page. There are four key elements to a reactor: fuel rods, coolant, control rods, and moderator.

Tubes packed with pellets of uranium oxide form the **fuel rods**, which are located in the reactor core. A typical commercial reactor contains fuel rods bundled in fuel assemblies of around 200 rods each. A fuel rod assembly is constructed so that a coolant can flow around the rods to remove the energy emitted from the fission chain reaction. The reactors used in the United States are light water reactors, which means that ordinary water is used as a coolant. However, the hydrogen nuclei of ordinary water have a tendency to capture neutrons. This removes neutrons from the chain reaction, so enriched uranium with 3 to 5% ^{235}U is used to achieve a critical mass. A nuclear reactor can run for 3 or 4 years before having to be refueled.

The chain reaction and energy output of a reactor are controlled by means of boron or cadmium **control rods**, which can be inserted into and withdrawn from the reactor core. Cadmium and boron have a high cross section for absorbing neutrons. By inserting the control rods between the fuel rod assemblies, neutrons are absorbed and removed from the chain reaction. The control rods can be adjusted so that energy is released at a relatively steady rate. If more energy is needed, the rods are further withdrawn from the core. When the control rods are fully inserted, enough neutrons are removed to curtail the chain reaction and shut down the reactor.

The water flowing through the fuel rod assemblies acts not only as a coolant, but also as a **moderator**. The fission cross section of ^{235}U is largest for slow neutrons (kinetic energies less than 1 eV). The neutrons emitted from a fission reaction are fast neutrons with energies on the average of 2 MeV. These are slowed down, or their speed moderated, by collisions. The hydrogen atoms in water are very effective in slowing down neutrons because their masses are

The reaction is

$$^1_0\text{n} + {}^1_1\text{H} \longrightarrow {}^2_1\text{H} + \gamma$$

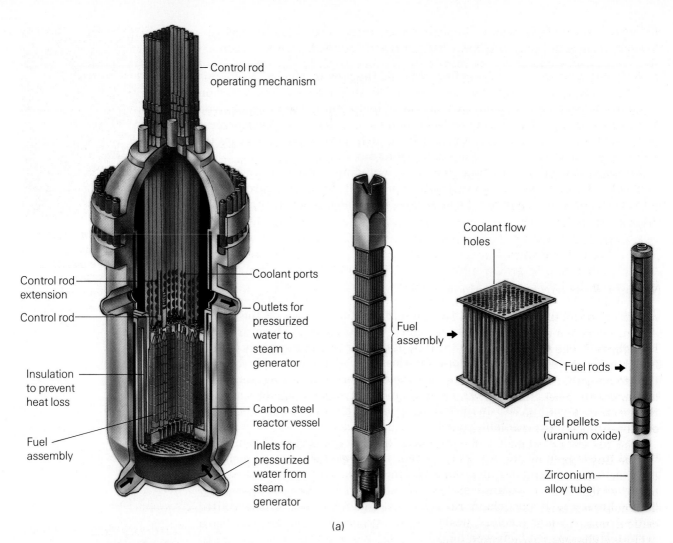

Control rod
operating mechanism

Control rod
extension

Control rod

Insulation
to prevent
heat loss

Fuel
assembly

Coolant ports

Outlets for
pressurized
water to
steam
generator

Carbon steel
reactor vessel

Inlets for
pressurized
water from
steam
generator

Fuel
assembly

Coolant flow
holes

Fuel rods

Fuel pellets
(uranium oxide)

Zirconium
alloy tube

(a)

Figure 30.5 Nuclear reactor
(a) A schematic diagram of
a reactor showing a fuel rod
assembly and a fuel rod.
(b) The core of a nuclear
reactor seen through shielding
water. The characteristic blue
glow is called Cerenkov
radiation, which is produced
by particles traveling at
speeds greater than the speed
of light *in a transparent
medium.*

nearly equal to the neutrons' masses. Recall from Chapter 6 that there is maximum energy transfer in a head-on collison of particles of equal mass. Not all collisions are head-on, but it takes only about 20 collisions to moderate fast neutrons to energies of less than 1 eV.

Heavy water (D_2O, where D stands for deuterium) can also be used as a moderator, eliminating the need for enriched uranium. Unlike ordinary hydrogen, deuterium does not readily absorb neutrons, so natural uranium (0.7% ^{235}U) can be used as fuel. Most of the reactors in Canada are heavy water reactors.

Other materials, such as graphite (carbon), may be used as moderators. Because of their relatively heavy nuclei, not as much energy is transferred per collision. On the average, about 120 collisions with carbon atoms are needed to produce slow or thermal neutrons. (A thermal neutron is one that has acquired the average thermal speed of the atoms of the moderator.)

Although not the best moderator, carbon in the form of graphite also permits a chain reaction to occur for natural (unenriched) uranium. The first experimental proof that a chain reaction was feasible was accomplished in such a reactor in 1942 by a team of scientists working with Enrico Fermi on the World War II Manhattan Project. The reactor was called a "pile" because it essentially consisted of a pile of graphite blocks (see Fig. 30.6). (A carbon reactor was involved in the 1986 accident at Chernobyl in the Soviet Union.)

In July 1945, the first nuclear device was exploded over the desert in New Mexico. Soon after, in August, the first atomic bomb was dropped on Hiroshima, Japan. To have a critical mass of practical size for a bomb, the uranium must be highly enriched (over 90% ^{235}U). In addition, the critical mass must be held together for about 1 μs for the chain to proceed to explosive proportions. The time to assemble the critical mass must be less than the average time for the appearance of a neutron by spontaneous fission that could prematurely initiate the reaction. For these reasons, a nuclear reactor used to generate electricity cannot explode like an atomic bomb.

The necessary conditions for explosion were realized in the Hiroshima "Little Boy" bomb by a gun-type assembly (Fig. 30.7). Interlocking pieces of uranium were brought together by an explosive charge. The second atomic bomb, which was dropped on Nagasaki, Japan, used plutonium-239 as the fissionable material. Because ^{239}Pu spontaneously fissions more rapidly, the critical mass had to be assembled more quickly. This was done by using explosive charges surrounding a sphere of material. When detonated, these charges produced an *implosion* on the fissionable material, bringing it quickly to critical mass.

In general, the ^{238}U in a reactor goes along for the ride, so to speak, without reacting. However, ^{238}U has an appreciable cross section for fast neutrons. Not all the neutrons are moderated, so there is some ^{238}U reaction, and there is conversion into ^{239}Pu via radioactive decay:

$$^{1}_{0}n + {}^{238}U \longrightarrow {}^{239}U \longrightarrow {}^{239}Np \longrightarrow {}^{239}Pu$$

Plutonium-239, with a half-life of 24,000 years, is fissionable and serves as additional fuel that prolongs the time to reactor refueling. It is possible to promote the reaction of ^{238}U in a reactor by reduced moderation. When, on the average, one or more neutrons from ^{235}U fission are absorbed by ^{238}U nuclei to produce ^{239}Pu, then the same amount or more of fissionable fuel is produced (^{239}Pu) than is consumed (^{235}U). This is the principle of the **breeder reactor**, which produces more fissionable fuel than it consumes.

Nuclear energy is used to generate a substantial amount of the electricity in the world [Fig. 30.8(a)]. Twenty-five countries now produce nuclear-generated electricity, and twelve more plan to do so in the 1990s. More than 430 nuclear

Figure 30.6 Chain reaction and nuclear energy The unveiling of a plaque at the University of Chicago commemorating the first self-sustained chain reaction. Enrico Fermi stands directly below the plaque.

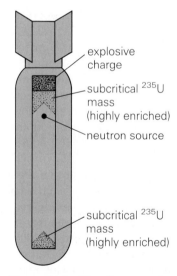

explosive charge

subcritical ^{235}U mass (highly enriched)

neutron source

subcritical ^{235}U mass (highly enriched)

Figure 30.7 The first atomic bomb A diagram of the principle of a uranium fission bomb. The subcritical masses are brought together by the explosive charge of a gun assembly.

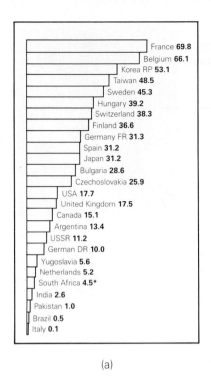

(a)

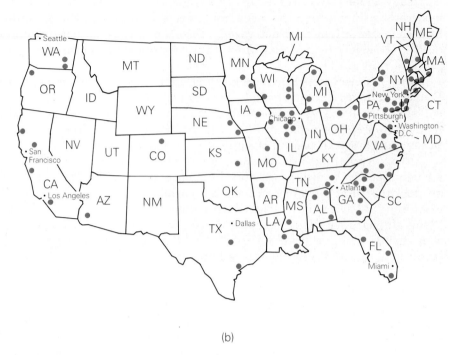

(b)

reactor units are in operation throughout the world, with about 110 operating units in the United States [Fig. 30.8(b)] and about a dozen under construction. With the increasing number of nuclear-generating facilities comes the fear of a possible nuclear accident and the release of radioactive materials into the environment. We now hear such terms as *LOCA* (loss-of-coolant accident) and *meltdown*. Should the reactor coolant be lost or stop flowing through the core, the reactor would shut down because of the loss of the moderator. However, residual reactions might cause the core to melt into a hot fissioning mass that could go through the floor of the containment building into the environment.

A partial meltdown occurred at the Three Mile Island (TMI) generating plant near Middletown, Pennsylvania. This was a LOCA in which fortunately only a relatively small amount of radioactive steam was vented into the atmosphere. The nuclear accident at Chernobyl was a meltdown caused by human error. The reactor involved was a graphite reactor, with some 1660 fuel assemblies encased in 1700 tons of graphite blocks (Fig. 30.9). Chemical explosions blew the top off the building, and when the fuel rods melted down, the graphite blocks burned like a massive coal pile. The radioactive smoke was carried over Europe by the wind. In the region near the plant, more than 30 people died. It is estimated that thousands will die of cancer related to the Chernobyl radiation.

Another problem with nuclear reactors is their leftovers, or radioactive waste. The by-products of the fission reactions are radioactive and have long half-lives. As a result, nuclear waste will be a continuing problem for many years. Currently, the waste in liquid form is stored in underground tanks. In the United States, legislation was passed to set up regional repositories. The Nuclear Waste Policy Act of 1982 specified that several sites be evaluated and

Figure 30.8 **Nuclear electrical generation**
(a) A comparison of the percentages of nuclear electrical generation in various countries. Keep in mind that 14% of the electrical generation in a large country such as the United States means more nuclear reactors than 51% of generation in a relatively small country such as Belgium. (b) Nuclear plants in operation in the United States.

Figure 30.9 **Nuclear accident**
The photo shows the damage to the reactor at Chernobyl in the USSR. This accident released large amounts of radioactive material into the environment with very dire consequences.

that two sites, one east of and one west of the Mississippi River, be chosen on the basis of suitability. However, in 1987, Congress amended the law, voting to evaluate only one site. This is an underground repository more than 300 m below the surface of Yucca Mountain in Nevada. The evaluation of this site for waste storage will take until 1995. If it is found to be suitable, the repository will not be opened until after the year 2000.

Fusion as an Energy Source

Fusion can be a source of nuclear energy, even though practical fusion technology is believed to be years away. The energy released in a single hydrogen fusion reaction is relatively small in comparison to the energy released in a single uranium fission reaction. However, the complete fusion of a quantity of hydrogen gives almost three times the energy released from the complete fission of an equivalent amount of uranium.

In a way, we see fusion as the ultimate energy source, since enough deuterium exists in the ocean in the form of heavy water to supply our energy needs for centuries. Also, radioactive waste would not be as great a problem with fusion as it is with fission. The light nuclei of fusion products have relatively short half-lives. For example, tritium has a half-life of 12.3 years.

Very high temperatures are needed to initiate fusion reactions. This reflects the energy needed to overcome the Coulomb repulsion of the fusing nuclei. Temperatures on the order of millions of degrees or kelvins are needed to initiate **thermonuclear reactions**, as they are called. Because of this, practical fusion reactors have not been achieved. The problem is in confining sufficient energy in a reaction region to maintain the necessary high temperatures. Uncontrolled fusion has been demonstrated in the form of the hydrogen (H) bomb. In this case, the fusion reaction was initiated by a small atomic bomb.

At such high temperatures, electrons are stripped from their nuclei, and a gas of positively charged ions and free, negatively charged electrons is obtained. Such a gas of charged particles is called a **plasma**. Plasmas have a number of special physical properties. As a result, they are sometimes referred to as the fourth phase of matter, a term used in 1879 by William Crookes, an English chemist, who generated plasmas in gas-discharge or Crookes tubes.

The problem of plasma confinement is being approached in two ways: magnetic confinement and inertial confinement. Since a plasma is a gas of charged particles, it can be controlled and manipulated by using electric and magnetic fields. In **magnetic confinement**, magnetic fields are used to hold the plasma in a confined space, a so-called magnetic bottle (see Fig. 19.22). Electric fields produce electric currents that raise the temperature of the plasma. Temperatures of 100 million kelvins have been achieved in magnetic confinement machines called tokamaks (Fig. 30.10). In addition, the plasma density and the confinement time are considerations. Putting these conditions together properly might produce a fusion reactor.

Inertial confinement depends on an implosion technique similar to that of the ^{239}Pu bomb mentioned earlier. Hydrogen fuel pellets would be either dropped or positioned in a reactor chamber (Fig. 30.11). Pulses of laser, electron, or ion beams would then be used to implode the pellet, producing compression and high temperatures. Fusion would occur if the pellet stayed together for a sufficient time, which depends on its inertia (hence the name, inertial confinement). At this time, lasers and particle beams are not powerful enough to induce fusion by this means.

However, research continues. The technological problems of controlled thermonuclear fusion are enormous, but so are the benefits. Practical energy production from such fusion is not expected until well into the twenty-first century.

A more recent approach to fusion, termed cold fusion, attempts to fuse nuclei without the high temperature requirements. One method involves muons, which are negatively charged particles about 200 times more massive than electrons (see Section 30.4). Muons are produced in particle accelerators and can take the place of electrons in muonic molecules of deuterium and tritium. Fusion becomes more likely in these molecules by quantum mechanical tunneling, whereby nuclei tunnel through the large energy barrier that normally keeps them apart. Researchers are investigating this so-called muon-catalyzed fusion.

Figure 30.10 **A tokomak**

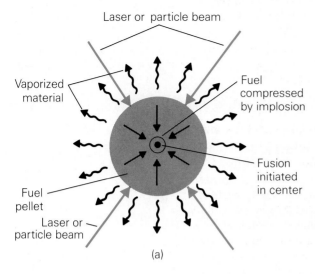

(a)

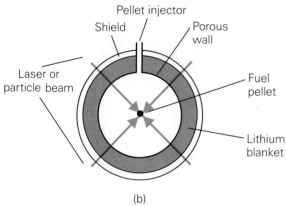

(b)

Figure 30.11 **Inertial confinement**
(a) An illustration of fuel pellet implosion. The compression is enhanced by the vaporization of the outer shell material of the pellet.
(b) A schematic diagram of a possible fusion reactor cavity. The lithium blanket captures neutrons from the fusion reactions, producing tritium, which may be used as a fuel.

In March 1989, a startling announcement was made that fusion had been achieved in the laboratory. This cold fusion process involved the electrolysis of heavy water (D_2O) using platinum-palladium electrodes. Scientists began looking closely at this process to see whether fusion was actually achieved, which now seems highly questionable.

30.3 Beta Decay and the Neutrino

As described in Section 29.2, beta decay appears to be a straightforward process. That is, certain radioactive nuclei naturally decay and emit an electron or a positron in the process. Examples of β^- and β^+ reactions are

$$^{14}_{6}C_8 \longrightarrow \; ^{14}_{7}N_7 \; + \; ^{0}_{-1}e$$

(14.003242 u)   (14.003074 u)

$$^{13}_{7}N_6 \longrightarrow \; ^{13}_{6}C_7 \; + \; ^{0}_{+1}e$$

(13.005739 u)   (13.003355 u)

However, more than this is going on in these beta decay processes. When they are analyzed in detail, there seem to be some violations of physical principles.

One of the difficulties arises with energy. The energy released in the above β^- process may be calculated from the mass difference in the nuclei:

$$\Delta m = 14.003242 \text{ u} - 14.003074 \text{ u} = 0.000168 \text{ u}$$

and

$$E = (0.000168 \text{ u})(931.5 \text{ MeV/u}) = 0.156 \text{ MeV}$$

Thus, the decay reaction has a Q value or disintegration energy of 0.156 MeV. The electron, being a light particle, would be expected to carry away virtually all of this energy or to have a $K_{max} = Q$. However, this is not what happens.

When the kinetic energies of the electrons from a given beta decay process are determined, a continuous spectrum of energies is observed up to a cutoff point of $K_{max} \cong Q$ (Fig. 30.12). That is, electrons are emitted that have less kinetic energy than expected; in fact, most of the beta particles have $K < Q$. This would appear to be a violation of the conservation of energy, since not all the energy is accounted for.

Nor is this the only difficulty. Observations of individual beta decay processes show that the emitted electron and the daughter nucleus from a stationary parent nucleus do not always leave the disintegration site in opposite directions. Thus, we have an apparent violation of the conservation of linear momentum. To top it off, there is also an apparent violation of the conservation of angular momentum. Nucleons and electrons have intrinsic (spin) quantum numbers of $\frac{1}{2}$. For a nucleus with an even number of nucleons (such as ^{14}C), the total angular momentum quantum number will be an integer, since an even number of spin $\frac{1}{2}$ particles is present. When an electron is created in the decay process, an odd number of spin $\frac{1}{2}$ particles is present, and the total angular momentum quantum number will be a half-integer. Therefore, the sums of the total angular momentum quantum numbers are not equal before and after spontaneous decay.

What, then, is the problem? We hope that our conservation laws are not invalid. The other alternative is that these apparent violations are telling us something about nature that we do not know or do not yet recognize. The difficulties would be resolved if we assumed that, in addition to the electron, another unobserved particle of intrinsic spin $\frac{1}{2}$ were created and emitted in beta decay.

This explanation and the existence of such a particle were first suggested by Wolfgang Pauli, a German physicist, in 1930. The particle was christened the **neutrino** (meaning "little neutral one"). The name suggests that, by the conservation of charge for beta decay, an additional particle must necessarily be electrically neutral. A charged particle would be easily detectable, and since the neutrino had not been observed, it must interact very weakly with matter, that is, by a *weak interaction* or a *weak nuclear force*.

The observations of beta decay suggest that the neutrino has zero rest mass and therefore travels with the speed of light like a photon. It also has linear momentum p with a total relativistic energy $E = pc$ and has an intrinsic spin quantum number of $\frac{1}{2}$. In 1956, a particle with these properties was detected experimentally, and the existence of the neutrino was established. Thus, we add the neutrino to the list of subatomic particles. The nuclear equations for some sample beta decay reactions become

$$^{14}_{6}\text{C} \longrightarrow {}^{14}_{7}\text{N} + {}^{0}_{-1}\text{e} + \bar{\nu}_e$$

Using the atomic mass of ^{14}N takes into account the emitted electron, since the daughter atom in beta decay would have only six electrons like the parent ^{14}C atom.

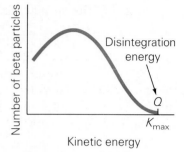

Figure 30.12 **Beta ray spectrum** For a typical beta decay process, most particles are emitted with $K < Q$, leaving energy unaccounted for.

The term *spin* has a historical origin. It was originally thought that observed splitting of spectral lines was due to the angular momentum associated with an electron spin (analogous to the Earth spinning on its axis). Quantum numbers of $\pm\frac{1}{2}$ were associated with different directions of rotation. However, electron spin is a purely nonclassical quantum mechanics effect caused by some intrinsic atomic property other than electron spin.

and

$$^{13}_{7}\text{N} \longrightarrow \text{ }^{13}_{6}\text{C} + \text{ }^{0}_{+1}\text{e} + \nu_e$$

where the Greek letter ν (nu) is the symbol used for the neutrino. The symbol with a bar over it represents an antineutrino. This bar method is a common way of indicating an antiparticle. In general, a neutrino is emitted in β^+ decay, and an antineutrino is emitted in β^- decay. The e subscript identifies the neutrinos as associated with beta decay. As you will learn in a later section, there is another type of neutrino.

Note that the antineutrino is associated with the decay of a neutron:

$$\text{n} \longrightarrow \text{p} + \text{e}^- + \bar{\nu}_e$$

The neutrino ν_e is associated with the decay of a proton into a neutron and a positron. A free proton cannot decay by positron emission because of the conservation of energy (rest masses). But because of binding energy effects, this does occur in the nucleus in β^+ decay with the emission of a neutrino. We will discuss neutrinos and their relationship to the weak nuclear force in a later section.

30.4 Fundamental Forces and Exchange Particles

All interactions take place by means of forces. The forces involved in everyday activities are complicated because of the large numbers of atoms or molecules that make up ordinary objects. Frictional forces, the forces holding this book together, and so on, are actually due to the electromagnetic forces between atoms. Looking at the fundamental interactions between particles makes things simpler. On this level, there are only four known **fundamental forces**: the *gravitational force*, the *electromagnetic force*, the *strong nuclear force*, and the *weak nuclear force*.

The most familiar of these four are the gravitational and electromagnetic forces. Gravity acts between all particles, while the electromagnetic force is restricted to charged particles. Both forces have an inverse square relationship for the separation distance between interacting particles, and an infinite range.

Classically, the action-at-a-distance of a force is described by using the concept of a field. For example, a charged particle is considered to be surrounded by an electric field that *interacts* with other charged particles. Quantum mechanics, on the other hand, provides an alternative description of how forces are transmitted. This process is pictured as an exchange of particles, analogous to you and another person interacting by tossing a ball back and forth.

The creation of such particles would classically violate the conservation of energy. However, within the limitations of the uncertainty principle, a particle can be created for a *short time* with no extra energy. For ordinary time intervals, the conservation of energy is obeyed. But for *extremely* short time intervals, the uncertainty principle requires a large uncertainty in energy ($\Delta E\,\Delta t \geq h/2\pi$), so energy conservation may be briefly violated. A particle created in such a manner and time is called a **virtual particle**.

The fundamental forces are considered to be carried by virtual **exchange particles**. The exchange particle for each force is different in mass. The greater the mass of the particle, the greater the amount of energy required to create

A fifth force has been postulated but has not been substantiated. It is said to be a repulsive force that depends on the chemical makeup of substances.

it and the shorter time it exists. Since a massive particle can exist for only a short time, the range of the interaction for the associated force would be small. That is, the range of an exchange particle is inversely proportional to its mass.

The exchange particle for the electromagnetic force is a (virtual) **photon**. As a "massless" particle, it has an infinite range, as does the electromagnetic force. The idea of a particle exchange is represented by the Feynman diagram of the collision of two electrons in Fig. 30.13. Such space-time diagrams are named after American scientist and Nobel Prize winner Richard Feynman (1918–1988), who used them to analyze electrodynamic interactions (look ahead to Fig. 30.15). The important points are the vertices of the diagram. One electron may be considered to create a virtual photon at point A and the other electron to absorb it at point B. Each of the two interacting particles undergoes a change in energy and momentum by virtue of the exchange photon and force interaction.

Using the idea of exchange particles, Japanese physicist H. Yukawa in 1935 proposed that the short-range strong nuclear force between two nucleons is associated with an exchange particle called the **meson**. An estimate of the mass of this particle may be gained from the uncertainty principle. If a nucleon were to create a meson, the conservation of energy would have to be violated by an amount of energy at least as great as the meson rest mass or

$$\Delta E = (\Delta m)c^2 = m_m c^2$$

where m_m is the rest mass of the meson.

By the uncertainty principle, the meson would be absorbed in the exchange process in a time on the order of

$$\Delta t = \frac{h}{2\pi \, \Delta E} = \frac{h}{2\pi m_m c^2}$$

In this time, the meson could travel a distance not greater than

$$R = c \, \Delta t = \frac{h}{2\pi m_m c} \tag{30.6}$$

Taking this distance to be the range of the nuclear force ($R \cong 1.4 \times 10^{-15}$ m) and solving for m_m give

$$m_m \cong 274 m_e$$

where m_e is the electron rest mass. Thus, if the meson existed, it would be expected to have a rest mass about 270 times that of an electron.

Of course, virtual mesons of an exchange process cannot be observed. But if sufficient energy were supplied to colliding nucleons, real mesons might be created through the energy made available in the collision process. The real mesons could then leave the nucleus and be detected. At the time of Yukawa's prediction, there were no known particles with masses between that of the electron (m_e) and that of the proton ($m_p = 1836 m_e$).

In 1936, Yukawa's prediction seemed to come true when a new particle with a mass of about $200 m_e$ was discovered in cosmic rays. Later called the **μ meson**, or **muon**, the particle was shown to have two charged varieties, positive and negative of electronic magnitude, with a rest mass of

$$m_{\mu_\pm} = 207 m_e$$

However, further investigations showed that the μ meson did not behave like the particle of Yukawa's theory. In particular, the interaction of μ mesons

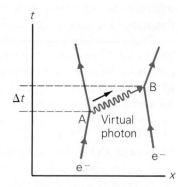

Figure 30.13 **Feynman diagram for electron-electron interaction** The interacting electrons have a change in energy and momentum due to the exchange of a virtual photon, which is created at A and absorbed at B in a time (t) that is consistent with the uncertainty principle.

with matter (nuclei) was very weak. The μ mesons from cosmic radiation could penetrate a large mass of material as evidenced by their detection in deep mines.

This situation was a source of controversy and confusion for several years. But in 1947, two more charged particles of the appropriate mass range were discovered in cosmic radiation. These particles were called π **mesons** (primary mesons), or **pions**, together with a less massive electrically neutral π meson. Measurement showed the masses of the pions to be

$$m_{\pi^{\pm}} = 274 m_e \qquad \text{and} \qquad m_{\pi^{\circ}} = 264 m_e$$

Moreover, the π meson interacted strongly with matter. The π mesons fulfilled the requirements of Yukawa's theory, and it has been generally accepted that this meson is the particle that transmits or mediates the strong nuclear force. The Feynman diagrams of nucleon-nucleon interactions are illustrated in Fig. 30.14.

Free mesons are unstable. For example, the π^+ meson quickly decays (in about 10^{-8} s) into a μ meson:

$$\pi^+ \longrightarrow \mu^+ + \nu_\mu$$

Here ν_μ is a μ neutrino different from that of beta decay. The μ mesons also decay into positrons and electrons with the emission of both types of neutrinos, for example,

$$\mu^+ \longrightarrow e^+ + \nu_e + \bar{\nu}_\mu$$

The discrepancies in beta decay discussed in the preceding section led to another discovery. Electrons and neutrinos are emitted from unstable nuclei, but there was evidence that they did not exist *inside* the nucleus. Enrico Fermi proposed that these particles did not exist before being emitted but were instantaneously created in the decaying radioactive nucleus. For β^- decay, this would mean that a single neutron was in some way transmuted into three particles:

$$n \longrightarrow p^+ + e^- + \bar{\nu}_e$$

That a neutron could do this was confirmed by the observation of free neutrons. Free neutrons disintegrate after a few minutes into a proton, an electron, and a neutrino. But the question arose as to what force could cause a neutron to disintegrate in this manner. None of the known forces was applicable, so some other force must be acting in beta decay. Decay rate mea-neutron to disintegrate in this manner. None of the known forces was ap-than the electromagnetic force, but still much stronger than the gravitational force. Thus, the **weak nuclear force** was discovered.

It was originally thought that the weak interaction was localized, without any range at all. However, we now know that the weak force has a range of about 10^{-17} m. In terms of an exchange particle, this means that the virtual carriers must be much more massive than the pions of the strong nuclear force. The virtual exchange particles for the weak nuclear force were called **W particles** (*w*eak). W particles have masses about 100 times that of a proton, which explains the extremely short range and weakness of the force. The existence of the W particle was confirmed in the 1980s when accelerators were built with enough energy to create the first real W particles.

The weak force acts between neutrinos, which explains why they are so difficult to detect. Research has shown that the weak force is involved in the

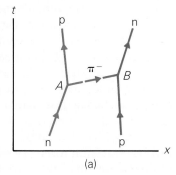

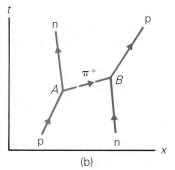

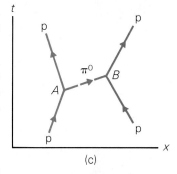

Figure 30.14 Feynman diagrams for nucleon-nucleon interactions Feynman diagrams for nucleon-nucleon interactions through the exchange of virtual pions. (a) n-p interaction. (b) p-n interaction. (c) p-p interaction.

transmutation of other subatomic particles. In general, the weak force is limited to transmuting the identities of particles within the nucleus. The only way it manifests its existence in the outside world is through the emitted neutrinos. One highly noticeable but infrequent announcement of the weak force at work is during the explosion of a supernova. The collapse of the core of an aging star gives rise to a huge release of neutrinos. The weak force of the neutrinos blasts the outer layers of the star into space in a cataclysmic explosion of a supernova.

Finally, let's not forget the gravitational force. The exchange particles of the gravitational force are called **gravitons**. There is still no firm evidence of the existence of this massless particle. Several searches have been made to detect the graviton or gravity waves, but the relative weakness of its interaction makes it a very elusive particle. A comparison of the relative strengths of the fundamental forces is given in Table 30.2.

30.5 Elementary Particles

The fundamental particles, the building blocks of atoms, are referred to as **elementary particles**. Simplicity reigned when it was thought that an atom was an indivisible particle and thus *the* elementary particle. However, scientists now know of a variety of subatomic particles. Many of which have not been mentioned here. Indeed, scientists are working to simplify and reduce a long list of elementary particles, since some may be variations of one type of particle.

There are several systems for classifying elementary particles on the basis of their various properties. One classification uses the distinction of nuclear force interactions. Particles that experience or interact by the strong nuclear force are called **hadrons**. These include the proton, neutron, and pion. Other particles, which interact via the weak nuclear force but not the strong force, are called **leptons** ("light ones"). The lepton family includes the electron, muon, and neutrino.

Note in Table 30.2 that gravity and the weak force are the only forces acting between *all* particles. But the force of gravity is so weak in elementary particle interactions that it can, in general, be neglected. This leaves the weak force as the only force that interacts with all particles.

Let's take a brief look at the lepton and hadron families.

Table 30.2
Fundamental Forces

Force	Relative strength	Action distance	Exchange particle	Particles with interaction
Strong nuclear	1	Short range ($\cong 10^{-15}$ m)	π meson (pion)	Hadrons*
Electromagnetic	10^{-3}	Inverse square (infinite)	Photon	Electrically charged
Weak nuclear	10^{-8}	Extremely short range ($\cong 10^{-17}$ m)	W particle†	All
Gravitational	10^{-45}	Inverse square (infinite)	Graviton	All

* Hadrons are elementary particles (see Section 30.5).
† Actually, three particles are involved: W^+, W^-, and Z. These are described in Section 30.5.

Leptons

The most familiar lepton is the electron. It is apparently the only lepton that exits naturally in atoms. There is no evidence of any internal structure, and, measured to be smaller than 10^{-17} m in size, the electron is considered to be a point particle.

Muons were first observed in cosmic rays. They are electrically charged (μ^-) and 200 times heavier than an electron. Since they appear not to have any internal structure, they are sometimes called heavy electrons. Muons are unstable and decay quickly (in about 10^{-6} s). This is the decay reaction:

$$\mu^- \longrightarrow e^- + \bar{v}_e + v_\mu$$

Recall that muon decay was used as evidence for relativistic time dilation in Chapter 26.

Only one other charged lepton has been discovered. Known as a **tau** (τ^-) or **tauon**, it has a mass twice that of a proton. The electron, muon, and tauon are all negatively charged and appear to have no internal structure. Their antiparticles are positively charged.

The only other known lepton is the neutrino, which is present in cosmic rays and is emitted in some radioactive decays. Neutrinos appear to have no mass and to travel at the speed of light. They feel neither the electromagnetic force nor the strong forces and so pass through matter as if it weren't there. Neutrinos are so weakly interacting that most neutrinos striking the Earth pass right through it.

Neutrinos come in several varieties. The electron (v_e) and muon (v_μ) neutrinos are well documented, and it is believed that a tau neutrino (v_τ) also exists. There is an antineutrino for each of these types.

This completes the list of leptons. With a total of six leptons plus antiparticles, there are twelve different leptons in all.

There has been a report that neutrinos actually possess a tiny mass. If this is true, the ocean of neutrinos that pervades the universe could provide enough mass to give rise to a closed universe in which the Big Bang expansion would be gravitationally halted and, in the future, collapse.

Hadrons

There are many more types of hadrons than there are leptons. All hadrons interact by the strong force, the weak force, and gravity. Also, some are electrically charged. The best known hadrons are nucleons: the proton and the neutron. All others are short-lived and decay via the weak force or more rapidly under the influence of the strong force.

The sheer number of hadrons suggests that they are perhaps composites of other elementary particles. Some help came to sorting out the hadron "zoo" in 1963 when Murray Gell-Mann and George Zweig of Caltech put forth the quark theory (Fig. 30.15). It suggested that **quarks** were elementary charged particles that made up hadrons. They could combine only in two possible ways, either in trios or in quark-antiquark pairs. Combinations of three quarks produce relatively heavy hadrons called **baryons** ("heavy ones"), which include the proton and neutron. Quark-antiquark pairs form lighter particles called **mesons**.

To account for the hadrons that were known at that time, the theory proposed three types or "flavors" of quarks. These were given the names "up" (u), "down" (d), and "strange" (s). In addition, the quarks carried fractional electronic charges. The u, d, and s quarks had charges of $+\frac{2}{3}e$, $-\frac{1}{3}e$, and $-\frac{1}{3}e$, respectively, antiquarks having charges of opposite signs (for example, $\bar{u}$ with

Taken from a line in James Joyce's novel *Finnegans Wake*: "Three quarks for Muster Mark!" The "three quarks" denote the children of Mr. Finn, who sometimes appears as Mister (Muster) Mark. This is bit strange, but it was originally thought that there were three kinds of quarks, one of which was "strange."

a charge of $-\frac{2}{3}$e). Thus, the quark combinations for the proton and neutron are *uud* and *udd*, respectively. A positive pion (π^+) is a $u\bar{d}$ combination.

The existence of quarks has been proven experimentally, but they are generally believed not to exist outside of the nucleus. This would explain why we do not observe fractional electronic charges in nature. Quarks can also exist in excited states, similar to the excited states of an atom. It is thought that many of the observed hadrons might be excited states of certain combinations of quarks.

Quarks interact by the strong force, but they are also subject to the weak force. A weak force acting on a quark changes its flavor and gives rise to the decay of hadrons.

The discovery of new elementary particles in the 1970s led to the addition of more quark flavors, which were called "charm" (*c*), "top" (or "truth," *t*), and "bottom" (or "beauty," *b*). A summary of the quark flavors is given in Table 30.3. There is an oppositely charged antiquark for each quark.

Since quarks interact, it is postulated that they too have interacting exchange particles. The exchange particle for quarks has been dubbed the **gluon**. Gluons bind or "glue" hadrons together, thus replacing the pion as the exchange particle.

The quark theory was further extended in terms of a force field. To give the strong force a field representation, each quark is said to possess the analog of electric charge that is the source of the gluon field. Instead of calling this property "charge," it was called "color," with no relationship to ordinary color. Each quark can come in one of three colors: red, green, and blue. There are corresponding anticolors for antiquarks.

When a quark emits or absorbs a gluon, it changes color. The effect is to change the identity of a quark—for example, a blue quark to a red quark. In this respect, the strong force resembles the weak force (for which there is a change of one particle into another with the exchange of a W particle).

Actually, the quark color scheme was developed after a major discovery was made concerning the weak force. Like Maxwell's combining the electric force and the magnetic force into a single electromagnetic force, the weak force and the electromagnetic force were combined, or shown to be two parts of a single **electroweak force**. This came about as the result of a theory put forth by Sheldon Glashow, Abdus Salam, and Steven Weinberg, for which they received a Nobel Prize in 1979. In this theory, weakly interacting particles such as electrons and neutrinos carry a weak charge that gives rise to a weak force field. The exchange particles for the weak force interaction are the heavy,

Figure 30.15 **Theoretical physicists** Murray Gell-Mann (left) and Richard Feynman (1918–1988). Gell-Mann was awarded a Nobel Prize in 1969 for his work on the quark model. Feynman shared the 1965 Nobel Prize for his work on quantum electrodynamics.

Table 30.3
Types of Quarks

Name	Symbol	Charge
Up	*u*	$+\frac{2}{3}$e
Down	*d*	$-\frac{1}{3}$e
Strange	*s*	$-\frac{1}{3}$e
Charm	*c*	$+\frac{2}{3}$e
Top (truth)	*t*	$+\frac{2}{3}$e
Bottom (beauty)	*b*	$-\frac{1}{3}$e

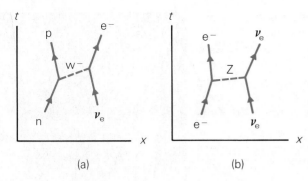

Figure 30.16 **Weak force interactions**
The Feynman diagrams for (a) the decay of a neutron into a proton and an electron through the exchange of a W^- particle and (b) the scattering of a neutrino by an electron through the exchange of a neutral Z particle.

electrically charged W^+ and W^- particles, along with a neutral Z particle for reactions in which there is no change in or transfer of charge. Two weak interactions are illustrated in Fig. 30.16.

In 1983, the existence of the Z was confirmed, and the four fundamental forces were reduced to three. Of course, scientists would like to reduce the list even further. A theory that would merge the strong nuclear force and the electroweak force into a grand unified force is the so-called **grand unified theory** (GUT). Some two dozen exchange particles are required, including the 12 already known particles.

Should the three fundamental forces be reduced to two by the GUT or a similar theory, there is the hope of a further reduction with the idea that all forces are part of a single superforce. The combining of a grand unified force with gravity is a real challenge. While the three components of a grand force may be represented as force fields in space and time, our current view is that gravity *is* space and time.

Important Formulas

Q value:
$$Q = (m_A + m_a - m_B - m_b)c^2 = (\Delta m)c^2$$

Threshold energy:
$$K_{min} = \left(1 + \frac{m_a}{M_A}\right)|Q|$$

Questions

Nuclear Reactions

1. How do chemical reactions and nuclear reactions differ?

2. How are nuclear reactions initiated?

3. What is the Q value of a reaction and how is it calculated?

4. What are the situations in which (a) $Q < 0$ and (b) $Q > 0$?

5. What is the sign of the Q value of a reaction in which (a) mass is converted to energy and (b) energy is converted to mass?

6. What is the interpretation of $Q = 0$?

7. If an incident particle striking a stationary nu-

cleus has a kinetic energy equal to the Q value of a possible reaction, will the reaction occur? Explain.

8. What is meant by the cross section of a nuclear reaction, and how does the cross section depend on whether the incident particles are charged particles or neutrons?

Nuclear Energy: Fission and Fusion

9. What makes nuclear fission energetically possible?

10. What is meant by a chain reaction?

11. Why is nuclear fusion energetically possible?

12. What is the difference between a reactor and an accelerator?

13. Give the conditions necessary for sustaining and controlling a fission chain reaction.

14. What is meant by a subcritical reaction?

15. Would lead (Pb) serve as an effective moderator for neutrons in a reactor? Explain.

16. Explain the meanings of the terms LOCA and meltdown.

17. Explain how a breeder reactor can produce more fuel than it consumes.

18. What is the critical mass for fusion?

19. What are some of the technical problems in developing a practical fusion reactor? How are these problems being approached?

Beta Decay and the Neutrino

20. If only an electron were emitted in beta decay, what would its kinetic energy be? Is this observed?

21. In beta decay, how does the neutrino resolve the apparent violation of the conservation of (a) linear momentum, (b) angular momentum, and (c) energy?

22. Is it possible for a free neutron to undergo β^- decay? Is it possible for a free proton to undergo β^+ decay? Explain.

23. Why is it so difficult to detect neutrinos experimentally?

Fundamental Forces and Exchange Particles

24. High-energy electrons entering the atmosphere give rise to cosmic ray showers. This process is illustrated in Fig. 30.17. Identify each of the labeled processes.

25. How is the photon viewed as taking part in electromagnetic interactions?

26. What is a virtual process? Is the conservation of energy violated during such a process? Explain.

27. If virtual exchange particles are unobservable, then how is their existence verified?

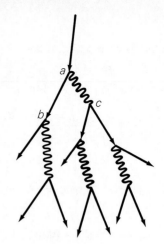

Figure 30.17 **Cosmic ray showers**
See Question 24.

28. Although the μ meson had the appropriate mass, why was it not believed to be the exchange particle predicted by Yukawa's theory?

29. According to exchange theory, on what does the strength of an interaction depend? How does this vary for the different forces?

30. Describe the exchange particle for each of the fundamental forces.

Elementary Particles

31. Distinguish among hadrons, leptons, baryons, and mesons.

32. Is the μ meson classified as a meson in quark theory? Explain.

33. What is meant by quark flavor and color? Can these be changed? Explain.

34. With so many types of hadrons, why aren't fractional electronic charges observed?

35. What are the electroweak force and the GUT?

Problems

30.1 and 30.2 Nuclear Reactions and Nuclear Energy

■■**1.** Complete the following nuclear reactions:

(a) ${}^{6}_{3}\text{Li} + {}^{1}_{1}\text{H} \longrightarrow {}^{3}_{2}\text{He} + \underline{\quad}\ \underline{\quad}$

(b) ${}^{58}_{28}\text{Ni} + {}^{2}_{1}\text{H} \longrightarrow {}^{59}_{28}\text{Ni} + \underline{\quad\quad}$

(c) ${}^{235}_{92}\text{U} + {}^{1}_{0}\text{n} \longrightarrow {}^{138}_{54}\text{Xe} + 5{}^{1}_{0}\text{n} + \underline{\quad\quad}$

(d) ${}^{9}_{4}\text{Be}(\alpha, \text{n})\underline{\quad\quad}$

(e) ${}^{16}_{8}\text{O}(\text{n}, \text{p})\underline{\quad\quad}$

■■**2.** Give the compound nuclei for the reactions in Problem 1.

■■**3.** Complete the following nuclear reactions:

(a) ${}^{13}_{6}\text{C} + {}^{1}_{1}\text{H} \longrightarrow \gamma + \underline{\quad\quad}$

(b) ${}^{10}_{5}\text{B} + {}^{4}_{2}\text{He} \longrightarrow {}^{12}_{6}\text{C} + \underline{\quad\quad}$

(c) ${}^{27}_{13}\text{Al}(\alpha, \text{n})\underline{\quad}\ \underline{\quad}$

(d) ${}^{14}_{7}\text{N}(\alpha, \text{p})\underline{\quad}\ \underline{\quad}$

(e) ${}^{13}_{6}\text{C}(\text{p}, \alpha)\underline{\quad\quad}$

■■**4.** Give the compound nuclei for the reactions in Problem 3.

5. Is the following reaction endoergic or exoergic?

$$\underset{(7.016005\ u)}{^{7}_{3}\text{Li}} + \underset{(1.007825\ u)}{^{1}_{1}\text{H}} \longrightarrow \underset{(4.002603\ u)}{^{4}_{2}\text{He}} + \underset{(4.002603\ u)}{^{4}_{2}\text{He}}$$

6. Determine the Q value of the following reaction:

$$\underset{(9.012183\ u)}{^{9}_{4}\text{Be}} + \underset{(4.002603\ u)}{^{4}_{2}\text{He}} \longrightarrow \underset{(12.000000\ u)}{^{12}_{6}\text{C}} + \underset{(1.008665\ u)}{^{1}_{0}\text{n}}$$

7. Is the reaction $^{200}\text{Hg}(p, \alpha)^{197}\text{Au}$ endoergic or exoergic? (See the equation in chapter for mass values.)

8. Find the threshold energy for the following reaction:

$$\underset{(15.994915\ u)}{^{16}_{8}\text{O}} + \underset{(1.008665\ u)}{^{1}_{0}\text{n}} \longrightarrow \underset{(13.003355\ u)}{^{13}_{6}\text{C}} + \underset{(4.002603\ u)}{^{4}_{2}\text{He}}$$

9. Find the threshold energy for the following reaction:

$$\underset{(3.016029\ u)}{^{3}_{2}\text{He}} + \underset{(1.008665\ u)}{^{1}_{0}\text{n}} \longrightarrow \underset{(2.014102\ u)}{^{2}_{1}\text{H}} + \underset{(2.014102\ u)}{^{2}_{1}\text{H}}$$

10. What is the minimum kinetic energy of a proton that will initiate the reaction $^{3}_{1}\text{H}(p, d)^{2}_{1}\text{H}$? (d stands for a deuterium nucleus.)

11. ^{226}Ra decays and emits a 4.706-MeV alpha particle. Find the velocity of the recoiling daughter nucleus from the decay of a stationary radium-226 nucleus.

12. Find the approximate energy released in the following fission reactions:

(a) $^{235}_{92}\text{U} + ^{1}_{0}\text{n} \longrightarrow$ with the release of 5 neutrons

(b) $^{239}_{94}\text{Pu} + ^{1}_{0}\text{n} \longrightarrow$ with the release of 3 neutrons

13. Calculate the amounts of energy released in the following fusion reactions:

(a) $^{2}_{1}\text{H} + ^{2}_{1}\text{H} \longrightarrow ^{3}_{2}\text{He} + ^{1}_{0}\text{n}$

(b) $^{3}_{2}\text{He} + ^{3}_{2}\text{He} \longrightarrow ^{4}_{2}\text{He} + 2^{1}_{1}\text{H}$

14. Compute the Q values of the following fusion reactions:

(a) $^{2}_{1}\text{H} + ^{3}_{1}\text{H} \longrightarrow ^{4}_{2}\text{He} + ^{1}_{0}\text{n}$

(b) $^{2}_{1}\text{H} + ^{2}_{1}\text{H} \longrightarrow ^{3}_{1}\text{H} + ^{1}_{1}\text{H}$

15. Assume that the average kinetic energy of ions in a plasma is given by the equation for the kinetic energy of the atoms in an ideal gas ($\frac{1}{2}mv^2 = 3/2kT$) and that fusion occurs when the ions approach each other within a distance of the upper limit for the nuclear diameter ($R = 10^{-12}$ cm). Calculate the temperature required for

fusion of two deuterium ions. (Boltzmann's constant is $k = 1.38 \times 10^{-23}$ J/K.)

30.3 Beta Decay and the Neutrino

16. Show that the disintegration energy for β^- is

$$Q = (m_p - m_d - m_e)c^2 = (M_p - M_d)c^2$$

where the m's represent the masses of the parent and daughter nuclei and the M's represent the masses of the neutral atoms.

17. What is the maximum kinetic energy of the electron emitted when a ^{12}B nucleus beta decays into a ^{12}C nucleus? (See Problem 16.)

18. The kinetic energy of an electron emitted from a ^{32}P nucleus that beta decays into a ^{32}S nucleus is observed to be 1 MeV. What is the energy of the accompanying neutrino of the decay process? (See Problem 16.)

19. Show that the disintegration energy for β^+ decay is

$$Q = (m_p - m_d - m_e)c^2 = (M_p - M_d - 2m_e)c^2$$

where the m's represent the masses of the parent and daughter nuclei and the M's represent the masses of the neutral atoms.

20. The kinetic energy of a positron emitted from the β^+ decay of a ^{13}N nucleus into a ^{13}C nucleus is measured to be 1.19 MeV. What is the energy of the accompanying neutrino in the process? (Neglect the recoil energy of the nucleus and see Problem 19.)

21. Show that the Q value for electron capture is given by

$$Q = (m_p + m_e - m_d)c^2 = (M_p - M_d)c^2$$

where the m's represent the masses of the parent and daughter nuclei and the M's represent the masses of the neutral atoms.

22. On the basis of the Q values given in Problems 16, 19, and 21, what are the mass requirements of the parent atoms for β^-, β^+, and electron capture processes?

23. (a) In the electron capture process of ^7Be nucleus converting into a ^7Li nucleus, what is the energy of the emitted neutrino? (See Problem 21.) (b) Is it energetically possible for ^7Be to β^+ decay into ^7Li?

30.4 Fundamental Forces and Exchange Particles

24. Draw the Feynman diagrams for (a) the Compton effect and (b) electron pair annihilation.

■■**25.** Assuming the range of the nuclear force to be on the order of the Bohr radius (about 0.050 nm), predict the mass of the exchange particle.

■■**26.** By how much energy is the conservation of energy violated in a meson exchange process?

■■**27.** How long is the conservation of energy violated in a meson exchange process?

■■**28.** Assuming the ratio of the ranges of the virtual W particle and the virtual pion to be equal to the ratio of the average lifetimes of the decay processes involving these interactions, estimate the mass of the W particle.

Additional Problems

29. Find the threshold energy of the following reaction:

$$^{13}_{6}C \quad + \quad ^{1}_{1}H \quad \longrightarrow \quad ^{1}_{0}n \quad + \quad ^{13}_{7}N$$
$$(13.003355\ u) \quad (1.007825\ u) \qquad (1.008665\ u) \quad (13.005739\ u)$$

30. Polonium-210 undergoes alpha decay. Find the energy released in the decay reaction.

31. Complete the following nuclear reactions:

(a) $^{27}_{13}Al + ^{1}_{0}n \longrightarrow ^{1}_{1}H +$ _____

(b) $^{13}_{7}N + ^{1}_{0}n \longrightarrow ^{4}_{2}He +$ _____

(c) $^{92}_{40}Zr(p, \alpha)$ ____ $+ ^{1}_{1}H \rightarrow$ ____ $+ ^{4}_{2}He$

(d) $^{25}_{12}Mg(\gamma, p)$ ____ $+ \gamma \rightarrow ^{1}_{1}H +$ ____

32. Write the compound nuclei for the reactions in Problem 31.

33. Lead-211 undergoes beta decay. Find the energy released in the decay reaction.

34. Determine the threshold energy of the following reaction:

$$^{16}_{8}O \quad + \quad ^{1}_{0}n \quad \longrightarrow \quad ^{13}_{6}C \quad + \quad ^{4}_{2}He$$
$$(15.994915\ u) \quad (1.008665\ u) \qquad (13.003355\ u) \quad (4.002603\ u)$$

35. Determine the minimum kinetic energy of an incident alpha particle that will initiate the following reaction:

$$^{14}_{7}N \quad + \quad ^{4}_{2}He \quad \longrightarrow \quad ^{17}_{8}O \quad + \quad ^{1}_{1}H$$
$$(14.003074\ u) \quad (4.002603\ u) \qquad (16.999131\ u) \quad (1.007825\ u)$$

Appendix

Appendix I **Tables**

    Table 1.   Physical Constants

    Table 2.   Conversion Factors

    Table 3.   Trigonometric Relationships

    Table 4.   Trigonometric Tables

    Table 5.   Areas and Volumes of Some Common Shapes

    Table 6.   Planetary Data

Appendix II **Kinetic Theory of Gases**

Appendix III **Einstein's Letter to President Roosevelt**

Appendix IV **Alphabetical Listing of the Chemical Elements**

Appendix V **Properties of Selected Isotopes**

Appendix I: Tables

Table 1
Physical Constants

Acceleration due to gravity:	$g = 9.80 \text{ m/s}^2 = 980 \text{ cm/s}^2 = 32.2 \text{ ft/s}^2$
Universal gravitational constant:	$G = 6.67 \times 10^{-11} \dfrac{\text{N-m}^2}{\text{kg}^2}$
Electron charge:	$e = 1.60 \times 10^{-19} \text{ C}$
Speed of light:	$c = 3.0 \times 10^8 \text{ m/s} = 3.0 \times 10^{10} \text{ cm/s} = 1.86 \times 10^5 \text{ mi/s}$
Boltzmann's constant:	$k = 1.38 \times 10^{-23} \text{ J/K}$
Planck's constant:	$h = 6.63 \times 10^{-34} \text{ J-s}$
	$\hbar = h/2\pi = 1.05 \times 10^{-34} \text{ J-s}$
Electron rest mass:	$m_e = 9.11 \times 10^{-31} \text{ kg} = 5.49 \times 10^{-4} \text{ u} \leftrightarrow 0.511 \text{ MeV}$
Proton rest mass:	$m_p = 1.672 \times 10^{-27} \text{ kg} = 1.007267 \text{ u} \leftrightarrow 938.28 \text{ MeV267/28}$
Neutron rest mass:	$m_n = 1.674 \times 10^{-27} \text{ kg} \times 1.008665 \text{ u} \leftrightarrow 939.57 \text{ MeV}$
Coulomb's law constant:	$k = 1/4\pi\varepsilon_o = 9.0 \times 10^9 \text{ N-m}^2/\text{C}^2$
Permittivity of free space:	$\varepsilon_o = 8.85 \times 10^{-12} \text{ C}^2/\text{N-m}^2$
Permeability of free space:	$\mu_o = 4\pi \times 10^{-7} = 1.26 \times 10^{-6} \text{ T-m/A}$

Astronomical and Earth data

Equatorial radius of Earth:	$6.378 \times 10^3 \text{ km} = 3963 \text{ mi}$
Polar radius of Earth:	$6.357 \times 10^3 \text{ km} = 3950 \text{ mi}$
Mass of Earth	$6.0 \times 10^{24} \text{ kg}$
Mass of Moon	$7.4 \times 10^{22} \text{ kg} = \frac{1}{81} \text{ mass of Earth}$
Mass of Sun	$2.0 \times 10^{30} \text{ kg}$
Average distance of Earth from Sun	$1.5 \times 10^8 \text{ km} = 93 \times 10^6 \text{ mi}$
Average distance of moon from Earth	$3.8 \times 10^5 \text{ km} = 2.4 \times 10^5 \text{ mi}$
Diameter of moon	$3500 \text{ km} \cong 2160 \text{ mi}$
Diameter of Sun	$1.4 \times 10^6 \text{ km} \cong 864{,}000 \text{ mi}$

Table 2
Conversion Factors

Mass	
	$1\ g = 10^{-3}\ kg = 6.85 \times 10^{-5}\ slug$
	$1\ kg = 10^3\ g = 6.85 \times 10^{-2}\ slug$
	$1\ slug = 1.46 \times 10^4\ g = 14.6\ kg$
	$1\ u = 1.66 \times 10^{-24}\ g = 1.66 \times 10^{-27}\ kg$
	$1\ metric\ ton = 1000\ kg$
Length	
	$1\ cm = 10^{-2}\ m = 0.394\ in.$
	$1\ m = 10^{-3}\ km = 3.28\ ft = 39.4\ in.$
	$1\ km = 10^3\ m = 0.62\ mi$
	$1\ in. = 2.54\ cm = 2.54 \times 10^{-2}\ m$
	$1\ ft = 12\ in. = 30.48\ cm = 0.3048\ m$
	$1\ mi = 5280\ ft = 1609\ m = 1.609\ km$
	$1\ \text{Å} = 10^{-10}\ m = 10^{-8}\ cm$
Area	
	$1\ cm^2 = 10^{-4}\ m^2 = 0.1550\ in.^2 = 1.08 \times 10^{-3}\ ft^2$
	$1\ m^2 = 10^4\ cm^2 = 10.76\ ft^2 = 1550\ in.^2$
	$1\ in.^2 = 6.94 \times 10^{-3}\ ft^2 = 6.45\ cm^2 = 6.45 \times 10^{-4}\ m^2$
	$1\ ft^2 = 144\ in.^2 = 9.29 \times 10^{-2}\ m^2 = 929\ cm^2$
Volume	
	$1\ cm^3 = 10^{-6}\ m^3 = 3.53 \times 10^{-5}\ ft^3 = 6.10 \times 10^{-2}\ in.^3$
	$1\ m^3 = 10^6\ cm^3 = 10^3\ L = 35.3\ ft^3 = 6.10 \times 10^3\ in.^3 = 264\ gal$
	$1\ liter = 10^3\ cm^3 = 10^{-3}\ m^3 = 1.056\ qt = 0.264\ gal$
	$1\ in.^3 = 5.79 \times 10^{-4}\ ft^3 = 16.4\ cm^3 = 1.64 \times 10^{-5}\ m^3$
	$1\ ft^3 = 1728\ in.^3 = 7.48\ gal = 0.0283\ m^3 = 28.3\ L$
	$1\ qt = 2\ pt = 946.5\ cm^3 = 0.946\ L$
	$1\ gal = 4\ qt = 231\ in.^3 = 3.785\ L$
Time	
	$1\ h = 60\ min = 3600\ s$
	$1\ day = 24\ h = 1440\ min = 8.64 \times 10^4\ s$
	$1\ year = 365\ days = 8.76 \times 10^3\ h = 5.26 \times 10^5\ min = 3.16 \times 10^7\ s$
Angle	
	$360° = 2\pi\ rad$
	$180° = \pi\ rad$
	$90° = \pi/2\ rad$
	$60° = \pi/3\ rad$
	$45° = \pi/4\ rad$
	$30° = \pi/6\ rad$
	$1\ rad = 57.3°$
	$1° = 0.0175\ rad$
Speed	
	$1\ m/s = 3.6\ km/h = 3.28\ ft/s = 2.24\ mi/h$
	$1\ km/h = 0.278\ m/s = 0.621\ mi/h = 0.911\ ft/s$
	$1\ ft/s = 0.682\ mi/h = 0.305\ m/s = 1.10\ km/h$
	$1\ mi/h = 1.467\ ft/s = 1.609\ km/h = 0.447\ m/s$
	$60\ mi/h = 88\ ft/s$
Force	
	$1\ N = 10^5\ dynes = 0.225\ lb$
	$1\ dyne = 10^{-5}\ N = 2.25 \times 10^{-6}\ lb$
	$1\ lb = 4.45 \times 10^5\ dynes = 4.45\ N$
	Equivalent weight of 1-kg mass $= 2.2\ lb = 9.8\ N$
Pressure	
	$1\ Pa\ (N/m^2) = 1.45 \times 10^{-4}\ lb/in.^2 = 7.5 \times 10^{-3}\ torr\ (mm\ Hg) = 10\ dynes/cm^2$
	$1\ torr\ (mm\ Hg) = 133\ Pa\ (N/m^2) = 0.02\ lb/in.^2 = 1333\ dynes/cm^2$
	$1\ atm = 14.7\ lb/in.^2 = 1.013 \times 10^5\ N/m^2 = 1.013 \times 10^6\ dynes/cm^2$
	$= 30\ in.\ Hg = 76\ cm\ Hg$
	$1\ bar = 10^6\ dynes/cm^2 = 10^5\ Pa$
	$1\ millibar = 10^3\ dynes/cm^2 = 10^2\ Pa$
Energy	
	$1\ J = 10^7\ ergs = 0.738\ ft\text{-}lb = 0.239\ cal = 9.48 \times 10^{-4}\ Btu = 6.24 \times 10^{18}\ eV$
	$1\ kcal = 4186\ J = 4.186 \times 10^{10}\ ergs = 3.968\ Btu$
	$1\ Btu = 1055\ J = 1.055 \times 10^{10}\ ergs = 778\ ft\text{-}lb = 0.252\ kcal$
	$1\ cal = 4.186\ J = 3.97 \times 10^{-3}\ Btu = 3.09\ ft\text{-}lb$
	$1\ ft\text{-}lb = 2.69 \times 10^6\ J = 1.36 \times 10^7\ ergs = 1.29 \times 10^{-3}\ Btu$
	$1\ eV = 1.60 \times 10^{-19}\ J = 1.60 \times 10^{-12}\ ergs$
	$1\ kWh = 3.6 \times 10^6\ J$

Table 2
Conversion Factors (cont.)

Power	1 W = 0.738 ft-lb/s = 1.34 × 10⁻³ hp = 3.41 Btu/h

Power $1\ \text{W} = 0.738\ \text{ft-lb/s} = 1.34 \times 10^{-3}\ \text{hp} = 3.41\ \text{Btu/h}$
             $1\ \text{ft-lb/s} = 1.36\ \text{W} = 1.82 \times 10^{-3}\ \text{hp}$
             $1\ \text{hp} = 550\ \text{ft-lb/s} = 745.7\ \text{W} = 2545\ \text{Btu/h}$

Mass- $1\ \text{u} = 1.66 \times 10^{-27}\ \text{kg} \leftrightarrow 931.5\ \text{MeV}$
Energy $1\ \text{electron mass} = 9.11 \times 10^{-31}\ \text{kg} = 5.49 \times 10^{4}\ \text{u} \leftrightarrow 0.511\ \text{MeV}$
Equivalents $1\ \text{proton mass} = 1.672 \times 10^{-27}\ \text{kg} = 1.007276\ \text{u} \leftrightarrow 938.28\ \text{MeV}$
(at rest) $1\ \text{neutron mass} = 1.674 \times 10^{-27}\ \text{kg} = 1.008665\ \text{u} \leftrightarrow 939.57\ \text{MeV}$

Table 3
Trigonometric Relationships

Definitions of Trigonometric Functions

$$\sin\theta = \frac{y}{r} \qquad \cos\theta = \frac{x}{r} \qquad \tan\theta = \frac{\sin\theta}{\cos\theta} = \frac{y}{x}$$

$\theta°$ (rad)	$\sin\theta$	$\cos\theta$	$\tan\theta$
0° (0)	0	1	0
30° ($\pi/6$)	0.500	0.866	0.577
45° ($\pi/4$)	0.707	0.707	1.00
60° ($\pi/3$)	0.866	0.500	1.73
90° ($\pi/2$)	1	0	$\to \infty$

See Table 4 for other angles.

The sign of a trigonometric function depends on the quadrant, or the signs of x and y; for example, in the second quadrant $(-x, y)$, $-x/r = \cos\theta$ and $y/r = \sin\theta$. The sign can also be assigned using the reduction formulas.

Reduction Formulas

	(θ in second quadrant)		(θ in third quadrant)		(θ in fourth quadrant)
$\sin\theta =$	$\cos(\theta - 90°)$	$=$	$-\sin(\theta - 180°)$	$=$	$-\cos(\theta - 270°)$
$\cos\theta =$	$-\sin(\theta - 90°)$	$=$	$-\cos(\theta - 180°)$	$=$	$\sin(\theta - 270°)$

Fundamental Identities

$$\sin^2\theta + \cos^2\theta = 1$$
$$\sin 2\theta = 2\sin\theta\cos\theta$$
$$\cos 2\theta = \cos^2\theta - \sin^2\theta = 2\cos^2\theta - 1 = 1 - 2\sin^2\theta$$
$$\sin^2\theta = \tfrac{1}{2}(1 - \cos 2\theta)$$
$$\cos^2\theta = \tfrac{1}{2}(1 + \cos 2\theta)$$

For half-angle ($\theta/2$) identities, replace θ with $\theta/2$, for example:

$$\sin^2\theta/2 = \tfrac{1}{2}(1 - \cos\theta) \qquad \cos^2\theta/2 = \tfrac{1}{2}(1 + \cos\theta)$$
$$\sin(\alpha \pm \beta) = \sin\alpha\cos\beta \pm \cos\alpha\sin\beta$$
$$\cos(\alpha \pm \beta) = \cos\alpha\cos\beta \mp \sin\alpha\sin\beta$$

For very small angles:

$$\cos\theta \cong 1 \qquad \sin\theta \cong \theta\ (\text{radians}) \qquad \tan\theta = \frac{\sin\theta}{\cos\theta} \cong \theta$$

Table 4
Trigonometric Tables

| Angle | | | | | Angle | | | | |
Degrees	Radians	Sine	Cosine	Tangent	Degrees	Radians	Sine	Cosine	Tangent
0°	.000	0.000	1.000	0.000					
1°	.017	.018	1.000	.018	46°	0.803	0.719	0.695	1.036
2°	.035	.035	0.999	.035	47°	.820	.731	.682	1.072
3°	.052	.052	.999	.052	48°	.838	.743	.669	1.111
4°	.070	.070	.998	.070	49°	.855	.755	.656	1.150
5°	.087	.087	.996	.088	50°	.873	.766	.643	1.192
6°	.105	.105	.995	.105	51°	.890	.777	.629	1.235
7°	.122	.122	.993	.123	52°	.908	.788	.616	1.280
8°	.140	.139	.990	.141	53°	.925	.799	.602	1.327
9°	.157	.156	.988	.158	54°	.942	.809	.588	1.376
10°	.175	.174	.985	.176	55°	.960	.819	.574	1.428
11°	.192	.191	.982	.194	56°	.977	.829	.559	1.483
12°	.209	.208	.978	.213	57°	.995	.839	.545	1.540
13°	.227	.225	.974	.231	58°	1.012	.848	.530	1.600
14°	.244	.242	.970	.249	59°	1.030	.857	.515	1.664
15°	.262	.259	.966	.268	60°	1.047	.866	.500	1.732
16°	.279	.276	.961	.287	61°	1.065	.875	.485	1.804
17°	.297	.292	.956	.306	62°	1.082	.883	.470	1.881
18°	.314	.309	.951	.325	63°	1.100	.891	.454	1.963
19°	.332	.326	.946	.344	64°	1.117	.899	.438	2.050
20°	.349	.342	.940	.364	65°	1.134	.906	.423	2.145
21°	.367	.358	.934	.384	66°	1.152	.914	.407	2.246
22°	.384	.375	.927	.404	67°	1.169	.921	.391	2.356
23°	.401	.391	.921	.425	68°	1.187	.927	.375	2.475
24°	.419	.407	.914	.445	69°	1.204	.934	.358	2.605
25°	.436	.423	.906	.466	70°	1.222	.940	.342	2.747
26°	.454	.438	.899	.488	71°	1.239	.946	.326	2.904
27°	.471	.454	.891	.510	72°	1.257	.951	.309	3.078
28°	.489	.470	.883	.532	73°	1.274	.956	.292	3.271
29°	.506	.485	.875	.554	74°	1.292	.961	.276	3.487
30°	.524	.500	.866	.577	75°	1.309	.966	.259	3.732
31°	.541	.515	.857	.601	76°	1.326	.970	.242	4.011
32°	.559	.530	.848	.625	77°	1.344	.974	.225	4.331
33°	.576	.545	.839	.649	78°	1.361	.978	.208	4.705
34°	.593	.559	.829	.675	79°	1.379	.982	.191	5.145
35°	.611	.574	.819	.700	80°	1.396	.985	.174	5.671
36°	.628	.588	.809	.727	81°	1.414	.988	.156	6.314
37°	.646	.602	.799	.754	82°	1.431	.990	.139	7.115
38°	.663	.616	.788	.781	83°	1.449	.993	.122	8.144
39°	.681	.629	.777	.810	84°	1.466	.995	.105	9.514
40°	.698	.643	.766	.839	85°	1.484	.996	.087	11.43
41°	.716	.658	.755	.869	86°	1.501	.998	.070	14.30
42°	.733	.669	.743	.900	87°	1.518	.999	.052	19.08
43°	.751	.682	.731	.933	88°	1.536	.999	.035	28.64
44°	.768	.695	.719	.966	89°	1.553	1.000	.018	57.29
45°	.785	.707	.707	1.000	90°	1.571	1.000	.000	∞

Table 5
Areas and Volumes of Some Common Shapes

Circle: $A = \pi r^2 = \dfrac{\pi d^2}{4}$ (area)

$c = 2\pi r = \pi d$ (circumference)

Triangle: $A = \frac{1}{2}ab$

Sphere: $A = 4\pi r^2$
$V = \frac{4}{3}\pi r^3$

Cylinder: $A = \pi r^2$ (end)
$A = 2\pi rh$ (body)
$A = 2(\pi r^2) + 2\pi rh$ (total)
$V = \pi r^2 h$

Table 6
Planetary Data

Name	Equatorial Radius (km)	Mass (Compared to Earth's)*	Mean Density ($\times 10^3 \ kg/m^3$)	Surface Gravity (Compared to Earth's)	Semimajor Axis $\times 10^6$ km	Semimajor Axis AU
Mercury	2,439	0.0553	5.43	0.378	57.9	0.3871
Venus	6,052	0.8150	5.24	0.894	108.2	0.7233
Earth	6,378.140	1	5.515	1	149.6	1
Mars	3,397.2	0.1074	3.93	0.379	227.9	1.5237
Jupiter	71,398	317.89	1.36	2.54	778.3	5.2028
Saturn	60,000	95.17	0.71	1.07	1427.0	9.5388
Uranus	26,145	14.56	1.30	0.8	2871.0	19.1914
Neptune	24,300	17.24	1.8	1.2	4497.1	30.0611
Pluto	1,500–1,800	0.02	0.5–0.8	~0.03	5913.5	39.5294

Name	Period Years	Period Days	Eccentricity	Inclination to Ecliptic
Mercury	0.24084	87.96	0.2056	7°00′26″
Venus	0.61515	224.68	0.0068	3°23′40″
Earth	1.00004	365.25	0.0167	0°00′14″
Mars	1.8808	686.95	0.0934	1°51′09″
Jupiter	11.862	4,337	0.0483	1°18′29″
Saturn	29.456	10,760	0.0560	2°29′17″
Uranus	84.07	30,700	0.0461	0°48′26″
Neptune	164.81	60,200	0.0100	1°46′27″
Pluto	248.53	90,780	0.2484	17°09′03″

* Planet's mass/Earth's mass, where $M_E = 6.0 \times 10^{24}$ kg.

Appendix II: Kinetic Theory of Gases

The basic assumptions are as follows:

1. The molecules of a pure gas all have the same mass (m) and are in continuous and completely random motion. (The mass of each molecule is so small that the effect of gravity on it is negligible.)
2. The gas molecules are separated by large distances and occupy a volume that is negligible compared to these distances; that is, they are point particles.
3. The molecules exert no forces on each other except when they collide.
4. The collisions of the molecules with one another and the walls of the container are perfectly elastic.

The magnitude of the force exerted on the wall of the container by a gas molecule colliding with it is $F = \Delta p / \Delta t$. Assuming that the direction of the velocity ($\mathbf{v}_x$) is normal to the wall, the magnitude of the average force is

$$F = \frac{\Delta(mv)}{\Delta t} = \frac{mv_x - (-mv_x)}{\Delta t} = \frac{2mv_x}{\Delta t} \tag{1}$$

After striking one wall of the container, which for convenience is assumed to be cubical with sides of dimensions L, the molecule recoils in a straight line. Suppose that the molecule reaches the opposite wall without colliding with any other molecules along the way. The molecule then travels the distance L in a time equal to L/v_x. After the collision with that wall, again assuming no collisions on the return trip, the round trip will take $\Delta t = 2L/v_x$. Thus, the number of collisions per unit time a molecule makes with a particular wall is $v_x/2L$, and the average force on the wall from successive collisions is

$$F = \frac{2mv_x}{\Delta t} = \frac{2mv_x}{2L/v_x} = \frac{mv_x^2}{L} \tag{2}$$

The random motions of the many molecules produce a relatively constant force on the walls, and the pressure (p) is the total force on a wall divided by its area:

$$p = \frac{\Sigma F_i}{L^2} = \frac{m(v_{x_1}^2 + v_{x_2}^2 + v_{x_3}^2 + \cdots)}{L^3} \tag{3}$$

The subscripts refer to individual molecules.

The average of the squares of the speeds is given by

$$\overline{v_x^2} = \frac{v_{x_1}^2 + v_{x_2}^2 + v_{x_3}^2 + \cdots}{N}$$

where N is the number of molecules in the container. In terms of this average, Eq. 3 may be written

$$p = \frac{Nm\overline{v_x^2}}{L^3} \qquad\qquad (4)$$

However, the molecules' motions occur with equal frequency along any one of the three axes, so $\overline{v_x^2} = \overline{v_y^2} = \overline{v_z^2}$, and $\overline{v^2} = \overline{v_x^2} + \overline{v_y^2} + \overline{v_z^2} = 3\overline{v_x^2}$. Then,

$$\sqrt{\overline{v^2}} = \bar{v}$$

where $\bar{v}$ is called the root-mean-square (rms) speed, and we write $(\bar{v})^2 = \bar{v}^2$. Substituting this result into Eq. 4 along with V for L^3 (since L^3 is the volume of the cubical container) gives

$$p = \frac{Nm\bar{v}^2}{3V} \qquad \text{or} \qquad pV = \tfrac{1}{3}Nm\bar{v}^2 \qquad\qquad (5)$$

This result is true even though collisions between molecules were ignored. Statistically, these collisions average out, so the number of collisions with each wall is as described. This result is also independent of the shape of the container. A cube merely simplifies the derivation.

Combining this result with the empirical perfect gas law gives

$$pV = NkT = \tfrac{1}{3}Nm\bar{v}^2$$

The average kinetic energy per gas molecule is thus proportional to the absolute temperature of the gas:

$$\tfrac{1}{2}m\bar{v}^2 = \tfrac{3}{2}kT \qquad\qquad (6)$$

The collision time is negligible compared with the time between collisions. Some kinetic energy will be momentarily converted to potential energy during a collision; however, this potential energy can be ignored because each molecule spends a negligible amount of time in collisions. Therefore, by this approximation, the total kinetic energy is the internal energy of the gas, and the internal energy of a perfect gas is directly proportional to its absolute temperature.

Appendix III: Einstein's Letter to President Roosevelt*

Albert Einstein
Old Grove Road
Nassau Point
Peconic, Long Island
August 2, 1939

F. D. Roosevelt
President of the United States
White House
Washington, D. C.

SIR:

Some recent work by E. Fermi and L. Szilard, which has been communicated to me in manuscript, leads me to expect that the element uranium may be turned into a new and important source of energy in the immediate future. Certain aspects of the situation seem to call for watchfulness and, if necessary, quick action on the part of the Administration. I believe, therefore, that it is my duty to bring to your attention the following facts and recommendations.

In the course of the last four months it has been made probable—through the work of Joliot in France as well as Fermi and Szilard in America—that it may become possible to set up nuclear chain reactions in a large mass of uranium, by which vast amounts of power and large quantities of new radium-like elements would be generated. Now it appears almost certain that this could be achieved in the immediate future.

This new phenomenon would also lead to the construction of bombs, and it is conceivable—though much less certain—that extremely powerful bombs of a new type may thus be constructed. A single bomb of this type, carried by boat or exploded in a port, might very well destroy the whole port together with some of the surrounding territory. However, such bombs might very well prove to be too heavy for transportation by air.

* Nathan, O and Noiden, H., *Einstein on Peace*, Avenel Books, NY (1960).

The United States has only very poor ores of uranium in moderate quantities. There is some good ore in Canada and the former Czechoslovakia, while the most important source of uranium is the Belgian Congo.

In view of this situation you may think it desirable to have some permanent contact maintained between the Administration and the group of physicists working on chain reactions in America. One possible way of achieving this might be for you to entrust with this task a person who has your confidence and who could perhaps serve in an unofficial capacity. His task might comprise the following:

a) To approach Government Departments, keep them informed of the further developments, and put forward recommendations for Government action, giving particular attention to the problem of securing a supply of uranium ore for the United States.

b) To speed up the experimental work which is at present being carried on within the limits of the budgets of University laboratories, by providing funds, if such funds be required, through his contacts with private persons who are willing to make contributions for this cause, and perhaps also by obtaining the cooperation of industrial laboratories which have the necessary equipment.

I understand that Germany has actually stopped the sale of uranium from the Czechoslovakian mines which she has taken over. That she should have taken such early action might perhaps be understood on the ground that the son of the German Under-Secretary of State, von Weizsäcker, is attached to the Kaiser Wilhelm Institut in Berlin, where some of the American work on uranium is now being repeated.

Yours very truly,

A. EINSTEIN

Appendix IV: Alphabetical Listing of the Chemical Elements

Element	Symbol	Atomic Number (Proton Number)	Atomic Mass*	Element	Symbol	Atomic Number (Proton Number)	Atomic Mass*
Actinium	Ac	89	227.0278	Gallium	Ga	31	69.72
Aluminum	Al	13	26.98154	Germanium	Ge	32	72.59
Americium	Am	95	[243]	Gold	Au	79	196.9665
Antimony	Sb	51	121.75	Hafnium	Hf	72	178.49
Argon	Ar	18	39.948	Hahnium	Ha	262	262.1138
Arsenic	As	33	74.9216	Helium	He	2	4.00260
Astatine	At	85	[210]	Holmium	Ho	67	164.9304
Barium	Ba	56	137.33	Hydrogen	H	1	1.0079
Berkelium	Bk	97	[247]	Indium	In	49	114.82
Beryllium	Be	4	9.01218	Iodine	I	53	126.9045
Bismuth	Bi	83	208.9804	Iridium	Ir	77	192.22
Boron	B	5	10.81	Iron	Fe	26	55.847
Bromine	Br	35	79.904	Krypton	Kr	36	83.80
Cadmium	Cd	48	112.41	Lanthanum	La	57	138.9055
Calcium	Ca	20	40.08	Lawrencium	Lr	103	[260]
Californium	Cf	98	[251]	Lead	Pb	82	207.2
Carbon	C	6	12.011	Lithium	Li	3	6.941
Cerium	Ce	58	140.12	Lutetium	Lu	71	174.967
Cesium	Cs	55	132.9054	Magnesium	Mg	12	24.305
Chlorine	Cl	17	35.453	Manganese	Mn	25	54.9380
Chromium	Cr	24	51.996	Mendelevium	Md	101	[258]
Cobalt	Co	27	58.9332	Mercury	Hg	80	200.59
Copper	Cu	29	63.546	Molybdenum	Mo	42	95.94
Curium	Cm	96	[247]	Neodymium	Nd	60	144.24
Dysprosium	Dy	66	162.50	Neon	Ne	10	20.179
Einsteinium	Es	99	[252]	Neptunium	Np	93	237.0482
Erbium	Er	68	167.26	Nickel	Ni	28	58.70
Europium	Eu	63	151.96	Niobium	Nb	41	92.9064
Fermium	Fm	100	[257]	Nitrogen	N	7	14.0067
Fluorine	F	9	18,998403	Nobelium	No	102	[259]
Francium	Fr	87	[223]	Osmium	Os	76	190.2
Gadolinium	Gd	64	157.25	Oxygen	O	8	15.9994

* Atomic masses given here are 1977 IUPAC values based on carbon-12. A value given in brackets is the mass number of the longest-lived or best-known isotope.

Element	Symbol	Atomic Number (Proton Number)	Atomic Mass*	Element	Symbol	Atomic Number (Proton Number)	Atomic Mass*
Palladium	Pd	46	106.4	Sodium	Na	11	22.98977
Phosphorus	P	15	30.97376	Strontium	Sr	38	87.62
Platinum	Pt	78	195.09	Sulfur	S	16	32.06
Plutonium	Pu	94	[244]	Tantalum	Ta	73	180.9479
Polonium	Po	84	[209]	Technetium	Tc	43	[98]
Potassium	K	19	39.0983	Tellurium	Te	52	127.60
Praseodymium	Pr	59	140.9077	Terbium	Tb	65	158.9254
Promethium	Pm	61	[145]	Thallium	Tl	81	204.37
Protactinium	Pa	91	231.0359	Thorium	Th	90	232.0381
Radium	Ra	88	226.0254	Thulium	Tm	69	168.9342
Radon	Rn	86	[222]	Tin	Sn	50	118.69
Rhenium	Re	75	186.207	Titanium	Ti	22	47.90
Rhodium	Rh	45	102.9055	Tungsten	W	74	183.85
Rubidium	Rb	37	85.4678	Uranium	U	92	238.029
Ruthenium	Ru	44	101.07	Vanadium	V	23	50.9415
Rutherfordium	Rf	261	261.1087	Xenon	Xe	54	131.30
Samarium	Sm	62	150.4	Ytterbium	Yb	70	173.04
Scandium	Sc	21	44.9559	Yttrium	Y	39	88.9059
Selenium	Se	34	78.96	Zinc	Zn	30	65.38
Silicon	Si	14	28.0855	Zirconium	Zr	40	91.22
Silver	Ag	47	107.868				

* Atomic masses given here are 1977 IUPAC values based on carbon-12. A value given in brackets is the mass number of the longest-lived or best-known isotope.

Appendix V: Properties of Selected Isotopes*

Atomic Number (Z)	Element	Symbol	Mass Number (A)	Atomic Mass[†]	Abundance (%) or Decay Mode (if radioactive)	Half-Life (if radioactive)
0	(Neutron)	n	1	1.008665	β^-	10.6 min
1	Hydrogen	H	1	1.007825	99.985	
	Deuterium	D	2	2.014102	0.015	
	Tritium	T	3	3.016049	β^-	12.33 years
2	Helium	He	3	3.016029	0.00014	
			4	4.002603	~100	
3	Lithium	Li	6	6.015123	7.5	
			7	7.016005	92.5	
4	Beryllium	Be	7	7.016930	EC[‡], γ	53.3 days
			8	8.005305	2 α	6.7×10^{-17} s
			9	9.012183	100	
5	Boron	B	10	10.012938	19.8	
			11	11.009305	80.2	
			12	12.014353	β^-	20.4 ms
6	Carbon	C	11	11.011433	β^+, EC	20.4 min
			12	12.000000	98.89	
			13	13.003355	1.11	
			14	14.003242	β^-	5730 years
7	Nitrogen	N	13	13.005739	β^+	9.96 min
			14	14.003074	99.63	
			15	15.000109	0.37	
8	Oxygen	O	15	15.003065	β^+, EC	122 s
			16	15.994915	99.76	
			18	17.999159	0.204	
9	Fluorine	F	19	18.998403	100	
10	Neon	Ne	20	19.992439	90.51	

* Data are taken from *Chart of the Nuclides*, 13th ed., General Electric, 1984, and from C. M. Lederer and V. S. Shirley, eds., *Table of Isotopes*, 8th ed., John Wiley & Sons, Inc., New York, 1986.

[†] The masses given throughout this table are those for the neutral atom, including the Z electrons.

[‡] EC stands for electron capture.

Atomic Number (Z)	Element	Symbol	Mass Number (A)	Atomic Mass	Abundance (%) or Decay Mode (*if radioactive*)	Half-Life (*if radioactive*)
			22	21.991384	9.22	
11	Sodium	Na	22	21.994435	β^+, EC, γ	2.602 years
			23	22.989770	100	
			24	23.990964	β^-, γ	15.0 h
12	Magnesium	Mg	24	23.985045	78.99	
13	Aluminum	Al	27	26.981541	100	
14	Silicon	Si	28	27.976928	92.23	
			31	30.975364	β^-, γ	2.62 h
15	Phosphorus	P	31	30.973763	100	
			32	31.973908	β^-	14.28 days
16	Sulfur	S	32	31.972072	95.0	
			35	34.969033	β^-	87.4 days
17	Chlorine	Cl	35	34.968853	75.77	
			37	36.965903	24.23	
18	Argon	Ar	40	39.962383	99.60	
19	Potassium	K	39	38.963708	93.26	
			40	39.964000	β^-, EC, γ, β^+	1.28×10^9 years
20	Calcium	Ca	40	39.962591	96.94	
21	Scandium	Sc	45	44.955914	100	
22	Titanium	Ti	48	47.947947	73.7	
23	Vanadium	V	51	50.943963	99.75	
24	Chromium	Cr	52	51.940510	83.79	
25	Manganese	Mn	55	54.938046	100	
26	Iron	Fe	56	55.934939	91.8	
27	Cobalt	Co	59	58.933198	100	
			60	59.933820	β^-, γ	5.271 years
28	Nickel	Ni	58	57.935347	68.3	
			60	59.930789	26.1	
			64	63.927968	0.91	
29	Copper	Cu	63	62.929599	69.2	
			64	63.929766	β^-, β^+	12.7 h
			65	64.927792	30.8	
30	Zinc	Zn	64	63.929145	48.6	
			66	65.926035	27.9	
31	Gallium	Ga	69	68.925581	60.1	
32	Germanium	Ge	72	71.922080	27.4	
			74	73.921179	36.5	
33	Arsenic	As	75	74.921596	100	
34	Selenium	Se	80	79.916521	49.8	
35	Bromine	Br	79	78.918336	50.69	
36	Krypton	Kr	84	83.911506	57.0	
			89	88.917563	β^-	3.2 min
37	Rubidium	Rb	85	84.911800	72.17	
38	Strontium	Sr	86	85.909273	9.8	
			88	87.905625	82.6	
			90	89.907746	β^-	28.8 years
39	Yttrium	Y	89	88.905856	100	
40	Zirconium	Zr	90	89.904708	51.5	

Atomic Number (Z)	Element	Symbol	Mass Number (A)	Atomic Mass	Abundance (%) or Decay Mode (if radioactive)	Half-Life (if radioactive)
41	Niobium	Nb	93	92.906378	100	
42	Molybdenum	Mo	98	97.905405	24.1	
43	Technetium	Tc	98	97.907210	β^-, γ	4.2×10^6 years
44	Ruthenium	Ru	102	101.904348	31.6	
45	Rhodium	Rh	103	102.90550	100	
46	Palladium	Pd	106	105.90348	27.3	
47	Silver	Ag	107	106.905095	51.83	
			109	108.904754	48.17	
48	Cadmium	Cd	114	113.903361	28.7	
49	Indium	In	115	114.90388	95.7; β^-	5.1×10^{14} years
50	Tin	Sn	120	119.902199	32.4	
51	Antimony	Sb	121	120.903824	57.3	
52	Tellurium	Te	130	129.90623	34.5; β^-	2×10^{21} years
53	Iodine	I	127	126.904477	100	
			131	130.906118	β^-, γ	8.04 days
54	Xenon	Xe	132	131.90415	26.9	
			136	135.90722	8.9	
55	Cesium	Cs	133	132.90543	100	
56	Barium	Ba	137	136.90582	11.2	
			138	137.90524	71.7	
			144	143.922673	β^-	11.9 s
57	Lanthanum	La	139	138.90636	99.911	
58	Cerium	Ce	140	139.90544	88.5	
59	Praseodymium	Pr	141	140.90766	100	
60	Neodymium	Nd	142	141.90773	27.2	
61	Promethium	Pm	145	144.91275	EC, α, γ	17.7 years
62	Samarium	Sm	152	151.91974	26.6	
63	Europium	Eu	153	152.92124	52.1	
64	Gadolinium	Gd	158	157.92411	24.8	
65	Terbium	Tb	159	158.92535	100	
66	Dysprosium	Dy	164	163.92918	28.1	
67	Holmium	Ho	165	164.93033	100	
68	Erbium	Er	166	165.93031	33.4	
69	Thulium	Tm	169	168.93423	100	
70	Ytterbium	Yb	174	173.93887	31.6	
71	Lutecium	Lu	175	174.94079	97.39	
72	Hafnium	Hf	180	179.94656	35.2	
73	Tantalum	Ta	181	180.94801	99.988	
74	Tungsten (wolfram)	W	184	183.95095	30.7	
75	Rhenium	Re	187	186.95577	62.60, β^-	4×10^{10} years
76	Osmium	Os	191	190.96094	β^-, γ	15.4 days
			192	191.96149	41.0	
77	Iridium	Ir	191	190.96060	37.3	
			193	192.96294	62.7	
78	Platinum	Pt	195	194.96479	33.8	

Atomic Number (Z)	Element	Symbol	Mass Number (A)	Atomic Mass	Abundance (%) or Decay Mode (if radioactive)	Half-Life (if radioactive)
79	Gold	Au	197	196.96656	100	
80	Mercury	Hg	202	201.97063	29.8	
81	Thallium	Tl	205	204.97441	70.5	
			210	209.990069	β^-	1.3 min
82	Lead	Pb	204	203.973044	β^-, 1.48	1.4×10^{17} years
			206	205.97446	24.1	
			207	206.97589	22.1	
			208	207.97664	52.3	
			210	209.98418	α, β^-, γ	22.3 years
			211	210.98874	β^-, γ	36.1 min
			212	211.99188	β^-, γ	10.64 h
			214	213.99980	β^-, γ	26.8 min
83	Bismuth	Bi	209	208.98039	100	
			211	210.98726	α, β^-, γ	2.15 min
84	Polonium	Po	210	209.98286	α, γ	138.38 days
			214	213.99519	α, γ	164 μs
85	Astatine	At	218	218.00870	α, β^-	~2 s
86	Radon	Rn	222	222.017574	α, β	3.8235 days
87	Francium	Fr	223	223.019734	α, β^-, γ	21.8 min
88	Radium	Ra	226	226.025406	α, γ	1.60×10^3 years
			228	228.031069	β^-	5.76 years
89	Actinium	Ac	227	227.027751	α, β^-, γ	21.773 years
90	Thorium	Th	228	228.02873	α, γ	1.9131 years
			232	232.038054	100, α, γ	1.41×10^{10} years
91	Protactinium	Pa	231	231.035881	α, γ	3.28×10^4 years
92	Uranium	U	232	232.03714	α, γ	72 years
			233	233.039629	α, γ	1.592×10^5 years
			235	235.043925	0.72; α, γ	7.038×10^8 years
			236	236.045563	α, γ	2.342×10^7 years
			238	238.050786	99.275; α, γ	4.468×10^9 years
			239	239.054291	β^-, γ	23.5 min
93	Neptunium	Np	239	239.052932	β^-, γ	2.35 days
94	Plutonium	Pu	239	239.052158	α, γ	2.41×10^4 years
95	Americium	Am	243	243.061374	α, γ	7.37×10^3 years
96	Curium	Cm	245	245.065487	α, γ	8.5×10^3 years
97	Berkelium	Bk	247	247.07003	α, γ	1.4×10^3 years
98	Californium	Cf	249	249.074849	α, γ	351 years
99	Einsteinium	Es	254	254.08802	α, γ, β^-	276 days
100	Fermium	Fm	253	253.08518	EC, α, γ	3.0 days
101	Mendelevium	Md	255	255.0911	EC, α	27 min
102	Nobelium	No	255	255.0933	EC, α	3.1 min
103	Lawrencium	Lr	257	257.0998	α	$\cong$ 35 s
104	Rutherfordium[§]	Rf	261	261.1087	α	1.1 min
105	Hahnium[§]	Ha	262	262.1138	α	0.7 min
106	No name adopted		263	263.1184	α	0.9 s
107	No name adopted		261	261	α	1–2 ms

[§] These are the unofficial names and symbols used in the United States. These elements were discovered at about the same time in the Soviet Union, where the proposed names and symbols are Kurchatovium (Ku) and Nielsbohrium (Nm), respectively. There has been no official adoption.

Answers to Odd-Numbered Problems

Note to the student: The answers given here were reached by solving the problems step-by-step and rounding the result at each step. If you work the problems fully and then round only your result, your correct answer may differ slightly from the answer you find here. Variations may be due either to rounding or to calculator differences.

CHAPTER 1

1.3 Dimensional Analysis

1. $[L^2] = [L^2]$
3. $[L^3] \neq [L^2]$, not dimensionally correct
5. $[L]/[T^2]$, e.g., m/s^2 in SI units
7. $[L]/[T] = [L]/[T] - [L]/[T]$
9. (a) yes, $s = \sqrt{m/(m/s^2)} = s$
   (b) yes, $m/s = \sqrt{(m/s^2)(m)} = m/s$
11. (a) yes, $[L^3] = [L^3]$
   (b) no, the dimensionless number should be $\frac{4}{3}$ rather than 4: $V = \frac{4}{3}\pi r^3$
13. (a) dimensionless   (b) m^{-1} or $1/m$ (c) m^2
15. m/s^2
17. (a) $F = ma$, $(N) = (kg)(m/s^2) = kg\text{-}m/s^2$
   (b) $Ft = mv$, $(kg\text{-}m/s^2)(s) = (kg)(m/s)$

1.4 Unit Conversion (Any number of significant figures assumed)

19. h (in.) $(2.54$ cm/in.$) = ?$
21. (a) 49.2 ft   (b) 30.5 cm   (c) 2.6×10^6 s
23. (a) 1 mi = 1.609 km, so 10 mi/h = 16 km/h, 20 mi/h = 32 km/h, and so on
   (b) 88 km/h (~90 km/h)
25. 27 mL more soda in 500-mL size
27. (a) 75 kg
   (b) weight $(1$ kg/2.2 lb$) = ?$
29. (a) 91.4 m $\times$ 48.8 m; $A = 4.46 \times 10^7$ cm^2
   (b) 27.9 to 28.6 cm
31. 137 m $\times$ 22.9 m $\times$ 13.7 m = 4.30×10^4 m^3
33. Hg: 6.8×10^3 g; H$_2$O: 5.0×10^2 g; 13.6 times greater.

1.5 Significant Figures

35. (a) 3   (b) 4   (c) 2   (d) 6
37. (a) 17.5 m   (b) 608 km   (c) 1.30×10^{-3} kg   (d) 9.30×10^7 mi
39. (a) 34.2 m
   (b) 93.3 m^2
41. 1.99 m^3
43. 1.12×10^{21} m^3

1.6 Problem Solving

45. 5.5×10^3 kg/m^3
47. 260 km
49. 9.5×10^{12} km = 9.5×10^{15} m
51. (a) 21 mi/gal; 8.9 km/L   (b) 5.5 gal/h; 21 L/h   (c) 1.1×10^2 mi/h; 1.8×10^2 km/h

Additional Problems

53. $[L/T]^2 = [L/T]^2$
55. 1.89 L < 2.0 L
57. (a) cm, since mm is estimated or doubtful figure
   (b) 1.24 m^2
59. 0.50 L
61. 2.98×10^3 = m/s
63. (s) = (s)
65. 1.3×10^7 m
67. 4.9×10^2 cm^3

CHAPTER 2

2.2 Speed and Velocity

1. 0.83 m/s
3. 32 km
5. (a) $v_{AC} = -75$ km/h; $v_{BC} = +15$ km/h
   (b) $v_{AB} = -90$ km/h; $v_{CB} = -15$ km/h
7. $v_{AC} = 1.90$ m/s, $v_{AB} = 2.00$ m/s, and $v_{BC} = 1.67$ m/s; $\bar{v}_{ABC} = 1.84$ m/s $\neq v_{AC}$
9. (a) (1) 1.0 m/s, (2) 0, (3) 0.80 m/s, (4) 4.0 m/s, (5) 0, (6) 1.3 m/s
   (b) (1) +1.0 m/s, (2) 0, (3) +0.80 m/s, (4) −4.0 m/s, (5) 0, (6) +1.3 m/s
   (c) at $t = 1.0$ s, +1.0 m/s; at $t = 2.5$ s, 0; at $t = 4.5$ s, 0; at $t = 6.5$ s, −4.0 m/s
   (d) −0.89 m/s
11. (a) 7.20 km/h (2.0 m/s) east, and 7.21 km/h (2.0 m/s) north
   (b) 7.2 km/h (2.0 m/s)
   (c) 7.2 km/h; zero
13. (a) +47.5 km/h   (b) −52.5 km/h

15. 11.1 s; 55.6 m from initial point of 5.0 m/s runner

2.3 Acceleration

17. 2.1 m/s²
19. (a) 6.28 m/s² (b) 100 km/h (27.8 m/s)
21. (a) 2.0 m/s², (2) 0 (3) −1.0 m/s²
23. (a) curve is horizontal straight line; area is $vt = d$
(b) triangle formed by line, area is $\frac{1}{2}vt = \bar{v}t$, where $\bar{v} = v/2$
(c) area of triangle plus rectangle beneath it is $v_\text{o}t + \frac{1}{2}(v - v_\text{o})t = v_\text{o}t + \frac{1}{2}at^2 = d$ (where $v = v_\text{o} + at$)
25. no; required deceleration is −5.0 m/s²
27. (a) (1) 0, (2) +5.0 m/s², (3) −4.4 m/s², (4) −4.4 m/s², (5) +8.0 m/s², (6) 0
(b) (1) at rest, (2) acceleration in + direction, (3) acceleration in − direction to a stop, (4) acceleration in − direction, followed by (5) a + acceleration, which is removed when (6) velocity is −4.0 m/s
(c) −11 m (d) 53.6 m

2.4 Kinematic Equations

29. (a) 81.3 m (b) 32.5 m/s
31. 6.3 s
33. (a) 4.0 s (b) +12 m (in the direction opposite to the acceleration)
35. 1.43×10^{-3} s
37. Initial velocity, reaction time distance, braking distance, and total stopping distance, respectively
(a) 10 m/s, 7.5 m, 6.3 m, 13.8 m
20 m/s, 15 m, 25 m, 40 m
25 m/s, 18.8 m, 39 m, 58 m
(b) 10 m/s, 7.5 m, 12.5 m, 20 m
20 m/s, 15 m, 50 m, 65 m
25 m/s, 18.8 m, 78.1 m, 97 m
39. (a) 536 m (b) 13.2 s
41. 11 m/s²
43. (a) 0.18 s (b) −3.4 ft
45. (a) 2.02 s (b) −19.8 m/s
47. 17.1 m/s
49. (a) −20.4 m (b) −20.1 m/s
51. (a) −49.6 m (b) −38.7 m/s
53. (a) 15.7 m/s (b) 12.6 m
55. (a) 43.1 m (b) −1.8 m/s (c) 3.8 s; −29.2 m/s
57. (a) parabola, $y = y_\text{o} - \frac{1}{2}gt^2$; straight line, $v = -gt$
(b) parabola, $y = v_\text{o}t - \frac{1}{2}g^2$; straight line, $v = v_\text{o} - gt$
(c) a versus t is horizontal straight line ($a = g$, a constant)
59. 5.03 s
61. (a) $t_\text{m} = 2.45 t_\text{E}$ (b) 169 m and 28.8 s
63. −3.2 m/s

65. Distance to water is between 51.1 and 51.6 m (time for fall and sound traveling back to person is 3.40 s; the stone would fall a distance of 51.6 m in this time, but the depth of the water is actually less; sound would travel this distance in 0.17 s ($v = 340$ m/s); assuming that the stone fell for 3.40 s − 0.17 s = 3.23 s, the distance would be 51.1 m); direct solution of the quadratic equation would be difficult because of values and significant figures

Additional Problems

67. 47.7 mi/h
69. 125 m
71. (a) 103 m (b) 9.2 s
73. (a) 175 km/h (b) 225 km/h
75. (a) 18 m/s (b) 16.2 m
77. (a) 5.6 m (b) $t = 0.71$ s (up) and $t = 3.6$ s (down)
79. 5.4 s
81. (a) 0.20 s (b) does not reach 3.5 m above starting point, goes to 3.2 m only (c) 0.20 s [In (a), roots correspond to 1.4 m going up and coming down; in (c), $-t$ has no meaning.]

CHAPTER 3

3.1 Components of Motion

1. $v_x = 4.3$ m/s, $v_y = 2.5$ m/s
3. (a) 37° (b) 6.0 m/s
5. 408 km
7. (a) $v_x = 26.0$ m/s, $v_y = 15.0$ m/s
(b) $x = 65$ m, $y = 38$ m
9. (a) $v_x = 3.0$ m/s, $v_y = 3.0$ m/s
(b) $x = 6.0$ m, $y = 13$ m
11. 1.9 m/s
13. $x = 11.1$ m, $y = 2.7$ m

3.2 Vector Addition and Subtraction

15. simple diagrams
17. $\mathbf{C}_x = 6.0$ m/s, $\mathbf{C}_y = 10.4$ m/s
19. (a) $-\mathbf{A} = (-3.0 \text{ cm})\mathbf{x}$ (b) $-\mathbf{B} = (10 \text{ cm})\mathbf{y}$
(c) $-\mathbf{C} = 8.0$ cm at an angle of −30° relative to the +x axis (or 330°)
21. simple diagrams (three ways to add)
23. $\mathbf{A} + \mathbf{B} = (14.4 \text{ m})\mathbf{y}$
25. (a) $\mathbf{C}_1 = 19.2$ m in +x direction
(b) $\mathbf{C}_2 = -19.2$ m in −x direction
27. (a) $(-1.9 \text{ m/s})\mathbf{x} + (4.9 \text{ m/s})\mathbf{y}$ (b) $(-5.2 \text{ m/s})\mathbf{y}$,
(c) $(-7.9 \text{ m/s})\mathbf{x} + (4.9 \text{ m/s})\mathbf{y}$
29. $v = 4.9$ m/s; $\theta = 3.5°$
31. $\mathbf{d} = (16.5 \text{ m})\mathbf{x} + (22.8 \text{ m})\mathbf{y}$

33. from a vector diagram, $\tan \theta = v_c/v_r$, so the greater v_c is, the greater θ will be (v_r is constant)

3.3 Projectile Motion

35. $t = 0.45$ s; $x = 2.0$ m
37. (a) 2.69 s   (b) 74.8 m
39. (a) 37°   (b) directly over the impact point
41. 34.6 m
43. (a) 32.5°   (b) no, $R_{max} = 3.3$ km
45. (a) parabola   (b) straight line   (c) horizontal straight line   (d) straight lines (positive and negative slopes)   (e) hyperbola, $v^2 = v_o^2 + (v_{y_o} - gt)^2$
47. $R_1 = 1.09R_2$
49. 25 m/s
51. 2.1 m/s
53. (a) 4.49 m   (b) 11 m
55. for all cases, $v^2 = v_o^2 + 2gh$, so all three have the same final speed
57. for the bullet, $t = x/(v_o \cos \theta)$ and $y = (v_o \sin \theta)t - \frac{1}{2}gt^2$; substituting for t yields $y = y_o - \frac{1}{2}gt^2$, so bullet will be at same place as falling object

3.4 Uniform Circular Motion and Centripetal Acceleration

59. 1.1 m/s²
61. no, $a_c = 1.33$ m/s², so friction is not enough
63. 2.2 s
65. (a) 2.99×10^4 m/s (6.7×10^4 mi/h)   (b) 6.0×10^{-3} m/s²; gravity
67. (a) 3.4×10^{-2} m/s²   (b) 2.9×10^{-2} m/s²   (c) zero
69. 1.01 m/s²

Additional Problems

71. 60°
73. $\mathbf{d} = (0.75 \text{ m})\mathbf{x} + (1.0 \text{ m})\mathbf{y}$
75. 8.7 m/s
77. 30.0 m from initial position at an angle of 25.9° relative to the road
79. $v_{up} = v_b(\sin \theta) = v_r$, but $v_r > v_b$, so boat cannot go directly across
81. (a) $a_t = 2.5$ m/s², $a_c = 9.5$ m/s²   (b) $a_c = g$ when $\theta = 0°$; $a_t = 0$
83. 0.90 m

CHAPTER 4

4.1 Newton's First Law of Motion

1. $m_2 = (2.6)m_1$
3. $\mathbf{F}_3 = (-1.5 \text{ N})\mathbf{x} + (2.5 \text{ N})\mathbf{y}$
5. $\mathbf{F}_3 = (-8.4 \text{ N})\mathbf{x} + (0.60 \text{ N})\mathbf{y}$

4.3 and 4.4 Newton's Second Law and Applications

7. 5.63 N
9. 20 m/s²
11. (a) 76 kg; 745 N   (b) Answer will vary.
13. 1.4 N
15. (a) 343 N   (b) 485 N   (c) it approaches infinity
17. zero
19. 25.3 m/s²
21. 139 m
23. (a) -7500 N   (b) -9375 N
25. -42.2 N
27. 1.1 m/s² up
29. 1.75 m/s² down the plane
31. (a) 3.0 m/s²   (b) $T_1 = 3.0$ N and $T_2 = 9.0$ N
33. (a) 9.9×10^3 N directed upward   (b) 8.7×10^3 N directed downward; yes
35. 1.0 m
37. 2.3 m/s², m_3 moving down
39. 1.3 m/s², m_2 moving down plane
41. (a) 637 N   (b) 671 N   (c) 603 N
43. (a) $T \sin \theta = mg$, $T \cos \theta = mv^2/(L \cos \theta)$
(b) when $\theta = 90°$, $v = 0$; as θ approaches zero, v approaches infinity

4.5 Newton's Third Law of Motion

45. equal and opposite forces add to zero, so table experiences a force equal to total weight of the blocks

4.6 Friction

47. 417 N
49. 0.40
51. (a) 30°   (b) 22°
53. no, because $\mu_s \leq 0.36$
55. (a) 75 N   (b) $F_2 = (0.097)F_1$
57. 0.59
59. 0.97
61. no, stopping distance = 24.2 m
63. (a) 26 N   (b) 21 N
65. $\theta < 20°$
67. (a) $m_3 = 0.18$ kg   (b) 0.86 m/s²
69. 0.24

Additional Problems

71. $\Sigma \mathbf{F} = 0$
73. (a) $\mathbf{F}_3 = (2.8 \text{ N})\mathbf{x} - (2.4 \text{ N})\mathbf{y}$ (b) at rest or moving with a constant velocity
75. 30°
77. $mv^2/r = N \sin \theta$ and $N = mg/\cos \theta$
79. $F_{Al} = (1.2)F_{st}$
81. 0.051

83. (a) 12 m/s = 43 km/h = 27 mi/h   (b) no, independent of mass
85. 5.87×10^3 N

CHAPTER 5

5.1 Work

1. 1.1×10^5 J
3. 6.0 m
5. 12 m
7. 11 J
9. (a) 1.5×10^5 J   (b) -1.5×10^5 J
11. (a) 9.0 ft-lb   (b) 12 J
13. 2.0×10^3 J

5.2 Work Done by Variable Forces

15. 160 N/m
17. 0.18 J
19. no, work is 2.5 J in both cases
21. (a) -0.020 J   (b) zero

5.3 The Work Energy Theorem: Kinetic Energy

23. (a) 3.1×10^5 J   (b) -3.1×10^5 J
25. (a) 45 J   (b) 21 m/s
27. 1.3×10^7 m/s
29. (a) 200 m   (b) 64 km/h
31. (a) 0.13 m   (b) 4.2 m/s   (c) 1.7 J; 4.75 m/s

5.4 Potential Energy

33. 2.9 J
35. parabola
37. (a) 8.8 J   (b) 6.5 J
39. $x_1 = (1.7)x_2$
41. 0.16 m
43. (a) 32 m   (b) 19 m
45. 6.3 m/s at lowest point of swing
47. (b) 0.14 J   (c) 1.3 m/s
49. (a) $U = 8.1$ J; $K = 19.9$ J   (b) 18.2 J
51. 13.6 m
53. (a) 0.18 m   (b) 4.7 m/s
55. (a) 1.0 m   (b) 0.82 m   (c) 2.1 m/s
57. (a) 2.9 m/s   (b) no

5.6 Power

59. 0.134 hp
61. (a) 200 J   (b) 40 W
63. (a) 8.2×10^5 J   (b) 3.9×10^3 W
65. 48.5 W
67. 5.3 W for Fig. 5.15; -0.375 W for Fig. 5.16

69. 0.34 hp
71. $0.30

Additional Problems

73. 30 W
75. (a) -137 J   (b) 578 J   (c) 578 J
77. 2.7×10^5 J
79. (a) 2.5 J   (b) 0.43
81. (a) 0.11 m   (b) 2.7 m/s
83. 300 N
85. $W = mgh$ for all paths

CHAPTER 6

6.1 Linear Momentum

1. (a) 4.0 kg-m/s   (b) 4.2×10^4 kg-m/s
3. 86 kg
5. 67 km/h
7. (a) (0.35 kg-m/s)**x**   (b) (0.092 kg-m/s)**y**
9. (a) $(-0.95$ kg-m/s)**x** $-$ (0.95 kg-m/s)**y**
     (b) not necessarily; total momentum is same as in (a)
11. (a) 420 kg-m/s   (b) yes; 450 kg-m/s
13. $(1.3 \times 10^{-2}$ kg-m/s)**x** $-$ $(2.5 \times 10^{-2}$ kg-m/s)**y**

6.2 The Conservation of Linear Momentum

15. -3.4 m/s
17. 1.7 m/s in direction of bullet's initial velocity
19. (a) 0.91 m/s toward lake   (b) same as in (a)
21. 545 s = 9.1 min
23. (a) 5.0 km/h   (b) 2.5 km/h   (c) 17.5 km/h
25. 7.6 m/s; 12°
27. 740 m

6.3 Impulse and Collisions

29. N-s = (kg-m/s²) = s = kg-m/s
31. 1.6×10^4 N
33. (a) 1.8 N-s   (b) 18 N
35. graphs
37. (a) 76.5 N-s   (b) 546 N   (c) 45 m/s
39. 13.5 N
41. 6.4×10^3 N = 1.4×10^3 lb
43. $v_1 = -0.48$ m/s; $v_2 = 0.02$ m/s
45. (a) 2.6 m/s; 45°, relative to $-x$ axis (third quadrant)
     (b) 0.50 m/s; 63°, relative to $+x$ axis (fourth quadrant)
47. 55% lost
49. no, $K_i \neq K_f$; energy goes into work to get grain moving horizontally
51. (a) $V = mv_o/(m + M) = v_o/90$ (b) 252 m/s   (c) 99% lost
53. 9.72 ms

55. $1 - K_f/K_i = M/(m + M)$

57. (a) add equations for conservation of momentum and conservation of kinetic energy
(b) for elastic collision, $e = 1$; for completely inelastic collision, $e = 0$

6.4 Center of Mass

59. $(-0.40 \text{ m}, 0)$

61. (a) 4.6×10^6 m relative to Earth's center
(b) 1.8×10^6 m below Earth's surface

63. $(8.0 \text{ m}, 0)$

65. $(5.3 \text{ m/s}^2)\mathbf{x} + (2.7 \text{ m/s}^2)\mathbf{y}$

67. (a) $x_1 = 3.2$ m and $x_2 = 4.8$ m (b) same distances as in (a)

69. 1.1×10^{13} m; near Sun's surface and ecliptic plane; moves

Additional Problems

71. 24 m/s

73. (a) 2.8 m/s in original direction of motion (b) 65% lost

75. Since racket does not slow down appreciably, ball has to have $2v_0$ to maintain same relative velocity.

77. 98.8 m from shore

79. 4.1×10^{-28} kg-m/s

81. (a) (2.0 m, 2.0 m) (b) (2.0 m, 2.0 m) (c) (2.8 m, 2.0 m)

83. -1500 N

CHAPTER 7

7.1 Angular Measure

1. (a) 0.262 rad (b) 4.71 rad (c) 0.872 rad (d) 9.43 rad

3. 4.7 cm

5. (a) 4.0 rad (b) 229°

7. $(r, \theta) = (2 \text{ m}, 45°)$

9. (a) 1.92 km (b) 12.6 km

11. 0.32 m

13. 6.1×10^6 m

7.2 Angular Speed and Velocity

15. 0.16 rad/s

17. 1.18 rad/s

19. (a) 0.026 rad/s, or 2.6×10^{-2} rad/s (b) 6.5 m/s

21. (a) 60 rad (b) 3600 m

23. (a) 3.16×10^{-8} Hz (b) 2.0×10^{-7} rad/s

7.3 Angular Acceleration

25. π rad/s² into table

27. (a) 0.30 m/s² (b) 1.95 m/s

29. $v/t = r(\omega/t)$

31. (a) 1.82 rad/s² (b) 28.7 rev

33. 0.09 m/s² toward center of circle

35. 8.62 m/s²; 86°

37. (a) 53 s (b) $(1.4 \text{ m/s}^2)\mathbf{t} + (8.4 \text{ m/s}^2)\mathbf{r}$

39. (a) frictional force not great enough to supply necessary centripetal force (b) 0.10

41. (a) zero (b) 1.1×10^3 N (c) 1.8×10^3 N

7.4 Newton's Law of Gravitation

43. 4.2×10^{-10} N toward the other sphere

45. 2.0×10^{30} kg

47. (a) 0.82 m (b) 3.4×10^3 kg/m³

49. (a) 5.1×10^{-10} N directed toward mass at opposite corner (b) zero

51. 3.72 m/s²

53. 690 N

55. 3.0 g's

57. (a) -2.5×10^{-11} J (b) zero

59. 2.7×10^{-3} m/s²

61. $g_m/g_e = 0.165 \cong \frac{1}{6}$

7.5 Kepler's Laws and Earth Satellites

63. (a) 3.0×10^{-19} s²/m³ (b) same as in (a)

65. 164 years

67. (a) 1.05×10^4 m/s (b) 95%

69. (a) satellite with smaller orbital radius has greater K
(b) at r_2, greater (less negative) U (c) $K_1K_2 = 1.02 = U_2/U_1$

71. 3.56×10^7 m $= 2.2 \times 10^3$ mi

73. 3.4×10^3 m/s

7.6 Frames of Reference and Inertial Forces

75. 9.77 m/s²

77. (a) 5.1×10^{-5} rad/s (b) Answers will vary

79. 30°

Additional Problems

81. $g_m/g_e = 0.38$

83. (a) 4.3×10^{-8} rad/s (b) 5.6×10^9 m

85. (a) 9.2×10^{-3} rad; 0.53° (b) 3.4×10^{-2} rad; 1.9°

87. 120 s = 2 min

89. 0.17 rad

91. 124 rpm

93. (a) 2 rad/s (b) 3 rad

95. 2.0×10^3 m

CHAPTER 8

8.1 Rigid Bodies, Translations, and Rotations

1. (a) 0.20 m/s (b) 0.40 m/s

3. 0.60 m/s

5. 4.0 rad
7. no, $0.45 \neq 0.465$

8.2–8.3 Torque, Moment of Inertia, and Rotational Dynamics

9. (a) 525 rad/s^2 (b) 600 rad/s^2 (c) 525 rad/s^2
11. $(r \sin \theta)F = r(F \sin \theta)$
13. zero
15. $I = \Sigma_i \, m_i r_i^2 = (\Sigma_1 + \Sigma_2) m_i r_i^2 = I_1 + I_2$
17. 2.2 N
19. (a) 9.2 m/s^2 (b) 8.9 m/s^2
21. 2.5×10^{38} kg-m^2
23. (a) 0.87 m/s^2 (b) 0.84 m/s^2
25. (a) $\mu_s = (m_2/m_1) - (\tau_s/m_1 g)$
    (b) $a = [g(m_2 - \mu_k m_1) - \tau_k/R]/(M/2 + m_1 + m_2)$

8.4 Rotational Work and Kinetic Energy

27. 1000 J
29. 2.6×10^{29} J
31. (a) 2.0×10^5 J (b) 71 m-N
33. 4.1 m/s
35. 52 J
37. 5.4 rad/s
39. K_{trans}/K for hoop = 1/2, for cylinder = 2/3, for sphere = 5/7; sphere has greatest translational kinetic energy
41. (a) 26.6 m (b) 37.5 m
43. (a) α from torque relationship; $a = r\alpha = L\alpha = (3g/2)\cos\theta$; $a_y = a \cos\theta$, $> g$; solve for θ (b) 0.18 m from end

8.5 Angular Momentum

45. 1.5×10^{-3} kg-m^2/s, up
47. 1.5 rad/s
49. $L_{orb}/L_{spin} = 3.8 \times 10^6$; not in same direction
51. 1.0 rad/s
53. 2.1 rad/s
55. (a) turntable rotates in opposite direction
    (b) 0.17 rad/s (c) no, angular difference between points will be 2.0 rad

Additional Problems

57. 0.18 J
59. (a) 6.5 rad/s (b) 1.1
61. (a) 1.3 m/s^2 (b) 9.8×10^{-3} m-N
63. 2.4 m/s
65. 3.0 N-s
67. 2.1×10^7 W
69. No, cylinder will be 7% higher

CHAPTER 9

9.1 Mechanical Equilibrium

1. (a) zero (b) -20 N
3. (a) 33.3 cm, or at 83.3-cm position (b) 83.3 g
5. $\mathbf{F}_3 = -(42 \text{ N})\mathbf{x} - (8.0 \text{ N})\mathbf{y}$
7. 21 N
9. $T_2 = 42$ N; $T_1 = 50$ N
11. (a) 436 N (b) -348 N
13. 3.0 N
15. as close as 0.33 m from the end
17. 1.2×10^4 N
19. 38 cm from the end where the weight is suspended

9.2 Stability and Center of Gravity

21. 9.8 J
23. (a) 9 books (b) 18 cm
25. (a) 2.5 cm (b) 36 cm
27. (a) 6.0 cm (b) 0.22 cm
29. (a) 1.3 (b) 0.75 m
31. 62.5%
33. 2.9
35. (a) 1.8 m from the input end (b) 3
37. 77%
39. 515 J
41. (a) 2.6 (b) 67%
43. 4.0 (b) 100 N
45. (a) downward force of 50 N (to maintain translational equilibrium)
    (b) 0.50; force multiplication sacrificed for distance multiplication
47. (a) 9.0 m (b) 333 N
49. (a) 8.6 N (b) AMA = 7.5; efficiency = 8.6%
51. TMA $= F_o/F_i = d_i/d_o = R/r$

9.4 Elastic Moduli

53. 3.1×10^6 N/m^2
55. 2.1×10^3 N/m^2
57. 1.2×10^{10} N/m^2
59. 9.5×10^{10} N/m^2
61. 6.7×10^3 N/m^2
63. 1.3×10^4 N
65. 2.3×10^6 N/m^2
67. 5.4×10^{-5}
69. 99.97 cm

Additional Problems

71. 7×10^4 N
73. 16°
75. (a) 5.0 (b) 0.20; distance multiplication

77. $k = AY/L_o = 1/m$
79. sketches with support strands equal to TMA's
81. (a) 2.7   (b) $F_o/F_i < 2.7$
83. 0.48 m

CHAPTER 10

10.1   Pressure and Pascal's Principle

1. $V_{Al}/V_{Pb} = 17.6$
3. 4.8×10^2 lb/in^2 = 33 atm
5. 33 cm^3
7. 3.0×10^3 N
9. (a) 1.1×10^8 Pa   (b) 1.9×10^6 N
11. 1.08×10^5 Pa
13. 1.36 m
15. (a) displacement of water   (b) no, 1.88×10^4 kg/m^3 is less than the density of gold
17. (a) 400 N   (b) 3.5×10^{-3} m^2
19. 3.4×10^3 kg/m^3
21. 0.38%
23. (a) -1.3×10^4 Pa   (b) 8.8×10^4 Pa

10.3   Buoyancy and Archimedes' Principle

25. no, cube's density is 1.06 g/cm^3
27. (a) 5.6×10^3 m^3 (b) 11 m
29. (a) 3.4×10^4 N   (b) 3.9×10^4 N
31. 11%
33. sp. gr. $= (w_s/V_s)/(w_w/V_w) = (m_s/V_s)/(m_w/V_w)$
35. ~2.0 m
37. 5.2 cm

10.3   Surface Tension and Capillary Action

39. (a) 4.7×10^{-3} N   (b) 8.5×10^{-2} N
41. 0.66×10^{-3} m
43. 6.0×10^{-4} m
45. 3.0 m
47. 0.053 N/m

10.4   Fluid Dynamics and Bernoulli's Equation

49. equation wtih $h_1 = h_2$
51. 0.049 m^3/s
53. 91 s
55. (a) V_s = 11 cm/s; V_n = 66 cm/s   (b) 1.2×10^3 Pa
57. 0.45 m

10.5   Viscosity Poiseuille's Law, and Reynold's Number

59. N-s/m^2 = Pa-s
61. 2.0×10^2 Pa
63. 200 s
65. 7.9×10^{-5} m^3/s
67. (a) 114 m/s   (b) 105 m/s

Additional Problems

69. 0.135 m^3/s
71. fluid B; 3 times greater
73. water; $h_w/h_{al} = 2.6$
75. zero
77. (b) 4.7×10^{-5} m^3/s
79. 0.71 m
81. 1.5×10^{-5} m^3; 6.0×10^3 kg/m^3

CHAPTER 11

11.2   The Celsius and Fahrenheit Temperature Scales

1. drops 18°F; rises 5.5°C
3. (a) 90°F   (b) 230°F   (c) 14°F   (d) -459.69°F
5. 136°F; -128°F
7. (a) drop = 55.7°C, from 6.7°C to -49°C; rise = 27.2°C, from -20°C to 7.2°C
     (b) -4.2°F/h, -2.3°C/h; 24.5°F/min, 13.6°C/min

11.3   Gas Laws and Absolute Zero

9. 2.026×10^5 Pa
11. (a) 273.16 K   (b) 373.16 K   (c) 293.16 K   (d) 248.16 K
13. (a) 10,340°F; 5727°C   (b) 4.6%
15. 313.16°C
17. 0.25 m^3
19. (a) (1) 44 g, (2) 98 g, (3) 36.5 g   (b) (1) 2 moles, (2) 1.5 moles, (3) 2.5 moles
21. (a) 0.033 mole   (b) 2.0×10^{22} molecules   (c) 0.92 g
23. ~7.8×10^4 °C, or ~78,000 K; no, would be hotter than the Sun's surface

11.4   Thermal Expansion

25. 5.1×10^{-4} m
27. 126 cm^3
29. (a) 60.06 cm   (b) 0.0037 cm^2; yes, because $v_1 A_1 = v_2 A_2$
31. 8.4×10^{-3} m
33. 0.99999 m
35. yes; its coefficient is 3.8×10^{-6} °C^{-1}, which is relatively low for a metal
37. (a) 1330°C   (b) no, the cross-sectional area of the hole would grow faster than that of rod
39. 7.8 cm
41. $L_2 = L_o(1 - \alpha^2 \Delta T^2)$ per cycle, and $L_2 \cong L_o$ as first-order approximation
43. 0.88 mm

11.5 Kinetic Theory of Gases

45. 9.5%
47. 313°C
49. (a) internal energies increase by 17% and 9.7%, respectively (b) 7.3% greater

Additional Problems

51. $T_K = 574.61 = T_F$
53. $\beta = [(p_o/p)(T/T_o) - 1]/[(T - T_o) + 273]$, with T's in kelvins
55. 13.40 g/cm³
57. (a) 6.21×10^{-21} J (b) 1.37×10^3 m/s
59. (a) 2.3×10^5 Pa (b) 114 K

CHAPTER 12

12.1 Units of Heat

1. 3.8×10^5 J
3. (a) 2.1×10^7 J (b) 5.8×10^3 W
5. 0.23°C
7. 2.1×10^2 kg
9. 0.2×10^8 J

12.1 Specific Heat

11. 29.8°C
13. 0.20 cal/g-°C
15. 33.7°C
17. aluminum, because it has a higher specific heat: $Q_{Al}/Q_{Fe} = 2$
19. 65°C
21. 1.4×10^2 s
23. 85°C
25. 400 m/s

12.3 Phase Changes and Latent Heat

27. 540 kcal
29. 153.8 kcal
31. 60.8 kcal
33. (a) 184.3 kcal (b) 182.4 kcal
35. 0.625 L
37. 0.15 L
39. 0.17 L

12.4 Heat Transfer

41. 200°C/m
43. 63 J/s
45. glass; $(\Delta Q/\Delta t)_{glass}/(\Delta Q/\Delta t)_{wood} = 84$

47. (a) 2.1×10^2 kcal/s $= 8.8 \times 10^5$ J/s (b) 120 L, not reasonable
49. zero (because water below ice is at 0°C)
51. (a) not affected (b) $P_2/P_1 = 1.3$ (c) similar effects, but $P_2/P_1 = 16$
53. 27°C
55. (a) 34 kcal (b) 0.43 kg (c) no, because ice is less dense than water
57. 94.5 kcal $= 3.9 \times 10^5$ J
59. 407°C

12.5 Evaporation and Relative Humidity

61. (a) 0.1 kg/h (b) 1.7 kg

Additional Problems

63. 3.7 kcal $= 1.6 \times 10^4$ J
65. 20.1°C
67. 62 kcal
69. (a) 5.1 in. (b) 5.7 in.
71. 37.55°C
73. 4888 kcal
75. tungsten; $L_W/L_{Pb} = 7.5$ and $L_W/L_{Au} = 2.9$

CHAPTER 13

13.1 Thermodynamic Coordinates

1. (a) vertical line (b) horizontal line (c) diagonal line (d) general curve
3. sketches

13.2 The First Law of Thermodynamics

5. (a) 200 J (b) $T_f = T_i$
7. (a) 640 J, (b) 198 J
9. (a) 3.0 kcal $= 1.3 \times 10^4$ J (b) 21.5°C
11. 6.1×10^2 J
13. 3.6×10^4 J
15. (a) isobaric (b) 260 J
17. 0.75×10^5 J

13.3 The Second Law of Thermodynamics

19. 0.29 kcal/K $= 1.2 \times 10^3$ J/K
21. $\Delta S_i = -0.22$ kcal/K $= -9.2 \times 10^2$ J/K
$\Delta S_{st} = 0.36$ kcal/K $= 1.5 \times 10^3$ J/K
ice has more order, or steam more disorder
23. 4.5×10^3 J
25. 1.2×10^3 J/K
27. zero; -5.0×10^3 J
29. 2.5 J/K
31. $Q/T_{cold} - Q/T_{hot} > 0$

13.4 Heat Engines and Heat Pumps

33. (a) 180 J (b) 420 J
35. (a) 83,720 J (b) 62,790 J/cycle
37. (a) 67% (b) no, $W_{out} = 1200$ J and $W_{in} = 5870$ J
39. $W = \Delta Q = (T_{hot} - T_{cold}) \Delta S$, and $Q_{in} = T_{hot} \Delta S$
41. (a) $Q_c \cong 71\%$ of output (b) $Q_c = 12{,}000$ J, $Q_h = 16{,}000$ J; Q_c from free energy source

13.5 The Carnot Cycle and Ideal Heat Engine

43. 37%; no
45. 273 K = 0°C
47. 1850 J; 498 K = 225°C
49. second engine has greater theoretical efficiency; 35% versus 67%
51. 100°C
53. (a) 6.7% (b) probably not, poor return compared to fuels
55. use $Q_h/Q_c = T_h/T_c$

Additional Problems

57. 2500 J
59. (a) 3.4×10^3 J (b) zero (isothermal process) (c) 12.5 J/K
61. 1.0×10^2 J/K
63. (a) 67% (b) $T_h = 3T_c$
65. 143 kcal = 6.0×10^5 J
67. change in entropies of reservoirs
69. 41 kW = 4.1×10^4 J/s

CHAPTER 14

14.1 Simple Harmonic Motion

1. 0.05 s
3. 0.60 Hz; 1.7 s
5. $E = \frac{1}{2}kA^2$ and $\omega = \sqrt{k/m}$
7. (a) 2.4 m/s (b) 2.4 m/s (c) 2.6 m/s at equilibrium position
9. 61 N/m
11. 8.0 m/s
13. $E_2/E_1 = 1/6$
15. 0.37 m/s

14.2 Equations of Motion

17. (a) $x = A \sin \omega t$
(b) $x = \pm A \cos \omega t$ (depending on sign convention used)
19. (a) 10 cm (b) 188 cm/s (c) 3.6×10^3 cm/s² (d) at top or bottom for (a) and (c); at equilibrium position for (b)
21. (a) 0.63 s (b) 90° (c) 2.4×10^{-3} J
23. (a) 4.8 cm (b) 4.4 cm/s (c) -1.2 cm/s²

25. (a) 14 cm (b) 89 cm/s (c) 0.16 J
27. $a = -kx/m = -\omega^2 x = -\omega^2 A \sin(\omega t + \delta)$
29. (a) 0.45 s (b) $y = (-10 \text{ cm}) \cos 14t$ (c) 0.20 J
(d) -5.0 cm
31. (a) zero (b) -94 m/s (c) zero
33. (a) Shorter L, clock will gain time (run fast) (b) 1.92 min (c) yes, because of linear expansion of the pendulum
35. (a) -5.6 cm (b) 8.1 cm

14.3 Wave Motion

37. 30 m/s
39. 10^5–10^2 m; 10^{-3}–10^{-6} s
41. 7.9×10^{-7} m
43. (a) 1.3×10^2 m (b) 88 m
45. (a) 1.5×10^3 m/s (b) zero (since fluids cannot support a shear and transverse waves do not propagate)
47. 1.0×10^4 m
49. (a) 0.4×10^3 s (b) depth = 1.9×10^6 m, so waves cross crust-mantle boundary (Moho discontinuity) (c) 1.6×10^3 s for P waves; S waves are not transmitted through liquid outer core

14.5 Standing Waves and Resonance

51. (a) 8.0 m (b) 2.7 m
53. 38 Hz, 76 Hz, 114 Hz, and 152 Hz
55. 210 Hz
57. 21.6 kg, 5.4 kg, 2.4 kg, and 1.35 kg

Additional Problems

59. $\sqrt{L/g} = \sqrt{m/k}$
61. (a) 2.5 J (b) 2.5 J
63. 17 m to 1.7×10^{-2} m
65. $k/m = 16$
67. (a) 12 cm (b) 3 cm, 9 cm, and 15 cm
69. 1.1 m/s
71. $T = 2\pi\sqrt{I/mgL}$ and $I = mL^2$, so $T = 2\pi\sqrt{L/g}$

CHAPTER 15

15.1 and 15.2 Sound Waves and the Speed of Sound

1. (a) 337.5 m/s (b) 349.5 m/s
3. (a) ≈2.7 km (b) ≈1.6 mi
5. 5.1×10^6 Hz
7. (a) 0.647 m (b) 0.671 m
9. 552.5°C
11. 171 m
13. 3.0 s
15. 38 vibrations

17. 45 m

19. (a) 337.9 m/s   (b) 10.7°C   (c) 3

15.3   Sound Intensity

21. $I_2/I_1 = 1/9$

23. (a) 100 dB   (b) 80 dB   (c) −10 dB

25. (a) $I_2/I_1 = 5.1$ (b) 5.7 m

27. $I_s/I_n = 6.3 \times 10^5$

29. 81 dB

31. (a) 73 dB   (b) (1) 76 dB, (2) 83 dB   (c) 3 dB

33. 10^3

35. 20 dB

37. 93 dB

39. $10[\log I_2/I_o - \log I_1/I_o] = 10 \log I_2/I_1$

41. 10^4

15.4   Sound Phenomena

43. (a) constructive   (b) destructive

45. 4 Hz; by making frequencies equal

47. (a) 435 Hz   (b) 370 Hz

49. 16 m/s

51. 42°

53. 2.88 m

55. (a) 17.7 m/s   (b) 6000 m

57. (a) 567 Hz   (b) 442 Hz

15.5   Sound Characteristics

59. (a) 0.39 m   (b) 4 Hz

61. no, because $2n \neq m$

63. 1.92×10^3 Hz

Additional Problems

65. 0.19 m, 0.58 m, and 0.97 m

67. (a) 10^6 (b) 10^{-2} W/m²; 10^{-8} W/m²

69. destructively

71. 7.5×10^{-3} J

73. (a) $v_w > v_{Hg}$ (b) $v_w/v_{Hg} = 1.1$

75. 10

CHAPTER 16

16.1   Electric Charge

1. (a) 50-cm mark (equal repulsive forces)   (b) 50-cm mark (equal attractive forces)

3. 6.3×10^{11}

5. (a) -8.0×10^{-10} C   (b) 5.0×10^9

16.2   Electric Force and Electric Field

7. 158 cm

9. 5.8×10^9 N/C

11. $q_2 = 3.0 \times 10^{-6}$ C, $+q_2$ at (0, −0.30 m) or $-q_2$ at (0, 0.30 m)

13. (a) 0.16 m from q_1 (b) same as (a)

15. (a) $\mathbf{E} = (74\ N)\mathbf{x} - (6.6\ N)\mathbf{y}$, or 96 N; 39.5° (fourth quadrant)

(b) $\mathbf{E} = -(6.6\ N)\mathbf{x} + (60.9\ N)\mathbf{y}$, or 61 N, 84° (second quadrant)

17. $(4.3 \times 10^6$ N/C$)\mathbf{x} - (2.5 \times 10^6$ N/C$)\mathbf{y}$, or 5.0×10^6 N, 30° (fourth quadrant)

19. $(3.8 \times 10^7$ N/C$)\mathbf{y}$

21. (a) zero   (b) 1.1×10^{12} N/C toward center of sphere

23. (a) 1.9×10^{-7} s   (b) (0.24 m, 0)

25. (a) 100 V   (b) 1.0×10^{-4} J, same

27. $(-4.2 \times 10^6$ N/C$)\mathbf{x} + (73.4 \times 10^6$ N/C$)\mathbf{y}$

16.4   Electrical Energy and Electric Potential

29. (a) -0.30 J   (b) no

31. (a) -8.6×10^{-14} J   (b) 8.6×10^{-14} J

33. -22.2 J

35. 2.1×10^5 V

37. -1.3×10^6 V

39. (a) 8.4×10^{-13} V   (b) 1.5×10^{-9} V   (c) top (+) and bottom (−) in (a); top (−) and bottom (+) in (b)

41. 0.52 m

43. (a) 5.9×10^7 m/s   (b) 5.9×10^{-9} s

45. 5.4×10^5 V

16.5   Capacitance and Dielectrics

47. 2.0×10^{-9} F

49. polyethylene (dielectric constant of 2.3)

51. 5.7×10^{-9} C; 3.4×10^{-8} J

53. 1.5

55. (a) 0.17 μF   (b) 2.0×10^{-6} C   (c) 5.3 V

57. 0.20 μF

59. 2.0 μF

61. (a) $V = V_o/K$ and $U = \frac{1}{2}QV_o/K = U_oK$

(b) $Q = KQ_o$ and $U = \frac{1}{2}QV = \frac{1}{2}KQ_oV = KU_o$

Additional Problems

63. (a) 8.8×10^9 m/s²   (b) $\theta = \tan^{-1}(d/L) \approx d/L$, where L is the length of the plates

65. (a) 2.4×10^{-9} C   (b) 1.4×10^{-20} kg

67. (a) 0.27 μF   (b) 1.2 μF

69. 4.2×10^6 m/s

71. 9×10^5 N/C toward -4-μC charge

73. 3.95×10^{-5} F

75. 6.0×10^{-19} J

77. (a) 1.3×10^{-9} C/m^2 (b) 8.1×10^9 e$^-$/m^2

CHAPTER 17

17.1   Batteries and Current

1. (a) 3.0 V   (b) 1.5 V
3. 0.33 A
5. 2.2×10^5 C
7. 1.1×10^{22} electrons
9. (a) $\Delta Q =$ (number of charges) (charge per particle) $=$
  $(nAx)(q)$
  (b) $x = v_d\Delta t$ in ΔQ, and $I = \Delta Q/\Delta t$

17.3   Ohm's Law and Resistance

11. 12 V
13. 0.12 A
15. 3.3×10^{-2} Ω
17. 2.5×10^{19} e$^-$/s
19. 0.25
21. 0.125 A
23. 1.4 Ω
25. 105 m
27. (a) 1.0×10^{-7} Ω-m   (b) platinum or iron (Table 17.1)
29. 1.7 Ω
31. 5.3×10^3 A
33. (a) copper, since carbon has $-\alpha$ (b) $R_{Cu}/R_C = 1.1$
35. heat copper wire to 108°C or cool aluminum wire to
  -67°C

17.4   Electric Power

37. 1.0 W
39. (a) 0.05 A   (b) 5.0 V
41. 1.8×10^6 J
43. (a) 0.15 A   (b) 1.4×10^{-4} Ω-m   (c) 2.3 W
45. (a) 400 Ω   (b) 4.0 W   (c) 480 J
47. 96 Ω
49. $P_2/P_1 = 9$
51. 36¢
53. 86 Ω
55. 73%
57. 413 s (6.9 min)
59. $120.27

Additional Problems

61. 34 C
63. 0.002 A = 2 mA
65. 2.7×10^{-2} m

67. (a) 0.42 A   (b) 14 Ω
69. no, $R_2/R_1 = 1.6$
71. 3.1 A less
73. $V^2/R = (W/q)^2[I/(W/q)] = (W/q)(q/t) = W/t$
75. $P_2/P_1 = 1.1$, or 10% power increase; 1650 W

CHAPTER 18

18.1   Resistances in Series, Parallel, and Series-Parallel Combinations

1. 90 Ω
3. 10 Ω
5. (1) series: 6 Ω, (2) parallel: 0.67 Ω, (3) parallel-series: 3
  Ω, (4) series-parallel: 1.3 Ω
7. 17 different combinations
9. (a) 11 A   (b) 6 A, 3 A, and 2 A, respectively   (c) 72 W,
  36 W, and 24 W, respectively   (d) 132 W in either
  case
11. (a) 0.5 A   (b) 120 V
13. (a) 0.34 A; 36 W   (b) 0.68 A; 150 W
15. 0.80 Ω
17. 6.9 Ω
19. 1.25 A; 2.5 V
21. (a) 0.42 A, 0.83 A, 1.25 A   (b) 290 Ω, 140 Ω, 96 Ω
23. (a) $I_1 = I_2 = 6$ A; $I_3 = I_4 = 3$ A   (b) $V_1 = V_2 = 12$ V;
  $V_3 = V_4 = 6$ V   (c) 180 W
25. (a) $I_1 = 1.0$ A; $I_2 = I_3 = I_4 = 0.60$ A   (b) 6.0 W; +
  3.6 W; + 2.2 W; + 1.4 W   (c) 13.2 W $\cong$ 13.1 W
  (rounding difference)
27. (a) 4.47 V   (b) 0.447 A
29. (a) R_1 and R_2 each dissipate 810 W   (b) 2160 W
31. 8.1 Ω
33. (a) $I_1 = 1.0$ A; $I_2 = I_4 = 0.40$ A; $I_3 = 0.20$ A
  (b) $P_1 = 100$ W; $P_2 = P_4 = 4.0$ W; $P_3 = 2.0$ W
35. $I_1 = 2.7$ A; $I_2 = 3.3$ A; $I_3 = I_4 = I_5 = 0.60$ A

18.2   Multiloop Circuits and Kirchhoff's Rules

37. (a) same equations with + and − changes   (b) solve or
  put in known solutions to show applicable
39. (a) $I_1 = 0.75$ A, $I_2 = 0.50$ A, $I_3 = 1.25$ A   (b) 12.5 W
41. $I_1 = 3.2$ A; $I_2 = 2.8$ A; $I_3 = 6.0$ A; $I_4 = I_5 = I_6 = 2.0$
  A
43. $I_1 = \frac{11}{12}$ A; $I_2 = \frac{9}{12}$ A, $I_3 = \frac{2}{12}$ A

18.3   RC Circuits

45. 5.0 s
47. 1.5×10^{-6} F and 1.5×10^{-6} Ω
49. (a) 0.38×10^6 Ω = 0.38 MΩ   (b) 11.4 V
51. (a) 4.7×10^{-4} C   (b) capacitor, 12 V; resistor, zero

18.4 Ammeters and Voltmeters

53. $6.0 \times 10^{-3}\ \Omega$
55. 0.59994 A
57. $R = V/I_R = V/(I - I_v) = V/[I - (V/R_v)]$

Additional Problems

59. (a) $I_1 = I_2 = 0.67$ A; $I_3 = 1.0$ A; $I_4 = I_5 = 0.40$ A
(b) $V_1 = 6.7$ V; $V_2 = 3.4$ V; $V_3 = 10$ V; $V_4 = 2.0$ V;
$V_5 = 8.0$ V
61. $0.0667\ \Omega$
63. $1.67\ \mu\Omega$
65. 0.75 A through each resistor; $V_{4\Omega} = 6.0$ V and $V_{8\Omega} = 3.0$ V
67. $I_1 R_x = I_2 R_2$ and $I_1 R_s = I_2 R_1$; form ratio
69. $I_1 = I_A = \frac{3}{11}$ A; $I_2 = I_A - I_B = \frac{15}{11}$ A; $I_3 = I_B = \frac{12}{11}$ A

CHAPTER 19

19.2 Electromagnetism and the Source of Magnetic Fields

1. 0.50 N, to left relative to v
3. 1.6×10^{-6} T, downward
5. 2.5×10^2 T, in $+z$ direction
7. (a) 3.3×10^{-12} N   (b) 2.3×10^{-12} N   (c) zero
9. 4.0 A
11. 6.3×10^{-5} T, toward north
13. 60 V
15. (a) 2.0×10^{-5} T, perpendicular to line joining wires (out of page)   (b) 0.096 m from wire 1
17. (a) 3.3×10^{-5} T, perpendicular to line joining wires (into page)   (b) nowhere
19. 2.4 A
21. 10 cm
23. 1.3×10^{-3} m
25. 7.7×10^{-21} J
27. $B_T = 1.4\mu_o I/\pi a$, at $\theta = 45°$ relative to $-x$ axis (third quadrant)
29. 0.44 T
31. (a) $f = qB/2\pi m$ (b) $T = 2\pi m/qB$ (c) 5.7×10^{-3} m; 2.8×10^6 Hz

19.3 Magnetic Materials

33. $m = IA = (e/t)\pi r^2 = e\pi r^2/(2\pi r/v) = evr/2$
35. 0.030 A-m²
37. (a) 38 T   (b) $B = (2000)B_o$
39. 1.3 N, upward
41. 4.0×10^{-3} T, north to south
43. (a) zero   (b) 400 N/m, $+z$ direction   (c) 400 N/m, $-y$ direction   (d) 400 N/m, $-z$ direction   (e) 400 N/m, $+y$ direction
45. 6.7×10^{-2} A, in the $-z$ direction

47. 1.8×10^{-5} N/m, repulsive
49. 3.0×10^{-4} N/m, repulsive
51. 2.7×10^{-5} N/m, toward other wire
53. 2.4×10^{-2} m-N
55. 0.26 m-N
57. 0.28 m

19.5 Electromagnetic Applications

59. (a) 2.4×10^3 V   (b) same as (a), since V is independent of the charge of the particle
61. (a) 1.6×10^4 N/C   (b) 1.6×10^3 V   (c) bottom plate
63. (a) 2.4×10^{-27} kg   (b) 1.2×10^{-18} J   (c) no, force is perpendicular to motion and $W = 0$

Additional Problems

65. (a) 0.090 A-m²   (b) plane of loop parallel to field
67. 5.8×10^{-7} m/s
69. 1.9 N, upward
71. short segments: 25 N, toward sides of page, equal and opposite; long segment: 75 N, toward top of page
73. (a) 1.3×10^{-2} m   (b) 7.5×10^6 V/m

CHAPTER 20

20.1 Induced emf's: Faraday's Law and Lenz's Law

1. 0.75 T-m²
3. -1.1 V
5. 5.2×10^{-2} T-m²
7. 1.1×10^{-6} T-m²
9. 1.4×10^{-7} T-m²
11. 12.5 V
13. 1.5 s
15. 0.27 V
17. -0.96 V
19. -5.0 T/s
21. 1.5×10^{-4} W
23. 0.13 V

20.3 Generators and Motor Back emf

25. 150 times or every $\frac{1}{150}$ s
27. (a) $\mathcal{E} = \mathcal{E}_o \sin(120\pi t)$ (b) 4.7 V
29. 169 loops (turns)
31. the second generator; 1.1 V
33. (a) 40 A   (b) 21 Ω
35. (a) 216 V   (b) 160 A   (c) 10.5 Ω

20.4 Transformers and Power Transmission

37. (a) step-down   (b) 21.3 A at 90 V
39. (a) 16 turns   (b) 500 A
41. (a) 840 V   (b) 0.36 A

43. (a) 300 V   (b) 320 turns
45. 40 turns
47. (a) 15 turns   (b) 3.0 V
49. 51 kWh
51. $I_s = 30$ A; $I_p = 0.33$ A

Additional Problems

53. 1.0; ideal conditions
55. 140 V; every $\frac{1}{120}$ s
57. (a) $N_s/N_p = 1/10$ (b) 0.05 A
59. (a) 3.0 W   (b) 1.1×10^5 V
61. (a) 5.0 A   (b) 30 A

CHAPTER 21

21.1 Resistance

1. 339 V
3. 0.35 A
5. 1.6 A
7. (a) 10 A   (b) 14 A   (c) 12 Ω
9. (a) $I_{eff} = 10$ A and $I_p = 14$ A   (b) $V_{eff} = 50$ V and $V_p = 71$ V
11. $\langle \sin^2 \omega t \rangle = \frac{1}{2} - \langle \cos \omega t \rangle \langle \cos \omega t \rangle = \frac{1}{2}$
13. (a) 3.3×10^2 A   (b) 311 V
15. 150 V
17. 0.042 T

21.2–21.3 Capacitive and Inductive Reactances

19. 5.3×10^{-8} F
21. 6.6×10^2 Ω
23. 0.25 H
25. (a) 91 V - (b) 90°
27. 0.70 H

21.4–21.5 Circuit Impedance and Resonance

29. 91 Hz
31. $-51°$
33. (a) 38 Ω (b) 55 Ω (c) 2.2 A   (d) zero
35. (a) 75 Ω (b) 92 W
37. (a) 465 Ω (b) 130 Hz
39. any resistance with a capacitance of 9.4 μF
41. 90 Ω
43. 51 W
45. 0.01×10^{-9} F $= 0.01$ nF
47. 27%

Additional Problems

49. $V_R = 12$ V; $V_C = V_L = 280$ V
51. 1.7×10^{-17} F

53. 3.1 A
55. 0.94 H
57. (a) 0.89   (b) 1.9×10^{-5} F; in parallel

CHAPTER 22

22.2 Reflection

1. 2.5 s
3. 50°
5. 68°
7. half the height of the person, or 0.90 m

22.3–22.4 Refraction and Total Internal Reflection

9. 1.24×10^8 m/s
11. 2.3×10^8 m/s
13. 29°
15. 20°
17. (a) 32°   (b) 28°
19. $\frac{16}{15} = 1.07$
21. 511 nm; 4.4×10^{14} Hz
23. 1.49
25. 35.9 mm
27. 8.6 cm
29. 6.0 cm
31. (a) 1.41   (b) 1.88
33. 2.6 cm
35. (a) 56°   (b) 30°

22.5 Dispersion

37. 0.10°
39. (a) 0.55°   (b) $\lambda_{red}/\lambda_{blue} = 1.73$

Additional Problems

41. 54°
43. 48.7°
45. 0.45°
47. 75%
49. 70°

CHAPTER 23

23.1 Plane Mirrors

1. (a) 6.0 m   (b) upright, virtual, same size
3. (a) 2.0 m   (b) 1.0 m/s
5. ray diagrams
7. 0.75 m

23.2 Spherical Mirrors

9. (a) $f = 10$ cm; $R = 20$ cm (b) 5.0 cm
11. 60 cm; 9.0 cm
13. $d_i = -10$ cm; $\frac{2}{3}$ of original height
15. (a) convex (b) $R = -10$ cm
17. $d_i = -10$ cm; $h_i = 0.67$ cm
19. $d_i = 60$ cm; real, inverted, magnified $(2\times)$, $h_i = 10$ cm
21. (a) $d_i = 37.5$ cm; real, inverted, magnified $(1.5\times)$
    (b) ray diagram not shown here
23. 4 times as large as the cavity
25. $d_o = 3$ cm
27. 7.5 m
29. graphs
31. yes, $d_o = 13.3$ cm and $d_o = 26.7$ cm

23.3 Lenses

33. (a) $d_i = -90$ cm; virtual, upright, magnified $(6\times)$
    (b) $d_i = 36$ cm; real, inverted, same size
35. (a) $d_i = -6.7$ cm; virtual, upright, reduced $(0.67\times)$
    (b) $d_i = -11.1$ cm; virtual, upright, reduced $(0.44\times)$
37. 95 mm
39. $d_o = 10$ cm, or midway between the screen and the object
41. (a) $d_i = -17$ cm (b) $d_i = -120$ cm
43. graphs
45. $d_i = 60$ cm; virtual, inverted, magnified $(8\times)$
47. by similar triangles, $y_o/d_o = y_i/d_i$, and $y_i/y_o = (d_i - f)/f$; combining these equations give (a) $1/d_o + 1/d_i = 1/f$ and (b) $y_i/y_o = d_i/d_o$
49. Sharp image on screen, rays parallel to axis of diverging lens. Rays reflected back through lens system to source. Rays incident on diverging lens transmitted parallel to axis if they apparently converge at image side focal point. Since rays originally converged at the mirror (screen), the mirror is at the focal point of the diverging lens.

23.5 The Lens Maker's Equation

51. R_1 and R_2 surfaces interchange
53. $+0.70$ D

Additional Problems

55. (a) $d_i = 24$ cm; real, inverted, magnified $(0.60\times)$
    (b) $d_i = 30$ cm; real, inverted, same size
    (c) $d_i = 60$ cm; real, inverted, magnified $(3\times)$
    (d) $d_i = \infty$; $M = \infty$
    (e) $d_i = -7.5$ cm; virtual, upright, magnified $(1.5\times)$
57. $f = -2.5$ cm; -40 D
59. 12 m/s
61. -6.0 cm
63. 5.0 cm and 15 cm
65. 0.50 m

CHAPTER 24

24.1 Young's Double Slit Experiment

1. (a) $\lambda/6$ (b) $\lambda/4$ (c) $\lambda/2$
3. 9.6×10^{-4} rad
5. 0.019 rad
7. (a) 6.0×10^{14} Hz (b) 0.20×10^{-2} m
9. 1.6×10^{-2} m
11. $y_n' = (0.75)y_n$
13. 4.5×10^{-5} cm = 450 nm

24.2 Thin Film Interference

15. 8.9×10^{-6} cm
17. (a) 115 nm (b) yes, $n_3 > n_2 > n_1$
19. (a) $d_2 - d_1 = \lambda/2$ (b) $d_2 - d_1 = \lambda/2n$

24.3 Diffraction

21. $\Delta y'/\Delta y = 0.25$
23. 2.9×10^{-8} m
25. $n \leq 3.6$, so $n = 3$
27. $6.7°$
29. diffraction minimum at $19.6°$

24.4 Polarization

31. 1.6
33. $32°$
35. graph

Additional Problems

37. (a) 2400 lines/cm (b) $n \leq 5.8$, so $n = 5$
39. $54°$ to $60°$
41. 0.91 m
43. (a) 1.12×10^{-3} m (b) would decrease to 7.5×10^{-4} m
45. 1.20×10^{-7} m
47. $39°$
49. $48.8°$
51. blue: $24°$ and $14.3°$; red: $53°$ and $30°$

CHAPTER 25

25.1 The Human Eye

1. $+2.0$ D
3. $+71$ D
5. (a) nearsighted (b) -0.20 D

7. (a) $+3.0$ D   (b) $d_i = 33.3$ cm, which is not inside
       corrected near point, so lenses would not have to
       be removed
9. $d_i = -67$ cm
11. $d_i = -6.7$ m
13. $+3.0$ D

25.2 Microscopes

15. 2.7
17. 12.5 cm
19. $1.7\times$
21. $375\times$
23. $300\times$
25. (a) 7.1 cm   (b) 3.5
27. by a small angle approximation, $m = \theta_i/\theta_o = (y_i/y_o)(25/D)$;
       by similar triangles, $y_i/y_o = -d_i/d_o = (D - d)/d_o$;
       This and the lens equation, $1/d_o = 1/f - 1/d_i$
       $= 1/f + 1/(D - d)$ give $m = (25/f)[(1 - d)/D]$
       $+ 25/D$
29. (a) $150\times$ (b) 7.2 cm

25.3 Telescopes

31. (a) $4.0\times$ (b) 75 cm
33. (a) $2.7\times$ (b) 135 cm
35. $m = \theta_i/\theta_o$, and $\theta_i \cong \tan\theta_i = y_i/f_e$ and $\theta_o \cong \tan\theta_o = y_i/f_o$

25.4 Diffraction and Resolution

37. 5.5×10^{-7} m
39. (a) 3.5×10^{-5} rad
       (b) 1.4×10^{-4} cm
41. (a) 1.6×10^{-6} rad
       (b) 9.6×10^{17} m

Additional Problems

43. $+3$ D
45. $3.7\times$
47. $560\times$
49. 1.1×10^{-2} rad
51. $+58$ D

CHAPTER 26

26.1–26.2 Classical Relativity and the Michelson–Morley Expt.

1. (a) 150 km/h   (b) -10 km/h
3. (a) 170 km/h   (b) 225 km/h
5. (a) 63 s   (b) 13 m   (c) 15° upstream
7. $t_2 = 2d/c[1 - (v/c)^2]$, but wth d contracted by $[1 - (v/c)^2]$, then $t_2 = t_1$
9. write $\Delta t = t_2 - t_1$ and $\Delta t' = t_1 - t_2$, then subtract Δt's

26.3 The Special Theory of Relativity

11. 45 min
13. 1.98×10^8 m/s
15. 7.5 years
17. 5.6 m
19. 1.98×10^8 m/s
21. (a) follows directly from figure   (b) with the length
       contraction, $x = x'(1 - v^2/c^2) + vt$, and $x' = \gamma(x - vt)$

26.4 Relativistic Mass and Energy

23. 938 MeV
25. 340 kg
27. (a) 1.8×10^8 m/s   (b) same height, smaller width
       (c) would increase by a factor of 1.56
29. 0.516 MeV
31. 2.90×10^8 m/s; 1.1×10^{-21} kg-m/s
33. (a) 1.53×10^{-10} J   (b) 3.0×10^{-12} J   (c) 1.0×10^{-19} N-s
35. (a) 2.6×10^8 m/s   (b) 2.9992×10^8 m/s

26.4 General Theory of Relativity

37. for an object with $R = 2GM/c^2$ or less, $v_e \geq c$, and not
       possible

Additional Problems

39. 0.588 MeV
41. 2.94×10^8 m/s; 2.511 MeV; 1.3×10^{-21} N-s
43. 6.0×10^{-7} kg
45. 2.29 years
47. (a) 0.42×10^8 m/s   (b) 1.3×10^8 m/s   (c) 2.9998×10^8 m/s

CHAPTER 27

27.1 Quantization: Planck's Hypothesis

1. 1.1×10^{-5} m, 2.7×10^{13} Hz
3. graphs
5. 5.2×10^{13} Hz
7. 3.1×10^{13} Hz
9. 1.4×10^{19} quanta/s/m²

27.2 Quanta of Light

11. infrared (6.0×10^{-6} m = 6000 nm)
13. 3.6×10^{18} quanta
15. doubled also
17. (a) 1.3×10^{-19} J   (b) 4.1×10^{14} Hz

19. yes; 1.3×10^{-19} J
21. no, wavelength of red light is greater than the threshold wavelength
23. $K_{max} = hf - \phi$, and plotting K_{max} versus f gives a line with slope h and intercept ϕ

27.3 Quantum "Particles": The Compton Effect

25. 1.0×10^{-27} N-s
27. $30°$
29. 2.5×10^{-10} m
31. 4.4×10^{-16} J

27.4 The Bohr Theory of the Hydrogen Atom

33. $R = 1.097 \times 10^{-2}$ nm^{-1}
35. (a) -1.51 eV (b) -0.378 eV (c) -0.136 eV
37. (b) and (c)
39. (a) 12.8 eV (b) 13.3 eV
41. (a) 433.6 nm; 121.6 nm (b) yes
43. four transitions

Additional Problems

45. 8.3×10^{20}
47. 0.850 eV
49. (a) 4.7×10^5 m/s (b) no emission
51. (a) 0.63 V (b) 8.4×10^{14} Hz
53. (a) 6.9×10^{14} Hz (b) 8.2×10^{14} Hz
55. (a) $E_1 = -2.82 \times 10^3$ eV and $E_2 = -0.705 \times 10^3$ eV; $r_1 = 2.56 \times 10^{-4}$ nm and $r_2 = 1.02 \times 10^{-3}$ nm
    (b) 0.585 nm
57. (a) slope $= h/e$, so $h = 6.4 \times 10^{-34}$ J-s
    (b) $\phi_o = hf_o = 2.91 \times 10^{-19}$ J $(= 1.82$ eV$)$

CHAPTER 28

28.1 Matter Waves

1. (a) 2.4×10^{-6} m (b) 1.3×10^{-9} m
3. 2.0×10^5 m/s
5. 4.7×10^{-35} m
7. (a) 1.7×10^{-46} N-s (b) 2.5×10^{-13} m
9. (a) 8.8×10^{-35} m (b) 1.7×10^{-38} m
11. 267 eV
13. 1.5 V

28.2 The Schrödinger Wave Equation

15. Function has required nodal points
17. maximum for first excited state, for example, at r^2 is 2.12 Å

28.3 The Heisenberg Uncertainty Principle

19. 11.5 m
21. 2.8×10^{-13} m
23. greater by a factor of 10^4

28.4 Particles and Antiparticles

25. 1.2×10^{20} Hz

Additional Problems

27. 680 nm
29. 43
31. 240 keV
33. -48%

CHAPTER 29

29.1 Nuclear Structure and the Nuclear Force

1. (a) 27 p, 33 n, 27 e$^-$ and 27 p, 32 n, 27 e$^-$
    (a) 27 p, 33 n, 29 e$^-$ and 27 p, 32 n, 29 e$^-$
    (a) 27 p, 33 n, 26 e$^-$ and 27 p, 32 n, 26 e$^-$
3. (a) $^1_1$H, $^2_1$D, $^3_1$T, $^3_1$T radioactive
    (b) H_2O, D_2O, T_2O, HDO, HTO, DTO
5. 35.452140 u

29.2 Radioactivity

7. alpha: $^{210}_{81}$Tl; beta: $^{214}_{84}$Po
9. (a) $_{-1}^0$e (b) $^{222}_{86}$Rn (c) $^{237}_{94}$Pu (d) γ (e) $_{+1}^0$e
11. (a) alpha to $^{223}_{87}$Fr, beta to $^{227}_{90}$Th (b) beta to $^{223}_{88}$Ra, alpha to $^{223}_{88}$Ra
    (c) both alpha to $^{219}_{86}$Rn (d) alpha to $^{215}_{84}$Po

29.3 Decay Rate and Halflife

13. (a) 6.76×10^{-6} Ci $= 6.76$ μCi (b) 2.50×10^5 Bq
15. (a) $1/4$ (b) $1/(2)^{24}$
17. (a) $^{28}_{12}$Mg (b) 9.1
19. 44 years
21. 4.6 μg $\cong 5$ μg
23. technetium-104
25. 1.9×10^6 kg
27. (a) -5.4×10^{17} decays/min (b) -8.9×10^{16} decays/min

29.4 Nuclear Stability

29. 2.013553 u
31. (a) 1.118×10^4 MeV (b) 3.728×10^3 MeV
33. (a) stable (b) unstable (c) stable (d) unstable
    (e) stable

35. (a) 7.72 MeV   (b) 2.57 MeV
37. 127.6 MeV
39. 10.1 MeV
41. 6.013470 u
43. (a) mass of two alpha particles is less   (b) Be: 56.5 MeV; from text example, He: 28.3 MeV   (c) expect it to be unstable

29.5   Radiation Detection and Applications

45. by observing beta emissions at a rate of about 2 emissions/min/g of carbon
47. (a) alpha particles   (b) X-rays: 2.0 rem; alpha source: 40 rem
49. yes, 3.7 rem exceeds 3 rem/3 months
51. ~2.5 μCi

Additional Problems

53. (a) 0.011 s^{-1} (b) 189 s
55. 7.57 MeV
57. 11,460 years
59. (a) $^{218}_{84}$Po   (b) $^{214}_{82}$Pb; $^{218}_{85}$At

CHAPTER 30

30.1–30.2   Nuclear Reactions and Nuclear Energy

1. (a) $^{4}_{2}$He   (b) $^{1}_{1}$H   (c) $^{93}_{38}$Sr   (d) $^{16}_{7}$N   (e) $^{12}_{6}$C
3. (a) $^{14}_{7}$N   (b) $^{2}_{1}$H   (c) $^{30}_{15}$P   (d) $^{17}_{8}$O   (e) $^{10}_{5}$B

5. exoergic
7. exoergic ($Q = 6.50$ MeV)
9. 4.36 MeV
11. 3.7×10^5 m/s
13. (a) 3.27 MeV   (b) 12.9 MeV
15. 5.6×10^8 K
17. 13.4 MeV
19. $Q = (m_p - m_d - m_e)c^2 = \{[M_p - Zm_e] - [M_d - (Z - 1)m_e] - m_e\}c^2$
21. $Q = (m_p + m_e - m_d)c^2 = \{[M_p - Zm_2] + m_e - [M_d - (Z - 1)m_e]c^2$
23. (a) 0.86 MeV   (b) no

30.4   Fundamental Particles

25. $M = (8.0 \times 10^{-3})m_e$
27. 4.8×10^{-24} s

Additional Problems

29. 3.23 MeV
31. (a) $^{27}_{12}$Mg   (b) $^{10}_{5}$B   (c) $^{89}_{39}$Y   (d) $^{24}_{11}$Na
33. 0.868 MeV
35. 1.53 MeV

INDEX

(Numbers followed by *f, m, p, g,* or *t* refer to pages with terms in footnotes, margins, problems, questions, or tables, respectively.)

A

aberration,
  lens,
    astimatism, 638
    chromatic, 637
    spherical, 636
  mirror,
    spherical, 629
absolute,
  pressure, 280
  temperature scale, 318
  zero, 316
    and third law, 381
ac,
  circuits, 581
  current, 560
accelerators, particle, 807
acceleration, 32
  angular, 184
  centripetal, 71
  due to gravity, 37, 190–93
  linear,
    average, 32
    instantaneous, 32
  in SHM, 399
  tangential, 74, 184
achromatic doublet, 638
activity, (radio-), 783
adhesive force, 289
adiabat, 368
adiabatic process, 367
air resistance, 107
alpha,
  decay, 780
  particle, 780
alternating current (ac), 560
alternator, 560
ampere (unit), 478
  definition of,
    ac, 584
    dc, 537
Ampere, Andre, 478
ammeter (dc), 510
Anderson, C.D., 769
angular
  acceleration, 184
  displacement, 182
  distance, 178
  magnification, 684
  measure, 177–80
  momentum, 232

  speed, 181
  velocity, 182
angle,
  Brewster's, 663
  contact, 289
  critical (internal reflection),
    607
  of incidence, 599, 601
  polarizing, 663
  radian (unit), 178
  of reflection, 599
  of refraction, 601
  of repose, 107
antinode, 409
antipartile, 769
antimatter, 772
anode, 476
Archimedes, 282
Archimedes' principle, 282
Aristotle, 38, 84
astigmatism,
  lens, 638
  vision defect, 681
astronomical unit, 214*p*
atmosphere (unit), 276
  pressure and density, 281
  standard, 280
atmospheric,
  pressure, 276
  scattering, 669
atom, model of,
    Rutherford-Bohr, 776
    solar system, 444
    Thomson, 776
atomic,
  bomb, 817
  mass unit, 790
  number, 778
  pile, 817
Atwood, George, 95
Atwood machine, 94, 224
aurora,
  australis, 545
  borealis, 545
Avogadro's number, 316
axis,
  optic,
    cyrstal, 664
    lens, 631
    mirror, 621
  transmission, 665

B

back (counter) emf, 563
Balmer series, 743
barometer,
  aneroid, 281
  mercury, 280
barrier penetration, 781
baryons, 827
battery, 476
  dry cell, 476
  storage, 477
beat(s), 429
  frequency, 429
becquerel (unit), 787
Becquerel, Henri, 779
bel (unit), 424
Bell, Alexander Graham,
    424*m*
Bernoulli, Daniel, 294
Bernoulli's equation, 295
beta,
  decay, 782
    and neutrino, 821
  particle, 780, 782
binding energy,
  of hydrogen electron, 746
  nucleon, 790
    average per nucleon, 791
    total, 791
binoculars, 689
birefringence, 664
black,
  body, 352, 733
  hole, 725
boiling, 355
  point, 343
    and pressure, 343
Bohr, Niels, 743
Bohr theory,
  and de Broglie waves,
    757
  of hydrogen atom, 742
  and probable radii, 765
Boyle, Robert, 316
Boyle's law, 316
branch (circuit), 503
Bragg, W.L., 662
Bragg's law, 662
Brahe, Tyco, 197
Brewster, David, 663
Brewster's angle, 663

British,
  system (of units), 6
  thermal unit (Btu), 335
Brown, Robert, 325
Brownian motion, 325
bubble chamber, 795
buoyancy, 281
buoyant force, 281

C
calorie (unit), 334
Calorie (kilocalorie), 335
calorimeter, 341
calorimetry, 341
cancer, 798
capacitance, 462
capacitor(s),
  in parallel, 466
  parallel plate, 462
  in series, 465
  variable air, 593
capillary action, 290
Carnot,
  cycle, 379
  efficiency, 380
Carnot, Sadi, 379
CAT scan, 574
cathode, 476
  ray tube, 541
CD (compact disc), 751
center
  of curvature,
    lens, 631
    mirror, 621
  of gravity, 166
    and stability, 251
  of mass, 163
centimeter, 7
  cubic, 7
centrifguge, 204
centrifugal force, 204
centripetal,
  acceleration, 71
  force, 89
Celsius, Anders, 315
Celsius temperature scale, 312
cgs system (of units), 6
chain reaction, 813
Chernobyl, 818
charge(s),
  conservation of, 445
  electric,
  law of, 444
  net, 444
  polarization of, 449
charging, electrostatic, 447
  by contact, 448
  by friction, 447
  by induction, 448
Charles, Jacques, 316
Charles' law, 316
chief ray,
  ,lens, 631
  mirror, 622
chromatic aberration, 637
circle of least confusion,
  637

circuit(s),
  ac, 581
    capacitive, 589
    inductive, 589
    resonance, 591
    series RC, 587
    series RL, 588
    series RLC, 588
  blinker, 509
  branch, 503
  breaker, 514
  household, 512
  junction, 503
  multiloop, 502
  neon-tube relaxation oscillator, 509
  parallel, 497
  RC (battery), 506
  series, 496
  series-parallel, 500
Clausius, Rudolf, 370, 378
cloud chamber, 795
coefficient,
  of diffusion, 327
  of expansion, temperature
    area, 321
    linear, 320
    volume, 321, 324
  of friction,
    kinetic, 102
    static, 101
    table of, 103
  of performance, 385p
  of resistivity, temperature, 481t,
    482
  of restitution, 175p
  of viscosity, 298
    table of, 299
coherent,
  light, 748
  sources, 647
cohesive force, 289
collision, 155
  elastic, 156
  inelastic, 156
    completely, 157
color(s)
  additive,
    method, 696
    primary, 696
  subtractive,
    method, 697
    primary pigments, 697
compact disk (CD), 751
Compton, Arthur H., 739
Comption,
  effect, 739
  wavelength, 741
compressibility, 264
condensation point, 343
condensations, 420
conductivity,
  electrical, 481
    super-, 485
  thermal, 348
conductor(s),
  electrical, 446
  thermal, 347

cones, 677
confinement,
  inertial, 820
  magnetic, 820
continuity, equation of, 293
constant,
  decay, 784
  gravitational, universal, 188
  phase, 396
  of proportionality, 88
  spring, 120
  Stefan-Boltzmann, 352
  time (RC), 507
conservation of,
  charge, 445
    in nuclear processes,
      780
  energy,
    mechanical, 129–36
    relativistic, 727
  momentum,
    angular, 233
    linear, 150
  nucleons, 780
contact angle, 289
  table of, 290
continental drift, 47p
control rods, 815
convection, 350
coordinates,
  Cartesian, 24, 178
  polar, 178
Coriolis, Gaspard, 206
Cariolis force, 206
coulomb (unit), 444
Coulomb, Charles A. de,
  444
Coulomb's law, 449
counter
  (back) emf, 563
  torque, 565
Cousteau, Jacques, 297
conversion factor, 12
critical,
  angle,
    internal reflection, 607
  mass, 813
Crookes, William, 820
Crookes', tube, 820
cross-section, reaction, 810
crystalline lens, 677
curie (unit), 787
Curie, Marie (Madam), 779
Curie, Pierre, 535m, 779
Curie temperature, 535
current(s),
  alternating (ac), 560
    ampere (unit), 584
  conventional, 478
  direct (dc), 476
    ampere (unit), 537
  eddy, 567
  peak, 582
  rms (effective), 582
cutoff frequency, 736
cyclotron, 550p, 721
  frequency, 550p

D

damped harmonic motion, 398
Davisson, C.J., 759
Davisson and Germer experiment, 759
de Broglie, Louis, 757
de Broglie hypothesis, 757
decay constant, 784
decibel (unit), 424
dew point, 355
density,
  mass, 11
    table of, 284
  of water vs. temperature, 324
deposition, 343
detectors, radiation,
  bubble chamber, 795
  cloud chamber, 795
  film badge, 793
  Geiger counter, 793
  scintillation counter, 794
  solid state, 794
  spark chamber, 795
deuterium, 779
deuteron, 779
dichrosim, 665
dielectric, 463
  constant, 463
diesel engine, 375
diffraction,
  crystal, 661
  electron, 759
  gratings
    reflection, 659
    transmission, 659
  and resolution, 693
  single slit, 657
  sound, 427
  wave, 408
diffusion, 326
  coefficient, 327
dimentional analysis, 9
diopter (unit), 640
Dirac, P.A.M., 766m, 769
disintegration energy, 809
diskette, computer, 558
dispersion, 611
displacement,
  angular, 182
  linear, 24
  oscillation, 389
distance,
  angular, 178
  image,
    lens, 631
    mirror, 619, 623
  linear, 24
  object,
    lens, 631
    mirror, 619, 623
  stopping, 36
Doppler, Christian, 430m
Doppler effect,
  red shift, 432
  sound, 430
drift velocity, 479
dual nature of light, 742
dynamics, 23

E

earthquakes, 404
eddy currents, 567
efficiency,
  Carnot, 380
  electrical, 488
  mechanical, 138
  thermal, 376
Einstein, Albert,
  and Brownian motion, 325
  letter to Roosevelt, 841
  and photoelectric effect, 735
  and relativity, 711
elastic,
  limit, 261
  modulus, 260
electric(al),
  charge(s), 443
    conservation of, 445
    law of, 444
    net, 444
    unit of, 444
  circuit breaker, 514
  conductor, 446
  conductivity, 482
  current, 478
    alternating, 560
    direct, 478
  efficiency, 488
  equipotentials, 459
  field, 452
  fuse, 513
  insulator, 446
  potential (voltage), 458
    and electric field, 459
  power, 484
  resistance, 479
  resistivity, 481
    temperature coefficient of, 482
  safety, 512–16
  semiconductor, 447
  voltage (potential), 458
electromagnetic,
  induction, 555
    mutual, 555
    self, 566
  force, 569, 823
  waves, 569
    types of, 571–74
electromagnetism, 527
  applications of, 539
electromotive force (emf), 476, 496m
  back (counter), 563
  motional, 560
electron,
  capture, 782
  electric charge, 443, 445t
  mass, 445t
  microscope, 760
    scanning, 760
    scanning tunneling, 760
    transmission, 760
  spin, 800
  volt (unit), 460
electroscope, 447
electrostatic,
  charging, 447–48

copier, 450
  potential energy, 456
electroweak force, 828
emf, 476, 496m
  back (counter), 563
  motional, 560
emission,
  spontaneous, 748
  stimulated, 748
emissivity, 352
endoergic reaction, 809
energy,
  binding,
    of hydrogen electron, 746
    nucleon, 790
      average per nucleon, 791
      total, 791
  conservation of, 129–36
    mechanical, 131
    relativistic, 727
  disintegration, 809
  internal, 311
  kinetic,
    linear, 124
    rotational, 227
  mechanical, 131
  nuclear, 8ll–21
    fission, 811
    fusion, 814
  potential,
    mechanical, 126
    electrostatic, 456
  relativistic, 727
  in SHM, 389
  threshold,
    pair production, 771
    reaction, 810
  and work, 124
entrophy, 370
equation(s),
  Bernoulli's,
  of continuity, 293
  flow rate, 294
  kinematic,
    linear, 35
    rotational, 187
  lens maker's, 638
  Maxwell's, 569, 707
  of motion,
    linear, 37
    SHM, 392
    rotational, 187
  pressure-depth, 276
  Schrödinger's wave, 762
  spherical mirror, 624
  thin lens, 632
equilibrium, 245
  mechanical, 245–46
    rotational, 246
    translational, 245
  neutral, 251
  stable, 250
  unstable, 251
equipotential, electric, 459
escape speed, 198
ether, 707
  wind, 709

evaporation, 354
exchange particles, 823
    gluon, 828
    graviton, 826
    pi meson (pion), 825
    photon, 824
excited state(s), 745
    lifetime, 746
exoergic reaction, 809
extraordinary ray, 664
expansion, thermal, 311
eye, human, 676
    crystalline lens, 677
    far point, 678
    near point, 678
    retina, 677
        cones, 677
        rods, 677

F
Fahrenheit, Daniel, 315
Fahrenheit temperature scale, 311
far point, 678
farsightedness, 679
farad (unit), 462
Faraday, Michael, 462m, 555
Faraday's law (of induction), 555
Fermi, Enrico, 817
ferromagnetism, 534
Feynman diagrams, 824
Feynman, Richard, 824
fiber optics, 608
fiberscope, 610
field,
    electric, 452
    gravitational, 193
    magnetic, 526
fifth force, 823m
film badge, 793
fission, nuclear, 792, 811
    chain reaction, 813
    critical mass, 813
    moderator, 815
FitzGerald, George F., 711
flow,
    incompressible, 292
    irrotational, 292
    laminar, 298
    nonviscous, 292
    rate, 299
        equation, 294
    steady, 292
    turbulent, 301
fluid, 274
    dynamics, 292
focal,
    length,
        lens, 631
        mirror, 622
    point,
        lens, 631
        mirror, 622
    ray,
        lens, 636
        mirror, 623m
force(s), 82, 86
    action-at-a-distance, 86

adhesive, 289
balanced, 82
buoyant, 281
centrifugal, 204
centripetal, 89
cohesive, 289
concurrent, 246
contact, 86
conservative, 130
Coriolis, 206
couple, 246
electromagnetic, 569, 823
electromotive (emf), 476
electroweak, 828
fifth, 823m
of friction, 100
    kinetic (sliding), 100
    rolling, 100
    static, 100
fundamental, 823
gravitational, 188, 823
inertial, 203
internal, 150
magnetic, 528
and momentum, 148
net, 82
normal, 99
nuclear,
    strong, 823
    weak, 823
spring, 120
unbalanced, 82
Foucault, Leon, 207
Foucault pendulum, 207
free fall, 37
free-body diagram, 97
freezing point, 342
frequency, 184, 389
    beat, 429
    cutoff, 736
    cyclotron, 550p
    fundamental, 410
    harmonic, 410
    of mass on a spring, 393
    natural, 409
        of organ pipes, 436
        of stretched string, 410
    resonant, 409
    threshold, 738
    wave, 402
Franklin, Benjamin, 447, 478
friction, 100–108
    air resistance, 107
    charging by, 447
    coefficient of,
        kinetic (sliding), 102
        static, 101
    force of, 100
    rolling, 101, 231
fuel rods, 815
fundamental,
    forces, 823
    quantities, 2
        electric charge, 443
        length, 2
        mass, 3
        time, 4

fuse, electrical
    Edison-base, 513
    type-S, 514
fusion, nuclear, 792, 814
    cold, 820
    reactions, 814

G
g's of force, 191
Galileo, Galilei, 37, 38, 83, 688m, 689
galvanometer, 539
    in ammeter and voltmeter, 509
gamma,
    decay, 783
    ray (particle), 780, 783
gauss (unit), 528m
Geiger counter, 793
Geiger, Hans, 776, 793
Gell-Mann, Murray, 827
Germer, L.H., 759
Gilbert, William, 543
Glashow, Sheldon, 828
Glazer, D.A., 795
gluon, 828
gradient,
    concentration, 327
    thermal, 347
grand unified theory (GUT), 829
gravitation(al),
    constant, universal, 188
    field, 193, 726
    force, 188, 823
    intensity, 193
    lensing, 726
    potential,
        energy, 127, 194
        well, 129, 194
    Newton's universal law of, 188
    red shift, 727
gray (unit), 798
Gray, Harold, 798m
grazer, 802p
greenhouse effect, 572
ground,
    electrical, 448m
    state, 745
grounding, dedicated, 514
GUT, 829

H
Hail Observatories, 691
half-life, radioactive, 785
    table of, 786
Hall effect, 547p
harmonic,
    frequencies, 410
    motion
        damped, 398
        simple (SHM), 389
heat, 311, 334
    of combustion, 336
    death of universe, 372
    engine, 374

diesel, 375
  four-stroke, 375
  two-stroke, 375
joule, 486
latent, 343
mechanical equivalent of, 337
pump, 377
reservoir, 363
specific, 338
  table of, 338
transfer,
  conduction, 347
  convection, 350
  radiation, 351
heavy water, 817
Heisenberg uncertainty principle, 776
  energy and time, 768
  position and momentum, 766
Heisenberg, Werner, 766
henry (unit), 586
Henry, Joseph, 555, 586m
Herapath, W., 665
hertz (unit), 184, 389
Hertz, Heinrich, 184m, 389f
Hooke, Robert, 261
Hooke's law, 262, 388
holography, 750
hologram, 750
horsepower (unit), 136
humidity, 355
  relative, 355
Huygen, Christian, 600
Huygen's principle, 600
hydraulic lift, 278
hydrometer, 286

I
ideal (perfect) gas law, 316
image,
  distance,
    lens, 631
    mirror, 619, 623
  lens,
    real, 631
    virtual, 631
  mirror,
    real, 619, 622
    virtual, 619, 622
impedance, 587
impulse, 154
inclined plane,
  as a machine, 258
index of refraction, 602
inductance, 586
induction,
  electromagnetic, 555
  mutual, 555
  self, 566
inertia, 85
  and mass, 86
inertial,
  confinement, 820
  force, 203
  frame of reference, 201, 706
infrared radiation, 571
intensity,
  decibel level, 424

gravitational, 193
  radiation, 353
  sound, 422
    level, 424
interference (wave), 405
  constructive, 406–207
  destructive, 406–407
  fringes, 648
  sound, 427
  thin film, 651
interferometer, 709
insulator,
  electrical, 446
  thermal, 347
I^2R loss, 486
isentropic process, 373
isobar, 365
isobaric process, 365
isomet, 366
isometric process, 366
isotherm, 366
isothermal process, 366
isotope(s), 779
  of hydrogen, 779

J
Jansky, Carl, 692
jet propulsion, 167
joule (unit), 118
joule heat, 486
Joule, James Prescott, 118
  free expansion experiment, 368
junction, 503
  theorem, 503

K
Kamerlingh Onnes, H., 485
K-capture, 782
kelvin (unit), 318
Kelvin, Lord, 318m, 377
Kelvin temperature scale, 318
Kepler, Johannes, 196
Kepler's laws,
  of areas (2nd), 197
  of orbits (1st), 196
  of periods (3rd), 197
kilocalorie (unit), 335
kilogram (unit), 3, 5t
kinematic(s)
  linear, 23
    equations, 35
  rotational, 187
    equations, 187
kinetic,
  energy,
    linear, 124
    and momentum, 155
    relativistic, 721
    rotational, 228
  theory of gases, 325, 839
Kirchhoff, Gustav, 503m
Kirchhoff's rules,
  first (junction theorem), 503
  second (loop theorem), 503
  methods of,
    branch current, 506
    loop, 524p

L
Land, Edwin H. 665, 697
latent heat, 343
  of fusion, 343
  of vaporization, 343
laser(s), 747
  and fusion, 820
  gamma ray, 802p
law(s),
  Boyle's, 316
  Bragg's, 662
  of charges, 444
  Charles', 316
  of cooling, Newton's, 357p
  Coulomb's, 449
  Faraday's, of induction, 556
  of gravitation, Newton's, 188
  Hooke's, 262, 388
  ideal (perfect) gas, 316
  Kepler's, 196–98
    of areas (2nd), 197
    of orbits (1st), 196
    of periods (3rd), 197
  Lenz's, 556
  of motion, Newton's, 82–100
  Ohm's, 479
  perfect (ideal) gas, 316
  Poiseuille's, 299
  of poles, 526
  of reflection, 599
  Snell's, 601
  Stefan's, 352
  of thermodynamics,
    first, 364
    second, 369
    third, 381
    zeroth, 311m
  Wien's displacement, 734
Laue, Max von, 661
LCD, 666
LED, 667
length, 2, 5t
  contraction, 719
  proper, 718
lens(es), 629
  biconcave, 630
  biconvex, 630
  combinations, 635
  converging, 630
  equation, thin, 632
  erecting, 689
  Fresnel, 636
  maker's equation, 638
  nonreflecting, 653
  oil-immersion, 695
  power, 640
Lenz, Heinrich, 556m
Lenz's law, 556
lever,
  (moment) arm, 220
  as a machine, 256
light,
  coherent, 748
  dispersion of, 611
  dual nature of, 742
  emitting diode (LED), 667
  ray, 599

light, *continued*
    reflection of, 599
    refraction of, 600
    visible, 572
lifetime,
    excited state, 746
lightning, 455
    rod, 455
limit,
    elastic, 261
    proportional, 261
line,
    broading, natural, 769
    spectra, 743
liquid crystal, 343
    display (LCD), 666
liter, 7
    milli-, 7
lodestone, 525
Lorentz, H.A., 711
Lorentz-FitzGerald contraction, 711, 718
loss-of-coolant accident (LOCA), 818
loudness, 434
loudspeaker, 575*g*

M
Mach, Ernst, 434*m*
Mach number, 434
machine, 254
    Atwood, 94, 224
    types of, 255
magnet, 525
magnetic,
    bottle, 545
    confinement, 545, 820
        tokamak, 820
    dipole, 526
    domains, 534
    field, 526–27
        around straight wire, 531
        at center of circular loop, 531
        of Earth, 543
        of solenoid, 532
        source of, 530
    flux, 555
    force, 528–30
        on current-carrying wires, 535
    levitation, 485
    materials, 533
    moment, 538
    poles, 525
    tape, 558
magneto, 562
magnification,
    angular, 684
    lateral, 620*m*
    lens, 632
    microscope,
        compound, 687
        simple, 685
    mirror, 620
    telescope, refracting, 688
    useful, 695
magnifying glass, 683
Magnus effect, 297

manometer,
    closed-tube, 280
    open-tube, 279
Marsden, Ernest, 776
maser, 748*m*
mass, 3–4
    center of, 163
    critical, 813
    and inertia, 86
    kilogram (unit), 3
    number, 778
    spectrometer, 542
    relativistic, 721
    rest, 722
    and weight, 4, 89
mass-energy,
    conservation,
    equivalence, 722
matter,
    and antimatter, 772
    waves, 757
Maxwell, James Clerk, 569, 707
Maxwell's equations, 569, 707
mechanical,
    advantage, 255
        actual (AMA), 255
        theoretical or ideal (TMA), 255
            inclined plane, 258
            lever, 256
            pulley, 258
    equilibrium, 245
    equivalent of heat, 337
mechanics, 23
melanin, 573
meltdown, reactor, 818
melting point, 342
meson, 824, 827
    mu (muon), 824
    pi (pion), 825
metastable state, 748
meter (unit), 2
methods of mixtures, 341
metric,
    prefixes, 6*t*
    system, 2
        cgs, 6
        mks, 6
        tonne, 8
Michelson, Albert A., 709
Michelson interferometer, 709
Michelson-Morley experiment, 708–711
microscope,
    compound, 686
        eyepiece (objective), 686
        objective, 686
    electron, 760
        scanning, 760
        scanning tunneling, 761
        transmission, 760
    simple, 683
microwave,
    oven, 352
    radiation, 352, 571
Millikan, Robert A., 711
mirror, 618
    plane, 618
    sperical, 621

    center of curvature, 621
    concave, 621
    converging, 622
    convex, 621
    diverging, 622
    focal point (length), 622
    radius of curvature, 621
    vertex, 621
mks system (of units), 6
moderator, 815
modulus,
    bulk, 264
    elastic, 260
    shear, 264
    Young's, 262
moment,
    arm (lever), 220
    of inertia, 221
momentum,
    angular, 232
        conservation of, 233
    and impulse, 154
    and kinetic energy, 155
    linear, 146,
        conservation of, 150
        and force, 148
    relativistic, 721
Morley, E.W., 710
motion, 23
    Brownian, 325
    circular, uniform, 70
    components of, 52
    curvilinear, 52
    equations of,
        linear, 36
        (SHM), 392
        rotational, 187
    linear, 24
    perpetual machine, 370
    rotational, 217
    translational, 217
    wave, 400
motor, dc, 540
multiplier resistor, 511
muon, 824
    in cold fusion, 820

N
natural line broading, 769
near point, 678
nearsightedness, 678
neon-tube relaxation oscillator, 509
neutrino, 822, 827
neutron, 445*t*, 777
    activation analysis, 800
    number, 778
newton (unit), 88
Newton, Isaac, 82, 84, 613, 655
    law of cooling, 357*p*
    laws of motion, 82–100
        first (law of inertia), 83
        second, 87
        third, 97
    Principia, 84
    rings, 655
node, 408
nonreflecting coating, 653

nuclear,
  electrical generation, 817
  force
    strong, 788
    weak, 778m, 822, 824
  magnetic resonance (NMR), 799
  notation, 778
  reactions, 806–11
    endoergic, 809
    exoergic, 809
    fission, 811
      chain, 813
    Q value, 808
    reactors, 815–19
nucleon, 777
  populations, 788
nucleus, atomic, 776
nuclide, 779
number(s),
  atomic, 778
  exact, 13
  mass, 778
  measured, 13
  neutron, 778
  proton, 778

O
object distance,
  lens, 631
  mirror, 619, 623
Oersted, Hans Christian, 530
ohm (unit), 479
Ohm, Georg, 479
Ohm's law, 479
optic axis,
  crystal, 664
  lens, 631
  mirror, 621
optics,
  fiber, 608
  geometrical, 599
  physical (wave), 647
optical,
  activity, 668
  flats, 655
  illusions, 698
ordinary ray, 664
oscillation, 389
  amplitude, 389
  frequency of, 389
  period of, 389
osmosis, 328
Otto cycle, 375
Otto, Nikolas, 375
ozone, 573
  layer, 573

P
pair,
  annihilation, 771
  production, 770
pairing effect, 789
parallax, 210p
parallel,
  axis theorem, 221
  ray,
    lens, 631
    mirror, 622

parsec (unit), 210p
pascal (unit), 275
Pascal, Blaise, 275, 277
Pascal's principle, 277
path difference,
  constructive interference, 649
  destructive interference, 649
particle(s),
  accelerators, 807
  and antiparticles, 769
  elementary, 826–29
    hadrons, 827
    leptons, 827
  exchange, 823
    gluon, 828
    graviton, 826
    pi meson (pion), 825
    photon, 824
    W particle, 825
  virtual, 823
Pauli, Wolfgang, 822
pendulum,
  Foucault, 207
  physical, 418p
  simple, period of, 394
PET, 799
perfect (ideal) gas law, 316
period, 389
  of curcular motion, 184
  of mass on spring, 393
  of pendulum, 394
permeability, 535
  of free space, 531
  relative, 535
permittivity, 464
  of free space, 462
perpetual motion machine, 370
  first kind, 370m
  second kind, 370m
phase(s)
  angle, 396, 588
  change, 343
  constant, 396
  diagram,
    electrical, 587
    thermodynamic, 346
  of matter,
    gaseous (vapor), 343
    liquid, 342
    plasma, 819
    solid, 342
phasor, 587
  diagram, 587
phosphorescence, 748
photoelectric effect, 735
photomultiplier tube, 794
photon, 735
  virtual, 824
physical (wave) optics, 647
physics, 1
  quantum, 732
pitch, 435
Planck, Max, 734
Planck's,
  constant, 734
  hypothesis, 734
plasma, 819

plate tectonics, 47p
plug, electrical,
  polarized, 515
  three-prong, grounded, 514
poise (unit), 298
poiseuille (unit), 298
Poiseuille, Jean, 298
Poiseuille's law, 299
polarization,
  of charge, 449
  direction, 664
  electrical, 515
  light, 622
    by double refraction, 664
    by reflection, 663
    by scattering, 669
polarized plug, 515
polarizing angle, 663
poles, magnetic, 525
  law of, 526
population inversion, 749
positron, 769
  emission tomography (PET), 799
positronium atom, 771
potential,
  energy, 126
    electrostatic, 456
    gravitational, 127, 193
    spring, 127
  electric (voltage), 458
    and field, 459
  stopping, 736
power, 136
  electric, 484
    transmission, 568
  factor, 590
  lens, 640
  resolving, 695
  rotational, 227
precession, 236
  of Earth's axis, 236
pressure, 274
  absolute, 280
  atmospheric, 276
  blood, 300
  -depth equation, 276
  gauge, 280
  radiation, 570
principal quantam number, 744
principle,
  Archimedes', 282
  of classical relativity, 706
  of equivalence, 724
  Huygen's, 600
  Pascal's, 277
  of relativity, 713
  of superposition, 405
  uncertainty, Heisenberg's, 766
probability density, 762, 766
process, thermodynamic, 364
  adiabatic, 367
  irreversible, 364
  isentropic, 373
  isobaric, 365
  isometric, 366
  isothermal, 366
  reversible, 364

projectile motion, 63–70
  at arbitrary angle, 65
  horizontal projection, 63
  range, 66
proper,
  length, 718
  mass, 720
  time, 716
proportional limit (elasticity), 261
proton,
  electric charge, 443, 445t
  mass, 445t
  number, 778
pulley, as a machine, 258
pumping, 749

Q
quality,
  factor (QF), 798
  sound, 435
quantum, 734
  physics, 732
  mechanics, 756, 762
  number, principal, 744
quark, 444f, 827
Q value, 808

R
rad (unit), 798
radar, 432
radian (unit), 178
radiation,
  blackbody, 733
  detectors, nuclear, 793–95
  electromagnetic, 569
    types of, 571–74
  heat transfer, 351
  infrared, 351, 571
  microwave, 352, 571
  pressure, 570
  thermal, 732
radio waves, 571
radioactive,
  dating, 795–97
  decay rate, 783
  half-life, 785
  isotopes, 779
radioactivity, 779–88
  alpha decay, 780
  beta decay, 782
  detectors of, 793–95
  domestic and industrial applications,
    800
  electron capture, 782
  gamma decay, 783
  medical applications, 797
radius,
  of curvature,
    lens, 631, 638
    mirror, 621
  of gyration, 240p
rainbow, 611
range, projectile, 66
  maximum, 67
Rankine temperature scale, 319f
Rankine, William, 319f
rarefactions, 420

ray, 599
  extraordinary, 664
  lens,
    chief, 631
    focal, 631
    parallel, 631
  mirror,
    chief (radial), 622
    focal, 622m
    parallel, 622
  ordinary, 664
Rayleigh,
  criterion, 693
  Lord, 670, 693
  scattering, 670
reactance,
  capacitive, 585
  inductive, 586
reaction(s),
  cross-section, 810
  endoergic, 809
  exoergic, 809
  fission, 811
    chain, 813
  fusion, 814
  nuclear, 806–11
  Q value, 808
  thermonuclear, 819
  time, 40
reactor(s), nuclear,
  breeder, 817
  fission, 815–19
  fusion, 820
real image,
  lens, 631
  mirror, 619, 622
red shift,
  Doppler, 432
  gravitational, 727
reference frame,
  inertial, 706
  proper, 716
reflection, 599
  internal (total), 607
  irregular (diffuse), 599
  law of, 599
  regular (specular), 599
  sound, 426
  wave, 407
refraction, 600
  angle of, 601
  index of, 602
  sound, 426
  wave, 408
refrigerator, 378
relative,
  biological effectiveness (RBE),
    798
  humidity, 355
  velocity, 30, 706
relativistic,
  total energy, 727
  kinetic energy, 721
  mass, 721
  momentum, 721
relativity, 705
  classical (Newtonian), 706

  general theory of, 724
  special theory of, 713
rem (unit), 798
resistance,
  air, 107
  electrical, 479
    in parallel, 497
    in series, 496
resistivity, 481
  temperature coefficient of, 482
resistor,
  multiplier, 511
  shunt, 499, 510
resolution, 693
resolving power, 695
resonance,
  electrical (circuit), 591
    frequency (resonant), 591
  mechanical, 413
    frequency (resonant), 409
rest mass, 722
retina, 677
revolution, 217f
Reynold's number, 301
rigid body, 216
rocket (jet propulsion), 167
rods, 677
roentgen (unit), 798
Roentgen, Wilhelm, 573
rolling,
  friction, 101, 231
  without slipping, 219
rotation(al), 217f
  instantaneous axis of, 217
  kinetic energy, 228
  motion, 217
  power, 227
  work, 227
Rutherford, Ernest (Lord), 776, 794m, 807
Rutherford-Bohr model of atom, 776
R-value, 361p
Rydberg constant, 743

S
Salam, Abdus, 828
satellites, Earth, 198–201
saturation, nuclear force, 793
scalar, 24
scattering,
  atmospheric, 669
  Compton, 739
  experiments, nuclear, 776
Shcrödinger, Erwin, 762
Schrödinger wave equation, 762
scintillation counter, 794
seismology, 404
shear,
  angle, 264
  modulus, 264
  strain, 264
  stress, 264
SHM, 389
  acceleration, 399
  energy, 389
  speed, 389
  velocity, 399
shunt resistor, 499, 510

SI system, 2
  base units, 5
significant figures (digits), 14
  rounding of, 14
simple harmonic motion (SHM), 389
  acceleration, 399
  energy, 389
  speed, 389
  velocity, 399
simultaneity, 714
single-photon emission tomography
      (SPET), 799
sievert (unit), 799
smoke detector, 800
Snell, Willebord, 601m
Snell's law, 601
solar system model (of atom), 444
solid state detector, 794
sonic boom, 433
sound, 419
  diffraction, 427
  reflection, 426
  refraction, 426
  spectrum, 420
    audible, 420
    infrasonic, 420
    ultrasonic, 420
  speed of, 421
  waves, 419
    interference, 427
spark chamber, 795
specific,
  gravity, 286
  heat, 338
    table of, 338
spectrum,
  continuous, 611, 743
  discrete (line), 743
    natural broadening, 769
speed,
  angular,
    average, 181
    instantaneous, 181
  escape, 198
  intensity, 422
  linear,
    average, 25
    constant, 28
    instantaneous, 26
  root-mean-square (rms), 326f
  in SHM, 389
  of sound, 421
SPET, 799
spherical,
  aberration,
    lens, 636
    mirror, 629
  mirror equation, 624
sphygmomanometer, 301
spin,
  electron, 534
  nuclear, 822
spring,
  constant, 120
  potential energy, 127
  force, 120
spyglass (telescope), 689

standard unit(s), 1
state(s),
  equation of, 363
  excited, 745
  ground, 745
  metastable, 748
  thermodynamic, 363
Stefan-Boltzmann constant, 352
Stefan's law, 352
stopping,
  distance, 36
  potential, 736
strain, 260
  shear, 264
  tensile (or compressional), 260
  volume, 264
streamline, 292
stress, 260
  shear, 264
  tensile (or compressional), 260
  volume, 264
strong nuclear force, 778
sublimation, 343
sunburn (tan), 573
sunglasses, polarizing, 667
sunscreen, 573
Superconducting Super Collider, (SSC),
      731p
superconductivity, 485
superposition, principle of, 405
surface tension, 287
  table of, 289
system(s),
  conservative, 131
  thermodynamic, 362
  of units, 1
    British (English), 6
    cgs, 6
    fps, 6
    metric, 2
    mks, 6
    SI, 2, 5

T
Tacoma Narrows bridge, 412
telephone, 575g
  fiber optics, 610
telescope,
  infrared, 692
  multiple mirror, 672
  radio, 692
  reflecting, 689
  refracting, 688
    astronomical, 688
    spyglass, 689
    terrestrial (Galilean), 689
television, 541
temperature, 310
  absolute, 316
  coefficient of,
    expansion, 320, 321t
    resistivity, 481t, 482
  Curie, 535
  and kinetic theory, 326
  scales,
    Celsius, 312
    Fahrenheit, 312

  Kelvin, 318
  Rankine, 319f
tesla (unit), 528
theorem,
  impulse-momentum, 154
  junction, 503
  loop, 503
  parallel axis, 221
  work-energy, 124
theory,
  of hydrogen atom, Bohr's, 742
  of gases, kinetic, 325, 839
  grand unified (GUT), 829
  of relativity,
    general, 724
    special, 713
thermal,
  conductivity, 348
    table of, 349
  conductor, 347
  contact, 311
  energy, 311f
  equilibrium, 311
  expansion, 311, 319
    of water, 324
  insulator, 347
  gradient, 347
  radiation, 732
thermodynamic(s), 362
  laws of,
    first, 364
    second, 369
    third, 381
    zeroth, 311m
  process, 364
    irreversible, 364
    reversible, 364
  system, 362
    closed, 362
    isolated,
      completely, 363
      thermally, 362
    open, 362
    state of, 363
thermometer, 311
  bimetallic, 312
  constant-volume gas, 316
  liquid crystal, 667
  liquid-in-glass, 312
thin film interference, 651
Three Mile Island (TMI), 818
3-D movies, 667
thermonuclear reactions, 819
thickness gauging, 801
Thomson, J.J., 776
Thomson, G.P., 760
Thomson model of atom, 776
Thomson, William, (Lord Kelvin),
      318m
threshold,
  frequency, 738
  of hearing, 423
  of pain, 423
time, 4
  constant (RC), 507
  dilation, 716
  proper, 716

time, *continued*
    reaction, 40
    second (unit), 4, 5*t*
time's arrow, 372
TMI, 818
tokamak, 820
ton (tonne), metric, 8
torque, 220
    and angular momentum, 232
    counter, 565
torr (unit), 281
Torricelli, Evangelista, 280
Tower of Pisa, 38–39
transformer, 565
    step-down, 566
    step-up, 566
transverse wave, 403
triple point (water), 319
tritium, 779
triton, 779
tunneling,
    in barrier penetration, 781
    scanning, electron microscope,
        761
twin paradox, 719

U
ultrasound, 420
ultraviolet,
    catastrophe, 734
    radiation, 573
unit(s)
    analysis, 10
    base, 2
    conversion, 11
    derived, 2
    determining, 11
    mixed, 10
    standard, 1

V
vector, 24
    addition, 57
        component method, 60–62
        parallelogram method, 58
        polygon method, 59
        triangle method, 58
    resultant, 58
        component notation, 60
        magnitude-angle form, 60
        subtraction, 59
velocity,
    drift, 479
    linear,
        average, 26
        constant, 28
        instantaneous, 28
        relative, 30

selector, 542
in SHM, 399
terminal, 107
wave, 402
Venturi meter, 309*p*
vertex, mirror, 621
vibration, 389
    amplitude, 389
    frequency of, 389
    period of, 389
virtual,
    image,
        lens, 631
        mirror, 619, 622
    particle, 823
    photon, 824
viscosity, 297
    coefficient of, 298
    table of, 299
vision,
    color, 695
    defects,
        astigmatism, ,681
        farsightedness, 679
        nearsigtedness, 678
    20/20, 682
volt (unit), 458
Volta, Allessandro, 458*m*, 476
voltage, 458
    peak, 581
    rms (effective), 582
voltmeter (dc), 511

W
W particle, 825
watt (unit), 136
Watt, James, 136
wave(s), 400
    amplitude, 402
    bow, 433
    de Broglie, 757
    diffraction, 408
    electromagnetic, 569
        gamma rays, 574
        infrared, 571
        micro-, 571
        power, 571
        radio, 571
        ultraviolet, 573
        visible, 572
        X-rays, 573
    equation, Schrödinger, 762
    frequency, 402
    front, 598
    function, 762
    interference, 405
        constructive, 406
        destructive, 406

length, 402
longitudinal, 403
matter, 757
mechanics, 762
motion, 400
optics, 647
P-, 405
periodic, 401
pulse, 401
radio, 571
    AM, 658
    FM, 658
reflection, 407
refraction, 408
S-, 405
sound, 419
speed, 402
standing, 408
superposition, 405
transverse, 402
water, 403
weak nuclear force, 778*u*, 822
wever/m^2 (unit), 528, 555
weber (unit), 555
weight, 4, 89
Weinberg, Steven, 828
wetting agent, 289
Wheatstone bridge, 523*p*
Wheatstone, Charles, 523*p*
Wien's displacement law, 734
Wilson, C.T.R., 795
work, 116–17
    energy theorem, 124
    function, 737
    rotational, 227
    in stretching spring, 122
    by variable force, 120

X
xerography, 450
X-rays, 573
    CAT scan, 574
    Compton effect, 739
    diffraction, 661
    tube, 573

Y
Yerkes Observatory, 690
Young, Thomas, 262, 647
Young's
    double slit experiment, 647
    modulus, 262
Yukawa, H., 824

Z
Z particle, 829
Zweig, George, 827

PHOTO CREDITS

Photo credits should be considered an extension of the copyright page.

CHAPTER ONE Figs. 1.1(b), 1.2(b), 1.3(c): National Institute of Standards and Technology. Figs. 1.5, 1.7, 1.12: John Smith.

CHAPTER TWO Fig. 2.2: John Smith. Insight, Fig. 1: Archiv/Photo Researchers, Inc.; Fig. 2: J. M. Charles-Rapho/Photo Researchers, Inc.; Fig. 3: PSSC Physics, 2nd edition, 1965; Educational Development Center, Inc., Newton, MA. Fig. 2.9: John Smith.

CHAPTER THREE Fig. 3.11(b) PSSC Physics, 2nd edition, 1965; Educational Development Center, Inc., Newton, MA.

CHAPTER FOUR Insight Fig. 1: The Bettmann Archive; Fig. 2: The Granger Collection, New York. Fig. 4.3: John Smith. Demonstration 1 (a–c): Michael Freeman. Demonstration 2 (a–c): Michael Freeman. Fig. 4.18: F. Rickard-Artdia, Agence Vandystadt/Photo Researchers, Inc. Fig. 4.19: John Smith.

CHAPTER FIVE Demonstration 3 (a–d): Arthur G. Schmidt, Northwestern University. Fig. 5.13: Greg Vaughn/TOM STACK & ASSOCIATES.

CHAPTER SIX Demonstration 4 (a, b): Michael Freeman. Figs. 6.6(a), 6.8 (a): H. E. Edgerton, MIT. Fig. 6.13 (a): John Smith. Fig.6.14(a): PSSC Physics, 2nd edition, 1965; Educational Development Center, Inc., Newton, MA. Fig. 6.14(b): WIDE WORLD PHOTOS. Fig. 6.16: John Smith. Fig. 6.18: Christian Petit, Agence, Vandystadt/Photo Researchers, Inc. Demonstration 5 (a, b): Michael Freeman. Fig. 6.21(a, b): NASA. Fig. 6.22(a): Runk Schoenberger/Grant Heilman Photography, Inc.

CHAPTER SEVEN Demonstration 6: Michael Freeman. Fig. 7.10: NASA. Demonstration 7 (a, b): Michael Freeman. Fig. 7.17: Biophoto Associates/Photo Researchers, Inc. Fig. 7.18 (b): The Museum of Modern Art/Film Stills Archive. Insight, Fig. 2: Richard Megna/FUNDAMENTAL PHOTOGRAPHS, NY.

CHAPTER EIGHT Fig. 8.2(b): H. E. Edgerton, MIT. Fig. 8.18(a, b): John Smith. Fig. 8.27(b): Tony Savino/The Image Works, Inc.

CHAPTER NINE Fig. 9.8(b): Taurus Photos. Fig. 9.10: California Institute of Technology Archives. Demonstration 8 (a–d): Michael Freeman. Insight, Fig. 1: WIDE WORLD PHOTOS. Demonstration 9 (a–c): Michael Freeman. Figs. 9.20(a, b), 9.21: John Smith.

CHAPTER TEN Fig. 10.6: George Haling/Photo Researchers, Inc. Demonstration 10 (a, b): Michael Freeman. Fig. 10.8: John Smith. Fig. 10.9: Biophoto Associates/Photo Reseachers, Inc. Fig. 10.14 (b): Richard Megna/FUNDAMENTAL PHOTOGRAPHS, NY. Demonstration 11 (a–c): Michael Freeman. Insight, Fig. 2: Blair Seitz/Photo Researchers, Inc. Figs. 10.20(a, b), 10.22(a, b), 10.24: John Smith.

CHAPTER ELEVEN Figs. 11.2, 11.3, 11.5(b): John Smith. Insight, Fig. 1: The Granger Collection, NY. Fig. 11.6(a–c): John Smith. Demonstration 12 (a–d): Michael Freeman. Fig. 11.11: Robert Mathena/FUNDAMENTAL PHOTOGRAPHS, NY. Figs. 11.15, 11.20(a, b): John Smith.

CHAPTER TWELVE Fig. 12.2: North Wind Picture Archives. Fig. 12.3(a): The Science Museum, London. Fig. 12.4: John Smith. Fig. 12.12: Dr. Ray Clark and Mervyn Goff/Photo Researchers, Inc. Figs. 12.13, 12.15, 12.16: John Smith.

CHAPTER THIRTEEN Fig. 13.9: John Smith.

CHAPTER FOURTEEN Demonstration 13 (a–d): Michael Freeman. Fig. 14.10: John Smith. Fig. 14.13(a) FUNDAMENTAL PHOTOGRAPHS, NY. Fig. 14.14(b): E. M. Ahart/TOM STACK & ASSOCIATES. Figs. 14.15, 14.18, 14.19, 14.20: PSSC Physics, 2nd edition, 1965; Educational Development Center, Inc., Newton, MA. Demonstration 14 (a, b): Michael Freeman. Insight, Fig. 1: WIDE WORLD PHOTOS.

CHAPTER FIFTEEN Fig. 15.1(b): John Smith. Fig. 15.3: E. R. Degginger, Color-Pic, Inc. Fig. 15.6: Guy Gillette/Photo Reseachers, Inc. Fig. 15.9 (b): PSSC Physics, 2nd edition, 1965; Educational Development Center, Inc., Newton, MA. Fig. 15.11(b): H. E. Edgerton, MIT. Fig. 15.14(b): John Smith. Fig. 15.15: E. R. Degginger, Color-Pic, Inc. Fig. 15.18: John Smith.

CHAPTER SIXTEEN Fig. 16.7(b): John Smith. Fig. 16.12(a): Gary Milburn/TOM STACK & ASSOCIATES. Fig. 16.12(b): Grant Heilman/Grant Heilman Photography, Inc. Fig. 16.18(c): Paul Silverman/FUNDAMENTAL PHOTOGRAPHS, NY. Fig. 16.25(a): Runk/Schoenberger/Grant Heilman Photography, Inc. Fig. 16.25(b): John Smith.

CHAPTER SEVENTEEN Fig. 17.7: Michael Dalton/FUNDAMENTAL PHOTOGRAPHS, NY. Insight, Fig. 1: IBM Research. Fig. 17.9: John Smith. Fig. 17.10: Reproduced from USAF DMSP (Defense Meteorological Satellite Program) film transparencies archived for NOAA/NESDIS at the University of Colorado, CIRES/National Snow and Ice Data Center, Campus Box 449, Boulder, Co, U.S.A. 80309. Fig. 17.11: John Smith.

CHAPTER EIGHTEEN Figs. 18.12(a), 18.17(b): John Smith. Fig. 18.18: Square D Company.

CHAPTER NINETEEN Figs. 19.1, 19.7(a), 19.8(a), 19.9(a): PSSC Physics, 2nd edition, 1965, Educational Development Center, Inc., Newton, MA. Fig. 19.11(b): Grant Heilman/Grant Heilman Photography, Inc. Fig. 19.18(b): Michael Freeman.

CHAPTER TWENTY Fig. 20.1(d): E. R. Degginger, Color-Pic, Inc. Demonstration 15 (a, b): Michael Freeman. Fig. 20.9: Mark Antman/The Image Works, Inc. Fig. 20.12(b): John Smith. Insight, Fig. 1: John Nettis/Photo Researchers, Inc.

CHAPTER TWENTY-TWO Fig. 22.4(a, b): PSSC Physics, 2nd edition, 1965; Educational Development Center, Inc., Newton, MA. Fig. 22.5(a): Scott Blackman/TOM STACK & ASSOCIATES. Fig. 22.5(b): John Smith. Fig. 22.6: PSSC Physics, 2nd edition, 1965; Educational Development Center, Inc., Newton, MA. Demonstration 16(a–c): Michael Freeman. Fig. 22.10(a): John Smith. Fig. 22.10(b): Runk/Schoenberger/Grant Heilman Photography, Inc. Fig. 22.13(b): PSSC Physics, 2nd edition, 1965; Educational Development Center, Inc., Newton, MA. Fig. 22.15: Charles Falco/Photo Researchers, Inc. Fig. 22.16: Courtesy of American ACMI, Division of American Hospital Supply Corporation, Stamford, CT. Fig. 22.17: Sugawa/Schuman, *Primer of Gastrointestinal Fiberoptic Endoscopy*, Little, Brown and Co., Boston, 1981. Courtesy of Dr. Hiromi Shinya. Fig. 22.18: David Parker/Photo Researchers, Inc. Fig. 22.20(a): Runk/Schoenberger/Grant Heilman Photography, Inc. Figs. 22.20(b), 22.21: John Smith.

CHAPTER TWENTY-THREE Demonstration 17 (a, b): Michael Freeman. Figs. 23.4(a, b), 23. 7: John Smith. Demonstration 18 (a, b): Michael Freeman. Fig. 23.16: John Smith. Demonstration 19 (a, b): Michael Freeman. Insight, Fig. 1(c): John Smith. Figs. 23.22(a, b), 23.23: John Smith. Fig. 23.24(b): Michael Freeman.

CHAPTER TWENTY-FOUR Fig. 24.1: PSSC Physics, 2nd edition, 1965; Educational Development Center, Inc., Newton, MA. Fig. 24.2(b): Courtesy of Addison-Wesley Publishing Company. Demonstration 20 (a, b): Michael Freeman. Fig. 24.8(b): E. R. Degginger, Color-Pic, Inc. Fig. 24.10: Courtesy of Addison-Wesley Publishing Company. Fig. 24.11(b): Eduquip-Macalaster Corporation. Fig. 24.13 (b): E. O. Wollan, Oak Ridge Laboratory. Fig. 24.16(b): E. R. Degginger, Color-Pic, Inc. Fig. 24.18(b): FUNDAMENTAL PHOTOGRAPHS, NY. Insight, Figure 2: John Smith. Figure 24.20(b): FUNDAMENTAL PHOTOGRAPHS, NY.

CHAPTER TWENTY-FIVE Fig. 25.3(b): John Smith. Fig. 25.7(b): American Optical. Fig. 29.9(c): John Smith. Fig. 25.10: BUSHNELL, Division of Bausch & Lomb. Fig. 25.11: Yerkes Observatory Photographic Services. Fig. 25.13(left and right): California Institute of Technology. Insight, Fig. 1: Grant Heilman Photography, Inc. Fig. 25.15(a, b): Michael Cagnet, *Atlas of Optical Phenomena*, Springer-Verlag, Berlin and Prentice-Hall, Englewood Cliffs, NJ, 1962. Fig. 25.17: Runk/Schoenberger, Grant Heilman Photography, Inc.

CHAPTER TWENTY-SIX Fig. 26.6: WIDE WORLD PHOTOS, INC.

CHAPTER TWENTY-SEVEN Fig. 27.3: AIP, Niels Bohr Library. Fig. 27.8(a): Courtesy of Bausch & Lomb. Fig. 27.17(a, b): Bob Hahn, Taurus Photos, Inc. Fig. 27.18(c): Ray Ellis, Photo Researchers, Inc.

CHAPTER TWENTY-EIGHT Fig. 28.1: AIP, Niels Bohr Library. Insight, Fig. 2: T. E. Adams; Fig. 3 (a–c): David M. Phillips, Taurus Photos, Inc.; Fig. 4: R. M. Feenstra, IBM Thomas J. Watson Research Center. Fig. 28.4(a, b): PSSC Physics, 2nd edition, 1965; Educational Development Center, Inc., Newton, MA. Figs. 28.5, 28.10: AIP, Niels Bohr Library. Fig. 28.13: G. D. Rochester and J. G. Watson, *Cloud Chamber Photographs of The Cosmic Radiation*. Pergamon Press, 1952.

CHAPTER TWENTY-NINE Fig. 29.5: AIP, Niels Bohr Library. Fig. 29.17: Brookhaven National Laboratory. Fig. 29.19: Paul Shambroom, Photo Researchers, Inc.

CHAPTER THIRTY Fig. 30.1(a): Fermilab National Accelerator Laboratory. Demonstration 22 (a–c): Michael Freeman. Fig. 30.5(b): Tom Bochsler. Fig. 30.6: University of Chicago. Fig. 30.9: WIDE WORLD PHOTOS, Fig. 30.10: Courtesy of the Princeton Plasma Physics Laboratory. Fig. 30.15: AIP, Niels Bohr Library.

Physical Constants

Acceleration due to gravity:	$g = 9.80$ m/s^2 = 980 cm/s^2 = 32.2 ft/s^2
Universal gravitational constant:	$G = 6.67 \times 10^{-11} \dfrac{\text{N-m}^2}{\text{kg}^2}$
Electron charge:	$e = 1.60 \times 10^{-19}$ C
Speed of light:	$c = 3.0 \times 10^8$ m/s = 3.0×10^{10} cm/s = 1.86×10^5 mi/s
Boltzmann's constant:	$k = 1.38 \times 10^{-23}$ J/K
Planck's constant:	$h = 6.63 \times 10^{-34}$ J-s
	$\hbar = h/2\pi = 1.05 \times 10^{-34}$ J-s
Electron rest mass:	$m_e = 9.11 \times 10^{-31}$ kg = 5.49×10^{-4} u $\leftrightarrow$ 0.511 MeV
Proton rest mass:	$m_p = 1.672 \times 10^{-27}$ kg = 1.007267 u $\leftrightarrow$ 938.28 MeV267/28
Neutron rest mass:	$m_n = 1.674 \times 10^{-27}$ kg × 1.008665 u $\leftrightarrow$ 939.57 MeV
Coulomb's law constant:	$k = 1/4\pi\varepsilon_o = 9.0 \times 10^9$ N-m^2/C^2
Permittivity of free space:	$\varepsilon_o = 8.85 \times 10^{-12}$ C^2/N-m^2
Permeability of free space:	$\mu_o = 4\pi \times 10^{-7} = 1.26 \times 10^{-6}$ T-m/A

Astronomical and Earth data

Equatorial radius of Earth:	6.378×10^3 km = 3963 mi
Polar radius of Earth:	6.357×10^3 km = 3950 mi
Mass of Earth	6.0×10^{24} kg
Mass of Moon	7.4×10^{22} kg $= \frac{1}{81}$ mass of Earth
Mass of Sun	2.0×10^{30} kg
Average distance of Earth from Sun	1.5×10^8 km = 93×10^6 mi
Average distance of moon from Earth	3.8×10^5 km = 2.4×10^5 mi
Diameter of moon	3500 km $\cong$ 2160 mi
Diameter of Sun	1.4×10^6 km $\cong$ 864,000 mi

Trigonometric Relationships

Definitions of Trigonometric Functions

$$\sin\theta = \frac{y}{r} \qquad \cos\theta = \frac{x}{r} \qquad \tan\theta = \frac{\sin\theta}{\cos\theta} = \frac{y}{x}$$

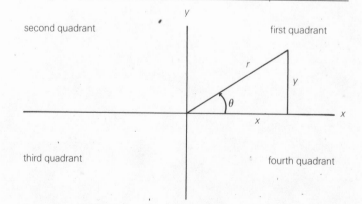

$\theta°$ (rad)	$\sin\theta$	$\cos\theta$	$\tan\theta$
0° (0)	0	1	0
30° ($\pi/6$)	0.500	0.866	0.577
45° ($\pi/4$)	0.707	0.707	1.00
60° ($\pi/3$)	0.866	0.500	1.73
90° ($\pi/2$)	1	0	$\rightarrow \infty$

See Appendix 1, Table 4, p. 837 for other angles.